HOUSE BUILDER'S REFERENCE BOOK

HOUSE BUILDER'S REFERENCE BOOK

Edited by

M. J. V. POWELL
MSc, MIOB

With specialist contributors

NEWNES–BUTTERWORTHS
LONDON - BOSTON
Sydney - Wellington - Durban - Toronto

The Butterworth Group

United Kingdom	**Butterworth & Co (Publishers) Ltd** London: 88 Kingsway, WC2B 6AB
Australia	**Butterworths Pty Ltd** Sydney: 586 Pacific Highway, Chatswood, NSW 2067 Also at Melbourne, Brisbane, Adelaide and Perth
Canada	**Butterworth & Co (Canada) Ltd** Toronto: 2265 Midland Avenue, Scarborough, Ontario M1P 4S1
New Zealand	**Butterworths of New Zealand Ltd** Wellington: T & W Young Building, 77–85 Customhouse Quay, 1, CPO Box 472
South Africa	**Butterworth & Co (South Africa) (Pty) Ltd** Durban: 152–154 Gale Street
USA	**Butterworth (Publishers) Inc** Boston: 19 Cummings Park, Woburn, Mass. 01801

First published 1979

British Library Cataloguing in Publication Data

House builder's reference book.
1. House construction
I. Powell, Michael John Vivian
690'.8'3 TH4811 78-40997

ISBN 0–408–00337–5

Typeset by Scribe Design, Chatham, Kent
Printed in England by Page Bros Ltd, Norwich
and bound by G & J Kitcat Ltd, London SE1

FOREWORD

by IVAN TOMLIN, JP, PPIOB, FAIB
Chairman of the Professional Practice Board of the Institute of Building

Builders have always found great challenge in working in the housing sector. In housing there is a direct contact between the builder and the consumer, whether he is an owner-occupier or a tenant. We all of us live in a dwelling and find the experience either a pleasure or a succession of problems. It is a pleasure if the house is well designed and convenient to live in, if the quality is such that adding fitments and carrying out redecoration and routine maintenance are easy and if, in the case of a new house, the builder is courteous and punctual in dealing with problems that have arisen in spite of his care. One is not dealing with the public at large or a committee or nameless workers in a factory, but with individuals. This is even more true in alteration, conversions and maintenance work to occupied houses.

In private development and in speculative improvement and conversion work the developer/builder is a creative entrepreneur, taking all the responsibilities and risks in masterminding the whole enterprise, including the assessment of the market, the securing of the finance, the buying of the land, the commissioning of the architect, and the physical process of building. Even though private development, like many aspects of modern life, is the subject of too much detailed legislation, it is nevertheless one area in which a person with building experience and much creative energy can establish a first-class business of his own.

Houses are complex structures. In terms of building technology, small certainly does not mean simple. The house builder has to know his way round the worlds of soil mechanics, timber technology, heating and insulation, to name but three areas. Technological change has been faster in the last twenty years than in any previous time and the pace is unlikely to slacken, although there may be some reduction from previous trends.

Customer consciousness, entrepreneurial flair and technological skill must be backed by sound financial management. There is no merit in being the best builder in Carey Street. Although profit does not necessarily have to be an end in itself, without adequate profit there can be no future for a business.

I believe that all engaged in the many-sided job of house building, whether as developers, architects, builders or site staff, will find this book a useful work of reference, sometimes to recap basic principles and sometimes to give signposts for the solution of problems that the particular reader has not experienced before. The detailed contributions from technical and other experts are well balanced by those from people directly engaged in the industry, forming judgements, making decisions, and creating jobs and houses for people.

PREFACE

This book is about building management, design and technology as they relate to new house building, house alteration and conversion work and housing maintenance. Much of the information is also relevant to other types of building.

It has been written primarily for those who run house building firms and sites, to provide a first source of information in areas in which they themselves are not specialists. It will also be relevant to surveyors, engineers, architects, valuers, building inspectors and students concerned with house building.

The team of contributors is a blend of practical and professional people involved in house building at grass-roots level with experts from the technical bodies and building colleges.

I thank the National House-Building Council for permitting me to undertake the editorship whilst in their employ, all the contributors for their enthusiasm and cooperation, the editorial staff at Newnes–Butterworths for their courtesy, and my wife for her administrative help.

M.J.V.P.

CONTENTS

MANAGEMENT

1 Financial management
2 Estimating and tendering
3 Purchasing
4 Marketing
5 Insurance
6 Building contracts
7 Quantity surveying
8 NHBC scheme
9 Plant
10 Site organisation
11 Production planning and control
12 Quality control
13 Human aspects of management

DESIGN

14 Site appreciation
15 Development design and landscape
16 Dwelling design
17 Convenience, safety and the user

CONSTRUCTION

18 Surveying and setting out
19 Earthworks
20 Estate roads, paths and drives
21 Foul sewerage and drainage
22 Surface water drainage
23 Soils
24 Site investigation
25 Standard foundations
26 Special foundations
27 Concrete technology
28 Reinforcement
29 Theory and design of structures
30 Temporary works
31 Concrete floors and roofs
32 Brickwork and blockwork
33 Protection against damp
34 Timber – the materials available to the builder
35 Timber floors and flat roofs
36 Pitched roofs
37 Timber-framed walls and partitions
38 Non-traditional construction

SERVICES

39 Water and sanitary services
40 Gas services
41 Electrical services
42 Heating services
43 Insulation

FINISHES

44 Joinery
45 Glazing
46 Roof tiling
47 Asphalt roofing
48 Built-up felt roofing
49 Plastering and rendering
50 Plasterboard dry linings
51 Plasterboard lightweight partitions and separating walls
52 Floor and roof screeds
53 Floor finishes
54 Painting and decorating

GENERAL

55 Defects in private house building
56 Diagnosis of structural failure
57 Diagnosis of damp conditions
58 Insulation improvements
59 Alterations and conversions (1)
60 Alterations and conversions (2)
61 Management of maintenance
62 Scottish practice
63 The metric system
64 Sources of building information

INDEX

CONTRIBUTORS

In this book, the views expressed by contributors are their own and not necessarily those of their firms or organisations.

B.H. ASPIN, BSc(Eng), CEng, MICE
Technical Director, New Ideal Homes Ltd (Sections 18, 19)

J.A. BAIRD, CEng, MIStructE, MWeldI, FIWSc
UK Managing Director, Swedish Finnish Timber Council (Section 34)

L.G. BAYLEY, BA(Com), MSc, FIOB, AMBIM, AIArb, FRSA
Lecturer in Building, University of Manchester Institute of Science and Technology (Section 13)

P. BRIGHAM, RIBA, MIOB, MBIM, MInstM
Construction Industry Marketing Group, Institute of Marketing (Section 4)

W.W.L. CHAN, BSc, PhD, DIC, CEng, FICE, FIStructE, FIWSc, MBIM
Consulting Engineer (Section 38)

J.K. CHAPMAN
Kent/Sussex Area Manager, National House-Building Council (Section 44)

C.R.I. CLAYTON, BSc(Hons), MSc, PhD, DIC, CEng, MICE, FGS
Senior Research Engineer, Ground Engineering Ltd (Sections 23, 24)

A.F. CONSTANTINE, CEng, MRAeS, HonMIAT
Director, Mastic Asphalt Council and Employers Federation Ltd (Section 47)

J.C. CORMACK, BSc(AppSc), CEng, MICE
Associate, Cottington Phillips and Associates, Consulting Engineers (Section 56)

J. COTTINGTON, CEng, FICE, MIMunE, FGS, FIArb, MConsE
Partner, Cottington Phillips and Associates, Consulting Engineers (Section 56)

E. DEARDEN, MIOB
North West Regional Director, National House-Building Council (Section 33)

D. FOSTER, DipArch, RIBA
Consulting Architect (Section 32)

J.F. GEORGE, BSc
Chief Executive, The Building Centre Group (Section 64)

D.A.S. GOODMAN, DipArch, FRIBA
Partner, Briars Goodman Associates, Chartered Architects and Industrial Designers (Section 45)

H. GRAHAM, FIMBM, FFB, MIOB
Formerly Assistant Director of Housing (Maintenance), Greater London Council (Section 61)

B. HARRISON, CEng, MICE, MIMC
Consultant, BAS Management Services (Section 2)

D.H. HAYES, CEng, FInstF, FIDHE, MIGasE, MCIBS
Consulting Engineer (Section 42)

J.D. HEALY, BSc, PhD
Senior Lecturer in Science, Vauxhall College of Building and Further Education (Section 63)

L. HILL, FIOB, FBIM
Chairman, R. Mansell Ltd (Section 59)

T.J. HOPPER, MInstPS
William Davis and Co (Leicester) Ltd (Section 3)

Mrs J.E. HUNTING
Librarian, The Building Centre Group (Section 64)

G.D. JOHNSON, BEng(Tech)
Structural Consultant (Section 32)

B.M. JONES, BSc, CEng, MIStructE
GKN Reinforcements Ltd (Section 28)

D.R. JONES, FCCA, FCIS, MIMC
Senior Consultant, BAS Management Services (Section 1)

N. KENT, MIOB
George Wimpey and Co. Ltd (Section 12)

C.D. KINLOCH, DFH, CEng, MIEE
Assistant General Manager, National Inspection Council for Electrical Installation Contracting (Section 41)

T.W. KIRKBRIDE, BSc, CEng, MICE
Technical Director, British Ready Mixed Concrete Association (Section 27)

J. LAWRIE, DSC
Vice-Chairman, Structural Insulation Association (Sections 43, 58)

K. LAWSON, CEng, MIMechE, MIArb
Construction Plant Consultant (Section 9)

F.J. McCANN, CEng, FIStructE, MICE, MIMinE, MSocCE(France)
Partner, McCann and Smith, Consulting Engineers (Section 29)

M. MANSER, DipArch, RIBA
Partner, Michael Manser Associates, Chartered Architects (Sections 14, 15)

J.H. MILAN, CEng, FIGasE
Sales Officer, Products and Utilisation, British Gas (Section 40)

M.R. MILLER, BArch(L'pool), MPhil, RIBA
Scottish Development Department (Section 62)

J. OLLIS, AADip, RIBA
Chief Architect, Timber Research and Development Association (Sections 35, 36)

M.I. OSPALA, BSc
Construction Industry Marketing Group, Institute of Marketing (Section 4)

P.H. PERKINS, CEng, FASCE, FIMunE, FIArb, FIPHE, MIWE
Senior Advisory Engineer, Cement and Concrete Association (Sections 31, 52)

M.J.V. POWELL, MSc, MIOB
Research Manager for Building, Construction Industry Research and Information Association (Section 55)

N.J. RAGSDALE, LIOB, FFB, AMBIM
Managing Director, H.A. Marks Ltd, Building Contractors (Section 60)

I. RANKIN, BSc(Hons), MPhil, MIOB
South East Regional Director, National House-Building Council (Section 48)

J.L. REYNOLDS
Lecturer in Painting and Decorating, Chelmer Institute of Higher Education (Section 54)

J.F. RYDER, BSc
Formerly Head of Limes and Plasters Section, Building Research Establishment (Section 49)

C.D.A. SCHOFIELD, MSc, AIQS, MIOB, AIArb
Consultant, BAS Management Services (Section 7)

J.G.A. SHEEHAN-DARE, DipArch, FIOB, MRSH
Director, Taylor Woodrow Homes Ltd (Section 11)

R.W. SMITH, MSAAT
Technical Literature Officer, British Gypsum Ltd (Sections 50, 51)

S. SMITH, CEng, MIStructE, MInstW
Partner, McCann and Smith, Consulting Engineers (Section 29)

D.A. STEPHENSON, BSc, CEng, FICE, FIStructE, FIArb
Consulting Engineer (Sections 20, 21, 22)

F.W. SUNTER, MRAIC, FRAIA, AIWSc
Architectural Consultant to the Council of Forest Industries of British Columbia (Section 37)

A.A. THOMPSON, MSAAT, LIOB
Lecturer in Construction Technology, Vauxhall College of Building and Further Education (Section 53)

E. THOMPSON, BSc
Director of Technical Services, National Federation of Building Trades Employers (Section 57)

M.J. TOMLINSON, CEng, FICE, FIStructE
Consulting Engineer; formerly Director, Wimpey Laboratories Ltd (Sections 25, 26)

T.A. TOMPSON, FRSH, FIPHE
Partner, T.A. Tompson and Associates, Consulting Engineers (Section 39)

J.K. TUNNICLIFFE
Director, Stewart Wrightson (Contractors Insurance Services) Ltd (Section 5)

D.F. TURNER, BA, FRICS, AIQS, MIOB
Senior Lecturer in charge of Building Economics Section,
Robert Gordon's Institute of Technology, Aberdeen (Section 6)

G. WHITE, FIOB, AMBIM
Principal Lecturer in Building Management Studies,
Guildford County College of Technology (Section 10)

A.J. WILKINS
Information Officer, The Building Centre Group (Section 64)

O.D. WILLMORE, DipArch, RIBA
Partner, Barton Willmore and Partners, Architect Planners (Sections 16, 17)

C.J. WILSHERE, BA, BAI, CEng, FICE
Manager, Temporary Works Department,
John Laing Design Associates Ltd (Section 30)

Miss R.I. WYLES, BA(Hons), Barrister
Legal Adviser, National House-Building Council (Section 8)

R.H. YOUNG
Redland Roof Tiles Ltd (Section 46)

1 FINANCIAL MANAGEMENT

BASIC KNOWLEDGE AND REQUIREMENTS 1–2

COST CONTROLS 1–13

ADMINISTRATION 1–33

TAXATION 1–36

EVALUATING A BUSINESS 1–39

FINANCIAL MANAGEMENT 1

1 FINANCIAL MANAGEMENT

D.R. JONES, FCCA, FCIS, MIME
Senior Consultant, BAS Management Services

This section outlines the basic accounting knowledge required by a builder, explains how to understand a set of annual accounts, gives details of typical cost control systems, and concludes with consideration of the funding of a company and taxation.

BASIC KNOWLEDGE AND REQUIREMENTS

Importance of financial management

House construction falls into two distinct classes:
- *contracts* to build for Local Authorities and other clients, on their land;
- *private development* carried out by a developer on his own land.

The difference between them is sufficiently substantial to cause many of the larger contractors to use separate departments, and separate personnel, for each of the two classes.

Contracts are obtained by pricing Bills of Quantity and tendering in open competition for the work. The margins included in tenders in good times for general overheads and profit are therefore small, and allow little room for error. In times of recession the margins are reduced further, even to a point at which profit is eliminated. Thus the emphasis in contracting is to carry out work within the planned (estimated) price. To this end techniques have been developed for:
- pre-tender planning and method study;
- analytical estimating, which shows separately the labour, materials, plant etc. cost for each Bill item;
- overall and short-term planning, and weekly programmes;
- productivity bonusing of labour;
- site cost controls over materials, labour, sub-contractors, plant, preliminaries, site overheads, etc.

In private development there is equal need for using the foregoing techniques in the interests of maximum cost efficiency and profits. However, private developers have frequently in the past adopted a more relaxed, less demanding attitude towards estimating, costing and cost controls, for two reasons. First, gross profit margins on private development have been substantial compared with those on contract housing (mainly due to the effects of inflation on old land stocks). Secondly, most emphasis has been placed on designing and marketing the product, on gearing the speed of production to the probable demands of the house-buying public, and on cash flow.

The situation is changing, mainly due to the Community Land Act and Development Land Tax. Gross profit margins in private development are likely to diminish progressively as old land banks become worked out. It will become as important in future to exercise tight estimating and production and cost-control procedures in private development as in contract housing.

The role of top management is concerned mainly with money and people. Professional, technical and trade skills can be recruited, and as a business grows in size it becomes increasingly necessary for top management to delegate to others more duties and responsibilities. It is a primary function of top management to select the right team in types, quality and numbers, to ensure that staff is properly integrated, motivated and remunerated, and to earn and retain the loyalty and cooperation of staff.

In return for the loyalty, integrity and efficient services of staff of all levels, top management should provide conditions of employment that are stable, continuous and progressive. Top management can do this only if it knows how to handle money in all its contexts, such as:

- wise borrowing, at reasonable rates of interest, and protected from sudden withdrawals of support by the lenders;
- obtaining up-to-date control information regularly and using it efficiently;
- turning over capital employed as many times each year as possible, whether the company's own funds or borrowed monies;
- forward financial planning in the short (monthly), medium (six-monthly) and long (several years ahead) terms.

Accountants within a business, or acting as external advisers, may be but are not necessarily entrepreneurs. Accountants should provide to top management essential information without which fully effective financial management would be impossible. It is a function of top management to interpret the results produced by accountants: they should do so in the spacious and forward-looking manner that is essential to business success, knowing that undue caution is inhibitive to that success, but carefully calculating and minimising the risks involved.

Basic requirements for financial control

The first requirement is that information should be provided easily and cheaply to top management each month, in order to reveal the overall profitability of each site and of the company as a whole. This applies to both contracting and private development, and includes:

- an overall cost/value comparison for each site, in which a cost within ½ per cent accuracy can be made available any day during a month it is required for comparison with a surveyor's valuation of the work done, completed that same day;
- a trading and profit and loss account for the business as a whole, which should become available in the first week of each month in respect of the preceding month's results.

With the aid of these, top management can each month:

- check the overall performance at each site, and see how actual results compare with those planned;
- see strengths to be exploited and weaknesses to be corrected;
- up-date its overall financial plans;
- and most importantly (as will be discussed later) keep bankers and other providers of funds fully informed each month about the current state of the business, and ensure their maximum cooperation.

Although they should take second place, other financial control requirements are also important. They are to identify:

- actual labour performances at sites, compared with those estimated;
- actual materials usage, wastage and prices compared with estimates;
- actual preliminaries and site overhead costs compared with estimates;
- actual sub-contract costs compared with estimates.

By these means the reliability of estimates and the performances of site management, planners, bonus surveyors, buyers etc. can be checked.

Cost of costing

Routine information of the kinds described above can be produced very cheaply and easily by unsophisticated office staff if the proper systems are used and simple routines are followed faithfully. However, some methods of costing can be expensive; it is important to decide on a costing system that will fulfil essential needs at the least possible cost in time, effort and money.

A clerk receiving £2500 per annum, plus labour overheads, working at 70 per cent efficiency, can cost about £2.46 per hour of 'worked time' (excluding general office overheads), i.e.

£2500 plus 13.5% National Insurance	= £2837.50
Hours worked for 47 weeks, after allowing for annual and public holidays, sickness, etc. (35 hour week)	= 1645
70 per cent efficiency to cover natural and other breaks (1645 × 0.7)	= 1152
Cost per worked hour equals $\frac{£2837.50}{1152}$	= £2.46

At these rates it is important to simplify essential clerical work, and to eliminate the unnecessary.

First principles of accounting

LEGAL NECESSITY FOR ACCOUNTS

An individual in business, a partnership or a company is required by the Inland Revenue to produce accounts each year so that tax due can be calculated from them. A company is also required to provide a set of accounts each year to the Registrar of Companies. These accounts consist mainly of:

- Trading Account
- Profit and Loss Account
- Appropriation Account
- Balance Sheet

see pages **1**–7 to **1**–13

Also, all businesses (individuals and companies) are required by law to keep book-keeping records for purposes of VAT, wages, and the new Tax Deduction Scheme.

CASH BOOKS AND PETTY CASH BOOKS

An accounts system commences with a Cash Book, which is virtually the same as a Bank Statement, showing payments into and withdrawals from the Bank Account. The balance at any date is the amount of money in the account, or the amount of the overdraft. An example is shown below.

Cash Book (i.e. Bank Receipts and Payments Book)

	Debit (Dr)			Credit (Cr)	
	RECEIPTS			PAYMENTS	
1979		£	1979		£
Jan 10	Tom	600	Jan 1	Balance (O/D)	1300
17	Dick	900	17	Smith	900
24	Harry	1000	24	Jones	500
31	Balance (O/D)	1500	25	Robinson	1100
			30	Brown	200
		£4000			£4000

Notes 1. A similar Petty Cash Book is usually maintained in addition, to record minor payments in cash rather than by cheque.
2. O/D means 'overdraft' or 'overdrawn'.

BANK STATEMENT

The Bank Statement for the above business might appear as follows:

		PAYMENTS	RECEIPTS	BALANCE	
		Debit	Credit		
1979		£	£		£
Jan 1	Balance			DR	1300
15	Tom		600	DR	700
22	Dick		900	CR	200
	Smith	900		DR	700
26	Harry		1000	CR	300
	Jones	500		DR	200
28	Robinson	1100		DR	1300

Note The debits and credits in the Bank Statement reflect entries in the bank's books, and are the reverse of the entries in the company's books.

BANK RECONCILIATION

The Cash Book shows an overdraft of £1500. The Bank Statement shows an overdraft of £1300. The difference between the two is the £200 cheque sent to Brown on 30 January, which had not been presented to the Bank by the end of the month. The adjustment of the Bank Statement to make it agree with the Cash Book is called a 'Bank Reconciliation'. As far as the business is concerned, the Cash Book (not the Bank Statement) is correct.

ANALYSIS OF RECEIPTS AND PAYMENTS

The next step in a firm or company's books is to analyse *receipts* to distinguish sales of properties, initial deposits on properties, receipts from contracts, sales of plant etc., and to analyse *payments* into four main categories:

- *prime costs of work carried out,* such as site labour, materials, sub-contractors, and plant and transport hire;
- *general overheads,* including directors' and staff's salaries and expenses, their car running expenses, rent, rates, printing, stationery, telephone, etc.,
- *capital items,* such as the purchase of land for development, the costs of new or second-hand plant, cars, lorries, vans, office furniture, office machines, etc.,
- *financial items* such as dividends, interest and Corporation Tax.

ADJUSTMENTS TO ACCOUNTS

The foregoing cash system is as far as the accounts go in local government (and central government). In private enterprise the accounts take an important step in the direction of realism. As at the closing date of the financial period concerned (year-end, month-end, etc. as the case may be):

- add to revenue received in cash the value of work done for which no cash had yet been received (work-in-progress) plus unsettled sales invoices;
- add to expenditure paid the value of goods received and services rendered, which had not yet been paid for to the suppliers, sub-contractors, etc.

Then deduct the corresponding revenue and expenditure provisions made in the previous set of accounts.

Finally, an adjustment is made in the accounts for the value used up (i.e. *depreciation*) in the accounting period by:

- major plant, lorries, buses, cars and vans,
- leaseholds,
- office machines, furniture and equipment.

For example, supposing a 5/3½ mixer costs £600, and has a working life of five years. The £600 when it is paid is not charged as expenditure against profits: it is placed instead into the capital account. Each year one-fifth of the cost of the mixer (£120) is entered as though it was expenditure, which thus reduces profits. The same £120 is taken from the value of the mixer in the Balance Sheet, which after the first year would appear as £480 instead of £600, £360 at the end of the second year, and so on. The same procedure is applied to all other items of plant, vehicles, etc. until the whole cost of each is written off.

Five years is a typical but not a standard period for writing-off assets. Top management decides the write-off period. The Inland Revenue is not concerned with top management's decisions: it uses its own rules, which are called 'capital allowances' (see page **1**–37)

ACCRUALS ACCOUNTING

Thus the accounts of a commercial undertaking do not merely show increases or decreases in bank balances. By reason of the adjustments made to them outlined above they reflect:

- true revenue earned in the accounting period (not merely the cash received in the period);
- true costs incurred in the accounting period (not only the cash paid out in the period);
- the difference between true revenue and true costs, i.e. a realistic profit or loss for the accounting period.

Increases or decreases in bank balances can mean many things quite unrelated to profitability, such as land purchases greater than land sales (or vice versa), increases or decreases in stocks of houses, the purchase of expensive equipment, and many other possibilities.

This system of accounting is called 'accruals accounting', and it is insisted upon by the Inland Revenue and all who are concerned with realistic accounting.

OTHER ACCOUNT BOOKS

A business, in addition to a Cash Book and Petty Cash Book, usually has other books of financial account, such as:

- *Purchases Journal* to record individual invoices from suppliers, and to show VAT paid to them;
- *Purchases Ledger*, for compiling an account of monies due to, and paid to, each supplier;
- *Sub-contractors Journal*, to show amounts due to sub-contractors, VAT paid to them, and tax deductions from them;
- *Sub-contractors Ledger,* for the individual account of each sub-contractor;
- *Sales Journal,* to record particulars of sales invoices rendered to clients, and the VAT included in those invoices;
- *Sales Ledger,* containing particulars of amounts due from, and received from, clients;
- *cost books*, showing the costs of each site (labour, materials, plant, haulage and sub-contractors).

Finally, and most importantly, there is a nominal ledger (also sometimes called a 'private' or 'general' ledger). This book contains a summary of all of the entries in all other books.

DOUBLE–ENTRY BOOKKEEPING, AND TRIAL BALANCE

Bookkeeping is 'double-entry', which means that all entries on the one side (the debit, or left-hand side of the nominal ledger) must have entries for corresponding amounts on the other side of the nominal ledger (the credit or right-hand side).

It follows that if all the balances of the ledger are extracted, the total of the right-hand (debit) balances must agree exactly with the left-hand total (the credit balances). This is called a Trial Balance. If the debit and credit sides of the Trial Balance do not coincide an arithmetical error has occurred, which must be traced and corrected.

Double-entry bookkeeping means that no alteration can be made to any figure on one side of the books without making a compensating adjustment to the other side of the books (or making a corresponding adjustment to another figure on the same side).

Producing a 'set of accounts'

A 'set of accounts' is produced, up to a defined date, from a final Trial Balance, i.e. after entries have been made in the books to adjust them as at that date for:

- payments made in advance (prepayments) of vehicle licences, general insurances, rates, deposits, etc.,
- amounts owing to suppliers and sub-contractors for goods provided and services rendered by them;
- depreciation (wear and tear for the month, quarter or year for the used-up portions of capital items such as plant and machinery, vehicles and office equipment and furniture);
- work done for clients, whether complete or incomplete, for which monies have not been received (work-in progress, and debtors for final accounts unpaid).

A set of accounts for a construction company consists of:

- Trading Account (*Figure 1.1*), which contains 'sales' and the 'prime costs of sales', leaving a balance of Gross Profit;
- Profit and Loss Account (*Figure 1.1*) containing overheads, which when deducted from Gross Profit leave a 'Net Profit before tax';
- Appropriation Account (in companies; *Figure 1.2*), which deducts from Net Profit the amounts of tax and dividends;
- Balance Sheet (*Figure 1.3*).

PERIOD AND SUMMARY OF ACCOUNTS

Trading, Profit and Loss and Appropriation Accounts relate to a period, such as a month, quarter or year. They show for the period of the accounts (taking *Figures 1.1 and 1.2* as our example):

Trading Account	£
Value of work carried out (whether completed or in progress)	2 000 000
less the *Prime Costs* of doing that work	1 740 000
equals the Gross Profit (*Gross Margin*) earned	260 000

		£
Profit and Loss Account		
less General Overheads		164 400
gives the *Net Margin* (Net Profit before tax and dividends)		95 600
Appropriation Account		
less *Corporation Tax*	43 800	
Dividends paid to shareholders	17 000	60 800
leaves 'after tax' net profits (*Retained Profits*)		34 800
Add retained profits accumulated over previous years		45 200
to give an accumulated *Profit and Loss Account balance* to date of		80 000

Trading and Profit and Loss A/C for the year ended 31 March 19..

Dr.				Cr.
Prime Costs:	£	*Sales (turnover):*	£	£
Wages (direct and 'labour only')	334 286	final A/Cs rendered in the year	1 800 000	
Materials/services	371 429	*Add* closing work-in-progress at valuation	1 400 000	
Sub-contractors	780 000			
Plant and lorry operating costs and depreciation	74 285		3 200 000	
Land consumed	180 000	*Deduct* opening work-in-progress at valuation	1 200 000	2 000 000
Sub-total	1 740 000			
Gross Profit c/d	260 000			
	2 000 000			2 000 000
General Overheads:		Gross Profit b/d		260 000
Salaries of directors and staff 105 462		Sundry revenue from rents, discounts, sales of scrap, etc.		3 000
Wages (storekeeper), small plant, stores losses, apprentices, etc. 20 088				
Car running costs and depreciation 11 718				
Premises, rent, rates & upkeep 8 370				
Printing, stationery, posts, telephone, audit, bank charges, subs and general expenses 21 762	167 400	(Overheads: £167 400 less £3000 equals £164 400 net)		
Net Profit before tax	95 600			
	263 000			263 000

Figure 1.1 Typical Trading and Profit and Loss Account (mixture of general building construction, private development and small works)

Appropriation Account for the year ended 31 March 19..		
		£
Net Profit for the year, brought forward from Profit and Loss Account (Note 1)		95 600
Deduct *Corporation Tax* (Note 2)		43 800
		51 800
Deduct *Dividends paid or payable:*	£	
£200 000 Ordinary Shares @ 5%	10 000	
£100 000 Preference Shares @ 7%	7 000	17 000
Balance		34 800
Add *Balance of Profit and Loss account* brought forward from last year		45 200
Balance as at 31 March 19.., as shown in Balance Sheet		80 000

Note 1. It is customary to show in the Appropriation Account, instead of in the Profit and Loss Account: Directors fees, Audit fees, Depreciation.
2. The amount of Corporation Tax payable is after deducting capital allowances made by the Inland Revenue, and Stock Relief.

Figure 1.2 Typical Appropriation Account

The final Profit and Loss Account (Appropriation Account) balance of £80 000 is entered in the Balance Sheet, which is a necessary part of the double-entry system.

After entering into the Trading, Profit and Loss and Appropriation Accounts all the items relating to them, the remaining items in the final Trial Balance (together with the Appropriation Account balance) must be shown in the Balance Sheet.

BALANCE SHEET PROOF

If the foregoing procedure is properly carried out, both sides of the Balance Sheet *must* add up to the same figure. This proves the arithmetical accuracy of the entire bookkeeping processes for the year (or shorter period). If the two Balance Sheet totals do not coincide, there will be an error to be traced.

BALANCE SHEET DEFINITION

Whereas Trading, Profit and Loss and Appropriation Accounts cover a period of a company's operations (e.g. month, quarter or year) the Balance Sheet reveals the financial position of the company as on a certain day (the date of the Balance Sheet). It shows (taking *Figure 1.3* as our example):

	£
Fixed Assets, at cost prices, such as land and buildings owned and occupied by the company, and plant and machinery, vehicles, office equipment and furniture, etc. owned by the company	220 000
Less *Depreciation* of the above, written-off against profits, not only for this accounting period but also in earlier accounting periods	40 000
Sub-total	180 000
Add the present value of small plant and loose tools owned by the company	20 000

to give the *net current value in use of Fixed Assets* owned by the company		200 000
Add *Investments* (shares owned by the company in other companies)		10 000
Sub-total		210 000
Add *Current Assets* (land bank, materials stocks, work-in-progress, debtors, cash balance, etc.)	450 000	
Deduct *Current Liabilities* (due to suppliers and sub-contractors, owing on bank overdraft, etc.)	300 000	
giving *Net Current Assets* (working capital)	150 000	150 000
Sub-total		360 000
Deduct *Monies Borrowed* from outsiders		60 000
Balance representing the *Net Worth* of the company, also known as Proprietors' Capital, or *Capital Employed* or *Net Assets*		300 000

Analysis of the Balance Sheet

The Balance Sheet shown in *Figure 1.3* has the same *content* as that described above, but the figures, although the same in amount, are displayed in a *horizontal* rather than a vertical form.

A Balance Sheet has several functions:

- It proves the arithmetical accuracy of bookkeeping processes, when both sides (Assets and Liabilities) total to exactly the same amounts (all figures having been extracted from the 'ledger').
- It is a statement of a company's affairs on a specific date, i.e. the date of the Balance Sheet. It does *not* cover a period such as a month, quarter or year. However, the differences between two Balance Sheets cover the movements of assets and liabilities over the period between the dates of those Balance Sheets.
- It is a necessary legal requirement (by the Inland Revenue and Company Registrar).
- It is a guide to management policy.

'Capital Employed' is susceptible to various definitions, e.g. it is used in the context of:

- Net Worth (£300 000), which is the Shareholders' Funds, or
- Shareholders' Funds (£300 000) plus borrowed money (£60 000) equals £360 000, or
- the total of the Balance Sheet (£660 000).

'Net Worth' is the most common definition of Capital Employed. Net Worth is the value of the business as a 'going concern' when all assets and liabilities are properly valued. The 'breakdown value' of a business could be much lower, i.e. stocks, vehicles, plant, etc. could fetch very little on the market in a forced sale, though their 'value-in-use' in a business may be considerable.

The book value of an ordinary share is calculated as follows:

Net Worth	£300 000
Deduct Preference Shares	£100 000
100 000 Ordinary Shares	£200 000

100 000 Shares @ £1 nominal value (£100 000) divided into £200 000 equals £2 each.

Balance sheet as at 31 March 19..

Liabilities (where the money came from)

		£ Authorised	£ Issued
A	*Share capital*		
	7% Cumulative Preference Shares @ £1 each	100 000	100 000
	Ordinary Shares @ £1 each	200 000	200 000
		300 000	200 000
	Capital Reserve		20 000
	Revenue Reserves & Profit and Loss Account		80 000
	Net Worth, Net Assets, or Capital Employed		300 000
B	*Long-Term Liabilities*		
	8% Mortgage Debenture	50 000	
	Medium-Term Liabilities		
	Loans received	10 000	60 000
	External Borrowings plus Net Worth		360 000
C	*Current Liabilities*		
	Bank overdraft	92 000	
	Trade creditors	52 000	
	Subcontractors	100 000	
	Sundry creditors, dividend, current taxation	56 000	300 000
	Total Capital Employed		£660 000

Assets (where the money is)

		£ Cost	£ Depreciation	£ Net
D	*Fixed Assets*			
	Freehold land and buildings	30 000	(+) 20 000	50 000
	Plant & machinery	100 000	26 000	74 000
	Motor vehicles	60 000	20 000	40 000
	Office equipment and furniture	30 000	14 000	16 000
		220 000	40 000	180 000
	Small plant and loose tools			20 000
	Total Fixed Capital			200 000
E	*Investments*			
	Outside the business			10 000
F	*Current Assets*			
	Stocks: land		40 000	
	timber	5 000		
	petrol & oil	1 000		
	general stores	14 000	20 000	
	Work in progress at valuation	1 200 000		
	Less cash received on account	900 000	300 000	
	Sundry debtors and prepayments		88 000	
	Cash-in-hand		2 000	450 000
				£660 000

Figure 1.3 Typical Balance Sheet

Stock-exchange share values are based on the dividend-earning capacity of a company (P/E ratio : see below). The 'yield' on a share is calculated as:

$$\frac{\text{nominal value} \times \text{dividend}}{\text{market price}}$$

$$\text{e.g.}\ \frac{100\text{p} \times 20\%}{250\text{p}} = 8\%$$

Preference Shares usually bear a fixed rate of interest. They may be:

- cumulative (interest accumulates as a debt if unpaid);
- non-cumulative;
- redeemable (the share money can be repaid to shareholders);
- non-redeemable;
- participating (providing a share of profits, as well as a fixed rate of dividend);
- or a combination of some of these.

Rules regarding Preference Shares are provided in either the Memorandum or Articles of Association, usually the latter. Unless otherwise stated, they are preferential over Ordinary Shares as to both repayment of capital and payment of their fixed dividend.

'Working capital' (£150 000) is the difference between current assets and current liabilities, i.e. £450 000 and £300 000 equals 1.5 to 1. This is a good ratio. A ratio of less than 1 to 1 indicates a possibility of trouble, or even disaster.

Working capital is required to pay for wages, materials, sub-contractors and other operating costs, plant running costs and general overheads. Money spent on fixed assets (£200 000 net), development land (£40 000) and stocks (£20 000) will deplete the liquid funds in working capital.

'Authorised Capital' (£300 000) is the amount on which stamp duty has been paid to the Company Registrar. 'Issued Capital' (£200 000) is the amount held by shareholders.

A Capital Reserve (£20 000) is usually created by uplifting the value of a fixed asset, such as land, or by creating an asset, such as Goodwill.

Long-term liabilities (£50 000) usually mean loans not repayable for three years or more. Medium-term liabilities (£10 000) are repayable in one to three years. Current liabilities (£300 000) are repayable within the coming year.

In a construction company Goodwill rarely has any value, and is usually omitted from Balance Sheets. 'Revenue Reserves and Profit and Loss Account' (£80 000) means much the same thing as a general rule. They indicate the profits after tax left in the business by shareholders over the years, to strengthen it rather than pay out those profits in the form of dividends.

Primary financial ratios

The primary financial ratios obtainable from the Trading and Profit and Loss Account are:

$$\%\ \text{Gross Profit on Turnover:}\quad \frac{\pounds 260\,000 \times 100}{\pounds 2\,000\,000} = 13\%$$

$$\left.\begin{array}{ll} \%\ \text{Overheads on Turnover:} & \dfrac{\pounds 164\,400 \times 100}{\pounds 2\,000\,000} = 8.22\% \\[2ex] \%\ \text{Net Profit on Turnover:} & \dfrac{\pounds 95\,600 \times 100}{\pounds 2\,000\,000} = 4.78\% \end{array}\right\}\ 13\%$$

These ratios are overall, for a company carrying out private development housing, general building contracts and small works. The separate ratios of gross and net profits, and of general overheads, would be very different for each of these three activities.

Price earnings (P/E) ratio

Stock exchange price of Ordinary Shares		£2.50 each
Number of Ordinary Shares (*Figure 1.3*)		100 000
Total worth of Ordinary Shares, therefore, is 100 000 @ £2.50		£250 000
Earnings available to ordinary shareholders (from *Figure 1.2*):	£	
Total earnings (Net Profit after tax)	51 800	
Less Preference Dividend	10 000	
Available for ordinary shareholders	41 800	£41 800

Therefore the P/E ratio $= \dfrac{250\,000}{41\,800} = 5.99/1$

COST CONTROLS

Standards for comparison

If it is proposed during the progress of the work to compare actual costs with those planned, it is axiomatic that there must be reliable predetermined standards (i.e. proper estimates) with which the actual costs, when they become known, can be compared.

Proposed selling price of a dwelling

Figure 1.4 shows a proposed selling price for each of two different types of bungalow, the 'Derby' and the 'Oaks', to be constructed as part of a group of six units at Epsom Downs, as Phase 2. These selling prices consist of a summary of the estimated costs of:

- land,
- roads, sewers, footpaths and kerbs,
- variable site overheads,
- site works,
- bungalow construction costs,
- garage construction costs,
- advertising and selling expenses,
- legal costs of sale,
- general overheads and profit.

Merged and separated costs

It is not uncommon to see all or most of the costs listed above merged into one conglomerate cost record. In such cases, if the final actual results differ substantially from the estimated costs, there is no means for determining where and how the discrepancies arose.

Schedule A: Pricing of six bungalows, Epsom Downs (Phase 2) Units Nos. 7–12 Contract No. 1096					
Schedule	*Cost codes*		*Total*	*Per plot, 'Derby' type*	*Per plot, 'Oaks' type*
			£	£	£
	99	Land, at cost	22 200	4 100	3 500
B	50/59	Roads, sewers, footpaths and kerbs	15 000	2 500	2 500
C	60/69	Variable site overheads	6 000	1 000	1 000
D	70/96	Site works	9 000	1 500	1 500
E	1/49	Bungalow construction:			
		Two 'Derby' @ £8832	17 664	8 832	
		Four 'Oaks' @ £8 300	33 200		8 300
F	1/49	Garages construction:			
		Six double @ £2 310	13 860	2 310	2 310
	98	Advertising and selling expenses	3 000	500	500
	97	Legal costs of sale	1 800	300	300
			121 724	21 042	19 910
		General overheads and profit	40 576	7 014	6 637
			162 300	28 056	26 547
				(56 112)	(106 188)
		Selling prices		28 000	26 500

Figure 1.4 Calculation of proposed selling price

Schedule B: Roads and Sewers (Bill of Quantities) *Site:* Epsom Downs, Phase 2: 6 Bungalows: Contract No. 1096				
Code	*Element*	*Quantity*	*Rate*	*Amount*
				£
50	Excavation for roads and sewers			
51	Kerbs			
52	Manholes			
53	Lean mix/hardcore			
54	Surface water drains			
55	Sewers			
56	Service crossings			
57	Footpaths			
58	Road surfacing			
			Total	15 000
			Per bungalow	2 500

Figure 1.5 Bill of Quantities for roads, sewers, footpaths and kerbs

It is recommended that, as far as practicable, each of the cost headings above should form a distinct cost centre. By this means it should become possible to isolate deviations between actual and planned expenditure.

Land Stock Account and land costs per unit

The calculation of land costs per plot is done by ascertaining the total cost of the land to the developer (including legal costs of acquirement, and the costs of any improvements carried out since purchase) and then dividing this total cost by the number of plots. On some sites the costs may be allocated to plots at different amounts, dependent on size, position, etc.

Most developers:

- maintain an asset account in the nominal ledger for the land cost of each site;
- do not take any profit on land sales (the profit is taken on construction work);
- at the time a plot is sold, *debit* the Cost of Sales Account: Land and *credit* the Land Stock Account with the cost value of the plot sold.

Thus at all times there remains in the nominal ledger for each site a balance of unsold land, at cost price, as below:

Land Stock Account: Phase 2, Epsom Downs

DR						CR
1978		£	1978			£
Jan 1	Balance in stock	22 200	Aug 20	Plot 7 Cost of sales		4 100
	(Plots 7–12)		Sep 12	Plot 8 Cost of sales		4 100
			26	Plot 9 Cost of sales		3 500
			30	Balance in stock c/d		
				Plot 10	3 500	
				Plot 11	3 500	
				Plot 12	3 500	10 500
		22 200				22 200

This system is straightforward where land is held in stock, and shown in the Balance Sheet, at 'cost or market value, whichever is the lower'. Where profit is added to land values, however, do not include this in the Land Stock Account, which should remain as above, but for all sales *debit* the Client (Purchaser's) Account and *credit* the Sales Account: Land with the selling price of the plot, whether with or without profit on land.

Roads, sewers, footpaths and kerbs

There should be a Bill of Quantities prepared for the above costs and others listed in *Figure 1.5*. The total costs in the Bill of Quantities should be allocated to plots.

A separate Works Order number (cost centre) should be given for roads, sewers, footpaths and kerbs, and this Works Order should be costed separately. Some builders will also wish to code costs into the categories 50 to 58 (see *Figure 1.5*).

Variable Site Overheads and their incidence

Variable Site Overheads are listed in *Figure 1.6*. A distinctive feature of these overheads is 'time'; the longer the site is in progress, the greater the cost per unit, e.g.

Number of units to be constructed	6
Cost of Variable Site Overheads for six months @ £1000 per month	£6000
Cost for each of six units over a six months construction period	£1000
Time extended to	9 months
Variable Costs at £1000 per month for nine months	£9000
Cost for each of six units over a nine months construction period	£1500

Thus an extension of construction time as in the foregoing example could reduce the profits on each unit by £500.

There should be a separate Works Order number (cost centre) for Variable Site Overheads for each site (or phase), and the costs might be coded in the series 61 to 69, as shown in *Figure 1.6*.

Site Works

Figure 1.7 is a sample list of Site Works (other than roads, sewers, footpaths, kerbs and variable site overheads). There should be for these:

- a Bill of Quantities;
- a Works Order Number, i.e. cost centre with code numbers 70 to 99, as shown in *Figure 1.7*.

Building construction: materials take-off schedules

The first step in arriving at construction estimates is to take off materials quantities from the drawings for each unit type in the fullest detail. When this is done a completed schedule can serve for all dwellings of the same type, for all sites. It thus forms a standard list, which can be stencilled or photocopied as many times as necessary, for the purposes of:

- estimating the materials cost of a unit (and subsequent revisions of estimates);
- preparing standard schedules for attaching to purchase orders (to save rewriting details on the orders).

An example of a take-off schedule is provided in *Figure 1.8*. Some builders shy away from the task of producing fully detailed materials schedules of this kind, but it is easier in the long run to prepare them, and avoids the risks inherent in sporadic take-offs at head office and/or sites, often undertaken as a rush job, and frequently involving duplicate, triplicate and more bites at the same task.

Order schedules

Once a detailed materials schedule is made available, standard ordering schedules can be prepared from it. These have the following advantages:

- They save the repeated work of detailing particulars of materials on order forms, and providing delivery schedules. (A standard schedule can be attached to a purchase order form.)
- They can be priced up at estimated rates, and again at order rates. The difference between the two will show a favourable or adverse variance between estimate and order prices, and those variances can be listed by the buyer.
- They can serve to check deliveries.
- They can be used for valuations of 'materials on site' at times of monthly valuations.

Figure 1.9 shows a procedure for carrying out these tasks.

Schedule C: Variable Site Overheads Budget *Site:* Epsom Downs, Phase 2: 6 Bungalows: Contract No. 1096							
	Total	*Jun*	*Jul*	*Aug*	*Sep*	*Oct*	*Nov*
	£	£	£	£	£	£	£
61 *Staff:* foreman	2000	100	400	400	400	400	300
site labour	1200	60	240	240	240	240	180
watchman	1200	60	240	240	240	240	180
62 *Transport:* van	500	25	100	100	100	100	75
63 *Hutting, fencing & notice boards:*							
internal hire charges	400	20	80	80	80	80	60
65 *Plant:* internal hire:							
Mixer	200	10	40	40	40	40	30
Scaffolding	100	20	40	40			
Dumpers, forklift and miscellaneous	220	5	45	45	45	45	35
66 *Telephone*	75	5	15	15	15	15	10
67 *Water*	30	5	5	5	5	5	5
68 *Power*	75	5	15	15	15	15	10
£	6000	315	1220	1220	1180	1180	885

Figure 1.6 *Variable site overheads budget*

Schedule D: Site Works (Bill of Quantities) *(other than roads, sewers, and site overheads)* *Site:* Epsom Downs, Phase 2: 6 Bungalows: Contract No. 1096				
Code	*Element*	*Quantity*	*Rate*	*Amount*
	Site : General			
70	Access and right of way			
71	Tree felling			
72	Site clearance			
73	Work to existing roads			
74	Demolition			
75	Land drains			
76	De-watering			
77	Gas main			
78	Water main			
79	Electricity supply main			
80	Street lighting			
81	Grassed areas			
82	Tree planting			
	Curtilages			
90	Retaining walls			
91	Fencing and gates			
92	Drives, paths and steps			
93	Gardens/landscaping			
			Total	£9000
			Per bungalow	£1500

Figure 1.7 Bill of Quantities for site works

Schedule 10: materials take-off				
Description	*Unit*	*Qty*	*Rate*	£ p
CARCASSING TIMBER (cont'd)				
Canopy (cont'd)				
50 mm × 50 mm softwood triangular fillet for felt 2.350 m long	No.	1		
75 mm × 50 mm studding				
50 mm × 38 mm grounds for pipe box casing 2 m long	No.	4		
250 mm × 35 mm window boards, bullnose one edge, rebated other edge:				
3 m long	No.	2		
2 m long	No.	2		
1.500 m long	No.	2		
750 mm long	No.	1		
50 mm × 20 mm architrave:				
3.050 m long	No.	2		
2.440 m long	No.	32		
1 m long	No.	10		
750 mm long	No.	6		
IRONMONGERY (cont'd)				
Mortice latch furniture	No.	8		
300 mm casement stays	No.	4		
250 mm casement stays	No.	5		
BS salt-glazed stoneware				
100 mm channel pipe	LM	14		
100 mm ½-section channel bend	No.	4		
100–150 mm channel taper pipe	No.	3		
100–150 mm channel taper bend	No.	7		
150 mm channel pipe	LM	23		
150 mm ½-section channel bend	No.	6		

Figure 1.8 Extracts from twenty pages of materials take-off schedule

Materials variance analysis

As explained above, standard order schedules can be used to disclose variances between 'order prices' and 'estimated prices'. Once a standard order has been placed, it prescribes the quantities of materials required for a particular unit. If an additional order becomes necessary for that unit, it will be a result of excessive wastage, or theft, or extra works. No orders should be placed by the buyer for such additional quantities without:

- prior consultation with the contracts manager;
- notification to the person responsible for raising charges for extra works.

The buyer should make a list of additional orders placed, and their cost, i.e. a Quantity Variance Schedule.

Summary of construction costs

Figure 1.10 shows a Summary of Construction Costs for garages for the 'Derby' type bungalow. The materials column is completed from the order schedules described above (priced-up at estimated prices); see also *Figure 1.9*.

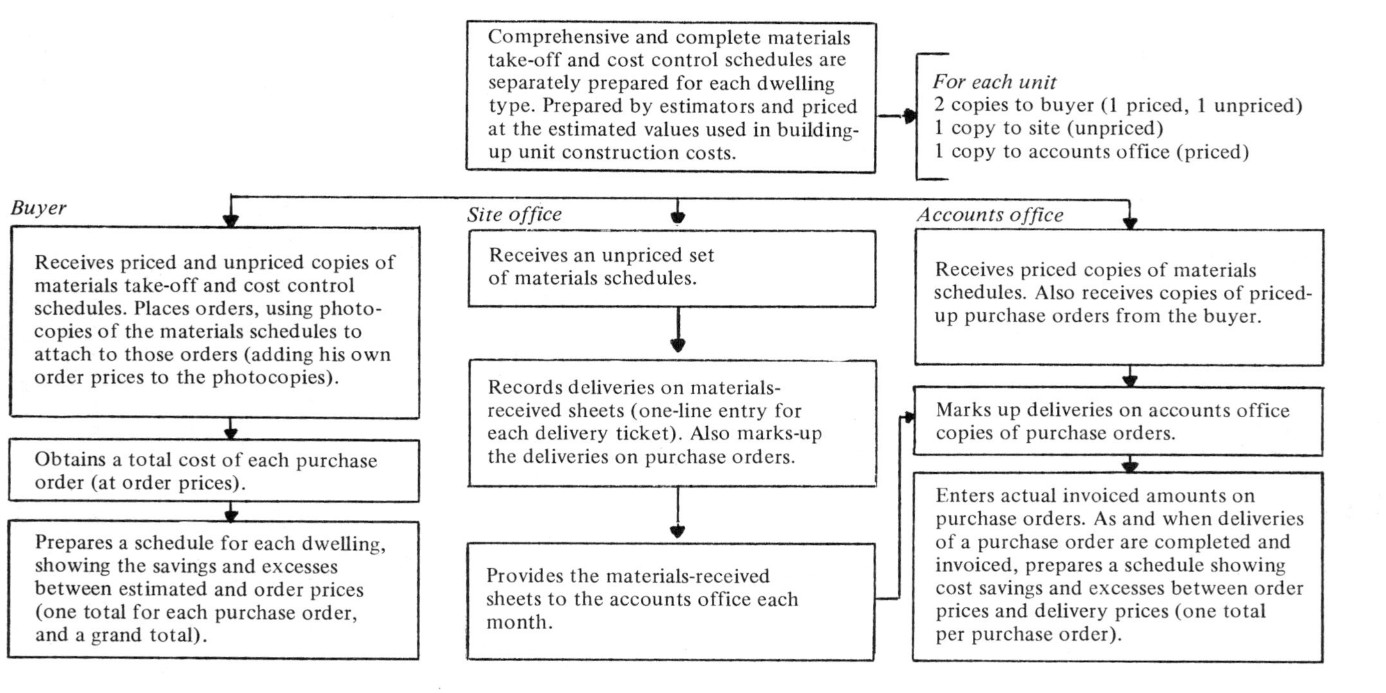

Figure 1.9 Work flow for materials take-off schedules and materials cost control

Schedule E: Summary of Construction Costs: six double garages

Code	Element	Labour	Materials	Sub-contractors	Total
01	Excavation	90	–	–	90
02	Concrete foundations	18	66	–	84
03	Brick footings	75	60	–	135
04	Hardcore	30	30	–	60
05	Concrete oversite	36	90	–	126
06	Brick superstructure	406	300	–	706
07	First-floor joists	–	–	–	–
08	Roof construction	135	342	–	477
09	Roof tiling	–	–	–	–
10	Flat roof covering	–	–	120	120
11	Flashings and rainwater goods	–	–	24	24
12	External cladding	–	–	–	–
13	Joinery first fix/door	42	274	–	316
14	Glazing	–	–	18	18
15	Internal partitions	–	–	–	–
16	Plastering	–	–	–	–
17	Floor screeds	–	–	–	–
18	Joinery second fix	–	–	–	–
19	Internal plumbing and heating	–	–	50	50
20	Electrical installation	–	–	50	50
21	Fireplace and slab flooring	–	–	–	–
22	Wall tiling	–	–	–	–
23	Ceiling finish	–	–	–	–
24	Floor finishes	–	–	–	–
25	External painting	–	–	36	36
26	Internal painting	–	–	18	18
27	Cleaning	–	–	–	–
28	Contingencies	–	–	–	–
		£832	£1162	£316	£2310

Figure 1.10 Summary of construction costs

DIRECT LABOUR

Having estimated the direct labour cost for each trade, for each unit type, the costs are shown in the Summary of Construction Costs (*Figure 1.10*).

Once construction commences, a card of the type shown in *Figure 1.11* can be prepared for each of the sub-headings for direct labour shown in the Summary of Construction Costs. The cards are in duplicate, one copy each for head office and the site.

The site foreman completes his card from Daily Allocation Sheets (*Figure 1.12*) and then sends it to the office weekly. At the office:

- the particulars of hours worked are copied from the foreman's copy into the office copy;
- the total hours on all copies are agreed with the total 'payroll hours' for the site;
- the foreman's copy is returned to him for re-use in the current week.

During the progress of the work it can be seen from these cards what hours, if any, have been paid for which there was no provision in the estimates.

Trade: Excavations *Site:* Epsom Downs Phase 2, Unit 7, Works Order No. 1096 *Weather:* (W) wet, (D) dry, (N) normal, (X) wintry					
Summary				*Hours*	£
Allowance in estimates				109½	
Actual costs as below				115	
Surplus					
Deficit				5½	11
Details Loam, chalk					
Date	*Name*	*Remarks*	*Weather conditions*	*Hours*	£
June 7	O'Reilly		N	40	
"	Moore		N	20	
"	Mahon		N	19	
June 14	O'Reilly		D	10	
"	Mahon		D	20	
June 21	Regan		W	6	
			Totals	115	

Figure 1.11 Direct-labour cost card

Labour: daily allocation *Site:* Epsom Downs Bungalows: Phase 2, Unit 7									*Date:* 10 July 1978
Contract		1096	1096	1096	1096	1097	1098		
Name	*Trade*	*Excavate trenches Unit 8*	*Unload bricks Unit 8*	*Brick work Unit 7*	*Concrete foundations Unit 8*	*Clearing -up site*	*Collecting Materials from merchants*		*Totals*
		Hours	Hours	Hours	Hours	Hours	Hours	Hours	Hours
Smith	BK			8					8
Jones	BK			8					8
Robinson	LAB		2	4		2			8
Brown	LAB	5		2			1		8
					8				8
			4	2	2				8
	Totals	5	6	24	10	2	1		48

Figure 1.12 Daily labour-allocation sheet

On completion of each operation, it can be seen from the cards the number of hours saved, compared with those estimated, or the number of hours excess. These can be priced-up at an average hourly rate applicable to the particular site, and listed in a Labour Variances schedule.

SUB–CONTRACTORS

Sub-contractors' costs are taken from the Summary of Construction Costs and Bills of Quantities; a separate card is prepared in duplicate (white and pink) for each sub-contractor, for each unit on which that sub-contractor is employed (see *Figure 1.13*).

The site foreman (or a visiting surveyor in consultation with the site foreman) should complete each card each week, then send the cards to the office. The office should enter particulars from site foremens' cards into the office copies, then:

- pay the sub-contractors;
- return each foreman's cards to him for further use.

These cards also have columns, and a summary section, for recording variances between estimated and actual costs. These variances should be listed.

It will be noted that these cards have provision for controls required under the Tax Deduction Scheme, in particular:

- certificate numbers,
- expiry dates,
- recovery of vouchers on form 715,
- tax deductions from those without a certificate.

Figure 1.14 shows a sub-contractor's payment certificate, which is a standard form obtainable from the NFBTE Publications Department. It is in quadruplicate:

- top: for sub-contractor's retention;
- duplicate: signed by the sub-contractor, serves as an Authenticated Tax Receipt for VAT purposes;
- cost office copy;
- surveyor's copy.

Monthly valuations

Monthly valuations for:

- Roads, Sewers, Footpaths and Kerbs
- Site Works

should be carried out by a measurement surveyor in the normal way, and made subject to cost/value comparison procedures.

The cost office should maintain a separate record of Variable Site Overheads, and each month show the 'planned' and 'actual' costs, and the differences between them (see *Figure 1.15*). These should be shown for the month and cumulatively.

Valuations of construction costs can be carried out on charts, an example of which is provided in *Figure 1.16*. This example relates to one property, and shows percentage completions each month. The cumulative percentage can be multiplied by the estimated cost, to obtain the value of work completed to date, e.g.

- completed to date 66%
- estimated total cost £8832
- cost value of work done to date 66% × £8832 equals £5829.

On large estates with similar house types, the valuation work is condensed by obtaining one percentage total to cover all dwellings. Similar calculations are made for free-standing garages.

A quick method of obtaining a monthly valuation of unfixed materials on site is shown in *Figure 1.17*.

Cost/value comparisons

No matter which other systems may be in operation, monthly cost/value comparisons of the work done at each site are of considerable value to top management by showing:

Sub-contractor's address 914 Langley Hill, Epsom Downs

Sub-contractor's name Joe Smith

Tax Certificate particulars

Trade Painter

No. I1234567 Expiry Date 9/79

Site: Epsom Downs Phase 2, Plot 7 Works Order No. 1096

Amounts due												*Amounts paid*					
Date	*Contract price*	*Agreed inc-reases*	*Deduc-tions*	*Total*	*Retent.*	*Balance*	*Disc.*	*Balance*	*Plus VAT*		*Total due*	*Date paid*	*Net amount paid*	*Tax deduct.*		*Total paid*	*Form 715 rec'd*
(1)	(2)	(3)	(4)	(5)	(6)	(7)	(8)	(9)	(10)	(11)	(12)	(13)	(14)	(15)	(16)	(17)	(18)
Sep 28	180.00	20.00	–	200.00	20.00	180.00	4.50	175.50	–	–	175.50	Sep 30	175.50	–	–	175.50	✓
Oct 5	193.00		10.00	180.00	18.00	162.00	4.05	157.95	–	–	157.95	Oct 8	157.95	–	–	157.95	✓
					(38.00)		(0.95)				(37.05)	Dec 12	37.05	–	–	37.05	✓
	370.00	20.00	10.00	380.00	–	380.00	9.50	370.50			370.50		370.50			370.50	

Final summary

Contract price	Agreed increases (+)	Agreed deductions (–)	Net	Discount	Balance	Estimates	Excess (+)	Saving (–)
£ 370	£ 20	£ 10	£ 380	£ 9.50	£370.50	£ 370.00	£ 0.50	£

Figure 1.13 Sub-contractor's ledger card

Vestry Series B.42A (Revised 1977)

SUB CONTRACTORS PAYMENT ADVICE AND AUTHENTICATED TAX RECEIPT

FROM

(Co. Registration No.:)

TAXED	I or P		C
		SUPPLY & FIX NOMINATED & OWN	
		HIRED PLANT/ TRANSPORT	
		LABOUR ONLY	

Tick box above as necessary

TO ....

V.A.T. Registration No.

Contract ....
Contract No ....
Certificate No ....
Payment No ....
Date of Valuation ....

Tax deduction certificate expiry date

SALE OF ....

CERTIFICATE	GROSS £	RETENTION %	RETENTION £	NET BALANCE £
PRESENT				
PREVIOUS				
BALANCE DUE				
LESS : DISCOUNT		← LESS : DISCOUNT →		
NET COST				

DEDUCT: Income Tax : Gross
@ ....% Less Mat.
Balance

ADD : VAT £..........@.....%
£..........@.....%

Total

p.p.

DEDUCT Contra Charges

Cheque £

We have pleasure in enclosing our cheque for the VAT exclusive amount in the above certificate, together with the VAT calculated thereon as shown. PLEASE COMPLETE THE APPENDED RECEIPT AND RETURN IT TO US IMMEDIATELY (Authenticated Receipt VAT (General) Regulations (1972) Reg.21(2)), TOGETHER WITH INLAND REVENUE FORM NO. 715 Signed ....

IMPORTANT. Do not detach. Please return the WHOLE OF THE DUPLICATE COPY OF THIS FORM to the main contractor (above) after the completion and signature of the authenticated receipt below.

AUTHENTICATED RECEIPT

Sub Contractor's Reference ....

Sub Contractor's VAT Registration No.

Value of Supply (5 above) £............ Rate of VAT (6 above) ....% ....%

(AMOUNT OF VAT (6 above)
(RECEIVED THIS AMOUNT OF VAT £..................

Date of Supply ....

Signed .... (Date when you received enclosed cheque)

PLEASE COMPLETE AND RETURN THE WHOLE DUPLICATE TO ENSURE FUTURE PAYMENT, TOGETHER WITH INLAND REVENUE FORM No. 715

Figure 1.14 Sub-contractor's payment certificate

- how actual costs compare with those planned, and if there are substantial deviations where these are arising;
- where action needs to be taken to counteract an adverse trend;
- the profitability of each site and of the company as a whole.

Regular information of this kind is not only anxiety-relieving, but invaluable when provided to those furnishing funds to the company.

Variable Site Overheads: costs, savings and excesses Epsom Downs Phase 2: Works Order No. 1096										
	June		*July*		*August*		*September*		*October*	
	Est.	*Actual*	*Est.*	*Actual*	*Est.*	*Actual*	*Est.*	*Actual*	*Est.*	*Actual*
61 Foreman	100	98	400	430	400	430	400		400	
61 Sundry site labour	60	66	240	260	240	270	240		240	
61 Watchman	60	30	240	200	240	240	240		240	
62 Site transport	25	30	100	90	100	120	100		100	
63 Hutting, fencing & notice b'ds	20	10	80	80	80	80	80		80	
65 Plant: Mixer	10	15	40	40	40	40	40		40	
Scaff'g	20	15	40	40	40	40				
Dumper	5		45	45	45	45	45		45	
Forklift										
Other										
66 Telephone	5		15	20	15	20	15		15	
67 Water	5	10	5	5	5	5	5		5	
68 Power	5	5	15	20	15	20	15		15	
69 Miscell.				10		30				
Monthly totals	315	279	1220	1240	1220	1340	1180		1180	
Monthly savings		36								
Monthly excess				20		120				
Cumulative		(–)36		(–)16		(+)104				

Figure 1.15 Site overheads chart

It is common practice in the industry to employ systems that enable a site cost to be obtained simply and easily any day during a month it is required. This spreading of cost/value comparison dates throughout a month is helpful to surveyors in private development, and is essential for contract work, in which valuation dates can arise at the dictation of the client's Quantity Surveyor.

The costs can be obtained within a half per cent, which is far more accurate than valuations. A half per cent on £100 000 worth of contract work is £500, but since this margin applies to only recent weeks' materials deliveries, it is in fact a margin of about 16.6 per cent (£500 on £3000, say).

Dwelling construction. valuation of work done
Site: Epsom Downs, Phase 2: Unit No. 7 (Derby type): Contract No. 1096

Code	*Element*	*Value* £	*Value* %	*Cont.* %	*Jun 30* %	*Jul 31* %	*Aug 31* %	*Sep 30* %
01	Excavation	219	2.50	2.50	2.50			
02	Concrete foundations	138	1.50	4.00		1.50		
03	Brick footings	525	6.00	10.00		6.00		
04	Hardcore	135	1.50	11.50		1.50		
05	Concrete oversite	345	4.00	15.50		4.00		
06	Brick superstructure:							
	DPC	448	5.00	20.50		5.00		
	1 metre above cills	805	9.00	29.50		9.00		
	F.F. joists }							
	Wall plate – top out }	1432	16.00	45.50			16.00	
07	First-floor joists	–	–	–				
08	Roof construction	621	7.00	52.50			7.00	
09	Roof tiling	645	7.25	59.75			7.25	
10	Flat roof covering	–	–	–				
11	Flashings and rainwater goods	180	2.00	61.75			2.00	
12	External cladding	–	–	–				
13	Joinery first fix	195	2.25	64.00			2.25	
14	Glazing	105	1.00	65.00				
15	Internal partitions	–	–	–				
16	Plastering	525	6.00	71.00			2.00 (Part)	
17	Floor screeds	150	1.75	72.75				
18	Joinery second fix	663	7.50	80.25				
19	Internal plumbing and heating:							
	first fix	260	6.00	86.25				
	second fix	400	1.50	87.75				
20	Electrical installation:							
	first fix	160	1.50	89.25				
	second fix	50	1.00	90.25				
21	Fireplace and slab flooring	111	1.25	91.50				
22	Wall tiling	72	1.00	92.50				
23	Ceiling finish	60	0.75	93.25				
24	Floor finishes	210	2.50	95.75				
25	External painting	120	1.25	97.00				
26	Internal painting	240	2.75	99.75				
27	Cleaning	18	0.25	100.00				
28	Contingencies	–	–	–				
	Totals	8832	100.00	100.00	2.50	27.00	36.50	
	Cumulative totals				2.50	29.50	66.00	
	Extras lists 111	140					70	
	124	80						
	Extras totals	220					70	

Figure 1.16 Valuation of work done

Dwelling construction: valuation of unfixed materials on site *Site:* Epsom Downs, Phase 2: Units nos. 7–12: Contract No. 1096						
Code	*Element*	*Value* £	*Jun 30* £	*Jul 31* £	*Aug 31* £	*Sep 30* £
01	Excavation					
02	Concrete foundations	108				
03	Brick footings	258	258			
04	Hardcore	66				
05	Concrete oversite	264				
06	Brick superstructure					
	DPC					
	1 metre above cills					
	F.F. joists	1629	1629	760		
	Wall plate – top out					
07	First floor joists	–				
08	Roof construction	447		447		
09	Roof tiling	–				
10	Flat roof covering	–				
11	Flashings and rainwater goods	–				
12	External cladding	–				
13	Joinery first fix	144	144	144		
14	Glazing	–				
15	Internal partitions	–				
16	Plastering	–				
17	Floor screeds	–				
18	Joinery second fix	453			453	
19	Internal plumbing and heating:					
	first fix	–				
	second fix	–				
20	Electrical installation:					
	first fix	–				
	second fix	–				
21	Fireplace and slab flooring	81		50	50	
22	Wall tiling	–				
23	Ceiling finish	–				
24	Floor finishes	–				
25	External painting	–				
26	Internal painting	–				
27	Cleaning	–				
28	Contingencies	–				
	Totals	3450	2031	1401	503	
	Cumulative totals	– –	–	–	–	– –
	Extras list 11	70			–	
	24	40			40	
	Extras totals	110			40	

Figure 1.17 Valuation of unfixed materials on site

Materials–received sheet			
Site: Epsom Downs, Phase 2: Contract No. 1096			
Date of receipt	*Supplier's name*	*Supplier's ticket no. (attached)*	If no supplier's ticket is attached, full details of deliveries must be entered here; if there is a delivery ticket attached, details of goods received are *not* to be entered below
April 3	Hall	XYZ 4321	
" "	Nichols & Clarke	B.4321	
" "	Jones Bros.	Missing	40 – 150 mm PIPES 2 – Y JUNCTIONS – 150 mm 1 INTERCEPTOR 3 – 150 mm SLOW BENDS 1 – 150 mm KNUCKLE BEND
" "	UBM	AB 982	
" "	Reeves	986421	
" "	Roberts Adlard	N4384	

Figure 1.18 Materials-received sheet

To achieve a cost on the day of a valuation:

- Sub-contract work is separated from own work in both costs and valuations.
- A purchase order must be issued for every purchase made, and a copy of each order must be provided to the cost office.
- Deliveries should be marked-up on cost office purchase orders regularly. An abbreviated Materials Return (*Figure 1.18*) should be provided by sites to the office. This return need contain only a one-line entry when a delivery ticket is attached to it.

Figure 1.19 shows a standard type of purchase order form, which is obtainable from the publications department of the National Federation of Building Trades Employers. This example shows:

- an order for 50 000 facing bricks worth £3000;
- of which 40 000 have been delivered;
- 16 000 have been invoiced for £960;
- 24 000 have been delivered but not yet invoiced, and are worth £1440;
- 10 000, worth £600, are not yet delivered.

Similar calculations can be made for all goods delivered but not yet invoiced.

Invoices checked but not yet costed are kept together under contracts, or sites, for ease of access for cost/value comparison purposes. The full cost/value comparison procedure is shown in:

- *Figure 1.20*, Cost/value comparison forms;
- *Figure 1.21*, Sub-contractors and services analysis;
- *Figure 1.22*, Cost adjustment sheet.

QUADRUPLICATE

ORDER

Telephone: 25-549999

DERBY CONSTRUCTION CO.LTD.
401-410 OAKS WAY,
EPSOM DOWNS, SURREY

No. 1096 / 1453

To:
EPSOM DOWNS BUILDERS
SUPPLIES LTD.
TATTENHAM CORNER STATION YARD

Date 1st December

Your Ref. X14298

Terms NET

QUANTITY	DESCRIPTION	PRICE	Del. Date	Del. Qty.	Inv. Date.	Inv. AM.
			Deliveries			
50,000	CORONATION FACING BRICKS	£60	3/1	8,000)	10/1	£960
			5/1	8,000)		
			8/1	8,000		
			12/1	8,000		
			18/1	8,000		

THE "Deatry" SERIES 872

Delivery Address and Date Required:

BEECH TREES
TATTENHAM CRESCENT
EPSOM DOWNS

INVOICES:
AMOUNT:

DATE:

J. Smith

DELIVERY DATE(S) As called up January

IT IS ESSENTIAL THAT DELIVERY IS MADE AS SHOWN

Figure 1.19 Standard purchase order form

In the case of sub-contractors it is known whether there are cost savings or excesses at the time an order is placed. For example:

Bill price		£1000
Sublet for	£960	
Less 2½% discount	£24	£936
Expected cost saving		£64 = 6.4%

Each time a sub-contract valuation is certified the profit or loss is calculated; e.g. a £200 certification at 6.4% equals £12.80 profit. This calculation is carried out each month separately for each sub-contract, and a grand total is obtained of savings and excesses, to add to or detract from the planned gross profit on own work.

Cost/value comparison

Contract . . . . Contract No . . . .

Valuation No . . . . To . . . . 19 . . . . .

Line no. *SECTION A (completed by Surveyors):* £

1. *Gross values of work done:*
   - 1.1 Main work
   - 1.2 PC sums
   - 1.3 Materials on site
   - 1.4 Increased cost claims
   - 1.5 Dayworks
   - 1.6 Other items
2. *Total gross valuation*
3. Add or deduct: Surveyors over-valuation (–) or under-valuation (+)
4. *Adjusted gross valuation*
5. *Deduct: sub-contractors & PC items* (from Figure 1.21): £
   - 5.1 Nominated sub-contractors at cost
   - 5.2 Profit on nominated sub-contractors
   - 5.3 Own sub-contractors at cost
   - 5.4 Profit on own sub-contractors
   - 5.5 PC items at cost
   - 5.6 Profit on PC items
6. *Gross value of own work*

SECTION B (completed by Accounts Office)

7. *Deduct: cost of own work* (from Figure 1.22):

Increase since last valuation	*Totals to last valuation*		*Totals to this valuation*
£	£		£
		7.1 Labour	
		7.2 Materials/services	
		7.3 Plant	
		7.4 Transport	
		7.5 Total	

8. *Gross Profit Margin to date on own work* (...% of Prime Cost to date in line 7.5)
9. Deduct: Gross Profit Margin, on previous valuation
10. Difference: Gross Profit Margin, this valuation
11. *Total Gross Profit Margin to date:*
    - 11.1 Own work (line 8 above)
    - 11.2 Nominated sub-contractors (line 5.2 above)
    - 11.3 Own sub-contractors (line 5.4 above)
    - 11.4 PC items (line 5.6 above)
12. *Total to date*

Figure 1.20 Cost/value comparison form

Analysis of sub-contracts & services (to be completed by the Surveyor)

Contract . . . . . . No. . . . . . .
Included in Valuation No . . . . to . . . .

Name	*Trade*	*Gross Value Recommended*		*Previous Certificate (Gross)*		*Builder's Profit*	
						%	£ p
Nominations							
Domestic (own)							
Services							
							£

Figure 1.21 Sub-contractors and services analysis form

Cost adjustment sheet: own work only (to nearest £) Contract . . . . Date				
(1)	*From Cost Sheet* (2) £	*From office documents* (3) £	*From information brought by surveyors from site* (4) £	*Extended (totals)* (5) £
1. *Labour (line 7.1)* Includes labour-only sub-contractors:				
1.1 Cost Sheet total	X			
1.2 Charges available in office, but not yet costed to Cost Sheets (at average hourly rates)		X		
1.3 Time worked on site, in current week (at average hourly rates)			X	X
2. *Materials purchased (line 7.2)*				
2.1 Cost Sheet total	X			
2.2 Invoices checked but not yet costed (in one folder per contract)		X		
2.3 Orders marked-up with deliveries but suppliers' invoices not yet received (priced at approximate rates), in one folder per contract		X		
2.4 Particulars of recent deliveries brought by surveyor from site (marked-up on office copies of orders)			X	X
3. *Stores issues/transfers (line 7.2)*				
3.1 Cost Sheet total	X			
3.2 Stores Notes priced but not yet entered on Cost Sheets (in one folder in contract no. sequence		X		X
4. *Plant (line 7.3)*				
4.1 Cost Sheet total	X			
4.2 Average per week for weeks not yet entered on Cost Sheet		X		X
5. *Transport (line 7.4)*				
5.1 Cost Sheet total	X			
5.2 Average per week for weeks not yet entered on Cost Sheet		X		X
6. *Own manufacturers joinery (line 7.2)*				
6.1 Cost Sheet total	X			
6.2 Value of joinery deliveries not yet entered on Cost Sheet		X		X
7. *'Own work' cost to date of valuation*	X	X	X	£ X

Notes 1. Reference nos. are to the cost/value comparison form
2. There would be sections for other internal departments, e.g. painting, unless treated as though an external sub-contractor.

Figure 1.22 Cost adjustment sheet

Monthly profit statement for the month of ... 19..		
Completed sites	£	£
Epsom Downs, Phase 1: completed sales		90 000
Deduct Cost of Sales:		
Land	12 000	
Roads, Sewers, etc.	9 000	
Variable Site Overheads	5 000	
Site Works	7 000	
Bungalow Construction	36 000	
Garages Construction		
Advertising Selling Expenses	1 000	
Legal Costs of Sales	1 000	71 000
Gross Profit on Sales		19 000
Deduct last month's profit on completed sales		17 000
This month's profit on completed sales		2 000
Work-in-progress		
Epsom Downs, Phase 2: work-in-progress, margin 33.3%		
33.3% of costs £45 000	15 000	
Deduct last month's work-in-progress	8 000	
	7 000	
Add savings made	500	
Less losses known	(–) 1 000	6 500
Total Gross Profit for the month		8 500
Less General Overheads ($^1/_{12}$ of year's budget of £60 000)		5 000
Net profit for the month		3 500

Figure 1.23 Monthly profit statement

Monthly profit statement

The gross profit margins depend for their validity on the valuations of work in progress, made in accordance with the cost/value comparisons procedures described in previous paragraphs.

From the gross profit margins is deducted one-twelfth of one year's general overheads. The method for preparing a general overheads budget is described in the booklet *Calculating builders' labour rates and general overheads*, published by the National Federation of Building Trades Employers. *Figure 1.23* is an example of a Monthly Profit Statement.

ADMINISTRATION

Companies, partnerships and sole traders

The advantage of a limited liability company is that the liability of any shareholder is restricted to the amount he has taken up (or agreed to take up) in shares. Thus if a shareholder has agreed to take shares with a face value of £1000, for which he has paid in full, he cannot be asked to pay anything further should the company fail.

On the other hand, a sole trader is liable to the extent of everything he owns, including his home and personal possessions, should his business fail. Thus if the assets of the business are inadequate to meet the debts of the business, the personal possessions of the trader will be attached to meet, or help meet, the deficit of the business.

Partnerships are even worse from the personal liability point of view, because the personal possessions of any partner can be attached to meet a partnership deficit. If the personal possessions of a partner are inadequate to meet his share of the deficit, then the other partner or partners are required to meet not only their own share of the deficit, but also the shares of defaulting partners.

A limited partner's liability is restricted to a stated amount, but such a partner must not be active in running the business: he is a sleeping partner.

There often are tax advantages for sole traders and partners. Where profits are low they are at present taxed at 33 per cent (up to £8000 at 33 per cent with a sliding scale thereafter). In the case of companies, the corporation tax rate is 42 per cent up to taxable profits of £50 000 and 52 per cent thereafter.

Tax considerations are sufficiently persuasive in some cases to cause sole traders and partners on the advice of their accountants to remain as such. In the writer's view the merits of limited liability, particularly in long term projects such as private housing, outweigh the tax advantages. The cost of forming a company can be relatively low (from £100, plus the costs of the involvement of legal advisers in each case).

Borrowing and other fund raising

The cheapest source of funds is usually the High Street bank, and the ability to persuade a banker to lend is dependent on the criteria he is obliged to use, chiefly:

- his confidence and faith in the managerial ability of the borrower, based on a proven profit record, and his fulfilment of past promises;
- the security offered;
- the ability of the borrower to provide a flow of up-to-date management information, such as monthly or quarterly profit statements, cash flow charts, etc.;
- the positive steps taken by the borrower to train adequate successors for top and middle management positions, particularly to meet eventualities.

The next best source of funds is often the Industrial and Commercial Finance Corporation (ICFC), which is owned by the Joint Stock Banks but concerned with longer-term lending. There are ICFC branches throughout the United Kingdom. This organisation carries out an in-company investigation of the proposed borrower, following which it will, if satisfied with its findings, lend funds by almost any arrangement, including:

- purchase of company shares or debentures;
- long-term loans repayable in instalments;
- loans for a specific project.

The many Merchant Banks will provide funds where the borrower's past record and future prospects are good, and there is sound security.

In general, borrowing should not be for too long a period ahead, as the borrowing company may find that through changed circumstances it has no use to which it can put the funds, but must nevertheless meet the regular interest payments.

Budgets and budgetary controls

Budgeting starts with a General Overheads Budget, and the reader is recommended to follow the overheads budgeting procedures detailed in the NFBTE booklet, *Calculating builders' labour rates and general overheads.*

Cash-flow programme
Site: Epsom Downs, Phase 2: Unit Nos. 7/12: Contract No. 1096 — Programme Date 1/1/78

	Type	*Apr*	*May*	*Jun*	*Jul*	*Aug*	*Sep*	*Oct*	*Nov*	*Summary*
Bungalow construction costs										
Unit 7	'Derby'			220	2385	3224	1500	1503		8832
8	"				1302	2805	2362	1500	863	8832
9	'Oaks'		2075	2075	2075	2075				8300
10	"		1038	2075	2075	2075	1037			8300
11	"			1038	2075	2075	2075	1037		8300
12	"				1038	2075	2075	2075	1037	8300
6 double garages construction costs					4620	4620	4620			13860
Variable Site Overheads				315	1220	1220	1180	1180	885	6000
Site Works		1000	1000	3000	1000	500	500	1000	1000	9000
Roads, Sewers, Kerbs and Footpaths		1000	2000	4000	5000			1000	2000	15000
Total construction costs		2000	6113	12723	22790	20669	15349	9295	5785	94724
Bungalow sales		–	–	–	–	26500	26500	54500	54500	162000
Cash-flow surplus (+) or deficit (–)		(–) 2000	(–) 6113	(–) 12723	(–) 22790	(+) 5831	(+) 11151	(+) 45205	(+) 48715	(+) 67276
Cumulative cash flow		(–) 2000	(–) 8113	(–) 20836	(–) 43626	(–) 37795	(–) 26644	(+) 18561	(+) 67276	

Note: Labour is paid in the month the work was carried out. Materials and supply-and-fix sub-contractors are payable a month after delivery.

Figure 1.24 Cash-flow chart

General overheads consist mainly of directors' and staff salaries (say 63%), directors' and staff cars (say 7%) and premises upkeep (say 5%). Thus general overheads are usually difficult to reduce in any substantial way unless staff is dismissed, but trained and integrated staff are usually the most valuable asset of a business. Thus general overheads must be regarded as a fixed future annual commitment, which a company can recover only by obtaining a sufficiently large turnover of profitable work.

A cash-flow chart should be prepared separately for each site. An example is provided in *Figure 1.24*. Then the cash-flow charts for every site should be amalgamated to provide an overall company cash-flow chart, in which will also be included commitments for:

- general overheads;
- capital purchases of land, plant, etc.;
- special payments, such as tax and dividends.

However, the most difficult part of cash-flow budgeting is the site budget, involving as it does projections of sales and construction programmes.

TAXATION

Taxation is now overwhelming, and must receive priority consideration, particularly in the case of CTT (Capital Transfer Tax).

In the context of *any family business*, CTT considerations are fundamental to the continued existence of the business. This is of importance not only to the proprietors of the business and their dependents, but to all others attached to the business (and their futures and families). It follows that tax planning should take place in every family business, failing which there could be disastrous results on the death of the proprietor.

Because taxes are interrelated they cannot be treated in isolation. This applies to all taxes, but especially to Capital Transfer Tax, Capital Gains Tax, and Development Land Tax. The Community Land Act also has far-reaching connotations for the construction industry. (Note that the rates and allowances quoted in these pages are those in force up to the 1978 Finance Act.)

Corporation Tax

Corporation Tax is based on a company's 'net profits before tax'. Adjustments to those profits are first made, particularly:

- *add* to the net profits: (a) the depreciation taken in the accounts, (b) disallowed expenditure, (c) benefits taken by directors and shareholders;
- *deduct* from the net profits: (a) capital allowances, (b) stock relief.

The current rate of Corporation Tax is 52 per cent, reduced to 42 per cent where the profits do not exceed £50 000 (there is marginal relief).

ACT means 'Advance Corporation Tax'. This is payable to the Inland Revenue at the time a dividend is paid (33/67 of the dividend). It is adjusted in the final tax settlement.

Close Companies are those under the control of five or less persons. For this purpose an entire family constitutes one person (grandparents to grandchildren, uncles, aunts, etc.). Such companies may be required to declare dividends so that the recipients can be charged higher rates of income tax on those dividends.

Stock Relief is a recognition by the Inland Revenue that because a business's stocks increase in amount there is not necessarily an increase in profit. Stock includes *work-in-progress*.

The system changed in April 1976. For example:

1976		£
Jan. 1	Opening stock and work-in-progress	160 000
Dec. 31	Closing stock and work-in-progress	200 000
	Increase	40 000
Deduct:	15% of Net Profits, less capital allowances (say £100 000)	15 000
Stock Relief		£25 000

For stock relief purposes, contracts are *not* complete at the practical completion stage, so can continue to be regarded as work-in-progress; they are complete when certified.

Reverse of Stock Relief. If a company's workload (work-in-progress) falls, this could have the reverse effect of Stock Relief, i.e. additional tax would be payable. This causes some construction companies to take work at low prices in order to maintain their turnover. (They also take work at low prices to avoid redundancy payments to staff and operatives.)

Capital allowances

Depreciation is a matter of company policy, and is ignored by the Inland Revenue. Instead they grant capital allowances, which are now (1978):

- *Plant* : 100 per cent for the year of purchase. Plant is not defined by the Revenue (there is some case law), and *includes* lorries, vans, office furniture and fittings and office machines. It *excludes* private motor cars bought by a company.
- *Motor cars* : 25 per cent per annum up to a maximum of £1250 per annum. Cars below £5000 cost price (new or second-hand) are aggregated by the Inland Revenue into a pool. Cars above £5000 cost price are treated individually by the Inland Revenue.
- *New industrial buildings:* initial allowance of 50 per cent; writing-down allowance of 4 per cent per annum.
- *Development Area Grants* : 20 per cent extra for industrial buildings and plant.
- *Special Development Area Grants* : 22 per cent.

Capital Gains Tax (CGT)

Capital Gains Tax applies to companies as well as to individuals, partners, etc. It arises when a sale or transfer takes place. The rate is 30 per cent on the gain since acquisition (or 5 April 1965, if the acquisition was before that date). The first £1000 is free of tax and the next £4000 is at 15 per cent. For example:

		£
Selling price		30 000
Value at 6 April 1965		18 000
Capital gain		12 000
Tax: £1000	nil	
£4000 @ 15 per cent	£600	
£7000 @ 30 per cent	£2100	2 700

Gains on the sale or transfer of company shares are subject to CGT. The main exemptions are: £1000 per annum; private cars and houses (one per taxpayer); gifts of less than £100 per grantor; life assurance; chattels worth under £2000.

It is to be noted especially that *transfers between husband and wife are exempt from CGT.* This is important in considering Capital Transfer Tax planning.

Development Land Tax (DLT)

Development Land Tax commenced on 1 August 1976. It replaced Development Gains Tax.

The rates are based on special and abstruse alternative methods of calculation, and are:
- until 31 March 1980: 66.6 per cent on the first £150 000 (in one year); 80 per cent thereafter.
- after 31 March 1980: 80 per cent.

The charge arises on the sale or part disposal of an interest in land, including leases and first lettings. Land held by a builder as *stock in trade on 12 September 1974* is exempt, provided it had planning permission then (or an appeal has been ante-dated to before that date).

The main exemptions are:
- £10 000;
- owner-occupiers (up to one acre, and after one year's occupation);
- one dwelling, on land owned before 12 September 1974 by the intended occupier of that dwelling;
- gifts at death (the DLT arises on the subsequent disposals);
- inter-company transfer where share holdings are 75 per cent or more;
- industrial buildings built by an industrialist for his own use on land he owns.

DLT is in lieu of and *not* additional to other taxes.

Capital Transfer Tax (CTT)

CTT is a birth-to-death tax on gifts; the rates are on a rising scale. Therefore a birth-to-death record of gifts should be maintained by everyone. The rates of CTT are different for *lifetime* and *death* gifts.

The tax is on the *diminution in the value of the donor's estate.* For example:

	%	*No.*	*Value*
A father holds shares having a net worth value of £14 each, but reduced to £7 each (see note on *majority holdings* below)	51	5 100	£ 35 700
His son holds shares having a net worth value of £6 each (minority holding value)	49	4 900	29 400
	100	10 000	6 300
Father transfers to son @ £7	2	200	1 400

However, the father has reduced the value of his estate:

from 51% @ £7 per share	£35 700
to 49% @ £6 per share	£29 400
	£6300

and pays CTT on the difference, i.e. £6300 at his current CTT rate. This is called the *diminution principle.* The diminution principle applies to everything – property, sets of chairs, pairs of antiques, etc., etc.

From October 1977 a new rule was introduced whereby the value of shares for a *majority holding* is reduced by 50 per cent. *Minority holdings* have always been valued at less than majority holdings. The Inland Revenue can introduce *goodwill* into the valuation.

Share valuation for CTT purposes is not defined by the law, except that it is to be at the *open market value.* In practice it is necessary to appraise book values and price earnings ratios in each case, and to take into account goodwill, and majority and minority holdings.

The most important CTT exemption is *gifts between husband and wife,* which permits some manipulation within the law. For example:

	Duty
Husband's estate is worth £100 000	
Wife's estate is worth £10 000	
Husband dies first and leaves £100 000 to wife	Nil
Wife dies first and leaves £10 000 to husband	Nil
Wife dies (after husband) and leaves £110 000 to children	£28 250
Husband in his lifetime equates his estate with his wife's. No CTT is payable at that time. Both leave their estates to their children:	
Husband £55 000	£6250
Wife £55 000	£6250

This shows a saving of £28 250 less £12 500, i.e. £15 570.

Other important *exemptions* are:

- the first £25 000 of each person's estate at a nil rate;
- gifts of £2000 per annum (husband and wife can each gift £2000, provided the wife has funds);
- gifts of £100 each to any number of separate individuals;
- £5000 to a child on *marriage,* by each of parents, and £2500 by each of the grandparents; also £1000 by any individual;
- gifts out of *savings from annual income* (provided the standard of living of the donor is not reduced).

Trusts are specially taxed, and no longer a good thing to have. Trusts for care, maintenance and education of children (up to age 25) are the exception.

EVALUATING A BUSINESS

The value of a business depends on the purpose of that valuation. A vendor will have his/her notions, and a buyer will think something quite different. The Share Valuation Division of the Inland Revenue will have its own ideas.

Under Capital Transfer Tax and Capital Gains Tax legislation the value of a share is what it would fetch in an open-market sale. For public companies this is the stock exchange price, but for private company share valuations the value is more difficult to determine.

The net asset value is a starting point, and in the example on page **1**–10 a figure of £2 per £1 ordinary share is calculated. The validity of this figure could be challenged if the asset values in the Balance Sheet are understated; for example the present values of business premises, the land bank and work-in-progress could materially affect the value per share.

The Inland Revenue then try to introduce goodwill into the valuation. Taking again the example on page **1**–10, supposing this were a quoted company and the shares are quoted at £2.50, against a £2 net asset value. The difference of 50p per share would be regarded as goodwill (£50 000 on 100 000 ordinary shares). However it is considered there is no goodwill in a construction business, because:

- although a successful private developer performed well in the past on specific sites, the circumstances surrounding this may never be repeated;
- a contractor's past successes have been attributable to the combined efforts of a team of estimators, planners, surveyors, contract managers, etc., who are not tied to the company; the loss of any member of the team could reverse the successes of the company. In any event, current and future tendering margins may be far lower than those obtainable formerly.

The Articles of Association of a company have a bearing on minority shareholdings. If they provide no right to a dividend, a minority shareholder is in the hands of the majority and the shares could be virtually valueless. On the other hand, a simple majority holding brings considerable power, and a 75 per cent holding even more.

Preference-shares valuations could be worth less than par. A preference shareholder may have a right to a low annual dividend, but has little chance of disposing of the shares for cash.

The purchaser of a company takes on liabilities for current contracts, and for redundancy pay.

In general, the sales value of a contracting/development business will usually be less than the net asset value shown in the Balance Sheet, and this should be put strongly to the Inland Revenue when share transfers are taking place.

2 ESTIMATING AND TENDERING

INTRODUCTION	2–2
ESTIMATING	2–2
TENDERING	2–18
FEEDBACK	2–24
TRAINING	2–24

ESTIMATING AND TENDERING 2

2 ESTIMATING AND TENDERING

B. HARRISON, CEng, MICE, MIMC
Consultant, BAS Management Services

This section demonstrates a logical method of estimating and tendering, and comments briefly on the use of feedback data and on the training of estimators.

INTRODUCTION

It is necessary to differentiate between the two very separate functions involved in estimating and tendering.

Estimating is forecasting or predicting the cost that a particular company will incur in carrying out a defined amount of work, and culminates in the production of a *net cost estimate.*

Tendering is converting the net cost estimate into a selling price that accommodates the company's requirements in relation to:

- general overheads;
- return on capital employed;
- risks from any source;
- the market situation.

Clearly, if a detailed and analytical approach is taken in the estimating stage, the decisions and judgements made in tendering will be easier and probably less risky.

The increasing tendency for companies to use estimating and tendering information for appraisal and control purposes throughout the duration of a project influences the degree of analysis required.

This trend towards additional use of the estimate has also influenced the increased involvement of different functions in the company. In a small firm, the person carrying out the estimating/tendering function may also be contracts manager and surveyor, while in the larger organisation there will be an estimating department responsible for producing net cost estimates so that senior management, in a formal manner, may apply certain criteria and produce a tender figure.

The variety of combinations between these two extremes is limitless, and the presentation of information reflects personal involvement and the eventual use to which the information will be put. The important factor for every company is to ensure that effort is being applied in the right direction at any point in time, and this is often very difficult to achieve.

ESTIMATING

Estimating can only be based on available facts. When these are not to hand, assumptions have to be made that in themselves may contain risk. Facts come from a variety of sources, and their collection and processing requires a systematic handling if the results are to be successful. Consequently, it is vital that a plan of campaign is produced with firm deadlines to ensure that facts are not only properly presented but also produced on time.

A comprehensive and logical collection and presentation of estimating techniques is provided in the *Code of estimating practice* produced by the Institute of Building, and used extensively by all sections of the construction industry.

Basic information

The main sources of basic information are:

- the Tender Documents, including:
  - (a) Conditions of Contract,
  - (b) Bills of Quantities,
  - (c) specifications,
  - (d) drawings;
- the site of the works;
- the client, or his representatives.

It is now common practice to develop checklists for the entire information-gathering process, and these are usually specific to a company's activities and requirements. Standard checklist forms for a site visit and a visit to the client or his representatives are given in *Figures 2.1 and 2.2.*

Project ______________________________ Tender no. __________
______________________________ Date __________
______________________________ Prepared by __________

Site investigation

1. Availability of site __________
2. Location:
   - (a) distance from head office/travelling time ______km ______hrs
   - (b) local transport services __________
3. Access to site
   - (a) road width __________
   - (b) existing access __________
   - (c) crossings __________
   - (d) temporary roads __________
   - (e) traffic intensity __________
4. Working space available for:
   - (a) offices, canteen, stores __________
   - (b) bulk material storage, handling __________
   - (c) construction purposes __________
   - (d) plant (and recovery) __________
5. Services/location:
   - (a) drainage: foul __________
     surface __________
   - (b) water __________
   - (c) electricity __________
   - (d) gas __________
   - (e) telephone __________
   - Are any available from client? __________
6. Nature of ground:
   - (a) surface __________
   - (b) boreholes __________
   - (c) levels: ground __________
     water __________
   - (d) piling __________
7. Site clearance:
   - (a) trees — Note difficulties, recommend methods __________
   - (b) streams __________

continued

(c) existing items ____________
(d) excavations: room to operate plant ____________
(e) tips: charges ____________
distance ____________

Local conditions

8. Availability of local labour:
   (a) type, quality ____________
   (b) inducement paid in area ____________
9. Labour Exchange
   (a) address ____________
   (b) telephone number ____________
10. Damage to existing properties:
   (a) adjacent buildings ____________
   (b) roads ____________
   (c) footpaths ____________
   (d) (photographs if necessary) ____________
11. Local resources
   (a) builders' merchants ____________
   (b) sub-contractors ____________
   (c) plant hire services ____________
   (d) others ____________
12. Other large contracts in area ____________
____
____
13. Name and address of adjacent property owners/occupiers ____________
____
____

Security

14. General security requirements:
   (a) fencing/hoarding ____________
   (b) fans ____________
   (c) lighting ____________
15. Watchman or patrols:
   (a) playgrounds ____________
   (b) schools ____________
   (c) general area ____________

Other conditions

16. Special conditions: implication of:
   (a) penalty/bonus clause ____________
   (b) work in phases ____________
   (c) special insurances ____________
17. Items requiring special attention when pricing ____________
18. Possibility of flooding ____________
19. Fire precautions ____________
20. Noise restrictions ____________
21. Work to be done by the client ____________
22. Other firms tendering ____________
23. Additional information ____________

Figure 2.1 Specimen checklist for pre-tender site survey

Project ______________________ Tender no. ________
______________________ Date ________
Prepared by ________

1. Client
2. Architect/consultant
3. Person in charge of project
4. Quantity surveyor
5. Brief description of works

6. Principal materials in:
   6.1 Substructure
   6.2 Superstructure
   6.3 Roofing
   6.4 Finishings
   6.5 External works
7. Drawings:
   (a) block plans
   (b) site plans
   (c) general locations
   (d) details
   (e) assess complexity
8. Discuss location of:
   (a) site accommodation
   (b) compounds
   (c) car parks
   (d) existing services
   (e) building services
   (f) spoil heaps
   (g) site boundaries
9. Phasing – client requirements:
10. Queries arising out of a study of the tender documents:
   (a) drawings
   (b) Bill of Quantities
   (c) Conditions of Contract
   (d) specification
   (e) other documents
11. *Materials reserved* — *Supplier*
12. *PC sums for:* — *Sub-contractor*

Figure 2.2 Specimen checklist for pre-tender visit to architect's/consultant's office

After assimilation of the basic information, requests for further specific information, including prices, are necessary to the providers of materials and/or services, who include:

- materials suppliers,
- sub-contractors,
- specialists from either outside, or within, the company.

Careful scrutiny of the tender documents is necessary if the enquiries and subsequent quotations are to reflect the conditions of contract included in the tender enquiry.

Additionally, if enquiries are to produce realistic quotations they must also include all relevant facts in regard to:

- quantities involved,
- specification,
- drawings or sketches,
- timing of the supply and/or service.

Much of the information can be provided by selective photocopying of the relevant contract documents, but special care is required to ensure that the information is both complete and relevant. When timing of materials and services is important, this assessment may require the assistance of either the company's planning department or a person delegated to carry out the planning function. The important aim is to provide sufficient accurate information in enquiries to ensure that quotations can be as realistic as possible, in order to obviate the need for qualification and/or amounts added for contingencies.

Estimating/tendering programme

No matter what tendering period is stated, there is never sufficient time to obtain all the information exactly as required, so an estimating/tendering programme is essential if the effort expended is to produce anything like a worthwhile result. Many of the operations overlap, and coordination of each individual's effort becomes more important as the build-up of the estimate proceeds.

In very broad terms this programme must include timings for:

- visit to the site of the works,
- visit to the architect's and/or consultant's offices,
- latest date for sending out enquiries,
- latest date for receipt of quotations,
- tender settlement date (preferably before submission date).

As already mentioned, the quality of information sent out to sub-contractors will influence the prices obtained; while this is rewarding, it is also time-consuming, and due allowance must be made for this fact in producing a realistic estimating/tendering programme.

Pre-tender planning

As with most management functions, planning depends upon making decisions using the facts that are available. In well run companies every activity is analysed to ensure that maximum return on effort is obtained, and pre-tender planning is no exception. It is possible to produce a pre-tender plan, expressed in some form of programme, that provides basic information on durations for:

- overall project;
- major items of work (and interdependencies);
- sub-contractors' visits.

There should be in-built, or specific, allowances for:

- initial planning period;

- setting-up site;
- agreed holidays;
- inclement weather.

However, many companies now look to pre-tender planning to provide information not only for estimating and tendering purposes, but also for general management control purposes if the tender is successful.

It is quite usual to find that pre-tender plans are being used to forecast many of the following:

Resource	*Planning provides extra information on:*	*Used to decide:*
Labour	Labour demand overall trade by trade	Labour employment policy
	Total labour content	Check on labour costs in the estimate
Materials	Hold-ups Bottlenecks	Procurement/storage/handling policy
Plant	Site set-up Plant demands	Preliminaries Hiring/buying policy
Small tools	Durations Costs Spread	Pricing policy
Scaffolding	Durations Timing	Sub-contract/own supply
Personnel	Site staffing Accommodation	Preliminaries Preliminaries
Finance	Payment systems	Overhead recovery Profit requirement
	Cash-flow projection	Cash loading of bills
Weather	Protection Start dates Performances	Risk items Programme Risk items

In most cases, this information can be provided by derivatives or extensions of the simple bar-chart programme.

No matter how basic the planning function in a company, it is essential that the following steps are taken:

- collecting facts;
- method study;
- sequence study;
- presentation (usually a programme).

Collecting facts The fact-collecting methods for estimating, already mentioned in this section, apply equally well to planning, and indeed most of this information will be used for both purposes.

Method study No two projects are exactly alike, and in a competitive situation it is often the correct choice of methods that will win the contract. Usually the choice of construction methods is left to the builder, and it may be possible to point out the benefit to the client of design changes that will lead to cheaper or quicker construction.

Clearly, all decisions on method should be made after proper consideration of possible alternatives and by balancing the advantages in time and/or cost. It is also important that the estimate reflects the methods incorporated in the pre-tender plan. Only the main operations can be studied at the pre-tender stage, and the decisions reached should be recorded for future reference. Standard forms, called *method statements*, are often used; these should indicate:

- details of the method,
- type of equipment or plant,
- outputs,
- size and composition of the gangs,
- estimated costs.

Sequence study Closely allied to method study is the consideration of sequence. In terms of the whole project, the sequence of operations may well dictate the methods that can be employed, and for this reason sequence and method are often studied simultaneously.

In addition, it is often necessary to determine the time taken for a cycle of operations, e.g. the floor cycle for a multi-storey block of flats. This type of study is called an *operation sequence study*, and the results of such studies are often indicated as a single bar of large duration on the pre-tender programme.

Presentation Although more sophisticated techniques are available, the standard presentation for most pre-tender plans is the simple bar chart. This is because of the universal understanding that it promotes, although for certain projects it has limitations. Nevertheless, it can be supplemented to provide the facilities mentioned earlier.

Computing costs

Most estimated costs are an amalgam of labour, materials and plant, so it is important first to establish basic costs for these items. This requires a series of calculations to establish built-up rates for:

- cost of labour per hour (trade by trade);
- cost of mechanical plant per hour (all main machines);
- cost of material per unit (all materials).

Having established these basic data, using performance standards, it is then possible to establish the cost of the items in the Bills of Quantities or Schedules of Rates. As a further consideration, it is also possible to calculate the cost of administering and servicing the entire project.

CALCULATION OF BUILT-UP RATES

Labour Companies must first decide, between departments, what should be included in the built-up labour rate and what should be catered for elsewhere. It is vital that items are neither missed nor included twice, and for this reason a labour-overheads checklist (*Figure 2.3*) is a very simple but useful insurance against mistakes.

For the majority of medium and small builders, the realistic period of time to consider when calculating the hourly rate is a full working year, although on certain projects of a special nature a similar exercise can be carried out over the planned duration of the project.

It is first necessary to establish the number of actual hours for which an employee is available for work, and to calculate the cost of employing a man for this period.

No.	Description	*Labour rates*	*General overheads*	*Prelims*	*Bill rates*
1	National Insurance				
2	Joint Board supplement				
3	Guaranteed minimum bonus				
4	1977 supplement				
5	Holiday pay stamps				
6	Unstamped holiday pay				
7	Tool money				
8	Sick pay				
9	Guaranteed (wet) time				
10	Daily travelling time				
11	Lodging allowances				
12	Fares allowances (daily)				
13	Expenses (daily)				
14	Overtime (non-productive)				
15	Plus rates (fixed bonus)				
16	Incentive bonuses earned				
17	Attraction money				
18	Company pensions				
19	Condition money				
20	Foremen				
21	Chargehands' additional pay				
22	Week-end and other abnormal overtime				
23	Redundancy pay pool				
24	Employer's and public-liability insurances				
25	CITB levy				

Figure 2.3 Labour-overheads checklist

Most companies have a standard form for carrying out this exercise, and an example of a typical format, giving a facility for both 'without overtime' and 'with overtime' situations, is shown in *Figures 2.4 and 2.5.* These forms do not include all the items shown on the labour-overheads checklist, and adjustments may have to be made to specific rates or sections of the project to cater for special circumstances.

Prime cost of daywork An additional calculation for an hourly rate for daywork prime cost, in accordance with the *Definitions of prime cost of daywork carried out under a building contract,* published by the NFBTE and RICS, requires slightly different treatment and is again best calculated using a standard format similar to *Figure 2.6,* which gives an example for a tradesman and a labourer.

To allow for overheads, profit and any other charges, additional percentages are added to these basic daywork rates in accordance with the officially recognised definitions mentioned in the last paragraph.

Materials Costs of basic materials are readily available from the particular suppliers, and the estimator's task is to relate these to the circumstances of the project. What is equally important is that additional factors that may affect material costs are assessed as accurately as possible. Some of these include:

- waste: (a) short lengths/laps
  (b) perishable items

Annual worked hours			Hours
40 hours × 52 weeks			2080
Deduct:			
Annual holidays, 3 weeks × 40	= 120		
8 days public holidays × 8 (Good Friday, Easter Monday, Spring, Late Summer, Christmas, Boxing, New Year and May Days)	= 64		
Sickness . . . days × 8	=		
Absenteeism . . . days × 8	=	Sub-total	
Deduct: inclement weather (guaranteed time) at . . . %			
Basic hours worked		(A)	

		Worked overtime, hours	Unworked (non-productive) overtime, hours
Overtime hours			
Summer working: 39 weeks of evenings at . . . hour per day = 39 × 5 × . . .	=		
Deduct: 2 weeks annual holiday and 5 days public holidays = 15 × 5 × . . .	=		
Add: 25% for time and a quarter			
Winter: 13 weeks of Saturday mornings less 1 week's annual holiday (not 3 days public holidays) = 12 × . . . hours			
Add: 50% for time and a half			
Summer: 39 weeks of Saturday mornings less 2 weeks' annual holiday (not 5 days public holidays) = 37 × . . . hours	=		
Add: 50% for time and a half			
Totals			
Deduct: . . .% inclement weather			
Final overtime totals			(B)
Add: basic hours worked	(A)		
Total hours worked			
Add: non-productive hours	(B)		
Total paid hours			

Note Sickness, absenteeism, and inclement weather are assessed from existing company records.

Figure 2.4 Calculation of annual worked hours

Note on Figure 2.5 Both guaranteed minimum and Joint Board supplement (lines 9 and 10) are payable only on:

- basic worked hours
- wet time hours
- public holidays

which are all obtainable from *Figure 2.4*.

	Without overtime, hours	With overtime, hours
Hours: 'worked' and 'paid'		
1. Standard annual hours after allowing for annual and public holidays, sickness, absenteeism, wet time, etc. (total 'A' from *Figure 2.4*)		
2. Add: 'worked' overtime	—	
3. Total 'worked' hours		
4. Add: non-productive hours (unworked overtime, or overtime premium hours) (total 'B' from *Figure 2.4*)	—	
5. Total annual 'paid' hours		
Annual taxable pay	£	£
6. Basic cost of 'worked' hours, on which labour overhead percentages are calculated (line 3)		
7. Basic cost of 'non-productive' hours, which are included in the labour overheads percentage (line 4)	—	
8. Cost of paid hours		
9. Add: Joint Board supplement on . . . hours at . . . (see note on page 2–10)		
10. Add: guaranteed minimum bonus on . . . hours at . . . (see note on page 2–10)		
11. Sub-totals		
12. Total annual taxable pay calculation:		

	Without overtime	With overtime
12.1 Sub-totals (line 11)		
12.2 Tool money (line 16)		
12.3 Sick pay (line 18)		
12.4 Public hols (line 19)		
12.5 Wet time (line 20)		
12.6 Travelling allowance (line 22)		
13. Total annual taxable pay		

	Without overtime	With overtime
13. Total annual taxable pay		
Labour overheads		
14. National Insurance (13.5% of line 13)		
15. Holiday pay stamps (49 weeks at . . .)		
16. Tool money (49 weeks at . . .)		
17. Sick pay (. . . days at £. . .)		
18. Public hols (. . . hrs at . . .p)		
19. Guaranteed wet time . . .hrs at . . .p)		
20. Fare allowances (. . .days at . . .p per day)		
21. Travelling all'ce (. . .days at . . .p per day)		
22. Employer's & public liability ins. (2¼% of line 13)		
23. Redundancy payment pool (. . .% of line 13)		
24. CITB levy		
25. Total labour costs		
26. Divide by: total worked hours (line 3)		
27. Labour cost per worked hour		

Figure 2.5 Labour rate build-up form

Line no.		Craftsmen		Labourers	
		Rate	Amount	Rate	Amount
	Standard working hours per annum		Hours		Hours
1	52 weeks of 40 hours		2080		2080
	Less				
2	3 weeks annual holidays = 120				
3	8 days public holidays = 64		184		184
4	*Standard hours used in calculation*		1896 (36.46)		1896
	Guaranteed minimum weekly earnings	£	£	£	£
5	Standard basic rate: 49 weeks	37.00	1813.00	31.40	1538.60
6	Joint Board supplement: 49 weeks	11.00	539.00	10.20	499.80
7	Guaranteed minimum bonus: 49 weeks	4.00	196.00	3.60	176.40
8	Sub-totals	52.00	2548.00	45.20	2214.80
	Employer's labour overheads				
9	Employer's National Insurance contribution	10.75% (5.59)	273.91	10.75% (4.86)	238.09
10	Employer's contributions for CITB levy	0.41	20.00	0.08	5.00
11	Annual holiday credits: 49 weeks at £4.20	4.20	205.80	4.20	205.80
12	Public holidays are included in guaranteed minimum weekly earnings, above				
	Totals	62.2	3047.71	54.34	2662.69
	Standard hourly base rate	$\frac{3047.71}{1896}$	= 1.61	$\frac{2662.69}{1896}$	= 1.40

Figure 2.6 Calculation of 'standard hourly base rate' as defined in Definition of prime cost of daywork carried out under a building contract *(1975)*

- delivery charges: (a) initial charges
  (b) cost of delays
  (c) demurrage
  (d) insurances/duties
  (e) small loads
  (f) special dates
  (g) return on empties
- credit values: (a) re-use, e.g. shuttering
  (b) resale, e.g. sheet-piles
- conversion factors: (a) bulking
  (b) compaction
  (c) consolidation

- handling and storage: (a) labour
  (b) plant and equipment
  (c) special facilities

Most companies have their own statistics relating to waste and conversion factors, but many of these are too arbitrary for the competitive situation. Companies that analyse actual results on the types of work that they carry out are able to be more precise when the need arises.

In an inflationary situation, the cost of materials can vary very quickly, and it is increasingly difficult to assess possible trends over a contract period.

In the case of 'specials', especially those which are imported, it is now quite common to find that these are treated separately in regard to price fluctuations; this eliminates the need to incorporate contingencies as a protection against this type of risk.

Mechanical plant When calculating rates for plant, it is important to comprehend the role that the items of plant are playing on the particular project. On a larger job, or one that has a substantial plant content, it is worthwhile preparing a plant programme using a simple bar chart but listing plant items rather than operations. The chart will thus show durations, and hence utilisations, of main plant items so that a decision to hire or buy can be made more sensibly. Whether the machine is bought or hired, the true costs of operating the plant must be decided by either:

- calculating a basic operating cost, which requires an understanding of cost accountancy and plant operation;
- using a published schedule of plant rates;
- obtaining a hirer's quotation.

In certain circumstances all three sources may be used, during which careful comparison is required before final selection is made. In addition, it is necessary to consider running costs such as:

- fuel;
- oil and grease;
- operator's wages and bonus;
- any other consumable stores.

Whether plant costs are applied to unit rates or to general project overheads depends upon their use on site for:

- specific operations or sections of work (e.g. excavation for foundations or trenches);
- a number of operations or trades (e.g. tower crane or forklift truck);
- general purposes (e.g. dumpers or hand tools).

It is always considered good practice to allow for specific operations related to particular items of plant (such as transport to and from site, and erection and dismantling) to be considered separately in the project overheads, or preliminaries, since these will be costed separately by the company.

SITE OVERHEADS AND PRELIMINARIES

The cost of administering and servicing a project requires staff, plant and equipment for varying periods of time, and these can be readily obtained from a properly prepared pre-tender programme. The groups of items most commonly considered are:

- site personnel;
- site accommodation;
- main mechanical plant } if not included elsewhere;
- small plant }
- site transport service;
- scaffolding;

- temporary services;
- safety, health and welfare;
- cleaning and clearing site;
- handover and defects liability costs.

These are usually produced in a standard company schedule. A comprehensive schedule is produced in the *Code of estimating practice* mentioned at the beginning of this section, and in many companies this schedule is multi-paged.

Presentation of net cost estimates

The fact that companies are becoming more analytical and questioning at the tendering stage, and use estimating information throughout the currency of a contract, has influenced the presentation of net cost estimates.

Estimates are used as a basis of control for:

- buying,
- planning,
- incentives,
- surveying,
- financial control,

as well as contributing to the production of the tender sum. Consequently, they are now presented in a form that can be readily understood and used by all involved functions within the company; this requires that the costs of the main resources are shown separately.

Figure 2.7 shows how the pages of the Bills of Quantities can be used for this purpose by utilising the gaps between the bill items. Using this system, any necessary calculations to substantiate the individual costs can be shown on the reverse of the preceding page, which is adjacent and to the left of the page being costed. Alternatively, lined sheets are affixed to the right-hand side of each Bill page and these are then priced in a similar manner.

	L	M	P	Net	*Quant*	*Unit*	*Rate*	*£*	*p*
2/46	Excavation in trench to suit 150 mm diameter porous concrete pipe to a depth not exceeding 1.5 m, average depth 300 mm, including grading, bottoming, compacting and disposing of surplus material.								
	0.88	0.05	–	0.93	100	1 m	1–02	102	00
2/47	Excavation in trench to suit 150 mm diameter porous concrete pipe to a depth exceeding 1.5 m but not exceeding 3 m, average depth 3 m, including grading, bottoming, compacting and disposing of surplus material.								
	6.60	0.75	–	7.35	20	1 m	8–09	161	80

Figure 2.7 Analytical build-up of unit rates using existing bill pages

Stuck on right-hand side of pages of Bills of Quantities	*Labour*		*Materials*		*Plant*		*Sub-contractors etc.*		*Totals*	
	Unit	*Total*	*Unit*	*Total*	*Unit*	*Total*	*Unit*	*Total*	*Unit*	*Total*
	0.88	88.00	0.05	5.00	–	–	–	–	0.93	93.00
	6.60	132.00	0.75	15.00	–	–	–	–	7.35	147.00

Figure 2.8 Analytical build-up of unit rates using five-column 'stick-on' sheets (direct-labour example)

Stuck on right-hand side of pages of Bills of Quantities	*Labour*		*Materials*		*Plant*		*Sub-contractors etc.*		*Totals*	
	Unit	*Total*	*Unit*	*Total*	*Unit*	*Total*	*Unit*	*Total*	*Unit*	*Total*
	–	–	–	–	–	–	0.93	93.00	0.93	93.00
	–	–	–	–	–	–	7.35	147.00	7.35	147.00

Figure 2.9 Analytical build-up of unit rates using five-column 'stick-on' sheets (labour and materials sub-contractor example)

Contract: Old peoples' dwellings, Halifax — *Date: 19.3.78* — *Tender No.: 1168*

Bill No.	Description	Own work			Domestic sub-contractors			Nominated amounts		Contingency provisions	Totals
		Labour	Material	Plant	Net	2½%	5%	Work	Sups		
1	Prelims. & on costs	48 000 –	22 000 –	40 000 –		30 000 –				40 000 –	180 000 –
2	Dwellings	160 000 –	300 000 –	10 000 –	10 000 –	50 000 –		80 000 –	10 000 –		620 000 –
3	Site works	20 000 –	20 000 –	10 000 –		30 000 –			10 000 –		90 000 –
	Totals (2) & (3)	180 000 –	320 000 –	20 000 –	10 000 –	80 000 –		80 000 –	20 000 –		710 000 –
	Totals	228 000 –	342 000 –	60 000 –	10 000 –	110 000 –	–	80 000 –	20 000 –	40 000 –	890 000 –
		£630 000			£120 000			£100 000		£40 000	

Figure 2.10 Summary of cost estimate analysis

Estimated costs for house type no . . .	date				
Stage	*Labour*	*Materials*	*Plant*	*Sub-contractors*	*Total*
1 Foundations excavated					
2 Foundations concreted					
3 Foundations brickwork					
4 Foundations hardcore					
5 Foundations o/s concrete					
6 Unload to eaves					
7 Garage founds. excavated					
8 Garage founds. concreted					
9 Garage founds. brickwork					
10 Garage floor laid					
11 Foul drains					
12 Storm drains					
13 1st lift brickwork					
14 Ground floor frames					
15 2nd lift brickwork					
16 First floor joints					
17 3rd lift brickwork					
18 First floor frames					
19 4th lift to wallplate brckwk					
20 Unload to 1st fix					
21 Roof timbers/frame fixed					
22 Brickwork topping out					
23 Roof tiled					
24 Garage brickwork					
25 Garage and porch roof					
26 Felt roofing					
27 Vertical tiling & shiplap					
28 Glazier					
29 Plumber 1st fix					
30 Central heating 1st fix					
31 Electrician 1st fix					
32 Flooring & staircase					
33 Joiner 1st fix					
34 Plastering					
35 Floor screeds					
36 Artex					
37 Unload to completion					
38 Plumber 2nd fix					
39 Central heating 2nd fix					
40 Electrician 2nd fix					
41 Joiner 2nd fix					
42 Joiner kitchen units					
43 Wall tiler					
44 External decoration					
45 Internal decoration					
46 Floor tiler					
47 Fireplace fixed					
48 Boundary walls & fences					
49 Paths & drives complete					
50 Services laid					
51 Turf laying					
52 Clean out					
53 Completion					

Figure 2.11 Estimated costs for house type

Examples of this format, often referred to as 'five-column sheets', are given in *Figures 2.8 and 2.9,* where the method has been used for directly employed labour and 'supply and fix' sub-contractors respectively. This format not only states the unit price for each item but also incorporates total elemental costs, which can be collected into a summary similar to *Figure 2.10.* In this form, costs can be used to give senior management a better indication of the disposition and importance of the various resources throughout the Bills of Quantities, and thus assist in the apportionment of risk, overhead and finance at the tendering stage.

The breaking down of costs into smaller elements, as already described, is referred to as *analytical estimating,* and provides the data for other functions in the company to use for control purposes.

In estimating for either smaller jobs or package deals, both of which may require the builder to produce his own Bills of Quantities or some other form of schedule, a similar format may be used but incorporating the facility to work up measurements as well to describe the works in more specific detail. This type of approach is especially used in those companies that believe in passing on estimating information to site management, and the document is set out in such a manner that the descriptive information can be passed on while more confidential financial information on the right-hand side can be blanked out or removed. In house building, standard schedules of the form compatible with the company's activities are used; these break down the construction into small elements, all of which can be priced separately in an analytical form. A typical schedule is shown in *Figure 2.11;* this may be supported by more specific build-ups relating to each element.

Estimating in this manner, house builders are able to quickly consider alternative combinations of specifications and detail when undertaking feasibility studies and calculations in relation to land values.

TENDERING

The commercial operation of converting a net cost estimate into a selling price involves the consideration of:

- company overhead recovery;
- profit requirement;
- allowances for risk;
- adjustment for market conditions.

Contractors involved in large projects employ a flexible yet systematic consideration of these factors, while smaller firms often ignore the latter two items and apply an overall percentage to cover overheads and profit. Indeed some companies incorporate this percentage in the build-up of their estimates, thus arriving at instant selling prices. Attempting to be both realistic and competitive calls for separate consideration of each of the foregoing factors.

Company overheads

Most companies produce annual budgets of their overhead costs and it is usual for these to be apportioned between the various operating functions of the organisation. The intention is to produce correct assessment of each department's expected costs, so that the right addition can be made to ensure full recovery.

A method of preparing budgets for companies with a variety of activities is shown in the NFBTE's book *Calculating builders' labour rates and general overheads,* and this type of document is used by every company but prepared in its own style.

In stable times, with a low rate of inflation and a steady flow of work, the practice of applying a fixed percentage to recover overheads can be reasonably successful. Especially when inflation is difficult to predict and workload is fluctuating and uncertain, good budgeting practice calls for a continuous review:

- to apply the most realistic amount at any point in time;
- to facilitate decision making should overheads have to be reduced in the event of workload or overhead recovery not being achievable.

In consequence, the consideration of budgeted overhead requirement is itself becoming more analytical so that each project may have differing overhead recovery rates, dependent on the company's circumstances and methods of operation.

Profit requirement

A common practice has been to apply a fixed percentage to net cost estimates to provide a profit requirement, often incorporating market considerations on a 'what we can get away with' basis. In these situations, this percentage may be combined with an overhead recovery percentage and is referred to as *margin* or *mark-up*. In the current economic climate, a more realistic and competitive approach is required to establish not only the amount of money required to finance the project, but also for how long. Factors that can influence this amount are:

- the frequency of interim or stage payments;
- the possibility of initial deposits or payments;
- the possibility of surcharging early items of work;
- the possibility of rapid payment on completion;
- demands for payment relating to items contained in the net cost.

Bearing in mind the need to use the same money as many times as possible in the same financial year, the profit requirement is governed by the need to achieve an acceptable return on the capital employed. Hence if an arrangement can be reached where clients' money is financing the entire operation (which is not uncommon), there is no need for a positive profit requirement, and indeed it may be possible to reduce the tender amount accordingly, without placing the company's interests in jeopardy.

Risks

In tendering for work, companies must assess the risks inherent in a project, and assess a potential cost, where appropriate.

There are two main categories of potential cost that require consideration:

- increased costs due to inflation, either nationally or internationally;
- possible costs due to uncertainties relating to the particular project and the company's capability to cope.

INCREASED COSTS DUE TO INFLATION

The variety of historical information on cost increases is considerable at both government and company level, and while some companies approach the whole question of assessing cost increases in a comprehensive and analytical fashion, others adopt a more global approach. Whichever method is chosen, it requires a blend of national information, local knowledge and management skill, and company results reflect the enormous risks that are often taken.

For many years, there have been facilities available within the JCT standard forms of contract to allow for fluctuations in the costs of labour, materials and certain taxes and levies. While cost increases were fairly small, the fact that there was a shortfall in recovery of increased costs was unimportant, and the calculation at the tendering stage was of little consequence and was often ignored. With high rates of inflation, the situation has changed dramatically and consideration of increased costs is now a vital function in the tendering process. Therefore, before submitting a tender, the contractor must forecast the cost of any increases that are likely to occur during the currency of a contract, irrespective of whether or not a fluctuation clause is included in the conditions of contract.

Increased cost calculation Increased cost calculations vary from rudimentary assessment on a very broad basis to methodical calculations using carefully prepared pre-tender programmes linked to specific increased-cost forecasts. The final figure should allow for all cost increases relating to:

- labour,
- materials,
- plant,
- sub-contractors,
- preliminaries and on-costs,
- company overheads,
- profit requirement.

It follows that this exercise can be more accurately completed if the costs have been presented in an analytical form.

Under-recovery of increased costs Although the JCT standard forms of contract incorporate a clause (31) divided into sub-clauses, none of these, as currently operated, gives a complete recovery of increased costs, and it is necessary to assess the amount of shortfall and to make a decision whether to include an allowance in the tender price.

'Traditional' sub-clauses The sub-clauses (31A, 31B, 31C, 31D) in the JCT conditions of contract provide a measure of reimbursement dependent upon the combination of clauses used, and can provide what is often described as:

- *full* fluctuations (labour and materials cost and tax fluctuations), or
- *partial* fluctuations (contribution levy and tax fluctuations),

to which a further sub-clause (31E) can provide a percentage addition which still falls short of the cost increases described and listed earlier.

An example of the recovery to be expected on a 'traditional clause' contract is set out in *Figure 2.12.* The following assumptions have been made:

- Analysis of the tender sum shows that:

35 per cent relates to labour
45 per cent relates to materials
5 per cent relates to plant
15 per cent relates to overheads, profit, preliminaries etc.

100 per cent total tender sum

- Average cost increases relating to these resources are:

Labour	15 per cent
Materials	20 per cent
Plant	15 per cent
Overheads etc.	15 per cent

- The department dealing with increased labour costs is not fully efficient and only recovers 90 per cent.
- The basic list of materials is only 85% complete.

Resource description	*Resource proportions in this project* %	*Average increased cost* %	*Actual cost increase (as percentage of tender sum)*	*Entitlement from clauses 31A to 31D* %	*Efficiency of company's increased cost system* %	*Cost increase paid (as percentage of tender sum)*
Labour	35	15	5.25	100	90	4.725
Materials	45	20	9.00	100	85	7.65
Plant	5	15	0.75	Nil	–	Nil
Overhead, profits, preliminaries, etc.	15	15	2.25	Nil	–	Nil
			17.25			12.375
If clause 31E incorporated at, say, 15%						1.856
						14.231

Under-recovery: (i) 31A to 31D = 4.875%
(ii) 31A to 31E = 3.019%

Figure 2.12 Calculation of under-recovery by traditional clauses 31A to 31E in the JCT standard forms of contract

Using these figures, it can be seen that the cost incurred by the contractor is 17.25 per cent of the tender sums, and his recovery will be either:

- clauses 31A to 31D = 12.375 per cent, or
- clauses 31A to 31E = 14.231 per cent,

an under-recovery of 4.875 or 3.019 per cent, depending on the conditions.

It is therefore necessary to decide whether to incorporate a contingency sum in the tender price to allow for the risk of under-recovery.

'Formula' sub-clauses Owing to the unwillingness of firms to quote firm-price tender sums, and to provide a system more relevant to a spiralling inflationary period, the formula method of calculating cost increase payments was introduced to the construction industry, initially to civil engineering (Baxter Formula in 1973) and then to building (Osborne Formula in 1974 – sub-clause 31F).

Further formulae concerned with specialist sub-contractors have also been introduced, but, while all these give a cost increase payment governed by the movements of indices specific to the types of work being carried out, there are factors that affect the level of recovery, including:

- a non-adjustable element (public sector only), currently 10 per cent;
- differences between national and local cost increases;
- differences between the proportions of labour, materials and plant used in the build-up of indices compared with those that apply on a particular project.

While in the vast majority of cases these factors result in under-recovery, it is possible to over-recover in particular categories; in either case the tender sum may require adjustment.

POSSIBLE COSTS DUE TO UNCERTAINTIES

In addition to the increased costs due to inflation, the company must consider particular circumstances that may affect predicted performance and thus incur cost. Again, some companies carry out this part of the tendering process in a very informal manner, while others have created a checklist of factors to be considered.

The following items indicate the type of list that can be prepared and that might reveal bonus, as well as risk factors affecting the final price.

Resource	*Possible risk*
Labour	Untrained for particular type of work Militant Scarce 'Local labour' clause in contract
Materials	Unknown or untried Difficult Scarce Liable to be stolen or damaged
Plant/equipment	New to the company Worn out, unreliable or unsuited Liable to be stolen or damaged
Management	Untried } company Unreliable } company Overworked } company Difficult – client or his representatives
Finance	Scarcity/abundance – company Poor payers } client Method of payment } client
Working methods	New method Restricted access Difficult site/location Ground conditions
Time	Time penalties Delay by others Shortage of time

Finalising the tender

The process of collecting together all the sums of money to be incorporated into the tender sum is usually carried out on a tender summary sheet similar to that shown in *Figure 2.13.* (A similar format is to be found in the *Code of estimating practice,* published by the Institute of Building.) Part of this summary has been completed to show where the monies from the cost estimated analysis are placed, and the following points are worth noting:

- Allowances in relation to discounts vary from company to company and should be treated with great caution.
- Contingencies, provisional sums etc. are deemed to include overheads and profit.

No attempt has been made to indicate the amounts relating to overheads, profit etc., since these will vary from company to company and contract to contract, and in certain projects from section to section. The need for a formal meeting to decide

Contract: Old peoples' dwellings, Halifax					Date: 22.3.78 Tender No.: 1168
	Labour	*Materials*	*Plant*	*Total*	*Disc.*
Own work	180 000	320 000	20 000		
Site on-costs	48 000	22 000	40 000		
	228 000	342 000	60 000		
± adjustments	–	–	–		
	228 000	342 000	60 000	630 000	
Domestic sub-contractors	*Amount*	*± Adjust*	*Total*		
Net	10 000	–	10 000		
2½% discount	110 000	– 2 750	107 250		
5% discount	–	–	–		
	120 000	– 2 750	117 250	117 250	
Nominated sub-contractors (2½%)				80 000	2000
Nominated suppliers (5%)				20 000	1000
			£	847 250	3000
Deduct discounts (2½%) = 2000 (5%) = 1000				3 000	
Net costs			£	844 250	
Add: company overhead 1. £. . . 2. £. . .					
Gross cost			£		
Add: profit £. . . £. . .			£		
Add: provisional sums (deemed to include contingencies, overheads and profit)			£	40 000	
Risks (see separate sheet)			£		
Firm-price risk			£		
Bonds and special insurances			£		
Tender price			£		

Total margin	
Overheads: Discounts: Profit:	Contract period as given: weeks Contract period required: weeks Allowance made for fixed price: yes/no
Total:	Tender prices due in:

Figure 2.13 Specimen tender summary sheet

these matters cannot be over-emphasised, and in common with other parts of the estimating/tendering process the trend is towards greater analysis and more careful consideration.

FEEDBACK

Estimators treat any feedback with considerable caution, mainly due to its unreliability in the past. Information on actual costs and performance can be misleading because, as mentioned earlier, estimates and costs are not always prepared on the same basis. If companies approach feedback in a systematic manner, however, it is possible to create a company library of relevant feedback that can be used for:

- estimating,
- planning,
- incentives,

as well as giving a series of management control parameters. In the short-term, unit labour costs can be regularly monitored and a checking system incorporated into the estimating function to ensure that estimated labour rates are, in fact, being achieved. Similarly, materials and plant costs can be monitored through a company's costing system; if this is so designed, it is possible to establish the effect of actual cost increases compared with the assessment at the tendering stage.

Feedback on performance has to be used on a more long-term basis since this information, which is usually derived from labour allocation sheets or bonus records, may have been 'adjusted' at site and may require additional research. For this reason most companies are very selective in the feedback they attempt to produce, putting their effort where there is the greatest need.

TRAINING

Most estimators have practical experience of the building industry, but there is no limit to the variety of backgrounds and disciplines to be found in estimating departments. In training and developing personnel to carry out the function, there is little doubt that a comprehensive and systematic approach is essential, and people of only average ability can give excellent results if this is the case. The trend is towards more pre-estimate analysis of the proposed tender to ensure that priority of attention is given to the important parts of the project, rather than that the estimator spends his time on the parts that interest him and that allow him to air his knowledge. A sound company policy is therefore necessary to discipline personnel so that the correct approach, from the company's point of view, is always taken.

Similarly, as the tendering process becomes more analytical, there is a need for persons involved in tendering to have a knowledge of basic finances and an awareness of the current economic, political and industrial-relations environment related to the area in which the project is to be carried out. Coupled with sound budgeting practices and a realistic presentation of estimating information, senior management quickly develops an effective approach.

FURTHER INFORMATION

Code of estimating practice, 3rd edition, Institute of Building (1973)

Amendment No. 1 to 3rd edition of *Code of estimating practice,* Institute of Building (Autumn 1975)

Code of practice for single-stage selective tendering 1977, RIBA Publications (Jan. 1977)

Davies, F.A.W., *Preparation and settlement of competitive tender for building works,* Institute of Building (April 1975)

Calculating builders' labour rates and general overheads, Building Advisory Service, NFBTE Publications (1977)

Definitions of prime cost of daywork carried out under a building contract, RICS/NFBTE (1975)

Estimating the prime costs of daywork under a building contract, Building Advisory Service, NFBTE Publications (1975)

In addition, the Estimating Information Service of the Institute of Building regularly publishes papers on a variety of matters related to specific aspects of estimating and tendering. Details can be obtained from the Sales Secretary, Institute of Building, Englemere, Kings Ride, Ascot, Berkshire SL5 8BJ.

3 PURCHASING

INTRODUCTION 3–2

PURCHASING OBJECTIVES 3–3

SOURCE SELECTION 3–6

3 PURCHASING

T.J. HOPPER, MInstPS
William Davis and Co (Leicester) Ltd

This section indicates the financial importance of good purchasing to a company, discusses the main objectives of a purchasing policy, and finally considers sources of goods and relationships with suppliers.

INTRODUCTION

It is only in recent years that the purchasing field within the construction industry has developed to become a specific responsibility of a member or a team within a company. Prior to this the responsibility for obtaining materials and sub-contract labour was generally that of the contracts manager, building surveyor or site foreman. As the industry developed and became more specialised, so therefore did the demands placed on the shoulders of individuals within the industry. Consequently those who has previously been expected to perform the purchasing task in conjunction with their more specific responsibilities found that they were unable to do so, both because of their own increased job demands and the broadening of the purchasing function.

Purchasing contribution to profits

Contractors, generally, have in the past paid too little attention to the subject of material and sub-contractor costs, which can amount to as much as fifty per cent of the total building cost. The potential of effective purchasing practice can easily be seen from the following example.

Assume that a company has an annual turnover of £1 million, out of which it makes £25 000. Sub-contract and material costs amount to fifty per cent of the total revenue. For just a two and a half per cent saving on sub-contractor and material costs, it would be possible to increase the amount of profit by fifty per cent to £37 500. To achieve the same profit improvement through turnover, it would be necessary to increase the turnover by fifty per cent to £1.5 million a year.

Price/cost

It is wrongly assumed that if a material is obtained at the lowest possible price, this will ultimately result in the lowest possible cost. This misconception is brought about because price is the most prominent feature of cost and therefore attracts most interest. The other features of cost tend to be ignored, and consequently these are the areas where the biggest losses occur.

Sub-contractors

In most organisations, the responsibility for the selection and subsequent control of sub-contractors is still that of the contracts manager. In the author's experience this appears to be an acceptable arrangement because of the contracts manager's responsibility for the total running of the job. It is therefore essential that he makes the final decision as to the labour that will be working under his control. It needs to be said, however, that this assumes that sound selection techniques are employed in the engagement of the sub-contractors; otherwise a monopoly situation might develop, which could lead to the ultimate detriment of the company. In this respect, therefore, it could be advantageous to employ the experience and professionalism of the purchasing department. Accepting that this is possibly the method adopted in selecting sub-contractors by the majority of companies, this section concentrates on purchasing as it is applied to materials and products.

PURCHASING OBJECTIVES

The objectives of effective purchasing practice can be summed up as obtaining goods of the right *quality* for the work in hand, in the right *quantity*, at the right *price*, from the right *source*, at the right *time* and in the right *place*.

It would be unrealistic to assume that all the factors mentioned are of equal importance in any given set of circumstances or that it would be possible to satisfy, to the maximum extent, all the factors simultaneously. In one instance it may be the quality of a material that is of the greatest importance, and in another instance it may be time that governs the eventual decision.

Quality

What do we mean when we talk of quality? We all have experience of buying items such as clothes and food, and can therefore appreciate that there is a wide and varied range to choose from when comparing similar items. This is true when we consider the range of materials and products offered to the construction industry. In terms of quality and price, when comparing materials and products of a similar functional ability, the adage 'you get what you pay for' is usually quite correct, and so-called bargains should be viewed with suspicion.

The problem of quality is twofold. Quality as defined in the Concise Oxford Dictionary means the 'degree of excellence'. This is the first and possibly the most difficult question that arises when deciding upon quality. What degree of excellence is required to meet the demands that will be put upon the material or product under consideration?

If a material or a product of an unnecessarily high quality is used, when in fact a lesser quality would have performed the same function to an acceptable degree, the difference in quality will invariably be reflected by a variance in price. Therefore, to avoid unnecessary expenditure, careful consideration should be given to the likely demands that a material or product is required to withstand for the purpose of the work in hand.

The second problem is that of quality assurance. Will the material or product that is chosen actually withstand the demands placed upon it? To satisfy oneself in this respect it may be necessary to visit a site where the particular material or product is being used to assess its suitability and performance. Alternatively, one can request copies of test reports that have been independently carried out, or compare the material or components specification with the standard called for by the British Standards Institution (approximately twenty-five per cent of current British Standards are devoted to materials or components used in the construction industry). New products may be initially assessed by referring to the Agrément Board Certificate, which is issued against specific products and states the uses that the product is recommended for.

Unfortunately, builders are not always responsible for writing the specification for a job. When a housing project is being built for a local authority, the specification is laid down by their own departments and it is not unusual for overspecifying in terms of quality to occur. Since local authorities are the largest single employer in the construction industry, it is necessary for building contractors to point out items of overspecification and thereby save the country what could be a considerable amount of otherwise wasted money and resources.

Quantity

It is a relatively simple matter to determine for most materials the quantity required for one house or one hundred houses, but how much extra is it necessary to add on to this quantity to cover damage, pilferage or vandalism?

The wastage allowance that is incorporated in the total ordered quantity has a direct relationship to the final cost of that material when fixed. Some materials and products require a greater wastage factor than other similar materials and products that are available for the same purpose. It is essential to have accurate information from within the company regarding the percentage wastage factor of commonly used materials and products so that overordering and underordering, both of which are costly, can be avoided.

The quantity of a material or product that is bought usually affects the unit price paid for it. To enable a potential supplier to offer a realistic price, it is necessary to present him with an accurate quantity. In some cases the supplier will specify the quantity required to make economic full-load deliveries, but when this is not the case it will be advantageous to find out exactly what he considers to be a full load, otherwise the contractor may find a fully loaded twenty-tonne articulated vehicle on his doorstep one morning.

For a contractor who has a number of small sites it may well be impossible for him to accept full loads of some materials on any one site. If this should be the case he will find himself paying small-load charges or additional transport charges or both. In order to eliminate this problem and take more economically sized loads, it may well be worthwhile considering a central location that could be used as a delivery point and stockyard. The cost of this type of operation may prove prohibitive, however, and close consideration will have to be given to its advantages and disadvantages, bearing in mind such intangible factors as the increased efficiency of the company.

As an alternative to this, the purchaser could examine the possibilities of *collecting* materials whenever uneconomical load sizes are involved. If one or two suppliers were approached with a view to supplying small but regular quantities of given materials, they might well be persuaded to offer attractive terms on a collected basis or even on a delivered basis, depending upon the location of the delivery point in respect to their regular delivery routes.

Time

This is a most important factor, but it is occasionally left to look after itself, with serious consequences. No matter how attractive an offer is made by a supplier in terms of price etc., if the material is not available or not delivered when required there is no point in considering the offer further. Promises may be made by the supplier, but it is up to the buyer to make a decision based on known facts and his own experience. If that decision is the wrong one, the possibility is that the site will be without material and the job will grind to a halt. When compared to the cost of a site standing idle through lack of material, the increased cost of the material from another source at a higher price is insignificant.

Place

The location of the work and the point of delivery have a direct bearing on the overall cost of a material. If the supplier has to travel a great distance to deliver his goods, it is quite likely that the price of the goods will be higher than that from a supplier who has only a short distance to travel. If the site is located on a regular delivery route for the supplier, there is the added benefit of less delay caused by transport holdups.

Most suppliers, these days, have terms and conditions of sale that the purchaser is bound by unless he revokes them with his own terms and conditions. One of the usual conditions of supply is that the supplier will undertake to deliver the goods to the appointed location to the extent of the nearest hard-surfaced road. This is to reduce possible delays, damage to vehicles that become bogged down in mud, and similar problems. Most suppliers will interpret this condition of sale quite freely, because they appreciate that the conditions on a building site are not ideal, but it does allow the drivers of the vehicles the option of refusing to deliver the goods to the ultimate location if they consider the conditions unsuitable. For this reason it is essential that any stores or compounds located on a site be as accessible by a hard-surfaced road as possible, and indeed that the estate roads be laid as soon as possible to allow access for delivery vehicles.

It must be appreciated that although this condition of sale is freely interpreted, it *may* be applied quite literally. The contractor could then find himself involved in the added expense of further handling to the place of work.

Price

Although price is obviously an important feature of cost, too much emphasis is usually placed upon it because it is easily understood and so prominent.

Source

The best source of supply can be defined as the source that offers the most economical service, taking into consideration all the other factors of material cost. There are three sources usually available to a purchaser: manufacturers, merchants and agents.

Manufacturer. By purchasing direct through a manufacturer, the contractor has a greater degree of control over the source because the lines of communication are shortened. It is usually necessary to purchase in large quantities if an economic price is to be achieved, and there are a number of manufacturers who discourage trading direct with a contractor by offering no discounts.

Merchant. A builders' merchant outlet is probably the most common source that building contractors use. Initially it may be difficult to appreciate why 'middlemen' such as builders' merchants are used so extensively, but there are several good reasons.

(a) They offer an off-the-shelf, take-away service of commonly used materials and products.
(b) The building contractor can buy small quantities of materials at attractive prices, because the merchant buys in bulk from his sources and passes on a proportion of the price advantage.
(c) The builders' merchant offers important credit facilities, which on average work out to be well over the two months maximum period available when dealing on a strict monthly-account basis.
(d) Manufacturers benefit from dealing with builders' merchants because they can appoint a number of merchants to cover an area much greater than they themselves could hope to.

(e) The possible financial loss is taken off the manufacturer's back by the merchants if a building contractor is declared bankrupt.

It must be remembered, however, that the builders' merchant is not a charitable institute, and someone has to pay for the services that are offered.

Agent. Invariably an agent does not have a stockholding of the manufacturer's products. Consequently he relies on the manufacturer totally to complete an order; if time is of the essence, dealing with an agent can just be an unnecessary extension of the communications network.

SOURCE SELECTION

Usually the best sources are companies that have offered their services and have taken the initiative, thus showing their willingness to trade with you. Experience on the buyer's part will show those companies who have proved their reliability in the past, and they are obviously the sources to be considered first. The important questions to consider when selecting a source are:

- What do I know of the company?
- Have they proved to be reliable on previous occasions?
- Are they financially sound, what are their resources?
- What are their existing commitments?
- What is their maximum capacity?
- Do they have the management capability to ensure that their promises can be met?

It may be necessary to answer these questions by visiting the supplier's office and works to satisfy oneself of their suitability. Another method of gathering information about a prospective supplier is to contact other contractors who have traded with him and enquire of the supplier's performance.

Pareto analysis

Pareto's Law states in essence that the significant items in a given group tend to be concentrated into a smaller part of that group. In purchasing, this can be taken to mean that eighty per cent of the money spent on materials tends to be concentrated with twenty per cent of the range of suppliers.

By referring to the purchase ledger and comparing it with the list of key materials, the buyer can isolate:

- the top ten suppliers – those with whom the most money is spent;
- the top ten materials – those on which he spends the most money.

This is a relatively simple tool of management for isolating the areas in which most effort should be directed.

Suppliers – how many?

There are two opposite dangers, having too many suppliers and having too few.

With too many suppliers it is likely that the purchasing power is spread too widely. This in effect means that the purchaser's degree of control, which is required to develop the source to the advantage of the company, is not great enough.

If a purchaser has too few suppliers of major key materials, the supplier may be tempted to increase the price of his products or services and take advantage of his position. Action to reduce this possibility can easily be taken by obtaining competitive offers from other sources and comparing them, but unless this is done regularly it could well be some time before it is noticed that an unnecessarily high price is being paid.

Registered No. 468397 England

N 1019

OFFICIAL ORDER

WILLIAM DAVIS & CO. (Leicester) LTD.

Registered Office: FOREST FIELD, FOREST ROAD

BUILDERS LOUGHBOROUGH, LEICS. LE11 3NS **CONTRACTORS**

Telephone: Loughborough 63404 (15 lines)

Date..

SUPPLIER	SITE

Please supply and deliver to the above site (carriage paid) the following:—

QUANTITY	DESCRIPTION	PRICE
	SPECIMEN COPY	

DELIVERY	TERMS

All Invoices must clearly state the following:—

NAME OF SITE . DELIVERY NOTE NUMBER . ORDER NUMBER

William Davis & Co. (Leicester) Ltd.

....

Buyer

Figure 3.1 Typical order form

DAVIS
WILLIAM DAVIS & CO. (LEICESTER) LTD.
FOREST FIELD · FOREST ROAD · LOUGHBOROUGH
LEICESTERSHIRE. LE11 3NS
Telephone: 63404 (15 lines)

Messrs.

Date..

ENQUIRY

CONTRACT..

Dear Sirs,

Will you please forward your most competitive quotation, in accordance with our terms and conditions printed overleaf, for that stated below.

Yours faithfully,
for WILLIAM DAVIS & CO. (LEICESTER) LTD.

Assistant Buyer

SUBJECT..

FIXED PRICE REQUIRED IN MONTHS:..

DATE QUOTATION TO BE RETURNED BY:..

DETAILS:

SPECIMEN COPY

TERMS ARE TO BE STATED ON QUOTATIONS

DIRECTORS. W. DAVIS · W. H. DICKINSON, F.C.A. · E. DAVIS, A.I.O.B. · T. HIGGINS, J.P., F.C.I.S.
EXECUTIVE DIRECTORS: E. R. HILL, F.C.A. · E. C. YEOMANS · J. R. MACDIARMID, A.C.W.A. · L. M. ORAM · M. SMALLMAN, A.I.O.B.

Figure 3.2 Typical enquiry form

The construction industry is traditionally reluctant to change, and it is perhaps for this reason that most small and medium size companies have a small number of suppliers with whom a working relationship has been built over a period of years. This situation has a definite advantage, in that the amount of work each supplier is getting is large as a proportion of the amount of work the contractor is able to offer. Because of this, the suppliers can be encouraged to offer the maximum price benefit together with reliable service.

When faced with equal offers from two suppliers for the same material, a method that can be adopted if one supplier is a regular supplier and the second a new one is to offer seventy per cent of the business with the regular source and thirty per cent with the other, letting each know of the arrangement and thereby creating competition between them to secure either the total amount of business or to increase their respective portion.

Relationship with suppliers

It must not be forgotten that suppliers as well as contractors are in business ultimately for profit, and without it neither can continue to operate. The job of the buyer should therefore not be to squeeze every last penny he can from his suppliers but to endeavour, by means of his training, experience and professionalism, to allow his suppliers to make a fair and reasonable profit for their services. It is, after all, the contractor who will suffer in the final analysis if the industry is governed by monopolistic concerns who are able to dictate to the buyer rather than allowing an element of negotiation to be available so that both parties can trade to their mutual benefit.

Typical stationery

Figure 3.1 shows an example of a typical official order form. This document is probably the most important of all those a buyer has at his disposal. It not only acts as a record of purchases but (far more important) it is an authorisation by the company for goods or services to be supplied and a form of assurance that they will be paid for if all the conditions of the purchase are fulfilled. For the latter reason in particular, it is essential that the official order form is used in all transactions and that a price is agreed upon and stated on the order form before it is issued.

Figure 3.2 shows an example of a typical enquiry form that can be used when obtaining quotations for goods or services. If a company wishes to impose its own terms and conditions, these can be detailed on the reverse of the form.

FURTHER INFORMATION

Dand, R. and Farmer, D., *Purchasing in the construction industry*, Gower Press
Baily, P. and Farmer, D., *Purchasing principles and techniques*, Pitman Press

4 MARKETING

INTRODUCTION 4–2

MARKETING ORGANISATION 4–4

MARKETING ACTION AND THE MARKETING MIX 4–6

CONCLUSION 4–9

4 MARKETING

P. BRIGHAM, RIBA, MIOB, MBIM, MInstM
M.I. OSPALA, BSc
Construction Industry Marketing Group, Institute of Marketing

This section discusses how a company can be oriented to its customers and markets, rather than be narrowly concerned with its production processes.

INTRODUCTION

Definitions of marketing vary widely in the choice of terminology, which is often biased towards a particular industry or need. It may be defined either from the customer's or the industry's viewpoint. Even within the building industry the differing interests of the parties lead to differing interpretations.

The definition offered by the Institute of Marketing reads: 'Marketing is the management function responsible for identifying, anticipating and satisfying customer requirements profitably.'

The place of marketing

It becomes clear after a moment's thought about the definition that the marketing function has sweeping coverage. The one phrase 'satisfying customer requirements' makes marketing responsible for the formulation, production, promotion and distribution of a company's products or propositions. Marketing is not an additional or separate activity. It is a concept that coordinates everything relating a company to its customers. Marketing means aligning all operating functions towards the evaluation and satisfaction of customer needs.

On the other hand, marketing is not the top policy-making activity in a company. That place belongs to the function labelled corporate planning, which sets the objectives towards which marketing action is to be directed.

In preparing a corporate plan the assets of a company are identified. These of course include plant and machinery, buildings and land, together with finance. These are the tangible assets, which represent the way in which the company's funds are disposed. In another, less tangible, category are the *skills* of the company, which rest largely in people.

Analysis of the factors that determine a company's future tends to bring recognition that present physical assets, such as special plant, need not dominate longer-term decisions. The tangible assets are best thought of as money that is currently invested in various items.

The corporate plan will aim to utilise the capital assets over an identified period of years. It will set objectives such as conserving the assets, causing them to grow, and generating increased levels of profitability from them. The analysis of the company's skills and strengths will be used to select the directions in which the company should move.

Where corporate planning leaves off, marketing takes over. Marketing is the means of implementing the corporate plan.

The principles outlined here are wholly valid for companies in general building and house building serving both the public and private sectors. Nevertheless the application of the principles to the building industry requires careful thought.

Building contracting is a service. Although the end result is an extremely finite object – a building – the building contractor is really providing the service of managing the erection process on site.

Contractors therefore have no product. Where marketing theory talks of products, it is best for builders to substitute the word 'propositions', and to think about the propositions that they are offering to their customers.

Speculative house builders, on the other hand, do have products and can relate their thinking much more directly to terms such as design, production and sales.

The product life-cycle concept

The objective of any company is to stay in business. As described earlier, there is usually also a desire to grow.

The 'product life-cycle theory' prognosticates that a commercial activity goes through a series of phases that form its life-cycle. The theory states that, following

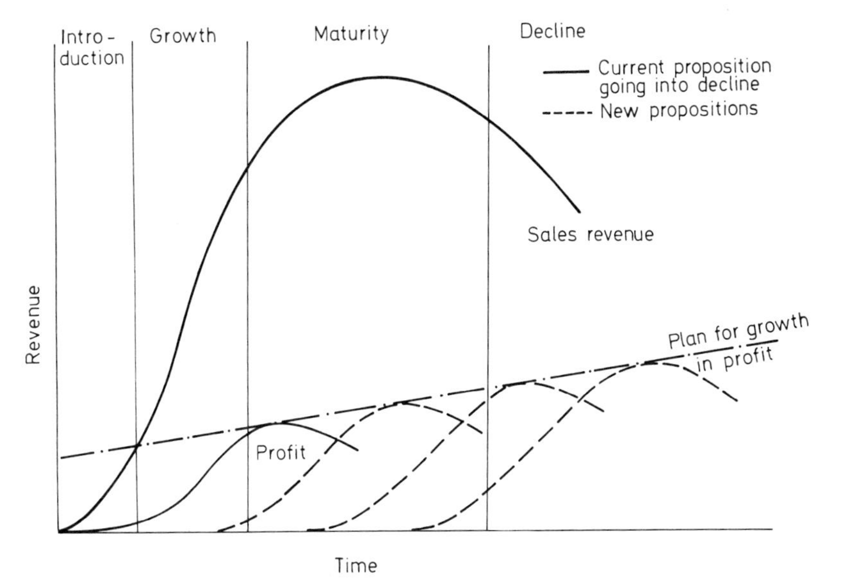

Figure 4.1 Product life-cycle

the successful introduction of a product or proposition, there will be phases of growth, maturity and then decline.

Figure 4.1 shows a hypothetical plotting of the life-cycle using two measures, sales turnover and profit. During the second stage, growth, the attractions of the market are likely to bring in competitors. The initiator will then have less opportunity to dictate price, and whilst sales continue to grow profits may fall away.

The most important inference of the life-cycle theory is that every product or proposition will in due course go into decline. The timescale is extremely variable – it may be weeks or decades – but the pattern is recognisable in almost all industrial and commercial sectors.

Life-cycle theory tells us that the things a company is doing today will not keep it prosperous in the future. If the company is to sustain its level of activity and hopefully achieve a growth target, new products or propositions must be brought forward to take over from those that go into decline. *Figure 4.1* attempts to show this graphically.

This view of the inevitable decline of today's activities provides one of the clearest rationales for marketing. The systematic identification of customer needs, followed by the formulation and promotion of new or updated propositions, is essential to survival.

MARKETING ORGANISATION

In a business context, organising means arranging activities, and the people engaged in them, in such a way as to achieve the best possible coordination and the highest possible output. In an organisation based on this precept the people involved would operate more efficiently as a group than they would as individuals.

The principles of marketing outlined above imply that organising a company to be effective in marketing is fundamental to its structure. This can be illustrated by the two contrasting manufacturing organisations shown in *Figure 4.2.* In (a), the production department (which is not in contact with customers) controls the design

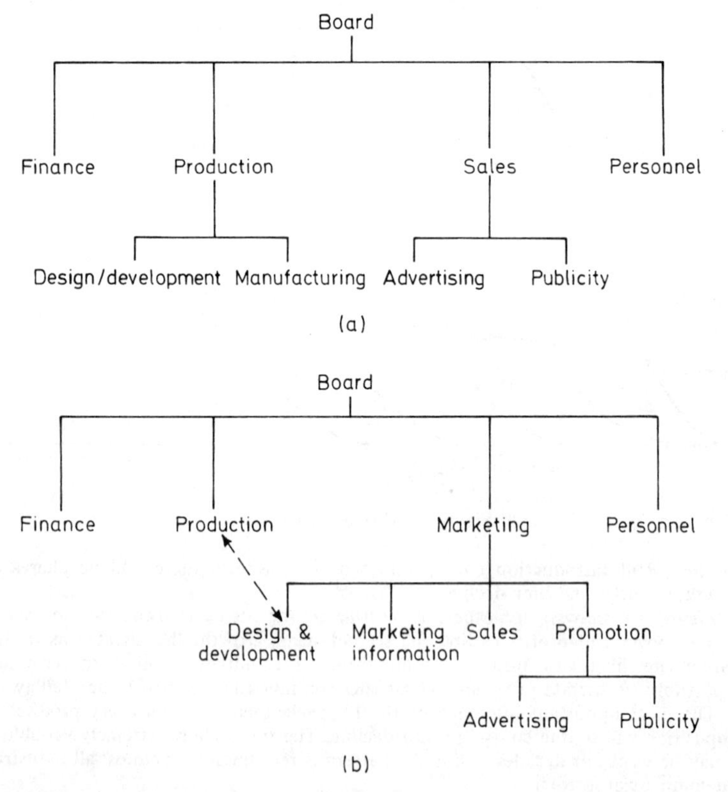

Figure 4.2 Manufacturing organisations: (a) production oriented, (b) marketing oriented

of products. The sales department, which is quite separate, has the task of selling the products devised and made by the *production men.*

The lower chart (b) shows an integrated marketing department in which design and development is combined with 'marketing information', sales and promotion. Here the designers are closely linked with those who have contact with customers. Advertising and publicity are also brought into a close relationship with product development.

The production department, in the example shown at (b), is still separate. Effective communication between marketing and production is, however, essential. The production men contribute their knowledge of how things can best be made, of manufacturing costs and of production capacity.

An integrated marketing department of the kind shown in chart (b) has the ability to forecast decline, identify new opportunities and develop and promote new propositions.

Marketing information system

A marketing information system may be defined as an interacting, continuing, future-oriented structure of people, equipment and procedures designed to generate and process an information flow that can aid business executives in the management of their marketing programmes. The essential elements of marketing information are:

- *marketing intelligence* – internal monitoring of the company's own performance, and monitoring of the external influences on the company;
- *marketing research* – specific searches for information about particular markets.

Internal monitoring Keeping a sensitive finger on the company's financial pulse is at the heart of business management. In a company structured for marketing, however, it will be particularly recognised that information is needed about each of the company's activities separately.

A company that plans its development in terms of increased profitability will be particularly interested to know which parts of the organisation are making the most contribution to profits, and which the least. For large companies performance in different geographical areas may be analysed in profitability terms.

Results also need to be sifted activity by activity or product by product. Analysis of the market performance of each product or activity can indicate the stage of the life cycle it has reached. In particular the onset of the decline stage needs to be predicted.

For a firm of building contractors the 'activities' requiring analysis might be open competitive tendering, selected tendering and variants on negotiated tendering. A firm of speculative house builders on the other hand might wish to analyse the sales and profit performance of each of its house designs.

External monitoring Businesses do not, of course, enjoy full self-determination. The performance of the company, referred to above, is influenced – even dominated – by external forces. In addition to analysing what is happening to the company, marketing intelligence has the even more important role of trying to understand *why* it is happening. Beyond this again is the task of predicting what will happen next.

The external influences on any one company are bewilderingly varied. Geographically they range from global to local happenings. In time-scale they vary from the gradual trend to the sudden impact.

Categorising these factors in the simplest way, they may be said to fall under the headings:

- economic and financial,
- political,
- social and demographic.

Changing external factors demand a response from each company. It is not possible to stand still and stay in business. Within each industry the responses of individual companies become the smaller-scale influences on their competitors. A further vital task within external monitoring, therefore, is that of collecting information about these competitors.

Search for opportunities The marketing research element of the marketing information system is mainly used in searching for opportunities. Marketing research is the term applied to special, one-off investigations, as against the continuous routine of market intelligence.

The phrase 'opportunity search' may be interpreted broadly and quite literally. It is, of course, associated with the development of new propositions, but it is equally applicable to a firm's existing propositions and markets.

A company that is already selling a proposition to a particular set of customers may wonder whether there is an opportunity to increase these customers' rate of consumption. Similarly, the company may seek ways of increasing its penetration of the existing market, either by attracting entirely new users or by winning sales from competitors.

Where marketing intelligence indicates that a proposition is going 'over the hump' and entering the decline phase, marketing research may look for ways of re-vitalising that proposition's appeal to customers – perhaps by making it more fashionable.

A common marketing research task is the exploration of new geographical territory for the company's products and services, to establish whether equivalent customers exist in these locations. Alternatively, a company may look for opportunities to sell its existing proposition to quite different types of customers.

Marketing research directed towards these aims starts with the product or service already defined and tries to gauge the reaction of segments of the total market to the proposition. The development of *new* propositions on the other hand requires an entirely different approach, as follows.

The definition of marketing set out at the beginning of this section tells us that the function starts with the identification of customer requirements. The satisfaction of these requirements, through the formulation of new propositions, is a subsequent step. Marketing research, which seeks opportunities for *new* propositions, should 'listen with an empty mind' to the many 'need signals' that come from the market. To identify the signals that are of interest to the company a filtering process is required, which eliminates by stages the 'need signals' that are too weak, too inaccessible or unrelated to the company's skills.

MARKETING ACTION AND THE MARKETING MIX

In the definition of marketing, 'satisfying customer requirements' follows the identification of the requirements. It was noted earlier that this three-word phrase, which slips off the tongue so easily, covers almost all of a company's activities. Satisfying customer requirements means devising the company's responses to identified needs, and activating these responses right through 'from factory to fireside'. This is marketing action.

The action of marketing needs to be seen as a unity, bringing maximum effort to bear on the achievement of objectives. In order to analyse and understand marketing action, however, it is necessary to distinguish its elements. These may be listed as:

- the proposition: product/service
  packaging
  brand
  price

- methods and tools: distribution channels
  personal selling
  advertising
  publicity
  sales promotion

In this context the term 'the marketing mix' is often used. The word mix, as in concrete mix, suggests proportioning the ingredients in order to find the optimum balance, giving the best results for the resources used. The last four items in the list are, in similar vein, referred to as 'the promotional mix'.

The proposition

The term proposition has already been explained. It is one identifiable offering by the company to its customers. Propositions are often complex, particularly in the construction industry, but the four subsidiary elements listed above help to categorise their basic ingredients.

Product/service A product such as a packet of washing powder is readily envisaged. Few propositions, however, are all product to the exclusion of service. For example, the washing powder manufacturer probably includes delivery in his proposition to his immediate customer, and thereby introduces the service element.

Propositions involving every relative balance of product and service will come to mind, through to those (such as the professions) that are almost purely service.

Decisions on the degree of service to offer with a product are extremely important in the formulation of propositions. Think, for example, of aluminium windows, where the proposition offered to the householder may include the services of designing, installing and glazing.

Packaging Packaging is an element of the proposition that may have three functions:

- decorative and promotional,
- protective,
- as a means of assembling units or diverse items into a 'package' that reflects the customer's needs.

Although packaging is a term usually associated with finite products, the third of these functions can readily be transposed to service propositions. The term 'package deal' in the construction industry, for example, suggest the bringing together of a group of items that it is thought the customer would like to buy in one 'deal'.

Brand Any one proposition in marketing may be supported by its association with a brand. The brand is usually common to a number of propositions – for example Heinz and the 57 varieties.

When a new proposition is being introduced to the market there are obvious advantages in its being associated with a familiar brand name. Similarly, when a proposition is being offered in competition the association with a trusted brand may prove to be the deciding factor in selection.

The adoption of the brand image is very familiar in building, both in general contracting and in house building.

Price Price is inseparable from the product or service. It is often the factor around which the rest of the proposition is developed.

Pricing may be approached in three ways. Cost-oriented pricing builds up the price on the basis of what it costs to provide the product or service, plus a mark-up for profit (cost-plus pricing). Demand-oriented pricing, on the other hand, sets the price in relation to what the market will bear. This approach produces, for example, the concept of price discrimination, where the same product may be priced differently according to the geographical location in which it is offered. It also introduces psychological pricing, where the price simply reflects the seller's assessment of the strength of the buyer's desire for the product. The third approach is that of competition-oriented pricing, where a seller's price is determined by his observation of his competitor's prices.

Clearly all three pricing attitudes occur in the building industry. Competitive tendering is a combination of cost-oriented and competition-oriented pricing. The pricing of speculative houses is demand-oriented.

Methods and tools

The elements of marketing action grouped under the phrase 'methods and tools' are the techniques in general use for distributing and promoting propositions.

Distribution channels Distribution aims to make the proposition available to all worthwhile customer groups. Devising and managing distribution systems for physical products is a major subject in its own right. It is certainly one of the most crucial factors in the success or failure of a marketing plan.

The distribution of a service such as building contracting is equally important but of a quite different nature. The things to be made available are ideas, techniques and know-how. The initiators of a superior proposition may seek to harvest increasing rewards from their brainwork by acquiring other companies or by franchising.

Personal selling This is the first item of the sub-group of four known as the promotional mix. Personal selling covers all uses of sales personnel, either as representatives or in retail establishments. It also covers the sales activities carried out by directors and senior management, face to face with their counterparts from customer companies.

Advertising Advertising consists of non-personal forms of communication conducted through paid media under clear sponsorship. By actually buying the time or space in the media the advertiser is able to completely determine the messsage he wishes to transmit to customers.

Publicity Publicity, in the sense of commercial promotion, means getting the company's name into the media without making direct payment for space. Basically, this means creating news-worthy events such as 'firsts', official openings or success stories.

Sales promotion Other promotional activities not so far defined may be grouped under the term 'sales promotion'. They include the use of exhibitions, displays, demonstrations, special offers and point-of-sale displays.

Marketing action uses all the elements of the marketing mix and should aim to bring them into balance and harmony, like the sections of a symphony orchestra. There is complete interaction between the elements. Each element should be considered in the development of a marketing proposition. The way in which a proposition is to be distributed, and the ways in which it is to be promoted, will materially alter its formulation or design.

CONCLUSION

In essence marketing means turning the whole of an organisation to face the market, tuning in to listen to the needs of customers and matching the organisation's offerings to those needs. To be effective and profitable a company should have a plan that, at least for a short period, sets its activities in specific directions. Marketing will therefore concentrate on the particular customer groups envisaged in the plan.

It should then be recognised that the needs of these selected customer groups will develop and change with time. Marketing must adapt and develop the company's propositions in response to these changes.

A sizable building company may have within it divisions engaged in design and build, competitive tender contracting, minor works and speculative house building. Other companies, particularly smaller companies, may be engaged in only one of these activities.

Building houses for sale can be fairly readily equated with the marketing concepts and terminology developed in the consumer goods industries. Product, product development, price and promotion are finite and identifiable. Marketing skills in house building need to centre on keeping the proposition matched to changing customer needs, with particular reference to finance.

General builders and building contractors, large or small, should recognise that their basic proposition is the provision of service to customers. Although builders are dealing in extremely tangible bricks and mortar, customers are looking to the firms for organisational and managerial skills. Most customers want a proposition that 'takes the trouble away'. Marketing for builders therefore means developing trading relationships that make it easier for customers to buy the service.

5 INSURANCE

WHAT IS INSURANCE? 5–2

INSURANCE OF WORK IN PROGRESS 5–3

INSURANCE AGAINST PUBLIC LIABILITY 5–5

INSURANCE AGAINST LIABILITY TO EMPLOYEES 5–6

PROFESSIONAL INDEMNITY INSURANCE 5–7

OTHER INSURANCES 5–7

5 INSURANCE

J. K. TUNNICLIFFE
Director, Stewart Wrightson (Contractors Insurance Services) Ltd

Insurance may be the only barrier between a builder and bankruptcy. This section indicates the main types of insurance that a builder may need, including principally work-in-progress, public-liability, employer's-liability and professional-indemnity insurance.

WHAT IS INSURANCE?

Many misconceptions exist about insurance. *The simple fact is that it may well be the only barrier between a builder and bankruptcy.* So it is an imprudent contractor who does not study it carefully or seek skilled professional advice. If underlining of this philosophy is needed, readers are recommended to consider the industry's bankruptcy statistics.

One of the most common misconceptions is that 'the insurance' is much the same whatever its source. In fact, whilst it is a fallacy to say that anything is capable of being insured at a price, the flexibility and non-standardisation of the British insurance market allows a wide degree of choice as to what may be covered. But with this goes the problem of selecting cover appropriate to a builder's needs.

Superficially the policies issued by most insurers look alike, and only careful scrutiny of the policy terms, conditions and (above all) exclusions will demonstrate what are often vitally important differences. At the end of the day, as in so many things, quality is rarely represented by the seemingly cheapest, but skilled marketing can produce the twin ideals of adequate cover at an acceptable price.

A difficulty often met is the intangible nature of insurance. Builders are used to physical things that they can buy via a reliable pre-known specification or measure, or can test before use. None of these things applies to insurance; it cannot be tested until the day when it is needed, at which time the builder must hope it will not be found wanting.

Insurance is (or should be) about the sensible and economic protection of the builder's assets and liabilities. Cover on such things as office typewriters and the tea-boy's bicycle is an irrelevance. The general object should be to pinpoint those risks that are capable of seriously threatening the profit from a job, or indeed the overall results of the company.

To some extent a line has to be drawn between builders engaged solely on private speculative development and those involved in commercial work for local authorities and others. In the former case, the builder has freedom to insure or not as he chooses. On commercial work, there will be a contract demanding that the builder insures against certain specified risks. But the principle remains as stated. It has to be remembered that contractually demanded insurance flows from the fear of employers that, without it, the builder may not be able to meet his obligations under the contract. The underlying *risk* to the builder remains, and often what is contractually demanded falls significantly short of what he should prudently be insuring to protect profits. And this *risk* applies just as much to speculative work as to that under commercial contract.

The scope of this section does not allow the comments that follow to do more than give general guidance and warnings on insurance policies designed to cover various aspects of the risks falling upon builders. They cannot possibly be exhaustive, and builders are urged to seek skilled professional advice, the obvious source of which must be an insurance broker with demonstrable experience of the industry. Direct dealing with an insurance company is entirely permissible, but builders are again reminded of the significant variations in cover between different insurers, who, for entirely understandable commercial reasons, are unlikely to offer totally impartial advice if what is required of them falls short of what the builder feels he needs. Governed no doubt by the sadly mistaken view that 'the insurance' is much the same whatever its source, many builders do deal with insurance companies primarily because of the 'own case agency' system, which gains them commission on their own business.

With regard to insurance premiums themselves, there is no 'tariff' followed by any insurer. Each makes up his own mind what he will charge, but inevitably the main influencing factor is the relationship between claims and premiums.

Significant premium economies are possible if the builder is prepared to accept a degree of self-risk. For instance, he may undertake to pay an agreed amount of each claim, the cost saving obviously rising in some proportion to the amount in question. There usually comes a point of optimum cost saving measured against the cost and incidence of claims. Each case has to be judged on its history and merits.

INSURANCE OF WORK IN PROGRESS

A sensible starting point is to consider what is represented by work in progress. In the case of a speculative house builder, it is a valuable company asset, from the eventual sale or perhaps letting of which profit will hopefully ensue. To the builder working on a commercial contract for an employer, it represents something that he has to produce and hand over in good order before he obtains final payment and again hopefully makes his profit.

In short, the works represent the potential profit that the business needs to survive and prosper. No prudent contractor therefore leaves it unprotected against physical damage. The decision really has to be how this is best and most economically achieved. That the work is on a commercial contract in no way affects the validity of this statement. The only difference is that the threat to the contractor's profit is represented by his irrecoverable rectification of his outlay following damage to the works for which he is contractually responsible.

Against what risks should the contractor insure? The answer will depend upon the commercial position of the contractor and upon those matters for which he finds himself with a liability to rectify.

Speculative developments

Since the builder is working for himself, obviously there will be no contractual *duty* to insure. But the potential threat to profit remains. Damage following fire, lightning, explosion, storm or water damage would have to be rectified at the builder's own cost. Many sites also suffer the 'running sore' of theft and vandalism.

For maximum reassurance, the speculative builder should insure his works against all risks of physical loss or damage under what is known as a 'Contractors All Risks' policy. Whether he does so is up to him. There is no compulsion to insure anything.

Experience shows practice to vary, with much depending upon the house builder's financial strength. If, as with so many of the small firms, this could not without embarrassment withstand a serious loss costing thousands of pounds, sensibly 'Contractors All Risks' cover should be the answer. The cost of insurance is infinitesimal compared to overall costs, and can be passed on in the sale price without rendering this uncompetitive.

Some builders compromise by insuring only against what they see as a source of major profit-affecting loss, e.g. fire, lighting and explosion, or perhaps even fire only. Some prefer to insure on an 'all risks' basis but accept a pre-agreed amount or percentage of each loss for their own account, so minimising premium outlay.

The nature of the work also has to be taken into account. If the builder erects only detached or, at most, semi-detached houses, a reasonable assessment can be made as to the worst he could lose in any given event. Perhaps it would be only £20 000 or less, and this the builder may feel he can stand. So, measuring the premium cost against the probabilities of the loss happening, it may be a worthwhile risk to run. On the other hand, the work may involve terraces of houses or low-rise flats, in which event the aggregation of value at risk would be vastly higher and a major fire would cause real financial problems.

Again, violent storms affecting a widespread area are not unknown in the United Kingdom, and half-completed houses are an obvious risk. The damage to each house may be only a few hundreds but repeated to greater or lesser degree across, say, twenty houses it may add up to a very significant total.

Every speculative house builder should prudently assess his own position against the risks he sees himself running. Optimism should never be allowed to rule over realism when not just profit but perhaps the builder's continued existence could be at stake.

Speculative builders should beware the naive assumption that because they sub-let much, if not all, of the work, their risk of financial loss is insignificant. This is particularly so where use is made of 'labour only' sub-contractors, who by and large will be unsophisticated as to their legal liabilities and even more so as to the need for insurance.

It may be argued that such sub-contractors can be tied up via sub-contract conditions. But these do not always exist, or somehow never get to be signed and returned. If they do exist, and are carefully drawn up, in many cases the liabilities so passed on are meaningless when the sub-contractor is a man of straw without equally well drawn up insurance protection. In addition, the practice usually is to sub-let the work between various trades, and it is most improbable that each trade, if any, will carry insurance on its particular part.

It is the builder-developer who stands in the front line and must face the loss. To rely totally upon sub-contractors, especially 'labour only' sub-contractors, as a source of redress is wishful thinking indeed.

Commercial developments

Most such work will be on the JCT Standard Form of Building Contract. A striking point about the JCT Form is that it contains no specific reference as to the contractor's *responsibility* for loss or damage to the works, yet in clause 20 there appears a fairly precise *insurance duty* requirement.

Builders under commercial contract must remember that, except to the extent that they may have contractual relief (e.g. via clauses 20(B) or 20(C) or the proven error of the employer, his representatives or agents), their contractual duty to make good is virtually absolute. Sensibly therefore they should effect a 'Contractors All Risks' policy rather than one restricted to the perils referred to in clause 20(A). They should also bear in mind that the escape provided by clauses 20(B) or 20(C) relates *only* to the 20(A) perils, which by no means represent all the sources of major loss.

It should also be noted that the JCT Form imposes no insurance duty in regard to loss or damage manifesting itself during the defects liability period. Nevertheless the builder remains responsible, although only for such loss or damage resulting from his work done prior to practical completion or what he may cause in the process of complying with his rectification responsibilities during the defects liability period. Any policy that he effects must necessarily reflect this situation, and in no circumstances must cover be allowed to terminate at practical completion.

Stress is also laid upon the comments made earlier (in connection with speculative builders) as to the possible folly of relying upon sub-contractors for escape. In commercial work this is even more relevant, since 'labour and materials' sub-contractors, nominated or direct, will normally be employed under the 'green' or 'blue' standard form of sub-contract, which absolves them from liability for the clause 20(A) perils and in other respects may fairly be said to be somewhat ambiguous and uncertain of interpretation.

A 'Contractors All Risks' policy can also be the most economical medium through which to obtain cover on constructional plant, equipment and temporary buildings. But it has to be remembered that, even under commercial contracts, there is no compulsion to insure such property. There may well be a case for non-insurance of low value and/or heavily depreciated plant, especially if it is used primarily on housing sites where the risk is statistically low. But equally it would be foolish for a builder to ignore the value tied up in high-value items such as cranes and excavators.

Both for reasons of economy and simplicity (but also as a precaution against inadvertent omission to insure), it is usual to effect a 'Contractors All Risks' policy on a 12-month annually renewable basis at a pre-fixed and standard premium rate automatically applicable to all work. Such a rate is normally applied to annual turnover, i.e. monies received or receivable in payment for work done.

INSURANCE AGAINST PUBLIC LIABILITY

Public-liability risks relate to the responsibility at law for death of or injury to persons, or loss of or damage to third-party property (including existing property of any employer).

Builders should always bear in mind three important factors:

(a) Theirs is an industry traditionally heavily productive of accidents on (or flowing from) sites or site work, and third-party claims not only represent probably the largest proportion of all insurance claims but frequently produce awards of quite staggering amounts.

(b) Consumer-orientated legislation has grown apace in recent years and, increasingly, Common Law decisions extend earlier case law to the disadvantage of the defendant. Equally, the public has become far more aware of its legal rights and has little cost deterrent in pursuing them.

(c) In this field, more than any other, what the insurance market offers varies widely between insurers, as to both cover and cost, and therefore the utmost care has to be exercised in the purchase of protection.

Special stress is laid upon the need for a realistic insurance amount. Builders must view this in relation to the nature of their work, but in today's circumstances a figure of anything less than £1 000 000 would be ill-advised. Premium cost does not increase *pro rata* with the amount insured, since insurers acknowledge the probability that catastrophic claims occur rarely.

Builders under commercial contract are reminded that, because the Bills may require a specific amount of public-liability cover, in no way does this limit or affect their true exposure at law, which will always be totally unlimited. The Bill item represents no more than the employer's view as to a 'safe' figure, and it is often grotesquely inadequate.

Space does not permit an exhaustive examination of the variations between the policies on offer from different insurers, but among the 'musts' for builders are the following.

(a) There must be no exclusion of liability specially assumed under contract or agreement. Such liability is imposed upon builders, both inside and often outside the formal building contract, and must automatically be covered.

(b) Any exclusion of damage to property in the builder's 'care, custody or control' must not apply to property or its contents worked in or upon.

(c) There must be no exclusion of 'technical or professional advice', these being matters in which builders are probably more involved than they realise.

(d) There must be no exclusion of damage to property caused by vibration, or removal or weakening of support.

(e) There must be no exclusion for defects in goods sold, supplied, repaired or treated.

It is emphasised once again that builders must have access to skilled, professional advice as to the acceptability of various aspects of policy wordings and that the confines of this section do not permit anything approaching a detailed examination of what is a complex matter. It is also important to examine public-liability policies as to their interaction with other policies held, e.g. employer's-liability and motor-insurance policies, which the mechanics of the insurance market dictate must be separately effected.

Public liability policies are normally issued on a 12-month annually renewable basis and premium-rated on annual wageroll.

INSURANCE AGAINST LIABILITY TO EMPLOYEES

A public-liability policy will exclude liability to persons arising out of or in the course of their employment by the insured, for which an employer's-liability insurance policy is required.

Employer's-liability insurance is now statutorily compulsory, but nevertheless builders still face the problem of a significant degree of non-standardisation in the cover offered by the insurance market. Superficially, employer's-liability policies appear more alike than public-liability policies, but the 'catch' frequently lies in restrictive endorsements affecting the type of work that is permitted without special reference to, and agreement with, insurers. Far too many builders fall into the trap of forgetting that policy exclusions perhaps acceptable when the policy was first effected may become highly dangerous as time passes and work becomes more diversified.

There is no substitute for skilled, professional advice and a constant review of the scope of insurance purchased against what the builder actually does or (even more important) may do. It makes much more sense to pay a little more premium and so avoid restrictions as to permitted work, rather than rely upon the memory or foresight of those responsible for insurance or of estimators whose familiarity with insurance must inevitably be questionable.

Normally, employer's-liability policies are issued for an amount totally unlimited, so the problems in this respect met on public-liability insurance do not arise. Care should also be taken to see that an employer's-liability policy includes liability for death or injury to 'labour only' sub-contractors and their employees. There is no reliable case law on this subject, and builders should prudently assume that they have similar liability to such people as they have to their own direct employees.

As in the case of public-liability insurance, employer's-liability insurance policies are normally issued on a 12-month renewable basis and premium-rated on salaries and/or wageroll.

PROFESSIONAL INDEMNITY INSURANCE

Particularly in the case of speculative house builders, but increasingly in commercial contracts, the builder may be required to exercise wholly or partially the function of the architect. In such circumstances, he will be concerned not just with the possibility of personal injury or physical damage to property but also with the possibility of the work simply failing to fulfil the purpose for which it was designed.

No more than general comment can be made, but this matter could involve heavy financial loss, so it is important that the builder seeks professional insurance advice as to where his financial exposure may lie and how far, and at what cost, this may be protected by insurance.

OTHER INSURANCES

In these areas, the needs of a builder are not markedly different from those of any other industry and space does not permit more than passing reference.

Fire and allied perils on buildings and/or contents Probably the most important point is to see that insured values are kept in line with current replacement cost.

Motor insurance With ever-increasing garage repair costs and inflationary third party awards causing all insurers constantly to increase their premiums, there is probably little point in constantly switching insurers. More important is an efficient claims service and the preservation of goodwill.

Personal accident and/or sickness insurance While this is no more than (at most) a moral obligation, there can be merit in attracting or retaining valued staff or site management if attractive insurance schemes can be offered.

Insurance of money In an increasingly criminal world, well-designed insurance is essential.

Engineering insurance/statutory inspections Particularly against the background of the Health and Safety at Work etc. Act, and earlier statutes, it is important that all types of plant and equipment are well maintained. Specialist engineering insurance companies can offer a valuable inspection service and, in some cases, an accident-prevention advisory service.

6 BUILDING CONTRACTS

SURVEY AND PRINCIPLES 6–2

JCT FORM WITH QUANTITIES, PRIVATE EDITION 6–3

OTHER JCT AND RELATED FORMS 6–17

6 BUILDING CONTRACTS

D.F. TURNER, BA, FRICS, AIQS MIOB
Senior Lecturer in charge of Building Economics Section, Robert Gordon's Institute of Technology, Aberdeen

This section outlines the main standard contract forms available for use between parties to building contracts, with emphasis on the practical procedural operations entailed. The discussion is restricted to those forms commonly occurring within the house building sector, and is preceded by a survey of the relationship of the forms to one another and to the law in general.

SURVEY AND PRINCIPLES

There are two broad approaches to the construction of a building by one person for ownership by another. The first is for the builder to erect it to his own design, or one that he has commissioned, and for the other person then to purchase the land and building from him. This is the approach used in the speculative housing market, for instance, whether the agreement is made before or after completion. The purchaser thus has to satisfy himself as to the adequacy of the intended purchase, either by inspection of the completed building, or by examination of a sample building (such as a show house), or by looking over plans and specifications prepared by the builder. In any case his later redress is governed by such legal matters as land law, misrepresentation and any specific terms embodied in the sale, such as the support of the National House-Building Council's warranty. This section is not concerned with these contracts of sale, beyond pointing out that they are quite distinct from the alternative approach here considered.

The second approach is that of a contract to build. While a speculative house builder may or may not agree to sell a house to a purchaser ahead of completion, in the case of a contract to build the agreement must precede the construction. There is also the difference that construction takes place on land provided by the client, and thus the land is not included in the price payable. Within this approach there are two variants. The first is that the contractor (as the builder is usually called) will provide the design service in accordance with the client's brief, and this is often referred to as a 'design and build' or 'package' service. This method is not covered by the group of contract forms to be considered, which are unsuitable in a number of respects. There are contracts available within the industry, but these have been prepared unilaterally by particular organisations.

The second variant of the contract to build is that in which the client's architect or other adviser provides the design service, and the contractor builds to the design and specification given. While the contractor is expected to exercise his skills and knowledge, he is not responsible for the adequacy of the design in terms of layout and the like. It is this variant arrangement that this section takes up by considering principally the family of contract forms produced by the Joint Contracts Tribunal (JCT), a special body constituted of members or representatives of such interests as the RIBA, the RICS, the NFBTE and local authorities and sub-contractors. The resultant contracts represent an agreed set of forms for several purposes. The main group can be divided in two ways:

1. By client: (a) private,
   (b) local authority.
2. By financial basis: (a) without quantities (a simple lump sum),
   (b) with quantities,
   (c) with approximate quantities.

Thus 1(a) and (b) are each available in three variants, 2(a), (b) and (c). In addition to this group of six, there are other forms produced by the JCT, and also a Scottish form and a sub-contract form that bear a relationship to them. The next part of this section looks in some detail at the JCT form in the private edition with quantities (1977), while the last part draws out some leading differences in the other versions.

In addition to all these forms, there are special forms used for central government contracts and for civil engineering work. These may thus be used for particular housing schemes or for site-works contracts respectively, but cannot be treated within the present compass.

Before passing on to the details of the forms, it must be emphasised that no building contract can safely be read in isolation when any important legal point is at issue. Proper legal advice should be sought if there is any doubt before entering into a binding contract, and also if any dispute arises over its interpretation during work or settling the account. Any legal document exists in the context of the wider law of the land, and must be read in the light of statute law and also case law (i.e. the results of legal decisions in the courts). It can only be said here that in such areas as insolvency, contra-charges and ownership of materials, for example, the law may be more complicated by far than a simple reading of a printed document might suggest. Indeed, this section simply provides an introduction to the subject, before the interested reader takes up the actual forms and more detailed books about them.

While the various forms do give precision to many of the intentions of the parties, it is true that a valid contract exists between them once there has been an unconditional offer made and acceptance given. This may be shown by the form of tender and a letter of acceptance, or even when there is no written evidence – provided intentions can be deduced from the actions of the parties, such as the client allowing the contractor to start work on the site. Usually there will be written evidence of the intention to use a particular form of contract and other documents.

JCT FORM WITH QUANTITIES, PRIVATE EDITION

The basis, parties and main obligations

The document is intended for use with contracts for several houses or a reasonably sized single building and upwards. It comprises three main elements: the conditions as the central bulk, preceded by articles of agreement and followed by an appendix. The conditions are intended to be used bodily as printed, subject to the deletion of several alternatives, although it is possible for the parties to the contract to amend them further. This needs care, since an innocuous-looking change may produce unexpected legal implications. The articles and the appendix contain a number of blank spaces so that those items may be inserted that give the contract its individuality. These include the names of the parties (i.e. of the contractor and of the client, who is referred to as the *employer* in the form and in the rest of this section), the title and site of the works, the contract sum and the contract period. All the items in the appendix are referred to in the conditions (which are individually called clauses), and several occur in this section.

The articles are the key to the contract, and they are the place where the parties sign or seal the contract. The most important difference in practice between *signing* and *sealing* is in the time that the contractor remains liable for any breach of his contract obligations. The period is respectively six and twelve years from the time of any breach, and thus at latest from the issue of the architect's final certificate under clause 30(6).

While only the contractor and the employer are parties to the contract, the architect and the quantity surveyor are also named in the articles. This is because they have specific powers and duties under the contract in dealing with issues between the parties, as will be considered.

Clause 1(1) states that the contract documents are the articles, the conditions, the contract drawings and the contract bills.

The articles and clause 1(1) establish the liability of the contractor to carry out the works to completion as an 'entire contract' obligation. On its own this is a very heavy assignment, so that he would be obliged to finish, no matter what catastrophe occurred, but it is substantially eased by other clauses that provide some protection against delays and damage beyond the contractor's control, for instance. While the articles state the sum payable between the parties, clause 13 states that this contract sum can be changed only for reasons set out in the conditions and in the ways specified. Despite the use of many detailed quantities, the contract is still on a lump sum basis with the quantities acting to analyse the sum, but not fragment it legally. Neither party can break off the contract part way through and settle for payment at the prices in the Bills of Quantities for only such items in the bills as happen to have been done. The individual prices in the bills are, however, used to arrive at any adjustment of the contract sum to allow for any design changes, considered later as 'variations'.

What the contractor is required to do in exchange for the payment he receives is covered by several rather scattered entries in the standard form. In addition to the title of the works and site, the articles list the contract drawings. These may be only the general arrangement drawings and site layout, since clause 3(4) provides for the detail drawings and other information to be issued during progress. The underlying idea is that the contract drawings show the type of work, while the Bills of Quantities contain the amount of work to be done, as is stated in clause 12(1). Any error in the bills is adjustable with an adjustment in the contract price, so that the quantity of work produced is matched by the payment eventually made. The implication of this pattern is that the detail drawings issued will not change the design from that covered by the contract drawings and the bills, but that they will simply 'explain and amplify' as clause 3(4) states. If there are changes they should be treated as variations under clause 11(1) so that the sum payable may be adjusted. Since the detail drawings may have been modified in the period between preparations of the bills and the issue of the drawings, or may actually have been drawn in that time, the contractor should check information carefully.

The list of contract documents excludes any specification as such, and clause 6(1) and clause 12(1) both set out that the contract bills give the quality of work required, this being in the item descriptions and also the trade preambles of the bills. Since neither drawings nor bills may show where a particular material is to be used, clause 3(3) allows the issue during progress of 'descriptive schedules', but these are not to introduce any variation and are thus on a similar footing to the detail drawings.

There remain matters over which the drawings and bills cannot be absolutely precise as to standard – such things as evenness of plaster are difficult to define, and what is acceptable on one type of building may be below standard on another. Clause 1(1) therefore gives the architect a right to 'reasonable satisfaction' in such matters as have not been defined. 'Reasonable' gives a right to arbitration to determine whether the standard sought is too demanding.

Should the contractor find any discrepancy between or within the various contract documents or any of the amplifying documents, clause 1(2) requires him to notify the architect so that the doubt may be resolved. What is not clear is what should happen if the contractor does not notice a discrepancy until he has performed work that has to be altered on site or at works. He is not absolutely obliged to find snags in advance, and the sequence and timing of information may make it difficult if not impossible to do so. On the other hand he is expected to possess the skill usual in the industry, and he should therefore look for discrepancies and, for that matter, for errors in design that are not discrepancies. He will be liable for carrying on rashly.

What the contractor may not do is to undertake any design work of his own, whether to correct errors or to save waiting for information. Even more, he must not change the architect's design as embodied in the contract documents and in any authorised variations. Only in any detail where the contract specifically requires him to undertake design must he do so, and even here he usually has to obtain the architect's approval to his proposals. These points are not explicit in the conditions, but are implied in the whole philosophy of the contract.

Under clause 4(1) the contractor is required to comply with statute, by-laws and the like – as legally he must. If again this uncovers any discrepancy, this time between the contract provisions and the statutory matters and so forth, the contractor is to seek instructions from the architect – unless it is necessary to act in an emergency, perhaps over safety. The clause is not easy to interpret and should be read very closely in any case of complexity.

These are the main provisions about carrying the works through in general. Clause 8 requires the contractor to keep a 'foreman in charge' on the job. The title is not significant: what is important is the everyday presence of someone to coordinate the site and to be available for receiving instructions from the architect. This means that most matters can be dealt with on a quite direct basis, while leaving just a few points of a serious nature that call for formal correspondence with the contractor's registered office.

Clause 5 puts the responsibility on the architect to give the contractor sufficient levels and other setting-out information to position the works correctly, while leaving the contractor liable for any errors in using this data.

Architect's instructions and approvals

Not only does the architect design the works before the contractor appears on the site, but he also has numerous rights and duties during and after construction. In view of the nature of the contract as an 'entire contract' it is necessary for these rights and duties to be made explicit, because otherwise the architect would have no right to affect events at all – in fact not even to go on the site. It follows that he can have no powers other than those spelt out in the conditions, extensive though these in fact are. By contrast, the employer has no powers in the matters reserved to the architect and has few formal points of contact with the contractor. In particular he is not entitled to give the contractor any instructions, and so he must act through the architect if he wishes to have the design modified, for instance.

INSTRUCTIONS

The key clause over the architect's instructions is clause 2. This gives him the power to instruct and requires the contractor to comply. If the contractor does not do so, there is a procedure for bringing in other persons after due warning to perform the subject of the instruction, but not to take over the rest of the works. The clause also contains a procedure for the contractor to query whether a particular purported instruction falls within the architect's powers. This means that the architect then has to quote the appropriate clause of the contract. As on nearly everything under the contract, there will remain a right to arbitration over the architect's powers here. An outline of the main subjects on which the architect may or must instruct is as follows:

Clause 1(2) Resolving discrepancies
Clause 4(1)(c) Statutory requirements
Clause 5 Information for setting out the works
Clause 6(3), (4) and (5) Inspecting and testing work, removal of defective work and materials, dismissal of persons
Clause 11(1) Varying the design of the works
Clause 11(3) Expenditure of prime cost and provisional sums in the contract bills
Clause 15(3) Defects after completion of the works
Clause 21(2) Postponement of work during progress
Clause 27 Nominated sub-contractors and their work
Clause 28 Nominated suppliers and their materials
Clause 32(2) Protective work during hostilities
Clause 33(1) Action over war damage
Clause 34(2) Action over antiquities and the like found while excavating

In addition there are several matters on which the contractor may have to act and which are then 'deemed to be variations required by the architect', the most important relating to restoring the works after damage covered in the insurance clauses of the conditions and the war damage clause.

In total these add up to a formidable list of powers – although they are counter-balanced where appropriate by a right to modified reimbursement for the contractor. A number of them are touched on elsewhere in this section. What none of them do is to give the architect any power to interfere with the contractor's site organisation or detailed operations to produce the works. The nearest approach to this is in the powers to require dismissal and to instruct postponement. But unless the technical parts of the contract documents specify in advance a particular sequence of work, a particular layout of site huts or a particular mode of mixing concrete (to give but a few examples), the contractor is entitled to proceed by any reasonable way that does not infringe any wider regulation. If he feels it to be expedient to comply with the architect's desires, however, he should ensure that his right to any extra payment is agreed in advance.

Since instructions during progress are notoriously difficult to track down at the time of final settlement, clause 2 requires these to be given in writing before they are effective and must be obeyed. There are two fall-back positions, however. Firstly, the contractor may confirm in writing any oral instruction, and his obligation to obey then runs from the end of a waiting period to see whether the architect wishes to dissent. Secondly, the architect may still confirm any instruction given orally right up to final settlement. This he should reasonably do, although there can be the possibility of argument over what was said or whether anything was said at all. But unless there is a written instruction the quantity surveyor has no power to include any extra payment in the final account (although he equally has no right to include any reduction of payment without a covering instruction).

Instructions under clause 11(1) and clause 11(3) listed above are among the most common. A variation is defined in such a way as to mean a change in the design of the works or of materials or workmanship used in them. The architect has discretion to instruct any variation before practical completion occurs under clause 15(1), although if it is awkwardly timed it may lead to additional payment under clause 11(6). He cannot instruct the contractor to introduce a change so drastic as to alter the scope of the contract, such as to halve or double its size or to substitute a building of different type or constructional nature. A variation is intended to be comparatively incidental, although a number of variations may add up to quite a significant difference and the contractor may be led into this a stage at a time without realising it. If this happens he may be entitled to some drastic revision of his payment, while if he sees it coming he may be entitled to decline to accept the instruction as beyond the architect's powers. On the other hand he may welcome the business and wish to accept, provided he can negotiate a suitably amended contract.

Clause 11(3) deals with instruction over the spending of prime cost sums and provisional sums as listed in the Bills of Quantites. These occur when sums of money are written in to cover work about which the architect wishes to reserve giving instructions until the contract is under way. Prime costs sums are used to deal with nominated work, about which more is said later. Provisional sums relate to work that the contractor may well perform in the end, but that is initially uncertain. In both cases the architect must give instructions as to what is to be done and by whom, while conversely the contractor must not press ahead with any work covered by one of these sums without obtaining instructions, even if it is fairly obvious what is to be done. Sometimes there may be provisional quantities in the bills; these should be treated as equivalent to provisional sums and instructions should be given.

APPROVALS

The other arm of the architect's rights and duties is the giving of approvals. These relate principally to the works as conforming to the contract documents in such things as accuracy of layout and quality. The contractor's obligation to produce the works correctly has been looked at already, and this applies whether anyone checks his work or not. He could thus face a call to redress or legal action within as much as six or twelve years of completion (as the case might be) if a defect came to light. This underlying position explains the way in which the conditions operate. They do not require the architect to give positive approval until the end of the defects liability period following practical completion (considered later) and when any defects have been righted. This is then regulated by clause 30(6) and (7) over the final certificate, but even here the architect is strictly approving only those items that are to be to his 'reasonable satisfaction'.

During progress the architect is not required to express any satisfaction at all, although he has power under clause 6(3) to inspect and test work and under clause 6(4) to instruct the removal of anything defective. No timing is specified, and the architect could defer inspecting the foundations until internal decorating was in hand. In this case he could ask to have them exposed, but would have to certify additional payment for the disturbance if they then turned out to be in order! In view of this a more moderate course of periodic inspection usually applies, but it cannot be claimed that work has been inspected and therefore the work cannot be condemned at a later stage.

In support of whatever extent of inspection the architect may decide to exercise, clause 9 gives him a right to enter the site and workshops of the contractor and sub-contractors. Further, a clerk of works may be appointed under clause 10 with similar powers of entry, to inspect but not to approve or condemn. Strictly he can only report to the architect on what he sees. In practice he is usually given a fair amount of delegation over these matters, and it is safer for the contractor to seek confirmation of the extent of this delegation before any misunderstanding arises over any direction that he receives from the clerk of works. Otherwise in theory the contractor could ignore him, which would be a poor way to treat someone whose help is usually of great value during progress.

Work by others on the site or for the works

Several cases modify the duty of the contractor to perform all the work in the contract and to have uninterrupted control of the site. At one end of the scale, clause 17 permits him to employ his own sub-contractors for work measured in the bills, provided the architect has no reasonable objection to the firms concerned. This may relate to work on site or to manufacture off-site, but will not apply to items of supply for which the manufacturer is exclusively named in the bills. The clause caters for the contractor wishing to assign execution of the whole of the works to another firm, but it is much more restrictive over this occurring. In any such case, the sum payable between the employer and the contractor and the responsibility of the one to the other in other ways both remain unaffected.

At the other end of the scale, clause 29 permits the employer to bring on to the site artists and others who are to perform work directly for him and who have no contractual relationship with the contractor over payment, coordination or any other matter. The contractor is obliged to allow them to operate, and usually the clause is only used to permit some restricted piece of work. There is some right in other clauses for the contractor to be reimbursed for any delay by such specialists, but the position is not entirely satisfactory.

Between these extremes come two other cases, variously sub-contractors and suppliers, covered by the procedure of nomination already referred to in connection with instructions under clause 11(3). The starting point is usually a prime cost sum (often abbreviated to 'PC sum') written into the bills. Occasionally there may be a PC rate embodied in a measured item, or a PC sum may occur when instructions are given about a provisional sum, but these cases need not be followed up here.

NOMINATION

The cardinal principle of nomination is that the architect has the right and duty to name the persons to perform work or supply materials as represented by each PC sum, while the contractor cannot choose such persons himself and place orders without instructions from the architect. Once the architect has given a nominating instruction and the contractor has accepted it and entered into a sub-contract with the firm, then the relationship is the usual one applying under a sub-contract. The employer has no contractual relationship with the firm, despite the action of the architect in selecting and putting forward the firm. The contractor has a right of objection to accepting a nomination, however: in the case of a supplier or a sub-contractor he may object to the terms offered if they are not in line with those given in the main contract; in the case of a sub-contractor only he may object to the firm as such. There is also the option under the clause for the contractor himself to be chosen by the architect to perform work under a PC sum, and thus to be his own sub-contractor.

The clauses recite a number of points that must be embodied in the sub-contracts or contracts of sale. In the main these affect in some way the relationship of employer and contractor, even though they are primarily between the contractor and the other firm concerned. Several of the most important are outlined below. There are standard documents embodying the appropriate clauses, which mirror the wording of the present contract, as well as giving other clauses that are not relevant here.

The detailed provisions over nominated sub-contractors are the more extensive and may be taken first. While the sub-contractor has to satisfy the architect over the standard of his work, he has also to carry it out in such a way as to fit in with the site operations of the whole contract, and he has to shoulder any liabilities due to his own lapses. There are therefore provisions that the sub-contractor is to work under the control of the contractor, and that he indemnifies the contractor against those matters for which the contractor is liable to the employer. This last is a very convenient provision to drop in (it is more precise in its wording, of course), but its effect is by no means certain at all points. In general it relates to such things as negligence. While the contractor has responsibility for controlling the sub-contractor, he cannot extend the time for the sub-contractor's work without the architect's written consent. If, however, the sub-contractor fails to meet his due date, his liability to pay the contractor damages for delay arises only upon a certificate of the architect.

The architect is cast in an active role over delay, but his role is even stronger over interim payments. (Final payments are considered later, under 'Financial matters and settlement'.) The contractor may not make any interim payment to the sub-contractor that is a sub-contract amount, as distinct from some other amount for (say) incidental work elsewhere on the job, unless the payment is authorised by an architect's certificate. When it is so authorised it becomes a mandatory payment except in very special circumstances, including set-off under certain precise rules and procedures. The gross amount due is subject to two deductions: a permanent one of 2½ per cent for cash discount and a temporary one for retention (discussed under 'Financial matters and settlement') at the contract percentage. The architect also has a discretionary power to seek proof of payment under one certificate before issuing the next. If the contractor has defaulted, the employer has an equally discretionary power to deduct all or part of the non-payment when honouring the next certificate. If the architect wishes to secure final payment to a sub-contractor ahead of that to the contractor, he also has power to authorise this under the clause.

In the case of suppliers the provisions are scaled down. Again the materials are to be to the satisfaction of the architect, but the contractor is protected against consequential damage as well as replacement of materials (but not against his own carelessness), if it is necessary to take out any items once they have been fixed. The requirement to fit the programme of the contractor relates simply to deliveries, and there are no statements about what happens if there is any delay: the contractor would have his usual remedies, but could obtain no special intervention by the architect. Payments are to be made on usual monthly terms and are not dependent on the architect's certificates for authority or amount, and equally the supplier has no special protection against the contractor failing to pay him. No retention can be deducted, although cash discount applies – this time at 5 per cent.

While the employer has no contractual relationship with nominated firms, there are documents that are frequently used alongisde the present conditions as between the employer and such firms. These agreements or warranties do strengthen the employer's hand in some problems that he otherwise faces, but do not affect the dual relationship of the contractor to those with whom he has contracts at different levels.

The programme and progress of the works

One of the basic concepts of the conditions is that the contractor is responsible for completing his work on time and by his own means, subject to meeting the specification and design. This is underlined by clause 21(1), which obliges him to finish 'on or before' the date given in the appendix. It also requires him to proceed 'regularly and diligently', and he could find himself in breach of contract if he fails in this respect. Minor lapses in programme are not what is envisaged, but such fitfulness as makes it almost impossible to finish anywhere near the contract date. How the contractor meets the date is his concern; not only does the architect have no power to instruct the contractor over sequence and working methods, but he cannot instruct him to speed up to regain lost time – whether by working overtime or other ways. This applies whatever the cause of delay, although the contractor may take steps voluntarily to make up his own deficiency and avoid the threatened consequences, or he may agree to an acceleration to make up other delays, with an extra payment to compensate. But on all of this the conditions are silent, and there need be no action taken. The architect's only power is to instruct a postponement under clause 21(2), ironically.

This rigid position is hedged about at both ends of the programme by clause 21(1). The employer has to give possession of the site on time, otherwise the contractor could modify his programme (perhaps by more than the length of the delay) or even repudiate the contract and sue, in an extreme case of legal frustration. At the other end of the programme the obligation to complete is eased by the right to an extension of time under clause 23 for causes given there. These in some cases are matters in which the architect has acted by way of instruction or has lapsed by way of delayed issue of information or in other ways. In other cases they are serious adverse happenings such as damage to the works arising from causes given in clause 20 (considered later), exceptionally inclement weather and the like. A third group covers matters that arise within the industry, and includes strikes, shortages of labour and materials that were unforeseen (although this provision is optional) and delays by nominated firms, provided in the last case that the contractor has progressed them adequately.

While the contractor is entitled to extension in these cases, he may lose his right if he does not write to the architect as soon as a threat of delay is 'reasonably apparent'. He must not wait until events have run their course and it is too late for the architect to assess the extent of the problem. The architect in turn has to grant an extension in terms of length as soon as he can to clarify the position for the contractor, although it is not always possible to do this ahead of the delay being suffered. Much will depend on its nature, but this subject is riddled with pitfalls and care is needed to work as closely to the letter as possible.

If the contractor overruns the original or amended completion date and the architect certifies in writing that the contractor is at fault, the employer is entitled, under clause 22, to recover liquidated damages from the contractor by sending him an account or by deducting the money progressively from interim payments. The damages are represented by a sum of money (usually expressed as so much per week) entered in the appendix. The sum should represent a reasonable estimate, at the time of entering into the contract, of the employer's anticipated loss if completion is delayed. This may bear no fixed relationship to the contract sum, but will depend on the use of the works or the return due from them. If it is assessed so high as to be a penalty (that is, a sum designed to intimidate the contractor rather than recoup the employer's loss), the contractor is entitled to repudiate the sum and plead for damages to be reassessed by the courts, even though the sum is written into the contract. But if circumstances change after entering into the contract so as to make the sum inappropriate when it is incurred, neither party has any redress; they are caught with what was reasonable initially.

The architect has to act under clause 15(1) to certify that the works are 'practically complete'. This odd expression does not mean 'nearly finished', but rather that the architect thinks that the works are complete and ready for occupation, although he may still need to check them over. Practical completion ushers in several events and states, including legal and financial ones, but it does not end the contract. It does mean, though, that the contractor need not accept any fresh work on the contract, that the employer may move in, and that the contractor must move out. It also begins the defects liability period (usually entered in the appendix at six months) and any defects found in this period are to be put right by the contractor himself, on receipt of a schedule from the architect at the end of the period or on receipt of earlier instructions if there is an urgent defect. After this the contractor is still liable for the cost of remedying defects, as already discussed, but is not obliged to return and do the work himself.

Superimposed on all these programme matters are two provisions dealing with phased completion. One is clause 16, which takes the case in which the employer, by agreement at the time with the contractor, is let into some part of the works ahead of the general completion. This is something that the contractor does not have to agree to, but it does have benefits for him. The other provision is a supplement to the main conditions to cover sectional completion that has been agreed to in the contract originally. Either provision could relate to the progressive handing-over of houses on a site, for example, but also to parts of a single building. Both provisions produce a scaled-down practical completion, and thus release the contractor from various obligations ahead of the final date for the rest of the works, and also set retention monies free earlier. There can be severe problems over physical interference and legal liability if the parts handed over are not distinct from those retained by the contractor, and any agreement should be viewed accordingly.

Disturbance and damage to the works

That all may not go well has been hinted at in the previous sub-section, and what follows ties closely with some of its matter. The starting point is clause 24, which deals with loss and expense to the contractor when regular progress has been disturbed. The causes that it allows are restricted to those that lead to extension of time and that broadly are due to the architect's fault, such as late information, unnecessary opening up of work and postponement. As with extension, the contractor has to give warning as soon as he can, so that again an assessment can be made on the basis of proper records. This will obviously be a progressive matter in most cases. Although the clause requires the architect to calculate the loss or pass the task to the quantity surveyor, it is often the contractor who can best draw up the first statement of what is usually termed a 'claim'.

A special case of such a claim can occur under clause 34, which treats the topic of antiquities. Owing to the value of such finds and the frequent need to have them inspected in position, the contractor must stop work and wait until he is instructed to proceed. This produces a postponement and he will be reimbursed for his loss.

DETERMINATION

A more drastic interruption occurs if there is a 'determination' under the contract by either party, that is if either is so aggrieved as to bring the contract in practical terms to an end, although it will continue in legal being until affairs have been wound up. Separate clauses deal with this according to which party determines, but they both proceed by securing 'determination of the employment of the contractor' (that is, the ending of work on site) and then settling up the financial consequences. Within this framework they differ considerably.

The case of the employer determining occurs under clause 25. There are two sets of provocations: default and insolvency, the consequences of which are somewhat different. Default breaks down into suspending the works without good reason, dilatory progress, not removing seriously defective work that would radically impair the finished works, and sub-letting without approval. Where default occurs, the architect has to give the contractor notice that it is held to have occurred so that he may cease it. If the contractor does not cease, or if he resumes the same type of default later, the employer may then give formal notice determining the contractor's employment. There is a precise timetable for these stages, with a rider that a minor technical infringement will not justify using the clause.

Insolvency, as a blanket title for several things, is said by the clause to lead to automatic determination, that is without notice. This provision is highly suspect under the legal requirements of liquidation in particular, and the liquidator or the like may well be able to elect to carry on. However, the mechanism of the clause is often accepted as a working basis.

To tidy things up, the clause then allows the employer to engage others to complete the works and to make certain payments to sub-contractor creditors of the outgoing contractor – this again is suspect where there is insolvency. The employer may also retain plant and the like until no longer required and without charge (although this does not allow him to retain hired plant); the contractor should then remove it, or the employer may sell it and hold the net proceeds for the contractor. The contractor receives no further payment until after completion of the works, not unexpectedly, and then only if the balance struck between what the employer has paid out and what he should have paid shows a sum due. Usually indebtedness is the other way round, and if there has been insolvency the employer lines up with other creditors for what settlement he can get.

The case of the contractor determining is covered by clause 26. Again there are the categories of default and insolvency, but with no attempt to make determination automatic. In general, all that is needed is a single notice from the contractor, although there are one or two sub-routines. In the case of default, the principal items are those leading to extension of time (except mainly those arising within the industry) provided that they cause nearly complete cessation of work for a period given in the appendix, commonly one month, and also matters concerning certificates. The latter involve the employer directly, and default occurs when he fails to pay an interim amount after notice or when he obstructs the issue of a certificate of any type and thus embroils the architect. If, however, the architect fails to issue a certificate by his own dilatoriness, the right of action does not apply.

When determination comes about, the contractor is to remove all his property from the site, as soon as he can leave this in a safe condition. He is to be paid for all work done and materials supplied, while having a temporary right of retaining materials pending settlement. In addition he is to be paid the costs of removal and direct loss caused by the determination, which in this case will include loss of profit on work not performed. On the whole the terms available to the contractor when he determines are better than those that the employer secures, and this is true of several subsidiary issues not discussed here.

Injury to or in connection with the works

The contract proceeds on the broad basis that the contractor is responsible for injury, but then softens this in detail. In clause 18 it deals with injury variously to persons and property, making each party liable if he is negligent and leaving some slight differences when neither is negligent. The contractor, but not the employer, is required by clause 19(1) to back up his liability by insurances, for which he must be prepared to produce substantiation that policies are operative. In the case of personal injury this means third-party and employer's-liability insurance, and the full statutory level applies to the latter. For third-party and also for property, the level of insurance is to be as given in the appendix; these risks may well be completely covered by the contractor's running policies, or he may wish to have higher cover, but the details should be checked. It is the contractor's responsibility to ensure that sub-contractors are maintaining similar cover. If there is any default, the employer can act to insure enough to cover the contract level.

Clause 19(2) deals with a special (and extremely costly) insurance, which is only to be taken out in peculiar circumstances to cover the employer's liability to adjoining owners. In any case the contractor would be reimbursed the premiums and given instructions about all the details. The clause need not therefore be explored further here. Clause 19A excludes nuclear and related risks from the contractor's responsibility, since these are always to be covered by the operating agency, if at all.

There are three variants of the provisions over insuring the works themselves during progress. Clause 20[A] relating to new buildings makes the contractor responsible for insuring against a fairly wide, but still limited, list of risks, which may be summarised as fire and explosion, natural hazards like flooding, aerial damage and rioting. The sum insured is to cover the works and professional fees until practical completion, when the employer assumes responsibility. The contractor may wish to use his all-risks policy (and thus cover additional risks such as theft), and can do this provided the employer's interest is endorsed. This time he is due to cover sub-contractors' work as well as his own. Again if there is any lapse the employer can insure, even though the contractor would still be liable under clause 1(1) to complete the works at his own cost if insurance monies were not forthcoming. The insurance is a back-up and not a substitute.

Under clause 20[B] and clause 20[C] this position is changed in some respects, even though the list of risks is the same. Clause 20[B] still relates to new buildings, but makes the employer responsible for insurance, while giving the contractor the right to insure on any default. However, the clause also transfers the underlying risk that the insurance covers to the employer, and this must reduce the interest of the contractor in insuring, to say the least. Only if the employer is his own insurer or can obtain special terms does there appear any advantage to him in considering this variant. Clause 20[C] deals with work that consists of alterations or extensions, when there will already be an existing structure for which the employer is responsible. It therefore more reasonably makes the employer responsible for insuring both old and new, with the contents of the old. Again the risk is transferred, and the contractor may insure on default. In both of these clauses, however, risks not in the list remain with the contractor, as in clause 20[A].

If there is damage, the way in which matters are settled varies. Under clause 20 [A], the contractor is to reinstate upon acceptance of the claim and be paid the settled amount. He will receive no more from the employer, but equally will not have to pay over any excess! He will, of course, have to pass on the amount for fees. Under the other clauses, the contractor is to be paid for reinstatement as a variation, that is specifically at contract terms and without regard to the settlement that the employer obtains. When there is damage to any existing building, the contractor may be prepared to repair this as well or even alone. He is not obliged to do this under his contract, which covers only reinstatement of the works themselves when damaged, and he is entitled to decline this extra work or negotiate a separate contract to cover it. As there may be a variety of consequences following damage to the old premises, the clause builds in a right to determination by either party if this is reasonable, with a counter-right of the other party to seek arbitration over this. There would be little point in continuing to build an extra storey on a building that had been burned down, for instance!

In none of these cases will the contractor be reimbursed his loss due to delay, under the terms of the required insurances, but simply the sum for reinstatement. He may therefore wish to review his insurances in this respect. If there is a prolonged delay, there is the possibility of determining under clause 26, as already discussed, and of being reimbursed for leaving the site and for lost profit, but still not for the delay costs.

Financial matters and settlement

There have been strong financial implications in issues already discussed, such as loss and expense, liquidated damages and insurances. These, however, leave aside the mainstream provisions about interim and final payments. It is easier to consider them in the reverse order in this section.

The starting point is the contract sum, defined in clause 13 as unalterable except as the conditions permit. In any case it is to be adjusted and not set aside, unless affairs deteriorate so severely that the whole contract has to be reconstructed – usually a matter for the courts. Three major areas of anticipated adjustment may be identified: design changes, market-price changes, and substitution of actual sums payable to nominated firms.

Variations have been considered as changes in the design introduced under clause 11(1) and defined under clause 11(2). Their evaluation as additions is provided for in clause 11(4). This gives a series of options to be taken in turn: to use the same prices as are in the contract bills for work of the same nature, to use proportionate prices for similar work, to use prices that are reasonable for work not like that in the bills but otherwise able to be measured and valued, or, lastly, to use daywork terms (also given in the bills) to value work according to the time and materials expended. While the quantity surveyor will usually undertake a number of further activities, the measurement and valuation of variations is specifically assigned to him. The contractor is entitled under the clause to be present when this is being done, and may well cooperate in some detail in it. He should provide any supporting information by way of obtaining previously unconfirmed instructions, and also invoices and other data for pricing.

While the quantity surveyor will expect to use the three options of measurement and pricing as the initial choices, the contractor should ensure that daywork records are produced for signature by the end of the week following the work being done. The clause requires this, but is not precise as to what happens if sheets are not signed in time. The quantity surveyor may set them aside as unsubstantiated and price the work without them, but he cannot refuse to include some payment for work authorised. If there is any doubt over whether daywork will be used as the method of valuation, the contractor should prudently have records signed. The mere fact of a sheet signed by the clerk of works, or even by the architect, does not settle the mode of payment, but without such a sheet there is no record to fall back on.

Variations often introduce additional work, but they also frequently lead to omissions and these will be measured and priced as in the bills. Sometimes an omission changes the circumstances of some remaining work, perhaps by making its quantity much less or by throwing on to it all the use of a particular piece of plant. In this case the clause allows for repricing the remaining work to reflect the change in cost.

It may be that the timing or other circumstance of the introduction of a variation causes expense that is not covered by any of the rules for valuation – even daywork does not allow for any extra cost of surrounding work not itself varied. In this case the contractor can seek additional payment under clause 11(6), using a similar procedure as for loss and expense under clause 24.

Changes in what the contractor pays for labour and materials are termed 'fluctuations' by clause 31, which has several alternatives within itself. The use or otherwise of these has a considerable effect on the calculation of the contract sum arising out of the tender, and the implications of the alternative named in the documents should be carefully weighed when tendering. Here only the principles can be outlined for what is the longest clause in the conditions. The main alternatives are as follows:

Clause 31A Calculation related to the hours worked and the quantities of materials purchased, making adjustment for changes in market rates and prices and changes in tax and other statutory payments bearing directly on them.

Clause 31B Similar calculation, but making adjustment for the statutory elements only.

Clause 31F Calculation related to the measured prices in the bills, adjusting these in relation to a national index published monthly, which shows changes in average costs of labour and materials caused by market and statutory elements.

The second division is therefore a reduced version of the first, whereas the third proceeds on a quite different basis and, as is shown by the clause number, is of more recent introduction. The first and second require wage sheets and invoices in the main, and the calculations can be performed at any time after expense is incurred. The third is independent of the precise expenditure of the individual contractor and must be done at the time of interim valuations. If either clause 31A or clause 31B applies in the contract, then clauses 31C, D and E apply also to fill up various details about giving warning of increases and decreases, provisions over sub-contractors and so on. Clause 31F, however, is used alone, if at all.

Clause 31A is so worded that the contractor, when tendering, should allow for any increases or decreases in labour expenses already agreed by the appropriate bodies, even though they have not come into effect. So far as materials are concerned, the contractor is not required to foresee any market change not actually current. Concerning statutory elements of labour and materials costs, under clause 31A or clause 31B, again he is not to be prophetic – and this is perhaps as well when the government is involved! With clause 31F he need do nothing at the time, since the index is automatically used; he must simply ensure that he tenders at the right level. There is a lot of fine detail in what is and is not allowable in clause 31A in particular, and bonus payments need care.

The method of adjustment under clause 31F in effect allows a profit element automatically in the calculation of changes. The other two methods are qualified by clause 31D as being net of profit. However, clause 31E allows an addition on a percentage basis to be stated in the contract appendix and to be made to all adjustments for fluctuations under clause 31A or clause 31B. This addition is unexplained, but is meant to cover changes in elements not allowed under the clauses, such as bonuses and supervisory staff. It may therefore be priceable so as to include a profit element, even though not so defined.

Value added tax is dealt with by clause 13A, and a supplemental agreement is attached to the conditions. The clause provides that no allowance has been made for the tax in the contract sum and that it will be charged by the contractor in parallel with the contract settlement under the terms of the supplemental agreement, which are related to matters of accountancy in the main and are based on the statutory requirements.

There will need to be adjustments to the contract sum in almost any case in which a prime cost sum is used for nominated work, even if the expenditure of the sum does not lead to any variation in the work as designed. These adjustments represent the difference between the quotation accepted and the sum in the bills, and also any later adjustment as work proceeds. In principle it is the contractor who is responsible for all matters relating to all his sub-contractors, but in the case of nominated firms the finance is of direct interest to the employer, since he pays the actual sums through the contractor. The architect therefore issues variation instructions and the like through the contractor, and the quantity surveyor agrees the financial effects – often most conveniently directly with the firms concerned. However, the contractor should keep in touch with what is happening, and can insist on acting in any negotiations, especially in any case of contra-charges or other items that may take on a 'three-cornered' aspect.

While work is proceeding, the contractor is to be paid each month for the value of work done and for materials on site, in accordance with clause 30(1). The certificate concerned is usually based on a valuation prepared by the quantity surveyor in agreement with the contractor, although the clause is not explicit on this point. The architect has a discretionary power to include for materials off-site, provided various conditions in clause 30(2A) have been met regarding completeness and proof of ownership. The total sum so calculated is obviously approximate, but is also subject to a deduction for retention; the deduction is a percentage to act as a buffer, and is given in clause 30(3) as 5 per cent unless some other percentage is given in the appendix. Amounts for fluctuations may be included in interim payments and those under clause 31F are subject to retention, but those under the alternative provisions are added or deducted in full. Some other sums, particularly those for loss and expense, are also to be paid in full. The value of variations and other changes to the contract sum are also to be included in interim payments, but these are subject to retention. The special position of nominated firms affects the requirements regarding interim payments, as has already been discussed.

The regular monthly issue of interim certificates for payment ends with the certificate of practical completion, but further certificates may be issued at not less than monthly intervals if progress towards settlement reveals further sums to be due. At practical completion a parallel certificate is issued to release half the retention fund; a further special certificate releases the other half, usually on the completion of remedying defects after the six months period. The final certificate includes a statement of the balance outstanding, and this can be in such a form as to recover any overpayment.

Bringing the contract to an end

Most of the matters concerned have already been considered and may be summarised here:

- The contractor's obligation to accept instructions for fresh work ends with the certificate of practical completion.
- The architect's power to instruct the contractor to remedy defects ends with the schedule that he issues during or after the defects liability period.
- The final certificate expresses satisfaction, pays off the final balance, and usually brings the contract to an end.
- The employer has the possibility of taking action against the contractor for up to six or twelve years from any breach of contract. This includes any defect coming to light, and from which the contractor is not exonerated by the final certificate.

This summary leaves only two elements to discuss. The first is the final certificate, which is dealt with by clause 30(6) and (7). It is to be issued within three months of the clearance of defects and of the contractor providing any papers needed for the quantity surveyor to complete the final account (such things as daywork sheets and invoices to permit pricing and nominated accounts). It expresses the architect's final satisfaction with those parts of the works that are to be to his reasonable satisfaction. It brings the contract to an end, unless there are outstanding disputes, and thus defines the limitations period during which a breach is actionable.

The second element is dispute; here the main provision of the conditions is in clause 35, which allows arbitration to occur, but allows it to get under way effectively in most cases only after practical completion. The exceptions are matters that need rapid resolution, the most important of which are disputes over the validity of architect's instructions and over certificate irregularities. A reference to arbitration can be initiated at any time up to fourteen days after the issue of the final certificate, and so can include the subject matter of the certificate, which otherwise becomes binding on both parties. Since arbitration can be a lengthy matter, the possibility is given of issuing the final certificate in a form conditional upon any reference still going on at the time.

There are only one or two relatively minor matters over which the conditions preclude the chance of going to arbitration. Although many matters of dispute arise from the action or inaction of the architect, arbitration is possible only between the two parties, but will usually arise when one party accepts the architect's stance and the other disagrees with him, thus starting the process. While the parties may agree on who the arbitrator is to be, this is not a necessary feature of the procedure, any more than is it necessary for the parties to agree to go to arbitration – one party may start things off on his own and without warning. However, a measure of agreement to disagree will often limit the extent of the reference and save time and money.

The presence of an arbitration clause in the contract does not remove the alternative of recourse to the courts, and this may be a better way if the point at issue is legal rather than, say, technical. Either way, affairs can become very involved and expensive and a compromise agreement is often more expedient.

OTHER JCT AND RELATED FORMS

Forms in the JCT series

The various forms available have been outlined, and the salient differences can be considered compactly by comparison with the private form with quantities, discussed above.

The differences resulting from using one of the local authority variants are quite small. The term 'architect' is replaced by 'architect/supervising officer', but the effect is the same. Slightly more menacing are two additional causes for determination available to the employer: an additional clause 17A requires the contractor to pay fair wages (hardly an onerous requirement, to say the least), and an additional clause 25(3) brings in corruption as a default. More pressing than either, clause 31F fluctuation payments are subject to a productivity deduction of up to 10 per cent, according to what is inserted in the appendix.

The approximate-quantities variants have a number of differences affecting the contract documents and the procedures for arriving at the final amount payable. The principle of approximate quantities is that the Bills of Quantities used for tendering purposes are not precise, but give quantities that are only a reasonable representation of the works required. As the works are finally designed and produced, so the quantities are completely remeasured accurately; the results are priced at the same prices as in the approximate bills, or prices are used that are deduced in the same way as variation prices under firm-quantities contracts. This procedure has a lot to be said for it, but there are at least two cautions to be expressed. One is that the contractor must take the quantities as given when tendering, and not try to reconcile them with the drawings issued to show the character of the works. He should therefore price the bills strictly for both the quantities given and the descriptions given, even though he may see something else on the drawings. The second caution flows from the first: the final quantities may be substantially different from the originals, if not in total value at least in the distribution of types of work. Either way the contractor, in agreeing the final account, may need to seek an adjustment of the level of measured prices or of preliminaries to reflect properly the changes. This the contract permits.

The without-quantities variants rely on a complete set of drawings (including all details) and a specification of quality as contract documents. They include no quantities in the contract, as their title indicates. They are intended for smaller projects of the scale of a single house, and the contractor will prepare his own quantities when tendering on his own basis. These do not become part of the contract, so any error in them remains the contractor's responsibility. It is common practice to ask that he reveals the prices that he used in tendering, so that these prices can be given as a schedule for valuing variations, but still without taking any account of the accuracy of the original quantities. The only other point to note in this outline consideration is that clause 31F fluctuations, i.e. the formula method, are omitted owing to the absence of quantities.

Also issued by the JCT is an agreement for minor building works. This is an extremely scaled-down version of the forms considered, and is only intended for something like the smallest of extensions to an existing house. It is a 'goodwill' contract, and while this is a fine sentiment it leaves many issues open. Only over matters like insurance is it reasonably full. It should not be expected in projects of any value, even when the parties know each other well. Precision is desirable to avoid awkward issues that may be the fault of neither.

Lastly, there is a fulsome form (very similar to the main forms) for use in prime cost contracts, i.e. contracts in which the contractor is paid on the basis of costs rather than of work produced. This is most likely to be used in conditions of uncertainty when prior estimating is impracticable. While it does rely on direct reimbursement of most site costs, it still requires the tendering of a fixed fee to cover overheads, supervision and a number of other elements – and the calculation of this fixed fee is obviously difficult if the work is uncertain. Calculation needs care too, because the fee must cover all items not expressly reimbursable at cost, including insurances. However, it is obviously a fairly safe financial proposition as a contract to the contractor when once he has got his initial sums right.

Forms related to the JCT series

There are several forms in common use that are clearly drafted to go with the JCT family but are not part of it. Two of these may be mentioned in closing.

The Scottish building contract is a short document designed to adapt the JCT forms (in their various versions) for use in Scotland, where there is a distinct legal system. The document is issued by a special joint body and consists of a section similar to the JCT articles of agreement, followed by three appendices. The second and third appendices deal with insertions, as in the JCT appendix, and with value added tax. The first appendix gives a number of interpretations of, and amendments to, the JCT conditions to make them suitable for use in Scotland. The most important of these cover the different arbitration law and the fact that the JCT clauses about payment for off-site materials are invalid in Scotland, the solution being for the employer to purchase the materials directly from the contractor and then have their value deducted from the contract. When all this is taken into account it is necessary to incorporate the JCT conditions, but not the articles and appendix, by reference; special 'conditions only' versions are obtainable for this purpose.

The other document is the standard form of nominated sub-contract issued by the NFBTE and sub-contractors' organisations. This is designed to fit the JCT conditions over provisions about sub-contractors, and also to regulate other matters not touched on in the main forms. It is useful to cover issues like payments (as do the main forms, and the points have been discussed earlier), the passing down to the sub-contractor of appropriate main contract obligations, and the use or otherwise of facilities like scaffolding. It is drafted to allow the sub-contract to be based on firm or approximate quantities, a schedule of rates, a simple lump sum, or daywork terms. While the form is entitled as for nominated sub-contracts, there is no reference to nomination in it and no provision that would make it unsuitable for a private sub-let arrangement that the contractor might enter into, although there is another related form available and labelled for this purpose.

FURTHER INFORMATION

Joint Contracts Tribunal, *Standard Form of Building Contract* in the appropriate version considered in this section and the latest edition (a revision is usually published each July, although there was not one in 1978); Royal Institute of British Architects hold copyright.

Robb, G.G.G., and Brookes, J.P., *Outline of the law of contract and tort*, Estates Gazette Ltd

Marshall, E.A., *General principles of Scots law*, Sweet and Maxwell

Building and Contract Journal Ltd, *Contractor's guide to RIBA forms of contract*
Jones, G.P., *A new approach to the standard forms of building contract*, Construction Press
Turner, D.F., *Building contracts: a practical guide*, George Godwin Ltd; contains a further and more advanced reading list
Building and other periodicals for articles and reports on current changes and cases
Joint Contracts Tribunal, *Practice notes*, an occasional series giving guidance over changes in the standard forms

7 QUANTITY SURVEYING

SURVEYOR'S RESPONSIBILITIES AND DUTIES 7–2

SURVEYING DOCUMENTS 7–3

CONTROL OF FINANCE AND PRODUCTION 7–5

FEEDBACK 7–14

TRAINING 7–14

QUANTITY SURVEYING 7

7 QUANTITY SURVEYING

C.D.A. SCHOFIELD, MSc, AIQS, MIOB, AIArb
Consultant, BAS Management Services

This section describes the surveying function in a building company, the preparation of Bills of Quantities and the financial control of projects. It comments briefly on feedback and on the training of surveyors.

SURVEYOR'S RESPONSIBILITIES AND DUTIES

The surveyor has three main areas of responsibility within the company, all of which involve liaison with other departments. These areas are:

- keeping the financial strength of the company at the highest level;
- protecting the company's interests;
- providing accurate information for control purposes, upon which management decisions can be made.

The duties required of a surveyor to perform his responsibilities are listed below, under the particular area of responsibility, but it should be noted that in many organisations some of the duties are delegated to other departments. This reinforces the need for the surveyor to coordinate and liaise with other departments as and when required.

Duties to keep the financial strength of the company at the highest level:

1. Preparing interim certificate valuations that represent the full value of work done and other recoverables in accordance with the conditions of contract.
2. Progressively building up information to facilitate the prompt preparation of final accounts.
3. Promptly preparing and submitting contra-charges to sub-contractors, and direct charges for work outside the scope of the contract.
4. Keeping a daily check on applications, certificates and receipts to ensure the earliest possible payments.

Duties to protect the company's interests:

1. Searching and reporting on contract documentation before completion by the company signatories.
2. Ensuring that the company obtain their full contractual entitlements, whilst at the same time discharging their full responsibilities and obligations.
3. Making application for and negotiating extensions of time to the contract period, where necessary, keeping all records in substantiation at the same time whilst the work is in progress.
4. Obtaining sub-contract quotations, negotiating, placing sub-contracts, certifying interim payments and subsequently agreeing the final accounts.
5. Checking invoices, where the surveyor's technical ability is required.
6. Liaising, when required, with the production department on targets and bonus.
7. Keeping running checks on quantities of the main materials used.
8. Preparing daywork returns.
9. Preparing and agreeing fluctuations on variation of price contracts.
10. Keeping measurements and records to facilitate the agreement of the value of any variation to the contract.

11. Giving all notices required under the condiitons of contract, and ensuring that all relevant communications are recorded in writing.
12. Maintaining a balance between the current interests of the company and the reputation and future standing with clients and their professional advisers.
13. Providing feedback information to estimators, bonus surveyors, managers and others.
14. Advising production managers on alternative construction methods and costs.

Duties to provide accurate information for control purposes:
1. Preparing forecasts and budgets, and updating these through the running of the contract.
2. Preparing cost/value comparisons at regular intervals as determined by company policy.
3. Preparing monthly reports, in conjunction with the cost/value comparisons, on the financial trend of current contracts.
4. Producing periodic reports to show the anticipated final profit/loss figures of contracts.

SURVEYING DOCUMENTS

Preparation of documents

The form and extent of documentation will depend on the person or organisation financing the contract, i.e. the building owner or contractor.

In the case of a contract being financed by the building owner and being tendered for in competition, there will be generally either
- drawings and Bills of Quantities, or
- drawings and specification,

according to the nature of the contract.

Where the contractor is financing the operation, as in speculative house building, it will be necessary for the contractor to prepare adequate documentation to facilitate the initial budgeting and pricing, and then cost control, purchasing and the other functions of the management of a building project as the contract proceeds.

Whether the documentation is provided with the invitation to tender, or is of necessity to be prepared by the contractor, the extent of such documentation can be classified under three headings:
- the Articles of Agreement and Conditions of Contract;
- drawings;
- Bills of Quantities and specifications.

The drawings and specifications will not be the responsibility of the quantity surveyor, although in some cases he may write the specification, but he will be responsible for the preparation of the Bills of Quantities. In the case where Bills of Quantities are part of the contract documents they will be prepared by the quantity surveyor appointed by the building owner, but otherwise they will be prepared by the contractor's quantity surveyor.

Principles of measurement for building works have been in existence since 1922, and the current edition is the Standard Method of Measurement (5th edition), but the principles are constantly under review, the 6th edition having been published in July 1978 for adoption into Bills of Quantities in 1979. Work is meanwhile proceeding on the 7th edition. The SMM, as it is generally referred to, is intended to provide uniform information to the contractors who are tendering, and consequent uniformity in the analysis of tenders received by the building owner.

However, the use of the SMM can be time-consuming, owing to the detail required, and contractors often prefer to use a more 'composite' approach when preparing Bills of Quantities for their own use.

If the contractor decides not to base his own Bills of Quantities on the SMM, the preparation of Bills based on composite items, trades, or elements of the construction process are all viable alternatives.

Bills of Quantities based on composite items would combine items required by the SMM to be given separately. An example of this would be a door lining, door stop, architrave, door and all ironmongery being combined into one item, and priced accordingly by the estimator.

Trade-based costs are an extension of the above in that, by further combining composite items, it is possible to price items based on the overall measurement of quantity. An illustration of this would be with brickwork, or concrete, measured and reduced to one unit of measurement and quantity, regardless of location in the building, and then priced by the estimator using historical records of compatible construction types on previous contracts. Such techniques have been used for many years for budgeting for construction works, but there is no reason why the technique should not be expanded to suit the more specific taking-off and pricing requirements of the contractor.

Quantities can also be prepared based on elements of the construction. Internal partitions are an example where this could be used: for example, lightweight block walls, plastering and decoration can be measured together, and one composite price then inserted by the estimator against the single superficial measurement. Where quantities are uncertain or the specification is not fully finalised, a Schedule of Rates can be prepared. When work is carried out it can be measured and valued at the relevant rates.

Whichever of the alternatives above is chosen by the company, and in some cases where a mix of the alternatives is used, it is paramount that Bills of Quantities be prepared for use throughout the contract, and it is essential that companies standardise the procedure for preparing Bills of Quantities. Only by such standardisation can worthwhile comparisons be made at the various stages of the construction process.

In addition to the continual development of the principles of measurement, as mentioned earlier, there is also development in the style of presentation of the information required by the Standard Method of Measurement. Two such developments are Operational Bills and Elemental Bills, which both follow the implications of their respective titles. Work is presented in sections following the operations involved in the construction process, or in sections representing the various elements used in the building. It should be emphasised that both these techniques are used mainly by the appointed quantity surveyor. Contractors when preparing bills generally follow the principles of the SMM, but reduce the work involved in preparing the Bills of Quantities by way of the methods outlined earlier.

Uses of documents at various stages of a project

Where the builder is contracting, the following documents will generally be involved, prepared by one or other of the parties, depending on the type of contract.

- Articles of Agreement and Contract Conditions;
- drawings;
- Bills of Quantities;
- specification;
- programme.

While the initial relevance of the documents to the building owner is to obtain tenders on an equal basis, to the builder this is not so, for he is concerned with using the documents to prepare his tender, and the information so compiled will be used throughout the project. The quantity surveyor will be using much of the information during the contract, and he needs to have knowledge of the relevance of the documents at the earliest possible stage.

The quantity surveyor should have a sound knowledge of the particular standard conditions of contract encountered by the company. He should check the conditions of contract and contract documents for the project before the project starts, and should highlight:

- significant deletions in the contract, for example regarding extensions of time;
- the fluctuation provisions, if these apply;
- the provisions for payment;
- clauses added to the standard conditions;
- discrepancies that may be found between the various contract documents;
- rates in the Bills of Quantities that are of advantage or disadvantage to the contractor;
- other areas of particular relevance to the contract.

If the builder is not contracting but building for sale, the above will not be necessary, for the builder will employ a solicitor to deal with the contracts of sale, and this does not affect the surveyor's role.

Before construction commences budgets should be prepared for each project, and this is dealt with later in the section.

When construction starts the Bills of Quantities will be used to prepare valuations for interim payments or stage payments, and also for the valuation of any variations. This will lead to the preparation of statements showing variances between estimated costs and value. The information then compiled will be used to prepare a final account for the work, both for the main contract and any sub-contracts.

CONTROL OF FINANCE AND PRODUCTION

Each company will set up its own financial and production controls, and procedures for so doing are set out elsewhere in this book. What follows is the part that the quantity surveyor must play in such procedures.

Materials scheduling

Material quantities will be required to be taken off from the Bills of Quantities, and must show:

- quantity required to be fixed,
- waste allowed in the estimate,
- the quantity to be ordered,
- dates of delivery,
- cost included in the estimate.

When materials are ordered, delivered, invoiced and paid for, any variances in quantity and cost should be abstracted and the financial effect shown on the cost/value comparisons; claims should be submitted for the value if fluctuations are allowed under the contract, and feedback given to the estimator and to the site.

Plant allocation

Schedules of the plant required for the contract must be prepared and should show:

- type of plant or transport,
- periods required,
- cost included in the estimate.

When plant is ordered, used, invoiced and paid for, any variances should be abstracted and the financial effect shown on the cost/value comparisons. Also, feedback should be given to the estimator, the site and the plant manager.

Labour allocation

In order that the maximum benefit is obtained from the labour resources, all the construction operations must be well planned, not just within the parameters of the overall contract programme, but by the use of much shorter time programming. Ideally this should be weekly, as the men are paid weekly, and reconciliations and comparisons can then be made. This weekly planning will enable schedules to be prepared for each gang showing:

- the operation(s) to be carried out,
- description of the work involved,
- quantities for the work involved,
- time allowed,
- cost allowed in the estimate.

It will be necessary to ensure that the operations are short-term, to allow for the switching of labour owing to unforeseen circumstances such as inclement weather, without the need to prepare revised schedules at uneconomic intervals.

Provision must be made within the schedule for the application of a bonus system if one is being operated by the company.

When the week's work is completed, the ganger or chargehand of each gang should fill in the worked hours of each operative who worked on an operation.

The calculation of bonus and wages will be carried out by the site clerk, bonus clerk or wages clerk, and the resultant information should be compiled to show:

- hours worked,
- non-productive hours,
- bonus hours saved,
- variance between hours paid and hours included in the estimate,
- variance between wages paid and cost included in the estimate,
- total wages cost.

This information can then be incorporated into the cost/value comparison to show production (hours) and cost (£) variances, and also given as feedback to estimators and site managers.

Incentive schemes

The quantity surveyor may be responsible for incentive schemes, but whether this is so or not he must still liaise with those carrying out the day-to-day work in running such schemes.

This is important because estimators do sometimes price earlier bill items more highly than later items to maximise cash-flow advantages, and it is important that the correct labour values are used, not what would otherwise be inflated values in the early stages of the contract, and deflated values towards the end.

Problems may also arise where the tender was on a fixed-price basis, and therefore included some allowance for future increases; if no increase has been suffered but is allowed in the tender after a certain point, again inflated values for labour may be used in bonus targets, unless adjustment is made.

The above problems should not arise if the estimating build-up is so prepared as to avoid it, but the quantity surveyor should be aware of the estimating build-up, and how allowances for increases in costs or for 'front-loading' have been made.

The quantity surveyor also needs to follow closely the labour outputs and variances, for possible use later: should variations or disturbance to the works give rise to claims for additional financial reimbursements, such information is often evidence of the facts as claimed.

Sub-contract selection, negotiation and placing

Own sub-contractors The selection of those sub-contractors who are to carry out sections of the work will have been carried out by the estimator, who will have based the tender on the prices so obtained.

Whether or not upon commencement of a contract it is decided to seek competitive tenders again from sub-contractors, rather than placing sub-contracts according to the tender enquiries, is a matter for the individual company. The placing of sub-contracts, in either situation, will often lie with the quantity surveyor. When placing sub-contracts, the quantity surveyor must consider:

- the sub-contractor's tender price;
- previous knowledge and performance of the sub-contractor;
- sub-contract contract conditions, especially if these are at variance with the main contract, or trading agreements of the company, in which case they should be evaluated before comparing tender submissions;
- the programme of the sub-contractor, to ensure that it is in accordance with the main programme.

Sub-contracts should, ideally, be placed using a standard form of contract as published by the NFBTE, since these have been agreed by the representatives of the associations of the main contractor and the sub-contractor, and serve to clarify the working contractual agreements between the parties. This is not always the case where special conditions are imposed by the builder upon the sub-contractor.

Nominated sub-contractors In situations where the contractor is, under the terms of the contract, to engage nominated sub-contractors, he must be aware of the contractual significance and associated problems. If this falls to the quantity surveyor, because of his contractual knowledge, then he must be sure to carry out the responsibilities of the contract, including recommending that the company exercise the right of objection to the nominated sub-contractor, where appropriate.

Allocation and recovery of overheads

Where the builder is not in a contractual situation, recovery of overheads becomes relevant when payment is made on exchange of contracts, except where stage payments have been agreed upon.

The position where the contractor is in contract is different, for he will generally be paid under certificates at intervals of one month.

Whichever system applies, control of overhead expenditure must be achieved through the costing system, and regular checks made, through the cost/value comparisons, to monitor estimated costs against actual costs for all items of overheads. For site overheads, this comparison must be made against the items priced in the budget or tender. Head office overheads will be composed by the accounts staff against the overhead budgets.

Analysis of the cost/value comparison results should be given to the estimator for feedback, and to the site manager for action, where appropriate.

Cash flow and budgetary forecasting

The quantity surveyor is responsible for ensuring that the inflow of finance for each site is full and prompt, and he is also reponsible for analysing the cost results against such recovery.

In order to gauge actual performance against estimated performance it is necessary to use a budget, and this has the added advantage of showing the expected capital commitment to the contract at any given stage. Such budgets are generally prepared by the accounts department with the assistance of the surveyor, and examples of these are shown elsewhere in this book.

Ongoing budget checks can be made more communicable to site staff by graphical presentation, and *Figure 7.1* is an example. This shows the relative estimated costs from the tender build-up, and each month the actual costs incurred will be plotted using cost/value results.

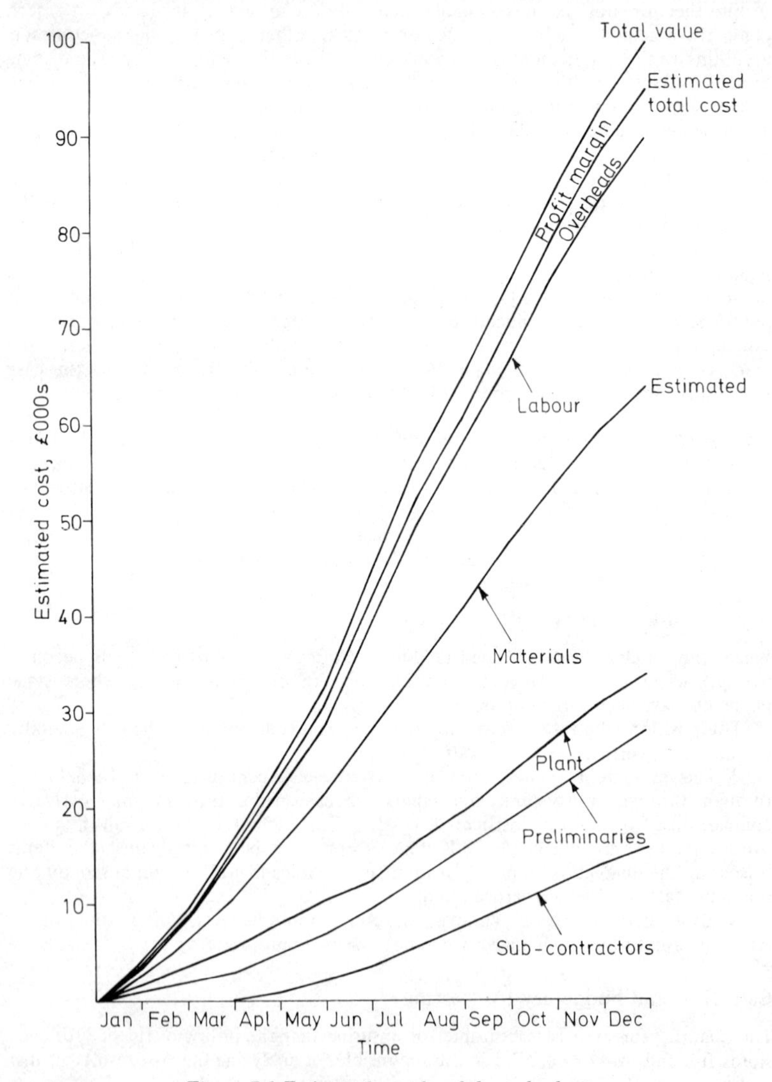

Figure 7.1 Estimated cost breakdown budget

Figure 7.2 shows a different form of presentation, this time using total costs. It is dependent upon the plotting of financial values of the contract and earliest and latest start points of each operation, using information from the Bills of Quantities and the site programme.

The graph is built up by plotting, as work proceeds, the cost and the values of work done, and is often useful in highlighting areas of significance to the quantity surveyor, other than just the flow and comparison of income and expenditure.

In *Figure 7.2* it can be seen that the value of work done exceeds the cost of work done until September, when it falls below cost. Up to that time progress had been within the values of the earliest/latest starts, but is now delayed and extra costs are

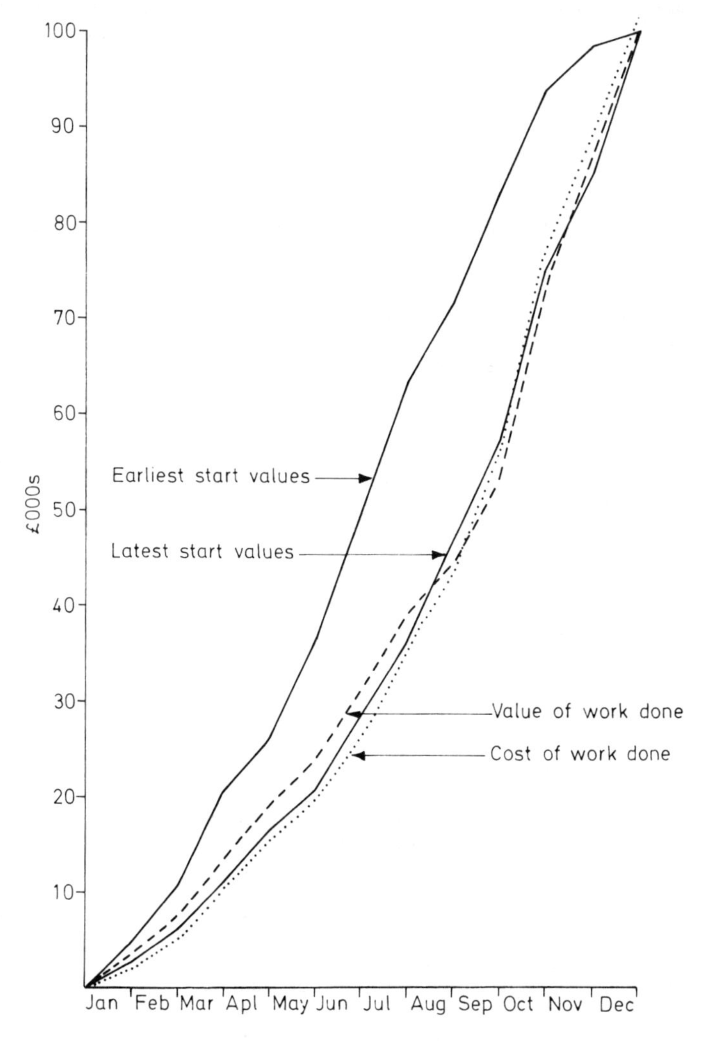

Figure 7.2 Cost/value envelope

showing. By the end of the contract, value is back within the envelope but costs are still greater than value. This may indicate that, despite completion according to programme, extra costs have been incurred that should be reimbursable, and if so may necessitate the preparation of a claim.

Whichever system of budgeting is adopted, it must be updated regularly if it is to have any significance, and the results should be disseminated for feedback.

Variations

Variations arise from omissions, additions, substitutions and changes in the quality or quantity of the works, and will require to be recovered and valued in order to make the necessary adjustments to interim applications/stage payments and final accounts.

This work is the responsibility of the quantity surveyor, but depends very much on the collection and communication of information concerning the changes in cost brought about by the variation. This will involve site staff, planning engineers and others.

Valuation of variations is not always a simple matter of substituting quantities and rates, for it can occur that the nature of the variation also causes the nature of the work to change, or an operation to be used out of sequence. These additional costs must be allocated to the variation and charged to the client, within the provisions of the contract.

The quantity surveyor is responsible for this, and must therefore ensure that the cost feedback comes to him for inclusion, and that he passes the agreed variation price to the estimator as feedback.

'Claims'

'Claims' is a word much used in recent years to signify protracted disputes concerning additional costs arising out of building contracts. It is too lengthy a subject to deal with here, save to say that success in a claim often depends on the strength of the information available; since that information is generally historical, this means the strength of the records that were kept. It is essential therefore that complete records are kept, and the quantity surveyor must take responsibility for controlling this in parallel with records on variations.

There are references to publications dealing with 'claims' at the end of the section.

Monthly valuations and certificates

'Cash flow is the life blood of the construction industry' were the words used by Lord Denning, Master of the Rolls, in a case dealing with building contracts.

Where housing or factories are being constructed for sale, the cash flow in to the company arises when payment is made to the builder upon the completion of the sale.

Where a building is being constructed for a building owner the contract will invariably allow for stage payments or interim certificates at regular intervals. The intervals are usually monthly, although the Government form of contract (GC Wks 1, Second Edition) provides for two certificates per month.

It is the responsibility of the quantity surveyor to ensure that full application is made for payment at the stipulated intervals. The following list shows the items to be considered when preparing a valuation.

1. Measured work executed by the builder and his sub-contractors
2. Preliminaries
3. Dayworks
4. Materials on site
5. Materials off site
6. Variations
7. 'Claims'
8. Work executed by nominated sub-contractors
9. Materials and goods supplied by nominated suppliers
10. Variations in price
11. Overheads and profit

Because builders prepare tenders in different ways it may be that overheads and profits are built into measured rates; if this is the case, the value would be recovered through item 1 when preparing a valuation.

When preparing a valuation it is not necessary to measure in detail every single item of work executed. Each item must be considered and included, but a little preparatory work by the quantity surveyor before the contract work commences can greatly reduce the volume of work to be done at each valuation.

Figure 7.3 illustrates the way this can operate on a housing site. The trades and operations are listed and values inserted against each of them using the Bill of Quantity values. Each valuation period, a percentage assessment is recorded of the work done against each operation. By applying the percentages to the values of the operations and totalling, a value of work executed at each period is compiled for each house for inclusion in the valuation.

The list contained in *Figure 7.3* is quite comprehensive, but many builders may choose to use a more condensed list, especially in cases where they have simplified Bills of Quantities prepared by themselves.

Use of a valuation progress chart will reduce much of the time involved in preparing the measured-work section of a valuation, but there will be other work that will require measuring, for example access roads, and to such items must also be added all relevant sums compiled in respect of the items to be considered in a valuation as earlier listed.

Having prepared and submitted a valuation the quantity surveyor must then ensure that payment in respect of the valuation is certified in full and made within the period stipulated in the contract.

Cost/value comparisons

At regular intervals actual progress and cost must be measured against estimated progress and cost to establish:

- where losses and profits are occurring;
- the reasons for the loss or profit;
- the action that should be taken, if any;
- feedback to estimators.

It must be emphasised that it is the estimated costs and values that are compared, i.e. net of profit.

The majority of companies choose to prepare cost/value comparisons in conjunction with the preparation of valuations. This generally therefore gives a monthly comparison, and the time for preparation of the cost/value comparison could mean that some of the information is five or six weeks out of date. Builders should therefore give due consideration to preparing such comparisons at shorter intervals, if they are to have information available within a time such that they are able to take action upon that information.

Trade and operation		*Value £*	%	£	%	£	%	£	%	£	%	£
Excavator	Oversite											
	House foundations											
	Drain trenches											
	Service trenches											
Concretor	Hardcore											
	Foundations											
	Oversite											
	Drain trenches											
Bricklayer	Up to DPC											
	DPC to 1st floor											
	1st floor to eaves											
	Above eaves											
	Partition walls											
	Fireplace											
Roof coverings and tiling	Felt and battens											
	Roof tiles											
	Vertical tiling											
Carpenter/joiner	1st floor joists											
	Roof timbers											
	Fascia and soffit											
	Floor boarding											
	Stairs											
	Partitions											
	1st fixings											
	2nd fixings											
	Kitchen units											
	Ironmongery											
Wall tiling	Kitchen											
	Bathroom											
	WC											
Electrician	Carcassing											
	Fittings											

Figure 7.3 Valuation progress chart

Whatever the periods decided upon for the preparation of cost/value comparisons, the periods should be equal so that comparisons can be made not only for the current period between costs and value, but also with previous periods.

Detailed workings of cost/value comparisons are covered elsewhere in this book under accounting procedures, but care should be taken to ensure that the bases of preparation of costs and values are compatible, so that comparison is worthwhile.

Costs will be reconciled to the date of comparison, with due allowance for:

- split of weekly wages, where the date of comparison falls within the week;
- costing of materials delivered to site, included in valuations, but not invoiced.
- hired plant on site the value of which is included in the valuation but which has not been invoiced.
- apportionment of preliminaries, overheads, and any other costs that are not compatible with the valuation at the date of comparison.

Likewise, it follows that the valuation should be reconciled to make comparisons viable, and the following are areas where this may be necessary:

- adjustment of total value to the date of reconciliation;

Trade and operation		*Value £*	%	£	%	£	%	£	%	£	%	£
Plumber	Roof and rainwater goods											
	Soil and wastes											
	Sanitary fittings											
	Water service											
	Central heating 1st fix											
	Central heating 2nd fix											
Gas fitter	Carcassing and fittings											
Telephone	Installation											
TV	Installation											
Plasterer	Ceilings											
	Walls											
	Floor screeds											
	Making good											
Glazier	Fixing											
Painter	Internal painting											
	External painting											
Floor tiler	Thermoplastic tiles											
Drainlayer	Foul drains											
	Stormwater drains											
	Testing											
Garage	Foundations excavated											
	Foundations concreted											
	Brickwork to DPC											
	Brickwork above DPC											
	Floor											
	Roof											
Services	To site											
Landscaping	Turfing and planting											
Roads and paths	Paths											
	Driveway											
Boundary walls and fences												
Handover	(insert date)											

Figure 7.3 continued

- adjustment of any over- or under-valuation, in respect of any of the areas of the valuation;
- adjustment for any over- or under-recovery of variation in price;
- adjustment for any 'front loading' of the tender;
- adjustment of any reserves taken by the surveyor so that a full and correct value is shown, *not* depreciated by the additional reserves that some surveyors have been known to take in a mistaken attempt to enhance their reputation at final account stage.

If the above guidelines are followed, in accordance with a formal procedure, then the costs and values, when compared, will be truly compatible; the information given, and the decisions taken, will accordingly be that much more meaningful.

Presentation of information

The comments above, concerning the compatibility of the information between costs and value, are also relevant to any comparison of information and run through the whole formal communications system of a company.

Estimates, programmes, valuations and cost/value comparisons must all be prepared so that any comparisons made are worthwhile and any feedback is of value. For example all too often feedback is given by surveyors to estimators in a form that is of little value because it is not compatible with the manner in which an estimator produces a build-up.

If a formal procedure is being set up, care must be taken to ensure that it is tailored to dovetail into other departmental systems, or that the other systems are adjusted accordingly. This may seem to be stating the obvious, but there have been instances of systems being introduced without due regard to the communications flow and information requirements of other departments. When this happens, all that is proved is that the system works, but in isolation, and consequently full benefit is not accorded to the company.

FEEDBACK

Information feedback to estimators, even when made available, suffers often from being too historical and is often presented in a format incompatible with the needs of the estimator.

Feedback should therefore be presented as soon as possible, and in a readily usable format. Only if this is done can an estimator hope to reflect the accurate net cost in the estimate.

Cost/value comparisons give only a general feedback on performance; more detailed information is available through the weekly labour, plant and sub-contract costs, which should be made available to the estimator as a matter of course.

The estimate and tender lay foundations for the surveyor to work on, and it is therefore in his interests to ensure that the correct feedback reaches the estimator. Surveyors are responsible for the preparation of cost/value comparisons, and should also follow this through by being responsible for feedback to the estimator. Before each contract starts the feedback should be reviewed, so that new construction processes, unusual conditions etc. are recorded separately for feedback. The areas to be covered should be decided by the estimating department, but it must remain the responsibility of the surveyor to prepare the information.

TRAINING

Surveyors in the construction industry generally follow a formal course leading to a qualification, the course being either full-time, part-time day release, or evening release. Such courses are available at technical colleges, polytechnics and universities, and the following bodies figure prominently in the field of qualifications sought by surveyors:

- Royal Institution of Chartered Surveyors,
- Institute of Quantity Surveyors,
- Institute of Building,
- Incorporated Association of Architects and Surveyors,
- Construction Surveyors Institute.

Courses are also run for ONC and HNC qualifications in quantity surveying.

If a company is to develop its surveying staff to the full, both for the benefit of the individual and the company, a formal pattern of development should be laid down. Periods of time should be spent in the departments of the company dealing with:

- land acquisition,
- planning applications,
- estimating and tendering,

- purchasing,
- planning,
- plant,
- bonus,
- contracts management.

Through such a training a surveyor will realise the significance and requirements of each of the departments, and should consequently provide a better service to the company.

FURTHER INFORMATION

Skoyles, E.R., *Preparation of operational bills,* Building Research Station (1967)

Skoyles, E.R., *Examples from bills of quantities (operational format),* Building Research Station (1969)

Clapp, M.A., *Cost comparisons in housing maintenance,* Building Research Station (1965)

Clapp, M.A., *Labour requirements for conventional houses,* Building Research Station (1965)

Skoyles, E.R., and Hussey, M.J., *Wastage of materials on building sites,* Building Research Station (1974)

Willis, A.J. and C.J., *Practice and procedure for the quantity surveyor,* Crosby Lockwood Staples (1976)

Seeley, I.M., *Building economics,* Macmillan (1972)

Stewart, E.P., *Techniques for preparing a bill of quantities,* E. & F.N. Spon (1971)

Wainwright, W.H., *Variation and final account procedure,* Hutchinson (1970)

Harlow, P.A., *Contractual claims – an annotated bibliography,* Institute of Building (1976)

Burke, H.J., *Claims and the standard form of building contract,* Institute of Building (1976)

Wood, R.D., *Builder's claims under the JCT form of contract,* Institute of Building (1976)

Robinson, T.H., *Establishing the validity of contractual claims,* Institute of Building (1977)

Powell-Smith, V., *Contractors' guide to the Joint Contracts Tribunal's standard forms of building contract,* IPC Building & Contract Journal Ltd (1977)

8 NHBC SCHEME

INTRODUCTION 8–2

HISTORY AND CONSTITUTION OF NHBC 8–3

THE REGISTER 8–4

TECHNICAL REQUIREMENTS 8–6

INSPECTION AND CERTIFICATION 8–6

BUILDERS' OBLIGATIONS TO PURCHASERS 8–7

NHBC'S UNDERTAKINGS TO PURCHASERS 8–11

DEFECTIVE PREMISES ACT 1972 8–15

CHANGES IN LAW AFFECTING BUILDERS' LIABILITIES 8–15

8 NHBC SCHEME

Miss R.I. WYLES, BA(Hons), Barrister
Legal Adviser, National House-Building Council

This section deals primarily with the National House-Building Council Scheme, outlining the obligations of builders and developers and the rights of purchasers under the Scheme, and concluding with a brief resumé of the major changes in the law relating to builders' liabilities at common law.

INTRODUCTION

The National House-Building Council (NHBC) is an independent, non-political, non-profit-making body approved by successive governments and now recognised under statute by virtue of Section 2 of the Defective Premises Act 1972.

Its main object is 'the improvement of the house-building industry throughout the United Kingdom (including in such industry the building of all forms of home accomodation)' – see Clause 3 of the Memorandum of Association. NHBC ensures, as far as possible, that this aim is achieved by:

- maintaining a Register of Builders and Developers whose financial and technical standards are vetted, both before and after registration;
- prescribing minimum building standards ('Technical Requirements') with which all builders must comply when constructing dwellings;
- carrying out spot-check inspections of all dwellings under construction, to ensure that, as far as can be seen, the builder complies with the Technical Requirements so that the Ten-Year Protection Certificate ('Certificate') may be issued;
- obliging vendors of dwellings to offer the current prescribed form of House Purchaser's Agreement to purchasers, which Agreement contains warranties relating to standards of workmanship and materials and the remedying of defects;
- providing an insurance scheme to indemnify purchasers in the event that there has been non-compliance with the Technical Requirements;
- promulgating Rules from time to time relating to the above matters, and taking disciplinary action against builders and developers who are in breach thereof.

The Rules, Certificate and House Purchaser's Agreement together constitute the NHBC Scheme. The Scheme, in its present form, has been in operation since 1965, although since that date a number of amendments have been made.

The description of the Scheme given in this section is limited to the current Scheme; however, its fundamental principles have not been substantially altered since 1965, so most of the text that follows is applicable to the obligations of builders and developers since that date. Essentially, the differences between the Scheme in 1977 and in 1965 are:

- tightening up of disciplinary procedures;
- improvement in Technical Requirements;
- increase in insurance benefits available to purchasers;
- since 1971, the developer has primary responsibility for dwellings under the Scheme; previously it was up to the builder and developer to choose which one of them offered the Agreement.

There is not sufficient space available to give more than a cursory description of the NHBC Scheme. However, it is hoped that this section will give those who have only heard of NHBC a working knowledge of all aspects of the Scheme, and will provide those who are already registered members with both background and detailed knowledge of the Scheme in a brief but comprehensive form.

HISTORY AND CONSTITUTION OF NHBC

The NHBC Scheme, in a limited form, was started in 1936 by a group of builders with the object of improving the reputation of the house building industry at a time when 'jerry building' was causing public concern. Thus, members were obliged to conform to prescribed technical standards and offer a two-year warranty to purchasers.

During the early 1960s consideration was given to the introduction of a compulsory registration scheme, and the matter was debated at some length by Parliament. As a result, NHBRC (National House-Builders Registration Council as it was then) held discussions with the Law Commission, the Law Society and official bodies concerned with house-building, to discuss what improvements were possible. In September 1965 the first ten-year Certificate was introduced, under which NHBC assumed primary responsibility for major structural defects appearing after the end of the builder's two-year warranty period. Consideration was also given to an improved minimum technical specification.

Consequently, Government, the Building Societies Association and other official bodies associated with the building industry agreed to support the NHBRC Scheme in place of a compulsory scheme, provided that more builders became registered. Once the Building Societies Association gave notice to their members that, as from 1 September 1968, no mortgage advances would be made on new dwellings without an NHBRC Certificate being available, the number of members registered increased rapidly, so that today approximately 21 000 builders and developers are registered (compared to 1700 in 1936).

Constitution

In 1967 a new Council consisting of 45 members in England and Wales was formed, again with builders as the largest single group, but as before without an overall majority. The various consumer bodies, and professional institutions associated with the house-building industry together with the Building Societies Association, all accepted invitations to nominate members. In 1968 the constitution of the Council was amended to include a Scottish Committee of 14 members, and in 1969 a Northern Ireland Committee of seven members was formed. These committees have a large measure of autonomy over the management of the Scheme in their own countries. All members of NHBRC are assigned to one or more committees, each responsible for a section of NHBC's work. Final authority rests with the full Council, but the day-to-day management and making of policy decisions are delegated to an *Executive Committee.* In addition, there is a *Finance Committee* responsible for the finances of NHBC and other related matters.

The Executive Committee has two sub-committees:

- *The Registration Committee* deals with the admission and removal of builders and developers from the Register. It also supervises the standards achieved by builder and developer members and takes disciplinary action where appropriate. It has a Disciplinary Sub-Committee before whom all builders accused of disciplinary offences have the right of appearance.
- *The Standards Committee* is responsible for considering and introducing improvements to the Technical Requirements.

Government observers from the Department of the Environment attend meetings of all committees, and the Secretary of State for the Environment appoints the Chairman of the Council and one member of the Finance Committee. All changes to the Rules have to be approved by the Secretary of State for the Environment (Rule 39), apart from the Scheme having been approved under the Defective Premises Act 1972.

THE REGISTER

Application

Any person falling within the definition of house builder or developer may apply to have his name entered on the Register (Rule 1). Builders are defined as those persons (including partners and corporations) who are engaged at any time in the construction of dwellings for others, and developers as persons (including partners and companies) who arrange for the construction of dwellings for sale directly or indirectly to the public (or who are, in the opinion of NBHC, concerned in or with such arrangements) (Rule 41).

Whilst builders, once registered, can also act as developers, the reverse is not possible. If a developer wishes to build, he must first apply to be transferred to the Builder's Register, and his building standards will be vetted as though he were a new applicant (Rule 4).

Housing Societies and Associations, such as self-build groups, are eligible for registration, provided that they are able to comply with the normal conditions of registration. Public Authorities (e.g. local authorities and development corporations) are ineligible for registration; however, special arrangements have been made to enable dwellings built by registered builders for sale by Public Authorities to carry the benefits of the Scheme.

When an application is made, the applicant must provide information concerning details of past, current and future building projects, the name and address of the applicant's bankers, and two other references. Limited companies must also give the names and addresses of directors and shareholders, and the amount of nominal and paid-up capital. The address of the registered office (if any) as well as a permanent address for correspondence (if different) must be given.

The current application fee is £100 plus VAT, and is non-returnable even if the applicant is refused registration.

Builders' applications are subject to vetting from both a technical and financial point of view; developers are subject only to financial investigation, but are obliged under NHBC's Rules to employ registered builders as main contractors unless exemption has been granted (Rule 26). Financial status is investigated by the taking up of references and obtaining credit status reports.

Technical competence is assessed by inspection of any dwelling under construction. Normally, a builder applicant is not accepted on the Register unless NHBC has inspected two dwellings from commencement to completion of construction, and his application is deferred until that time. An application is only deferred further if the builder's standards are below NHBC's Technical Requirements and he shows no sign of improvement. If the applicant is accepted on the Register, the dwellings that have been inspected for the purpose of the application may be covered by the Scheme if the applicant wishes; in any event, the applicant must pay the normal inspection fees for those dwellings once he is registered.

Developers' applications are unlikely to be deferred, since their applications are assessed on a financial basis only.

Conditions

Once NHBC has decided that the applicant is suitable, registration is offered, normally subject to conditions. These ensure as far as possible that NHBC has some financial security in the event that the builder or developer, once registered, fails to fulfil any of his obligations under the Scheme. Conditions are normally imposed for the first three years registration or for the first, say, 20 dwellings.

Conditions that may be offered are as follows:

- To pay the sum of up to £500 (normally £100 or £200) per dwelling in respect of such dwellings that the applicant submits for inspection. Such sums are normally repaid after three years if no claims have been made and none are anticipated, and interest is payable at the normal Building Society rate for deposits, less 1 per cent.
- To provide directors' or shareholders' personal guarantees, where the applicant is a company. Such guarantees are unlimited in time, but again are normally discharged after three years. Parent companies may be asked to provide guarantees in respect of their subsidiaries. Guarantees limited financially may be acceptable, although they are unusual.
- To provide a bond, backed by an insurance company or bank, normally limited to two or three years, the financial limits on the bond being related to the first year's building programme (usually calculated at £500 × the number of dwellings the applicant intends to build).
- To increase the paid-up capital of a company, where in the opinion of NHBC the company is under-capitalised in relation to its commitments.
- Where the applicant is associated with a previously registered builder or developer, the applicant may be asked to assume responsibility for dwellings built or sold by the builder or developer. Where no claims have already been met, the applicant will be given the alternative of paying a non-returnable insurance surcharge, usually in respect of all dwellings registered for inspection purposes for a period of up to five years. The degree of responsibility, or the surcharge, that the applicant will be asked to pay will be based on the degree of association or control (i.e. as director or shareholder) that the applicant exercised over the previously registered builder or developer.

Most applicants nowadays are subject to one or more of the conditions of registration set out above, unless the applicant is a sole trader and his building standards are acceptable and the number of dwellings he intends to build is small (i.e. below 20), or unless the applicant is associated with another registered member whose reputation is good.

Mention should be made here of the introduction of 'premium rating', which NHBC intends to introduce in 1979. Premium rating will effectively operate as a no-claims bonus for good builders, with 'bad' builders, or builders previously unknown to NHBC, paying increased insurance premiums. As details have not yet been published it is impossible to give further details at this time. However, suffice to say that 'premium rating' should go some way towards meeting the criticism made by many builders, that the good builders pay for the defaults of the bad. Presumably the introduction of premium rating will also result in some alteration of NHBC's policy of imposing conditions of registration.

Where an application is refused, or the applicant objects to the conditions offered, he may appeal to the Registration Committee, where the evidence given is in written form only. Where an applicant wishes to appear in person or through his solicitors, the case may be referred to the Disciplinary Sub-Committee. Final appeal lies to an independent Appeals Tribunal whose three members are nominated by the Law Society (which nominates the Chairman), the NFBTE and the RIBA with the RICS.

Discipline

Mention has already been made of the Registration Committee and Disciplinary Sub-Committee, which deal with disciplinary matters as well as the imposition of conditions of registration and the refusal of registration. Disciplinary offences include failure to build in accordance with NHBC's Technical Requirements, failure to remedy defects, failure to satisfy purchasers' complaints, and other breaches of the Rules.

Automatic deletion, without reference to those Committees, is provided for in Rule 11, (a) when liquidation or bankruptcy occurs, (b) when a Receiver is appointed, (c) when NHBC satisfies an arbitration award or court judgement, or (d) when there is no reply within 28 days to Recorded Delivery correspondence. Whenever a decision to delete is made, whether by a Committee or by NHBC officials as provided for in the Rules, final appeal lies ultimately to the independent Appeals Tribunal referred to above, which has the same power to award costs as a judge of the High Court (see Rules 8–18).

TECHNICAL REQUIREMENTS

Rule 26 obliges all builders to build in compliance with NHBC's minimum building standards. These minimum building standards are known as NHBC's Technical Requirements and are contained in Part 2 of the Registered House Builder's Handbook.

There is no space for detailed consideration of these Technical Requirements. Briefly, the Requirements set down minimum standards to ensure soundly constructed dwellings, and incorporate the Building Regulations. They cover all trades involved in house building and include some design matters. Special provisions cover timber-frame dwellings, design details of which must be undertaken by a 'competent person' and submitted to NHBC before commencement of work.

Mention should be made here of NHBC's power to require a structural engineer's report where a builder intends building on a hazardous site (e.g. made-up ground or a deep-fill site). The builder must notify NHBC two months (see Rule 23(c)) before he intends to start work on site, and if NHBC thinks it necessary the builder will have to employ a structural engineer to design the foundations and/or supervise the works. Where NHBC is uncertain as to the competence of the design (insofar as it relates to the structural stability of the dwellings), NHBC may require the structural engineer to sign a document confirming his satisfaction with the design and the works carried out. In the event that the dwelling later failed, NHBC might have recourse to the engineer's professional indemnity insurance.

In the event that there is any dispute between a builder and NHBC as to whether the Requirements have been complied with, NHBC can refuse to issue its Ten-Year Protection Certificate and the builder has the right to refer the dispute to independent arbitration (see Rule 33).

INSPECTION AND CERTIFICATION

All dwellings that are being built for sale or for letting directly or indirectly to members of the public must be registered with NHBC for inspection purposes, and a fee (calculated according to the sale price of the dwelling) must be paid at that time. This fee covers both the inspection costs and the insurance premium.

Rule 23 provides that the application must be made 14 days (shortly to be increased to 21 days) before commencement of work on site (the two-month rule having already been complied with wherever necessary). Failure to give proper notice will result in NHBC requiring the vendor (or builder, if he is also the developer) to pay an increased premium and special survey fee in order for a Certificate to be issued once the dwelling is completed to the satisfaction of NHBC. The cost of obtaining a Certificate in such circumstances is approximately £200 per dwelling (the maximum fee payable normally is currently £90).

A plan of the site must be provided at the same time as the application for inspection. Builders must also provide samples of materials for testing purposes when called upon to do so (Rule 27).

Once the application for inspection is received, NHBC arranges for inspections to be carried out during the course of construction, normally once every three weeks. Defects noted by the inspectorate are reported to the vendor by way of entries in a Site Record Book provided for that purpose. Defects so entered must be remedied within, normally, 28 days, otherwise disciplinary action will follow, which may result in the builder's and/or developer's name being deleted from the Register (see Rules 8, 9, and 10). Once a dwelling is completed to the satisfaction of the NHBC inspectorate, the Certificate is issued directly to the vendor who applied for inspection, and should be handed to the purchaser once the dwelling is sold.

BUILDERS' OBLIGATIONS TO PURCHASERS

These are contained in the House Purchaser's Agreement, which every registered developer is obliged to make an irrevocable offer to enter into, where a dwelling is to be sold to a member of the public who is buying it for his own or his family's occupation (Rule 28(a)).

Builders are obliged to offer the Agreement only where:

- the builder is also the vendor (i.e. he is acting as both builder and developer); or
- he is constructing a dwelling under contract on the purchaser's own land; or
- he is constructing dwellings under contract for a Public Authority to sell, and the Public Authority wishes the dwellings to be covered by the NHBC Scheme; in those circumstances, the builder is authorised to enter into the HB5B Agreement directly with the purchasers (i.e. not the Public Authority);
- in exceptional circumstances where NHBC allows the builder to offer the Agreement and 'the developer' is not registered.

Public Authorities may themselves enter into the HB5C Agreement with the first purchasers, where the dwellings have been constructed by a registered house builder. This they would do normally only where they wish the Scheme to apply, but have not required the builder in the contract to offer the Agreement.

The HB5, HB5B and HB5C Agreements are in all fundamental respects identical, although the HB5C Agreement has no pre-certification bankruptcy cover (since it is hardly conceivable that a Public Authority would go bankrupt).

For the purpose of the rest of this section, the word 'Vendor' should be construed as also meaning 'builder' or 'Public Authority', i.e. the person authorised by NHBC to enter into the appropriate form of House Purchaser's Agreement. The main clauses of the Agreement are as follows:

(a) In Clause 2 the Vendor warrants:

(1) *that his name is entered on the Register*

If an Agreement were entered into by an unregistered Vendor so that NHBC's undertakings did not apply, an action for damages would lie.

(2) *that he has undertaken to abide by the Rules of the Council and that he has submitted or will submit an application for periodic inspection of the dwelling by the Council during construction*

The first part of this warranty is self-explanatory. In the event that the Vendor is in breach of the second part, NHBC would be unable to issue its Certificate, so that an action for breach of the warranty contained in Clause 3(3) (see paragraph (b) (3) below) should, in appropriate circumstances, be coupled with an action for breach of this warranty. As a purchaser might have difficulty in selling his house without the Certificate, damages could be considerable. However, in appropriate circumstances, NHBC does allow the purchaser to make the application for inspection and the purchaser can recover the fees from the Vendor as damages for breach of this warranty.

(b) In Clause 3 the Vendor warrants that the dwelling has been built, or agrees that it will be built:

(1) *in an efficient and workmanlike manner and of proper materials and so as to be fit for habitation*
This is the warranty normally implied at common law where dwellings are built or sold, and is now imposed as a statutory duty under the Defective Premises Act 1972. It should be noted that this warranty (together with the other warranties contained in Clauses 2 and 3) is a six-year warranty in accordance with Section 2 of the Limitation Act 1939. Case law has established that this warranty should be read disjunctively, i.e. that if the dwelling is not fit for habitation, the Vendor is still in breach of the warranty even if he has built in an efficient and workmanlike manner and of proper materials – see *Hancock v Brazier (1966)* and *Batty v Metropolitan Realisations Limited (1977)*, both decisions of the Court of Appeal. Breach of this warranty would give a purchaser a right to damages, but this warranty should be considered in relation to the builder's two-year warranty and NHBC's obligations. Normally, a purchaser would only bring an action for breach of this warranty where NHBC's indemnity is insufficient to cover his loss or is, for other reasons, not applicable.

(2) *so as to comply in all respects with the Council's Requirements*
This warranty is self-explanatory. If breaches occur and are manifest during the 'initial guarantee period' (broadly speaking, defined as two years from the date of the Certificate) the purchaser would be entitled to call upon the Vendor in accordance with Clause 6 of the Agreement to carry out the remedial works. After expiry of the two-year period, the purchaser could take action against the Vendor under the warranty given in Clause 3(1) and/or this warranty, provided that NHBC has no liability under Clause B(2) of the First Schedule to the Agreement, or its indemnity is sufficient to meet the purchaser's claim.

(3) *so as to qualify for the Full Certificate*
If for some reason NHBC is unable to issue its Full Certificate a purchaser would be entitled to damages. The value of the Certificate (given that it brings into operation the major benefits of the NHBC Scheme – see 'NHBC's undertakings to purchasers', page 8–11) is difficult to quantify; damages as high as £750 were awarded in an unreported County Court case. Given that Building Societies are advised not to grant mortgages in the absence of the Certificate, it can be seen that breach of this warranty can mean that the dwelling would be difficult to resell.

Normally a Certificate would not be issued by NHBC in circumstances where either the Vendor failed to make the necessary application (in such circumstances the Vendor could apply for the special indemnity Certificate – see 'Inspection and certification', page 8–6) or there are outstanding defects that prevent its issue. In the latter case, an action for breach of the warranty could also be coupled with a claim for damages for breach of the warranties contained in Clause 3(1) and (2).

(c) *Hand-over of Certificate*
The Vendor is also obliged under Clause 4 of the Agreement to deliver the Full Certificate (or the endorsed or structural certificate as the case may be) to the purchaser as soon as it is received. A note at the foot of the Agreement makes it clear that this may be several weeks after the completion of the dwelling, and the purchaser should not be unduly concerned in the event that the Certificate is not immediately forthcoming after legal completion.

(d) *Two-Year Warranty*
Under Clause 6 of the Agreement, the Vendor is obliged to make good within a reasonable time, and at his own expense, any defects in the dwelling, and any damage to the dwelling caused thereby, that are consequent upon a breach of NHBC's

Technical Requirements and that have been reported in writing within the initial guarantee period as soon as practicable after they appear. In the event that defects appear in an adjoining or adjacent dwelling that could cause damage in the purchaser's own dwelling, the Vendor is obliged to remedy the defects provided that he is given access to the adjoining or adjacent dwelling. The Vendor is not obliged to make good any defects that are caused by wear and tear or by normal shrinkage arising from the drying out of the dampness normally to be expected in any dwelling after construction, or by defective design where the purchaser has provided the structural or installation design details. In the event that the defects are in, or are caused by, anything not originally built in to the dwelling pursuant to the contract between the purchasers and the Vendor, the Vendor is not obliged to make good the defects.

This two-year warranty is expressed to be without prejudice to the Vendor's liability for breach of warranty or agreement under Clause 3(1). A purchaser might, therefore, have a concurrent claim under both Clause 6 and Clause 3(1). After the initial guarantee period has expired, his remedies under Clause 3(1) would still be available.

It should be noted that the liability of the Vendor is in respect of defects reported in writing during the initial guarantee period. Clause 5 of the Agreement provides that the purchaser shall report to the Vendor any defects in the dwelling that are consequent upon any breach by the Vendor or the Council's Requirements as soon as practicable after the defects appear or ought reasonably to have become apparent to the purchaser. It has been a vexed question in many disputes as to whether or not the Vendor's obligation remains where a defect has not been reported in writing although it has been brought to the attention of the Vendor by other means. It seems clear from the general law of estoppel that if a purchaser does give verbal notice to a Vendor, and the Vendor does not require the purchaser to then put it in writing (which is a common situation where a builder still has workmen on site and a foreman authorised to deal with purchaser complaints) the Vendor cannot object at a later date if he has not received notice strictly in accordance with Clause 5 of the Agreement. In relation to the NHBC scheme, there is authority on this point; see the decision of the Official Referee in the case of *Marchant v Caswell & Redgrave (1976).*

If, on the other hand, the purchaser does not give any notice during the initial guarantee period where defects have appeared during that period, he may still have his remedies under Clauses 3(1) and (2). However, it is considered that in those circumstances he would not be able to recover the full extent of his loss, as the builder could plead that he had been prejudiced by the purchaser's delay in notifying the defects, particularly if the purchaser had had the works carried out before he made a claim against the builder. Furthermore, it is arguable that the builder might have an absolute defence or counterclaim, as he could claim breach by the purchaser of Clause 5 of the Agreement.

(e) *Successors in title*

Clause 7 of the Agreement makes it clear that, whilst successors in title are covered by the Scheme in precisely the same way as first purchasers (see Clause 1 of the Agreement and Rule 29(b)), a successor in title of the first purchaser would not have a claim against either the Vendor or NHBC, as the case may be, in respect of defects in the dwelling that have not, but that should have been, reported to either the Vendor or NHBC by the prevous purchaser. The reason for this limitation in respect of successors in title is that, if defects are apparent to a prospective purchaser, it is for him to negotiate a reduction in the contract price, leaving the vendor who is selling the dwelling to make a claim against the Vendor or NHBC (as appropriate). Alternatively the successor in title could pay the full purchase price, after the vendor has made his claim, and the vendor could then assign the benefits of that claim to him.

(f) *Authority to bind the Council*
Clause 8 of the Agreement provides that the vendor, by virtue of his registration with NHBC, undertakes on behalf of NHBC that NHBC will perform the undertakings set out in the First Schedule. It is important to note that it is solely *by virtue of the Vendor's registration* that NHBC is bound by its undertakings. In the event that the Vendor is not registered at the date of signing the Agreement (which is the relevant date for the purposes of this clause) NHBC would not be bound by its undertakings and the purchaser has the right to claim damages in respect of the Vendor's breach of warranty contained in Clause 2(1).

Clause 8(b) of the Agreement provides that, in the event that defects or damage appear in the dwelling after the initial guarantee period, the purchaser must pursue his remedies against NHBC, and that any relief obtained from NHBC shall be taken into account in mitigation of his damages against the Vendor. The effect of this clause is that, if a purchaser started proceedings against the Vendor for breach of the warranty contained in Clauses 3(1) and 3(2) in respect of damage appearing in the structure of the dwelling, it would be open to the Vendor to argue that the purchaser has not pursued his remedies against NHBC and that, until such time as he does so, he is not entitled to any remedy against the Vendor. Of course, if the purchaser can show that he would be unable to obtain a remedy against NHBC (because of the effect of any of the exclusion clauses contained in the First Schedule to the Agreement), the action would proceed.

(g) *Agreement additional to any other rights*
Paragraphs 9 and 10 of the Agreement make it an express condition of the Agreement that rights under the Agreement are additional to any other rights conferred on the purchaser by any other contract between himself and the Vendor, and that nothing contained in any other contract can in any way restrict or override the provisions of the House Purchaser's Agreement (except where the contract provides for higher standards). These two clauses are to ensure that the Vendor cannot, as a matter of contract, vary the terms of the House Purchaser's Agreement in any way detrimental to the purchaser. This is an important provision, given that it is a usual conveyancing procedure for the Agreement to be signed on the same day as the contract of sale or building. Such a provision makes it clear that, on reading the two documents together, the House Purchaser's Agreement is the document later in time, and therefore overrides the sale or building contract if there is any conflict between the Agreement and the sale/building contract. Section 10 of the Unfair Contract Terms Act would probably prevent a Vendor from evading his responsibilities under the Agreement.

(h) *Arbitration clause*
Paragraph 11 of the (current) Agreement contains a binding arbitration clause between the purchaser and Vendor. There is an exception in relation to disputes arising under Clause 3(1), which may be referred to arbitration only if both parties agree. Otherwise, any dispute under this Clause would have to be referred to the courts. A proviso to Clause 11 also provides that, if there are any contractual disputes arising at the same time as disputes under the Agreement, the matter need not be referred to arbitration unless both parties agree.

It is worth noting that the case of *Purser & Co (Hillingdon) Limited v Jackson (1976)* is authority for the proposition that the NHBC Scheme allows for a number of arbitrations to be undertaken in relation to different defects in the same dwelling. This means that a purchaser can take the Vendor to arbitration in respect of defects, say, in his roof, and having received an award in respect of these defects, could then commence arbitration proceedings in respect of defects, say, in the foundations in his dwelling.

The purchaser cannot, of course, take the builder to arbitration twice in respect of the same defects, whether under the Agreement each time, or under the

Agreement in the first instance and the contract in the second; see the case of *Willday v Taylor (1976)*, a decision of the Court of Appeal. In that case, the arbitrator did not in the first arbitration (under the Agreement) find that the defects did not exist; he made no decision as to the existence of the defects or otherwise, as he felt that the purchaser's claim must fail for other reasons. The purchaser then sued the builder in the County Court under his contract, and the County Court Judge found that he was estopped from doing so. The Court of Appeal confirmed the decision of the County Court Judge on the grounds that the matter was 'res judicata', as the same defects had already been the subject of the arbitration award. In terms of strict justice this was probably unfair to the purchaser; but any other decision would have been contrary to existing law, which prevents the same matter from being litigated twice.

NHBC'S UNDERTAKINGS TO PURCHASERS

These are contained in the First Schedule to the Agreement, and are divided into three parts:

(a) prior to certification, limited to (currently) £1500 (this limit has in fact been increased to £5000 since April 1978 for all dwellings registered with NHBC for inspection purposes after this date, although the Agreements have not yet been altered);

(b) post-certification, where the vendor fails to honour his obligations during the initial guarantee period (which is normally defined as two years from the date of the certificate);

(c) for years three to ten from certification, to indemnify purchasers against major damage caused by structural defects.

NHBC's indemnity under (b) and (c) jointly is currently £20 000. In relation to any one builder or developer, there is a limit of £250 000 relating to all applications for inspection received in any one insurance year. This limit increases to £500 000 when more than 5000 applications for inspection are received. The total limit of NHBC's liability is £5 000 000 for a generation of dwellings, i.e. all dwellings applied for in one insurance year.

Prior to certification: Clause A

NHBC indemnifies purchasers against:

(a) loss of deposit where the dwelling has not been started; or

(b) the cost of remedying defects or any excess cost of completing the contract where the dwelling has been partly built (or in NHBC's discretion, reimbursing the deposit); or

(c) remedying defects where the dwelling has been substantially completed.

The indemnity applies where the loss has been caused by the bankruptcy, liquidation or fraud of the builder or developer and contracts have been exchanged by the purchaser for the building or sale of the dwelling. Notice of the claim must be given to NHBC within three months of the bankruptcy, liquidation or fraud as the case may be.

Post-certification: where the Vendor defaults during the Initial Guarantee Period: Clause B(1)

Where the Vendor fails to honour an arbitration award or court judgment made in respect of the Vendor's liability to remedy defects appearing during the initial guarantee period, NHBC will do so on his behalf, provided that the application to

appoint the arbitrator or service of the Writ has been made within twelve months after the expiry of the initial guarantee period. It should be noted, however, that the indemnity does not cover:
(a) consequential losses: e.g. removal expenses, alternative accomodation, legal or other professional fees;
(b) defects or breaches of the contract specification, unless they are also caused by, or are breaches of, NHBC's Technical Requirements; or
(c) the first £20 of the cost of remedying any defect, subject to a maximum deduction of £90 from the claim;
(d) professional fees incurred in making the claim against NHBC;
(e) any items endorsed on the Certificate (the Certificate is usually endorsed to exclude NHBC's liability in respect of minor breaches of NHBC's Technical Requirements where the purchaser has agreed in writing or there are other exceptional circumstances);
(f) non-structural items where NHBC has issued a 'Structural Only Certificate' (NHBC usually does this only where the Vendor has contracted to provide such a certificate, or the Vendor goes into liquidation before a certificate is issued and the dwelling is completed by another builder following a claim against NHBC).

The indemnity arises as soon as the award or judgment is made, and the time for appeal has expired, or an appeal has been dismissed.

In the event that the purchaser could have instituted proceedings but the Vendor is in liquidation or without assets, NHBC will pay the cost of remedying any defects caused by breaches of the Technical Requirements and any damage to the dwelling caused thereby, subject to the exclusion clauses referred to in (a) to (f) above (Clause B(1)(b) refers). NHBC will not meet any claim unless it has been given a reasonable opportunity to inspect the defects before remedial works are carried out.

Major structural defect liability: Clause B(2)

In Clause B(2) of the First Schedule to the Agreement, NHBC undertakes to make good, or to pay the cost of making good:
(a) any major damage to the dwelling that has been caused by any defect in the structure. The defect must have been caused by breach of NHBC's Technical Requirements, and the indemnity covers the cost of remedying the defect.
(b) any major damage to the structure of the dwelling consequent upon subsidence, settlement or heave, and to remedy the defect causing such damage. No breach of the Requirements need be proved.

The purchaser must notify NHBC in writing as soon as practicable after the appearance of such damage, and in any event not later than the expiration of the tenth year from the date of the Certificate. Having notified NHBC, the purchaser will be required to pay an investigation fee (currently £40), which is returnable if the claim is valid or if it is reasonable for the purchaser to make a claim.

The investigation is carried out by a member of the RIBA, RICS or other appropriate institution, who draws up a schedule of remedial works. NHBC will pay the claim after the purchaser has obtained estimates (usually two) on the basis of that schedule. NHBC does not usually employ the remedial contractors itself, unless there are special circumstances (e.g. a number of dwellings on the same site with similar defects).

In the event that the purchaser disagrees with NHBC's schedule of remedial works or that the claim is rejected by NHBC, the purchaser has the right of arbitration. The arbitrator is appointed by the Institute of Arbitrators, from a panel jointly nominated by the RIBA and RICS.

There are a number of exclusion clauses applicable to this indemnity, the most important of which are:

(a) *negligence or nuisance of a person other than the Vendor or his sub-contractors* (note that, although architects and engineers are not usually referred to as 'sub-contractors', it is probable that the exclusion clause would not apply where the claim arises out of the negligence of the builder's architect or engineer, whether 'in house' or independent; however, there would appear to be no reason why NHBC, having met the claim, could not recover from an independent architect or engineer, as the case may be);
(b) *loss, damage or liability that at the time of the claim is covered by insurance, or for which compensation is provided for by legislation* (this excludes liability where subsidence has occurred, and the purchaser has recourse to either his household insurers – most household insurance policies nowadays cover damage caused by subsidence – or to the National Coal Board, in respect of mining subsidence);
(c) *defective design where the purchaser has provided his own structural or installation design details;*
(d) loss or damage consequent upon the presence of a lift or swimming pool;
(e) damage to, or caused by, anything not built into the dwelling pursuant to the building or the sale contract;
(f) loss or damage caused by dampness or normal condensation, or by wear and tear, or by gradual deterioration caused by neglect;
(g) defects that have been, or should have been, reported to the builder during the initial guarantee period, or any claim payable arising out of the builder's default during that period;
(h) subsidence where the claim is by the Public Authority that employed the builder to build on the Public Authority's own land.

Top Up Cover

Top Up Cover was introduced by NHBC on 1 October 1975, and is applicable to all Agreements exchanged on or after that date. A distinction must be drawn, however, between Top Up Cover available in respect of dwellings for which application for inspection was made on or after that date, and dwellings that were applied for prior to 1 October 1975, but that were not sold (i.e. contracts exchanged) until after that date.

Dwellings where application for inspection was made on or after 1 October 1975
From 1 October 1975 until and including 31 March 1977, all dwellings applied for were allocated a Certificate number prefixed with the letter 'G'. Since 1 April 1977, the Certificate reference number is 'H'.

In relation to these dwellings, NHBC's liability for defects arising during the initial guarantee period, where the Vendor is in liquidation or without assets (Clause B(1) (b) refers), or during the major structural defect period (Clause B(2)) is limited to the cost that would have been incurred if the claim had arisen and been met at the date of issue of the Certificate (defined as the 'notional cost' in Clause B(3)(a)). This cost is calculated by reference to the housing cost index published monthly by the RICS. Thus, if a major structural defect claim is made, say, in November 1982, and the dwelling was certified in January 1977, the cost of the claim of (say) £8000 is backdated to the price prevailing in January 1977 of (say) £4000, and this is the sum NHBC will pay to the purchaser. The purchaser must then look to his legal remedies against third parties, e.g. the builder, the Vendor (where different from the builder) or the local authority, in order to recover the shortfall.

However, the purchaser has the option of buying Top Up Cover for a once-and-for-all premium of £20. This is additional insurance to cover the inflation of building costs between certification and the date the claim is made.

(a) In relation to 'G' Certificate dwellings, the notional cost is increased at the rate of 25 per cent per annum compound from the date of the Certificate until the date of the claim. If the notional cost exceeds £20 000, the sum of £20 000 is substituted and projected forward until the date of the claim. Under no circumstances, of course, can the purchaser recover a sum greater than the cost of the claim.

(b) In relation to 'H' Certificate dwellings, the notional cost is increased at the rate of 20 per cent per annum compound from the date of the Certificate until 31 December 1978, and 15 per cent per annum compound thereafter. Otherwise, the method of calculation and limitations are the same as for 'G' Certificate dwellings.

Dwellings where application for inspection was made between 30 September 1971 and 30 September 1975 (both dates inclusive) but contracts are not exchanged until or after 1 October 1975 These dwellings will have Certificates prefixed with the letters 'C', 'D', 'E' or 'F', and the Agreement exchanged will be marked 'F' on the top righthand side to distinguish it from the Agreements marked 'G' or 'H', which have different application forms for Top Up Cover attached to them.

The Top Up Cover available to these purchasers is different. Claims made in respect of these dwellings are met by NHBC up to the appropriate limit of indemnity (£10 000 in respect of 'C' and 'D' Certificate dwellings; £20 000 for 'E' and 'F' Certificate dwellings) and are not backdated to the date of the Certificate.

However, for a once-and-for-all premium of £12.50, the purchaser has the opportunity of purchasing additional insurance cover, which is also called Top Up Cover. In the event that the claim costs more than NHBC's limit relating to that particular Certificate, the notional cost is increased at the rate of 25 per cent per annum compound from the date of the Certificate until the date of the claim. Thus, when a claim is made in respect of an 'F' Certificate (£20 000 limit) and the claim is made for £25 000, provided that the average rate of inflation does not exceed 25 per cent per annum compound from the date of the Certificate, the claim will be met in full.

Advantages for builder if Top Up Cover is purchased

Where claims are made during the years three to ten from the date of Certification, NHBC accepts primary responsibility for any structural defects. If NHBC's indemnity is insufficient to meet a claim, the purchaser will wish to recover the shortfall from the builder amongst others, by pursuing his rights under contract, statute, or at common law. Prior to the cases of *Sparham-Souter and Ryan v Town & Country Developments and Benfleet UDC (1976)* and *Anns v London Borough of Merton (1977)*, the latter a decision of the House of Lords, it was considered that such rights would be statute-barred under Section 2 of the Limitation Act 1939 after six years from the date of construction of the dwelling. The Anns decision confirmed that actions may be brought within six years from the date on which damage to the dwelling becomes apparent, and that the builder will be liable if he has been negligent when constructing the dwelling.

It follows, therefore, that the greater NHBC's liability is, the less likelihood there will be of a purchaser having to sue the builder. For that reason, NHBC has always given builders and developers the opportunity of purchasing Top Up Cover on behalf of the purchaser, at the date the application for inspection is made.

Future development

NHBC intends to introduce a number of fundamental alterations to its Scheme sometime in 1979. Unfortunately, at the time of writing, the precise details of the amendments are still under consideration.

DEFECTIVE PREMISES ACT 1972

Mention has been made previously of this Act, which came into operation on 1 January 1974. The Act imposes a statutory duty on all persons engaged in the construction of dwellings, or conversion or extension or repairs to dwellings, to build or design in an efficient and workmanlike manner and of proper materials (or in a professional manner, in the case of architects and engineers) and so that the dwelling will be fit for habitation.

The statutory duty under the Act is in similar terms to the warranty given by the vendor in Clause 3(1) of the Agreement. Because of this, and the fact that the NHBC Scheme indemnifies purchasers against loss where the Vendor is bankrupt or in liquidation, the NHBC Scheme has been given exemption from the Act under Section 2. This means that no action will lie under the Act where the first purchaser entered into the House Purchaser's Agreement and a full Certificate was issued by NHBC in relation to the dwelling.

In order to obtain approval under the Act, the NHBC's Agreement, Rules and Certificate have to be approved by the Secretary of State for the Environment, and a Statutory Instrument has to be laid before Parliament. To date three successive Schemes have been approved by virtue of Statutory Instruments 1973 No 1843, 1975 No 1462 and 1977 No 642. No change to the NHBC's documentation can therefore be made without such approval.

The limitation period under the Act is six years from the date of construction of the dwelling, which is the same limitation period as applies to the Vendor's general warranties under the House Purchaser's Agreement.

CHANGES IN LAW AFFECTING BUILDERS' LIABILITIES

The NHBC Scheme in its present form was introduced in the mid-1960s in order to provide purchasers with remedies, in respect of defective building, that were not otherwise available. However, during the last ten years the law relating to builders' liabilities has undergone fundamental changes, so that builders' legal liabilities (outside the NHBC Scheme) are very different from what they were in, say, 1965. It seems appropriate, therefore, to conclude this section on the NHBC Scheme with a brief discussion on the changes brought about by the courts, and to explain how, indirectly, these changes have also resulted in the NHBC Scheme becoming more valuable to builders.

Changes in common law position of builders/vendors

Prior to 1972, the common law position of builders who were also vendors was that they were immune from actions in negligence brought by purchasers or other persons. This immunity had been established by the case of *Bottomley v Bannister (1932)* and confirmed in the 1936 case of *Otto v Bolton and Norris.* Although there had been some discussion of the Donoghue v Stevenson principles in that case, it was held that the common law duty applicable to manufacturers had no relevance to the law of property.

However, by the 1960s it had been held that a builder building on someone else's land (whether that of the purchaser or a third party) held a duty of care to anyone who suffered personal injury or property damage as the direct and reasonably forseeable consequence of careless workmanship and/or design (insofar as the builder, as opposed to a third party, had designed the works). Authority for this proposition is to be found in the case of *Sharpe v Sweeting (1963).*

As far as the builder's contractual position was concerned, a builder who was also the vendor owed no implied duty of care under the contract, although a builder building under contract for a third party was under an implied duty 'to build with reasonable workmanship...', the same duty, in fact, that is expressly included in Clause 3(1) of the NHBC's House Purchaser's Agreement; see the case of *Hancock v W.R.Brazier (Anerley) Limited (1966).*

It was largely because of the immunity of builder/vendors at common law that it was decided to expressly include the warranty normally implied in contracts of building in the House Purchaser's Agreement. Furthermore, the fact that the Agreement contained this warranty was a material factor in Parliament's decision to exempt the NHBC Scheme from the provisions of the Defective Premises Act 1972.

However, at the same time as Parliament was passing the Defective Premises Act, the Court of Appeal was considering the duty of builders/vendors in the case of *Dutton v Bognor Regis UDC (1972).* It was decided in that case that the immunity conferred on builder/vendors was illogical and should be removed. (The case itself dealt with the liabilities of local authority building control inspectors when carrying out their functions under Public Health Acts.) As from 1972, therefore, builders, whether vendors of premises or merely building under contract on land owned by a third party, owed the same duty at common law as they owed under the Defective Premises Act or under the NHBC Scheme.

The common law duty owed was, of course, a contractual duty, and any liabilities in tort could only be owed to third parties not in a contractual relationship with the builder, as dual liability in contract and in tort did not exist: see the case of *Bagot v Stevens Scanlan (1966)* (a case principally concerned with the liability of architects and the limitation periods relating to actions in contract and in tort).

It was always possible, of course, for a builder to exclude this implied duty at common law by inserting exclusion clauses in the building or sale contract. However, such exclusion clauses would not prevent a purchaser's successor in title from suing in circumstances where a contracting party could not; see the cases of *Hazeldine v Daw & Son Limited (1941)* and the recent case of *Sutherland v C.R.Maton Limited (1976).*

Whilst a builder could not be held liable both in contract and in tort to a contracting party, builders had the protection of Section 2 of the Limitation Act 1939, so that any action in tort (i.e. negligence) brought after the expiry of the limitation period (six years from the date of the breach of the contract) could be successfully resisted on the grounds that the remedy was statute-barred. The importance of this principle was highlighted in the case of *Sparham-Souter and Ryan v Town and Country Developments and Benfleet UDC (1976),* another decision of the Court of Appeal. In that case, it was held that a cause of action in negligence accrues not at the date of the negligent act or omission, but at the date when damage is sustained by the plaintiff; thus in a case involving the defective foundations of a house, the damage is done not when the foundations are laid, or approved under the Public Health Acts, but when the defect first appears, and for the purposes of Section 2(1) of the Limitation Act 1939 time begins to run only when a plaintiff knows of the defective foundations, or could with reasonable diligence have discovered them.

It follows that, after *Sparham-Souter,* builders could be held liable in negligence for a period of six years from the date a plaintiff became aware of the damage to his dwelling. However, whilst the decision of Lord Diplock in *Bagot v Stevens Scanlan (1966)* still held good, purchasers in a contractual position with a builder or builder/vendor could not avail themselves of this extended limitation period. However, the case of *Esso v Mardon (1976)* (followed in the cases of *Midland Bank Trust Company Limited v Hett, Stubbs & Kemp (1977)* and *Batty v Metropolitan Realisations and Others (1977)*) decided that, in the case of builders, architects and engineers, the duty to use reasonable care not only arises in contract, but is also imposed by law apart from contract and is therefore actionable in tort. It follows from that, therefore, that any purchaser could rely on the *Sparham-Souter* decision where defects appeared after the expiry of the contractual limitation period.

The House of Lords confirmed the *Sparham-Souter* decision in the case of *Anns v London Borough of Merton (1977)*, which decided that a cause of action in negligence did not accrue until the state of the building was such that there was 'present or imminent danger to the health or safety of the occupants', and that the limitation period did not start running until the plaintiffs became aware of the danger. Since *Anns*, the most important case to reach the courts has been the case of *Batty v Metropolitan Realisations Limited and Others (1977)*, a decision of the Court of Appeal.

It is worth setting out the facts of the case briefly, because, although the facts were unusual, the principles enunciated by their Lordships have relevance to all building contracts. The builder (second defendant) purchased land from the local authority in the late 1960s. The builder sold the land to the developer (first defendant), who agreed to finance the development and employ the builder to construct the dwellings. The site was on a plateau on top of a hill; the land that sloped away down to the valley below was at all times in the ownership of the local authority, apart from a small part that formed the plaintiffs' garden. A representative of the builder and of the developer inspected the site and walked over the surrounding terrain before the decision to build was taken. The plaintiffs' house was built on the edge of the plateau, with the garden on the slope of the hill. The plaintiffs purchased the house in 1971 and the developer exchanged the NHBC standard form House Purchaser's Agreement. In 1974 there was a severe slip to the garden, although there was no damage to the plaintiffs' house or to its foundations, either then or since. As a result of the damage to the garden, however, proceedings were started against the developer, builder and the local authority, which had inspected the foundations in accordance with its statutory duty. It was held by the Court of Appeal that:

(a) The warranty given in Clause 3(1) of the House Purchaser's Agreement should be read disjunctively, i.e. the dwelling must be built so as to be fit for habitation, and it is no defence if the vendor has complied with the first part of the warranty 'to build in an efficient and workmanlike manner and of proper materials'. The fact that the ground adjoining the house was unsuitable, and proper inspection would have proved it to be so, meant that the vendor was in breach of this warranty.
(b) There is no limitation on the right of a plaintiff to have a finding made in contract and in tort, where the facts giving rise to the breach in contract would also constitute a breach of the common law duty apart from that imposed by the contract. This is important because of the limitation period for negligence as laid down by the cases of *Sparham-Souter* and *Anns.*
(c) The fact that the defects were on land owned by a third party did not negative the builder's duty of care when constructing the dwelling.
(d) The fact that the house itself was undamaged did not mean that there was no present imminent danger to the occupants; as expert evidence showed that the house would inevitably slide down the slope, the plaintiffs were entitled to judgment under the *Anns* decision.

Conclusion: the builder's position

Following the decisions referred to above, culminating in the *Batty* case, the legal position of builders appears to be:

(a) Contractors, vendors and architects have a dual liability in contract and in tort. A plaintiff may sue, therefore, for breach of contract and/or in negligence. The damages recoverable in either case would be the cost of remedial works and any consequential losses.
(b) Breach of the common law duty of care may give rise to actions by both purchasers and their successors in title.

(c) Liability may be incurred even if there is no breach of the Building Regulations. The ground conditions of neighbouring land (even if not in ownership of the builder or developer) must be taken into consideration.
(d) Economic loss alone is recoverable; the fact that a dwelling is still 'habitable' is not relevant if the dwelling is unsaleable, and likely to become uninhabitable in the future.
(e) 'Present or imminent danger to the health or safety of the occupants', referred to in *Anns,* does not mean that damage must have occurred in order to found an action; it is enough that there is a probability that damage will occur in the future. This is important if one considers the 'live shale' cases, or even dwellings build with high-alumina cement.
(f) The warranty to build a dwelling so as to be fit for habitation is virtually an absolute warranty; the fact that the workmanship has been reasonable and proper materials have been used will not afford a builder any defence if the dwelling is not fit for habitation.
(g) Whilst the builder can expressly exclude his common law warranty (except where he is a registered member of the NHBC and enters into the House Purchaser's Agreement), this does not prevent a subsequent purchaser from suing. Furthermore, if the builder wishes to exclude negligence he must specifically use the word 'negligence' in any exclusion clause (see the House of Lords decision in *Smith v South Wales Switchgear Limited (1977).* In any event, such exclusion clauses would probably be void under the Unfair Contract Terms Act 1977.
(h) Builders are not entitled to expect a purchaser to employ a surveyor when purchasing a house, although if the purchaser did so the chain of causation in an action in negligence by a second or subsequent purchaser might be broken and the builder might have a good defence: see *Sutherland v C.R.Maton Limited (1976).*

To sum up, builders and developers owe a duty at common law to build dwellings in an efficient and workmanlike manner and of proper materials and so as to be fit for habitation, and this duty is imposed both in contract and in tort. Tortious liability may remain for several years after the construction of the dwelling, even if breach of the duty does not manifest itself until after expiry of the contractual limitation period. Whilst liability may be excluded under contract, this duty is still owed to subsequent purchasers. Builders are also under a duty to carry out a proper site investigation, both of the land upon which they intend to construct dwellings and also of the neighbouring land if appropriate.

Under the NHBC Scheme, NHBC accepts primary liability for any major damage caused by defects in the structure that appear (generally) between three and ten years after the date of completion of construction of the dwelling. NHBC's liability becomes operative where the defect in the structure is consequent upon breaches of the Council's Requirements or is caused by subsidence, settlement or heave. In most cases, breaches of the Council's Requirements would also constitute a breach of the warranty given in Clause 3(1) of the Agreement. It can be seen, therefore, that if NHBC accepts primary responsibility to the purchaser (subject to its limit of indemnity in the Agreement), NHBC is also effectively, if indirectly, insuring builders against the consequences of their own mistakes in workmanship and materials. For that reason, NHBC is of value not only to purchasers but also to builders, in particular because insurance for defective workmanship and materials (commonly called 'product guarantee' insurance) is virtually impossible to obtain nowadays.

FURTHER INFORMATION

It is hoped that the information given above will provide a useful working knowledge of the NHBC Scheme. For those who wish to read more, a book called *Guarantees for new homes* by D. Marten and P. Luff, Oyez, 1973 (together with a supplement,

1975) is a useful practical book on the more detailed workings of the Scheme. NHBC itself publishes a number of information booklets, most of which are available upon request at no charge.

9 PLANT

GENERAL USE OF PLANT 9–2

PLANT FOR EARTHWORKS 9–4

CONCRETING PLANT 9–7

SITE HAULAGE 9–9

LIFTING PLANT 9–11

PUMPS 9--14

BREAKING AND DRILLING 9–15

MISCELLANEOUS 9–16

9 PLANT

K. LAWSON, CEng, MIMechE, MIArb
Construction Plant Consultant

This section begins with brief observations on the place of plant in the builder's business and on plant management. It then discusses the types of plant available for earthworks, concreting, site haulage, lifting, pumping, breaking and drilling, and other common operations.

GENERAL USE OF PLANT

Every building contract will require the use of some item of mechanical plant or an electric hand-tool during its construction period. Their use will always be labour-saving, but the economic value is less certain, unless considerable thought has been given to choosing the correct item and comparing its cost-effectiveness with the alternatives available to the builder.

A machine will generally be chosen for one or several reasons:

- speed up of work;
- replacing manpower, which may be expensive or scarce;
- providing a better finish or quality to the work.

It will be the aim to reduce costs, but this requires a careful analysis of all factors, and will not necessarily result in the selection of plant for every operation.

Planning for plant at time of tender

Although the time for tendering allowed to the contractor is very limited, it is advisable to study the alternatives for the use of equipment as far as it is practical. It is not necessary to select specific makes at this stage (since the builder will have a wide choice of similar plant), but a matching of related equipments should be carried out before the tender is finished.

The choice of a correct size of forklift, selected to carry concrete, will demand a matching size of mixer output and also a scaffold, suitably designed to accept the load. Terrain and access have to be considered for all stages of the construction process, to ensure that the machines can travel without difficulty or delay. At the estimating stage problems of plant storage, erection and dismantling should be clarified: a tower crane marooned in the centre of its own construction can become an expensive exercise in dismantling and removal.

Plant selection

When the contract has been awarded, a reappraisal of the plant tender is advisable coupled with the production of a simple bar-chart of required plant resources, which will match the site progress chart.

Too many small and large companies claim that they cannot spare the time for some initial planning; the result is a series of *ad hoc* decisions, panic orders for plant and a far from economic operation. Construction has sufficient inherent problems of changing weather, clients changing demands and other difficulties without compounding these by the unplanned use of expensive plant.

Correct selection demands a study of:

- the different types of work that can be carried out by a particular machine;
- machine output, with a view to matching equipment to practical circumstances;
- the availability of suitable operator and service back-up;
- the effect of weather on operation;
- costs.

It will often be necessary to compromise in the final selection. For instance, a small dumper fitted with a concrete skip will be most advantageous when discharging concrete, but will be wrong when muck has to be carried by the same machine; a forklift truck may be an ideal solution in combining the function of dumper and hoist, but will not be safe to operate in poor ground conditions.

Too little attention is paid to the planning of access roads, and a hardcore or metalled road can often be a means of reducing plant breakdown and achieving satisfactory output. The power of a wheeled machine has to be increased by a factor of eight, if one compares travel on a good, hard road and travel on a wet track with 15 cm rutting.

Plant hire

The contractor in the UK is fortunate in having at his command a comprehensive range of plant and tools that can be hired for short or long periods. Unlike his counterparts in Europe, where plant hire is just emerging, he can therefore restrict his capital commitment to such plant where he is reasonably certain of continuity of use. He is also able to select his plant to match his particular needs, which may change as a contract progresses.

The builder will tend to restrict his plant fleet – and thereby his commitment of capital – to the equipment most commonly used by him. It is often assumed that a decision whether to buy or hire is a simple comparison of relative costs; in practice it is important to include in the consideration other factors that may influence ownership, including:

- future utilisation of the equipment;
- availability of skilled operators and service staff;
- depot or yard facilities;
- ability to raise the necessary capital or loan facilities.

The conditions under which plant is hired in the UK are well established, and generally conform to the terms of the Contractors Plant Association in England and Wales, and to the Scottish Plant Owners Association in Scotland. Users of hired equipment should take care to acquaint themselves with the hire conditions, which clearly define the user's responsibilities.

Plant maintenance

Machines will not continue to function effectively unless they have been maintained to a good standard. Neglect will be costly in lost production and will result in heavy repair expenditure. All mechanical plant should be checked daily for:

- correct oil level;
- cooling water (unless air-cooled);
- conditions and inflations of tyres (where fitted);
- effectiveness of brakes (where fitted);
- loose components;
- safety hazards.

The manufacturer's instruction book will lay down further service requirements to be carried out at periodic intervals, based on the number of weeks or hours worked by the machine. On plant supplied with permanent operators, e.g. excavators, loading shovels, etc., the driver will generally carry out all maintenance and receive an additional payment for this. Other machines should be attended by a mobile maintenance mechanic supplied by the owners of hired plant, or by the builder's own maintenance staff based at a depot or, for a major contract, on site.

Routine lubrication is perhaps the most vital factor. Care must be taken to use the correct oils recommended by the plant manufacturer or to ensure that an acceptable alternative can be provided by a reputable lubricant supplier. It will assist site control if the actual varieties of lubricating oil and greases are kept to a minimum to avoid wastage and error. The provision of small marker plates on the oil filler cap or adjacent to it, giving details of the correct lubricant, helps in preventing incorrect oils being used, which can cause serious damage.

The effective control of maintenance requires a simple schedule or wall chart; this will ensure systematic attention to the plant and avoid costly breakdowns. Cleanliness of a machine is important in preventing excessive wear, and attention to the paintwork will prevent rusting. Both help in giving operatives a sense of pride in the machinery and should be conducive to a higher standard of maintenance.

Safety

The Health and Safety at Work etc. Act of 1974 laid the foundation for a new approach to safety in the UK. It does not in itself alter existing legislation on safety, but it makes all concerned with the operation, both management and operatives, personally responsible for ensuring that correct safety measures are observed at all times. New safety legislation and codes of conduct are being progressively introduced to tighten up and improve on the existing law.

The Act gives particular emphasis to the importance of training the operative in the safe use of plant and tools, and of communicating to him any safety hazards. Many companies now find it advisable to issue their operatives with written instructions on the safe use of plant, and to issue a certificate of competence before a driver is permitted to operate a machine.

All construction machinery is now fitted with guards, to prevent accidents caused by moving gears, shafts or chains. Care should be taken that these are always fitted when a machine is at work. Other legislation demands the periodic inspection and testing of all lifting appliances, lifting gear and compressed air vessels, and the maintenance of suitable records covering these items. The specific statutory requirements are changing progressively, so it is not practical to provide details in a reference book, which cannot be updated at short notice.

PLANT FOR EARTHWORKS

The builder will require plant for a variety of different earth-moving tasks. These may include:

- stripping top soil,
- general excavation,
- trenching for footings, drains, etc.,
- loading soil,
- moving soil on the ground,
- compacting.

It is possible to use either single-purpose machines, designed solely for the particular job, or multi-purpose plant. In general the latter will not be as efficient for a particular earthworks, but may gain by their all-round adaptability and prove more economical on the smaller contract, where they can undertake a variety of tasks.

Rope excavators

These are generally track-mounted machines, less commonly used on building sites since the introduction of hydraulic excavators. They can be fitted with different equipments to perform various duties:

- *backacter* (or *back-hoe*): trenching work or mass excavation where the machine has to remain above the level of dig;
- *dragline:* excavation below ground level and loading, where a considerable outreach is required;
- *forward shovel:* mass excavation above ground level; loading from spoil-heap;
- *grab:* excavating to greater depth, especially where limitations in the geometry of other equipments make this the only alternative. Also loading from stock-piles.

Hydraulic excavators (360°)

This type of excavator is fully slewing and is able to dig at greater speed in normal soil than the rope-operated machine.

- *back-hoe*
- *forward shovel*
- *grab*

} Uses are similar to those for the rope-operated machines, but in the hydraulic excavator a simple bucket change or its re-positioning on the boom converts the machine from one operation to the other.

- *dragline:* some of the larger hydraulic excavators fitted with long boom sections enable the machine to perform similarly to draglines.

In the UK the majority of hydraulic 360° machines are track-mounted and not on rubber tyres. In other countries the smaller machines are often the wheeled variety, for preference.

Hydraulic excavators (180°)

The tractor-based excavator/loading shovel is the most commonly used machine in the UK. It is particularly attractive to the small builder by virtue of its dual ability to dig trenches and back-fill or load with the front shovel. This shovel can also be used for site stripping, but only in favourable soil conditions. See *Figure 9.1.*

It is a major advantage that these machines can be driven to site on the highway, while the other excavators or loaders have to be transported by low-loader.

Trenchers

A variety of small trenchers are available to the builder for drainage or pipe-laying work. They permit an operation where the bucket-scoop digs continuously, forming the trench and side-casting the material.

Loading shovels and tractor dozers

Most reduced-level work is undertaken by tractor loading shovels. These are generally track-mounted, although the more mobile wheeled machines are often equally suitable, and frequently used on the Continent. Loaders move extremely fast and can achieve high outputs. Many are fitted with special '4 in 1' buckets, which allow the machines to be used for dozing, back-blading or grabbing. They are also able to provide a good output for mass excavation above ground level, but their need to travel

Figure 9.1 Hymac 370C excavator loader, showing back-hoe equipment working on site clearance

from the excavation face to the loading vehicle requires more space and demands reasonable ground conditions.

Dozers may be fitted with a dozing or tilting blade, depending on the operation. They are designed to push the spoil ahead assisted by grouser-plated tracks, which provide good traction. They can also be fitted with attachments to give greater versatility (e.g. rippers, tow-bar, winch, side-boom crane, pusher, etc.). Their primary purpose on a building site is the levelling of soil movement or the back-filling of trenches.

Graders, scrapers and compaction equipments

This type of plant is seldom seen on any but the largest building sites. Graders have the ability to level and 'grade' the ground to very fine limits, and are also used for road maintenance. Scrapers may be tractor-towed or mobile. They permit the movement of large volumes of soil over a longer distance at an economical rate. They cut the soil as they travel and fill the body of the scraper to its capacity. The load is then carried to the dumping area where it is discharged. A 'self-elevating' scraper cuts the soil, similar in operation to a lawn mower, but it is restricted to lighter soils.

Compaction may be achieved by a small rammer (petrol or air driven), a vibrating or deadweight roller, or a tractor-towed compactor. Hand-held rammers are generally used for consolidating the ground in small trenches or excavations.

A deadweight roller will compact the ground by its own weight, while the vibrating roller can achieve a similar efficiency (in suitable ground) with a much lighter machine whose impact is assisted by a vibratory pressure. The double-drum vibratory unit is a more efficient variety, and for the larger contract a heavy towed vibratory roller may be necessary. Major earthworks may call for a mobile compactor or rubber-wheeled, sheeps-foot or grid rollers, which are tractor towed.

In all compacting operations the type of soil and weather conditions will be the important criteria, together with the final compaction that is required.

CONCRETING PLANT

The choice of concreting plant will depend on many factors, e.g.
- volume of concrete requirement (average and peak),
- time available,
- quality of concrete,
- use of special cement or aggregates,
- site location,
- site layout and haulage conditions,
- expected weather conditions,
- availability of plant (for mixing and placing),
- availability of labour,
- power supply,
- costs.

In practice the conditions will not be consistent and the correct selection of plant will be difficult, except for the simplest operations. The plant required on site can be divided into these groups:
- storage of aggregates and cement,
- mixing,
- conveyance and lifting,
- placing.

Mixers and batching plants

Depending upon the location of a site and the demand, ready-mixed concrete delivered to the site in truck mixers will be accepted as a simple (though possibly expensive) means of providing concrete on most smaller sites. Ready-mix does, however, require an ability to accept fairly large quantities at a time, and the site will be dependent upon the punctual arrival of the concrete transport from a distant mixing plant.

A tilting-drum mixer is used for smaller volumes of concrete, e.g. for bricklaying. It will have an output of between 85 and 170 litres.

The reversing-drum mixer is used for greater outputs between 200 to 400 litres. It rotates the drum in one direction when mxing, and discharges the concrete when the rotation is reversed. It has replaced the closed-drum mixer previously found on site.

For a faster mix or for concretes of a particularly high quality the pan mixer is most suitable. This may operate as a stationary circular drum in which the horizontally rotating blades mix the concrete, or the drum may also revolve in an opposite direction to the blades. The size of machine may be as large as 4000 litres or as small as 200 litres. Special pan mixers may be used for plaster or grout, where revolving rollers crush the lime as they revolve.

Batching plants are generally units that have an integral system for the storage of aggregates and cement in bins, and a mechanical or electronic means of weighing and delivering the constituents into the mixer, which then discharges the mixed concrete to a receiving hopper or transportation unit. They are only introduced on site when large quantities of concrete are needed for a prolonged period. Automatic controls may enable one or two operators to produce concrete volumes in excess of 200 m^3/h.

Concrete distribution

Whether ready-mixed or site-made concrete is supplied it will be necessary to distribute it to the working area. While this may involve only a hod-carrier or a barrow, the larger quantities require a less labour-intensive, and perhaps quicker, conveyance.

Typical plant that may carry concrete includes:

- dumper,
- truck,
- hoist,
- pump or placer,
- conveyor,
- crane,
- forklift.

It will be seen that the first two alternatives only permit the horizontal movement of the concrete, while the hoist can only move it vertically. All the other machines are able to combine the functions to some extent.

Dumpers and trucks demand a good access for travel. On dumpers a hydraulically operated discharge can provide a better performance, especially if this allows a 180° slew or hydraulic tip for side discharge.

Figure 9.2 Schwing lorry-mounted concrete pump with 45 m placing boom, model BPL.600.HD.KVM.45 (courtesy Burlington Engineers Ltd)

Forklift trucks have become increasingly popular for moving concrete, which they can place at a reasonable height from ground level. They do, however, require a good road surface for safe performance.

Cranes, and in particular tower cranes, are widely accepted for the efficient distribution of concrete. They have the additional advantage of being able to lift shutters, reinforcing and other materials without difficulty. The use of a crane will demand a suitable skip for holding the concrete, and these can be side- or bottom-discharge or roll-over types, depending on the situation of the concrete.

Hydraulic concrete pumps, which displace the concrete by a reciprocating piston action along a pipeline, have become popular to an increasing extent; see *Figure 9.2*. They are frequently hired in conjunction with ready-mix supplies. The modern hydraulic pump may be fitted with a distribution boom, and the very largest units can reach a height of 250 m or convey it horizontally over 1000 m. Successful pumping will, however, require the ability to accept considerable quantities of concrete at a time and demands a strict control of concrete quality designed for pumping.

Conveyors are less frequently used for concrete distribution because of problems with weather, moving the conveyors, and washing down. Even so they can be effective and economical for certain applications below ground level and to a limited height.

Placing concrete

The efficient vibration of concrete is important for the final quality of the work, and this is normally achieved by:

- plate, tamper or float,
- internal poker,
- clamp-on.

Vibrator plates and poker vibrators may be engine, electricity or air driven, while clamp-on vibrators, affixed to the outside of a shutter, are generally electrically powered.

Different sizes of vibrator head, varying in weight and vibration frequency, provide different results and their choice depends upon the required quality and constituents of the concrete.

SITE HAULAGE

Apart from transporting muck or concrete on the site, numerous other materials and components have to be moved from a stockyard or stores to the building. There is a wide choice of plant for transporting on a building site, including:

- dumpers,
- dumptrucks,
- lorries and tippers,
- forklifts,
- tractors and trailers,
- railway, monorail.

Dumpers

The small dumper was evolved from the mechanical barrow, which is still used to some extent. It is the accepted all-round haulage unit on site for muck, concrete, bricks, etc. The simplest unit is rear-axle driven with a gravity-discharge skip situated in front of the driver. In poor ground a four-wheel-drive unit is preferable, and reference has already been made to special hydraulic discharge and purpose-made designs.

The carrying capacity of dumpers ranges between half a tonne and five tonnes. They require a road fund licence and have to be equipped with lights, horn, mirror and number plate, if used on the public roads.

Dumptrucks

For major earthworks the spoil may be carted away by dumptrucks. Although these may be manufactured in capacities exceeding 100 tonnes, even a major building site is unlikely to require units larger than 25 tonnes capacity. Not all dumptrucks are permitted on the highway with their full payload, since their axle loading and width may exceed the legal limits. Articulated dumptrucks are increasingly used for the smaller capacities, and another alternative consists of dump-trailers towed by tractors.

Lorries and tippers

The tipper is still the most common vehicle for hauling muck from a building site with payloads ranging from 6 to 20 tonnes. Some specially adapted bodies, fitted with paddles to agitate the concrete, may be used for concrete conveyance on site. Poor site conditions may not permit the ordinary tipper to travel, and four-wheel-drive units can then be used.

Forklifts

Forklifts have been developed as the specialist site machine for mechanical handling; see *Figure 9.3*. They may be equipped with forks, concrete skip, crane hook, clamps or grabs, and they are generally easily converted from one attachment to another. They are based on either a tractor or a dumper-type chassis and have the ability to raise the load. Very large units can have a capacity of over five tonnes and might lift lesser loads to a height of 10 m. Special features on machines include:

- articulated steering,
- side shift of forks,
- rotation of forks,

Figure 9.3 Benford 'Liftmate' rough-terrain forklift, model MA2300H; lift 2300 kg to 3.6 m and 1270 kg to 6.4 m, or 1000 kg to 6.4 m with 1.2 m forward reach using load extender as illustrated

- forward reach of mast,
- side tilt of mast.

Although forklifts have the capability to traverse rough terrain, care must be taken to avoid accidents by providing the best possible road conditions. Overturning and loads falling off the forks are a very real hazard.

Tractors and trailers

The agricultural-type tractor and trailer can be effectively used on site for many materials, including reinforcing and pipes. The articulated version, similar to the dumptruck, will give a better performance in poor travel conditions.

Railway and monorail

Rail-based haulage is seldom used on building sites, but it may have an application in some industrial construction as an economical alternative. It will generally demand the laying of Jubilee track, perhaps 18 inch or 24 inch gauge, on which bogey-mounted trucks can travel.

Monorail was used as a conveyance, but it is seldom encountered in modern construction. It utilises a self-propelled vehicle truck travelling along a single rail, made up of short sections elevated on stands.

LIFTING PLANT

Some reference has already been made to plant that combines horizontal travel with a limited lifting capacity. All building construction requires some lifting, whether it involves a bungalow or a multi-storey office block. Such lifts invariably require a mechanical appliance, i.e.

- wheeled or lorry mounted cranes,
- tracked cranes,
- tower cranes,
- portal cranes,
- goods hoists,
- passenger hoists.

Selection of the correct plant will depend upon many factors, including:

- load details,
- lifting height,
- ground condition on site (including access),
- positioning of loads,
- time available,
- cost.

When selecting a crane, care must be taken to check its capacity for the required load. This must take into account the radius at which the lift takes place, the height under the hook and the bulk of the load. It is also important to determine whether the crane may move with the load or is required to remain static for stability.

Wheeled and lorry mounted cranes

Crane developments have eroded the clear distinction between wheeled and lorry-mounted units, and the trend has moved towards a chassis that is able to travel at speed along the highway.

The simple wheeled crane is a relatively slow-moving unit, which requires a good road surface for travel. The 'rough-terrain' crane was developed to permit wider use of cranes on a building site, and its traction on both axles enables it to negotiate poor site conditions. It is rare for either type of crane to have a capacity greater than 20 tonnes, and most building sites would use a 7 to 10 tonne capacity.

The use of telescopic jibs, although more expensive, has the advantage of allowing the cranes to travel on the roads and to extend their telescoping jib within seconds, as against the need to assemble a required jib length on site.

Lorry-mounted cranes are invariably hired for short-term lifting requirements. They too are likely to have telescopic jibs, except for the very largest cranes.

'Fly jibs', available for some cranes, permit an improved outreach; an auxiliary jib is fixed at an angle to the jib-end of the normal boom. With such an attachment it may be easier to lift or place a load inside a building without extending the main jib excessively.

Tracked cranes

The rope excavators described earlier have been developed into superior cranes with a high lifting capacity and the ability to negotiate poor ground. They have a particular application on building sites where mobility must be combined with indifferent ground conditions, e.g. for placing precast concrete beams, etc. Tracked cranes with capacities ranging from 6 tonnes to over 100 tonnes can also be used with a 'fly jib'.

Tower cranes

Tower cranes have for many years been widely accepted mechanical handling equipment on the larger building sites. The use of the smaller units for housing or other small construction has been less common in the UK than on the continent of Europe, however.

Tower cranes have a vertical mast, generally assembled from sections, and a horizontal jib mounted at the top of the mast. When this jib is in a fixed position at 90° to the mast it is described as a 'horizontal jib', and the lifting hook will be suspended from a trolley travelling along the jib. Other jibs are designed to pivot upwards from the top of the mast, and are described as 'luffing' jibs, where the loading hook will be held by a rope vertically beneath the jib-end sheave.

Tower cranes also differ in other respects. They may be:

- rail-mounted
- static
- climbing
- wheeled (self-propelled or trailer type)
- lorry mounted
- tracked

Rail-mounted tower cranes provide a better cover for a building by their ability to travel on the rails. Considerable care has to be taken in the preparation and maintenance of the rail base to ensure that it remains level. The cranes are mounted on bogeys and (depending upon design) they may be able to negotiate slight curves. There is a limitation to the crane height if it is to be allowed to travel; where this height has to be exceeded, the crane must be static and tied into the building by steel supports.

Static tower cranes are held on the ground by weights ballasting the bogey base, or they may be fixed to steel angles held in a specially prepared concrete base. Climbing tower cranes are sometimes used on high construction to reduce the number of mast sections required. They are generally positioned inside a lift-shaft, where it may be central to the building operation. Sections of the crane covering only the top three or four floors of the building are secured by steel sections to these floors, which then support the crane.

The wheeled tower crane may be self-propelled or a trailer type. It is particularly suitable for the smaller building operation where its lightness and mobility are an asset. These cranes generally have a 'luffing' jib, and their erection and dismantling may be a matter of minutes rather than hours. See *Figure 9.4*.

Figure 9.4 Small 'Record' trailer-type tower crane as used in house construction

Lorry-mounted tower cranes are of particular value where there are heavy lifts for a short period, and used at a height and radius not easily covered by conventional lorry-mounted cranes.

Tracked tower cranes are rarely seen on site. They have a limited use where ground conditions demand tracks rather than wheels; even so, the crane has to be completely level when a lift is involved and the deflection of the mast and jib does not permit travel over really bad ground in an erected state.

Portal cranes

These electrically driven cranes are rarely used for construction in the UK, where they are more often found in stockyards for concrete or steel. A horizontal lifting beam is mounted on legs that are bogey-mounted on rails, perhaps as much as 30 m apart. The lifting hook travels along the beam, which may also 'cantilever' beyond the rail legs. The excellent coverage of the portal crane can be used to the best advantage on long, low-rise buildings.

Goods hoists

Platform hoists have been the traditional plant for raising materials on a building site. The smaller types (half tonne capacity) may be termed 'mobile', i.e. they are mounted on wheels and can be easily pushed along the ground when the mast is lowered.

'Static' hoists, with capacities of up to two tonnes, can be used to greater heights than mobile units. The platform is raised by a rope up a vertical, sectional mast or guided between steel scaffold poles. Rack-and-pinion drive has been introduced on hoists in recent years to provide a more positive travel with greater speed and safety. Concrete skips can be fitted in place of the platform.

A variety of small-engined ladder-hoists can assist the small builder; these have a platform or skips travelling along a ladder-frame on an incline, and may feature a hinged portion for laying on a roof for tiling.

Passenger hoists

High buildings may necessitate the facility to move men up and down. Passenger hoists are subject to very stringent safety conditions and are generally rack-and-pinion driven.

PUMPS

The use of water pumps on a site is frequently the result of changes in weather or unexpected ground conditions. It is therefore not always easy to select the correct pump for a particular operation. Different types of pumps are available to the builder, each suitable for particular operations. They include the following pumps:

- lift and force (or diaphragm),
- induced flow,
- centrifugal (self-priming or normal),
- submersible (electric, air or hydraulic).

The selection of a pump will depend on several factors, including:

- volume of water to be pumped in a given time,
- height and distance,
- condition of the water, i.e. clean, muddy, etc.

The total volume that can be moved by a particular pump will be dependent on the 'total head', i.e. the sum of the suction and delivery heights plus frictional losses in the pipeline.

Lift and force pumps

This type of pump is particularly suitable for limited quantities of water in an excavation or trench, which may have up to 15 per cent of mud in suspension. The pumps do not require priming and can be allowed to 'snore' (i.e. run idle) for long periods. The *induced flow pump* is also diaphragm-operated and can operate in similar circumstances, providing an improved performance.

Centrifugal pumps

Larger volumes of water are more generally pumped by centrifugal units, where a rotary impeller displaces the water. The simpler units of this type require priming, and care has to be taken that the water is clean and does not damage the impeller blades. Self-priming pumps are able to pump continuously even if the excavation runs dry for a while.

The engine-driven lift and force and centrifugal pumps require inlet and delivery hoses. It is important that the reinforced inlet hose is treated with care and that it is not damaged by site traffic, which destroys the reinforcing. A filter basket must be fitted at the inlet end, to prevent stones and other foreign materials entering the pump body.

Submersible pumps

In submersible pumps, the pump body is lowered into the water so that no suction hose is needed. These pumps are generally electrically or air driven, and in rare cases hydraulically operated via an engine-driven power unit.

The need for a separate power supply is a disadvantage in these pumps, but they are extremely compact, easily handled and more simply maintained than the engine-driven pumps. An electric flow switch may be incorporated with some pumps to permit starting and stopping of pumping at a predetermined water level.

Dewatering

Where it is necessary to lower the water table over a given area, 'dewatering' will be carried out. This involves a series of 'well points' (perforated pipes), which are lowered into the ground. Water from a jetting pump delivered at great pressure is forced through the ground, which then settles more solidly and allows the water to be pumped out by a centrifugal pump aided by a vacuum pump, which removes air from the system and allows the water to rise.

BREAKING AND DRILLING

At some phase of a building construction it is necessary to break up or drill into a hard material; this may be rock or stone in the foundation or concrete, bricks or plaster in the superstructure. This work will demand a machine or tool with sufficient energy and will be powered by:

- internal combustion engine,
- compressed air,
- hydraulics,
- electricity.

Engine-driven breakers

These are powered by a two-stroke petrol engine mounted as an integral part of the breaker. It is a hand-held unit, somewhat heavy and only suitable for limited breaking requirements.

Compressed-air breakers

Although the conversion of mechanical or electrical energy into compressed air is very inefficient, this is the commonly accepted means for most large-scale breaking or drilling operations. Compressors, diesel or electrically driven, may be stationary, portable or mobile, and their output is measured in cubic metres of air delivered per minute at a given pressure. The usual pressure is about seven atmospheres, but there is a tendency to improve efficiency by using higher pressures. Compressed air is generated by reciprocating pistons, a rotary vane or oil-immersed screws.

Compressor tools may be percussive, rotary or a combination of both. Breakers or picks are termed 'light', 'medium' or 'heavy' depending upon their weight and penetrating power. Pneumatic drills also come in various sizes ranging from a small 'pistol' sized drill to one weighing about 25 kg.

To overcome the high noise level, 'silenced' compressors are now common where noise-absorbing linings dampen the sound. Muffled breakers reduce noise, and special nylon adaptors for the steels serve a similar purpose.

Other tools operated by compressed air can be used for grinding, chasing or scabbling.

Hydraulic breakers

The same desire to reduce noise levels has led to the introduction of hydraulic breakers, where the power is generated by an internal combustion engine via a hydraulic pump. Similar breakers can be connected to the hydraulic pump on a vehicle or 180° excavator.

Electric breakers and drills

A variety of electric tools are available to the builder for breaking, drilling, grinding or chasing. The power supply should be transformed to 110 volts to avoid serious accidents.

MISCELLANEOUS

Saws

Electrically or diesel driven saw-benches with blade diameters from 20 cm to 65 cm are used on sites. A coarser blade is used for 'ripping', and one with finer teeth is preferred for cross-cutting timber. The safe operation of the saw demands a frequent inspection of the blade, and a 'riving knife', which parts the timber, must be fitted at all times.

Chain saws are a portable alternative for smaller cuts and have petrol, air or electric drive.

Bar bender and cutters

The cutting and bending of steel reinforcing requires suitable machinery, which can be hand-operated for the smaller diameters. Diesel or electric plants are available to suit different capacities and operating speeds. It is, however, common for sites to buy-in steel 'cut and bent', which creates a problem of storage but makes it unnecessary to carry out the operation at the contract.

Welding plants

Small diesel or electrically driven welding sets are frequently needed in construction as well as for plant maintenance. The smallest units will be about 150 amp capacity, although major welding will require sets of 200 to 400 amp output.

Welding can also be carried out by oxygen/acetylene gas equipment, which is especially useful for cutting metals. An alternative to the use of these is propane gas, which is cheaper than acetylene. The gas is supplied in bottles, and is fed through a gauge to pipes that combine in a 'tool' fitted with special nozzles and controls to regulate the flow of the ignited gas.

Generators and electrics

Where no mains power is available, diesel or petrol generators can provide electric power. They are normally rated by their output in kW or kVA, and may be 110, 230 or 440 volts, either three-phase or single-phase supply (440 volts in three-phase only).

Transformers can reduce the voltage as required and provide a safer level for electric tools (110 volts). Special site equipments are now available that distribute the electric power throughout a building during its construction, with safe outlet points for plugging in.

Cartridge hammers

The speedy fixing of timbers and other components has led to the increased use of cartridge hammers. A percussion cartridge is fired and drives the nail or pin into the material similar to a hammer blow. Although the guns have inbuilt safety devices, they must be used with great care to avoid accidents. Different gun sizes and cartridge strengths make it possible to fix timber or metal to similar materials or to bricks, concrete and stone.

Concrete cutting

A number of specialist equipments on the market can cut concrete or bricks with suitable nylon blades. Other machines can groove a concrete roadway, scabble or grind concrete walls, ceilings or floors. In all these units the relative costs of the consumable blades or tool head will be a deciding factor.

10 SITE ORGANISATION

POLICY **10**–2

SITE LAYOUT **10**–2

SAFETY ON SITE **10**–9

SECURITY ON SITE **10**–14

10 SITE ORGANISATION

G. WHITE, FIOB, AMBIM
Principal Lecturer in Building Management Studies,
Guildford County College of Technology

This section discusses the decisions that need to be made regarding site layout, site accommodation and materials control; how to establish safe methods of working (with examples of safety checklists); and some measures that can be taken to improve site security.

POLICY

It is up to the contractor to develop policies concerning site organisation. The first question is to decide which matters require consideration and then discuss these with the site manager. The following are typical of the matters that require consideration:

- the policy regarding site layout,
- the policy regarding site safety,
- the policy regarding site security.

It is worthwhile committing the findings of the discussion to paper in the form of a brief statement, thus providing a reference throughout the job.

SITE LAYOUT

The factors that go to make an orderly development of the site installation can be considered under the following headings:

- sequence of working,
- site roads,
- site accommodation,
- siting the mess hut,
- the site office,
- control of incoming materials,
- bulk material location,
- mixing-point location,
- security storage,
- the compound.

Sequence of working

Assuming a private development, the contractor will consider the sequence in which he proposes to develop the site. If there is only one outfall to the local authority sewer, the general rule would be to commence at the outfall point so that the first houses handed over or sold are those operating on the shortest run of private sewer. A drainage system designed for one hundred houses will not function satisfactorily for, say, six houses at the farthest point from the outfall to the sewer. Having established what is the desirable sequence of the work, the following matters require thought.

Site roads

Temporary roads are wherever possible laid in the position of the permanent road, and are laid after the services have been completed. However, the designed road may not be suitable in layout for some of the vehicles that will be used to supply materials to site. For example, certain vehicles used by suppliers of timber-framed dwellings cannot negotiate sharp bends. The question to be answered is: which delivery vehicle's minimum turning circle will have the largest radius? Find out before you lay the road, not afterwards! This factor might require temporary roads to be increased in width at bends beyond the width of the eventual estate road.

If the specification calls for a scrape 150 mm thick with a hardcore bed of 150 mm on top, it is generally advisable to scrape only 100 mm. The site traffic will punch the hardcore into the earth about 50 mm. By making this allowance, one avoids providing a hardcore bed with thickness in excess of the specification. The amount allowed for the settlement and consolidation will vary with the nature and condition of the soil and the volume of site traffic; a 50 mm allowance will suffice in most cases.

SITE ROADS ADDITIONAL TO ESTATE ROAD LOCATIONS

Frequently, installations such as compounds and site hutments will be sited in areas that will ultimately become back gardens to the properties being erected. Materials such as ex-railway sleepers are often favoured for the purpose of connecting the site road (laid in the estate road position) to the compound area. Sleepers are easily laid and easily removed, leaving the ground clean. If sleepers are laid separated by bricks, one at either end, the 'road' surface will be self-draining. The bricks should be stood on end so that they bear on the full thickness of the sleepers.

If sleepers are unavailable and conventional hardcore is to be used, an effort should be made to grub up the hardcore and re-use it before the last concreting operation.

In certain circumstances, a lean-mix concrete road, or part thereof, may be desirable. Examples are:

- if it is planned that a ready-mix truck will regularly use one location on the road when discharging its load;
- if bricks are to be unloaded at one point;
- the section of the road adjacent to the site mixing point;
- the road and visitors' parking area adjacent to the site offices;
- a section outside the sales office;
- a section adjacent to the mess hut.

Site accommodation

Two very important questions that must be considered at the outset are:

- How much site accommodation will be required?
- Will the accommodation be in one position throughout the project?

For purposes of illustration, assume a development of 100 houses to be completed in 18 months and a sale price of £25 000 each. The cost breakdown for one house might be as follows:

	£
Land	6 000.00
Contribution to head office overheads (including interest charges)	1 500.00
Site supervision	300.00
Contribution to external works	1 800.00
Defects – maintenance period	200.00
Gross labour costs	6 000.00
Material costs	7 000.00
Site overheads, Plant, etc.	1 200.00
Profit	1 000.00
	£25 000.00

Total labour costs (excluding labour in external work)
= £6 000 × 100 houses = £600 000

The *average* weekly wage bill $= \dfrac{\pounds 600\,000}{78 \text{ weeks}}$

= £7 692

Assuming a gross average wage of £200, the *average* number of men employed

$= \dfrac{\pounds 7\,692}{\pounds 200}$

= 38 men

Note the stress on the word *average*. On a housing site proceeding for eighteen months it would be at least six months before labour strength peaked, and a fall-off of labour employed would commence after 12 months. An empirical rule with a high degree of reliability is that 50 per cent of the activity takes place in the middle third of the contract period. Hence:

1st third	2nd third	last third
25 per cent	50 per cent	25 per cent
29 men	58 men	29 men

These figures are obtained as follows:

38 × 3 = 114
114 ÷ 4 = 29 men (first third and last third)
114 ÷ 2 = 58 men (middle third)

The provision for workmen must therefore be based on a peak labour force of 60 men. The following are affected:

- car parking space required,
- mess hut size,
- canteen facilities,
- drying rooms,
- toilets,
- protective clothing,
- first-aid arrangements,
- amount of supervision provided.

Some of the factors that must be considered when siting accommodation and installations are:

1. There should be a minimum amount of walking for operatives.
2. The walk should be downhill from site stores, etc. in preference to uphill (i.e. since goods move outwards from site stores, take advantage of gravity).
3. Minimise expenditure on temporary services by grouping installations that require the same service.

4. Take advantage of local authority street lighting to light your hutment area at night as a security measure.
5. Site workmen's parking area away from the materials stores and the work place.
6. Arrange the minimum walking distance from the visitors' car park to the site office.
7. Arrange stores, mixing points, and compounds adjacent to site roads, avoiding creating a noise nuisance by placing mixing points near neighbours.
8. Toilets for female workers should be screened to be out of view of male workers and convenient to the place where the females are employed.
9. Male toilets should be near the work area.
10. The site office should be in view of the work face but if possible away from noisy operations, i.e. fixed-location mixing points, etc.
11. There should only be one single entry and exit point for vehicles to a site, and this should be in the view of the materials checker.

Siting the mess hut

The correct siting of the mess hut is critically important. The mess hut is the focal point of a lot of walking activity. If we use the example of an average labour strength of 38 men employed for eighteen months at a gross wage of £5.00 per hour, a site mess hut incorrectly sited by 50 metres could run up a bill as follows:

Activity	*Walking distance* (m)
From clothes-drying shed to work face at 08.00	50
From work place to mess hut and return for mid-morning break	100
Midday meal break	100
Mid-afternoon mug of tea	100
Return to drying shed at 17.25	50
	400

Working days in 18 months:	78 weeks – 6 weeks holiday	= 72 weeks
	72 weeks × 5 days	= 360 days
Then: 400 metres × 360 days × 39 men		= 5 616 000 m
		= 5 616 km

At 5 kilometres walking pace per hour, this takes 1 123 hours
At £5.00 per hour, this represents a cost of £5 616

Washrooms, toilets and mess hut(s) should be in close proximity on a compact site where they are to occupy a once-and-for-all location. On widespread sites, mobile toilets can considerably reduce walking time.

Figure 10.1 shows a good way of working out the position of the mess hut.

Having determined ideal locations for the mess hut, the following should be considered:

- Are there any drains or other service 'runs' passing under the mess hut position?
- What is the floor finish (if any) specified under the area occupied by the mess hut?
- At what stage in the sequence of hand-over will the properties adjacent to the proposed position be handed over?
- If a site is to be treated as two phases, there is obvious advantage in having all hutments *just* within phase 2 and arranging the sequence of erection for phase 2 houses to terminate adjacent to site hutments, if compatible from all other organisational aspects.

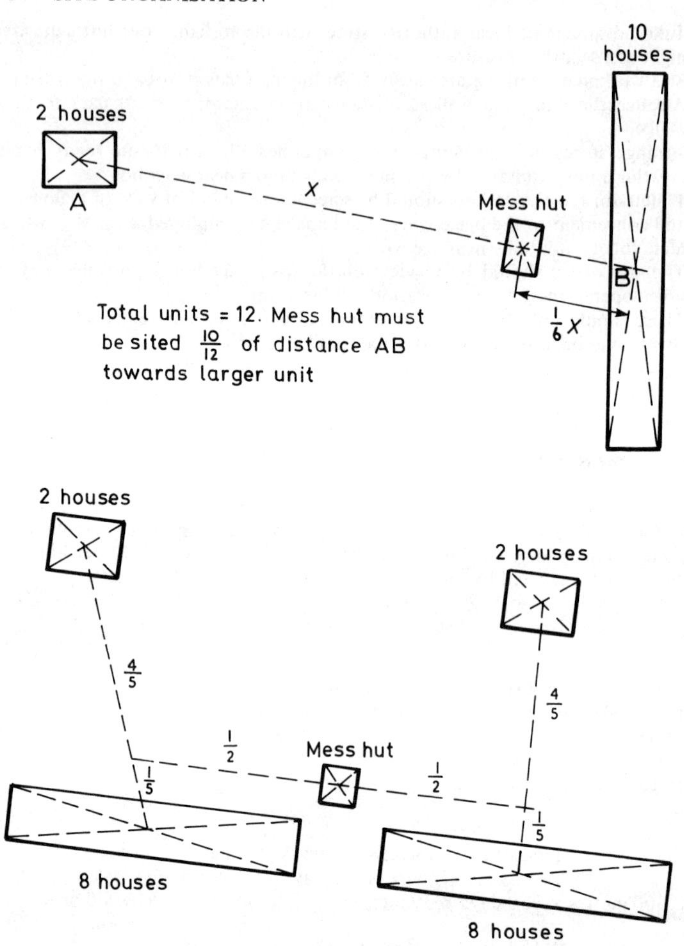

Figure 10.1 Locating the mess hut: two examples

The site office

Some of the questions that arise are:

- Will the building be used purely as an office or will some storage of special equipment, etc. be required? High-risk items such as holidays-with-pay cards, wages held for some hours, petty cash, raise the question of whether the site needs a safe. Where will high-risk equipment such as engineers' levels and staffs be kept on site overnight? If in the office, will the insurance cover be adequate if they are not stored in a separate lock-up metal cupboard?
- Will prospective purchasers be invited here to discuss matters concerning their possible choice? If so, clearly a separate table with three or four chairs is necessary, preferably partitioned off from the area where others might be working.

• Will site meetings be held here? If so, how many people will attend? If this is a local-authority housing site, a surprisingly large number of people from the various sections of the authority might wish to attend. There should also be due allowance of space for the drawings that will need to be viewed at the meeting.
• Is it possible to provide parking space adjacent to the office?
It is desirable that bona fide visitors to the site should not be allowed to wander freely about the site without the manager's knowledge. Therefore there should be conspicuous signs directing the visitor to the office.

On larger sites where a materials checker is employed, a sign should direct the visitor to the checker's hut; the checker can then direct the visitor to the office and also note what is in the visitor's vehicle when arriving and leaving.

Control of incoming materials

The materials checker should be informed of the materials that have been called forward for any given week. The checker should be informed of the quantity, quality, size, colour, type and specification clauses of the material he is checking. If he is not in possession of this information, he does not have a chance to do his job properly.

If a vehicle commences its stay on site at one point and returns to that point on leaving, it can be seen by the same person when entering and leaving. There is then a chance that control can be exercised over the many 'fiddles' that develop around delivery vehicles.

If the checker can look into the back of a vehicle (either from an elevated checker's office, or from a raised platform accessible by a ladder), he has a better chance of knowing what, and how much, goes on to the site and what comes off.

THE CHECKING PROCESS

1. All drivers should report to one and the same person. They should surrender their delivery ticket to him.
2. The checker will ascertain whether the materials were expected and if they conform to the order in description.
3. The checker should then give the driver a sheet of company-headed paper with a description of the load and tell the driver to whom he should report, e.g. Bill Jones, foreman carpenter, working on block 3.
4. Bill Jones should arrange the off-loading in the right place, minimising double handling. Jones should sign the company-headed note and give it back to the driver.
5. The driver must return to the checker, present the note signed by Jones and receive from the checker his delivery ticket signed.
6. The last act for the checker is to ensure that there is nothing on the back of the vehicle belonging to his company.

The reader may feel that this procedure only applies to larger sites. In this case, the following questions are posed:
• Have you ever unloaded unsuitable, unwanted materials?
• Have you, on carrying out a materials reconciliation, suspected that deliveries have been short in weight or quantity?
• Have you ever suspected that small plant, equipment or materials have left the site in unauthorised vehicles?

From the foregoing, it will be seen that one access/exit point is recommended. Any other possible entry/exit points should be adequately signed and barricaded to prevent their use.

The checker's hut should be sufficiently raised from the ground to allow vision into the back of a delivery vehicle. The hut should be located at the site entrance.

Bulk material location

The best location of material stacks depends very much on the intended mode of operation. On large sites, where forklift trucks are in use, fewer material stacks may be planned; the materials will be brought forward as required by the forklift.

Palleted materials require to be stacked on a level spot. On sloping sites, this may involve a 'scrape' for material stacks.

Material positions will be changed as the work moves forward. Size of stack, and the quantity of materials in it, will be relative to the number of houses being served from that point. It is very simple to build up check figures of space required. A stack of standard size bricks 1.5 m high × 1.0 m run × 3 bricks thick contains 675 bricks. This is very useful when considering how much ground should be cleared, and also when checking whether enough bricks are available to complete the job, etc.

Mixing-point location

This is a point that often gets scant attention when planning site layouts, yet the amount of sand that is wasted through being deposited in an unplanned, and therefore unprepared, position is considerable. It also often results in further wastage by having more mixing points than are necessary.

In the light of the Control of Pollution Act 1974, neighbours should not be exposed unnecessarily to noise; thus sound screens may be necessary at some mixing points.

If concrete, mortar, or plaster is transported from the mixing point by hand labour, obviously the machine must be as near the work face as possible. When site dumpers are used to transport the mixed materials, the distance becomes less critical and attention can be focused upon avoiding congestion of working space and site roads. Naturally, they will be sited away from face brickwork to avoid splashing; where this is not possible, the brickwork should be protected with polythene sheeting or other suitable means.

Security storage for small high-value items

Some special storage will be required to house joinery timbers, sanitary fittings, piping and fitments, paints, etc. With a wastage figure of around 20 per cent for these materials, the old method of storing them (in houses in course of completion, and accessible to all workmen, if not to other people) is unsuitable. Where garages are being erected, these can frequently be used for storage of the materials mentioned. In a battery of garages, if alternate division walls between garages can be left out temporarily, a better opportunity is provided for accommodating some of the longer materials, the missing division walls being inserted at the end of the storage periods. In many cases this suggestion will involve temporary support to purlins.

The compound

It is desirable that nothing other than bricks, blocks, aggregate and the like is left around the site. Other things (such as window and door frames, crated glass, soffit boarding, tilers' laths, gypsum plasterboard, plaster, cements, etc.) should be placed in a central compound, or at least an equally secure location. Apart from the aspect of security from theft, there is the question of protection from damage and deterioration of condition due to dampness.

There may be enough central garage space, as mentioned earlier, to deal with these materials in the same way as for the more costly sanitary fittings, kitchen units, etc.

If there is inadequate garage space, a central compound should be created. The compound fence may be either wire mesh with steel supporting stakes, approximately 2 metres high, or a similar construction formed of chestnut paling. The compound should be entered through an opening which, after working hours, is secured by a sound pair of gates fitted with a hasp and staple with concealed fixing, and a well selected lock. Ideally, the compound should be large enough to accommodate site dumper tracks, compressors, scaffold-tube fitments and boards not in use. Ladders not in use should be padlocked to the inside of the compound fence.

Conclusion

After a programme of work has been prepared, which should be before operations start on site, it is worthwhile to consider what will be suitable site layouts at four stages during the contract. Four pieces of perspex can be used, each large enough to cover the site plan drawing. For each stage, mark on the perspex by means of chinagraph pencil the houses that will have been completed and therefore will not require 'service'. Mark in those that will not yet have been started, and those that will have been completed in carcase. Use different coloured pencils for each stage. An examination of each will help to develop the best site layout for that stage of the contract.

SAFETY ON SITE

- Safety on site is very much a matter of attitude. Despite a comprehensive legislative system and an easy-to-follow Highway Code, we have an appalling number of road accidents in the United Kingdom. This proves that legislation does not necessarily ensure results. Correct attitudes can.
- The need for tidiness is emphasised.
- Almost every operation can be considered against a 'safety checklist' before commencing.
- Vehicles entering and leaving the site need special consideration.

Moulding attitudes towards safety

Workmen's attitude to safe working will be influenced most when management is seen to have a total commitment to safety. Some managers think that they have discharged their responsibility if they wear safety helmet and suitable footwear. Fostering and developing safe work practices goes much deeper than this. Setting an example is the most potent method of forming habit and tradition. In the site manager's work this includes:

- Being seen to plan the layout of work places, scaffolds, etc., with safety in mind.
- Developing and using safe methods of working, which includes care in the selection of personnel for the responsible posts, i.e. banksman, timber man, etc. There is nothing more likely to undermine morale than some idiot driving a site dumper around the site.
- Being seen to carry out site inspections and examinations in an efficient manner; the manager who forgets the inspection but remembers the clerical procedure of signing the register is undermining morale.
- Setting up proper systems for the maintenance of plant and equipment, with arrangements for spares, storage of fuel, etc.
- Providing an effective system for the issue, collection and cleaning of protective clothing.

ENVIRONMENT

A further responsibility of a site manager is the maintenance of a clean, well-ordered environment; this will tend to give rise to constructive, helpful, and cooperative attitudes towards safety. Examples of this type of environment are:

- clean canteens, toilets, etc.;
- clean walkways about the site and in the partially completed houses;
- good artificial lighting when artificial lighting is necessary.

TRAINING AND INVOLVEMENT

Managers should train operatives in operations that are new to them, and not leave the matter to chance. During training, the hazards of the job should be explained and safeguards that nullify the hazards should be explained.

Operatives should be encouraged to get involved in developing safe methods. Group loyalty can be harnessed towards group safety.

PAYING FOR SAFETY

A further way of emphasising the site manager's total commitment to safety is to include in bonusable tasks details of the safety measures that the operative is expected to comply with. A time allowance should be set against the safety measure. When operatives fail to carry out the safety measure, the safety allowance should be disallowed.

RAMMING THE MESSAGE HOME

There are available many effective, vivid posters from the various safety organisations. The site manager should make it his business to get the posters and display them, preferably apart from other notices; they should also be changed frequently. If a site manager equips himself with twenty such posters, they can then be moved about frequently to gain maximum effect.

A further aspect of 'ramming the message home' should be a willingness to reprimand workers who blatantly ignore the safe practices that the site manager is attempting to get established.

Tidiness on site

A large percentage of site accidents are caused through operatives having their feet injured by projecting nails and the like. Three factors can be considered:

- *Footwear.* Many contractors now have arrangements with suppliers of safety footwear so that workmen can purchase, through their firm, suitable site footwear. This idea would seem to be worth fostering. Few workmen are likely to go out on Saturday and buy safety boots themselves, but men may buy if they see their comrades buying on site, especially if a hire-purchase scheme is offered by the contractor.
- *General tidiness.* The present practice of sub-contracting almost all trades on many housing sites has undoubtedly improved productivity. It has also, in many instances, increased the amount of waste and untidiness. Many sub-contractors leave material stacks in an untidy condition and, because of zest to get on and create measured work, operate in an untidy workplace. In order to control this situation, the sub-contractor must be financially penalised for any material waste in excess of

the previously agreed percentage. The sub-contractor should be given every facility to permit him to get on with the job in a tidy and safe manner. Considerations include freedom from interference from other gangs working in too close proximity, materials stacked conveniently to the work place, and also stacked safely. Any equipment provided by the main contractor should be in sound condition.

- *Tidy walkways.* Occasionally, men take short cuts across areas where, for various reasons, the ground is not clear. Such areas should be cordoned off by means of coloured plastic markers attached to a line and secured to pickets. The areas to be used as walkways should be 'maintained', i.e.

(a) free from obstruction by objects,
(b) free from obstruction by bad ground conditions,
(c) artificially lit when work is proceeding during the hours of darkness.

Safety checklists

Before any task is commenced, it is useful to have available a checklist as an aid to memory. The checklist may be prepared for use in three stages:

- before the work commences,
- whilst the work is proceeding,
- on completion of the task.

An example of a checklist for excavation work is as follows.

BEFORE EXCAVATION WORK COMMENCES

1. Have details of the position of all services (water, gas, electricity, telephone, foul sewers, stormwater, etc.) been obtained?
2. Have the positions of these services been made known to all concerned?
3. Are the necessary barriers, warning notices, lighting equipment (including generator, fuel for generator and, where required, dry batteries) readily available?
4. Has a permit been obtained from the local authority
   (a) for road closure (semi-permanent),
   (b) for part road closure (semi-permanent),
   (c) to place skips on the Highway?
5. Has Form 10 been completed and sent to the Factory Inspector?
6. Have the local weather conditions for the period in question been ascertained from the Meteorological Station?
7. Has a method statement been prepared?
8. What is the proposed method of support for the trenches?
9. Is there adequate trench-sheeting equipment available, including
   (a) steel sheeting,
   (b) adjustable struts,
   (c) timber walings?
10. Is there a pump available on site? What are the means for disposing of pumped water?
11. Is the excavating machinery available and in good condition? Are the required sizes of bucket available? Are they all fitted with good teeth?
12. Will spoil be deposited alongside trenches? Is there enough space to allow 0.6 metre clear space between trench and spoil heap?
13. If there is a need for a spoil heap away from immediate excavating area, where will it be located? Is it accessible for site dumpers from all parts of the excavation area?
14. Are trench ladders available, i.e. with square-section rungs?
15. Will a trench bridge be required for the public? An additional one for workmen? One adjacent to the site office?

DURING EXCAVATION WORK

1. Is the timbering work being carried out satisfactorily? Are steel sheets placed correctly?
2. Is site transport operating too close to the trench? Have 'stops' been provided to prevent dumpers being overturned?
3. Are exposed services properly supported with timber and wire bonds? Are cover tiles being retained for re-use when back filling?
4. Are there proper sumps for collecting seepage? Is pumping drawing material from behind the timbers?
5. Are spoil heaps being properly controlled? Is the method adequate in the event of an adverse change in the weather?
6. Are there any pipes, bricks, tools or loose equipment on the inside of the spoil heaps?
7. Are trench ladders and bridges being used? Are walkways clear of mud, etc?
8. Is the labour being properly supervised?
9. Is the work properly fenced off and signed during the day?
10. Is the work properly fenced off, signed and lighted during the night?
11. Are tools and equipment being properly collected in at the end of the working day?
12. Is mobile plant being brought into plant compound at night and being immobilised? Is the plant parked in such a way that access to fuel tanks is denied to vandals?

ON COMPLETION OF EXCAVATION WORK

1. Have all the signs been removed?
2. Trench backfill properly carried out?
3. Trench overfilled to allow for subsidence?
4. Trench position marked on both sides to prevent vehicles getting bogged down?
5. Bridging provided at vehicle crossover points?
6. Plant and equipment checked over and returned to yard or moved forward on site?

SUGGESTED CHECKLIST FOR BRICKLAYERS AT WORK ON HOUSING

Working at ground level

1. Has the ground been cleared and levelled to make a proper standing?
2. Are bricks properly stacked, i.e. plumb and not too high (assuming bricks are being manhandled)?
3. Are mortar boards positioned properly, i.e. 2 metres apart and 0.75 m clear space between wall face and mortar board?
4. Is there an adequate amount of wall run for each man to work safely?
5. Are window and door frames properly 'stayed' whilst being built in?
6. Are the labourers serving the bricklayers without risk to the bricklayers?
7. Are the bricklayers keeping the walkway clean of brick waste from cutting, and also rejected bricks?

Loading and working on scaffolding

1. Are scaffold standards properly supported on sole plates?
2. Is the spacing of standards right for the method of loading the scaffold, e.g. forklift truck with palleted load?

3. Is the scaffold platform at the right level for the work, with the gap between boards and wall not more than 50 mm?
4. Has adequate bracing been provided?
5. Are the boards adequately supported and lying flat on the supports, with no loose end binding to trip anyone up?
6. Are boards overhanging the end support too little? too much? just right?
7. Have wedge boards been fitted at unavoidable step points?
8. Are toeboards and handrails OK?
9. Are toeboard and handrail properly terminated at access to scaffold platform from the ladder?
10. Is the ladder properly selected for the work?
11. Is the ladder in good condition, with no rough timber on poles to cause hand damage? Are all rungs intact and in good condition?
12. Is the ladder properly secured top and bottom? Is the base positioned in soft ground and likely to 'settle'? Is the angle of inclination to the horizontal all right for hod carriers? All right for ordinary usage? Is the rung nearest to the platform at the right level for stepping off on to the platform? Is there any way in which the user may trap his feet under a scaffold member?
13. Are the brick stacks laid out in right relationship to the work to minimise walking distances? Are brick stacks on the scaffold stable, of the right height, and not encroaching upon the walkway? Are they kept clear of the ladder alighting point?
14. Are mortar boards overloaded, thus creating the danger of mortar falling from scaffold to ground level?
15. Are loose bricks being left on the top of walls or on the scaffold platform?
16. Is brick waste being cleaned up as the work proceeds and carried down?
17. Is anything being thrown from scaffold to ground level?
18. Is the hoistway clear of projections from the scaffold?
19. Is the hoistman exposed to any risk?
20. Is the hoist properly secured into the scaffolding?
21. Are the areas of access to the hoist platform at ground and scaffold levels in a tidy condition?
22. Is the hoist being operated in a proper manner by a competent person? Are the hoist gates correctly fitted and kept closed while the hoist is in motion?
23. Is the 'green' brickwork stable? Is any operative standing on 'green' brickwork?
24. Is there any section of work being raised to too great a height in any one day?
25. Are free-standing walls being properly supported by 'staying' until they are incorporated into the fabric of the building?
26. Are warning notices fixed to uncompleted scaffolding?
27. Are ladders made unusable at the close of the day's work by having a scaffold board roped to them?

It is recommended that checklists are developed for all work operations and that the use of these lists by trade supervisors, whether they be directly employed or sub-contractors' supervisors, should be developed and encouraged by every means possible.

Vehicles entering and leaving the site

Whenever possible the entrance gates should be wide and set back from the back edge of the pavement, even if the pavement crossover is single, i.e. 2.30 metres wide. This enables the driver to get a better turn in to the site and use only one lane on the public highway. When the vehicle leaves the site, other road users and pedestrians get a better view of what the driver intends to do.

General aspects of safety on site

The 1970s will be remembered for, amongst other things, the great amount of employment legislation passed by Parliament. This legislation is still being added to at the time of publication, and the reader is therefore recommended to keep informed of changes through such bodies and organisations as:

The Department of Employment
HM Factory Inspectorate
Her Majesty's Stationery Office
The Federation of Master Builders
The National Federation of Building Trade Employers
The Building Advisory Service

It is advisable that at least one person in the organisation is kept up to date with current legislation by attending courses, reading the appropriate literature and (perhaps most important) keeping his managerial colleagues informed of his findings.

CHECKLIST FOR GENERAL SAFETY MATTERS

1. Are the First Aid boxes supplied with all necessary medical goods and equipment, pharmaceuticals, etc. relevant to the kind of operations being carried out on this particular site?
2. Are the First Aid duties being carried out by a competent person?
3. Is proper control exercised over all First Aid goods, equipment, ambulance room, all protective clothing and equipment?
4. Are the accident book, all registers of inspections and examinations, all returns to the Factory Inspectorate and other official bodies being properly administered by a competent person?
5. Has the competent person been fully briefed as to the extent of his duties, responsibilities, authority?
6. Are the location of the First Aid box and the name of the competent First Aid person known to all site workers?
7. Are sharing arrangements with other contractors understood and properly communicated to all concerned?
8. Is there readily available on site all necessary safety equipment, including helmets, goggles, eye shields, nets, belts, dust masks, protective clothing and footwear?
9. Does the storage of petrol and other fuel oils conform to Regulations?
10. Are fuel storage areas properly signed and is the 'no smoking' rule being observed?
11. Is mobile dust-extraction equipment available for use with portable power tools?
12. Is a vacuum cleaner available to free protective clothing from dust before its removal?
13. Are all saws, abrasive wheels, etc. not only fitted with proper guards but also being used with the guards in position?
14. Is a proper policy being implemented towards worn equipment, i.e. recognising the need to destroy equipment before it becomes dangerous?

SECURITY ON SITE

Perhaps the easiest way to improve profit performance is to take better care of the materials of construction and also the plant and equipment used.

In recent years there has been a considerable increase in the value of goods stolen from building sites. Housing sites probably suffer more than commercial and industrial sites. There is a large, ready market for all materials and fittings used in house construction. Housing sites are often large in area, and construction activity of one kind and another may be taking place on fifty houses at the same time; under these conditions a loss may not be detected for some days. It is perhaps worth categorising time zones and groups of people who may at some time be involved in thefts from building sites.

		Nights between ordinary working days (B)	*Weekends*			
Group	*Ordinary working days* (A)		*Summer* (C)	*Winter* (D)	*Wakes week* (E)	*Christmas holiday* (F)
1 Work force	√					
2 Member of work force with outside agent	√					
3 Occupiers of houses adjacent to work area		√	√	√	√	√
4 Public at large		√	√	√	√	√
5 Professional thieves				√	√	√

Group 1 – time A

If workers observe that there is a disregard by supervisors in setting up proper systems of control on materials, fitments, etc., temptation is increased. Examples of this may be:

1. Obvious over-ordering of goods, and, the mistake having been made, no effort made to get the excess removed to where it can be used for the purpose intended.
2. No effort made to control waste when using materials.
3. No arrangements made for the storage of materials, e.g. joinery materials left lying on the ground without even protective cover.
4. Fitments such as baths, hand basins, kitchen cabinets stored in unsecured houses.
5. No perimeter fence or gates.
6. No 'watching' arrangements at night-time or weekends.
7. No stage-by-stage clearing up as the work progresses.

Minimising theft in this area is largely a question of supervisors planning needs, coordinating supply with construction, and establishing methods that minimise waste in use. In addition, certain stock should be stored in secure conditions. There is a case for a supervisor being the first person on site in the morning and the last person to leave the site in the evening; amongst other advantages a good control can be kept on stock-room keys.

Group 2 – time A

Collusion in theft between someone employed on the site and a regularly visiting transport vehicle and driver can be extremely difficult to spot. It is largely a matter of good general supervision, good entry and exit control, and care in noting when 'losses' have occurred if they are coincidental with visits to site by outsiders.

Group 3 – times B to F

Stealing materials from a builder, with his widespread site and many stock piles, in the minds of many people is neither tremendously risky nor tremendously immoral. It is generally known that builders do not have involved security arrangements, especially on a Sunday morning, to deter the enthusiast who needs twenty bricks to form a base for his greenhouse. The risk of detection cannot be compared to the supermarket with its closed-circuit television cameras.

Possible counter measures: pay a regular retainer to some senior citizen resident in the area to 'watch' at vulnerable times and make an example of the first offender caught, i.e. maximum publicity in the local press. Publicise your intentions by means of a notice that reads: 'If you see anyone removing materials from this site between 17.30 and 08.00 (week-ends 17.30 Friday to 08.00 Monday), they are being stolen. Please notify the police immediately and you will be rewarded. A. Rodgers. Builder.'

Group 4 – times B to F

If you have a trustworthy employee who is a caravan dweller during the working week, make arrangements for him to live on site. Recognise the problem created when he goes home.

Group 5 – times D, E and F

Dealing with professional thieves who are obviously interested in large-scale returns is extremely difficult. Clearly, however, times like the Christmas holiday (when a theft could go unnoticed for ten days, and when the hours of darkness are at their longest), or a fortnight's 'Wakes' when the whole town goes dead, require special consideration. Plan the work at this period to emphasise external doors (or temporary external doors), glazing, etc. Scale down stock levels; deliveries immediately preceding the stoppage should, whenever possible, be unloaded in the compound. If garden fencing can be erected, it makes an extra job for thieves to get a vehicle up to the compound. Park vehicles and plant awkwardly in the compound to make access difficult for another vehicle.

FURTHER INFORMATION

The practice of site management, Institute of Building

11 PRODUCTION PLANNING AND CONTROL

INTRODUCTION 11–2

PRODUCTION PLANNING 11–2

PRODUCTION CONTROL 11–6

CONCLUSION 11–8

PRODUCTION PLANNING AND CONTROL 11

11 PRODUCTION PLANNING AND CONTROL

J.G.A. SHEEHAN-DARE, DipArch, FIOB, MRSH
Director, Taylor Woodrow Homes Ltd

This section considers production planning and control as applied to contract and private development work; while the techniques used are the same, the constraints are different.

INTRODUCTION

The last part of the title of this section is the key to logical and effective production, that is 'planning and control'. As much care and consideration has to be devoted to this aspect of the operation as to the initial layout design, the design of the dwelling units and the practical execution of the work.

What is being commented upon in this section applies equally to housing in the public sector as to the private sector, but each has particular facets that have to be carefully considered. The fact that a dwelling unit is a small structure in comparison with a factory, school or warehouse, for example, should not lead one to suppose that scant attention can be paid to the site organisation, ordering of materials, sequence of work and overall supervision, all of which, if overlooked or discounted, will result in confusion of many kinds, unnecessary expenditure and, at the best, an indifferent end product.

PRODUCTION PLANNING

Construction sequence

What then is the simplest and most satisfactory way to approach this very important aspect in the production of a house, whether it be a single unit or many hundreds extending over a considerable site area?

First, consideration must be given to the planning, irrespective of the number of dwellings, their type, number of storeys and configuration, in order to create a logical sequence for their construction. Avoid the classic situation where someone who is staining the floorboards in a room works all round the perimeter and progressively towards the centre, only to find himself with no means of escape and no room to manoeuvre. It is more than likely that the greater the number of units to be constructed the greater will be the chance of dividing the whole development into smaller groups, or phases; this will enable each to be dealt with individually, but with due consideration for the whole of which it is a part.

Above all else, there must be a reasoned approach to the operation, which will result in a practical sequence with a realistic timescale to produce continuity and avoid leap-frogging from unit to unit or from one part of the site, or phase, to another. The foregoing is a discipline that must be observed, however tempting and expedient the 'ad hoc' approach may appear at the time; once the correct sequence has been broken, there is every chance that the economics will be adversely affected to a greater or lesser degree.

Contract work

There is no denying that a site being developed on a 'contract' basis, within the public sector, is somewhat simpler to consider in planning terms, because the total number of units (or in the case of a large site an appreciable proportion of the total) will be required to be constructed within a specified, and usually short, period. This creates the ideal situation. In a case such as this, it is possible to construct all the carriageways and related works and to lay the whole of the foul and surface-water sewer networks, therby providing hard surfacing upon which to deposit bricks, roof framing, carcassing timber, roof covering and other 'heavy' materials. The ideal approach is to programme the work, and with this must be correlated the ordering and delivery of materials, so that when these are received on site, they can be unloaded in the most convenient position in relation to each unit, or group of units. Unnecessary handling of materials on site causes many problems, not least of which is the time involved and the additional expenditure resulting from this double operation.

It is accepted that the ideal situation is difficult to achieve, and in certain cases virtually impossible, because of delays resulting from adverse weather conditions, shortages of labour and late delivery of materials. However, provided these and any similar factors are recognised and some positive action is taken, the effect on the overall programme should not be in any way disastrous. Problems, and possibly chaos, only seem to occur when situations such as these are not recognised and acted upon, or are ignored completely for whatever reason.

There does, however, appear to be a trend towards another method of operating a site, which could find more favour in the future. This is the establishment of a much more extensive 'compound' area, to enable all materials to be received and stored there and eventually distributed throughout the site by means of a forklift vehicle. This method does constitute double handling, which tends to increase costs, but it is quite possible that this can be compensated for by a reduction in the wastage of materials (a certain amount of which is unfortunately unavoidable even in ideal situations), total control, and better security against loss by other agencies.

Private development

Private-sector developments are not, in general terms, as simple to operate as those in the public sector. The ultimate goal in the latter case is completion of the whole of the development, or a specified part, as one operation within a set (and frequently limited) period of time. In the former case, on the other hand, the demand for the houses will often have considerable influence on the rate of production. Those who can afford, or who are prepared to commit, an inordinate amount of capital for work in progress will be few, and this is one of the most important factors in the private sector: the fine balance of the work in progress against the demand for houses at any particular time. There may be occasions when the demand for houses on a particular site, or sites, diminishes or possibly ceases altogether. A decision then has to be taken, not whether to continue the development, but whether merely to complete the work in progress and remove all, or any, surplus materials from the site.

Assessment of these situations is not easy, and there has to be a well considered judgement as to whether the demand will recover in a reasonable period of time, or not. Situations can change radically in a relatively short period and it is not unknown for a site, where demand has diminished or ceased altogether, to become active again virtually overnight. What are the factors responsible for this? They could be economic, e.g. dwelling construction costs cause the total cost to rise above the current selling price, or social situations may change, e.g. employment patterns alter, resulting in population movements. However, these changes cannot always be predicted far enough in advance (or, indeed, be predicted at all), and it is considered that they have to be regarded as an occupational hazard.

Planning techniques

What is the most satisfactory and easily understood method of portraying the planning, or programme, for a particular project? Critical Path Networks appear to enjoy some popularity for this purpose, and it is not denied that in certain circumstances, usually the larger commercial or industrial projects, they are extremely useful, although there are those who experience some difficulty in appreciating them.

It is considered that the 'bar chart', with which all in the building industry are familiar, is eminently suited to housing development, although not capable of conveying as much detailed information as a Critical Path Network diagram. The bar chart must be of a size sufficient to contain all the basic information required, to depict all trade categories, and to extend over the whole period of development of the site in question, with sufficient additional space to allow for any further time that may be required as the result of delays, however caused. It is not practical, nor is it considered desirable, to produce a bar chart to take account of dwelling units on an individual basis, since this would be far too unwieldy, and possibly difficult to read. The chart must embrace the whole operation, whether it be for six units or for sixty, and in sufficient detail to indicate the various operations in connection with the sub-structures and super-structures, together with all the ancillary operations that have a bearing on the total site operation. As previously commented upon, it is not felt advisable to attempt to show individual dwellings as a reference point, but if such reference is required, it is better that small groups of units are used. The site agent, together with other supervisory staff, will be in possession of all relevant information in respect of completions and, if it is considered necessary, a supplementary chart indicating only the projected completion of units can be prepared and read in conjunction with the construction bar chart.

To illustrate the principles contained in the preceding paragraph, *Figure 11.1* shows a bar chart for a small hypothetical site consisting of eighty-one two-storey dwelling houses on a site (*Figure 11.2*) that could be considered either as an isolated one or, as in this instance, part of a larger development. It has been assumed that it is possible to construct all the roads, sewers and dwelling units on a continuous and uninterrupted programme. To achieve the target completion in this programme it would be necessary to have ideal meteorological conditions, a naturally well drained site, regular and prompt material deliveries, and a constant and adequate supply of labour.

It is not always possible to commence work at an ideal period of the year (considered by some to be between the middle of March and the middle of May, for obvious reasons). Where a late or early start occurs, the fifty-two week programme could well become extended for various reasons and adjustments will have to be made accordingly. The fact that this programme indicates commencement at Week 1 does not necessarily mean that this is coincident with the first week of January.

It is accepted that where a project is being carried out as a 'contract' for an authority, i.e. other than for individual direct sale to the public, the terms of the contract may require a specific date of completion and handing over, or may require that a certain section of the site be constructed and completed before another. Conditions of this nature could vary the period of the programme and have an effect on the siting of the compound; the latter therefore requires due consideration, although where possible it should not be positioned so as to obstruct general progress by occupying the site of eventual dwelling units. It should be pointed out that this situation may, on certain restricted and high-density developments, be unavoidable. The final units will then have to be constructed and completed on a 'kerbside' basis, where materials and components have to be placed or stored in any available and not

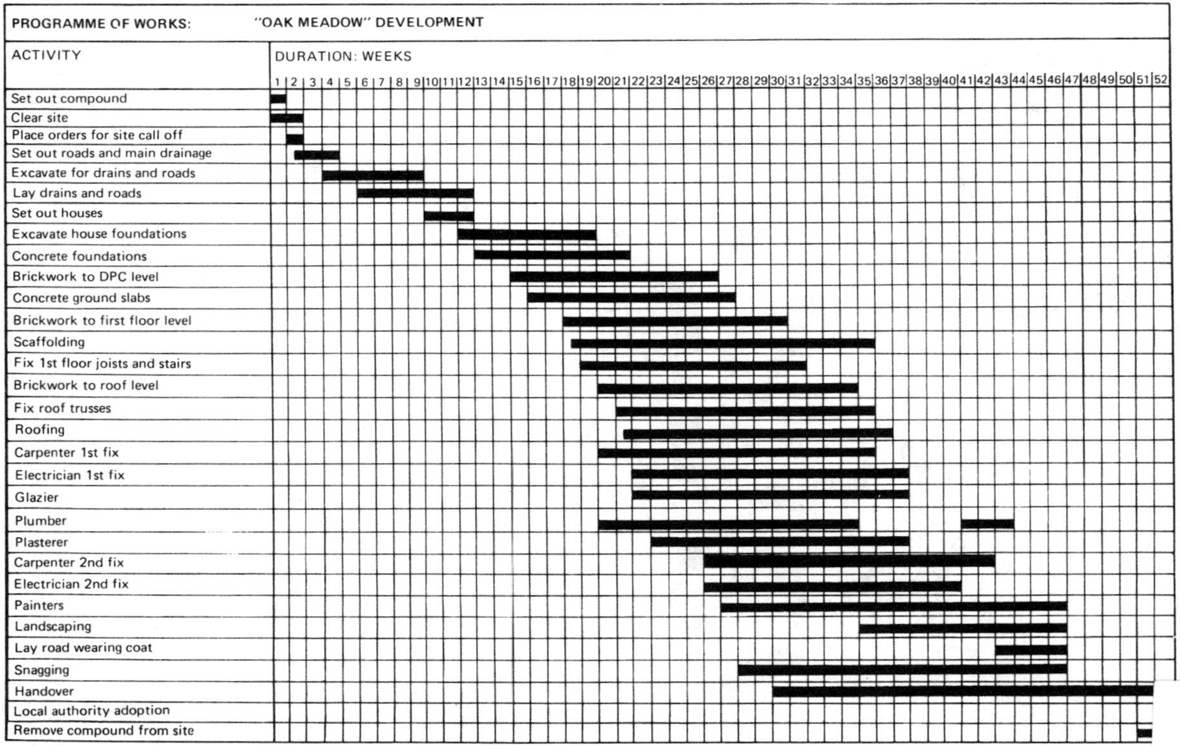
PROGRAMME OF WORKS:
"OAK MEADOW" DEVELOPMENT
ACTIVITY
DURATION: WEEKS
1 2 3 4 5 6 7 8 9 10 11 12 13 14 15 16 17 18 19 20 21 22 23 24 25 26 27 28 29 30 31 32 33 34 35 36 37 38 39 40 41 42 43 44 45 46 47 48 49 50 51 52
Set out compound
Clear site
Place orders for site call off
Set out roads and main drainage
Excavate for drains and roads
Lay drains and roads
Set out houses
Excavate house foundations
Concrete foundations
Brickwork to DPC level
Concrete ground slabs
Brickwork to first floor level
Scaffolding
Fix 1st floor joists and stairs
Brickwork to roof level
Fix roof trusses
Roofing
Carpenter 1st fix
Electrician 1st fix
Glazier
Plumber
Plasterer
Carpenter 2nd fix
Electrician 2nd fix
Painters
Landscaping
Lay road wearing coat
Snagging
Handover
Local authority adoption
Remove compound from site

Figure 11.1 Typical bar chart programme, for the site shown in Figure 11.2

Based on 260 working days

Figure 11.2 Estate layout: 'Oak Meadow', a low-rise residential development of 81 two-storey units (stage 1); scale 1:2500

too inconvenient a position: the footway, small areas of the site that will eventually be landscaped, or a garage or garages of unsold or unoccupied properties. The foregoing circumstances do require very careful consideration at the planning stage; the programme must be prepared accordingly so that the operation can proceed unhindered by inconveniences, either major or minor, which can cause delay and increase the cost.

PRODUCTION CONTROL

Organisation of site

Once the layout has been prepared, the house types designed, working drawings produced, all the necessary applications under the current Planning and Building Control legislation made and the related consents obtained, materials schedules completed and orders for materials and sub-contract work placed, some might suppose that the arduous work is finished and that what follows presents no difficulties or problems. Nothing could be further from the truth. There are other practical considerations that many would regard as not essential to efficient production, or even as pedantic. One of these is the intelligent distribution of 'heavy' materials throughout the operating area of the site in the most convenient situations in relation to the work in progress. However, it is accepted that certain materials and components, until they are incorporated in the structure, must be held in a secure place, namely the site compound, because of the possibility of theft (although this is regarded by some as an acceptable risk, provided its incidence is very infrequent). It would be an ideal situation if, for example, staircases could be received on site, checked and deposited outside each dwelling and then taken inside and fixed immediately, but this seldom occurs.

Production planning, which has already been discussed, plays an extremely important role in the construction and completion of each dwelling, as does production control, the latter being the practical application of the former theory. It must not be assumed that, because one has programmed and organised a projected development thoroughly, it will work automatically and without further attention. In fact, control is possibly the more demanding of the two aspects, planning and control, inasmuch as it requires constant monitoring and must be well coordinated with the general management of the site.

In order to control production – i.e. to regulate the rate of completions, to take account of changes in demand or such unforeseen circumstances as may occur – there must be good organisation from the start. However, it must be remembered that a flexible approach must be maintained, since it is not always possible, even with the ideal situation and every circumstance in one's favour, to adhere strictly to the planning as originally conceived. In order to be fully aware, at all times during the period of the work and at any time after completion, it is recommended that a Day Book, or Site Diary, be maintained and entered up at the conclusion of each day's work. The entries need not be lengthy, but should include all matters regarded as important, i.e. meteorological conditions, labour on site and related problems, delays due to late delivery of materials, and finally a general comment on progress. This document will prove invaluable at all times, either to recapitulate on progress or to resolve any differences of opinion on recorded matters which may arise either during, or after, completion of the work.

Supervision

How much supervision is required, or how much supervision should be provided? Without effective and constant inspection there will be no control, and it follows that standards can deteriorate, resulting in work having to be re-executed, consequent delays, and a reduced rate of production. Some may believe that intensive supervision is the determining factor in the control of a multi-unit project; this is not so. Saturation supervision achieves very little. It should be regular and at predetermined stages during the course of the work – thus ensuring that each stage of the work, in each unit, is satisfactory before the next stage of the work is proceeded with. This situation is not at all difficult to achieve where the project has been intelligently programmed and the discipline of regular inspection is adhered to. Local authority inspections in pursuance of the Building Regulations must not be allowed to take the place of those carried out by the developer as part of his administration of the project.

Disruptions

During the course of a project there is bound to be one occasion, if not more, when progress is interrupted for one reason or another, possibly one of those previously referred to. It is then that the appropriate remedial action must be taken without delay. Little purpose is served by procrastinating at this time and hoping that a situation will correct itself. This seldom if ever happens, and any delay while hoping that the situation does correct itself can cause greater aggravation than making a wrong decision. With a flexible and readily adaptable approach to supervision, these problems could still occur, but are unlikely to have any unduly damaging results.

A complicated system of control must be avoided, as this tends to result in errors because of the likelihood of duplication of documentation. The simplest system is, without doubt, the easiest to devise and to operate, and there is also less chance of errors and omissions. There is also the possibility that too much control can cause far more problems than inadequate control, and it should be clearly understood that the agent, or foreman, must be in complete control of the operation and be aware of the state of the work on any part of the site at any time. Any decision concerning the redisposition of materials or redeployment of labour can then be taken immediately, without having to spend time making a complete reappraisal of the work in progress.

CONCLUSION

The building operation, whether for a single unit or for a multi-unit scheme, is not an easy or automatic one. Without the necessary forethought and subsequent vigilance, the benefit or 'profit' that will result only from a well organised and administered development can be lost very easily and in a very short time. Never lose sight of the fact that the developer or house builder is judged ultimately by what he produces – the end product – and not by the method whereby he achieves it. But it must follow that without planning and control, adequate supervision and constant attention a good product cannot result. It is quite possible to devise and implement a very satisfactory system of planning and control and for a bad, or at best an indifferent, product to result. Conversely, however, a bad or indifferent system that is well administered can never produce an acceptable product.

To summarise planning and control, let us consider once again the purpose of this part of the building operation. To plan and control is not something peculiar to the building of a house, it is an essential part of every manufacturing operation, whether it be the production of biscuits or motor vehicles. Without it, the whole operation will be adversely affected, as will also the end product. There is nothing extraordinary about this aspect of the building process (not only in house building, but in the whole building industry), and one does not have to search long or far to witness the results of building operations where it is so apparent that something has not gone according to plan – if in fact there ever was a plan.

12 QUALITY CONTROL

INTRODUCTION 12–2

ACHIEVING THE OBJECTIVES 12–3

TRAINING 12–7

12 QUALITY CONTROL

N. KENT, MIOB
George Wimpey and Co Ltd

Quality control can contribute to more profit and is a sound investment. To achieve the correct level of control, construction staff should display a depth of knowledge in construction technology and be aware of how control is applied. This can be successfully achieved by training courses essentially of a practical nature.

INTRODUCTION

Too often the builder (that is, everyone engaged in the practice of building design, construction and marketing) needs convincing that better profits can be achieved through the application of quality control to their operations. Similarly, even when measures are adopted by the contractor to set up a system of quality control, there is uncertainty that money can be saved. This section indicates the area of work where much can be done not only to enhance a construction firm's reputation but also to establish sound customer-contractor relationships time and time again. Furthermore, most people in the construction industry find dealing with excessive remedial work to be neither a particularly satisfying task nor a rewarding one.

Much has been written on the subject of quality control and quality assurance, and readers can familiarise themselves with its terminology. Two publications readily available for initial reading are BS 4778[1] and BS 4891[2]. The following quote from BS 4891 is worthy of note:

> 'A major task of management is that of persuading personnel to have the correct attitude of mind towards quality; to behave in a manner which is not always 'the easiest' and to remember that 'prevention' is usually less costly than 'cure'. Management for their part can assist this process by ensuring adequate definition and communication of their objectives, leading to the establishment of the programme for quality assurance which can provide the basis upon which periodic reviews and evaluations may be made after taking account of essential feedback.
>
> 'In the majority of small companies overall responsibility for quality assurance rests with the chief executive, but, irrespective of whether quality specialists or quality departments are established, executive management is ultimately responsible for making balanced judgements, assessing the significance of variations in this sphere and taking decisions. In arriving at such decisions the quality and personal integrity of staff are of fundamental importance and management should ensure that each person in the organisation:
> (a) understands that quality assurance is important to his future;
> (b) knows how he can assist in the achievement of adequate quality;
> (c) is stimulated and encouraged to do so (incentives).'

The inference is that management are held responsible for the setting up of a quality-control system within a company. Management need to be aware of the objectives. Simply stated, the objectives of quality control are:

- to give the customer satisfaction
- as cheaply as possible
- within the allotted time scale.

Once these objectives are known, it is the purpose of those engaged in quality control to pursue them in the easiest, simplest and most practical ways.

ACHIEVING THE OBJECTIVES

Selecting the various methods required to achieve the objectives can be considered along the following lines.

An experienced member of the organisation should draft the requirements to be adopted, although in a larger firm various team leaders may want to discuss their own particular contributions towards setting up quality control. It must be remembered, however, that success may not be immediately apparent and that what may prove to be a satisfactory and worthwhile system on one project may completely fail on another scheme. Questions must be asked and answers given to set up even the most basic of systems.

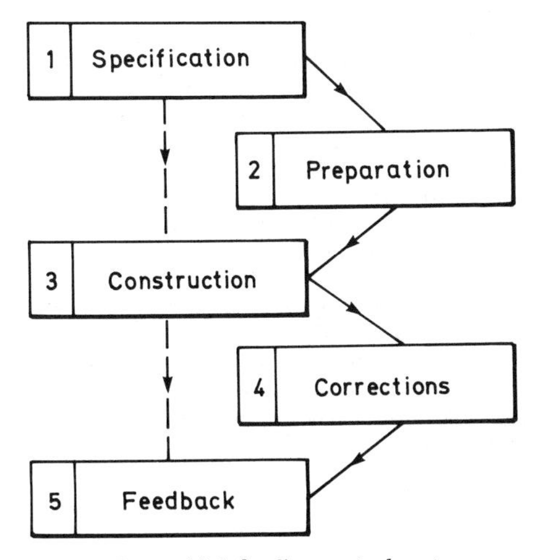

Figure 12.1 Quality-control system

However much quality controls seem to differ, there are many basic similarities. A simple quality control system might be set out as shown in *Figure 12.1*. All stages 1 to 5 are essential, although it will be seen that 2 and 4 can often be overlooked or, at best, skimped. Feedback (5) is then redundant.

Specification

An ideal specification should adequately describe the quality of the materials to be used in the project. The specification should lay down the good practices to be followed in the use of those materials. This is not easy, indeed volumes have tended to be written even for the smallest contracts. It is intended that specifications should be descriptive of how the work is to be carried out to the acceptable standard and finish. What is *not* intended is that doubts should be created in the minds of the contractor and his work force.

To be successful in construction quality control, it is necessary for the person implementing a system of control to have a full understanding of the specification and what is intended. When standards are quoted these should be fully consulted. Where descriptions of works are given, including the finishes to be obtained, agreement on the finishes may mean seeking an early understanding with the client's representative. Although beyond the scope of this section, it should not be forgotten that builders are often consulted on the drawing up of specifications. Great care is needed, however, and much thoughtful work must be done in the production of specifications so that they can be easily interpreted.

Preparation

Planning is given due weight by all engaged in the construction industry. What is not so obvious is that quality-control thinking must play a large part in planning.

Every project can be considered a new business operation, and those called upon to lead the work force must have available all the information on the materials/components/standards. This is the time to examine in detail all that the specification demands, to seek out the various standards, codes of practice, workmanship clauses and the like. In practice, in other than the smallest businesses, this will mean calling on the expertise of many. In fact all who have a contribution to make from their own disciplines should be encouraged to be quality conscious in their own field.

Buyers, planners, engineers and others should be made aware of the system of management necessary for total quality control to be operative at all times. Clear definitions should be given so that no mistake is made in who should coordinate the functions of quality assurance on the site. Similarly, some thought must be given to the deployment of the right operative to work on the project, and there may even be some consideration of new recruitment. Training of both new and long-serving members of the company must be considered in order to obtain the desired standards. Competence in anyone is difficult to detect, the more so with new members; consequently, selection of workforce and training is important in the build-up to any contract. It is not unreasonable to suggest that contractors should have considerable knowledge of their sub-contractors and those whom they directly employ. Such a picture could have been built up over many months and years of association. This is an area, indeed, that will form a basis for sound quality control throughout a project.

Construction

The control of the actual construction should include properly functioning quality control, so that the required standards are achieved. Prevention of errors that can make remedial work costly at a later date is desirable. The advantages of careful preparation, as indicated above, cannot be overstressed.

Decisions will have been made on the various aspects of the work to which control must be applied, and how wide a system is to be adopted. Ignoring design, the following areas can be looked at separately: materials; components; workmanship; total concept.

MATERIALS

Although the quality of all materials will be specified by the relevant British Standards, the number of site tests that can be carried out on materials will be strictly limited. The greatest need will be for accurate sampling and inspection of the various materials that will be received in the course of a contract. Sampling must be followed by skilled packaging of the materials, which must be sent either to the test house of one's choice or to a nominated establishment. Considerable attention must also be given to the labelling and despatch of samples, and to the keeping of true records on site. It is in the interest of the builder to make sure that sampling is carried out by properly trained persons. Many believe that sampling is best done by either the laboratory/test house representative or some technical assistant in the employ of the suppliers. This is wrong thinking: the builder should be just as proficient and should have experience of the collecting and despatch of samples correctly labelled and recorded. What is not acceptable is the taking of any samples without the full knowledge of the contractor's site management.

Sampling may be on a limited scale on the smaller site. Nonetheless it is an important aspect of control, even if the only sampling is of ready-mixed concrete being delivered to the site. It is essential that sampling should reflect no bias; hence the method specified in a particular standard should be accurately followed.

Materials that are commonly sampled on site to comply with the contract specification and also for control are aggregates, concrete, plasters and paints. Other materials that have to be examined for certified markings as having been examined to a requisite standard include stress-graded timber and drainage goods. Note should be taken of any trade association's authorisation scheme that will help in the demand for a material to be of uniform quality. The authorisation scheme[3] set up by the British Ready Mixed Concrete Association provides standards for all aspects of ready-mixed concrete production, including quality-control procedures. Chartered engineers engaged by the BRMCA carry out regular inspections of all depots of the member companies of the trade association. Following the issuing of a satisfactory report upon the inspection, an approval certificate will be granted. The withholding of an approval certificate from any depot failing to maintain an adequate standard for the supply and delivery of quality-controlled concrete may ultimately lead to the deletion of the concrete depot from the approved list of the association.

Notwithstanding these remarks about trade associations and approval schemes, of which the builder should always take advantage, such a scheme is only an agreement on delivery of materials of acceptable standard. Although agreement may have been made to purchase and supply materials from companies within such associations, abuse can still be met on site. In the case of concrete, the material still has to be handled, placed in position and compacted in a satisfactory manner. Similarly with all other materials received on site, examination and testing is only part of the quality control to be exercised. Hence a high degree of cooperation is required between those who supply materials on the strength of an association's approval and those who receive and use the material, albeit in good faith. Remember that, whatever the source of any material, checking to confirm its fitness for the purpose intended is only part of total quality control at site level.

COMPONENTS

Practical difficulties are likely to be experienced in component quality control. Dimensional tolerances are often specified in the design. Such matters are, however, also within the sphere of quality control, and consideration should be given not only to the components themselves but also to the tolerances for fixing. Components can be inspected on arrival and can be physically examined by measurement, weighing and finish, but the matter of dimensional tolerances calls for the experienced site man, who should have had all the facilities of prior negotiations with manufacturers. Knowing the intended purpose of the component, as well as measuring the component against the specification, is essential. Facilities are sometimes available to visit the manufacturer's workshops, although it is doubtful if any useful purpose can be served unless a specific item being fabricated is to be examined because of an in-built fault.

Component control on site demands not only scrutiny on arrival but also adequate storage to prevent further problems. For example, it is well known that excess moisture content in timber components results in movement and distortion. Similarly, plasterboard partitions are frequently found unfit for use through bad practice in storage, much to the embarrassment of site management.

A useful publication that highlights many of these problems is produced by the NHBC[4] and should be read by all site personnel who have to control quality.

WORKMANSHIP

This is probably the most difficult of all aspects of quality control to exercise. An architect will make certain assumptions in the design, and the workmanship should be consistent with the demands of the design. The specification will go some way to describing the requirements, but in the end considerate and careful supervision is required. For good workmanship *strong and positive leadership* is necessary. If samples, such as brickwork panels, have to be constructed for approval, careful particulars should be recorded to make sure that errors are eliminated before the main construction begins. This may seem a small point, but often a sample of work is presented, approved and totally ignored when reproduction is required on the building proper. Although positive leadership is regarded as a major factor in workmanship control, a sound basic knowledge of the trade being supervised is advisable to obtain the best results.

TOTAL CONCEPT

This is often regarded as an inspection for completeness, covering a building in its entirety. Alternatively it can be considered as inspection for the completion of trades in a section of a building, or even on a single floor in the case of flats. It is not intended to be a handover procedure but simply confirmation of the completion of trades, so that the workforce can have clearance and be released on satisfactory conclusion of the work.

Corrections

Complete quality control will include built-in correction procedures. Through the proper channels tradesmen, sub-contractors and others can be alerted to work that needs to be corrected. It is essential for the person carrying out the task of corrections and snagging to have a diary to keep an accurate record of work to be corrected. Such diaries are a valuable asset to any site records.

Typical inspection sheets are issued by the NHBC. Such sheets record who inspected and the date of inspection, and give notes on what to do after completion of the form. They also indicate that works should be inspected systematically and that all should be checked on completion. It does not matter whether the snagging strictly follows the pattern laid down by, for example, the NHBC. What is important is the systematic way in which correction work is highlighted and carried out. Inspecting work will not dramatically improve the quality of the work but it should effectively lead to the final part of a quality-control system, namely feedback.

Feedback

Feedback provides data for the analysis one must make in quality control to see whether certain men or firms have any faults that are repeated time and time again – known as *trends*. An analysis of feedback leads to further decisions in quality control: is it worth it? what steps must be eradicated?

It is essential for management to make a careful study of feedback, since this is a basic requirement of efficient quality control. Among the tasks resulting from efficient feedback, high priority should be given to considering the need for training and the type of training required. No feedback (and many supervisors are loath to emphasise any site problem) will lead only to no quality control, with lowering of standards throughout the site. Every person on site should be encouraged to feed back information on anything likely to impede the attainment of satisfactory standards. The obvious person to coordinate this information is the one who has been given the task of setting up and advising on quality control. This should be the site manager/subagent/general-foreman.

The cost of quality

It is obvious that quality-control systems *cost* money, but not so obvious that a good system can *save* money. The graph in *Figure 12.2* gives some indication of where costs are incurred in quality control. Poor sites can be envisaged as those not applying any quality control or only a minimum amount.

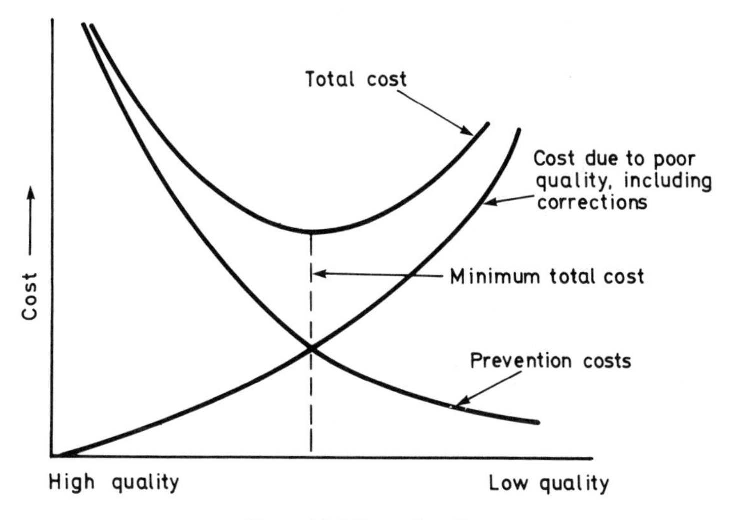

Figure 12.2 Cost of quality

The more prevention applied the greater the cost, but total costs can be shown by adding the prevention curve to the poor-quality curve. This gives a minimum total cost to be aimed at in effective quality control.

TRAINING

To obtain sound quality control, training is essential. This is necessary to reduce incompetence, ignorance and errors, which cause the biggest problems resulting in poor quality on a construction site. Key areas of quality control, where concentration should be focused to improve control on site, can be readily identified. Perhaps the biggest efforts over the last decade have been made in concrete quality control, because the material lends itself particularly well to testing and statistical method. However, many other areas have also proved to be equally adaptable to a system of control.

It is important that training courses, while aiming to cover the whole spectrum of quality control, should concentrate on key elements that can be covered effectively and make a positive contribution. Short courses are available in quality control in the following:

- concrete,
- mortars and renders,
- timber acceptance,
- finishes,
- bricks and brickwork,
- painting.

Many other specialisations can be added to this list. Laboratories may be necessary in teaching the subjects contributing to quality control, but it is important to keep the work both practical and simple.

Ultimately everyone engaged in construction should have participated in some form of quality-control work, the aim being to make everyone responsible for the maintaining of construction standards. Quality control is not static but is a constantly changing process, and personnel should be encouraged to take short updating courses to keep abreast of new methods, materials and standards.

REFERENCES

1. BS 4778, *Glossary of general terms used in quality assurance*, British Standards Institution (1971)
2. BS 4891, *A guide to quality assurance*, British Standards Institution (1972)
3. *Code for ready-mixed concrete, Part 4: Authorisation scheme*, 3rd edition, British Ready Mixed Concrete Association (1974)
4. *Registered house builder's site manual*, National House-Building Council (1974)

FURTHER INFORMATION

Powell, M.J.V., *Quality control in speculative house building*, Occasional paper 11, The Institute of Building

Caplen, R.H., *A practical approach to quality control*, 2nd edition, Business Books Ltd

13 HUMAN ASPECTS OF MANAGEMENT

INTRODUCTION	13–2
RELATIONSHIPS WITH CLIENTS, DESIGNERS AND OTHERS	13–4
MANAGING SUB-CONTRACTORS	13–4
SELECTION AND ENGAGEMENT	13–5
TRADE UNION MEMBERSHIP AND FACILITIES	13–7
HEALTH AND SAFETY	13–9
COMMUNICATING AND DELEGATING	13–11
MOTIVATING EMPLOYEES	13–12
INCENTIVES	13–15
DISCIPLINE AND DISMISSAL	13–16
CONSULTATION: SAFETY OR WORKS COMMITTEES	13–20
FINAL CHECKLIST	13–21

HUMAN ASPECTS OF MANAGEMENT 13

13 HUMAN ASPECTS OF MANAGEMENT

L.G. BAYLEY, BA(Com), MSc, FIOB, AMBIM, AIArb, FRSA
Lecturer in Building, University of Manchester Institute of Science and Technology

This section combines a review of current industrial relations and health and safety legislation with a discussion of the fundamental man-management problems of motivation, incentives and communication.

INTRODUCTION

Human relationships are experienced by each one of us in industry. We believe that we know what motivates us and therefore we are inclined to expect others to mirror our thoughts and aspirations. In consequence we are often surprised, not to say alarmed, when others take actions that are contrary to our own perceptions of industrial life or that indicate a misinterpretation of our thoughts and actions. This situation arises because our own experience is small compared to the totality of relationships. Care and thought are therefore needed to ensure that the interdependent relationships of those at work can flourish to ensure the success of the enterprise. It is worth recalling that while the starter of a motor car may 'kick back', a human being, unlike other resources a company uses, can answer back!

The first step needed to establish healthy industrial relations is to analyse the forces that impinge upon these relationships. *Figure 13.1*, developed in conjunction with Peter Samuel of Rohrer, Hibler and Replogle Inc., shows that every activity affecting a company's operation, even the national government's policies (e.g. pay restraint), affects attitudes and possibly behaviour within a company. The company therefore has two major tasks. First it must provide the motivation, in those matters within its control, to bring about a pattern of behaviour that at worst reduces aggravation and at best aids the achievement of the firm's objectives. The second task arises from forces (shown on the left of the diagram) external to the company. The often harmful effects of external pressures are not easy to offset, but careful explanation through effective communication can reduce the possible disincentive to effort, provided some positive motivation is directed towards those matters that are within the control of the company.

Money is not the only factor that motivates people. There are a number of factors, often neglected, to which importance should be given by management. These are set out later, in the section on motivation.

Every company should set out clearly its industrial relations policy. Hard as it is to do so, committing it to paper provides not only the opportunity to think the problem through but also the means whereby all staff know the objectives. There are therefore certain basic premises upon which the policy must be developed.

1. In view of the wide differences in background, aspirations, knowledge and experience of each person, senior managers must not assume that their own aspirations etc. are shared by their subordinates.
2. Employees' attitudes do not start from the basis that management is to be trusted. Management has to earn trust.
3. The first task of management at all levels is to earn trust through fairness in all its dealings, coupled with strict honesty particularly in respect of information supplied. Deviousness or any sense of duplicity will destroy any feeling of trust.

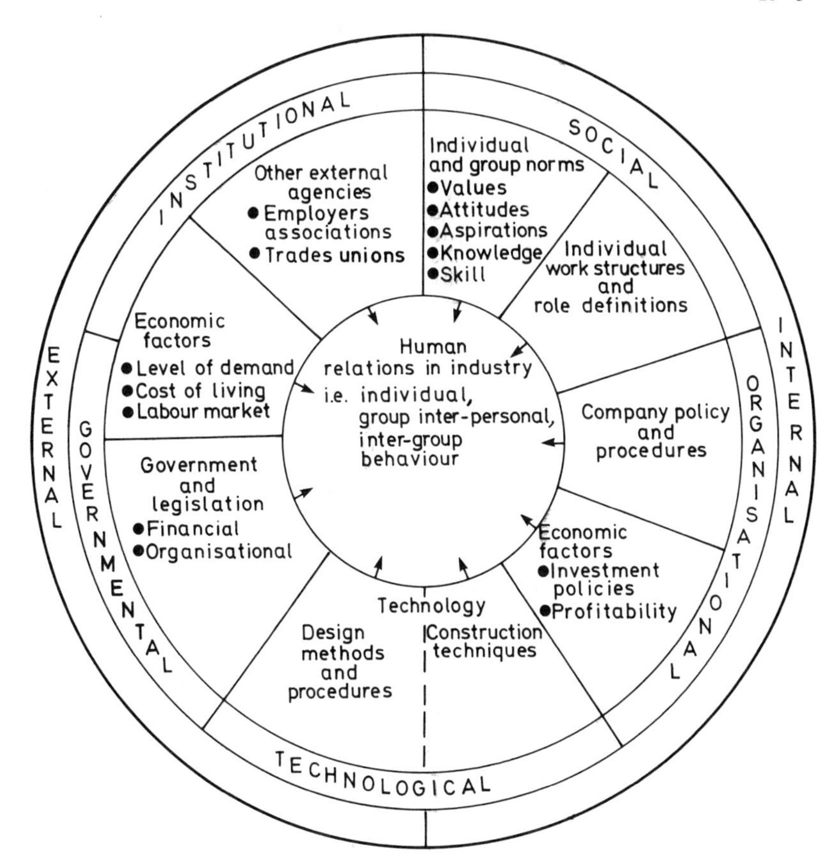

Figure 8.1 Influence of industrial relations in the building industry

4. Consequently management must devote as much time to creating good industrial relations as it spends on such activities as tendering, estimating, production control etc. The Industrial Relations Code of Practice lays the prime responsibility upon management.
5. Management must be prepared to train all supervisors who are in the front line of industrial relations as well as more senior managers.
6. Management must be able and willing at all levels to communicate with, as well as provide information to, its employees.
7. Management must be prepared at times to enter into meaningful consultation with representatives of, or directly with, some of its employees.
8. Management must have clear and precise procedures for exercising discipline and must be consistent in enforcing it.
9. Management must regard the employment of sub-contractors as an extension of its relationships with its own direct employees.
10. Contracting companies cannot ignore their relationships with clients and members of the design team, together with their effect upon site industrial relations.

11. Management must have on hand the relevant Codes of Practice and ensure that every manager is refreshed as to their contents at least annually.
12. Finally, the achievements of any company cannot exceed the aspirations of the person with the greatest power. This is very true of industrial relations.

It is impossible to condense the mass of legislation and Codes of Practice into a small section. In the rest of the chapter these Acts and Codes are borne in mind together with the experiences of successful companies, but in case of doubt reference must be made to the relevant Acts, Codes and (if necessary) such organisations as the various employers' associations, the Department of Employment, and the Advisory, Conciliation and Arbitration Service (ACAS).

RELATIONSHIPS WITH CLIENTS, DESIGNERS AND OTHERS

Formal contracts set out the rights and duties of the parties. Human relationships, however, often determine whether the project proceeds smoothly. To establish sound and healthy relationships certain matters must be borne in mind.

1. It is the client who pays. His needs are paramount.
2. If a promise is made to do something by or on a certain date, do it if humanly possible. If impossible, inform those affected immediately.
3. Architects and other designers do not like asking a client for more money. Avoid being pressurised into footing the bill yourself by warning of extra costs or indicating alternative solutions.
4. Remember that the local authority's and other bodies' inspectorates have a job to do. Give them as much warning as possible of any alterations or of the need for their presence for inspection purposes.
5. Exchange information regularly to help keep each other up-to-date.
6. In the event of conflict with any party, keep the client always in mind.
7. Above all, use all the company's technical knowledge, skill and experience positively for the benefit of the client.
8. Ensure that other parties take an interest in good work by your employees, and get them to express it directly to them.

MANAGING SUB-CONTRACTORS

It must be emphasised that the only differences between nominated and non-nominated sub-contractors is that the former are selected by another party and have special provisions as to payment procedure. There is no difference in management and control. The main contractor must accordingly exercise that control through the powers confirmed in the appropriate contract, and must resist at all costs the temptation to permit special links with architects or others, whereby he is by-passed, even for temporary administrative convenience. A temporary easement of control can turn into a long-term financial noose.

There are certain rules that must be followed.

1. All correspondence, instructions, etc. must pass through the main contractor.
2. Written contracts should be used, preferably using standard forms. If the main contract is not one of the standard forms, it is essential to ensure that the sub-contract is compatible with, and comprehensive in respect of, the conditions of the main contract.
3. Ensure that, when the master plan is prepared, all the operational programmes of the sub-contractor fit in with the master plan. This requires the sub-contractor's agreement, preferably attested by his signature. It is important to commit every party to its fulfilment.

4. It is essential that all revisions to the plans should be notified to all concerned. There is no more damaging effect to relationships than for sub-contractors to arrive and find that the work is not advanced sufficiently for them to carry out their work efficiently. Disputes, wrangles and claims are liable to flow fast.

Employment of 'labour only' sub-contractors Company policy must be clear about the problems of employing 'labour only' sub-contractors, especially when there is a possibility of friction between them and direct employees, whether on a single site or within the company. Should a company decide to use 'labour only' contractors instead of its own direct employees, redundancy is very likely to arise with the extra costs involved. If, nevertheless, 'labour only' sub-contractors are engaged, a written contract in which the rights of both parties are clearly defined is essential. Some sub-contractors (especially self-employed ones) dislike legal documents. Pressure to dispense with written contracts should be strenuously resisted.

The contract must be explicit on the following points.

1. The work to be done.
2. The standard of workmanship to be maintained, and who is to adjudicate.
3. Who is to be responsible for rectifying unsatisfactory work, and how the cost is to be recovered or offset.
4. The amount to be paid, how and when. If it is measured work, then the person responsible for measuring.
5. The amount to be retained and the timing of its release.
6. The basis of payment for variations.
7. The obligation to comply with site safety policy and procedures, and the penalties for breach.
8. Power of the main contractor to exclude any employee of the sub-contractor from the site.

This last point may be essential in the case of a known subversive, after frequent or dangerous breach of the Health and Safety at Work etc. Act (and its associated legislation) or of the Codes of Practice, or because of conduct disruptive to the site.

In managing sub-contractors there are a number of points to watch.

1. Possible causes of friction are differences in take-home pay or hours of working between directly employed and sub-contracted labour.
2. A regular (at least daily) check on standards is essential, together with a check on the possible wastage of materials.
3. The retention money must be sufficient to cover the cost of possible remedial work. But it should not be unreasonably withheld, and all payments should be regular and prompt, otherwise relationships will sour.
4. Avoid arrangements whereby directly employed operatives 'clean up' after sub-contractors. It will cause friction.
5. Avoid at all costs different standards of procedure, discipline and safety controls as between the two types of labour.

To assist contractors with preparing contracts there are standard forms available through the National Federation of Building Trades Employers.

SELECTION AND ENGAGEMENT

No manager should underestimate the importance of a selection procedure for all posts, however humble, within a company. Great care is taken to purchase plant and equipment. The same attention should be paid to personnel selection for the following reasons:

1. There is a substantial cost in terms of time as well as money in engaging personnel.
2. Mistakes in selection, especially when undetected for twenty-six weeks, can be very costly, besides being disruptive to the deployment of labour resources.

3. The failure to screen personnel may lead to disruptive elements gaining a foothold. This is particularly true of large projects, or those sensitive to the national economy or security.
4. The right choice contributes quicker to the success of the business.

Before selection, every post (even for operatives) should have:

1. An appropriate title (not too specific, e.g. carpenter, second fixer, otherwise redundancy could arise later).
2. An accurate job description.
3. A statement of the appropriate working conditions (wages, salary, fringe benefits, etc.).
4. The required qualifications, skills and experience.
5. The appropriate age bracket.
6. Any special requirements (e.g. health).

The above should be included in any advertisement or be readily available should application be made direct to the office or site.

An application form should be used for all posts and contain:

1. The title of the post.
2. Name, address and telephone number (if any).
3. National Insurance number.
4. Date and place of birth, marital status.
5. Name and address of next-of-kin.
6. Education.
7. Training and qualifications.
8. Medical history (serious illness, any disability or anything likely to affect employment).
9. Previous employment – names of the last three employers, jobs held with dates and reasons for leaving.
10. Other information the applicant wishes to record.
11. Whether the applicant has made a previous application. If so, when.
12. A space for applicant's signature below the words 'This information is correct to the best of my knowledge.'
13. Date.

All application forms should be filed with the reasons for rejection, if appropriate, since they may be required to rebut claims for discrimination on grounds of race or sex.

No applicant should be engaged without:

1. An interview.
2. Checking details, particularly items 3, 9 and 11 above.

Item 3 is likely to be a more accurate check on the identity of a former applicant or employee. When checking with previous employers, check (in addition to other information) that the reason for leaving set out on the form is the same as that recorded by the previous employer.

The purpose of the interview is:

1. To clarify any point arising from the application form.
2. To probe in greater depth the applicant's experience, potential aptitude and motivation.
3. To identify possible favourable or unfavourable traits likely to affect performance.
4. To enable the applicant to query any point in the proposed conditions of service.

The draft terms and conditions of employment should be available at this stage (omitting the employee's name).

The interviewer should give the applicant a rating between 1 and 10 for the following points and then make a judgment between the various applicants.

1. Experience required for the post.
2. Motivation.
3. Potential for promotion.

4. General intelligence.
5. Personal circumstances (domestic or travelling, which might affect performance).
6. Disposition – whether he will fit in well with his colleagues.
7. Other interests (some interests, such as trade union officer, civic duties, etc. will require absences from time to time).

Having decided to offer the post, and before doing so, the following should be shown to the applicant:

1. The 'Statement of Conditions of Employment' including appropriate documents such as working rules.
2. The disciplinary code of the company with the grievance procedure.
3. The company's safety policy.

If the offer based on the above is accepted, hand the requisite documents, duly completed, to the new employee for his retention.

Although the law requires that the 'Statement of Conditions of Employment' must be given to the employee before the end of thirteen weeks from the time of engagement, the above procedure minimises the chance of any contention as to the substance of the Contract of Employment, besides ensuring that the Statement is not overlooked. It establishes a fixed procedure for dealing with all applicants.

Immediately the applicant is engaged, arrange for any occupational and/or safety training as required.

The 'Statement of Conditions of Employment' must contain the following items, some of which may be covered by reference to supporting documents.

1. The names of the employer and employee.
2. The date when employment began.
3. Whether any employment with a previous employer counts as continuous employment, and if so the date when continuous employment began.
4. The title of the job the employee is employed to do.
5. The scale or rate of pay and the method of calculating it.
6. The intervals at which payment is made (i.e. weekly or monthly).
7. (a) hours of work (including normal working hours);
   (b) holiday pay including public holidays – showing how it is calculated and any accrued payment due on termination of employment;
   (c) sickness or injury, including any provision for payment;
   (d) pensions and pension schemes.
8. Length of notice on each side.
9. The date of termination if it is for a fixed term.
10. A note specifying:
    (a) the disciplinary rules applicable to the employee;
    (b) the person (preferably by designation) to whom the employee can apply to redress a grievance (including dissatisfaction with a disciplinary decision) and how the application should be made,
    (c) the subsequent steps in the grievance procedure.

Organisations such as the National Federation of Building Trades Employers have standard forms available.

There are two dont's: *don't* engage at the gate, and *don't* skimp the selection procedure.

TRADE UNION MEMBERSHIP AND FACILITIES

The building industry has very few companies with closed-shop agreements. Most trade unionists are found in the larger companies. However, discrimination between union members and non-unionists must be avoided, otherwise a company exposes itself to considerable penalties.

There are occasions, however, in non-union companies when group representation is appropriate. Management should make arrangements in such cases for formal discussions. When a union official makes representations on behalf of a company's employees, and the management considers that there is a discrepancy between the view expressed and their understanding of their own employees' attitudes etc., every step must be taken to establish the true facts, because misunderstandings can have serious consequences.

However, should a company have employees who are union members, there is a procedure for trade-union representation laid down on the NJCBI agreement, which should be followed (Rule 7). These provisions (marked *), together with the relevant Code of Practice, are:

1. A full-time official should be allowed access to site etc.,
   (a) to carry out his union duties,
   (b) to ensure observance of the Working Rule Agreement.
2.* The employer should not unreasonably withhold recognition of a steward appointed and issued with credentials by the union. The steward should have at least four weeks service with the firm.
3.* Meetings of stewards or operatives need prior permission of management if they are to be held within working hours.
4. Stewards should be notified of written warnings and dismissal notices issued to their members.
5.* The steward should be permitted to accompany his member in making representations of any kind to management.
6. Every unionist is entitled to be fully active in his union outside working hours and any dismissal arising from his right to do so is unfair.
7. If an employee refuses to join the appropriate union specified in a closed-shop agreement he may be fairly dismissed unless there are genuine religious objections held by that employee.
8. Discrimination between union and non-union members because of membership of a union can lead to complaints to an industrial tribunal. Successful complaints in the case of dismissal can be very punitive to a company, including a stay of dismissal until a tribunal hearing.

If a firm is having difficulties in assessing what action should be taken, advice should be sought immediately from the firm's employers' association or ACAS. It is worth remembering that only a few trade unionists and their officials are subversive, and until there is clear evidence that they are they should be treated in a manner similar to any other employee or person making representations. If, however, there is strong evidence that the motive is disruption, firms must act according to the book to avoid trouble. It should be remembered that once a genuine or potential cause of grievance is removed, the ground is removed from under the antagonists' feet. There are two golden rules:

1. Management is at all times entitled to talk to, discuss with and directly inform employees of its decisions, proposals and intentions. Since there is a contract between the two, it should be used more frequently than is customary.
2. Management should never use stewards to convey management decisions etc., however strongly stewards may suggest otherwise. Stewards are not bound by any constraints to pass on management's views in the manner that management desires. They are responsible only to their members. They form the part of the representative system whereby the views of their constituents are collectively presented or whereby negotiations take place on their behalf.

If trade unions are recognised and stewards appointed, certain measures need to be taken as a matter of policy by the management in the light of the Codes of Practice.

1. There should be an agreed arrangement for time off with pay to undertake specific trade-union duties within the company, such as recruitment and main-

taining membership, subscription collection, handling members' grievances, negotiation and consultation.
2. There should be a procedure for obtaining permission to do these duties, and a special procedure in respect of trade union duties outside the company (e.g. membership of industrial conciliation panels etc.).
3. There should be an agreement in respect of meeting place and times, access to telephone, the use of notice boards and, where there is a volume of work for the steward, use of office facilities.

Some companies with a large number of employees have made a practice of having regular meetings (often half-yearly) with leading Regional Officials of the union relevant to their employees, with the prime object of informing the unions, sometimes in confidence, of future programmes and development, besides obtaining feedback from the unions. Firms with many years experience of consulting at this level find it smooths the way for instituting procedures and practices that without consultation could lead to industrial action. There is evidence that it encourages a cooperative attitude by the unions towards the company.

HEALTH AND SAFETY

There are two human aspects of this subject: the first is the pastoral one of caring for all employees in such a way that they do not expose themselves to either illness or injury, and the second is to ensure that all levels of management, in particular supervisors, are sufficiently equipped with the requisite skills on all safety matters. Already a number of individual foremen have been fined because employees under their charge have been injured.

The first duty of management is therefore to ensure that all equipment is used solely for the purpose for which it was designed and in the prescribed manner with the appropriate protective equipment. The second is to ensure that all supervisors have an effective back-up system to help them discharge their duties in respect of safety.

One fact is paramount. The appointment of a safety officer in no way diminishes any manager's, supervisor's or employee's duties in law for these matters. Further, no delegation can absolve the chief executive of his ultimate responsibility in law. The Health and Safety at Work etc. Act 1974 extends responsibility for those outside the immediate control of a contractor, such as passers-by, future occupants, and visitors to the site (not forgetting the client, who might find at a much later date a defect in workmanship or materials that could give rise to danger, injury or ill-health). Clear procedures are therefore needed.

Every sub-contractor has exactly the same duties and responsibilities as the main contractor in his capacity as an employer; however, the main contractor has a responsibility in regard to all his sub-contractors under the Occupier's Liability Act 1957, in addition to his express function as a supervisor of construction work under the Health and Safety at Work etc. Act 1974 (HASAWA). It is therefore imperative that the safety policies and procedures, including the appropriate disciplinary codes, should apply throughout the site. This may be difficult, but differences in treatment for different employees will be liable to cause friction. A typical example is the different policies towards the use of safety helmets by various contractors on many sites.

Building employees could be exposed to health risks due to the use of materials such as lead paint or radioactive substances (there are others specified in various regulations). Such an employee may have to be suspended on medical grounds if there is no other suitable employment free of risk. Such suspension will entitle the employee to be paid for up to twenty-six weeks, provided he has been continuously employed for four full weeks. In any case of doubt, consult ACAS. During such suspension it will be fair to dismiss a replacement on termination of the suspension, provided this arrangement was incorporated in the contract of employment.

The regulation governing *safety representatives and safety committees* comes into effect on 1 October 1978. These regulations form part of the HASAWA. It is possible that a trade union might appoint a safety representative from amongst the employees. If the union gives written notification of its appointee, that appointee must have been employed in his occupation for two years with one or more employers. If this arises, the company should consider the following, bearing in mind that a safety representative must be paid his normal pay when undertaking such duties.

1. Whether the representative should represent more than one group of employees.
2. Whether the representative should cover more than one site, and if so how many.
3. The provisions and arrangements for consultation with representatives on promoting and developing safe and healthy methods of working.
4. The procedure for the investigation by the representative of potential hazards and dangerous events.
5. The procedure by which the representative makes representations concerning complaints by employees regarding health and safety.
6. The procedure for the representative to inform management of his desire to exercise his right of inspection of the workplace, whether routine (normally not more frequently than every three months), or following a notifiable accident, etc., as well as management's arrangements to accompany him on such inspections.
7. Management's own attitude towards safety representation. If the attitude is one of hostility or even indifference, management may produce an unfavourable reaction from the representative; due to the ever-changing nature of building work, this could dislocate site works from time to time through frequent inspections. A cooperative attitude (particularly at supervisory level and immediately above) is more likely to bring a reasonable response.

It is not the purpose of this section to provide a complete and detailed guide to various aspects of safety procedures. However, the 'Further Information' list at the end of the section includes those publications, current and anticipated, that every company should possess. A list of action areas is set out below so that management can double-check whether it is fulfilling those tasks necessary to fulfil its duty in law. Reference is made in the sub-section on discipline to the need to ensure that the non-observance of safety matters is appropriately included in disciplinary codes and procedures.

The safety action areas include:

1. Safety Policy Statement.
2. Appointment of Safety Officer(s).
3. Thorough inspection of plant, vehicles, excavations, protective clothing as prescribed under the regulations.
4. Safe storage and labelling of inflammable and explosive substances.
5. Provision of safety and health information to employees.
6. A training programme for new employees and all supervisors either in-company or at special centres.
7. Provision of sufficient protective goggles, clothing and headgear, and rules for their use.
8. Disciplinary rules in relation to safety.
9. Welfare – changing rooms. canteen facilities.
10. Health – latrine, washing facilities, first-aid boxes and first-aid training.
11. Reporting procedures.
12. Health records of employees – reasons for sickness.
13. Visitors – obeying rules, adequate protection.
14. Consultation scheme as required by the regulations.
15. Observed safe systems of working.
16. Cleanliness and tidiness on site and in workshops.

17. Common safety policy procedures with all sub-contractors on site.
18. Location of and access by employees to safety literature.
19. Checks on medical hazards.

COMMUNICATING AND DELEGATING

There is a difference between communication and information, and between being able to communicate and the system of communicating. *Information* presupposes that communication is functioning, and is an aid in the task of communicating.

Communication in business at any time has one of two objectives.

1. To provide the recipient with directives as to the action required by the sender.
2. To provide the recipient with information to enable him to make decisions or to aid him in adopting a particular attitude. This can arise where there is:
   (a) delegation of responsibility and/or authority, or
   (b) a desire to inform one's colleagues either as a warning or for counteracting gossip, or
   (c) a desire to inform one's superior.

There is, regardless of the objective, an intention by the sender to influence the recipient. Success therefore depends upon whether the sender gets the message across to the receiver in such a way that the latter acts in accordance with these intentions, which may be clear or partly hidden in the spoken or written word. No communication can operate without a listener or reader, so it is the receiver who is the key to the communication process.

The effectiveness of any communication depends upon factors that can differ substantially between sender and receiver. They are:

1. Individual needs, hopes and fears.
2. Attitudes and values both at work and outside.
3. Individual's background and experience.
4. The environment of the communication – whether in front of others, etc.
5. The knowledge of the subject matter.
6. Built-in prejudices.

The biggest obstacles to communication in business arise from gaps in experience and different expectations. Only when these two differences are narrowed during the communication process can there be a real understanding of the purpose of the communication. Accordingly it is the receiver who is the effective communicator, the sender merely utters!

There is an assumption that downward communication will be accurate. If it isn't the junior will have the excuse of being misled. But in the case of upward communication, the whole truth can be distorted or even withheld as a result of a number of considerations.

1. Does the boss need to know?
2. What does he need to know?
3. Might he hear from someone else?
4. What effect will it have on our relationship, my job, etc?

An effective communication system depends upon the state of inter-personal trust. Like all trust, it is not a gift, it has to be earned. Effective communication is therefore bound up with relationships, and these differ as between a superior and his subordinates. It is easy for a single sentence or instruction to lead to industrial action, however minor! This can arise because the factors of experience, perception and expectation of the parties have not been reconciled within a framework of trust or through the medium of communication.

Delegation

To effect delegation, the communication must be both precise and clear. Every task has a *prescribed* content as well as a *discretionary* one, and these two components must be clearly defined otherwise delegation becomes inadequate. Delegation can break down very easily, especially when:

1. A senior manager wittingly, or more often unwittingly, himself exercises the authority that he believes he has delegated. This is a common experience at contract and site-management level.
2. A junior, unsure of his ground, may refer a matter upwards for advice or information, whereupon the senior may frame the advice/information as an instruction, with the result that the junior for reasons of personal security ceases to exercise such discretion himself.

These situations are common when senior management withholds information from a junior, typified by the senior who asks 'Is there anything you want to know?' when the question is best framed by the junior as 'Is there anything you consider I need to know?'

Remember that delegation is a powerful motivating factor especially when the discretionary element is considerable. It should always be the aim of senior management to extend the discretionary element through planned experience and training.

Checklist when communicating

1. Have I given all the appropriate facts?
2. Have I differentiated between facts and opinions?
3. Is there anything in our respective *backgrounds and attitudes, knowledge, skills, expectations, experience or perceptions* that is a barrier to understanding the communication? Identify and explain.
4. As a receiver, is there a *clear* reason behind it and a *clear* objective, as opposed to conjecture? If not, clarify.
5. Is the right method of communicating being used – verbal, written, notice, letter, etc? Check.
6. What provisions have been made for employees to communicate upwards? Provide channels wherever possible.
7. Has delegation been precise in terms of prescribed duties and discretionary tasks?
8. Having delegated, have the discretionary tasks remained effectively with the junior employee?
9. If discretion has been clawed back, what steps have been taken to re-delegate?
10. Can the area of discretion be enlarged?
11. In the event of a misjudgment, has someone checked whether the decision maker was in possession of all the facts? If not, why not?
12. Never use shop stewards for communicating management decisions. Use management's chain of command.

MOTIVATING EMPLOYEES

There is one major difference between money as a motivating force in a company and other forms of motivation. It is that money is only useful outside the workplace whereas the other factors that influence industrial behaviour are in-company, at the work interface. Money is important, of course, and the place and value of money incentive as an inducement to effort are dealt with later.

There are two distinct, but often overlapping, areas of industrial life that tend to condition behaviour. The first includes those actions or inactions that tend to stultify performance in some way; the second includes those that lead to a sense of well-being leading to enhanced performance. They sometimes affect individuals, a gang, a whole site or even a whole company. Setting aside money, they are:

1. Those likely to cause aggravation:
   (a) company policy, procedures, administration;
   (b) the quality of supervision;
   (c) working conditions and associated facilities;
   (d) relationships – with others on the site or in the office.
2. Those likely to motivate towards greater effort:
   (a) achievement of personal or group goals;
   (b) recognition as an individual;
   (c) work itself – how it is carried out;
   (d) responsibility – no matter how small.

It is easier to tackle the aggravation problems than the second group. They are the chief causes of frustration, and once tackled they remove some of the brakes on production; wisely handled they can smooth the way for other more positive action to supplement any effect monetary inducements may have.

Company policy, procedures and administration Companies seldom spell out their policies. Management often assumes universal knowledge for some reason or other! How many companies explain their procedures, especially when new ones are introduced? Often such innovations change work patterns and call for additional work. Why not tell juniors the reasons? It will do two things: show juniors that management has an objective in mind, and provide management with a yardstick against which they can test the achievement of the stated objective. Further, if there is a reliable and respected feedback, juniors might suggest improvements, which they could not do if they did not know the reason for it.

Quality of supervision Management must always remember that they are judged by the quality of their appointments, especially the immediate supervisors of work groups. No-one likes bad or indifferent supervision. Research shows that the quality of supervision at foreman level is a major determinant of morale and that poor morale is closely related to absenteeism, whether voluntary or due to short-term sickness.

Inter-personal relationships Work is a social activity for most people, where likes and dislikes of colleagues arise. People work better with those whom they respect or like. Positive efforts are needed to avoid placing socially unacceptable persons in certain gangs or management teams.

Working conditions Measurement in productive terms of improvements in working conditions, amenities etc. is difficult, but a happy and contented workforce is likely to work better than a disgruntled one. The site is never as congenial in its surroundings as an enclosed factory protected from the elements, so attention has to be paid to providing as much protection from the elements as possible and countering adverse conditions with suitable facilities for change of clothing and welfare. It was once stated that there was a correlation between the number of times nurses smiled and the speed of recovery of their patients. How happy are the company's sites?

Achievement A sense of achievement presupposes personal or group goals. 'Management by objectives' provides the philosophy for creating a sense of achievement for both an individual and the company. The more the capacity of an individual is stretched, the greater the sense of achievement. The use of monetary targets as a measure of achievement can lead to pressure to reduce targets without increasing performance.

The completion of a project can bring a sense of achievement, but it must not be coupled with a sense of insecurity of employment. Management must counter this potentially disruptive effect by having a positive policy for transferring employees at the end of their stint on a project. This requires early notification of the next place of work; leaving the situation open to rumour or surmise is inadequate.

Pride in achievement is provided in some firms by visits to completed projects in conjunction with the clients.

Recognition Many have experienced being shunned and ignored, yet positive efforts are often needed to recognise people and their worth. Formerly the boss knew everyone by his or her Christian name, yet today bosses or middle management seldom take the same trouble. In the building industry the design team seem content to leave comment to others. Site managers as a matter of policy should request architects and consultants who comment favourably on workmanship to make them face to face with those involved. Recognition of good workmanship enables the workforce to accept criticism when due. It often reassures workers that management knows the difference!

It is not only praise or blame that gives rise to the motivating effect of recognition. Most people at one time or another like to be asked for advice. It is sometimes appropriate to ask for advice from juniors even when the answer is known, if by doing so the person asked is spurred to greater effort. It is also true that management may not always know the practical problems or the best way of doing things. If more effort were directed towards developing workers' skills and their contribution to the overall success of the business through recognition, a powerful motivating force would be released. This is closely associated with responsibility.

Responsibility Responsibility is normally associated with managerial function which should be delegated down the line of command to assist in providing greater job satisfaction to junior staff. However the hallmark of a craftsman is his acceptance of responsibility for the quality of his work. This can also be true of semi-skilled operatives and labourers. Management must therefore instil such a sense of responsibility; delegation, which was discussed earlier, with its discretionary element is a powerful motivator.

Checklist of steps

1. When changing company policies and procedures, tell those affected the objectives and reasons for the change.
2. Check administrative procedures for simplification.
3. What use is made of the various returns? Are they vital for decision-making? If so, keep. Are they only interesting? If so, cancel.
4. Can something be delegated? Does someone need training before delegation can take place?
5. When did you last check the rate of absenteeism? Is it increasing? Where? Why? Is it too high already? Investigate.
6. Is there an odd-man-out in a team or gang? Can he be moved elsewhere?
7. Do team leaders and gangers have a sufficient say on the composition of their team or gang?
8. Does the site appear contented or sullen? If sullen, what steps are being taken to find the cause?
9. Is management proud of its projects? Does management share its pride with its employees?

10. Is there a positive policy on transferring employees on completion of projects? Does management let the employees know where their next job is as early as possible?
11. When was a worker last praised for good work by management, by the architect?
12. Can employees be encouraged to show how improvements can be made?
13. Does management ever ask for advice from employees on how things might be done better and more efficiently?
14. Do your craftsmen require strict supervision in respect of the quality of their work? What is management doing to ensure that each employee takes responsibility for the quality of his or her workmanship?

INCENTIVES

Monetary incentives are normally assumed to be desirable for manual workers, but they can be provided for non-manual workers as well. Some companies operate a productivity bonus for the site and some form of profit-sharing for others. The crucial question every company must ask itself is: Is an incentive scheme a bribe or a reward? The two are different, reflecting the attitude of management to its employees. The NJCBI lays down in its working rules a statement of general principles for incentive schemes, the basis upon which they should operate as well as machinery for settling disputes about such schemes.

The first important thing to remember about every incentive scheme is that its motivational effect will diminish over time unless it is changed. The lifetime of every scheme is limited to between five and seven years, after which it becomes the norm in terms of the earnings level expected. If the bonus earnings fluctuate excessively they cause friction. So every company, whether operating an incentive scheme or proposing to do so, should review the situation in the light of a number of considerations.

Aims of an incentive scheme

1. To provide a simple and comprehensible scheme whereby employees can identify feasible targets with the payment due on attainment.
2. To provide a company with a system that reduces unit costs.
3. To provide increased earnings for increased effort.
4. To encourage new and improved methods of working.
5. To integrate management and employee goals.

The success therefore depends upon the achievement of these aims, which is most likely to occur when the level of output at inception tends to be low.

It is essential that management takes care during the currency of an incentive scheme to watch out for a number of factors that could undermine its overall effectiveness. Warnings of potential or actual failure arise when:

1. There is a lowering of the standard of workmanship. This is indicated by the amount of remedial work undertaken by non-bonused employees.
2. Bonus earnings fluctuate wildly.
3. Absenteeism increases.
4. Management uses the system to signal possible management action, e.g. shortage of supplies etc.
5. Extra earnings reach 40 per cent of the basic rate.
6. Special bonuses have to be introduced to ensure completion etc. This is especially true when 'standing bonuses' are paid.
7. It destroys the corporate activity needed to overcome mutual difficulties.
8. Disputes on bonus payments become frequent.
9. Management no longer controls output, it being in the hands of employees.

10. Wastage of materials runs at a high level or increases.
11. There is increasing friction between bonused and non-bonused employees.
12. Supervisors and junior managers consider that their differentials are being or have been eroded.

Once these signals of danger appear, management must review the scheme before it destroys a healthy industrial relations climate. The action required is:

1. A complete revision of the scheme (possibly its abandonment), as opposed to minor adjustment.
2. A review of management's action in operating the scheme. Management could have become lazy and left too much of the organisation to the employees themselves.

Some companies turn to 'labour only' sub-contracting as an alternative to introducing an incentive scheme themselves for normal workers. The operation of 'labour only' sub-contracting is discussed separately. However, whatever scheme is adopted, particular attention needs to be drawn to (a) the maintenance of standards of workmanship, and (b) the provision for rectification of shoddy or unacceptable workmanship.

Incentives for staff

Staff incentive schemes usually take the form of annual bonuses. However, unless the staff know the details of the company's aims and its achievements, they will not be motivated and soon the bonus will be expected as a right. There are two basic types of scheme: (a) based on earned profits, (b) based on value added.

Both types require that management is open and frank about the company's finances. It is possible to use value added as a basis for earned bonus. It can be distributed monthly. Some such schemes operate in manufacturing industries, not only for the staff but for the whole company or plant. For such a scheme to be effective there is a need to create a participation structure to direct the goals of the employees towards a common aim – namely the progress of the company – from which all derive benefit. No company should introduce any incentive without thorough preparation.

DISCIPLINE AND DISMISSAL

The law provides protection against unfair dismissal after twenty-six weeks of employment; it is therefore unsound to provide two standards, one for those with less than twenty-six weeks employment and another for those with longer service. If you do, someone could slip up and cost the company money. With sites widely dispersed and a greater decentralised management structure than in most industries, all who have the duty to exercise discipline should be aware of the extent of their powers and the company's standards. Aids to a good industrial relations climate are:

1. A clear statement of company policy and objectives in relation to the employee's role. This is a valuable aid to management in clarifying its thoughts and attitude.
2. A clear statement of employees' tasks, targets and standards.
3. A positive manpower policy covering transfer on project completion.
4. A clearly defined and understood disciplinary code of procedure.
5. An effective appeals system associated with it.
6. A thorough understanding of the appropriate Working Rule Agreement.

The essential features of a disciplinary procedure are contained in the appropriate Code of Practice issued by ACAS.

Disciplinary procedures should:

1. Be in writing.
2. Specify to whom they apply.

3. Provide for matters to be dealt with quickly.
4. Indicate the disciplinary actions that may be taken.
5. Specify the levels of management that have the authority to take various forms of disciplinary action, ensuring that immediate superiors do not normally have the power to dismiss without reference to senior management.
6. Provide for individuals to be informed of the complaints against them and to be given an opportunity to state their case before decisions are reached.
7. Give individuals the right to be accompanied by a trade union representative or by a fellow employee of their choice.
8. Ensure that, except for gross misconduct, employees are not dismissed for a first breach of discipline.
9. Ensure that disciplinary action is not taken until the case has been carefully investigated.
10. Ensure that individuals are given an explanation for any penalty imposed.
11. Provide a right of appeal and specify the procedure to be followed.

The disciplinary code should cover certain standards and sanctions and be known to all. It should cover:

1. Failure to obey an executive instruction.
2. Breaches of Construction Regulations and company safety policies.
3. Breaches of company procedure.
4. Poor behaviour and conduct in relation to clients, professional design staff, clerk of works, local-authority inspectors and others.
5. Poor standards of attendance.
6. Abuse of arrangements for intermittent sickness, absences, etc.
7. Providing false information.
8. Failure to perform satisfactorily.
9. Misuse of company plant and materials.

Some offences are graver than others, but where summary dismissal is appropriate (e.g. action that endangers life) it should be clearly indicated, as well as in those cases where a single final written warning is appropriate.

This means that management must lay down:

1. Who has the power to warn.
2. Who issues the final warning.
3. Who has the power to dismiss.
4. The keeping of adequate records of disciplinary action.
5. Who should hear appeals.

Remember:

1. It is in order to give verbal reprimands.
2. Warnings may be expunged over time.
3. Final warnings should be in writing, with a copy to the employee's shop steward or union if appropriate.
4. It is dangerous to create precedents. If it is imperative, the reasons should be carefully recorded in case they are queried later.

Dismissal

The period of notice by either party will be set out in the appropriate Statement of Employment or associated documents.

Dismissal is the most serious sanction available to management. Tribunals look carefully at the following.

1. Was the reason justified in law?
2. Was the manner in which the employee was dismissed fair?

It is easy for personal bias or prejudice to enter into consideration in a one-to-one situation, so every care should be taken to ensure that a person senior to the affected manager should hear the case and adjudicate upon it. Decisions must be referred in this way, if the supervisor/manager is not to have his authority undermined.

There are two important aspects of dismissal not referred to earlier, namely constructive dismissal and dismissal for 'an inadmissible reason'. Constructive dismissal, regarded as unfair, occurs when the conduct of the employer leaves his employee with little or no alternative but to resign. Dismissal for an inadmissible reason is likewise unfair, but there is no qualifying period of continuous service in this case. It occurs when the reason for dismissal has been determined to be:

1. That the employee is or proposes to become a member of an independent trade union, or is engaged or proposes to engage in its activities.
2. That the employee refuses to belong to a non-independent trade union.
3. That the employee refuses to belong to an appropriate union in a closed shop because of genuine religious objections.

The procedure leading up to dismissal is set out in the Code of Practice. If it arises through conduct or performance, the disciplinary procedure should enable the matter to be dealt with fairly. However, care must be taken in the event of ill-health or redundancy.

Dismissal arising from sickness or injury

Certain tasks in the building industry should not be undertaken if an employee suffers from certain sickness or injuries. Hence it is important that the applicant, prior to engagement, should make a written declaration on these matters. Once engaged the law (HASAWA) places a heavy duty on employers to safeguard the health of employees.

The Building Industry's National Working Agreement contains provisions for payments in the event of absence through sickness or injury in Rule 8. Some companies supplement such payments to encourage a sense of loyalty. In the case of other employees some companies provide benefits often diminishing with extended absence.

A person absent for whatever reason is still regarded as being in the company's employment unless there is a serious breach of contract of employment. However, it is essential to consider the employment position in the case of long absence. Considerations include the following:

1. How long is the absence likely to last? If a return to work is possible within the foreseeable future, a tribunal may regard it as unfair to dismiss.
2. If the employee is able to return to work, will he or she be able to fulfil his or her original duties? If not, is there another task available either temporarily or permanently? It might necessitate a revised contract of employment.
3. Is there a query over possible fitness? Consider a separate medical examination by the company's doctor.
4. Is the employee near retiring age? Would it be possible to effect early retirement?
5. Is there redundancy in the company? If it is company policy to consider health as a factor in selection for redundancy, this might be appropriate.
6. Is the job being filled when there is dismissal for reasons of ill-health? If it is not filled, there could be a claim for redundancy.
7. Tribunals are concerned that dismissal for ill-health is not premature; therefore, if a sickness scheme still has some time to run when dismissal takes effect, they are likely to question the reasoning very closely.

Lay-offs: shortage of work, exceptional weather, etc.

Management must take one of three courses once normal working is no longer possible:

1. To provide alternative jobs.
2. To put employees on short-time or lay them off.
3. To dismiss employees, i.e. make them redundant.

The power to lay off employees or put them on short time must be contained in their Contract of Employment. Any provision for guaranteed payments during short-time or lay-offs must be honoured; any relaxation during a period of suspension or lay-off will be viewed with extreme scepticism by a tribunal. The Building Industry's National Working Rules provide for lay-offs in Rule 2A 2.3.

If the decision to lay off or introduce short-time is made, an employee may under certain circumstances decide to resign and claim redundancy. To effect this, the following timetable is applicable.

Stage	*Event*	*Time limit*
1	If at the end of 4 weeks consecutive employment, or at the end of 6 weeks within a 13 week period, the employee has been laid off or on short-time or a mixture of both, Stage 2 can operate.	–
2	Employee gives written notice of intention to claim redundancy.	Within 4 weeks of Stage 1.
3	Employee gives written notice of termination of employment.	Within 4 weeks of Stage 2.
4	Employer entitled to submit counter-notice in writing stating intention to contest liability, if he can envisage a period of 13 weeks continuous employment starting from 4 weeks after the serving of the notice in Stage 2.	Within 7 days of Stage 3.
5	If, however, work does not materialise within the 4 weeks referred to in Stage 4, the employee is redundant and (if qualified by age, service, etc.) can claim redundancy payment.	

Before accepting the redundancy claim, check that every stage requirement has been met.

Dismissal due to redundancy

Redundancy can arise from the completion of any project where a person is dismissed and not replaced. However, it is important to distinguish between the fairness or otherwise of dismissal for reasons of redundancy and an entitlement to payment for redundancy. The latter procedure should now be well known, but if in doubt consult the Department of Employment.

Redundancy, if new to a company, can harm industrial relations, but it is important to give as much advance notice as possible of transfer, in order to provide a feeling of security and to prevent the falling-off of production at the end of a contract.

A positive manpower policy, including effective selection procedure and a sound industrial relations programme, should reduce the incidence of redundancy. If it arises there are two questions to be answered.

1. Can the employee be transferred to another site or depot, etc? To do this, there must be a clause in the Contract of Employment (viz. the Building Industry's National Working Rule 2B) or by agreement. If it is in the contract, quote the clause when making transfer arrangements.
2. Is the selection procedure for redundancy fair? This poses additional questions. (a) Is there an agreement on redundancy selection? If so, follow it.

(b) Has every consideration been taken into account, and would a third party accept that the criteria chosen are both fair and reasonable in the circumstances?

'Last in, first out' is not the only choice (unless in an agreement). It is possible to consider: conduct, attendance record, age, potential for advancement, performance, health, length of service, attitude to the company, and other matters.

Procedure in the likelihood of redundancy

The law requires notification to the appropriate unions and the Department of Employment (90 days notice before first dismissal if 100 persons or more are involved on any site, 60 days notice if 10 or more are involved). There are standard forms available from employers' organisations. However, to regularise the situation a monthly return setting out possible redundancies seems to be the only administrative way to avoid action against the company. It must be remembered that, if representations are made by the unions concerned, a reply in writing is necessary setting out the reasons for rejection, if applicable. For further points consult with the Department and ACAS.

Once dismissal for redundancy occurs, the employee concerned, if qualifying for redundancy payment (two years service) is entitled to time off to look for another post with payment up to two-fifths of one week's pay for the whole notice period.

Checklist following dismissal

1. Record the reason for dismissal.
2. Make sure that it is the true reason – it can be challenged before an industrial tribunal.
3. Use that recorded reason in all correspondence and communications.
4. If requested by the dismissed person, supply the reason in writing unless special circumstances exist, such as those highly damaging to the person concerned.

CONSULTATION: SAFETY OR WORKS COMMITTEES

Safety committees may have to be established by a company: see the Health and Safety sub-section. Some companies may have works committees. Whatever purpose such a committee has, it is dealing with the whole problem of consultation. Consultation can be a traumatic experience and requires very considerable patience by management. It should be developed and not rushed, since consultation (as opposed to negotiation) is a new experience to many.

The following points can be made regardless of the type of consultation proposed.

1. Consultation does not mean management by committee.
2. The top management must believe in it, must be seen to believe in it, and therefore must demonstrate that it has real respect for the individual's opinions, besides having a capacity for fair dealing.
3. Management must have a willingness to accept criticism and a readiness to learn from such criticism.
4. Consultation should provide a framework within which management is free to manage.

Once these facts are accepted and it is proposed to establish some form of consultative machinery, certain steps are needed.

1. The aims of the consultation process must be set out with the topics to be covered. The number of topics can be increased from time to time.

2. Management's representative must be the person who has the power to make decisions on the subject matter. This invariably means the managing director. Unless this happens all consultation will fade away. This is the basic reason why so many works councils have been ineffective. However, the managing director need not be the chairman.
3. Adequate representation must be effected at all levels within the company in clearly defined constituencies with an election procedure.
4. A procedure must be set up for recording minutes, publicising them widely, and ensuring a thorough follow-up of all decisions. Once decisions are made, delaying tactics will create suspicions of management's real intentions.
5. Such a machinery must not be used to promulgate management decisions arising from the consultation. Management must still use its own executive line of command for this purpose.
6. Meetings must be formal and frequent at fixed intervals.

Against this background certain experiences are bound to arise.

1. There will be suspicions of management's intentions and these will only be dispelled in time.
2. Success will be expedited if management consults as early as possible about changes. It must not be used for instant decision-making.
3. Discussions that management requires, for business reasons, to be kept confidential must invariably be kept so.
4. Some resistance must be expected to certain consultations by some representatives. Only time and experience will break down reticence on these matters.
5. Management must expect opposition from time to time and be prepared to accept it. A decision to override opposition may generate reaction to the concept of consultation, and, if the feeling is strong, could lead to industrial action.
6. Management may hear unpleasant comments about itself. These must be taken on the chin.

FINAL CHECKLIST

1. Know the Codes of Practice backwards.
2. Provide a positive plan to motivate employees other than with money.
3. Don't be devious. Tribunals can catch you out.
4. Don't use union stewards for conveying management decisions. Use your executive line of command.
5. Provide means for your employees to consult you.
6. Communicate effectively. The other party must understand your message as you intended.
7. Explain instructions and give reasons wherever possible.
8. Consider regular meetings with union officials.
9. Make sure that there is a clear procedure on sites for identifying possible sources of dispute and for notifying senior management.
10. Ensure that there is machinery on site to prevent or adjudicate on complaints.
11. Ensure that there is adequate provision for recording decisions reached in the event of settlement of disputes, grievances, etc.
12. Ensure that foremen/agents know which of the officials or elected union representatives with whom you deal have authority to call for industrial action. If there is doubt consult with the unions to clarify the position.
13. Examine the circumstances and events surrounding recent disputes and take remedial action where appropriate.
14. If there is a stoppage of work or a threat of stoppage, ensure that there is a swift procedure for notifying senior management, and if necessary the appropriate employers' organisation official.

15. Make a practice of carrying out an annual review of industrial relations, procedures, all work rules and rules of conduct, using the Code of Practice as the basis.
16. Finally, remember that you are not employing robots, but human beings like yourself.

FURTHER INFORMATION

Acts of Parliament and Regulations (available from HMSO)

Industrial Relations:

Redundancy Payment Act 1965
Equal Pay Act 1970
Contracts of Employment Act 1972
Trade Union and Labour Relations Act 1974
Health and Safety at Work etc. Act 1974
Sex Discrimination Act 1975
Employment Protection Act 1975
Trade Union and Labour Relations (Amendment) Act 1976
Race Relations Act 1976

Health and Safety:

Health and Safety at Work etc. Act 1974
Factories Act 1961 (especially Section 127)
The Construction (General Provisions) Regulations 1961
The Construction (Lifting Operations) Regulations 1961
The Construction (Working Places) Regulations 1966
The Construction (Health & Welfare) Regulations 1966

Codes of Practice (available from HMSO)

Industrial Relations Codes of Practice Nos. 1–3:
No. 1 *Disciplinary practice and procedures in employment*
No. 2 *Disclosure of information to trade unions for collective bargaining purposes*
No. 3 *Time off for trade-union duties and activities*

Code of Practice for Safety Representatives and Safety Committees

Books

Barrel, B., et al., *Industrial relations and the wider society,* Open University (1975)
Bayley, L.G., *Building – teamwork or conflict?*, Godwin (1973)
Brown, Wilfred, *Organisation*, Heinemann (1971)
Drucker, Peter F., *Management*, Heinemann (1973)
Graham, H.T., *Human resources management,* M & E Handbook (1977)
Humble, John W., *Improving business results*, McGraw Hill (1967)
McGregor, Douglas, *Human side of enterprise*, McGraw Hill (1960)
McGregor, Douglas, *Leadership and motivation*, MIT Press (1968)
Mitchell, Ewan, *The employer's guide to the law on health, safety and welfare at work,* Business Books (1977)
Payne, Desmond, *Employment law manual*, Gower Press (1977)
Porter, R.W., *Guide to employment conditions*, 2nd edition, Godwin (1977)

Publications by specialist bodies and institutions

Advisory, Conciliation and Arbitration Service (ACAS):

This is ACAS
An industrial relations service to industry
Advice on personnel management and industrial relations practice
Trade union recognition – how ACAS can help resolve recognition problems
Conciliation between individuals and their employers
Arbitration in trade disputes

Department of Employment:

Employment Protection Act 1975
No. 1, *Outline*
No. 2, *Procedure for handling redundancies*
No. 3, *Employees' rights in insolvency of employer*
No. 4, *New rights for the expectant mother*
No. 5, *Suspension on medical grounds under Health and Safety regulations*
No. 6, *Facing redundancy? Time off for job hunting and to arrange training*
No. 7, *Trade union membership and activities*
No. 8, *Itemised pay statement*
No. 9, *Guarantee payments*
No. 10, *Terms and conditions of employment*
No. 11, *Continuous employment and a week's pay*
No. 12, *Time off for public duties*

Contracts of Employment Act 1972: a guide (April 1976)
Trade Union and Labour Relations Acts 1974 and 1976: a guide (July 1976)
Dismissal – employees' rights (amended March 1977)
The Redundancy Payments scheme, 10th revision (June 1976)

The Home Office:

Racial discrimination – guide to the Race Relations Act 1976
Sex discrimination – guide to the Sex Discrimination Act 1975

The National Federation of Building Trades Employers, 82 New Cavendish Street, London W1 (*available to members only):

National Joint Council for the Building Industry – National Working Rule Agreement, together with regional variations to the agreement
Guaranteed minimum bonus and Joint Board Supplement
Travelling and lodging, explanatory notes and examples
Road Haulage Workers Working Rule Agreement
Annual holidays with pay, two explanatory booklets
London region travelling map
*Employment and the law
*Practical employment procedures, guidance notes
Construction safety, an up-dating manual

The Institute of Building, Englemere, Kings Ride, Ascot, Berkshire:

Occasional papers:
No. 6, *Personnel management* by M.J. Dowham
No. 10, *Financial incentives – do they work?* by P.J. Talbot
No. 11, *Quality control in speculative house-building* by M.J.V. Powell

Chapters in *The practice of site management:* 'Hiring and firing in the building industry' by L.G. Bayley, and 'Disputes and conciliation machinery' by R.E. Stenning

The Industrial Society, Robert Hyde House, 48 Bryanston Square, London W1M 1BQ:

Notes for managers:
No. 2, *The manager's responsibility for communication* (January 1976)
No. 8, *A guide to employment practices* (February 1976)
No. 9, *A practical guide to joint consultation* (July 1976)
No. 11, *Industrial relations* (April 1976)
No. 12, *The manager's responsibility for safety* (February 1976)
No. 15, *Absenteeism – causes and control* (November 1975)
No. 19, *Delegation* (July 1976)

Guides to legislation, and related books:
A guide to the Employment Protection Act (July 1976)
A guide to the Health and Safety at Work Act (Revised May 1977)
A guide to the Race Relations Act 1976 (November 1976)
A guide to the Sex Discrimination Act (January 1976)
A guide to the Trade Union and Labour Relations Acts 1974 and 1976 (March 1976)
The closed shop (January 1977)
Legal problems of employment, 6th edition (August 1976)
Model procedural agreements (July 1976)
Unfair dismissal (July 1976)
The work challenge (Revised October 1977)

14 SITE APPRECIATION

CHOOSING A SITE 14–2

CHECKLIST 14–4

14 SITE APPRECIATION

M. MANSER, DipArch, RIBA
Partner, Michael Manser Associates, Chartered Architects

This section gives guidance on the assessment of a site before it is bought, to ensure its suitability for the type of development envisaged.

CHOOSING A SITE

Design is where you begin and end. If the design is not right, then, regardless of the quality of execution, the buildings will not sell. Remember that time at design is never wasted. It is easy to tear up a sheet of paper and start again, but cast concrete and laid bricks take a lot of moving.

At the beginning is site appreciation. You must be sure, if there is a choice, that you have the best site, and if there is no choice, that the one available is suitable at all. There are certain elementary rules about searching for a housing site. Always take the best you can find, best in the sense that it is in an acknowledgeably desirable area, i.e. an area popular in the price bracket in which you contemplate building. It is also important to pitch your investment at the same bracket. It is no use buying a site in a low-cost area and building medium-price homes. Neither will you reap the maximum profit from medium-price houses in a high-price area. Always match investment to sale price.

Look for some topographical features that can give your development character. Building is instant, but a group of trees or a mature hedge take twenty or thirty years to grow. Likewise, the existence of an old wall or restorable building, which can be incorporated with the development by an imaginative architect, will show a cash reward when sales time comes. It hardly needs spelling out that the proximity of an old church or other historical feature of which a view can be obtained from the site is another sales bonus.

There are other fundamental considerations: transport to work, to school, to leisure activities. Even though cars are now universal, it seems likely that congestion and the cost of running cars are going to drive people back to public transport. Even Los Angeles, the ultimate suburban city, seventy miles square, entirely dependent upon motor transport, using an elaborate motorway system laid over a street layout, is showing signs of strain. There are more small European and Japanese cars and bicycles in this city than in any other part of America, and the reason is the increasing cost of motoring.

Even in low-density high-price development owners are looking for public transport, bicycles and amenity shops reasonably close by. In future developments, it would be prudent to bear this in mind.

The existence of a park or golf course nearby, or a canal or river, are attractions, as are sports centres and clubs and any other entertainment facilities.

Physical characteristics

The physical characteristics of the site are important. Low-lying land or even just flat land is suspect for flooding, and sloping land, apart from increasing building costs, can contain slips and faults. Underneath can be sewers or gas, electricity or water mains, which will inhibit development; also overhead cables, depending upon the electric load carried, create a zone beneath in which it is not possible to build. The land may contain footpaths, the rights of which cannot be extinguished, although they can usually be re-routed, or disused burial grounds which are difficult to move and de-consecrate. In an area known to be of archeological interest, there may even be undiscovered relics, the revealing of which can cause extreme delay and cost.

Always bear in mind that the generally prevailing wind over the British Isles comes from the south-west. In most towns and cities the best residential areas are found on the south-west side of the town, i.e. the downwind side where the smoke and fumes are least apparent. To a certain extent this is a legacy from the real smoke and grime of the industrial revolution. Nevertheless, even in these clean-air times, the view from a high-flying aeroplane on a hot day shows dirt and haze pluming off any conurbation in a north-easterly direction.

Planning considerations

Having assessed the proposed site for its physical advantages and limitations, for its neighbourhood and character possibility, and for its communication and amenities, it is necessary to check its planning potential.

This is done in two stages. First, obtain from the local town planning office an initial confirmation that the site is zoned for housing, so that it can be decided whether the possible development justifies the asking price; second, if the price seems right, check the broader planning future of the area: see the county structure plans, find out about proposals for motorways, airport developments, industrial developments, reservoirs or national parks. All these will have more local effects – some good, some bad – on the marketability of the houses you propose to build.

If the broader plans for the area prove satisfactory and the site is zoned for housing, it is worthwhile researching more detailed planning and cost considerations.

In the first instance you need to obtain from the local town planning department an indication of the density they will readily permit. From this it will be possible to calculate, by dividing the number of housing units into the asking price, the land cost attributable to each unit of accommodation. Add to this the construction cost, including the relevant proportions of roads and services, the fees of architects, engineers and agents, and, of course, the interest charges on the capital laid out. The resultant total cost of the development per unit is then compared with the estimated market price per unit, and the difference between is the notional profit – or, if it is a minus figure, the loss. This is an elementary calculation which you make in order to ascertain if the venture is worth taking a step further, by entering into more sophisticated calculations and more specific purchasing negotiations. There are many refinements to be made in the calculations. On a rising market, the timescale between construction cost and selling price may show an additional profit. Similarly, cost inflation between initial calculation and actual construction, after perhaps protracted town planning negotiations, could play havoc with what had initially been a profitable forecast.

Assuming the initial calculations show a favourable profit forecast, an immediate outline planning application should be made on the basis of the density already indicated by the planning authority, and the contract to purchase should be subject to this application being successful. With the outline permission secured, it is possible to proceed with the purchase and embark upon a detailed design layout. At this stage, by good design, it may be possible to increase the density, thereby increasing the profit margin and almost certainly the quality of the project. It is fairly easy for an indifferent designer to cobble together a conventional layout. A more imaginative and painstakingly devised plan, perhaps at a higher density, by a good architect, will obviously have cost and environmental benefits.

CHECKLIST

The person who makes the initial site appreciation would do well to work to a checklist, and one is set out below as a suggestion. Using a camera on any kind of survey is invaluable. Accurate and clear elevational photographs of a building can often be used to count bricks and thus make surprisingly accurate drawings without specific vertical measurements being taken. They can also be used to check horizontal plan measurements.

Using a camera in an initial appraisal is particularly important. Often there is a long journey to a prospective site, and time actually on the spot may be restricted. A short visit can leave a somewhat distorted and misleading impression. However, if a good photographic record has been made, the prints can be studied at leisure later and will help to form a more measured appraisal.

1. *Access to site*
   (a) Condition and location of serving roads.
   (b) Width of roads.
   (c) Classification of roads.
   (d) Position of bus stops, if any.
   (e) Adjacent intersections that may be a potential hazard.
   (f) Footpaths and rights of way.
   (g) Discharge points nearby, e.g. factory gates, bus station, car park, school.

2. *Services*
   (a) Power cables (above or below ground – look for Electricity Board ground signs or manholes).
   (b) Gas mains (inspection covers).
   (c) Water mains (inspection covers).
   (d) Sewers (inspection covers).
   (e) Telephone lines above or below ground.

3. *Boundary features*
   (a) Fenced, walled, hedged, ditched or undefined.
   (b) Adjacent buildings, whether overhanging or posing underpinning hazards or involving rights of light.
   (c) Proximity of undesirable features, chimneys, flues, scrap yards, noise sources.

4. *Topography*
   (a) Orientation relative to north point and access points.
   (b) General slope of site.
   (c) Nature of top soil and vegetation.
   (d) Mounds or hillocks – check whether natural or made up.
   (e) Signs of flooding, positions of ponds or water courses.
   (f) Cuttings or excavations or signs of ground having been worked and then made up.
   (g) Types, sizes and positions of main trees.
   (h) Degree of exposure or shelter and climatic condition.
   (i) Views, orientation and nature.

5. *Underground hazards*
   Check with local authority engineers for:
   (a) Mine working.
   (b) Railway tunnels or underground trains.
   (c) Springs, watertable, flooding, ground movement.

6. *Neighbouring properties*
   Check for:
   (a) Rights of light or air.

(b) Rights of support.
(c) Rights of access.
(d) Fire risks.
(e) Drainage or services.
(f) Noise or smells.

7. *Legal and statutory*

(a) Obtain the names of the vendor's solicitor and agent.
(b) Find if any earlier survey drawings are available.
(c) Enquire into tenure offered – freehold or leasehold – and whether there are any title rents or restrictive covenants.
(d) Establish ownership of fences, rights of way, rateable values.
(e) Check if there have been previous town planning applications made for the site.
(f) Obtain the name of the local authority and county authority, or the metropolitan authority if a large city is concerned. Obtain the names and addresses of the relevant town Planning Offices, Building Control Offices, District Surveyors, Fire Officers, Highways Officer and the appropriate area representatives of the main utilities, water, gas, electricity and drainage.

8. *Photography*

(a) Either obtain an ordnance map or make a rough sketch. Mark viewpoints of photographs and note time of day.
(b) General panoramic views (useful panoramas can be obtained by taking four or five overlapping shots across 180° from a fixed position; these can later be pasted together in a strip).
(c) A series of views from opposite viewpoints, to give an impression of the general nature of the site.
(d) Specific photographs of:
   (i) adjacent buildings,
   (ii) buildings already on site,
   (iii) particularly good or bad views,
   (iv) dominant features or pylons etc.,
   (v) mature trees.

The care with which the site appraisal is made can eliminate costly discoveries at a late stage in development plans. There are many instances where enthusiasm for a development and elation over a profit forecast blind an otherwise prudent eye. Salutary is the story of a very famous name in property who purchased a wooded site hoping to obtain permission for housing. He did. The vendor then pointed out that the standing timber had not been included in the sale and obtained an injunction to prevent felling. The ransom proved to be another twenty-five thousand pounds. More care with the legal aspect of the site appraisal might have revealed this problem earlier.

FURTHER INFORMATION

Further information on this subject can be obtained from the RIBA Practice Handbook.

15 DEVELOPMENT DESIGN AND LANDSCAPE

PRINCIPLES 15–2

DENSITY 15–4

ACCESS AND CIRCULATION 15–6

PRIVACY 15–9

LANDSCAPING 15–10

DEVELOPMENT DESIGN AND LANDSCAPE 15

15 DEVELOPMENT DESIGN AND LANDSCAPE

M. MANSER, DipArch, RIBA
Partner, Michael Manser Associates, Chartered Architects

This section discusses the main principles to be followed in development design, and comments in detail on density, access, privacy and landscaping.

PRINCIPLES

Good development design and landscape is not just a question of taste and preference; it is the practical arrangement of houses, and the public and private spaces between them, to give maximum privacy and convenience in a visually agreeable way. It is not necessary to be an academic to know the ingredients of a well designed housing estate. Examination in detail of any small town or village whose core was constructed in the eighteenth century or earlier yields the answers. In the past, people lived and worked close to nature and they developed an instinct for scale and proportion. They understood about shape and the size of a human being and how to set a house in its site, to be sheltered and sunny and have a sense of belonging. They seldom made a mistake, even though they were without the benefits of Building Regulations or Town Planning Acts.

One of the most destructive factors in more recent town planning and estate lay-out has been the assumption that the road pattern was of paramount importance, i.e. that the convenience of motor vehicles took precedence over that of pedestrians. The point has been missed that motors are at their most effective *between* towns. *Within* towns pedestrians are more efficient, and it is easier and more practical to inconvenience vehicles in order to ease pedestrian routes. Making the road networks paramount led to rigid building lines (unheard of before this century) and the by-law road and minimum turning radii, all of which destroy the scale and character of any development. The establishment of rigid building lines alone was the birth of the boring suburb. Now more modern thinking is going away from the fixed rules, and local authorities are being encouraged to accept more intimate layouts and to accept restrictions on the use of cars. See *Figure 15.1*. It does not, after all, inconvenience a car-borne traveller very much if he has to travel another two hundred yards to his destination. To a housewife carrying a shopping basket, another two hundred yards is too far.

Landscape means more than a strip of grass and a flowering tree. Skilfully planned landscaping can provide at least half the success of a good housing development; this again was well understood by our forebears, and it is something that is only slowly being re-exploited and re-understood.

Each development site is unique. There is no other spot on earth that is identical. Even the flattest, most featureless site has its own character and micro-climate, which will indicate a line to be taken in its development. At the worst, a bleak site may have to be shaped in its development to protect the future inhabitants from a scouring wind; at the best, a site may be dotted with magnificent forest trees and enclosed by an old brick wall that will give instant warmth and character. Most sites will lie somewhere between the two extremes.

In architectural terms the area of the development may have strong local characteristics, for example a local material, or a particular brick or stone. Parts of Sussex are characterised by flint and red brick, parts of Kent by clapboarding, and parts of East Anglia by decorated rendering. In the north, window surrounds are often emphasised by heavy rendered frames. These characteristics can be taken up in

Figure 15.1 (a) A highway-dominated layout; (b) layout not dominated by highways, showing architectural use of trees

a modern manner and in modern materials, and can be incorporated in designs to give a regional flavour. So much recent housing development has a boring monotony, not only because it is poor in design and layout but because the same design and materials are used throughout the country in an arrogant disregard of traditional local variations.

Purchasers of houses are motivated mainly by cost. However, offered a choice within a price bracket, once they are satisfied that the space and equipment are compatible they will choose between two developments on the basis of which is the most distinctive, which gives the most sense of individuality, which inspires the least feeling of living in a hen battery. Accepting the character of the site and developing it in a sympathetic and perhaps local way will produce those qualities of identity. See *Figure 15.2,* for example.

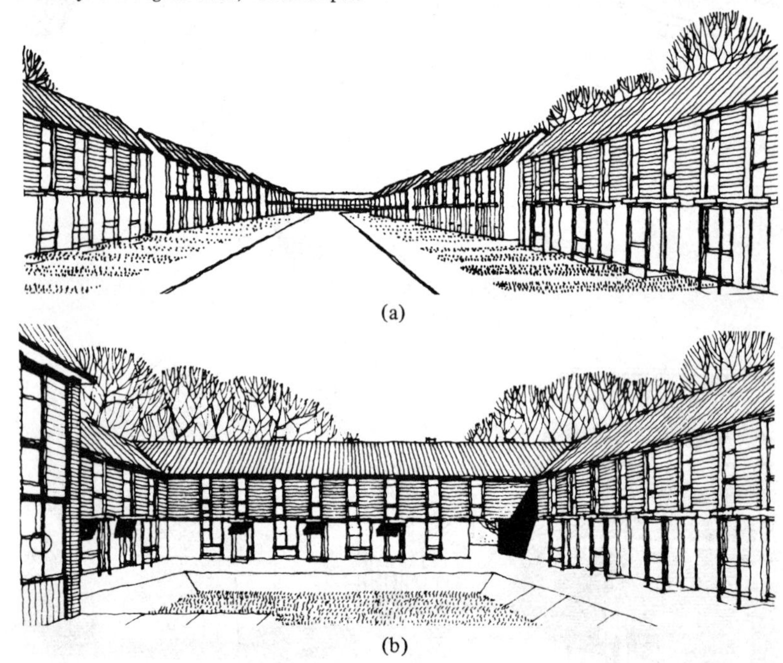

(a)

(b)

Figure 15.2 (a) Visually exhausting, no sense of identity; (b) visually resolved, a place with which to identify

The design of a good development layout is a skilled and specialised job and it cannot be emphasised too much that it is not a task for an amateur. An architect must be appointed. The success of the development will depend on the skills put into the layout, not only for its visual appearance but also, bearing in mind the high cost of land, for the best revenue.

DENSITY

There are probably four basic types of housing development:
- low-density suburban,
- high-density rural,
- high-density urban,
- super-high-density urban.

Low-density suburban is, inevitably, high-cost housing on the outer fringes of conurbations, providing substantial houses in upwards of half-acre (0.16 hectare) plots. Here the market force is the owner who wants the illusion of a country home close to the city. In this case the layout should aim to further the illusion, and the existing vegetation must be used and fortified to shield each house from its neighbours. Hedges are preferable to fences. Roads should look as much like lanes as possible. Houses should be individualistic and set back from the road and in the middle of their plots. See *Figure 15.3.* This category is a small corner of the housing

Figure 15.3 Traditional low-density privacy

market, but profitable for a small operator prepared to give a very personal service. (There is a huge subdivision of this category, *high-density suburban,* which accounted for most of the between-wars and immediately postwar sprawl. It consists of detached or semi-detached houses in small plots. Constrained now by the diminishing availability of land, this area of development is being forced to take the form of the next two types of housing because the waste of providing small gaps of land between the houses is reducing each curtilage to derisory proportions. It was the growth of population and the limitations of horse transport that brought the eighteenth-century designers to urban terraced housing and close-knit villages. The shortage of land and the cost and congestion of motor travelling have forced us now to similar housing solutions.)

High-density rural housing is informal cottage-scale grouping. Houses of various types and sizes on footways, interlocking and laid out in an apparently random way (*Figure 15.4*).

High-density urban housing is formal terracing of similar houses in various squares or crescents.

Super-high-density housing is apartment blocks. Although in recent years there has been hysterical rejection of this kind of housing, it suits many people at some periods in their life. Most urban dwellings in continental Europe are flats, and commercial flat developments are successful. The complaints have come from the public sector of housing where occupants have no choice. If you have no children and hate gardening you will not be happy in a house and garden; on the other hand, if

Figure 15.4 Traditional high-density privacy, achieved by design

you have six children and suffer from vertigo the tenth floor of a tower block will be a nightmare. The failure of high-rise housing is also mainly one of management; there has to be a concierge who has strict control of entrances and stairs.

The two-room flat market has a large unmet demand, and is one that the smaller builder on the smaller development could well consider.

ACCESS AND CIRCULATION

The starting point for any layout is the access and circulation. These are for:

- residents and visitors by car or on foot,
- trades deliveries,
- removals and heavy deliveries,
- refuse collections,
- emergency services – fire engines and ambulances.

Ideally the pedestrian should be able to reach each house separately from the vehicle route. There are two reasons: one is safety, the other is a more subtle reason. It is a question of scale. A car is a large bulky object designed to travel at speed. It needs wide spaces and gentle curves. Pedestrians are small and travel slowly. A long wide straight road, ideal for a swiftly moving vehicle, will be desperately boring and psychologically exhausting for a pedestrian. He needs a narrow path with frequent changes of direction, so that his interest is always to see what is around the next bend; his view constantly changed and refreshed.

In a small village-type development it may be possible to keep the main access road short, with several cul-de-sac mews-type courtyards off it, each serving a few houses – see *Figure 15.5.* These could be approached over ramps through narrow entrances – the object being to restrict the speed of vehicles to walking pace. It

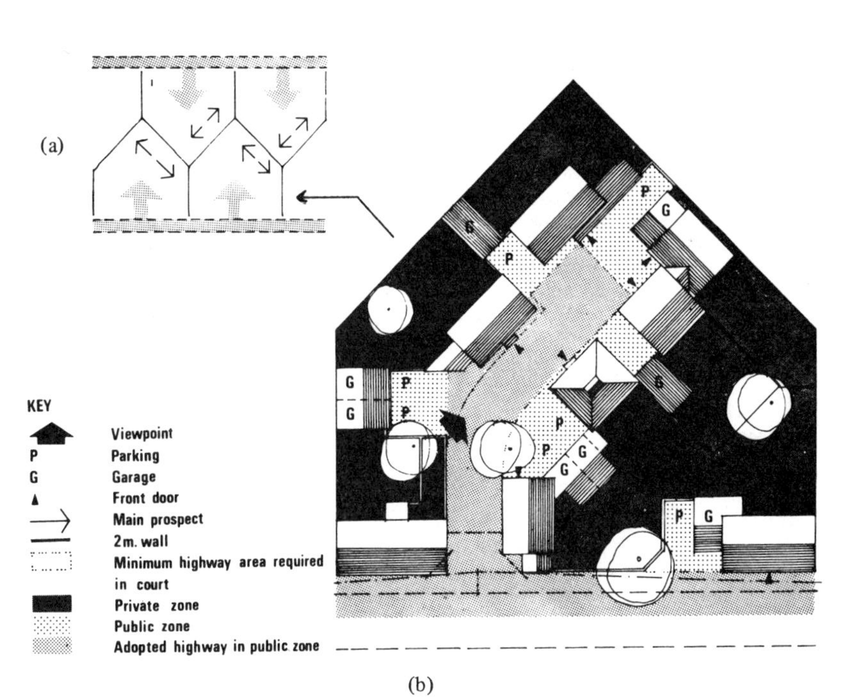

(b)

(c)

Figure 15.5 (a) Interlocking courtyards; (b) layout of individual courtyard; (c) view of courtyard from black arrow-point

would be assumed that furniture removal vans would back into such courts and heavy deliveries or refuse trucks would park outside. To make this practical, no dwelling must be more than 36 m from the entrance to the cul-de-sac; this is the maximum reasonable carrying distance for dustbins etc. In a suburban development, the roads themselves can be for the unrestricted use of cars – apart from speed. It is assumed that a car will serve all essential travelling, and strategic transverse footpaths may only be for pleasure walking or visiting neighbours.

Avoid useless public spaces in all types of development. Grass verges become dog lavatories and are a maintenance liability. A good development has the minimum 'amenity' land. What cannot be conveyed to a particular curtilage is better paved. It is also an illusion, except in low-density areas, that open space around houses creates privacy – rather the reverse. Houses that are close together and interlocking can shield each other to create immensely private zones between. Deep front gardens

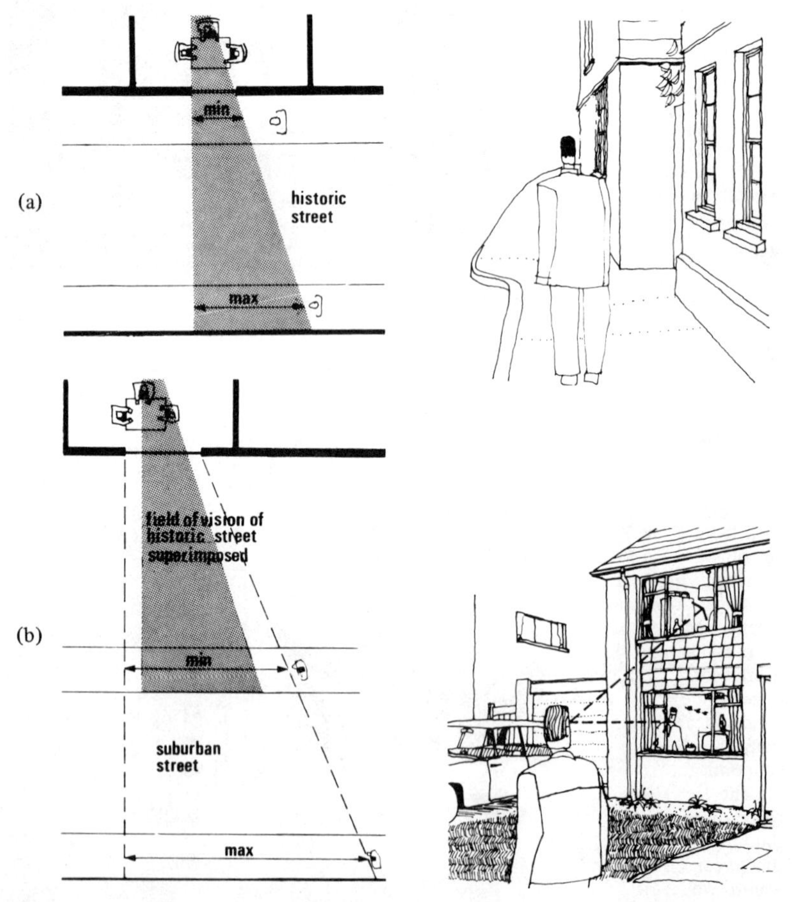

Figure 15.6 (a) Narrow field of vision of the historic street, with small windows and 'set forward' buildings; buildings that were set back had walls or planting to provide a physical barrier. (b) Wider field of vision of 'set back' suburban street provides less privacy

and picture windows are another delusion. They merely produce 'goldfish bowl' conditions for the occupants, offering a long leisurely look in for every passer-by as he traverses the frontage, whereas a smaller window close against the pavement shows only the briefest glimpse into the interior, unless the passer-by actually stops and peers in, and few people are uninhibited enough to do that. See *Figure 15.6.* Houses should be close to the footway, or else they should be set back and shielded by a fence or bank and hedge to the footpath.

Adequate provision, usually in the ratio of one or two per dwelling, has to be made for car parking. Again, because of their scale, cars are better parked between buildings rather than in front, where they tend to dominate a small house. Ideally, every house should have an adjacent parking space as well as a garage, because in practice a garage is often used as a general store and workshop, rather than for car storage. In high-density schemes it is usually inevitable that garages are in blocks away from the houses they serve. In this case, enclosed garage cul-de-sac courts should be avoided. They are vandal-prone and always deteriorate into neglected and unsightly enclaves. It is better to arrange the garages in banks discharging on to the access roads, where they will be properly lit by street lights and constantly in the public eye.

PRIVACY

One of the major problems of layout is providing privacy between houses. In a suburban layout the size of the plots is usually sufficient, and owners are free to increase their planting or fencing at will within their own curtilage. In high-density schemes, however, the designer has to be sure to build-in adequate sight and sound screening, because a high-density scheme is less suitable to later modification by occupants. One man's screen wall, for instance, will steal another's daylight.

The most effective high-density privacy can be achieved in high-density rural layouts, the village-type development where the flank wall of one dwelling can be the screen wall of another. In an urban high-density layout, which will almost inevitably be a terraced formation, there has, in the author's view, been no improvement yet on eighteenth-century housing. The basics are strictly repetitive house plans facing small front gardens, tall narrow windows, and back gardens screened by walls or fences. The houses have to be strictly repetitive so that each front door is a house width away from the next front door. There is an enormous degree of privacy created by a front door that seems to be sheltered from the neighbours. Houses should never be laid out so that their front doors are adjacent, unless each house is stepped back to screen each door.

After front doors, windows can give the greatest impression of lost privacy. In recent years windows in mass housing have been far too large. Wall-to-wall glazing is a luxury for those with land. Nothing is more magnificent than a wall of window overlooking a two-acre lawn against a beech wood. Nothing is more squalid than a huge window that has to be obscured with net because it overlooks a busy street. Tall windows are preferable to wide windows. Tall windows let in the light and their shape restricts the view in. From a curtaining point of view, they are also far more economical than wide shallow windows.

The last point where loss of privacy is most noticed is in the garden sitting-out space immediately behind the house. Here it is absolutely essential to build in flanking screen walls that project at least 3 metres down each garden. Having formed the screens, this is then the one place in the house at ground-floor level where a large wall-to-wall and floor-to-ceiling window could be successfully installed. Here the window is screened from a sideways view and looks out on the length of the garden: there are no passers-by, and from the inside the view out into a flank-walled terrace will create an impression that will appear to enlarge the room. See *Figure 15.7.*

Figure 15.7 Present-day privacy by design

LANDSCAPING

Landscaping, like estate layout, is a specialist job and needs to be integrated with the layout. Grass verges and rows of flowering cherries are no delight to anyone but the local dogs, and are a positive burden to the local authority lumbered with their maintenance.

As a principle, the layout will incorporate existing trees, as far as is practical, and the landscaping will develop from these. See *Figures 15.8 to 15.12.* The point has already been made that as much as possible of the terrain should be conveyed to individual curtilages; indeed except for specific comparatively large open spaces (for example, a green that could be combined with a children's play area), landscaping should be hard paving planted with specimen trees. These should be chosen to contribute to the layout of the buildings. A forest tree might be planted where bulk is required to fill a gap between buildings; something delicate and light like a silver birch will add to the scale of a small court, without overpowering it. Always plan for easy maintenance and, if a green is provided, ensure that a path is laid where people will want to take short cuts. However delightful the appearance, few people are ever prepared to walk around two sides of a square; they will take the diagonal, certainly when pressed for time.

Locate the green itself at a central point where people and traffic occur, especially if it is combined with a children's play area. Tucked-away amenity areas are a molester's joy, and an unused amenity area soon becomes a rubbish dump.

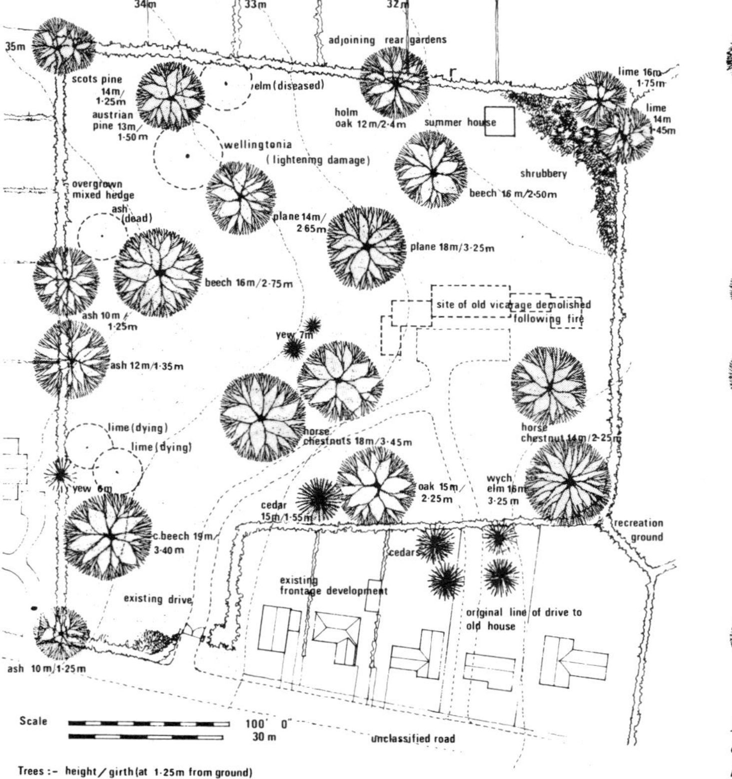

Figure 15.8 Site survey

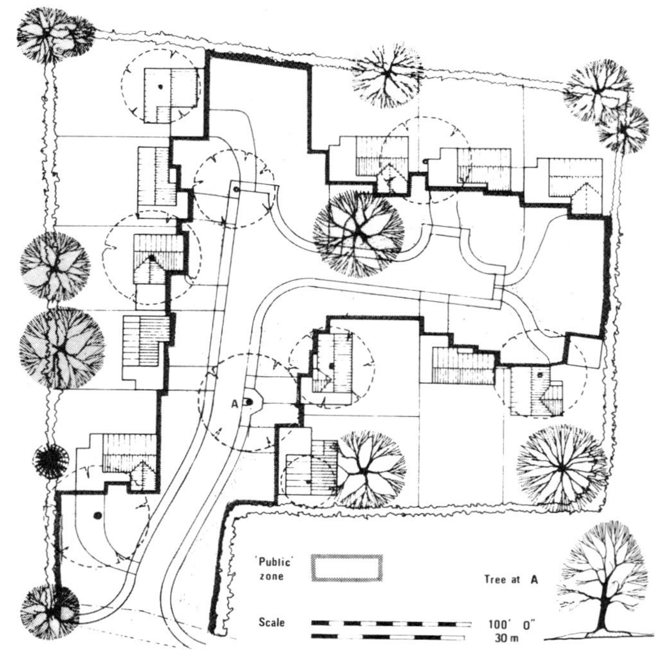

Figure 15.9 Poor layout on the site in Figure 15.8: unacceptable amount of public areas to maintain; individual houses lack privacy; neither a low-density suburban nor a high-density rural scheme

Figure 15.10 Good layout on the site in Figure 15.8: conveys most of land to individual owners. A low-density suburban layout, its nature would make it an up-market sales scheme. By linking some of the houses with an infill house (e.g. inserting house between 1 and 2, and between 3 and 4), density could be much increased without seriously diminishing amenity

Figure 15.11 View from arrow on Figure 15.10

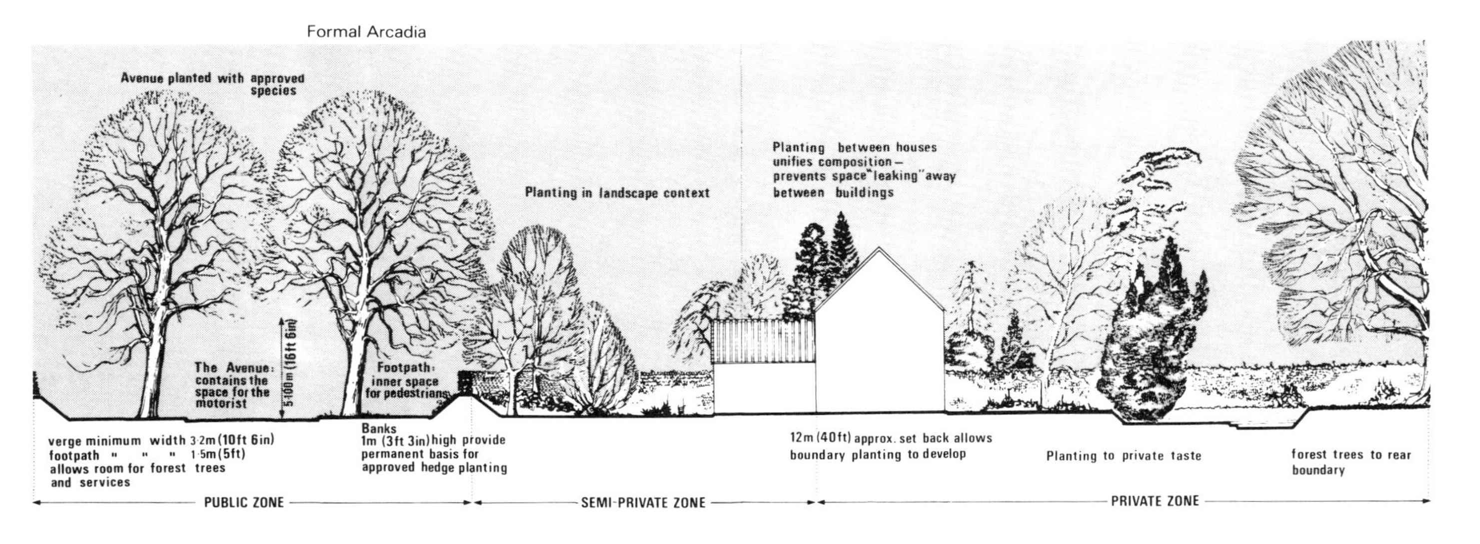

Figure 15.12 Planting: typical section

Except for large robust shrubs like laurel and may, the landscape designer should stick to grass and trees in an amenity area and, to give it shape, can use some building spoil to make a mound at one end or side. If maintenance is reduced to grass cutting and annual tree and shrub trimming, most local authorities will do a reasonable job. Paving does not have to be the ubiquitous 50 mm concrete paving slabs. They can be interspersed with bricks or cobbles or sawn logs treated with preservative and set end grain up to look like wooden discs – moss will soon form in the interstices. A cheap and effective paving, often used in America, is concrete poured between a pattern of treated timber slats and then brushed just before it hardens so that the aggregate is exposed. Never provide sand pits or ponds. Local animals soon make sand unusable, and ponds collect old tins and rubbish or become slimy. Ponds can only be contemplated if they are lake-sized or under constant supervision. Most parents of young children will see them as a drowning hazard.

Planting in place of fences is practical, cheap and effective. Thorn bushes or shrub roses planted as hedging are good to look at and (as any crime prevention officer will tell you) almost impenetrable. To preserve boundaries initially, a simple black plastic-covered chain-link fence can be used. Once the planting gathers strength, the black plastic becomes almost invisible. Never use green plastic. The green always clashes with the greens of the plants and is always prominently visible.

Finally, always ensure that adequate watering space is left around trees that stand in paved areas. It is quite a good plan to dish the paving so that the rainwater turns towards the trunk. Many trees preserved in the design are subsequently lost because their roots dry out owing to the paving.

FURTHER INFORMATION

Those needing recommendations to architects or landscape architects should contact the Clients' Advisory Service of the Royal Institute of British Architects, 66 Portland Place, London W1, or the Institute of Landscape Architects, Nash House, 12 Carlton House Terrace, London SW1. In each case, if the project is described, a selection of suitable names will be suggested from which a choice can be made.

ACKNOWLEDGEMENT

The figures in this section are reproduced from *A design guide for residential areas,* by courtesy of Essex County Council.

16 DWELLING DESIGN

INTRODUCTION	16–2
BASIC DESIGN CONSIDERATIONS	16–3
REGULATIONS AND STANDARDS	16–12
ASPECTS OF DETAILED DESIGN	16–13
KITCHEN DESIGN	16–19
BATHROOM DESIGN	16–27
STORAGE DESIGN	16–31
APPEARANCE	16–33

DWELLING DESIGN 16

16 DWELLING DESIGN

O.D. WILLMORE, DipArch, RIBA
Barton Willmore and Partners, Architect Planners

This section examines the context in which design has to be carried out – the market, purchasers' desires, and regulations. It then considers basic design choices such as type and mix of dwelling, and finally gives further detailed guidance on the design of kitchens, bathrooms and storage.

INTRODUCTION

The market

The number of households in England and Wales has grown steadily in the past, and recent projections prepared by Government show that this trend is expected to continue, at least until the end of the century. The statistics[1] show that, from a total of approximately 14.8 million households in 1961, the total number of households had grown to about 17.4 million by 1976 and that by 1991 this total is expected to grow to about 19.6 million. By the end of the century the projection is that there will be about 19.8 million households and potential households in England and Wales.

Though such demographic projections must be treated with caution, they indicate not only a continuing increase in the number of households headed by a married couple, but also a proportionately higher increase in the number of one-person potential households, at least until 1991.

The proportion of householders who own the home in which they live has also grown steadily, and the encouragement of this trend is the policy of all main political parties. By 1975 55 per cent of homes were owner-occupied, and there seems to be no reason why this proportion should not grow in Britain, as in other countries, to, say, 70 per cent of dwellings.

Though the growth in the total number of households is expected to be rather smaller from now on than it has been in recent years, there will nevertheless be about two million additional households in existence in 1991. There is also every likelihood that personal expectations of living standards and expectations of environmental quality will continue to rise. Many dwellings now deemed acceptable will no doubt cease to be regarded as satisfactory.

All this led, in June 1977, to the Government suggesting[2] that a substantial level of new house-building will be necessary to cope with the increase in the number of households, to meet the needs of those households that are currently overcrowded or sharing involuntarily, and to replace houses that have to be demolished. A table in one of the supporting Technical Volumes suggests that the total output of the house building industry might be 290 000–310 000 in 1981 and 305 000–325 000 in 1986. It suggests too that the owner-occupied share of these projected outputs might be 170 000–190 000 in 1981 and 195 000–215 000 in 1986.

All the foregoing, in the context of a future national economic climate that, at best, seems likely to call for stringency in non-productive expenditures, points to a continuing and increasing emphasis on the provision of value in meeting the demand for the supply of dwellings, and particularly in the provision of houses for private sale.

Constraints

There are, however, also increasing numbers of constraints on the house builder and on his designer. Costs continue to rise and in recent years have risen at a substantially greater rate than the selling price of houses, while the supply of land, a strictly limited and finite resource, is a matter of continuous concern and increasing difficulty. Notwithstanding Government pressure[3], some planning authorities continue to resist applications aimed at gaining planning consent for economic densities and higher proportions of smaller dwellings.

National economic stringency has led to constraints on the provision of infrastructure, and in turn to difficulties with statutory authorities over the financing of the necessary extensions to their distribution networks. National economic difficulties have also, from time to time, led to fluctuations in the supply of funds to building societies and other finance sources, and this has been reflected in constraints on the availability of mortgages.

Individual dwelling design has therefore become, more than ever, a matter of reconciling complex and changing needs on the part of the public with onerous constraints, frequently financial but sometimes imposed by legislation.

Purchasers' desires

There has, until recently, been little research into the needs and wishes of householders and therefore, by extension, into the desires of potential purchasers. A report from the Building Research Establishment[4] shows widespread dissatisfaction with a large proportion of our present housing. Table 16.1, reproduced from this report, illustrates the importance that householders attach to features of a house and its surroundings.

Care is necessary in the interpretation of this table because of what has become known as 'the escalator theory of housing desires'. In essence, this is founded on the fact that people who enjoy fewest facilities expect least, while those who have achieved higher standards tend to regard more items as being basic necessities. Housing for people with low incomes may therefore have to provide facilities in a priority founded on a subjective view on the part of the *provider* as to the ranking order of importance of the various requirements. There can be no simple set of rules that will produce quantitatively satisfactory new housing solutions.

BASIC DESIGN CONSIDERATIONS

Dwelling types and mix

Reference has already been made to recent projections concerning the numbers and sizes of households between the present time and the end of the century. The latest information currently available is that published by the Department of the Environment in August 1977, which again confirmed that there will be an increasing preponderance of single people and childless couples demanding their own homes. The average household size is expected to decline from a mean of 2.78 persons in 1976 to 2.57 persons in 1991. One-parent and single-person households are expected to increase at a substantially faster rate than households consisting of married couples.

The estimates also show that except in Berkshire, East Sussex, Greater London, Oxfordshire and Avon there was, by 1976, a surplus of empty dwellings over households before making allowance for vacant, unfit or otherwise inadequate dwellings or for second homes. A recent interpretation of the statistics by Christine Whitehead[5], however, indicated that this 'surplus' disappears if account is taken of the need to re-house households that are sharing dwellings involuntarily.

Table 16.1 ITEMS RANKED AS BASIC NECESSITY, DESIRABLE, OR UNNECESSARY

	Percentage of respondents classing this item as:		
	Basic necessity	*Desirable*	*Unnecessary*
Electric light	94.6	4.5	0.9
Hot water supply	92.5	6.5	1.0
Bathroom	90.2	8.2	1.6
Refuse disposal	86.0	12.2	1.8
Internal WC	84.0	14.1	1.8
Living-room heating	83.2	14.8	2.1
Power points	82.6	15.1	2.3
Free from damp	81.0	17.1	2.0
Ventilation	77.9	19.1	3.0
Daylight	71.3	26.8	1.9
Safe stairs	65.6	26.4	8.0
Refrigerator	61.3	30.7	8.0
Gas supply	61.2	19.2	19.7
Street lighting	59.4	35.3	5.3
Airing cupboard	58.3	35.6	6.1
Clean air	56.1	40.3	3.6
Ventilated food store	53.8	34.4	11.8
General storage	48.6	44.9	6.5
Convenient for shops etc.	45.9	47.9	6.2
Back garden	44.8	43.8	11.4
Safe play space	41.9	34.8	23.3
Thermal insulation	38.7	48.4	12.9
Sound insulation	37.0	47.4	15.5
Good neighbours	33.7	60.1	6.2
Privacy from neighbours	33.7	53.1	13.2
Bedroom heating	31.4	47.7	20.9
WC separate from bathroom	30.2	47.6	22.3
Well maintained property	30.3	59.6	10.1
Garage or parking space	29.8	41.0	29.1
Free from heavy traffic	28.7	53.2	18.1
Privacy in back garden	27.3	53.4	19.3
Front garden	26.9	54.1	19.0
Informal eating area	21.6	49.6	28.8
Greenery or trees	21.2	61.7	17.1
Privacy from family	21.1	43.4	35.5
Quiet surroundings	20.9	63.0	16.1
Two WCs	19.0	44.3	36.7
Level access	16.2	45.8	38.0
Pram space	13.3	33.3	53.4
Free from parked cars	12.7	50.7	36.6
All on one level	11.1	33.6	55.3
Shaver socket	8.4	32.5	59.1

In 1963 Harris and Clausen carried out research into the duration of occupancies and from their findings the Building Research Establishment has calculated that the median period of occupancy was slightly under 10 years. For a dwelling with a life of 60 years this indicated that there will be rather more than six occupations, but the Establishment makes the point that mobility is known to vary from one part of the country to another and that there are signs that mobility will increase in future. The Department therefore suggests that, in a life of 60 years, each dwelling is likely to be occupied eight times. The implications for dwelling design are examined subsequently, but it should be noted here that a substantial degree of flexibility should be provided in the uses to which the various spaces within the dwelling can be put, in order to cope with the needs of changing occupiers as well as the changing space requirements of the individual family in the course of each occupation.

Consideration is given to the possibility of extensions to the home, perhaps made more likely by the increasing availability of proprietary products, as well as to the increasing incidence of 'do-it-yourself' building.

Several factors, not least an appreciation of the economic and productivity benefits of repetition, have led to the production of numerous ranges of preferred dwelling types. It is perhaps sufficient to refer here to the Bulletin of generic house plans produced by the National Building Agency[6], the GLC preferred dwelling plans[7], the range of house shells produced by the National Building Agency[8] and the study on their use by the Housing Development Directorate of the Department of the Environment[9]. In addition, almost all house builders will have produced at one time or another their own range of 'standard plans', and the reasons why these expand until they are frequently indistinguishable from 'ad hoc' designs, site by site, are manifold. In addition, the Government has weighed in with well-intentioned advice[10] aimed at variety reduction in components and therefore better value for money.

Reverting now to the changing use of space throughout the family cycle, it is evident that even in the course of an occupancy of no more than average duration, say six to eight years, the needs of the household both for the uses of space and for the amounts of space in each use may change substantially. A newly married couple are likely to add two or three, perhaps even four or more, children in their household. A middle-aged couple might well see all their children married within this time span, and the trend for some time past has been for increasing numbers of newly married couples to leave their parents' home immediately and to set up separate households. No doubt this trend will continue into the future.

In all these circumstances it is not surprising that considerable thought has been given to the provision of 'extendible homes'. These fall into two categories. In the first, extensions and annexes to the main built fabric are planned and detailed but not built. Frequently, planning and building regulation approval is obtained for the extensions at the time of the original construction, and the subsequent additions are thereby integrated more satisfactorily into the visual and structural framework provided by the original dwellings. Difficulties can arise, however, particularly visually, if such annexes are not carried out in accordance with a carefully considered original design or if they are not subsequently constructed to the same standard as the overall development. Leaving space for subsequent extensions can, unless considerable care is taken in layout design, detract substantially from the benefits in land economy (and therefore cost) resulting from the original construction of small dwellings.

An interesting variation is the construction of houses that are extendible in various phases within the original building envelope. In a typical example a terrace of dormer bungalows with integral garages might have only the ground floor fitted out in phase one. The accommodation might be a hall, kitchen, bathroom and bed-sitting room, resulting in a dwelling suitable for a single person or a childless couple. In phase two a staircase is added, and access gained thereby to a bedroom in the roof, while the ground floor reception room reverts to living-room use only. In phases three and four, second and third bedrooms are added in the roof.

While such schemes have a place and some have found ready markets, there is perhaps a balance of opinion among house builders that small households are probably best catered for by the provision of small but complete dwellings, the currently fashionable 'starter homes'. This is the second broad category normally covered by the term 'extendible homes'. Should the family grow in size it is necessary for them to move to another dwelling, but this in itself tends to produce a healthy future housing market, and there will no doubt be small potential households anxious to acquire the original 'starter' dwelling in their turn.

Dwelling form

The balance of opinion perhaps favouring this alternative category is no doubt accounted for, in large part, by the difficulty frequently encountered in designing extendible dwellings with economic building forms.

When considering the matter of economic building form, leaving on one side the factors affecting the design of blocks of flats or maisonettes, and dealing specifically with houses and bungalows only, it is necessary to distinguish between single detached dwellings on the one hand and groups of dwellings such as terraces on the other. It is a commonplace that the most economic shape for the plan of a detached dwelling would be a perfect square. Rectangles on plan and re-entrants involve the construction of more walling for each given unit of floor area. Other design considerations apply, however, and it is necessary to strike a balance between these and the achievement of a minimum wall-to-floor ratio.

In terrace houses, where front and back walls (including doors and windows) have a higher cost of construction per unit area than party walls, this rule no longer applies. In general terms, the cost of each unit of area of front and back walls is likely to be twice that of the same area of walling in a party wall, perhaps even three times. In these two cases the most economic plan shape would be to have the house depth either twice the frontage or three times the frontage approximately. In paying regard to this principle, account should be taken of the higher cost of pine-end walls at the end of terraces, particularly where the number of houses in each terrace is small.

Repetition

No review of basic design considerations would be complete without reference to the productivity, and therefore economy, benefits accruing from the utilisation of designs that take account of 'the learning effect'. It is common experience that a degree of repetition or habit in the performance of tasks can result in dramatic decreases in the time and effort required to complete them. This principle applies to the construction of housing, as elsewhere.

The principle does not, however, lead inevitably to the production of large numbers of identical dwellings with all the problems of monotony (and loss of saleability) that would result. Those parts of the individual house that are most complicated and time-consuming to construct and install can be designed with repetition as a vital criterion. In one form, this approach involves the design of varying types and sizes of dwellings around a single unchanging 'core'. This core can consist of the main services elements of the building (kitchen fittings and sanitary appliances with plumbing and drainage connections, hot water storage, heating installation for space warming and main electrical harness), all grouped and disposed in a fixed relationship to each other. The groups of elements would include the staircase in the case of a two-storey dwelling. The remainder of the elements of the house can then be designed and built to a variety of different sizes, with a variety of

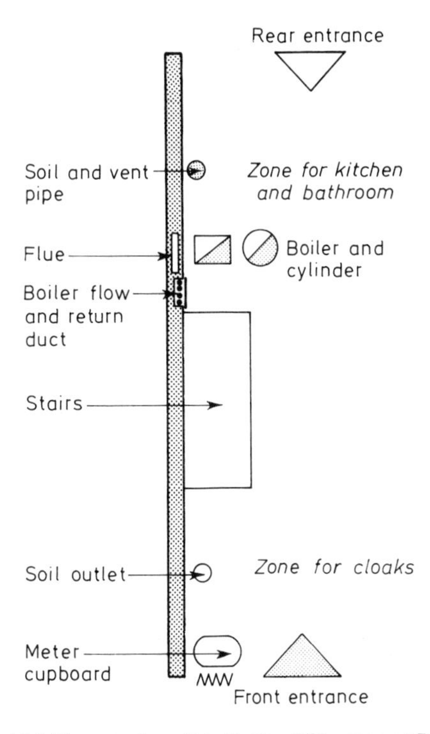

Figure 16.1 The core (courtesy Barton Willmore and Partners)

materials and with individual form where appropriate, e.g. where called for by site conditions (see *Figures 16.1 to 16.4*).

A core of this type, by rationalised use of conventional elements and techniques, is directed at improving on-site productivity and is not to be confused with prefabricated (off-site manufactured) cores from suppliers.

Flexibility

A final basic consideration in the design of dwellings should be the shaping and disposition of the accommodation so that as many spaces as possible have the potential for dual use. Apart from the obvious instances, such as bedrooms acting as studies for the use of children who bring work home from school, and the inclusion of breakfast rooms with glazed vision panels so that they can act as toddlers' play spaces under the parents' supervision during the day, there are other possibilities that should be borne in mind in this respect.

For example, it is often possible, at minimal additional cost, to size and locate a garage in such a way that there is space for a deep-freeze unit and/or a work bench with ready access to the working part of the house itself. Utility rooms can frequently, with advantage, act as storm porches by confining the 'rear' access (for refuse disposal and to the garden) to this unit. Finally, the waste of enclosed building volume in roof spaces has been very substantial in generations of British housing,

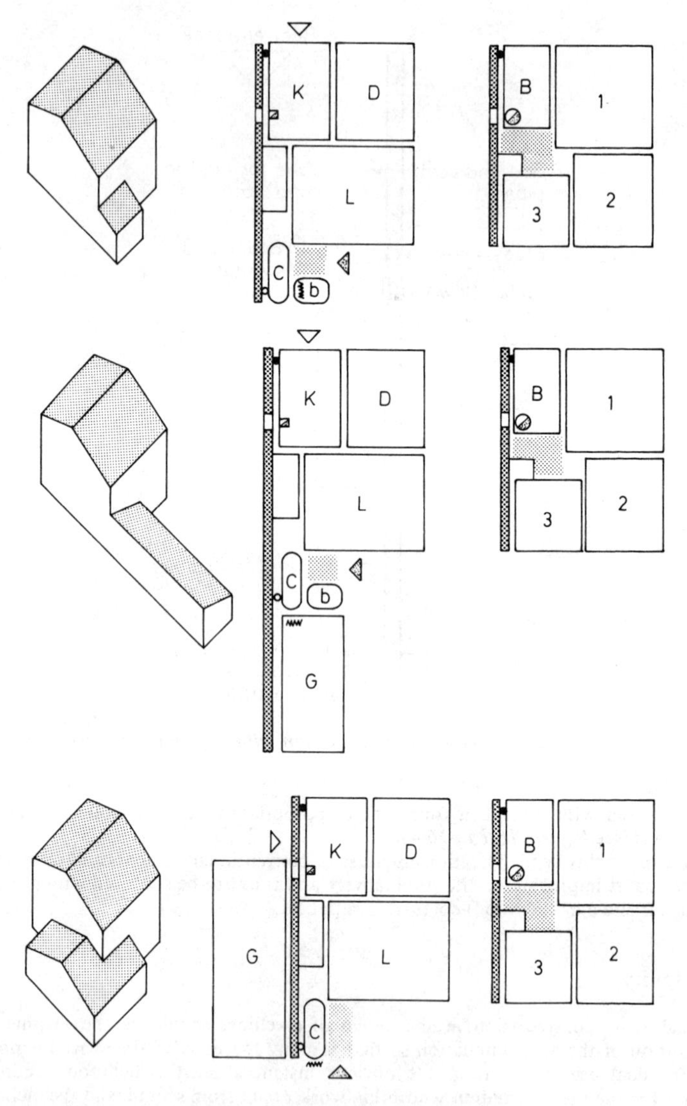

Figure 16.2 Variety in form and content, based on a common core: A (courtesy Barton Willmore and Partners)

though recently there has been much advice directed at reducing this, for example in *Better use of space in housing,* BRE CP 52/75; see *Figure 16.5*. Although costs may in many cases preclude the construction of roofs with attic trusses (giving the internal extendibility referred to above), wherever a pitched roof is employed an access point in a convenient location and of a sufficient size to permit the passage of large items for storage should be an essential feature. In addition to the gangway boarding from

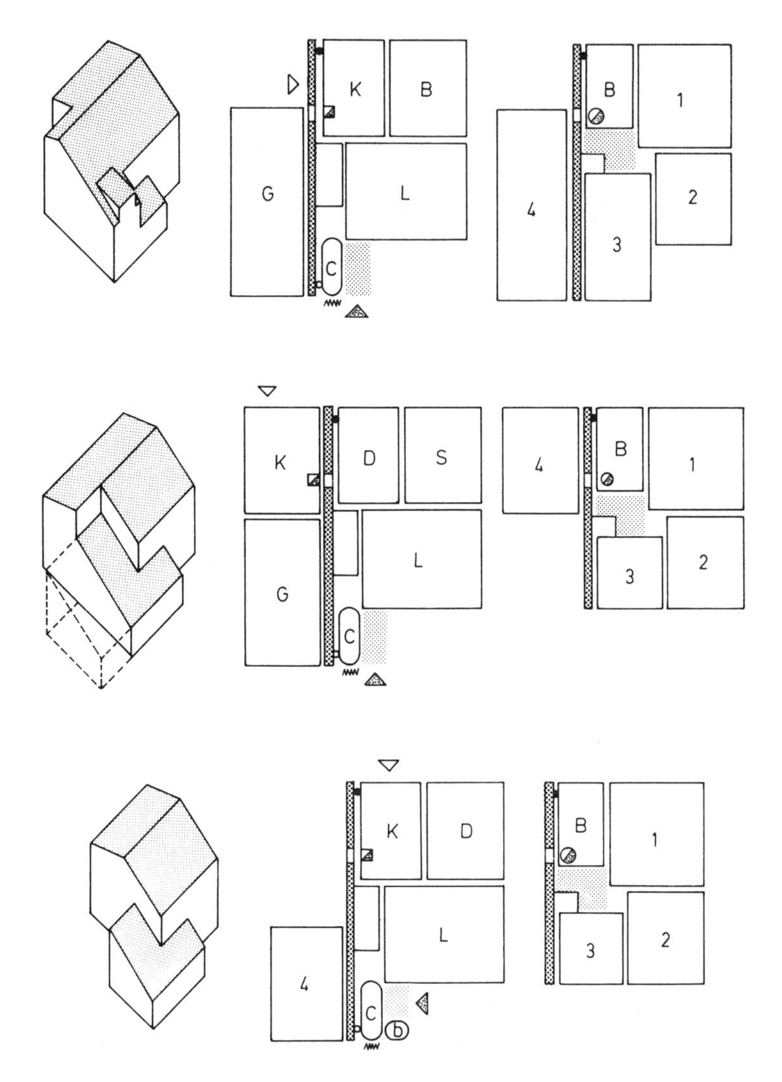

Figure 16.3 Variety in form and content, based on a common core: B (courtesy Barton Willmore and Partners)

the access opening to each cistern and round the cistern called for in the NHBC Schedule, there is benefit in providing a boarded area for storage purposes.

Some house builders are prepared to market 'shells' leaving most (if not all) internal partitions to be added, either as optional extras at the time of purchase, or subsequently, as in *Figure 16.6*.

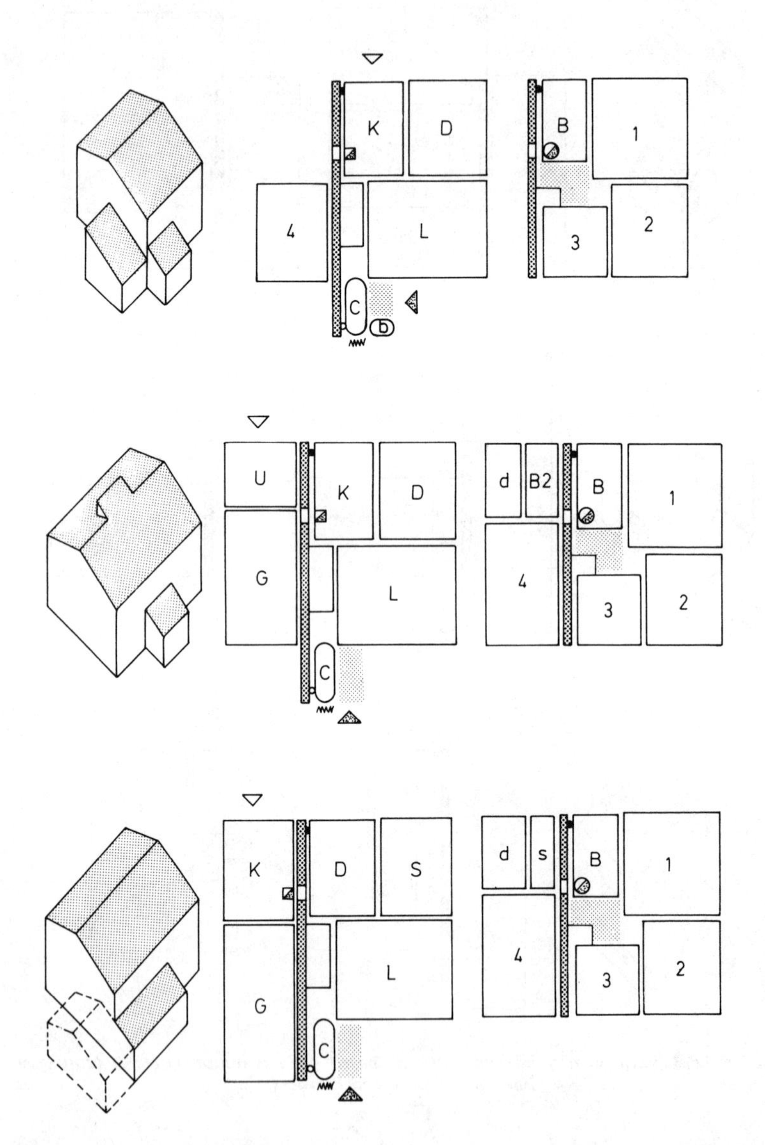

Figure 16.4 Variety in form and content, based on a common core: C (courtesy Barton Willmore and Partners)

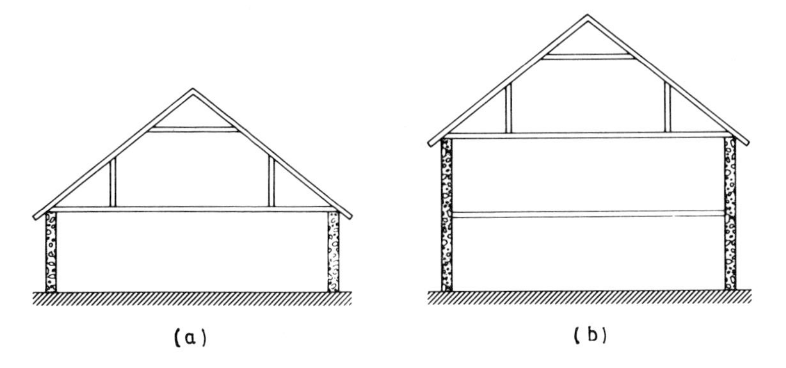

Figure 16.5 Possibilities of using roof spaces can give (a) more than 50 per cent extra usable space in some single-storey houses, (b) up to 29 per cent more for some two-storey houses (from BRE CP52/75: Crown copyright, reproduced by permission of the Controller, HMSO)

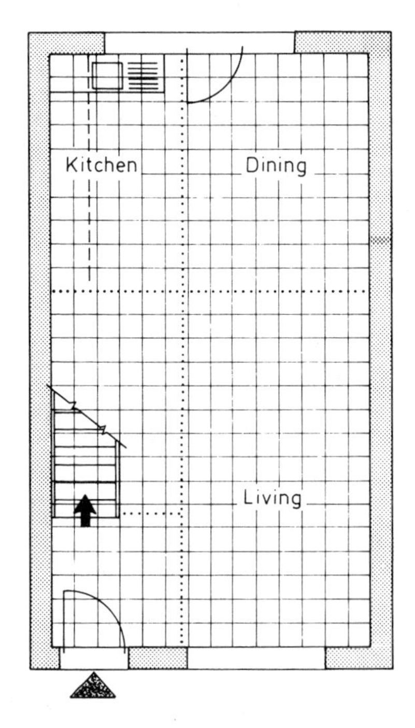

Figure 16.6 Typical shell (courtesy Barton Willmore and Partners)

REGULATIONS AND STANDARDS

Mandatory

All house building in the United Kingdom takes place within a complex framework of constraints and guidance. Some constraints are mandatory and of national application; for example, the Building Regulations[11] bear upon the construction of new dwellings as follows:

Thermal insulation and sound	Parts F and G
Structural stability	Part D
Fire precautions	Part E
Stairs	Part H
Chimneys	Part L
Drainage	Part N
Ventilation and height of rooms	Part K
Materials generally	Schedules

Scotland has its own national regulations in the form of the Building Standards (Scotland) Regulations 1971–1975.

Some mandatory regulations are of local application, for example the London Building (Construction) By-Laws.

Other parts of the framework have a mixture of mandatory and advisory status, for example the Codes of Practice covering such matters as fire (CP 3 Chapter VI) and glass (CP 152). British Standards (generally the lowest common denominator of acceptability from amongst a range of alternative products available to the industry) and Agrément Certificates are frequently adopted to provide a specification basis for the dwelling design.

The great majority of builders, who are on the register of the National House-Building Council, will regard the Handbook and Rules of the NHBC, as supplemented and updated from time to time, as a mandatory document. Builders providing housing for local authorities or Government-sponsored and supported agencies such as housing associations will frequently find that the client bodies have their own standards and specifications and that they require these to be complied with. An example is the GLC Housing Manual, and there are many others.

Advisory

The Building Research Establishment produces a series of digests and a range of current papers, most, if not all, bearing on the design and specification of dwellings and many giving sound guidance on sensible practice for the avoidance of future defects.

In the past the Ministry of Housing and Local Government, and nowadays its successor the Department of the Environment, has issued design bulletins, many with special application to the house-building field, for example:

Design Bulletin 6, *Space in the home*
Design Bulletin 13, *Safety in the home*
Design Bulletin 14, *House planning (a guide to user needs with checklist)*
Design Bulletin 16, *Coordination of components*
Design Bulletin 24, Part 1, *Spaces in the home (bathrooms and WCs)*
Design Bulletin 24, Part 2, *Spaces in the home (kitchens and laundering spaces)*

The Housing Development Directorate of the Department of the Environment issues valuable occasional papers, for example *Starter and extendible homes* (1975), *Monitoring provisions for small households* (April 1976), etc. Though advisory in nature, these documents form an important part of the framework of standards within which the house-building industry operates.

ASPECTS OF DETAILED DESIGN

General

'People demand a lot of different things from their home, and the designer's task is to satisfy as many of these demands as possible within the limitations imposed by density and cost, site conditions, technical requirements, and the need to increase productivity. This means that a balance has to be struck between many competing priorities.'[12] This paragraph, still perhaps the best and simplest statement of the house designer's job, is followed in the Design Bulletin that it introduces by a detailed checklist designed to assist in appraising the advantages and disadvantages of various dwelling plans. It is never possible for the designer to incorporate all the features he would wish to see within the given cost ceiling. The balancing of priorities is a matter for judgement, but that judgement can be aided by such studies as the BRE current paper 26/77 with its table of relative desirabilities (Table 16.1).

Floor areas

The NHBC Handbook contains in Part II, *Technical requirements for the design and construction of dwellings*, mandatory requirements (printed in red ink) that every dwelling shall comply with all relevant building legislation and, in the mandatory (red) schedule of facilities and services, requirements are set out for facilities related to dwellings of varying floor areas.

In general, housing provided for the public sector is required to conform to the standards set down in the Parker Morris Report, most recently set out in DoE Circular 27/70, though Circular 24/75 refers to the possibility of reductions in these mandatory areas where private house builders are engaged in constructing dwellings to their own design for the public sector.

Access, privacy

The majority of house plans and layouts nowadays allow for both front and rear access, but if this is not provided Design Bulletin 14 contains a mandatory requirement for public sector housing that there should be 'a way through the house from front to back . . . not through the living room. In such cases the dustbin compartment shall be on the front.' NHBC Advisory Note 1 has diagrams illustrating the point; see *Figure 16.7.*

The desirability of giving access to the house for the motor car, and making provision for garaging or parking within the curtilage, is commonly accepted. There should ideally be covered access from the house to the garage, to a separate ventilated enclosed store for refuse and to any external fuel store provided. Callers at the house should be able to stand under cover for protection from rain; this is particularly important for doors in west or south-west facing walls, becoming increasingly vital where houses are built on exposed sites. (In general, height above sea level and closeness to the western coasts are the main determining factors in increasing degrees of exposure. Information on the degree of exposure likely to be experienced in

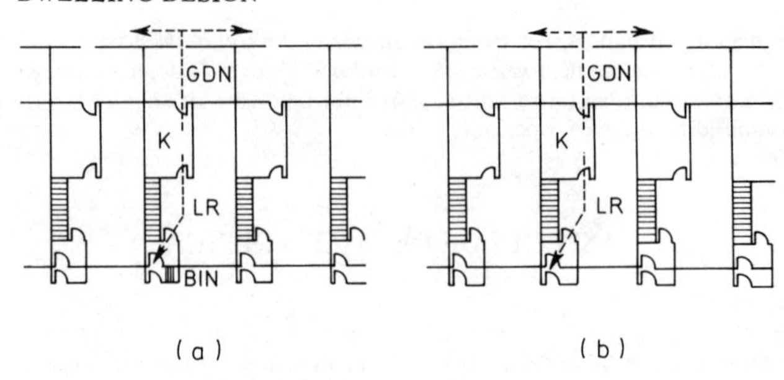

Figure 16.7 (a) Taking dustbins through living room is prevented by locating them at the front or back access road; (b) taking garden waste through living room is prevented by back access road (courtesy NHBC)

various localities is available from the Meteorological Office, but local site factors also have a substantial bearing and must be taken into account.)

There is nowadays a desirable and increasing tendency to fix meters so that they can be read from outside the house. Information is obtainable from the statutory authorities as to their requirements and recommendations for meters and prepayment meters.

In so far as privacy is a matter for the design of the individual dwelling within its curtilage, rather than for the layout design of the site as a whole, the design should give reasonable privacy to living rooms, and particularly to bedrooms, from callers and passers by. Its garden should have at least part of its area that is not directly overlooked from adjoining houses and where any overlooking from houses opposite is ameliorated by distance. The design guides produced by Essex County Council[13] and others contain specific provisions designed to secure this benefit. The guides are

Figure 16.8 Privacy by vertical window shape (courtesy Essex County Council)

also useful in pointing to the various ways in which room privacy can be provided. In brief, as well as by keeping a minimum distance between callers and the windows of rooms, it is possible to tackle this problem by care in the design of the shape and size of the windows themselves, as well as by the siting of screens and/or outbuildings. See *Figure 16.8*. Even net curtains may have a place, though their use by more than a small proportion of occupiers (i.e. those who place a specially high value on privacy) should be regarded as an indication of a degree of failure in design.

Daylight and sunlight

Apart from the Building Regulations (Part K), which require minimum zones of open space outside certain windows and which therefore have an effect on the daylight and sunlight entering the rooms behind, there is no mandatory statutory standard for the provision of these benefits. The British Standard Code of Practice (1945) had among its recommendations that 'in living rooms and, where practicable, in kitchens and bedrooms also, one of the windows forming the main source of daylight should be so placed that sunlight can enter for at least one hour at some time of the day, during not less than the ten months of the year from February to November. It is preferable for sunlight to enter living rooms in the afternoon and kitchens and bedrooms in the morning. Larders should be sited so that they are protected from the heat of the sun. Sunlight should not be considered to enter a room if the horizontal angle between the sun's rays and the plane of the window is less than 22½°.'

The British Standard Code of Practice CP 3: Chapter 1, Part 1 (1964) has among its recommendations that 'in domestic kitchens . . . a daylight factor of 2 per cent be provided over 50 per cent of the total floor area within a minimum of 50 square feet and with the cooker, sink and preparation table sited well within these zones. The sink is usually placed underneath the window but, if for planning reasons it is desired to place the sink in any other position, particular care is required to avoid the housewife being obliged to stand in her own light. Where the working surface is placed directly under the window the cill should not be more than 6 inches above it.

'In living rooms a minimum daylight factor of 1 per cent should cover at least 75 square feet and the penetration of the 1 per cent zone should extend at least half the depth of the room facing the main window. For rooms of unusual shape, or where the main window is sited at one end of the room, provision of light from windows in more than one wall assists a satisfactory distribution of daylight, and is a definite aid towards good quality lighting.

'For bedrooms a good distribution of daylight with no dark corners is more important than a high level of daylight. A minimum daylight factor of 0.5 per cent should cover at least 60 square feet, with the penetration not less than three-quarters the depth of the room facing the window.

'Rooms which are expected to be used for more than one main purpose (e.g. kitchen-living rooms) should be lighted to the more exacting of the relevant recommendations.'

The reflection factors recommended to be used were as follows:

Walls = 40 per cent reflection factor
Floors = 15 per cent reflection factor
Ceiling = 70 per cent reflection factor

and these values were recommended for rooms of all the above uses. The provision of sunlight and daylight by design of the layout is dealt with elsewhere, and is fully covered in a DoE study[14].

Orientation, aspect and view

In so far as the above factors bear on housing layout design they are covered elsewhere, but it is perhaps worth noting that CP 3[15] gives desirable orientations for the various rooms in dwellings. In broad terms, no habitable room should have its windows facing due north, bedrooms are best facing between north-east and south-east and living rooms between west and south. The requirement for food-storage spaces to be placed on a north wall continues to have validity, notwithstanding the increased use of refrigerators and deep-freeze units.

Regarding the provision of a view, windows in individual dwellings should be so disposed as to take advantage of any available view over a distance (this is more important than 'openness' of view). The placement of windows to take advantage of such opportunities as are available in this respect can have important visual benefits in the appearance of the house, by making it evident that it has been individually designed and 'tailor-made' for the space and place that it occupies. Research Report 6 issued free by the Department of the Environment, entitled *The value of standards for the external residential environment* (1976) is of particular benefit in aiding the house designer to assess the usefulness of commonly accepted standards in this and a dozen other respects varying on housing and housing-layout design.

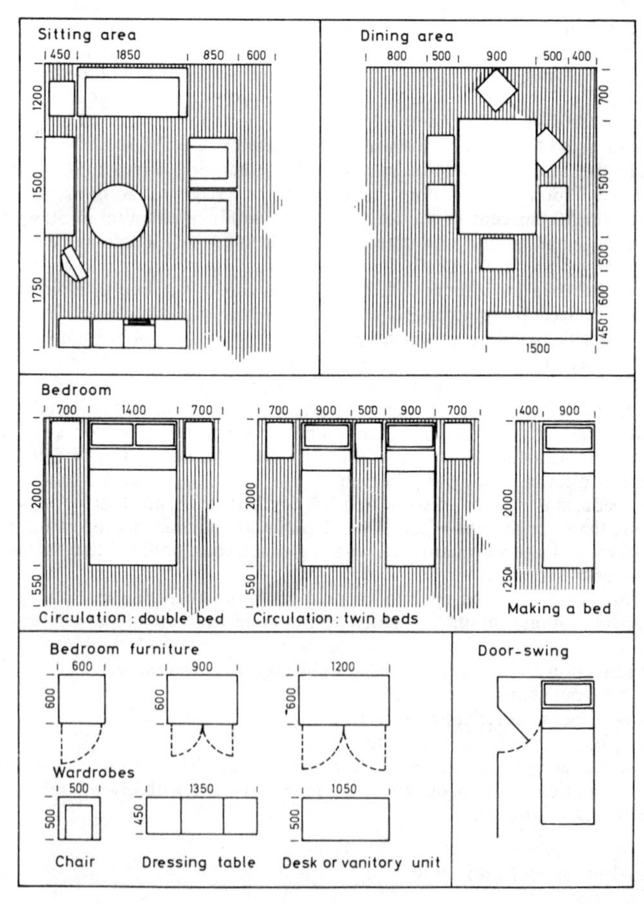

Figure 16.9 Furniture sizes: specimen layouts showing typical space allowances (courtesy NHBC)

Circulation

The NHBC Advisory Note 1, *Guide to internal planning*, illustrates the typical sizes of furniture in various rooms and the space required for circulation about them; see *Figure 16.9.* This note also contains diagrams illustrating desirable and undesirable relationships between the principal spaces in terraced houses; see *Figure 16.10.*

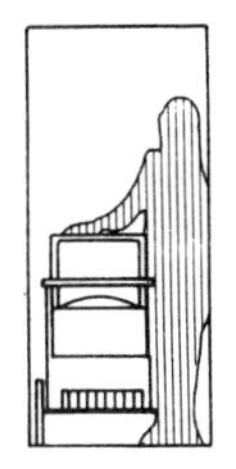

(a) Nowhere to put pram

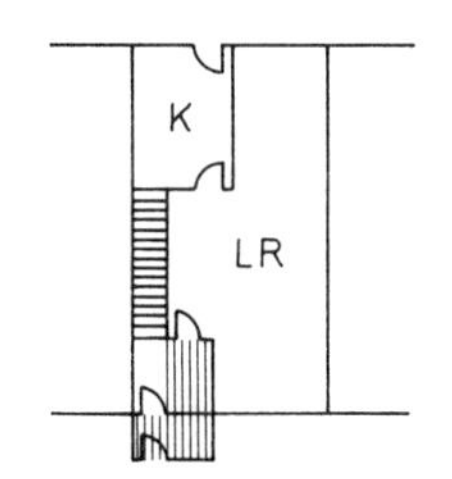

(b) Outdoor clothing in living room

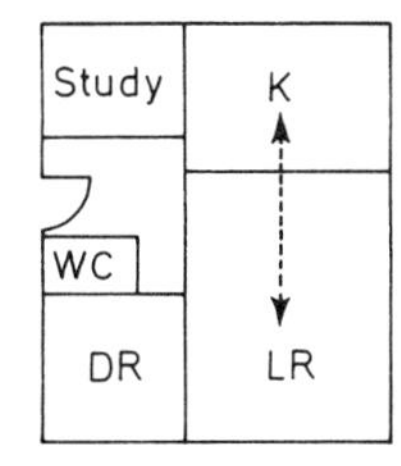

(c) Long way from kitchen to dining room

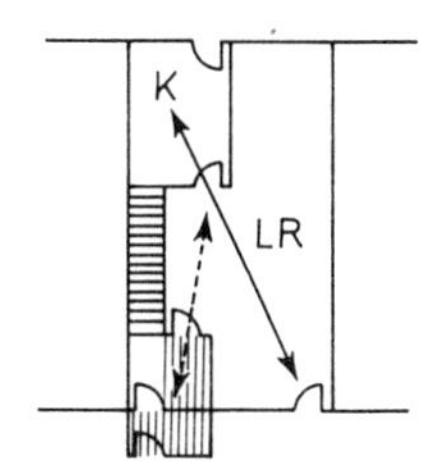

(d) Crossing living room from kitchen to front door

Figure 16.10 Some planning faults to be avoided in terrace housing: (a) and (b) can be prevented by adding porch or extending hall as shown, (c) by locating kitchen next to dining room, (d) by relocating front door as shown (courtesy NHBC)

In general, it is important if possible to be able to pass through the house without entering the main living area. It is important that the kitchen is directly accessible from the dining room, and that there is a reasonably direct route between the kitchen and the main living area. Visitors should be able to pass from the front door to the main living area without passing through the kitchen and, if possible, without crossing the dining space. Members of the household should be able to get from the front door to their bedrooms without crossing the living room (or more than a corner of it, if the house is open-plan).

The circulation spaces of the house should be sufficient in size to permit furniture of normal size to be moved about internally, from room to room and floor to floor.

Room sizes

Typical space allowances around furniture in sitting areas, dining areas and bedrooms are shown in the NHBC Advisory Note 1. These should be regarded as desirable minima, and only the provision of strong compensating advantages in some other form should justify any reduction in them. They are reproduced in *Figure 16.9.*

Sheet 4 of that note shows how, even with limited space, bathrooms and cloakrooms can be well planned. This sheet also shows the normal limit of overlap of user space requirements in the use of bathroom fittings (*Figure 16.11*).

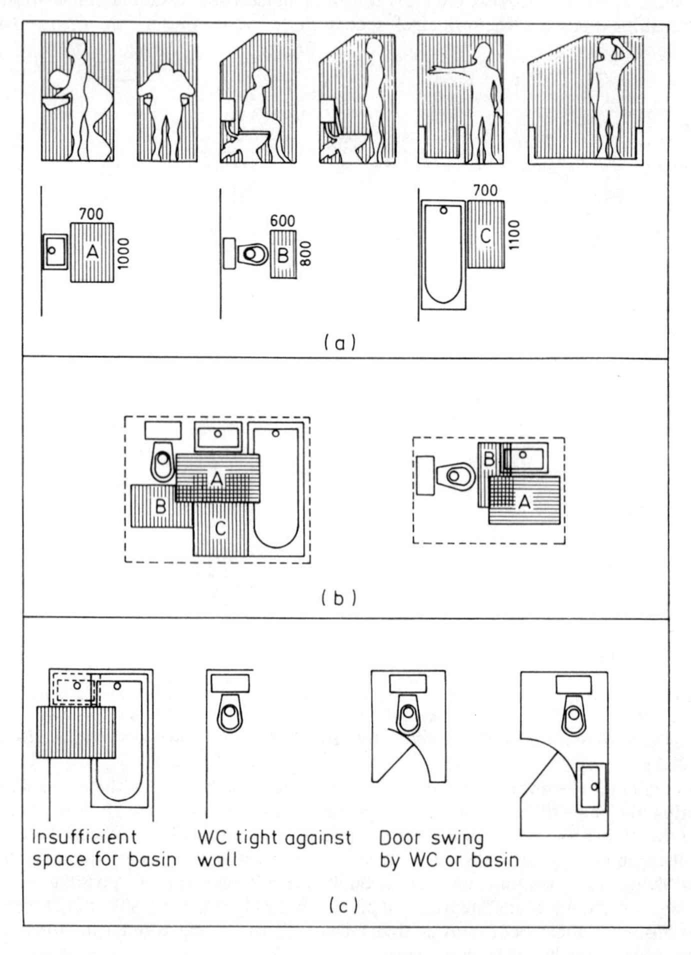

Figure 16.11 Bathroom planning: (a) requirements; (b) normally up to 100 mm overlap of space is acceptable; (c) three common faults can be prevented by adjustment of room size or more careful planning (courtesy NHBC)

For further information DoE Design Bulletin 6, *Space in the home*, should be consulted, and Design Bulletin 14 gives further planning guidance on desirable room sizes. Perhaps the most clear-cut desire on the part of occupiers to be highlighted by this and other work[16] is the very strong preference that many households have that provision should be made in the design for them to eat at least some of their meals in the kitchen space itself.

Fitments, fittings and services

The NHBC Handbook has mandatory schedules of facilities and services covering the number, size and disposition of kitchen fittings and spaces for additional appliances; see *Figure 16.12*. The schedules also call for storage accommodation (half within the kitchen at least) and an airing cupboard.

Other points to be respected and incorporated as far as possible in the house design are the provision of spaces for dishwasher, deep freeze and tumble drier in the working part of the house; sufficient space in the dining area to seat visitors as

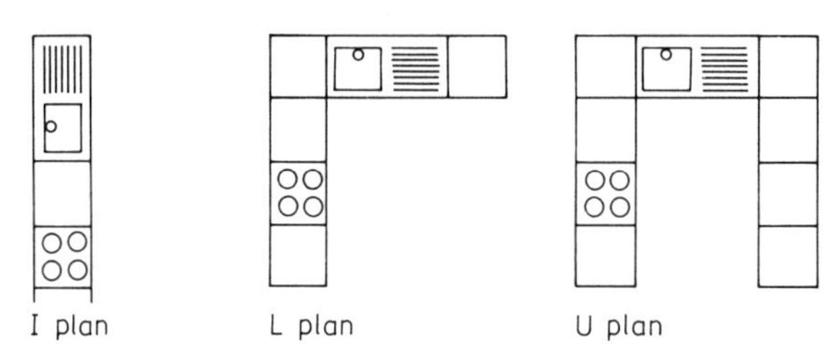

Figure 16.12 Kitchen planning: three plans that comply with best sequence (work surface/cooker/work surface/sink/work surface), unbroken by doors or trafficways (courtesy NHBC)

well as members of the family; and space for wardrobes, chests of drawers, dressing tables and cupboards in bedrooms, all with rooms of a shape that permits alternative arrangements. Attention should also be directed to the size of the linen cupboard, the size and location of any storage accommodation for fuel that is to be provided, and the possibility of locating a work bench within the dwelling.

KITCHEN DESIGN

Background

Kitchens are the most complex spaces in the house, and perhaps the aspect of British housing design at present leaving greatest scope for improvement. They have to cater for complex uses and they therefore call for a degree of complexity in fixtures, fittings and services.

There is evidence[16,17] that householders and potential house purchasers have become increasingly aware of the criteria by which kitchen design should be judged, and also some evidence that this awareness is affecting the choice of purchase and therefore saleability of the house as a product. Since studies have shown that the housewife spends on average about a quarter of her waking hours in the kitchen this is, to say the least, understandable.

The principal activities in the kitchen area itself are food preparation, cooking, serving, eating (reference has already been made to the general preference for an eating facility in the kitchen for casual meals), and washing-up. In support of these activities there is required (in addition to the fixtures, fittings and services referred to above) a provision for food storage (both dry store and cold store), services for an increasing range of fittings, work surfaces and storage space for utensils, crockery, cutlery and appliances.

In addition, and as part of this general area of the house, consideration must be given to the clothes washing, drying and ironing activities within the laundering process.

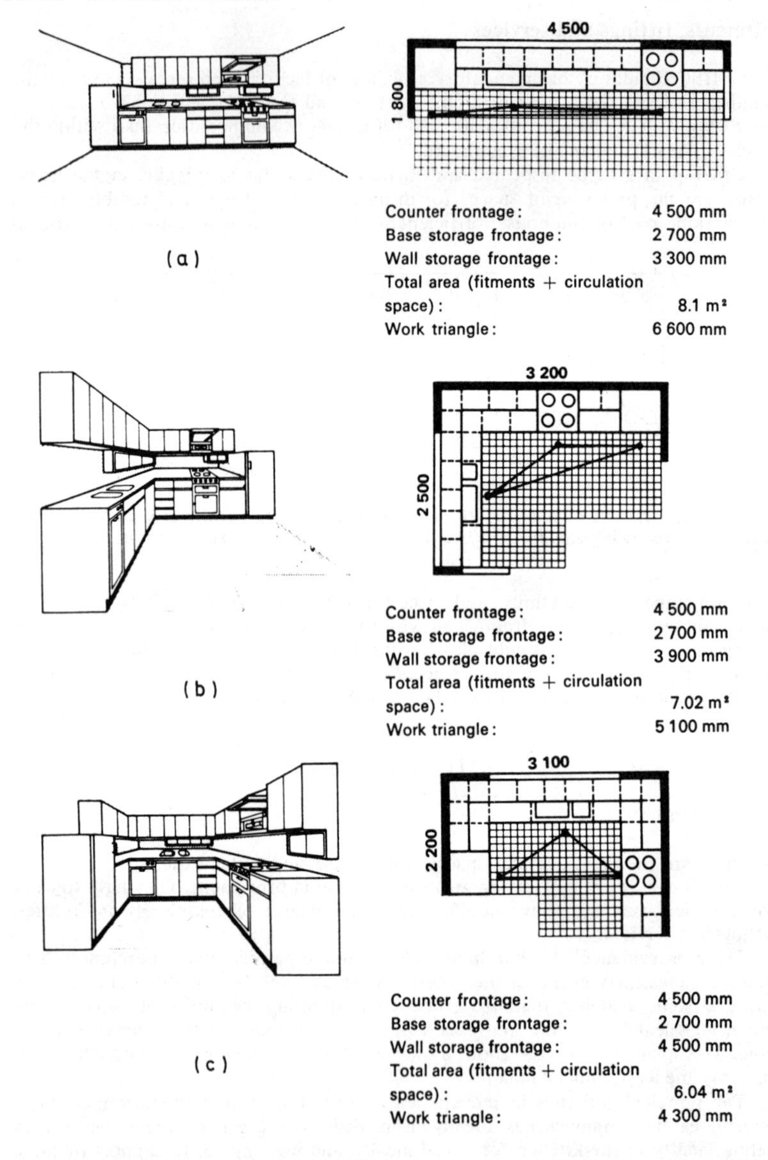

Design Bulletin No 24 Part 2, *Spaces in the home (kitchens and laundering spaces)*, gives comprehensive guidance on user requirements and the design of both kitchens and laundering facilities. The main sections include criteria and dimensions for fitments and appliances; activity zones, their design and sequential arrangements (see *Figure 16.13*); the room as a whole, its location and character; how household type affects requirements; services installations – gas, electricity, water and plumbing, heating, ventilation and waste disposal. The main sections also include an

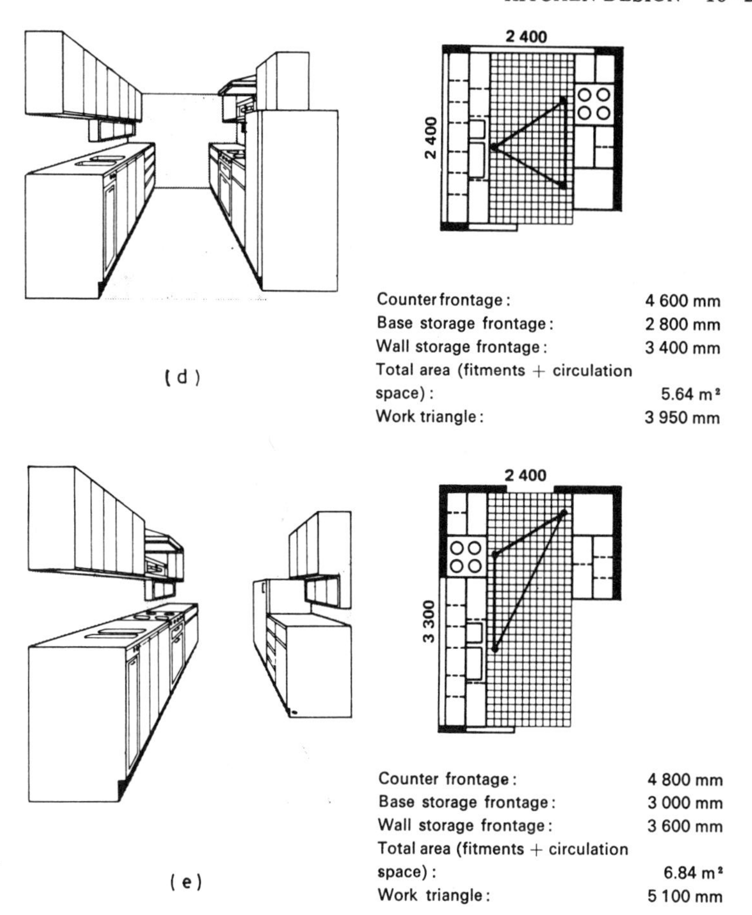

2 400
2 400
(d)
Counter frontage: 4 600 mm
Base storage frontage: 2 800 mm
Wall storage frontage: 3 400 mm
Total area (fitments + circulation space): 5.64 m²
Work triangle: 3 950 mm
2 400
3 300
(e)
Counter frontage: 4 800 mm
Base storage frontage: 3 000 mm
Wall storage frontage: 3 600 mm
Total area (fitments + circulation space): 6.84 m²
Work triangle: 5 100 mm

Figure 16.13 Diagrammatic meal-preparation sequence, showing not actual room shapes, but minimum dimensions capable of containing assembly of essential units: (a) single-line assembly; (b) 'L' assembly; (c) 'U' assembly; (d) parallel assembly without through circulation; (e) parallel assembly with through circulation (from Design Bulletin 24 Pt 2: Crown copyright, reproduced by permission of the Controller HMSO)

analysis of the activity sequences in the laundering process; space, appliances, and storage requirements; and consideration of suitable locations (see *Figure 16.14*).

The appendices to the bulletin include checklists for the designer, a summary of the current standards, and a list of relevant regulations and codes of practice for services installations. The final part of the bulletin contains a review of people's reactions to their kitchens, draws some general conclusions and goes on to illustrate kitchens and analyse the housewives' comments on each.

House builders will know the mandatory requirements in the schedule of facilities and services printed in red in the NHBC Handbook. These cannot be regarded as more than essential minima, which are helpfully illustrated at item 17 of the NHBC Advisory Note 1 (*Figure 16.12*).

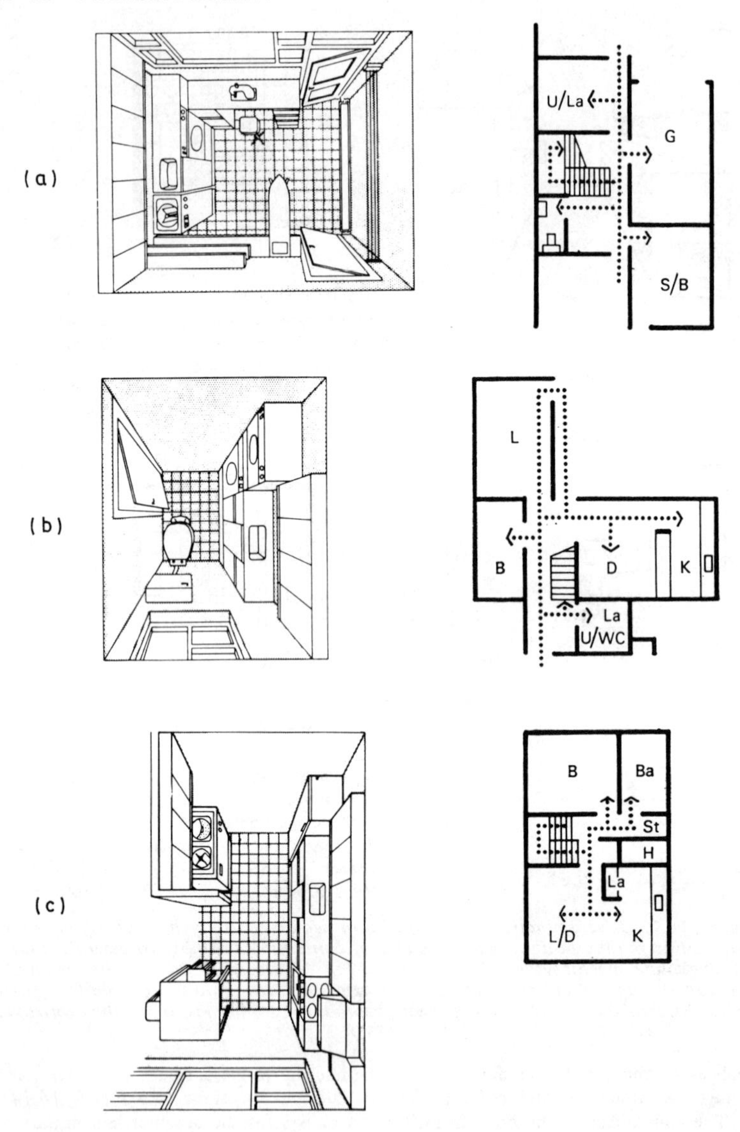

Services

Services are dealt with in detail in other chapters, but it should be noted here that a high level of illumination is required on working surfaces. The Illuminating Engineering Society (IES) Code, *Recommendations for lighting building interiors*, recommends 200 lux (= lumens per square metre) as good practice. This is best provided by lighting beneath wall units shining downwards on to work surfaces, with any central

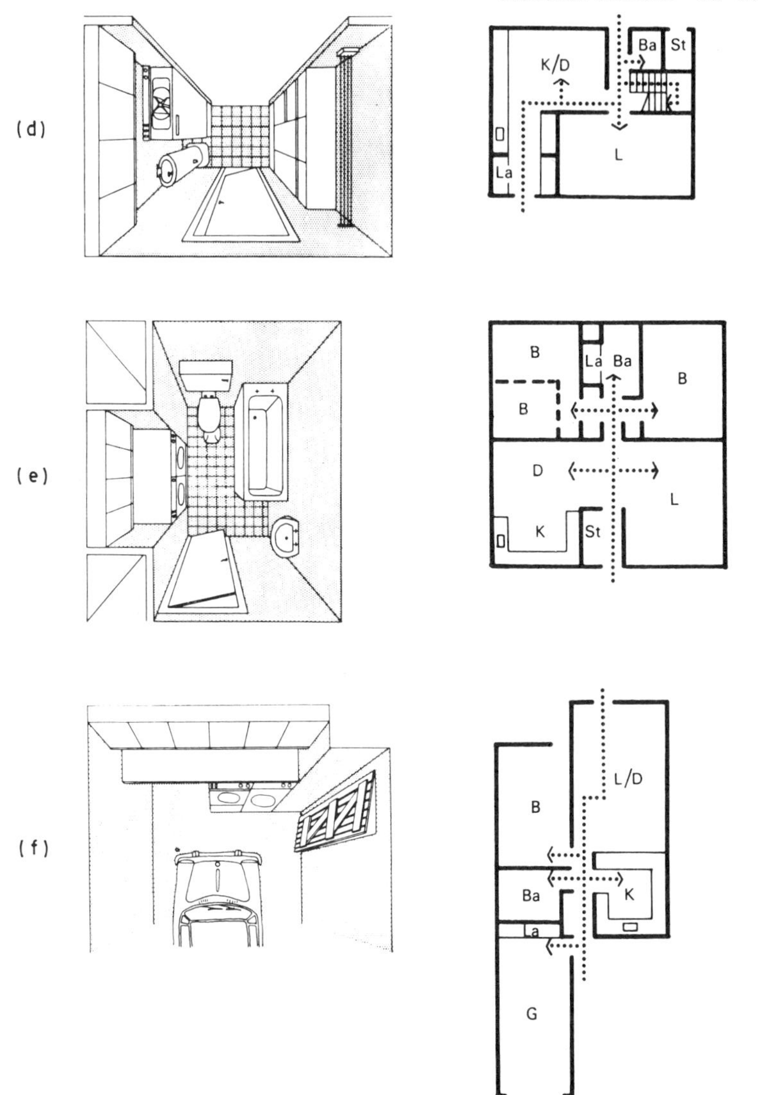

Figure 16.14 Laundering sequence diagrams showing relationship between appliance location, storage and drying; (a) utility room, (b) washroom, (c) kitchen; (d) back porch, (e) bathroom, (f) garage (from Design Bulletin 24 Pt 2: Crown copyright, reproduced by permission of the Controller, HMSO)

fitting, if one is employed, being used as background illumination only. Where there is an occasional eating area in the kitchen or, more important, where the kitchen is *en suite* with a dining space it is important that there should be separate switching to the working-area lights so that this zone can be left unlit while the meal is in progress.

As an alternative to lighting beneath wall units, ceiling-mounted lighting can be satisfactory if it located directly above the work surface so that it avoids the housewife working in her own shadow.

Heating in this general zone should be to living-room standards, i.e. 18°C when the outside ambient temperature is –1°C. This standard can be reduced in purely working kitchens, and the NHBC Handbook has a mandatory requirement for kitchen temperatures in the range from 12°C to 15°C.

If the kitchen area is independently heated it should be by an appliance that gives a rapid heat build-up. If, as is frequently the case, the central-heating appliance for the house is located in this zone, care is necessary to ensure that the casing of the appliance is well insulated in order to avoid over-heating the space.

Balanced flue appliances require an external wall, appliances with conventional flues are also available. The location of appliances for convenience and safety will be discussed in Section 17.

Food preparation

The fitments in connection with this activity are units to provide dry storage for foods and storage for utensils. This is normally provided in conjunction with work surfaces, with the storage being disposed above and below worktops in base units, tall units or wall units. The relevant British Standards are BS 1195: 1972, *Kitchen fitments and equipment*, and BS 3705: 1972, *Recommendations for provision of space for domestic kitchen equipment*. BS 3705: 1964 gave information on a pull-out working surface at which smaller-than-average women can work comfortably standing up.

The convenience and safety of the user of these fitments are considered in Section 17. It is sufficient here to note that provision should always be made in the disposition of fitments for the appliances mentioned below to be incorporated, and that continuous easy-clean surfaces are to be preferred, with junctions between units and at walls, floor and ceiling being sealed. In the absence of generally available adjustable fitments, table/counter heights of about 900 mm for stool seating arrangements and about 720 mm for chair seating arrangements are normal. A height to sink-rim and worktop of 975 mm can be combined with work surfaces at 900 mm, provided that care is taken not to employ a change in level immediately alongside a cooker hob, or to break up useful lengths of continuous worktop with such steps.

Widths and depths of base units are normally 600 mm; wall cupboards, which are normally 300 mm deep, are usually placed about 400 mm above the worktop. Hinged doors are preferable for access reasons.

The storage capacity of cupboards of all types can be greatly improved by the adoption of adjustable shelving. If this is used in conjunction with pull-out base unit shelves and baskets and the adoption of rack fitments on the rear face of doors, maximum benefit can be obtained.

Other fittings in the food-preparation zone are the sink and its associated draining boards. In this connection, it is worth noting that the usefulness of the sink is greatly enhanced, during both food preparation and cleaning up activities, if a double-bowl arrangement is adopted.

Other appliances in this zone are refrigerators and deep freezes, and there is a current tendency for these two units to be incorporated into a single tall cabinet with two side-hung front opening doors. The freezer in this case has a relatively small capacity, and freezers of this type are increasingly being supplemented for long-term storage of cold food by a chest-type freezer located either here or elsewhere.

If brooms, brushes, vacuum cleaner and other cleaning equipment and materials are not stored in the laundering area they are frequently kept in a tall cupboard in the food preparation zone, though this is not ideal.

Services to the food preparation area consist of hot and cold water to sinks, and electricity to power an ever-increasing range of utensils and appliances. These range from power-driven carving knives through liquidisers to power tin-openers and mixing machines. The location of service outlets is discussed in Section 17.

Cooking

Fitments for accommodating the needs of this activity take the form of worktops into which either single or multiple hob units can be recessed, and these can be powered by gas or electricity. Other fitments are cupboards into part of which the oven for 'roasting' can be let.

Cooking is in fact carried out by a number of different methods in the majority of households. The fittings to permit this can include, apart from boiling rings and ovens referred to above, fixed or portable grill units, rotary spits, frying and deep frying pans and, particularly in rural districts, solid-fuel ranges incorporating perhaps a griddle and hotplate warmer. Appliance sizes vary, but many are based on the recommendations in PD 6444: Part 2 issued by the British Standards Institution. Again, the services to power these appliances require convenient and safe location in close proximity to the anticipated cooking area, but on no account should an electric power socket outlet be placed above the cooker. Extract fans and cooker hoods require electrical supply in the vicinity of the cooker.

Serving

Reference has been made previously to the strong desire of many families that some provision should be made for eating food in the kitchen. Serving in these circumstances is simple and direct, provided that sufficient heat-proof standing area is available for hot pans and that the space is not too cramped to permit the housewife to move about while the family are seated.

Fitments on which the family can eat occasional meals are frequently designed to do double duty as sideboards beneath serving hatches; above such a unit, high-level cupboards are conveniently arranged to open at both front and back, so that crockery can be stored and taken either into the kitchen or into the dining area. Fitments in the serving zone could well with benefit also include storage space for trays, table linen, bread board and cheese board.

Fittings in this part of the kitchen include electrically driven carving knives, a hotplate and, not infrequently, an electric kettle, a coffee percolator and a toaster.

Eating zone

In so far as this falls within the kitchen, it may be either a fitment permanently located, a hinged or sliding table top, or a separate table. In all cases it is desirable that every member of the household should have a seating place for such meals, generally informal, as are taken in this part of the house. For two people the table size should not be less than 800 mm × 800 mm, for three people 800 mm × 1000 mm, for four 800 mm × 1200 mm, for five 800 mm × 1350 mm, and for six 800 mm × 1500 mm. Wherever possible there should be at least 550 mm behind each chair back for passing space, and where there are storage cupboards this should be increased to 1000 mm. Considerable space savings can, however, be achieved by bench seating arrangement in alcoves, where the passing space needs to be only at one end.

Washing-up

The activity of washing-up and cleaning both eating and cooking utensils centres again on the sink. In connection with this fitting there has been a slow tendency, which still continues, towards the provision of waste disposal units to deal at least with soft rubbish.

Extract ventilation by wall- or window-mounted variable-speed fans is frequently desirable, particularly where provision has not been made for extract ventilation over or adjoining the cooker, where the volume of space enclosed is small, and particularly where laundering facilities are incoporated in the kitchen or in a non-divisible alcove to it.

Apart from electricity supplies to the above appliances, there may well be the call for electricity or gas to a water heater mounted above the sink.

Laundering

This is an activity which, by convention up to the present time, has been considered as part of the kitchen zone of the house. As standards rise there is an increasing tendency to separate the laundering activity from the kitchen and to provide a second sink to serve it. This is a much more satisfactory arrangement, since it greatly reduces the risk of cross-contamination between clean and unclean material. The location of the laundry fitments can be in a separate utility room, in a combined utility/store room or in a utility/circulation lobby. Alternatively they may be associated with the garage and workshop zones referred to earlier.

The fitments associated with this use may, when dirty clothes are not kept either in the bathroom or bedrooms, include a ventilated cupboard in the laundry space.

The fittings associated with this activity are semi-automatic or automatic washing machines, twin-tub washing machines, tumble dryers, spin dryers, drying cabinets, airing racks, ironing boards, rotary ironers and storage shelving. Tumble dryers are now increasingly frequently mounted on top of automatic washing machines, making a single tall unit with both component parts being front-loaded. The size of semi-automatic washing machines is generally 600 mm × 600 mm with a 100 mm zone at the rear for services. The size of twin-tub machines is generally 500 mm deep × 800–900 mm wide, again with a 100 mm zone at the rear for services. Automatic washing machines are generally 600 mm × 600 mm on plan, though some may be as wide as 700 mm. Tumble dryers are generally 600 mm × 600 mm on plan with 100 mm at the rear for services. Drying cabinets are generally 600 mm × 600 mm. Where collapsible racks or retractable lines are used over a shower base, the base is the normal size of 900 mm × 900 mm on plan. Spin dryers are generally in circular casings, having a diameter of approximately 400 mm. The space required in which to iron, standing up at an ironing board, is a total overall of approximately 1650 mm × 1200 mm. This latter dimension requires to be increased to 1400 mm if the ironing board is mounted at a lower level, in order to permit the work to be carried out sitting down.

The airing space required is usually associated with general linen storage, but for airing purposes alone the volume provided should be 0.2 cubic metre per person per week.

Services required to laundry fittings are water and hot water (though an increasing proportion of washing machines are plumbed-in and heat their own supply), electricity and possibly gas.

The future

It is perhaps appropriate here to examine briefly the general directions in which the design of kitchens appears to be evolving. While it may not be appropriate for the individual house builder to incorporate 'experimental' plan dispositions, fitments or fittings, particularly when he has to pay regard to the demands of the mass market, the kitchen is, above all others, one area of the house where the prospective purchaser expects to see that full account has been taken of current technology.

If, in addition, the kitchen arrangements can be shown to have sufficient flexibility to be able to cope with currently evolving trends this is readily appreciated. Experience in other, currently more well-to-do, countries such as the United States and West Germany indicates that ideas adopted there at the present time will flow into the UK by way of the communications media and be adopted in the relatively near future.

Among the ideas currently gaining full acceptance in this country are adjustable shelving and the pull-out trays, racks and baskets referred to earlier. Separate 'split-level' cookers are another well known case in point. German manufacturers of cookers, fridges and other fittings have for some time been producing fully coordinated ranges of appliances that are visually integrated. Cooperation with fitment manufacturers has produced visually unified kitchen designs. These are now being made by British firms also and are becoming widely available in the United Kingdom.

The variety and number of small hand-held and medium-sized portable appliances are likely to continue to increase. The scale of provision of electrical socket outlets thought sufficient only a few years ago is already less than adequate. Double socket outlets provide a most worthwhile increase in the convenience of this facility at minimal cost.

Above all, it should be possible to show that the kitchen area, no matter what the total size of the dwelling, has been planned with an eye to the incorporation of all the fittings and appliances likely to be in common use in the near future. Care in design can achieve this end within limited areas and tight construction budgets.

BATHROOM DESIGN

Background

House builders will be aware of the NHBC mandatory standard for the provision of sanitary fittings in Section S7 of the Schedule of Facilities and Services in the NHBC Handbook. These standards do not apply to Scotland because the matters are covered by Scottish statutory requirements. In Northern Ireland, dwellings for which statutory subsidy is to be paid are exempt from the mandatory schedule but must comply with the Housing (Owner Occupation) Regulations (Northern Ireland) 1973. Non-subsidy dwellings may comply either with those regulations or with the NHBC Schedule.

Part S8 of the NHBC Handbook mandatory schedule covers the provision of water services.

Further criteria were set by the NHBC, originally in Practice Note 4, *Kitchens, WCs and storage*, but these have since been made mandatory in the red printed section S2 of the NHBC Handbook. The NHBC requirement is currently that every dwelling must have at least one sink, one bath, one washbasin and one WC, but present standards in both the private and public sector generally are such that, in all save small dwellings, these standards are frequently exceeded. In addition, all WC compartments must contain a washbasin unless they are alongside a bathroom that has a washbasin in it. The final mandatory requirement in the Schedule is that, in a dwelling with only one WC, it must be in a separate compartment unless the dwelling

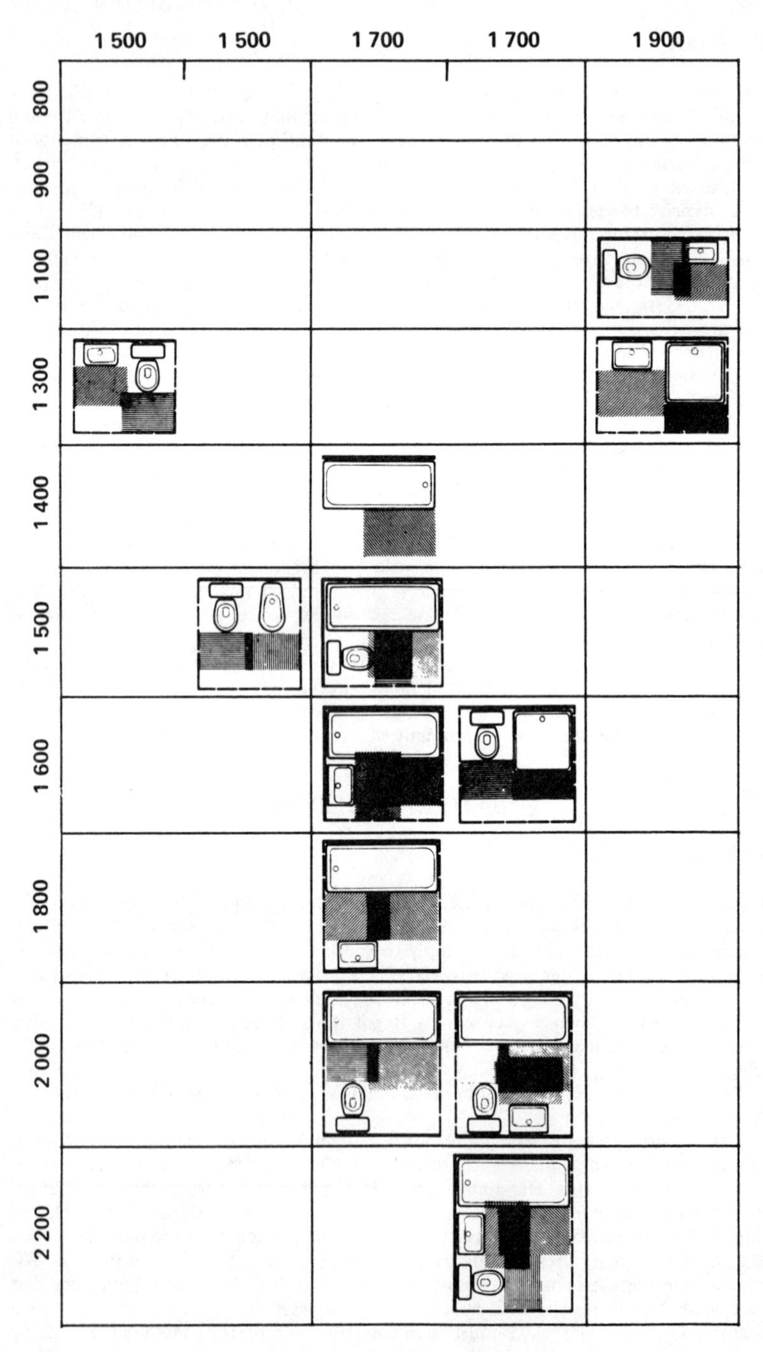

Figure 16.15 Bathroom layouts: combined dimensions of appliances and their required activity spaces (from Design Bulletin 24 Pt 1: Crown copyright, reproduced by permission of the Controller, HMSO)

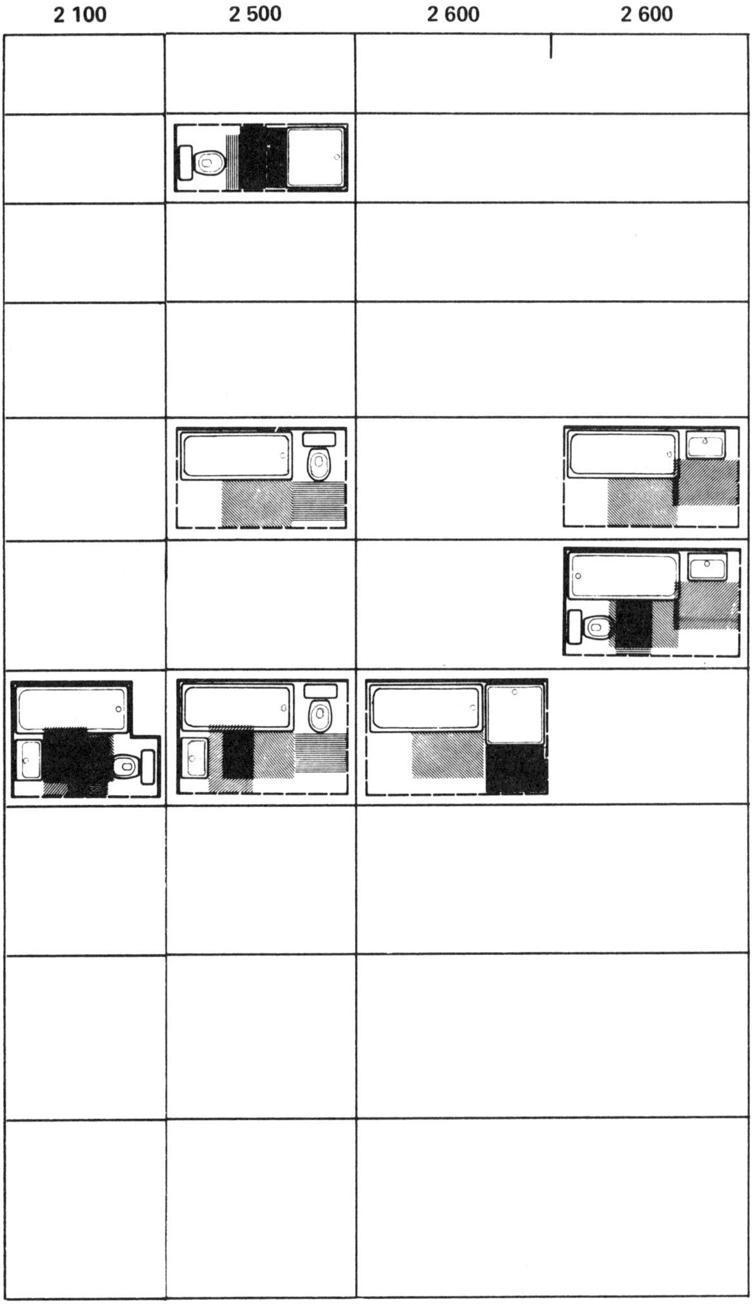

2 100 2 500 2 600 2 600

16.15 continued

is a single-storey one of less than 75 m², or, if on two or more storeys, of less than 80 m². For houses built to Parker Morris standards the provisions required are as follows:

'The WC and washbasin provision shall be as set out below.

(a) In one-, two- and three-person dwellings, one WC is required, and it may be in the bathroom.

(b) In four-person, two- or three-storey houses and in four- and five-person single-storey houses, one WC is required in a separate compartment.

(c) In two- or three-storey houses at or above the minimum floor area for five persons, and in single-storey houses at or above the minimum floor area for six persons, two WCs are required, one of which may be in the bathroom.

(d) Where a separate WC does not adjoin a bathroom it must contain a wash-basin.'

User requirement and attitude surveys carried out by Government sociologists show that in the private sector about one fifth of householders are 'very satisfied' and about a further two-thirds are 'fairly satisfied' with their bathroom and WC facilities. The proportion 'very satisfied' in the public sector was a good deal higher.

The size of the bathroom is the most important single criterion, and in order to satisfy more than half the housewives a bathroom size of about four square metres is necessary. Above this size most users are satisfied with the size of the room and are also generally satisfied with the accessibility of the fittings in it. The most frequent reason for dissatisfaction in 'tight' bathrooms is that there is insufficient space around the wash-basin for it to be used conveniently. There is also a demand for space in which to keep extra items in the bathroom, for example a stool or laundry box.

Design Bulletin 24 Part 1, *Spaces in the home (bathrooms and WCs)*, issued by the Department of the Environment contains, in addition to information on the background to current mandatory standards and the findings of a survey of user needs and attitudes, recommendations on the planning of bathrooms and the provision of a satisfactory internal environment. There are diagrams of fitting sizes, the activity space required for their use, and the degree of overlap of activity spaces normally permissible in various configurations of fitment groupings. Part of the key matrix is shown in *Figure 16.15*.

The general factors bearing on the optimum provision are generally regarded as being (a) the best disposition and grouping of fittings into compartments in order to suit the family for whom the dwelling is designed, (b) economy in the provision of services and drainage, and (c) locations to minimise noise and to secure satisfactory privacy.

Uses

Dealing with the first of these, there is no doubt that, if costs and other factors permit, the most important guiding principle should be a wash-basin in each compartment in which there is a WC. The training of children is a constant responsibility in many homes, and ensuring that they wash their hands is onerous enough, even when both fittings are in the same compartment.

Probably the next most important consideration is that visitors should have access to toilet facilities, including a WC, without the necessity for them to enter the more private parts of the house, if possible. The provision of a separate WC (with its attendant wash-basin) therefore assumes high priority. This space should be large enough to be used by an adult for washing, as an alternative to the bathroom.

There is already a tendency to incorporate shower baths, sometimes in lieu of a second bath, and some evidence of the increasing employment of bidets. Both these fittings can be expected to become increasingly popular in time to come, if only because of an increasingly wide awareness of the need for energy conservation, perhaps chiefly by way of dissatisfaction with rising fuel bills.

Whichever group of fittings is adopted in any given house type, it is important that the compartment size is adequate and that it allows sufficient activity space, room for the door-swing free of fittings and space for a stool and/or storage box for linen, if possible.

Airing cupboards opening off bathrooms, though acceptable if made necessary by the requirement of the plan form, are not desirable because of the occasional difficulty of access and also because of the frequently high moisture content of the bathroom atmosphere.

Economy

All builders will be well aware of the desirability of avoiding a disposition of bathroom, cloakroom and kitchen fittings that involves waste drainage outlets on both front and back faces of the house, or excessive lengths of drain under the ground floor slab. The ideal disposition from the economy point of view would permit all waste pipes to be connected into a single soil stack.

Nuisances

Finally, the disposition of the elements of the plan should, if possible, insulate (from the noise of fittings in use, the passage of effluent along waste pipes and the refilling of cisterns) not only the main entrance and the living areas of the house, but also, if possible, the bedrooms as well. The most recent amendment to the Building Regulations, permitting soil stacks to be either internal or external, can be expected to reduce the incidence of complaints due to soil stacks passing through living areas.

Internal bathrooms, i.e. bathrooms with no window in an external wall, are not generally liked by purchasers. Though this may be due to past experience of this space so placed but with inadequate ventilation, it is quite probably also due to the common dislike of being in a totally enclosed space which, in its more extreme form, is the unbearable medical condition of claustrophobia. The Building Research Station Digest 78, *Ventilation of internal bathrooms and WCs in dwellings*, contains useful information on this matter. The noise of extract fans is another source of potential nuisance to occupiers.

STORAGE DESIGN

General

The mandatory provisions in the NHBC Schedule of Facilities and Services are as follows:

‘(a) In every dwelling, enclosed domestic storage accommodation shall be provided as follows:

Area of dwelling (m^2)	*Minimum volume of storage* (m^3)
less than 60	1.3
60–80	1.7
over 80	2.3

(b) At least half the above volume shall be in the kitchen. The remainder shall be easily accessible to the kitchen.

(c) Part of the storage accommodation shall be suitable for brooms and similar equipment.’

These requirements do not apply to Scotland, which has statutory requirements.

Circular 27/70 issued by the Department of the Environment sets out, in metric dimensions, the mandatory provisions for housing in the public sector built in accordance with the Parker Morris Report. These provisions show separately the minimum general storage space for dwellings designed for a range of households between one and seven bed spaces. Circular 24/75 sets out certain relaxations in these mandatory requirements where a private builder is providing his own dwelling types for the public sector. Wherever possible, space should be provided for a workbench and tools, extra to the general storage.

The particular considerations that apply to the provision of storage space in addition to the above are as follows.

Fuel

The current growing awareness of the value of Britain's coal reserves, and the fact that with present technology our coal supplies can be expected to last for three hundred years, will undoubtedly lead again in future towards a higher incidence of use of solid-fuel fires, enclosed fires and heating appliances. There will therefore be a requirement for adequate covered storage space for solid fuel, and this should be accessible from the kitchen area, under cover, with adequate artificial lighting. The minimum areas given in Design Bulletin 6 are 1.5 m square where there is only one appliance, 2 m square where there are two, or in rural areas, and 1 m square in flats if there is no auxiliary storage. Hot ashes do not go well into plastic refuse containers and special provision may be necessary for them.

Refuse

The general principles involved in the design of facilities for the storage and collection of refuse are set out in British Standards Institution CP 306: Part 1: 1972. In general, the factors to be considered are (a) adequacy of the storage provision, (b) maximum convenience for the householder and refuse collector, and (c) the highest practicable standard of hygiene, amenity, safety from fire risk, and smoke and sound insulation.

The weight of refuse produced per dwelling has risen in the past and is expected to continue to increase in the future. The *density* of refuse has fallen, however, and the consequence of this is that the *volume* of refuse has risen substantially more sharply than its weight.

Systems of refuse storage and collection other than individual containers (bins or disposable sacks) in private-sector housing, apart from flats and maisonettes, have in the past been very rare. There will perhaps be some increase in the employment of communal refuse containers where houses are built in tight clusters, but in general refuse storage seems likely to continue to be best effected by the provision of individual storage containers for each dwelling. The containers should be located in the open air, if not in freely ventilated compartments. Disposable sacks should be attached to holders or used as liners to dustbins. Containers should be on a hard drained surface and, where a shelter is constructed, it should be of sufficient size to take two containers and be of adequate height to permit the lid to be removed or fully opened without the withdrawal of the container. The recommended cubic capacity of the shelter is approximately 0.7 m^3.

This space should be accessible for collection without passing through the living accommodation. It should also enjoy covered access to the kitchen part of the dwelling with provision for adequate artificial light.

Dwellings in three- or four-storey blocks should have communal refuse storage containers with chutes where practicable. The occupier should not be required to carry refuse more than 30 m to the nearest chute.

APPEARANCE

The creation of a satisfying living place out of mere housing depends in part on the achievement of an optimum balance of solutions to meet the detailed requirements set out above. It depends also on the qualities of imagination and flair brought by the builder and his architect to the complex issues of layout design described in the next section. Neither of these, alone or together, is enough. The happy blend that results in the whole being of greater value than the sum of its constituent parts comes about from the exercise of an intangible skill in dealing with the appearance of the relationship of things; this is a gift given to relatively few people. That gift, plus training and experience, can be exercised to create a sense of the rightness of what has been done which is deeply satisfying to the providers of the building, the passers by, the purchasers and the users. Some ingredients can be identified, however, and these paragraphs are intended to touch upon the more important of them.

The house builder, in creating a building such as a single house, or a group of buildings forming a cluster or a whole neighbourhood, will have several objectives and concerns in the forefront of his mind. Above all he will wish to create maximum value (and therefore saleability) at minimum cost. In pursuit of this proper objective he will be seeking to establish and to maintain a feeling of quality, both inside and out. It is vital to recognise that higher-than-average quality does *not* necessarily involve higher-than-average cost. It *does* involve much higher-than-average care, and concern not only for the workability of the spaces created, but also for their appearance.

Other proper concerns of the house builder (such as minimising delay in obtaining planning approval, or designing for maximum productivity) are facets of his overall concern to minimise cost. Much opposition has been engendered by the preparation and adoption of design guides by numerous local planning authorities, the first and best known of them undoubtedly being the one published by Essex County Council[13]. On analysis it will be found that almost all builders and architects will subscribe enthusiastically to the overall principles that motivated the authorities (a good statement of these, as far as they concern the design of building, is given on page 15 of the Essex guide). Adverse reaction has generally been either against the apparent condescension with which design principles are set out and seemingly reduced to rules of thumb, or against the too rigid determination of 'standards' which, when applied inflexibly to particular cases, tend to allow no opportunity for balancing achieved benefits against a degree of infringement on the margin of the 'standard'. There are encouraging signs that this is being realised in both local and central government and grounds for hope that house builders' architects can do good work in the new climate.

From the point of view of the purchaser, various pieces of evidence, such as a report from the National Council of Women of Great Britain[17], indicate that the purchaser seeks some measure of individual identity in the dwelling. It is, to say the least, understandable that he should wish to have some identifiable entity, on which, when he completes the purchase, he could well be promising to spend a major part of a lifetime's effort in order to discharge the mortgage. A large part of the architect's skill lies in assembling a variety of built forms in such a way that they make a unified overall composition. The National Council of Women are quite clear that their members do not want a neighbourhood to look like 'an estate', and the enclosure of spaces by the design and disposition of individual dwellings in order to create a sense of 'place' in any one of a hundred different ways can meet their point.

Some degree of individual identity is not necessarily in conflict with the objective frequently stated hopefully in planning authorities' development briefs and design guides, namely the retention or re-creation of a regional character. The choice of building forms and of the various elements, the choice of materials and their colour, can all be directed towards this end. Materials and elements may well be those that are nationally available, but careful choice from standard ranges can generally maintain reasonable sympathy with the indigenous building of the region. The choice of regional forms (for example, steeper pitches, gabled dormers, even jettied first floors) can give great visual satisfaction without necessarily descending to pastiche, let alone the creation of 'mock-'.

At the heart of successful design is care in choice, not only in the main features but also, perhaps most important of all, in the design and construction of the details. Fascias and soffits, verges, copings and flashings, cills and lintols, retaining walls and steps, screen walls and fences – how each is designed, and above all how each matches and marries with its neighbours, is a major part (perhaps *the* major part) of the difference between the so-called 'developer's pot-boiler' and a product that enhances the company's reputation. House builders will not need to be reminded of the long-term advantage of a high reputation, and the design of their product, in all its senses, makes a very large contribution towards producing that desirable end.

REFERENCES

1. Supplementary table 15, in *Housing and construction statistics*, Issue No. 17, HMSO
2. *Consultative document on housing policy* (Command 6851), HMSO
3. *Housing: needs and action*, Department of the Environment Circular No 24/75 (42/75 Welsh Office) (March 1975) and *Residential density: meeting changing needs*, Department of the Environment draft Circular RB76.258 (September 1976)
4. Britten, J.R., *What is a satisfactory house? A report of some householders' views*, BRE current paper 26/77
5. *Centre for Environmental Studies Review*, Centre for Environmental Studies, 60 Chandos Place, London WC2 (July 1977)
6. *Generic plans: two and three storey houses*, National Building Agency, Arundel Street, London WC2 (October 1965)
7. Department of Architecture and Civic Design of the Greater London Council (Housing Branch), *GLC preferred dwelling plans*, Architectural Press, London (1977)
8. *Metric house shells – two-storey plans*, National Building Agency (1969)
9. *House shells – part 1*, Housing Development Note VI, Department of the Environment – Housing Development Directorate
10. *Coordination of components in housing*, Design Bulletin 16, Ministry of Housing and Local Government, HMSO (1968)
11. *The Building Regulations 1976*, HMSO (1976)
12. *House planning – a guide to user needs, with a checklist*, Ministry of Housing and Local Government Design Bulletin 14
13. *A design guide for residential areas*, County Council of Essex (1973)
14. *Sunlight and daylight: planning criteria and design of buildings*, Department of the Environment, HMSO (1971)
15. CP 3: Chapter 1(B), *Sunlight (houses, flats and schools only)*, British Standards Institution (1945)
16. *Houses and people*, a review of user studies at the Building Research Station, Ministry of Technology, HMSO (1966)
17. *Houses of today and tomorrow*, a report of the Housing Committee of the National Council of Women of Great Britain (April 1974)

17 CONVENIENCE, SAFETY AND THE USER

INTRODUCTION 17–2

BASIC SAFETY CONSIDERATIONS 17–2

REGULATIONS AND STANDARDS 17–3

DESIGN FOR SAFETY 17–4

CONVENIENCE, SAFETY AND THE USER 17

17 CONVENIENCE, SAFETY AND THE USER

O.D. WILLMORE, DipArch, RIBA
Barton Willmore and Partners, Architect Planners

This section brings together those matters that the designer needs to consider in relation to convenience and safety in the home. It refers to both formal regulations and good practice.

INTRODUCTION

About 7 000–8 000 people in England and Wales are killed in accidents in their own homes *every year*. Probably ten times as many are seriously injured *every year*.

Government studies[1] indicate that rather less than a tenth of these are *solely* attributable to bad design. There are strong grounds for believing, however, that bad design contributes to a large proportion of the balance. It is a commonplace observation that tiredness, irritability and bad temper are frequently exacerbated, if not originated, by an inconvenient and cramped working and living environment. In this sense, the homes people live in contribute to safety by the convenience of their design, and any inconveniences resulting from design contribute to accident, injury and death.

Design Bulletin 13 issued by the Department of the Environment[2] summarises the design objective excellently as follows: 'The aim of designing for safety is not merely to prevent accidents by the use of elaborate precautions and gadgets. It is to prevent strain which leads to accidents long before they happen, by providing an environment really in tune with human beings and their activities'.

While it would be going too far to say that all that is convenient is safe (it might even be convenient at times to have an electric socket outlet within reach of the bath) it is certainly true in broad terms to say that the design considerations discussed in the preceding section, the majority of which are directed at enhancing the convenience of the dwelling as a home, do reduce the tendency to accidents that lies in the unconsidered use of ordinary everyday things.

In this section no special consideration will be given to the particular problems of designing for old people. They and young children are, it is true, the groups most at risk, but designing for maximum convenience and greatest safety for the ordinary user will go far to reducing the incidence of accidents to special categories of people also. There is a great deal of sound guidance available on designing specifically for the old, notably in the Ministry of Housing and Local Government Design Bulletin 1, *Some aspects of designing for old people* (metric edition), 1968. The special problems of designing dwellings for the disabled are well covered in Selwyn Goldsmith's *Designing for the disabled*, published by the RIBA.

BASIC SAFETY CONSIDERATIONS

It is worth remembering that many aspects and refinements of the basic act of constructing a building to provide warmth and shelter to its occupants are considerations that stem from a concern for the occupants' safety in all its aspects. Thus structural stability, covered elsewhere, is an obvious prerequisite for the avoidance of injury both to occupants and to passers by; the requirements for waste disposal, in the form of both refuse and water-borne sewage, are precautions against the risk of disease and indeed epidemics common in times past up to and including the nineteenth century. The codes and regulations governing the installation and distribution of services have as their underlying motive the minimisation of risk of injury and accident, and these are referred to below.

Turning now to factors more commonly regarded as safety risks nowadays, there is the ever-present concern in buildings of all types to take precautions against fire. Fire is an everyday hazard but nonetheless a terrifying one, particularly in its ability to injure and kill numbers of people as the result of any one incident. While modern techniques and procedures of fire-fighting help to reduce this risk, as do many of the structural and services provisions of modern buildings, there are factors that increase fire hazards. Not least among these is the propensity for certain furnishing materials to give off noxious fumes on combustion.

The most common individual accidents are falls, burns, scalds, suffocation, poisoning and electric shock. To these must be added lacerations from glass and sharp metal objects, puncture wounds resulting from human impact with hard objects, and physical damage with largely internal lesions such as result from crushing. Each of these is considered below in relation to the particular locality where injury is most likely to take place.

REGULATIONS AND STANDARDS

Mandatory

The main statutory instrument controlling the form and nature of the construction of dwellings is the Building Regulations[3], the current instrument having come into operation on 31 January 1977. Compliance with the Regulations is mandatory except for those types of building that are exempt (see Section III – Application, A5 Exemptions), and dwellings of all types are subject to the regulations.

Chapter 37 of the Health and Safety at Work etc. Act 1974 has, in Schedule 5, a comprehensive list of the subject-matter of Building Regulations. Included within this subject-matter is a great deal with implications for the builder, and ultimately for the occupier or user of houses. For example, Part B of the Regulations deals with the fitness of materials (including surface finishes), and Part D includes, *inter alia*, the requirement for the structural stability of strip foundations and structure above foundations. There are, for example, deemed-to-satisfy provisions for chimneys of bricks, blocks or plain concrete where the underlying concern is the safe dispersal of exhaust fumes as well as structural stability.

In Part E, *Safety in fire*, Section 1 deals with structural fire precautions and (after certain preliminary definitions and interpretations) the table to Regulation E2 records that private dwelling-houses are brought within the orbit of the Regulation under purpose group I while flats and/or maisonettes are within purpose group III.

Schedule 5 of the Health and Safety at Work etc. Act 1974 makes it plain that among the purposes of the subject-matter of Building Regulations are fire precautions, including:

(a) structural measures to resist the outbreak and spread of fire and to mitigate its effects;
(b) services, fittings and equipment designed to mitigate the effects of fire or to facilitate fire-fighting;
(c) means of escape in case of fire and means for securing that such means of escape can be safely and effectively used at all material times.

While structural fire precautions are covered in Section 1 of Part E, Section 2 deals with means of escape. Specific means of escape provision in the Regulations for domestic premises are at present limited to flats and maisonettes. However, in addition to the provisions for staircases required under E13 (stairways, ramps, balustrades and vehicle barriers), Part H contains, in addition to the general requirements under H2, further requirements under H3 and the table to Regulation H3. These requirements are founded, it is plain, on a concern for prevention of falls and other accidents, as well as a concern to ensure the provision of proper means of escape.

Non-mandatory

House builders providing houses for the public sector under the requirements of the Parker Morris report[4] may come to regard Appendix 4 of that report as having virtually mandatory effect. In this Appendix, entitled Safety in the Home, there is a list of design points 'to be considered'.

Other publications setting standards that should be part of the house-designer's library of information include the Building Research Station Digest 43[1] referred to earlier; the Regulations for the Electrical Equipment of Buildings[5]; British Standard BS 3922: 1976, *Domestic medicine cabinets*; Building Research Station Digest 44 (second series), *Safety in domestic buildings – 2*, available from HMSO.

In addition, *Building* magazine has carried a series of articles on 'Designing Against Fire'; of particular relevance are articles published on 11 June 1976 (*External fire spread – two-storey housing*), 10 September 1976 (*The choice of interior materials (1) – ceilings*), and 8 October 1976 (*The choice of interior materials (2) – surface finishes for walls and ceilings*).

DESIGN FOR SAFETY

Immediate environs and entrance

The immediate environs of the house constitute one of the most prolific sources of accident and injury. Single steps are a frequent case of falls: if such a step cannot be avoided, its presence should be signalled by other features, for example a flower box, dwarf walls, or a clearly marked change of material, preferably with a change in colour also. The dimensions of external steps need to be larger generally than they would be if the steps were internal, but excessively shallow or excessively steep steps themselves constitute an additional hazard.

A sudden ramp is even harder to mark and to see and can be the more dangerous on that account. A handrail serves as a visual warning as well as a practical physical aid, and a rough surface to the ground finish is highly desirable. Artificial lighting should be placed to illuminate all such hazards.

Paths around the dwelling should be separated from the external walls by a strip with planting or with a surface too rough to walk upon (large cobbles); this will substantially reduce the risk of accidents caused by passers-by walking into open ground-floor casement windows (see *Figure 17.6*). Even a low curb at the rear of parking spaces can at times be instrumental in preventing accidents while cars are reversing.

A step is expected at entrances, but the change in level should if possible be such that only a single step in this position is required. The threshold should form the nosing to the step. The door is safer if opening inwards; see *Figure 17.1*.

External play spaces for small children should be securely enclosed and shut off from vehicles. External fencing should be large enough to be clearly seen, while vertical boarding is preferable to horizontal, which can be climbed by small children (see *Figure 17.2*).

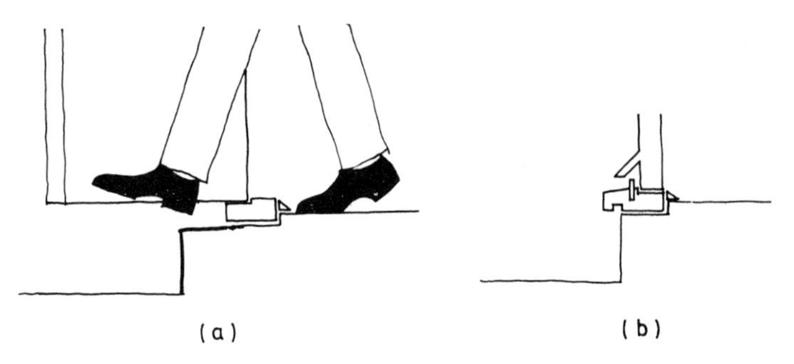

Figure 17.1 External doors: a threshold forming the nosing of the step is safer than one set back; (a) bad – door opens outwards; (b) good – door opens inwards (from Design Bulletin 13: Crown copyright, reproduced by permission of the Controller, HMSO)

Figure 17.2 Fencing: horizontal is easy to climb, vertical is hard to climb (from Design Bulletin 13: Crown copyright, reproduced by permission of the Controller, HMSO)

Stairs

Staircases (see *Figure 17.3*) are the source of the greater number of accidents in the home. Risers and tread sizes will be governed by the Building Regulations but winders, which are permissible, should be avoided if possible, particularly at the top of the stairs. There should, if possible, be a handrail on each side of the stair. The handrails should follow the pitch of the stair throughout, be continuous, and be carried well beyond the top and bottom risers. Single steps should be avoided.

On landings, doors (including cupboard doors if possible) must not open out into the circulation area near the head of the stairs, and balustrading should have no gap wider than 90 mm (75 mm is even better). Top and bottom steps should not encroach into landing or hall spaces; see *Figure 17.4*.

Staircases with open risers can be confusing and can cause vertigo. Building Regulations now require that there should be no open riser, or opening in a riser, larger than would permit the passage of a 100 mm diameter sphere.

The lighting in the vicinity of stairways should cast illumination on to the risers, and this applies to both natural and artificial light. Windows in stair wells should not,

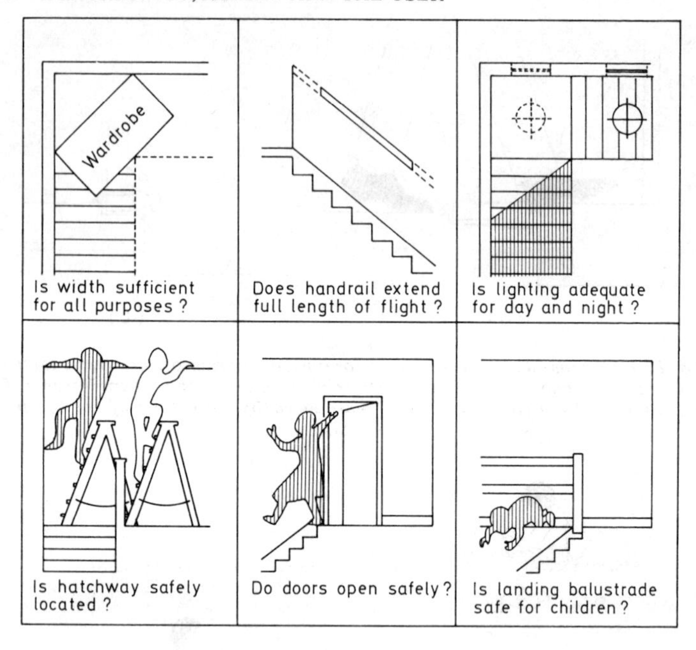

Figure 17.3 Points to consider in relation to staircases (courtesy NHBC)

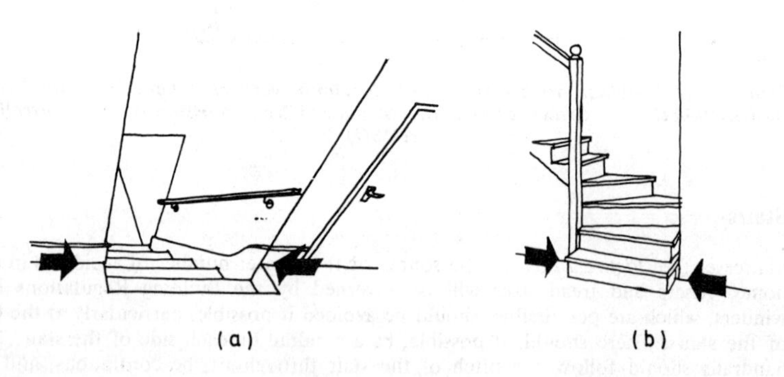

***Figure 17.4** (a) Top step of a flight encoraching on to a landing space is a hazard; (b) projecting step at bottom of a flight may also cause a fall (from Design Bulletin 13: Crown copyright, reproduced by permission of the Controller, HMSO)*

however, be located at high level above the foot of the stairs, where access for opening and cleaning presents particular risks of a fall.

Access hatches into roof spaces should be located away from the staircase, while if possible there should be ample headroom at the hatchway within the roof itself, and a pull-string light-switch within easy reach.

Doors

The placing of doors near the head of staircases has been mentioned above, but the same point applies with almost equal force to any door opening into a circulation area. Doors mounted on double-swing hinges should have sufficient glazing to permit a person approaching to see another person moving towards the door on the other side. On the other hand, doors should never be so fully glazed that they are difficult to see. Glazing below lock-rail level should never be of ordinary-quality glass. The danger of over-glazed doors is compounded if the glass, both above and below lock rail level (800 mm) is not suitably toughened by treatment; see *Safer glazing at home*, produced by the Safer Glazing Information Service in conjunction with the Royal Society for the Prevention of Accidents.

Windows

The above comments regarding doors apply with equal force to windows carried down to or close to floor level, whether they be french windows or fixed lights. In addition, openable windows at first floor level and above present the hazard of the temptation to lean too far out of an open casement or to sit too far out on the cill in order to clean the exterior surfaces; see *Figure 17.5.* Some types of window have frames that project internally as well as externally when they are standing open. Window-opening limiting devices should be used on all wide-opening lights above ground floor level. Cills of wide-opening lights need to be high enough to prevent small children climbing over. See BS Code of Practice 153 Part 1, *Windows and roof lights – cleaning and safety* for further information.

Floors

Floor finishes, where they are provided by the builder, should be chosen with care to avoid materials that are excessively slippery when wet. A case in point is rubber in porches or utility rooms. Mats at entrances should be recessed into mat wells so that they are flush with the remainder of the floor.

Services

Permanent fixing points for a fireguard to BS 2788 should be incorporated wherever a solid-fuel open fire is installed.

The safety aspects of services installations will be safeguarded by installation in accordance with the requirements of the statutory authorities, but it is worth making a few general points. Gas and electricity meters should be in separate cupboards and in fire-resisting boxes. Light switches should be located so that *the way ahead* can be illuminated. Illuminated switches have a useful role to play, particularly in circulation areas. The staircase lighting should invariably be switched from both top and bottom. Switches and socket outlets in bedrooms should be placed so that the light can be switched on from the door and also from the bed. Pull switches are generally in use in bathrooms and WCs, and the string should be long enough to permit a child to use them.

Gas appliances with pilot lights should not be located in positions where there is likely to be a draught, for example alongside an open window.

Other points of general application on services are shown in *Figure 17.6,* while electricity, and the safe use of it in the home, is covered in guidance issued by the Electricity Council, Trafalgar Buildings, Charing Cross Road, SW1.

Figure 17.5 Importance of placing and design of windows: (a) maximum reach for cleaning windows; (b) fixed equipment below window restricts reach for cleaning; (c) difficult to clean extending window if it is too deep; (d) difficult to clean horizontal-pivot window if it is not reversible through 180° (from Design Bulletin 13: Crown copyright, reproduced by permission of the Controller, HMSO)

Kitchens

Government research shows that 40 per cent of personal accidents in the home occur in the kitchen/eating area. In the course of food preparation falls, burns and scalds are prevalent, but their incidence can be substantially reduced by care in design.

The sequence of work first identified in *Homes for today and tomorrow*[4], i.e. work-surface/cooker/work-surface/sink/work-surface (all of which should be continuous and uninterrupted), greatly reduces the risks of collisions and spillages. NHBC Advisory Note 1 shows three arrangements, the principle common to all of them being that no access or door (even a cupboard) should be allowed to interpose between the fittings. In addition, access to the refrigerator should be in close juxtaposition (completing the well-known triangle of fittings) and there should be no through circulation route crossing between these points; see *Figure 17.7*.

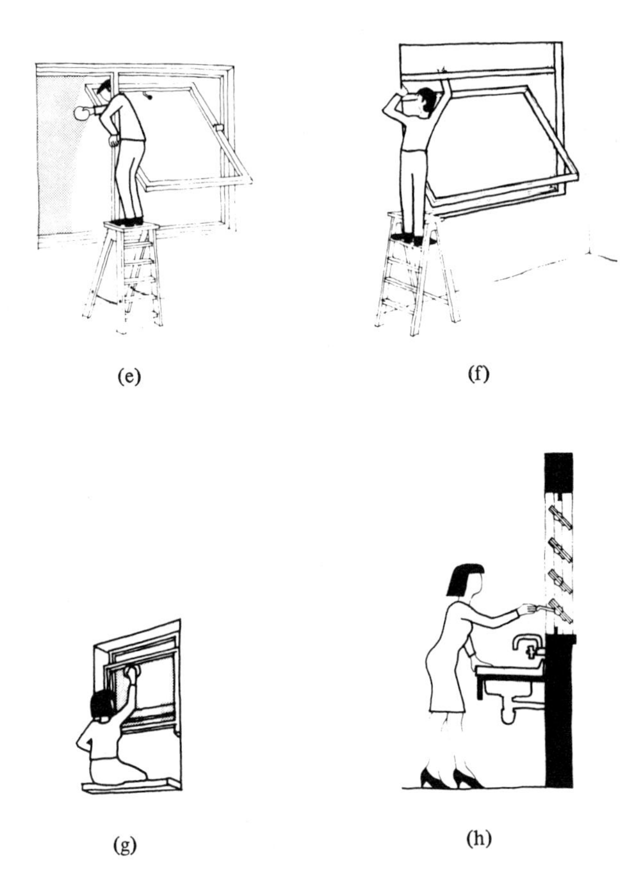

Figure 17.5 Continued: (e) difficult to clean fixed light next to horizontal-pivot window; (f) difficult to clean fixed light above pivot window; (g) easiest way to clean vertical sash windows, by sitting on cill, has obvious risks; (h) louvre windows with controls readily accessible over sink (from Design Bulletin 13: Crown copyright, reproduced by permission of the Controller, HMSO)

It is impossible to select a single worktop height that is equally convenient for all people in the normal range of heights, say between five feet and six feet tall. The current compromise is about 900 mm high. Until adjustable worktops at heights ranging between, say, 850 mm and 1050 mm are available, it is possible to incorporate worktops at two levels. If this is done, however, particular care must be taken to avoid abrupt changes in level partway along a worktop run; above all, the worktop on each side of the hob plate of a cooker must be at the same level as the hobs themselves.

Cookers must never be placed in the corner of a room where access is inconvenient and where pan handles will inevitably project over the front face of the appliance. Doors should never be positioned so that a door-swing passes through the standing space in front of a cooker, for obvious reasons.

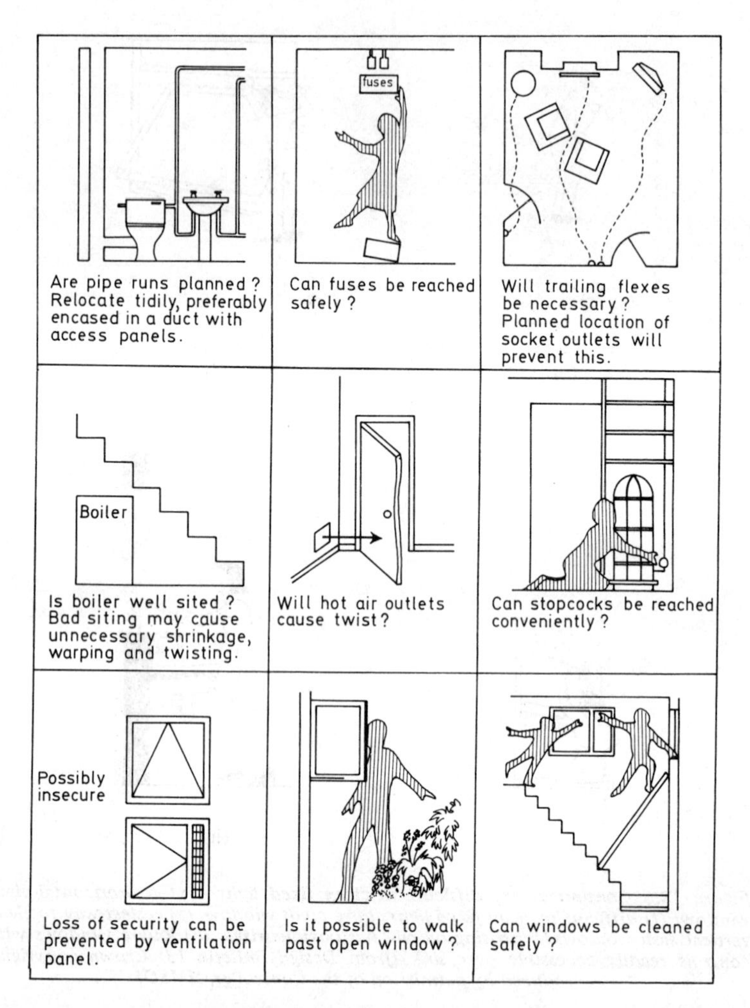

Figure 17.6 Design and placing of services and components affect occupants' safety and convenience (courtesy NHBC)

If individual hob units are incorporated in worktop surfaces they should be set towards the rear of the fitments so that handles can not protrude beyond the face.

In connection with the serving of food, it is also important to ensure that there is no through traffic passing between the cooking zone and the serving zone.

Mention has been made earlier of the desirability, from the convenience viewpoint, of arranging that the mother at work in the kitchen can oversee the play of young children both in an enclosed outdoor space and in a separate and secure area within the house. In the interest of safety it is vital that the mother is not forced to have young children playing under her feet during any of the processes of work in which she is engaged in the kitchen.

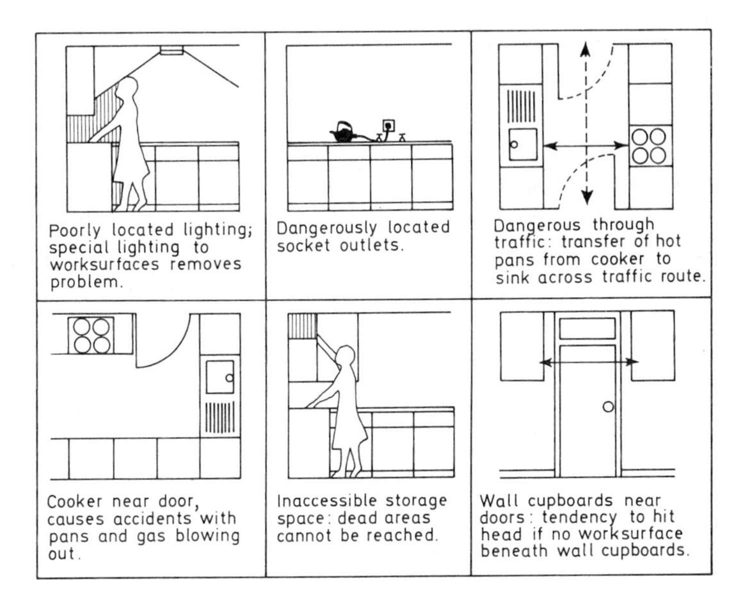

Figure 17.7 Some avoidable mistakes in kitchen planning (courtesy NHBC)

Wall storage units should not be mounted in positions where people can walk into the corners of them. This generally means locating them above base units, but when they are hung in this position care must be taken to avoid placing them inconveniently high.

Whether the storage units are for food or for utensils, it is important to ensure that they are not located above hob plates and it is desirable that the maximum shelf height should be about 1800 mm. There is a school of thought that urges that only sliding doors should be used in cupboards mounted on the wall above work surfaces. Sliding doors, however, are generally held to be rather less convenient in that they restrict access to the whole of the cupboard at any given time, and channel guides are frequently criticised for the difficulty they present in cleaning, particularly after minor spillages.

Perhaps the measure best designed to relieve strain and to reduce the incidence of burns and cuts is the provision of adequate natural and artificial light directly on to the working surface. This is discussed in Section 16.

The laundering activity is frequently considered in conjunction with the kitchen, although as mentioned elsewhere there is great benefit if it can be catered for in a separate but adjoining space. Laundry work is associated with a relatively high incidence of burns and scalds, particularly if the space available is cramped and ill-lit.

It is important that both hot water and space heating are available in this locality, in order to avoid both the carrying of pans of boiling water and also the use of portable heating appliances in a wet condition and steamy atmosphere. Laundry work is particularly tiring, and ventilation will assist in reducing accidents due to fatigue. As mentioned above, the floor finish must be non-slip.

Throughout all the activity zones of the kitchen, the provision of an adequate number of electrical socket outlets is an important safety measure. Only by this means will the use of long trailing leads be avoided. While long trailing leads are undesirable in any location (see *Figure 17.6*) they can be particularly dangerous in the kitchen, where, for example, they may be connected to a pan of boiling fat.

Bathrooms

The chief risk in the bathroom is of falls due to slipping in the bath and on entering and leaving it. The chance of this type of accident can be minimised if a securely fixed grab rail is provided, if the bath specified has a flat bottom, and if the floor finish of the room itself is of a non-slip variety, even when wet.

Medicine cabinets, if provided, should be fixed at a high enough level to be out of the reach of small children (say a minimum of 1400 mm to the lowest point) and they should preferably be of the lockable variety.

Locks that can be opened in emergency from outside the compartment are of benefit on both bathrooms and WC doors, as there are circumstances where quick access to a child or person taken ill is vital.

The electrical installation permissible in bathrooms is governed very closely by the IEE Regulations[5] because of special danger from electricity, when wet hands and feet may provide a good return circuit to earth. Specially safeguarded shaver sockets should be installed, and the light fitting should be of an enclosed type. The latter point will reduce, if not entirely eliminate, the risk of people plugging appliances into the lighting circuit. Heating should be of an adequate standard to avoid any temptation to bring in portable appliances, while if an overhead electric heater is installed it must be fixed at high level and well away from the bath. It must be operated by a pull-cord.

The British Standard Codes of Practice for gas installations stipulate the safeguards necessary with this service. In general, meters, main switches and fuse boxes should all be fixed high enough to be out of the reach of small children. Switches and fuse boxes should be close to an entrance, but they should not be fixed so high that falls result from their use.

That purchasers and users are aware of the importance of many of the points made above is made plain by reports by the National Council of Women of Great Britain[6] and others.

REFERENCES

1. *Safety in domestic buildings – 1*, Building Research Station Digest 43 (second series), HMSO (1964)
2. *Safety in the home*, Department of the Environment Design Bulletin 13, metric edition, HMSO (1971)
3. *The Building Regulations 1976*, HMSO (1976)
4. *Homes for today and tomorrow*, Ministry of Housing and Local Government/ Department of the Environment, HMSO (1961)
5. *Regulations for the electrical equipment of buildings*, 14th edition, The Institution of Electrical Engineers (1966); reprinted in metric units incorporating amendments (1970)
6. *Houses of today and tomorrow*, a report of the Housing Committee of the National Council of Women of Great Britain (April 1974)

18 SURVEYING AND SETTING OUT

INTRODUCTION 18–2

SURVEYING 18–3

SETTING OUT 18–8

SURVEYING AND SETTING OUT 18

18 SURVEYING AND SETTING OUT

B.H. ASPIN, BSc(Eng), CEng, MICE
Technical Director, New Ideal Homes Ltd

This section provides a practical summary of the surveying and setting-out procedures used in house building.

INTRODUCTION

Whatever the size of a building project, whether it be the erection of a small porch on an existing house or a large multi-storey office block, at some stage it will be necessary to do (a) some surveying and (b) some setting out.

Simple methods can be used for both surveying and setting out, and will produce results of adequate accuracy. Such simple setting out and surveying should be well within the competence of the average builder.

Available information

Before starting to survey a site it is well worth checking the extent of the available information. The first and most readily available source in the UK is the Ordnance Survey. Most built-up areas of the country are available in sheets plotted to 1 : 2500 or 1 : 1250 scale. These are regularly updated and show in great detail all roads, buildings, trees, etc. Up-to-date revision information for particular areas can be obtained from local Ordnance Survey agents or from the headquarters of the Ordnance Survey in Southampton.

The Ordnance Survey sheets are also essential for establishing data such as benchmarks for levelling and National Grid references.

It is also worth checking with a local authority office, as they often keep up-to-date revisions of Ordnance Survey sheets. Some care is necessary in using any revisions that are hand plotted, as these can be somewhat inaccurate.

A further check with the legal transfer plan for the site is always worthwhile. This should always be available to the surveyor and should show the apparent position of the boundaries of the site, but it is known from experience that very often these plans are not drawn particularly accurately or to any particular scale. They are generally to be regarded as indicative only.

Site reconnaissance

It is essential to carry out a detailed site reconnaissance, firstly to check the available material against the actual conditions on the ground and secondly so as to be able to assess the actual nature of the site. Very often such a reconnaissance will reveal site encroachment, new building or alteration to boundaries etc. Site reconnaissance will generally emphasise the need for an accurate survey of the site.

Instruments and equipment

Every builder should have directly available to him the following basic surveying and setting-out equipment:

- two 30 or 50 metre steel tapes;
- five ranging rods;
- a supply of 50 × 50 mm wooden pegs, approximately 450 mm long;
- a quick-set level with metric levelling staff;
- a Cowley or similar site level for setting out.

A useful addition to the above list is a Site Square, used for quickly setting out right angles and especially useful for setting out foundations etc.

It may well be that, in larger organisations, builders have access to a theodolite (possibly of the microptic direct-reading type) and perhaps an electronic distance-measuring device. Distance-measuring devices, although comparatively simple to use, are really the tools of a professional surveyor. They require a considerable amount of care and maintenance, and in particular require very careful setting up. Their great value is when setting out extensive base lines, but such base lines are normally only met when carrying out surveys of a comparatively large site.

However, using the first few items of equipment, the builder should be able to survey and set out to a suitably high degree of accuracy for most normal buildings and structures.

SURVEYING

Before starting or commissioning any survey, consideration must be given to the ultimate use of that survey. Will it be necessary to set out buildings to the nearest millimetre, or will the nearest half metre be sufficiently accurate? What accuracy is required in terms of levels? What information is required about the surroundings of the site? Do we need to know the position of manholes on adjoining sewers? Is information required about the types and position of trees? And so on.

The dimensions of a site in a highly built-up area will need to be known with a great deal of accuracy, whereas the dimensions of a site for houses in comparatively open country will probably only need to be known to a low degree of accuracy.

What scale will the survey be plotted at? This again reflects the accuracy necessary in the actual surveying. For most residential developments, a scale of 1 : 500 is sufficiently accurate for normal purposes. However, for a small site, especially in a built-up area or where the density of development of the site is very high, a scale of 1 : 200 or even larger may well be necessary.

Whatever scale is selected, a high level of accuracy can be achieved by very simple methods of surveying, and it is not necessary to use expensive and advanced instruments. The golden rule is to keep the survey as simple as the area and use of the land and the accuracy required dictate.

Triangulation

The basic principle behind surveying is known as *triangulation*. Consider the triangle shown in *Figure 18.1*. Although geometrically this can be drawn if the length of all three sides is known, this information in surveying terms is not considered to be sufficiently accurate to establish the exact location of a particular point. In surveying it is considered that point B (known as a *station*) can only be truly fixed if a third line is established to prove it, such as line D–B.

Similarly, consider a quadrilateral (*Figure 18.2*). Geometrically a quadrilateral can be drawn by knowing the length of one of the diagonals in addition to the lengths of the sides and perhaps one of the angles. For surveying purposes, however, it is necessary to known the length of both diagonals.

It can be shown that the greatest accuracy is obtained with triangles that are equilateral. This of course is not always possible, but it is always worthwhile to include an extra leg in a survey to ensure that the triangulation consists of as nearly equilateral triangles as possible. If the lengths of the sides of triangles and the 'third line' is known, it is not necessary to know the angles between the lines and a survey can be plotted using drawing compasses only. However, when surveying larger areas and a theodolite is available, angles as well as distances can be measured; this enables

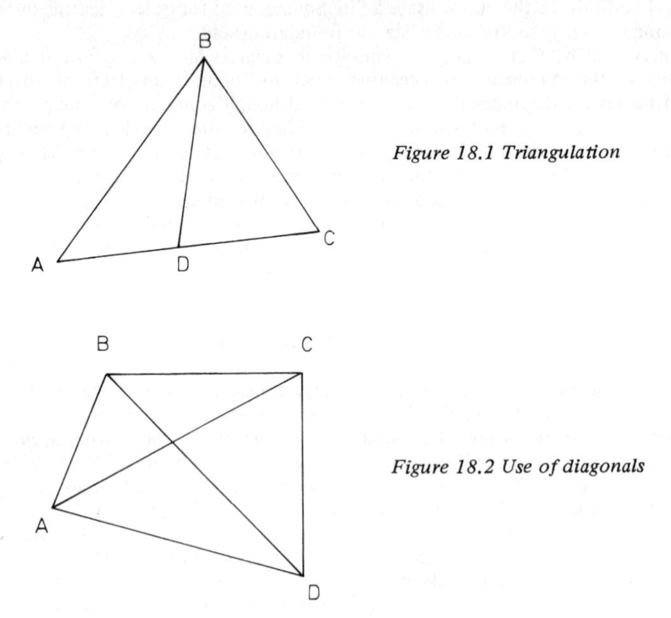

Figure 18.1 Triangulation

Figure 18.2 Use of diagonals

greater accuracy to be obtained. A typical perfect triangulation situation is shown in *Figure 18.3*. However, on other than fully cleared sites there are usually extensive obstacles, which will then dictate the survey-line pattern and the location of stations.

Survey stations

The first consideration when on the site is to consider the location and then the establishment of the survey stations. In general terms they should be placed as close to the boundaries of the site and the limit of the survey as possible, so that unobstructed views can be obtained in as many directions as necessary. The station should be marked by a wooden peg, well driven into the ground and if necessary concreted in. It should be noted that, where reference to stations will be required over long periods, the construction of more permanent markers will be worthwhile. A station should always be triangulated by measurements to the nearest permanent objects so that it can be readily relocated if removed by vandals or by unfortunate action on the site! Permanent stations will always be of use at a later date when setting out. The setting out is then directly related to the survey of the site.

When measuring between stations, it is useful to set up a ranging rod immediately behind each station and to drive a nail into the centre of the peg to form the reference point. Even on a small site it is unusual to be able to measure from one station to another without an intermediate reference on the ground, and it will be necessary to range in 'by eye' an intermediate marking point on a straight line between the stations. This normally will be at the maximum limit of the tape. The ranging in, of course, must be done standing at one station and looking directly at the other, as it is impossible to range in accurately from an intermediate point. Survey details are picked up from the survey lines by means of right-angled offsets from the stretched-out tape, and should be plotted into a survey book as shown in *Figure 18.4*. More important points are best located by tie lines, as shown in *Figure 18.4*. Note that it is not considered necessary to use a third line to form a fixed triangle in this particular operation.

Use of theodolite

As mentioned previously, a theodolite may well be available. Modern theodolites are comparatively robust, but nevertheless do have to be handled with care as they are precision instruments. A complicated theodolite is not necessary for the simple

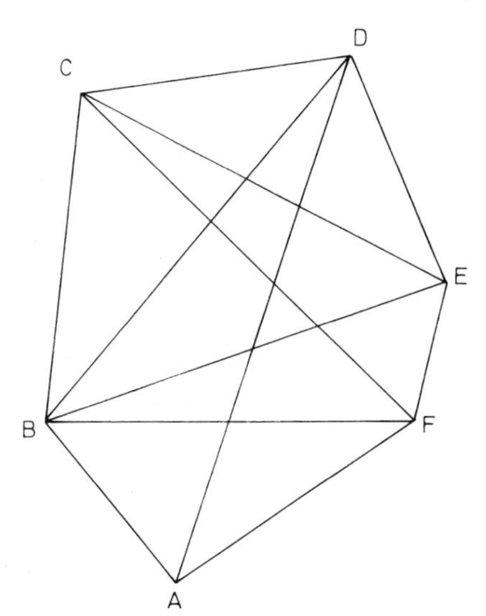

Figure 18.3 Perfect triangulation

survey work being considered here, but any theodolite used should be capable of being read easily and directly.

When the theodolite has been set out at a station, it should always be remembered that it is useful to read off as many angles from that setting as possible so that work is not duplicated. It is always easier to read off and book more angles than are perhaps actually necessary than to have to reset the theodolite at that station, with a risk of not precisely coinciding with the first setting up. Great care is needed with the booking to ensure that it is clear and precise and that the correct sighting is recorded against a particular reading. All reputable survey textbooks show various methods of booking. If a theodolite were available to carry out the traverse ABCDEF shown in *Figure 18.3*, it would not be considered so essential to check the cross ties AD, BE etc.; the theodolite thus facilitates surveying large sites or around obstacles.

Details of the survey are again plotted between stations using offsets and triangulation, but it is also often worthwhile taking specific readings to principal objects from survey stations rather than leave these to be picked up from survey lines.

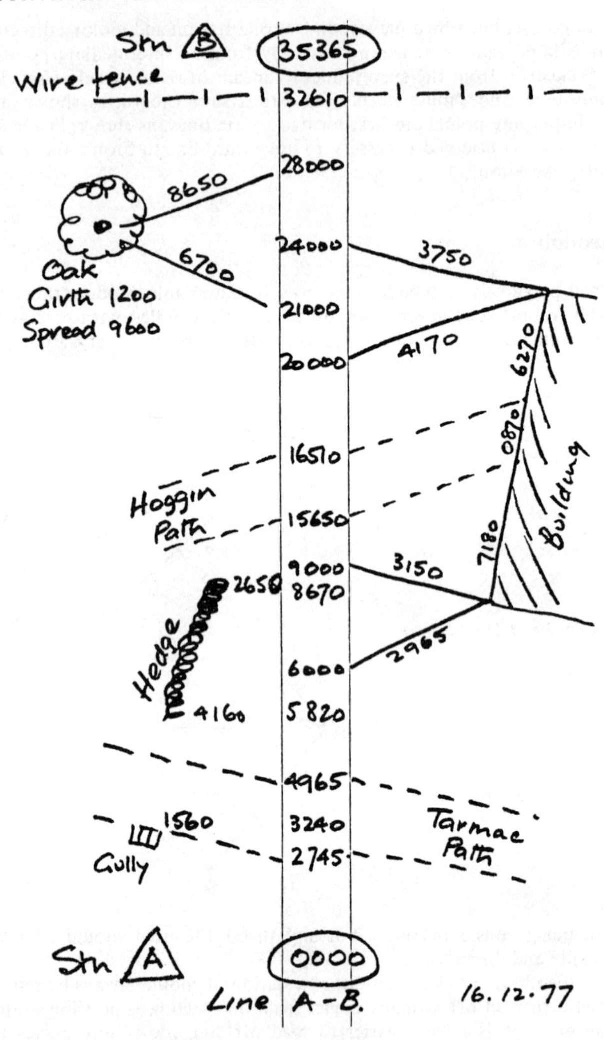

Figure 18.4 Typical page from a survey book

In the writer's opinion, the use of a theodolite needs fairly constant practice. If the survey is such as to really require the use of such an instrument, serious consideration should be given to the employment of professional surveyors to carry it out.

Aerial surveys

For large areas of land, aerial surveying becomes distinctly economical. In this type of survey, overlapping photographs are taken of the ground from an aircraft flying at a suitable fixed altitude and on fixed courses. The photographs are then processed through a stereographic scanner and the plotting of the survey is virtually automatic. However, aerial surveys do require extensive ground control work to establish a base line and therefore the accuracy. They also require extensive ground work to check around such objects as trees (if these are to be located accurately, and identified), and to check around buildings to ensure that there are no overhangs and other hidden work. Aerial photographs do, of course, only show the extent of the spread of a tree, and the position of the trunk cannot normally be identified.

The cost of taking the photographs is high and requires expert flying. There are, however, numerous companies operating over the whole of the country who can produce excellent results. The plotting of these surveys is quite fast, but it should be remembered that if flying is required anywhere near a large civil airport, particular permission has to be obtained and there could be some delay in obtaining results. Local authorities regularly use aerial surveys in the preparation of trunk-road improvement and other schemes, and it is therefore worthwhile checking with them to see if any company has recently flown over the area. If this is the case, they may already have a set of photographs that can be fairly quickly translated into an excellent survey.

Recording and plotting

One of the secrets of good, accurate survey work is neat and tidy field recording. This is sometimes very difficult in inclement conditions, as all builders must know! The corollary to neat and tidy recording is neat and accurate plotting in the office. All surveys should be plotted on stout paper using a sharp pencil. The plot should be kept as a master for reference, and if further copies are required a tracing can of course be taken. One should always remember that virtually all tracing media move with changes of temperature and over time; if exceptional accuracy is required, therefore, reference must continually be made to the original plot. Ordnance Survey field work is plotted directly on to metal plates with a minimal coefficient of expansion.

Levelling

This is of course an integral part of the survey operation. In some instances only spot levels will be required, and these are very often obtained after the survey has been plotted. However, when extensive field work is required levels are generally taken as the survey proceeds. Again specification of the survey is most important in this respect.

On large open sites contours are generally required. These are normally obtained by taking a grid of levels over the whole site, plotting them, and then interpolating to produce the run of the contours. The contours therefore cannot be taken as precise levels, but can only give a general guide as to the nature of the fall or rise of the ground.

All levels should, of course, wherever possible be related to Ordnance Survey datum. The Ordnance Survey again produce up-to-date lists of the height of bench-marks. All levelling should start and finish on a known datum point so that the levelling accuracy can be checked. All surveyors have their own preference for recording levels; in general, for checking long runs of levels, the rise-and-fall method of recording is more easily checked, but for setting-out purposes the height of collimation method could be preferred.

SETTING OUT

The first step in setting out is to consider the material actually available to produce the information that is required on the site. Before leaving the office, check:

- that all drawings are available;
- that all information is available – i.e. the necessary angles or dimensions marked on the drawings; the value of the data for levelling; all the information about the surroundings of the site. A setting-out drawing may well be available showing all this information, and it is always worth checking this drawing most carefully so that total reliance can be placed on it when out in field.

The best type of setting-out drawing shows the necessary outlines of the project only and gives considerable detail as to the necessary dimensions. For instance, a setting-out drawing for estate housing roads should clearly indicate all the centre lines, where curves start and end, the length of any transition curves, and of course all necessary dimensions. The drawings will be prepared bearing in mind the type of setting out that is to be used. Again it is emphasised that all setting out should be kept to the simplest possible methods.

A convenient way of seeting out for the professional surveyor is to use coordinates. In other words this is the plotting of all information referred back to a grid system. The grid system is very often based on the National Grid. It is particularly easy to set out such things as estate roads using coordinates, because the plotting and calculation is made very much more simple. As a builder would generally only have comparatively simple instruments available to him, this method is really best left, as stated before, to the professional.

All structures that are set out should be checked, and to this end it is always necessary to calculate diagonals of rectangular buildings and any other check measurements. The principle of surveying (i.e. triangulation) is applied to setting out, so that the correct diagonals and check lines can easily be calculated in the office; it is then only a moment's work to check the squareness etc. on site.

Setting out curves is not normally a particularly complicated job, and the calculation of offsets is quite simple. If a theodolite is available the curves can be more readily set out with a minimum of labour, but nevertheless both operations should be well within the capability of the average person.

As mentioned previously, it is common to use transition curves, especially on road works. These can be comparatively complicated to calculate, and as long as the total offset is easily known they can readily be set by eye. This step may horrify some professional surveyors, but in the author's experience it has always been necessary to ease transition curves, even on major road works, by eye, and there seems to be no reason why they could not be eyed in from the very start of the operation. Do not forget that transition curves are also used vertically.

Very often projects are set out in stages and at every stage it is important to make a running check. It is rather inconvenient to find, for example, that when you have set out eight houses on a small estate of twelve there is only room left for three because the first house had been set out incorrectly! Running checks need not be made to the accuracy of the actual setting out, but they are so easy that they are sometimes forgotten. It is not difficult to run one's eye along the front of a line of houses to see if they correspond, even roughly, to the plan. Simple pacing on a small project is very often sufficient as a check.

Simple setting-out reminders

Straight lines Quite long straight lines can be established by eye using ranging rods, although of course if a theodolite is available this eases the task and produces a more accurate result.

Offsets Offsets from known centre lines can be established using simple right-angled triangles. Don't forget the 3–4–5 triangle and the 5–12–13 triangle. The 3–4–5 triangle, being more nearly equilateral, gives a more accurate result, but if a shorter base line is available requiring a longer offset the 5–12–13 triangle has its value; conversely, of course, it can be used to establish more accurately a shorter offset with a longer base line. An optical square or a Site Square is of immense value in this operation.

Triangulation Remember the theory of triangulation, i.e. a third check line is always required for important points and both diagonals of a quadrilateral should always be checked.

Setting-out levels Level pegs should always be clearly marked so that there is no confusion about their meaning. It is easy to write on the side of the peg the level to which it has been set, and if there are offsets or any other instructions these can be written on the peg as well. Level pegs very often have to be somewhat offset from the work to which they refer, especially during drainage runs, and there is always room for confusion.

Curves and radii Do not forget that the simplest way of setting out a true radius is to establish the centre and swing an arc using a tape. Curves can be set from a straight line using offsets.

A final reminder: use the *simplest* method suitable for the job.

19 EARTHWORKS

INTRODUCTION 19–2

SITE INVESTIGATION 19–2

EXECUTION OF EARTHWORKS 19–4

19 EARTHWORKS

B.H. ASPIN, BSc(Eng), CEng, MICE
Technical Director, New Ideal Homes Ltd

This short section provides a practical aide-memoire to the house builder working on a site requiring major earthworks. Site investigation, prevention of defects and excavation are considered.

INTRODUCTION

To the average house builder, the mention of the word earthworks immediately suggests a vast expensive operation that interrupts all other operations on site, an operation that is messy and always causes the greatest inconvenience. However, earthworks carried out correctly and at the right time need not be too expensive, although, like all other building operations, they do require much previous thought and detailed planning. Earthworks should always be carried out on the largest scale possible; by this means costs are generally reduced. On occasions it is possible to improve a difficult, sloping or uneven site by extensive earth-moving and filling operations, making it far more economical in terms of building and access – this point should always be borne in mind.

It is suggested that the following guidelines should help to reduce costs and to improve site efficiency.

SITE INVESTIGATION

Site reconnaissance

A reconnaissance of the site and immediate surroundings will often reveal details of the subsoil. A close consideration of the existing vegetation is often worthwhile: the presence of marsh-type grasses, for instance, is often indicative of water-retaining ground, and if vegetation as a whole is sparse, questions should be raised in the mind as to why. There could be very little topsoil, or immediately below the topsoil could lie hard chalk, rock or particularly infertile and probably difficult ground.

Do not forget to look at the immediate surroundings to the site. With some luck it may be possible to see disturbance such as earth-moving in progress, and this can give a very good guide as to the general nature of the subsoil. Have a good look at all the buildings that may surround the site: if there are signs of cracking or settlement, further investigation is obviously necessary. This reconnaissance of the site will also help the builder to get a feel for the actual physical shape so that he can assess in general terms, if earth moving is being considered, how this can best be achieved. It will also help him to consider what further information is required.

Existing information

It is always worth checking with local authorities to see if they have any existing knowledge or records of the site. Building control officers often have quite good memories about the type of ground that has been met in the locality, and a quick chat with them is always worthwhile. A final source to check is of course the

Geological Survey. Unfortunately, the survey is printed on maps that are generally well out of date, and great care is often necessary to locate a site accurately. However, these maps give very good indications of the general type of ground to be met. It must always be remembered that in certain types of substrata – Bagshot and Reading Beds for instance – conditions are very variable, thus necessitating very extensive site investigation.

Sub-soil investigation

It is always worthwhile carrying out some form of subsoil investigation on the site. It may well be that detailed investigations have already been made. The type of earthworks that are envisaged will affect the type of site investigation necessary. If only small amounts of earth are to be moved and from a small depth, it may simply be necessary to dig a few trial pits in the area of the proposed works. For more extensive operations trial bore-holes carried out by a specialist would be a particularly good investment. Where trial pits are dug, and the builder does not have the knowledge to assess the conditions found himself, he should employ a specialist to advise. Always arrange for some samples to be taken and tested in a specialised laboratory; in particular, wherever clay is to be dug and moved it is essential that moisture content and strength tests are made. It has been known for clay when compacted to occupy less space than it did before excavation!

The investigation will also reveal the depth of the standing water table on the site; the water table level is probably one of the most important matters to be considered when disturbing natural ground in any way.

Appreciation of possible failure

Whenever naturally formed ground is disturbed by any building work at all, there is always a risk that the natural geological balance of the site will be seriously disturbed. Even the removal of vegetation can seriously affect the characteristics of the subsoil. There are specific areas of the country where it is known that the subsoil is unstable even in its undisturbed state; this of course should be revealed by the site investigation. Hillsides in clay districts are particularly treacherous, and it is not often appreciated that disturbance on one part of the hillside can result in further movements on the hillside well away from the disturbance. It cannot be emphasised too strongly that where any work is contemplated in such areas, expert advice is essential. The cost of this advice will always be repaid.

All materials have a natural 'angle of repose', i.e. they will always tend to assume a particular slope when placed without any compaction. Ultimately, any slope that is man-made will tend to stabilise at this natural angle. The effect of rain and wind will be to flatten this angle by erosion, and careful consideration should be given to the design of slopes bearing this in mind.

A further important point to remember when considering the scope of earthworks is the possibility of failure due to the surcharge of a tipped material on to natural earth that is already very close to the limit of its natural equilibrium. Yet again, expert advice should be sought. It is not unknown for the weight of a pile of bricks to start a downward movement of a slope because the natural balance of the slope has been disturbed by the weight.

Soil mechanics is a particular speciality whose importance has been appreciated only during the last fifty years or so. Knowledge is rapidly widening, and the importance of obtaining specialist advice in areas where difficulties could be encountered cannot be stressed too highly. If you employ an expert, do not forget that you must give him all the facts, otherwise his conclusions and advice could be wrong.

EXECUTION OF EARTHWORKS

Balancing of cut and fill

To minimise the cost of earthwork, the general object must be to endeavour to balance the amount of excavation and the amount of filling, i.e. to achieve cut and fill balance. Generally carting to tip off the site is the most expensive operation. As every builder knows, not only is it expensive, but often very difficult, to dispose of surplus material!

On small sites, it is often difficult to achieve a balance of cut and fill, as it is usually not possible to carry out the operation in one and there is no room to hold a balancing stockpile. Local conditions may make carting off the site so expensive as to make it necessary to consider an adjustment of the design (such as raising floor levels etc.) to take advantage of the possibilities that are then open for disposing of surplus materials on the site.

Equipment

Obviously the execution of earthworks depends on numerous factors. If the builder wishes to carry out the work himself, he will wish to utilise to the best advantage the plant that he already has. However, if extensive work is involved, it is always worthwhile to consult a specialist firm. The quotation obtained from such a firm will almost certainly be very economic and a far better proposition than endeavouring to carry out work using the wrong type of plant, even though it is available.

Modern earth-moving plant is available in such a range of sizes that a specialist firm can easily arrange to carry out the operation at an optimum level. However, if it is decided to carry out the work directly, great thought must be given to the type of plant to be used. It is no use trying to excavate 1000 cubic metres of material and cart it away in one-cubic-metre dumpers.

Great care and thought must be given to the type of compaction equipment used where filling is concerned. It is no use placing fill without consolidating it efficiently, and again this is where the expert can help. There are numerous types of compaction plant available, some of which are suitable for one job rather than another. Badly filled and compacted ground can only lead to trouble, and most careful site supervision is most necessary.

Earth-retaining structures

Earthworks by their very nature are often associated with earth-retaining structures such as retaining walls, which in themselves are more often than not sources of considerable expenditure. These earth-retaining structures are often a considerable hindrance during the execution of the earthworks themselves, and great care has to be taken in the planning of the site to ensure that carrying out one aspect of the works does not interfere with another. Filling has already been mentioned in a previous paragraph, but filling behind retaining walls needs the most careful execution and supervision. It must not be forgotten that concrete does take time to cure. This means that some days' wait is essential before the filling can actually take place.

It must also be mentioned that retaining walls and similar structures affect the environment considerably. Consideration should therefore be given to whether these structures can be eliminated from the site by the use of plateaux and slopes; such alternatives do of course take up a considerable amount of space, and the physical restrictions may rule them out.

Temporary works

If any temporary works are involved during earth-moving operations or during any other operation it is essential that they are properly designed. The present legislation on safety on sites lays great emphasis on the responsibility for the design of temporary works, and it cannot be emphasised too strongly that expert advice must be obtained.

20 ESTATE ROADS, PATHS AND DRIVES

ESTATE ROADS 20–2

PATHS AND DRIVES 20–15

ESTATE ROADS, PATHS AND DRIVES 20

20 ESTATE ROADS, PATHS AND DRIVES

D.A. STEPHENSON, BSc, CEng, FICE, FIStructE, FIArb
Consulting Engineer

This section relates to estate roads constructed as part of a development, to footpaths within and without the curtilage, and to drives. It considers the statutory requirements, layout, geometric design, structural design, and accommodation of public utilities.

ESTATE ROADS

Statutory powers

Section 40 (2) of the Highways Act 1959[1] empowers local highway authorities – generally County Councils – to agree to undertake the maintenance of private or occupation roads, whether existing or proposed. Section 40 (5) of the Act empowers the authority when entering into such an agreement to impose such conditions as it may think fit, having regard to the expense of construction, maintenance or improvement of the road.

This effectively gives the highway authority power to construct estate roads itself, at the cost of the developer, or to require that the developer shall construct the roads to the authority's standards. Such standards are aimed at ensuring that estate roads are safe, both for pedestrian and for vehicular traffic; that they are of suitable geometric design for the traffic to be carried; and that they are of adequate strength to carry that traffic without needing excessive expenditure on maintenance.

Many highway authorities publish notes setting out their requirements for the design and construction of estate roads, for the guidance of developers. Such requirements may of course vary from one authority to another, so the guidance notes may need to be consulted for any specific development. Nevertheless the information given in this section represents a standard that is believed to be generally acceptable to many authorities.

Sections 192 to 199 of the Act – the 'Advance Payments Code' – empower highway authorities to require payments to be made, or security to be given, in respect of road construction costs, before building is commenced. Where the authority itself is to construct the roads, the sum required may be the full estimated cost of the roads; while in cases where an agreement has been entered into under section 40 of the Act, for roads to be constructed by the developer, a bond for a sum being a more detailed estimate of the road works costs, based on the approved plans, will probably be acceptable.

Road layout

Economy in the use of land and in the cost per dwelling of roads and services is achieved when the number of dwellings per hectare is the maximum that planning and other constraints may permit, and when the length of road per dwelling is kept to a minimum.

Until the first few decades of the present century both these objectives resulted in the construction of long terraces of houses of narrow frontage – perhaps as little as 3–4 metres – with a minimum of links between frontage roads. Allowing for building on both sides of the road, such an arrangement could give an average road length per dwelling of little more than half of the frontage.

More recently the objectives have led to other forms of construction, notably flats and maisonettes. These may facilitate the elimination of one of the objectionable features of long terraces (the absence of external access to the back door, for refuse removal etc.) while still requiring relatively little land and road per dwelling. But although such types of building are still acceptable in large towns, where land values are high, recent years have seen an almost complete rejection of the more extreme forms of housing designed to achieve economy in land use, namely flats in 'high rise' blocks of perhaps 20 or 30 storeys. For most ordinary housing projects, both private and local authority, opinion has swung in favour of individual dwellings. But, unlike their Victorian predecessors, such dwellings must now provide external access to the back door, and must make provision for the motor car.

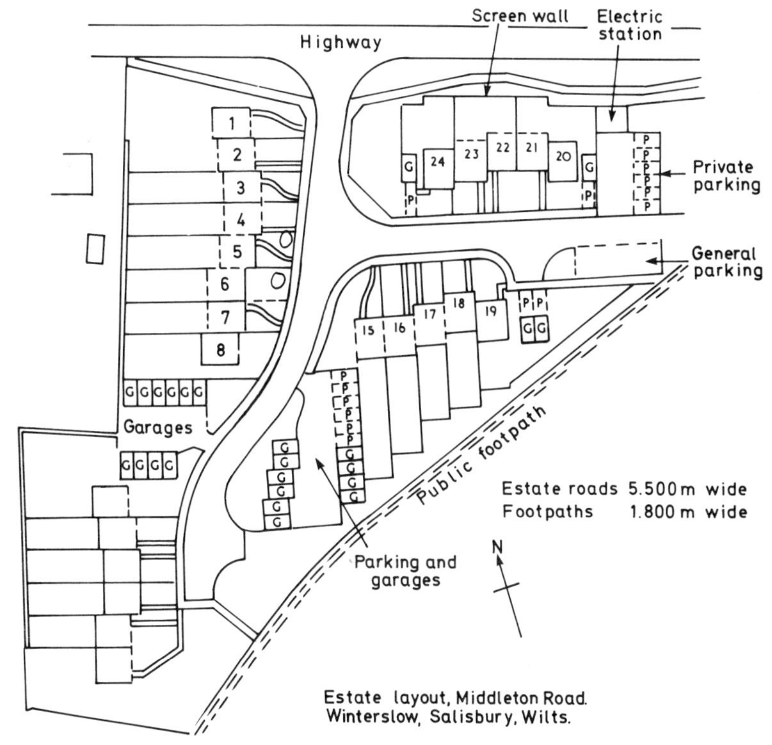

Figure 20.1 Example of modern housing layout designed to achieve economy in land use and heating costs. Scale 1 : 1500 approx. Housing construction and layout at this site gained a Medallion Award of the Electricity Council; architect Victor Nicholas, RIBA (courtesy Dellshore Properties Ltd, Salisbury)

These two requirements inevitably increase the cost per dwelling of both land and site works; in consequence many developers have looked for alternatives to the conventional solution of semi-detached houses each with its own garage, car port or parking space.

One modern solution is the town house, in which much of the ground floor area is devoted to the garage; living accommodation is mainly on the first and second floors.

Where economy in site costs is not so vital, attractive layouts have been achieved by deviating from the conventions that garages must be adjacent to their respective dwellings, and that each house must necessarily front onto a road (*Figure 20.1*). Such layouts must however make adequate provision not only for daily traffic, but for heavy delivery vehicles and for the emergency services. The requirements of road safety are of especial importance, often requiring that special provision be made for school-children and for the aged and infirm.

Again, planning authorities may require the provision of open spaces within housing layouts, as recreational or amenity areas; while the highway authority may require roads within a housing estate to provide links between two or more access points, and as such to provide for heavier vehicles, or for greater traffic density, than would otherwise be needed.

These possible requirements indicate the need for early consultation with both planning and highway authorities, in order that any constraints that may be imposed may be taken into account at an early stage in planning the site layout.

Geometric design

Under this heading are considered the four aspects of the physical shape of the road surface: horizontal alignment, vertical alignment, road and footpath widths, and lateral falls. Although these subjects are interdependent, they may conveniently be dealt with separately.

HORIZONTAL ALIGNMENT

It is desirable that the road layout, having been decided in sketch form in the light of considerations dealt with in the preceding paragraphs, should be defined precisely in mathematical terms. Most layouts comprise a number of straight lengths linked together through curves of uniform radius; in such cases the main geometry of the layout may be defined by fixing:

(a) the location, on a site or on the national grid, of each intersection point between adjacent lengths of straight road, and

(b) the radius of the curves linking these straight lengths; see *Figure 20.2.*

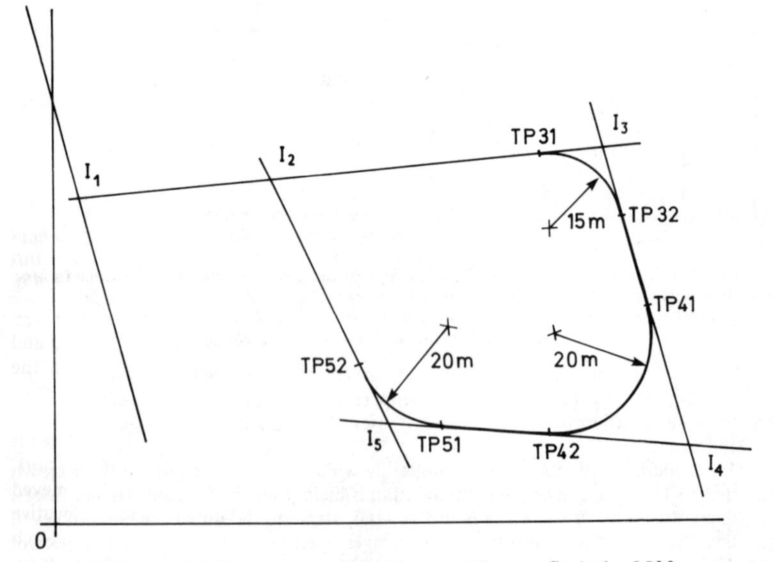

Figure 20.2 Framework of an estate-road layout. Scale 1 : 1500

From this basic information may be determined:
(c) the length between intersection points,
(d) the angle at each intersection,
(e) the length of each curve, and
(f) the distance from each intersection point to the adjacent tangent points.
If the grid coordinates of two adjacent intersection points are (E_1, N_1) and (E_2, N_2), the distance between them is:

$$\sqrt{[(E_1 - E_2)^2 + (N_1 - N_2)^2]}$$

The direction of the line joining the intersection points is:

$$\tan^{-1} \frac{E_2 - E_1}{N_2 - N_1}$$

and the angle between two such lines is the difference between the directions calculated in this way for each line.

Having calculated the angle θ between the two lines (which is the direction change, not the internal angle), the length of the curve is that angle expressed in radians, multiplied by the radius of the curve, i.e.

$$\text{Length of curve} = \text{angle } \theta \text{ in degrees} \times \frac{\pi}{180} \times \text{radius of curve}$$

The distance from the intersection point to the adjacent tangent points is given by

$$\tan\frac{\theta}{2} \times \text{radius of curve}$$

For example, the basic data for the road layout shown in *Figure 20.2* is:

Point	*Eastings*	*Northings*	*Curve radius*
I_1	5.400	65.000	-
I_2	45.000	68.613	-
I_3	115.000	75.000	15.000
I_4	132.800	15.000	20.000
I_5	70.000	20.000	20.000

From this data may be derived the information needed in determining chainages along the road and for accurate setting-out, which is:

Section	*Length*	*Direction*
I_1–I_2	39.765	84.787°
I_2–I_3	70.291	84.787°
I_3–I_4	62.585	163.476°
I_4–I_5	62.999	274.552°
I_5–I_2	54.665	332.785°

Curves:

At	*Angle* (deg)	*Angle* (rad)	*Arc length* (m)	*I – TP* (m)
I_3	78.689	1.3734	20.601	12.297
I_4	111.076	1.9386	38.773	29.142
I_5	58.223	1.0164	20.328	11.137

Chainages:

I_1	0
I_2	39.765
TP31	97.759
TP32	118.360
TP41	139.506
TP42	178.279
TP51	200.999
TP52	221.327
I_2	264.855

The equipment needed for setting out the horizontal alignment of estate roads is a steel tape and a theodolite. Having calculated data as described above, it is usual to start by setting out the main framework of the roads (I_1–I_2–I_3–I_4–I_5 in the example) and then to fill in the details of the curves.

The first step in setting out the curves is to locate the tangent points, again using the data calculated as described above. Where radii are small and the site is unobstructed, it may be possible also to locate the centre of each curve (simply by measuring the curve radius in a direction perpendicular to the straight line, at the tangent point), and to plot intermediate points on the curve by measuring radii from the centre.

More usually, however, these conditions do not apply, and it becomes necessary to set out points on the curve from the tangent point at which it starts. This is done by choosing suitable intervals between the setting-out points on the curve, and calculating a deviation angle from the tangent direction, for each point. Provided that the interval chosen is small in relation to the radius, the difference between an arc of the curve and the chord length actually measured is small enough to be ignored. It then follows from simple geometry that the deviation angle, in radians, is one half of the chord length divided by the radius of the curve.

Thus, for example, if intermediate points are required at 10 m intervals on a curve of radius 100 m, the deviation angle for each point is:

$$\frac{1}{2} \times \frac{10}{100} \times \frac{180}{\pi} = 2.865°$$

In this example the inaccuracy resulting from the assumption that the chord and the arc are of equal length is about 4 mm in each 10 m length, an amount that would affect the deviation angle by 0.0012 of a degree, or about 4 seconds. Such a minute angle is not measurable on the ordinary type of theodolite used in work of this type.

It must, however, be noted that errors of this type are cumulative, and could become material where a large number of points are set out in this way.

VERTICAL ALIGNMENT

Economy demands that in general estate roads should conform reasonably closely to natural ground levels, in order that large volumes of excavation or filling may be avoided. This does not, however, imply that roads should rise and fall at each minor hump and depression; the road should follow a smooth vertical alignment, consisting of sections of uniform gradient linked together by curves in the vertical plane of uniform radius.

Gradients should not be so steep as to impair road safety or to create undue inconvenience to any form of road user. Conversely, they should not be so flat as to create drainage problems. Generally a maximum gradient of 6–7 per cent may be adopted as a suitable limit for estate roads, and a minimum of 0.7 per cent. Limits may, however, be imposed by the highway authority, whose specifications should be consulted at an early stage in the design.

Vertical curves linking such gradients should be long enough to provide a smooth and comfortable transition, and (where necessary) to provide adequate sight distance over summit curves. On the other hand, vertical curves should not be so long as to create drainage problems at summits or in valleys, and it may be desirable to reduce the length of the curve where a steeply sloping road joins another.

With these provisos, a useful rule-of-thumb is that the length in metres of the vertical curve should be not less than four times the algebraic difference between the gradients linked, expressed as percentages. Thus, for example, a change from a downward grade of 3 per cent to an upward grade of 4 per cent would require a curve length of $(3 + 4) \times 4 = 28$ m.

It is usual to design vertical curves to be of parabolic shape rather than circular arcs. This is because the parabolic curve greatly simplifies the calculation of levels and gradients. The difference between the two shapes in practical terms, where gradients are within normal limits, is negligible.

The general form of the parabola is $y = ax^2$, where y is the offset at any point at distance x from the tangent point. The value of a, which is constant for any given curve, depends upon the rate of change of gradient. It is equal to the change of gradient (expressed as a fraction) divided by twice the distance in which that change takes place. Thus, for a vertical curve designed to give a change of gradient of 1 per cent in 4 m,

$$a = \frac{1}{100} \times \frac{1}{2 \times 4} = \frac{1}{800}$$

and hence the equation of the vertical curve is

$$y = \frac{x^2}{800}$$

For this curve, which may be adopted as standard for all vertical curves on estate roads, values of y for given values of x are given in Table 20.1.

Table 20.1 OFFSETS y FROM A TANGENT AT x FROM TANGENT POINT

x =	*0*	*1*	*2*	*3*	*4*	*5*	*6*	*7*	*8*	*9*
0	0	0.001	0.005	0.011	0.020	0.031	0.045	0.061	0.080	0.101
10	0.125	0.151	0.180	0.211	0.245	0.281	0.320	0.361	0.405	0.451
20	0.500	0.551	0.605	0.661	0.720	0.781	0.845	0.911	0.980	1.051
30	1.125	1.201	1.280	1.361	1.445	1.531	1.620	1.711	1.805	1.901
40	2.000	2.101	2.205	2.311	2.420	2.531	2.645	2.761	2.880	3.001
50	3.125	3.251	3.380	3.511	3.645	3.781	3.920	4.061	4.205	4.351

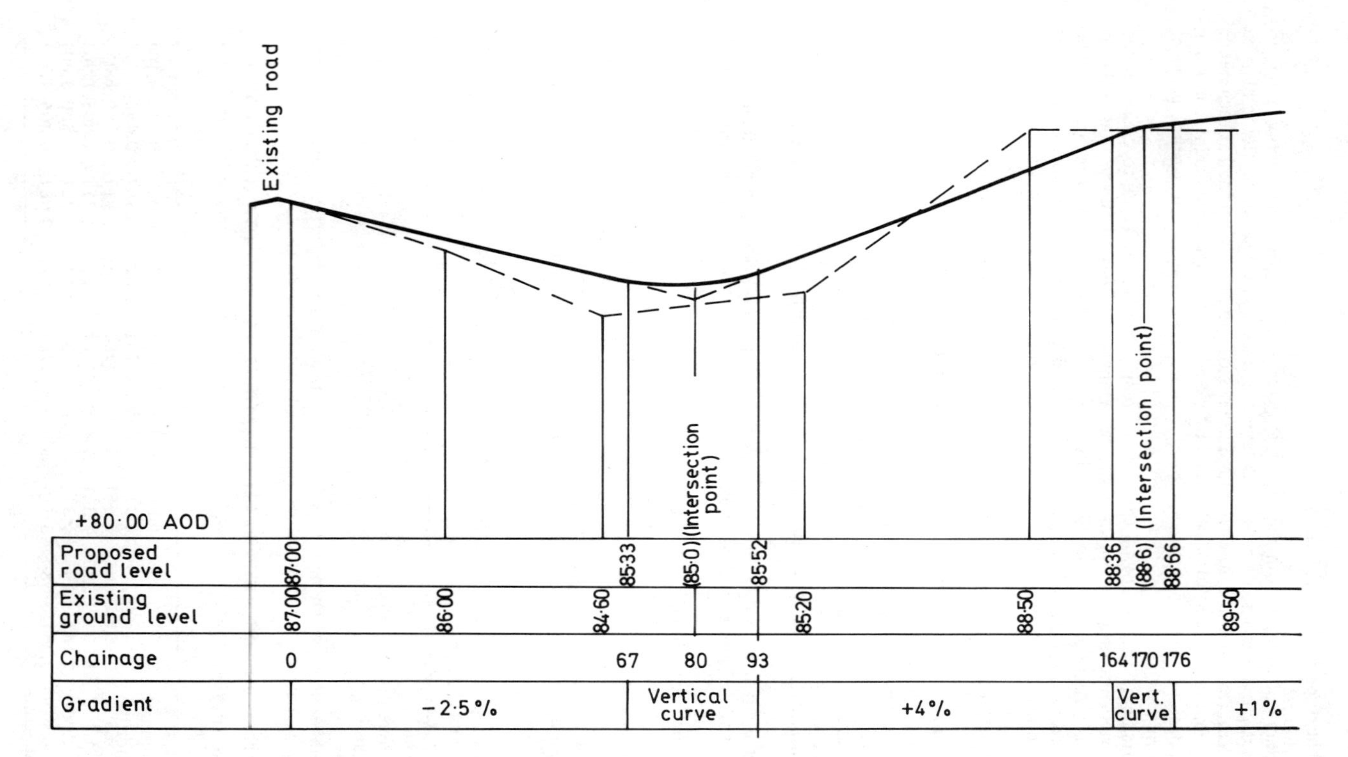

Figure 20.3 Longitudinal section of an estate road. Vertical scale 1 : 150, horizontal scale 1 : 1500

The proposed vertical alignment of a new road is conventionally shown on a longitudinal section drawing in which the vertical scale is exaggerated. In the typical drawing shown in *Figure 20.3,* a vertical scale of 1:150 is used in conjunction with a horizontal scale of 1:1500, giving a ten-times vertical exaggeration. This enables gradients and curves to be plotted more clearly than would be the case with a natural scale. Such section drawings usually include tabulated information as to existing ground levels, proposed road levels, chainages, and gradient and vertical curve details. Additionally, other related information, such as sewer levels, may also be included in the level information and plotted in the section.

ROAD WIDTHS

The minimum width between kerbs, even in the case of minor estate roads, is usually taken to be 5.5 m. This is sufficient just to allow two heavy vehicles to pass; the clearance would, however, be such as to require the drivers to move with great caution. On the other hand, the more normal traffic on such roads – private cars and small delivery vans – can generally move at more normal speeds when passing other similar vehicles.

Where roads are used both for access to the dwellings they serve and as a link between more important routes, or where they are on bus routes, the width between kerbs is usually required to be 6.7 m or more.

LATERAL FALLS

The minimum lateral fall for estate roads is usually taken to be 2.5 per cent. This may be provided either as the camber on each side of the road centre line, or as a crossfall over the full width of the road. Generally straight roads are cambered, while those on a curve are given a crossfall to provide a small degree of superelevation.

It may, however, be convenient and economic to provide a crossfall on certain lengths of straight road, for example where the natural fall of the ground lies across the direction of the road. By adopting an arrangement of this type, road gullies are all located on the side of the road adjacent to the lower ground, so that drainage lengths are reduced; additionally there may be a saving in earthwork quantities.

It is not usual, for vehicles moving at the slow speeds expected in estate roads, to provide a crossfall in excess of 2.5 per cent, even when the crossfall needed to counterbalance centrifugal force may be much in excess of this figure. The purpose of the crossfall or camber is simply to facilitate drainage; the absence of anything more than token superelevation – where a crossfall is provided – is considered to discourage undesirably high speeds.

Structural design

Almost all road construction in the UK is initiated or controlled by national or local governments or their agencies. Structural design standards required by these authorities are based on research, mostly of a practical nature, carried out over many years by such bodies as the Transport and Road Research Laboratory of the Department of the Environment. These design standards are incorporated in the DoE publication *Road Note 29*[2]; the following design recommendations are based on that publication insofar as it relates to estate roads.

ESTIMATION OF TRAFFIC

Roads should be designed to perform their intended function, without undue distortion or damage, during a period termed the 'design life' of the road. Research has shown that damage to roads is caused almost entirely by heavy vehicles; private cars do not contribute significantly to the damage.

Estate roads, which carry mostly private cars, with relatively few commercial vehicles, suffer least, and even the commercial vehicles using such roads are less heavy than those on main roads and motorways. Thus for most estate roads the volume of potentially damaging traffic is far less than for other types of road. Where the estate road also forms a bus route the difference is less marked, because even the standard figure of 25 public-service vehicles per day in each direction represents a substantial increase in the volume of heavy vehicles using the road.

On the other hand, the design life of an estate road should be longer than the corresponding period for other types of road. This is because the likelihood of a change being required to the road layout is less in the case of roads serving established developments than for roads carrying through traffic, in which changing patterns of numbers and types of vehicle may involve alterations to roads at relatively frequent intervals. Furthermore, the light construction of estate roads allows life expectancy to be increased at modest cost, while strengthening of heavy road construction is a much more costly operation.

These considerations have led to a recommendation that estate roads should be designed for a life of 40 years. During this period the number of commercial vehicles expected to use such roads, allowing for expected increases in traffic, is 0.36 million where estate traffic only is carried, and 2.7 million where the road also serves as a bus route.

Each of these figures is converted into a number of 'standard axles' by applying factors that take into account the average number of axles for various classes of commercial vehicle, and the average load on each axle. Both of these factors are smaller for the two types of estate road considered than for other roads, and they result in design requirements of 0.16 million standard axles for roads to carry estate traffic only, and 1.22 million where the road is also a bus route.

GROUND CONDITIONS

The second factor to be taken into account in road design is the strength of the ground on which it is to be constructed, and the possibility that damage may be caused by frost action.

The recognised basis of assessment of ground strength is the California Bearing Ratio (CBR) test. This is an empirical test, usually carried out in the laboratory, and originating from work of the California State Highways Department. The test is carried out on a prepared sample of the subsoil on which the road is to be built, which is placed in a machine and subjected to loading through a plunger having an area of 1935 mm². The load required to produce a penetration of 2.54 mm is measured, and this is compared with the load required to produce the same penetration when the sample is of crushed rock to a standard degree of compaction. The result is expressed as a percentage, which may be taken to be a measure of the strength of the subsoil in comparison with that of an ideal formation.

One of the defects of this test is that sample preparation may influence the result, especially where moisture content has a strong influence on ground strengths, as is often the case with the wet cohesive soils commonly found in the UK. One way in which attempts have been made to overcome this problem is to modify the test so that it may be carried out *in situ;* but here again the results may be affected by variations in the subsoil moisture content, which depends on the season of the year, the weather conditions before the test is carried out, and the way in which the ground is prepared for testing.

For these reasons the reliability of CBR tests of either type is open to question, and road designs based on CBR values assumed to apply to various types of subsoil are often acceptable instead of, or perhaps even in preference to, actual test results.

The level of subsoil water has an important influence on ground bearing capacity. Where necessary and feasible, drainage should be provided to prevent the water table from rising above a level of 600 mm below formation level, or alternatively the same objective should be achieved by constructing the road on an embankment. Where neither course is practicable, allowance should be made for the high water table in designing the road.

For the same reason water must be prevented from penetrating through the road surface and weakening the subsoil. This is achieved, in modern road construction, by either of two means:

(a) In flexible road construction, tar or bitumen is used to coat roadstones of various gradings, which are laid hot and compacted to form an impervious membrane.

(b) In rigid road construction, a slab of concrete, which may or may not be reinforced, depending on the traffic it is to carry, is used as the impervious membrane.

FLEXIBLE ROAD CONSTRUCTION

Typical sections of flexible roads are shown in *Figure 20.4*. The wearing course and base course are often termed the 'surfacing' and may in some cases be of similar materials, enabling them to be laid as a single course. The thickness of the sub-base depends on the traffic to be carried by the road, the type of subsoil, and the moisture present in the subsoil. Where ground conditions are favourable there may be no need for a sub-base – see Table 20.2.

Table 20.2 THICKNESS OF SUB-BASE

Type of subsoil	*Water table**	*Estimated CBR*	*Sub-base thickness* (mm)** *Estate traffic*	*Bus routes*
Sandy gravel: well-graded	low	60	Not reqd	Not reqd
	high	20	Not reqd	150
Sand: well-graded	low	40	Not reqd	Not reqd
	high	15	80	150
Sand: poorly graded	low	20	Not reqd	150
	high	10	80	150
Sandy clay	low	6–7	140	170
	high	4–5	230	260
Silty clay	low	5	180	220
	high	3	300	340
Heavy clay	low	2–3	400	450
	high	1–2	550	600
Silt	low	2	400	450
	high	1	550	600

* Where the water table is less than 600 mm below formation level the water table is 'high'; where more than 600 mm below formation level it is 'low'.

** Where the subgrade is susceptible to frost action the sub-base depth should be increased where necessary to give a total construction depth of 450 mm, except where local experience during severe winters indicates that this precaution is unnecessary. Susceptible soils include sandy clays, granular soils with more than 10 per cent dust, chalks, limestones, burnt colliery shales and certain types of pulverised fuel ash (pfa).

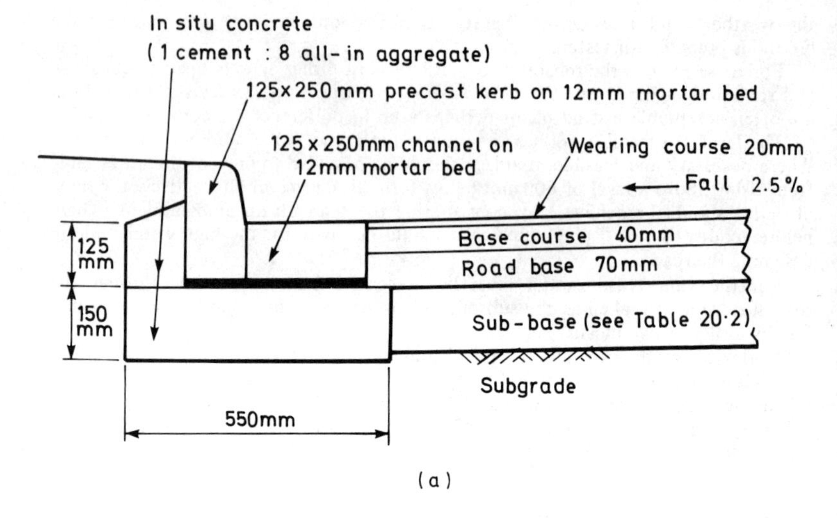

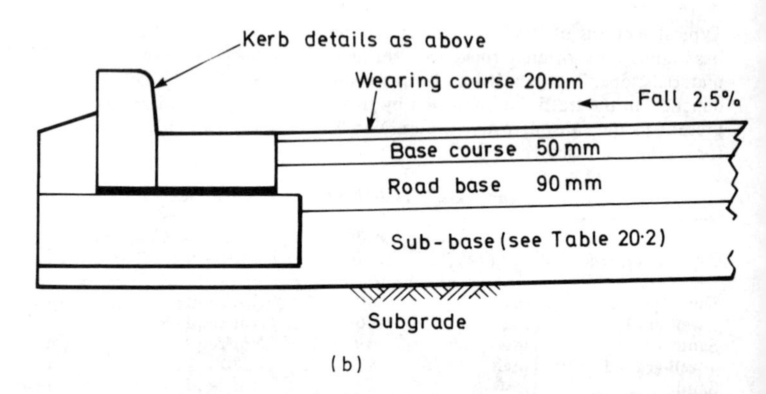

Figure 20.4 Typical sections of estate roads of flexible construction: (a) estate traffic only; (b) estate traffic and bus route

Materials for the wearing course, base course and roadbase of flexible roads may be either coated macadam to BS 4987 or rolled asphalt to BS 594. In both cases the British Standard specifies different grades for use in the several courses.

For the sub-base, where needed, the material should be well-graded natural sand, gravel, crushed rock, crushed slag, crushed concrete or well-burnt non-plastic shale complying with a grading envelope specified in the DoE *Specification for road and bridge works*[3], clause 804, which is:

Passing 75 mm	100 per cent
37 mm	85–100 per cent
10 mm	45–100 per cent
5 mm	25–85 per cent
1 mm	8–45 per cent
125 μm	0–10 per cent

The sub-base material must also have a CBR value not less than 20 per cent for ordinary estate roads, and not less than 30 per cent where the road also forms a bus route. Where the subgrade has a CBR that itself meets this requirement, no sub-base is needed.

Care must be taken, in constructing flexible roads, to obviate any risk of damage, either during construction where traffic flow has to be maintained, or in service, by the ingress of water through an unsealed surface. This applies especially where open or medium textured macadam or cold asphalt is used in construction. In such cases a suitable surface dressing material should be applied, either to the completed road or at any intermediate stage of construction during which traffic is to be carried.

RIGID ROAD CONSTRUCTION

A rigid road consists of a concrete slab with or without reinforcement, laid on a suitable sub-base. Where the ground is suitable to meet the requirements of the sub-base, the concrete slab may be laid directly on to the prepared subgrade, i.e. after it has been trimmed and compacted. The thickness of the concrete and of the sub-grade, where needed, depend on the traffic to be carried – including traffic during the construction period of the estate – and on the type of subsoil and its water content.

When reinforcement is provided in the concrete slab, the slab thickness may be reduced by a small amount. In the case of estate roads, however, the saving in cost resulting from the reduced thickness of concrete is usually much less than the cost of the reinforcement, and therefore the use of reinforced concrete is not justified for reasons of economy.

The thickness of the sub-base and of the unreinforced concrete slab needed to conform to the requirements of the DoE are given in Table 20.3. Where it is intended that the sub-base only should be used as a road for construction traffic, the depth of the sub-base should be increased by 80 mm, or by 150 mm in cases where the CBR of the subgrade is less than 4 per cent. In such cases the top 150 mm of the sub-base

Table 20.3 THICKNESSES OF SUB-BASE AND OF CONCRETE SLABS

Type of subsoil	*CBR*	*Sub-base thickness** (mm)	*Concrete slab thickness* (mm)	
			Estate traffic	*Bus routes***
Very stable (well-graded sands and gravels)	15 or more	Not reqd	130	150
Normal (poorly-graded sands, clays and silty clays)	2–15	80	160	180
Weak (poorly-drained heavy clays, silts etc.)	2 or less	150	180	200

* Where the subgrade is susceptible to frost action, the sub-base depth should be increased where necessary to give a total construction depth of 450 mm, except where local experience during severe winters indicates that this precaution is unnecessary. Susceptible soils include sandy clays, granular soils with more than 10 per cent dust, chalks, limestones, burnt colliery shales and certain types of pulverised fuel ash (pfa).

** The higher concrete slab thicknesses applicable to estate roads that are also bus routes should be used where estate roads are to carry the heavy construction traffic resulting from the construction of 100 houses or more, or equivalent buildings.

should be of lean concrete or of a cement-stabilised material, except where the construction work is to be limited to the summer months, when normal sub-base material, as recommended for flexible road construction, may be used.

The concrete used for unreinforced road construction should have a minimum crushing strength at 28 days of 28 N/mm² using ordinary Portland or Portland blast-furnace cement. The top 50 mm of the slab, or the whole of the slab, should be of air-entrained concrete.

Public utility services

Provision must be made in the construction of estate roads for the public utility services: sewerage (foul and surface water), water supply, gas, electricity and GPO telephones. In some areas where radio reception is bad, it may also be necessary to provide for 'piped' television and radio programmes.

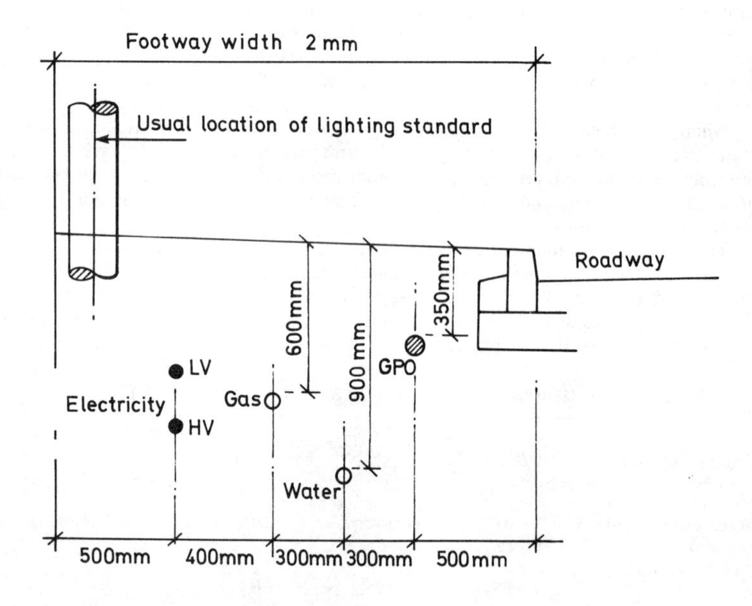

Figure 20.5 Standard locations and depths of public utility services

In addition to the pipe or cable providing the main service, connections are required to each property and to certain street apparatus or equipment associated with the service, namely:

Drainage:	Road gullies, manholes, and in some cases pumping or ejector stations.
Water supply:	Sluice and washout valves; fire hydrants.
Gas:	Control valves.
Electricity:	Junction boxes, street lighting standards, lights for traffic control and direction signs.
Telephones:	Junction boxes; public callboxes.

Drainage, or more accurately sewerage, is generally regarded as being part of the road construction, and may well be undertaken by the roadworks contractor.

Other services are generally installed by the responsible authority or by contractors employed by the authority, who should be advised of the proposed road construction at an early stage and invited to indicate on copies of the road plans the mains and apparatus they require to instal. It then usually devolves on the roadworks designer to coordinate the installation of mains and equipment, in order to avoid conflict both as to the location of the various authorities' apparatus and as to the programme for the installation work.

Wherever possible main services should be located in accordance with recommendations contained in a report by the Institution of Civil Engineers[4], which has become accepted nationally as a standard. Details of the locations and depths recommended in this report are shown in *Figure 20.5.*

No standard location or depth can be given in the case of sewers, because the sewerage system must be designed to provide the falls necessary to ensure suitable rates of flow (see Section 21). For reasons of economy, sewer depths should be kept to the minimum consistent with adequate protection against damage and with serving all points of discharge within the estate; these two requirements usually result in a sewer layout following the natural drainage channels of the area it serves.

Where it is possible to locate sewers within the overall width of the road (including verges etc.), it is desirable for ease of maintenance that this should be done. In some cases, however, and in particular where building sites are substantially below the level of the road serving them, sewers may have to be located within the curtilage of the properties served by them.

PATHS AND DRIVES

Statutory powers

Where footpaths are to be constructed adjacent to estate roads, they are treated by the highway authority as being part of the road for purposes of advance payments or security, standards of construction, adoption by the authority and future maintenance (see the beginning of this section).

Footpaths and drives within the curtilage of a dwelling are not usually the concern of the highway authority, except as regards safety at the point of access to the road. They may, however, be of interest to the planning authority who, besides requiring to be satisfied as to road safety, may wish to ensure that adequate access is provided to all dwellings, for the occupants, for commercial vehicles, and for the emergency services.

Layout

The layout of footpaths flanking estate roads is generally determined by the road layout, although minor deviations may be needed where, for example, a wide verge is provided at a bend in the road for reasons of visibility. In such cases it is desirable that the footpaths should follow the route likely to be used by pedestrians – namely the shortest route – since otherwise a path will be worn into the surface of the grass verge.

Similar principles apply to paths giving access to dwellings where no road is provided, i.e. the layout should attempt to anticipate the routes likely to be followed by pedestrians. Special attention should be given to the need to provide safe access between dwellings, schools, recreation areas and shops.

Within the curtilage of each dwelling, paths should be provided from the road to the main entrance to the dwelling and to the kitchen entrance (except where the latter is not accessible from the road, as in mid-terrace houses). Paths should also be provided from the dwelling to the garage, carport or paved standing area where provided.

In cases where part of a garden is made inaccessible by the presence of a retaining wall, suitable access in the form of a path or steps should be provided.

A drive should be provided from the highway to the garage, car port or parking space. In some cases, where access from the dwelling is to a busy road, provision for turning the car within the dwelling curtilage may be needed.

Geometric design

Where footpaths are adjacent to roads or are separated from them only by a narrow verge, both horizontal and vertical alignments may be related to the road; the path should be of fixed width with a standard crossfall (usually 3 per cent) to the kerb.

In other cases, the principles used in the geometric design of roads (see earlier) may be applied to the design of footpaths and drives, although it is not usually necessary to prepare designs in such elaborate detail. Both horizontal and vertical curves can be adequately set out by eye by an experienced worker, such as a kerb layer experienced in providing a smooth curve linking quite widely separated setting-out points.

Footpaths to be used by the public should be at least 1.8 m in width, to allow sufficient space for perambulators, invalids' wheel chairs and childrens' tricycles to pass. Short paths providing access only to single properties should preferably be not less than 1 m in width.

Within the dwelling curtilage, footpaths may be as little as 600 mm in width, or 700 mm where against the dwelling. Gradients of footpaths should not exceed 16 per cent without steps; and where steps are provided they should be designed to give suitable rises consistent with the going, both of which should as far as possible be uniform. Generally a satisfactory balance between rise and going is achieved when the sum of the going and twice the rise is between 550 and 700 mm, as required by the Building Regulations.

Drives should preferably be not less than 2 m in width, and should have a gradient no steeper than may be negotiated by an ordinary family saloon. The limiting gradient may be taken to be 20 per cent, although a flatter gradient is desirable wherever possible.

Paths and drives should have a crossfall or camber not less than 3 per cent where required for purposes of drainage. This, a somewhat higher fall than that required for roads, recognises the fact that the surface of a road is usually more accurately laid than the surface of a path or drive. In cases where the drive or footpath is constructed of a permeable material such as gravel, surface water drains through the material and no crossfall is needed.

Where a path is constructed against the dwelling, the crossfall should be away from the building. The level of the path should be at least 150 mm below dpc level.

Structural design

Drives and footpaths should have a clean, smooth, regular, stable and adequately drained non-slip surface. Drives should be capable of bearing the weight of a normal motor car without undue distortion.

Factors that may result in damage to the surface include inadequate foundations or drainage, settlement caused by frost heave, and the growth of weeds through the material of the path or drive.

Where an impermeable surface is used these dangers are greatly reduced. Permeable materials, such as gravel, are generally only satisfactory within the curtilage of the dwelling, where the routine maintenance work of clearing weeds and forking to improve drainage is acceptable.

In many cases a foundation will be required, depending on the type of surfacing material to be used and on the type of subsoil. Where the ground is soft, such as in silty or clayey subsoils, at least 150 mm thickness of hardcore, consisting of broken stone or brick or other inert material, should be provided. Such a foundation should always be provided, irrespective of the ground type, where the drive or path is to be of gravel surfacing.

Suitable materials for surfacing drives are listed in Table 20.4. Foundations of hardcore or other material should be thoroughly consolidated by rolling, as should the surfaces of asphalt, macadam and gravel drives. Edge strips, of precast concrete or treated timber, should be provided for all types of drive except those of *in-situ* or precast concrete.

Table 20.4 SURFACING MATERIALS FOR DRIVES

Type of construction	*Material*	*Minimum thickness (mm)*
Single-course asphalt	Rolled asphalt	40
Single-course macadam	Bitumen macadam	50
Two-course macadam	Cold asphalt or macadam	20
	on	
	Base course (20 mm) macadam	30
In-situ concrete	Concrete 1:3:6 or 1:8	100 (note 1)
Precast concrete	Precast slabs	50 (note 2)
Permeable	Gravel or broken stone	50 (note 3)

Notes (1) Where *in-situ* concrete is used, movement joints 10 mm wide should be provided at intervals not exceeding 4 m and at points where the drive abuts against a wall. Where necessary a non-slip finish should be provided (i.e. on sloping drives).

(2) Where precast slabs are used, they should be laid on a base of 150 mm hardcore, or on 75 mm of 1:10 lean-mix concrete.

(3) Gravel drives should always have a foundation of not less than 150 mm of hardcore.

Paths should be constructed of similar materials to those recommended for drives, but the thicknesses may be reduced. In the case of *in-situ* concrete the thickness may be 75 mm, and in general foundations of hardcore may be laid to 75 mm thickness where required.

REFERENCES

1. The Highways Act 1959, HMSO
2. Department of the Environment, *Road Note 29*, 3rd edn, HMSO
3. Department of the Environment, *Specification for road and bridge works*, HMSO
4. *Report of the Joint Committee on Location of Underground Services*, Institution of Civil Engineers

FURTHER INFORMATION

Further information on the design and construction of estate roads should be obtained from the highway authority in whose area the works are situated, as well as from the above references.

21 FOUL SEWERAGE AND DRAINAGE

INTRODUCTION	21–2
SEWERAGE SYSTEMS	21–2
FLOW CALCULATIONS	21–2
SEWER LAYOUT	21–4
MATERIALS	21–6
SEWER CONSTRUCTION	21–7

FOUL SEWERAGE AND DRAINAGE 21

21 FOUL SEWERAGE AND DRAINAGE

D.A. STEPHENSON, BSc, CEng, FICE, FIStructE, FIArb
Consulting Engineer

This section explains separate and combined systems of drainage. In relation to foul sewerage and drainage it considers flow calculations, layout, materials and construction.

INTRODUCTION

Drains are the pipes through which liquid wastes (sewage or surface water, as the case may be) are conveyed from individual properties to the *sewers* that serve them. Drains are usually vested in the property owner, while sewers are always the property of the public authority in whose area they are situated. Since local government reorganisation in 1974 the authority responsible for sewerage has been the Water Authority for the area concerned.

SEWERAGE SYSTEMS

Sewerage systems may be separate, partially separate or combined, depending upon the extent to which surface water is allowed to enter the foul sewers. Modern practice favours the separate system, notwithstanding that this requires additional drains and sewers to be laid to take surface water. This is because the maximum rate of flow of surface water is usually vastly in excess of the flow of sewage, and it therefore causes problems at the sewage treatment works, and more particularly at any pumping stations there may be between the area served and the treatment works. By eliminating surface water, a system of comparatively small diameter sewers may be used to serve a large number of houses.

There is, however, some benefit to be derived from allowing a small quantity of surface water to enter foul sewers: namely that it provides a natural means of flushing the sewers from time to time, and may therefore reduce the need for cleaning by rodding. Furthermore, the larger diameters of sewer needed where part of the surface water is admitted allow sewers to be laid at flatter gradients; this is of particular advantage in areas where there is little natural fall in the ground, where sewer depths tend to become greater than is desirable or economic.

Again, it may not always be practicable or desirable to exclude surface water entirely from the foul sewerage system, especially where it may be necessary to take flow from older properties served by combined drains.

FLOW CALCULATIONS

The first step in the design of a sewerage system is to calculate the flow to be carried. In this section flow of sewage only will be considered; the calculation of surface water flow is dealt with separately in Section 22, and must be taken into consideration in the design of combined or partially separate systems.

The flow of foul sewage from an estate may generally be equated with the water consumption in the same area. In the case of new estates, however, no consumption figures are available, and therefore average values must be assumed.

The BS Code of Practice for sewerage[1] recommends that the allowance be from 136 to 227 litres per head of population per day, depending upon the type of property. For new estates, a figure at or near the upper end of this range should be assumed, because new houses will be fully equipped with modern sanitation and bathrooms. It is also necessary to make an assumption as to the number of occupants per dwelling, and this does of course depend upon the type and size of the dwellings. It is suggested that a reasonable figure to assume for purposes of sewer design is 1.5 persons per bedroom in each dwelling.

Using these two assumptions the total flow from the dwellings served is, in round figures, 350 litres per day for every bedroom. The total flow calculated in this way is termed the dry weather flow (DWF), and is expressed in litres or cubic metres per day, one cubic metre being equal to 1000 litres.

Another factor to be taken into account in designing sewers is that the rate of flow of sewage varies during the day, and, to a lesser extent, from one day to another. Flow during the night and the small hours of the morning is negligible, while during the day, and especially at times when many people may be using toilet facilities, the rate of flow may be several times the average. To allow for these fluctuations sewers should be designed to carry at least four times, and preferably six times, DWF.

For purposes of sewer design the flow of sewage calculated in this way is usually expressed in litres per second except where quantities are very large, when cubic metres per second may be used.

Flow capacity of pipes

The flow of water or sewage in a pipe is the product of the cross-sectional area of the pipe and the velocity. Velocity depends upon three variables: the gradient and the diameter of the pipe, and its surface 'roughness'.

Many formulae for the flow in a pipe, knowing the values of these three variables, have been developed during the past two hundred years, from the early work of Chézy in about 1775. The most authoritative current work on the subject is probably that of the Hydraulics Research Station, whose research paper[2] published in 1969 includes tables of velocities and discharges for pipes of varying diameters, each table taking a different value for the roughness coefficient of the pipe.

Recommendations are made in the paper as to the roughness coefficients to be assumed for various types of pipes, generally when clean and new, although figures are also given for pipes that have deteriorated with the passage of time. From these values it may readily be seen that roughness depends much more on the condition of the pipe than on its material; and because sewer designs must take into account conditions likely to obtain after many years of service, the justification for assuming different roughness values for different materials has been questioned.

For the practical purpose of designing small-diameter sewers to serve housing estates, flow velocities calculated from the Crimp and Bruges formula are often used, even though the values so obtained may in some cases be conservative. The formula has the merit of simplicity. Expressed in metric units it is:

$$V = 83.33m^{2/3}i^{1/2}$$

where V is the flow velocity in metres per second, m is the hydraulic mean depth (equal to one-quarter of the pipe diameter in the case of a pipe running full) in metres, and i is the gradient of the pipe.

From this equation the discharge from the pipe may be calculated by multiplying the velocity by the pipe area. Expressing both the area and the hydraulic mean depth in terms of the pipe diameter, the expression for discharge becomes:

$$Q = 25\,976\, D^{8/3} i^{1/2}$$

where Q is the discharge in litres per second, D is the pipe diameter in metres, and i is the gradient.

Sewers should be designed to give velocities of flow falling between lower and upper limits. Where the gradient is very flat, slow rates of flow may allow solid matter to settle to the pipe invert, causing deposits of silt to build up. A minimum velocity of 0.76 metre per second is generally taken as the lower limit in order that the sewer shall be 'self cleansing'.

At the other extreme, excessively high rates of flow may cause scour, which will in time damage the surface of the pipe. This upper limit may be taken to be 3 metres per second.

The flow capacities of pipes of diameters normally used for drains and sewers for housing sites are given in Table 21.1. Flows above the upper zig-zag line give conditions in which scour may occur. Below the lower zig-zag line, flows are not sufficient to give self-cleansing conditions.

Sewers should never be less than 150 mm, and drains never less than 100 mm, in diameter.

Example of a flow calculation

To find the diameter of sewer needed to serve an estate of 190 houses, each of three bedrooms, assuming the separate system of sewerage.

Estimated population: $190 \times 3 \times 1.5 = 855$ persons

Daily flow (DWF): $855 \times 227 = 194\,000$ l/day

Six × DWF: $6 \times 194\,000 = 1\,164\,000$ l/day

Flow per second: $\dfrac{1\,164\,000}{24 \times 60 \times 60} = 13.47$ l/sec

Hence, from Table 21.1, a pipe of 150 mm diameter laid at a gradient of 1 in 150 is adequate for this flow.

This is the minimum gradient to provide the flow capacity required to serve the estate; it also happens to be the minimum gradient necessary to ensure that the sewer is self-cleansing. It follows therefore that a 150 mm diameter sewer, laid to gradients sufficient to ensure self-cleansing, will always serve a minimum population of 855 when the separate system of sewerage is adopted. Where the topography of the ground is such as to provide steeper gradients, a greater population can be served.

SEWER LAYOUT

Sewers must be laid on routes and at depths sufficient to allow all properties within the estate to be connnected; provision should be made, where appropriate, to ensure that future developments may be served without the need to relay sewers. Economy demands that the lengths of sewers and of house connections, and the depths to which they must be laid, should be kept to the minima necessary to achieve their purpose, while giving sufficient protection to the pipes against damage being caused by loads on the ground, by frost action, or by movements in clay subsoils caused by variations in the moisture content.

Table 21.1 FLOW CAPACITIES IN LITRES PER SECOND FOR PIPES OF VARIOUS DIAMETERS, AT VARIOUS GRADIENTS, RUNNING FULL

Gradient	*Pipe diameter (mm)* 100	150	225	300	375	450	525
1 in 10	17.7	52.2	153.8	331.3	600.6	976.7	1473.3
20	12.5	36.9	108.8	234.3	424.7	690.6	1041.8
25	11.2	33.0	97.3	209.5	379.9	617.7	931.8
30	10.2	30.1	88.8	191.3	346.8	563.9	850.6
35	9.5	27.9	82.2	177.1	321.1	522.1	787.5
40	8.9	26.1	76.9	165.6	300.3	488.4	736.7
45	8.3	24.6	72.5	156.2	283.2	460.4	694.5
50	7.9	23.3	68.8	148.2	268.6	436.8	658.9
55	7.6	22.3	65.6	141.3	256.1	416.5	628.2
60	7.2	21.3	62.8	135.2	245.2	398.7	601.5
65	6.9	20.5	60.3	129.9	235.6	383.1	577.9
70	6.7	19.7	58.1	125.2	227.0	369.2	556.9
75	6.5	19.1	56.2	121.0	219.3	356.6	538.0
80	6.3	18.5	54.4	117.1	212.4	345.3	520.9
85	6.1	17.9	52.8	113.6	206.0	335.0	505.3
90	5.9	17.4	51.3	110.4	200.2	325.6	491.1
95	5.7	16.9	49.9	107.5	194.9	316.9	478.0
100	5.6	16.5	48.6	104.8	189.9	308.9	465.9
110	5.3	15.7	46.4	99.9	181.1	294.5	444.2
120	5.1	15.1	44.4	95.6	173.4	282.0	425.3
130		14.5	42.7	91.9	166.6	270.9	408.6
140		13.9	41.1	88.5	160.5	261.0	393.8
150		13.5	39.7	85.5	155.1	252.2	380.4
160		13.0	38.5	82.8	150.2	244.2	368.3
170		12.7	37.3	80.4	145.7	236.9	357.3
180		12.3	36.3	78.1	141.6	230.2	347.3
190		12.0	35.3	76.0	137.8	224.1	338.0
200		11.7	34.4	74.1	134.3	218.4	329.4
220		11.1	32.8	70.6	128.1	208.2	314.1
240			31.4	67.6	122.6	199.4	300.7
260			30.2	65.0	117.8	191.5	288.9
280			29.1	62.6	113.5	184.6	278.4
300			28.1	60.5	109.7	178.3	269.0
350			26.0	56.0	101.5	165.1	249.0
400				52.4	95.0	154.4	232.9
450				49.4	89.5	145.6	219.6
500				46.8	84.9	138.1	208.4
550				44.7	81.0	131.7	198.7
600				42.8	77.5	126.1	190.2
650					74.5	121.1	182.7
700						116.7	176.1
750						112.8	170.1
800							164.7

The minimum depth to avoid such damage is generally taken to be 1.2 m to the crown of the pipe where it is laid beneath roads or footpaths, or 0.9 m to the crown where beneath fields or gardens. These figures need not, however, be taken as absolute minima, because where necessary protection (such as encasement in concrete) may be provided. Where such protection is needed only for a short length of pipe it may be much less costly than laying the sewer (perhaps considerable lengths on the downstream side) at greater depths.

Where sewers have to be laid beneath the beds of streams or other watercourses, the authority responsible for drainage of the area may require a depth of cover over the pipes sufficient to allow for future deepening or widening of the watercourse. Details of such crossings, and of any other crossing of underground pipes, cables or other apparatus, should be agreed with the responsible authority.

Special precautions may be needed when sewers are to be laid in rocky ground, or in ground liable to mining subsidence. Where appropriate a bed of soft material, such as pea gravel, may be needed to safeguard the pipes against damage by angular rocky projections; where ground movement is to be expected the use of flexibly jointed pipes may be necessary.

The general rule for designing sewer layouts is that the pipes should follow the natural falls of the ground. A study of the natural drainage channels of surface water will indicate appropriate sewer routings, and although there may be times when some other route is appropriate (for example, when a small part of an estate lies just beyond the highest ground) it is not usually possible to deviate substantially from the natural drainage paths without incurring the greatly increased costs resulting from excessive sewer depths.

Sewers should always be laid in straight lengths between manholes, to facilitate maintenance and the location of obstructions. Manholes are required wherever there is a change of direction, gradient or size of the sewer, or wherever a junction occurs between two or more sewers. Lengths between manholes should in any case not exceed about 100 m.

Provision for connections from dwellings under construction or from future dwellings should be made by the inclusion of capped branches at appropriate points. These enable the sewer to be tested for watertightness before house connections are to be made.

MATERIALS

Pipes of the small diameters needed for housing site sewerage are manufactured in several different materials, the choice lying between the following:

Glazed stoneware to BS 65 & 540
Asbestos cement to BS 3656
Concrete to BS 556
Pitch fibre to BS 2760
Cast iron to BS 436
Spun iron to BS 1211

The main factors influencing choice of pipe material are durability and economy, but sometimes there are special factors, such as the type of ground, that may override these considerations. The durability of stoneware, asbestos cement, concrete and pitch fibre generally exceeds that of iron, because of corrosion of the latter. Cost comparisons should take account not only of pipe costs but also of the costs of laying and jointing. Pipes that are light in weight and are supplied in long lengths with easily formed joints cost much less to lay than those supplied, for example, in lengths of 1 m, and requiring a mortar or similar packed joint to be formed.

Where subsoil conditions are aggressive, because of acidity or the presence of sulphates in subsoil water, stoneware will generally prove most durable. Where strength is needed, to resist stresses created by ground movements in mining areas or to span across streams etc, spun or cast iron pipes are likely to be most suitable.

Where pipes have to be laid in very soft ground the risk of settlement may be reduced by providing a concrete bed, or bed-and-haunch. In rocky ground a bed of pea gravel may be used to protect the pipes against damage from the hard, uneven surface of the rock. Protection against damage from loads on the ground surface may be given by a concrete bed-and-surround. See *Figure 21.1*.

Whenever concrete is used to bed or to protect the pipe additional difficulties arise in the event of unforeseen connections having to be made, because of the danger that the pipe may be damaged in breaking out the concrete. The use of concrete, especially as a surround to the pipes, should not be called for unnecessarily.

Manholes are almost invariably constructed from precast concrete units manufactured to BS 556: Part 2 Section 3C[3]. Units detailed in this standard include manhole rings and covers, and, for deep manholes, taper rings and shaft rings, with the smaller diameter covers needed where access is down a shaft. Step irons are built into the rings at suitable intervals. Base units, complete with channels and benching, may generally be obtained from the precasting works, to suit the requirements of each individual manhole.

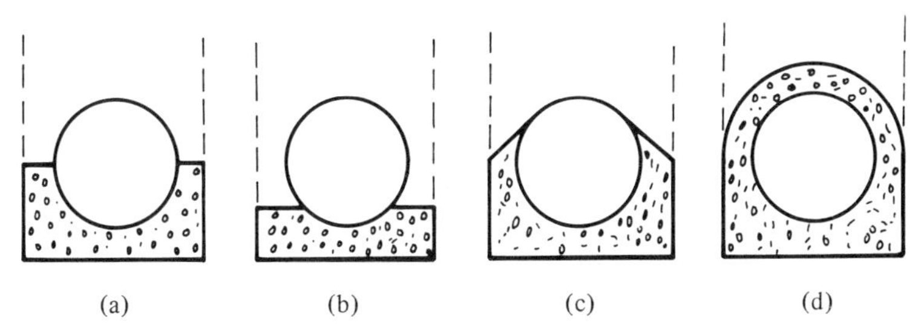

Figure 21.1 Protection of sewers against damage: (a) gravel bed, (b) concrete bed, (c) concrete bed-and-haunch, (d) concrete bed-and-surround

Joints between precast units are usually of the ogee type, except at the joint between the highest ring and the cover slab, where a straight joint is provided, the cover slab projecting below the joint on the outside of the ring, to ensure correct alignment of the two components.

Similar but lighter precast sections, manufactured to the same British Standard, are also available for use in constructing inspection chambers for domestic drainage. These units should not be used under public roads.

Cast iron manhole covers are manufactured to BS 497[4], which specifies three grades:

Grade A	Heavy-duty double triangular or single triangular, for use in roads.
Grade B	Medium-duty circular or rectangular, for use in verges etc., and generally where it is unlikely that heavy commercial vehicles will pass over them.
Grade C	Light-duty rectangular or square, for use in locations inaccessible to wheeled vehicles.

SEWER CONSTRUCTION

The first step in sewer construction is to set out lines and levels. The centre of each manhole should be marked by a peg; in long sewer lengths, intermediate pegs between manholes may be advantageous. Offset pegs should also be driven at a recorded distance from the pipe centre line, so that this line can still be located after the trench has been excavated.

Sight rails, consisting of accurately planed timbers supported on substantial posts, and set at a fixed distance above the sewer invert level, should be fixed at each manhole location and at one or more intermediate points in the length between manholes. In this way a check is available against accidental displacement of the sight rails, in that the three or more rails above each sewer length should always be in alignment. The centre line of the sewer may also be marked on the sight rails as an additional reference point, and as a check against displacement of any of the setting-out points.

A traveller, consisting of a short horizontal sight rail firmly fixed to a vertical timber to form a 'T', and with a steel angle fixed to the base, is constructed to an overall length equal to the height of the fixed sight rails above sewer invert level. See *Figure 21.2.*

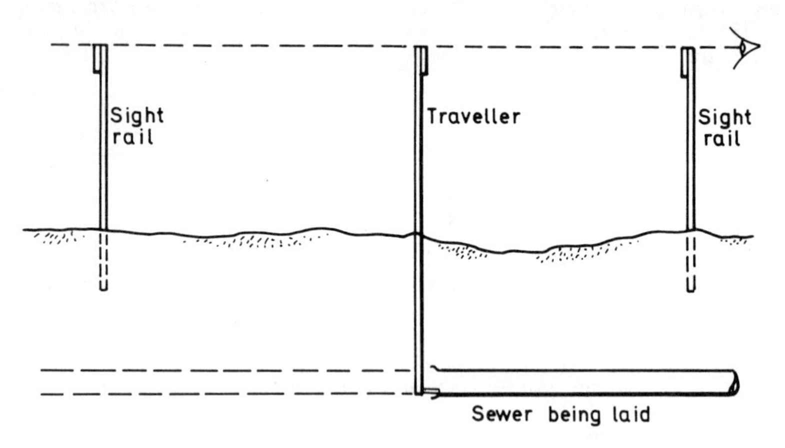

Figure 21.2 Use of sight rails and traveller for levelling pipes

Excavation of sewer trenches is usually started at the lowest point, in order that rain and any subsoil water that may flow into the trench drains away from the working point. Trenches should be at least 300 mm wider than the sewer diameter, with an additional width allowance for timbering in the trench. The method of timbering necessary to support the sides of the sewer trench is a matter of the greatest importance, on which a decision should be made when sewer depths and types of subsoil are known. It should be borne in mind by those responsible for deciding on the working method and of the trench supports needed that the collapse of the sides of such trenches on to men working in them is one of the most prevalent causes of fatalities in the construction industry, which itself is hazardous in comparison with most other industries.

Guidance as to the need for, and the design of, trench supports is given CP 2003[5]. In general, it is recommended that support should be provided to the sides of any trench, irrespective of depth, in soft peat, soft clay or silt, in loose gravel and in any gravel below the water table. In compact, or slightly cemented, sands and gravels and in firm peat, supports should be provided to trenches more than 1.5 m in depth; while in firm or stiff clays and in fissured rock support should be provided where the depth exceeds 4.5 m.

Attention should be given to the fact that ground conditions may vary as the work proceeds. A change (for example, from firm to soft clay) may tend to be ignored when the aim is rapid progress, and changes resulting from climatic conditions – in particular heavy rainfall or frost – may cause a previously safe excavation to become unstable.

When designing supports for trench excavation it is necessary first to evaluate the lateral pressure of the ground, known as the 'active' pressure. It is convenient to consider this pressure as being that of an 'equivalent fluid', the density of which depends upon the actual density of the soil and its mechanical strength. For most granular soils the equivalent fluid density is unlikely to exceed one third of its actual density. It is therefore usually safe to take sands and gravels to have an equivalent fluid density of 0.7 tonne per cubic metre.

For cohesive soils the active pressure depends upon the shear strength of the soil, and the concept of equivalent fluid is not strictly applicable. However in most cases, excepting only trenches in very soft clays and silts to depths in excess of 3 m, an equivalent fluid density of 0.7 tonne per cubic metre may safely be assumed.

Greatly increased loading will, however, be caused if ground water pressure is allowed to build up behind the trench supports. In general this condition is unlikely to apply, because groundwater can pass through the trench supports into the trench, thereby relieving the pressure. However, where trench sheeting or steel sheet piling is used as the means of excluding groundwater from the trench, allowance must be made for water pressure in addition to the pressure of the saturated ground. In such cases the equivalent fluid density is likely to be not less than 1.3 tonnes per cubic metre.

More accurate determination of active pressures requires a knowledge of the ground data: density (γ) and angle of internal friction (ϕ) in the case of granular soils, and density and shear strength (c) for cohesive soils. Given this information the active pressure at a depth h in granular soil is:

$$\gamma \times h \times \frac{1 - \sin\phi}{1 + \sin\phi}$$

and in cohesive soils:

$$(\gamma \times h) - (2 \times c)$$

Allowance in both calculations must be made for any loading on the ground surface due to plant etc. that may be needed near the trench, and this can be done by converting the average ground load or 'surcharge' into an equivalent additional depth of trench.

Example of ground-pressure calculation

To calculate the active ground pressure at the bottom of a trench 3 m deep in granular soil having a density of 2 tonne/m^3 and an angle of internal friction of 30°, when the ground surcharge is 1 tonne/m^2.

Trench depth	3 m
Add for surcharge ½	0.5
Effective depth	3.5 m

Active pressure $= \gamma \times h \times \dfrac{1 - \sin\phi}{1 + \sin\phi}$

$$= 2 \times 3.5 \times \frac{1 - \frac{1}{2}}{1 + \frac{1}{2}} = 2.33 \text{ tonne/m}^2$$

$$= 22.9 \text{ kN/m}^2$$

Had the soil in the example been clay, of the same density but with a shear strength of 30 kN/m^2, the active pressure would have been:

$$(\gamma \times h) - (2 \times c) = (2 \times 9.806 \times 3.5) - (2 \times 30) = 8.64 \text{ kN/m}^2$$

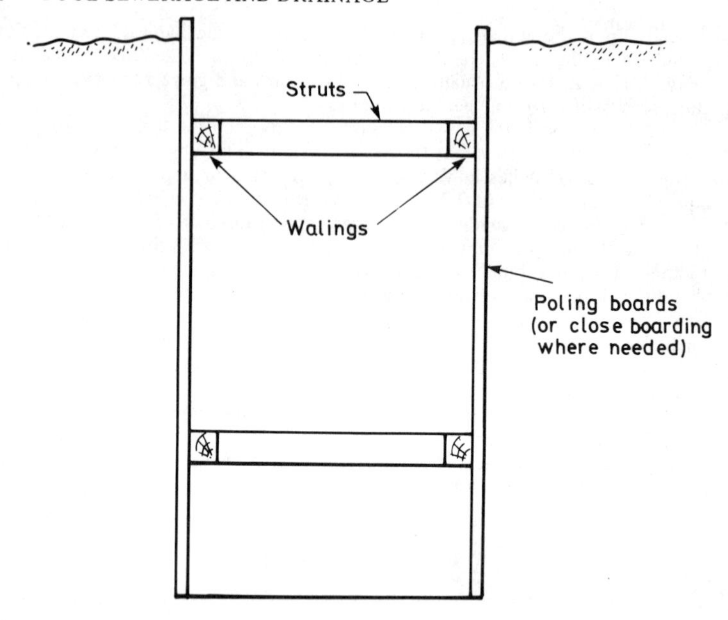

Figure 21.3 Typical timbering to sewer trench

Having calculated the active pressure of the ground, stresses in the trench timbering may be calculated, checks usually being needed of:
- bending stresses in the poling boards,
- bending stresses in the walings,
- axial stresses in the struts,
- bearing stresses of struts against walings.

Of these, bending stresses in the walings are usually most critical, and therefore determine the maximum spacing of the struts. See *Figure 21.3*.

Pipe laying and jointing

Having checked that the pipes are sound and undamaged (and the fact that they may have been examined by the manufacturer is no safeguard against damage having been caused in transit to the site), pipes should be placed in position and accurately aligned both horizontally and vertically. Alignment may be checked by looking through the sewer line, when a full circle of light should be seen. Where it is not intended to use imported material for bedding, the bottom of the trench should be accurately formed to a flat 'V' section, with additional pockets at the end of each pipe to accommodate the pipe sockets. In this way the pipes should each be firmly supported by their barrels resting in the shaped bottom of the trench. Where gravel or concrete beds are to be provided, the trench must be excavated to the additional depth needed to accommodate them, and the bedding material should be carefully packed beneath the pipes, ensuring that the correct alignment is maintained.

Joints should be formed carefully to the maker's instructions. Many modern pipes incorporate a rubber ring or similar type of seal, and require only that the ring shall be placed in position on the spigot, and the spigot pushed into the socket to form a seal. In some cases a mortar or grout joint may be provided instead of or in addition to the rubber joint, but where this system is used the advantage of flexibility to accommodate minor ground movements is lost.

Provision for house connections should be made by inserting capped branches at suitable points along the line of each sewer length, and recording the location and direction of each branch.

Testing

Testing for watertightness is usually carried out before backfilling is commenced, so that any leaks can be found and rectified without undue difficulty. Tests are carried out using water or air.

In the water test a plug is firmly fixed at the lowest end of the pipe run to be tested, and on the upstream side of the upstream manhole if that manhole is complete and ready for inclusion in the test. If it is not, a bend is inserted in the socket of the upstream pipe and jointed temporarily using puddled clay or similar material, the bend pointing upwards. If necessary a further short length of pipe may be needed, so that the open end is at least 1.2 m above the crown of the pipes to be tested. The pipe length is then filled with water to the required head, and the loss of water is measured after a period of 30 minutes, by pouring in the quantity of water needed to restore the level to that at the beginning of the test.

Assuming that there is no air trapped in the system (and there should be none if the pipes are laid to a uniform fall), any fall in water level indicates either leakage or absorption by the pipes or by the joints. The latter cause may be eliminated or at least reduced by repeating the test if necessary, but to allow for inevitable losses it is usual to permit wastage at a rate of 1 litre per hour per lineal metre of pipe per metre of pipe diameter. Thus for example a sewer of 150 mm diameter, 100 m long should, in a 30-minute test, show a loss not exceeding:

$$1 \times 0.5 \times 100 \times 0.15 = 7.5 \text{ litres}$$

In cases where pipes are excessively absorbent, difficulty may be experienced in obtaining a satisfactory test result, and in such cases information as to the actual degree of absorption of the pipes should be obtained from the pipe manufacturer. Any leaks that may be visible during testing should be made good.

In the air test both ends of the pipe run to be tested are plugged, with one or both plugs having small-diameter outlet pipes, to one of which is fitted an air pump, and to the other a glass U-tube to form a pressure gauge. Air is then pumped in until the pressure reaches 100 mm water gauge. After allowing conditions to stabilise, there should be no more than a drop to 75 mm water gauge during a period of 5 minutes.

The air test is convenient in that it requires little equipment, and obviates the need to obtain and dispose of the substantial quantity of water needed for the alternative test. It does not, however, provide any indication of the location of any leaks there may be, and it may be misleading because unsatisfactory results are often found to be caused by leaking plugs or pipes from the plugs. In a scheme involving many lengths of sewer it may be worth using the air test initially, and then reverting to the water test for any lengths that fail to satisfy the air test.

Backfilling

Pipe trenches are usually backfilled using excavated material where this is suitable. Care should be taken to keep topsoil separate from subsoil in cultivated land, or to preserve any paving or other surfacing materials that may be suitable for re-use when working in roads or footpaths. Selected material should be carefully packed around the pipes to ensure that there is no danger of their being displaced by large lumps of subsoil or by pieces of rock: such large pieces of soil are used only at levels well above the pipes. Backfilling should be well consolidated by hand or by mechanical ramming, again using care to prevent any damage to the pipes.

Where the surface layers of topsoil etc. have been kept separately, it may be desirable to defer replacing them until the main trench has consolidated to its final level; alternatively the topsoil may be left a few inches above the ground level to allow for settlement.

Sewage outfalls

In most cases sewage from new housing estates is discharged into the nearest available and suitable public sewer, and the absence of such a sewer may well result in the local authority withholding planning permission for development of the site.

There may, however, be occasions when some other means of disposal of sewage may be acceptable, although generally as an interim measure only, until main sewerage becomes available. In such circumstances consideration may be given to the construction of a small sewage treatment plant to the recommendations of CP302[6] or of a cesspool to CP302.200[7].

Components for either of these types of installation may be obtained from specialist manufacturers, many of whom offer advisory and design services to assist developers in complying with local authority requirements, which are usually related to codes of practice.

House connections

The laying of house connections should generally be deferred until after the sewers into which they flow have been laid and tested. Connections should as far as possible be made only to branches left in the main sewer for the purpose, and should be arranged to enter the sewer from the side, with the crown of the connection level with the crown of the sewer.

House connections usually serve a secondary purpose of providing ventilation to the main sewer. The requirement is to prevent the formation of air locks in the sewer, and at one time this was usually achieved by the provision of vent pipes connected to the sewer and discharging to the atmosphere at a height, usually some 6 m above road level. However this type of installation is now rarely used, the need for ventilation being met by house soil/vent pipes, which provide for the need through each house connection.

REFERENCES

1. CP 2005, *Sewerage*, British Standards Institution (1968)
2. Ministry of Technology, Hydraulics Research Paper No. 4, HMSO
3. BS 556: Part 2, *Concrete cylindrical pipes and fittings including manholes, inspection chambers and street gullies*, British Standards Institution (1972)
4. BS 497, *Cast manhole covers, road gully gratings and frames, for drainage purposes*, British Standards Institution (1967)
5. CP 2003, *Earthworks*, British Standards Institution (1959)
6. CP 302, *Small sewage treatment works,* British Standards Institution (1972)
7. CP 302.200, *Cesspools*, British Standards Institution (1949)
8. *The Building Regulations 1976*, HMSO

FURTHER INFORMATION

Further information on the design and construction of foul sewers and drains may be obtained from the relevant British Standards and Codes of Practice, from the Concrete Pipe Association of Great Britain, and from the Building Regulations[8].

22 SURFACE WATER DRAINAGE

INTRODUCTION 22–2

FLOW CALCULATIONS 22–2

SEWER LAYOUT 22–6

SEWER CONSTRUCTION, PIPE LAYING AND TESTING 22–7

SURFACE WATER DRAINAGE 22

22 SURFACE WATER DRAINAGE

D.A. STEPHENSON, BSc, CEng, FICE, FIStructE, FIArb
Consulting Engineer

This section relates to separate systems of surface water drainage and considers flow calculations, layout, materials and construction.

INTRODUCTION

The term *drainage* when used in connection with surface water usually denotes the whole system by which rainwater is conveyed from the site to the point of discharge. It therefore includes both drains and sewers.

In this section consideration is given primarily to surface water drainage by the separate system. Section 21 refers to the other systems of drainage that may sometimes be used, namely the combined and partially separate systems. Where either of these latter systems is to be used the total flow in the sewers is the sum of foul flow (calculated in accordance with the principles set out in Section 21) and the surface water flow (calculated in the manner set out in this section). Problems that arise from excessive quantities of flow, and the need for stormwater overflows, are also dealt with in this section, being due mainly to large volumes of surface water.

FLOW CALCULATIONS

The flow of surface water from an area of land to be drained is the product of three factors: intensity of rainfall, permeability of the ground, and the area of ground on which it falls.

Intensity of rainfall

It is well known that very intense rainfall occurs only in storms of short duration, being associated generally with short but violent rainstorms, usually in hot weather. During such storms intensities of rainfall as high as 200 mm per hour have been recorded at the peak of the storm, which may have lasted one or two minutes.

It would be both uneconomic and unnecessary to design a system of surface water sewers to cope with such exceptional conditions, for several reasons. Firstly, the likelihood of a recurrence – termed the 'return period' by statisticians – is so remote that the risk of flooding for a short period of time can be tolerated. Secondly, a storm of short duration cannot result in the corresponding rate of flow in a sewer serving anything but a small area, because of the varying times taken for flows from different parts of the area to arrive at any given point in the sytem. Thirdly, the sewers themselves, including manholes etc., have a sufficient capacity to form a reservoir, mitigating the most severe effects of the storm.

Until the early 1960s it was usual to allow for a rainfall intensity related to the duration of the storm by a simple formula adopted by the then Ministry of Health, which gave:

For storms of 5–20 minutes duration, $$R = \frac{30}{T+10} \times 25.4$$

For storms of 20–100 minutes duration, $R = \dfrac{40}{T + 20} \times 25.4$

where R is the intensity of rainfall in millimetres per hour, and T is the duration of the storm in minutes.

This formula gives a maximum intensity of slightly more than 50 mm per hour for the most intense storm of 5 minutes duration, reducing to less than 10 mm per hour for a storm lasting 100 minutes: see Table 22.1.

Table 22.1 INTENSITIES OF RAINFALL IN MM PER HOUR FOR VARYING STORM DURATIONS

Storm duration, T (min)	*Intensity of rainfall, R* (mm/hour)
5	50.8
6	47.6
7	44.8
8	42.3
9	40.1
10	38.1
11	36.3
12	34.6
13	33.1
14	31.8
15	30.5
16	29.3
17	28.2
18	27.2
19	26.3
20	25.4
21	24.8
22	24.2
23	23.6
24	23.1
25	22.6
26	22.1
27	21.6
28	21.2
29	20.7
30	20.3
40	16.9
50	14.5
60	12.7
70	11.3
80	10.2
90	9.2
100	8.5

Much research, including extensive recordings of rainfall intensities, has since been carried out, the results of which are given in Road Note 35[1]. Recommendations for the design of stormwater sewers in that document are based on the use of rainfall statistics for the exact location of the scheme, which may be obtained at a small charge from the Meteorological Office, Met O8c, London Road, Bracknell, Berkshire.

The information available from this source, computed for the location of the scheme, includes rainfall intensities for storms varying from 2 to 20 minutes in duration, for return periods of 1, 2, 5, 10, 20, 50 and 100 years.

Whether or not the high degree of precision implied by the use of these statistical values is of any practical value is open to question, especially in the case of relatively small schemes for the sewerage of estates. For the designer is still left with the arbitrary choice of a return period on which to base his calculations; also the design capacity of the sewers themselves contains a large element of conjecture, in that it depends heavily upon the degree of silting or encrustation that may be present during the useful life of the sewers.

The recommended return period for use for separate systems of sewers in modern estates is given in the document as one year. An example of rainfall statistics included in the document, for Crowthorne, Berkshire (an area having rainfall intensities near the upper limit of those plotted for the whole of the United Kingdom) shows the figures for storms of one year return period, for durations between 5 and 20 minutes, to be remarkably close to those given by the earlier Ministry of Health formula. The table also gives figures for storms of shorter duration, down to 2 minutes, and the intensity corresponding to that duration is about 50 per cent more than that at 5 minutes duration. Storms of such short duration are however unlikely to be of practical significance, for reasons that are explained later in this section.

It is therefore the author's contention that, for preliminary design purposes at least, the Ministry of Health formula may still be relied upon to give reasonably accurate figures for rainfall intensity.

The storm duration to be used in calculating rainfall intensity (and hence the rate of flow at any point in a sewerage system) is equal to the 'time of concentration' for that point, i.e. the time taken for rainfall on the most remote part of the area draining to the sewer under consideration to reach that sewer. The time of concentration is made up of two parts: time of entry and time of flow.

Time of entry is the time taken for surface water to flow over the area to be drained and to enter the sewer. It is usually taken to be 2 minutes in normal urban areas; a larger figure may be used in exceptional situations, such as where large paved areas having very little fall are drained, when a value up to 4 minutes may be used.

Time of flow is calculated from the length of the sewers from the point of entry to the sewer under consideration, and the velocity of flow in the sewers, asssuming that they are running full. Such velocities may be determined from the formula developed by Crimp and Bruges, and referred to in Section 21:

$$V = 83.33m^{2/3}i^{1/2}$$

where V is the velocity in metres per second, m is one-quarter of the diameter of the pipe in metres, and i is the gradient. Velocities of flow calculated from this formula are given in Table 22.2. Velocities above the upper zig-zag line may cause damage by scour. Velocities below the lower zig-zag line are not self-cleansing.

Permeability and area

For purposes of design of surface water sewerage it is usual to assume that paved areas directly connected to the system contribute to the flow, and that unpaved areas, or paved areas not directly linked to the system, make no contribution. The justification for this assumption is that the critical consideration in design is intense rainfall of short duration, which usually occurs in hot weather, when the ground is likely to be relatively dry. In such conditions unpaved areas have considerable capacity to absorb rainfall, either by its soaking into the ground or by the formation of pools in local depressions.

Paved areas are taken to include roads and footpaths that adjoin them directly or have their own system of gullies connected to the sewer, plus any roofs or paved terraces that may be so connected. Such areas are taken to be completely impermeable, the run-off being the full volume of rainfall upon them.

Table 22.2 FLOW VELOCITIES IN METRES PER SECOND FOR PIPES OF VARIOUS DIAMETERS, AT VARIOUS GRADIENTS, RUNNING FULL

Gradient	*Pipe diameter* (mm)						
	100	150	225	300	375	450	525
1 in 10	2.25	2.95	3.87	4.69	5.44	6.14	6.81
20	1.59	2.09	2.74	3.31	3.85	4.34	4.81
25	1.42	1.87	2.45	2.96	3.44	3.88	4.30
30	1.30	1.70	2.23	2.71	3.14	3.55	3.93
35	1.20	1.58	2.07	2.51	2.91	3.28	3.64
40	1.13	1.48	1.93	2.34	2.72	3.07	3.40
45	1.06	1.39	1.82	2.21	2.56	2.98	3.21
50	1.01	1.32	1.73	2.10	2.43	2.75	3.04
55	0.96	1.26	1.65	2.00	2.32	2.62	2.90
60	0.92	1.21	1.58	1.91	2.22	2.51	2.78
65	0.88	1.16	1.52	1.84	2.13	2.41	2.67
70	0.85	1.12	1.46	1.77	2.06	2.32	2.57
75	0.82	1.08	1.41	1.71	1.99	2.24	2.49
80	0.80	1.04	1.37	1.66	1.92	2.17	2.41
85	0.77	1.01	1.33	1.61	1.87	2.11	2.33
90	0.75	0.98	1.29	1.56	1.81	2.05	2.27
95	0.73	0.96	1.26	1.52	1.76	1.99	2.21
100	0.71	0.93	1.22	1.48	1.72	1.94	2.15
110	0.68	0.89	1.17	1.41	1.64	1.85	2.05
120	0.65	0.85	1.12	1.35	1.57	1.77	1.96
130		0.82	1.07	1.30	1.51	1.70	1.89
140		0.79	1.03	1.25	1.45	1.64	1.82
150		0.76	1.00	1.21	1.40	1.59	1.76
160		0.74	0.97	1.17	1.36	1.54	1.70
170		0.72	0.94	1.14	1.32	1.49	1.65
180		0.70	0.91	1.10	1.28	1.45	1.60
190		0.68	0.89	1.08	1.25	1.41	1.56
200		0.66	0.87	1.05	1.22	1.37	1.52
220		0.63	0.82	1.00	1.16	1.31	1.45
240			0.79	0.96	1.11	1.25	1.39
260			0.76	0.92	1.07	1.20	1.33
280			0.73	0.89	1.03	1.16	1.29
300			0.71	0.86	0.99	1.12	1.24
350			0.65	0.79	0.92	1.04	1.15
400				0.74	0.86	0.97	1.08
450				0.70	0.81	0.92	1.01
500				0.66	0.77	0.87	0.96
550				0.63	0.73	0.83	0.92
600				0.61	0.70	0.79	0.88
650					0.67	0.76	0.84
700						0.73	0.81
750						0.71	0.79
800							0.76

Units

Rainfall intensity, denoted by R, is expressed in millimetres per hour, and impermeable area A in square metres, although in the case of very large areas hectares (10 000 m^2) may be used. Run-off, Q, is expressed in litres per second, or in cubic metres per second where quantities are very large.

Using the smaller units in each case, the run-off from an impermeable area of 1 square metre resulting from rainfall of 1 mm is 1/1000 of a cubic metre, i.e. 1 litre.

Hence the run-off Q in litres per second from an area of A m² resulting from rainfall of R mm per hour is:

$$Q = \frac{R \times A}{3600} \text{ litres per second}$$

Example of a flow calculation

To determine the size and gradient of a surface water sewer to drain a 110 m length of road 5.5 m wide, with footpaths 2 m wide on both sides.

Assume that a pipe of 150 mm diameter, falling at 1 in 150, will be adequate. Then:

Velocity of flow in pipe (from Table 22.2) = 0.76 m/s
Time of flow = 110/0.76 = 145 seconds
Time of entry = 2 minutes

Time of concentration = 4.5 minutes (approx.)
Allow for the rainfall intensity of a 5-minute storm, i.e. (from Table 22.1): R = 50.8 mm/hr

$$\text{Then } Q = \frac{50.8 \times 110 \times (5.5 + 4)}{3600} = 14.75 \text{ l/sec}$$

Increase the gradient of the sewer to 1 in 125, to give a flow capacity of: Q = 14.8 l/sec

In this example the maximum intensity of rainfall has been taken to be 50.8 mm per hour, the figure given in Table 22.1 for a 5-minute storm. In consequence no adjustment is needed for the fact that the time of concentration is slightly reduced because of the increased gradient from that assumed originally. In many cases it is necessary to re-check a sewer design to take into account the increased value of R that will result from a reduced time of concentration, and to arrive at a final design by successive approximations.

Flow capacity of pipes

The flow capacities given in Table 21.1 for sewers of various diameters and gradients are equally applicable to the design of surface water sewers.

SEWER LAYOUT

The principles stated under the corresponding heading in Section 21 are equally applicable to the design of surface water sewerage systems.

Generally the flow of surface water to be allowed for in design is likely to be many times greater than the flow of sewage from the same development. It was seen, for example, in Section 21 that a 150 mm diameter sewer in a separate system would take the flow from at least 855 persons. If it is assumed that those persons occupy some 250 houses, the size of the estate on which they are built may be of the order of 10 hectares, of which it may be further assumed that at least 10 per cent is impermeable area of roads etc. connected directly to the stormwater sewerage system.

Making these assumptions, the flow of surface water to be allowed for in the design of separate sewers, assuming the time of concentration to be 10 minutes, is

$$Q = \frac{38.1 \times 100\,000}{3600 \times 10} = 105.8 \text{ litres per second}$$

This is nearly eight times the flow of foul sewage to be allowed for in design, and requires a sewer of 300 mm or 375 mm diameter, depending upon the gradient.

Because of the relatively large quantities of stormwater flow in comparison with foul sewage, both the combined and the partially separate systems of sewerage have in general been superseded by the separate system. Even when a separate system is provided, it is desirable in the interests of economy that drainage from paved areas of domestic properties, roofs etc. should be excluded from the surface water sewers, which are then provided only for road and footpath drainage.

Materials

Materials for the construction of surface water sewerage systems are generally similar to those described in Section 21 for foul sewers. It is, however, recognised that the requirements of surface water sewerage are not so demanding as those of foul sewerage, and in consequence lower qualities of materials are generally acceptable. The British Standards for drain and sewer pipes, BS 65 and 540, for example, allow somewhat wider tolerances for pipes intended for use in surface water drainage.

Other materials required for surface water sewerage are covered by the following British Standards:

Street gullies (concrete and plastic) to BS 556
Street and yard gullies (stoneware) to BS 539
Gully gratings and frames (cast iron) to BS497

SEWER CONSTRUCTION, PIPE LAYING AND TESTING

The construction of surface water sewers should be carried out in accordance with principles described in Section 21 for the construction of foul sewers. Reference should also be made to Section 21 for notes on pipe laying and jointing, and for testing of sewer lengths.

Surface water outfalls

The outfall from a surface water sewerage system is normally discharged into the nearest available watercourse, subject to the approval of the drainage or river authority responsible for the watercourse. Special requirements may, however, be imposed where the sewerage system serves a large area in relation to the existing watercourse, because of the danger that there may be an unacceptable degree of pollution in the discharge, especially after long periods of dry weather. In such cases the authority may require that the flow be conveyed in the sewer to a point further downstream where the flow in the watercourse is sufficient to absorb the polluted flow without unduly affecting the purity of the watercourse.

A headwall and apron should be constructed at the point of discharge of the surface water sewer into the watercourse, to prevent damage to the sewer and scour to the stream bed. In the case of large-diameter outfalls, where there may be a danger of debris from the stream entering the pipe, or unauthorised entry by children etc., a screeen of bars at not more than 150 mm centres should be provided. A typical headwall is illustrated in *Figure 22.1*.

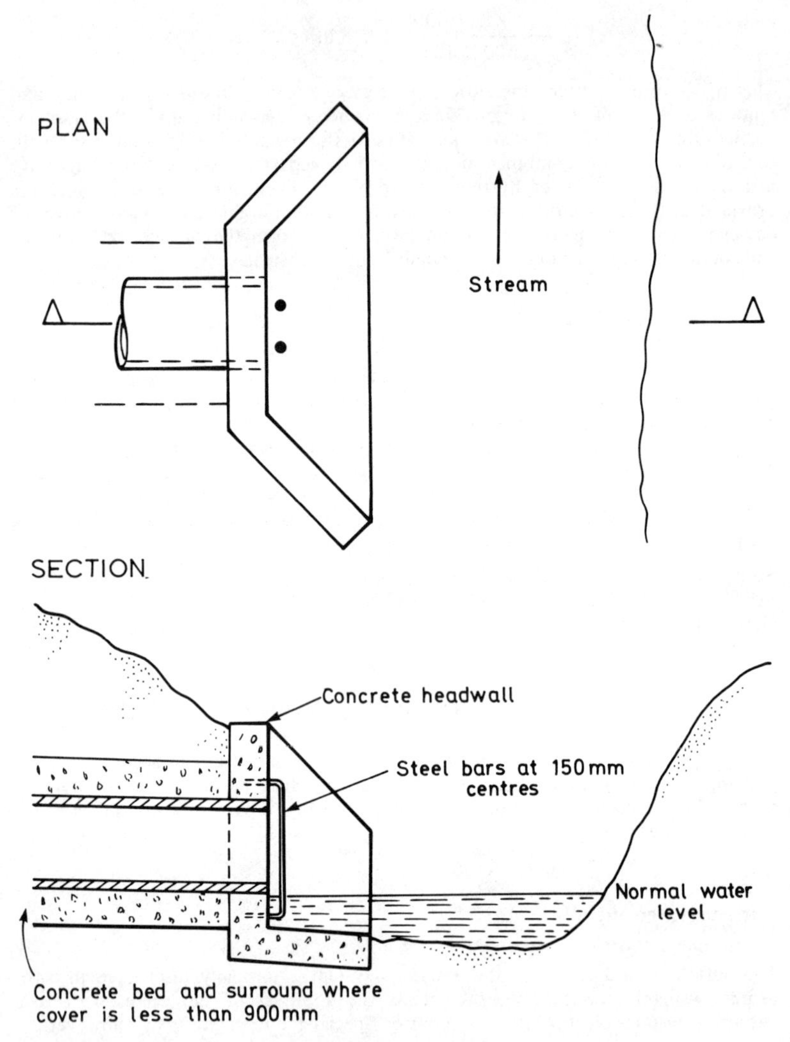

Figure 22.1 Typical headwall and apron at outfall of surface-water sewer

Stormwater overflows

In order to limit the maximum flow that may be taken into sewers designed on the combined or partially separate system, provision is sometimes made for flows in excess of the capacity of the downstream sewers, pumping stations or treatment plants to be discharged to a suitable watercourse.

This is done by means of a special stormwater overflow manhole constructed on the main sewer and having a weir on one or both sides of the channel. Flow in excess of the predetermined figure passes over the weirs, and is led to a separate pipe that discharges into the watercourse.

It was at one time acceptable in principle that flows in excess of six times DWF could be discharged in this way, even though this involves the discharge of diluted but untreated sewage. However, this principle has in recent years met with growing opposition, and such overflows are now likely to be permitted by the drainage authority only where they are for practical purposes unavoidable. In many cases it will be argued against the use of an overflow that surface water should be excluded by use of a separate system of sewers; and that where this is not feasible the full flow should be carried to the treatment works.

The code of practice for sewerage[2] recommends that no overflow should be provided on sewers of less than 460 mm diameter: a recommendation that is likely to apply to most estate sewerage systems.

Soakaways

The volume of surface water for which provision is needed in the sewerage system may in suitable types of ground be substantially reduced by the use of soakaways for drainage from roofs, terraces and similar paved areas.

The principle used in the design of a soakaway is that it should provide storage capacity for the volume of water that may flow from the drained area during a period of intense rainfall, allowing it to be dispersed into the ground over a much longer period.

A soakaway can only be effective in ground of sufficient permeability to allow the water it contains to be absorbed, and clearly this cannot apply to any part of the soakaway that may be below the water table.

To check that the ground is sufficiently permeable, the Building Research Station Digest[3] on this subject recommends a test using a shallow borehole. By measuring the rate at which a known volume of water in the borehole is absorbed into the ground, a measure of the size of the soakaway needed for a given area to be drained is obtained.

The Digest includes recommendations as to the construction of a soakaway, which usually takes the form of a cylindrical excavation of roughly equal diameter and depth, filled with coarse granular material, and covered with granular material of gradually reducing size, followed by topsoil and grass.

REFERENCES

1. Road Note 35, Second edition, HMSO
2. CP 2005, *Sewerage,* British Standards Institution (1968)
3. Building Research Establishment Digest 151, HMSO

FURTHER INFORMATION

Further information on the design and construction of surface water sewerage systems may be obtained from the highway authority in whose area the works are situated, as well as from the references given above.

23 SOILS

INTRODUCTION 23–2

BEARING CAPACITY 23–2

SETTLEMENT 23–5

OTHER CAUSES OF FOUNDATION PROBLEMS 23–16

SOILS 23

23 SOILS

C.R.I. CLAYTON, BSc(Hons), MSc, PhD, DIC, CEng, MICE, FGS
Senior Research Engineer, Ground Engineering Ltd

This section begins with background information on how soils behave. It goes on to consider particular types of problem ground-conditions. The general reader can pass over the mathematics without disadvantage.

INTRODUCTION

To the builder and civil engineer, soils comprise not only 'topsoil' (the agriculturalist's soil) but all the relatively loose or 'soft' deposits that occur near the earth's surface and cover the harder rock deposits. Soils may be readily recognised because the bonding between their particles is considerably weaker than that of the minerals of which they are composed.

Inevitably building must be founded on either soil or rock. In practice rock is usually less troublesome as a founding material, so this section concentrates on soil mechanics, a branch of engineering devoted to understanding the behaviour of soil under changing load conditions.

Soil is conventionally divided into cohesive and non-cohesive types, which are approximately equivalent to:

(a) slow-draining materials, such as clay soils;
(b) free-draining materials, such as sands, gravels, etc.

When soil is loaded, for example by a footing, two processes occur simultaneously: the soil compresses, and during this process its strength increases. Therefore, provided that the amount of settlement that occurs is small, if the foundation does not fail immediately it is loaded it will normally be safe for all time. Two criteria are therefore normally considered when designing foundations:

(a) *allowable settlement* – the settlement that the building can tolerate without cracking of brickwork or overstressing of the structure;
(b) *safe bearing capacity* – the maximum stress level that can be applied to the soil by the structure while ensuring that failure of the soil does not occur.

BEARING CAPACITY

Bearing-capacity failure of soil beneath footings is always accompanied by large vertical and rotational movements. *Figures 23.1* and *23.2* show a typical footing being progressively loaded until failure occurs.

When settlement is large, differences between the settlement at different positions around the structure (known as differential settlements) are likely to be large, because inevitably there will be some differences in soil properties and bearing pressure from place to place. It is differential movements of structures that lead to problems of structural overstressing and unsightly cracking, and in order to avoid these problems the foundation designer aims to keep his applied load in the 'elastic' phase (*Figure 23.1(a)*). To ensure this, the applied foundation pressure must not exceed between one-third and one-half of the ultimate bearing capacity. This reduction factor is known as the *factor of safety*, and therefore

$$\text{Allowable applied foundation stress} = \frac{\text{ultimate bearing capacity}}{2 \text{ to } 3}$$

In order to choose a suitable material for foundations to rest on, a calculation of ultimate bearing capacity is necessary.

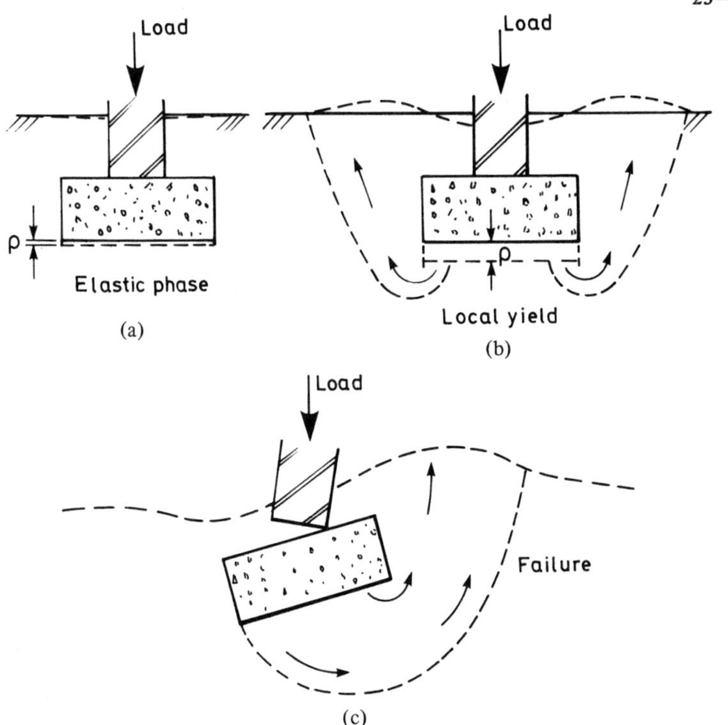

Figure 23.1 Behaviour of typical footing under increasing load. (a) When applied stress ≪ ultimate bearing capacity, settlement (ρ) is small and proportional to applied stress. (b) When applied stress > half the ultimate bearing capacity, local failure occurs at edge of footing; settlement (ρ) is large but total failure does not occur. (c) When applied stress = ultimate bearing capacity, total failure of footing occurs; settlement is very large and significant rotation of footing occurs

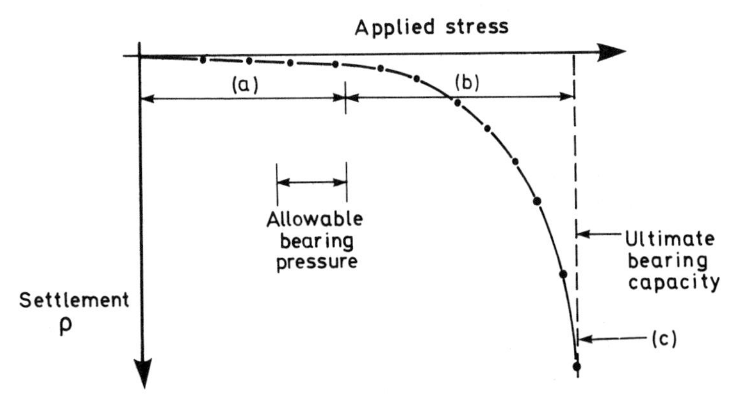

Figure 23.2 Stress-settlement curve for typical footing; (a), (b), and (c) refer to load conditions in Figure 23.1

Bearing capacity of cohesive soils

Since clay soils become stronger under loads that do not induce local-yield or bearing-capacity failure, the bearing capacity of the material under load will increase with time. The bearing capacity used in design is therefore calculated using the strength of the soil before loading occurs, which in clays is known as the 'undrained shear strength'.

Skempton[1] has provided a series of 'bearing capacity factors' by which the undrained shear strength is multiplied in order to obtain the ultimate bearing capacity

Table 23.1 BEARING CAPACITY FACTORS[1]

	Depth/width ratio (D/B)				
	0	0.5	1.0	2.0	4.0
N_c (circle, square)	6.2	7.1	7.7	8.4	9.0
N_c (strip)	5.1	5.9	6.4	7.0	7.5

of the soil. These factors are determined according to the shape of the foundation; see Table 23.1. It can be seen that the bearing capacity factors range from 5.1 to 9.0. The total allowable bearing pressure on a footing (*Figure 23.3*) is

$$q\ (\text{allowable}) = \frac{N_c \times C_u}{F} + p_o$$

where N_c = bearing capacity factor (above)
C_u = average undrained shear strength for a depth below the foundation of two-thirds width
p_o = vertical pressure originally applied by the soil at the founding level
F = factor of safety

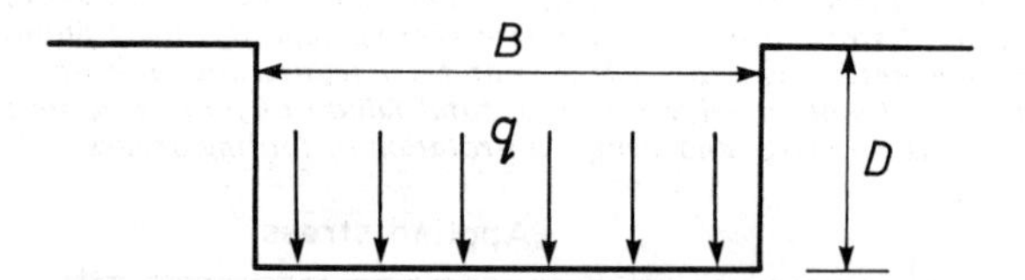

Figure 23.3 Bearing pressure on footing

If, as is common, the structure does not contain a basement but starts at ground level, the *net* allowable bearing pressure is more useful. The net allowable bearing pressure is the allowable stress increase at foundation level (i.e. above that existing at proposed foundation level before the start of construction):

$$q\ (\text{net allowable}) = \frac{N_c \times C_u}{F}$$

The net allowable bearing pressure is particularly convenient where foundation excavations are completely replaced by brick or concrete. In this case the load applied at foundation level due to the substructure (i.e. that part of the structure below ground level) will approximately equal the preexisting soil load, and the additional stress at foundation level due to the superstructure and other loads must not exceed the net allowable bearing pressure.

In the case shown in *Figure 23.4*, the superstructure load divided by the foundation area must be less than the net allowable bearing capacity.

$$\text{Net allowable bearing pressure} = \frac{\text{net ultimate bearing capacity}}{\text{factor of safety}}$$

Since the bearing capacity factor, N_c, is never less than between five and six, a reasonable net allowable bearing pressure for small structures can be obtained by multiplying the undrained shear strength of the soil by two.

$$\text{Net allowable bearing pressure} = N_c \frac{(5 \text{ to } 6) \times C_u}{F \times (2 \text{ to } 3)}$$

$$\approx 2 \times C_u$$

Bearing capacity of granular soils

For sands and gravels, the net ultimate bearing capacity can be calculated from the equation:

$$q \text{ (net ult.)} = \left[\frac{N\gamma}{2} + (N_q - 1)\frac{D_f}{B}\right]\gamma B$$

where $N\gamma$, N_q = bearing capacity factors
D_f = depth of foundation below ground level
B = width of foundation
γ = bulk density of soil below foundation level

The bearing-capacity factors $N\gamma$ and N_q can be found either from the angle of friction of the soil or from its density (*Figure 23.5*), which may be either determined by the standard penetration test, or estimated.

SETTLEMENT

Settlements of soil beneath structures are usually calculated by combining soil compressibility with vertical stress distribution in the soil beneath the foundation, as found using the theory of elasticity. The derivation of these stress distributions is outside the scope of this section, but since they are of such use in the calculation of settlements, *Figure 23.6* reproduces values by previous authors. From these figures it can be seen that:

(a) The distribution of vertical stress beneath a foundation is proportional to its width (B) or radius (R). If the footing width is doubled and the applied stress is kept constant, the stressed depth beneath the foundation will double. If the soil compressibility is constant with depth, the settlement of the footing will double.

(b) The vertical stress distribution depends on the shape of the footing. A strip footing will stress the soil to a much greater depth than a square or circular footing.

The accurate prediction of the settlement of structures is both complex and expensive. Solutions to such problems generally require the acquisition of good-quality soil-compressibility data, which can only be achieved after careful laboratory testing of good-quality soil samples. In addition such calculations are thrown into doubt by factors such as foundation rigidity, variation of soil compressibility (both laterally and with depth), the effects of foundation depth, and unknowns such as pore pressure coefficients and the relationship between vertical and horizontal soil compressibility during loading.

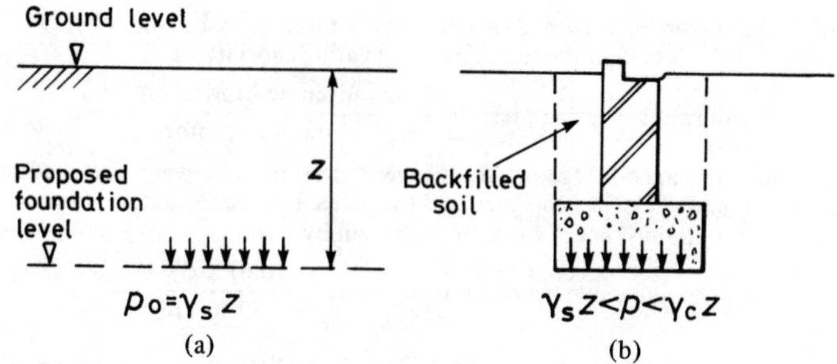

Figure 23.4 Net allowable bearing pressure: (a) at foundation level, before construction, (b) with substructure complete
γ_S = density of soil (≈ 2.0 Mg/m³)
γ_C = density of concrete and/or brick (≈ 2.3 Mg/m³)

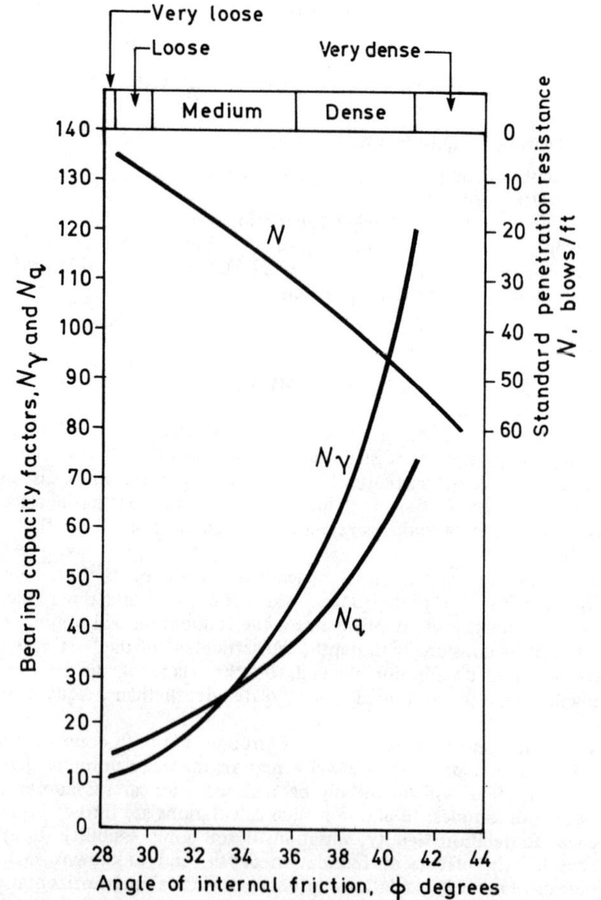

Figure 23.5 Bearing capacity factors N_γ and N_q as a function of angle of friction ϕ and SPT N value (after Peck, Hanson and Thornburn[2])

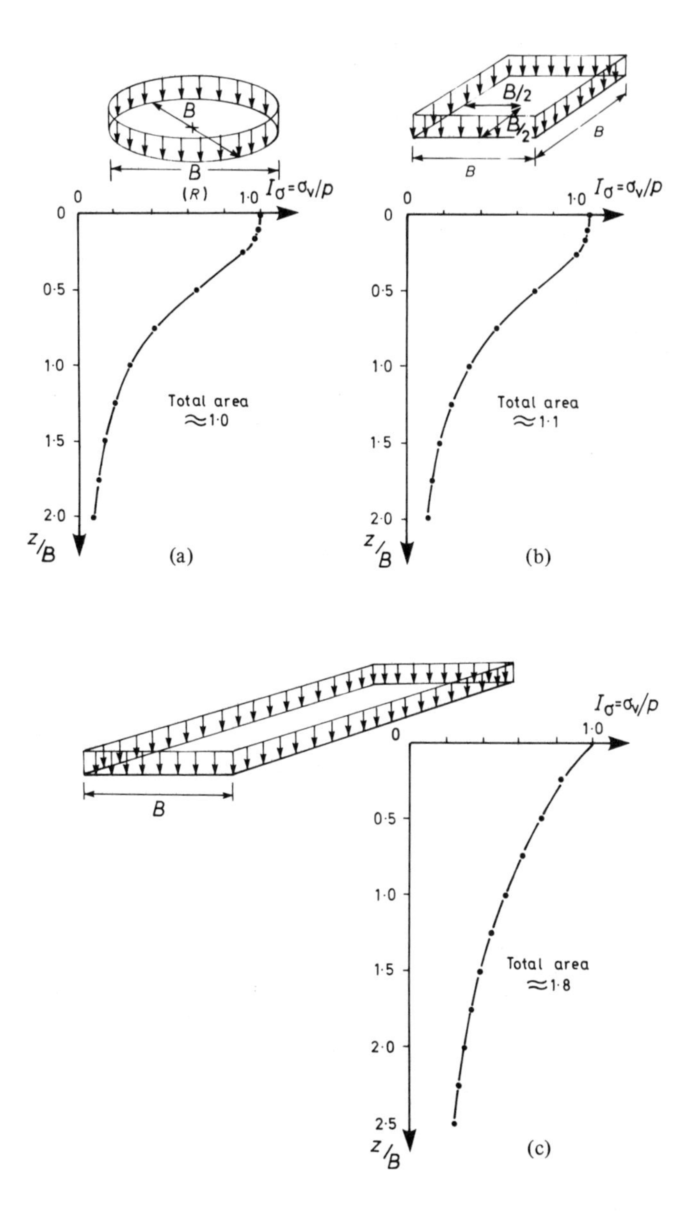

Figure 23.6 Vertical stress distribution beneath centre of different footing shapes, derived from elastic theory: (a) uniformly loaded flexible circular footing; (b) uniformly loaded flexible square footing; (c) uniformly loaded flexible, infinitely long, strip footing

Low-rise housing and other lightly loaded structures suffer from additional disadvantages. Their foundations are usually placed as close as possible to the original ground level, whereas larger buildings might be piled or founded on basements, and the soil they encounter is likely to be more variable. Because of the relatively low cost of each structure, smaller funds are available to investigate this variability.

Sophisticated calculations are pointless when carried out with poor-quality data, and therefore the best course is to attempt a conservative estimate of settlement. The following paragraphs detail methods that, although technically incomplete, give a rapid calculation method suited to the problem. When large structures are encountered these methods are not recommended, and the reader is referred to the review papers by Sutherland[3], Simons[4], Butler[5] and Hobbs[6] in the Cambridge conference on settlement of structures (1975).

Stress-distribution charts can be used in a fairly simple way to obtain a distribution of pressure increase with depth below a footing. Example 1 below shows how the stress at any depth beneath a loaded footing can be calculated.

Having obtained the distribution of vertical stress that will occur as a result of the load increase given by the structure, the settlement can be calculated by combining stress with compressibility.

The modulus of compressibility, E, is an expression of the stiffness of the soil, and the vertical strain ϵ (i.e. displacement per unit length of stressed material) of a column of soil would equal the applied stress divided by the modulus of compressibility, i.e.

$$\epsilon = \sigma/E_V'$$

The actual settlement ρ of the top of such a column of soil would be the strain multiplied by the length of the column, i.e.

$$\rho = \epsilon z = \sigma z/E_V'$$

In a layered system such as shown in Example 2 below, the soil can be imagined as a series of soil columns piled on top of each other, each contributing a portion of the settlement. However, unlike a series of columns, the stress level in each layer of the soil decreases with distance from the load. In this case the total settlement can be calculated as the sum of the settlements of each layer. The settlement of each layer

$$\rho = \Delta\sigma' \times \Delta z/E_V'$$

where ρ = settlement of individual layer
$\Delta\sigma'$ = average vertical stress increase in the layer as a result of the structure (this can be taken as the stress increase at the middle of the layer)
Δz = layer thickness
E_V' = modulus of compressibility of the soil

Where ground conditions are uniform with depth, the settlement can be calculated from the total area under the graph in *Figure 23.6*, i.e. for a circular foundation:

$$\rho = \frac{1.0pB}{E_V'}$$

For a square foundation

$$\rho = \frac{1.1pB}{E_V'}$$

For a strip foundation

$$\rho = \frac{1.8pB}{E_V'}$$

where B is the footing width and p is the applied foundation pressure.

Example 1 – Calculation of vertical stress beneath a footing

A proposed strip footing for a domestic dwelling is to be 450 mm wide. The structure is a semi-detached house, and the party wall is to apply a load of 50 kN/lineal metre. What will be the vertical stress 0.5 m below the bottom of the footing?

(a) Stress applied to soil

$$p = \frac{50}{450} \times 1000 = 110 \text{ kN/m}^2$$

(b) Since the footing is long compared with its width, it may be assumed to be an 'infinite strip' load. In *Figure 23.6*, B = footing width = 0.45 m. At 0.5 m depth below the bottom of the foundation:

$$z/B = \frac{0.5}{0.45} = 1.11$$

From *Figure 23.6(c)*, when $z/B = 1.11$:

$$\sigma_V/p = 0.48$$
$$\sigma_V = 0.48 \times 110 = 53 \text{ kN/m}^2$$

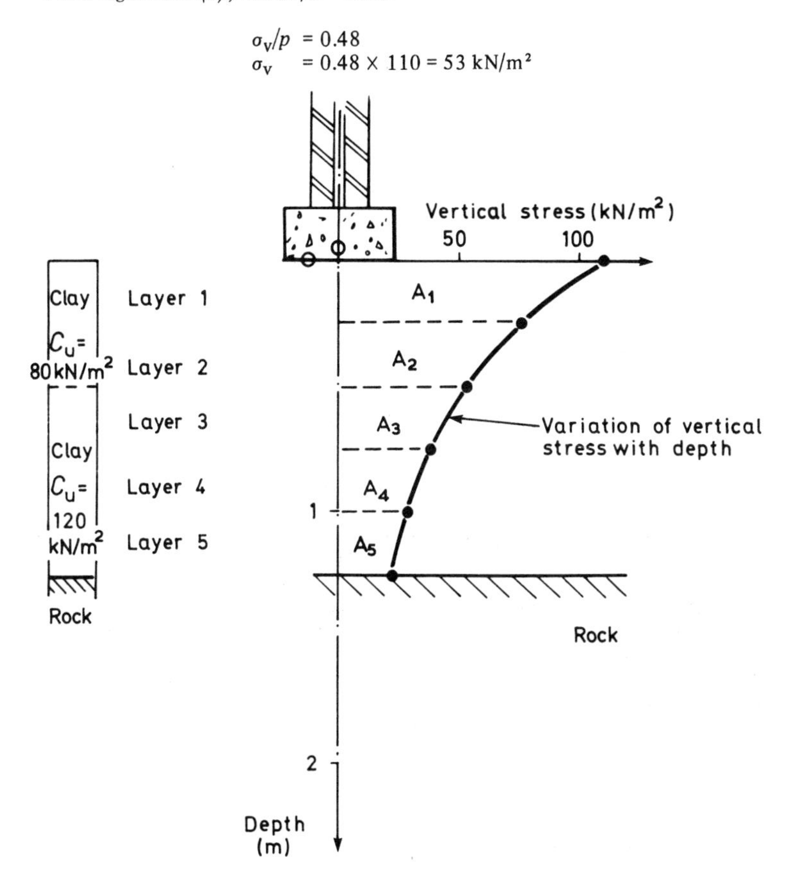

Figure 23.7 Stress distribution and soil profile for Example 2

Example 2 – Calculation of settlement

The strip footing used in Example 1 is to be founded on a layer of clay 1.25 m thick, beneath which lies incompressible rock. The clay is of stiff consistency with an undrained shear strength derived from undrained compression tests on 100 mm diameter samples of about 80 kN/m² to a depth of 0.5 m below the foundations. Below this the average strength is higher, at 120 kN/m². How much will the footing settle?

(a) Approximately, $E_V' = 130 \times C_u$ for the clay layer.
  0–0.5 m: $E_V' = 130 \times 80 = 10\,400$ kN/m²
  0.5–1.25 m: $E_V' = 130 \times 120 = 15\,600$ kN/m²

(b) The stress distribution for the clay is calculated as in Example 1, but for various depths below the footing. The settlement contribution of each layer (*Figure 23.7*) is:

$$\rho_1 = \frac{A_1}{E_1} \approx \frac{110 + 76}{2} \times \frac{0.25}{10\,400} \times 10^3 = 2.2 \text{ mm}$$

$$\rho_2 = \frac{A_2}{E_2} \approx \frac{76 + 53}{2} \times \frac{0.25}{10\,400} \times 10^3 = 1.6 \text{ mm}$$

$$\rho_3 = \frac{A_3}{E_3} \approx \frac{53 + 39}{2} \times \frac{0.25}{15\,600} \times 10^3 = 0.7 \text{ mm}$$

$$\rho_4 = \frac{A_4}{E_4} \approx \frac{39 + 29}{2} \times \frac{0.25}{15\,600} \times 10^3 = 0.5 \text{ mm}$$

$$\rho_5 = \frac{A_5}{E_5} \approx \frac{29 + 23}{2} \times \frac{0.25}{15\,600} \times 10^3 = 0.4 \text{ mm}$$

Total = 5.4 mm

Settlement of cohesive soils

The simplest method of calculating settlements of foundations on cohesive soils is to estimate a modulus of compressibility (E_V') value.

Three methods are available, based on properties derived from:
(a) the laboratory triaxial compression test (Butler[5]);
(b) the laboratory oedometer test (Skempton and Bjerrum[7] and Simons and Som[8]);
(c) the *in situ* SPT test (Stroud and Butler[9]).

Butler[5] gives $E_V = 130 \times C_{u100}$ for stiff fissured clays, where C_{u100} is the undrained shear strength obtained from laboratory triaxial tests on 100 mm diameter samples.

It has been shown by Agarwal[10] that the shear strength of stiff fissured clays varies with the size of sample tested, because fissures act as planes of weakness and are more likely to occur in the direction of a shear plane in a large sample than in a small one. Large samples give lower undrained shear strengths than small samples taken in the same clay. The following reduction factors are suggested to apply to undrained shear strength results to give an equivalent strength that would be expected from a 100 mm diameter sample in the triaxial test.

Pocket penetrometer	$\times \frac{1}{3}$ to $\frac{1}{5}$
Hand vane	$\times \frac{1}{2}$ to $\frac{1}{3}$
38 mm triaxial compression test	$\times \frac{2}{3}$ to $\frac{1}{2}$

When clays are not fissured a correction should not be applied.

The laboratory oedometer test is commonly used to determine soil compressibility, which is expressed as the coefficient of compressibility, m_v, in m²/MN.

$$m_v = \frac{\Delta h}{h_o} \times \frac{1}{\Delta p}$$

where Δh = change in height of specimen for a pressure increase of Δp
h_o = original height of specimen before the pressure increase
The sample tested is a disc of soil, usually 19 mm high with a 76 mm diameter, which is prevented from squeezing laterally when loaded by a ring placed around it. Because of this, only vertical strain occurs in the oedometer, while in the field the soil is able to move both laterally and vertically. The oedometer compressibility results are therefore not directly applicable to the field situation. This problem has been considered by Skempton and Bjerrum[7] and also by Simons and Som[8], who have both proposed complex solutions. For the purpose of estimating the settlement of small structures, it is sufficient to use

$$E_v' = 1/m_v$$

which, although not theoretically correct, will usually yield conservatively high settlements.

Stroud and Butler[9] have investigated the use of the standard penetration test (SPT) to give values of modulus E_v' for settlement calculations. The standard penetration test (BS 1377: 1975[11] test 19) involves driving a 50 mm diameter hollow tube into the ground using repeated blows of a 65 kg hammer dropping freely through 760 mm. The SPT N value is the number of blows required to advance the tube 300 mm into the soil, after an initial 150 mm of penetration has been achieved. Stroud and Butler have proposed that for clays

$$E_v' = 130 F_i N$$

where F_i = a factor approximately between 4 and 6
N = SPT value of the soil
It is sufficiently accurate to assume $F_i = 5$ for small structures.

Settlement of granular soils

The settlement analysis of foundations on granular soil has been developed from analysis of field experience of small-diameter plate-bearing tests and actual footings (Bazaraa[12], Peck, Hanson and Thornburn[2]). These investigations have led to the conclusion that, for footings on sand,

$$\rho = \frac{p}{0.47N}$$

where p = applied bearing pressure (kN/m²)
N = SPT N value (blows/300 mm)
ρ = settlement (mm)
The SPT N values used in this calculation should be values taken from the test and then corrected for groundwater position and depth of test (*Figures 23.8* and *23.9*).

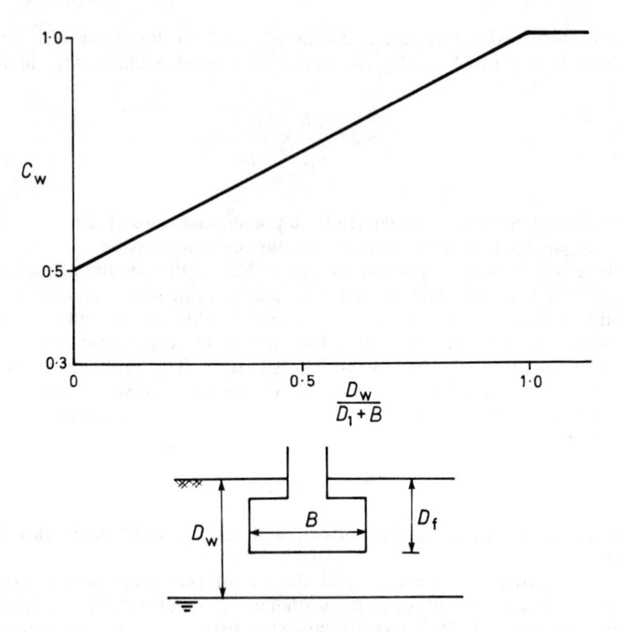

Figure 23.8 Corrections for SPT N values for depth of water table (after Peck, Hanson and Thornburn[2])

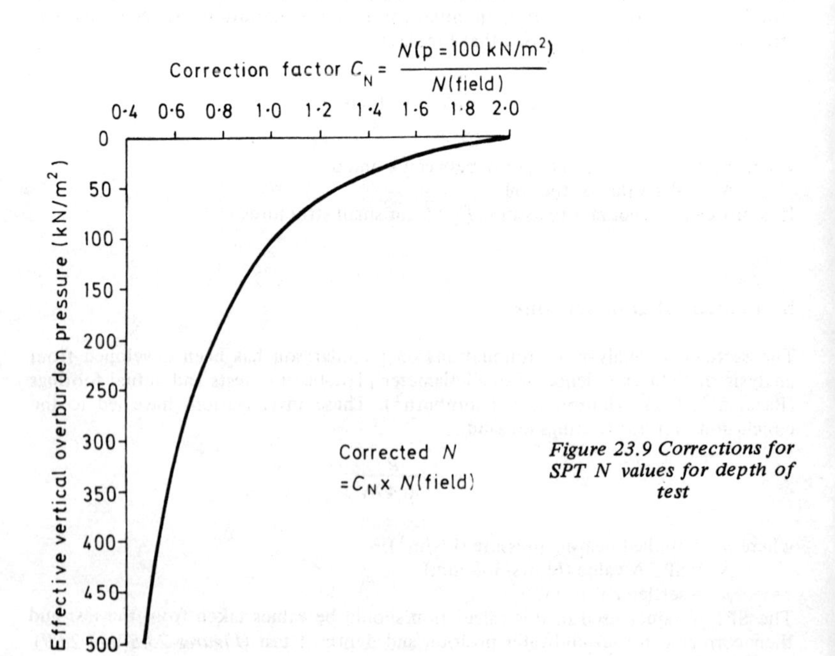

Figure 23.9 Corrections for SPT N values for depth of test

Differential settlement

Footing design charts by Terzaghi and Peck[13] and Peck, Hanson and Thornburn[2] for sand have both considered that a 25 mm total settlement would restrict differential settlements of square footings to such a low level as to prevent structural damage. However, their charts were not specifically designed for brick-clad structures founded on strip footings, but are more applicable to framed structures founded on a series of square-pad footings. Burland and Wroth[14] have shown that Skempton's and MacDonald's[15] criterion that angular distortions should be restricted to less than 1/500 is satisfactory for frame building but may be unsafe for load-bearing brick walls, especially those with a length to height ratio (L/H) greater than four. According to Burland and Wroth[14] the 1955 Building Code of the USSR gives the maximum allowable relative deflections shown in Table 23.2. *Figure 23.10* shows the definition of the terms 'angular distortion' and 'relative deflection'. Approximately, an angular distortion of 1/1000 is equivalent to a relative deflection of 1/2500.

Table 23.2 MAXIMUM ALLOWABLE RELATIVE DEFLECTIONS (USSR)[14]

	Maximum relative deflection	
Length/height ratio	*on sand*	*on soft clay*
< 3	1/3300	1/2500
> 5	1/2000	1/1430

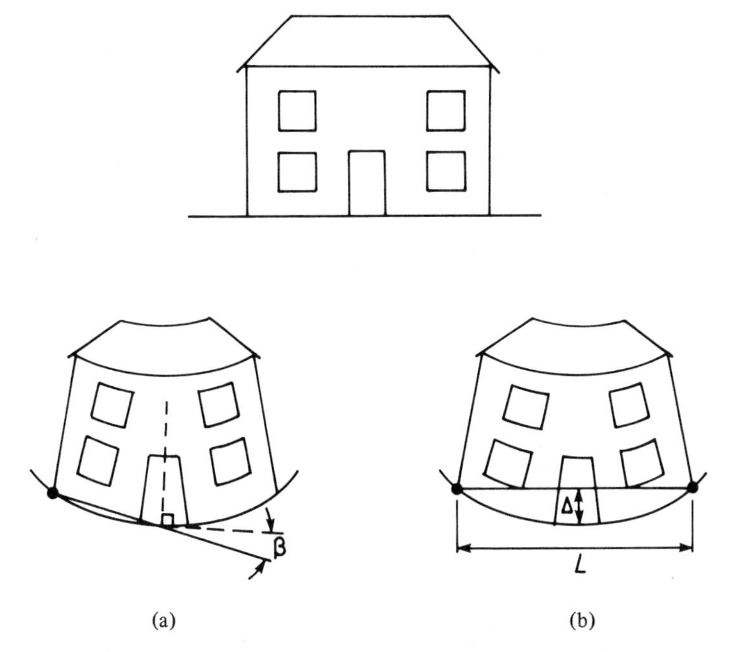

Figure 23.10 Definitions of differential settlement: (a) angular distortion β (after Skempton and MacDonald[15]); (b) relative deflection ΔL (after Polshin and Tokar[16])

Table 23.3 MINIMUM WIDTH OF STRIP FOUNDATIONS (table to Regulation D7 of Building Regulations 1976[17])

Type of subsoil		*Conditions of subsoil*	*Field test applicable*	*Minimum width in millimetres for total load in kilonewtons per lineal metre of load-bearing walling of not more than:* 20 kN/m	30 kN/m	40 kN/m	50 kN/m	60 kN/m	70 kN/m
I	Rock	Not inferior to sandstone, limestone or firm chalk	Requires at least a pneumatic or other mechanically operated pick for excavation.	In each case equal to the width of wall					
II	Gravel Sand	Compact Compact	Requires pick for excavation. Wooden peg 50 mm square in cross-section hard to drive beyond 150 mm.	250	300	400	500	600	650
III	Clay Sandy clay	Stiff Stiff	Cannot be moulded with the fingers and requires a pick or pneumatic or other mechanically operated spade for its removal.	250	300	400	500	600	650
IV	Clay Sandy clay	Firm Firm	Can be moulded by substantial pressure with the fingers and can be excavated with graft or spade.	300	350	450	600	750	850
V	Sand Silty sand	Loose Loose	Can be excavated with a spade. Wooden peg 50 mm square in cross-section can be easily driven.	400	600	*Note.* In relation to types V, VI and VII, foundations do not fall within the provisions of regulations D7 if the total load exceeds 30 kN/m			
VI	Silt Clay Sandy clay Silty clay	Soft Soft Soft Soft	Fairly easily moulded in the fingers and readily excavated.	450	650				
VII	Silt Clay Sandy clay Silty clay	Very soft Very soft Very soft Very soft	Natural sample in winter conditions exudes between fingers when squeezed in fist.	600	850				

It is not uncommon for brick-built dwellings to have wall lengths of 5–8 m, with L/H ratios well below three. In this case it appears that differential settlements should be limited to about 2–4 mm. Where L/H ratios exceed five those values can be increased to 4–6 mm. The relationship between total settlement and differential settlements of a structure remains a major unknown, depending on many factors such as soil variability, structural rigidity and loading patterns. Differential settlements equal to 50 per cent of the maximum total settlement are certainly common, and if this figure is used a maximum total settlement of 8–12 mm is obtained.

Building regulations

The Building Regulations[17] give minimum widths for strip footings in table D7 (see Table 23.3). These dimensions are deemed to satisfy the regulations, but do they give satisfactory results?

According to the Building Research Establishment's Digest 67[18], a typical two-storey semi-detached house of about 85 m^2 area, in cavity brickwork, with light-weight concrete or clay-block partitions, timber floors and a tiled roof has a mass of about 100 tonnes. The loads at ground level, in kN/m, are approximately:

Party wall	50
Gable end wall	40
Front and back wall	25
Internal partitions	15

Clearly the Building Regulations are designed to reduce the probability both of bearing-capacity failure and, theoretically, of achieving high settlements. Using the approximations in the previous paragraphs the typical values shown in Table 23.4 are obtained.

Table 23.4 TYPICAL VALUES FOR SETTLEMENT OF FOOTINGS DESIGNED TO BUILDING REGULATION D7

	Type of subsoil	*Settlement* (mm) *for width of footing specified in table to Regulation D7 for total load per lineal metre of not more than:*					
		20 kN/m	30 kN/m	40 kN/m	50 kN/m	60 kN/m	70 kN/m
I	Rock	≈ 0	≈ 0	≈ 0	≈ 0	≈ 0	≈ 0
II	Compact gravel (N = 30 blows/300 mm)	5.7	7.1	7.1	7.1	7.1	7.6
III	Stiff clay (C_u = 120 kN/m^2)	2.2	3.4	4.5	5.6	6.7	7.8
IV	Firm clay (C_u = 60 kN/m^2)	4.5	6.7	*9.0*	*11.3*	*13.5*	*15.7*
V	Loose sand (N = 15 blows/300 mm)	7.1	7.1	–	–	–	–
VI	Soft clay (C_u = 30 kN/m^2)	*9.0*	*13.5*	–	–	–	–
VII	Very soft clay (C_u = 10 kN/m^2)	*27.0*	*40.3*	–	–	–	–

Using Skempton's bearing-capacity factor for a strip footing at ground surface, and a factor of safety of 2.5, the allowable net bearing pressures shown in Table 23.5 are obtained.

Table 23.5 ALLOWABLE NET BEARING PRESSURES FOR FOOTINGS DESIGNED TO BUILDING REGULATION D7

	Type of subsoil	*Allowable net bearing capacity* (kN/m^2) *and value from table D7 for the width of footing specified in the table to Regulation D7 for a total load per lineal metre of not more than:*					
		20 kN/m	30 kN/m	40 kN/m	50 kN/m	60 kN/m	70 kN/m
I	Rock	Probably greater than the structural stresses allowed					
II	Compact gravel (N = 30 blows/300 mm)	110/80	125/100	145/100	165/100	190/100	200/100
III	Stiff clay (C_u = 120 kN/m^2)	240/80	240/100	240/100	240/100	240/100	240/100
IV	Firm clay (C_u = 60 kN/m^2)	120/67	120/86	120/89	120/83	120/80	120/82
V	Loose sand (N = 15 blows/300 mm)	70/50	90/50	–	–	–	–
VI	Soft clay (C_u = 30 kN/m^2)	60/44	60/46	–	–	–	–
VII	Very soft clay (C_u = 10 kN/m^2)	*20/33*	*20/35*	–	–	–	–

a/b: a = net allowable bearing capacity, b = value applied by Table D7

The Building Regulations 1976 do not appear satisfactory for foundations on soft and very soft clays, or for foundations supporting a load in excess of 30 kN/m on firms clays (see figures in italic type in Table 23.4). Indeed it is bad practice to found on any clay with soft or very soft consistency, since these materials are usually not only weak but have properties, such as compressibility, that vary considerably over very short distances.

OTHER CAUSES OF FOUNDATION PROBLEMS

Soil variability

Soil variability is a major problem in foundation design, and can arise as a result of several factors.

Soft and organic soils are highly compressible, and also generally have a high level of variability of compressibility. Unless a building is to be constructed on a reinforced raft, its foundations should be placed on material that is consistently firm (i.e. with an undrained shear strength of >40 kN/m^2).

In addition, house foundations are generally constructed at shallow depth, and shallow soils tend to be much more variable than those at depth. This presents difficulties to the designer, since funds available for site investigation normally restrict the amount of work and the quality of information on which foundation designs are to be based.

Therefore, even though site investigation has been carried out by a competent engineer, it is important to recognise that changes in design may still be necessary when construction of the footings is started. Site-investigation recommendations can only be based on the subsoil encountered at the locations of trial pits and borings. At all times during the construction of footings the contractor should be examining the founding strata, and any changes in soil properties should be considered, in order to determine if foundation design should be altered.

Softening of soil

When trenches are opened for the construction of footings they should be left exposed to the effects of the weather and ground water for the minimum possible period. Excavations should be covered during periods of rain, at night and at week-ends, if concreting has not been completed. Whenever possible, foundation concrete should be placed within 24 hours of excavation to foundation level. Clay soils, which may have originally been very good bearing material, will take in any water allowed to reach the base of the trench, and will then swell. As a result of this swelling they lose shear strength (and hence bearing capacity) and increase their compressibility.

Granular soils, and in particular sands and silts, are badly affected by water flowing out of them. If water flows into the sides of footing trenches then they will rapidly lose stability and collapse. Similarly, water contact with sand in the base of the trench can cause problems of loss of fines and, consequently, weakening. Where strip foundations are to be constructed below the water table, the use of well-pointing should be considered.

Frost heave

Certain soils, notably chalk, suffer from heaving under heavy frost action. When temperatures at the ground surface are maintained below freezing for some time, the frost gradually penetrates the soil. As a result of the freezing of the water held in the soil, a small increase in volume (of the order of 5 per cent) occurs. However, the major cause of frost heave occurs as a result of the build-up of ice lenses in the soil. Initially the whole volume of soil does not freeze, but a small crystal of ice forms at some depth below the surface. As a result of this freezing further water is attracted to this crystal and, provided temperature conditions at the surface are low enough, the crystal grows to form an ice lens in the soil.

The growth of the ice lens leads to heave at the surface, and this would clearly damage any structure above it. Further, when thawing occurs the release of water generally leads to a loss of bearing capacity, and large settlement occurs due to collapse of the void left by the melted lens.

Clearly, structures should not be founded in the freezing zone of frost-susceptible soils. For practical purposes the freezing zone may be taken as up to 0.5 m below ground level. Frost-susceptible soils can either be determined using the Frost Susceptibility test (Croney and Jacobs[19]) or from Casagrande's empiricisms set out below. In general clay soils are not affected by frost.

Casagrande's criteria indicate that soils susceptible to frost heave are of the following categories:

(a) fairly uniform soils ($D60/D10 < 5$) having 10 per cent or more of particles finer than 0.02 mm;
(b) well graded soils ($D60/D10 > 5$) having 3 per cent or more of particles finer than 0.02 mm;
(c) perfectly uniform soils ($D60/D10 = 1$) having all particles equal to or smaller than 0.01 mm;

D60 is the hole diameter through which 60 per cent by weight of a soil passes in a sieve analysis; D10 is the hole diameter through which 10 per cent by weight of a soil passes in a sieve analysis.

Seasonal swell and shrink

Seasonal swell and shrink will primarily affect structures on shallow foundations on plastic clays. Seasonal fluctuation of pore-water pressures in the clay leads to increases

and decreases in effective stress, causing the clay surface to rise and fall. The amount of this rise and fall and the depth of soil affected are a function of:
(a) plasticity of the soil,
(b) climatic conditions at the site,
(c) groundwater level and, to a limited extent,
(d) type of construction.
In Britain the depth of significant movement below open grassland is generally less than 900 mm (Ward[20]) but in arid environments or after long periods with reduced rainfall it may be much larger. Footings should not be founded within the top 900 mm in heavy shrinkable clay soils, such as the Gault Clay, the London Clay, and the Oxford Clay. Any clay with a liquid limit (as defined in BS 1377: 1975[11] test 2a) in excess of 50 per cent is almost certainly shrinkable.

Significant swelling and shrinkage will only occur above the water table, so provided plentiful water is found a short distance below the foundations, they should not be affected. However, swelling clays will also cause uplift on the sides of footings or piles in the active zone, and may apply very high upward loads to the underside of reinforced concrete ground beams spanning over the top of piles. Piles may be protected from the upward forces by the inclusion of a steel cage capable of withstanding the tensile forces due to upward movement of the clay, but the base of the ground beam must also be protected since the piles will not stop soil heave. It is suggested that expanded polystyrene around the ground beam should allow soil movement without causing the application of load. This method, or the use of a polythene sheath or thick bitumen coating, may also be used on the sides of footings or piles in the active zone.

Vegetation effects

In the same way that climate can induce swelling or consolidation in clay soil, changes in vegetation lead to changes in ground level. The main difference in the two effects is that in the temperate British climate seasonal effects are relatively small compared with those due to vegetation changes. These effects can lead to significant soil movements to a depth of 4.5 m below ground level. Trees and shrubs may cause foundation movements in several different ways.
(a) Where trees are planted close to existing structures their roots may penetrate the brickwork below ground level, and the growth of roots may have a jacking effect on parts of the building, leading to differential settlement and cracking.
(b) Trees and shrubs planted close to existing structures will year by year extend their foliage, and because of their increasing need for water will therefore extend their roots. Usually the water required by the root system will be taken from water percolating through the soil, but in a dry summer it will have to be taken from the water bonded to the clay. When this happens severe soil shrinkage will occur in the area of the tree roots, and may affect foundations several metres deep.
(c) The removal of mature trees or shrubs immediately before construction will have a reverse effect. Soil may swell for a period of years, particularly when the water table is very low.

National House-Building Council requirements restrict the position of foundations relative to trees, as shown in *Figure 23.11* and Table 23.6. The mature heights (*H*) of some common trees are listed in Table 23.7.

In view of the fact that ground swells of up to 5.5 inches (140 mm) have been reported, beams between piles should be kept well above the soil, and floors should be suspended. A specialist soils engineer should be consulted to determine the probable soil swell. In broad terms, the degree of swell to be expected can be obtained from a comparison of moisture contents versus depth profiles in and away from the area desiccated by vegetation and from the use of oedometer swelling tests on undisturbed samples.

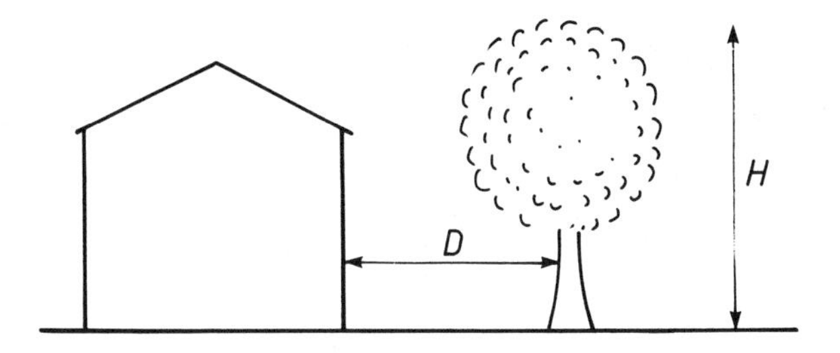

Figure 23.11 Trees and buildings (D = distance from house to centre of tree, H = expected mature height of tree)

Table 23.6 DISTANCES OF TREES FROM DWELLINGS

Species of tree	*Depth of foundation trench* (m) *for the following values of ratio D/H (see Figure 23.11):* 1/10	1/4	1/3	1/2	2/3	3/4	1
Poplar, elm and willow, sycamore	Not acceptable	2.8	2.6	2.3	2.1	1.9	1.5
All others	Ditto	2.4	2.1	1.5	1.5	1.2	1.0

Note. Up to 50 per cent extra depth may be necessary where trees are in rows or groups.
Intermediate cut-off concreted trenches may sometimes enable these distances to be reduced.

Table 23.7 MATURE HEIGHTS OF TREES

Species	*Mature height* (m)
Ash	27
Beech	27
Birch	21
Cedar	30
Chestnut	27
Elm	27
Larch	30
Lime	30
Oak	24
Pine	35
Poplar	30
Sequoia	30
Sycamore	24
Willow (weeping)	15
Willow (other)	24
Yew	14

Mining subsidence

Many northern towns of Britain owe their origins and development to the presence of coal or other minerals. During the 19th century exploitation occurred in a haphazard manner, with many deposits being worked at both shallow and deep levels. The roofs of underground coal workings may stand for a very long time before collapse, and so subsidence is very often unexpected.

It should be remembered that other minerals have also been extensively mined in Britain, for example limestone (e.g. Worcestershire), iron ore (e.g. Northamptonshire, Lincolnshire and Cleveland Hills), salt (e.g. Cheshire, south Lancashire, Staffordshire, Worcestershire), oil shale (e.g. Lothians, south Fife), gypsum (e.g. Staffordshire, Nottinghamshire, Leicestershire, Derbyshire), tin (Cornwall) and lead (Derbyshire).

The best way of determining whether mining subsidence will occur is, initially, to study available records. Maps and memoirs of the Institute of Geological Sciences will give information on the position of mineral sources, and large-scale Ordnance Survey maps (new and old issues) will show the positions of old and existing shafts and refuse tips. The National Coal Board will investigate past and projected workings in a particular area if requested, at no expense to the enquirer.

Generally local builders' experience, and an examination of adjacent structures for damage and method of construction, will give a good idea of the need for precautions. Relevant foundation designs are discussed in Section 26.

Collapse of fill

Fill materials provide one of the commonest causes of foundation and ground-slab problems. Spread or strip footings should not be constructed on pre-existing, untreated fill, and floor slabs may only be constructed directly on up to 600 mm of well-compacted fill. If ground conditions necessitate building on pre-existing fill, a more rigid foundation (such as the raft – Section 26) should be used, or the foundation should be carried through the fill to solid ground, perhaps by using piles or a basement.

If any point in a self-contained area of fill is more than 600 mm deep, a suspended floor of timber, precast beams or reinforced concrete slabs should be used, according to the National House-Building Council requirements.

Where fill is to be placed during construction in order to provide a level base for floor slabs, it should consist of a well-graded material that should not self-settle if it subsequently wets up. To achieve this, use one of the following.

(a) Stable well-graded fill (sand and gravel) may be used, compacted to the DOE Specification for Road and Bridge Works[21], clauses 601 and 609.

(b) A less suitable material (such as brick rubble hardcore) may be used, provided:
  (i) the maximum size of the material is restricted to 75 mm;
  (ii) the material is compacted with a heavy vibrating roller and is well-graded after compaction;
  (iii) a check on the percentage air voids at various parts of the site is undertaken and the fill recompacted when more than 10 per cent air voids is found.

For further details see BRS Digest 142[22].

Generally it is better to use a good-quality well-graded aggregate, since site control can then be reduced. Where significant depths of fill are present on the site vibroflotation may provide an economical foundation (see Rogers[23]).

Frequently, a proposed site will have small depths of fill over parts of its area. If this fill is similar to the subsoil it may be very difficult to detect. Look for man-made inclusions such as pieces of metal or small fragments of brick.

Hillside soil creep

Soil creep occurs on the sides of relatively steep clay slopes. It is associated with seasonal swell and shrink, because as the clay swells each winter it does so perpendicular to the soil surface, but during the summer it shrinks approximately vertically. In this way a slow movement of the soil in the upper layer occurs.

It has been observed that soil tends to creep where clay slopes are steeper than about 10 per cent. Under these conditions foundations should go to below the zone of movement, and not less than 1 m below ground level.

Pre-existing slope instability

Old slope failures can exist on slopes as flat as 3–4°, but generally they will be associated with steeper ground. The best ways of locating slipped areas are by the examination of aerial photographs and geological maps, and by a close examination of the site on foot. Old landslips generally take the form of hummocky ground, which may well become boggy in the winter despite being on the side of a hill.

Sites of old landslips may be covered with trees that will obscure the problems. Where landslips exist foundations should be carefully designed. It may be possible to use footings or piles founded below the slipped mass, but a geotechnical engineer should always be consulted. A detailed site investigation will be necessary.

Chemical attack on concrete

Many of the natural clay deposits in the UK contain aqueous solutions of sulphates that may attack concrete, destroying its strength. In addition, the presence of acid in groundwater will lead to attack of the lime portion of cement.

Samples of soil and groundwater from relevant depths and locations must be tested for acidity and sulphate content on all sites where concrete is to be placed on the ground. Tests should be in accordance with, and interpretation should be carried out on the basis of, Digest 174[24].

In addition, other types of chemical attack may occur in industrial infill sites, as a result of previous usage.

Swallow holes

Swallow holes exist where highly water-erodable materials, such as chalk or limestone, are covered by a thin layer of impermeable material, such as clay. In addition they may have been formed when the surface of the chalk was permanently frozen, during the last ice age, which would have made the surface relatively impermeable.

Under these conditions, when the surface water managed to find a path through the impermeable material all the flow in an area would flow down this 'sink hole', which would be progressively enlarged to form a hollow shaft. At later times the tops of these holes became filled, leaving voids below. Small changes of load or drainage patterns may trigger the collapse of the surface capping. This collapse is nearly always very rapid and unexpected.

Swallow holes, particularly in chalk, may often be detected on aerial photographs where they show up as circular features or depressions (see Dumbleton and West[25]).

REFERENCES

1. Skempton, A.W., 'The bearing capacity of clays', *Building Research Congress*, ICE, Div. 1, 180 (1951)
2. Peck, R.B., Hanson, W.E. and Thornburn, T.H., *Foundation Engineering*, 2nd edn, Wiley (1974)
3. Sutherland, H.B., 'Granular materials', *Conf. on Settlement of Structures*, Cambridge, 473 (1975)
4. Simons, N.E., 'Normally consolidated and lightly over-consolidated cohesive materials', *Conf. on Settlement of Structures*, Cambridge, 500 (1975)
5. Butler, F.G., 'Heavily over-consolidated cohesive materials', *Conf. on Settlement of Structures*, Cambridge, 531 (1975)
6. Hobbs, N.B., 'Rocks', *Conf. on Settlement of Structures*, Cambridge, 579 (1975)
7. Skempton, A.W. and Bjerrum, L., 'A contribution to the settlement analysis of foundations on clay', *Geotechnique*, 7 No. 4, 168 (1957)
8. Simons, N.E. and Som, N.N., 'The influence of lateral stresses on stress deformation characteristics of London Clay', *Proc. 7th Int. Conf. on Soil Mech. and Found. Engineering*, Mexico, 369 (1969)
9. Stroud, M.A. and Butler, F.G., 'The Standard Penetration Test and the engineering properties of glacial materials', *Symposium on the Engineering Behaviour of Glacial Materials*, Birmingham (1975)
10. Agarwal, K.B., 'The influence of size and orientation of sample on the undrained strength of London Clay', *PhD thesis*, Univ. of London (1968)
11. BS 1377, *Methods of test for soils for civil engineering purposes*, British Standards Institution (1975)
12. Bazaraa, A., 'Use of the standard penetration test for estimating settlements on shallow foundations on sand', *PhD thesis*, Univ. of Illinois, Dept. of Civil Engineering (1967)
13. Terzaghi, K. and Peck, R.B., *Soil mechanics in engineering practice*, 2nd edn, Wiley (1968)
14. Burland, J.B. and Wroth, C.P., 'Settlement of buildings and associated damage', *Conf. on Settlement of Structures*, Cambridge, 611 (1975)
15. Skempton, A.W. and MacDonald, D.M., 'The allowable settlement of buildings', *Proc. ICE*, 5, Part 3 (1956)
16. Polshin, D.E. and Tokar, R.A., 'Maximum allowable non-uniform settlement of structures', *4th Int. Conf. on Soil Mech. and Found. Engineering*, 1, 402 (1957)
17. Building Regulations, HMSO (1976)
18. *Soils and Foundations*, Building Research Establishment Digest 67 (1966)
19. Croney, D. and Jacobs, J.C., *The frost susceptibility of soils and road materials*, Road Research Laboratory Report LR90 (1967)
20. Ward, W.H., 'Soil movement and weather', *Proc. 3rd Int. Conf. on Soil Mech. and Found. Engineering*, 477 (1953)
21. *Specification for road and bridge works*, Department of the Environment, HMSO (1976)
22. *Fill and hardcore*, Building Research Station Digest 142 (1972)
23. Rogers, T.B., 'Soils and foundations', *Building Trades Journal*, 7 June, 5 July, 2 August, 6 September, 4 October, 1 November, 29 November (1974)
24. *Concrete in sulphate-bearing soils and groundwaters*, Building Research Establishment Digest 174 (1975)
25. Dumbleton, M.J. and West, G., *Air-photograph interpretation for road engineers in Britain*, Road Research Laboratory Report LR369 (1970)

24 SITE INVESTIGATION

INTRODUCTION 24–2

PRELIMINARY INVESTIGATION 24–2

SUBSURFACE INVESTIGATION 24–5

LABORATORY TESTING 24–12

SITE INVESTIGATION 24

24 SITE INVESTIGATION

C.R.I. CLAYTON, BSc(Hons), MSc, PhD, DIC, CEng, MICE, FGS
Senior Research Engineer, Ground Engineering Ltd

This section gives a logical and practical procedure for assembling information about a site. Subsurface investigation should not be undertaken until preliminary investigations have been completed; the final stage is laboratory testing.

INTRODUCTION

CP 2001[1] gives the primary objects of site investigations as:
(a) to assess the general suitability of the site for the proposed works;
(b) to enable an adequate and economic design to be prepared;
(c) to foresee and provide against difficulties and delays that may arise during construction due to ground and other local conditions;
(d) to investigate the occurrence or causes of all natural or created changes of conditions and the results arising therefrom.

Site investigation is a combination of widely differing processes, such as desk studies of existing information, site visits, land survey, and subsurface exploration. It is important to recognise that the small local builder can gain considerable advantage in this field, because by careful study of his area over a period of years he can obtain a knowledge and understanding of the factors likely to affect his work.

The wide definition of site investigation given by CP 2001 in most cases reduces to avoidance of bearing-capacity failure, excess settlement, and the many other problems that tend to affect lightly loaded structures such as housing (see Section 23).

Do not rush out to the site and carry out borings. There are much cheaper ways of gathering information. A good site investigation should consist of three stages:

- preliminary investigation,
- subsurface investigation,
- testing, and interpretation.

A simple site investigation should be carried out before site purchase in order to ensure the land is suitable for the envisaged development.

PRELIMINARY INVESTIGATION

A preliminary investigation can provide a broad guide to the types of problem that may be found both during and after construction. This part of the investigation may consist of several parts:

- desk studies of published information,
- assessment of local experience,
- site visit,
- study of aerial photographs.

All these activities can be carried out at relatively low cost and, with the possible exception of air-photograph interpretation, by non-specialists. This preliminary work will often indicate the type and amount of subsequent subsurface investigation required.

Published information

It is very unusual to find a site where no documentary information is available. A comprehensive list of sources is given in Dumbleton and West [2], and the following selections from this list are of most general use.

TOPOGRAPHICAL MAPS

Ordnance survey maps are currently widely available at 1 : 50 000 scale and 6 inch and 25 inch to the mile. Old editions of large-scale maps are particularly valuable in urban areas since they may show the positions of structures that have subsequently been demolished.

GEOLOGICAL MAPS

The Institute of Geological Sciences issues 1 inch to 1 mile geological maps of most parts of the country. In addition some coalfield areas, and the London area, are covered by 6 inch to 1 mile maps. All these maps are available from The Geological Museum, Exhibition Road, London, SW7.

Geological maps are published in three varieties:

(a) *Solid:* give only the details of the solid, recognisable strata below any transported material such as alluvium or boulder clay. The extent or occurrence of surface deposits of these materials is not given.

(b) *Drift*: chart the outcrops of all known deposits at ground level. The extent of different types of 'solid' deposit beneath any surface deposits is not given.

(c) *Solid and Drift*: are essentially similar to drift maps, but also show the boundaries of solid deposits beneath any drift. Where possible 'solid and drift' maps should be obtained.

GEOLOGICAL REGIONAL GUIDES AND SHEET MEMOIRS

Regional guides are available from the Geological Museum and provide descriptions of the strata reported on 1 inch to 1 mile geological maps. Sheet memoirs provide specific details of the geology of a particular area covered by one 1 inch to 1 mile geological map.

MINING HISTORY

Records of mining from 1860 are kept by the National Coal Board. Their regional contacts are given in the *Guide to the coalfields*, and they should be contacted whenever risk of subsidence is suspected. Other sources of information are:

- *Catalogue of abandoned mines* published by HMSO, available at National Coal Board Area Offices.
- Plans of coal mines, in the Plans Records Office of the NCB Area in which the mine was located.
- Plans of oil-shale mines, in the Plans Record Office of the NCB Edinburgh.
- Plans of all other mines, in the Plans Record Office of the Safety and Health Commission, London, SW1 (*except* plans of Cornish mines, which are in the County Records Office, Truro).

Current mining and quarrying operations are listed in the following annual publications:

1. *Guide to the coalfields*, published by Colliery Guardian, John Adam House, 17/19 John Adam Street, London, SW1. Lists coal, stratified-ironstone, shale and fireclay mines.

2. *Miscellaneous mines in Great Britain* published by the Department of Energy, Thames House, Millbank, London, SW1.
3. *Directory of quarries and pits* published by the Quarry Managers Journal Ltd, 62-64 Baker Street, London, W1. Lists currently operating quarries.

HISTORIC DATA

Records of landsides, subsidence, refuse tipping, etc. may be held in a number of uncoordinated places. It is worth consulting local Universities, Libraries, Museums, County archives, Natural History Societies and Newspaper offices. In addition local residents may be able to offer significant information.

PREVIOUS SITE INVESTIGATION REPORTS

Occasionally site investigation reports may be available for adjacent sites; however this source of information may be difficult to track down. Try contacting a local site investigation company.

Assessment of local experience

The builder who operates in a fairly restricted area can obtain a major advantage over the national or international construction company by using his observations of his own and other builders methods and problems in that area. Much additional information can sometimes be gained from local residents.

Site visit

After the collection of published information a site visit is essential. Some of the information previously collected can be checked, and new information can usually be obtained. Take the oldest and youngest topographical maps, of the largest scale, to the site, together with the geological map. Look for the following:

- types and differences in vegetation (may indicate changes in soil conditions),
- marshy ground,
- evidence of subsidence (particularly look for evidence of cracking in nearby structures or precautions taken by other builders),
- slope instability (hummocky or marshy ground or evidence of movement on hillslopes),
- infilled quarries, pits or old basements,
- lines of ditches,
- outcrops of rock or soil (for example in quarries or adjacent excavations),
- springs or other evidence of the position of the groundwater table,
- positions, type and size of trees.

If there is evidence of subsidence in any adjacent structures check with the owners, builders or occupiers the type of foundations used.

Air photographs

At a fairly modest cost (compared with the cost of land and its development) stereo air photographs can be purchased from a number of sources. By applying to the relevant authority (see below) giving a six-figure map reference, a list of the photographs covering a site can be obtained. The list will also give the address from which the photographs are purchasable.

England	Air Photographs Unit Department of the Environment Prince Consort House Albert Embankment London, SE1
Wales	Air Photographs Officer Welsh Office Summit House Windsor Place Cardiff, CF1 3BX
Scotland	Air Photographs Officer Scottish Development Department York Buildings Queen Street Edinburgh, 2
Northern Ireland	The Deputy Keeper of Public Records Public Records Office of Northern Ireland Law Courts Building May Street Belfast, 1
Ordnance Survey	The Air Photographs Officer Air Survey Branch Ordnance Survey Romsey Road Maybush Southampton, SO9 4DH

Dumbleton and West[3] give examples of the interpretation of air photographs. Overlapping vertical photographs at about 1 : 5000 to 1 : 10 000 scale are normally used for site-investigation purposes. They are particularly valuable when infilled quarries or swallow holes are suspected. The detection of landslips and the interpretation of geology usually require a more experienced eye, and it may then be better to seek expert advice.

SUBSURFACE INVESTIGATION

The preliminary investigation should have provided sufficient information to highlight any problems that will need to be investigated before design and construction proceed. These problems should then be at the centre of the subsurface investigation.

For most small structures, and particularly for low-rise housing, trial pits provide the best method of investigating the subsoil since they allow a very detailed examination of near-surface deposits. In many instances a hand auger or power auger will allow sufficient examination of deeper deposits. Only in special circumstances are shell-and-auger boring rigs likely to be necessary in the investigation of low-rise housing projects.

The depth of the borings or pits should be planned (a) to establish a suitable depth for foundations, (b) to go through any drift or near-surface soils and find the position of the solid geology beneath. It is most important to find the position of the solid geology and to ensure that foundations will not be placed on a thin crust, which lies over much softer, more compressible soil.

During subsurface investigation:

- describe the soil in detail,
- determine the level of groundwater,

- check for tree-root penetration,
- take sufficient samples for laboratory tests,
- carry out *in-situ* strength and density tests.

All these activities should be carried out with a view to avoiding the problems described in Section 23.

Excavating trial pits and drilling boreholes

Shallow trial pits provide a cheap method of examining shallow deposits *in-situ*, but the cost increases dramatically with depth, since pits over about 1.2 m deep must be supported. (Factories Act (1961) No. 1580 – Construction (General Provision) Regulations.)

Shallow pits are usually excavated by JCB 3C or MF50 type backacters, and these machines have a reach of up to about 4.0 m. Deeper pits may be excavated by Hymac 580 or similar machines, to a depth of about 6.0m.

Deep pits usually require much heavier support, and are conventionally excavated by hand. It is sometimes worthwhile to excavate with a 22–RB with a heavy double-rope grab, since hand excavation is extremely expensive.

The advantages of trial pits are that a thorough examination of soil structure can be made, and that high-quality block samples can be taken. However, the excavation of pits is normally limited to dry or cohesive soils.

The hand auger provides a light, portable method of sampling soft to firm soils near the surface. No special access is required, and the method is cheap. However, there are a large number of disadvantages associated with the method.

Firstly, the depth of boring is very limited. In stiff clays it may be possible to advance a hole to about 6–7 m. However, the presence of gravel makes progress almost impossible. In granular soils and below the water table, progress is difficult since the hole cannot be cased and tends to collapse.

Secondly, sampling is usually limited to small (1 kg) disturbed samples. Small diameter (38 mm) undisturbed samples may also be taken, but this is not normal practice.

At least five types of auger are in use (see *Figure 24.1*) but the most common is the posthole or 'Iwan' auger. This is available at diameters between 100 mm and 200 mm. Screw augers provide a very quick method of sampling stiff clays, but become almost impossible to use when water is encountered.

Figure 24.2(a) shows a typical 'shell and auger' rig. In fact, augering is seldom now done with such rigs, since progress can be made by claycutter (*Figure 24.2(b)*). The rig is light and is commonly towed by Landrover. In difficult locations it can be winched into position, and in very tight spaces it can be partially dismantled, although this is not common. This rig is the most commonly used for soft ground boring in the United Kingdom.

The one-ton shell and auger rig commonly uses 200 mm and 150 mm tools and casing. In a stiff clay such a rig will have little difficulty in boring to depths of 45 m. However, in sandy soils more frequent reductions of casing will be necessary, and casing sizes of 300 mm and 250 mm are sometimes used, when depths of up to 60 m may be achieved. Because the standard undisturbed sampling tool is 102 mm i.d., hole sizes less than 150 mm are not used.

It is good practice on any site to sink at least one deep boring to establish the geology. On extended sites several of these may be necessary, partly in order to establish the depth of weathering, which may be irregular, and also to establish the depth to which cavernous or mined areas descend.

Hvorslev[4] has suggested a number of general rules:

'The borings should be extended to strata of adequate bearing capacity and should penetrate all deposits which are unsuitable for foundation purposes – such as unconsolidated fill, peat, organic silt, and very soft compressible clay. The soft strata

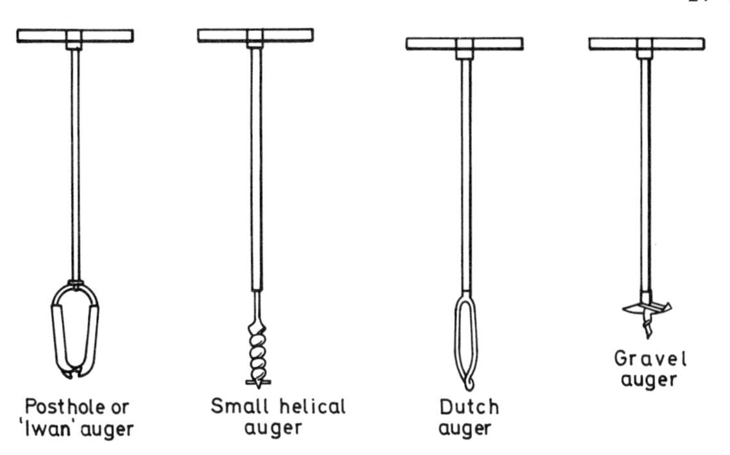

Figure 24.1 Common types of hand auger

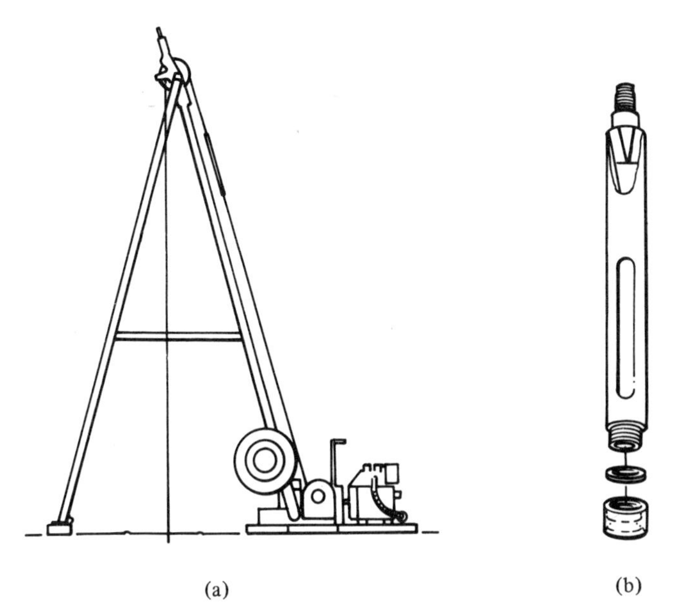

Figure 24.2 Soft ground boring equipment: (a) typical shell and auger rig: (b) claycutter

should be penetrated even when they are covered with a surface layer of higher bearing capacity.

When structures are to be founded on clay and other materials with adequate strength to support the structure but subject to considerable consolidation by an increase in the load, the borings should penetrate the compressible strata or be extended to such a depth that the stress increase for still deeper strata is reduced to values so small that the corresponding consolidation of these strata will not materially influence the settlement of the proposed structure.'

Therefore exploration should extend at least 1.5 times the width of the proposed foundation below proposed foundation level, and where strip footings are considered this figure should be increased to three times the footing width. Where a structure is to be founded as a pile group, a notional 'equivalent raft' should be used to determine the depth of exploration. See *Figure 24.3*.

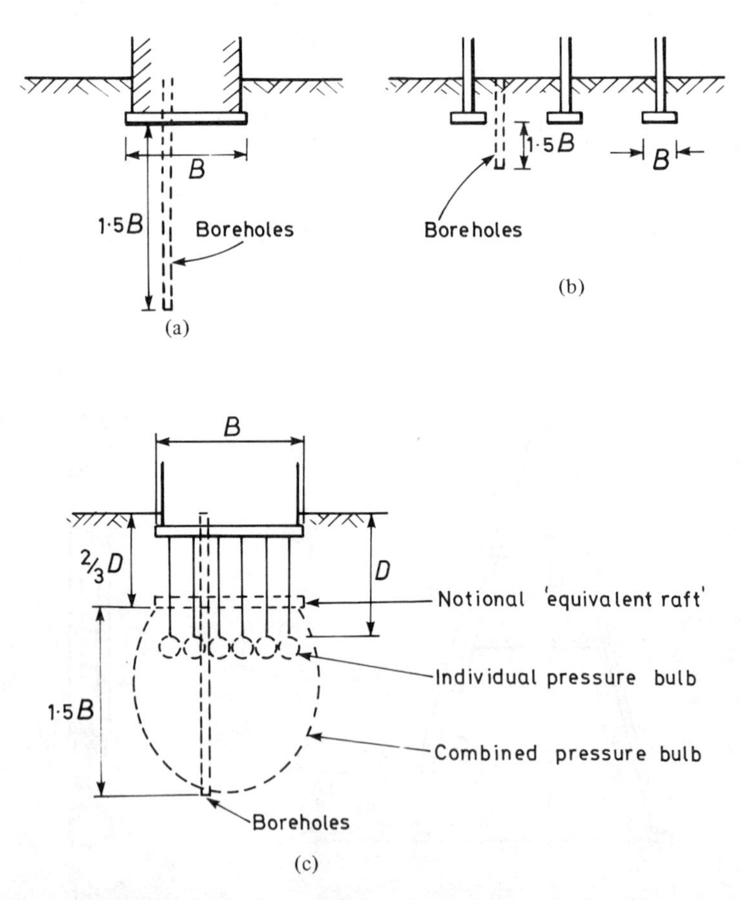

Figure 24.3 Boring depths for foundations: (a) structure on raft; (b) light structure on footings (if spacing is less than 5B, consider as a raft); (c) heavy structure on friction piles

Soil description

Soil description is an art, but it is based on a fairly rigid set of rules that should always be adhered to since they allow a description by one man to be interpreted by another. The following sequence of wording should be observed:
(a) consistency, or relative density,
(b) structure or texture,
(c) colour,
(d) minor constituents,
(e) PRINCIPAL SOIL TYPE,
(f) any additional description.

For example: Very stiff fissured dark grey silty CLAY (London Clay)
(a) (b) (c) (d) (e) (f)
Loose brown fine to coarse angular flint GRAVEL
(a) (c) (d) (e)

CONSISTENCY

The consistency of a clay soil is the shear strength of the material, ignoring any effects of fissuring. This may therefore be much higher than the undrained shear strength of the material when tested as large samples, or as it performs in the field. The consistency of a soil can be determined in a number of ways:

- manual classification,
- the pocket penetrometer test,
- hand vane testing.

Manual strength determination. The approximate consistency of a clay can be determined by handling a clay. The criteria listed in Table 24.1 are used.

Table 24.1 TESTS FOR CONSISTENCY OF CLAY

Description of consistency	*Method of test*	*Approximate undrained shear strength* (kN/m²)
Very soft	Exudes between fingers when squeezed in hand	<20
Soft	Moulded by light finger pressure	20–40
Firm	Moulded by strong finger pressure	40–80
Stiff	Indented by thumb	80–150
Very stiff	Indented by thumbnail	150–300
Hard	Difficult to indent by thumbnail	>300

The pocket penetrometer. Figure 24.4 shows a pocket penetrometer that is suitable for the approximate determination of consistency. The device is manufactured by Wykeham Farrance Engineering Ltd, Western Road Trading Estate, Slough, Berkshire, and cost about £10 (1977). The penetrometer consists of a spring-loaded point that is pushed into the soil up to the calibration groove. The reading from the penetrometer is given as unconfined compressive strength in ton/ft² or kg/cm². The undrained shear strength $C_u \approx 50 \times$ (penetrometer reading) kN/m².

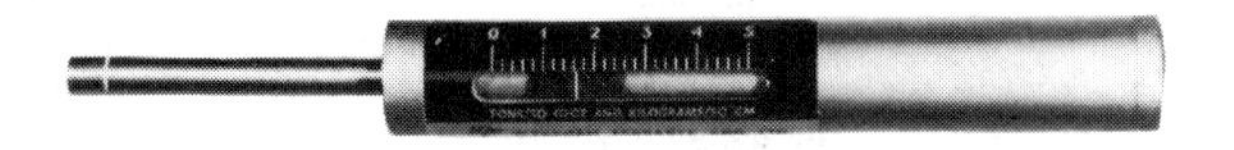

Figure 24.4 Pocket penetrometer

The hand vane. More accurate determinations of undrained shear strength can be made using a hand vane tester, such as the one shown in *Figure 24.5*. The vane is pushed into the soil a distance of about twice its length, and the handle, which incorporates a torque-measuring device, is turned slowly until the vane rod is seen to rotate. As the vane rotates in the ground it shears a cylinder of soil, and the device is calibrated to read directly the average undrained shear strength on the cylindrical shear surface.

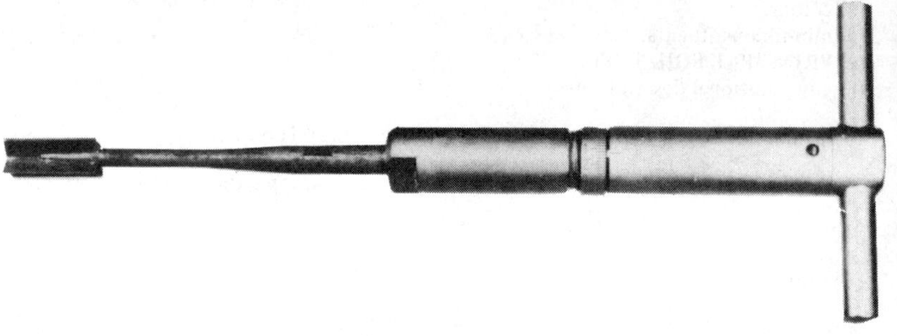

Figure 24.5 Hand vane tester

RELATIVE DENSITY

The relative density of granular soils is difficult to determine accurately. Once again, several methods can be used:

- visual classification, simple hand methods,
- the Perth penetrometer,
- the standard penetration test.

Visual classification and simple hand tests. Table 24.2 gives simple methods of determining the relative density of granular soils as described in CP 2001[1], the Building Regulations 1976, and Teng[5].

Table 24.2 TESTS FOR DENSITY OF GRANULAR SOILS

Relative density	*Field tests*
Loose	Easily excavated with a shovel, a wooden peg 50 mm square in cross-section can be easily driven, and a reinforcing rod can be pushed several feet into the ground.
Compact	A pick is required for excavation, and it is hard to drive a wooden peg of 50 mm square cross-section more than 150 mm.

These methods are by their nature inaccurate, and results should be viewed conservatively.

The Perth penetrometer. Glick and Clegg[6] proposed a penetrometer for use in sands; it is much more portable than the standard apparatus described in the next paragraph and gives similar values to the standard test when used in sand. It consists of a 16 mm diameter square-ended hardened steel probe, which is driven into the soil by repeated blows of a 9 kg weight falling through a distance of 600 mm. The SPT N value is roughly equal to the number of blows required to drive the tip 300 mm into the soil. According to Scott[7] the device gives conservative values in loose sand and optimistic values in dense sand. As might be expected, because of the small diameter of the driven probe, the results are unreliable in coarse granular soils.

The standard penetration test (SPT). The standard penetration test consists of driving a 50 mm outside-diameter tube into the ground using repeated blows of a 65 kg weight in a free-fall automatic trip-hammer. Because of the weight of the apparatus this test can only be carried out using a boring rig. The SPT *N* value is obtained by first driving the tube a distance of 150 mm into the ground (the 'seating drive'), and then measuring the number of blows to drive it a further 300 mm. The relative density of the soil is related to the SPT *N* value as shown in Table 24.3.

Table 24.3 RELATIVE DENSITY OF SOIL RELATED TO SPT *N* VALUE

Relative density	*Blows/300 mm penetration*
Very loose	<4
Loose	4–10
Medium dense	10–30
Dense	30–50
Very dense	>50

It is very difficult to carry out the SPT test on very shallow soils (<1.0 m deep) because of the size of the drilling equipment, and when carried out at these depths a considerable correction factor has to be applied to allow for the low overburden pressure. This correction has been a matter of considerable dispute in the past. Peck, Hanson and Thornburn[8] proposed the correction shown in *Figure 23.9*, which should be applied before the *N* value is used in calculations.

SOIL SIZES

The names of the various types of soil (for example 'gravel') are precisely linked to the size of the individual particles making up that soil type; see Table 24.4.

Table 24.4 SOIL PARTICLE DIAMETERS

Soil name		*Particle diameter (mm)*
Boulders		>200
Cobbles		60–200
Gravel:	coarse	20–60
	medium	6–20
	fine	2–6
Sand:	coarse	0.6–2
	medium	0.2–0.6
	fine	0.06–0.2
Silt		0.002–0.06
Clay		<0.002

Groundwater levels

During the excavation of trial pits or the formation of boreholes, particular care should be taken to find groundwater level, since a groundwater level above foundation level can cause problems during construction.

During trial pit excavation the level of every seepage into the pit, however small, should be noted, together with some idea of the rate at which the water is entering the excavation.

Since groundwater levels tend to be higher in the early spring than in the summer, it is a good idea to install a simple device to allow measurement of the groundwater position after trial pits or boreholes are backfilled. A standpipe can be made from small-diameter plastic (25–75 mm diameter), with holes drilled in the lower section. These holes can be wrapped in a water-permeable cloth in order to prevent the soil entering the base of the tube. The standpipe is buried in the trial pit or borehole, preferably with some sand around the base of the pipe at the level of the holes.

The level of the water in the ground can then be checked after the investigation is complete by lowering a 'dipmeter' down the standpipe. The dipmeter consists of a probe, sensitive to water, mounted on the end of a cable. When the water level is reached, the dipmeter emits a high-pitched tone. After the installation of a standpipe the water level will usually rise for some time. Water levels should be taken until the level stabilises.

Tree-root penetration

The depth to which tree roots penetrate will give a guide to the minimum depth that foundations on shrinkable clays should be taken. It is necessary to find the very finest 'hair' roots, which are the deepest, and these can be detected by breaking open lumps of clay taken from various depths in a trial pit.

LABORATORY TESTING

In small-scale site investigation for foundations, several types of laboratory tests are likely to be used:

- chemical tests,
- plasticity tests,
- strength tests,
- compressibility tests.

Chemical tests should be carried out on soil wherever concrete foundations are to be placed. Where groundwater is present at foundation level, samples of both groundwater and soil should be tested. BRE Digest 174[9] recommends precautions for different sulphate contents, as given in Table 24.5.

Plasticity tests, the determination of the Atterberg limits, are useful to give warning of highly shrinkable clays. The Atterberg limits are:

(a) the liquid limit (LL, w_L), the moisture content at which a remoulded clay changes from a plastic material to a liquid;

(b) the plastic limit (PL, w_P), the moisture content at which a remoulded clay changes from a brittle material to a plastic one.

The plasticity index (PI) is the difference between the liquid and plastic limit.

When shallow trial pit investigation is carried out, an approximate value of undrained shear strength can be obtained in the field using one of the methods detailed earlier. However, when deep soil investigation is required, borings will have to be made with a drilling rig, and under these circumstances tests will need to be carried out on samples obtained from these holes. Under these conditions the best method of test available is the undrained triaxial compression test.

The triaxial test is carried out on a cylindrical soil specimen, which may be either 38 mm diameter × 76 mm high or 102 mm diameter × 204 mm high. *Figure 24.6(a)* shows a diagram of the test apparatus. The soil sample is sealed in a rubber membrane and confined by water pressure inside the cell. A vertical load is then applied to the

Table 24.5 SULPHATE CONTENT OF SOIL[9]

	Concentration of sulphates expressed as SO_3				*Requirements for dense, fully compacted concrete made with aggregates meeting the requirements of BS 882 or BS 1047*			
	In soil				*Minimum cement content (kg/m^3)*			
					Nominal maximum size of aggregate			
Class	*Total SO_3 (%)*	*SO_3 in 2:1 water:soil extract (g/litre)*	*In ground-water (parts per 100 000)*	*Type of cement*	*40 mm*	*20 mm*	*10 mm*	*Maximum free water cement ratio*
1	<0.2	–	<30	Ordinary Portland or Portland-blastfurnace	240	280	330	0.55
2	0.2–0.5	–	30–120	Ordinary Portland or Portland-blastfurnace	290	330	380	0.50
				Sulphate-resisting Portland	240	280	330	0.55
				Supersulphated	270	310	360	0.50
3	0.5–1.0	1.9–3.1	120–250	Sulphate-resisting Portland or supersulphated	290	330	380	0.50
4	1.0–2.0	3.1–5.6	250–500	Sulphate-resisting Portland or supersulphated	330	370	420	0.45
5	>2	>5.6	>500	As for Class 4, but with the addition of adequate protective coatings of inert material such as asphalt or bituminous emulsions reinforced with glass-fibre membranes				

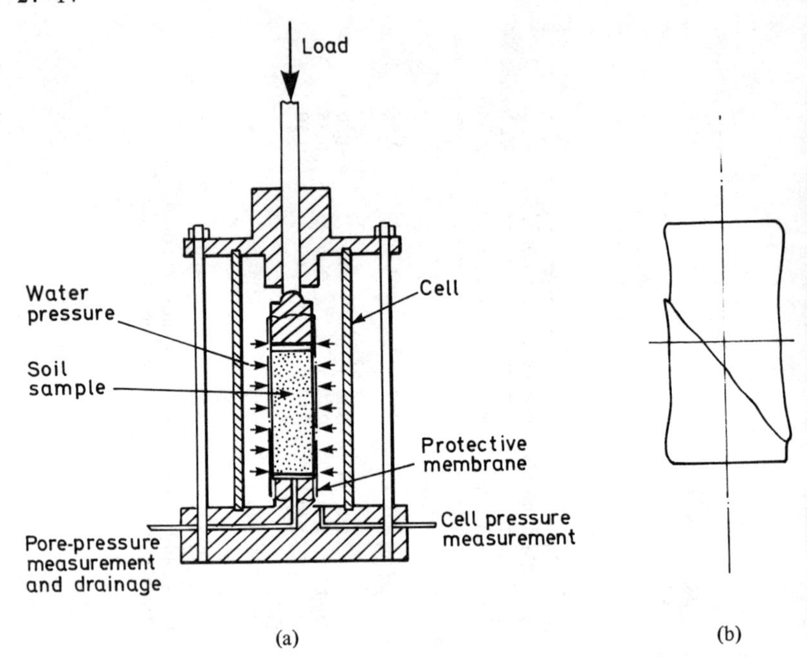

Figure 24.6 Triaxial test: (a) typical triaxial compression cell; (b) triaxial specimen after failure

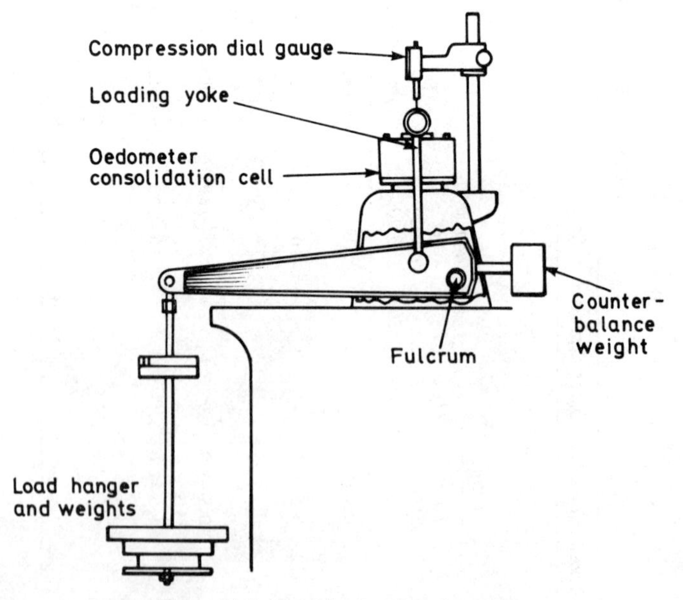

Figure 24.7 Front-load oedometer press

top cap and increased until failure of the soil takes place, as in *Figure 24.6(b)*. The maximum value of vertical load is termed the deviator stress, and the undrained shear strength is half this value.

In a typical test three specimens are tested at three different confining pressures, the middle pressure being approximately equal to the vertical soil pressure at the depth from which the sample was obtained. Under these conditions the average undrained shear strength is required for bearing capacity calculations.

Soil compressibility is conventionally measured in the oedometer test (*Figure 24.7*) where the soil specimen is confined so that it compresses in one direction only. The vertical load on the specimen is increased, and its change in height is measured together with the rate at which this change takes place.

The results may be shown in two ways: (a) as a plot of change in height divided by original height, as a function of the pressure applied to the sample, or (b) as the void ratio of the soil as a function of pressure (*Figure 24.8*). Soil is a mixture of

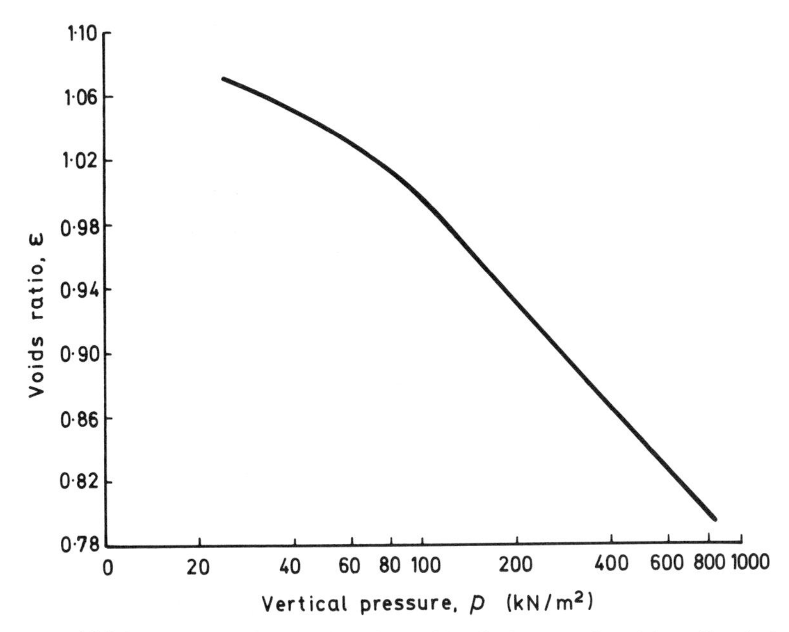

Figure 24.8 Oedometer test result plotted as void ratio (ϵ) versus logarithm of vertical pressure

solid particles and water (and sometimes air), and the void ratio (ϵ) is the volume of water and air in the voids divided by the volume taken up by solid soil particles.

The compression of the soil beneath a foundation load may be calculated using the coefficient of compressibility (m_v) values of the soil at various depths. The value of coefficient of compressibility used in calculation should be obtained from the height or void-ratio change from *existing* vertical effective overburden pressure up to the *expected* vertical overburden pressure (at the level the sample was taken) after the foundation is placed and loaded. Over this pressure increase:

$$m_v = \frac{\Delta h}{h_o} \times \frac{1}{\Delta p} = \frac{\Delta \epsilon}{1+\epsilon_o} \times \frac{1}{\Delta p}$$

where Δh = change in height of specimen over pressure increment Δp
h_o = height of specimen before pressure increment Δp
Δp = increase of pressure
$\Delta\epsilon$ = void-ratio change due to pressure increment Δp
ϵ_o = void ratio before pressure increment Δp

h_o and ϵ_o are the height and void ratio of the soil at the existing effective overburden pressure, σ_v'. This can be found if the soil density and the groundwater level are known.

$$\sigma_v' = \gamma_s . z - \gamma_w . h_w$$

where γ_s = soil density
γ_w = density of water (1 Mg/m³)
z = sample depth
h_w = height of water table above sampling position (*Figure 24.9*)

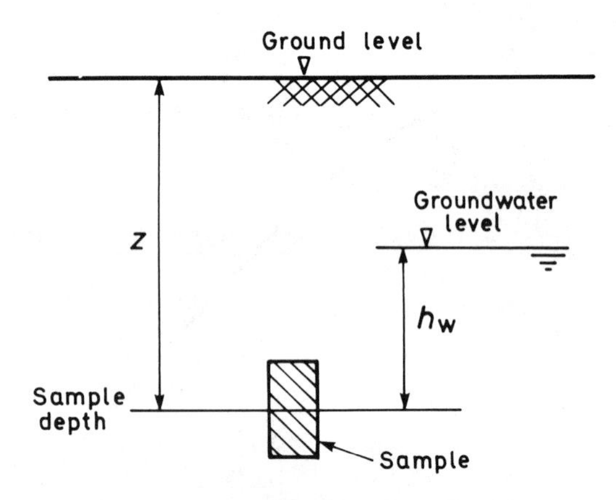

Figure 24.9 Coefficient of compressibility

REFERENCES

1. CP 2001, *Site investigations*, British Standards Institution (1957)
2. Dumbleton, M.J. and West, G., 'Preliminary sources of information for site investigations in Britain', *Transport and Road Research Laboratory Report* **403**, (Revised 1976) (1971)
3. Dumbleton, M.J., and West, G., *Air-photograph interpretation for road engineers in Britain*, Road Research Laboratory Report LR369 (1970)
4. Hvorslev, M.J., *Subsurface exploration and sampling of soils for civil engineering purposes,* Waterways Experiment Station, Vicksburg, Mississippi (November, 1949)
5. Teng, W.C., *Foundation design,* Prentice-Hall Inc., New Jersey (1962)
6. Glick, G.L. and Clegg, B. 'Use of a penetrometer for site investigation and compaction control at Perth, W.A.' *Perth Engineering Conf.,* Inst. of Engineers, Australia (1965)
7. Scott, M.B., private communication (1977)
8. Peck, R.B., Hanson, W.E., and Thornburn, T.H., *Foundation engineering*, 2nd edn, Wiley (1974)
9. *Concrete in sulphate-bearing soils and groundwaters,* Building Research Establishment Digest 174 (1975)

25 STANDARD FOUNDATIONS

TYPES OF STRIP FOUNDATION 25–2

FOUNDATION DEPTHS 25–9

ALLOWABLE BEARING PRESSURES 25–10

FOUNDATIONS ON SHRINKABLE CLAYS 25–11

STRIP FOUNDATIONS ON MADE GROUND 25–13

MATERIALS FOR FOUNDATIONS 25–14

STANDARD FOUNDATIONS 25

25 STANDARD FOUNDATIONS

M.J. TOMLINSON, CEng, FICE, FIStructE
Consulting Engineer
Formerly Director, Wimpey Laboratories Ltd

This section is concerned with traditional and narrow strip foundations and concrete beam foundations. It considers their use on normal sites, sloping, clay and filled sites.

TYPES OF STRIP FOUNDATION

Strip foundations are a suitable form of construction for the foundations of buildings with load-bearing walls, and for structures supported by rows of close-spaced columns where a combined foundation in strip form is more economical than individual pads beneath the columns.

Traditional strip foundations

The design of strip foundations traditionally adopted for the brick load-bearing cavity walls of dwellings up to four or five storeys in height is shown typically in

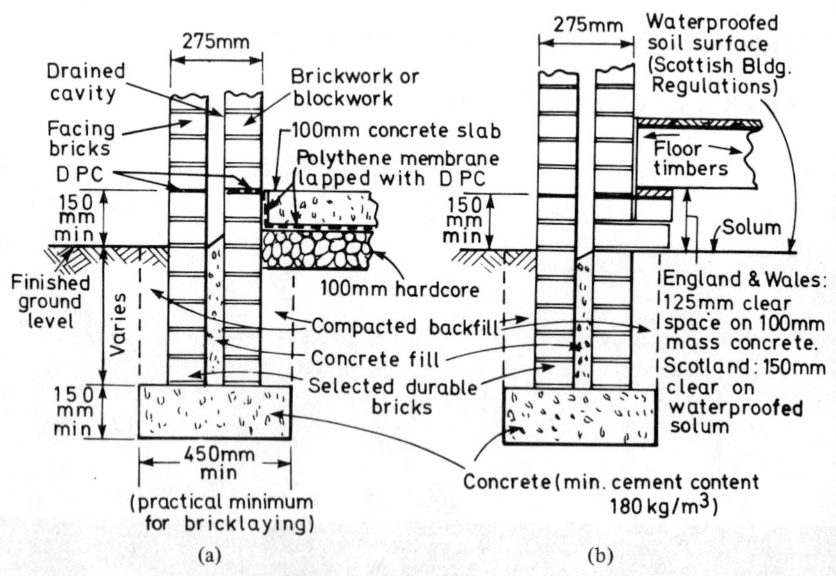

Figure 25.1 Traditional strip foundation: (a) with ground-bearing floor slab; (b) with suspended timber floor

Figure 25.1. The required depth from ground level to the base of the foundation is discussed later. The base has the function of spreading the load from the footing walls on to the ground, and its width is dependent on the bearing pressure permitted by the particular ground conditions (see page 25–10). A minimum width of about 0.45 m is required in order to provide sufficient working space for operatives to place and level the concrete and for bricklayers to construct the footing walls within the trench.

The thickness of an unreinforced concrete base is determined by the depth required to spread the load at an angle of 45° from the base of the footing walls on to the ground. A thickness of not less than the projection of the base from the wall

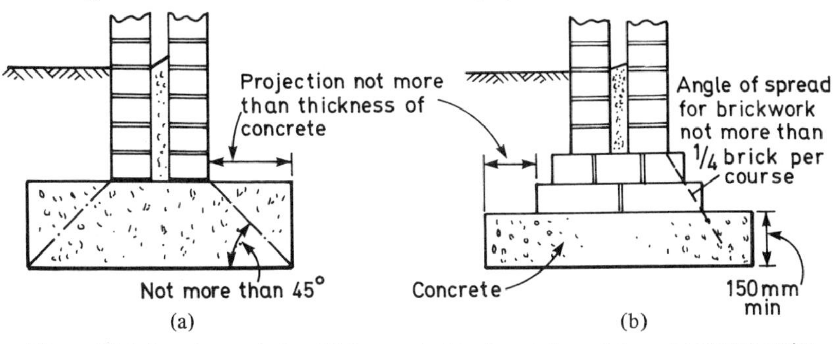

Figure 25.2 Requirements for thickness of strip foundation: (a) mass concrete strip; (b) stepped brick footings with mass concrete strip

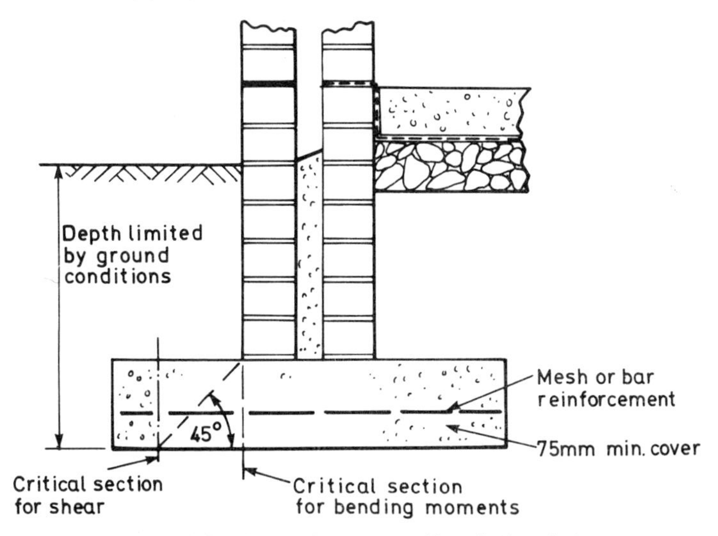

Figure 25.3 Reinforced-concrete wide strip foundation

or footing, or a minimum thickness of 150 mm, is required by the Building Regulations[1] and by CP 101[2]. Spreading the load through a concrete base in this way (*Figure 25.2(a)*) is usually more economical than stepping out the lower courses of brickwork (*Figure 25.2(b)*). Where limitations are placed on the depth of the foundation, say by the presence of groundwater, it may be necessary to construct a wide strip foundation with steel reinforcement placed transversely to enable the base to cantilever from each side of the footing wall (*Figure 25.3*). The materials used in the construction of the foundation are discussed at the end of this section.

Where strip foundations are used to support the cavity walls of buildings, the cavity should be filled with fine concrete up to the level of the underside of the ground bearing slab (*Figure 25.1(a)*), or in the case of houses with suspended timber floors up to the level of the solum or sub-base (*Figure 25.1(b)*). The object of filling the cavity is to provide structural strength where the footing walls are required to act as retaining walls, and also to prevent infestation or penetration of roots to the foundation system. The cavity above the infill should be drained by providing vertical open joints in the brickwork at suitable spacings. The polythene membrane beneath the ground-floor slab is turned up at the edge of the slab and bonded into the dpc to prevent transmission of moisture to the slab across the infilled portion of the footings.

On sloping sites where the ground floor on the downhill side of the building is elevated above ground level, the footing walls may have to withstand substantial lateral pressures from the retained soil or filling material (*Figure 25.4*). In these conditions the walls should be designed as retaining walls; the required width of the brickwork and concrete base should then be considered in relation to the stage of partial construction of the building (i.e. with the foundations at dpc level), when the superstructure load may not act to keep the resultant of the vertical and horizontal forces within the middle third of the base. The surcharge of the plant used to compact the underfloor fill should also be taken into consideration. Guidance on the design of earth-retaining structures is given in CP 2002[3].

Clay should not be used as a backfill material behind the footing walls, because it will usually gain moisture in this situation, with the resulting development of considerable swelling pressures against the confining structure. Granular materials such as sand, gravel, crushed rock, clean hardcore or lean concrete are suitable as backfilling materials. In the case of private dwelling houses covered by the structural-protection provisions of the National House-Building Council, the depth of filling that is permitted to support a ground-bearing floor slab is limited to 600 mm[4]. A suspended floor slab is required for greater depths. Instead of leaving a void beneath the suspended floor, it may be economical for the builder to construct a reinforced concrete slab cast-in-place on lightly compacted fill. The latter will remain in place and will exert lateral pressure on the footing walls.

Drainage should be provided on the uphill side of buildings on sloping sites, and also on the uphill side of internal walls placed transversely to the slope where these may act to impede the flow of groundwater down the slope.

On sloping sites, the concrete base to the strip foundation is stepped at intervals to avoid excessive depths of construction of the horizontal base. The Building Regulations and CP 101 require the steps to be of a height not greater than the thickness of the foundation (unless special precautions are taken), and the base should be lapped at the steps over a distance at least equal to the thickness of the foundation or twice the height of the step, whichever is greater. The Building Regulations require a minimum lap distance of 300 mm (*Figure 25.4*). Where the foundation is stepped at a transverse wall it is necessary to provide a vertical dpc in the footing wall. The dpc can be constructed from lapped slates, two layers of bitumen-impregnated hessian, or self-adhesive plastic sheeting (*Figure 25.5*).

The traditional strip foundation is constructed first by excavating by backacter machine to within 25 mm or so of the final level. A line of cement or lime dust is placed along the edge of the trench to guide the machine operator. Pegs are then driven into the bottom of the trench to the level of the top of the concrete, after which the final excavation is taken out by hand, measuring down from the top of the pegs to give the desired thickness of concrete. After loose soil or water has been removed from the trench, the concrete is placed and levelled by eye or by straight edge between the pegs.

A high order of accuracy is not required in the line and level of concrete strip, because the five or more courses of footing brickwork between ground level and the

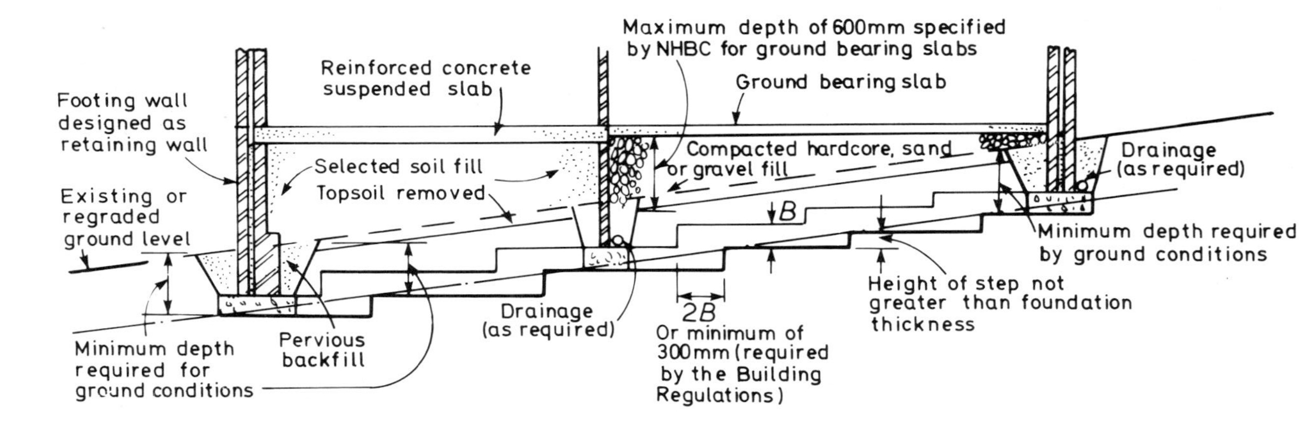

Figure 25.4 Strip foundations on sloping ground

top of the concrete permit the correction of any but gross inaccuracies in setting-out and construction. The backfilling of the foundation trench should be compacted to prevent settlement of the ground surface or of the underfloor fill within the footing walls. Points for entry of services should be made in the footing walls above the level of the concrete. The bricks at these points can be laid with sand joints to facilitate removal of the bricks, when the pipework or ducts are built-in. If the foundations are expected to settle under the superstructure loading, provision should be made for relative movement between the footing walls and the service ducts or pipework, in order to prevent their fracture when settlement takes place. For the same reason it is not advisable to construct services beneath the base of the foundations.

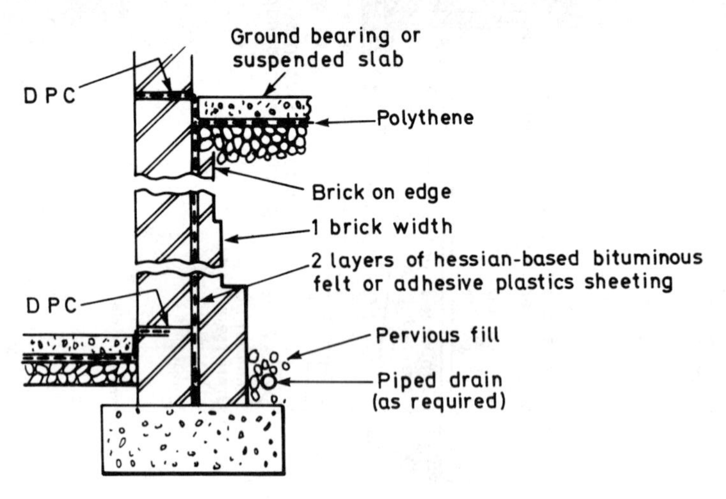

Figure 25.5 Arrangement of vertical damp course in footing wall

The traditional strip foundation is best suited to excavation depths not exceeding 750 mm. Greater depths involve difficulties for bricklayers working in the confined space of the foundation trench, and also problems with the disposal of the large volume of surplus soil. Clay heaped around foundation trenches can cause difficulties of access for construction, and also impedes the natural drainage of surface water, particularly when construction is continued through the winter months. Where trenches are deeper than 1.2 m, supports to the sides are necessary to maintain safe working conditions for the operatives. In weak soils such as soft clay, or in cohesionless granular soils, it may be necessary to support the trench sides by close sheeting, particularly if it is necessary to take the excavations below groundwater level.

Narrow strip foundations

The narrow strip or trench-fill foundation was evolved to overcome some of the difficulties described in the preceding paragraph. The method is suitable for soils such as firm to stiff clays or weak and weathered rocks such as chalk, Keuper Marl or clay-shales. These will stand unsupported for the full depth of the trench, which may be excavated to depths of 2 m or more. The essential feature of this design is that operatives are not required to work within the trench, since the unsupported sides are required to remain stable only over the time necessary to trim the bottom of the excavation and place the concrete to within 150 mm of ground level, or above ground level on sloping or undulating sites (*Figure 25.6*). The width of the trench is governed only by considerations of allowable bearing pressure, though 375 mm is a practical minimum for machine excavation.

The narrow strip foundation is a suitable type for use in shrinkable clays. These clays are usually firm or stiff in consistency, and will stand unsupported to the depth required to get below the zones of significant moisture-content changes (see page 25–11).

The need for formwork is avoided on level sites by terminating the concrete 100–150 mm below ground level. The brickwork is seen to be extending below the ground surface, thus preserving the architectural appearance of the building. On sloping or undulating sites, formwork is needed where the concrete is elevated above ground level. The formwork need only be rough, since the concrete is hidden when soil fill is placed to the regraded ground surface.

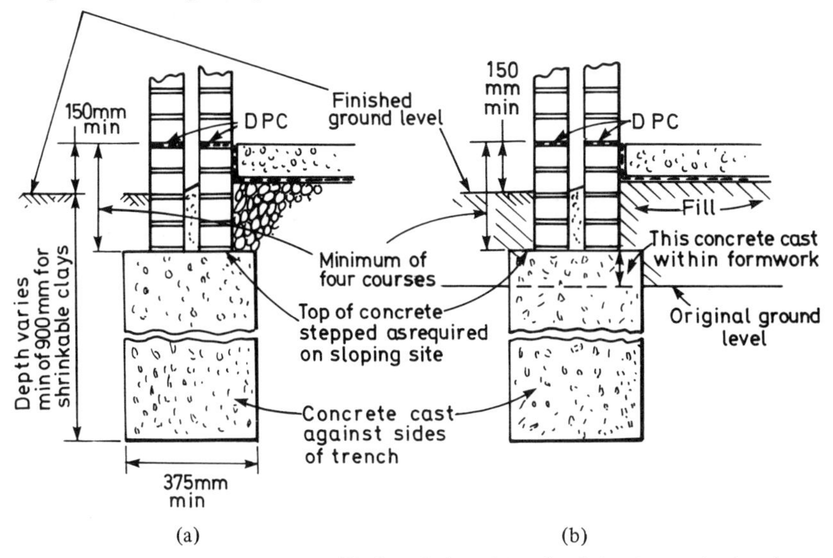

Figure 25.6 Narrow strip (trench-fill) foundation: (a) on level site; (b) on sloping site

The narrow strip foundation requires careful attention to accuracy in setting-out and construction. It is difficult for the operator of a small trenching machine to cut the trench sides to within 50 mm of the correct position. Thus, when a 275 mm wide cavity wall is constructed on a 375 mm wide concrete strip, any appreciable error in setting-out may result in the brickwork oversailing the edge of the concrete. Also, there may be no more than four courses of brickwork between the top of the concrete strip and the dpc. The upper two courses and part of the third course will be exposed on the outside face of the building. Thus any gross inaccuracy in the level of the top of the concrete cannot be corrected in a single course of brickwork without excessive labour in cutting the bricks. Cavity infilling is necessary with the narrow strip foundation, as described earlier for the traditional strip.

The concrete should be placed as quickly as possible after completing the excavation. If placing is delayed, soil may fall into the bottom of the trench, where it is not easily seen, or if seen is not easily removed. The concrete can be discharged directly from the chute of a truck-mixer, or if the area is inaccessible to vehicles it can be placed by barrow or dumped from the bucket of a long arm backacter excavator.

Concrete beam foundations

Load-bearing walls or rows of columns at close spacing carrying heavy superstructure loading can be supported by a continuous concrete beam foundation. This is a suitable type for weak soils of varying compressibility, where strength is required in a longitudinal direction to enable the foundation to bridge over the weaker zones. The beam can be a simple rectangular section (*Figure 25.7(a)*) or, where a wide foundation is needed, an inverted T-beam can be provided (*Figure 25.7(b)*). The

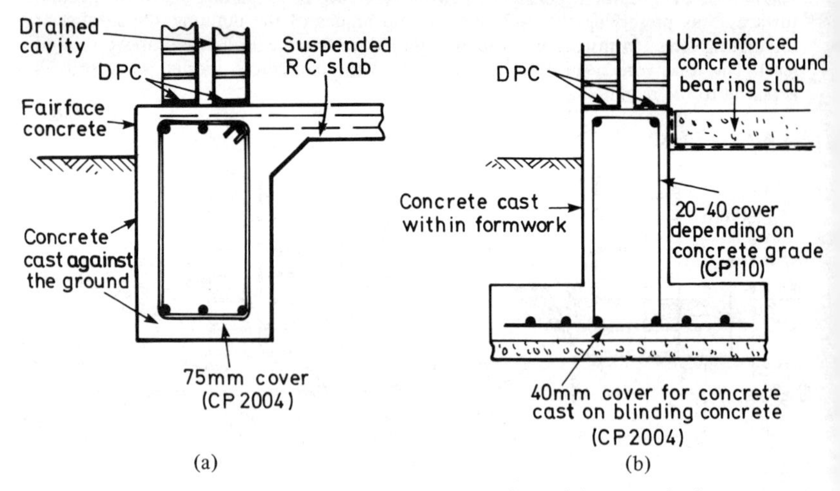

Figure 25.7 Reinforced concrete beam foundations: (a) rectangular beam; (b) tee-beam

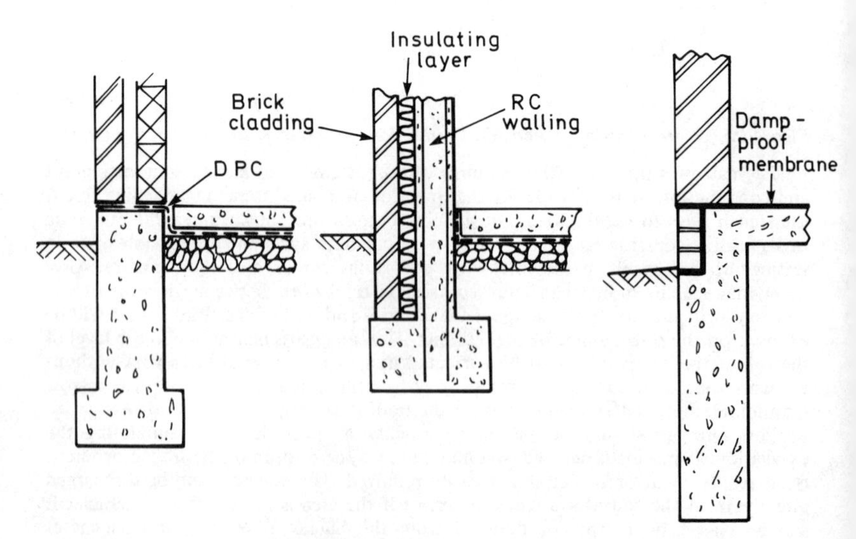

Figure 25.8 Unreinforced concrete beam foundations

reinforcement in the beam is designed to resist the bending moments due to the estimated differential settlements along the length of the member.

In situations where bricks of suitable quality for foundation work are unobtainable, consideration can be given to an unreinforced concrete beam or strip foundation commencing at dpc level. This type of construction may be economical where brickwork is not used in the superstructure, for example in dwelling houses constructed of no-fines concrete, or for timber buildings.

Three typical designs for unreinforced concrete beam foundations are shown in *Figure 25.8.* While these are simple in outline they can be costly in comparison with the traditional or narrow strip foundation. This is because of the need to provide formwork on both sides of the portion of the substructure projecting above ground level. The exterior formwork must be set accurately to line and level, and the exposed concrete should have a fair face free from honeycombed patches or pronounced board-marks. Thus the formwork can be expensive to fabricate and erect, particularly on sloping sides where the depth of the forms must be varied to suit the profile of the ground surface. The potential savings of repetitive formwork can rarely be achieved where reinforced or unreinforced concrete beam foundations are used for domestic buildings. This is because of the variations in the plan of the buildings and variations in ground-floor levels typical of most housing schemes, whether for private or municipal development. Repetition of formwork is most likely to be achieved in industrial projects such as low-rise factory or warehouse buildings.

FOUNDATION DEPTHS

The Building Regulations require the foundations of a building 'to be taken down to such a depth, or be so constructed, as to safeguard the building against damage by swelling, shrinking or freezing of the subsoil'. The requirements for swelling or shrinking of soils are discussed later. In soils not susceptible to these movements, the required foundation depth is governed either by the level of a suitable bearing stratum or by the need to take foundations below the level at which the soil may heave at times of severe frost (see Section 23, page 23–17). In the case of single or two-storey domestic buildings, considerations of frost heave are usually the determining factor since satisfactory bearing conditions can generally be obtained immediately below the zone of the disturbed topsoil. Only where buildings are sited on filled ground or in peaty or marshy soils is it necessary to consider deep foundations for these lighter buildings.

For the heavier structures, such as three- or four-storey dwellings or industrial buildings, it may be economically advantageous to take the foundations down to a stratum of soil or rock of low compressibility, and hence to achieve savings in foundation width by adopting high bearing pressures. However, as a general rule it is more economical to construct foundations at a constant, relatively shallow, depth and to accept the low bearing pressures that may be needed on the weaker shallow soils. This is because the soil strata capable of accepting high bearing pressures frequently lie at varying depths below the ground surface and may be of variable quality. It is rarely possible at the site investigation stage to explore the depth and consistency of the soil strata in sufficient detail to enable a contractor to quote a fixed price for a predetermined variable foundation depth. Thus the foundation work will have to be undertaken on a measurement basis, requiring an inspection of the ground and a decision on the varying construction depths for every building on the site. This will inevitably lead to arguments and delays awaiting decisions in ground of variable quality, and the contractor will allow for these delays in fixing his unit price for the work. Also, delays in placing concrete in the trench may result in collapse of the sides. Excavation below groundwater level to reach a suitable bearing stratum can be costly, because it will involve pumping and close-sheeting in granular soils.

The Building Regulations state that the requirements will be deemed to be satisfied if there is no 'wide variation in the type of subsoil within the loaded area... within such a depth as may impair the stability of the structure'. This could be construed to read that the regulations would not be satisfied if the type of soil changed, say, from a clay to a sand along the base of the foundation trench. This has led to much unnecessary deepening of foundations to obtain uniform conditions, especially in variable soils such as glacial till (boulder clay). There is no reason why the soil should not vary widely in type, *provided* that there is no wide difference in compressibility. For example, a stiff clay would have roughly the same order of compressibility as a medium-dense sand, and the stability of the structure would not be impaired in any way if the foundations were bearing on these two different soil types within the same trench.

CP 101 states that for frost-susceptible soils a foundation depth of 450 mm will be satisfactory as a precaution against frost-heave in most parts of the British Isles. Frost-susceptible soils include silts, chalk, fine sands and some clays. The Department of the Environment publication *Winter building*[5] states that 'only once or twice a century when the mean daily air temperature is below freezing for a month or more, the greatest depth of frozen ground is not likely to be more than 0.6 m and this depth will only be reached in bare well-drained sands or gravels'. Thus when the soil is covered by the foundation substructure it is hardly conceivable that frost will penetrate below a foundation constructed at a depth of 450 mm, i.e. the minimum depth recommended by CP 101 for frost-susceptible soils. In the case of soils that are not liable to frost heave, such as free-draining sands or gravels, there is no reason why foundations should not be constructed immediately below the topsoil. However, for practical reasons (to allow for the entry of services and for the disturbance or cultivation of soil around buildings, and as a safeguard against surface erosion by wind or water), a minimum foundation depth of 350 mm is desirable.

In frost-susceptible soils there is no need to take the foundations of internal walls such as party or partition walls to the same depth as the exterior wall foundations. Frost will not penetrate beneath the ground floor slab after the building has been roofed and glazed. However, problems can arise when garages or other outbuildings are attached to the main structure and are left with their doors open in severe frosty weather. Frost heave beneath the ground-bearing slabs may then be transmitted to the walls of the outbuildings, and possibly to the main structure as well.

Material that is susceptible to frost heave, such as chalk or some other types of limestone, may be used as fill beneath ground floor slabs. Frost penetrating below the slabs may then cause uplift of the slabs, which may in turn lift the peripheral walls of the building. Frost-susceptible material need not necessarily be rejected for under-floor fill, but the possibility of trouble in frosty weather should be recognised, and exposed floor slabs should be covered with insulating mats or warmed by braziers in partly completed buildings during spells of severe weather.

The requirements for foundation depths in swelling and shrinking clays are dealt with later.

ALLOWABLE BEARING PRESSURES

Domestic buildings up to three storeys in height impose quite a low order of loading on strip foundations. For single- and two-storey houses, the more heavily loaded party walls may impose loadings in the range of 20–25 and 40–50 kN/metre run respectively. Therefore there is no need to base allowable bearing pressures on the results of field or laboratory tests using the methods described in Section 24, provided an adequate site investigation has been made to identify the soil type at and below foundation level, and to obtain information on all other relevant factors that can influence the stability of the structure.

It is the usual practice to assess allowable bearing pressures for these lighter structures from 'deemed-to-satisfy' tabulated values, which appear in the Building Regulations and CP 101. See Section 23.

For buildings of five storeys or more in height, or for industrial buildings having heavy wall loadings from cranes or plant, the most economical design of foundations will be obtained by basing allowable bearing pressures on field or laboratory testing of soils or rocks using the methods described in Section 24. Allowable bearing pressures of strip foundations are likely to be governed by considerations of the total and relative settlements that can be tolerated by the superstructure. Therefore attention should be given to obtaining sufficient information on the compressibility of the ground, so that reliable estimates of settlement can be made.

Criteria for the maximum permissible relative settlement of structures are given in Section 23 (page 23–13). It should be noted that the degree of damage to which the criteria apply can be taken as very slight to slight. 'Very slight' damage implies crack widths not exceeding 1 mm. In external brickwork these cracks are visible only on close inspection, and internal cracks are easily dealt with during normal decoration works. 'Slight' cracking implies crack widths not exceeding 5 mm. Repointing of external cracks is required, and internal cracks require filling before redecoration.

Building structures can be designed to accommodate larger values of deflection ratio than those given in Section 23. Reinforcement can be incorporated in load-bearing walls, or vertical joints can be provided to reduce the panel length/height ratio of the walls. The theory of structure-soil interaction and a discussion of design methods are given in a report published by the Institution of Structural Engineers[6].

FOUNDATIONS ON SHRINKABLE CLAYS

Where buildings are sited on shrinkable clays, strip foundations must be taken to a depth sufficient to avoid movements due to seasonal moisture-content changes in the clay, or to depths below the zones affected by the drying action of the roots of trees, bushes or hedges. Alternatively the foundations can be stiffened and reinforced to accommodate the ground movements caused by moisture in the clay. The effect of seasonal moisture-content changes is to cause shrinkage of the clays with an accompanying fall in the ground surface in the summer months, followed by swelling of the clays and a rise in the ground surface during the winter months. The depths of soil moisture variations and the recognition of potentially shrinkable clays are discussed in Sections 23 and 24.

CP 101 recommends a foundation depth of 1 m as satisfactory for foundations in shrinkable clays in British climatic conditions. This does not mean that seasonal moisture-content changes do not occur at and below this depth, but the resulting vertical movements of the soil should not be such as to cause significant damage to ordinary building structures. For structures that are sensitive to small differential movements, foundation depths greater than 1 m should be considered. Narrow strip (trench-fill) foundations of the type shown in *Figure 25.6* are suitable for depths of 1 m or greater, since the clay affected by the drying action of surface vegetation is usually firm or stiff in consistency and will stand unsupported over considerable depths of excavation. However, the narrow strip foundation can be vulnerable to damage caused by soil movements while it is in an unloaded condition. At this stage uplift may occur on the unloaded strip by the adhesion of the clay to the sides of the foundation at the time of year when soil swelling occurs. Therefore narrow strip foundations constructed in the summer months should not be left unloaded during the following winter. Similar uplift effects can take place on an unloaded ground floor slab constructed on a desiccated clay.

A foundation depth of 1 m is inadequate where buildings are sited closer to growing trees, bushes or hedges than the mature height of these growths (*Figure 25.9*). The drying action of the roots, accompanied by shrinkage and settlement of the ground, extends over the full depth and lateral extent of the root system, affecting clays to depths of 5 m or more. Narrow strip foundations extending below the desiccated zone can be used, but bored piles are likely to be a more economical solution (see Section 26). Where trees are in rows, CP 101 states that buildings should not be sited closer to the trees than 1½ to 2 times their mature height unless special precautions are taken. These precautions should be observed when trees or bushes are planted as part of an estate development scheme or by individual owners.

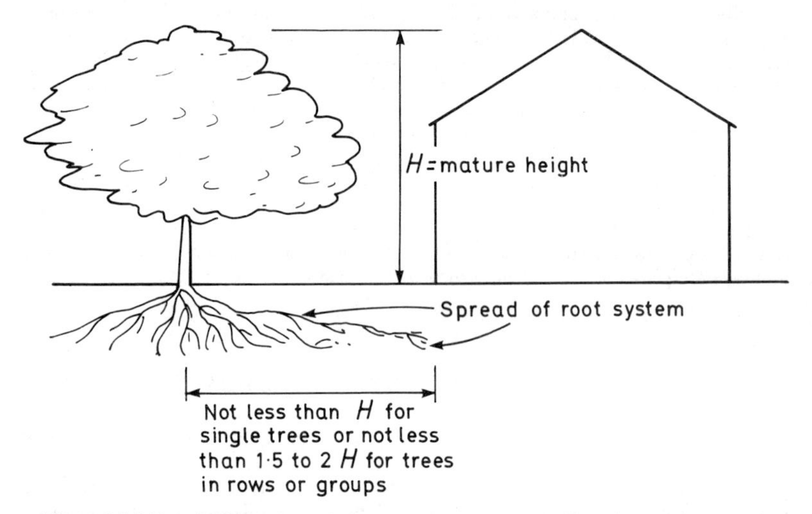

Figure 25.9 Required clearance between a structure on shallow foundations on clay soil and a growing tree

In any soil conditions, buildings should not be sited closer to trees than 4 m or one-third of their mature height, because of the possible damage to the foundations caused by the growth of large roots close to the trunk.

Where trees and similar deep-rooted vegetation have been cut down before building commences, the desiccated clay takes up moisture from the surrounding ground, and swelling of the clay occurs until the stage of equilibrium moisture is reached. Soil swelling can take place over a period of 10 to 20 years, and the resulting uplift of the ground surface caused by the removal of a large tree can be as much as 100–150 mm. Swelling takes place both in vertical and horizontal directions, causing severe damage to any building sited within the zone of the root spread.

Deep narrow strip foundations are not entirely suitable as a precautionary measure for buildings sited within this zone. Although the foundations may be taken below the zone of soil swelling, the horizontal pressures exerted by the swelling soils on the internal faces of the foundations may be sufficient to split the building apart. These lateral forces can be absorbed by providing a vertical layer of expanded polystyrene (*Figure 25.10*) on the sides of the foundations, but this method involves large quantities of excavation, particularly for party walls where the compressible layer is required on both sides of the wall. Uplift also occurs on the underside of the ground floor slab, requiring either a suspended floor with a 150 mm void beneath it, or a suspended floor cast on a layer of compressible material.

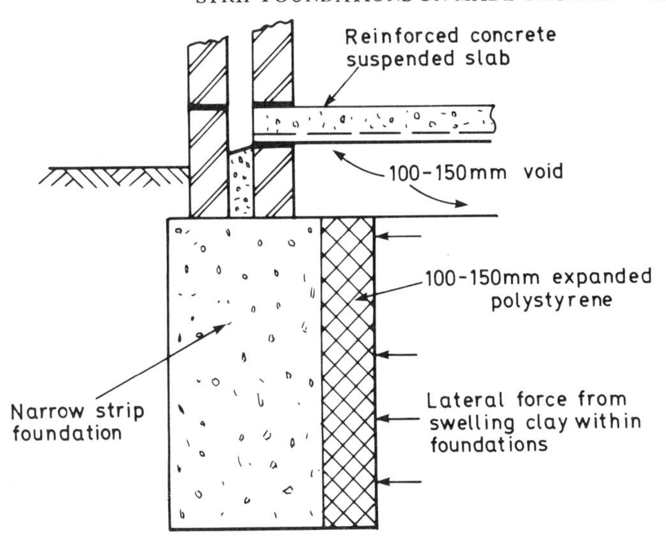

Figure 25.10 Horizontal swelling of soil against a deep strip foundation

As in the case of growing trees, the bored pile foundation in conjunction with a suspended ground floor slab is technically and economically the best solution to the problem of deep-seated soil swelling (see *Figure 26.24*), but in some circumstances a stiff raft foundation can be an economical alternative (see Section 26).

STRIP FOUNDATIONS ON MADE GROUND

Where fills are placed in layers, each layer being adequately compacted, the strength and compressibility of the soil will not be inferior to that of the same ground in its natural state. Therefore traditional strip or narrow strip foundations of the types illustrated at the beginning of this section can be used on these compacted fills. However, where the made ground has been formed by uncontrolled tipping without compaction, the resulting settlements of the fill can be substantial and can take place over a long period of years. Where buildings are sited on uncompacted granular fills, the total settlements may be fairly small and virtually complete within a few years of completing the fill placement. In these conditions settlement of the structure will be due only to consolidation of the fill under the load from the foundations. This will be quite small for granular fills, and traditional strip foundations with bearing pressures of about 50 kN/m^2 will be satisfactory. It may be advisable to provide a layer of mesh reinforcement in the concrete base as a precaution against cracking due to small differential settlements caused by variations in density of the fill.

In the case of uncompacted clay or weak rock fills where substantial long-term settlements of the fill are likely, strip foundations will not be satisfactory and either reinforced concrete beam foundations (see above) or raft foundations (Section 26) should be adopted.

It is inadvisable to build on domestic refuse tips, even with raft foundations, because of the likely large long-term settlements caused by biological degradation of substances within the fill, and the possibility of deep-seated combustion.

Loose granular fills can be compacted by inserting large vibrating units at close-spaced intervals along the length of the foundations. The vibrators may sink under their own weight, or penetration can be assisted by water jetting. As the units are withdrawn, clean graded stone or slag is fed into the void; the units are then lowered again to compact the infill material, thus forming close-spaced columns of compacted stone. Strip foundations reinforced by a layer of steel mesh are constructed over the heads of the columns. Compaction depths of up to 10 m can be achieved. These deep vibration processes are known as 'vibroflotation' or 'vibro-replacement' depending on the type of equipment used. Another method of achieving deep compaction involves dropping a heavy weight from a crane or tripod on to the ground from a considerable height. The various processes, including their applications and limitations of use, are described in a publication of the Institution of Civil Engineers[7].

The deep vibration process has been used extensively for the foundations of houses sited on loose brick rubble resulting from the demolition of buildings in urban redevelopment schemes.

Where building on recently-placed fill is undertaken, consideration should be given to long-term consolidation of any compressible natural soil layers beneath the filled area. Precautions may be necessary against attack on concrete in foundations and ground bearing slabs by aggressive substances present in industrial waste tips and some types of rock used as fill material (see below).

MATERIALS FOR FOUNDATIONS

Foundations are exposed to the influences of wind, driving rain, and frost over the portion between ground level and the damp course. Frost action also extends below ground level. On wet and exposed sites the effects of alternate wetting and drying or freezing and thawing can be severe, causing disintegration of porous or poor-quality bricks. In addition the foundations below ground level may be in contact with sulphate-bearing soils or groundwaters, which can cause deleterious expansion of brickwork mortar and concrete made with Portland cement. Some soils or groundwaters in marshy or peaty areas contain organic acids, which can cause leaching of concrete or mortar to a damaging extent if there is a flow of groundwater to carry away the products of leaching. Where foundations are constructed on made ground consisting of industrial wastes, there is the possibility of attack on concrete or brickwork by acids or alkalis present in the fill.

A thorough investigation of the ground is necessary wherever there is a possibility of attack on foundation structures by aggressive substances. The methods described in Section 24 should be used, and it is important that the considerations of precautionary measures should be based on an adequate number of test results. Too few tests may result in over-conservative precautions being adopted. Where industrial waste materials are present it may be necessary to make detailed chemical analyses, to determine the constituents of the material. Attention should be given to the possibility of soluble material being present in the made ground, which could be subject to long-term leaching by the percolation of surface or subsoil water.

Precautions to be adopted for foundations in sulphate-bearing soils and ground waters are shown in Table 25.1. These are based on the BRE Digest 174[8], but they have been modified to take account of the type of concrete used in building foundations and the level of the groundwater table. It should be noted that sulphate-resisting cements are ineffective for this purpose if they are used in lean mixes.

Precautions necessary for foundations in acidic marshy soils or peats are shown in Table 25.2. These have been based on Technical Note 69 published by the Construction Industry Research and Information Association[9]. The table does not apply to acidic conditions resulting from the presence of industrial wastes. Specialist advice based on detailed chemical analyses should be sought for these conditions.

Table 25.1 PRECAUTIONS AGAINST SULPHATE ATTACK ON BURIED CONCRETE IN BUILDING FOUNDATIONS

SO_3 in ground-water (parts per 100 000)	*SO_3 in soil*		*Precautions*	
	Total (%)	*In 2:1 aqueous extract (g/l)*	*Foundations wholly above groundwater table*	*Foundations in contact with fluctuating water table*
<30	<0.2	–	No special precautions necessary	Use ordinary Portland cement at $\nless$310 kg/m^3, max. w/c ratio 0.55
30 to 120	0.2 to 0.5	–	Use ordinary Portland cement at $\nless$330 kg/m^3, max. w/c ratio 0.50	Use ordinary Portland cement at $\nless$350 kg/m^3 *or* use sulphate-resisting cement at $\nless$310 kg/m^3, max. w/c ratio 0.50
120 to 250	0.5 to 1.0	1.9 to 3.1	Use ordinary Portland cement at $\nless$400 kg/m^3 *or* sulphate-resisting cement at $\nless$350 kg/m^3, max. w/c ratio 0.50	Use sulphate-resisting cement at $\nless$390 kg/m^3 max. w/c ratio 0.50
250 to 500	1.0 to 2.0	3.1 to 5.6	Use ordinary Portland cement at $\nless$400 kg/m^3, w/c ratio 0.50, *or* use sulphate-resisting cement at $\nless$350 kg/m^3, max. w/c ratio 0.45 (see note 2)	Determine the cations, then seek specialist advice on suitable precautions, e.g.special cements or membrane sheathing
>500	>2.0	>5.6	Use ordinary Portland cement at $\nless$400 kg/m^3, *or* use sulphate-resisting cement at $\nless$350 kg/m^3, max. w/c ratio 0.40 (see note 2)	Use sulphate-resisting cement at $\nless$390 kg/m^3, max. w/c ratio 0.40, and *in addition* protect concrete surfaces by asphalt or plastics sheathing

Notes (1) The cement contents recommended in the above table are suitable for low workability concrete (10–25 mm slump). Lower cement contents can be used if placing conditions are satisfactory for concrete mixes of lower workability provided that the concrete can be compacted to produce a dense impermeable mass.

(2) In this case the soil must remain 'dry'.

(3) No admixtures of any kind must be used with sulphate-resisting cement.

Table 25.2 PRECAUTIONS AGAINST ATTACK BY ORGANIC ACIDS IN PEATY OR MARSHY SOILS ON CONCRETE IN BUILDING FOUNDATIONS

pH value	*Precautions for foundations above groundwater table in any soil and below groundwater level in impervious clay*	*Precautions for foundations in contact with flowing groundwater in permeable soil*
>6	None necessary	None necessary
6 to 5	Use ordinary Portland cement at $\nless 370$ kg/m^3, max. w/c ratio 0.50	Use ordinary Portland cement at $\nless 380$ kg/m^3 or sulphate-resisting cement at $\nless 350$ kg/m^3, max. w/c ratio 0.5
<5	Use ordinary Portland cement at $\nless 400$ kg/m^3, or sulphate-resisting cement at $\nless 390$ kg/m^3, max. w/c ratio 0.5	Use supersulphate cement, or ordinary Portland cement at $\nless 400$ kg/m^3, or sulphate-resisting cement at $\nless 390$ kg/m^3, max. w/c ratio 0.5; *in addition* protect by external sheathing

Note (1) The cement contents recommended in the table are suitable for medium-workability concrete (50–75 mm slump). Lower cement contents can be used if placing conditions are satisfactory for concrete mixes of lower workability, provided that the concrete can be compacted to produce a dense impermeable mass.

(2) The precautions apply only to concrete exposed to organic acids in peaty and marshy soils. Special advice should be sought where mineral acids or acid industrial wastes are present.

Where sulphates or acid conditions are not present, CP 101 permits mass concrete in strip foundations with a minimum cement content of 180 kg/m^3, which is approximately 1 : 9 cement/aggregate by volume. In these conditions the Code of Practice for walling[10] permits the use of ordinary-quality facing bricks for foundations, i.e. below dpc level. Suitable bricks are defined in BS 3921[11] as 'less durable than special quality but normally durable in the external face of a building'. Thus the facing bricks used in the superstructure can be carried down below ground level, provided that experience has shown that they can withstand conditions of freezing and thawing and wetting and drying without damage to the facings. Where bricks may remain wet for long periods and may be subjected to frost action when in this condition, CP 121 recommends special-quality bricks. These are defined in BS 3921 as being 'durable even when used in situations of external exposure'. Engineering bricks normally attain this standard of quality, but BS 3921 permits the use of facing or common bricks if there is evidence of ability to withstand frost action over a period of years.

Past evidence is not always reliable in this respect, since modern houses can have higher standards of insulation than in the past; this means that there is less heat transmission through the walls, with correspondingly longer periods in which the bricks are subjected to freezing temperatures.

Calcium silicate bricks to BS 187[12] and concrete bricks to BS 1180[13] are suitable for foundation work. Solid concrete blocks to BS 1364[14] (type B) made with natural aggregate may be suitable, provided that precautions are taken with the type of cement and the cement content in sulphate-bearing ground.

Engineering bricks to Class B are necessary where they are exposed to sulphate-bearing groundwater. However, where the brickwork is above groundwater level and it is not exposed to such water seeping through the pervious backfilled zone, no special precautions are necessary.

Where the sulphate content of the groundwater exceeds 30 parts per 100 000 but is less than 300 parts per 100 000, sulphate-resisting cement should be used in the brickwork mortar in the proportion of one part of cement to three of sand. If the workability of the mortar needs to be increased a plasticiser can be used with this mix or lime can be used in the proportions of 1 : ¼ : 3 cement/lime/sand. Special advice must be sought for the precautions necessary when the sulphate content of the groundwater exceeds 300 parts per 100 000.

Although bricks on the inner leaf of a footing wall are not subjected to frost action, and therefore common bricks can be used in non-aggressive ground conditions, the brickwork can be exposed to wetting and drying and frost action during the construction stage. Therefore some attention to the quality of the bricks for the inner leaf is required.

The polythene membrane beneath the ground floor slab normally protects this concrete against attack by sulphates in the soil or in the material used as underfloor fill. In any case it is bad practice to use fill material containing sulphates or other chemically active substances. This is because of the risk of attack on the mortar in the footing brickwork, and of expansion of the fill caused by chemical reactions induced by conditions of warmth and dampness beneath the ground floor slab. Cases have occurred of expansion of ironstone shales, used as underfloor fill, due to the oxidation of pyritic minerals held within the laminations of the shale. An expansive movement of only a few centimetres of fill was sufficient to cause severe damage to the buildings consequent on heave of the ground floor slab.

REFERENCES

1. *The Building Regulations 1976*, HMSO
2. CP 101, *Foundations and substructures for non-industrial buildings of not more than four storeys*, British Standards Institution (1972)
3. CP 2002, *Earth-retaining structures*, British Standards Institution (revised code in preparation)
4. *Registered house-builders's handbook*, National House-Building Council (1974)
5. *Winter building*, Department of the Environment, HMSO, 23 (1971)
6. *Structure-soil interaction, a state-of-the-art report,* Institution of Structural Engineers (1978)
7. *Ground treatment by deep compaction,* Institution of Civil Engineers, London (1976)
8. *Concrete in sulphate-bearing soils and groundwaters*, Building Research Establishment Digest No 174, HMSO (Feb. 1975)
9. Eglinton, M.S., *Review of concrete behaviour in acidic soils and groundwaters*, CIRIA Technical Note No 69 (Sept. 1975)
10. CP 121, *Walling:* Part I, *Brick and block masonry,* British Standards Institution (1973)
11. BS 3921, *Clay bricks and blocks,* British Standards Institution (1974)
12. BS 187: Part 2, *Calcium silicate (sandlime and flintlime) bricks,* British Standards Institution (1970)
13. BS 1180, *Concrete bricks and fixing bricks*, British Standards Institution (1972)
14. BS 2028, 1364, *Precast concrete blocks*, British Standards Institution (1968)

[illegible]

Where the sulphate content of the ground water exceeds 50 parts per 100 000 [illegible] [illegible]

[illegible]

[illegible]

REFERENCES

1 [illegible]
2 [illegible] Station [illegible]
3 CP [illegible]
4 [illegible] National House-Building [illegible]
5 [illegible]
6 [illegible] Structural Engineer [illegible]
7 [illegible]
8 [illegible] Digest No. [illegible]
9 [illegible] Note No. [illegible]
10 [illegible] British Standards Institution [illegible]
11 BS [illegible] British Standards Institution [illegible]
12 [illegible] British Standard Institution (19[illegible])
13 BS [illegible] British Standards Institution (19[illegible])
14 BS [illegible] British Standards Institution (19[illegible])

26 SPECIAL FOUNDATIONS

RAFT FOUNDATIONS 26–2

PILED FOUNDATIONS 26–16

FOUNDATIONS FOR MULTI-STOREY BUILDINGS 26–28

SPECIAL FOUNDATIONS 26

26 SPECIAL FOUNDATIONS

M.J. TOMLINSON, CEng, FICE, FIStructE
Consulting Engineer
Formerly Director, Wimpey Laboratories Ltd

This section describes the various types of raft and piled foundations that may be used in house construction, the conditions for which each type is appropriate, and precautions against attack by sulphates, acids etc. The choice of foundation for multi-storey buildings is also discussed, together with basement design.

RAFT FOUNDATIONS

A raft foundation is a means whereby concentrated loads from building columns or line loadings from walls can be distributed over a large area, thus reducing bearing pressures to a minimum. This makes it possible for heavy foundation loads to be carried by weak compressible soils or loose fill materials without risk of general shear failure of the ground. Settlements cannot be avoided when rafts transmit bearing pressures to these compressible materials, but, by providing a stiff (not rigid) substructure, the differential movements due to variations in load distribution or compressibility of the ground can be reduced to a degree that can be tolerated by the superstructure.

Rafts are also adopted as a matter of structural convenience where building columns are located at a close spacing in two directions. The column bases can then be linked together to form a continuous slab, which avoids the necessity of constructing separate formwork units for each column base and of cutting and bending reinforcement for each base.

A raft combined with peripheral retaining walls is a suitable form of construction for a basement, when the continuity of the base slab with the walls provides an effective means of preventing the ingress of groundwater to the substructure. By providing a membrane of asphalt or plastics sheeting beneath the base slab, and extending this up the exterior face of the walls, a substantially watertight structure can be achieved.

Where mining subsidence occurs as a result of extraction of coal on a long advancing face, horizontal tensile and compressive strains are induced in the ground. A raft foundation is an effective means of accommodating these strains and the changes in curvature of the ground surface.

Types of raft foundations

Plain **slab rafts** (*Figure 26.1*) can be used in ground conditions where large settlements are not anticipated and hence a high degree of stiffness is not required. A layer of mesh reinforcement in the top and bottom of the slab is usually provided to resist bending moments due to hogging or sagging deflections at any point in the slab.

For architectural reasons it is usual to conceal the whole of the slab below finished ground level. Thus the top of the slab is placed at a depth of 100–150 mm to allow for minor variations in ground surface level.. The slab can be cast either directly on to the soil or on to a thin layer of blinding concrete. The waterproof membrane is provided above the slab and is lapped with the dpc. In addition a layer of polythene sheeting is provided beneath the slab as a waterproof membrane where soils aggressive to concrete are present. For a raft slab 200 mm thick suitable for a light building, the resulting foundation depth of 300–350 mm below ground level

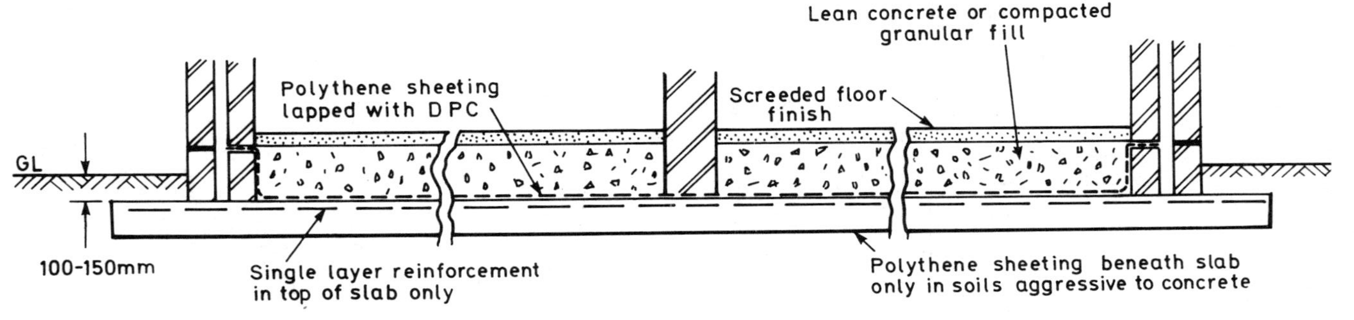

Figure 26.1 Flat slab raft

will be satisfactory for soils that are not susceptible to frost heave. In frost-susceptible soils the periphery of the slab can be constructed on a layer of lean concrete or compacted granular fill to provide an exterior foundation depth of 450 mm as a safeguard against frost heave (*Figure 26.2*). Somewhat greater depths may be needed in areas subjected to severe frost (see Section 25). Where a flat slab raft has its upper surface below ground level, it is necessary to provide a separate ground floor slab on a layer of lean concrete or compacted fill as shown in *Figure 26.1*. Alternatively a timber floor can be constructed on brick sleeper walls or timber bearers.

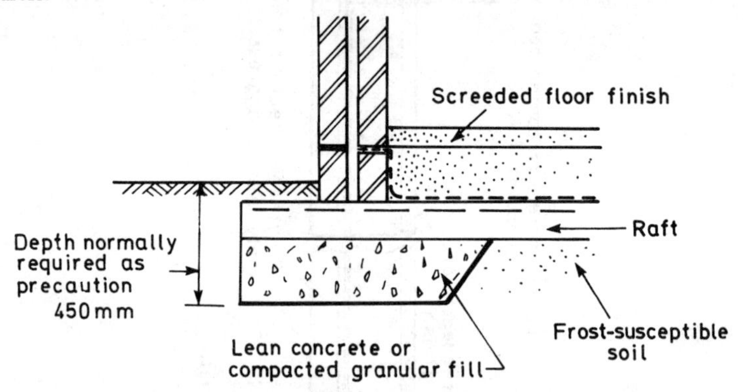

Figure 26.2 Treatment of periphery of flat slab raft in frost-susceptible soil

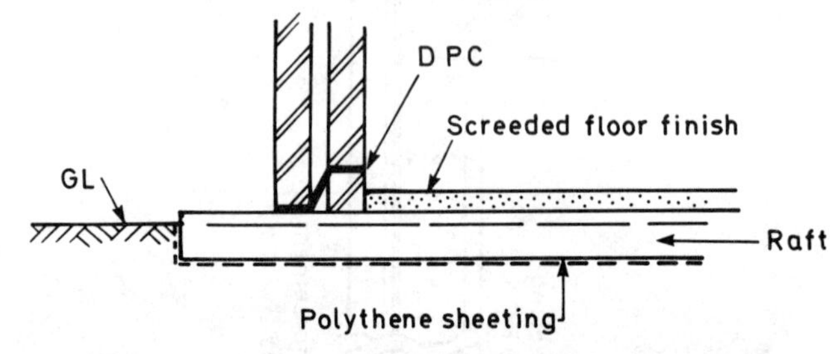

Figure 26.3 Flat slab raft forming ground floor of building

Where architectural considerations do not necessitate concealing the exposed edge of the slab the raft slab can also form the ground floor (*Figure 26.3*). Precautions are then necessary against rain driving beneath the walls or the external cladding of the structure.

The **stiffened-edge raft** (*Figure 26.4*) is suitable for the foundations of single- or two-storey houses on weak compressible soils or loose granular fill materials. The ground floor slab is made an integral part of the raft, and by stepping down the peripheral part of the slab a cut-off is provided against the ingress of water to the ground floor. A stiff beam can be formed integrally with the stepped down portion of the raft and its projecting toe (*Figure 26.4(a)*). In a construction of lesser stiffness the edge beam is cast separately from the slab, in which case it performs structurally as a wide strip foundation (*Figure 26.4(b)*). In soils of high to very high compressibility, such as soft peaty clays or loose recently deposited fill material, a more massive thickening should be provided to accommodate a stiffer beam with a more rigid connection to the slab as shown in *Figure 26.5*.

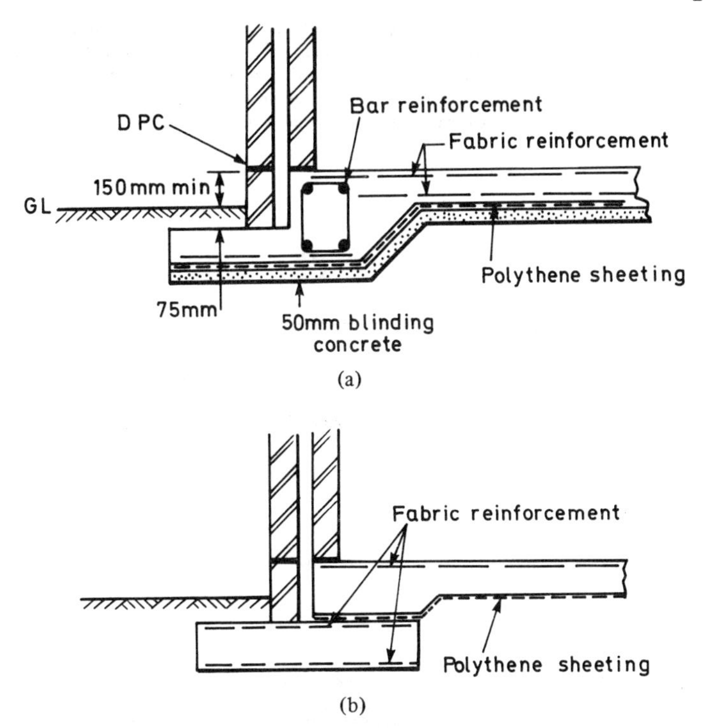

Figure 26.4 Stiffened-edge raft: (a) with integral beam, (b) with separate edge beam

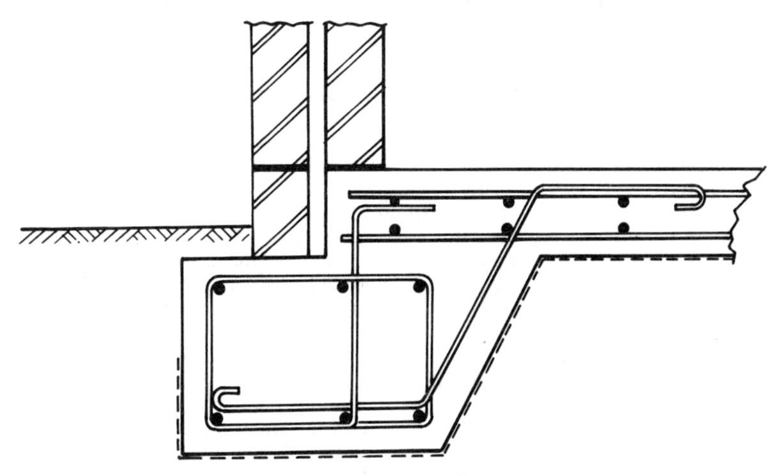

Figure 26.5 Stiffened-edge raft for highly compressible soils and fill material

It is not always necessary to thicken the slab (*Figure 26.6(a)*) beneath party walls or other load-bearing internal walls. A strip of transverse reinforcement in the bottom of the slab (*Figure 26.6(b)*), when combined with the stiffness of the wall above, may give sufficient resistance to bending moments induced in the slab by variations in compressibility of the soil or in the superimposed loading.

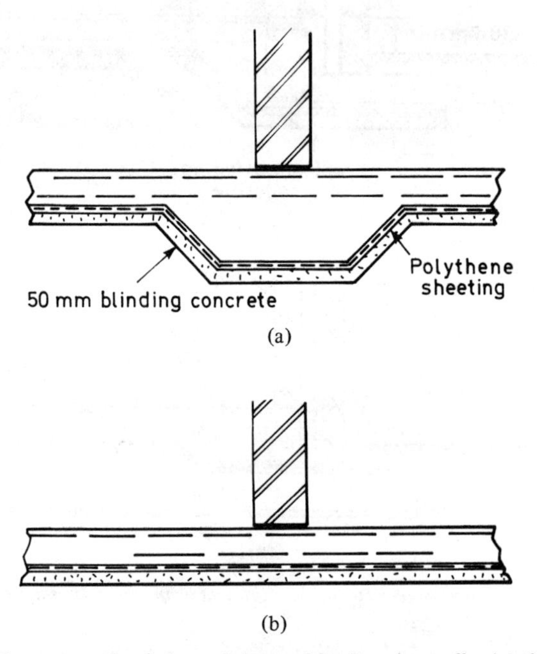

Figure 26.6 Treatment of raft beneath internal loadbearing walls: (a) thickening of slab on weak compressible ground; (b) flat slab on ground of low to moderate compressibility

The **slab and beam raft** (*Figure 26.7*) is used as a foundation for heavy buildings where stiffness is the principal requirement in order to avoid excessive distortion of the superstructure as a result of variations in the load distribution over the raft or in the compressibility of the supporting soil. These stiff rafts can be designed either as 'downstand-beam' (*Figure 26.7(a)*) or 'upstand-beam' types (*Figure 26.7(b)*). The downstand-beam raft has the advantage of providing a level surface slab that can form the ground floor of the structure. However, it is necessary to construct the beams in a trench, which can cause difficulties in soft or loose ground when the soil requires continuous support by sheeting and strutting. Construction may also be difficult when the trenches are excavated in water-bearing soil requiring additional space in the excavation for subsoil drainage and sumps. The downstand-beam raft is suitable for firm to stiff clays, which can stand for limited periods either without any lateral support or with widely spaced timbering.

The upstand-beam design ensures that the beams are constructed in clean dry conditions above the base slab. Where excavation for the base slab has to be undertaken in water-bearing soil, it is easier to deal with the water in a large open excavation than in the confines of narrow trenches. However, the upstand beam design requires the provision of an upper slab to form the ground floor of the structure. This involves the construction and removal of soffit formwork for the upper slab, or the alternative of filling the spaces between the beams with granular material to provide a surface on which the slab can be cast. Precast concrete slabs can be used for the top decking, but these require the addition of an *in-situ* concrete screed to receive the floor finish.

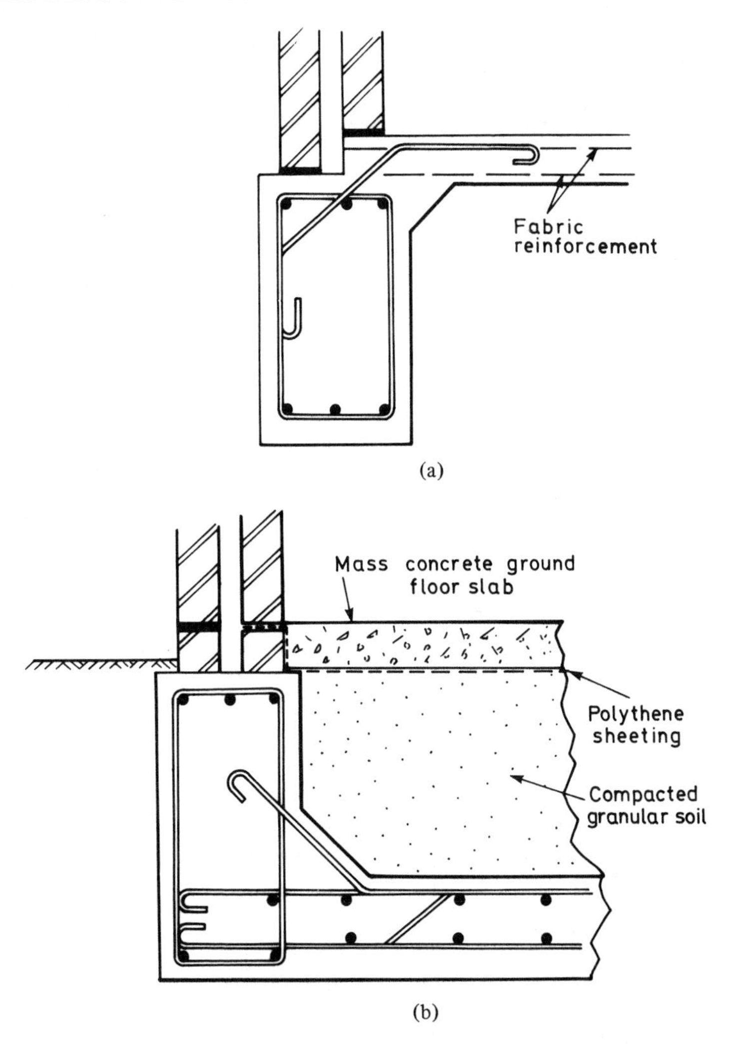

Figure 26.7 Slab and beam raft: (a) downstand beam, (b) upstand beam

The **buoyancy raft** or **cellular raft** is an expedient by which the gross loading from a heavy building can be reduced to a comparatively small net value by supporting it on a deep buoyant substructure. The soil beneath the substructure is relieved of the weight of material removed in excavation. This weight is deducted from the combined gross loading of the building and its foundation to give the net bearing pressure. As a guide to the reduction in gross load that can be achieved, the dead load combined with 60 per cent of the maximum design live load on the plan area of a conventionally designed multi-storey dwelling or office block is about 12.5 kN/m^2 for each storey. Excavation to a depth of 6 m relieves the ground of a vertical pressure of about 125 kN/m^2. Thus it is theoretically possible to balance the loading from a multi-storey building by the weight of soil removed in excavating for the cellular raft foundation. In practice it is not usually feasible to achieve this state of balance, since the deeper the required depth of excavation the heavier the substructure becomes in order to resist earth pressure and hydrostatic pressure from the surrounding ground. Also, settlements cannot be eliminated by balancing the load so that there is zero net pressure on the soil. This is because the ground at the base of the excavation will heave due to elastic rebound on removal of the overburden soil, with some additional swelling when the pore pressure in the soil is reduced. The amount of heave will depend on the length of time the ground at the base of the excavation remains in an unloaded condition[1].

Buoyancy rafts can be constructed within an excavation formed with sloping or supported sides (*Figure 26.8*). For this type of construction the raft is designed as a

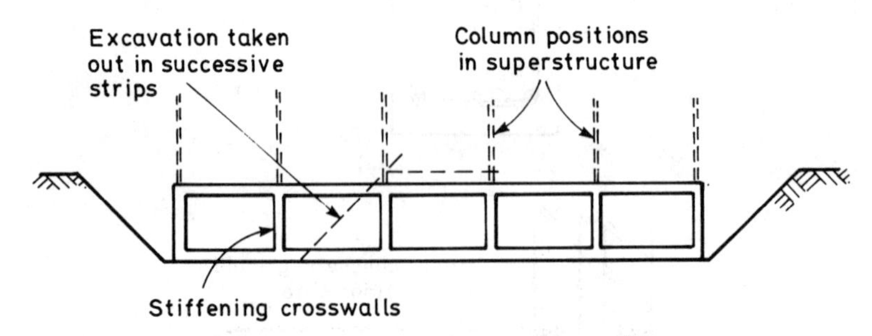

Figure 26.8 Buoyancy raft constructed in open excavation

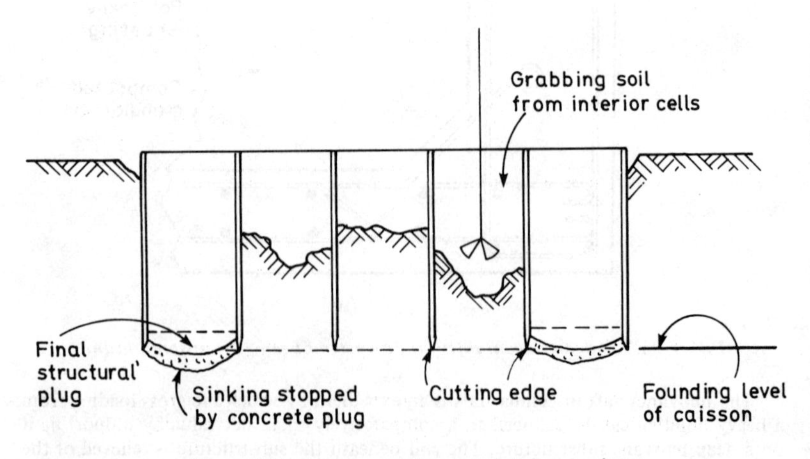

Figure 26.9 Buoyancy raft constructed as a caisson

cellular double-slab structure with vertical crosswalls to provide stiffness in order to resist a tendency to dishing settlements, which are typical of structures on soft clays. Heave of the soil can be minimised by limiting the area at the base of the excavation exposed at any given time. The excavation is taken out in successive strips working from one side to the other across the area of the raft.

Alternatively the buoyancy raft can be constructed as a caisson (*Figure 26.9*), in which case the soil within the cells is removed by grabbing and the whole structure sinks down as the soil is removed from beneath the level of the 'cutting edges' of the crosswalls and exterior walls. To control the verticality of sinking it is usual to grab from the exterior cells, keeping the soil in the interior cells only partially excavated until the caisson reaches its final level. At this stage the bottoms of the exterior cells are plugged with concrete to stop further sinking, then the interior cells are excavated, plugged and sealed.

Grabbing is only effective in loose granular soils or soft clays, which will flow from beneath the cutting edges. Thus the soil beneath the caisson is in a disturbed condition and it will reconsolidate as the superstructure load is built up, causing settlement of the building.

Much buoyancy is lost if the cells of the substructure become flooded. It is difficult to achieve watertightness in a light cellular structure, so it is usually necessary to connect the cells by drainage holes and to provide a sump with automatically-controlled pumps. It is also necessary to seal the cellular structure against the entry of gas, as a precaution against internal explosions if gas leakages were to reach dangerous concentrations.

The main purpose of the **mining-subsidence raft** is to prevent a structure from cracking as a result of tensile strains induced in the ground by the wave of subsidence as it crosses the site ahead of the line of advance of the coal face in the mine below (*Figure 26.10*). The tensile forces in the raft caused by the frictional contact with the soil can be reduced by casting the raft slab (*Figure 26.11*) on a sheet of polythene laid on 150 mm of compacted sand. The underside of the slab should be completely flat, since any downstand beams or thickened edges will key the structure to the ground, resulting in the development of very high tensile forces in the raft.

Reinforcement is provided to resist the residual tensile stress in the raft, but the aim should be to achieve flexibility rather than rigidity in this structure. The raft should be capable of bending without fracture while conforming to the changing curvature of the subsiding ground. Any cracks that open in the superstructure (due to hogging movement as the tension wave crosses the site) will partially close when the stage of the following compression wave is experienced. Final repairs are made when subsidence is practically complete.

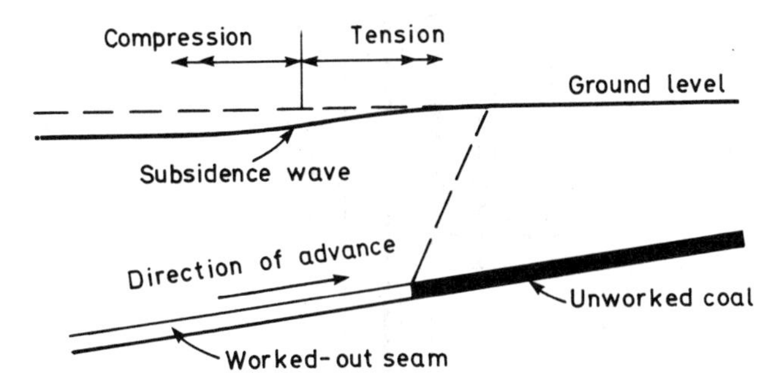

Figure 26.10 Subsidence caused by advance of mine workings on long face 'longwall mining'

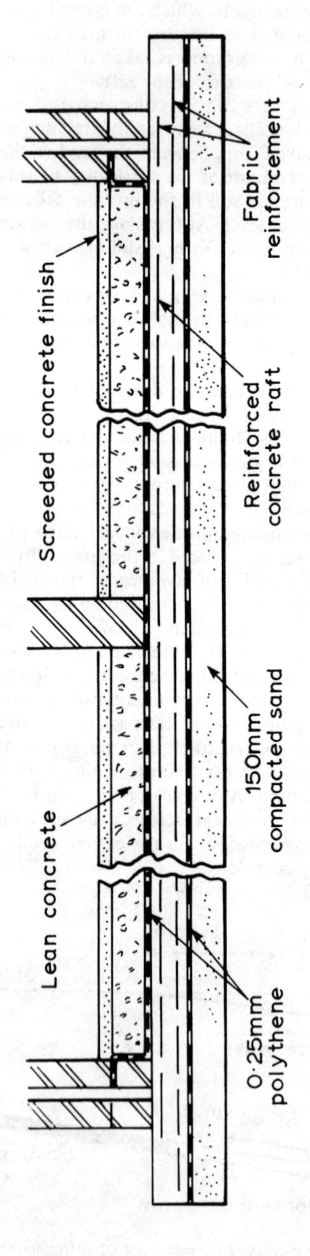

Figure 26.11 Mining-subsidence raft suitable for two-storey building

Service pipes should not be taken through the raft, and connections to services outside the building should be designed to accept relative movements between the building and the ground.

Rigid connections between adjacent buildings such as screen walls, fences or outbuildings must be avoided, since these transmit horizontal forces during the passage of the compression wave which are sufficient to crush the walls of a building.

General guidance on precautions against mining subsidence are given in a National Coal Board publication[2].

Estimation of settlements

Reliable estimates of settlement are essential to the safe and economical design of a raft foundation. It is usual to adopt rafts on sites where the ground is weak and compressible; in such cases the principal aim should be to estimate the total and differential settlements, allowing for variations in the distribution of superstructure loading and in the compressibility of the supporting ground. Guidance on methods of calculating settlements from the results of field or laboratory tests is given in Section 24.

The first approach to the problem is to assume that the superstructure and raft are completely flexible and then to calculate the total and differential settlements for the maximum and minimum distributed loads, basing these calculations on the maximum, minimum and average coefficients of compressibility of the soil or rock. This 'parametric study' will give an indication of the likely range of settlements, and will enable the engineer to decide on the need or otherwise for a sophisticated method of settlement analysis that takes into account variations in ground compressibility and the combined stiffnesses of the structure, foundation and soil. An analysis of this type can be made using finite element techniques[3].

If the calculated range of total and differential settlements, as estimated on the assumption of fully flexible conditions, is of no great significance in relation to the design and function of the structure, then such sophisticated methods of analysis are not justified and the structural design can be based on the estimates of settlement made on the assumption of the flexible foundation.

Structural design of rafts

There is no 'exact' method of calculating bending moments and shear forces in a raft because the supporting ground is never completely uniform in compressibility, and the distribution of dead and live load cannot be predicted over the life of the structure. Any realistic design method must take into account the interaction of the structure with its foundation and the supporting ground, since their combined stiffnesses act to redistribute stresses in the soil and structure and hence to reduce differential movements[4].

Sophisticated design methods based on finite element techniques are available for raft foundation design[5], but the need to adopt them depends on the settlements predicted from fully flexible conditions (see above), the sensitivity of the structure to differential movements, and the effects of such movements on the function and architectural appearance of the building. Guidance on the criteria that can be adopted to assess the possible damaging effects of relative settlements on load-bearing wall structure is given in Section 23 (page 23–13).

The Institution of Structural Engineers report[4] quotes damage criteria for framed buildings and reinforced load-bearing walls as shown in Table 26.1. If application of the criteria shown in the table shows that the differential settlements for a fully flexible raft and superstructure will not be damaging to the structure, simple methods

Table 26.1 LIMITING VALUES OF ANGULAR DISTORTION (RELATIVE ROTATION)

	Authority			
	Skempton and Macdonald	*Meyerhof*	*Polshin and Tokar*	*Bjerrum*
Structural damage	1/150	1/250	1/200	1/150
Cracking in walls and partitions	1/300 (but 1/500 recommended)	1/500	1/500 (0.7/1000 to 1/1000 for end bays)	1/500

Note For definition of relative rotation see *Figure 26.12*. The limiting values are for structural members of average dimensions. They do not apply to exceptionally large and stiff beams or columns, where the limiting values may be less and should be evaluated by structural analysis.

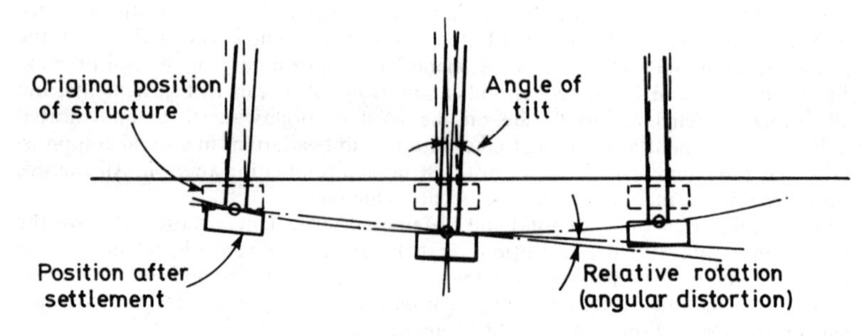

Figure 26.12 Definition of relative rotation (angular distortion) in framed structure

of design can be used for the raft. In the case of the flat slab and stiffened slab rafts (*Figures 26.1* and *26.4*), the strips beneath the load-bearing walls or rows of columns can be assumed to act as wide rectangular beams. Bending moments are calculated on the assumption that the beams will span or cantilever over local areas of weak soil (*Figure 26.13*) up to a maximum effective span length L, beyond which the structure would tilt and redistribute bearing pressures as the soil yields under the concentrated loading at the supports of the beams. The estimation of maximum span lengths requires some engineering judgement. The slab in the interior of the structure is reinforced, with the mesh steel linked to the reinforcement beneath the peripheral and interior loadbearing walls. This reinforcement is of nominal size and serves to prevent fracture at the junction of the interior slab and stiffened portions of the raft.

In the case of the slab and beam raft (*Figure 26.7*), a simple method of design is to assume that the raft acts as an inverted deck structure supported by columns. Uniform bearing pressure is assumed over the area of the deck, and the design follows conventional methods for reinforced structures with or without allowance for increase in bending moments and shear forces due to differential yielding of the supports. The assumption of uniform bearing pressures and unyielding supports can lead to unsafe designs where differential settlements occur between portions of the raft.

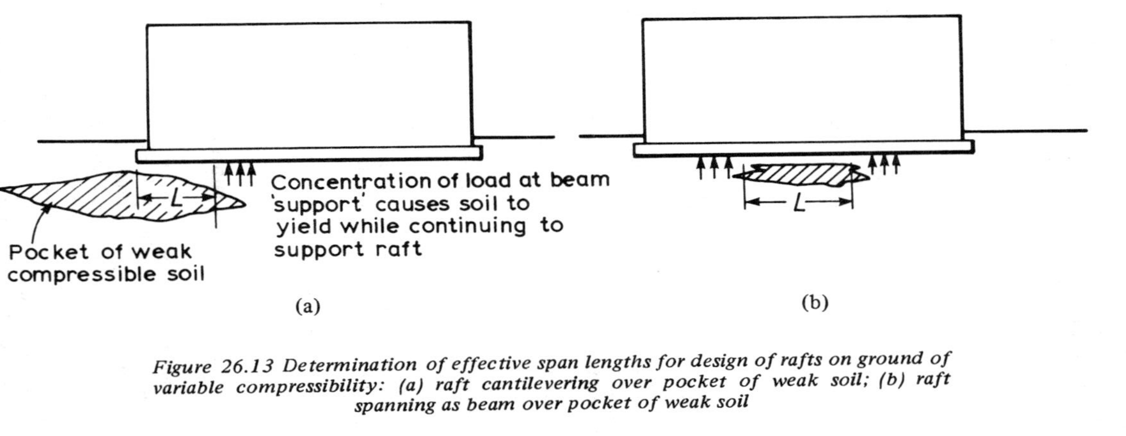

Figure 26.13 *Determination of effective span lengths for design of rafts on ground of variable compressibility: (a) raft cantilevering over pocket of weak soil; (b) raft spanning as beam over pocket of weak soil*

Rafts covering large areas should be articulated by means of joints in the slab. These joints should be continued upwards to form vertical-movement joints in the superstructure.

Sophisticated design methods based on finite element techniques[5] are available to take into account the relative stiffnesses of the superstructure, raft and soil. These methods can be applied to soils of varying compressibility and to cases of varying load distribution, and are justified in cases of heavily loaded rafts on weak compressible soils.

Rafts on sloping ground

Where rafts are sited on sloping ground, it is preferable to construct them on a layer of compacted granular fill (*Figure 26.14*) rather than cutting into the slope to place the raft wholly on natural ground. The total and differential settlements due to compression of the compacted fill will be insignificant compared with those caused

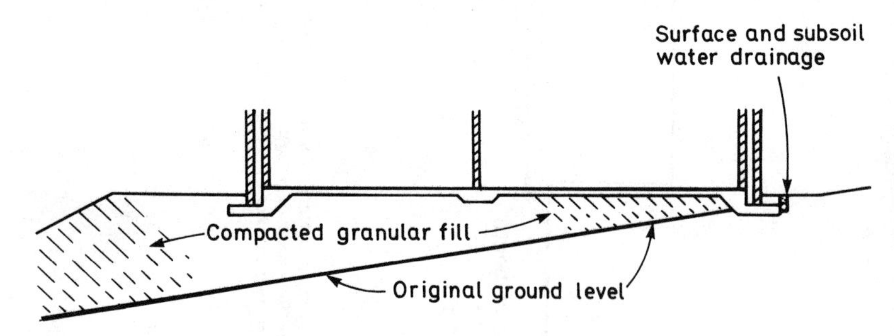

Figure 26.14 Construction of raft on sloping site

by consolidation of the compressible natural soil, and the reinforcement necessary to accommodate movements due to bearing-pressure variations on the natural soil can be achieved economically. On steeply sloping sites it is necessary to step down the raft to avoid excessive depths of fill.

Drainage is required on the uphill side of the raft to intercept surface and subsoil water flow.

Precautions against attack by aggressive substances in the ground

Adequate protection against the action of sulphates, acids or other aggressive substances in the soil or groundwater can be obtained by constructing simple slab or stiffened slab rafts on a layer of heavy-duty polythene sheet. The upstand beam type of raft can be protected in a similar manner. However, in the case of the downstand beam raft it is not practicable to lay the sheeting in a narrow trench, particularly at the intersections of the beams, and it is necessary to design the concrete mix to resist the aggressive action from the ground.

Precautions suitable for various concentrations of sulphates in soils and groundwater are shown in Table 25.1. Precautions against attack by organic acids present in the ground are shown in Table 25.2.

Basement rafts and buoyancy rafts constructed at the base of an excavation can be protected by a layer of asphalt beneath the base of the raft and carried up on the outside face of the exterior retaining walls. Alternatively heavy-duty polythene sheeting can be used on the underside of the raft. The sheeting is turned up around the wall foundations and lapped with self-adhesive plastics sheeting applied to the walls. Buoyancy rafts constructed as caissons should be protected by adoption of the concrete mixes shown in Tables 26.2 and 25.2. These tables have been based on Building Research Establishment Digest No 174 and CIRIA Technical Note 69 respectively (references 8 and 9 in Section 25), with modifications to take account of the ground-water conditions and type of concrete.

Table 26.2 PRECAUTIONS AGAINST SULPHATE ATTACK ON THIN REINFORCED CONCRETE SECTIONS IN BASEMENTS AND CAISSONS WHERE UNPROTECTED BY EXTERNAL MEMBRANES

SO_3 in ground waters (parts/100 000)	*SO_3 in soil*		*Precautions*	
	Total (%)	*In 2:1 aqueous extract (g/l)*	*Structures wholly above groundwater level (see note 2)*	*Structures subject to external water pressure*
<30	<0.2	–	Use ordinary Portland cement at $\nless$ 310 kg/m^3, max. w/c ratio 0.55	Use ordinary Portland cement at $\nless$ 370 kg/m^3, max. w/c ratio 0.55
30 to 120	0.2 to 0.5	–	Use ordinary Portland cement at $\nless$ 370 kg/m^3, max. w/c ratio 0.50	Use ordinary Portland cement at $\nless$ 380 kg/m^3 or sulphate-resisting cement at $\nless$ 340 kg/m^3, max. w/c ratio 0.50
120 to 250	0.5 to 1.0	1.9 to 3.1	Use ordinary Portland cement at $\nless$ 400 kg/m^3 or sulphate-resisting cement at $\nless$ 350 kg/m^3, max. w/c ratio 0.50	Use sulphate-resisting cement at $\nless$ 380 kg/m^3, max. w/c ratio 0.50
250 to 500	1.0 to 2.0	3.1 to 5.6	Use ordinary Portland cement at $\nless$ 400 kg/m^3 or sulphate-resisting cement at $\nless$ 350 kg/m^3, max. w/c ratio 0.45 (see note 3)	Determine cations, then decide whether sulphate-resisting or super-sulphated cements can be used or external sheathing essential
>500	>0.2	>5.6	Use ordinary Portland cement at $\nless$ 400 kg/m^3 or sulphate-resisting cement at $\nless$ 350 kg/m^3, max. w/c ratio 0.45 (see note 3)	Use sulphate-resisting cement at $\nless$ 390 kg/m^3, max. w/c ratio 0.40, and *in addition* protect by external sheathing

Notes
(1) The cement contents in the above table are suitable for medium workability concrete (50–75 mm slump). Lower cement contents can be used if placing conditions permit lower workability concrete, but concrete must be capable of being compacted to produce dense impermeable mass.
(2) If surface water is likely to collect in backfilled zone around structure, adopt precautions as for external water pressure.
(3) In this case the soil must remain 'dry'.
(4) No admixtures of any kind must be used with sulphate-resisting or supersulphated cements.

PILED FOUNDATIONS

Piling, as used to support building structures, has the function of transmitting loads through weak compressible soils or fill materials to denser, less compressible soils or rocks at depth. Piling is used to support heavy structures in ground conditions where excessive total and differential settlements would result from shallow spread or raft foundations, particularly in the case of framed structures with heavy and variable column loading, where the differential settlements would cause distortions and cracking of the structure. Piling is also used for light structures in shrinkable clays, as a means of transferring loads to soil layers that are unaffected by seasonal moisture-content changes or by the desiccation and swelling effects associated with the growth or removal of trees in clays (see Section 25).

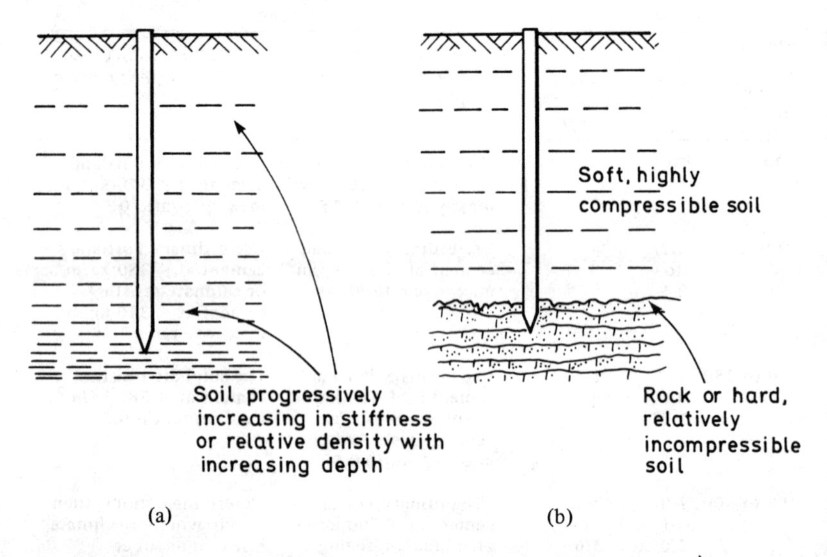

Figure 26.15 Types of bearing piles: (a) friction piles, (b) end-bearing piles

Where piles are supported mainly by skin friction from the soil surrounding the pile shaft, and to a lesser degree by end bearing, they are known as *friction piles* (*Figure 26.15(a)*). Piles supported almost wholly by a hard or dense soil, or a strong rock, at toe level are referred to as *end-bearing piles* (*Figure 26.15(b)*).

Types of piling

Piles are grouped into two principal classes related to their method of installation: *displacement* piles and *non-displacement* piles. There are also *composite* types, combining the features of both classes.

Displacement piles are installed by hammering or vibrating a pre-formed element, usually precast concrete or steel sections, into the ground. The soil is displaced by the entry of the pile; in some soil types, notably in clays, there will be a heave of the ground surface around the pile. In loose granular soils, the vibrations caused by the entry of the pile will densify the soil, resulting in a subsidence of the ground surface.

Non-displacement piles are installed by removing the soil by a drilling process, then placing concrete in the drilled hole to form the shaft. Precast concrete or steel sections can also be installed in drilled holes to form a non-displacement pile, although this is not an economical method of using these materials.

The method of installation in either of the above two classifications has an important effect on the carrying capacity of the pile. In the case of displacement piles, a soft clay can be weakened by the remoulding action of the pile entry, thus reducing the potential skin friction. On the other hand a loose sand will be densified, with an increase in both skin friction and end bearing. Drilling for non-displacement piles does not materially weaken firm or stiff clays, but it can loosen a dense granular soil, with a marked reduction in both skin friction and end-bearing.

TYPES OF DISPLACEMENT PILES

Timber piles are used only in countries where this material is cheap and plentiful. At the time when large logs were available it was usual to cut square-section sawn timbers for the piles, but the most economical way to use the timber is to drive the piles as round unbarked logs with their butt uppermost. The bark needs to be removed only when uplift loads are carried. The head of the pile is protected from splitting or brooming by a steel hoop, and the toe is protected from similar damage by a pointed metal shoe (*Figure 26.16*).

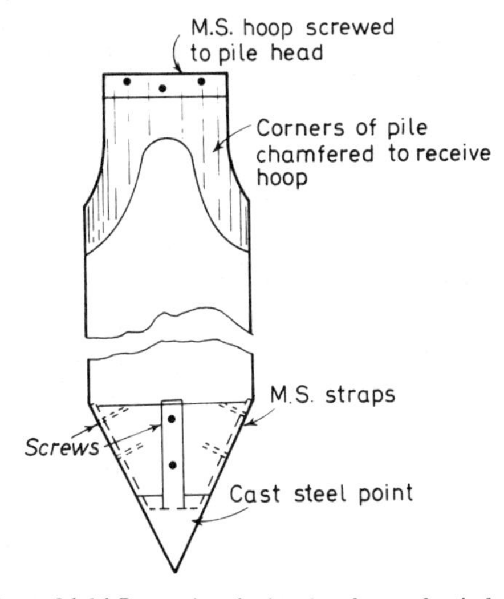

Figure 26.16 Protecting the head and toe of a timber pile

Although timber piles are light and easy to handle, they are susceptible to unseen damage by splitting at the toe when they are driven hard to develop a high bearing capacity. For this reason working loads on timber piles are usually limited to about 200–300 kN. Another disadvantage is the decay of the timber that can occur above groundwater level. This can be overcome by the methods described later.

Precast concrete piles are manufactured in solid units of square, hexagonal or octagonal cross-section and as hollow cylindrical or octagonal sections. The piles need to be fairly heavily reinforced in order to withstand the bending stresses that are caused when the piles are lifted from their casting bed for transport to a stacking area and from there to the piling rig. Some maximum lengths for typical cross-sections of piles made with concrete of 25 N/mm² characteristic strength and mild-steel reinforcement with 40 mm cover to the link steel are shown in Table 26.3.

Table 26.3 PRECAST CONCRETE PILES

Pile size (mm)	*Main reinforcement (mm)*	*Max. length for pick-up at:*		
		Head and toe	*One-third length from head*	*One-fifth length from head and toe*
300 × 300	4 × 20	9.0	13.5	20.0
	4 × 25	11.0	16.0	24.0
350 × 350	4 × 20	8.5	13.0	19.5
	4 × 25	10.5	16.0	23.5
	4 × 32	12.5	19.0	28.5
400 × 400	4 × 25	10.0	15.0	22.5
	4 × 32	12.5	18.5	28.0
	4 × 40	15.0	22.0	33.0

Closely-spaced link steel is required at the head and toe of the pile to prevent shattering of the concrete during driving. A typical design for a reinforced concrete pile is shown in *Figure 26.17*. The steel required to withstand handling and driving stresses is usually greatly in excess of that required for normal compression loading in the service life of the foundations. This can put the precast pile at an economic disadvantage compared with cast-in-place types. Another disadvantage is that the

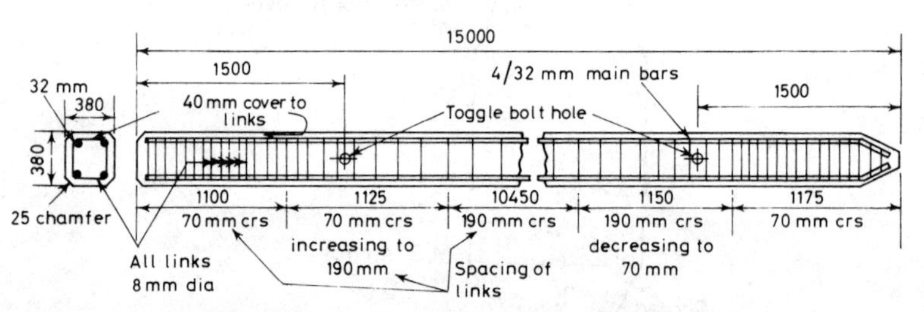

Figure 26.17 Design of a precast concrete pile suitable for fairly easy driving conditions

required length of pile cannot be predicted accurately in advance of construction. This means that piles driven to a lesser depth than predicted must be cut down, with consequent waste of the cut-off portions. Piles may need to be driven to greater depths than predicted. This leads to delays while additional lengths are cast-on and left to harden to a state ready for driving. These difficulties with predicting the required lengths can be overcome by using jointed piles. The 'West's shell pile' consists of hollow cylindrical units, which are threaded on a mandrel (*Figure 26.18(a)*). The mandrel is withdrawn on completion of driving and the interior of the pile is filled with concrete. The 'Herkules pile' (*Figure 26.18(b)*) consists of solid hexagonal precast concrete sections with bayonet-type locking joints that can withstand bending and tension forces. The piles are cast in single lengths to suit the predicted penetrations, and additional short lengths are locked-on if greater driving depths are required.

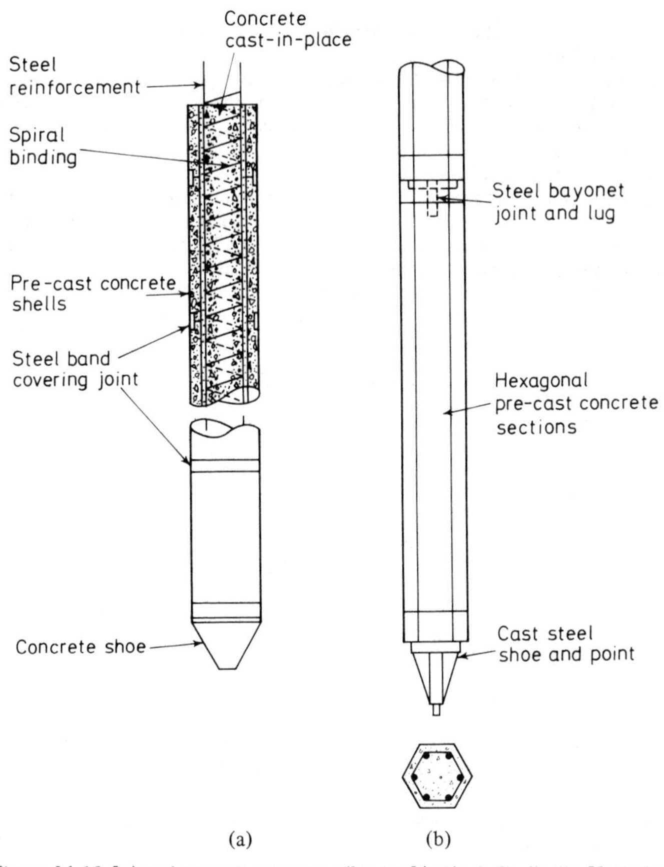

Figure 26.18 Jointed precast concrete piles (a) West's shell pile (b) Herkules pile

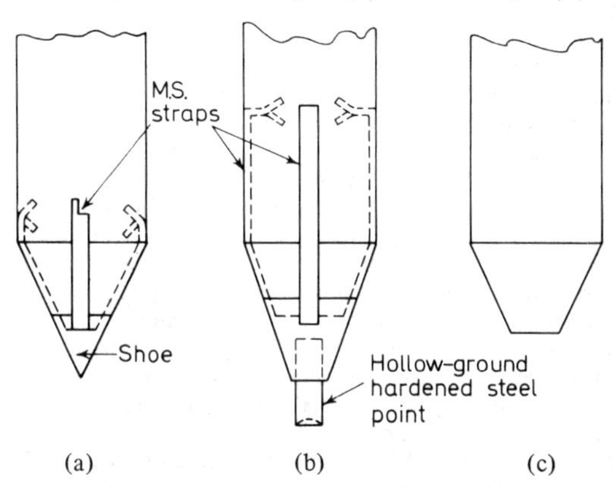

Figure 26.19 Design of toe for precast concrete piles

The West's 'Hardrive' and the GKN-Keller and Balken piles are similar types of jointed solid sections.

Where piles are driven through thin layers of compact material or into clays interbedded with thin seams of rock, the pointed toe (*Figure 26.19(a)*) is suitable. The rock shoe (*Figure 26.19(b)*) is designed for seating a pile on to a sloping hard rock surface, and the butt end (*Figure 26.19(c)*) is suitable for a pile driven into firm to stiff clays or granular soils.

Steel piles are manufactured in H-sections (Universal bearing piles) and in tubular or box sections (*Figure 26.20*). They cost more delivered to site than precast concrete piles, but they can carry heavy loads and can withstand hard driving, which makes them suitable for use in difficult ground conditions (e.g. in soils containing boulders or in rubbly fill material). Steel piles can be readily extended by welding on additional lengths, and cut-off portions have some value as scrap. Steel piles have a high resistance to lateral loads.

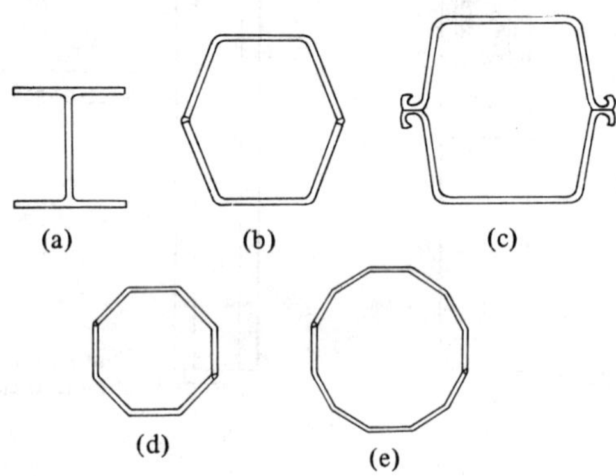

Figure 26.20 Steel bearing piles of various types: (a) Universal bearing pile; (b) Rendhex box; (c) Larssen Box; (d) Frodingham octagonal; (e) Frodingham duodecagonal

H-sections and tubular or box piles driven with open ends are referred to as 'small-displacement piles'. They have a useful application in avoiding excessive vertical and lateral soil displacement when driving piles close to existing structures. When open-ended hollow piles are driven to a deep penetration in soil or weak rock a plug will form in the interior of the pile. The plug can be drilled out, and the drilling continued below the pile toe to assist in penetrating obstructions to driving.

'Raymond piles' consist of thin tapered steel shells driven by a mandrel; this is withdrawn and the hollow interior is then filled with concrete. Raymond steel piles are readily adaptable to a varying penetration depth.

Driven and cast-in-place piles are installed by driving a tube with a closed lower end to the desired penetration. A reinforcing cage is then lowered down the tube, and concrete is placed in stages and rammed as the tube is withdrawn. Alternatively, the tube can be filled completely with high-slump concrete before it is withdrawn. Proprietary driven and cast-in-place types include the Franki and Dowsett piles, in which the tube is forced down by blows of a heavy rammer acting on a plug of gravel at the base of the tube. On reaching the required depth, the gravel and some added concrete are hammered out to form an enlarged base, which gives a useful increase to the bearing capacity of the pile. The shaft is then concreted in stages using a dry concrete compacted by rammer (*Figure 26.21*).

The GKN, Simplex, Alpha and Vibro piles are formed by driving the tube with its end closed by a shoe. The concrete is compacted while withdrawing the tube in stages, or a self-compacting concrete is used to fill the tube completely before withdrawal (*Figure 26.22*).

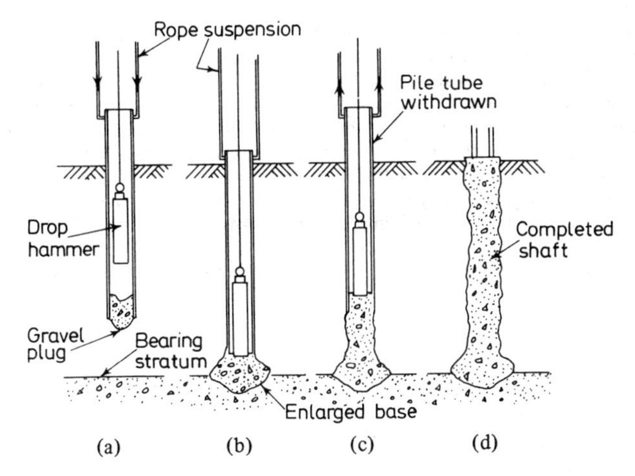

Figure 26.21 Franki pile

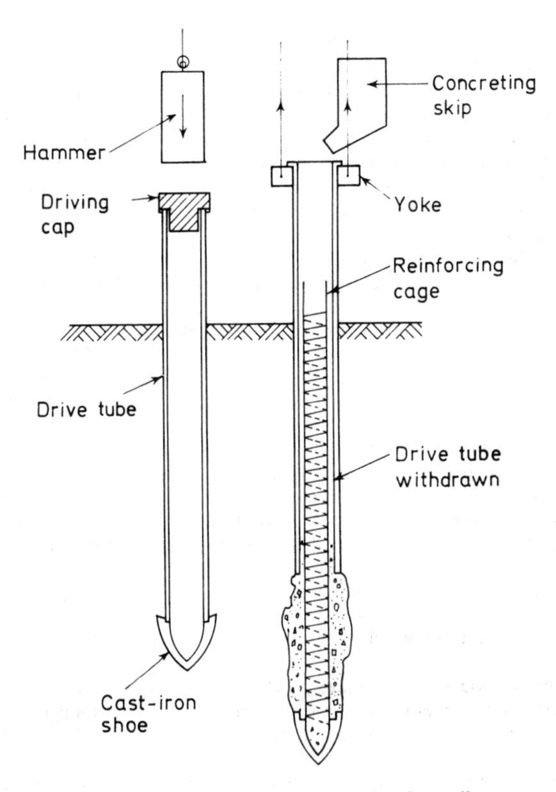

Figure 26.22 Driven and cast-in-place pile

Driven and cast-in-place piles have the advantage of avoiding the need to cut-down or extend the shaft in conditions of variable penetration depth. Also the reinforcing cage is needed only to provide for stresses due to eccentric loading on the pile head or uplift caused by ground heave. Care is needed in forming the shaft to avoid lifting the concrete while withdrawing the tube, and to avoid 'waisting' or 'necking' of the shaft due to squeezing of the unset concrete by soft clays or peaty soils.

There is a limit to the penetration depth of driven and cast-in-place piles, governed by the pulling capacity of the piling rig to extract the driving tube.

Some typical maximum working loads on driven and cast-in-place piles are:

Nominal shaft diameter (mm)	*Nominal maximum working load (kN)*
275	400
350	500
400	700
450	900
500	1100
600	1400

TYPES OF NON-DISPLACEMENT PILES

Non-displacement piles are installed by drilling a hole to the required depth; concrete, with or without a reinforcing cage, is then placed in the drilled hole to form the pile shaft. In firm to stiff or hard clays the hole can be drilled by a rotary plate or spiral auger. Alternatively, a cable percussion rig can be used in conjunction with a claycutting tool. In these cohesive soils, support to the sides of the borehole is not required for short or medium-length piles. However, a length of temporary steel casing should be lowered down the drilled hole of long piles in order to avoid the risk of collapse of loose or fissured clay from the sides while concrete is being placed. Under-reaming tools can be used to form an enlarged base in clays. In the case of large-diameter piles the enlarged base can be formed by hand excavation.

Continuous support is necessary to the sides of the pile borehole when installing piles in granular soils. A cable percussion rig in conjunction with a 'shell' or baler is used for drilling in these soils. Large-diameter piles can be drilled by mechanical grabbing rigs. The casing is forced down hydraulically to follow the drilling, with the assistance of a semi-rotary movement of the casing actuated by clamps mounted on the piling rig. Alternatively, pile boreholes in granular soils can be drilled by a rotary auger with support to the sides of the borehole by a bentonite slurry. Concrete is placed through a tremie pipe to displace the slurry.

As in the case of driven and cast-in-place piles, care is needed when forming the pile shaft to avoid lifting the concrete while withdrawing the temporary casing tubes, and to avoid squeezing of the unset concrete in soft or peaty clays. In severe conditions, squeezing can be prevented by placing the concrete in a permanent thin steel lining tube.

Where groundwater cannot be pumped or baled-out of the pile borehole, the concrete should be placed under water by tremie pipe. If pumping or baling is possible, the rate of inflow should not be such as to wash the cement out of the unset concrete.

The 'Prestcore' pile is a proprietary type in which hollow cylindrical concrete units are lowered on a mandrel down the borehole. The mandrel is then withdrawn, and cement grout is pumped down to fill the voids within and surrounding the concrete units.

The Dowsett 'Prepakt' pile is formed by drilling the pile borehole by a continuous-flight spiral auger. When drilling has reached the required depth, cement grout is pumped down the hollow stem of the auger as it is withdrawn, to form the pile shaft. This type of 'grout-intruded' pile has useful applications for installation in water-bearing granular soils close to existing structures where it is necessary to avoid loss of ground caused by drilling.

Calculation of the carrying capacity of piles

The method of installation of piles has a considerable effect on the calculation of their bearing capacity. Thus driving a pile into a firm or stiff clay can have the effect of compacting the soil, which increases in strength in a zone around the pile shaft, whereas drilling for bored piles can cause softening due to swelling of stiff clays. Similarly, drilling for bored piles in granular soils can cause loosening of the soil beneath the pile toe, with consequent loss of end bearing resistance, whereas pile driving will densify the soil.

In view of the effects of installation, the calculation of the carrying capacity of driven and bored piles is a matter of practical experience allied to soil mechanics theory, and the reader is referred to the 'further information' list at the end of this section for guidance on design methods.

Negative skin friction

Where piles are installed through recently deposited fill material, negative skin friction or 'dragdown' will act on the pile shaft due to downward movements of the fill as it consolidates under its own weight. This dragdown force must be added to the working load on the pile. When the consolidation of the fill is completed there may be some residual downdrag stresses remaining in the pile shaft. If the ground level is then raised by placing further fill, the negative skin friction will be re-activated, both in the portion of the shaft within the old fill (which will consolidate further under the new surcharge load) and within the new fill. Negative skin friction also occurs where piles are installed in soft compressible clays and a surcharge load of fill is applied to the surface of these clays. The negative skin friction acts on the portion of the shaft within the soft clay (which consolidates under load) and on the portion within the new fill.

The downdrag forces can be calculated by the same methods that are used for calculating skin friction support to a pile shaft, but the magnitude of these forces and the length of shaft over which they are activated depends on the amount of yielding of the pile under load. The dragdown force is at a maximum when the pile is bearing on a hard incompressible stratum.

Where precast concrete or steel piles are used, the negative skin friction can be reduced by coating the pile shaft within the consolidating material by a thick layer of soft bitumen. However, it is usually more economical to regard the dragdown as an addition to the working load. A somewhat lower safety factor may be permissible for the combined dragdown and normal working load on a pile.

Groups of piles

If piles are installed at too close a spacing there will be interference between the respective zones of transmittal of load from the pile to the soil, with the result of excessive load concentration and possible yielding of the piles. This is particularly liable to happen where driven piles deviate from their intended position such that the pile toes may be very close together. In cohesive soils piles should not be installed closer together than three times their least diameter or width, with a minimum spacing centre-to-centre of 1 m. The minimum spacing for piles end-bearing in a granular soil is twice their least diameter or width, but a minimum spacing of three times the width would be preferred for long driven piles, to allow for deviation from their true line.

Piles in small groups can be considered to act as individual units, but in the case of large groups the effect of interaction of the stresses transmitted to the ground must be considered in relation to the settlement of the group. The settlement of the group is likely to be more than that of the single pile under the same working load as each pile in the group.

The interactive effects can be considered in a simple way by assuming that the group acts as a block foundation with the load transmitted to the ground, as shown in *Figure 26.23*. The safety factor against general shear failure and the settlement of the group can then be calculated by the methods described in Section 23.

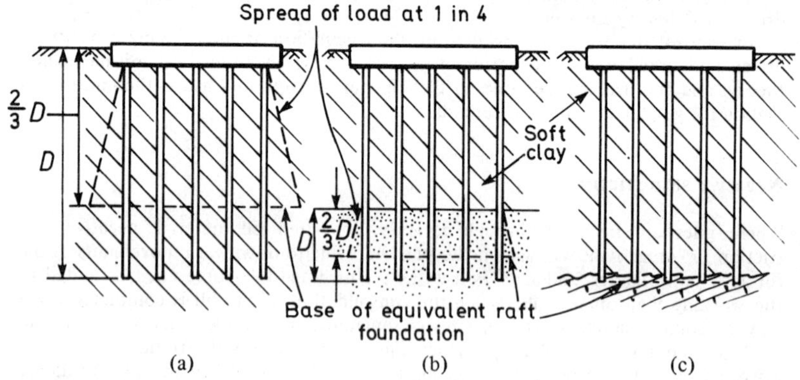

Figure 26.23 Spread of load from pile group considered as a block foundation: (a) group of piles supported predominantly by skin friction; (b) group of piles driven through soft clay to combined skin friction and end-bearing in a dense granular soil stratum; (c) group of piles end-bearing on a rock stratum

Interactive effects must also be considered at the construction stage. For example, driving piles at close spacing in clays can cause heave of the ground surface and uplift of piles already driven. Boring for non-displacement piles in water-bearing granular soils and in soft clays can cause withdrawal of soil from the area surrounding the pile groups, with consequent settlement of the ground surface.

The durability of piles

Timber piles need to be protected against decay in the portion of the shaft above the lowest groundwater level on the site. This can be effected either by taking the pile cap down to groundwater level or by driving a composite concrete and timber pile.

A limited life can be given to piles by treating the portion of the shaft above groundwater level with creosote or other preservative applied by pressure-treatment.

Concrete piles are susceptible to attack by any sulphates that may be present in the soil or groundwater. Precast concrete piles have a high resistance to attack since they can be made with a good-quality, well-compacted and well-cured concrete. Protective coatings on the shaft of precast concrete piles are liable to be stripped off by contact with the soil into which they are driven. However, resin or bituminous coatings are satisfactory for piles in soft clays or peats.

It is not always possible to ensure well-compacted concrete in some types of driven and cast-in-place or bored and cast-in-place piles. In severely aggressive conditions it may be necessary to surround the pile shaft with a protective sheath of metal or plastics.

Recommendations for protective measures suitable for various concentrations of sulphates are shown in Table 26.4.

Acids in industrial wastes and organic acids in peaty or swampy soils can attack concrete piles. Specialist advice should be sought where piles are installed in ground contaminated with industrial wastes. Precautions suitable for natural soils or groundwater containing organic acids are shown in Table 25.2.

Steel piles are unlikely to suffer significant corrosion if they are wholly embedded in natural soil. Oxygen is necessary to initiate and sustain the process of corrosion, and this is unlikely to be present if the pile is driven in close contact with the soil. Therefore no special protective measures are necessary for piles driven into natural soil deposits. Specialist advice should be sought if piles are installed in disturbed ground, particularly in industrial wastes or in ground polluted by sewage or other decomposing organic material.

Pile testing

Loading tests on piles may be needed at two stages: in the first stage to verify the carrying capacity of the piles either in compression, uplift or lateral loading, and in the second stage to act as a proof load to verify the soundness of workmanship or adequacy of penetration of working piles.

For first-stage testing either the constant rate of penetration (CRP) test or the maintained load (ML) test may be used. The latter is to be preferred if it is desired to obtain information on the deflection of the pile under the working load, or at some multiple of this load.

For proof loading of working piles the maintained load test should be made. It is not usual to apply a load of more than 1.5 times the working load in order to avoid overstressing the pile. The procedures for the CRP and ML tests are described in CP 2004.

Pile caps and ground beams

A pile cap is necessary to distribute loading from a structural member, e.g. a building column, on to the heads of a group of bearing piles. The cap should be generous in dimensions to accommodate deviation in the true position of the pile heads. It is

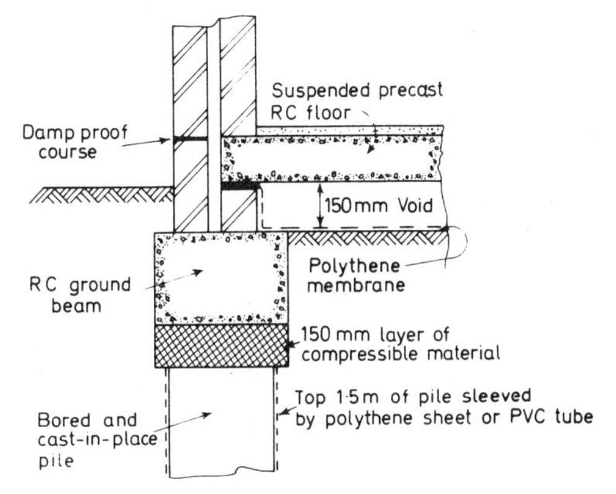

Figure 26.24 Ground beam for piles carrying a load-bearing wall in conditions where swelling of a soil can occur

Table 26.4 CLASSIFICATION OF SULPHATES IN SOILS AFFECTING CONCRETE IN PILED FOUNDATIONS AND RECOMMENDED PRECAUTIONS

Concentration of sulphate expressed as SO_3				*Types of cement and limiting mix proportions (note 1) for dense fully compacted concrete and other protective measures for:*		
Class	*In soil*		*In ground water (parts/100 000)*	*Precast concrete piles and pile shells; precast concrete caps and ground beams*	*Concrete placed in thin steel shells in dry conditions (note 2); reinforced concrete in pile caps and ground beams (note 3)*	*Concrete in driven and cast-in-place and bored and cast-in-place piles (note 4)*
	Total SO_3 (%)	*SO_3 in 2:1 aqueous extract (g/l)*				
1	<0.2	–	<30	OPC min. 300 kg/m^3, max. w/c ratio 0.55	(a) Above highest water level: OPC min. 300 kg/m^3, max. w/c ratio 0.55 (b) In contact with fluctuating water level: OPC min. 310 kg/m^3, max. w/c ratio 0.55	(a) Above highest water level: OPC min. 330 kg/m^3, max. w/c ratio 0.55 (b) In contact with fluctuating water level: OPC min. 370 kg/m^3, max. w/c ratio 0.55
2	0.2–0.5	–	30–120	(a) Above highest water level: OPC min. 310 kg/m^3, max. w/c ratio 0.50 (b) In contact with fluctuating water level: OPC min. 350 kg/m^3 or SRPC min. 310 kg/m^3, max. w/c ratio 0.50	(a) Above highest water level: OPC min. 330 kg/m^3, max. w/c ratio 0.50 (b) In contact with fluctuating water level: OPC min. 350 kg/m^3 or SRPC min. 310 kg/m^3, max. w/c ratio 0.50	(a) Above highest water level: OPC min. 370 kg/m^3, max. w/c ratio 0.50 (b) In contact with fluctuating water level: OPC min. 380 kg/m^3 or SRPC min. 340 kg/m^3, max. w/c ratio 0.50
3	0.5–1.0	1.9–3.1	120–250	(a) Above highest water level: OPC min. 380 kg/m^3 or SRPC min. 340 kg/m^3, max. w/c ratio 0.50 (b) In contact with fluctuating water level: SRPC min. 350 kg/m^3, max. w/c ratio 0.50	(a) Above highest water level: OPC min. 400 kg/m^3 or SRPC min. 350 kg/m^3, max. w/c ratio 0.50 (b) In contact with fluctuating water level: SRPC min. 390 kg/m^3, max. w/c ratio 0.50	(a) Above highest water level: OPC min. 400 kg/m^3 or SRPC min. 350 kg/m^3, max. w/c ratio 0.50 (b) In contact with fluctuating water level: SRPC min. 390 kg/m^3, max. w/c ratio 0.50 (note 5)

4	1.0–2.0	3.1–5.6	250–500	(a) Above highest water level: SRPC min. 380 kg/m^3, max w/c ratio 0.45 (b) In contact with fluctuating water level in lower range of SO_3 and favourable cations: SRPC min. 390 kg/m^3, max. w/c ratio 0.45. For higher range of SO_3 and unfavourable cations: SSC min. 370 kg/m^3, or provide permanent sheathing in metal or plastics	(a) Above highest water level: OPC min. 400 kg/m^3 or SRPC min. 350 kg/m^3, max. w/c ratio 0.45 (b) In contact with fluctuating water level: as for precast concrete, but external sheathing to consist of polythene, hot bitumen spray, bituminous paint, trowel-applied mastic asphalt or adhesive plastic sheet.	(a) Above highest water level and soil free from seepage water: SRPC min. 400 kg/m^3, max. w/c ratio 0.45 (b) In contact with fluctuating water level in lower range SO_3 and favourable cations: SRPC min. 390 kg/m^3, max. w/c ratio 0.45. For higher range of SO_3 and unfavourable cations, place concrete in durable metal or plastics sleeve left in place.
5	>2.0	>5.6	>500	(a) Above highest water level: SRPC min. 390 kg/m^3, max. w/c ratio 0.40 (b) In contact with fluctuating water level: permanent sheathing in metal or plastics over SRPC min. 390 kg/m^3, max. w/c ratio 0.40	(a) Above highest water level: OPC min. 400 kg/m^3 or SRPC min. 350 kg/m^3, max. w/c ratio 0.40 (b) In contact with fluctuating water level: SRPC min. 390 kg/m^3, max. w/c ratio 0.40 with permanent external sheathing as above	(a) Above highest water level and soil free from seepage water: SRPC min. 390 kg/m^3, max. w/c ratio 0.40 (b) Cast-in-place piles are unsuitable for installation below the water table

OPC ordinary Portland cement
SRPC sulphate-resisting Portland cement
SSC supersulphated cement
HAC high alumina cement

(1) The minimum cement contents recommended above are suitable for:

(a) Precast concrete – low workability (12–25 mm slump)
(b) *In-situ* concrete in pile shells, pile caps and ground beams – medium workability (50–75 mm slump)
(c) Cast-in-place piles – high workability (100 mm slump)

If concrete of lower workability than (c) is required by specialist piling contractor, the cement content may be reduced provided that concrete can be compacted to dense impermeable mass. In no case must the maximum water/cement ratio be exceeded.

(2) Precautions assume that shells may be ruptured or lost by corrosion.
(3) Precautions assume fairly massive sections where concrete can be vibrated.
(4) Precautions assume that drive tube or borehole casing is extracted during or after placing the concrete.
(5) Higher cement content may be required if concrete is placed under water by tremie pipe.

usual to allow piles to be driven out of position by up to 75 mm, and the positioning of reinforcement that ties into the projecting bars from the pile heads should allow for this deviation. Caps are designed to beams spanning between the pile heads and carrying concentrated loads from the superimposed structural member. The heads of concrete piles should be broken down to expose the reinforcing steel, which should be bonded into the pile cap reinforcement. The loading on to steel piles can be spread into the cap by welding capping plates to the pile heads or by welding on projecting bars or lugs as shear keys.

Piles placed in rows beneath load-bearing walls are connected by a continuous cap in the form of a ground beam (*Figure 26.24*). In the illustration the ground beam is shown as constructed over a compressible layer designed to prevent uplift on the beam due to swelling of the soil, and the pile is sleeved over its upper part to prevent uplift within the zone of swelling.

FOUNDATIONS FOR MULTI-STOREY BUILDINGS

Influence of type of structure on type of foundation

The structural form of the building has a considerable influence on the choice of a particular method of foundation design. Multi-storey buildings of *loadbearing wall* construction are normally supported on strip foundations of the types described in Section 25. In the case of high buildings, heavy loads per metre run of wall may need to be carried by the ground, necessitating the adoption of reinforced-concrete wide strip foundations (*Figure 25.3*).

For loadbearing wall structures of the *crosswall type* (*Figure 26.25*), the infill between the loadbearing walls may be of a light panelled form of construction that

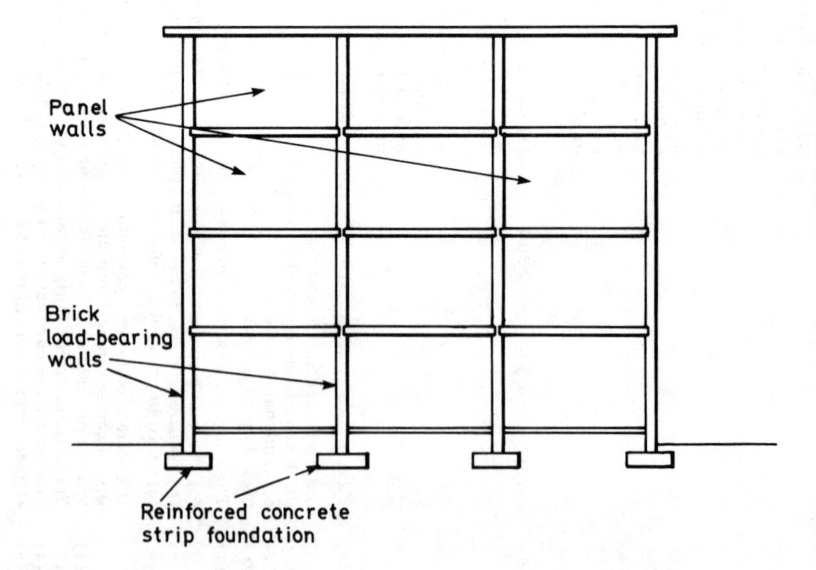

Figure 26.25 Foundations of crosswall type multi-storey building

provides no stiffening to the structure in a plane at right-angles to the crosswalls. In such cases care is needed to avoid excessive eccentric loading on the crosswall foundations, or differential settlement between adjacent foundations.

For buildings of framed construction, pad foundations (*Figure 26.26*) are suitable for supporting columns at normal spacings in domestic and office buildings. Where columns are closely spaced it may be convenient to combine rows of closely adjacent pads into single strip foundations, thus avoiding costs of fabricating individual sets of

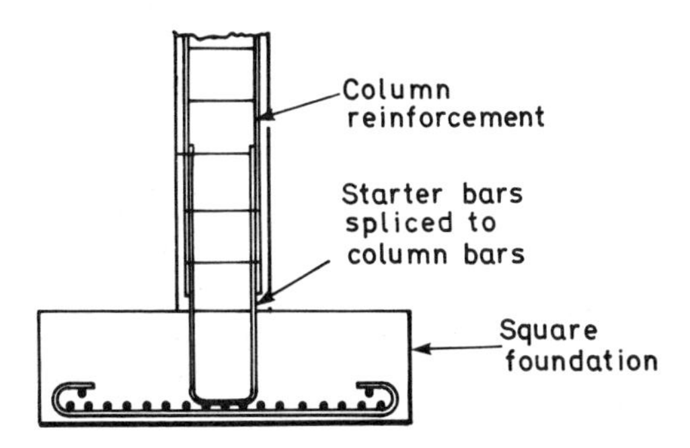

Figure 26.26 Pad foundations for framed multi-storey building

foundation formwork and also avoiding the problems of dealing with individual close-spaced excavations. Where differential settlements, calculated on the assumption of a flexible foundation, are not excessive the combined strip foundations can take the form of abutting pads with a simple butt joint to allow relative vertical movement between them. This involves cutting reinforcing bars for each pad, but steel will be saved since the foundations are not required to act as beams spanning between column positions. However, where estimates of settlement show high values of relative rotation (*Figure 26.12*) between adjacent columns, it is necessary to prevent excessive differential movement in the structure by designing the strip foundation to act as a stiff continuous beam (*Figure 26.27*).

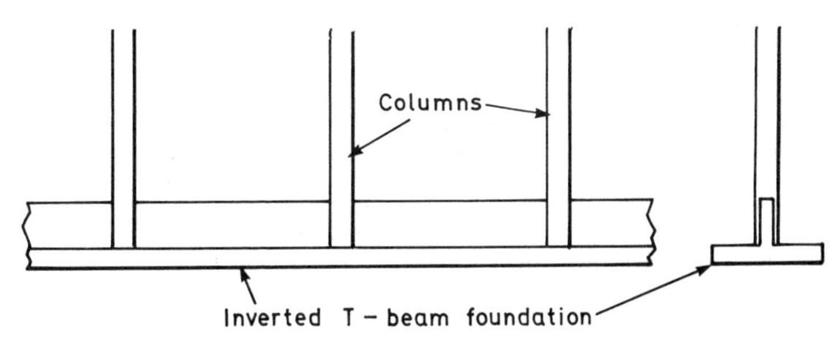

Figure 26.27 Continuous beam foundation for row of close-spaced columns

There may be a wide variation in loading from adjacent columns in a framed building. An example of this is the variation between lightly loaded exterior columns and more heavily loaded adjacent interior columns of a building. While the majority of the interior columns carrying fairly equal loading can be satisfactorily carried by pad foundations, it may be necessary to combine the foundations of pairs of columns in the outer two rows in order to minimise differential settlements between them

(*Figure 26.28*). This form of construction is sometimes known as a *strapped foundation*. The need for strapping may also arise where there is insufficient space around the exterior column to provide a symmetrical pad foundation (due, for example, to the proximity of an existing building). In such cases the required bearing area can be provided by combining the exterior column foundation with that of the adjacent interior column. Reinforcement is necessary to prevent rotation due to eccentric loading on the exterior column support.

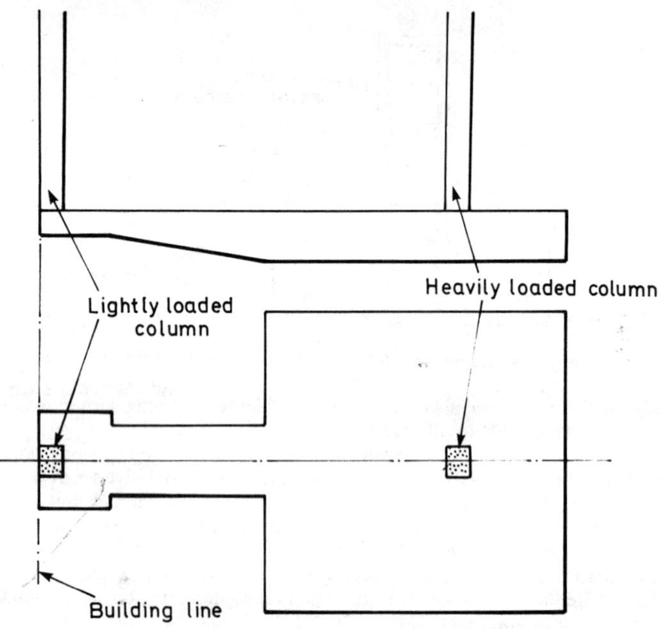

Figure 26.28 Combined (strapped) foundation for adjacent columns carrying unequal loading

A **basement** may be necessary beneath a multi-storey building for some functional purpose, for example as an underground car park or to provide space for the storage of goods or to accommodate building services. It is usual to design the base slab as a raft (see earlier), since the alternative of pad foundations linked by a relatively thin slab raises problems of providing sufficient stiffness in the slab to prevent cracking from bearing pressures caused by external earth pressure. Pad foundations in conjunction with a thin slab can be used where the basement is founded on an incompressible stratum, e.g. rock or dense granular soil, as shown in *Figure 26.29*.

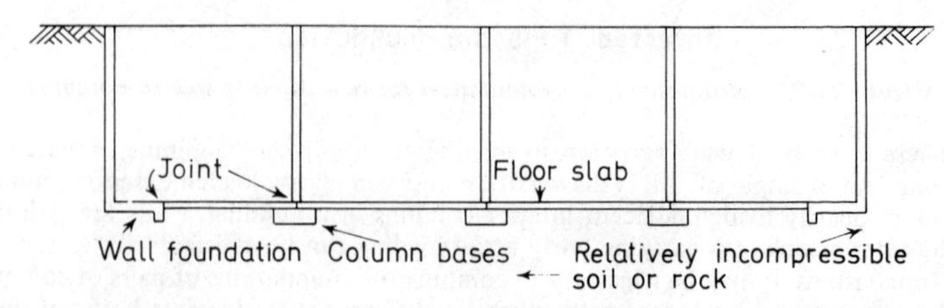

Figure 26.29 Basement foundations suitable for relatively incompressible ground

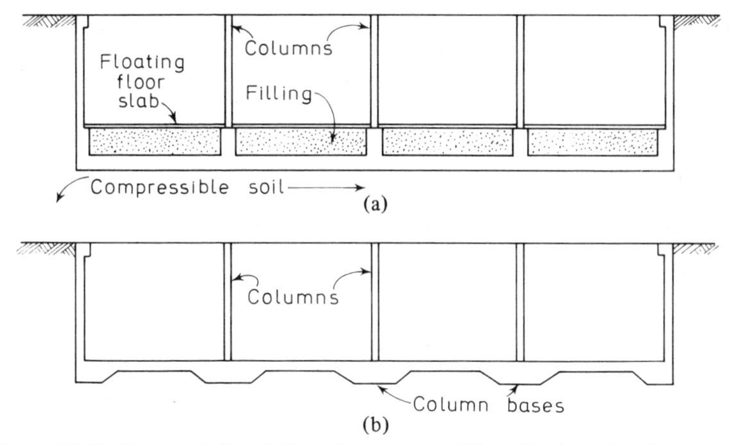

Figure 26.30 Basement foundations for compressible soils: (a) raft with upstand beams; (b) raft with thickened slab

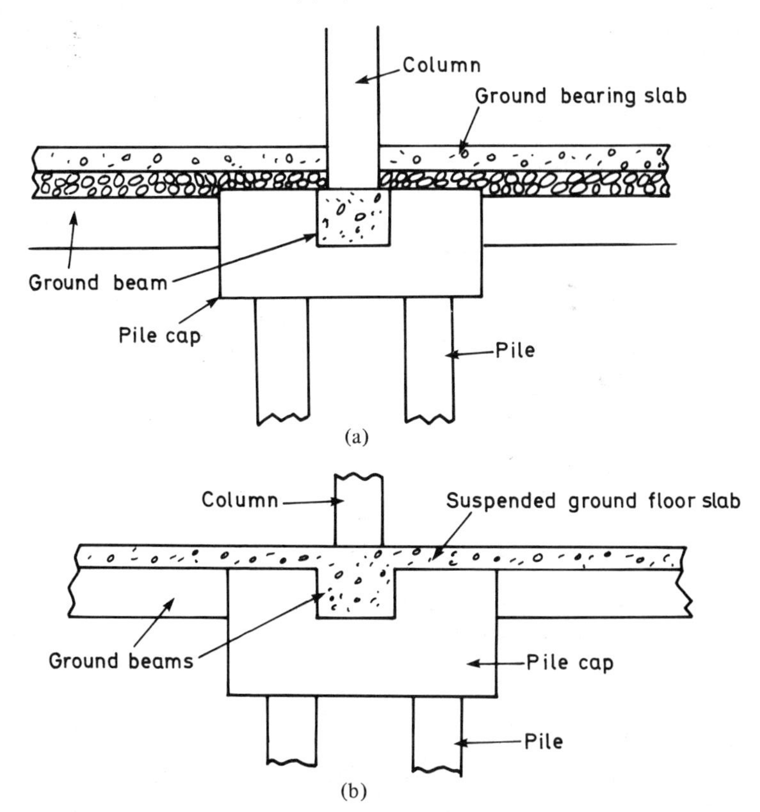

Figure 26.31 Piled foundation for multi-storey buildings: (a) in conjunction with a ground bearing slab; (b) in conjunction with a suspended ground floor

Waterproofing by asphalt tanking or plastics sheeting is more readily achieved on the underside of a slab raft stiffened by upstand beams (*Figure 26.30(a)*) or a slab thickened by column positions (*Figure 26.30(b)*). The design and construction of basements is dealt with below.

Influence of ground conditions on type of foundation

Where the ground is too weak to support strip or pad foundations of the types described above, either *raft* or *piled* foundations will need to be adopted. Usually pile foundations will be more economical since the piles can be placed directly beneath the points of loading, for example beneath column positions or load-bearing walls. Thus excessive bending moments are avoided in the substructure, which can consist of simple pile caps or capping beams, with the ground floor constructed either as a ground bearing slab or as a suspended slab (*Figure 26.31*).

Ground conditions may make piling difficult or costly, in which case the alternative of the raft foundation needs to be considered. Considerable stiffness is required for rafts beneath multi-storey buildings because of the bending moments induced by the soil pressures acting on the wide span lengths between column positions. Either a heavy uniform slab is required or a slab stiffened by upstand beams (*Figure 26.30(a)*).

Where piling is adopted, there are economic advantages in supporting column loads by single piles of large diameter rather than by clusters of small- or medium-diameter piles. Large reinforced-concrete pile caps are avoided, only simple trimming and capping of the pile head is needed, and overall construction time is saved since the installation of a single large-diameter pile takes little longer than a small-diameter pile of the same shaft length.

Design and construction of basements

It is presumed in the following paragraphs that the basements are required for some functional purpose in the building, such as underground car parking or storage space. Thus clear spaces are required between the columns, which extend upwards through the basement to support the superstructure of the building. The cellular form of construction, where crosswalls are used to provide stiffness combined with lightness of the substructure, as in the design of buoyancy rafts (*Figure 26.8*), cannot be used for basements where working space is required.

General guidance on basement construction is given in a publication of the Institution of Structural Engineers[6].

CONVENTIONAL BASEMENT DESIGN

The basement structure can be designed either as a slab stiffened by upstand beams or with the slab spanning between columns and exterior walls (*Figure 26.30*). The exterior walls are designed as conventional retaining walls using the method described in CP 2002[7]. These walls should normally be designed as free-standing structures, since propping by the ground floor slab is not usually provided until all the below-ground work has been completed.

Basements can be constructed in open excavations with the sides sloped back to a stable angle of repose. Where space is not available for sloping back the sides of the excavation, or where it is desirable to exclude groundwater from the excavation, the sides should be cut vertically and supported by sheeting and strutting or shoring.

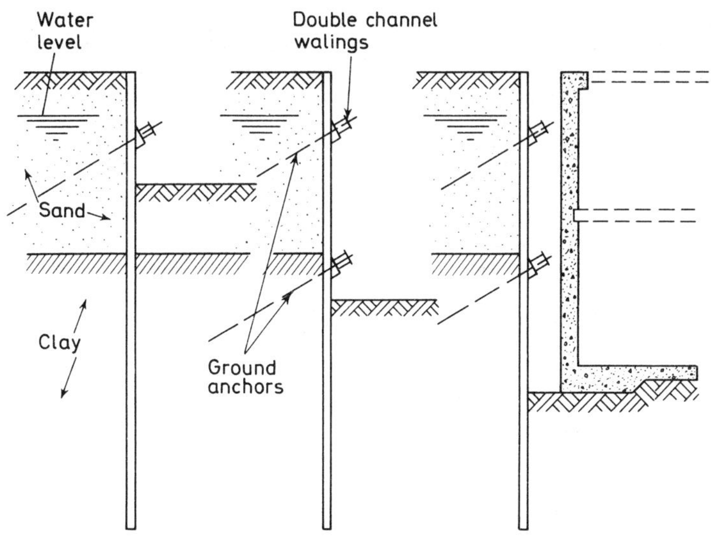

Figure 26.32 Stages of excavation for basement with support by tied-back sheet piling

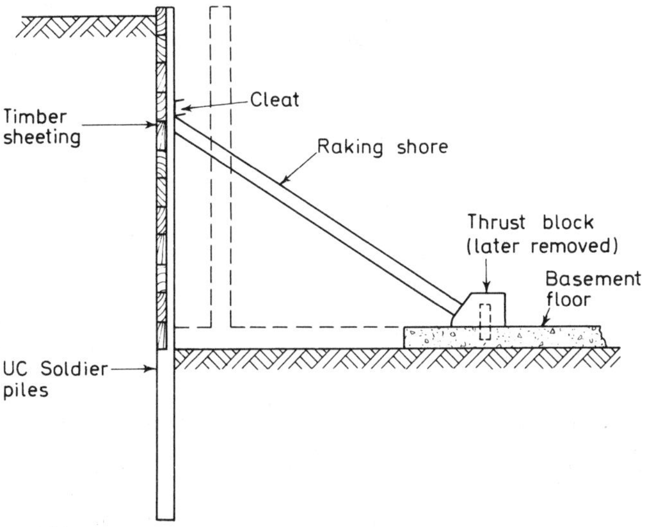

Figure 26.33 Basement excavation supported by horizontal timber sheeting and H-section soldier piles

Interlocking steel sheet piles (*Figure 26.32*) are required as a support where it is required to exclude groundwater from an excavation, e.g. in water-bearing sands or gravels. Where groundwater is not present, or if the soil type is such that some inflow can be permitted without causing instability, the sheeting can consist of vertical timber runners supported by horizontal walings, or by horizontal timber sheeting supported by vertical H-section steel soldier piles (*Figure 26.33*).

The sheeting to the sides of excavations of narrow to medium width can be supported by horizontal struts spanning across the width of the excavation (*Figure 26.34*). Restraint against buckling of long struts is provided by king piles. These should be bolted or welded to the struts as a restraint against either upward or downward buckling and against horizontal buckling in either direction. In deep

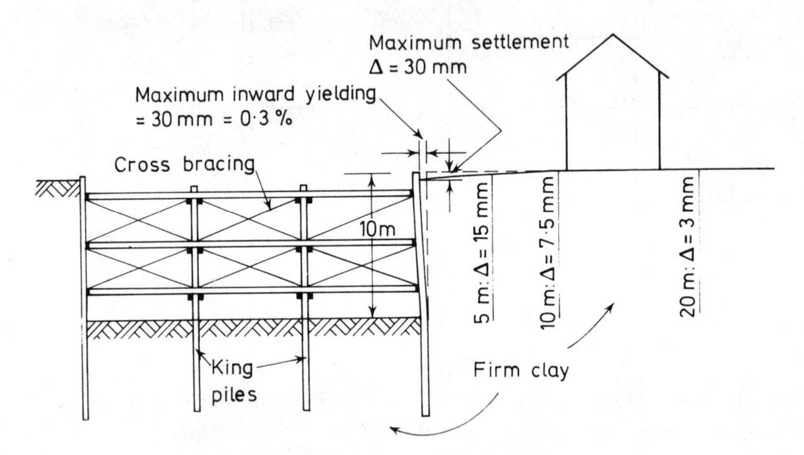

Figure 26.34 Support of medium width excavation by horizontal strutting (expected inward movement of sides of excavation also shown)

excavations, additional cross-bracing is desirable as a safeguard against progressive collapse of the support system due to the failure of one member. Sheeting to wide excavations can be supported by raking shores as shown in *Figure 26.33*, or they can be tied back by ground anchors as shown in *Figure 26.32*.

General guidance on methods of supporting excavations is given in CP 2003[8] and CP 2004[9].

DIAPHRAGM WALL CONSTRUCTION

Where basements are sited close to existing structures, such that sloping back the sides of the excavation is not feasible, or where noise and vibrations due to driving steel sheet piles would cause damage to the buildings or annoyance to their occupants, the sides of the excavation can be supported by a diaphragm wall. This system of support consists of excavating a trench in alternate panel lengths of 3–4 m, the sides of the trench being supported by a bentonite slurry[10]. Concrete is placed in the trenches by a tremie pipe, and then the intermediate trenches are excavated and concreted in a similar manner to provide a continuous vertical wall (*Figure 26.35*). The concrete is reinforced against bending stresses, and joints between panels are keyed to prevent the entry of water. Diaphragm wall construction can be used for all types of supported excavations where obstructions such as large cobbles or boulders in the soil would prevent full penetration of driven sheet piling. The diaphragm wall constructed in this way acts as the permanent retaining wall surrounding the structure. Alternatively, the permanent support can take the form of a row of close-spaced bored piles ('contiguous piling'). In water-bearing soils the sides of alternate bored piles can be grooved by chiselling them. When the intermediate piles are concreted a continuous watertight wall is obtained ('secant piling').

Excavation within the ground enclosed by the diaphragm wall or bored-pile walling is taken out in stages, and support to the wall is provided by means of strutting or shoring as described above, or by ground anchors as shown in *Figure 26.35*. The ground anchor or 'tied-back' method of support has the advantage that a clear working space is provided within the excavation, but it is necessary to obtain wayleaves from owners of the adjoining property to install the anchors beneath their land, and the effects of drilling and pressure grouting on the foundations of existing buildings must be considered.

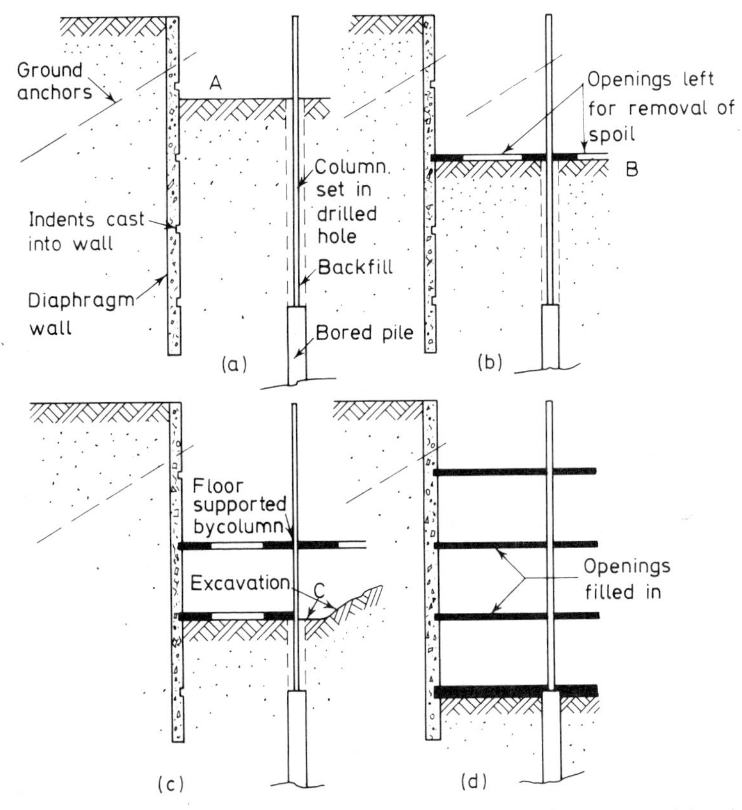

Figure 26.35 Construction of deep basement with excavation supported by tied-back diaphragm wall: (a) excavation to level A and ground anchors installed; (b) excavation to level B and floor slab cast; (c) excavation to level C and further floor slab cast; (d) completed excavation with all floor slabs cast

Any deep excavation is accompanied by inward yielding of the sides, resulting in settlement of the surface of the surrounding ground (*Figure 26.34*). The amount of yielding is proportional to the depth of excavation, and it is higher in cohesive than in cohesionless soils[11]. It can be kept to a minimum, but not entirely prevented, by a rigid form of horizontal support. One method of providing this support, which is sometimes practised in conjunction with diaphragm wall construction, is to use the ground floor slab and any intermediate basement floors as rigid permanent strutting. Excavation is taken out beneath the floors, the spoil being removed through openings left for the purpose and later filled in. Restraint to buckling of the floor slabs is provided by the building columns, which are installed in holes drilled below ground level in advance of the excavation (*Figure 26.35*).

Diaphragm walling is generally more costly than sheet pile support, particularly where the sheet piles can be withdrawn and re-used. The guide walls can be a significant proportion of the cost of diaphragm wall construction in cases where the guide walls have to be constructed in a timbered trench to a sufficient depth to prevent the bentonite escaping through drains or voids in filled ground.

PILED BASEMENTS

Where basements are constructed in a compressible soil such that settlement of the building will be excessive under the net loading (after deducting the weight of soil removed in the excavation from the gross load), the basement will have to be supported by piles taken down to a relatively incompressible stratum. As described earlier in this section, the soil at the base of the excavation will heave, and in a clay soil there will be some long-term swelling after completing the excavation. Thus swelling pressures will be exerted on the underside of the basement slab, which will relieve the piles of some, if not all, of their load. With time there will be a partial transfer of load to the piles [12] and they may eventually carry the full load of the building. It is the usual practice to design the base slab to resist a ground pressure equal to that imposed by the full net load of the building, but also to design the piles to carry the same total load.

Where driven piles are installed in clay, the ground heave due to piling displacement will initially carry the building load, but the heaved ground will then subside accompanied by full load transfer to the piles.

Some building structures consist of a tower block with adjoining low-rise wings. Where basements are provided beneath both the tower and the wings there will be relative vertical movement between the tower (which will settle) and the wings (which will rise due to relief of overburden pressure on the soil). In such cases it may be necessary to pile both the tower and the wings to reduce or eliminate the relative vertical movement. The piles beneath the wings will act as tension members to prevent uplift movement of these parts of the structure.

REFERENCES

1. May, J., 'Heave on a deep basement in the London Clay', *Proc. Conf. on settlement of structures, Cambridge*, Pentech Press, 177–182 (1975)
2. *Subsidence engineer's handbook*, National Coal Board, Mining Dept. (1975)
3. Zienkiewicz, O.C., and Cheung, Y.K., *The finite element method in structural and continuum mechanics*, McGraw-Hill (1967)
4. *Structure-soil interaction – a state-of-the-art report*, Institution of Structural Engineers (1978)
5. Wood, L.A., and Larnach, W.J., 'The effects of soil-structure interaction on raft foundations', *Proc. conf. on settlement of structures, Cambridge*, Pentech Press, 460–470 (1975)
6. *The design and construction of deep basements*, Institution of Structural Engineers, London (1975)
7. CP 2002, *Earth-retaining structures*, British Standards Institution (under revision)
8. CP 2003, *Earthworks*, British Standards Institution (under revision)
9. CP 2004, *Foundations*, British Standards Institution (1972)
10. Various authors, *Proceedings of the Conference on diaphragm walls and anchorages*, Institution of Civil Engineers, London (1975)
11. Tomlinson, M.J., *Foundation design and construction*, 3rd edn, Pitman (1975)
12. Hooper, J.A., 'Observations on the behaviour of a piled raft foundation on London Clay', *Proc. Instn Civil Engineers*, **55** Part 2, 855–877 (Dec. 1973)

FURTHER INFORMATION ON PILING

Cornfield, G.M., *Steel bearing piles*, Publication No 2/74, Constrado, London (1974)

Choosing a type of pile, Building Research Establishment Digest 95, 2nd edn (1968)

Thorburn, S., *Review of problems associated with the construction of cast-in-place concrete piles*, CIRIA/Property Services Agency, Report No PG 2 (1977)

Tomlinson, M.J., *Pile design and construction practice,* Viewpoint Publications (1977)

Weltman, A.J., and Little, J.A., *A review of bearing pipe types,* CIRIA/Property Services Agency, Report No PG a (1977)

Whitaker, T., *The design of pile foundations,* 2nd edn. Pergamon Press (Oxford) (1974)

CP 2004, *Foundations,* British Standards Institution (1972)

27 CONCRETE TECHNOLOGY

WHAT IS CONCRETE? 27–2

MATERIALS FOR CONCRETE 27–3

PROPERTIES OF FRESH CONCRETE 27–6

PROPERTIES OF HARDENED CONCRETE 27–7

INTERPRETING SPECIFICATIONS 27–10

TESTING CONCRETE 27–11

SAFE WORKING WITH CONCRETE 27–12

CONCRETE TECHNOLOGY 27

27 CONCRETE TECHNOLOGY

T.W. KIRKBRIDE, BSc, CEng, MICE
Technical Director, British Ready Mixed Concrete Association

This section gives background information on concrete materials (cements, aggregates, admixtures), the properties of concrete, and concrete practice (specifications, testing).

WHAT IS CONCRETE?

Concrete should be considered in two stages of its life cycle: in the *fresh* condition, when it must be fit for use by the builder, and in the *fully hardened* condition, when it must satisfy the client's requirements for strength and durability.

Fresh concrete is a mixture of cement, water and aggregate. The cement and water make up a paste that surrounds the particles of aggregate (normally a mixture of gravel and sand). If properly proportioned, this mixture will be workable and suitable in every way for placing, compacting and finishing to the required specification.

An hour or two after mixing, the paste sets through chemical action between the cement and water, binding the aggregate particles into a solid mass, which hardens, gaining in strength as the chemical reaction proceeds. About half the ultimate strength is reached in three days, and at 28 days almost the full strength is achieved. The chemical reaction (called 'hydration') also produces heat, which can be beneficial in cold weather.

The rate of strength gain depends mainly on the type and quantity of cement in the mix and the temperature of the concrete during the hardening period: higher temperatures result in more rapid gains of strength. The ultimate strength achievable depends not only on the mix, but also on there being enough water to react with the cement; premature drying of concrete must not be permitted.

The most commonly used cement is ordinary Portland cement, which is the best choice for most jobs, and the aggregates are usually sharp sand and gravel or crushed rock. There are different types of cement for different purposes, e.g. for higher early strength or for sulphate resistance. Admixtures, normally chemicals or bulk powders, are sometimes used to improve certain properties of the concrete in the fresh or hardened condition.

Specifications for concrete are normally in terms of *workability*, measured by slump test, and *one* of the following:

- *mix proportions*, e.g. 1:2:4 (1 volume cement, 2 volumes sand, 4 volumes gravel);
- *cement content*, in kilograms of cement per cubic metre of concrete,
- *compressive strength*, measured by the crushing strength of 100 mm or 150 mm cubes.

The majority of the concrete (70–75 per cent) used in building work is ready mixed, available at short notice (1 day or less) from over 1100 depots throughout the United Kingdom. Over 95 per cent of ready mixed concrete production is from depots that comply with the requirements of the British Ready Mixed Concrete Association. Under this scheme depots are classified as 'BRMCA Approved Depot' or 'BRMCA Approved Depot with Quality Control Procedures', the latter including requirements for quality control of strength mixes. Approved depots of both types are checked regularly by Chartered Engineers. Depots complying with the requirements hold certificates. The scheme provides an assurance that the depots maintain the standards of production and control laid down.

Handling, placing, compacting, finishing and curing of concrete are described by the single word 'workmanship'. *Handling and placing* involve transport from the mixer and pouring into place. This can often be done direct from the truckmixer when ready mixed concrete is used. When concrete has to be placed quickly, or access for barrows, dumpers etc. is difficult, it is often advantageous to pump the concrete into place. *Compaction* involves rodding, tamping or vibration, the purpose being to work the air out of the fresh concrete and obtain the maximum density and strength. *Curing* is the term used for the protection of concrete during the hardening period, and has three main objectives: to retain moisture so that the cement will not be deprived of the water it requires for the chemical reaction; to prevent cracking due to premature drying or rapid cooling at night; and to protect the concrete from damage by frost in winter weather. Curing normally involves covering exposed concrete with plastic sheeting for the first 7–14 days, including insulation in cold weather.

MATERIALS FOR CONCRETE

Cement

Ordinary Portland cement (OPC) is most commonly used for making concrete for all general purposes where no special properties are required. It develops strength quickly enough for the general run of concrete work.

Concrete and mortars made with ordinary Portland cement can be attacked by acids and sulphates. Sulphates may be present in soils and in groundwaters, for example. Acids may occur in soils and groundwaters as a result of industrial processes or from organic matter in the soil. Special cements may have to be used or other measures taken in these situations.

Rapid-hardening Portland cement, by modified chemical composition and/or finer grinding, gives increased strength at early age compared with OPC. At later age, e.g. 28 days, the strengths of RHPC and OPC are roughly the same.

The cement produces more early heat than ordinary Portland cement, and so can be used to advantage in cold weather to offset the effects of low temperature. It should be stored and used in the same way as ordinary Portland cement. It is slightly more expensive than ordinary Portland.

Sulphate-resisting Portland cement (SRPC), by modified chemical composition, improves the resistance of the concrete to attack by sulphates. The cement is usually a little darker in colour than most other Portland cements. Sulphate-resisting cement is slightly more expensive than ordinary Portland.

Other cements, e.g. Portland blastfurnace cement, masonry cements and coloured cements, are sometimes used. The manufacturers should be consulted for details.

Delivery and storage of cement

If the concrete is mixed on site, bagged cement will normally be used, but for large jobs bulk cement may be advantageous. Separate storage must be provided for each type and class of cement, and the contents of all containers should be clearly marked at the inlet point. The storage silos or bins must be proof against moisture; they must also be provided with vents and filters or other means of dust control, sufficiently dimensioned to prevent build-up of pressure or vacuum, to avoid interference with the batching system. Failure to provide adequate venting may result in excessive build up of pressure when filling, risk of damage, incorrect operation of discharge mechanism due to pressure, and deterioration of cement by condensation.

When bagged cement is used, weatherproof covered storage must be provided to protect the cement from moisture; different types and classes of cement should be separately stored to minimise the risk of using the wrong cement.

Aggregates

Natural dense aggregates are obtained from land and marine sources and are normally classified as follows:

- *Uncrushed material* – sands and gravels that are used 'as dug' or graded into suitable sizes, but without further processing. Sea dredged aggregates may be washed to remove detrimental salts, but no other processing of this material normally takes place. (The 'as dug' aggregate is coarse and fine mixed, usually called 'all-in'.)
- *Crushed material* – naturally occurring rocks or gravels, which are crushed and graded into suitable sizes.
- *Lightweight aggregates* – available as naturally occurring volcanic material or as processed materials, e.g. pumice, sintered materials or clinker. The lighter materials are often used for roof screeds. Some of the harder lightweight aggregates are used structurally, e.g. for floors, walls and roofs, to reduce dead weight and improve thermal insulation.

Coarse aggregates are designated by maximum size, e.g. 40 mm, 20 mm etc. The 20 mm size is used for the majority of concrete building work, e.g. for foundations, ground slabs and structural members.

Fine aggregate is material the bulk of which passes a 5 mm sieve, and is normally considered in two types : coarse sand (usually called 'concreting sand') for making concrete; and fine/soft sand for mortars and renderings. In the British Standard for aggregate (BS 882), fine aggregate is sub-divided into four numbered zones according to predominant size of particle. Zone 1 is the coarse material, with Zones 2, 3 and 4 being progressively finer. In practice, it is usually costly for the aggregate producer and uneconomic for the builder to insist on material within a particular zone. For general building work, the use of a 'concreting sand' of unspecified zone is satisfactory. For special work, e.g. structural concrete, a ready mixed concrete supplier will provide a consistent concrete by maintaining his own control over the fine aggregate.

Aggregate impurities

Aggregates may contain small quantities of impurities such as clay, silt, chloride salts, sulphate salts, coal and lignite according to location. BS 882 places limits on the allowable quantities of clay, silt and dust.

Strict limits are now imposed by certain Regulations and Codes of Practice on the introduction of chloride salts into concrete. In the case of marine aggregates, which contain chlorides from sea water, there will normally be no problem when the concrete is made with ordinary Portland cement and does not contain embedded metal, e.g. reinforcing steel. In all other cases, e.g. when sulphate-resisting cement is used, or the concrete is reinforced concrete, chlorides present in marine aggregates may be above the permitted limits. In these cases specialist advice should be obtained.

Storage of aggregates

Storage of aggregates poses special problems when concrete has to be mixed on site. To comply with current Codes of Practice there should be separate storage for each different size of aggregate used, with adequate provision for drainage. It is also a recognised standard of good practice to provide solid bases for aggregate storage, and to take special precautions to prevent intermingling of different sizes and types by using bins or constructing ground storage bays.

Water

The simplest guide is 'water that is fit to drink is suitable for making concrete'.

Admixtures

Admixtures are useful when it is necessary to modify the characteristics of a mix to correct for deficiencies in cement or aggregates, or to impart special properties for better placing, compaction or finishing. They are substances that are added to concrete and that, by chemical or physical action or both, change properties of the concrete, e.g. its workability, hardening or setting. They normally provide negligible bulk. They are available as liquids, pastes or powders and perform the following functions either singly or in some cases in combination:

Accelerators increase the initial rate of reaction between cement and water and thereby accelerate the setting and early strength development of concrete. These materials are sometimes incorrectly called antifreezes. There are now severe restrictions on the use of chloride-based accelerators, but non-chloride accelerators are available.

Retarders decrease the initial rate of reaction between cement and water and thereby retard the setting of concrete. These can be useful when ready mixed concrete is obtained from long distances. Retardation of several hours is possible, after which the setting and hardening continue at the normal rate.

Water-reducing agents (plasticisers) increase the fluidity of the cement paste and increase the workability of the concrete, or permit concrete to be made at the same workability with the consequent increase in strength. These can provide useful economies in some cases.

Air-entraining agents alter the properties of fresh and hardened concrete by propagating minute bubbles of air in the cement paste. Normally, segregation and bleeding (i.e. water rising and standing on the surface of concrete) are reduced, impermeability is increased, and durability (in particular, resistance to frost and de-icing salts) is greatly improved.

Superplasticisers when added to normal concrete impart very high workability, or allow a large water reduction beyond the limits of normal plasticising admixtures. Superplasticised concretes are sometimes called 'flowing' concretes.

Admixtures should be stored in clearly labelled containers, and any special storage conditions recommended by the vendor must be followed.

Tests on materials

When ready mixed concrete is used, the concrete producer is responsible for checking that materials meet specified requirements. BRMCA Approved Depots maintain regular checks on cements and aggregates, and these are confirmed by BRMCA inspections.

For site production the following procedures should be followed:

Cements. Proof of compliance should be obtained by the supplier's certificates in accordance with the relevant British Standard. In cases of doubt, cement tests should be carried out by specialist laboratories.

Aggregates. Proof of compliance should be obtained by supplier's certificate in accordance with the relevant British Standard. In the case of marine aggregates, certificates of chloride content (in per cent chloride ion by weight) should be obtained. Specialist laboratories are available for independent tests when required.

Admixtures. Before any admixture is used the admixture supplier should provide data on its suitability from acceptance tests, in accordance with BS 5075, with the proposed materials and mixes; and certificates of test with recommended dosage should be obtained. Generally, the approval of the Architect, Engineer or Building Control Officer will be required before using admixtures.

Water In cases of doubt, water should be laboratory tested in accordance with BS 3148, *Tests for water for making concrete.*

PROPERTIES OF FRESH CONCRETE

The main requirements for fresh concrete are that the concrete should be easy to transport, place and consolidate into its final position without segregation (i.e. without the concrete changing from a mixed into an 'unmixed' condition), and that the desired finish should be readily achieved, either 'off the form' or by trowelling or other means of finishing. These characteristics cannot be specified by performance tests to cover all possible combinations of materials and workmanship, and the choice of the 'right' mix depends mainly on the experience and skill of the concrete technologist. In the ready mixed industry, repeated use and observation of familiar materials has lead to an understanding that goes beyond the normal techniques of mix design. Therefore when ready mixed concrete is used the builder should state his proposed procedures for handling, placing and finishing the concrete, rather than attempt to define mix proportions in detail.

Table 27.1 SPECIFICATION OF SLUMP

Nominal slump value (mm)	*Allowable tolerance (mm)*	*Workability classification*	*Typical use*
25	±20	Low	Unreinforced concrete with powerful vibration.
75	±25	Medium	Unreinforced and reinforced concrete with internal vibration or tamping. Pumped concrete.
150	±50	High	Heavily reinforced concrete, vibration difficult or impossible. Cast *in-situ* piling.

Normally only one property of fresh concrete is defined, i.e. workability. The term 'workability' is used to described the ease with which the concrete can be compacted. This depends on cement content, overall grading of aggregate, shape of the aggregate particles, and amount of water in the mix. Workability is measured by standard tests, normally the slump test.

Figure 27.1 Slump is the distance between the highest point on the concrete and the underside of the rod

Slump is specified as a nominal value in millimetres. 'Nominal' means 'in name only', as defined in dictionaries, and is used because in practice the actual measured slump varies over a range due to differences from batch to batch and because of variations in the test itself. Thus tolerances must be allowed in practice: a 75 mm slump has an allowable tolerance of ±25 mm, i.e. should be within the range 50 mm to 100 mm. See *Figure 27.1*. Recommended nominal slumps for different methods of placing, with the corresponding allowable tolerance range, are indicated in Table 27.1 (the allowable tolerances conform to current standards, e.g. BS 5328).

PROPERTIES OF HARDENED CONCRETE

For hardened concrete, strength and durability are the main requirements. These are considered in the following paragraphs.

To ensure that the best possible strength and durability are obtained, the concrete must be cured to prevent premature drying and to protect exposed concrete from extremes of temperature. In house building and associated works, concrete should be cured by covering with plastic sheets as soon as it is finished. See *Figure 27.2*. In frosty weather additional protection by insulating blankets should be provided. Curing should normally continue for 7 days, but this should be increased to 14 days in hot summer weather or frosty winter weather.

(a)

(b)

Figure 27.2 (a) Protecting and curing freshly placed concrete with temporary materials. (b) Curing slab after removal of forms: plastic sheeting placed directly on concrete and weighted along edges (courtesy Cement and Concrete Association)

Figure 27.3 Compacting concrete in a cube mould by hand (courtesy Cement and Concrete Association)

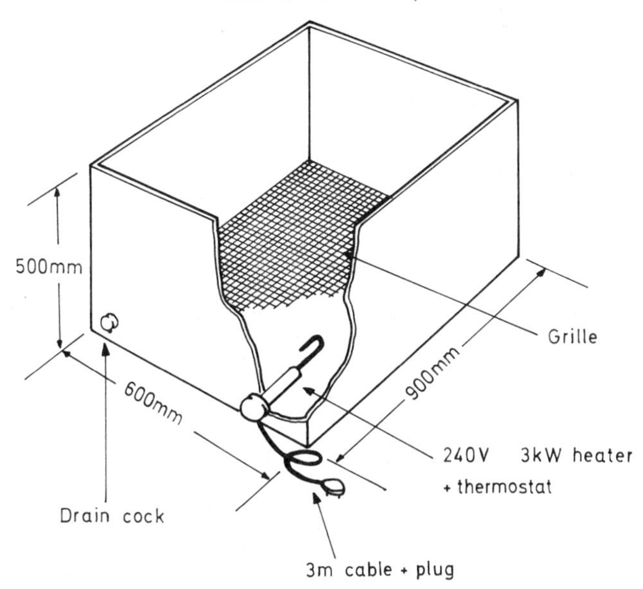

Figure 27.4 Warm-water tank, for cube curing on site, can be hired and will hold thirty 150 × 150 mm cubes. The cubes are maintained at 18–22°C by thermostatically controlled immersion heater, in accordance with BS 1881

Concrete strength

The strength of a concrete mix is normally measured by the cube test in accordance with BS 1881. The preparation of cubes is shown in *Figure 27.3*. The cube test is carried out under standard conditions of compaction and curing (*Figure 27.4*), and is not intended to be a direct measure of the strength of the concrete in the finished job. The difference between the 'standard strength' and the 'structure strength' is allowed for by the designer of the structure when following Codes of Practice. Strength is linked to the other main property of concrete, i.e. durability. Higher strength means more durable concrete, including resistance to chemical and frost attack and abrasion.

Durability against sulphates

Sulphates are salts that are present in some soils and groundwaters; over a period of years, they can destroy concrete unless the proper precautions are taken. The presence of these salts cannot be detected by eye or by simple test, but guidance on possible soil problems can be obtained from the local Building Control Officer or the Building Research Establishment. In cases of doubt, soil specialists should be called in to sample and analyse the soil and make a report on sulphate concentration. When the sulphate concentration has been obtained, the correct concrete mix to use should be obtained from Building Research Establishment Digest No. 174[2]. This will not always involve the use of SRPC cement unless the sulphate concentration is moderately high, but stronger mixes are usually necessary in all cases.

Durability against frost

After a few days of hardening, good-quality concrete is with few exceptions immune to frost. In the fresh condition, however, the water in the concrete must be prevented from freezing, otherwise the concrete may be permanently damaged and will not harden even when the temperature rises later. In severe weather, fresh concrete should be preheated to 10°C or more and protected from frost after placing by insulating sheets or blankets for 3–7 days depending on the severity of the weather.

Hardened concrete exposed to the weather will normally resist frost attack, provided it is of sufficient strength (cube strength 25 N/mm^2 or over) and has been fully compacted by vibration followed by proper curing. Concrete that is treated with de-icing salt (e.g. garage drives, which are sometimes salted by householders or receive salt from underneath a parked car following a winter journey) is more susceptible to frost attack and should be increased in strength or an air-entraining agent used.

INTERPRETING SPECIFICATIONS

Normally the specification for the concrete will be included in the contract between the builder and his client, to satisfy the requirements of the job and to comply with Building Regulations and any relevant local requirements or Codes of Practice. Before concrete can be ordered from a ready mixed supplier or produced on site, the full details of each mix should be scheduled. A typical schedule for a designed mix is given below.

Typical schedule for a designed mix from a main specification

Item	*Specified requirement*
1. Grade (i.e. strength at 28 days)	20 N/mm^2
2. Type of cement	OPC
3. Type of aggregate	BS 882
4. Nominal max. size of aggregate	20 mm
5. Nominal slump	75 mm
6. Tests required	Slump test Cube test (1 each day of concreting)
7. Other items	None

Notes

(1) If any of items 1–4 is missing in the main specification, the builder should decide this in consultation with the client's architect or engineer as appropriate.

(2) Slump (item 5) should be chosen by the builder according to the use and method of placing the concrete (see Table 27.1).

Mix proportions for designed mixes

For designed mixes, ready mixed concrete suppliers will provide a suitable mix based on previous quality-control data. In the absence of previous data from control tests, the Ordinary Prescribed Mixes in Table 1 of BS 5328 may be used, and these will normally be the most convenient for site use.

Quotations and ordering – ready mixed concrete

When inviting quotations or ordering ready mixed concrete, the supplier should be given the following information:

Name of builder
Name and location of job
Starting date (approx.)
Estimated construction period
Quantities per day, m^3 (approx.)
Total quantity of concrete, m^3
Proposed method of placing (e.g. skip, pump)
Mix description (i.e. mix schedule – see above)

TESTING CONCRETE

Normally the builder will be required to carry out only two tests on concrete : the slump test and the cube test. These tests are carried out on samples that must be representative of the batch of concrete (e.g. ready mixed truck load) that is being tested.

The procedures for sampling and testing are fully detailed in BS 1881, and these should be followed in detail. Cards[3] for use by operatives, giving the equipment required and step-by-step procedures for each test, are available from the British Ready Mixed Concrete Association or from ready mixed suppliers.

Cube crushing is normally carried out by specialist test houses. It is important to check the competence of test houses. The BRMCA publishes a Register[4], which gives the names of laboratories that have demonstrated their ability and intention to test 100 mm and 150 mm concrete cubes in accordance with BS 1881 and to record and make available data to confirm the validity of results.

SAFE WORKING WITH CONCRETE

Concrete is one of the safest building materials in use, but the usual precautions in dealing with heavy lifting (e.g. lifting bags of cement) and protection of eyes from splashes of cement paste should be observed. Body contact with fresh concrete should be avoided because it contains alkali, to which some skins may be sensitive, causing irritation or burning through prolonged exposure. When working with fresh concrete, waterproof gloves, long-sleeved shirts, full-length trousers and rubber boots are recommended. Waterproof kneeling pads should be used when trowelling fresh concrete.

REFERENCES

1. *Code of practice for ready mixed concrete: specification, ordering and production,* British Ready Mixed Concrete Association (1975)
2. *Concrete in sulphate-bearing soils and groundwaters,* Building Research Establishment Digest 174, HMSO (1975)
3. *Better cube making cards,* British Ready Mixed Concrete Association (1975)
4. *Register of commercial test houses (cube testing),* 4th edn, British Ready Mixed Concrete Association (1978)

FURTHER INFORMATION

British standards

BS 12, *Portland cement (ordinary and rapid hardening)*
  Part 2: 1971, *Metric units*
BS 146, *Portland blastfurnace cement*
  Part 2: 1973, *Metric units*
BS 812, *Methods for sampling and testing of mineral aggregates, sands and fillers*
  Part 1: 1975, *Sampling, size, shape and classification*
  Part 2: 1975, *Physical properties*
  Part 3: 1975, *Mechanical properties*
BS 882, 1201, *Aggregates from natural sources for concrete (including granolithic)*
  Part 2: 1973, *Metric units*
BS 881, *Methods of testing concrete*
  Part 1: 1970, *Methods of sampling fresh concrete*
  Part 2: 1970, *Methods of testing fresh concrete*
  Part 3: 1970, *Methods of making and curing test specimens*
  Part 4: 1970, *Methods of testing concrete for strength*
BS 3148: 1959, *Tests for water for making concrete*
BS 3681, *Methods for the sampling and testing of lightweight aggregates for concrete*
  Part 2: 1973, *Metric units*

BS 3797, *Specification for lightweight aggregates for concrete*
Part 2: 1976, *Metric units*
BS 3892: 1965, *Pulverised fuel ash for use in concrete*
BS 4027, *Sulphate-resisting Portland cement*
Part 2: 1972, *Metric units*
BS 4550, *Methods of testing cement*
Part 2: 1970, *Chemical tests*
BS 5075, *Concrete admixtures*
Part 1: 1974, *Accelerating admixtures, retarding admixtures and water-reducing admixtures*
BS 5328: 1976, *Methods for specifying concrete*

Other publications

Man on the job leaflets, Cement and Concrete Association
Effective use of admixtures in concrete, Cement Admixtures Association (1977)
Working safely with concrete, Portland Cement Association, Skokie, Illinois (1977)

28 REINFORCEMENT

SPECIFICATION	28–2
PROPERTIES	28–4
BAR REINFORCEMENT	28–8
FABRIC REINFORCEMENT	28–13
DETAILING	28–15
ORDERING	28–17
FIXING	28–17

REINFORCEMENT 28

28 REINFORCEMENT

B.M. JONES, BSc, CEng, MIStructE
GKN Reinforcements Ltd

This section describes how both bar and fabric reinforcement for concrete is specified, and how its properties are defined; it gives guidance on reinforcement detailing and fixing.

SPECIFICATION

In the UK it is usual to specify reinforcement by reference to the relevant British Standard specifications. Typical clauses from a contract document might be as follows.

MATERIALS

REINFORCEMENT

The reinforcement shall consist of:
(a) Hot rolled mild steel bars complying with the requirements of BS 4449, *Hot rolled steel bars for the reinforcement of concrete.*
(b) Cold worked high-yield bars complying with the requirements of BS 4461, *Cold worked steel bars for the reinforcement of concrete,* and having Type 2 bond characteristics.
(c) Steel fabric reinforcement complying with the requirements of BS 4483, *Steel fabric for the reinforcement of concrete.*

WORKMANSHIP

REINFORCEMENT

(a) The reinforcement shall be clean of oil and grease and free of all loose mill scale or rust.
(b) The reinforcement shall be cut and bent in accordance with the schedules provided.
(c) The cutting and bending of the reinforcement shall comply with the requirements of BS 4466, *Bending dimensions and scheduling of bars for the reinforcement of concrete.*

In addition, clauses prohibiting the re-bending or the hot bending of reinforcement, requiring prior approval of any welding, specifying approved methods of tying and spacing and dealing with any special requirements for a particular contract might also be expected.

In general the Bill of Quantities section of the contract documents will be prepared in accordance with the Standard Method of Measurement for Building Works, but this is not always adhered to quite strictly. In particular, the separate measurement of special spacers or chairs is usually ignored, and fabric reinforcement specified on schedules is usually measured by weight rather than area.

Table 28.1 BRITISH STANDARD REINFORCEMENT: SUMMARY OF MATERIAL PROPERTIES

Material	*Standard*	*Description*	*Characteristic values*				
			Yield (N/mm^2)	*Margin (%)*	*Elong. (%)*	*Site Bend Radius (mm)*	*Bond*
Mild steel bars	BS 4449	Mild steel billets are hot rolled to a plain round section and are sold in their as-rolled condition	250	15	22	2Φ	n/a
Hot rolled, high-yield, ribbed bars	BS 4449	High-yield steel billets, produced by an improved chemical composition during melting, are hot rolled to a round ribbed section and are sold in their as-rolled condition.	Up to 16Φ: 460	15	12	3Φ	Type 2
		Identification note: the longitudinal rib runs straight along the length of the bar.	Over 16Φ: 425	15	14	3Φ	Type 2
Cold worked, high-yield, ribbed bars	BS 4461	Mild steel billets are hot rolled to a round ribbed section and its strength is then increased by twisting.	Up to 16Φ: 460	10	12	3Φ	Type 2
		Identification note: the longitudinal rib spirals along the length of the bar	Over 16Φ: 425	10	14	3Φ	Type 2
Cold worked, high-yield, square twisted bars	BS 4461	Mild steel billets are hot rolled to a plain square or chamfered square section and its strength is then increased by twisting.	Up to 16Φ: 460	10	12	3Φ	Type 1
			Over 16Φ: 425	10	14	3Φ	Type 1
Hard drawn, high-yield, plain wire	BS 4482	Mild steel billets are hot rolled to a plain round section and its strength is then increased by cold drawing to a reduced section size.	485	10	n/a	3Φ	n/a
Hard drawn, high-yield, indented wire	BS 4482	Plain hard drawn wire, produced as above, is then passed through indenting rollers	485	10	n/a	3Φ	Type 1

British Standards for reinforcement

The relevant British Standards for reinforcement are:

BS 4449 *Hot rolled steel bars for the reinforcement of concrete*
BS 4461 *Cold worked steel bars for the reinforcement of concrete*
BS 4466 *Bending dimensions and scheduling of bars for the reinforcement of concrete*
BS 4482 *Hard drawn mild steel wire for the reinforcement of concrete*
BS 4483 *Steel fabric for the reinforcement of concrete*

Until these standards are all revised to include clauses on bond classification, reference may be necessary to CP 110, *The structural use of concrete*, Appendix E.

The principal requirements of these standards are summarised in Table 28.1; a more detailed consideration of the significant material properties is given below.

Other reinforcement

From time to time other materials are used for the reinforcement of concrete. Higher-strength steels or stainless steels may be used when specialised applications favour their adoption and, although these materials are not covered by British Standards, they sometimes have the benefit of an Agrément Board Certificate for their use in particular applications.

In unusual circumstances even old railway track, structural steel sections, expanded metal, wire rope, etc. have been pressed into service as reinforcement.

Consideration of these materials is not within our scope, and attention is subsequently confined to the range of British Standard reinforcement. Prestressing wires and strand, together with steel or other fibres for use as reinforcement, are also excluded.

PROPERTIES

Characteristic strength

The characteristic strength of reinforcement is defined in terms of its yield stress. Typical stress–strain curves are shown in *Figure 28.1;* for mild and high-yield hot rolled steel complying with BS 4449 the yield stress corresponds with the visible yield point on their curves. Cold worked reinforcement complying with either BS 4461 or BS 4482 has no visible yield point, and its yield stress is defined as the 0.2 per cent proof stress (i.e. the point where a line from 0.2 per cent strain, drawn parallel to the start of the stress-strain diagram, intersects the curve).

Reinforcement suppliers are required to take tensile test samples at a specified frequency for routine testing and to maintain the test records for inspection. In these tests not more than two results in every forty may fall below the specified characteristic strength, and none may fall more than 7 per cent below the specified value.

Ultimate strength and margin

The ultimate strength is the maximum strength shown by the tensile test before the test piece fractures. British Standards do not specify a value for the ultimate strength, but require a specified margin between the actual yield stress measured in the tensile

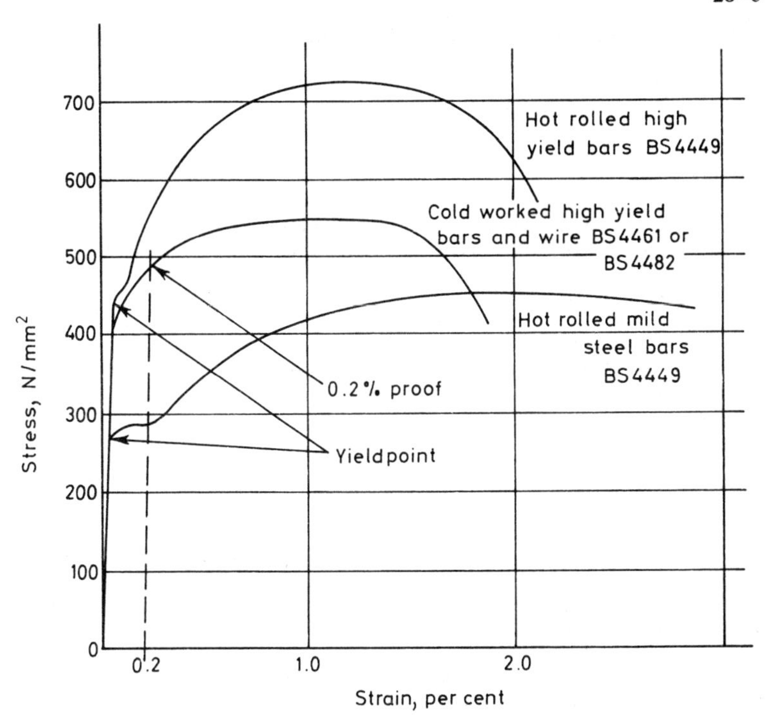

Figure 28.1 Typical stress–strain curves for reinforcement

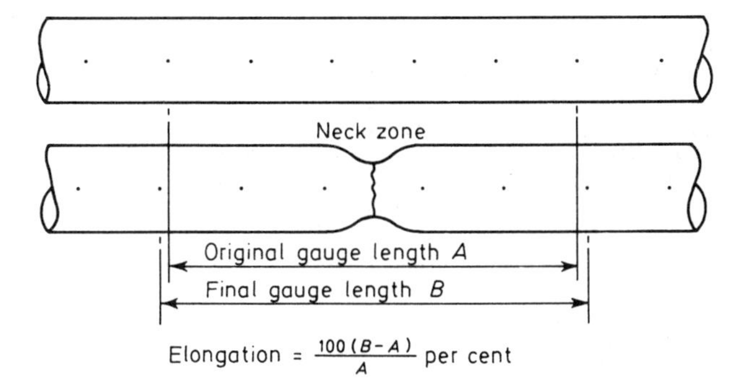

Figure 28.2 Measurement of elongation

test and the ultimate stress measured. The specified margin differs in each Standard but ranges from 5 to 15 per cent.

The margin is considered to be an important measure of ductility, and too small a margin could indicate that the steel is brittle.

Elongation

The elongation of tensile test specimens is measured over a gauge length of approximately five times the diameter. The elongation is expressed as a percentage; see *Figure 28.2*.

The elongation is not uniform and it will be fairly obvious that more strain has occurred in the neck zone than elsewhere. The British Standard test therefore measures the average elongation over the chosen gauge length. It is important to appreciate this relationship when considering non-British materials. If the elongation is measured over a longer gauge length, the influence of the neck zone is reduced and the average elongation appears a lower percentage figure.

Bending tests

The reinforcement should possess adequate ductility to withstand the bend test and the re-bend test without fracture. The re-bend is the more severe of the two tests.

When a bar is bent it is subjected to a mild form of cold working. A consequence of this is a locally higher strength, a reduction of the margin between yield and ultimate strengths, and a slightly more brittle steel. The re-bending of this steel is a very severe trial of the material.

Both the bend test and the re-bend test are carried out using bending formers half the diameter of those recommended for normal bending in accordance with BS 4466, and provide some guarantee against site and production failures. It is not an absolute guarantee against bending failures. The laboratory tests are at room temperature, using a smoothly powered bender and correctly sized formers. Site bending may be at very low temperatures, around a scaffold pole or other improvised former, and may involve the re-bending of bars that have been accidentally or deliberately misplaced.

Ductility

Reference has been made to the margin between yield and ultimate strengths, to elongation and to the bending tests. All of these in different ways measure the ductility of the steel; as a property of reinforcement, adequate ductility is as important as adequate strength.

The opposite of a ductile material is a brittle one like concrete or china. The type of failure that occurs in a brittle material is sudden, complete and may occur without any prior warning. This is exactly the type of failure that the design philosophy of reinforced concrete seeks to avoid. It is important that, when overloaded, a reinforced concrete member should show its distress with excessive deflection and extensive cracking, thus providing adequate warning of its likely failure.

Bond characteristics

In reinforced concrete construction the reinforcement must provide the tensile strength that the concrete lacks. The steel must remain bonded within the concrete, carry all the tensile force across the cracks that occur, and prevent the cracks growing wider. It is the composite behaviour of the two materials that is all-important. It is easy to imagine the situation in a beam if the reinforcement rested loosely in a preformed duct: the first concrete crack would merely get wider and wider as the reinforcement slid along the duct to accommodate the movement (*Figure 28.3*).

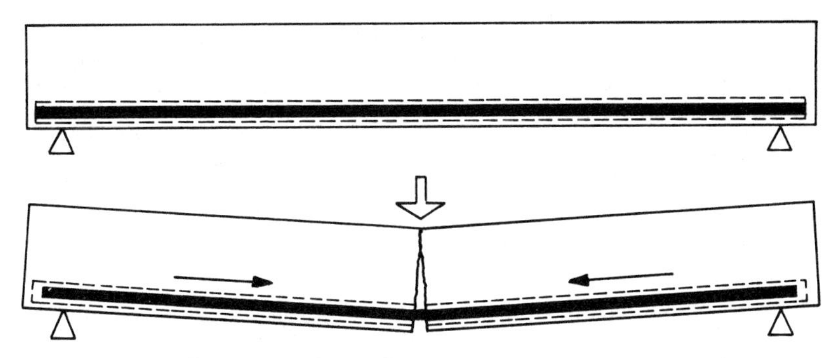

Figure 28.3 Result of lack of bond between steel and concrete

Bond between the steel and the concrete is therefore an essential characteristic of reinforced concrete, and every effort is taken to ensure that the bond is not impaired. Thus specifications stress the need to remove loose rust, guard against mould oil on the bars, etc.

Bond is achieved in two ways. The first is simple adhesion and the second is mechanical bond, which becomes the more important when the simple adhesion bond breaks down. Plain surface reinforcement relies primarily upon adhesion bond, supplemented by the mechanical anchorage provide by hooks and bends when necessary.

With deformed bars the surface geometry of the bar is designed to provide a mechanical bond with the concrete that will continue after the adhesion bond has failed.

Until recently all deformed bars were uniformly recognised by design code CP 114 to have an allowable bond stress 40 per cent greater than that for plain bars. Further investigations into bond have shown that deformed bars with a ribbed surface are significantly better than square twisted bars, and are approximately 80 per cent better than plain bars. This fact was recognised by the new design code CP 110 and is being incorporated in the reinforcement standards as they become due for revision. CP 110 introduced two categories of high bond bar:

- *High Bond, Type 1*, which covers square twisted bars and indented wires;
- *High Bond, Type 2*, which covers ribbed bars.

Some recent advertising has used the terms 'medium bond' for Type 1 and 'high bond' for Type 2. This is incorrect, and it is important that Type 2 be specified when it is required, since the specification of High Bond does allow the supply of Type 1 material.

It should be noted too that bond characteristic relates only to the surface geometry of the bar and has nothing to do with the steel strength. Thus if ribbed mild steel bars became available in the UK they would be High Bond Type 2, but would remain mild steel bars with the same low strength as the ordinary plain mild steel bars.

Weldability

Basically all steels can be welded if the appropriate techniques are used, but some techniques are too specialised, or require such a high level of skill and supervision that they are unsuitable for normal site application. Reinforcement classified as 'weldable' is therefore limited to those steels that allow normally skilled welders, following the supplier's recommended procedures, to achieve sound welds without the use of specialised techniques.

In general terms, hot rolled mild steel complying with BS 4449, cold worked bars complying with BS 4461 and steel fabric complying with BS 4483 will automatically fall into the 'weldable' classification. Hot rolled high-yield bars complying with BS 4449 are the only category of reinforcement covered by the British Standards where there is the possibility of a weldability problem. In all cases where 'weldability' is an essential requirement the customer must specify this as an additional 'special' requirement at the time of placing his order, since it is not a requirement that suppliers are obliged to meet to satisfy the standards.

Other properties

The reinforcement properties considered above are all referred to within the British Standards and cover the properties relevant for the vast majority of structural applications. Occasionally the designer of a specialised structure may find the need for additional properties relating to performance under extreme temperatures, dynamic loading, aggressive environments, etc., and it is important that the extent of the standards is properly appreciated so that any such additional requirements are fully specified when seeking tenders and placing orders.

BAR REINFORCEMENT

Size range

Since metrication in 1969 the term 'size' denotes the diameter in millimetres of a circle having the same cross-sectional area as the reinforcement. Obviously for plain round mild steel bars the actual diameter will correspond with the nominal size, but

Table 28.2 REINFORCEMENT SIZE RANGE

Bar sizes: BS 4449 and BS 4461	*Wire sizes: BS 4482*	*Cross-sectional area (mm²)*	*Nominal mass (kg/m)*
–	5	19.6	0.154
6*	6	28.3	0.222
–	7	38.5	0.302
8	8	50.3	0.395
–	9		
10	10	78.5	0.616
12	12	113.1	0.888
16	–	201.1	1.579
20	–	314.2	2.466
25	–	490.9	3.854
32	–	804.2	6.313
40	–	1256.6	9.864
50*	–	1963.5	15.413

* Bar sizes 6 and 50 are not preferred sizes and may not be readily available.

GKN Reinforcements Ltd

Metric bar schedule

Ref. 123 01 ☐

Date 12 03 78

T 45 M

Job ref. A.B.C. Housing

Member	Bar mark	Type and size	No. of mbrs	No. in each	Total no.	Length of each bar mm †	Shape code	A mm *	B mm *	C mm *	D mm *	E/r mm *
	01	Y10	1	164	164	1200	54	300	600			
	02	Y8	1	35	35	4000	20					

All bending dimensions are in accordance with BS 4466   * specified to the nearest 5 mm   † specified to the nearest 25 mm

Figure 28.4 Typical bar schedule

PREFERRED SHAPES

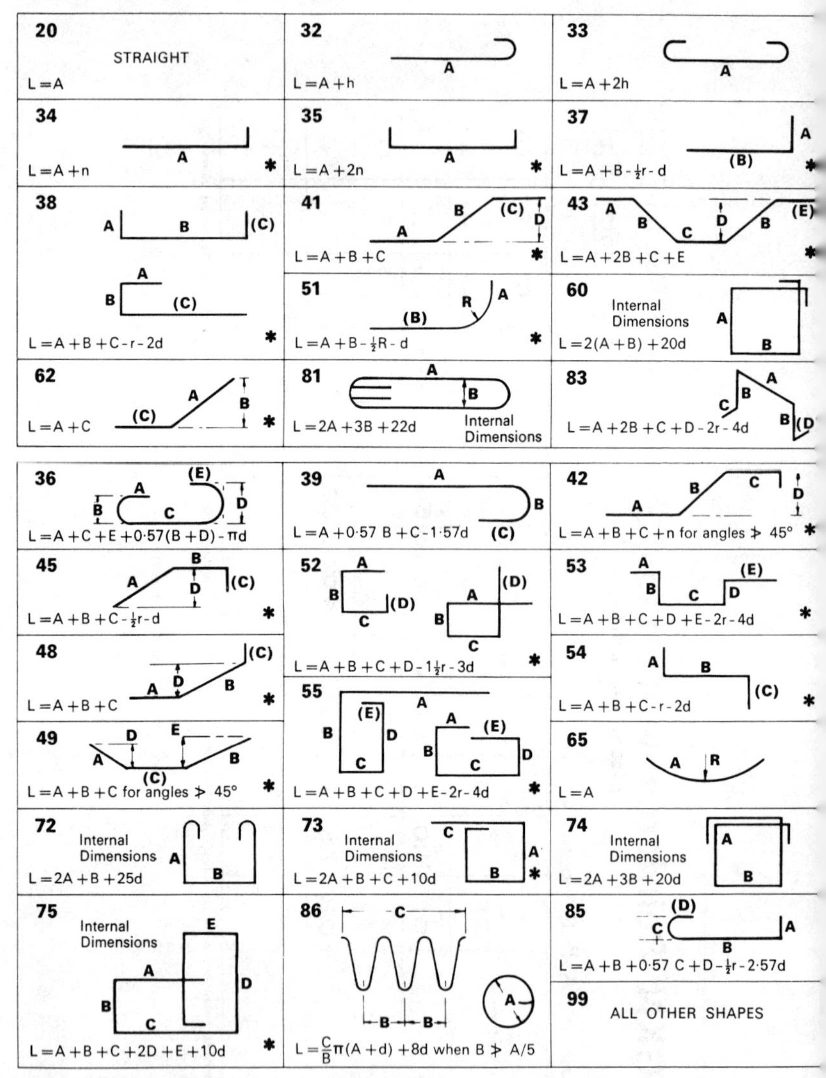

*For these shapes in particular the effect of a positive cutting tolerance increasing the actual length of the "free" leg o legs by up to 25 mm should be considered.

Bracketted dimensions are run-out dimensions and may be omitted from the final schedule.

For shape codes 60, 81, 72, 73, 74 and 75 internal dimensions, **I.D.**, are assumed and for all other shape codes outside dimensions, **O.D.**, are assumed unless otherwise marked.

Figure 28.5 Bar shape codes and dimensions (courtesy GKN Reinforcements Ltd)

ribbed bars or square twisted bars can have a maximum dimension up to 10 per cent greater than their nominal sizes.

The preferred size ranges included in the British Standards are shown in Table 28.2, together with the first choices in non-preferred sizes, but it is important to check the likely supply situation before making extensive use of any non-preferred size.

Cutting and bending

The cutting and bending of reinforcement is covered by BS 4466, which not only specifies methods of scheduling one's requirements but also specifies the quality of workmanship in terms of dimensional tolerances.

The headings of a typical bar schedule are shown in *Figure 28.4;* these are designed for use in conjunction with the range of shape codes and related dimensions shown in *Figure 28.5*. The formulae given allow calculation of the required cut length for each shape code, where the *n* and *h* values for standard hooks and bends should be not less than the values given in Table 28.3. CP 110 recommends that when calculating these schedule dimensions, for reinforcement between two concrete faces, the nominal dimensions should be reduced by the deductions shown in Table 28.4 to allow for constructional variations.

It is always assumed that these deductions have been made during the preparation of the schedules. The actual reinforcement dimensions should lie within the plus/minus tolerance range allowed by BS 4466 and shown in Table 28.4.

Most shape codes leave one dimension unspecified to serve as a run-out. In some cases, particularly for those shape codes marked with an asterisk in *Figure 28.5*, consideration should be given to the possible variation of the actual run-out due to the cumulative effect of individual tolerances, as illustrated by the following example.

EXAMPLE: NORMAL TOLERANCES

	A	*B*	*C*	*D*	*E*	*Cut length*
Nominal dimensions	200	1200	350	850	200	*
CP 110 deductions	−10	−15	−10	−10	−10	*
Scheduled dimensions	190	1185	340	840	*	2650†
BS 4466 tolerances	± 5	+ 5 −10	± 5	± 5	*	± 25
Possible range of dimensions: Min.	185	1175	335	835	(150)‡	2625
Max.	195	1190	345	845	(245)	2675

* Assumes that E will be the run-out dimension and omits it from the schedule.
† Assumes the cut length is rounded to a multiple of 25 mm.
‡ Note that the run-out accommodates the cumulative variances and may vary significantly from the required dimension.

Table 28.3 MINIMUM HOOK AND BEND ALLOWANCES

For mild steel bars complying with BS 4449

Bar size, d (mm)	6	8	10	12	16	20	25	32	40
Hook allowance, h	100	100	100	110	150	180	230	290	360
Bend allowance, n	100	100	100	100	100	100	130	160	200

For high-yield bars complying with BS 4449 or BS 4461

Bar size, d (mm)	6	8	10	12	16	20	25	32	40
Hook allowance, h	100	100	110	140	180	220	280	360	440
Bend allowance, n	100	100	100	100	100	110	140	180	220

Note. These dimensions are nominal minima but may vary due to cutting tolerances.

Table 28.4 DEDUCTIONS AND TOLERANCES

	Not greater than 1.0 m	*Between* 1.0 m *and* 2.0 m	*Greater than* 2.0 m
CP 110: required deductions from nominal dimensions	– 10 mm	– 15 mm	– 20 mm
BS 4466: allowed tolerances on cutting	± 25 mm	± 25 mm	± 25 mm
allowed tolerances on bending dimensions	± 5 mm	+ 5 mm – 10 mm	+ 5 mm – 25 mm

Table 28.5 BARS BENT TO A RADIUS

Bar size (mm)	6	8	10	12	16	20	25	32	40
Critical radius (m)	2.5	2.75	3.5	4.25	7.5	14	30	43	58

Shape code 65 referring to radius reinforcement need only be specified when the required radius is less than the critical values shown in Table 28.5. When the required radius is equal to or greater than the critical value, straight bars are used and pulled to shape during fixing.

FABRIC REINFORCEMENT

Fabric reinforcement complying with BS 4483 is produced from reinforcement arranged in a square or rectangular mesh and resistance welded at each intersection to produce a rigid sheet. The reinforcement used may be cold worked bars complying with BS 4461 or, more commonly, hard drawn wire complying with BS 4482. With the possible exception of wrapping fabric, which may use steel with a lower characteristic strength, all fabric is therefore high-yield reinforcement.

Hard drawn wire may be plain surfaced or indented to provide improved bond performance, but in most cases the mechanical anchorage provided by the welded cross-wires is of greater practical significance than the bonding of the individual wires. The range of preferred sizes for hard drawn wire is shown in Table 28.2.

BS 4483 includes the designated range of meshes listed in Table 28.6 which are generally available in flat, stock-size sheets 4.8 m long and 2.4 m wide.

Table 28.6 DESIGNATED RANGE OF STOCK FABRIC

Fabric reference	*Longitudinal wires*			*Cross-wires*			Nominal mass
	Size	*Spacing (mm)*	*Area (mm^2/m)*	*Size*	*Spacing (mm)*	*Area (mm^2/m)*	*(kg/m^2)*
A 98	5	200	98	5	200	98	1.54
A 142	6	200	142	6	200	142	2.22
A 193	7	200	193	7	200	193	3.02
A 252	8	200	252	8	200	252	3.95
A 393	10	200	393	10	200	393	6.16
B 196	5	100	196	7	200	193	3.05
B 283	6	100	283	7	200	193	3.73
B 385	7	100	385	7	200	193	4.53
B 503	8	100	503	8	200	252	5.93
B 785	10	100	785	8	200	252	8.14
B1131	12	100	1131	8	200	252	10.9
C 283	6	100	283	5	400	49	2.61
C 385	7	100	385	5	400	49	3.41
C 503	8	100	503	5	400	49	4.34
C 636	9	100	636	6	400	70.8	5.55
C 785	10	100	785	6	400	70.8	6.72
D 49	2.5	100	49	2.5	100	49.1	0.77
D 98	5	200	98	5	200	98	1.54

For the appropriate application the benefits offered by fabric lie in the reduced level of steel-fixing skills required, faster steel fixing, increased accuracy and easier checking. The designated range of sheets offers early availability ex-stock, but at the cost of on-site cutting and increased waste when the stock size is not a multiple of

Schedule

Designer: A. N. Other

Contract: A.B.C. Housing

Schedule reference: 1 23 01

Date: 12 03 78

Job number: 77111

No. of sheets	Length	Width	GK Fabrik reference or sheet details						Special details and/or bending dimensions							Mark No.
			No. of wires	Size	Pitch	Length	Overhangs O_1 O_3	O_2 O_4	BS4466 shape code	Bending instruction	A	B	C	D	E/R	
	metres	metres		mm	mm	mm	mm	mm			mm	mm	mm	mm	mm	
19	1.2	2.4		C 283					54		300	600				01
16	4.8	0.6		A 193					20							02
11	4.8	2.4		A 193					20	Stock sheets						03

Figure 28.6 Typical fabric schedule (courtesy GKN Reinforcements Ltd)

the structural dimensions. Sheets in lengths made to order can provide greater economy by eliminating off-cut wastage and reducing the frequency of laps. Made-to-order sheets are readily available, but are not ex-stock and may be specified on a schedule; see *Figure 28.6*.

Cutting and bending

Although BS 4466 refers only to the scheduling of bar reinforcement, its requirements should also be observed when scheduling fabric reinforcement in other than flat stock-size sheets.

The typical schedule shown in *Figure 28.6* is very similar to the standard bar schedule, and allows for the specification of flat sheets in made-to-order sizes and for the specification of bent sheets using the usual bending shape codes. Clearly certain shape codes, such as chairs 83 and spirals 86, are not relevant for fabric; others such as 55 and 75 may only be used when the cross-wires are carefully located to avoid clashing on the final bend, forming a closed shape. In addition the dimensions should avoid bends in a weld zone, and the direction of bending needs to be indicated to show which side has the cross-wires. The example in *Figure 28.6* shows the scheduling of strips and bent sheets for a typical semi-raft foundation of a house.

DETAILING

There is no British Standard for the method of detailing reinforced concrete drawings, but the industry took a major step forward with the publication of a joint report by the Concrete Society and the Institution of Structural Engineers entitled *Standard method of detailing reinforced concrete*. In addition to guiding detailers on the choice of drawing scales, line thicknesses, etc., the report also codified the notation for describing bars that is now generally adopted.

Number of bars	*Steel type and size*	*Mark number*	*Spacing and position*	*Notation*
4	Y20	01	Bottom	4Y20–01B
52	R8	02	150 mm centres in near face	52R802–150NF

The abbreviations for steel type are:

Y for high-yield bars; either hot rolled high-yield bars complying with BS 4449 or cold worked high-yield bars complying with BS 4461. It is assumed that only one type of high-yield bar will be used in the job; which one is used may be determined by the contractor with freedom of choice or may be determined by the contract specification.
R for hot rolled mild steel bars complying with BS 4449.
X for any other grade of steel to be used as reinforcement and specified in the contract documents.

Generally reinforcement spacing dimensions are only a guide to ensure that the correct number of bars are more or less uniformly distributed throughout the specified length; the actual dimensions of individual spaces are not really critical. Commonly, therefore, the steelfixer would measure the location of perhaps every fourth bar and place the intermediates by halving and quartering the gap by eye. In detailing, spacings are stated in millimetres and should preferably be a multiple of 25 mm for ease of setting out.

Abbreviations used in the joint report to describe the location of reinforcement are:

EW	each way	NF	near face
EF	each face	B	bottom
FF	far face	T	top

Any other abbreviations used should be defined on the drawing on which they occur. Some that are fairly common, but not necessarily recommended, are:

OF	outer face	AS	alternate bars staggered
IF	inner face	AR	alternate bars reversed
EE	each end	ASR	alternate bars staggered and reversed
LKS	links	AW	alternate with
ER	each rib	ES	equally spaced
ALT	alternate	LL	lower layer
STG	staggered	UL	upper layer

Reinforcement detail drawings show the concrete outlines within which the steel should be placed, in the form of a plan (for slabs, rafts and foundations) or an elevation (for walls, beams and columns). Whenever possible the detailer will group together members having similar reinforcement: this is highly desirable both to reduce the detailer's work and to reduce the volume of data that is communicated to the site. Seeking similarity and repetition is an important detailing skill, the essence of which is to convey the maximum information in a simple, clear and unambiguous manner.

The plan or elevation drawing should contain all the descriptive data concerning the reinforcement, identifying the number, type, size, mark and location. Further information on the location of the reinforcement, and showing how the steel fits together, may be contained on appropriately selected sections, where the reinforcement is identified only by its mark number so that the reinforcement is fully described just once on each drawing. See *Figure 28.7(a)*.

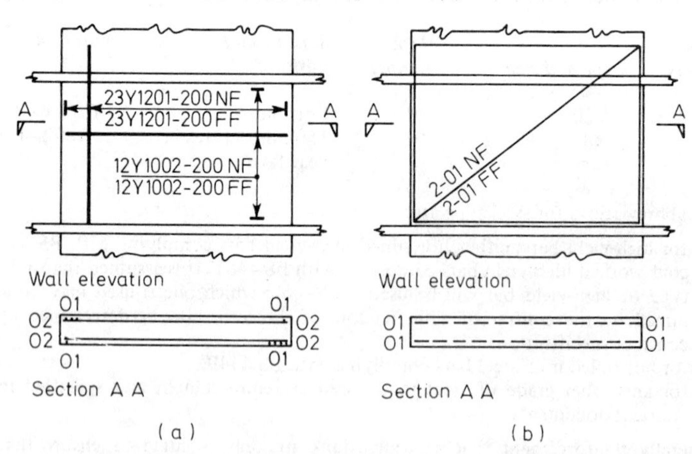

Figure 28.7 Detailing notation: (a) bar reinforcement, (b) fabric reinforcement

The detailing of fabric reinforcement follows the same convention except that the details tend to be simpler. The sheets are located and described on the plan or elevation view, and further detailed on sections where the sheets are normally shown as a broken line. See *Figure 28.7(b)*.

Economic pressures have encouraged the development of alternative methods of communication between the designer and the site, leading to efforts at reducing the detailing skills required. The 'standard' drawing is one well known technique and, with the addition of computers, lineprinters and plotters, techniques are being further refined. Most techniques rely upon some degree of standardisation and the use of not-to-scale drawings, but the ability of computers to turn out typescript and schedules at a great rate is employed to reduce the quantity of line-work drawing.

It is impossible to review the range of 'automated' methods available. They start with methods that mechanically duplicate the traditional, progressing to the methods of those who argue that the problem is merely one of communication and, if a requirement can be clearly described in words, a drawing is unnecessary. The important thing, when encountering any of these techniques for the first time, is to receive a full briefing from the originators: never attempt to struggle through under one's own efforts.

ORDERING

When ordering reinforcement that complies with a British Standard, the requirements to be specified are:

- the number of the British Standard,
- the material type,
- any additional requirements relating to material properties,
- the bar sizes or fabric references,
- the quantities and dimensions.

In most cases the ordering procedure does not follow the recommended practice. Commonly an enquiry for a quotation is accompanied with photocopied extracts of the contract specification and bill of quantities, to be followed later, when the order is placed, by copies of the bar and fabric schedules. In some ways this simplified procedure has the merit of ensuring that the supplier's information exactly corresponds with the specification, but occasionally the system breaks down when the client has additional requirements that are embodied in another part of the specification or in the contract drawings.

FIXING

The range of activities included within the steel fixer's duties can vary enormously from job to job. In the extreme the steel fixer may be responsible for:

- producing schedules from the detailed drawings,
- checking the accuracy of schedules,
- editing the schedules into batches for production or delivery in phase with the construction programme,
- checking material deliveries,
- unloading, sorting and stocking delivered material,
- sorting, bundling and transport of material from stock,
- on-site cutting and bending of bars and fabric,
- prefabricating reinforcement assemblies,
- hoisting reinforcement to shutter,
- placing, marking out and tying reinforcement in position,
- blocking out and fixing spacers,
- final check and post-inspection rectification.

Producing schedules from a detailed drawing is an activity that should be undertaken by the detailer producing the drawing, and that should be checked by the designer with overall responsibility for the project. Anthing less than this is a very poor second. The production of schedules from a strange drawing often calls for a high degree of design appreciation; except for the simplest of structures, the task is better sublet to specialists.

Checking the accuracy of schedules is a task that most specifications assign to the contractor, and in many ways it is a most unfair imposition. Thorough checking calls for the same skills as those required to prepare the schedule, and failure to detect an error could result in expense to the contractor without any redress against those who originated the error. In general the letter of the law is rarely observed; the schedule producer tends to accept responsibility for his own errors, but will invoke the law to ensure that the contractor cannot entertain the idea of claiming for consequential losses.

Editing schedules into production batches, with the appropriate lead-time to phase with the construction programme, is clearly an important activity. It requires close liaison with the material suppliers to prevent hold-ups due to shortage, inefficient working due to excess stocks on site, or loss of economy due to small deliveries attracting a transport surcharge. Also, if the editing is carried out in conjunction with the detail drawings, it may provide the skilled steelfixer with an opportunity to modify some of the detailing for increased economy. Commonly, when the detailer has scheduled shorter bars with laps, the steel fixer can see the opportunity for replacing two bars with one longer bar and saving the lap length.

Checking material deliveries is an important activity. Basically, it aims to establish that the reinforcement is in acceptable condition and clearly labelled, and that the correct number of bundles are delivered in accordance with both the delivery notes and the schedule. Too often the checking is limited to a comparison of the material with the supplier's own delivery documentation, and the possibility of the documents not being in accordance with requirements is overlooked until a material shortage becomes critical.

Sorting the delivered material into a well-ordered stock area is the hallmark of a skilled steel fixing operation. It is true, even on the most congested of sites where sorting space is at a premium, that the care taken in properly organising stocks is always compensated by the subsequent time savings. Unfortunately at least once something will take a higher priority at the time of unloading and, with even one load dumped at random, it is very difficult to restore a previously well organised system. An important key to any system is the quality of labelling on the bundles. BS 4466 specifies the minimum content that should appear on the label: *Figure 28.8*. The

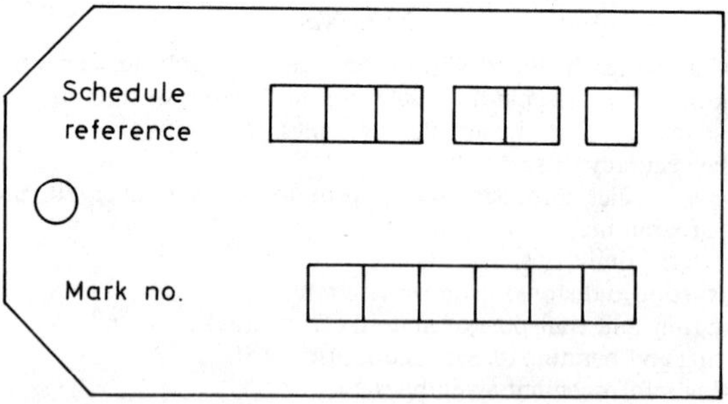

Figure 28.8 Reinforcement label: minimum data

first three figures of the schedule reference are the detail drawing number, and are the most important aid to identification. The second two digits number the schedule pages, often starting from 01 for each drawing, and the reinforcement mark number often starts from 01 on each drawing. Several Mark 01s on Schedule 01s may occur on the job; so, although all the label information is vital, the key identifier is usually the drawing number.

The high scrap levels usually associated with stock lengths for on-site cutting and bending make it unpopular except for very large civil engineering contracts, where the flexibility provided by the on-site facility may outweight its cost. Most building contracts receive their bar reinforcement already cut and bent in accordance with the schedules, although many persist in ordering their fabric reinforcement in stock sheets.

The skilled steel fixer will seek to prefabricate as much of the reinforcement as practical. The advantages of prefabrication are that part of the fixing work can be carried out before or during the period that the formwork is being prepared, giving continuity of work and reducing the more critical fixing time on the formwork to allow earlier concreting. The prefabrication can usually be undertaken in more favourable working conditions and, by spreading the work load, fewer steel fixers may be able to cope. The emphasis, however, must be on the word *practical*. It is customary nowadays to see column cages being prefabricated and hoisted into position as completed assemblies. The tying of the steel in such cages must be even more secure to withstand handling than would be strictly necessary to hold the reinforcement in position during concreting, and a crane must be available. The weight of the cages is one limitation on the practicality of prefabrication. Others are the size of the unit, the stiffness of the assembly, and the availability of hoist or crane capacity. Frequently with bar assemblies the stiffness of the unit is a major limitation, and additional temporary bracing bars are needed to prevent distortion during handling.

Welded fabric is an example of factory-prefabricated reinforcement. The designated range of meshes in BS 4483, using sizes in the range 5 to 12, was developed to eliminate the bar-by-bar assembly of reinforcement grids for a range of common structural elements.

Considerable scope exists for the greater use of welded fabric, particularly when detailers appreciate the availability of a wider range of meshes, fabric in made-to-order sheet sizes, and the possibility of bending sheets in accordance with BS 4466 to most of the available shape codes.

Fabric is inherently stiff, so prefabricated assemblies incorporating bent sheets reduce the need for ties and bracing, speed up prefabrication, and (more important) extend the practical scope for prefabrication into the critical areas where bar-by-bar assembly on the formwork is the only practical alternative.

The contractor is specifically responsible for the correct location of the reinforcement and for securely holding it in position during the placing and compaction of the concrete. Spacers of the correct sizes, in the form of either cement blocks or proprietary plastic spacers, are obviously an essential requirement, but so too are chairs and ties.

Chairs are required to correctly locate the top layers of reinforcement in slabs, and may be one of the proprietary ranges available or may be produced from reinforcement bars. However, chairs are not usually detailed on the reinforcement drawings or included on the schedules or measured as part of the reinforcement content of the job. Even when they are indicated on drawings or specified as being required at a certain frequency, it remains the contractor's responsibility to securely locate the reinforcement, and to provide whatever number of chairs is required for this purpose. The Standard Method of Measurement lists chairs as special spacers to be measured separately, but this rarely occurs.

The tying of reinforcement may be undertaken with proprietary clips and ties or with fixing guns, but the usual material is 16 gauge soft iron wire. Again, tying wire is not measured with the reinforcement (it is deemed to be included), but as a rough rule of thumb it is usual to allow about 10 kg per tonne of reinforcement. The usual soft iron wire is suitable for normal reinforcement, but should not be used with galvanised steel or stainless steel reinforcement.

The overall organisation of a well-ordered steel fixing activity that phases into the construction programmes, and uses such common services as transport and craneage without disrupting other trade operations, requires from the steel fixer more than mere skill in the tying together of reinforcement.

29 THEORY AND DESIGN OF STRUCTURES

INTRODUCTION 29–2

LOADING 29–2

METHODS USED IN DESIGN OF STRUCTURAL ELEMENTS 29–3

THEORY OF STRUCTURES 29–4

DESIGN OF STRUCTURES 29–12

THEORY AND DESIGN OF STRUCTURES 29

29 THEORY AND DESIGN OF STRUCTURES

F.J. McCANN, CEng, FIStructE, MICE, MIMinE, MSocCE(France)
S.SMITH, CEng, MIStructE, MInstW
Partners, McCann and Smith, Consulting Engineers

This section explains the types of loading that occur in buildings, and the effects that loads produce in structural elements; the design of simple beams and retaining walls is demonstrated. Readers unfamiliar with mathematics will find the reasoning easy to follow.

INTRODUCTION

Theory of structures is concerned with the methods by which the effects produced in a structure by the applied loading can be analysed and quantified, and from which actual stress conditions can be calculated. *Design of structures* is the process by which a suitable and economical arrangement of structural members is obtained which are of a sectional size adequate to ensure that the permissible conditions (specified in the relevant Codes of Practice and British Standards) are not exceeded.

It is the structural function of brick and masonry house walls to support roof, floors and lintols, and to transfer to the foundations (and hence to the sub-soil) all the load to which these elements of the construction are subjected. A minority of houses are framed in steel or timber, or sometimes reinforced concrete. In such cases the walls function mainly as a cladding agent, and all loading is applied through the framing to the foundations.

The most common structural element used in house construction is the member that principally resists bending. This is generally called a *beam*, and can take the form of purlin, floor joist, lintol, or reinforced concrete flat roof or floor constructed in precast units or as an *in-situ* slab.

Outside the actual house construction it is the retaining wall that is of extreme importance. Whether this retains a fill of half a metre or several metres depth, it is essential that it remains stable under normal conditions of loading.

Both the above areas of construction are considered in this section. Consideration is given to the source and effect of applied loading, the current theoretical structural assumptions made regarding the behaviour of the items described above, and finally some typical design calculations.

LOADING

In house construction the loading applied to the various constructional elements includes:

- dead load,
- imposed load,
- wind load.

Dead load

This can generally be described as the self-weight of the various constructional elements, such as roof covering and its supports, floors, partitions and walls, etc.

Information concerning self-weight can normally be obtained either from manufacturers of the materials or from BS 648, *Schedule of weights of building materials.*

Imposed load

This is given in CP 3: Chapter V: Part 1 for a wide range of buildings and situations. It is the loading assumed to be produced by the intended occupancy or use, and includes snow loading. A great deal of work is still being undertaken in the assessment of such loading, including the statistical probability of specified values of loading being exceeded during a specific period of time in which the construction is expected to stand. No doubt further information on this subject will be available in the future.

It should be noted that Table 1 of CP 3: Chapter V: Part 1 contains two main columns, the first specifying the 'intensity of *distributed* load' and the second the '*concentrated* load to be applied over any square with a 300 mm side'. These are alternatives and must be considered, when provision for effective lateral distribution of load is not made, when designing floor slabs and beams and joists spaced not more than one metre apart. The alternative that produces the most critical condition should be used in the design of the structural element concerned.

With larger constructions, imposed loading can actually be reduced in multi-storey buildings when considering the design of a member such as a column or pier that supports a number of floors. Imposed loading can also be reduced in the case of a beam that supports an area of floor not less than 46 m^2.

Information concerning permitted reduction is given in Table 2 and clause 5.2 of CP 3: Chapter V: Part 1.

Wind load

Recommendations concerning the calculation of wind loading are given in CP 3: Chapter V: Part 2. This is a most comprehensive document and is based on the results of extensive investigations and study. It specifies wind velocities for the whole of the British Isles. Calculation of wind pressure takes into account local variations in the siting of the building relative to elevation and exposure, roughness of ground, height and protection from surrounding objects, and the statistical probability of extra-severe conditions arising during the expected life of the building.

Because of the comprehensive nature of CP 3: Chapter V: Part 2, some difficulty may be experienced in its application by those not normally accustomed to wind calculations. Under these circumstances it is suggested that the services of a structural engineer be obtained to make these calculations.

However, it is extremely important to realise two basic points. First, wind pressure is not uniformly applied over the areas of a building; the greatest concentration is around the edges. Secondly, most surfaces of a building, particularly roofs, are subjected to a *negative* pressure, i.e. suction; this is increased by conditions inside the building if air is allowed to freely enter and leave through door and window openings or through joints in the cladding. Because of this, great care should be exercised in securing roof cladding, particularly at verges, and in anchoring flat and lightweight roofs to the supporting structure.

METHODS USED IN DESIGN OF STRUCTURAL ELEMENTS

With few exceptions, the individual structural elements in a house will be constructed in timber, steel or reinforced concrete. Generally these take the form of purlins, rafters, roof frames, floor joists, beams, lintols and slabs. The basic principle used in the design of structural members of timber, steel and reinforced concrete is in course of change. Briefly, at present when designing in timber and steel, sectional sizes of members are calculated so that the appropriate values of permissible stress specified in CP 112 and BS 449, respectively, are not exceeded.

Permissible stresses can be regarded as 'safe stresses' that incorporate a 'factor of safety' against ultimate failure and are within the elastic range. Therefore the fundamental on which the design principle is based is that, when the normal expected loadings are applied to the structural element in question, the appropriate permissible stress will not be exceeded.

This used to be the case also when designing in reinforced concrete, and in fact still is if the obsolescent CP 114 is used. In 1972, however, a new Code of Practice (CP 110, *The structural use of concrete*) was published. The principle adopted in CP 110 is that of 'limit state', which is a relatively new term. Very briefly and simply, what is envisaged by limit state is that a member of reinforced concrete must be designed to satisfy certain specified conditions when functioning as part of a structure. These include strength, stability, deflection, cracking and vibration. Durability and fire resistance must also be provided for.

All design calculations are based on what are called 'design loads', and as far as strength calculations are concerned these are 'ultimate loads'. In the limit-state method, ultimate loads are obtained by taking the normal working loads described in the paragraph headed 'loading' and multiplying them by certain factors as follows:

Combination of dead and imposed loads	= 1.4 × dead + 1.6 × imposed
Combination of dead and wind loads	= 0.9 × dead + 1.4 × wind
Combination of dead and imposed and wind loads	= 1.2 × dead + 1.2 × imposed + 1.2 × wind

The loading that is applicable (or, if all three types of loading *can* apply, the one that produces the *critical* case) is used in the design of the member.

The reader must now be wondering why the actual loads (which, incidentally, in this method should be referred to as 'characteristic loads') are modified in this way and, in most cases, increased in value. The answer is that the stresses in the materials used, steel and concrete, are based on yield stress and crushing stress respectively, each being modified by a partial safety factor. It is therefore an 'ultimate' design method.

The reason behind the limit-state method of design is that it produces a more uniform strength in the building structure as a whole, and this leads to uniform safety and economy of the structural members.

In the near future a revised British Standard for the design of structural steelwork will be issued, and this will also be based on the above method. It is envisaged that ultimately structural design in all materials will be similarly carried out, and it is therefore a method to be learned and used by all concerned with design. The method is illustrated in the paragraph headed 'Reinforced concrete beam' (page 29–16).

THEORY OF STRUCTURES

Effects produced by loading applied to beams

REACTIONS

The purpose of a beam, whatever form it takes, is to support a given system of loading, and to transfer it by means of the rigid nature of the beam to points on the beam span where it can be supported, and eventually transferred to the foundations. The support given to the beam at these points is usually referred to as a *reaction.* That part of the structure giving the support must have sufficient stability to produce this reaction force without itself collapsing or moving by any appreciable amount.

BENDING

In order to function as required, the sectional dimensions of a beam must be adequate to resist the moments from the applied loads and forces. These are called *bending moments*. The bending moment at a given point on a beam span is determined by calculating the algebraic sum of moments of loads and forces to *one* side of that point.

SHEARING

The sectional dimensions of the beam must also be adequate to provide resistance against shear; in the case of reinforced concrete, special reinforcement may have to be introduced to do this. Shearing force at a given point on a beam span is determined by calculating the algebraic sum of loads and forces acting to *one* side of the point.

Note 'Algebraic' means that a positive or negative sign is allocated in order to signify direction.

EXAMPLE 1

For the beam shown in *Figure 29.1(a)*, determine the reaction force at each support, and draw diagrams that show the variation in the value of shearing force and bending moment across the length of the beam.

To determine the reaction at B take moments about A. Since moments balance, their algebraic sum will be zero.

$$20 \times 2 + 15 \times 4 + 10 \times 7 - R_B \times 8 = 0$$
$$8R_B = 40 + 60 + 70$$
$$\therefore R_B = 170/8 = 21.25 \text{ kN}$$

Since the algebraic sum of loads and forces applied to the beam must be zero, for equilibrium, then

$$20 + 15 + 10 - 21.25 - R_A = 0$$
$$\therefore R_A = 23.75 \text{ kN}$$

The shear-force diagram – *Figure 29.1(b)* – is drawn by plotting loads and forces according to their magnitude and direction.

Bending moment at C	$= 23.75 \times 2$	$= 47.50$ kN m
Bending moment at D	$= 23.75 \times 4 - 20 \times 2$	
	$= 95 - 40$	$= 55$ kN m
Bending moment at E	$= 21.25 \times 1$	$= 21.25$ kN m

All the bending moments will cause the beam to sag and therefore they are plotted

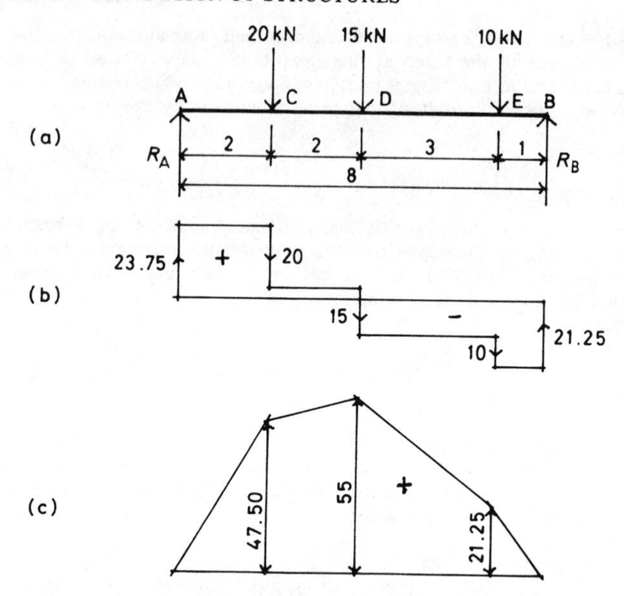

Figure 29.1 Example 1: (a) beam, (b) shear-force diagram, (c) bending moment diagram

on the same side of the base line: *Figure 29.1(c)*. Usually this is called *positive-type* bending.

Note The maximum bending moment always coincides with the point of zero shearing force.

EXAMPLE 2

Repeat Example 1 for the beam shown in *Figure 29.2(a)*.

The majority of loads in buildings (e.g. imposed loading on floors or roofs, self-weight of floors and roofs, brickwork supported by a beam, etc.) are not concentrated but distributed. The moment of a distributed load about a given point is calculated as the product of total load and distance to its centre of gravity.

To determine reaction at B, take moments about A.

$$(3 \times 5)\frac{5}{2} + 12 \times 10 - R_B \times 8 = 0$$

$$8R_B = 37.5 + 120$$

$$\therefore R_B = 19.69 \text{ kN}$$

$$\text{Also } (3 \times 5) + 12 - 19.69 - R_A = 0$$

$$\therefore R_A = 15 + 12 - 19.69 = 7.31 \text{ kN}$$

Before calculating bending moments in order to draw the bending-moment diagram, the position on the beam span where maximum bending moment occurs must be located, as follows.

Let the position of maximum bending moment (which is also the position of zero shear force) be distance x from A. At this position,

$$\text{Shear force} = 7.31 - 3x = 0$$
$$3x = 7.31$$
$$x = 7.31/3 = 2.44\ \text{m}$$

$$\text{Maximum bending moment} = 7.31 \times 2.44 - (3 \times 2.44)\frac{2.44}{2}$$
$$= 17.84 - 8.93 = 8.91\ \text{kN m}$$
$$\text{Bending moment at C} = 12 \times 5 - 19.69 \times 3$$
$$= 60 - 59.07 = 0.93\ \text{kN m}$$
$$\text{Bending moment at B} = 12 \times 2 = 24\ \text{kN m}$$

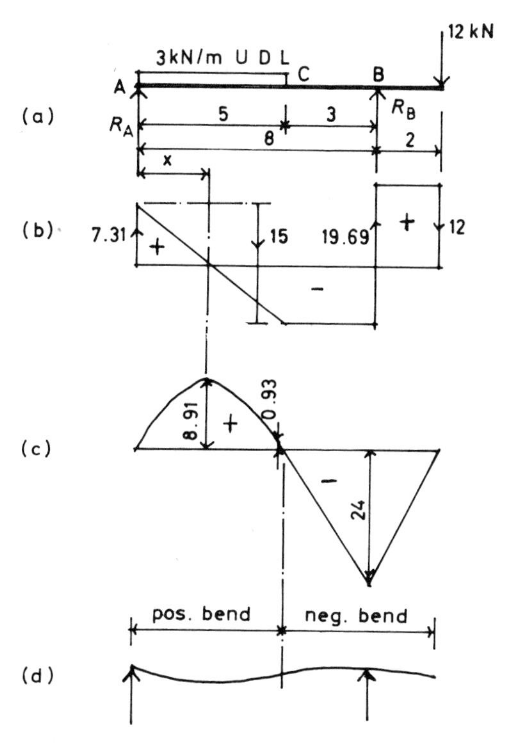

Figure 29.2 Example 2: (a) beam, (b) shear-force diagram, (c) bending-moment diagram, (d) curved profile of beam

The maximum bending moment will cause the beam to sag. The moment at B will cause it to hog, and this is called *negative-type* bending. These bending moments therefore must be plotted on different sides of the base line: *Figure 29.2(c)*. The profile of the bending moment diagram under a uniformly distributed load is parabolic. If it is required to plot the curve accurately, values of bending moment should be calculated at close intervals along the length of the load.

Stresses in beam sections

It can be seen from the above examples that bending moment and shearing force co-exist and that they are the direct cause of loading applied to a beam. It is now necessary to examine the stress from both bending and shear, and its distribution within the beam section.

STRESS DUE TO BENDING

It has been mentioned earlier that the design method applied to structural members of timber and steel is different from that applied to reinforced concrete. The former, based on permissible stresses, is called the *elastic method* and this will now be described.

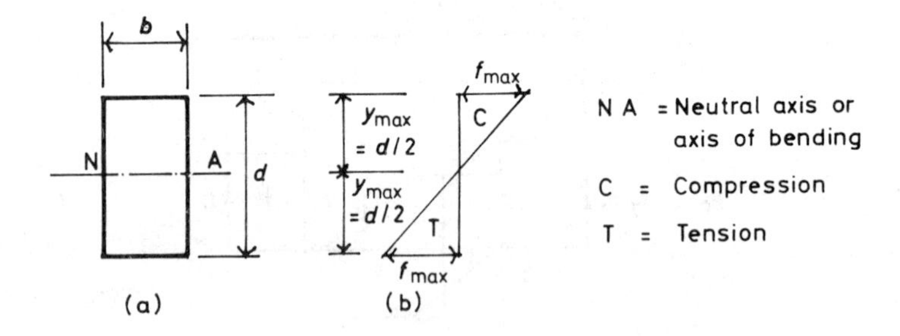

Figure 29.3 Bending stress: (a) beam section, (b) stress diagram

Without entering into a mathematical proof, it should be appreciated that when a beam is in positive bending the upper part is in compression and the lower part in tension. Maximum stress occurs at the extreme edge of the section, and the stress distribution is as shown in *Figure 29.3.*

The expression derived from the elastic theory is:

$$\frac{M}{I} = \frac{f}{y} = \frac{E}{R}$$

and the meaning of the terms is no doubt understood by most readers. If, however, the particular problem deals with maximum stress in a beam section (and this is usually the case) applied to either design or analysis, the above expression can be simplified to:

$$M = fZ$$

where M = bending moment
f = maximum stress at extreme edge of beam section
Z = section modulus, which in general terms $= \dfrac{I}{y}$

where I = second moment of area of beam section
and y = distance from centroid of beam section to outer edge

Formulae for Z of variously shaped sections can be obtained from standard information tables. For a rectangular section, which is the most common section for timber beams, Z about the major axis of bending = $bd^2/6$. The Z value for standard steel beam sections is listed in steelwork section tables as published by CONSTRADO.

STRESS DUE TO SHEAR

The effects of shearing force in a beam are not always fully appreciated, and therefore will be dealt with in some detail.

Consider a beam that supports any system of vertical downward loading as shown in *Figure 29.4*. If a particle 'X' is arbitrarily chosen within the depth of the beam, a

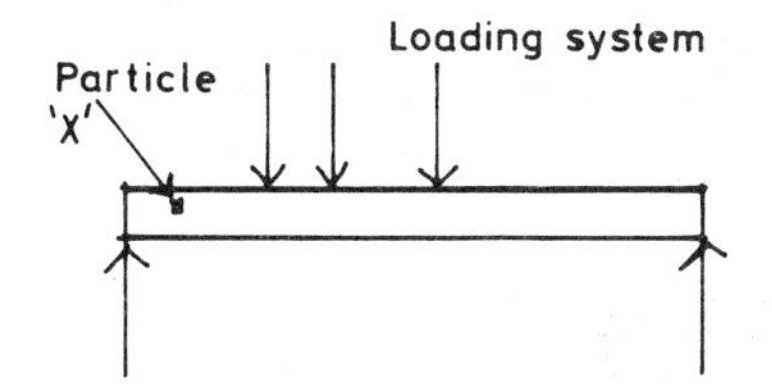

Figure 29.4 Arbitary particle 'x' in a beam

shearing force S_1 as shown in *Figure 29.5(a)* will be present. Since this would obviously cause clockwise rotation, it must be concluded that a horizontal shearing force S_2 must also be present in order to cause equal but opposite rotation for equilibrium.

If the particle is equal sided, $S_1 = S_2$. Also, if the particle is of constant thickness, the area of each face that is under shear will be equal and the shearing stress both vertically and horizontally will be of the same value. *At any point within the depth of a beam, therefore, there exists equal vertical and horizontal shear stress.*

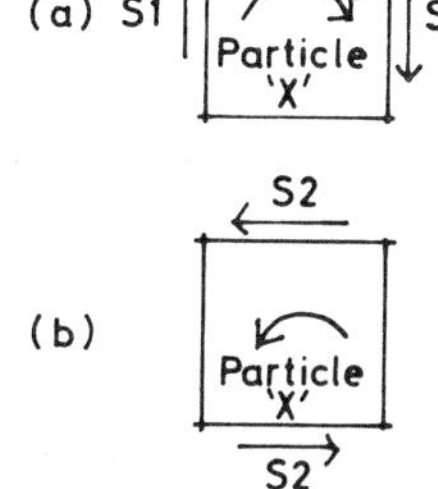

Figure 29.5 Enlargement of particle 'x'

The forces S_1 and S_2 can be resolved at each corner of the particle into a single force F, as shown in *Figure 29.6(a)*. This will tend to cause distortion, as shown in *Figure 29.6(b)*.

It would appear therefore that shear stress is rather complex. It can, however, be simplified by remembering that both vertical and horizontal shear are *separate* components of diagonal force. The material from which a beam is constructed, i.e. timber, steel or concrete, will have its inherent weakness exploited by *either* diagonal force *or* its components, as illustrated in *Figure 29.7*.

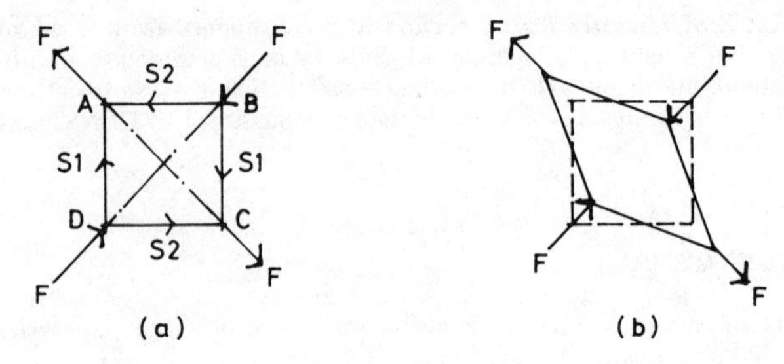

Figure 29.6 Result of shear forces in particle 'x'

Part (a) of the figure shows a typical shear failure in a timber beam by splitting along the grain at mid-depth. Part (b) shows a deep universal beam or plate girder. The web in the diagonal plane is very slender and so tends to buckle due to the diagonal *compressive* forces. Stiffeners are usually welded on to the face of the web

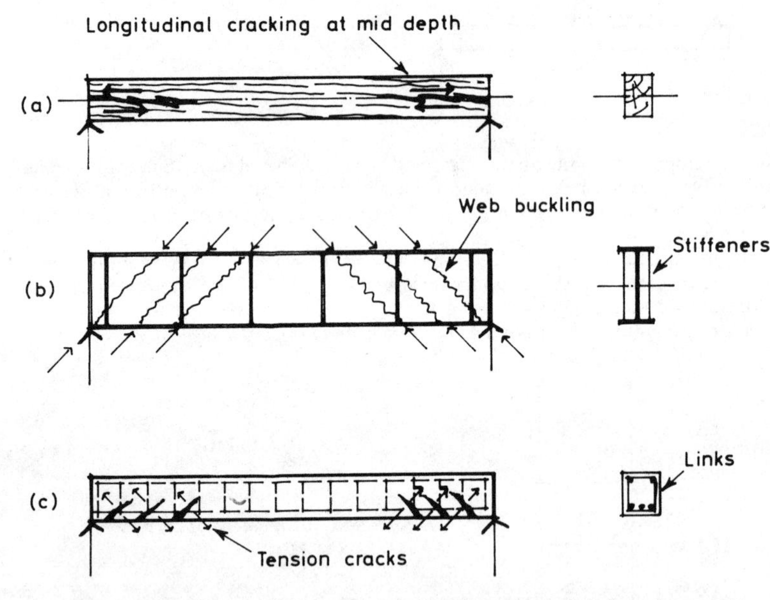

Figure 29.7 Shear stress in beams: (a) timber, (b) steel, (c) reinforced concrete

in order to stiffen it against buckling. Part (c) shows a reinforced concrete beam. Concrete is relatively weak in tension, and would therefore tend to crack due to the diagonal *tension* forces. Links are included, which pass across the plane of the cracks and hence provide the necessary shear reinforcement.

CALCULATION OF STRESS DUE TO SHEAR

Exact method. The value of stress due to shear varies within the depth of a beam. Unlike stress due to bending, its maximum value is always at the level of the neutral axis (hence the reason for the position of the shear failure crack in timber beams).

The diagram that shows the variation in the value of shear stress in the depth of a beam is parabolic in profile when the beam is rectangular in section – *Figure 29.8(a).* In beams where the breadth of section varies, as in universal steel beams, the shear-stress diagram is basically parabolic but discontinous – *Figure 29.8(b).*

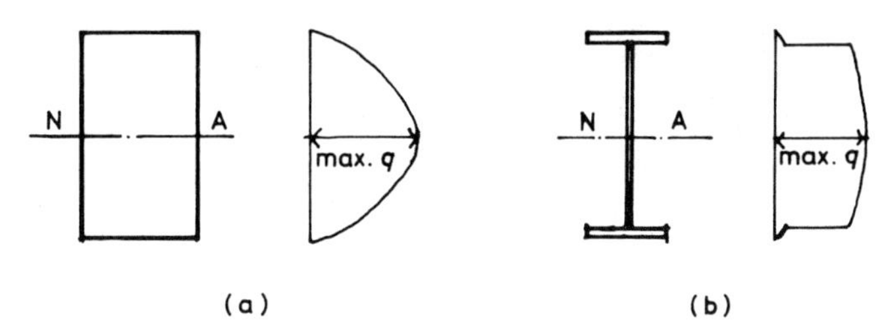

Figure 29.8 Beam sections and shear-stress diagrams (exact method): (a) timber, (b) steel

The value of shear stress can be calculated at any level within the depth of a beam section from the following formula. Usually it is the maximum value of shear stress that is required, and this will occur at the neutral axis at a point on the beam span where *shear force* is a maximum.

$$\text{Shear stress, } q = \frac{SA\bar{y}}{Ib}$$

where S = shear force
A = area of part of beam section above *or* below the level considered
$\bar{y}$ = distance from axis NA to centroid of area A
I = second moment of area of beam section about axis NA
b = breadth of beam section at level considered

Easier method. When the beam section is rectangular the maximum value of shear stress can alternatively be determined by assuming the stress to be uniformly distributed over the beam section, thus calculating the mean stress. Multiply this value by 3/2 to obtain maximum shear stress. See *Figure 29.9(a).*

$$\text{Mean shear stress, } q' = \frac{\text{shear force}}{\text{sectional area}} = \frac{S}{bd}$$

$$\text{Maximum shear stress, } q = \frac{3}{2} \cdot \frac{S}{bd}$$

In the case of a universal beam type of section, the greatest proportion of shear is resisted by the web, and hence the mean shear stress in the web is calculated. Since this value is slightly less than the maximum value at the NA position, the permissible shear stress value in BS 449 is downgraded accordingly for the purpose of comparison. See *Figure 29.9(b).* At a point on the beam where shear force = S,

$$\text{Mean shear stress in web, } q' = \frac{S}{Dt}$$

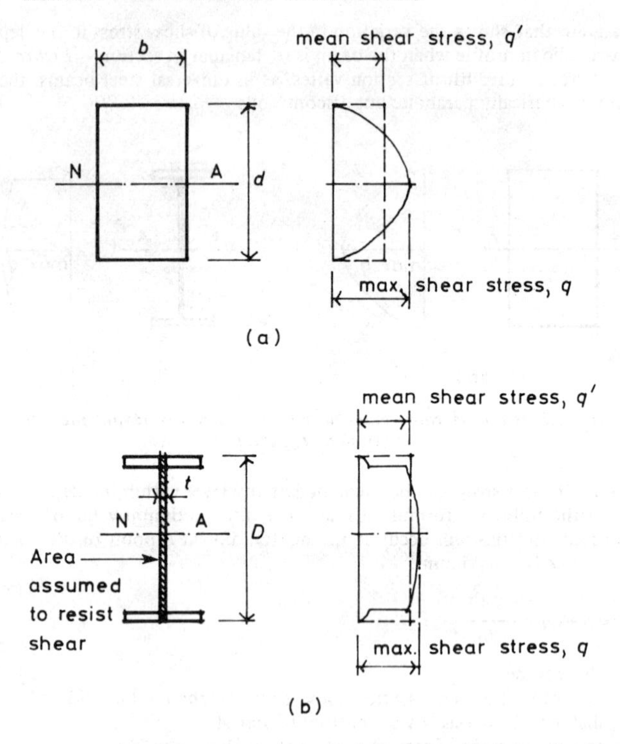

Figure 29.9 Beam sections and shear-stress diagrams (easier method): (a) rectangular, (b) universal type

DESIGN OF STRUCTURES

Design of beam sections

Alternative designs of beams in timber, steel and reinforced concrete will be prepared for the example of a lintol supporting brickwork and a first floor, as shown in *Figure 29.10*. Roof and ceiling are supported by trussed rafters, and the load from them is transferred to the side walls.

Load from brickwork over lintol, including self-weight of beam	= 8.70 kN/m
Self-weight of first floor supported by lintol	= 2.70 kN/m
Imposed load on first floor	= 6.00 kN/m
	17.40 kN/m

TIMBER BEAM

Effective span of beam = distance between centres of support, say 2.80 m (see *Figure 29.11*).

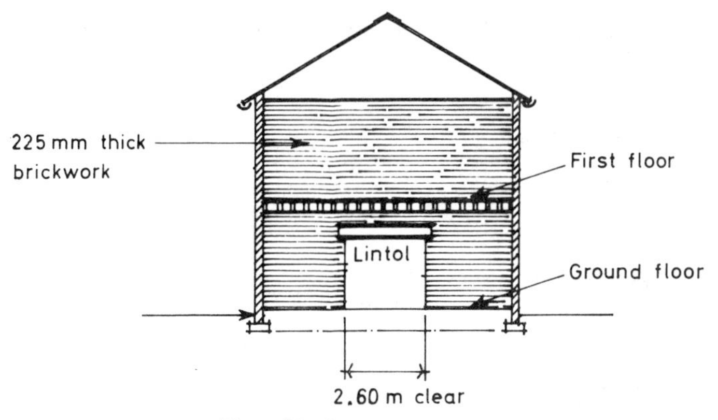

Figure 29.10 Example of lintol

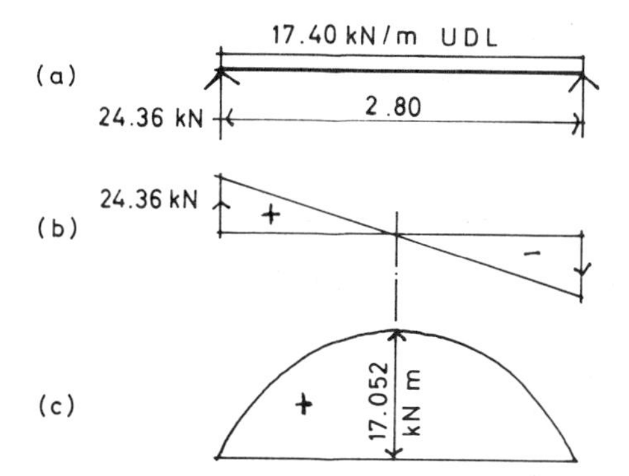

Figure 29.11 Lintol example: (a) timber beam, (b) shear-force diagram, (c) bending-moment diagram

$$\text{Reaction at each end of beam} = \frac{17.40 \times 2.80}{2} = 24.36\ \text{kN}$$

$$\therefore\ \text{Maximum shearing force} = 24.36\ \text{kN}$$

$$\text{Maximum bending moment} = 24.36 \times \frac{2.8}{2} - \frac{17.4 \times 2.8}{2} \times \frac{2.8}{4}$$

$$= 34.104 - 17.052 = 17.052\ \text{kN m}$$

Timber species S2, grade SS will be used. From CP 112, permissible stresses for this timber, in dry condition, are as follows:

Bending = 6.9 N/mm^2
Compression perpendicular to grain = 1.55 N/mm^2
Shear parallel to grain = 0.79 N/mm^2
Mean value of modulus of elasticity, E = 8900 N/mm^2

Two separate sections each 100 mm in breadth will be used side by side, as shown in *Figure 29.12.*

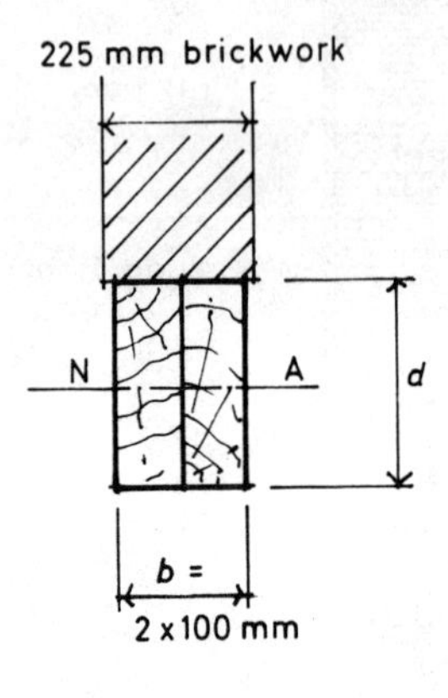

Figure 29.12 Two timber sections used side by side for lintol

$$M = fZ$$

$$\text{i.e. } Z = M/f$$

$$\therefore\ Z \text{ required to be provided } = \frac{17.052 \times 10^6}{6.9}$$

$$= 2.471 \times 10^6 \text{ mm}^3$$

$$\text{But } Z = bd^2/6$$

$$\therefore\ \frac{2 \times 100 \times d^2}{6} = 2.471 \times 10^6$$

$$d = \sqrt{\frac{2.471 \times 10^6 \times 6}{2 \times 100}} = 272 \text{ mm}$$

Therefore use two 100 × 300 mm sections. Check shear stress:

$$\text{Maximum shear force, } S = 24.36 \text{ kN}$$

$$\text{Maximum shear stress, } q = \frac{3}{2} \times \text{mean stress}$$

$$\therefore\ q = \frac{3}{2} \times \frac{24.36 \times 10^3}{2 \times 100 \times 300} = 0.61 \text{ N/mm}^2$$

Permissible shear stress parallel to grain = 0.79 N/mm²; therefore the beam section is suitable in shear.

Check deflection. According to CP 112, maximum deflection is not to exceed 0.003 of span.

$$\therefore\ \text{Permitted deflection} = 0.003 \times 2.80 \times 10^3 = 8.40 \text{ mm}$$

$$\text{Maximum deflection due to UDL} = \frac{5WL^3}{384EI}$$

where W = total load = (17.40 × 2.80) kN
L = span = 2.80 m
E = modulus of elasticity of timber = 8900 N/mm²
I = second moment of area of beam section

$$= \frac{bd^3}{12} = \frac{2 \times 100 \times 300^3}{12} \text{ mm}^4$$

$$\therefore \text{ Maximum deflection } = \frac{5 \times (17.40 \times 2.80 \times 10^3) \times (2.80 \times 10^3)^3 \times 12}{384 \times 8900 \times (2 \times 100 \times 300^3)}$$

$$= 3.50 \text{ mm}$$

Therefore the beam deflection is not excessive.

Calculate the length of bearing at each end of the beam. Let the length at each end be l mm.

Permissible compressive stress perpendicular to grain = 1.55 N/mm²
Compression on beam at each support = 24.36 kN

$$2 \times 100 \times l \times 1.55 = 24.36 \times 10^3$$

$$l = \frac{24.36 \times 10^3}{2 \times 100 \times 1.55} = 78.60 \text{ mm}$$

say 100 mm at each end, as originally assumed.

STEEL BEAM

Maximum bending moment = 17.052 kN m
Maximum shearing force = 24.36 kN
(Both as calculated in previous design example.)

Using grade 43 steel, permissible bending stress (from BS 449) = 165 N/mm²

$$M = fZ$$

$$\text{i.e. } Z = M/f$$

$$\therefore Z \text{ required to be provided } = \frac{17.052 \times 10^6}{165}$$

$$= 103.345 \times 10^3 \text{ mm}^3 \text{ or } 103.345 \text{ cm}^3$$

An appropriate steel beam section can now be chosen from structural steelwork tables. Use 152 × 89 mm RSJ; the Z_x provided = 115.6 cm³. It will be necessary to weld a 200 × 8 mm plate to the top flange, as shown in *Figure 29.13(a)*, in order to support the brickwork. As an alternative, two 127 × 64 mm RSCs could be used, as shown in *Figure 29.13(b)*; the Z_x provided = 151.98 cm³.

Figure 29.13 Alternative steel beam sections for lintol

It is not necessary to check either shearing stress or deflection. Only under exceptional circumstances would these be critical for a small-span beam.

REINFORCED CONCRETE BEAM

It is necessary first of all to identify the *type* of loading to be supported by the beam, since, to obtain the 'design load', factors of 1.4 and 1.6 are to be applied to dead load and imposed load respectively. From the previous design example:

Dead load, G_k = 8.70 + 2.70 = 11.40 kN/m
Imposed load, Q_k = 6 kN/m
Design load = $1.4G_k + 1.6Q_k$ = 1.4 × 11.40 + 1.6 × 6
= 15.96 + 9.6
= 25.56 kN/m

Reaction at each end of beam $= \dfrac{25.56 \times 2.80}{2} = 35.78$ kN

Design bending moment at centre of span

$$= 35.78 \times \frac{2.80}{2} - \frac{25.56 \times 2.80}{2} \times \frac{2.80}{4}$$
$$= 50.098 - 25.049$$
$$= 25.049 \text{ kN m}$$

Figure 29.14 shows the reinforced concrete beam.

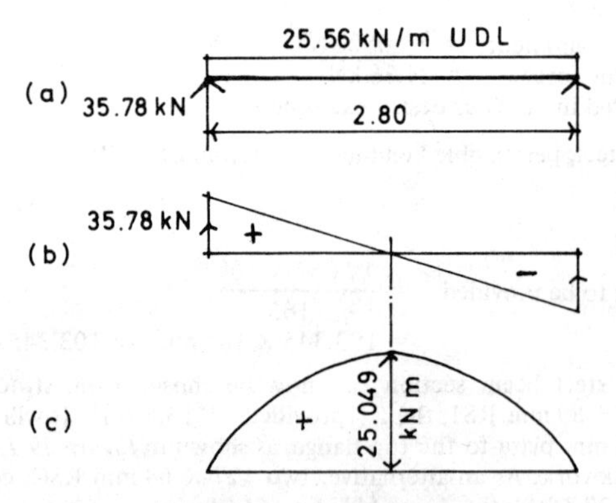

Figure 29.14 Lintol example: (a) reinforced concrete beam, (b) shear-force diagram, (c) bending-moment diagram

As already described on page 29–4, the design load calculated above is that which will theoretically cause failure of concrete and steel to occur, in as much as both materials will approach their critical stress. The limiting stress of each material is modified by a partial factor of safety.

Limiting stress in concrete $= \dfrac{0.67 \times f_{cu}}{1.5}$
$= 0.45f_{cu}$

where f_{cu} = characteristic strength. This value is reduced to $0.4f_{cu}$ in order to simplify the profile of the stress-distribution diagram into rectangular form.

$$\text{Limiting stress in steel} = \frac{f_y}{1.15} = 0.87f_y$$

where f_y = yield stress. The simplified distribution of maximum stress in the beam is as indicated in *Figure 29.15*, it being assumed that no tension is resisted by the concrete.

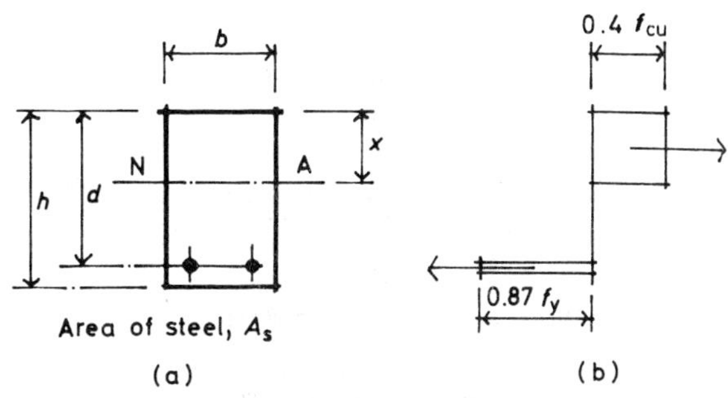

Figure 29.15 Concrete beam: (a) section, (b) simplified stress diagram

Without going too deeply into the derivation of design principles, two formulae from CP 110 can be used in order to proportion the section to resist bending, as follows:

$$M_u = 0.15f_{cu} \cdot bd^2 \qquad \text{...(1)}$$

from which the sectional dimensions can be determined, and

$$M_u = 0.87f_y \cdot A_s \cdot 0.75d \qquad \text{...(2)}$$

from which the area of steel can be obtained. M_u is the moment of resistance of the beam at maximum stress, and is numerically equal to bending moment.

Concrete with a characteristic strength f_{cu} of 20 N/mm² and steel with a yield stress f_y of 250 N/mm² will be used in the design. Let breadth of beam b = 225 mm to suit brickwork. From (1):

$$25.049 \times 10^6 = 0.15 \times 20 \times 225 \times d^2$$

$$\therefore d = \sqrt{\frac{25.049 \times 10^6}{0.15 \times 20 \times 225}} = 193 \text{ mm}$$

From (2):

$$25.049 \times 10^6 = 0.87 \times 250 \times A_s \times 0.75 \times 193$$

$$\therefore A_s = \frac{25.049 \times 10^6}{0.87 \times 250 \times 0.75 \times 193} = 795 \text{ mm}^2$$

Use three 20 mm diameter bars; the A_s provided = 943 mm².

The beam must also be designed to resist shearing, and this should be undertaken using the recommendations in CP 110. For relatively small spans, nominal shear reinforcement in the form of links of, say, 8 mm diameter spaced at centres not greater than $0.75d$ can be used. See *Figure 29.16.*

$$\text{Overall depth of beam, } h = d + \tfrac{1}{2} \text{ dia. of bar} + \text{cover}$$
$$= 193 + \frac{20}{2} + 25$$
$$= 228 \text{ mm, say } 250 \text{ mm}$$

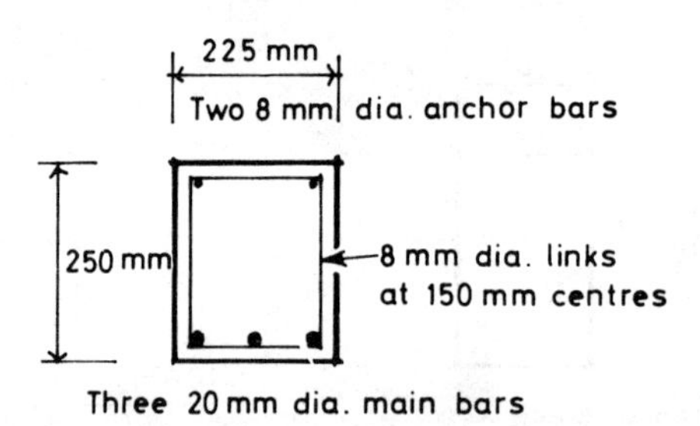

Figure 29.16 Concrete beam section showing position of reinforcement

It must be emphasised that the design of reinforced concrete is rather complex, and anything other than a small beam should be entrusted to a structural engineer.

Design of gravity retaining walls

Such walls are usually required to resist a combination of earth, live and hydrostatic loadings. The basic requirement is that the wall is capable of holding the retained material in place without deflection, tilting, overturning or sliding.

Tensile failure Where a wall has insufficient weight, and very high tensile stresses try to develop due to the overturning moment, it will fail in tension and overturn; see *Figure 29.17(a).* The mass of the wall must therefore be increased to counteract the overturning moment, or reinforcement must be added to resist the tensile stresses due to bending.

Ground pressure failure. To avoid this type of failure – *Figure 29.17(b)* – the actual ground pressure below the base must not exceed the permissible ground pressure.

Overturning Where a wall and its foundation have insufficient weight to counteract the overturning moment from horizontal forces about the toe, the wall will overturn – *Figure 29.17(c).* In gravity walls the resultant thrust should not fall outside the middle third of the base of the wall (*Figure 29.18*). In other types of wall, a factor of safety of at least 2 is required against overturning.

Sliding Where the total active horizontal force exceeds the total horizontal resistance the wall will slide – *Figure 29.17(d).* A factor of safety of approximately 2 should be obtained against sliding.

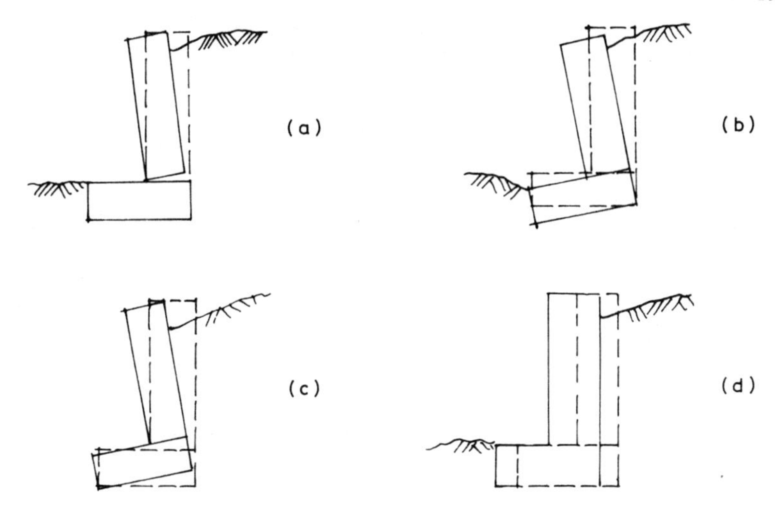

Figure 29.17 Gravity retaining wall: (a) tensile failure resulting in excessive deflection; (b) ground failure resulting in tilting; (c) overturning; (d) sliding

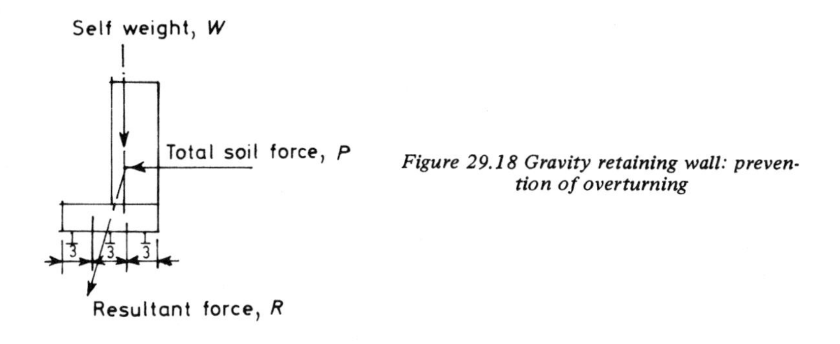

Figure 29.18 Gravity retaining wall: prevention of overturning

COHESIONLESS SOILS, VERTICAL WALLS WITH HORIZONTAL GROUND PRESSURE DUE TO EARTH

For gravity walls retaining cohesionless soils, the distribution of active pressure may be taken as varying linearly in any one stratum. The total horizontal pressure acting normal to the back of the wall may be taken as:

$$P_{an} = K_a \gamma \frac{H^2}{2}$$

where K_a = coefficient of active earth pressure = $\dfrac{1 - \sin\phi}{1 + \sin\phi}$

γ = density of the soil
H = height of wall
ϕ = angle of shearing resistance of the soil

The centroid of this pressure acts at $H/3$ from the base of the wall (*Figure 29.19*). Values of K_a are given in Table 29.1 for various values of ϕ.

Table 29.1

	ϕ			
	25°	30°	35°	40°
K_a	0.41	0.33	0.27	0.22
K_p	2.5	3.0	3.7	4.6

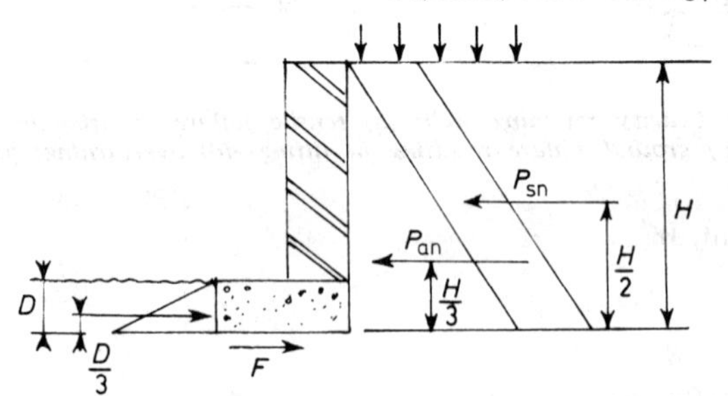

Figure 29.19 Gravity wall retaining dry cohesionless soil: distribution of pressure

PRESSURE DUE TO SUPERIMPOSED LOADING

The active pressure on the wall acting horizontally from superimposed loading may be taken as constant in any one stratum. The total horizontal pressure acting normal to the back of the wall may be taken as:

$$P_{sn} = K_a W_s H$$

where $K_a = \dfrac{1 - \sin \phi}{1 + \sin \phi}$

W_s = superimposed loading
H = height of wall
ϕ = angle of shearing resistance of the soil

The centroid of this pressure acts at $H/2$ from the base of the wall, and the overturning moment from superimposed loading is added to the overturning moment from the earth (*Figure 29.19*).

FRICTIONAL RESISTANCE

The coefficient of friction between a concrete foundation cast *in situ* and a cohesionless soil may be taken as the tangent of the angle of shearing resistance of the soil.

$$\text{Coefficient of friction} = \tan\phi$$

The product of the total vertical load and the coefficient of friction gives the total frictional resistance against sliding, F (*Figure 29.19*).

PASSIVE RESISTANCE

The total passive resistance of the earth in front of the wall may be taken as

$$P_{pn} = K_p\gamma\frac{D^2}{2}$$

where K_p = coefficient of passive earth pressure = $\dfrac{1+\sin\phi}{1-\sin\phi}$

γ = density of the soil

D = depth of earth in front of wall

ϕ = angle of shearing resistance of the soil

The centroid of this pressure acts at $D/3$ from the base of the wall (*Figure 29.19*). Table 29.1 gives values of K_p.

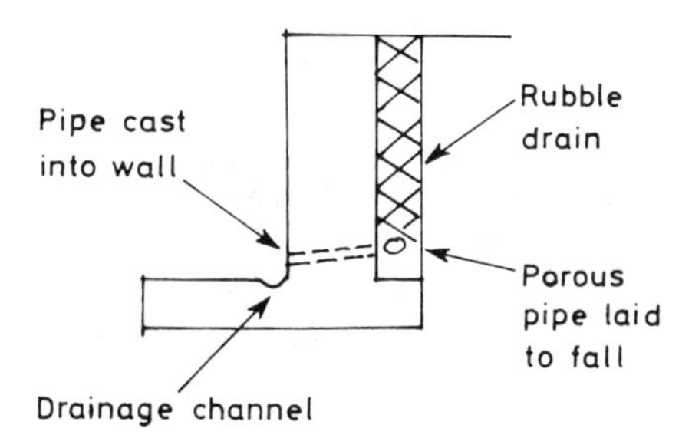

Figure 29.20 Reduction of hydrostatic pressure by means of drainage

POSSIBLE HYDROSTATIC PRESSURE

The back faces of retaining walls will usually be subject to hydrostatic forces from groundwater. This may be reduced by the provision of a drainage path at the face of the wall. It is usual practice to provide such a drain by a layer of rubble or porous blocks as shown in *Figure 29.20*.

DESIGN EXAMPLE 1

Gravity wall with vertical back, retaining cohesionless soils (Figure 29.21)

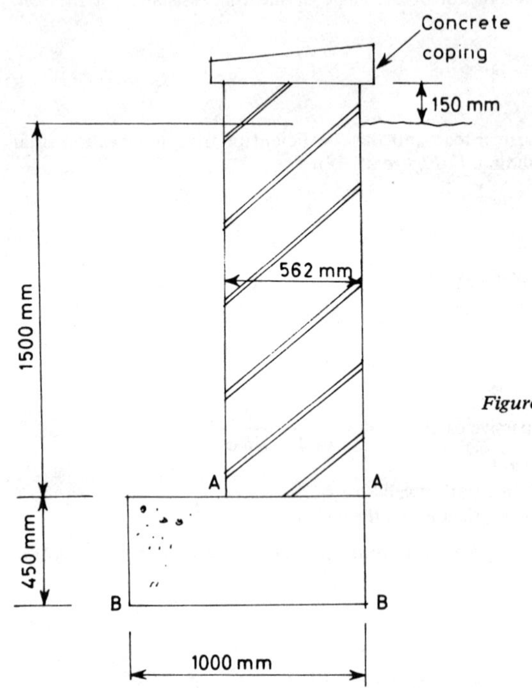

Figure 29.21 Design example 1

Data: Earth at 1600 kg/m³, ϕ = 30°
Brickwork at 2000 kg/m³ laid in cement mortar (1:6, with plasticiser)
Mass concrete base with a weight of 24 kN/m³
Permissible ground pressure 200kN/m²

Check stress at level A–A. Consider a 1 m run of wall.

$$H = 1.5\,\text{m} \quad \gamma = 1600\,\text{kg/m}^3 \quad \phi = 30° \quad K_a = 0.33$$

Total active horizontal force, $P_{an} = K_a \gamma \dfrac{H^2}{2}$

$$= 0.33 \times 1600 \times \frac{1.5^2}{2} = 594\,\text{kg}$$

$$= (594 \times 10)\text{N} = \frac{594 \times 10}{1 \times 10^3}\,\text{kN} = 5.94\,\text{kN}$$

OTM from earth pressure $= 5.94 \times \dfrac{1.5}{3} = 2.97$ kN m

Self-weight of wall $= 2000 \times 1.65 \times 0.562$

$= 1854$ kg $= 18\,540$ N $= 18.54$ kN

Area of wall at level A–A $= 562 \times 1 \times 10^3 = 562 \times 10^3\ \text{mm}^2$

$$Z = \frac{1 \times 10^3 \times 562^2}{6} = 52.64 \times 10^6\ \text{mm}^3$$

Stresses at extremities of section are given by:

$$\frac{W}{A} \pm \frac{M}{Z} = \frac{18.54 \times 10^3}{562 \times 10^3} \pm \frac{2.97 \times 10^6}{52.64 \times 10^6}$$

$$= 0.0329 \pm 0.0564$$

$$= 0.0893 \text{ and } -0.0235\ \text{N/mm}^2$$

A tensile stress of $-0.07\ \text{N/mm}^2$ could be accepted (reference Clause 316 of CP 111).

$$\text{Average shear stress at bedface} = \frac{5.94 \times 10^3}{1 \times 10^3 \times 562} = 0.01\ \text{N/mm}^2$$

This is a very low value; the permissible shear stress is given in CP 111 as $0.1\ \text{N/mm}^2$.

Check stress at level B–B, where $H = 1.95$ m, as follows.

$$\text{Total active horizontal force} = 0.33 \times 1600 \times \frac{1.95^2}{2} = 1003.86\ \text{kg}$$

$$= 10\,038\ \text{N} \qquad = 10.04\ \text{kN}$$

$$\text{Eccentricity from centroid of wall to centroid of base} = 0.5 - \frac{0.562}{2}$$

$$= 0.5 - 0.281 = 0.219\ \text{m}$$

$$\text{Net moment on base} = 10.04 \times \frac{1.95}{3} - 18.54 \times 0.219$$

$$= 6.526 - 4.06 = 2.466\ \text{kN m}$$

Weight of concrete base $= 0.450 \times 1 \times 24 = 10.8$ kN

Total vertical load $= 18.54 + 10.8 = 29.34$ kN

Area of base $= 1 \times 1 = 1\ \text{m}^2$

$$Z = \frac{1 \times 1^2}{6} = 0.167\ \text{m}^3$$

$$\frac{W}{A} \pm \frac{M}{Z} = \frac{29.34}{1} \pm \frac{2.466}{0.167}$$

$$= 29.34 \pm 14.76 = 44.1\ \text{kN/m}^2 \text{ and } 14.58\ \text{kN/m}^2$$

Maximum stress is less than $200\ \text{kN/m}^2$ and is therefore satisfactory.

Design of diaphragm walls

Diaphragm walls offer an economic solution to gravity retaining walls constructed using brickwork. The mass required for stability can be obtained by filling the voids with gravel or *in-situ* concrete (see *Figure 29.22*). The filling process is usually carried out in stages after the brickwork has been constructed to heights of 600–900 mm.

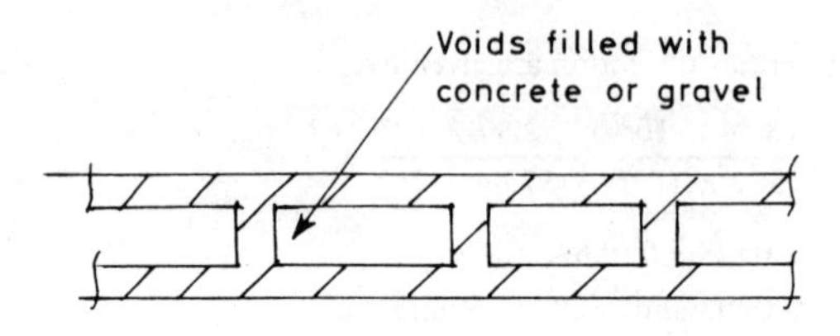

Figure 29.22 Typical section through diaphragm wall

DESIGN EXAMPLE 2

Gravity wall with vertical back using diaphragm wall construction (Figure 29.23)

Data: Earth at 1700 kg/m³, ϕ = 30°
Brickwork at 2000 kg/m³ laid in cement mortar (1:6, with plasticiser)
Concrete to be placed in layers not exceeding 600 mm in depth

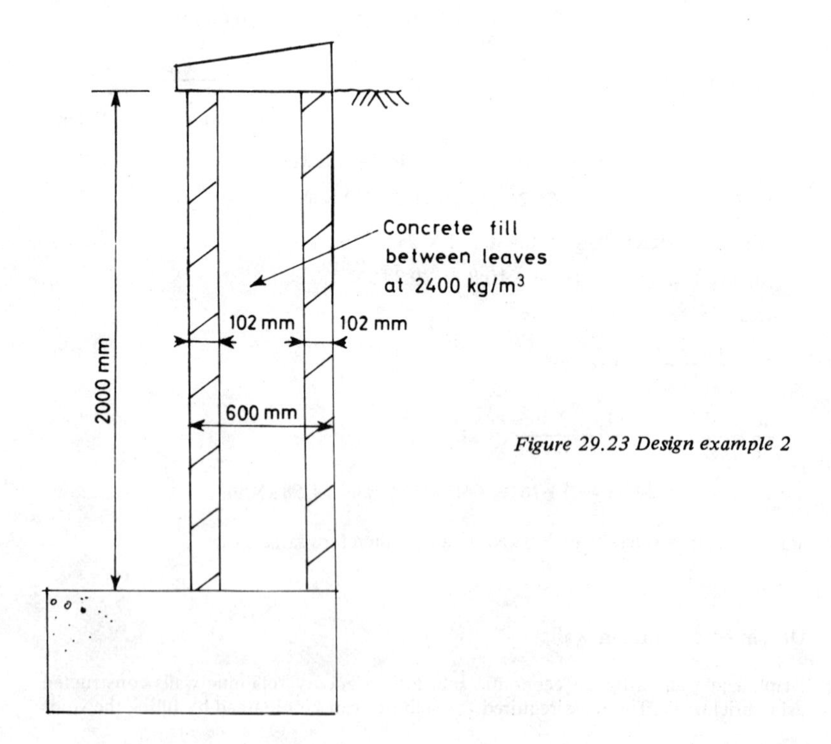

Figure 29.23 Design example 2

Consider a 1 m run of wall.

$$\text{Total active horizontal force, } P_{an} = K_a\gamma\frac{H^2}{2}$$

$$= 0.33 \times 1700 \times \frac{2^2}{2} = 1122\,\text{kg} = 11220\,\text{N}$$

$$= 11.22\,\text{kN}$$

$$\text{OTM from earth pressure} = 11.22 \times \frac{2}{3} = 7.48\,\text{kN m}$$

Self-weight of wall = $0.204 \times 1 \times 2 \times 2000$ = 816 kg
Concrete infill = $0.396 \times 1 \times 2400 \times 2$ = 1901 kg

2717 kg = 27.17 kN

$$\text{Area of wall at level A–A} = 600 \times 1 \times 10^3 = 6 \times 10^5\,\text{mm}^2$$

$$Z = \frac{1 \times 10^3 \times 600^2}{6} = 6 \times 10^7\,\text{mm}^3$$

Stresses at extremities of section are given by:

$$\frac{W}{A} \pm \frac{M}{Z} = \frac{27.17 \times 10^3}{6 \times 10^5} \pm \frac{7.48 \times 10^6}{6 \times 10^7}$$

$$= 0.04 \pm 0.12 = 0.16\,\text{N/mm}^2 \text{ and } -0.08\,\text{N/mm}^2$$

The tensile stress is slightly in excess of 0.07 N/mm², and the wall would therefore have to be increased in thickness to prevent an excessive tensile stress developing.

Design of reinforced retaining walls

A reinforced brickwork retaining wall requires less brickwork than a gravity retaining wall by virtue of the reinforcement developing all the tensile stresses. During the construction the brickwork acts as formwork for the concrete, thus eliminating special shuttering, which is always necessary for a reinforced concrete wall.

DESIGN EXAMPLE 3

Reinforced brick retaining wall (Figure 29.24)

Data: Earth at 1600 kg/m³, ϕ = 30°

Level A–A

$H = 2.2\,\text{m} \quad \gamma = 1600\,\text{kg/m}^3 \quad \phi = 30° \quad K_a = 0.33$

$$\text{Active horizontal force from earth, } P_{an} = K_a\gamma\frac{H^2}{2}$$

$$= 0.33 \times 1600 \times \frac{2.2^2}{2} = 1277.76\,\text{kg}$$

$$= 12.77\,\text{kN}$$

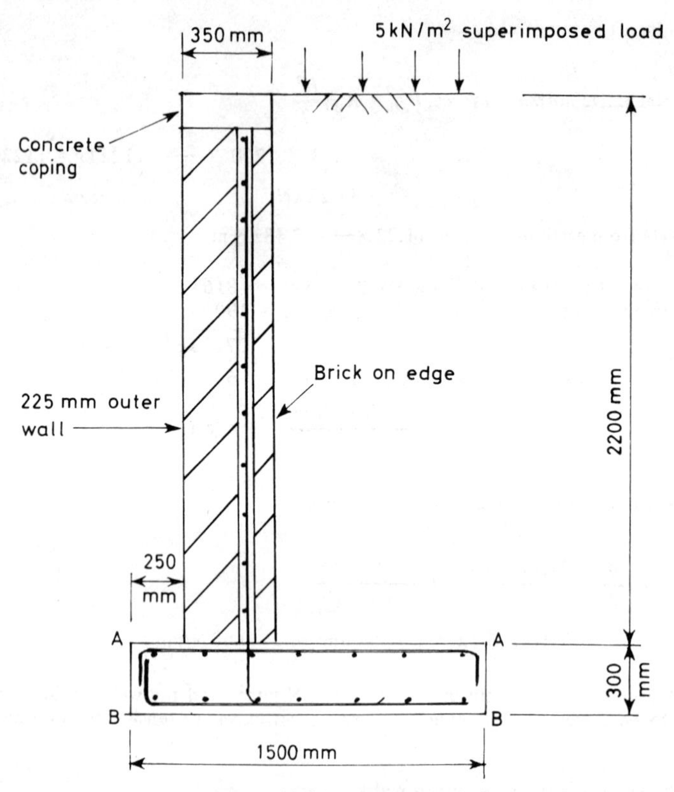

Figure 29.24 Design example 3

OTM from earth pressure $= 12.77 \times \frac{2.2}{3} = 9.36$ kN m

Active horizontal force from superimposed loading $= 0.33 \times 5 \times 2.2 = 3.63$ kN

OTM from superimposed loading $= 3.63 \times \frac{2.2}{2} = 3.993$ kN m

Total OTM $= 9.36 + 3.993 = 13.353$ kN m

Total horizontal force $= 12.77 + 3.63 = 16.4$ kN

Crushing strength of brick unit $= 27.5$ MN/m²

Permissible flexual stress = 2.8 N/mm²

Modular ratio, $m = 24$ (reference Table 6a of CP 111)

Permissible tensile stress in steel reinforcement $= 140$ N/mm²

Neutral axis

$$\frac{f_t}{f_{bc}} = \frac{m(d_1 - n)}{n}$$

$$\frac{140}{2.8} = \frac{24(250 - n)}{n}$$

$140n = 67.2(250 - n)$
$140n = 16\,800 - 67.2n$
$207.2n = 16\,800$

$$n = \frac{16\,800}{207.2} = 81 \text{ mm}$$

$$\text{Lever arm} = 250 - \frac{81}{3} = 223 \text{ mm}$$

$$\text{M of R (brickwork)} = \frac{2.8}{2} \times 81 \times 1 \times 10^3 \times 223$$
$$= 25.288 \times 10^6 \text{ N mm} = 25.288 \text{ kN m}$$

This is greater than 13.353 kNm, and therefore satisfactory.

$$A_{st} = \frac{13.353 \times 10^6}{140 \times 223} = 428 \text{ mm}^2\text{; use R10 at 150 mm spacing}$$

Shear force = 12.77 + 3.63 = 16.40 kN

$$\text{Shear stress} = \frac{16.4 \times 10^3}{1000 \times 223} = 0.07 \text{ N/mm}^2$$

This is less than 0.21 N/mm^2, and therefore satisfactory.

Level B–B

$H = 2.5$ m

$$\text{Active horizontal force from earth} = 0.33 \times 1600 \times \frac{2.5^2}{2}$$
$$= 1650 \text{ kg} = 16\,500 \text{ N} = 16.5 \text{ kN}$$

$$\text{OTM from earth pressure} = 16.5 \times \frac{2.5}{3} = 13.75 \text{ kN m}$$

Active horizontal force from superimposed loading = 0.33 × 5 × 2.5 = 4.125 kN

$$\text{OTM from superimposed loading} = 4.125 \times \frac{2.5}{2} = 5.15 \text{ kN m}$$

Total OTM = 13.75 + 5.15 = 18.9 kN m
Total horizontal force = 16.5 + 4.125 = 20.625 kN

Take moments about the toe to find the position of the resultant force on the base (*Figure 29.25*).

Weight of earth (1)	= 0.9 × 2.2 × 16	= 32 kN
Weight of wall (2)	= 0.35 × 2.2 × 20	= 15.4 kN
Weight of base (3)	= 1.5 × 0.3 × 24	= 10.8 kN

Moments about point X:

Total downward force = 15.4 + 32 + 10.8 + 4.5 = 62.7 kN
Σ(moments about X) = 0
$(15.4 \times 0.425) + (10.8 \times 0.75) + (36.5 \times 1.05) - (62.7 \times \bar{x}) - 18.9 = 0$
$6.545 + 8.1 + 38.325 - 62.7\bar{x} - 18.9 = 0$
$34.07 = 62.7\bar{x}$

$$\therefore \bar{x} = \frac{34.07}{62.7} = 0.543 \text{ m } (>0.5)$$

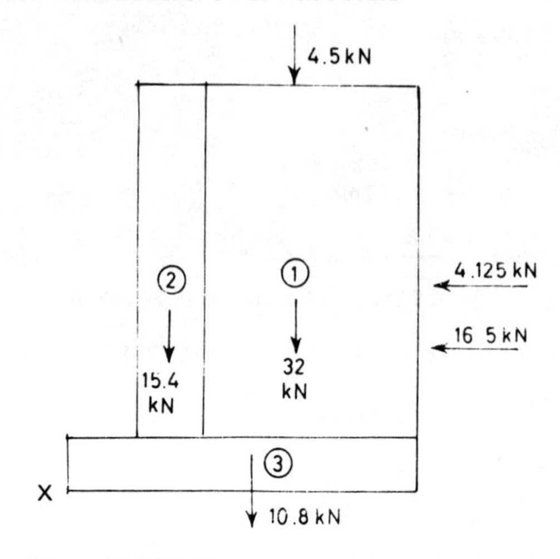

Figure 29.25 Taking moments about toe (point A)

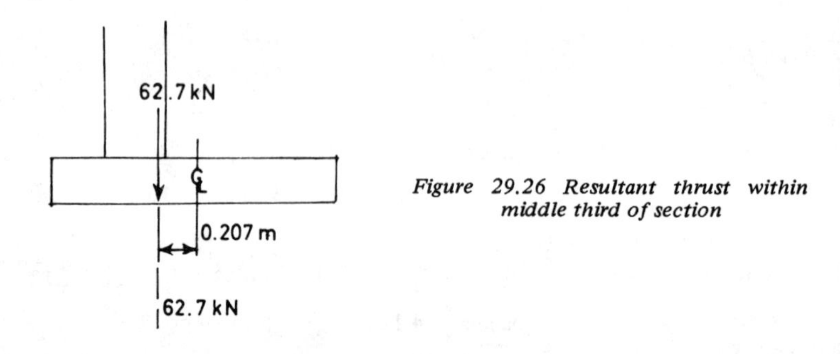

Figure 29.26 Resultant thrust within middle third of section

The resultant thrust is located within the middle third of the section (*Figure 29.26*).

$$e = 0.75 - 0.543 = 0.207 \text{ m}$$

$$\text{Base area} = 1 \times 1.5 = 1.5 \text{ m}^2$$

$$Z \text{ of base} = 1 \times \frac{1.5^2}{6} = 0.375 \text{ m}^3$$

$$\text{BM} = 62.7 \times 0.207 = 12.97 \text{ kN m}$$

Maximum ground bearing stress is given by:

$$\frac{W}{A} + \frac{M}{Z} = \frac{62.7}{1.5} + \frac{12.97}{0.375}$$

$$= 41.8 + 34.58 = 76.38 \text{ kN/m}^2 \quad (<100 \text{ kN/m}^2)$$

Minimum ground pressure is given by:

$$\frac{W}{A} - \frac{M}{Z} = 41.8 - 34.58 = 7.22\,\text{kN/m}^2$$

Check for sliding:

Total active horizontal force = 16.5 + 4.125 = 20.625 kN
Total vertical force = 62.7 kN

Frictional resistance = 62.7 × μ, where μ = tan 30° = 0.577
= 62.7 × 0.577 = 36.2 kN

$$\text{Factor for sliding} = \frac{36.2}{20.625} = 1.75$$

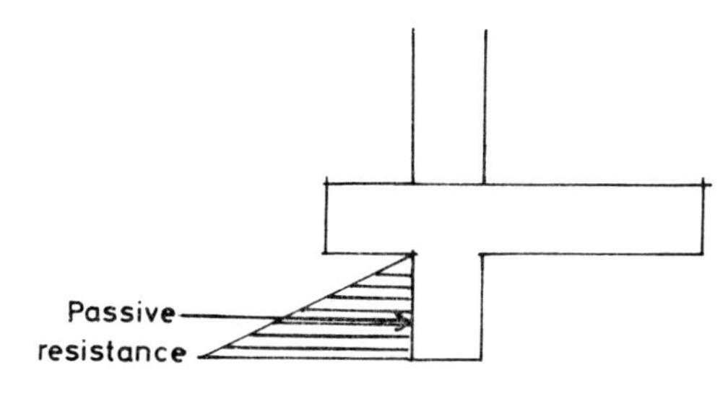

Figure 29.27 Projection below base

This is less than 2, therefore develop passive resistance on a projection below the base (see *Figure 29.27*). Passive resistance plus frictional force gives a factor for sliding of 2.

30 TEMPORARY WORKS

INTRODUCTION	**30**–2
ACCESS SCAFFOLD	**30**–3
FORMWORK	**30**–8
FALSEWORK	**30**–15
EXCAVATIONS	**30**–16
SHORING	**30**–17
UNDERPINNING	**30**–18
HANDLING STRUCTURAL ELEMENTS	**30**–18

30 TEMPORARY WORKS

C.J. WILSHERE, BA, BAI, CEng, FICE
Manager, Temporary Works Department, John Laing Design Associates Ltd

This section is concerned with scaffolding, formwork, falsework and other types of temporary works that the house builder has to carry out. Legal and contractual, as well as technical, aspects are discussed.

INTRODUCTION

While the architect (and engineer) will provide adequate information to define clearly the final structure, there are frequently problems for the builder in providing various sorts of equipment to enable the work to be done. For example, in almost all structures it is necessary to provide access for those who are actually going to do the work. This access may be for the builder's own employees or for the employees of other contractors. At the beginning of a job or when drainage is required, where a shallow hole is needed, sloping the sides of the excavation may make it safe. But if the ground is bad or the excavation is of any depth, it may be essential for reasons of safety and economy to provide a support. For concrete, formwork is required, and for many other items as well, falsework is needed. Much building work is connected with repairs and alterations, and for this shoring and underpinning can be needed.

Relation to contract

In some cases a builder will be his own client, for example if he decides to build houses for sale. This makes any question of responsibility during construction very simple, as the builder will carry it all. However most work is done contractually.

The main contract in use in building today is the SFB, which developed from the RIBA contract. In it the contractor undertakes to provide the structure as described in the drawings and specification, and to safeguard the client from the consequences of any mishaps during the contract.

This means that he must provide and operate all the relevant equipment and techniques suggested above, although they are not usually described in detail. The preliminaries to the contract will require that the construction is carried out safely and that proper access is provided. It will specify that this access is made available as necessary to sub-contractors. Formwork is an item that is normally billed. In excavation, it is normal to bill excavation items that require shoring either as planked and strutted or to specify a particular size of sheet piling. In either case the contractor must ensure that excavation is carried out satisfactorily.

Generally the contract will not specify methods, but there will be occasions when the architect dictates the method by which he wants something done, and will then set this down in detail. Because under this form of contract the builder takes full responsibility, the architect is not normally concerned in approving the method. However what follows below should be noted.

Common law

The builder is bound like everybody else by common law. This applies to him as a corporate body, and to his staff as individuals. Briefly this means that, if a man is connected with construction and has some knowledge and experience that he is able to bring to that construction, failure to exercise that judgement would render him liable to paying damages should ill befall. In a normal management situation, naturally a manager will be organising the work of those he is managing. However, failure to do this is a breach not merely of his duty as an employee, but of common law.

This duty to one's neighbour has been supplemented in the last two centuries by statute law. This started with the Morals of Apprentices Act of 1801, and the first regulations specifically relating to building were made in 1926. There are currently three main sets of Regulations, which were made under the various Acts that relate to building [1,2,3].

Health and Safety at Work etc. Act 1974

This Act is based on the Robens Report. It enshrines the concept of the common-law duty, and the more detailed concepts of earlier Factories Acts. It requires that all work shall be undertaken with a safe method and that the place be safe. It makes everyone, including employees, responsible for their actions. It requires that manufacturers of products and equipment shall investigate these sufficiently and then they shall issue instructions for their safe use. It has a series of sanctions that the Health and Safety at Work Act Inspectors can apply, and sets down the legal penalties should a successful criminal prosecution be undertaken.

Sub-contracting

Where a firm specialises, for example in scaffolding, greater experience can be built up, and it will be better able to deal with both the legal and practical problems. The 1974 Act makes both main and sub-contractors responsible for the safety of their operations. Should a sub-contractor fall foul of the 1974 Act, it is likely the main contractor would be prosecuted too.

It is normal for the main contractor to be responsible for the provision of access scaffolding for all the sub-contractors. In all types of sub-contract, it is essential both for safety and commercial reasons to make quite clear exactly where the split of responsibility lies. Detailed information and advice should be obtained from the Health and Safety Executive.

General

It will be seen that the need to provide adequate temporary works is important for the safety and well-being of all concerned. It is also most important from an economic point of view to assess the need correctly, since the cost of temporary works can be very large.

ACCESS SCAFFOLD

The fundamental purpose of access scaffold is to provide a working place and a means of access for operatives, materials and equipment. This may be in initial construction or it may be in maintenance, repair or modification of an existing structure. No matter which purpose is to be served it is important that there are no hazards, and that working on it will be easy.

Safety is required specifically by law, and the Regulations set down a number of parameters that give guidance on the minimum that must be achieved. It is important, with access scaffold, to have as a main consideration the ease of working. The right access scaffold will enable the quality of work needed to be achieved, which is sometimes difficult from more rudimentary scaffolds, and in addition the work will be done more rapidly, thus saving cost.

What is needed

A platform must be provided adjacent to the work; it must be strong enough for the loads that will be placed upon it, so that it does not break or get bent, and sufficiently rigid to provide an adequate sense of security to those using it. Certain minimum widths are required depending on the use to which it will be put. These are set out in Table 30.1. It must have a handrail or guard-rail where there is a possibility of people falling off the edge. The Regulations require this if it is more than 6.5 ft (1.98 m) above the level on to which people might fall, but it is obviously sensible to provide such a handrail at all times. This handrail must be at least 3 ft (915 mm) above the level of the platform. In addition, there must be a toe-board of minimum height 6 inches (152 mm) to prevent small items being kicked off the side of the scaffold. It also reduces the risk of someone being able to slip below the handrail. The resulting gap below the lower side of the handrail and the top of the toe-board must not exceed 2.5 ft (762 mm).

The use of wire grids called brick guards provides additional safety. The platform should be level, and there should be no steps likely to cause people to trip.

This access has been provided in the UK for many years by scaffold made with tube and fittings, although a lot of proprietary equipment is used today. For smaller jobs various types of trestles and towers are used.

Putlog scaffold

This is used primarily for brick construction. Standards are erected about 1200 mm from the building to be constructed, and are joined parallel to the building by horizontal ledger tubes. To these ledgers are fixed putlogs whose inner ends are supported on the brickwork being constructed, and which support scaffold boards forming platforms. It is usual, because it suits bricklaying, to have these platforms about every 1350 mm in height.

The end of a putlog is flattened and it is usual for it to be in the horizontal joint between two brick courses. The weight of brickwork built above each putlog end tends to grip it, thus reducing the risk of the scaffold falling away from the building. But it is very important to remove them carefully when the time comes, without chipping the edges of the bricks. In some cases a putlog scaffold may not be possible as the thickness of the joints is insufficient to accommodate a putlog end.

In some putlog scaffolds, the putlog is turned so that the flat of the end is in a vertical joint. This may be convenient at one side of a corner of the building where the scaffold boards cannot be made to fit neatly and have to overlap, as putlogs on edge and on flat are at different heights. However, weak bricks may be split by the edge of the putlog. Putlog scaffolds require careful tying in to the building so that it does not fall away from it.

Table 30.1 LOAD ON SCAFFOLD PLATFORMS (based on CP 97: Part 1: 1967, courtesy British Standards Institution)

Type of scaffold	*Spacing of standards* (m)	*Width of platform*		*Platform area per bay* (m^2)	*Permissible loading* (kgf/m^2)	*Maximum load per bay* (kgf)	*No. of working platforms in use at one time*	*Example of maximum load per bay (see Note 2)*
		No. of 225 mm boards	*Nominal width* (m)					
Putlog	1.8	5	1.2	2.22	275	610	1	1 man, mortar on spot board and 180 bricks
Putlog	2.4	5	1.2	2.97	180	537	1	1 man, mortar on spot board and 160 bricks
Light independent tied	2.7	3	0.6	1.67	73	122	1	1 man, and hand tools, paint or plaster
Light independent tied	2.7	4	0.9	2.5	73	184	1	2 men and hand tools, paint or plaster
General-purpose independent tied	2.1	5	1.2	2.6	180	470	4	1 man, mortar on spot board and 130 bricks
General-purpose independent tied	2.4	5	1.2	2.97	145	435	4	2 men and 280 kgf materials
Heavy-duty independent tied	1.8	5	1.2	2.22	290	653	See Note 1	1 man, mortar on spot board and 200 bricks
Heavy-duty independent tied	1.8	6	1.37	2.5	290	735		2 men and 580 kgf of materials
Heavy-duty independent tied	1.8	5	1.2	2.22	180	403		2 men and 250 kgf of materials

Note 1. Information as to the number of platforms that may be in use at any one time is given in the appropriate section of CP 97.

2. Assumptions made: weight of a man = 75 kgf
   weight of spot board and mortar = 32 kgf
   weight of a brick = 2.75 kgf
3. If bricks weighing more than 2.75 kgf each are used, the total number of bricks per bay should be proportionately reduced.
4. Scaffolds are designed for uniform loading; point and impact loads should be avoided. If point and impact loads cannot be avoided, strengthening of the scaffold is required
5. Material that has a mass of 100 kilograms (kg) causes a load of 100 kilograms-force (kgf) on the platform.

Independent scaffold

In this case the scaffold has pairs of standards adjacent to the building, usually about 1200 mm apart. They are connected similarly with ledgers parallel to the building. The inner and outer grids so formed are joined by horizontal transoms, which also serve to support scaffold boards. The rigidity and strength of this scaffold depends on its being attached to the structure that it serves. In addition, diagonal bracing to alternate pairs of standards is required, and scaffolds of any size should be braced diagonally up the face parallel to that of the building. The requirements for handrails and toe-boards are the same as for putlog scaffolds. Despite its name, the scaffold requires full tying in to the building it is serving.

Freestanding towers

A scaffold can be erected without tying it to the structure. Such scaffolds are frequently provided with castors so that they can be moved easily. Both proprietary scaffolding and tube and fittings can be used, or alternatively special components are available, many in aluminium.

A tower is very useful, but must be used with care. The rules for construction normally adopted are to limit the height-to-base ratio. The base is the smaller of the two dimensions, while the height is the height to the top platform on which operatives will work. For outdoor applications the ratio should not exceed 3 : 1, whereas indoors, where there is no risk of wind, 3½ : 1 is acceptable. No tower should ever be moved with people on top of it, nor should it be used unless the wheels have been locked.

Birdcage scaffold

A birdcage scaffold covers an area, and when used for access will be fully boarded, for example, to work on a large ceiling. A grid of vertical tubes is assembled with horizontal tubes in the two main directions, and diagonal bracing is added.

Vertical access

Whatever the type of scaffold, it is necessary for people to reach the working level. This will normally be achieved by the use of ladders. For towers, a special ladder component may be specifically provided as part of the construction. In general, however, it is necessary to provide a separate ladder. This should rise at least 1060 mm above the level which it serves, or alternatively should have a satisfactory hand-hold for getting on and off. The height that one climbs should not exceed 10.3 metres without a landing on which to rest. The ladder should normally be at a slope of 1 horizontal to 4 vertical. Where it is arranged to come through the platform, care should be taken to keep this hole as small as possible, and it should preferably be outside the normal area of the scaffold deck. The ladder must be tied at the top.

Smaller equipment

As well as the larger items of equipment mentioned above, there are a number of smaller components of considerable use. These start with trestles, which provide a painter's access, and there are various types of trestle of which *Figure 30.1* is typical.

Details

All access scaffolding should be set up on adequate foundations. A concrete slab or a pavement will normally be satisfactory. Inside a structure, if it is necessary to erect a scaffold on a timber floor, it will be prudent to check that this will stand the weight if the scaffold is of any size. If the scaffold has to be erected on earth or clay, the surface should be inspected carefully first. This is to ensure that there will be no settlement, particularly if the soil has been dug over recently, perhaps to put in

Figure 30.1 Timber trestles and lightweight batten staging (courtesy Stephens and Carter)

drains. It is also most important that the scaffold remains safe, and that excavations do not take place close to it.

In scaffolds built of tube and fittings, it is important that the correct materials are used. Most scaffolding is made with 48 mm steel tubes, but it is very necessary that steel and aluminium are not mixed. The fittings normally used are of steel, and these can be used with aluminium tube. Aluminium fittings are also sometimes used.

It is important that the correct types of fittings are used. For a putlog scaffold, right-angle fittings are used to connect ledgers to standards, and putlog couplers to connect the putlogs to the ledgers. In a similar way, in independent scaffolds the standards are connected to the ledgers with right-angle couplers, and the transoms or board bearers are connected to the ledgers with putlog couplers. Wherever possible the bracing is also connected with right-angle couplers. Where tubes have to be connected end to end it is best to use sleeve couplers. However, for the standards a spigot coupler is usually satisfactory.

It is frequently necessary to tie the scaffold effectively to the structure. This is often done by using tubes. In one method, a reveal screw and short length of tube are tightened between the reveals of the window, if these are parallel to each other; a tube is then attached between this anchorage and the scaffold. A more effective method is to connect to the inside face of the outer wall of the structure; for example, a tube is placed horizontally behind the window, and the tie tube attached to it. Care should always be taken to assess ties to make sure that they are in fact effective.

In some cases threaded anchorages are provided in the face of the structure: an eye bolt can be screwed into each of these and attachment made to that. Such anchorages may be cast in, where the structure is of *in-situ* concrete, or may be drilled in later. There is at present no standard arrangement for them, so should they be found on an existing building it will be necessary to investigate them carefully to see if they can be reused.

The Regulations

The Regulations[3] have been mentioned with reference to particular points, and form the minimum requirements for scaffolds. A number of exceptions to the general rules are set down, which are appropriate for specific circumstances.

In particular they require that a record be kept of the inspections that the scaffold undergoes. Inspections should be carried out once a week, and the results entered in the register that must be kept on site.

Sub-contracting

Where the main contractor decides to subcontract his scaffolding it is most important that he works out carefully beforehand what service he wants, and that he clarifies exactly where any split of responsibility comes. Once the scaffold is erected and complete, there is frequently a need to modify it. It is normally desirable that this should be clarified, as responsibilities can get very blurred in the absence of such an agreement.

Other problems

There are numerous situations where the common methods outlined above do not answer the problem. Reference should be made to the literature [1,3,4,5] or to a specialist scaffolding firm.

FORMWORK

Formwork enables concrete to be compacted and to set in the required shape. The quality of surface finish may be important simply to show that compaction is effective, or it may be the ultimate visible finish. Most formwork falls into the first category. Unless it is actually going to be totally hidden, the basic minimum quality is usually referred to as 'fair face', but there is no standard definition. The terms formwork and shuttering are virtually interchangeable. (Falsework is a temporary structural support, and in this section it is taken to mean the structural support below soffit formwork, and similar supports to things that are not made of concrete.)

Basic materials

The sheeting in contact with concrete is normally plywood, from 12 to 19 mm thick. There are several types, although for many purposes these are interchangeable and the question of price is the one that guides a choice. Plywoods should have a WBP glue, which is in practice waterproof and will stand up to normal site conditions without delaminating.

Where the quality of finish is important, the choice is less straightforward. Finnish birch plywood gives a smooth finish until the surface is damaged and moisture enters the timber at one point, causing it to swell locally. It leaves little impression of grain on the concrete. Douglas fir gives a much stronger grain impression, which is much coarser. It is relatively more durable as it does not suffer from the same problem with moisture. In addition, there are now various hardwood types coming from Africa and south-east Asia, and these are satisfactory in most cases. However, one or two have given rise to retardation of the concrete, which makes them quite unsuitable. The worst species found is Meranti.

It is also possible to use timber boards, but they are not suitable for a fair-faced finish. They move with changes in moisture content, and the joints between board and board print out on the concrete.

For framing formwork, timber is the most commonly used material. For most purposes, and particularly for the smaller items, the lower grades of timber are perfectly satisfactory. However, it is possible, by purchasing stress-graded timber and carrying out appropriate design, to produce formwork that is both lighter and more economical than that produced with the lower-grade timber. In all cases an examination of the timber should be made before use, so that the knots, which could effect strength, are placed in positions where they are not highly stressed. Where this is done, grades such as Scandinavian redwood and whitewood fifths are satisfactory. Timber can be used both for the smaller members directly behind the sheeting and in larger sizes as soldiers or walings. It is best to adopt twin timbers for these heavier members. Note that, where shutters are built up of more than one layer, timbers should be 'thicknessed', i.e. the timbers in any one layer should all be the same thickness. Without this, structural difficulties can occur in the shutter and the line of the face is unlikely to be straight.

It is possible to use steel as a backing member to plywood, but this is seldom done and it is more usual to adopt it for soldiers and walings. These may take the form of structural steel sections, either unfabricated or built up into reusable components. At present the use of aluminium is virtually non-existent, but with present trends in the relative prices of steel and aluminium there may be a change.

A number of specialist suppliers will make column shutters in glass reinforced plastic (GRP). This is particularly suitable for circular and elaborately shaped columns.

Proprietary equipment

An alternative to the use of basic material is ready-made equipment. Complete systems can be obtained that will enable most shapes to be constructed without any additional material. These proprietary systems are in the main based on a panel, to form the concrete face, made of plywood with a metal surround to give it durability and strength. They are used in connection with their own accessories such as ties, and are generally not interchangeable with other systems; some components from systems, however, can be used with advantage in conjunction with traditional materials. Other systems are designed for use with plywood sheets, although providing everything else from standard reusable components.

With all wall formwork systems some means of supporting the side panels is required. In a typical shutter, this will be done by ties that pass through the concrete, connecting one face to the other to resist the pressure of the concrete. It is possible to use conventional steel items such as threaded rods and nuts in conjunction with plastic sleeves, but the proprietary items available today are usually so much quicker in use that the slightly greater initial cost is amply saved in the labour. Ties are of several types: those that remain in the concrete, and those that can be extracted for re-use; those that space the shutters apart as well as tying them together, and those

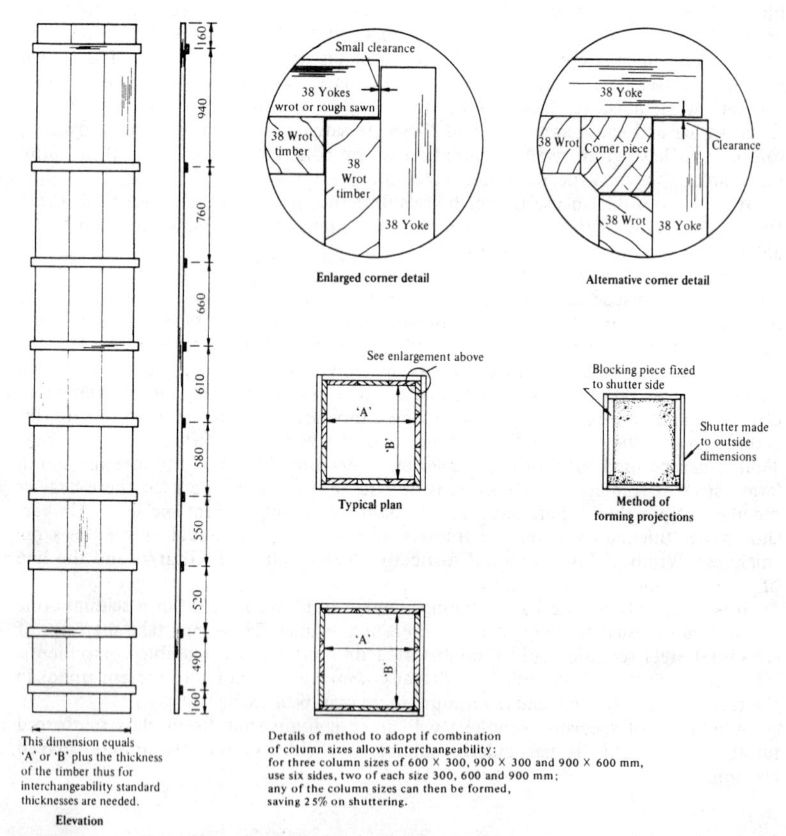

Figure 30.2 Rectangular-column formwork using timber sheeting, for columns up to 550 mm wide; not suitable for fair-faced work. Spacing of yokes is commenced from the top; where this would leave an unsupported length below the lowest yoke of more than 170 mm, an extra yoke is necessary fixed flush at the bottom. Column clamps to be handed at alternate levels and if necessary a template used to ensure a square column

that only do the latter; those that provide a long path against water leakage – as needed, for example, in the walls of tanks. There are also those that are needed when a tie is wanted at one face of the concrete only, to form an anchorage to construct the next section higher up.

For beams and columns, there are clamps available that serve the same purpose as ties, but are advantageous in that they support shutters very much more effectively. The traditional column clamp, which has been in use for many years, appears likely to continue for a long time. For beams, most of the main manufacturers produce beam clamps to their own particular design. The cost of these is justified where they can be reused rapidly, so designs that enable them to stripped out at an early stage are to be preferred.

How to shutter common items

A wall will normally be constructed using shuttering panels, possibly 2439 × 1219 mm. This is as large a panel as can conveniently be handled without a crane. The proprietary systems with metal-edged plyfaced shutters all use smaller panels, which are lighter, although considerably more have to be handled to construct an equivalent area of formwork. On the other hand they can be erected by less skilled people as the system is more like Meccano. Their use may enable joiners to deploy their skill to better advantage elsewhere. For large walls that repeat a number of times, shutters the size of the wall handled by crane are likely to be most economic.

Traditionally timber has been used to construct columns, largely because the length of timber that can be obtained is as long as a column, whereas it is almost always necessary to join plywood. Nonetheless plywood is frequently used, with vertical timbers put up outside the ply panels to make a straight panel. *Figures 30.2 and 30.3* show methods of constructing column formwork. Columns of this size are almost invariably constructed using column clamps.

The traditional approach for soffits of slabs is to produce a grid of supports with larger timber beams at the top and smaller beams above, directly supporting the sheeting. This has provided a satisfactory method, particularly since the timber prop has given way to the telescopic prop. The grid may also be of tube and fittings, or one of the various proprietary systems. These proprietary systems are often allied with soffit systems. It is sometimes desirable to strip out the majority of the equipment, leaving only a limited number of props. This can be done using a 'quick strip' system. In this the various proprietary components are supported from a special head device so that everything can be released from the concrete except a support area about 150 mm square at the head of the prop itself.

Beams are frequently constructed on their own, even when they are part of a beam and slab construction. It is essential that the support is stable, and thus it is almost always necessary to have pairs of supports. These are spread out along the beam, depending on the strength of the soffit member. The sides are supported from the soffit and are held in a vertical position either by small timber braces, by beam clamps as described above, or by small proprietary adjustable braces. If the slab is being cast with the beam, it is usual to arrange that the beam sides can be stripped in advance of the beam soffit. Where a beam is deep, it will be necessary to treat it as a wall and arrange the formwork accordingly. See *Figure 30.4*.

Stairs

For stairs, a sloping soffit shutter is needed. It is important to set this up securely. Shaped packers should be put in where the supports connect with the sloping members. The side is constructed with a string member cut to the profile of the

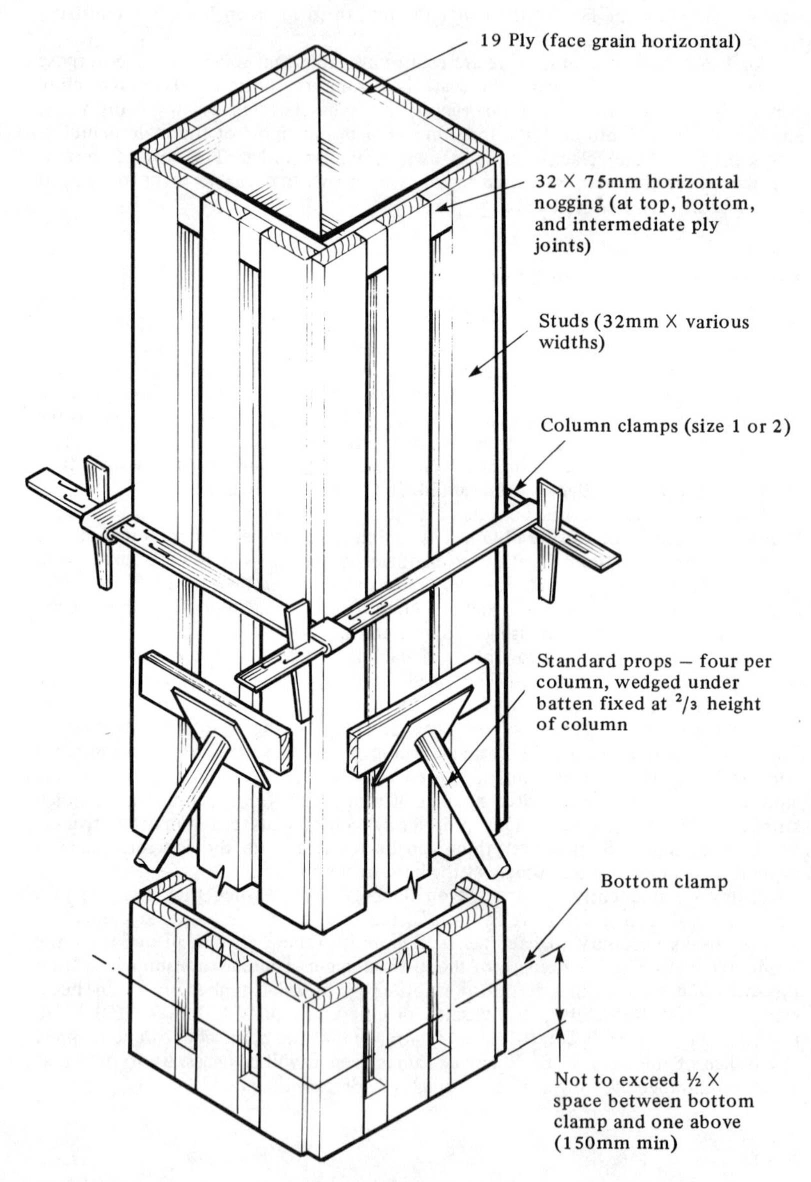

Figure 30.3 Rectangular-column formwork using plywood face, suitable for fair-faced work

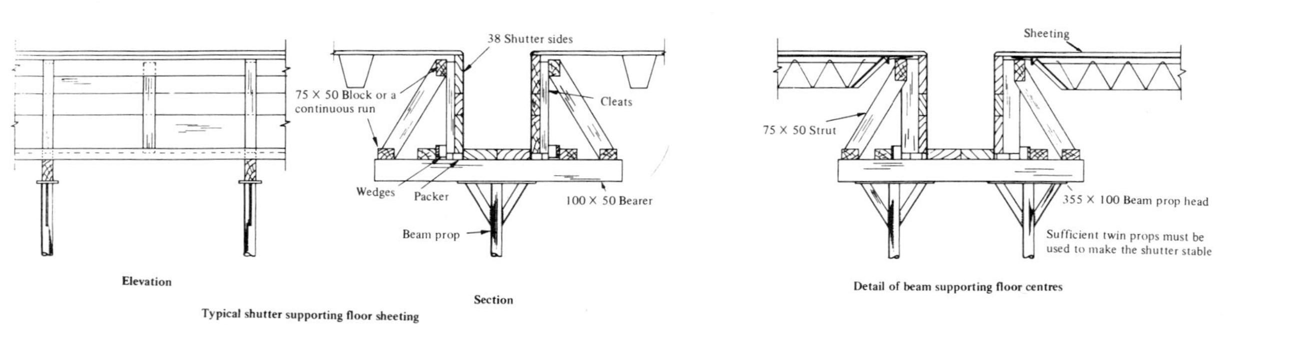

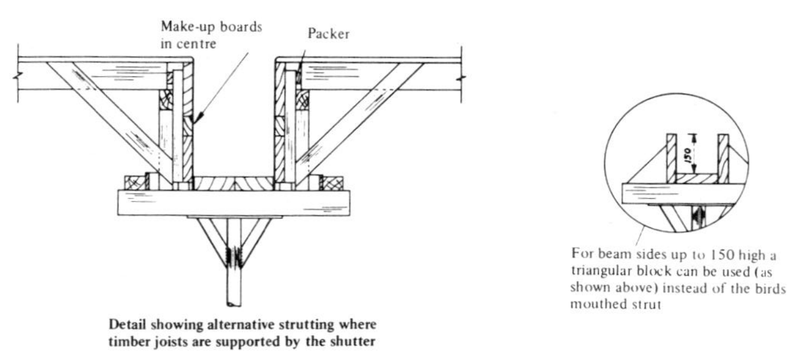

Figure 30.4 Small-beam formwork; where stripping of side shutter will not precede stripping of soffit, omit packers below side seams

treads and risers to be constructed. If the stair is against a wall an inverted string should be nailed to it. To the strings are fixed the riser boards, which form the front of the stairs. It is important that the top edges of the riser boards and the string enable the concrete surface to be finished to the correct level.

Where there is significant repetition and a crane is available, consideration should always be given to pre-casting flights, which can be done on edge, enabling the concreting to be carried out more simply.

Finished concrete

In some buildings, the concrete forms the finished visible surface and is made a positive feature, either plain or in some cases with a texture or pattern. There may be some treatment after casting to expose some of the aggregate within the concrete. To provide a surface of this nature it is necessary to consider carefully the composition of the concrete. The main points to watch when constructing formwork are as follows.

Dimensional errors A shutter may deflect unduly, giving a wavy surface that will be noticeable where the light strikes the concrete at a narrow angle. Individual panels or boards of the shutter may not be precisely level, creating a step in the concrete face. The joint between concrete in one pour and the previous pour may not be made accurately, and again will create a step. All these faults must be carefully avoided by satisfactory design and care in workmanship.

Colour faults If concrete is carefully placed against a uniform shutter surface of uniform quality, there will be little colour variation. But problems can arise from a number of causes. Sugary substances, almost always present to some extent in plywood, give a darkening retardation effect. Black marks may be caused if plywood with a very high gloss plastic face is used. Incorrect application of release agent, particularly too much or the wrong type, will also cause some discoloration. If the shutter is made with joints that are not completely watertight, there is a risk of minor leakage at these joints, and this will show as dark irregular marks along the line of the joint. If the joint is really badly made, there may actually be gaps in the concrete due to the concrete fines escaping through the formwork. Further information can be obtained from the C & CA 'Yellow Book'[6], which discusses the importance of choosing an appropriate release agent. A release agent should always be used.

Stripping

It is essential that stripping does not take place until the concrete is strong enough. For walls and columns (where it is normally sufficient that the surface is not damaged, nor the corners knocked off) the concrete's strength will normally be adequate for stripping (2 kN/mm^2) the next day, unless the weather is very cold. For soffits, a considerably greater strength is needed, normally from half to two-thirds the design strength of the concrete.

Where the strength of the concrete is critical, and it is desired to strip as early as possible, reference can be made to the CIRIA research[7,8]. If the appearance of several pours of concrete is important and they can be seen together, it is very desirable that the procedures of stripping and curing be as similar as possible. If it is necessary to do any remedial work or rubbing up, this is best done as soon as possible.

There are a number of documents on formwork [10,11,12].

FALSEWORK

The primary consideration with falsework is the risk of collapse. It is of concern with soffits and beams and in the assembly of precast lintols and the like, which have to be temporarily supported. It is always possible to use a lot of material to increase safety, but this is uneconomical.

Telescopic props

The most ubiquitous piece of support equipment in the industry is the telescopic prop, commonly called an 'acrow'. It comes in four main sizes, covering the range of soffit heights normally met with. While it is a very adequate piece of equipment, certain points about using it are important. Its strength is significantly impaired if it is not erected vertically; and similarly the load should be arranged over the centre of the head. There will be many occasions on which the full capacity of the prop is not required, and these points are then not so critical, but good practice should be followed on all occasions so that when the prop is heavily loaded it has been put up properly. The pin provided with the prop should always be used, and should be replaced from the manufacturer if it is lost. Alternatives are weaker. Any prop that shows visible signs of damage should not be used but returned to the depot. Data on strength is given in CIRIA research data[9].

Telescopic centers

Telescopic centers can be adjusted to suit most of the spans in buildings, and are a convenient support for plywood. However, their use requires care, and experience shows that a significant number of mishaps occur in their use.

The end bearing of telescopic centers is very critical. When they are heavily loaded it is generally unsatisfactory to support them from timber. Instead some form of steel plate should be used. Where there is a timber beam supporting them it is most important that it is supported in such a way that it cannot turn sideways, and that the support to the beam is centred underneath it and is vertical. Centers should be locked to the appropriate span so that they do not move under vibration. In some cases there is adjustment for camber. This type of equipment is fairly flexible and the shape changes appreciably when the wet concrete is put on, and it will then set in that shape.

Others

There are various other possibilities for support involving scaffolding of various types and the structure itself. It is important in all cases that there is no risk of the support-work collapsing sideways. This instability happens with very small loads along the deck, and will be more likely in windy situations.

Procedures

Because of the risks in falsework, careful supervision and independent checking are desirable. Each problem should be examined systematically and an appropriate method chosen. Its erection should be carefully carried out. Just before loading an independent check should be made that the method has taken account of all aspects of the problem, and that the quality and quantity of the intended materials are correct.

Further information is available in the Falsework Report[13], the Bragg Report[14], and the draft Code of Practice[15].

EXCAVATIONS

In many situations it is practical and safe to excavate shallow holes and trenches without providing any support to them. But if the excavation is of any depth, or if the ground is in any way treacherous, some support work is essential. There is a risk to life, as those working in a trench may well be buried should there be a fall of the bank.

The problems that arise stem from the fact that one has very seldom got a complete picture of the ground in which the trench or other excavation is to be formed. If there have been previous excavations, the ground will have much less strength. If there are planes of weakness, a little rain may cause a section of bank to slide in. In particular, where a previous trench has been dug alongside, the bank may become extremely unstable. Whatever support is provided, it is important to see that it is on a scale appropriate to the problem, and not something that is so flimsy as to be of no practical use.

It should be noted that on most occasions trench supports are an insurance, and it is only in a small number of cases that the full load is placed on the support-work. Unfortunately it is impossible to predict which cases these will be.

Trenches

Trenches in rock normally do not require support-work. Where there are cleavage planes, however, it is possible for large sections of rock to slide in, and these should be specifically supported.

In the more common trench in soft ground, the traditional approach is to use timber poling boards or trench sheeting, wales and struts. This provides a fully sheeted support, and is necessary if the ground is wet and sandy. Where the ground is not as poor as that, it may be adequate to space out the boards on a hit-and-miss basis behind walings. Where the ground is considerably better, it may be satisfactory to use trench sheets spaced apart with trench struts. It is important that everything is placed so that the pipe-laying or subsequent operation can be completed without moving the struts.

There are several proprietary devices available today. For example, frames with hydraulic jacks can be obtained for supporting the sheeting to the side of the trench.

Wider excavations

Where the excavation is too wide to permit strutting straight across it, but the sides must be supported, there are various approaches. The sheeting may be similar to trenching, or sheet piling may be used. It may be practical to use raking struts to a suitable support in the floor of the excavations. Alternatively, it may be possible to carry the struts from one side to the other using intermediate vertical supports called king posts. Such design is beyond the scope of this section.

When strutting is used, the working space is always obstructed. It is frequently possible to use ground anchors as an alternative. In this case the waling is tied back by a tie, which is drilled in at an angle. Provided that this does not need to go under neighbouring property, it is fairly straightforward, although even if it does it may be possible to negotiate for such a facility. Should this approach look appropriate, further information should be obtained from the soil-mechanics firms who provide this service. Details of timbering may be obtained from *Mitchell's building construction*[16].

SHORING

Raking shores

Older buildings may crack, and external walls tend to fall outwards. This is particularly likely if an adjacent building is demolished. It may be practical to use ties right through the building, but it is usually more practical to use raking shores from the ground. The principle involved is to provide a horizontal restraint. The raking shore is connected to the wall by placing short struts right through it. At its lower end a foundation in the ground must be made so that any force carried by the strut does not merely cause it to bury itself in the ground.

Such a shore will be at a rake of 60 to 70°. It should be designed to carry about 10 per cent of the dead weight of the structure that it is to support. It can be constructed in timber, or (as is becoming more common nowadays) in steel scaffold tubing or even in purpose-made steelwork. Timber strutting is normally the work of specialist timbermen, and tubular steel shoring is provided by scaffolding specialists.

Figure 30.5 Typical needling: falsework or centering to a brick arch

Flying shores

Where a house in a row has to be demolished for some reason, the support to two party walls is removed. If there is any doubt about their stability, it is desirable to provide an alternative support with a flying shore. This is a horizontal shore reaching from one wall across the gap to the other; as with a raking shore, it should be designed to take about 10 per cent of the possible vertical load. It can be made of similar materials.

Needling

Many alterations require the replacement of windows by doors, or the insertion of new doorways. To construct these in a brick wall requires temporary support above the lintol that will be put in in the modification. Needling involves making holes above the level of the new lintol, putting a horizontal beam through and supporting both ends. This should be designed to carry the load that could come on to it from directly above the lintol.

It is important that the support at the bottom of the props is quite firm and will not sink, and it is important that the whole is well tightened up before starting to cut the main hole. When the new lintol is in position, it is essential that it is packed from the brickwork so that the load is actually taken by it. In some cases it is possible to use a hydraulic jack to pre-load the lintol and then to pack the remainder of the space.

Traditionally needling has been done with timber, but it is frequently practical to use steel horizontal beams of rolled steel section supported on proprietary props – see *Figure 30.5.* Some of the specialist firms will provide this service using their own equipment.

Some further information on these topics will be found in *Mitchell's building construction*[16].

UNDERPINNING

When foundations are constructed, it may be that they later prove to be inadequate. They may not have been taken to a sufficient depth, so that moisture changes in the ground cause the foundation to settle, or it may be that subsequent construction alongside has to go down lower. In either case it is desirable to remedy the situation by extending the foundation downwards, and this is called underpinning.

It is normally practical to excavate a short section below the foundation and fill it with concrete, packing it solid to the underside of the foundation, and then to repeat the process a number of times until the entire wall has been supported at a lower level. The length to be dealt with at any one time should not exceed 2 m and should often be less. No other point should be opened at the same time that is closer than five times the width of the hole being dug. Timbering is likely to be necessary.

Underpinning is normally carried out by specialist sub-contractors. For further information on this subject the reader is referred to *Mitchell's building construction*[16].

HANDLING STRUCTURAL ELEMENTS

With the advent of cranes as common site equipment, there is a need to handle safely the large number of heavier items such as precast units. It is important that they are picked up correctly. Failure to do this will cause damage to the item, and it may be difficult to place in position. When it is positioned, it may be necessary to provide temporary supports or jigs, to position it correctly and securely until such time as it is finally built into the structure. As this is not part of traditional building, it is essential that full consideration be given to it, and that proper equipment is provided.

Lifting gear is subject to the requirements of the Regulations. It must be tested before initial use and inspected at intervals. Where an item such as a concrete unit has sockets or other fixings cast into it, while these are not lifting gear as defined in the Regulations, it is important that they are appropriately designed to be strong enough to withstand the likely site maltreatment, and are properly fixed into the unit. *Construction safety*[5] gives some information.

REFERENCES

1. *The Construction (General Provisions) Regulations 1961*, HMSO No. 1580 (1961)
2. *The Construction (Lifting Operations) Regulations 1961*, HMSO No. 1581 (1961)
3. *The Construction (Working Places) Regulations 1966*, HMSO No. 94 (1966)
4. CP 97, *Metal Scaffolding*, British Standards Institution
   Part 1: 1967, *Common scaffolds in steel*
   Part 2: 1970, *Suspended scaffolds*
   Part 3: 1972, *Special scaffold structures in steel*
5. *Construction safety*, Sections 7 to 9, *Working places, etc.*, Building Advisory Service (1969–76)
6. Blake, L.S., *Recommendations for the production of high quality concrete*, Cement and Concrete Association (1967)
7. *Tables of minimum striking times for soffit and vertical formwork*, Report No 67, CIRIA (1977)
8. *Formwork striking times – methods of assessment*, Report No 73, CIRIA (1977)
9. *Safe working loads for adjustable props: the influence of prop conditions and site workmanship*, Technical Note No 79, CIRIA (1977)
10. Wynn, A.E., and Manning, G.P., *Design and construction of formwork for concrete structures*, 6th revision, Cement and Concrete Association (1974)
11. Richardson, J.G., *The formwork notebook*, Cement and Concrete Association (1972)
12. *Formwork Report of the Joint Committee of The Concrete Society and the Institution of Structural Engineers*, The Concrete Society (1977)
13. *Falsework Report of the Joint Committee of The Concrete Society and the Institution of Structural Engineers*, The Concrete Society (1971)
14. Advisory Committee on Falsework (the Bragg Report), *Falsework: Final Report of the Advisory Committee*, HMSO (1976)
15. Draft Code of Practice, *Falsework*, British Standards Institution (1975)
16. Foster, J.S., and Harington, R., *Mitchell's building construction: Structure and fabric*, Parts 1 and 2, Batsford (1972 and 1975)

31 CONCRETE FLOORS AND ROOFS

CONCRETE FLOORS 31–2

CONCRETE ROOFS 31–9

CONCRETE FLOORS AND ROOFS 31

31 CONCRETE FLOORS AND ROOFS

P.H. PERKINS, CEng, FASCE, FIMunE, FIArb, FIPHE, MIWE
Senior Advisory Engineer, Cement and Concrete Association

This section deals with concrete floors that are uniformly supported on the ground and suspended floors; domestic conditions and medium/heavy loading conditions are considered. Concrete roofs are considered in relation to structure, fire, moisture penetration and condensation.

CONCRETE FLOORS

Structurally there are two categories of concrete floors:
- floors uniformly supported on the ground;
- suspended floors, where there is a void beneath the structural slab, which is supported on beams, columns or walls.

While the majority of suspended floors are upper floors, there is a growing tendency for ground floor slabs also to be suspended. The use of suspended slabs for ground floors was given considerable impetus by Practice Note 6 issued by the National House-Building Council. Reference should be made to the Note itself for complete details, but in brief it requires that, where the depth of fill beneath a ground floor exceeds 600 mm, the floor should be suspended, which means that it should be carried on independent supports and not supported on the fill material.

Reference is made in the text of this section to the Building Regulations 1976; these apply to England and Wales (apart from the Inner London Boroughs) but do not apply in Scotland. For building work in Scotland, the Building Standards (Scotland) Regulations must be complied with. They are similar to corresponding Regulations in England but not identical, and therefore require careful study.

Floors uniformly supported on the ground

Floors in this category can be divided into three groups:
- lightly loaded (domestic),
- medium loaded (light commerical and light industrial),
- heavily loaded (heavy commerical and heavy industrial).

LIGHTLY LOADED FLOORS (DOMESTIC)

These floors are usually not designed in the normal meaning of the word. However, the use of a solid concrete ground floor slab should not be confused with the laying of oversite concrete, since the latter is intended to seal off the ground and provide a reasonably dry surface over which a suspended timber floor is constructed.

Construction of floors next to the ground is dealt with in Part C of the Building Regulations 1976 (Clauses C3 to C5). Clause C3 requires that any floor that is next to the ground shall be so constructed that it is not adversely affected by moisture or water vapour from the ground. Clause C4 details 'deemed to satisfy' provisions for suspended timber floors. Clause C5 details 'deemed to satisfy' provisions for floors of solid construction incorporating timber (*Figure 31.1*). It should be noted that there is

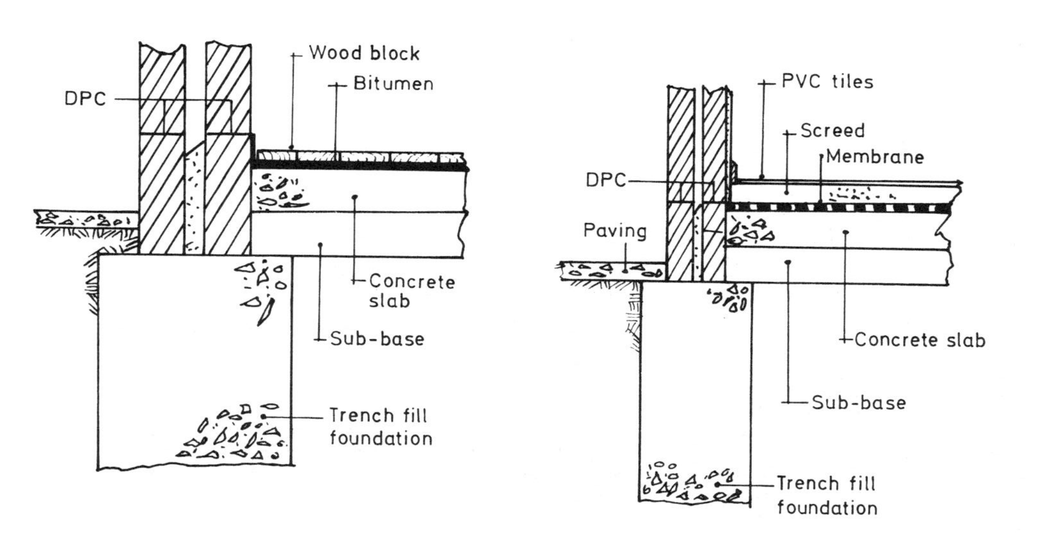

Figure 31.1 Concrete ground floor construction

no specific 'deemed to satisfy' clause for solid concrete ground floor slabs that do not incorporate timber, e.g. a concrete slab finished with linoleum, vinyl, carpet, ceramic tiles etc.

A properly constructed solid concrete slab laid on the ground would comply with Clause C3 of the Building Regulations because the concrete would not be adversely affected by moisture from the ground. This type of floor, usually incorporating a vapour-resistant membrane, is accepted by Building Control Officers throughout the country.

An important point to remember is that ordinary Portland cement concrete is liable to be attacked by sulphates in solution in the subsoil and groundwater. Sulphates can also be present in shale and in imported fill material such as clinker and old bricks. Therefore, if there is any doubt about the possible presence of sulphates, sulphate-resisting Portland cement should be specified and used. The minimum cement content should be as for OPC, which is given below (300 kg/m^3).

Unless the floor can be laid direct on a natural, inert and stable subsoil such as well drained gravel, chalk etc., a layer of inert material (hoggin, broken brick, hardcore etc.) should be used as a sub-base. The sub-base should be laid on the excavated ground and compacted to a thickness of 100 mm. Hardcore should be broken up into pieces not exceeding 50 mm and blinded with fine sand. The structural slab can be laid directly on this unless a vapour-resistant membrane is required as described below.

Where the structural concrete slab is not separated from the ground by an air space it is often necessary to provide a vapour-resistant membrane on or under the slab; the membrane should be carried up to connect with the damp-proof course in the walls (see *Figures 31.1 and 31.2*). The reason for the provision of the membrane

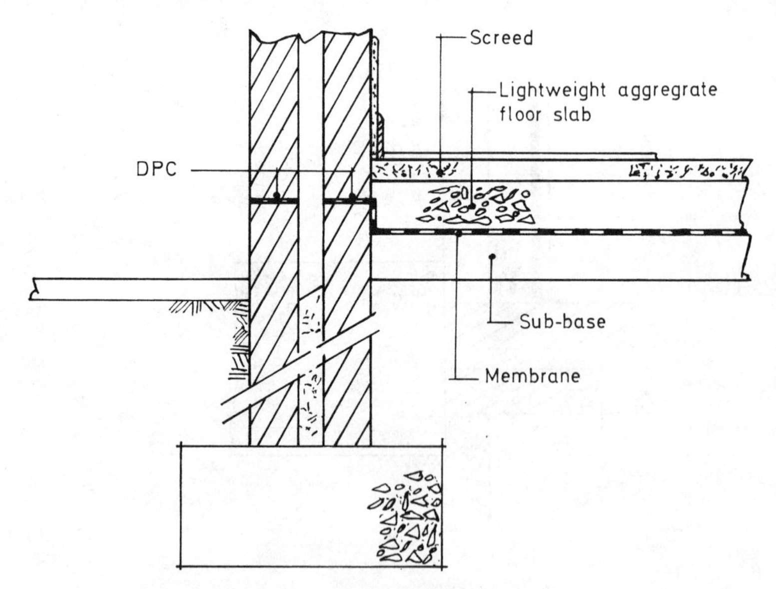

Figure 31.2 Concrete ground floor with membrane below the slab

is that good-quality concrete is watertight but is not vapour tight. Unless the membrane is provided, water vapour will pass through the concrete (even though it does so very slowly), and the vapour will not be able to escape (evaporate) if the top surface of the slab is sealed by a floor covering such as linoleum, vinyl, carpet etc. The result will be an accumulation of moisure beneath the floor covering, with all the trouble that this can cause.

The **concrete mix** for a solid floor of this type should contain not less than 300 kg of ordinary Portland cement per cubic metre of compacted concrete. In terms of BS 5328, *Methods of specifying concrete*, this should be a C20P mix, i.e. a 'prescribed' mix. It is also advisable for the water/cement ratio not to exceed about 0.5, and for the concrete to be well compacted and properly cured for a minimum period of four days. Curing can best be carried out by the provision of polythene sheets well lapped and held down around the edges by scaffold boards. The thickness of the concrete should not be less than 100 mm. The workability of the concrete is important, and for this type of work an 'average' slump of 75 mm is often found satisfactory.

For floors of domestic room size, the slab can be laid in one operation without joints; for larger areas the concrete should be laid between side forms and compacted with a vibrating beam tamper. Reinforcement is not required unless the maximum dimension of the bay exceeds about 4.0 m.

Considerably more attention is now being paid to **thermal insulation** in buildings. The U-value of floors is not easy to calculate, as many factors have to be taken into account. A considerable percentage of heat loss occurs around the perimeter of the slab when this abuts on external walls. Therefore the larger the floor area the less influence the perimeter has on the overall heat-loss and the lower the average U-value is for that particular floor. This is shown by the following figures.

The U-value of a slab 7.5 m × 7.5 m would be about 0.76 W/m^2 °C whereas the U-value for a slab 3 m × 3 m would be about 1.4 W/m^2 °C, on the assumption that both slabs have all four edges abutting on external walls. The thickness of the concrete slab has little effect on the U-value, because the thermal conductivity of the concrete is comparable with that of the ground. A U-value of about 0.5 W/m^2 °C is a reasonable one for a slab with two edges abutting on external walls.

The U-value of a concrete floor can be improved by the use of lightweight aggregate instead of normal aggregates and/or the provision of a thermal insulating screed, or thermal insulation above or below the structural slab. It is important that the thermal insulation should be maintained in a dry condition; this means either using a material that is impervious to moisture or providing a membrane below the insulation. See *Figure 31.2*.

MEDIUM AND HEAVILY LOADED FLOORS

For the lower range of loading, where the live load approaches that used for floors of dwelling houses, schools, hospitals etc., there is no need to consider a mathematical design of the floor slab.

However, as the loading increases, and if the use of the floor involves forklift trucks, heavy trolleys and heavy static loads, it is necessary to take a close look at the design. There are well known and accepted design methods for suspended floor slabs and road slabs, but the same cannot be said for floor slabs supported on the ground. The author recommends the use of Road Note 29 (third edition 1970) with certain modifications to allow for the less arduous condition of use of a floor slab compared with a road. This idea is discussed in detail in *Floors – construction and finishes* (see list at end of section).

A natural question is: to what extent should such floor slabs be reinforced? This depends largely on the size of the slab and its thickness. For domestic buildings there is no need for reinforcement, but for larger areas with a slab thickness of 150 mm it is advisable either for the slab to be laid in bays not exceeding 4.5 m long and about 4.0 m wide, or for reinforcement to be introduced. If reinforcement is used, the bay length can be considerably increased. The length will depend largely on the weight of reinforcement in the longitudinal direction. The reinforcement should be in the top of the slab, about 40 mm from the top surface, and should consist of a high-tensile fabric to BS 4483, *Steel fabric for reinforcement of concrete*. For rectangular slabs the fabric should be rectangular, and for square slabs it should be square. The methods used for calculating the thickness and the amount of reinforcement is beyond the scope of this section, and reference should be made to the 'further information' list at the end of the section.

Except for floors of a very large area, full movement joints (sometimes known as expansion joints) are not required. They are normally only used in floors to coincide with full movement joints in the structure of which the floor forms part. Therefore joints between bays are looked upon as stress-relief joints or contraction joints, and for most floors these can safely be sealed with an epoxy resin as late in the construction process as possible. It is unlikely that there will be any significant thermal movement in the slab once the building is completed, and therefore these joints can be locked. For the slabs that are considered in this section, a minimum thickness of 150 mm is recommended, and the minimum cement content should be 330 kg/m^3. If BS 5328 is used to specify the concrete, the mix should be C25P for the medium loaded floor slabs, and C30P for the heavily loaded ones. If a design mix is used, this should have a characteristic strength of 30 N/m^2 and a minimum cement content of 330 kg/m^3. For directly finished heavy industrial floors a minimum cement content of 360–400 kg/m^3 and a maximum water/cement ratio of 0.5 should be specified. The average slump is usually 50 mm.

DIRECT FINISHING OF FLOOR SLABS

Floor screeds can cause a great deal of trouble from the point of view of poor quality, cracking, curling, hollowness etc. Therefore in recent years there has been a move towards the direct finishing of floor slabs, so that with a minimum amount of 'skim-coat' the floor surface is fit to receive direct, thin-sheet and tile coverings. The Cement and Concrete Association has done a considerable amount of development work in this field, and reference should be made to the 'further information' list at the end of this section if more details are required.

If the floor slab can be finished directly without the addition of a screed there will be a saving in both time and money. However, it must be realised that while most contractors can, if they wish, produce a floor screed with a reasonable finish, there are far fewer who have had the necessary experience to provide an acceptable finish direct to the structural slab. If the surface of the concrete is rough and uneven after it has hardened, it is very difficult and expensive to put things right.

A high-quality concrete floor that has been correctly finished has good wearing properties and can be used for a wide range of light industrial purposes with few maintenance problems. Careful hand or power floating by an experienced floor layer should result in a very durable finish, and in many cases the use of a granolithic or proprietary topping is unnecessary.

Floor sealants such as silico-fluorides and low-viscosity epoxide resins can be very useful in sealing the floor surface and reducing oil penetration etc., but they will not convert a poor-quality floor into a durable one.

PRECAST REINFORCED CONCRETE RAFT SLABS

This method of construction can be very effective for heavily loaded floors on sites subject to continuous and long-term settlement. A well known proprietary raft unit is that made by Stelcon. The units measure about 2.0 m × 2.0 m × 114 mm thick and are made of high-strength reinforced concrete with a steel angle frame around the perimeter. Each unit has two lifting holes and a crane is needed to move them and place them in position, as each raft weighs about 1100 kg. More information on this method is given in the book *Floors – construction and finishes.*

Suspended floors

Suspended floors can be *in-situ*, precast, reinforced or prestressed concrete, with or without an *in-situ* concrete topping, or of composite construction consisting of steel beams with an *in-situ* or precast concrete deck. There are a large number of proprietary flooring systems on the market, and reference should be made to the National Building Agency's Commodity File, where comprehensive information on suppliers and systems is given. The National Building Agency based their classification of floors on five main groups, which can equally well be used for concrete flat roofs:

1. Solid units
2. Hollow units
3. Units of trough or 'T' section
4. Assemblies of beams and blocks
5. Concrete planks used as permanent shuttering

In fact, many of types 1 to 4 require a structural topping in cases where the span and/or live load exceeds certain limits laid down in the manufacturer's literature. Some types of precast unit are illustrated in *Figure 31.3.*

It must be remembered that certain characteristics of precast prestressed floor units differ from those of *in-situ* reinforced concrete. Deflection tends to be somewhat greater, although recovery is satisfactory. When loading tests are carried out, it is often found that deflection under the first loading is greater than under subsequent loading, and there is a more noticeable 'permanent set'.

The concrete is usually very high strength (50 N/mm^2) with low suction; since scabbling of the units to expose the coarse aggregate is not permitted, this can cause problems when trying to bond with subsequent layers of concrete or mortar. The drilling of holes for fixings can be very time-consuming. While the individual precast units may be prestressed, the floor as a whole is not completely prestressed. To obtain this, post-tensioning the whole floor after erection is required. Post-tensioned flat slab floors are used extensively in other countries, particularly in the US, and their virtual absence in the UK is difficult to explain. In 1971 a working party was set up by the Concrete Society to investigate this problem. The working party made considerable use of American design and construction practice, and in 1974 issued an excellent report entitled *The design of post-tensioned concrete flat slabs in buildings.* One of the advantages of this method is that the floor is permanently in compression, and therefore trouble from thermal contraction and shrinkage cracks is eliminated. This is particularly important in those buildings where the floor must be watertight, e.g. multi-storey car parks. It is also obviously a great advantage in roof slabs, which should also be watertight, even though a waterproof membrane is provided.

During the past fifteen years there has been intensive research in many countries on the reinforcement of concrete with fibres as a replacement for conventional steel reinforcement. Numerous types of fibres have been investigated, including special alkali-resistant glass, polypropalene, nylon and steel wire (short steel fibres). Generally, the advantage of random fibre reinforcement is that it increases impact resistance and flexural strength. There is no reason to suppose that concrete containing random steel fibres will be any less durable than ordinary reinforced concrete. Steel wire fibres (Wirand) were used as a reinforcement for precast concrete floor units in a multi-storey car park at London Airport in 1971. Recent information indicates that the floor units are still in good condition, although it was originally intended as a temporary structure.

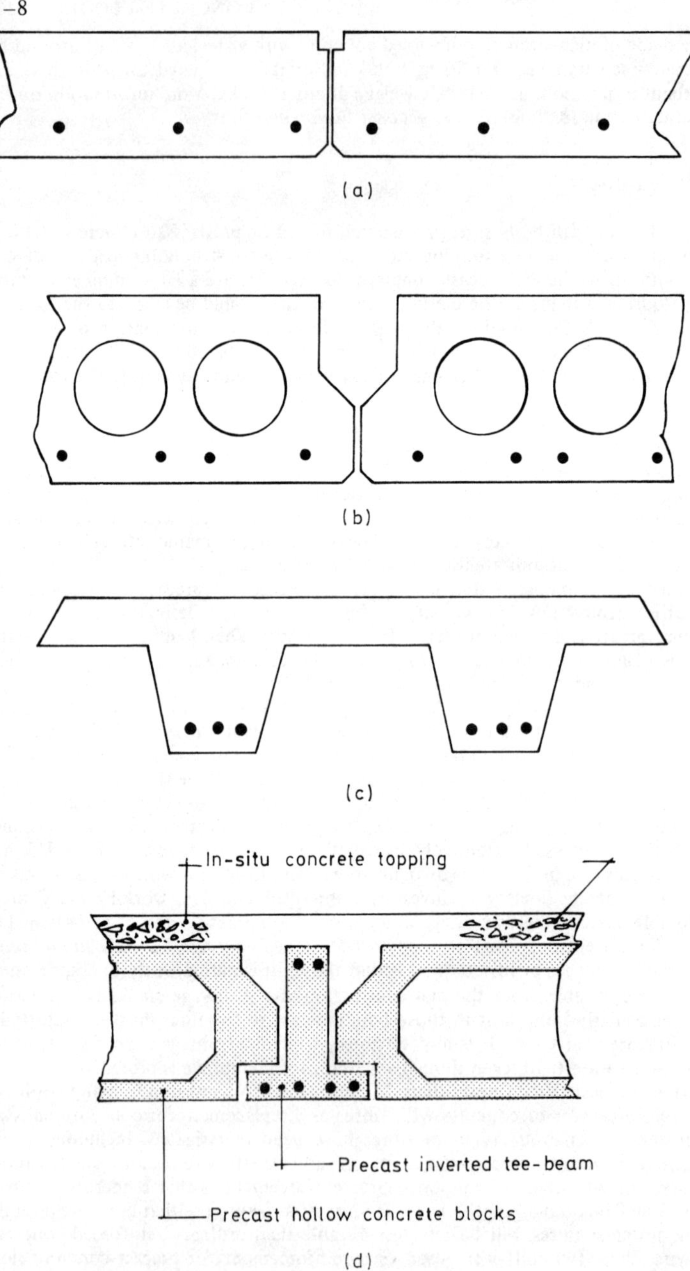

Figure 31.3 Precast reinforced concrete floor/roof units: (a) solid slab, (b) hollow or voided slab, (c) double-tee beam, (d) hollow blocks with inverted tee beam

DESIGN AND SPECIFICATION

Suspended concrete floors should be designed and constructed in accordance with one of the following Codes of Practice:
CP 114, *The structural use of reinforced concrete in buildings*
CP 115, *The structural use of prestressed concrete in buildings*
CP 116, *The structural use of precast concrete*
CP 117, *Composite construction in structural steel and concrete*
CP 110, *The structural use of concrete*
CP 3: Chapter V: Part 1, *Dead and imposed loads*
CP 110 was intended to replace CP 114, 115 and 116, but at the present time and into the foreseeable future these codes are likely to run in parallel. The relevant clauses in the Building Regulations must also be complied with, and these are contained in the following parts of the Building Regulations 1976:
Part C, *Preparation of site and resistance to moisture*
Part D, *Structural stability*
Part E, *Structural fire precautions*
Part F, *Thermal insulation*
Part G, *Sound insulation*
Details of the concrete mix, location and amount of reinforcement, thickness of the concrete and similar matters are covered by the design requirements, and there are numerous textbooks on the subject. In all cases, the basic principles of concrete technology should be applied, such as proper compaction of the concrete, and adequate curing by means of polythene sheets or a sprayed-on resin based membrane. The membrane must be completely removed if subsequent layers or finishes are to be bonded to the concrete.

For this type of structural concrete, **quality control** is mainly based on cube tests and slump. Detailed recommendations for sampling, making the cubes and testing them are given in BS 1881, *Methods for testing concrete*. The BS requirements must be followed exactly, otherwise the results of the tests will be invalid. Any action against the contractor or supplier of the concrete that is based on incorrect testing will not stand up in court nor in arbitration. Another point is that the strength of the concrete in the structure obtained from core tests is likely to be about 70 per cent of that obtained from test cubes, after adjustment for age. This is due largely to the differences in compaction and curing of the concrete in the cube compared with the concrete in the structure.

CONCRETE ROOFS

Concrete roofs are either reinforced or prestressed, *in situ* or precast, and may be divided into three main categories:
- flat roofs (*in-situ* or precast),
- folded plate roofs (*in-situ* or precast),
- shell roofs (invariably *in-situ* reinforced concrete or reinforced gunite).

Folded plate and shell roofs are generally the economic answer in cases where large unobstructed floor areas are required. Flat roofs represent well over 90 per cent of the total annual area roofed in concrete.

The Codes of Practice relevant to the design of concrete roofs are the same as for those mentioned above for floors. The Building Regulations 1976 must also be complied with, and the relevant Parts are:
Part D, *Structural stability*
Part E, *Structural fire precautions*
Part F, *Thermal insulation*

The three most serious problems with concrete roofs arise from:

- water penetration,
- condensation,
- thermal insulation and vapour barriers.

Water moisture penetration through concrete roofs

A concrete roof should be reasonably watertight in its own right without having to depend on waterproofing by a layer of asphalt, roofing felt etc. The usual sources of water penetrating through a roof are:

- rain and snow,
- water in the screed and/or in the thermal insulation when this is located on the top of the roof deck and below the waterproof membrane,
- interstitial condensation in the thermal insulation and/or the concrete deck.

The normal thickness and quality of concrete required for structural purposes is adequate for watertightness, provided there is no honeycombing and no cracking that penetrates the slab; generally this type of cracking is due to thermal contraction, sometimes aggravated by drying shrinkage. It is seldom that drying shrinkage is sufficiently severe to cause cracks penetrating through the structural slab, but thermal contraction cracks invariably do penetrate through the slab itself. The possible occurrence and effect of thermal-contraction cracking, particularly in large roof slabs, are often overlooked.

When water is added to cement a chemical reaction is started; this releases a considerable amount of heat and raises the temperature of the concrete. As the concrete cools it contracts, and any restraint against this contraction will result in the build-up of stress (usually tensile) within the concrete member. If the stress exceeds the strength of the concrete in tension, cracks will develop. These stresses will be increased by any increase in restraint at points of support of the slab and by an increase in the area of the slab that is cast in one operation. The characteristics of these cracks, i.e. their width, length, depth and spacing, depend on many factors, but adequate crack control can be exercised by the provision of properly designed and located reinforcement and stress-relief joints.

There is pressure from contractors to cast floors and roofs in larger and larger pours, and these casting techniques are facilitated by the use of ready mixed concrete and modern concrete pumps. This is a matter that requires careful consideration by the designers, who should provide adequate reinforcement in the top of the slab to ensure proper crack control, thus eliminating a potential source of leakage. The location and detailing of joints are also very important.

Condensation and thermal insulation

In recent years a considerable amount of information has been made available on interstitial condensation. This can be eliminated by proper design and detailing, but to do this the designer must understand the basic principles involved.

Due to vapour pressure, warm humid air inside a building diffuses into the roof slab and then into the screed. If the temperature of the slab or screed containing this humid air falls below the dew point, condensation will occur within the material itself, and this is known as interstitial condensation (*Figure 31.4*). The provision of an adequate vapour barrier, either immediately below the roof slab or between the slab and the screed, should eliminate interstitial condensation.

In the UK, it is usual to provide thermal insulation for concrete flat roofs where the building is to be used for human habitation or occupation. The normal method of construction is for the thermal insulation to be laid on the concrete slab and then

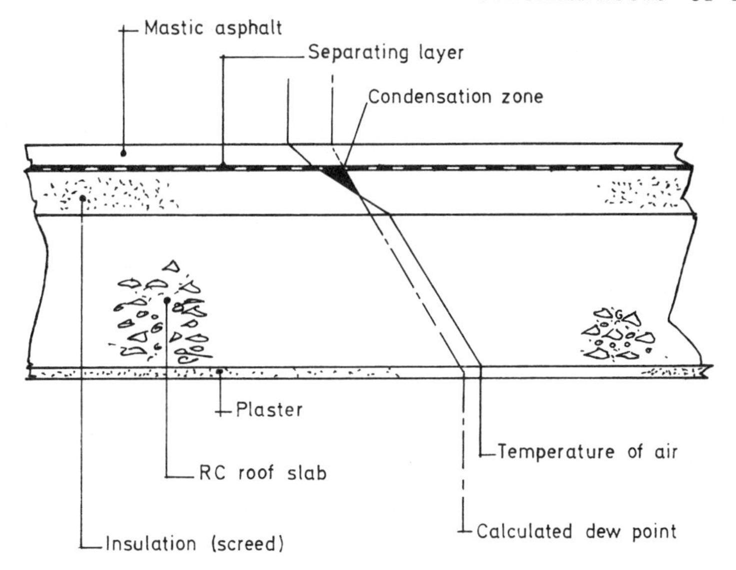

Figure 31.4 Formation of interstitial condensation in roof screed

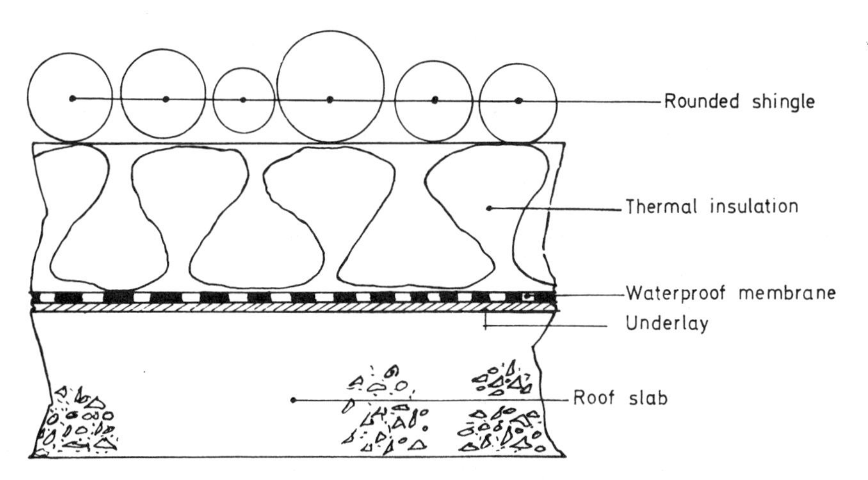

Figure 31.5 The 'inverted' roof

for the waterproof membrane to be laid on the insulation. The use of thermal insulating screeds has been mentioned earlier. However, it is a sufficiently important matter to repeat that any form of thermal insulation that is not impermeable to water and is laid on the concrete slab is liable to become partly or fully saturated by rain during the period that the roof is under construction. This can only be completely avoided by enclosing the whole of the roof structure in a waterproof covering (cocooning) so that the screed is laid under completely dry conditions. Needless to say, this is very seldom done. Any form of thermal insulation that becomes saturated or partly saturated loses a great deal of its thermal insulating capacity. In addition, it is virtually impossible to dry out the insulation before the waterproof membrane is placed on top. The result is that, in the course of time (it may even be years), the trapped moisture in the thermal insulation drains down on to the roof slab, and any cracks or honeycombed areas in it will show up as damp spots in the ceilings of the rooms below.

A method of construction that is very popular on the continent, particularly in Germany, and that has been introduced into this country in recent years is what is known as the 'upside down' or 'inverted' roof (*Figure 31.5*). In this type of construction the thermal insulation must be virtually impervious to moisture, and it protects the waterproof membrane from the deteriorating effects of sunlight, heat and cold.

Fire regulations

In considering the overall design and detailing of the roof, particularly with regard to the type of finish on the soffit of the slab and the type of materials that will go on top of the slab, reference must be made to the Building Regulations Part E, *Safety in fire*, and to Schedule 9.

Clause E1(b) of the Building Regulations requires that the roof shall be capable of satisfying the relevant test criteria for that particular designation of roof. The criteria are laid down in BS 476, *Fire tests on building materials and structures*, Part 3, *External fire exposure roof tests*. It should be noted that the Building Regulations stipulate only the fire designation of the roof from the point of view of *external* fire and the spread-of-flame characteristics of the surface of the roof deck. The standard fire tests in Part 3 of BS 476 are designed to give information on the hazard that exists of fire spreading to the roof from a nearby fire outside the building itself, and are not concerned with the behaviour of the roof when subjected to the effects of fire from the inside. In the tests, roofs are graded according to the length of time they resist fire from the outside and the distance travelled by the spread of flame on the external surface of the roof. The tests require that the material should be complete with any lining that is an integral part of the roof construction. The tests are detailed in Appendix A of Part 3 of BS 476.

The grading of roofs referred to above uses designating letters A, B, C and D in two sets, and the grading can give the result in any combination of pairs of letters such as AA, AC, BC, BA, etc. The first letter refers to the penetration of fire, the second to the spread of flame on the external surface. The meaning of the letters is as follows.

First letter:
A Specimens that have not been penetrated within 1 hour.
B Specimens that have been penetrated in not less than ½ hour.
C Specimens that have been penetrated in less than ½ hour.
D Specimens that were penetrated in the preliminary fire test.

Second letter:
A Specimens on which there was no spread of flame.
B Specimens on which there was not more than 21 in (533 mm) spread of flame.
C Specimens on which there was more than 21 in (533 mm) spread of flame.
D Specimens that continued to burn for 5 minutes after the withdrawal of the test flame, or spread more than 15 in (355 mm) across the region of burning during the preliminary test.

The test specimens are tested either sloping at 45°C (prefix S) or flat (prefix F).

An example of the use of the designating letters is: EXT SBA. This means that, for the external fire exposure test, the specimen was tested at a slope of 45° (S), it was penetrated by fire in a period of between 30 and 60 minutes (B), but there was no spread of flame on the upper surface (A).

It should be noted that the formation of holes, mechanical failure or molten drips is indicated by the suffix X added to the designation, but this is not detrimental from the point of view of compliance with the Building Regulations. Regulation E17 sets out the various acceptable designations, and the conditions attached to them. Designations AA, AB and AC carry no restrictions on use, but roofs with a lower designation than AC are subject to the restrictions set out in the Regulations.

Fire tests are carried out by the Fire Research Station at Boreham Wood, which is part of the Building Research Establishment. Information on testing is given in Fire Note No. 4, *Tests on roof constructions subject to external fire*.

From the above it can be seen that *all* the materials used in the roof construction can influence the final result of the test. For example, a roof covering may achieve an AA designation on a concrete deck, but only BC on a timber deck.

The waterproofing of concrete roof slabs

The following are the principal materials used for the waterproofing of flat roofs:

- mastic asphalt,
- built-up roofing felts,
- *in-situ* materials based on bitumen or polymers,
- special sheet material, such as PVC, butyl rubber, polyisobutylene, etc.

While there are no statistics available on the quantities of the above materials used annually, it appears that there has been a shift from mastic asphalt to built-up roofing felts in recent years. Also there has been some penetration of the market by new *in-situ* materials and special sheetings such as butyl rubber, hypalon (based on neoprene), polyisobutylene and PVC.

Mastic asphalt laid on a concrete deck has an AA fire grade, and two layers of built-up roofing felt (to CP 144) finished with white spar chips on a concrete deck will achieve AB grade.

There are a large number of special sheet materials mentioned above that are available on the market, and of these a number have obtained Agrément Certificates.

An important fact to remember in connection with a flat concrete roof is that, unless there is thermal insulation on the top of the roof slab, the concrete slab will be subjected to a significant range of temperature. The usual practice in the UK is for the waterproof surfacing to be laid as the final surface, i.e. it is laid on top of any thermal insulation. This means that the waterproof surfacing (asphalt, roofing felt, special sheeting material etc.) is subjected to a very wide temperature range, and the effects of frost, thaw, and ultra-violet light from the sun. All these tend to cause deterioration in whatever material is used, and for this reason the continental practice of the 'inverted' roof (*Figure 31.5*) whereby the waterproof layer is placed below the thermal insulation has a definite advantage. In this type of construction the thermal insulation should be virtually impermeable to moisture.

In this section it is not possible to give details of methods of laying of all the various types of roof covering, and all that can be said is that the recommendations of the suppliers should be carefully followed. Information on general methods of laying such materials as mastic asphalt and roofing felt can be obtained from CP 144, and from trade organisations such as the Mastic Asphalt Council.

With sheet material there are the alternatives of bonding it to the substrate (whether this be thermal insulation or the concrete deck) or laying it loose and only securing it around the edges and possibly at a few points on the surface of the roof. This is a method recommended by such firms as Dynamit Noble Ltd, for their Trocal

PVC sheeting, and by Alkor Plastics (UK) Ltd, for their range of PVC 'Alkor-plan' sheeting. Other materials such as polyisobutylene and hypalon are normally fully bonded to the substrate.

A new membrane material imported from Germany is a proprietary chlorinated polyethylene sheeting. Although this is claimed to be superior in a number of respects to PVC, all unbonded membranes have the serious disadvantage that, if they are damaged, water penetrating the damaged area can travel a considerable distance below the membrane. It may then pass through cracks or porous areas in the concrete roof slab. The chance of being able to locate the defect in the membrane is therefore very small indeed.

FURTHER INFORMATION

CP 204: Part 2: 1970, *In-situ floor finishes*

CP 110: 1972, *The structural use of concrete* (including Amendment Slip No. 3, May 1977)

CP 114:1969, *Structural use of reinforced concrete in buildings*

CP 115:1969, *The structural use of prestressed concrete in buildings*

CP 116:1969, *The structural use of precast concrete*

CP 117, *Composite construciton in structural steel and concrete.* Part 1:1965, *Simply-supported beams in buildings*

CP 3:Chapter V, *Loading*. Part 1:1967, *Dead and imposed loads*

CP 102:1973, *Protection of buildings against water from the ground*

CP 144: Parts 1–4, *Roof coverings*

CP 3: Chapter VIII: 1949, *Heating and thermal insulation*

BS 5328:1976, *Methods for specifying concrete*

BS 476:Parts 1–8, *Fire tests on building materials and structures*

Perkins, P.H., *Floors – construction and finishes*, Cement and Concrete Association (1973)

Perkins, P.H., *Concrete ground floors for domestic buildings*, Cement and Concrete Association (1977)

Barnbrook, G., *Concrete ground floor construction for the man-on-the-site*, Parts 1 and 2, Cement and Concrete Association (1975)

Chaplin, R.G., *The abrasive resistance of concrete floors*, Cement and Concrete Association (1972)

Chaplin, R.G., *The regularity of concrete floor surfaces*, Report 48, Construction Industry Research and Information Assoc. (1974)

Deacon, R.C., *Concrete ground floors, their design, construction and finish,* Cement and Concrete Association (1976)

Suspended floor construction for dwellings on deep-fill sites, Practice Note No. 6, National House-Building Council (1974)

The Building Regulations 1976, HMSO

The Building Standards (Scotland) Regulations 1972, HMSO

Tests on roof constructions subject to external fire, Fire Note No. 4, HMSO (1970)

Built-up felt roofing, Building Research Establishment Digest 8, HMSO (1970)

Developments in roofing, Building Research Establishment Digest 51, HMSO (1964)

NBA + Building Commodity File, The Builder Ltd

Condensation in dwellings, Part 1: A design guide, HMSO (1970)

32 BRICKWORK AND BLOCKWORK

CLAY BRICKS 32–2

CALCIUM SILICATE BRICKS 32–9

CONCRETE BRICKS 32–11

CONCRETE BLOCKS 32–12

MORTARS 32–14

DURABILITY 32–17

DIMENSIONAL STABILITY 32–22

PERFORMANCE OF WALLS 32–26

CONCLUSION 32–40

BRICKWORK AND BLOCKWORK 32

32 BRICKWORK AND BLOCKWORK

D. FOSTER, DipArch, RIBA
Consulting Architect
G.D. JOHNSON, BEng(Tech)
Structural Consultant

This section examines two specific functions of brickwork and blockwork walls: durability, and dimensional stability (i.e. movement). The properties of materials (both units and mortar) and of methods of manufacture, where these affect properties, are considered. The likely performance of different types of unit and hence of walls is then examined in relation to their positions in different types of housing in different degrees of exposure.

CLAY BRICKS

Sources of clay

The properties of clay bricks depend on the properties of the clay from which they are made. The UK – primarily England – is rich in brick clays and at one time there were hundreds of local brickworks. Many of these small works have now closed and the vast bulk of production is made from two main types of clay:

- the Oxford clay;
- carboniferous clays – effectively those associated with the coal measures.

The Oxford clay is found chiefly in Buckinghamshire, Bedfordshire, Cambridgeshire and Huntingdonshire, and the bricks made from it are called Flettons. Carboniferous clays are found mainly in the north of the UK: Lancashire, Yorkshire, North Wales, some parts of the Midlands, and of course north-eastern England and Scotland, although those from Scotland are of far less importance than elsewhere.

The Oxford clay and the carboniferous are the source of perhaps 80 per cent of all bricks made. A further 10 per cent comes from the Keuper beds, a Triassic clay appearing at the surface over a large part of central England and extending north-easterly to the mouth of the Tees, north-westerly into Lancashire, Cumberland and Northern Ireland, and south-westerly into Gloucestershire, South Wales and Somerset – even down as far as Devon. The bulk production lies in the Midlands.

The remaining clays that are of some importance are:

- the glacial clays (boulder clay), mostly in the north of England;
- the brick earth (silts) from which London Stock bricks are made;
- the Weald clays, in Sussex;
- old rocks such as Devonian Shales.

Manufacture

There are three main processes:

- *pressing,* which can be subdivided into stiff plastic pressing and semi-dry pressing;
- the *extrusion* or *wire cut* process;
- the *soft mud* process.

PRESSING

Pressing accounts for perhaps two-thirds of the total national output of bricks, although the percentage is changing in favour of extrusion with practically every new plant that is built. However, the whole of the Fletton brick output (approximately half the total national output) is made by the semi-dry pressing process, and this is unlikely to change. Here the water content of the clay as it is won from the pit is such that no water needs to be added in the making process. It is this relative simplicity (together with mass production, and the fact that the clay as won contains a large amount of natural fuel) that makes Fletton bricks as cheap as they are. In pressing, the clay is rammed into the mould; this action produces a depression or 'frog', which in Fletton bricks is quite deep but only on one face. As all builders know, this means that quite a lot of mortar is required to fill it when the brick is laid frog up in the normal way. The argument of 'frog up or frog down' is a consequence.

Carboniferous shale bricks are made by the stiff plastic process and by the extrusion process. The carbonaceous clays require a fair amount of breaking, grinding and mixing with water to bring them to a suitable consistency for manufacture. The stiff plastic process is used mainly in the north of the country – Durham, Northumberland, Lancashire, West Yorkshire, Staffordshire and South Wales. In pressing processes also, frogs are formed in the brick. Sometimes these are shallow and sometimes they occur on both faces. In both pressing processes another characteristic is that the brick in the unfired state has sufficient strength and stiffness to enable it to be put in the kiln without a prior drying process.

EXTRUSION OR WIRE CUT PROCESS

The simplest analogy to the wire cut process is the squeezing of toothpaste out of a tube. A clay extruder obviously requires considerable force, and the greater the water content of the clay as extruded, the less the power required.

In the process of extrusion the size of the mouthpiece, or 'die', is the bed face of the brick, with an additional allowance for drying and firing shrinkage. As the ribbon of clay is extruded it is chopped into individual bricks; many of them are solid (hence the term solid wire cuts) but by far the greater proportion are perforated. It is a relatively simple matter to arrange for the clay to flow around pieces of steel in the die, thus forming holes in the bricks. As the green ribbon is extruded these holes lie in a horizontal plane, but in the wall, when the brick is laid on its bed, they are vertical.

The extrusion process produces planes of weakness in the green body if the differential rate of flow of the clay between, say, the middle of the brick and the edge of the clay is excessive. Other parts, such as the metal used to form holes, inevitably distort the ideal, even flow. These particular weaknesses are not present in green pressed bricks; sometimes, therefore, in the search for more cohesion and accuracy of shape and size, pressing succeeds extrusion to form 'pressed wirecuts'.

SOFT MUD PROCESS

This method, as its name implies, uses clays made so malleable with water that they can be very easily moulded. The machines used were devised to simulate the centuries-old hand process. Bricks made by soft mud processes are almost inevitably less accurate in size than their counterparts made by pressing, or indeed extrusion processes, but of course to many their attraction lies in this unevenness of shape and size. Hand-made bricks have frogs, and can be recognised by the uneven creasing of the face caused by sand in the mould before the clay is thrown or pushed into the moulds.

Classification of clay bricks

Clay bricks and blocks are classified in BS 3921: 1974[1]. The somewhat bewildering combination of three Varieties, three Qualities and four Types, excluding special shapes, is perhaps best understood by reference to *Figure 32.1*.

The house builder will be concerned primarily with two Types:

- solid
- perforated

Somewhat strangely, a solid brick is defined as one 'in which holes passing through, or nearly passing through, do not exceed 25 per cent of the volume of the brick or block, or in which a frog does not exceed 20 per cent of the volume'. In other words BS 3921 does not contain a separate classification of a brick without any

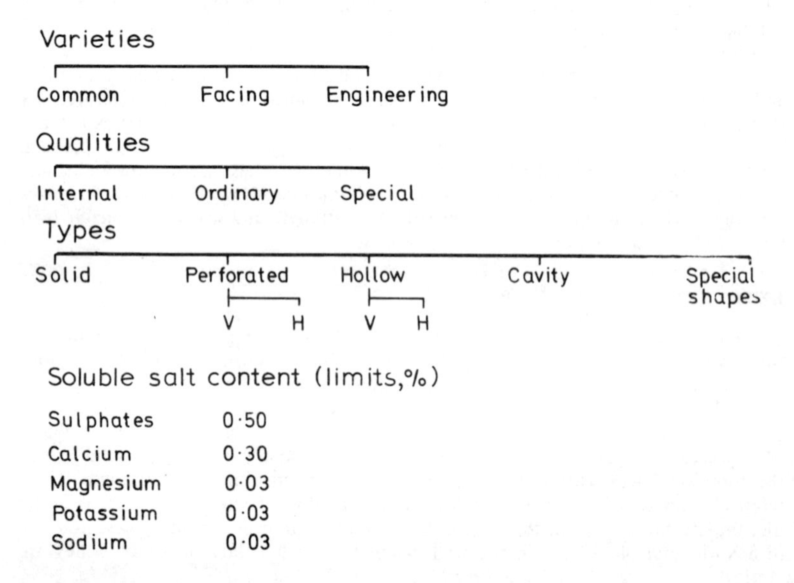

Figure 32.1 Clay brick classification to BS 3921: 1974

perforations or frogs – a 'solid' solid, so to speak. The reason is that the performance of brickwork of bricks with a percentage of perforation not exceeding the defined figure is held to be equal to that of brickwork of equivalent 'solid' solid bricks of the same material, when bricks are crushed on bed. Strengths on edge and on end can be startlingly different.

As for Varieties, the builder will be mainly concerned with two of the three:

- facing
- common

Of these, facings will predominate because today the inner leaf and walls below DPC of the external wall are usually in blockwork. Occasionally builders will use the third Variety, engineering bricks, but it must be realised that an 'engineer' can also be a facing or a common, since the difference between the last two is simply one of appearance. A facing is defined as 'specially made or selected to give an attractive appearance when used without rendering or plaster or other surface treatment of the wall', whereas no such claim is made for common bricks.

Of the three Qualities, the builder usually will be interested only in two:

- Ordinary Quality
- Special Quality

The other, Internal Quality, is not (to the best of the authors' knowledge) manufactured as such but is appropriate for rejected commons or facings. If such rejection has been on grounds of accuracy of shape or faulty appearance, the bricks may prove highly suitable and economic for plastered internal work, but if rejection has been because the bricks were underfired the builder should beware. Extra care is required with Internal Quality bricks, to avoid them becoming wet and subjected to frost action during construction or they will most likely spall or 'shale', i.e. lose part or the whole of their faces. The footnote to para 3.3.1 of BS 3921 reads: 'bricks of internal quality may need protection on site during winter'.

It should be realised that all three Varieties, common, facing or engineering, can be of Ordinary or Special Quality and be either solid or perforated. As examples, one may have an engineering perforated brick of Special Quality, or a facing solid brick of Ordinary Quality.

Ordinary Quality and Special Quality are of particular importance to the builder. Special Quality is defined as 'durable even when used in situations of extreme exposure where the structure may become saturated and be frozen – e.g. retaining walls, sewage plants, or pavings'. Ordinary Quality is defined as 'less durable than Special Quality but normally durable in the external face of a building'. Special Quality bricks are frost resistant and have stringent limits on content of soluble salts, the latter being determined by recognised analytical methods laid down in BS 3921. However, there is as yet no satisfactory test that is universally accepted for the determination of frost resistance. The Standard therefore requires that a brick or block of Special Quality shall satisfy one of three different criteria, namely:

1. 'The manufacturer shall provide evidence that bricks and blocks of the quality to be offered have been in service under conditions of exposure at least as severe as those proposed for not less than three years in the area or locality in which their use is now to be considered, and that their performance, by inspection, has been satisfactory.
2. Alternatively, in the absence of structures built from the bricks or blocks in question in the area concerned, manufacturers should arrange for sample panels of their products to be built in an exposed position under the supervision of some independent authority, e.g. the local authority of the area in which the works are situated, so that independent testimony to their quality shall be available. Details of construction of sample panels and for determination of the strength plus water absorption are given in an Appendix.

   Bricks or blocks which give a satisfactory performance after not less than three years' exposure in panels built in accordance with the above paragraph are deemed to satisfy the frost requirements for bricks or blocks of Special Quality.
3. Where it is not possible to provide evidence of frost resistance by either of the above methods, for example if the brick to be offered is essentially a new product, or if it is to be used for the first time in a certain area, a brick or block with either a strength not less than 48.5 N/mm^2 or a water absorption not greater than 7 per cent shall be deemed to be frost resistant.'

The requirements in (3) above are those for Class B engineering bricks. Usually these bricks and, of course, Class A (the corresponding figures for which are 70 N/mm^2 and 4 per cent) also meet the limits of soluble salts, but they are *not* automatically classified as of Special Quality. Compressive strength is not necessarily a guarantee of frost resistance, because bricks of startlingly different strength can be frost resistant. For example, London Stocks, made by the soft mud process, and most hand-made bricks, are not only very porous but have unusually large pores, so that water drains out of them and the brick does not become fully saturated in the work.

The authors have seen ice extruding from such bricks without any subsequent damage, yet they are quite weak in terms of compressive strength, being of the order of 14 MN/m^2 or less. Porosity is high (water absorption of the order of 25–30 per cent by weight), and thus the bricks are light in weight.

By contrast, bricks of high compressive strength are heavy, have low water absorption and achieve frost resistance largely because of their high internal cohesion (and hence resistance to the bursting forces of ice), owing to the formation of glass-like bodies.

By and large, the higher the firing temperature the more likely it is, for a given clay, that compressive strength and frost resistance will be increased, and water absorption and soluble salts content kept as low as possible. But there are limits to firing temperatures. The higher the firing temperature the greater is the firing shrinkage, and hence the more likely the 'setting' (the arrangement of the bricks in the kiln) is to tilt and collapse, with disastrous results. Therefore there is an optimum firing temperature for each clay, which determines the finished properties – strength, water absorption and limits of size. The first two have been discussed to some extent, and the picture would be incomplete without some comment on limits of size.

Limits of size

There is nothing in BS 3921 that specifies limits of size on individual bricks, although it is commonplace for users to think that there is, or (when they find to their cost that this is not so) that any reasonable person would expect that there should be! But the very nature of the raw material and methods of manufacture make this impossible. There are variations in the clay, differences in firing temperature across the width and height of the kiln, wear in the die and fluctuations in the water content, to name a few factors. Some of these will cause short-term, others long-term, variation. The only fair and reasonable method of assessing the differences caused is to measure a number of bricks jointly, not individual bricks. The Standard does this, as shown in Table 32.1.

Table 32.1 LIMITS OF SIZE (24 BRICKS)

Size (mm)		*Overall measurement of 24 bricks* (mm)	*Limits of size* (mm)
Length	215	5160	± 75
Width	102.5	2460	± 45
Thickness	65	1560	± 45

It will be seen that the range of allowable total variation in the length of 24 bricks laid dry and touching is ± 75 mm. This does *not* mean that the maximum and minimum length of an individual brick is 215 ± (75 ÷ 24), i.e. 215 ± 3.125 mm. More sensibly it means that there are some that are longer or shorter than others.

With perfect production the distribution would be as shown in *Figure 32.2*, which is an idealised distribution of the length of 100 bricks. In practice the results from different sets of 100 will vary, even if they are from the same batch of production, but the typical form of the histogram will be the same. In *Figure 32.2* all bricks are within 5 mm each side of the average length, and the further away from the average the greater is this difference. The dotted line shows what would happen if many bricks were measured and the lengths plotted in this way but with much smaller differences than 1 mm. A continuous curve would result, which statisticians call the *Normal distribution.*

For Normal curves it is usually accepted that 95 per cent of values will lie within ± 1.96 times the standard deviation from the mean. The standard deviation on length derived during the development of BS 3921 is 1.9 mm, so one can conjecture that, within a batch where the mean is 215 mm, 95 per cent of lengths will be within the range of 215 ± (1.96 × 1.9), or roughly 211–219 mm. A much smaller percentage (0.1 per cent) will be about ± 6 mm, which means that one standard brick in every 1000 could be 209 or 221 mm long.

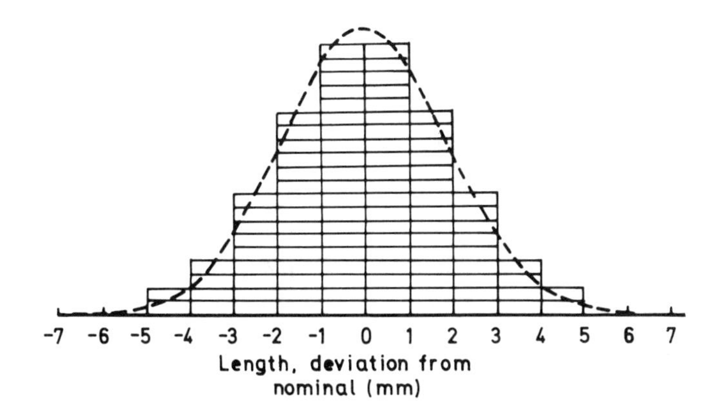

Figure 32.2 Idealised distribution of lengths of 100 bricks

There must therefore be a method of sampling that selects the 24 bricks to be measured so that there is a fair representation of the batch. BS 3921 lays this down very carefully. Manufacturers must not only be on the lookout for short-term variations, they must also guard against long-term variation. Wear of the die is an example from the extrusion process. They do this, of course, by quality control methods at the works, where the aim is to keep the mean from 'drifting'. Only small amounts can be tolerated. In a Normal distribution where the average is the nominal 215 mm, 99 per cent of all bricks could be expected to be within ± 4.9 mm. However, if the batch average is only 1 mm greater than the nominal 215 mm, two out of every 100 bricks could be expected to be greater than that upper limit of 4.9 mm, but only 1 in 1000 would be less than the lower limit.

The builder's difficulties are worse when bricks are longer, so whenever his design or what he is asked to build includes very short lengths of brickwork (such as narrow piers) he should price for selection if it has to be done on site, or arrange for the specification to include for selection before delivery, or at least draw somebody's attention to the problem before there is a dispute!

Modular clay bricks

According to BS 3921 a clay brick should not exceed 337.5, 225 and 112.5 mm in length, width and height respectively. Thus the Standard embraces the sizes of the biggest metric modular bricks currently in vogue as well as the standard metric bricks. A Draft for Development, DD 34[2], published separately, lists four formats of metric modular bricks and their respective limits of size, as shown in Tables 32.2 and 32.3.

The builder should realise that all the statistical background outlined above for standard metric bricks applies equally to clay metric modular formats, and in particular that the 288 work size has a range (between limits) of 200 mm, compared with the 150 mm range of the 215 length of the standard metric brick, i.e. almost exactly pro rata. Therefore, it must be expected that there could be considerable variation in length in a batch that still meets the requirements of BS 3921. It follows

Table 32.2 MODULAR BRICKS

Format	*Size with joint*		*Height*
	Length	*Width*	
1	300	100	100
2	300	100	75
3	200	100	100
4	200	100	75

Table 32.3 LIMITS OF MODULAR SIZE (24 BRICKS)

Work size (mm)	*Overall measurement of 24 bricks* (mm)
65	as Table 32.1
90	± 45
190	± 66
288	± 100

that the earlier remarks about tolerance problems with narrow piers of brickwork will apply with even greater force to work in 300 mm metric modular bricks. When these are used for a pier nominally 300 mm wide, either accurately gauged bricks will be required or else a considerable amount of selection will be entailed if the job is to look well. This applies equally to 215 mm piers in standard metric bricks. The wider a pier or column, therefore, the less is the problem of size because the greater is the number of joints to absorb variations.

Strength of clay bricks to BS 3921

There is no difference in Table 32.4 between engineering and loadbearing Class 10 and 7 in terms of compressive strength, and the builder may be forgiven for wondering why they are classed separately. The reason is that there are also specific requirements for level of water absorption for engineering bricks, namely 4.0 and 7.0 per cent by weight for Classes A and B respectively.

Table 32.4 AVERAGE COMPRESSIVE STRENGTH ON BED, MN/m^2

Kind of brick	*Class*									
	15	10	A	B	7	5	4	3	2	1
Engineering	–	–	69.0	48.5	–	–	–	–	–	–
Loadbearing	103.5	69.0	–	–	48.5	34.5	27.5	20.5	14.0	7.0

Efflorescence

The soluble salts limits in Special Quality bricks are still enough to cause efflorescence in certain circumstances. Only very small percentages of salt are necessary, and all clay bricks contain more than enough. Hence the liability to efflorescence, i.e. moderate as rated by the test, is the same for bricks of Ordinary and Special Quality. This does not mean much to the builder, who would be best advised to avoid the problem by the methods described in various ways later; in short, the bricks should be kept as dry as possible.

Moisture expansion of clay brick

The short-term wetting and drying movement of clay bricks is very small, but there is a long-term expansion caused by the glass-like compounds within the brick slowly adsorbing water from the atmosphere over a long period from the time they leave the kiln. No limits or method of test are given in the body of BS 3921 but the phenomenon is briefly discussed in the Foreword, where an amount up to 0.2 per cent linear is postulated. This is high, although it will be less in actual brickwork and it varies with different clays and firing temperatures. Broadly, the higher-strength, more vitrified dense bricks have the highest long-term expansions; the weaker and more porous the brick, the less the expansion. In the authors' opinion the greatest difference is not less than 3 to 1, with Flettons and Keupers in the middle range. At the time of writing these are at best suggestions based on experience, but there is no doubt that substantial differences exist and that builders need to be aware of them when designing to accommodate movement.

CALCIUM SILICATE BRICKS

Raw materials and manufacture

The raw materials of bricks to BS 187: 1978[3] are a mixture, largely, of sand and about 5–9 per cent of lime. If there is a substantial proportion (the amount is not defined) of crushed flint in the sand, the bricks may be described as flint-lime bricks, but practically the distinction does not mean much. The essential bond is formed when the lime and silica in the compacted mix are treated with steam at high pressure for several hours in sealed chambers, or autoclaves.

Calcium silicate bricks are made throughout the UK and form perhaps 6 per cent of all bricks made.

Classification to BS 187: 1978

Three designations are given:

- loadbearing
- facing
- common

The buyer should specify that the bricks he requires be one of these.

Strengths are classified as in Table 32.5. If no strength is specified, BS 187 decrees that the lowest strength in that class is required.

Table 32.5 CLASSES OF COMPRESSIVE STRENGTH

Kind of brick	*Class*					
	Green 7	*Blue* 6	*Yellow* 5	*Red* 4	*Black* 3	– 2
	(a) Mean compressive strength on bed (MN/m²)					
Loadbearing and/or facing	48.5	41.5	34.5	27.5	20.5	–
Facing or common	–	–	–	–	–	14.0
	(b) Predicted lower limit compressive strength on bed (MN/m²)					
	40.5	34.5	28.0	21.5	15.5	10.0

It will be seen that the classes follow closely those in BS 3921 but the maximum is much lower, although it will be adequate for most loadbearing situations. Class 2 is reserved for commons, but some bricks of this strength are of sufficiently good appearance to be deemed facings, which in these bricks is related to evenness of the generally smooth texture produced, and to accuracy, shape and colour.

The mean strengths are an average of ten, as in BS 3921, but the predicted lower limit strengths, determined statistically from the ten individual results, are included to demonstrate uniformity of production. A specified class must meet both requirements, and the practical effect of meeting the second is that there is only a small chance (about 1 in 40) of the average of any further ten bricks falling below it, in other words a guaranteed lower limit for the mean. This is unlikely to matter very much to the house builder concerned only with simple low-rise construction, but it has importance for engineered structures.

No 'Qualities' of durability are specified as in BS 3921. Instead, durability relates directly to strength, in that the more arduous the exposure the higher the strength requirement, as will be seen later.

An interesting and useful addition to BS 187: 1978 is the sensible use of colour marking to denote strength classes.

Limits of size

Here the nature of the material and method of manufacture are closely reflected in the degree of uniformity to be achieved. Limits of size have been set for individual bricks:

Length	215	+2 −3
Width	102.5	+2.5 −1.5
Height	65	±2

The deemed-to-satisfy requirements when the bricks are measured to the nearest millimetre are clearly stringent in comparison with clay bricks, and accurate brickwork in small lengths presents no problems.

Modular calcium silicate bricks

The same four modular formats are made in calcium silicate bricks as in clay. They are listed, with limits of size, in DD 59:1978[4], and their accuracy of manufacture gives them clear advantages in purely modular terms. With them there will be no difficulty in fitting into allotted modular spaces whatever the wall or pier size. In this respect it is significant to note that the work size for the length of the 300 formats is 290, compared with 288 for clay bricks. The limit on this work size is ± 3 mm, which range is easily absorbed by the 'standard' 10 mm mortar joints.

Drying shrinkage

This part of BS 187 is of particular importance to the house builder. Calcium silicate bricks have a high reversible moisture movement compared with clay bricks, but in practical building it is the drying shrinkage part of this movement that is important, particularly the initial drying shrinkage occurring from the time of building to the relatively steady state in use. BS 187 sets a maximum of 0.040 per cent, except by agreement between purchaser and manufacturer. This figure is a relaxation from earlier editions, where the highest standard was 0.025 per cent. The measurements are made, of course, on individual bricks not walls, and the relation between the two, even in best practice, is not well understood, primarily because of the effects of restraints, discussed later.

CONCRETE BRICKS

Raw materials and manufacture

The manufacture of concrete bricks needs no explanation, being simply that of small units of precast concrete by moulding. They are classified in BS 1180: 1972[5]. Aggregates can be dense or lightweight, and range from stabilised wood particles and all kinds of lightweight aggregate, to gravels. There are differences in the definitions of solid, perforated, hollow and cellular between BS 1180, BS 3921 and BS 187, which may affect volumes of mortar used per m^2 of wall if used in the normal way, i.e. on bed. As with calcium silicate bricks, strength is tabularly related directly to durability in different degrees of exposure (see Table 32.6), the format of the table being derived from CP 121: 1973[6]. It will be noted that again there is a degree of uniformity of production required. The drying shrinkage figures exceed those of BS 187.

Table 32.6 PHYSICAL REQUIREMENTS (Table 2 of BS 1180, reproduced by permission of the British Standards Institution)

Physical property	*Compressive strength category*					
	7.0	10.0	15.0	20.0	30.0	40.0
Compressive strength (wet) average of 10 bricks to be not less than MN/m^2	7.0	10.0	15.0	20.0	30.0	40.0
Coefficient of variation of compressive strength not to exceed %	30	30	30	20	20	16
Drying shrinkage not to exceed %	0.06	0.04	0.04	0.04	0.04	0.04

Limits of size – standard brick

Ranges are greater than in BS 187. For the standard metric brick they are:

Length 215 $^{+4}_{-2}$

Width 103 ±2

Height 65 ±2

Modular formats

Formats are those of DD 34, except for the 300 × 100 × 75. The limits on the 290 work size are +4, –2. These give the same range as the calcium silicate equivalent (290 ± 3) but it will not be so easy to accommodate concrete brick length variation in standard 10 mm joints.

CONCRETE BLOCKS

Materials of manufacture

Aggregates are listed in BS 2028, 1364: 1968[7] and have the same range as for concrete bricks. They can be dense or lightweight, with the emphasis on the latter because of thermal insulation and ease of handling. Aerated concrete, which is a mixture of cement and sand with pfa foamed by mixing with aluminium powder[8], is another technique that accounts for a large volume of production.

Aerated blocks are not subjected to any pressing or moulding process (being cut from a large cake) but aggregate blocks are moulded and compacted in various degrees. Curing is by air, steam at atmospheric pressure, or autoclaving. The last has the advantage of forming hydrated calcium silicate, with a reduction in cement content.

Classification to BS 2028

Three types of blocks are listed:

- *Type A* (density 1500 kg/m^3 or greater). For general use, including below DPC at ground level.
- *Type B* (density less than 1500 kg/m^3). Type B can be used as Type A apart from

use below DPC in an external wall or leaf, but some Type B blocks can also be used in that situation of high exposure to frost action. The restriction obviously reflects the durability of different blocks, with stronger (>7.0 MN/m²) blocks of denser aggregates being more suitable. However, if a manufacturer can show authoritative evidence of suitability, any other Type B block may be used; e.g. some aerated blocks have been shown to be highly resistant to frost action.

- *Type C* (density less than 1500 kg/m³ – usually much less). Type C blocks are intended for internal non-loadbearing walls and partitions.

Table 32.7 SUMMARY OF COMPRESSIVE STRENGTH REQUIREMENT

Type	*Designation*	*Minimum average compressive strength* (MN/m²)	*Lowest individual strength* (MN/m²)
A	(3.5)	3.5	2.8
	(7)	7.0	5.6
	(10.5)	10.5	8.4
	(14)	14.0	11.2
	(21)	21.0	16.8
	(28)	28.0	22.4
	(35)	35.0	28.0
B	(2.8)	2.8	2.25
	(7)	7.0	5.0

Compressive strength requirements are specified in BS 2028 for Types A and B and transverse strengths for Type C. The former is an average of ten blocks, the latter of five blocks. See Table 32.7.

Form

All types may be solid, hollow or cellular, with the obvious exception of aerated blocks. 'Solid' blocks have at least 75 per cent of solid material by volume. Hollow blocks have one or more holes right through and the total volume of solids is 50–75 per cent of the gross volume. Cellular blocks are like hollow ones, but the holes are closed at one end.

All kinds of faces are available from manufacturers (profiled, marble and stone chippings, coloured sands) and are acceptable in BS 2028 provided that the blocks they are on meet the requirements of their Type.

Formats

A wide range of formats, e.g. 450 × 225 × 75, is made. A 'standard' joint thickness of 10 mm is assumed for length and height in the work. Thickness is obviously unaffected by this. Thus the work size of the above example is 440 × 215 × 75. See Table 32.8.

Limits of size

Length and height +3 mm
–5 mm

Thickness ±4 mm (individual measurements)

Variation in the average thickness of any one block of a sample shall not be greater than the specified thickness ±2 mm.

Table 32.8 DIMENSIONS OF BLOCK (based on Table 1 of BS 2028, by permission of the British Standards Institution)

Type	*Length*	*Height*	*Thickness*
A	390	90	75, 90, 100,
	390	190	140, 190
	440	215	75, 90, 100, 140, 190, 215
B	390	90	75, 90, 100, 140,
	390	190	190
	440	190	75, 90, 100,
	440	215	140, 190,
	440	290	215
	590	190	
	590	215	
C	390	190	60, 75
	440	190	
	440	215	
	440	290	
	590	190	
	590	215	

Moisture movement

Change in dimension due to variation in moisture content is one of the prime considerations when considering movement in blockwork. Drying shrinkage from the delivered or stored state to that finally in use is the main criterion in this respect.

Table 32.9 DRYING SHRINKAGE OF TYPE A BLOCKS

Strength	*Drying shrinkage*
3.5	0.05
7.0	0.05
10.5	0.06
14.0	0.06
21.0	0.06
28.0	0.06
35.0	0.06

The maximum permitted drying shrinkage of Type B blocks ranges from 0.07–0.09 per cent, but for Type A the percentage varies according to the strength as shown in Table 32.9. For Type C the maximum is 0.09 per cent.

MORTARS

Mortars form more than 16 per cent of the volume of a half brick wall of standard bricks or about 7 per cent of a block wall of 450 × 225 blocks. Although manufacturers control properties of blocks, properties of walls are affected greatly by the builder's understanding of mortar, and its importance is therefore obvious.

Cements for mortars

Four types of cement are used in bricklaying mortars. They are:
- Portland cement (ordinary and rapid hardening) to BS 12: 1958, or Portland blast-furnace cement to BS 146.
- Sulphate-resisting cement to BS 4027. Useful for certain highly exposed situations where sulphates are present.
- Masonry cement to BS 5224: 1976. Consists of Portland cement mixed with about 25 per cent its volume of finely ground inert mineral fillers and an air entraining agent.

Limes for mortars

- Non-hydraulic limes (high calcium or white lime) to BS 890.
- Semi-hydraulic limes (grey limes are semi-hydraulic) to BS 890.
- Magnesian limes.

Lime in mortars is usually non-hydraulic or semi-hydraulic. Hydraulicity of a lime denotes its capacity to harden and develop strength *in a wet state*. The degree of hydraulicity depends on the proportion of clay-like ingredients in the raw materials and on the production in consequence of compounds akin to those in Portland cement. Magnesian lime is successfully used in some areas of the country where dolomitic limestone occurs, and can give higher strengths than high calcium limes in straight lime : sand mixes.

Limes are usually delivered to the site nowadays in the dry hydrated form rather than as quicklime. They can be used directly from the bag, but it is better to soak them first for sixteen hours before mixing with the sand and cement.

Sands for mortars

Sands for mortars are usually from local sources to avoid the costs of transport. They should comply with BS 1200: 1976. Because about one volume of binder is required for three volumes of sand, the sand forms the main bulk of any mortar; therefore as far as possible it must be physically and chemically stable. For example, it should not contain large amounts of materials such as loam or clay, because these are moisture sensitive and will shrink considerably when drying out. Sand such as unwashed sea sand containing magnesium chloride and sodium chloride (common salt) is also unsuitable because these salts are deliquescent, that is they attract moisture.

Siliceous sands (i.e. most ordinary sands) are practically immune, when clean, from physical attacks of these kinds and are therefore highly suitable for mortars. Where they are not obtainable, they may be replaced by crushings from hard rocks like granites, sandstones and hard limestones. Crushed brick is also useful when it can be obtained from brickyards, but not if made from bricks that have been salvaged from old buildings. These may be contaminated with gypsum from plasters.

Proprietary additives to mortars include air entraining agents. The effects of air entrainment on mortar properties are discussed below. Air entraining agents should comply with BS 4807: 1963.

Water

Water must be clean. Usually drinking water supplies, which are almost always suitable, are readily available.

Functional requirements of the mortar

A mortar joint forms both a sealant and a bearing pad. It sticks the units together yet keeps them apart, and in this sense it performs as a 'gap filling' glue. If it is to function properly in this dual role the bond at the interface between the unit and the mortar must be complete. Bond – and other properties of a mortar – can be affected by factors such as the water/binder ratio, the suction of the unit, the ability of the mortar to retain water against the suction of the unit (i.e. its water retentivity), the amount of lime present, the pressure in forming the joint (part of the craftsmanship), the textural quality of the bed surfaces of the unit, and finally the air content of the mortar.

An ideal mortar:

- adheres completely and permanently to the unit;
- remains workable long enough to set the brick right to line and level; this implies good water retentivity;
- stiffens sufficiently quickly to permit the laying of units to proceed smoothly, and gives rapid development of strength and adequate strength when hardened;
- is resistant to the action of frost, abrasion and chemical attack such as sulphate attack;
- resists the penetration of rain;
- accommodates early movement of the structure;
- has good appearance;
- is cheap.

Mortar mixes

The builder will use any of the following:

- Portland cement : lime : sand mixes
- Masonry cement : sand
- Cement : sand with air entraining agent

Because of the fact that sand contains about 30 per cent voids, Portland cement : lime mortars are always proportioned 1 : 3 binder : aggregate, as shown in Table 32.10.

Table 32.10 PROPORTIONING CEMENT : LIME MORTARS

Type	*Cement*	*Lime*	*Sand*
1	1	0 (¼ part is added for workability)	3
2	1	½	4½
3	1	1	6
4	1	2	9

It is a common error to assume that the same ratios apply with masonry cements, or in air entrained mortars. The equivalent mixes are given in Table 32.11. Differences in proportions are due to effects of air entrainment, which produces minute bubbles that set to form a skeletal matrix, with air replacing lime.

Today there is an increasing use of ready mixed lime : sand 'coarse stuff'; this is delivered wet to the site, where it should be mixed with the cement unless a pure lime : sand mortar is required. Care should be exercised to ensure that the cement is first mixed to a slurry, otherwise 'balling' occurs, with subsequent staining of the work. Reputable producers of ready mixed mortar ensure that this does not happen

Table 32.11 SELECTION OF MORTARS

Grade	*Hydraulic lime: sand*	*Cement: lime: sand*	*Masonry cement: sand*	*Cement: sand + plasticiser*
I	–	1 : 0–¼ : 3	–	–
II	–	1 : ½ : 4–4½	–	–
III	–	1 : 1 : 5–6	1 : 4½	1 : 5–6
IV	1 : 2 : 3	1 : 2 : 8–9	1 : 6	1 : 7–8
V	1 : 3	1 : 3 : 10–12	1 : 7	1 : 8

resistance to frost →

← increasing bond

↓ accommodation of movement

↑ increasing strength

with the lime : sand mix before delivery, but others are not so careful. The builder should be wary of them.

DURABILITY

By durability is meant resistance to deterioration. Agents of destruction are illustrated in *Figure 32.3*. However, no chemical or physical attack on walls can take place unless water is present[9]. It must therefore be excluded as much as possible by good detailing.

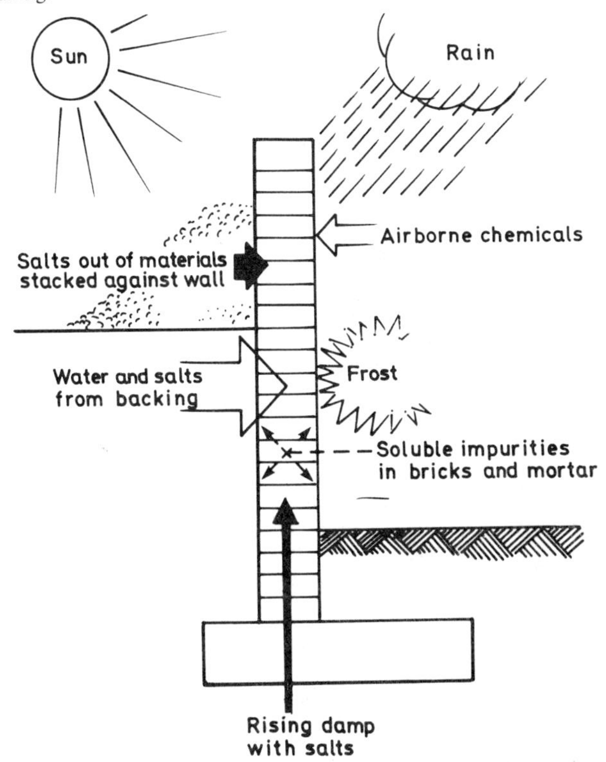

Figure 32.3 Agents of destruction of brickwork

Before considering aspects of good detailing the house builder must recognise that one primary consideration in choosing units and mortar is the degree of exposure. This has two aspects:

- the geographical exposure to which the building or buildings is to be subjected;
- the location of the element or wall within the particular building – e.g. whether the wall is a normal external wall, or a parapet or free standing wall.

Exposure is classified in CP 121: Part 1: 1973 as sheltered, moderate, or severe. These areas of exposure are modified by height and distance from the sea coast or the coast of large estuaries. Thus within five miles of the sea the geographical exposure in a sheltered area is upgraded to moderate. Similarly within five miles of the sea, moderate exposure is upgraded to severe.

As can be imagined, the classification of exposure is determined in the first instance by the amount of driving rain and the speed of the wind, the notation of which is m^2/s. Simply, the classification is that sheltered conditions occur where the driving rain index is 3 m^2/s or less, moderate conditions lie in areas where the index is between 3 and 7 m^2/s, and severe exposure is where the driving rain index is greater than 7 m^2/s. Superimposed on this is the requirement of distance from the coast as described above, and a commonsense further requirement is that buildings that stand above their surroundings or on top of a hill in any given area of exposure are likely to be more severely exposed and hence must be upgraded to the next grade.

Only a small area of the UK is sheltered exposure (*Figure 32.4*). Much the greatest area is moderate, and the whole of the west side is severe. It is essential to realise

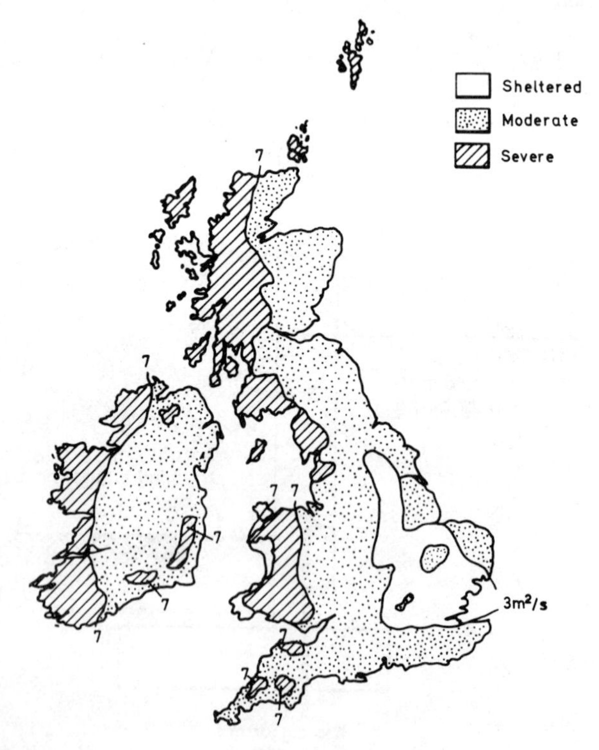

Figure 32.4 Degrees of exposure: geographical (based on BRE Digest 127)

that units suitable for use in one area of the UK may not prove to be suitable in another. For example, there are varieties of brick that should only be used in the west UK with considerable care. BS 3921 implies this of course, but does not draw attention to the different geographical areas of exposure. However, it does imply that there are differences that occur in a natural building complex. Thus even if the building itself is in an area of sheltered exposure, free standing walls, retaining walls, parapets and the like must be regarded as being severely exposed; see *Figure 32.5*. This apparent paradox has misled many builders and designers.

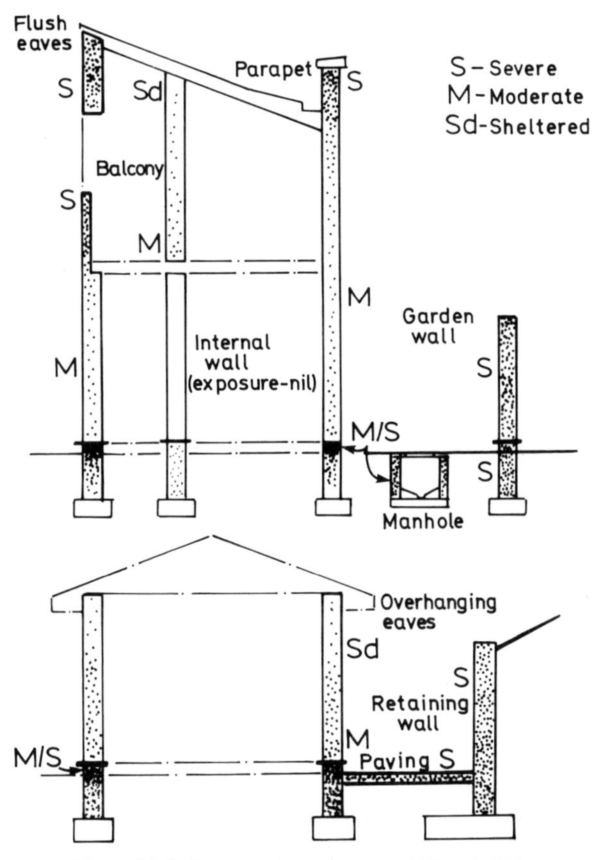

Figure 32.5 Comparative exposure within a building

In general the reader is referred to Table 4.1 and 4.2 of CP 121: Part 1: 1973 for specific guidance on choice of unit for a given situation, and also to the illustrations of building types and appropriate details discussed below.

Specific aspects of durability

Frost action (which is readily understood) and sulphate attack (which is not) are the main destructive agents.

FROST ACTION

The better qualities of concrete blocks and bricks (i.e. broadly those with higher compressive strengths) are frost resistant, as are calcium silicate bricks. For clay bricks, specification is more complicated, but by and large the stronger bricks will be more frost resistant. Much frost attack could be avoided by good detailing to prevent ingress of water, but where this cannot happen the best approach is to ask the manufacturers to show evidence of frost resistance or to specify bricks of Special Quality.

SULPHATE ATTACK

When clay bricks containing soluble sulphates become and remain wet for long periods, salts can be dissolved and migrate in solution to the mortar, where they can react with a constituent of cement called tri-calcium aluminate (C_3A for short). This reaction, which happens readily in the presence of the highly soluble salts of magnesium, sodium and potassium, creates calcium-sulpho-aluminate and is accompanied by significant expansion of the mortar; this leads to crumbling and disintegration of the joints and, in the more severe cases, to substantial growth and possibly subsequent disruption of the brickwork. The same reaction can occur where the source of the salts is other than the bricks. The sand in the mortar can be the source, although fortunately rarely so. Even an industrial atmosphere, if highly aggressive, can give rise to sulphate attack. More common is the presence of sulphates in soils, either in foundations or in retained earth, particularly where these are associated with a flow of groundwater. Attack on ordinary Portland cement in mortars, and also in concrete bricks and blocks themselves, is certain in such conditions. Clay bricks themselves are usually unaffected in any situation, even though they may be the source of the sulphates. Calcium silicate bricks are not affected.

Remembering first that it takes about one year for the attack to manifest itself, significant signs are as follows:

- If attack is taking place in an outer skin of brickwork, horizontal cracking of interior surfaces (plastered or otherwise) may occur, perhaps coinciding with lines of headers in a solid wall or even, perhaps, of wall ties in a lightly loaded cavity wall. If attack is on an inner skin the cracks will be on the outer.
- If cracking of the kind described above is present, clearly expansion has taken place and its effects will show at critical points such as junctions of ceiling and wall. Expansion without the other signs noted in the following lines is not likely to be sulphate attack.
- The mortar will whiten, and later may show a definite line of lamination roughly in the middle of the joint. Eventually it will be completely friable and resemble a weak lime : sand mix. It is easy to confuse sulphate attack with frost attack on mortar. Indeed, each weakens the resistance of the mortar against the other.

Other causes of cracking such as thermal or moisture expansion are physical phenomena and do not cause mortar to disintegrate in this way. However, insofar as they can lead to ingress of water through cracks and hence to saturation and possible sulphate attack, they are potential agents of destruction. They are described below under 'Dimensional stability'.

Vertical expansion due to sulphate attack can be as much as 2 per cent in a lightly loaded wall, such as the wall of ordinary two-storey housing. Horizontal expansion from any cause depends very much on the degree of restraint due to vertical load, friction against the earth (below DPC), friction on the DPC itself, and the number and stiffness of return walls. Expansions in clay brickwork commonly ascribed to sulphate attack a decade ago are now recognised to have been caused by irreversible moisture expansion of the bricks. Sulphate attack can be prevented by keeping the brickwork dry, by ensuring that sulphates are not present (or are only so in insignificant amounts), or by using cement of low C_3A content.

It is clear that only internal walls can be kept completely dry. Hence BS 3921 does not lay down any requirements for bricks for internal work. Some bricks of Internal Quality would fail in frost or give rise to sulphate attack if exposed to the weather.

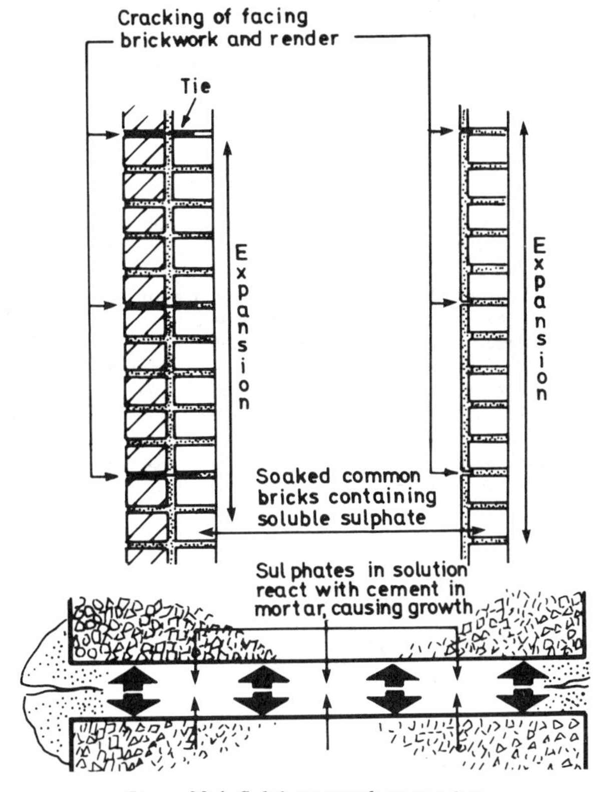

Figure 32.6 Sulphate attack on mortars

Clay bricks in external walls (which normally would be of Ordinary Quality to BS 3921) clearly cannot be kept dry. However, they rarely become saturated sufficiently, and for a sufficient length of time, for the sulphates they contain to give rise to sulphate attack. This undoubtedly is because they are being constantly dried by heat loss from the building they enclose. Many machine-made clay bricks of Ordinary Quality would fail in frost or eventually give rise to sulphate attack were it not for this fortuitous drying process (which is effective, incidentally, even below the base damp proof course).

Clay bricks in parapets, free-standing walls and retaining walls obviously will be thoroughly wet for much of the year. In these positions sulphate attack is likely, and the designer must resort to more stringent control. Wherever possible he should specify bricks of low soluble salt content such as Special Quality to BS 3921, which contains severe and strict limits. Where it is not possible to use Special Quality bricks, sulphation of the mortar must be resisted by making it as impermeable as possible and/or by using cements of known low C_3A content. The action of sulphation is illustrated diagrammatically in *Figure 32.6*.

Sulphate-resisting cement to BS 4027 contains not more than 3.5 per cent C_3A and hence is lower than the average Portland cement in this respect. However, a weak and hence porous mix containing sulphate-resisting cement is unlikely to be as resistant to a sulphate solution as a dense and hence less permeable mix using ordinary Portland cement. Clearly a combination of Special Quality bricks with a dense mortar containing low C_3A cement represents the upper end of an imaginary scale of resistance, whereas nondescript bricks in a weak mortar with ordinary Portland cement form the lower bound.

DIMENSIONAL STABILITY[10]

This somewhat grandiose title embraces the problems associated with movements of materials. The dimensions of building components such as bricks or blocks, or of complete elements of a building such as a wall or floor, or in fact of a whole building, are continually changing. Despite widespread dissemination of knowledge it is still not uncommon to see buildings over thirty metres long without any means of absorbing the differential movements that are bound to occur. In consequence there is cracking – far too much of it – in both brickwork and other materials.

The greater incidence of cracking of modern brickwork in comparison with earlier work is still widely queried, yet it is surprising that this should be so because the reasons for it have long been set out in the literature. By far the most important has been the change during this century from straight lime : sand mortar mixes to those gauged with cement. Accompanying this has been the inevitable decrease in massivity deriving from the better understanding of the properties of the material and its

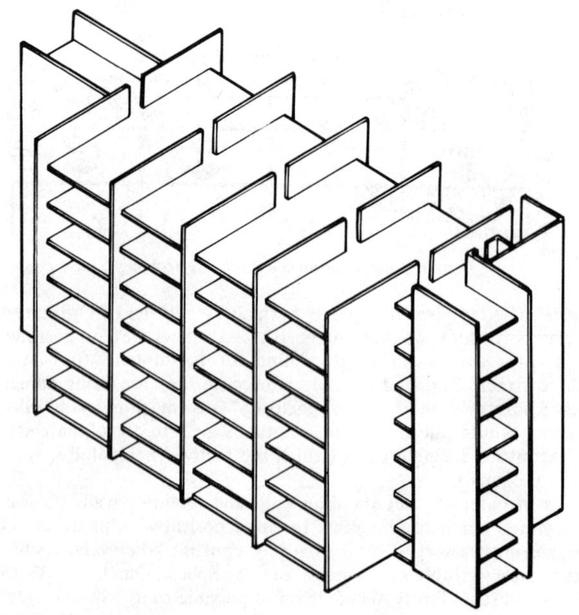

Figure 32.7 Modern brickwork: strong bricks and strong mortar in thin walls

associated structural performance. As a result brickwork is now used in thin plates (*Figure 32.7*) attached to and *receiving* essential support from other building elements, instead of being the virtually independent and relatively massive solid walls that gave support and stability, and that were tolerant of movement because of the long-term plasticity of the lime mortars. Economics and greater knowledge have forced these changes, so ways must be found of accommodating them and their side effects.

Most contemporary cracking can be avoided by understanding the causes and by correct detailing, which is often the builder's responsibility. It is hoped that this note will help him when his building is characterised by brickwork or blockwork.

Causes of movement

The four main causes of dimensional change are:

- deformation due to loads;
- movements due to temperature change;
- movements due to changes in moisture content;
- movements due to chemical action.

All these can cause cracking of walling. Another cause is foundation settlement, which occurs most frequently in smaller buildings where the foundations are not properly designed.

DEFORMATION DUE TO LOADS

When a structural member – say a brickwork pier under a vertical compressive load – is subjected to a load it shortens a tiny amount. If after removal of the load the member recovers fully, i.e. returns to its original length, it is said to have deformed and recovered *elastically*. If, on the other hand, there is any residual deformation, such deformation is said to be *inelastic* or *plastic*. Creep is a more common name for this. The ratio between any change in dimension and the original dimension is known as *strain* and, as might be expected, its amount is related directly to stress. The greater the stress the greater the strain so long as the material behaves elastically.

The elastic deformations that are caused by stresses permitted in normal building design are very small and are unlikely to give trouble, but inelastic or plastic deformations can be much greater and will certainly do so if not accommodated. Concrete in particular is known to 'creep' or 'flow' plastically to a marked extent under sustained load – so much so that concrete columns and loaded block walls under normal, i.e. working, compressive stresses can shorten up to 3 mm in a storey height.

By contrast, clay brickwork creeps very little because clay bricks themselves virtually do not creep under normal loads. (Mortar, of course, can creep like concrete, but it forms much the smaller part of the member – only about 13 per cent of any height, assuming 65 mm bricks and 10 mm joints).

MOVEMENT DUE TO CHANGE IN TEMPERATURE

All building materials alter their dimensions with variation in temperature. Coefficients of expansion, i.e. increase in length per unit length per degree change in temperature, when materials or components are in their unrestrained state have long been determined. Clay brickwork, for example, has a coefficient of about 6×10^{-6}

(six millionths) per degree C, which means that for every degree rise in temperature a 10 m length of brickwork will expand the following amount:

$$10 \times 6 \times 10^{-6} = \frac{60 \times 1}{1\,000\,000} \text{ m}$$
$$= 0.00006 \text{ m}$$
$$= 0.06 \text{ mm}$$

It follows that a temperature rise of 50 degrees C would cause *unrestrained* expansion of 50 × 0.06 = 3.0 mm, a surprisingly high figure for such a relatively short length of wall. If the wall was fully restrained, any increase in temperature would subject the wall to compressive stress rather than cause it to expand linearly. In real structures full restraint or full freedom to move rarely exists. Temperature rise is therefore most likely to cause a combination of some stress and some movement.

Movement due to temperature is of course not one way only. Contraction occurs with cooling, and cracking results from restraint because the wall as a whole is prevented from returning to its original position.

Movement will be different in the vertical direction from the horizontal. It will vary with aspect, being obviously greater on a south-facing facade, and it will be greater in an infill wall in a framed structure than in a loadbearing wall because of difference in restraint. Any assessment of temperature change in a wall must take into account differences between external and internal surface temperatures, and it must not be forgotten that these are not necessarily the ambient air temperatures. The surface temperature of a south-facing brick wall can be as high as 49°C (120°F). These facts are of particular importance in brickwork cavity walls infilling framed structures, where the inner leaf is bounded and restrained by the frame but the outer may be largely free from it.

MOVEMENTS DUE TO CHANGE IN MOISTURE CONTENT

These have been discussed briefly earlier, but will be mentioned again later when discussing buildings. Broadly, the builder needs to remember that clay brickwork *expands* and calcium silicate and concrete brickwork and blockwork *shrink*.

MOVEMENT DUE TO CHEMICAL ACTION

Here we are concerned with sulphate attack, which has also been discussed above. Growth of the C_3A causes expansion and sometimes entire disruption of brickwork. It can be avoided entirely, unlike movement from physical causes. An actual example is shown in *Figure 32.8.* Here a half brick skin of handmade facing bricks was tied to common brickwork of bricks high in soluble salt content. The tower was used for hose exercises, was sprayed with water once or twice a week and had no roof water protection. Sulphate attack on the mortar of the inner brickwork caused cracks in the outer at the tie intervals.

FOUNDATION MOVEMENT

Many foundations for small houses suffered badly in recent hot summers due to shrinkage of clay soils. Now deeper foundations are demanded by Building Regulations[11] and there should be less trouble. Cracking due to hogging and sagging is easily recognised (*Figure 32.9*).

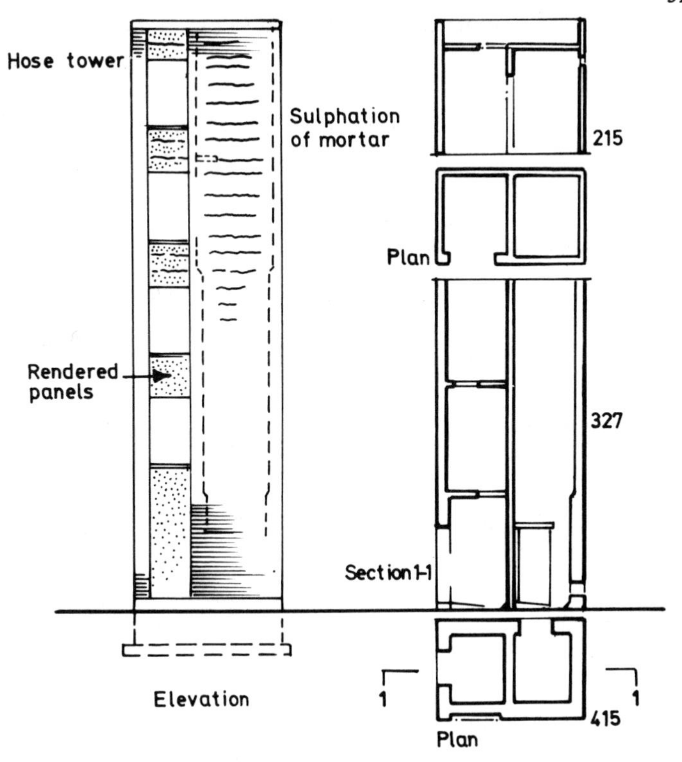

Figure 32.8 Vertical expansion: sulphate attack on mortars

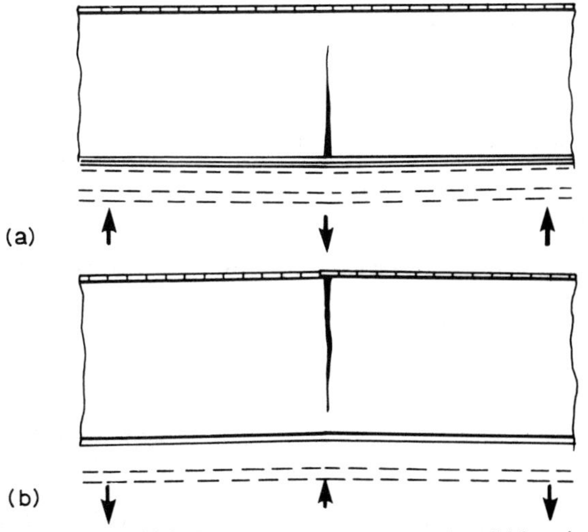

Figure 32.9 Settlement cracks: (a) sagging, (b) hogging

PERFORMANCE OF WALLS

Wall functions

An external wall has to be strong, stable, durable, rain resistant and free from the effects of movement due to changes in temperature, moisture and chemical action. Aspects of durability and dimensional stability are best considered by examining some forms of housing and possible defects that can occur due to:

- ignorance of the properties of materials,
- faulty design and detailing.

Perhaps the greatest single difference between modern and older brickwork, which certainly had less faults, is thickness. Walls today are simply thin veneers much more

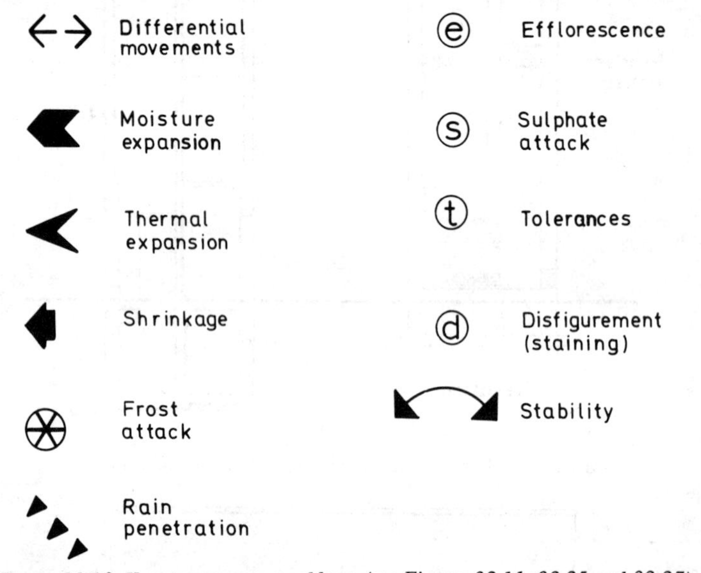

Figure 32.10 Key to masonry problems (see Figures 32.11, 32.25 and 32.27)

vulnerable to damage. The range of problems is listed symbolically in *Figure 32.10*, which applies to all types of building considered.

Semi-detached houses

The somewhat stark design shown in *Figure 32.11* is deliberately full of many faults, but all are unfortunately too often evident.

Severe exposure exists in these situations:

- parapet walls to garages;
- garden walls, which are part free-standing and part retaining walls.

In these areas care in choice of materials is essential.

CLAY BRICKS

These should be of Special Quality, or be of Ordinary Quality with the frost resistance of Special Quality. If the latter, they should be set in a mortar made with

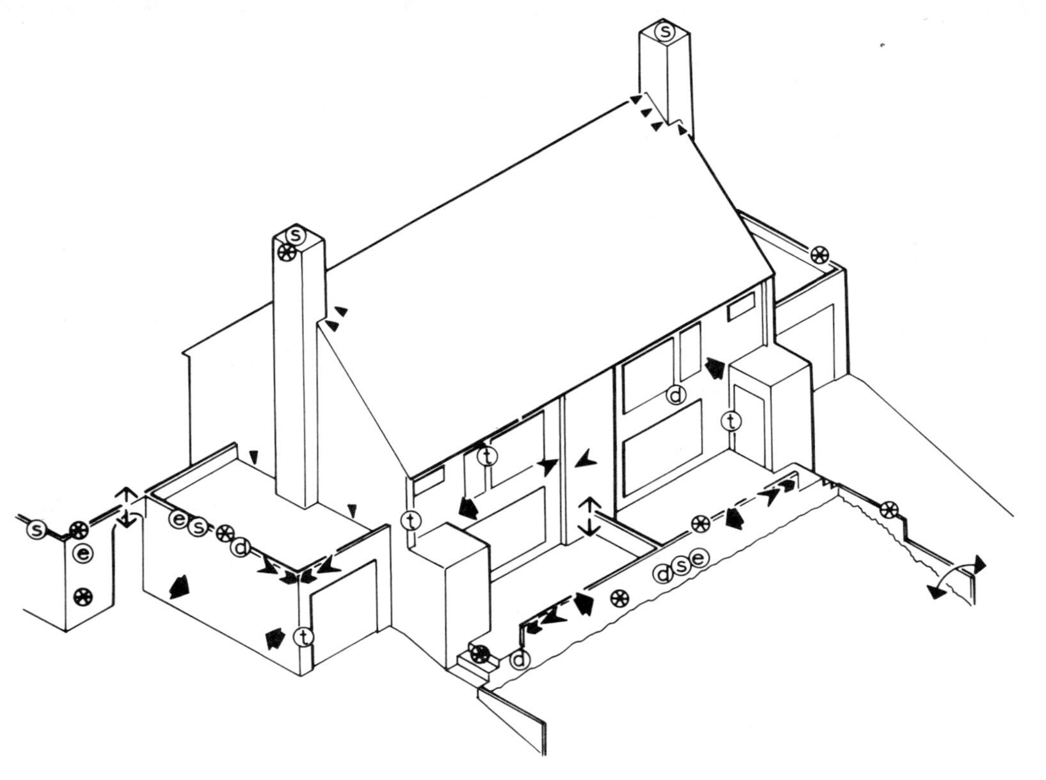

Figure 32.11 Masonry problems in pair of semi-detached houses (see Figure 32.10 for key)

sulphate-resisting cement. As this is darker than Ordinary Portland a problem of appearance due to differences in colour of finished brickwork will arise if sulphate-resisting cement is not used throughout. However, the builder should remember that in time some difference in colour is bound to occur, if only because parapets and garden walls are wetter – and colder – for longer than the parent brickwork.

CALCIUM SILICATE BRICKS AND CONCRETE BRICKS

The minimum quality of calcium silicate bricks for these situations is Class 3 for parapets and external free standing walls and Class 4 for retaining walls. For concrete bricks the categories are 20.0, 15.0 and 30.0 respectively. In addition, for concrete bricks (and for the mortar) there is need to consider the possible presence of sulphates in soils behind retaining walls. It will be recalled that calcium silicate bricks themselves are not affected by sulphates.

CONCRETE BLOCKS

For concrete blocks the requirements are generally for Type A, but some Type B may be used for parapets and external free-standing walls. For retaining walls or external free-standing walls below DPC, one must again ensure that sulphates in the groundwater are low.

COPINGS

All the walls discussed above need protection from downward penetration of water. External free-standing and retaining walls need to be protected from rising damp due

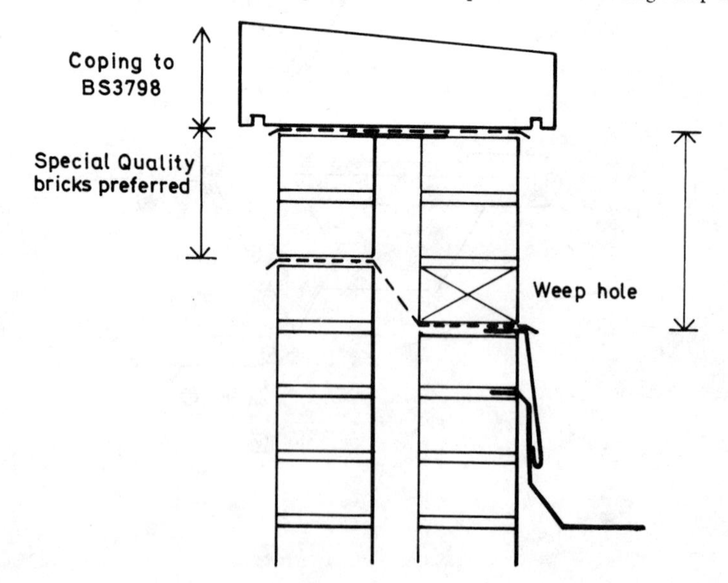

Figure 32.12 Cavity parapet: note DPC projection

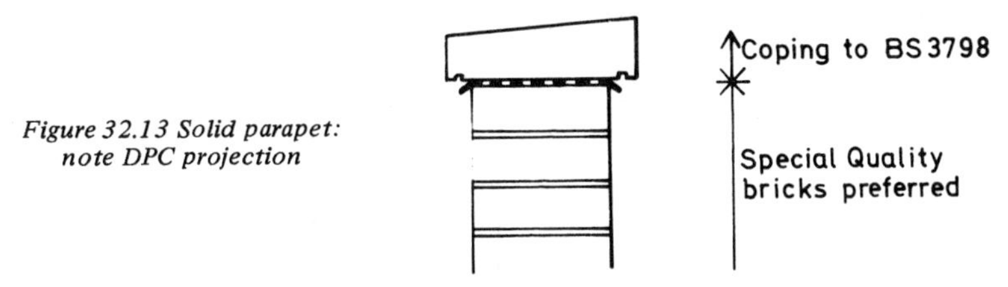

Figure 32.13 Solid parapet: note DPC projection

to capillary action. Damp-proof courses are dealt with elsewhere in this book, but are mentioned here insofar as they relate to durability. The builder must realise that all damp-proof courses to BS 743 may be used to check rising damp but only flexible, jointless materials should be used to check downward penetration of water.

When such DPCs are used they must *totally* cover the top of the wall. In the authors' opinion this means they should *project* from the face (*Figures 32.12 and 32.13*), but others argue against this, on the grounds of appearance, and say that they should be in line with the face but not project.

In practice this means that bricklayers (who in general dislike projections) keep DPCs back from the face. Consequently the DPC is by-passed all too often, and staining, efflorescence and frost attack may ensue on bricks below. It is usually necessary to protect the damp-proof membrane itself, and often a concrete coping is used (*Figure 32.12*). This should be of adequate weight per unit length to ensure that it is not easily dislodged by, say, a ladder. Thin concrete sections are useless.

In BS 3798 the minimum weight of 'flat bottomed copings not intended to be fixed by cramps' is 23 lb/ft run (≈ 34 kg/m). This is by no means too liberal; in the authors' opinion it is too low. Clay brick copings should be of Special Quality to BS 3921 and again of adequate weight. In practice one course of brick-on-edge above the DPC, which could be achieved easily by many bricks, is very vulnerable to dislodging and movement. Bricks, mortar and flexible DPCs do not stick together with any degree of permanence. Brick-on-edge plus at least one, and preferably two, courses of bricks below – which should also be of Special Quality – would be sensible. See *Figures 32.14 to 32.17*.

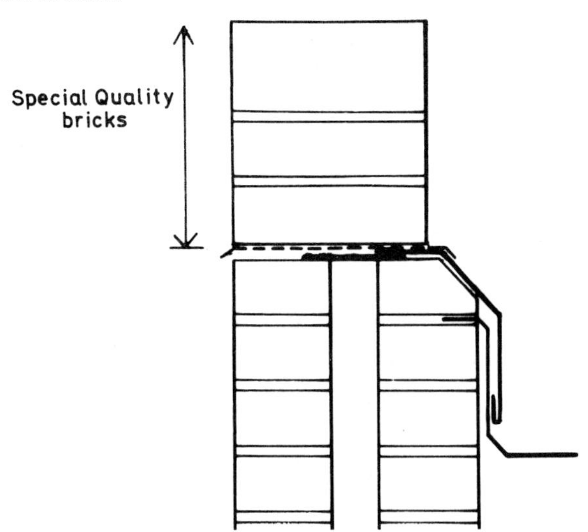

Figure 32.14 Solid parapet: note DPC projection

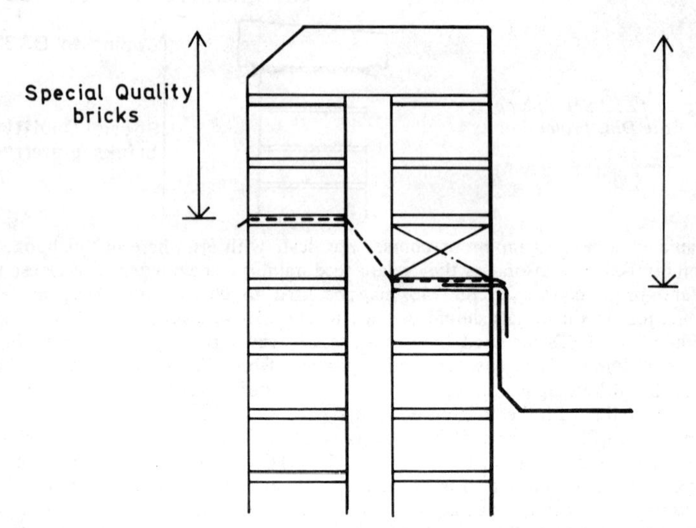

Figure 32.15 Cavity parapet: note DPC projection

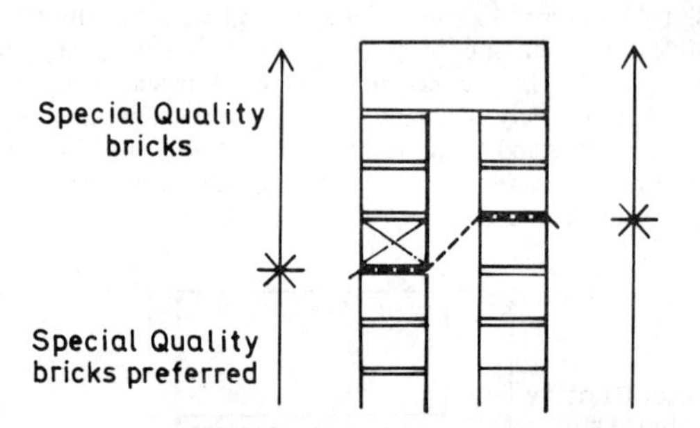

Figure 32.16 Cavity parapet: note DPC projection

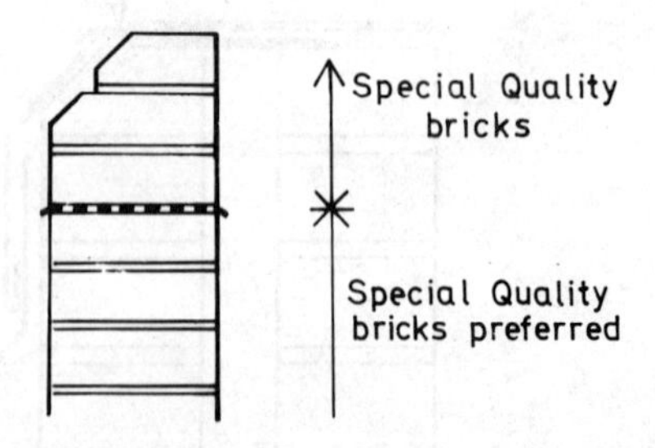

Figure 32.17 Solid parapet: note DPC projection

CHIMNEY STACKS

Chimney stacks are vulnerable to frost and sulphate attack. Wherever possible they should not be part of external walls. If they are, the cavity should continue around and not stop at the stack. The likelihood that the use of solid fuel will become widespread again brings with it the dangers from the use of slow burning stoves. Fortunately flues now have to be lined so attack from sulphates internally is eliminated, but there is still the problem of exposure externally. Good cappings to prevent downward rain penetration, and preferably sulphate-resisting cement in mortars, are the main defences against saturation of the work. Brick-on-edge tops imply treatment similar to copings just described.

WINDOW CILLS

Water running down glass of windows should not be allowed to soak into the body of a wall. The logical way to avoid this is to use an impermeable projecting cill. Modern wooden windows achieve this by being built into the outer leaf so that the throated cill projects about 38 mm, although originally they were intended to be set on a sub-cill so that the jamb could be behind the line of the vertical DPC. Nevertheless the current arrangement protects the brickwork below the cill, which can then be of the usual quality (for the different materials) required for the normal outer leaf. See *Figure 32.18.*

Figure 32.18 Standard timber frame in outer leaf

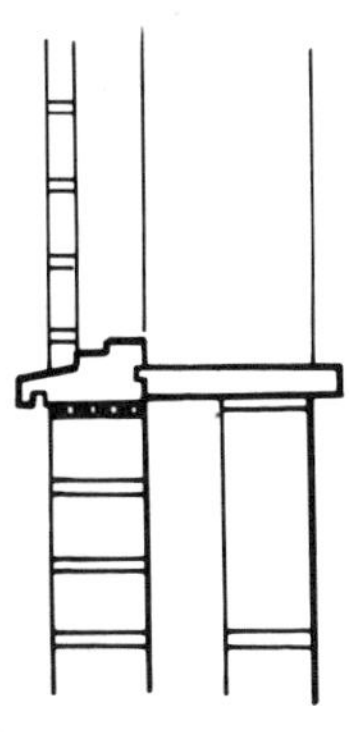

However, if the designer wishes to use a brick-on-edge sub-cill or recess the window by the use of plinth or cant bricks (to BS 4729), these must be regarded as having the same exposure as copings and thus must be of the appropriate quality. They must also have a flexible membrane DPC under to protect the wall below. See *Figures 32.19 and 32.20*. 'DPC' quality bricks are not adequate to prevent downward penetration of water, nor for that matter is a course (even a double course) of plain tiles, even though tiles themselves may be virtually impervious.

NARROW PIERS

Narrow piers between windows or projecting ends of gable walls need to have at least one perpend per course in clay brickwork. Thus they must be one and a half bricks (not one brick) wide as a minimum, to enable variations in brick lengths to be

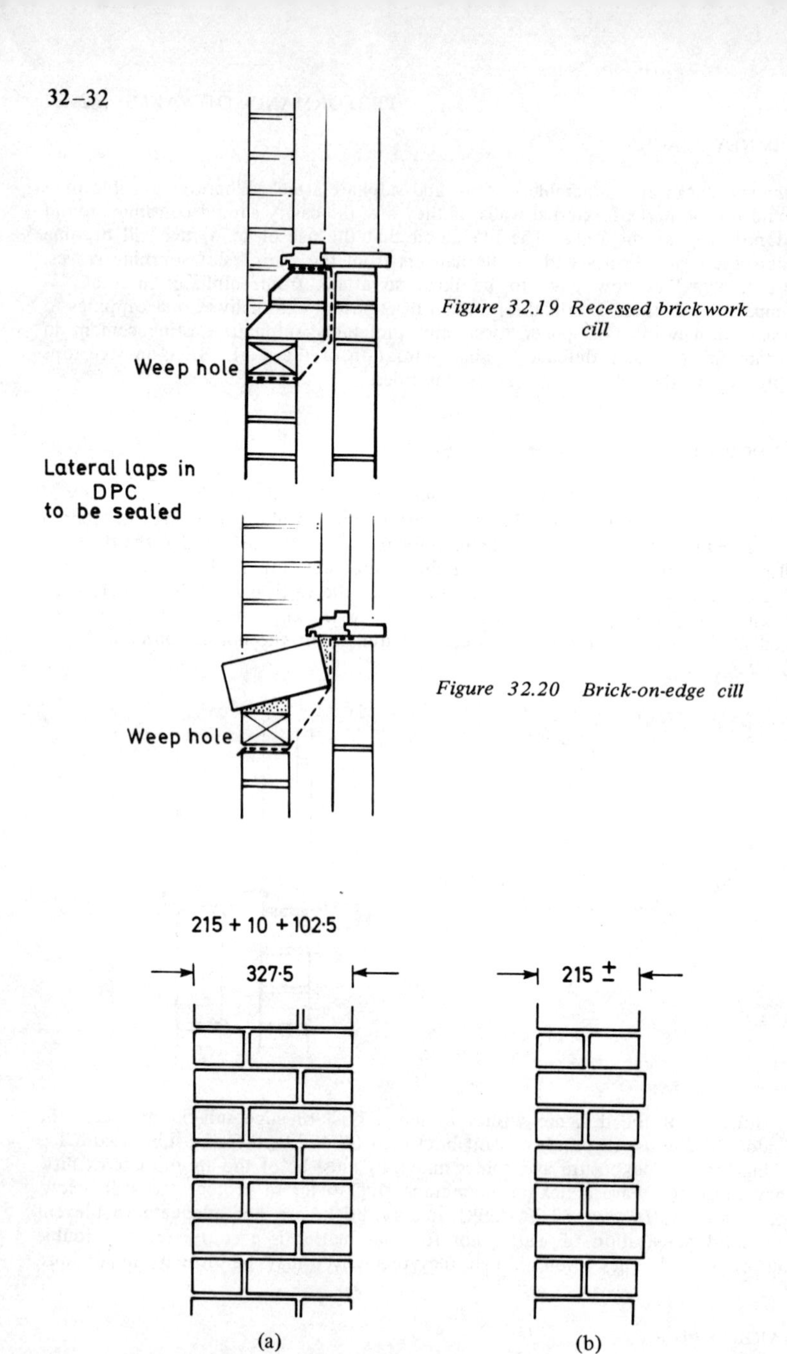

Figure 32.19 Recessed brickwork cill

Figure 32.20 Brick-on-edge cill

Figure 32.21 (a) 1½ brick pier: fair-face possible both sides. (b) Single brick pier, showing exaggerated effect of tolerance on length; fair-face possible on one side only

absorbed so that the two sides of the pier can be truly vertical and parallel (see *Figure 32.21*), otherwise the bricks will have to be selected at extra cost.

In calcium silicate brickwork the much smaller variations in length of individual bricks may permit one brick wide piers without selection. Concrete bricks and blocks may need more attention to ensure good work in this respect.

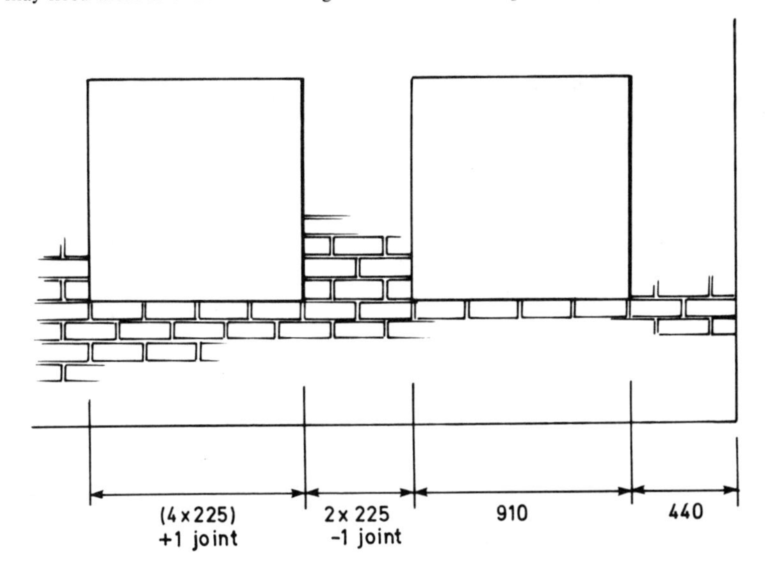

Figure 32.22 Openings and piers: sizes

It is a common error to assume that piers are multiples of brick plus joint. They are not, nor are openings. Piers are multiples *minus* a joint, openings are multiples *plus* a joint. See *Figure 32.22.*

FOUNDATION MOVEMENT

Even the small complex like the pair of semis in *Figure 32.11* contains many potential positions where cracking of walling could occur. Possible differential foundation movement makes the archway next to the garage very vulnerable whatever the material; so is a garden wall fully attached to the parent wall. It is better to leave a slot from DPC level upwards (*Figure 32.25*).

EXPANSION AND CONTRACTION OF WALLING

Very little guidance on position and frequency of movement joints in clay brickwork exists. As long ago as 1965, BRE issued a Digest[12] which said that total expansion in long walls (moisture plus thermal) could amount to as much as 3/8 inch in 40 feet (now 10 mm in 12 m). CP 121[6] published in 1973, recommended a 10 mm movement joint every 12 m. The difference is startling and incompatible. If 10 mm of movement occurs, it cannot possibly be accommodated by a 10 mm wide joint. It *might* by one 20 mm wide, assuming an allowable compressibility of the filler and sealant of 50 per cent, which is unlikely. In practice, restraints reduce the frequency

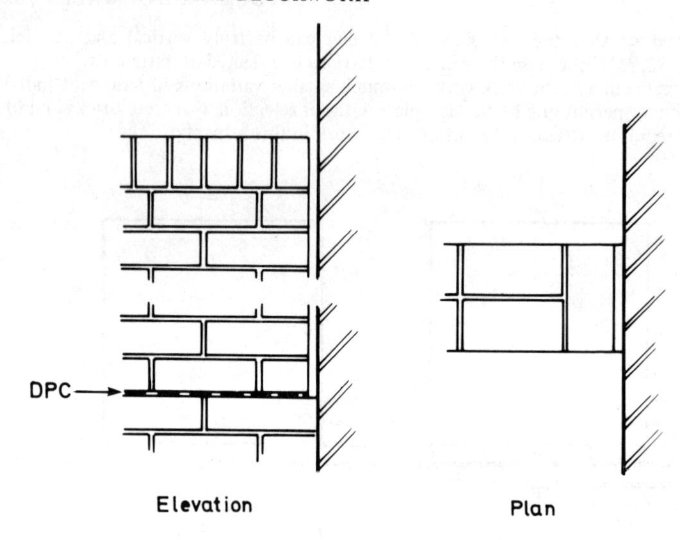

Figure 32.23 Junction of free-standing wall and main wall

of joints in clay brickwork, and in any event it is known that the basis of assessment was mostly measurement on bricks[13] not brickwork. Nevertheless it behoves the builder to think carefully about joint spacing, particularly if he is using high-expansion bricks.

CLAY BRICK WALLS

Materials of the external walls are not shown in *Figure 32.11*. If we assume that the gables, garages and porches (at each end) and the centre facade panel are in clay brick, and that the infill walling is, say, rendering on timber framing, then there is probably no need to incorporate movement joints to accommodate expansion. There may be a need in the long garden wall, otherwise cracking at the returns after the steps is possible. The maximum restraint in such a wall is at the centre, where there is a T junction. Restraint from the ground, by friction, also occurs, so walls tend to expand more above ground – or DPC if this is included. The best DPC in the free-standing wall would be a rigid type: slates in cement or three courses of DPC bricks in cement mortar. Membrane-type DPCs form slip planes.

If the infilling walling of the main facade were also clay brick, cracks might occur at the short returns where the recessing of the infill occurs if a high moisture expansion brick is used. Short returns are certainly vulnerable when there is a relatively long length of brickwork at either side. The action is shown in *Figure 32.24*.

WALLS OF CALCIUM SILICATE BRICKS, CONCRETE BRICKS AND CONCRETE BLOCKS

Cracking due to contraction – initial drying shrinkage – is likely in the walling of these semis where changes of section occur, i.e. in brickwork between heads of ground-floor and cills of first-floor windows. Guidance in CP 121 ('subject to the disposition of any openings') is that movement joints should be provided at intervals of 7.5–9 m (25–30 ft). In other words, joints should only be at these relatively large intervals if the panels are free from openings, and have an aspect ratio (height to length) of about 1 : 2 or less. Where there have to be openings, as in house walls, it is necessary either to reinforce the panels across the line of change of section, or incorporate movement joints – more correctly, joints that allow for contraction

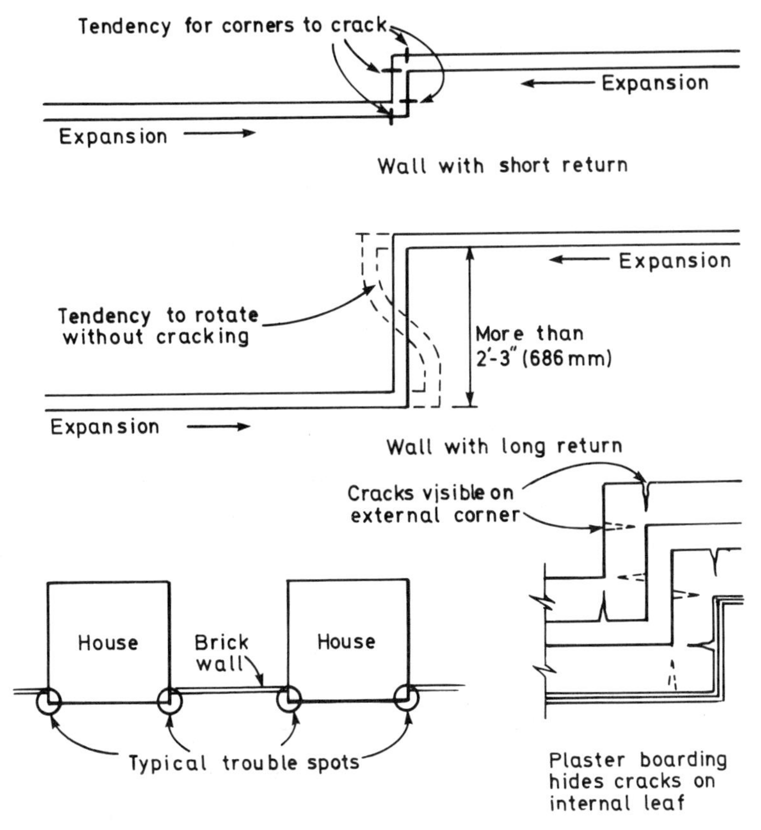

Figure 32.24 Short returns: moisture and thermal movement

without cracking. Sometimes contraction joints can be incorporated that follow the line of the bond (i.e. toothing) and that are simply strips of clear polythene sheet. These prevent adhesion of mortar to brick, but allow the same mortar to be used as in the rest of the wall[14].

For long unbroken lengths of walling of aerated concrete blocks, joints at about 6 m intervals are recommended[15]. It would appear from the literature that all these recommendations of spacing of joints assume that the units have been correctly cured and are kept dry during delivery and on site. The builder who does not take these elementary precautions on site will suffer more cracking due to initial drying shrinkage.

There are other, long-term, effects such as carbonation shrinkage (due to the effect of carbon dioxide from the atmosphere), which is non-reversible. Movement of

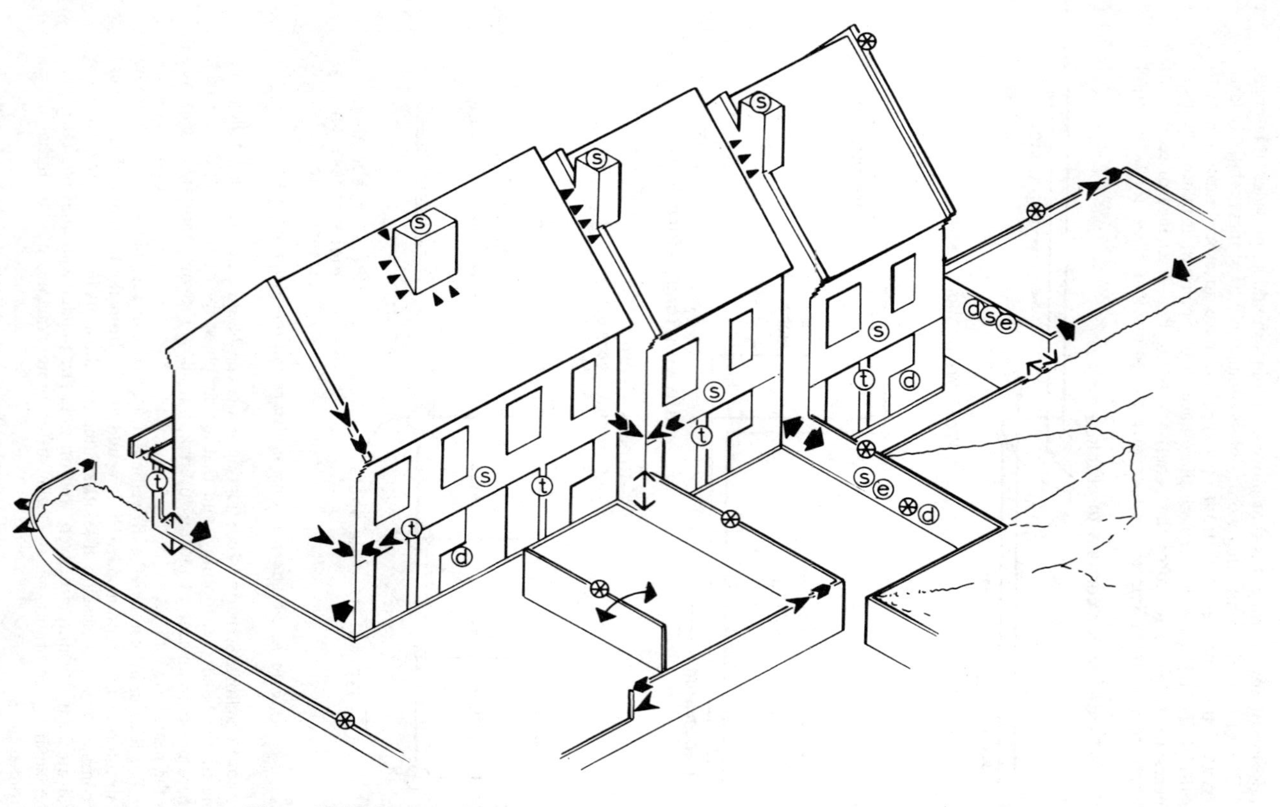

Figure 32.25 Masonry problems in two-storey terrace (see Figure 32.10 for key)

blockwork due to wetting expansion is rarely troublesome because it is less than drying shrinkage. With concrete blocks used internally – by far the majority – it is obviously of no importance once the work is enclosed.

Terraced two-storey housing

The discussion above (on both durability and dimensional stability) also applies to terraces, but in these the effects of horizontal expansion movement will be increased. In *Figure 32.25* some staggering is shown and hence the cumulative effects of clay brick expansion will be reduced, though perhaps at the expense of cracking at the external returns.

A relatively new phenomenon – to the authors at least – has been the expansion of coping bricks. This is indicated symbolically in *Figure 32.25* and an example is shown in *Figure 32.26*. Coping bricks of dense high glass content (blue brick) are

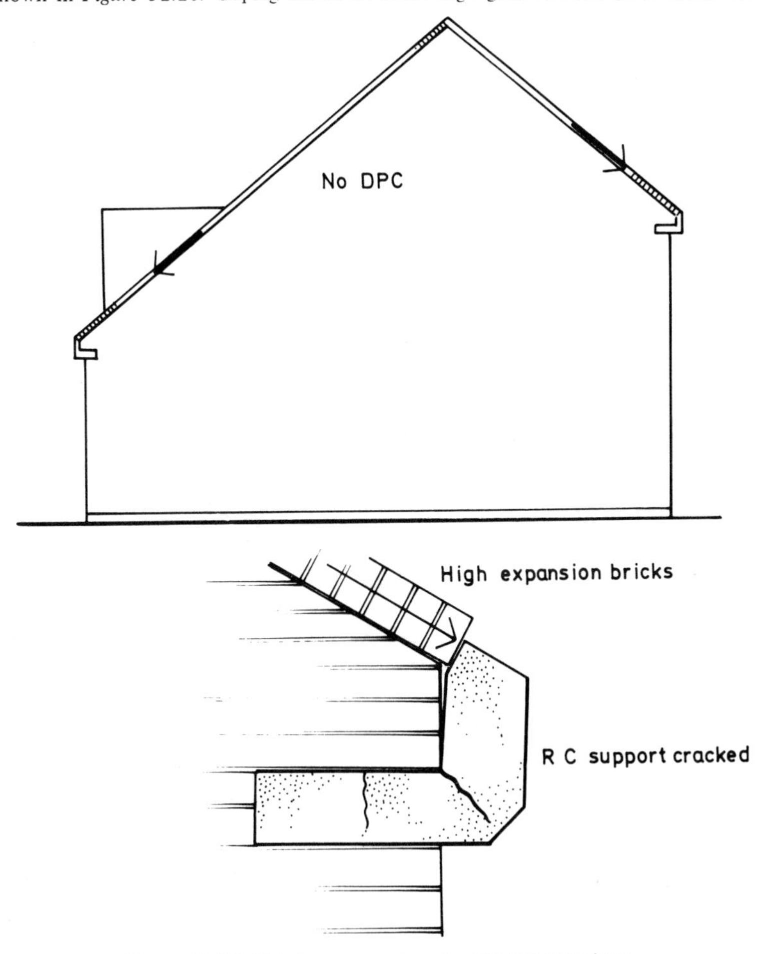

Figure 32.26 Moisture expansion of gable brick copings

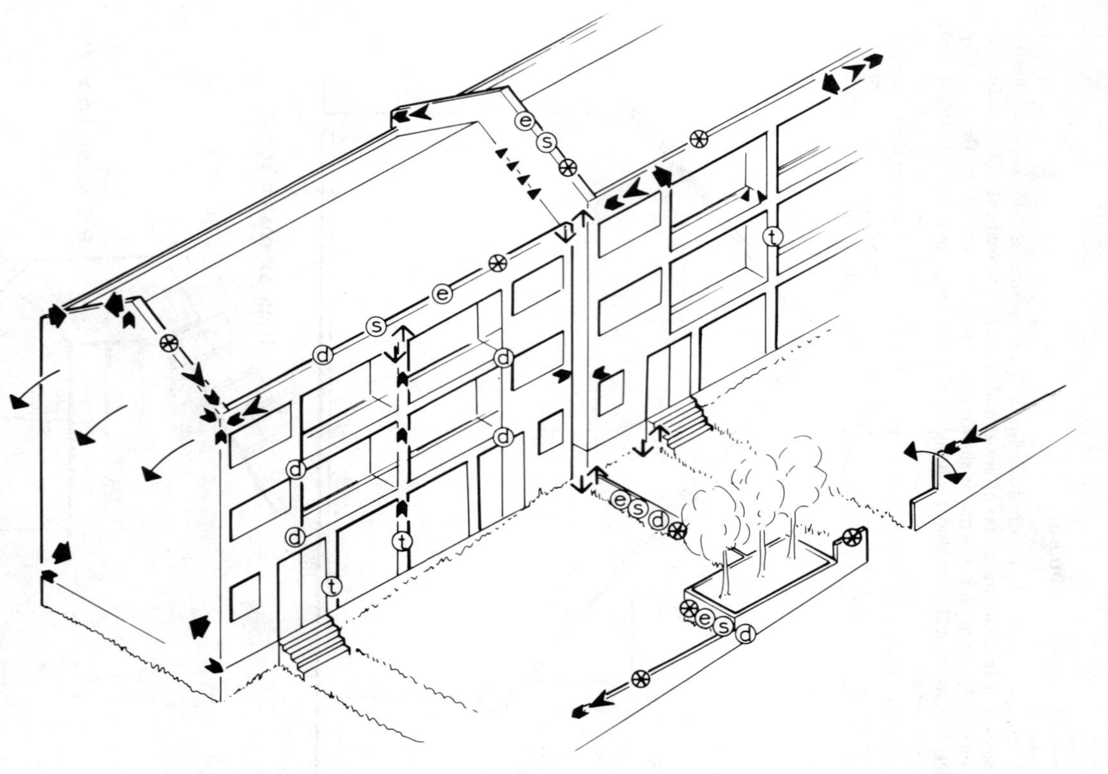

Figure 32.27 Masonry problems in three/four-storey terrace (see Figure 32.10 for key)

supported on the foot of each slope by a reinforced concrete corbel, which has been cracked by the force of expansion.

Terraced three- and four-storey housing

In two-storey housing, differential vertical movement seems to be insufficient to cause trouble. This is not so in three- or four- storey blocks (*Figure 32.27*) where the concrete blockwork is the structural support and where it has to carry reinforced concrete floor slabs. The extra weight is such that some creep and shrinkage under load (maybe the latter) is likely, and this, allied to the vertical expansion of clay brickwork, can cause cracking in the latter where the inner and outer leaf are connected – e.g. by lintels over openings. See *Figure 32.28*.

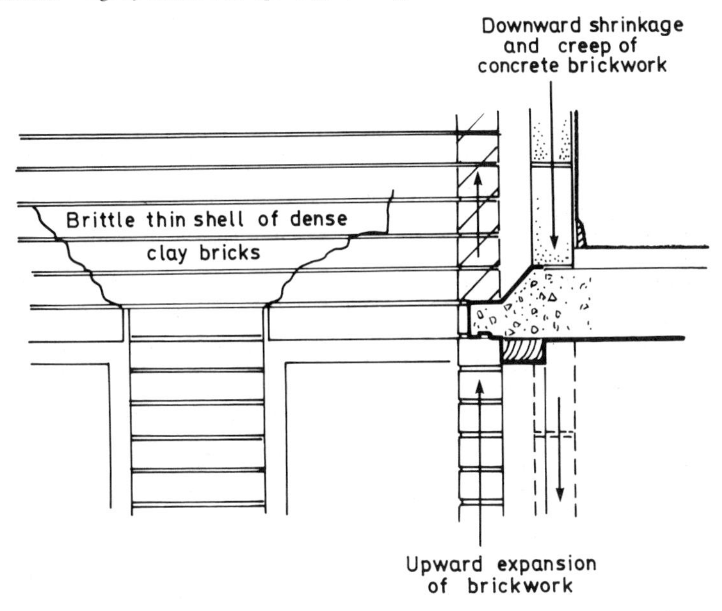

Figure 32.28 Differential vertical movement between support and cladding

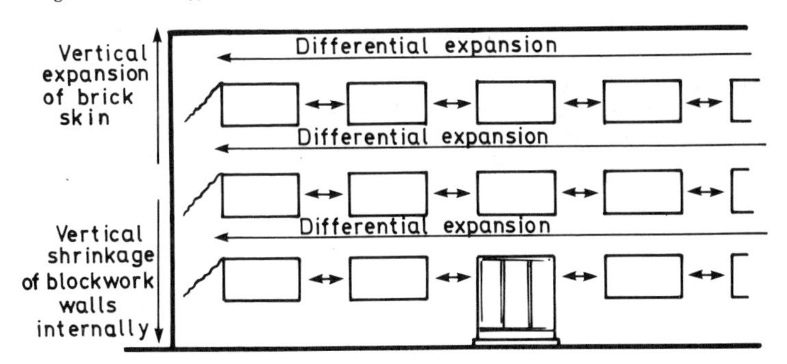

Figure 32.29 Differential vertical and horizontal movement

One answer to this problem is to avoid the use of materials of greatly differing properties, i.e. use one only for both support and facing. Another solution would be to accept the different materials but avoid rigid connections.

Vertical differential movement of the kind described can of course be combined with differential horizontal movement (*Figure 32.29*).

CONCLUSION

Marked differences in properties of walling units lead to equally marked differences in performance of walls. The functions of durability and dimensional stability have been studied to some small depth to illustrate these differences. Problems of durability are fairly consistent in all kinds and size of building, but dimensional stability problems are greatly affected by size and degree of restraint.

REFERENCES

1. BS 3921, *Clay bricks and blocks*, British Standards Institution (1974)
2. DD 34, *Clay bricks with modular dimensions*, British Standards Institution (1974)
3. BS 187, *Calcium silicate (sandlime and flintlime) bricks*, British Standards Institution (1978)
4. DD 59, *Calcium silicate bricks with modular dimensions*, British Standards Institution (1978)
5. BS 1180, *Concrete bricks and fixing bricks*, British Standards Institution (1972)
6. CP 121, *Walling*, Part 1, *Brick and block masonry*, British Standards Institution (1973)
7. BS 2028, 1364, *Precast concrete blocks*, British Standards Institution (1968)
8. Gage, M., and Kirkbride, T.W., *Design in blockwork*, Architectural Press, Cement and Concrete Association, London (1972)
9. Foster, D., *Brickwork: durability*, SCP tn3, Structural Clay Products Ltd, Hertford (1971)
10. Foster, D., *Brickwork: dimensional stability*, SCP tn 4, Structural Clay Products Ltd, Hertford (1971)
11. *Building Regulations 1976*, HMSO, London
12. Building Research Station Digest 65, 2nd Series, HMSO, London (1965); superseded by Digest 165
13. Smith, R.G., *Transactions British Ceramic Society*, 72(1), 1(1973)
14. Building Research Establishment Digest 157, HMSO, London (Sept. 1973)
15. Building Research Establishment Digest 178, HMSO, London (June 1975)

ACKNOWLEDGEMENT

Special thanks are due to Structural Clay Products Ltd for permission to reproduce freely from their booklets. The discussion on limits of size of clay bricks is derived from SCP tn7, *Brickwork: tolerances*, by D.G. Beech.

Extracts from British Standards are reproduced by permission of the British Standards Institution, from whom copies of complete standards can be obtained.

33 PROTECTION AGAINST DAMP

INTRODUCTION 33–2

PREVENTION OF PENETRATION FROM GROUND 33–2

PREVENTION OF PENETRATION THROUGH EXTERNAL WALLS 33–9

PREVENTION OF PENETRATION THROUGH ROOF 33–16

PROTECTION AGAINST DAMP 33

33 PROTECTION AGAINST DAMP

E. DEARDEN, MIOB
North West Regional Director, National House-Building Council

Emphasising that protection against damp is cost-effective to the builder in the long term, this section describes techniques and details suitable for preventing damp penetration from the ground, through the walls and through the roof.

INTRODUCTION

The rules governing the prevention of damp penetration in dwelling houses are described in Part C of the Building Regulations 1976[1] and in the NHBC Handbook[2]. The applicable sections of the latter publication are those concerned with foundations, concrete superstructure, brickwork superstructure and roof covering. These rules permit various methods of construction to be adopted to achieve the desired appearance and plan layout. It is important that the choice of detail be dictated more by full consideration of the site location and degree of exposure to which a dwelling will be subjected, rather than cost. While cost must of necessity play a crucial part in design, logic dictates that the most effective method of preventing penetration must be the most cost-effective to the builder in the long term, particularly as in most cases a builder's liability extends for a period of at least two years from satisfactory completion of a dwelling.

Sources of penetration

This section is sub-divided according to the areas of possible damp entry: from the ground, through external walls, and through the roof. A fourth source of dampness is condensation; this is dealt with in Section 57, so will not be discussed here.

PREVENTION OF PENETRATION FROM GROUND

To prevent the penetration of moisture from the ground, one must construct a continuous barrier in the form of linked membranes and damp-proof courses. The method is determined by the foundation design, the topography of the site and the groundwater conditions: all vary from site to site, in some cases even from dwelling to dwelling. Primarily due to the scarcity of land, foundation design now encompasses many different options – traditional concrete strip foundations, trench fill concrete (both of which incorporate ground bearing floor slabs), timber-framed hollow vented floors, prestressed or cast *in-situ* reinforced concrete floors, coupled in some cases with construction on two or more levels (split level). Other common forms are the semi-raft with downstand beams, the flat slab raft, pier and beam vibroflotation, and piling. All require the adoption of differing methods of preventing penetration from the ground. Examples are illustrated in *Figures 33.1 to 33.6.* The technique for piling, though not illustrated, is similar to that shown for semi-raft foundations with perimeter and intermediate downstand beams, unless the linking floor slab is at ground level and a sub-floor is incorporated to provide the finished floor, in which case the technique will be that illustrated in *Figure 33.6(a).* However, the same basic rules apply whatever the design.

The cavity

The cavity of the external walls must extend 225 mm below the lowest horizontal damp-proof course around the external cavity wall, primarily to prevent mortar droppings building up in the cavity and creating a bridge across the damp-proof course. If the horizontal damp-proof course is wider than the brick or block comprising the wall, it must be trimmed. Some bricklayers permit the excess width to extend into the cavity from each skin, thus reducing the width of the cavity; mortar droppings are then retained by the damp-proof course and form a bridge.

The membrane

The horizontal membrane (usually polythene of minimum 500 gauge) must be laid under the floor slab in traditional solid ground floor slab construction. The membrane *must* be continuous and must link with the horizontal damp-proof course in the inner skin of the external wall. This is achieved by turning the membrane up the face of the wall, then over the inner skin brickwork at finished floor level to unite with the horizontal damp-proof course, laid directly on top; see *Figure 33.1.* It is

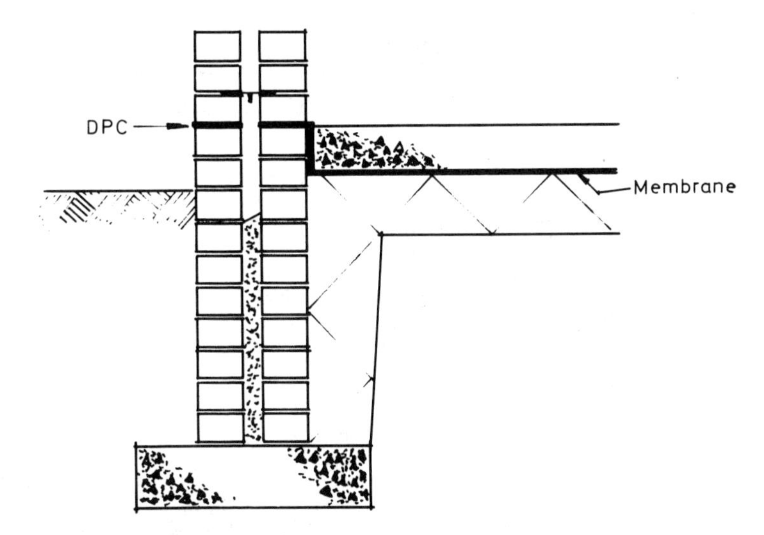

Figure 33.1 Damp-proofing in traditional solid ground floor slab construction

important to ensure that the membrane is not stretched taut at the perimeter, since this can cause feather-edging of the floor slabs. This problem can easily be avoided by spotting the membrane with concrete in the angle of the wall, around the perimeter, immediately before placing the concrete slab. It is permissible to omit the polythene membrane if an asphalt floor finish is used, providing the horizontal damp-proof courses in the inner skin of the perimeter wall and internal partitions extend beyond the wall faces to provide continuity. It is not advisable to omit the underslab membrane when building on a sloping site, or in conditions of high water table or where groundwater is subject to hydrostatic pressure.

Timber-framed ground floors

In hollow vented floor construction, horizontal damp-proof courses in the inner skin of the perimeter walls and internal partition walls must be positioned below the bottom edge of the floor joists (*Figure 33.2*). A polythene membrane beneath the oversite slab is not required. Adequate ventilation of the underfloor space is essential; specific rules relating to this are given in Clause Fo37 in the Foundations section of the NHBC Handbook[2], which stipulates that air bricks should be located at 2.0 m

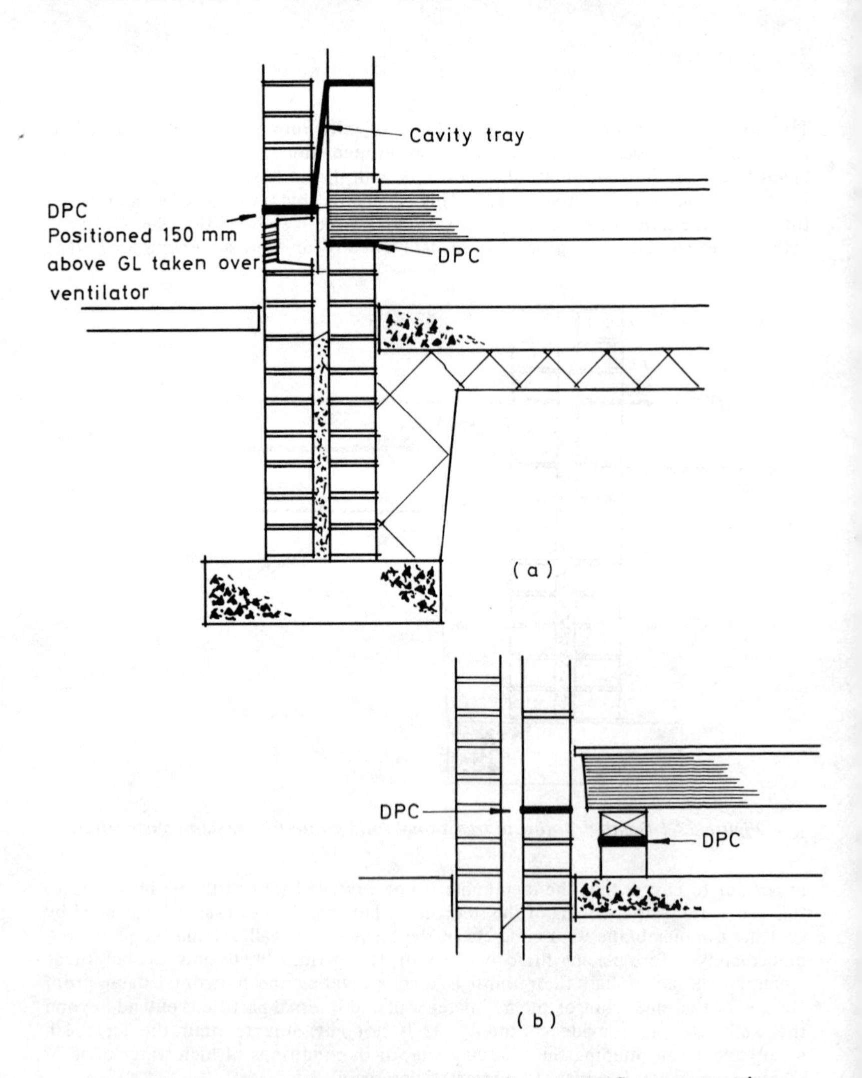

Figure 33.2 Damp-proofing in two types of hollow vented floor construction

centres and not more than 450 mm from the end of any wall or any intersecting foundation wall. The air bricks should be 225 mm × 150 mm; they should be ducted through the inner skin, which necessitates the incorporation of a cavity damp-proof course tray over the duct, the tray to extend approximately 150 mm either side of the duct.

In-situ concrete floors

If the ground floor slab is suspended, the polythene membrane can be laid across the foundation walls and be turned over the edge of the slab around the perimeter inner skin of the external wall (*Figure 33.3*). The superstructure work can then be raised directly off the horizontal damp-proof course laid over the slab membrane.

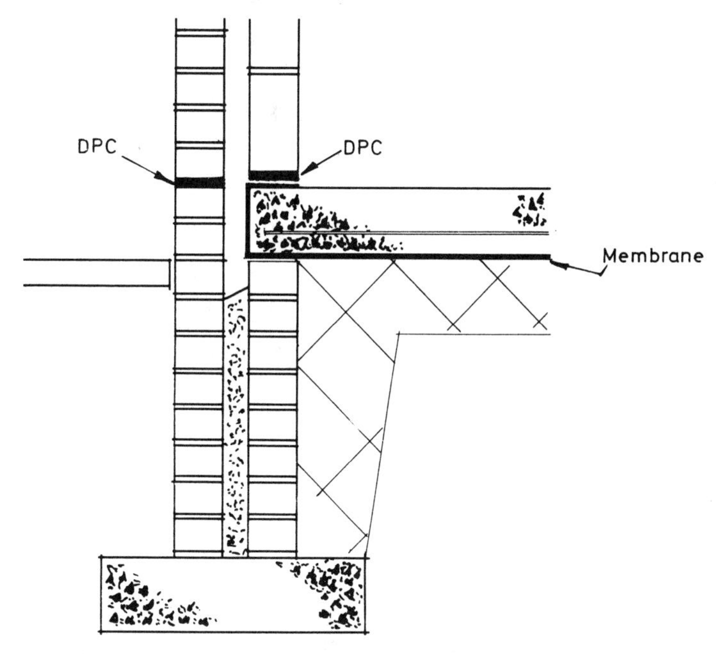

Figure 33.3 Damp-proofing in suspended slab construction

In split-level construction, it is more practical (and advisable) to use rubber bituminous polythene in preference to 500 or better gauge polythene. This is because the membrane, not being self-supporting, must be laid as work progresses through the various stages. A membrane that can be positively bonded provides a more effective means of avoiding subsequent failure. Although the Building Regulations and National House-Building Council requirements permit the use of polythene in split-level construction, it is not sufficient to lap the polythene; it must be jointed by welting or welding. Great care msut be taken to ensure that joints are not obstructed by mortar droppings, brick and block debris, and also that sufficient projection of polythene is allowed at upstands and offsets to form the joints. The membrane should preferably be located toward the external faces of walls in contact with the ground; then that part of the load-bearing structure below ground level remains dry and any water pressure cannot give rise to failure or breakdown, which could occur if the membrane were located on the inside face. Most basement or sub-basement walls act as retaining structures in addition to supporting the superstructure. The membrane must not be so located that it interferes with the former of the two functions. The membrane must be continuous throughout floor and walls, and must link with the horizontal damp-proof courses at general ground level.

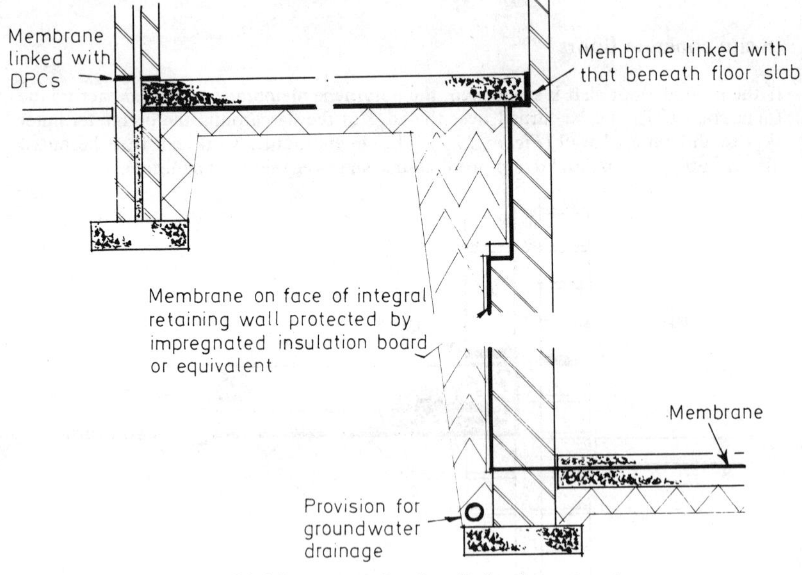

Figure 33.4 Damp-proofing in split-level construction

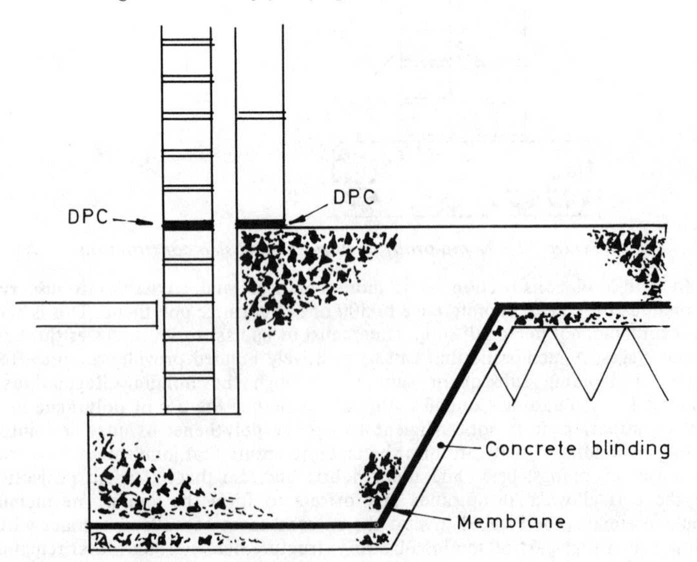

Figure 33.5 Damp-proofing in semi-raft foundation

In some situations retaining-wall design is such that the membrane cannot be located on or near the external face. In such cases the membrane should be laid internally with an adequate weight of concrete, brickwork or blockwork backing in floor and wall construction, coupled with adequate external perimeter groundwater drainage. See *Figure 33.4.*

Rafts

Of the three types of raft foundation commonly adopted, the simplest to construct with effective damp-proofing is the semi-raft illustrated in *Figure 33.5.* The membrane is laid over the full plane area, on the hardcore or preferably concrete

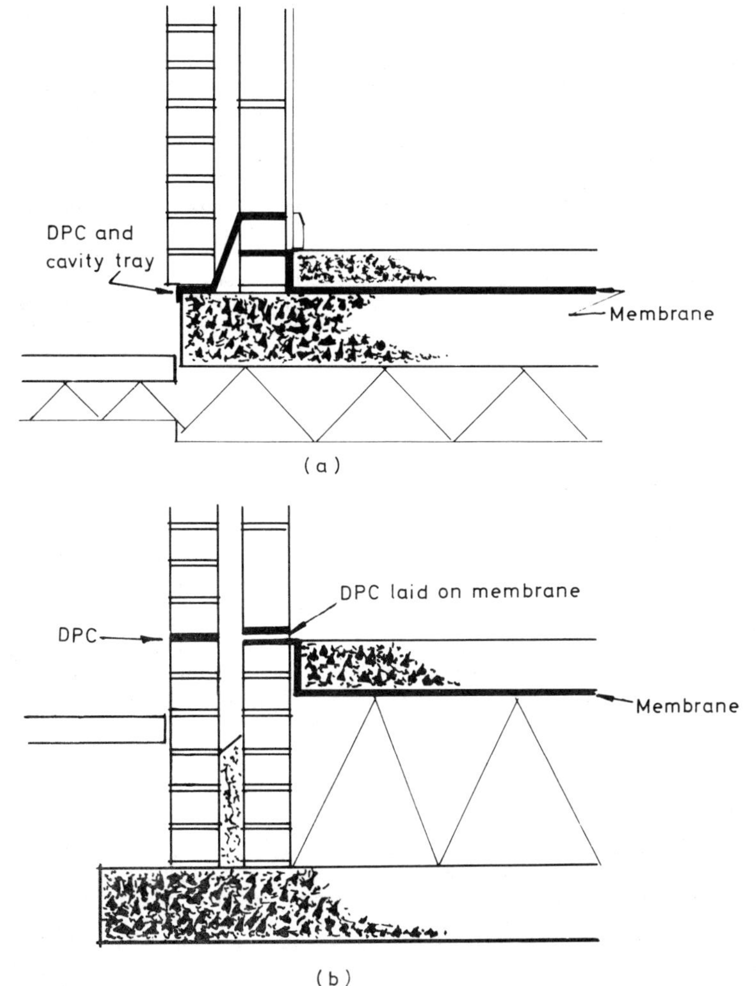

Figure 33.6 Damp-proofing in flat slab raft: (a) at ground level, (b) below ground level

blinding, terminating at the external face of the ground beam. The raft is poured with the edge of the tied slab lineable with the cavity side of the inner skin of the external wall; the cavity between the outer brick skin and the concrete extends (as in the case of traditional construction) 225 mm below the lowest of the horizontal damp-proof courses, which are laid in the traditional position.

In flat slab raft construction, at the discretion of the designer, the raft can be constructed either at or below ground level. In the former case – *Figure 33.6(a)* – the raft serves two functions, providing both the foundation and the floor slab. To be effectively damp-proof, the membrane must be laid on top of the slab and be overlaid with a 75 mm thickness of fine concrete or of sand-cement or pea-gravel floor screed. The membrane is turned up at the external wall perimeter. A cavity tray must be constructed around the full perimeter of the dwelling, care being taken at internal and external angles to incorporate pre-formed cavity trays to overcome the otherwise impossible task of endeavouring to work a polythene or bitumen-base material around an angle at two levels. To prevent mortar droppings bridging the cavity tray, the cavities in superstructure construction must be kept clean, as debris cannot be removed without risking perforations of the tray. To avoid any build-up of water in the cavity in severe weather conditions, weep holes should be incorporated in the perpends of the first course of brickwork in the outer brick skin, at approximately 1.0 m centres. A patent slotted plastic component is marketed for this purpose.

If the flat slab raft is located below ground level – *Figure 33.6(b)* – the void between sub-floor and raft must be damp-proofed, or filled, to comply with the Building Regulations. The usual practice is to construct an *in-situ* concrete sub-floor on consolidated hardcore fill sandwiched between the raft and sub-floor. The membrane and horizontal damp-proof courses are then formed in the same way as described above for the *in-situ* concrete ground bearing slab (*Figure 33.1*). Any load-bearing internal partition walls must be built off the raft.

Pile foundations

In the case of pile foundations, the method of damp-proofing is the same as that described for the semi-raft foundation or the flat slab raft at ground level, depending on the particular design. In some cases, however, ground levels and conditions may dictate the construction of the suspended ground floor slab at a level much higher than that of the perimeter and intermediate ground beams; damp-proofing at floor level would then be the same as that described for *in-situ* concrete suspended-slab construction.

No mention need be made of the vibroflotation or the stone pile technique, since the type of foundation or sub-structure construction used in conjunction with these systems is usually traditional strip or the semi-raft. The method used with the pier and beam system is similar to that for pile foundations.

Heating in the ground floor

Occasionally electric underfloor heating installations are incorporated in ground floor slab construction. Full constructional detail in diagrammatic form can be obtained through the appropriate Electricity Board. Lapped, welted or welded polythene membranes are not recommended. Claims experience has established that the practice of embedding closed-circuit central-heating pipes in floor slabs can lead to failures in the circuit, the symptoms of which are sometimes mistaken for damp penetration. If the problem is to be avoided, it is advisable to ensure that such services are laid in properly constructed ducts, which permit free movement of the pipes and do not transmit stress due to movement of the concrete floor. The best answer is not to embed the pipes, but this is not always aesthetically acceptable, although some Water Authorities may insist on pipes being accessible throughout their length.

Land drainage

The success of damp-proofing to prevent penetration from the ground depends in large measure on skill and care in installation. There is considerable variation in the exercise of these virtues. In some cases the margin of safety is dictated by the effectiveness of land drainage, particularly on the sloping site. It is of the first importance to properly collect and divert any field or land drainage encountered during the course of excavation, and also to incorporate drainage systems to relieve any water pressure and prevent the build-up of groundwater around split-level and semi-basement excavations.

PREVENTION OF PENETRATION THROUGH EXTERNAL WALLS

The criteria governing the construction of external walls to prevent damp entry are specifically described in Clauses C8 and C9 of the Building Regulations 1976. The building construction and design options are open, provided the wall does not transmit moisture due to rain or snow to the inside of the building or any part of the building that would be adversely affected. The options can be divided into two categories: solid walling with an impervious outer skin, and cavity wall construction. A third option is provided by patent building systems, which can loosely be placed in the first of the two categories.

In traditional construction the choice is generally the cavity wall, since it provides an effective tried and tested barrier whatever the degree of exposure to which a dwelling may be subjected. In addition, the cost of construction is probably lower than that of any alternative, and permits a choice of materials, design and dimensional variation that no other method can match.

The cavity

The wall should be built so that the width of the cavity is not less than 50 mm or more than 75 mm. Subject to certain conditions, however, CP 111[3] permits a cavity width of up to 150 mm.

The cavity must be kept clean. The usual technique to achieve this, currently frowned upon by the bricklayer, is to raise a lath to protect the cavity as work proceeds. Another is to leave apertures at the base of the wall and wash mortar droppings out before they harden and cause a bridge. A third is to ensure sufficient depth at the base of the cavity to prevent any mortar droppings building up above the horizontal damp-proof course; this is coupled with care in building, to ensure that bed joints are not struck off on the cavity side in such a way that mortar is allowed to drop down the cavity. Any mortar adhering to ties or damp-proof course trays must be cleaned off as work proceeds.

Rainwater will permeate the outer brick skin. The degree to which this occurs can be limited by taking into account the exposure conditions of the site when choosing the type of facing brick to be used or, in the case of blockwork, when choosing the type of block and applied finish. It is also important to ensure that perpends are filled solid with mortar; some bricklayers merely tip the end of the brick, with the result that joints are open on the cavity side. Other factors that promote a higher permeation rate through the outer brick skin are the adoption of recessed pointing and the use of reclaimed bricks; the latter is not an uncommon practice in country areas. Features are often incorporated in cavity wall construction, e.g. rendered, tile-hung or weather-boarded panels and the like. These will be referred to in greater detail later. Suffice it to say now that, in designing for or forming any such feature, it is preferable to retain the cavity, thus avoiding the necessity of flashing at abutments, and at top and bottom edges.

Cavity ties can be the butterfly or galvanised twisted fishtail type or stainless steel, to comply with BS 1243[4]. Their general spacing should be 450 mm vertically and 900 mm horizontally, subject to design modification for reduced or increased block thickness and cavity width. The ties must be embedded to a minimum of 50 mm in each leaf and must be fixed in a horizontal position. If a variation in bed-joint alignment occurs, necessitating slight tilt of the tie, the tilt must be outward; under no circumstances should the ties be permitted to tilt inward. Some publications dealing with timber frame show, in illustration, the cavity reduced to a minimum of 25 mm. This is not recommended, unless very special means are adopted to ensure that surplus mortar, on the cavity side of the facing brickwork, is cleaned off and prevented from dropping as the work rises. If this is not done the cavity becomes filled solid with mortar; even if the brickwork is separated from the timber-framed structure by a breather membrane, the latter will not prevent penetration, with the resulting unseen deterioration and eventual failure of the timber-framed structure.

To achieve better thermal insulation and structural stability in the external wall, it is preferable to close the cavity at the top edge; in conjunction with this, provision should be made at the soffit to permit through ventilation of the roof space in order to avoid the massive condensation that can and does occur in roof spaces.

Openings

In constructing openings in cavity walls there are three areas that must be damp-proofed or otherwise rendered imperforate: the cill or threshold, the jamb, and the head.

CILL AND/OR THRESHOLD

The design of the threshold must be determined, as stated previously, having regard to the degree of exposure to which the dwelling will be subjected. One must also consider the prevailing weather conditions on the elevation in which the opening occurs. The NHBC Handbook[2] (Clause Br14, Brickwork Superstructure, headed Prevention of Damp Entry) stipulates that detailing must be appropriate to the exposure conditions. The threshold (whether precast or *in-situ* concrete, stone, terracotta, block or brick) must be separated, at every point of contact with the structure, by a damp-proof course linked with the horizontal damp-proof course in the outer brick skin. To prevent penetration across the threshold it may be necessary to insert a weather bar or fix a patent threshold; examples are illustrated in *Figures 33.7 to 33.9.* Wind-driven rain will sometimes by-pass the threshold by running around the edge of the door; insertion of a deep throating in the edge of the frame rebate will prevent this. A weather mould, splayed and drip-grooved, can also be introduced on the bottom edge of the door to prevent dripping water being blown over the top edge of the weather bar or threshold strip.

Cills are now incorporated as part of the window frame, the frames being located in the outer brick skin. Other methods of cill construction include the use of natural, artificial or reconstructed stone, precast concrete, brick, tile, slate, asbestos, aluminium and ferrous metal pressings. In each case the basic principle is the same: to

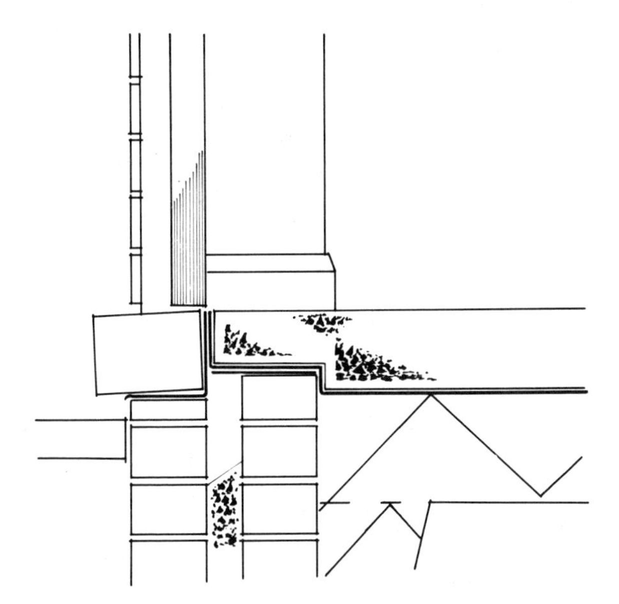

Figure 33.7 Threshold detail: sheltered exposure

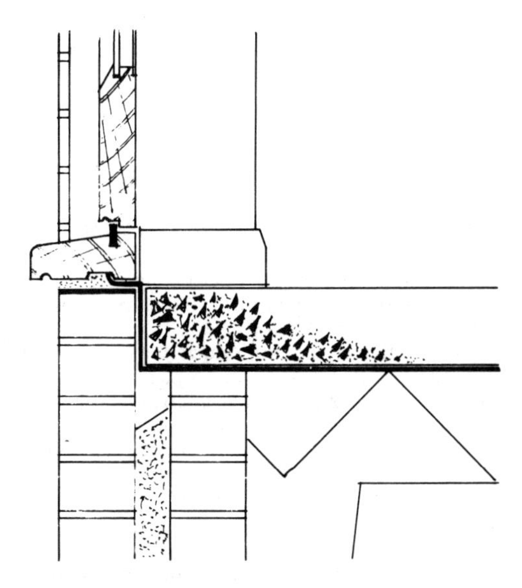

Figure 33.8 Threshold detail: moderate exposure

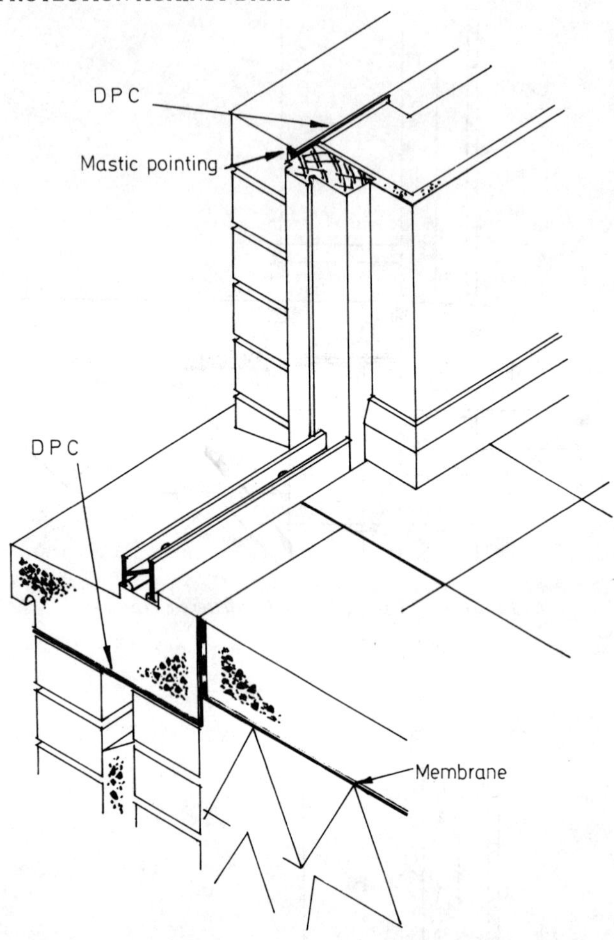

Figure 33.9 Threhold detail: severe exposure

provide adequate weatherings, to discharge water running off the frame, clear of the wall face below, and to prevent any moisture penetrating across the cavity. Some examples are illustrated in *Figures 33.10 and 33.11.* In the case of one-piece cill components (for example timber, asbestos, metal) a damp-proof course is not necessary, unless the cavity is closed to provide a platform for a window board, in which case a vertical damp-proof course must be inserted immediately behind the cill. In areas of severe exposure, a precast concrete or equivalent cill should be installed, sunk-weathered and drip-grooved, with water bar and continuous damp-proof course on bed and back edge.

JAMB

The frame, whether door or window, can be set in three positions: in the outer skin, centrally, or with the external face in contact with the cavity side of the outer skin.

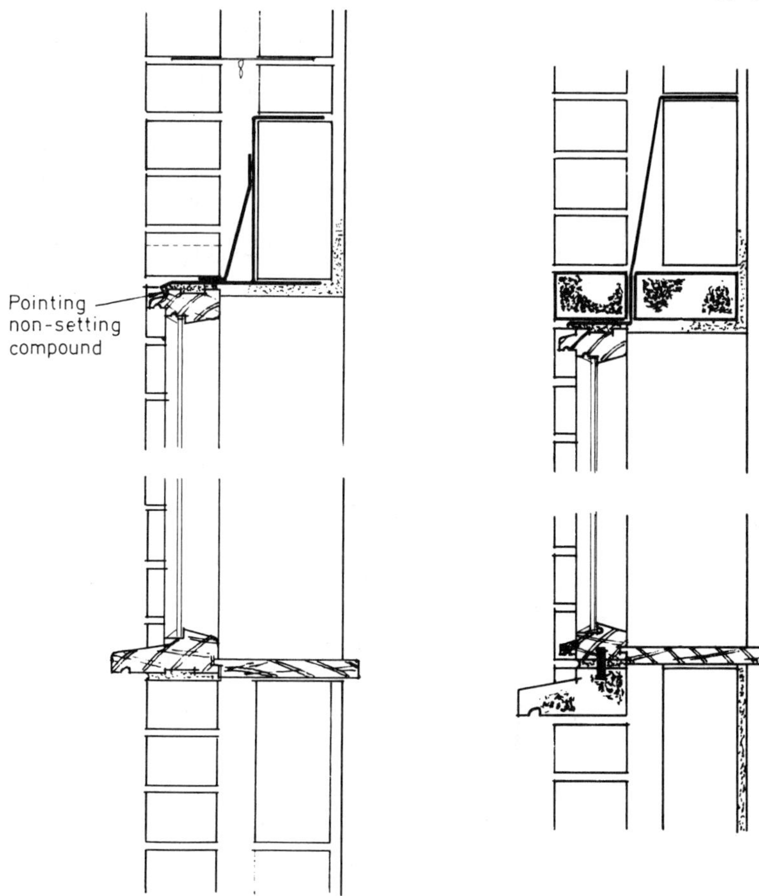

Figure 33.10 Cill and head detail

Figure 33.11 Cill and head detail, concrete

The reveals of the inner and outer skin are generally finished flush, one with the other; this is satisfactory in sheltered or moderate exposure zones, but in areas of severe exposure it is preferable to form a rebated reveal. The frame could still remain in the outer brick skin, with the inside face in contact with the edge of the rebate and damp-proof course. The three alternatives are illustrated in *Figures 33.12 and 33.13.* There are, of course, other ways of preventing damp penetration at reveals – for example, by not closing the cavity, or by inserting patent plastic sections; the latter method has been used with success by many builders. However, the methods discussed and illustrated provide the recommended answer. In some areas the tradition is to build the external wall in two skins of blockwork and to finish the outer skin with sand-cement render, spar dash or equivalent. This finish can provide the protection obtained by constructing a rebated reveal if there is sufficient margin of recess to the face of the frame.

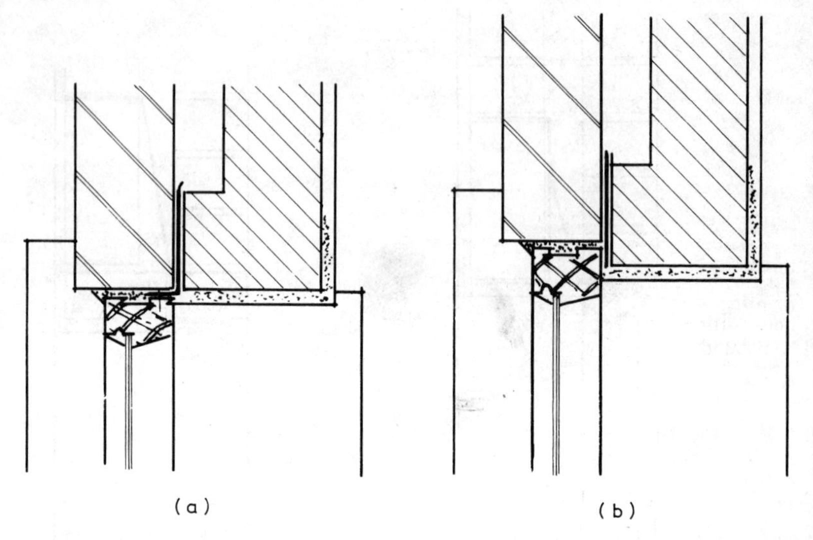

Figure 33.12 Jamb detail: (a) sheltered/moderate exposure, (b) moderate/severe exposure

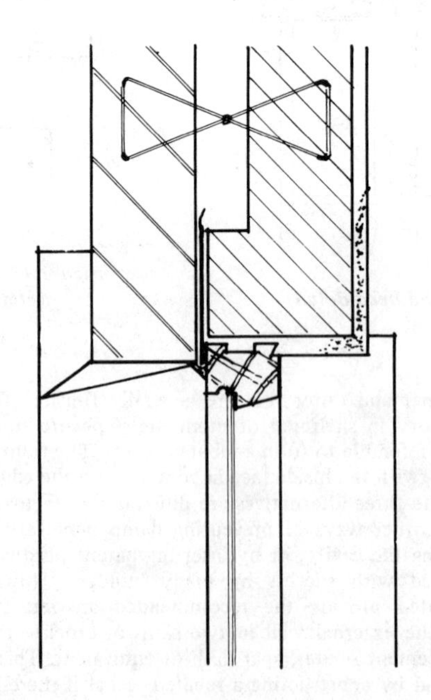

Figure 33.13 Jamb detail: severe exposure

HEAD

The head of the opening is formed by the lintol. The traditional methods of lintol support to the inner and outer brick skin, i.e. timber, angle iron and precast concrete, have been superseded by patent pressed-steel and pre-stressed concrete lintols, other than for exceptional spans.

Generally patent pressed-steel lintols serve two functions: they provide support for work over the opening, and they act as a cavity tray to prevent water that permeates the outer skin, over the opening, from running down the cavity and penetrating to the interior. When fixing steel lintols intended to serve the dual function, the end bearing must be at least 150 mm to ensure that any permeating water is discharged over the end of the lintol into the cavity and not down the reveal. Some types of patent lintol have insufficient splay across the cavity, inside to outside, to prevent a build-up of mortar droppings, causing a bridge. If such lintols are used, a separate damp-proof course cavity tray must be provided; in this case, of course, the length of end bearing is no longer important in providing protection against penetration, although very important structurally. The insertion of weep holes in the perpends of the brick course, immediately above the lintol, although rarely seen nowadays, is advantageous in preventing saturation of the brickwork. Care is needed in the positioning of pressed-steel lintols to ensure that the external edge of the lintol projects beyond the face of the frame; the gap between the head of the frame and the underside of the lintol should be filled with a non-setting compound. Pointing from the frame to the arris of the facing brick, across the edge of the lintol, is not recommended as the pointing can lead moisture from the cavity around the edge of the lintol over the top of the frame, causing penetration on the plaster soffit. Weep holes mitigate the severity of such penetration.

Pre-stressed plank lintols can be used in different ways, depending on span and loading. Generally when used in the external wall they should be two-piece, the wider part supporting the inner skin, the narrower the outer. The damp-proof course tray should be taken across the cavity between the two lintols and be turned over the head of the frame. The edge of the tray must project clear of the frame to prevent capillary attraction between tray and timber head.

Other openings or breaks in the effective barrier provided by cavity wall construction occur when panels and features are incorporated in the outer brick skin, without retaining the cavity. In the case of rendered panels the backing may be 150 mm or 205 mm blockwork; in the case of tile-hanging or boarding, it may be either blockwork of the above thickness, or timber studding. In the former case, care is needed to ensure that the outer brick skin and the blockwork backing are separated, at the point of abutment, by a vertical damp-proof course that extends beyond the point of abutment into the cavity below. Rendering should be finished with a drip-grooved bellcast at the base of the panel or over any opening below. At the top edge, the panel must either finish at the soffit or other overhanging projection, or, if under a cill, should be protected by a flashing turned up behind the cill. Tile-hanging and boarding must be flashed top and bottom with metal and at the abutment with the outer brick skin by a continuous stepped flashing, preferably lead. If the backing is timber-framed, continuous flashings must also be provided at abutments, top and bottom edge, so positioned (in conjunction with the breather membrane) that any water penetrating the cladding is collected and discharged beyond the external face of the work below.

Features and projections

Chimney stacks and other features built into, or projecting beyond, an external wall can be adequately protected by continuing the cavity behind. If this is not possible,

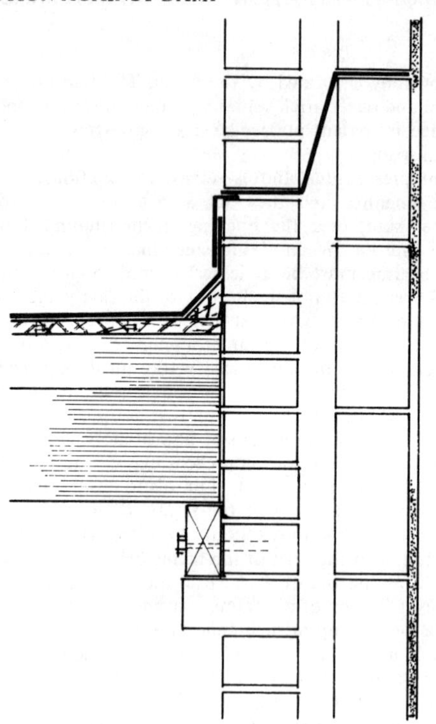

Figure 33.14 Flat roof abutment detail

an imperforate membrane must be inserted, terminating in the cavity at the perimeter edges.

Abutting structures (porches, garages, single-storey projections etc.) require the insertion of a damp-proof course tray across the cavity, inside to outside; the tray must be so positioned that it terminates in the outer brick skin immediately above the apron flashing of the abutting roof. See *Figure 33.14*.

Cavity insulation

Some see the cavity in hollow wall construction as a means of providing insulation. There are a number of insulation materials, including urea formaldehyde foam, glass fibre resin bonded slabs, and polystyrene pellets. If the cavity is to be filled in this way, it is imperative that no obstructions exist and that such installations are carried out in strict accordance with the recommendations set out in the appropriate Agrément Certificate.

PREVENTION OF PENETRATION THROUGH ROOF

The different types of roof construction are dealt with in other sections of this volume. It is sufficient, therefore, to discuss the few aspects that complete the installation of an impervious barrier. These relate to the discharge of rainwater and provision of flashings at abutments, projections, parapets and intersections.

Rainwater discharge

In locating rainwater pipes, care must be taken to ensure that these discharge directly into a drain or over a gulley, the latter in such a manner that the surrounding structure is not saturated. When rainwater discharges from one roof to another, as in the case of chalet construction, this should be contrived so that in severe weather conditions eaves gutters are not loaded to the point of overflow, and so that the part of the roof taking the discharge is protected by a lead slate, or equivalent, at the point of discharge.

Undercloak

Felt or polythene undercloaks should terminate into the eaves gutter and extend over the barge board on to the undercloak at the verge to prevent penetration down the cavity or into the verge or eaves-box framing. The latter would result in wet rot and penetration over window heads at points where the head coincides with the soffit. Support for the undercloak at the eaves position should preferably be provided by a tilting fillet, rather than the fascia board.

At the intersection of flat and pitched roofs, it is important to ensure that the bottom edge of the tiling is not in contact with the flat roof surface, as otherwise the tiles will cut or wear through the covering, causing penetration. Verge bracketing must be solidly built in to prevent sag, which would result in rainwater running around the edge of the verge tiling and ultimately discharging into the eaves boxing or the roof space through shrinkage cracks or gaps in verge pointing.

Projections

Projections through the roof, and abutments, must be properly flashed with lead (or equivalent) stepped flashings, soakers, gutter backs and apron flashings. Where pipes penetrate a roof the aperture must be sealed with a sleeved lead slate, or a patent plastic or metal flashing; there are many types, usually marketed by the pipe manufacturer.

Projections such as dormer cheeks and the like should be flashed in the same way as the chimney stack. It is important to ensure that at the flat roof/pitched roof intersection the felt is dressed over the layer board and the tilting fillet, and taken sufficiently high up the roof slope to prevent a snow load by-passing the top edge.

Stepped roofs

If dwellings are constructed in terraces with stepped roofs, stepped damp-proof course cavity trays must be incorporated to prevent rainwater or snow from permeating that part of the external face of the flank wall projecting above the lower roof slope, and then running down the cavity side of the wall to penetrate above or below ceiling level. The tray can be constructed in short lengths following the line of the stepped flashings, or alternatively the tray can be preformed in a continuous length. A third alternative would be to form the tray as a trough inset above the lower ceiling level. The preferred method is the continuous tray. The point of discharge in each case is the cavity, front and rear elevation.

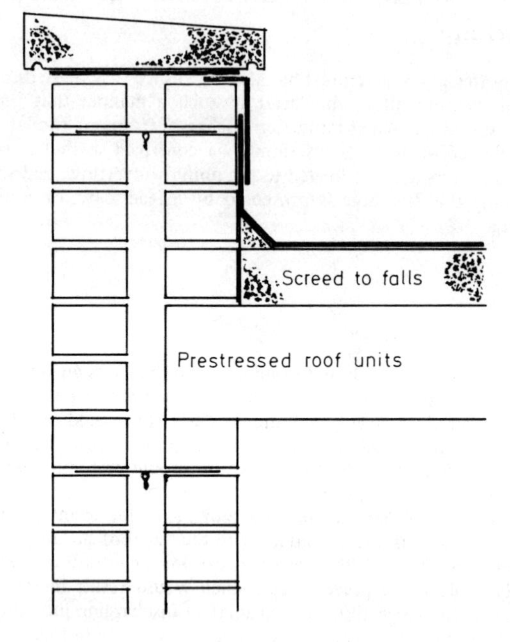

Figure 33.15 Parapet detail

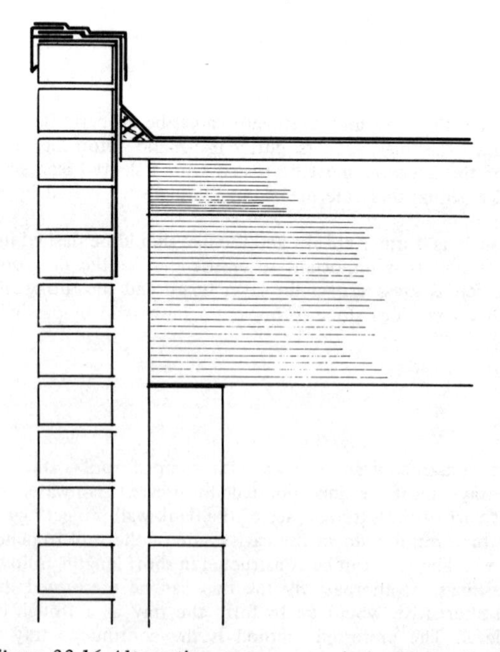

Figure 33.16 Alternative, more economical, parapet detail

Parapets

In flat development the design of roof often chosen is the flat roof with different perimeter configurations, depending on the desired architectural effect. Generally, this involves a form of parapet construction, as shown in *Figures 33.15 and 33.16*. The same principles should be followed as described for other forms of cavity construction. Copings should be weathered, project beyond wall faces, and be laid on a continuous damp-proof course. If the inner skin of the external wall extends above roof level, a damp-proof course tray must be built in at apron flashing level, extending down across the cavity to terminate in the outer skin of the cavity wall. In some designs only the outer brick skin is taken above roof level and the roofing is taken up and over the wall terminating in an extruded aluminium flashing. In this method of construction a movement joint must be incorporated in the roofing, at the upstand, to prevent breakdown due to thermal movement and shrinkage.

REFERENCES

1. *The Building Regulations 1976*, HMSO
2. *National House-Building Council Handbook*, 1974 Edition
3. CP 111, *Structural recommendations for loadbearing walls*, Part 2, *Metric units*, British Standards Institution (1970)
4. BS 1243, *Metal ties for cavity wall construction*, British Standards Institution (1972)

34 TIMBER – THE MATERIALS AVAILABLE TO THE BUILDER

SOFTWOODS	34–2
GRADING OF TIMBER	34–5
HARDWOODS	34–9
PLYWOODS AND OTHER SHEET MATERIALS	34–9
TIMBER HAZARDS, PRESERVATION AND TREATMENT	34–12
TIMBER FIXINGS	34–14
TIMBER MOVEMENT	34–16

TIMBER – THE MATERIALS AVAILABLE TO THE BUILDER

34 TIMBER – THE MATERIALS AVAILABLE TO THE BUILDER

J.A. BAIRD, CEng, MIStructE, MWeldI, FIWSc
UK Managing Director, Swedish Finnish Timber Council

This section gives guidance on the availability of sawn timber, board materials and fixings, and gives sources where further information can be obtained. The accent is on those materials more likely to be used regularly by builders (e.g. softwoods more than hardwoods).

SOFTWOODS

The most common softwoods available in the UK are European redwood (pine) and whitewood (spruce), which are imported primarily from Sweden, Finland and Russia. Canada provides Hem-fir (a mixture of Western hemlock and fir), spruce-pine-fir, a limited quantity of Douglas fir-larch and commercial hemlock. There are also certain UK home-grown species (e.g. Sitka spruce).

Sizes of sawn softwoods

In 1970 the European timber trade adopted metric measure, although Canada is still using Imperial measure and therefore quoting metric equivalents. It is important to realise that the previous tolerance of plus or minus one eighth of an inch on width and thickness no longer applies. The tolerances as laid down in BS 4471, *Dimensions for softwood*, are given below, but the whole accent is on eliminating minus tolerance as far as possible. The trade and users are also encouraged to specify actual sizes rather than 'nominal' sizes, although 'nominal' is still being used by some. Table 1 of BS 4471 (reproduced here as Table 34.1) is a good guide to those sizes of European whitewood and redwood that are available (those shown within the dotted lines). The larger sizes are usually available only in Canadian timbers. Table 34.1 has been presented to take account of modifications recently agreed to BS 4471: Part 1: 1969.

These sawn sizes are measured at 20 per cent moisture content and may have tolerances of minus 1 mm, plus 3 mm on dimensions not exceeding 100 mm, and minus 2 mm, plus 6 mm on large dimensions. BS 4471 permits minus tolerance to occur only on 10 per cent of the pieces. For further information on Swedish or Finnish timber, contact the Swedish Finnish Timber Council, 21 Carolgate, Retford, Nottinghamshire (0777 706616). For information on West Coast Canadian timber, contact the Council of Forest Industries of British Columbia, 81 High Holborn, London WC1 (01 405 0467). For information on other softwoods, contact the Timber Research and Development Association, Stocking Lane, Hughenden Valley, High Wycombe, Buckinghamshire (Naphill 3091).

Sizes of processed softwood

For timber used in joinery it is usual to plane the timber all round, and the reductions for planing from basic sizes are given in items 6 and 7 and Table 3 of BS

Table 34.1 BASIC CROSS-SECTIONAL SIZES OF SAWN SOFTWOOD (from BS 4471:Part 1, reproduced by permission of the British Standards Institution)

	Width (mm)								
Thickness (mm)	75	100	125	150	175	200	225	250	300
16	*	*	*	*					
19	*	*	*	*					
22	*	*	*	*					
25	*	*	*	*	*	*	*	*	*
32	*	*	*	*	*	*	*	*	*
36	*	*	*	*					
38	*	*	*	*	*	*	*		
44	*	*	*	*	*	*	*	*	*
47†	*	*	*	*	*	*	*	*	*
50	*	*	*	*	*	*	*	*	*
63		*	*	*	*	*	*		
75		*	*	*	*	*	*	*	*
100		*		*		*		*	*
150				*		*			*
200						*			
250								*	
300									*

†This range of widths for 47 mm thicknesses will usually be found to be available in constructional quality only.

Note. The smaller sizes contained within the dotted lines are normally but not exclusively of European origin. The larger sizes outside the dotted lines are normally but not exclusively of North and South American origin.

Table 34.2 REDUCTIONS FROM BASIC SIZE TO FINISHED SIZE BY PLANING OF TWO OPPOSED FACES (from BS 4471: Part 1, reproduced by permission of the British Standards Institution)

	Figures (mm) to subtract from basic sizes			
Class	15 to and including 35 mm	Over 35 to and including 100 mm	Over 100 to and including 150 mm	Over 150 mm
(a) Constructional timber	3	3	5	6
(b) Matching*; interlocking boards	4	4	6	6
(c) Wood trim not specified in BS 584	5	7	7	9
(d) Joinery and cabinet work	7	9	11	13

*The reduction of width is overall the extreme size and is exclusive of any reduction of the face by the machining of a tongue or lap joint.

Regularising. The following reductions will apply to timber to be regularised: 3 mm off the width only for timber up to and including 150 mm width, and 5 mm off the width for timber over 150 mm width.

4471 (reproduced here as Table 34.2). On the finished sizes a tolerance of plus or minus 0.5 mm is permitted. It may be acceptable to use stress-graded or construction timber in the sawn state, but if closer tolerance is required the timber can be surfaced all round (see item 1 of Table 34.2) or the timber can be 'regularised'. Whereas surfacing would normally be a machining operation that would surface both opposing faces, 'regularising' can be carried out by removing timber from only one face, as would occur by using one straight edge and one circular saw set from the straight edge. It is quite normal to regularise on one dimension only. For example, to obtain uniformity of floor joists one could regularise 50 × 225 joists to 50 × 220, specifying them as '50 × 220 regularised' or 'regularised to 50 × 220'. With regularising it is normal to remove 3 mm regardless of dimension, although merchants do occasionally remove up to 5 mm from the larger widths, as now stated by BS 4471. Where size is important it is prudent for the builder to specify the actual size required rather than the basic size from which he wants the processing to be carried out.

Certain Canadian Lumber Standard (CLS) sizes are imported that have already been surfaced. These sizes when dried to 19 per cent moisture content are:

38 × 89, 38 × 139, 38 × 185, 38 × 235, 38 × 285 mm

The surfacing is permitted to be carried out when the timber is unseasoned and is intended to give a surface suitable for construction rather than joinery. CLS spruce-pine-fir is usually kiln-dried before being surfaced, CLS Hem-fir is usually surfaced unseasoned. Minus tolerance is not permitted but oversize is not limited. Corners are 'eased' to about 3 mm radius.

RESAWING

When timber is resawn the saw will remove approximately 3 mm. For example, two pieces resawn from 50 × 200 will not be 50 × 100 but 50 × 98. BS 4471 requires that such timber be specified as 50 × 100 're-sawn ex larger'. Alternatively the actual size can be specified. If size is important it is always safer for the builder to specify the actual size.

LENGTHS

Standard lengths are quoted in BS 4471 in increments of 0.3 m as follows, and on these there should be no minus tolerance.

1.80	2.10	3.00	4.20	5.10	6.00	7.20
	2.40	3.30	4.50	5.40	6.30	
	2.70	3.60	4.80	5.70	6.60	
		3.90			6.90	

These are the lengths that the importer or merchant will normally buy. It will normally be more difficult to obtain lengths over 5.70 m except with Canadian timber. A builder can of course specify exact length or cut to length himself.

Dryness

Moisture content (MC) is defined as the amount of moisture in timber expressed as a proportion of its oven-dry weight. With timber it is usually expressed as a percentage.

$$MC = \frac{\text{total weight minus weight of oven-dry timber}}{\text{weight of oven-dry timber}} \times 100 \text{ per cent}$$

It is normally acceptable to measure moisture content by electronic moisture meter.

Most imported European redwood or whitewood is dried to 22 per cent or less before being shipped. Some 80 per cent of Swedish and Finnish timber is kiln dried to 18 per cent (+4% −2%) before shipping, all but a few per cent of the remainder being air dried. About 40 per cent of Russian timber is kiln dried, the remainder mostly air dried. A small percentage of redwood and whitewood is imported unseasoned. The majority of timber imported from Canada is imported unseasoned, the exception being CLS spruce-pine-fir, which is imported dried, usually kiln dried.

All British Standards, regulations and authorities emphasise that timber must have been dried to a suitable moisture content before being manufactured into components or built into the work. The following values give a guide to the moisture contents attained in service by timber:

Internal floorings, doors and linings:	
intermittent heating	12–15%
continuous heating	9–11%
underfloor heating	6–10%
External doors and windows	12–18%
External boarding	16–20%
Pitched and flat roof timbers	15–16%
Upper floor joists	15%
Ground floor joists	18%

Costing

The price of timber may be quoted (or required) as a rate per metre length or a rate per cubic metre. If quoted per metre length there is no room for confusion. There is rarely confusion if sawn timber is quoted per cubic metre; however there may be confusion if the price for processed timber is being quoted by cubic measure. The builder should clarify whether the price per cubic metre is being quoted on the basis of the actual finished dry size of the processed timber or on the basic size from which the piece is processed. The rate can vary by 10–50 per cent.

GRADING OF TIMBER

Quality or commercial grading

The commercial grading of Swedish and Finnish timber is based on rules for

- unsorted (u/s),
- fifth (Vth) and
- sixth (VIth) qualities.

The commercial grading of Russian, Polish and Czechoslovakian timber is based on rules for

- unsorted (u/s),
- fourth (IVth) and
- fifth (Vth) qualities

} roughly equivalent to Swedish or Finnish fifth and sixth qualities.

The commercial grading of Canadian timber is based on rules for

- No. 2 Clears and Select Merchantable,
- a mixture of No. 1 Merchantable and No. 2 Merchantable, and
- No. 3 Commons.

Factory 'flitches' are also available: these are large-section pieces that contain few knots and are sold for reconversion by manufacturers.

None of the grades listed above are end-use grades, although for certain uses (e.g. wallplates, ridge boards) it may be sufficiently precise to specify one of the commercial grades (e.g. Scandinavian fifth quality for wallplates and ridge boards). However, where a more precise specification is necessary, BS 4978, *Timber grades for structural use* and BS 1186, *Quality of timber and workmanship in joinery*, are to be recommended. Stress grading and joinery grading are explained below.

Stress grading

For all practical purposes a builder should now consider it mandatory to provide correctly stress-graded timber for structural use, including structural timber in housing. The principal British Standard on stress grading is BS 4978, *Timber grades for structural use*. One notable improvement for builders in this standard over previous British Standards on stress grading is that each individual piece of timber has to be marked to identify the grade. It is also worth noting that Sweden, Finland and Poland have taken steps to stress grade at source to BS 4978 rather than to export their domestic stress grades. The current tables in Schedule 6 of the Building Regulations 1976 are based primarily on BS 4978 stress grades.

BS 4978 permits stress grading to be carried out visually or by machine. Machine stress grades are identified by the letter M in front of the grade marking. There may also be colour splash marks but these are not essential. (There is no letter V in front of the mark of visually graded timber.) The primary stress grades of BS 4978 and the identification marks are:

General Structural	GS	MGS
Special Structural	SS	MSS

Builders will also probably encounter two additional machine grades: M50 and M75.

Further information on stress grading to BS 4978 is available from the Swedish Finnish Timber Council, the Princes Risborough Laboratory and the Timber Research and Development Association. Permissible stresses are given in CP 112, *The structural use of timber*, including amendment slips AMD 1265, AMD 1846 and AMD 1887.

Table 34.3 CATEGORIES AND GRADES FOR CANADIAN TIMBER

Grade category	*Grades*	*Sizes*
Light framing	*Construction* *Standard* Utility	63–100 mm wide, 38–60 mm thick
Structural light framing	Select No. 1 No. 2 No. 3	63–100 mm wide, 38–50 mm thick
Stud	*Stud*	38 X 63 mm 38 X 89 mm
Structural joists and planks	Select *No. 1* *No. 2* No. 3	150–300 mm wide, 38–50 mm thick

Timber stress-graded in Canada to the NLGA Rules is available in the UK; the allowable stresses are tabulated in CP 112: Part 2: 1971, and its amendments and the grades are included in the Schedule 6 Span Tables. Each piece carries a stamp that includes the grade and the species grouping. The timber is usually shipped in packages containing two and sometimes three grades. The timber may be grade sorted, but if it is incorporated into the building as a mixed grade, the stresses for the lower grade must be used in the structural design.

Canadian timber is stress graded into the categories and grades shown in Table 34.3. As a general rule only those grades in *italic* are available in the UK. Further information on Canadian stress grades is available from the Council of Forest Industries of British Columbia, Templar House, 81 High Holborn, London WC1, and the Canadian High Commission, Canada House, Trafalgar Square, London SW1.

Joinery grading

If a builder wishes to specify a joinery grade he can refer to BS 1186, *Quality of timber and workmanship in joinery*. Part 1 covers the quality of timber, Part 2 covers the quality of workmanship.

The standard describes three classes: Class 1, 2 and 3. There is also a Class IS, which is selected for clear finishes and permits larger knot sizes than Class 1. The standard differentiates between surfaces fully exposed, concealed all the time, and semi-concealed (e.g. exposed only when a window is open). BS 1186 is regarded as a base standard, and it is the intention that component standards should refer to it. It was written to avoid the necessity of designers writing their own (often unworkable) specifications ... 'best unsorted Scandinavian north of a line from ...'. NHBC now require opening windows to comply with Class 2, and fixed frames to comply with Class 3. Class 1 is really written for hardwood joinery or extremely high quality softwood joinery requiring a high degree of selection. Class 2 is for good quality joinery. Class 3 is not a particularly low quality.

Although BS 1186 is due for revision it is to be preferred for joinery specification rather than individual one-off specifications.

Specifying for non-stress graded and non-joinery uses

There are cases of timber use (e.g. wallplates, blockings, ridge boards) that are not joinery (in the BS 1186 sense) and where the timber does not have to be stress graded. In these cases the builder should simply state what he requires: size, length, limited wane, straightness, or whatever he does require.

BS 584 covers a limited number of wood trims (softwood).
BS 1297 covers the grading and sizing of softwood flooring.
BS 2482 covers timber scaffold boards.

Home-grown softwoods

A certain amount of the available softwood is home-grown, mainly Sitka spruce with some Scots pine. Part of the production is suitable to be stress graded to BS 4978, but little is suitable for joinery. Certainly the production is adequate for certain uses: fencing, pallets etc. When using or considering using home-grown timber for the first time, the builder is recommended to obtain specialist advice. For example, although Sitka spruce can be stress graded, for it to be used to satisfy the GS tables for imported timber in Schedule 6 of the Building Regulations it would have to be machine stress graded to M75.

Table 34.4 CHARACTERISTICS OF HARDWOODS

Species	*Colour guide*	*Density*	*Stresses quoted in CP 112*	*Normal uses*			
				External joinery	*Interior joinery & fittings*	*Flooring*	*Structural*
Afrormosia	Yellowish-brown darkens in light	Heavy	No	✓	✓	✓	✓
Afzelia	Reddish-brown	Very heavy	No	✓	✓	✓	✓
Ash	White to light brown	Heavy	Yes (home-grown)		✓	✓	✓
Beech	Pinkish-brown to red brown	Heavy	Yes (home-grown)		✓	✓	✓
Elm	Dull brown	Medium	No	✓*	✓	✓	✓
Iroko	Yellowish-brown with lighter markings	Medium	Yes	✓	✓	✓	✓
Jarrah	Pinkish to dark red	Very heavy	Yes	✓	✓	✓	✓
Keruing	Pinkish brown to dark brown	Heavy	No			✓	✓
Mahogany (African)	Red brown	Medium	Yes	✓	✓	✓	✓
Makoré	Pinkish to dark red	Medium	No	✓	✓	✓	✓
Oak (Japanese)	Light brown	Heavy	No	✓	✓	✓	✓
Opepe	Orange yellow	Heavy	No	✓		✓	✓
Ramin	Pale straw colour	Heavy	No		✓	✓	
Sapele	Fairly dark red brown	Medium	Yes	✓	✓	✓	✓
Teak	Golden brown darkens in light	Medium	Yes	✓	✓	✓	✓
Utile	Red to purplish brown	Heavy	No	✓	✓	✓	✓

* If preserved

HARDWOODS

The use of hardwoods is more specialist than the use of softwoods and is rather beyond the scope of a single section of this reference book. The notes given below really only touch the surface of the subject, and before using, costing or specifying a hardwood that he has not used before it would be prudent for the builder to seek advice from the Timber Research and Development Association, the Princes Risborough Laboratory or the supplier.

Characteristics of a limited number of more readily available hardwoods are given in Table 34.4. Standard sizes at 15 per cent moisture content taken from BS 5450 are given in Table 34.5.

Table 34.5 BASIC CROSS-SECTIONAL SIZES OF SAWN HARDWOODS
(from BS 5450, reproduced by permission of the British Standards Institution)

Thickness (mm)	*Width* (mm)										
	50	63	75	100	125	150	175	200	225	250	300
19			*	*	*	*	*				
25	*	*	*	*	*	*	*	*	*	*	*
32			*	*	*	*	*	*	*	*	*
38			*	*	*	*	*	*	*	*	*
50				*	*	*	*	*	*	*	*
63						*	*	*	*	*	*
75						*	*	*	*	*	*
100						*	*	*	*	*	*

Note. Designers and users should check the availability of specified sizes in any particular species.

PLYWOODS AND OTHER SHEET MATERIALS

The specifications of plywoods used in the UK have undergone a considerable amount of change during the last few years, and builders should question the validity of any codes, standards or company literature printed before 1972. Plywood is imported from at least forty countries and also is manufactured in the UK. Only the principal plywoods are mentioned below. It is convenient to consider those that are suitable for structural use and those that are not.

Four categories of plywood are used in the UK related to the type of adhesive used (as defined in BS 1203). The categories of adhesive are:

WBP Weather- and boil-proof
BR Boil resistant
MR Moisture resistant

There is also an ‘Interior’ category. Builders should be quite clear that it is the *adhesive* that is WBP, BR, MR or ‘I’, not the actual veneers from which the plywood is made.

For a plywood to be used for structural purposes in the UK and to be included in CP 112, it is the intention of the CP 112 committee that it must be made with WBP adhesive and its manufacture must be inspected regularly by a control organisation that satisfies the CP 112 committee.

Canadian Douglas fir plywood

This plywood has face veneers of Douglas Fir and inner veneers of West Coast softwood species. It is suitable for structural use and is normally available in thicknesses of 7.5–20.5 mm for unsanded grades and 6–19 mm for sanded grades in panel sizes 2440 × 1220 mm and 2400 × 1200 mm. The grain of the face veneer runs parallel to the longer side. The glue used is WBP. The sanded grades available are 'Good 2 sides' and 'Good 1 side'. The unsanded grades available are 'Select sheathing tight face', 'Select sheathing', and 'Sheathing'.

Canadian softwood plywood

This plywood is made entirely from softwoods and is suitable for structural use. The range of sizes and thicknesses is similar to that for Douglas fir plywood. Canadian softwood plywood is available in unsanded grades only, and is classified in the same way as Douglas fir plywood.

For further information on Canadian Plywood, contact the Council of Forest Industries of British Columbia, Templar House, 81 High Holborn, London WC1.

Finnish all-birch plywood

This plywood is suitable for structural use and is normally available in thicknesses of 6.5–27 mm and in panel sizes 1220 mm and 1525 mm wide up to 3660 mm long. The grain of the face veneer runs parallel to the shorter side. The glue used is WBP. The face qualities are B, S, BB and WG. B face quality is suitable for top class painting and clear finishing, S for good quality paint finish, BB for interior paint finishes and overlaying, and WG where appearance is not important.

Panels are available in combinations of face qualities. Those commonly stocked are B/BB, BB/BB, BB/WG and B/WG in tongued and grooved flooring panels.

Finnish combi plywood

This plywood has face veneers of birch and inner veneers of birch and softwood; otherwise the specification is as for the Finnish all-birch plywood. It is suitable for structural use, is usually more readily available than all-birch, but has lower permissible stresses.

Finnish softwood plywood

The plywood is made entirely from softwood, is suitable for structural use, and the range of sizes is similar to the birch plywoods. For further information on Finnish plywoods, contact the Finnish Plywood Development Association, Broadmead House, 21 Panton Street, London SW1.

Swedish spruce plywood

This plywood is available in thicknesses of 8–19 mm in panel sizes 2440 × 1220 mm and 2400 × 1200 mm. The grain of the face veneer runs parallel to the longer side. The glue used is WBP. The face qualities readily available in the UK for structural use are: C plugged C, and CC. For structural use the mark P30 should also appear on the plywood; this is a reference to the minimum failure stress in bending (30 N/mm^2). Further information is available from the Swedish Finnish Timber Council, 21 Carolgate, Retford, Nottinghamshire.

British plywood manufactured from tropical hardwoods

There are a few British plywoods manufactured from tropical hardwoods to BP 101 Specification (Association of British Plywood and Veneer Manufacturers) that were considered suitable for structural use when CP 112 was written in 1967.

Other structural plywoods

Before using any other plywoods for structural use the builder should be certain that they are approved, preferably making reference to the Timber Research and Development Association or to the Princes Risborough Laboratory. The builder should be particularly careful before using any plywood with a 'D' rated veneer for structural use.

Decorative and general-purpose plywoods

There are many types of decorative plywoods (Gaboon, Lauan, Mahogany, Sapele etc.), and several general-purpose plywoods (American, Russian birch etc.) suitable for hoardings, packaging, flooring etc. Details cannot be given in this section, and the builder should refer to the supplier and a publication of the Timber Research and Development Association: *Plywood: its manufacture and uses*. For current information, refer to the *Timber Year Book and Diary*. For information on American plywood, refer to the American Plywood Association, Westwood House, South Street, Great Waltham, Chelmsford, Essex.

Particle board (chipboard)

There are several specifications for particle board: high, medium or low density boards, boards of the same density throughout, and those with a denser layer near the surface. Certain types are moisture-resistant or flame-retardant, or have surfaces specially prepared for painting, or are veneered with plastic, metal or wood.

One large use for chipboard is in flooring. The builder should ensure that he is using a flooring-grade board, which will be stamped 'Flooring grade to BS 1181', and must ensure that boards do not become wet during storage or construction. Unless a board is specially treated or covered against moisture, should it become wet it would be prudent not to use it for flooring. Further information on chipboard is available from the Chipboard Promotion Association, 7a Church Street, Esher, Surrey (Esher 66468).

Blockboard and laminboard

For data on blockboard and laminboard, contact the Finnish Plywood Development Association.

Fibre building board

As well as the basic fibre building boards there is a comprehensive range of boards having special insulation, decorative or flame-retardant properties. A list of the types is given below, taken from the Brand List of the Fibre Building Board Development Organisation Ltd, 6 Buckingham Street, London WC2N 6BZ, from whom further information can be obtained.

Standard hardboard
Tempered hardboard
Medium board
Duo-faced hardboard and medium board
Perforated hardboard
Surfaced boards (painted, enamelled etc.)
Surfaced boards (surface laminated, including plastic and textile facings)
Moulded, embossed and similar hardboards
Insulating board (sheets)
Insulating board (tiles and planks)
Sarking or sheathing grade insulating board
Acoustic board (sheets)
Acoustic board (tiles and planks)
Boards supplied to Class 0 standard
Boards achieving Class 0 with site-applied surface coating
Boards with Class 1 surface spread of flame

TIMBER HAZARDS, PRESERVATION AND TREATMENT

Insect attack

In the UK there are two major pests that infest timber in buildings, the common furniture beetle and the death-watch beetle. The house-longhorn beetle also occurs, but is confined to a small area near London. The powder post beetle may occur in the sapwood of certain hardwoods.

The common furniture beetle is by far the most common cause of any damage attributed to 'woodworm'. The larvae of the beetle may spend three years or longer tunnelling and growing in size until they emerge as adult beetles through exit holes usually less than 2 mm diameter. Adult beetles do not again bore into timber.

The death-watch beetle occurs mainly in large-sized timbers in old buildings, such as churches, where timbers have been allowed to become damp. The beetles emerge after the larvae have tunnelled for a number of years leaving exit holes 3 mm diameter. Attack is usually confined to partly decayed timbers, and keeping the timber dry is usually the best preventative. Since the beetle tends to attack large *in-situ* timbers, insecticide treatments may well be ineffective and gas fumigation is the only reasonable treatment.

Softwoods or hardwoods may show signs of small insect holes caused by the pinhole borer (Ambrosia beetle). This is an insect of the forest which dies when the logs are sawn and does not re-infest the timber. Dried timber with signs of this attack can safely be used, and is permitted by BS 1186 for joinery timber and BS 4978 for structural timber. Pinhole borer is usually recognised by hole diameters of 0.5–3 mm surrounded by a dark stain and no dust, with holes usually at right angles to the grain.

Fungal attack

There are two main types of wood-destroying fungi, known commonly as wet rot and dry rot, that can attack timber if the moisture content rises above 20 per cent for any length of time.

Wet rot is much less virulent than dry rot and will only continue to flourish while there is a source of moisture. Wet rot will not spread to dry timber, and the most effective remedial treatment is to locate and remove the source of moisture. Affected timbers should, where necessary, be removed and replaced by sound, preservative-treated timber.

Dry rot is much more difficult to deal with, and remedial treatment is really a job for a specialist contractor. The fungus can spread to otherwise dry, sound timber, and strands can even find new timber by spreading over plaster and brickwork. All traces of the fungus must be removed by cutting out all affected timber and any sound timber within 600 mm of the last visible signs of decay. Surrounding masonry must be sterilised, and it is a wise precaution to preservative-treat all sound timber left within the affected area and any new replacement timber.

Sap stain

Sap stain or blue stain is caused by the timber, once sawn, having been allowed to remain wet (over 20–22 per cent MC) for several weeks. In itself sap stain is not a structural defect (a certain amount is permitted in BS 4978 stress grades). It can be covered by paint once the timber is dry. Once the timber has dried to below 20 per cent MC the sap stain is no longer active.

Timber preservation

Providing timber is maintained at a moisture content of 20–22 per cent or below, there is rarely any necessity to preserve. There are, however, cases where it is considered essential to preserve most softwoods (such as timbers permanently in contact with the ground or set into concrete or masonry that is likely to be damp for long periods of time), and also cases where building regulations call for preservation (such as external timber boarding). BS 5268:Part 5 (the new numbering for CP 112:Part 5) details the various risk conditions and gives examples of suitable preservatives. Regulation B3 of the Building Regulations 1976 requires roof timbers in certain areas around London to be preserved against the longhorn beetle by one of the treatments given in Schedule 5 (Table 5) of the Regulations.

The most effective method of preservation is to impregnate under pressure or vacuum. The two primary types are water-borne (with copper chrome salts or similar) and organic-solvent preservatives.

TAR-OIL PRESERVATIVES

There are a number of tar-oil preservatives, of which creosote is the most common. The effectiveness of creosote against insect and fungal attack is well established. Creosote can be applied by brushing, spraying or steeping where a surface application is adequate; however, to be fully effective in high-risk situations it is normal to pressure impregnate.

Creosote suffers from the disadvantage that it is pungent, taints foodstuffs, kills plant life and leaves a strong odour. It also tends to 'bleed' from the timber, leaving an oily, sticky surface, not suitable for overpainting by normal paints.

WATER-BORNE PRESERVATIVES

With this type of preservative metallic salts are dissolved in water. The water carries the salts into the timber by pressure-impregnation methods. The water is dried off leaving salts fixed in the timber. The most common forms of water-borne preservatives are copper-chrome-arsenate types. CCA treatments have the advantage that they are equally effective against insect or fungal attack, they are usually odourless, and the timber can be painted once it is dry. The disadvantage is that the treatment adds considerably to the moisture content of the timber, which must then be redried before being used.

ORGANIC-SOLVENT PRESERVATIVES

With these types of preservative the toxic content is dissolved in solvent, usually a light petrochemical solvent. These solvents tend to penetrate timber rather more than water and therefore brushing, spraying or dipping can be effective, but vacuum impregnation is required for the best results.

Because the moisture content of the timber is not increased there is no need for drying after treatment. After residual solvent has evaporated the surface is usually suitable for painting. Water-repellent oils or colouring can be incorporated. Remedial *in-situ* treatments are usually of the organic solvent type because they can be applied by brush or spray. Care must be taken of the fire risk during application. Preferably the electricity supply should be switched off and electrical appliances protected.

DIFFUSION PROCESSES

It is possible to buy timber that has been preserved at the sawmill by a diffusion process using boron salts. This must be carried out on unseasoned timber. The timber must not be used where it can be affected by rain because the preservative can be leached out by water.

Flame-retardant treatments

Both impregnation and surface-coating treatments are available that can upgrade the surface spread-of-flame characteristics of timber from the usual level of Class 3 to Class 1, and even to Class 0 in exceptional cases. Surface coatings may not always be permanent and may require maintaining as would a paint finish. Certain fire-retardant treatments are claimed to increase the fire resistance of a timber structure, but each case requires critical consideration.

TIMBER FIXINGS

Fixings readily available are nails, improved nails, screws, bolts, coach bolts, tooth-plate connector units, split-ring connector units, shear-plate connector units, punched-metal plate fasteners, and glues. Perhaps the most comprehensive notes on structural fixings including glue are contained in Chapters 1, 18 and 19 of the *Timber designers' manual* published by Crosby Lockwood Staples. Permissible loads for structural uses are given in CP 112: Part 2, *The structural use of timber*, with details of edge and end distances.

Nails, screws and bolts

Nails are satisfactory for lightly loaded connections where the nails are in shear. Nails can be obtained unprotected or treated against corrosion. 'Improved' nails can be obtained that will take higher shear loads and can resist tension and vibration. The 'improvement' usually takes the form of annular rings. Such nails can be obtained unprotected or treated against corrosion. Most softwoods nail easily. Care (perhaps preboring) is necessary with many hardwoods and hemlock or Hem-fir.

Screws will take shear or tension, although screws driven into end grain will not take tension. Usually predrilling is necessary. Untreated, treated or non-ferrous screws are available.

Bolts or coach bolts used with timber are usually ordinary mild steel, either 'black' or galvanised. It is usual to use fairly large washers under the head and the nut to minimise indentation.

Tooth-plate connector units

Tooth-plate connectors can be either single- or double-sided, square or round. Each connector is used with a bolt with a washer under the head and the nut. Split-ring and shear-plate connector units should be used only by builders or manufacturers with the correct cutting tools. Punched-metal plate fasteners should only be inserted by manufacturers with the correct pressing equipment (as with trussed rafter manufacture). Any plate dislodged on site should not be replaced without reference to the manufacturer of the component. However, there are variations (pre-holed metal plates), suitable for site nailing, in which the plate or connector is used like a small gusset plate.

Glue

WEATHER- AND BOIL-PROOF GLUES

When a component or joint is likely to be directly exposed to weather, either during its life or for lengthy periods during erection, it is necessary to use a glue that can match the 'weather- and boil-proof' (WBP) requirements of BS 1204:Part 1: 1964. These glues are normally gap-filling resorcinol-formaldehyde or phenol-formaldehyde resins. They are also usually used for structural finger-jointing and for conditions of high humidity or for components in corrosive atmospheres. These glue types require careful quality-control during storage, mixing, application and curing, and are therefore normally only intended for factory application.

BOIL-RESISTANT OR MOISTURE-RESISTANT GLUES

When a component is unlikely to be subjected to any serious atmospheric conditions once in place, and is only likely to receive slight wettings during transit and erection, it should be possible to use a glue that matches the 'boil resistant' (BR) or 'moisture resistant' (MR) requirements of BS 1204:Part 1:1964. These glues are normally gap-filling melamine/urea-formaldehyde or urea-formaldehyde, and are usually cheaper than the WBP types. They also require careful quality-control during storage, mixing, application and curing, and are therefore normally intended for factory application.

INTERIOR GLUES

Only interior non-structural components are normally glued with a glue that matches only the 'Interior' requirements of BS 1204:Part 1:1964. There are exceptions, usually where a site joint is required or where control of the gluing can not be guaranteed to a sufficient degree to enable a WBP, BR or MR glue to be used. An 'Interior' glue should be used only when the builder can be certain that the glue joint will not be subjected to moisture in place. A cold-setting casein adhesive to BS 1444:1970 Type A would normally be used.

GLUING

Any form of gluing requires quality control. The quality-control requirements to obtain a sound glue joint are correct storage, mixing and application of the glue, correct surface conditions, moisture content and temperature of the timber, and correct temperature of the air during application of the glue and during curing. The instructions of the glue manufacturer must be followed. The joints must be held together during gluing, either by clamps or by nails (glue/nailing, i.e. nails left in although their only duty is to hold the surfaces together during curing, design being based on glue stresses only).

Several sources point out that the glue must be compatible with any preservative used. This is not usually a problem, but should a treatment (such as application of a fire retardant) emit ammonia, then gluing by resorcinol types (and perhaps others) should never take place after treatment, nor should treatment take place until at least seven days after gluing.

TIMBER MOVEMENT

Moisture movement

Timber is a hygroscopic material and will move with changes of moisture content. Once timber has been dried correctly, subsequent movements with moisture content change are significantly less than movement during the initial drying. The guide values tabulated below are those for subsequent moisture change BS 4471 states that, for softwood, the actual size at moisture contents between 20 and 30 per cent shall be greater by 1 per cent for every 5 per cent of MC, and may be smaller by 1 per cent for every 5 per cent of MC below 20 per cent.

	Softwood (guidance values)	*Hardwood*
Tangential movement expressed as a value per 1% of MC change:	0.28%	
Radial movement expressed as a value per 1% of MC change:	0.13%	varies with species
Longitudinal movement expressed as a value per 1% of MC change:	0.005%	

Plywood	Up to around 0.022% along the grain per 1% of MC change. Up to around 0.015% across the grain per 1% of MC change.
Particle board	BS 2604 specifies a mean increase of not greater than 0.35% in length and 7% in thickness when relative humidity is increased from 65% to 90%.
Fibre building board	BS 1142:Part 2 specifies limits of change in size of standard hardboard and medium board after changes in relative humidity from 33% to 90%. BS 1142:Part 3 specifies maximum absorption levels for sarking and sheathing boards.

Storage of timber and components

To minimise movement, it is essential for timber, boards and components to be stored carefully on site. Storage must be clear of the ground, under cover, but ventilated. Supports must be levelled and positioned to limit distortion. During construction the timber and components should be protected from water, cement splashes etc.

35 TIMBER FLOORS AND FLAT ROOFS

TIMBER–JOISTED FLOORS 35–2

TIMBER–JOISTED FLAT ROOFS 35–12

TIMBER FLOORS AND FLAT ROOFS 35

35 TIMBER FLOORS AND FLAT ROOFS

J. OLLIS, AADip, RIBA
Chief Architect, Timber Research and Development Association

After discussing the structural and non-structural aspects of timber floors and flat roofs, this section considers those parts of design and construction that require particular care (including economical design, and acoustic and thermal insulation).

TIMBER–JOISTED FLOORS

Fire performance

Fire resistance involves the capacity of a complete floor element to contain and withstand the effects of a standardised test fire, supporting the design loads for a given period of time, remaining imperforate to flame and not getting too hot on the surface furthest from the fire.

The ground floor of a house is not required to provide any fire resistance unless it is over a basement or a basement garage, when it is required to provide fire resistance in all respects for 30 minutes.

The first floor of a two-storey house is required by regulations to provide 30 minutes resistance regarding load-carrying capacity but only 15 minutes resistance to flame penetration and excess surface temperatures. This is referred to as 'modified ½ hour'. First floors in two-storey flats, intermediate floors in maisonettes and all upper floors in three-storey houses are required to provide 30 minutes fire resistance in all respects. Floors between flats and/or maisonettes in three- or four-storey blocks are required to provide 60 minutes fire resistance.

Deemed-to-satisfy floor constructions with appropriate fire resistances are given in Schedule 8 of the Building Regulations 1965. Many of these constructions depend on the fact that joists can withstand, when fully loaded and after fire has penetrated the ceiling, between 6 and 12 minutes fire exposure, depending on the breadth of the joist.

There are accepted methods of calculating the fire resistance of exposed timber beams by which it can be shown that beams of 75 mm width will in most cases provide 30 minutes fire resistance in themselves. Some references to further information on methods of calculating fire resistance of timber sections and on tested constructions other than those in Schedule 8 are given at the end of this section.

Ceilings on the undersides of floors have to meet surface spread-of-flame test ratings as follows:

- Class 3 : rooms in two-storey houses.
- Class 1 : rooms in three-storey houses and in all blocks of flats and/or maisonettes and in the circulation areas of two-storey houses.
- Class 0 : circulation areas in all dwelling units and blocks (apart from one- and two-storey houses).

Timber in its untreated state generally provides Class 3 performance. It can be raised to Class 1 by impregnation or with special clear varnishes or intumescent paints. Thin treated-timber veneers on non-combustible backings can provide Class 0 performance.

Structural behaviour

The main structural duties of a floor are performed by individual joists, but acting together in taking live loads and heavy point loads without excessive vibration or local deflection, and in partly shedding loads from weaker joists on to stronger neighbouring ones. This load-sharing characteristic of a timber floor is assumed in the structural design when taking mean values for the modulus of elasticity of a particular grade and species of timber, if the joists are no more than 600 mm apart, and increasing the permissible bending stress by 10 per cent. Load sharing is achieved by the stiffness of flooring spanning across the joists and, to a lesser extent, by the ceiling in spreading the load sideways. With long spans, load sharing can be assisted by herringbone strutting or solid blocking between joists, provided that these are well fixed and tightly fitted when the timber is relatively dry, and that there is end anchorage to the walls for what can amount to a transverse spreader beam.

Timber joists can vary much in their stiffness, even within a stress-graded batch, and to get an even floor effective load-sharing is important. The stiffness of a floor may in practice also be improved slightly by the presence of ceilings and sheet-material flooring acting as a stressed skin in the longitudinal direction, although this is not normally allowed for in calculations (unless floor panels are deliberately designed with plywood stressed skins, which is rarely an economical proposition).

Floors without ceilings attached to the undersides of joists can occur in two main situations: with suspended timber ground floors and with certain types of sound-resistant party floors that have independent or resiliently fixed ceilings, described later. Other types of party floor have structurally detached 'floating' floor surfaces. In all these cases the floors are likely to be ones with 'living' rather than sleeping activities, and therefore they will in practice have to take more dynamic and heavier point loads – parties, washing machines and deep freezers, for example.

In the case of such floors spanning more than 3600 mm, it may be advisable to compensate for the lack of a load-sharing feature either by designing comfortably within the normal deflection limit or by increasing the assumed normal domestic superimposed design load by about 25 per cent, which would correspond with Canadian design practice. On the other hand, principal members such as beams and trimmers are normally designed on the basis of minimum modulus of elasticity values for a particular species and grade and without increased design stresses, because they have to take more concentrated loads virtually on their own. However, certain joists in a house plan may also be located where they are liable to take more concentrated loads, although perhaps not acting as principal members. This should be remembered when selecting joist material for different parts of the floor on the site.

Operations and workmanship affecting quality

For a timber-joisted floor to be an effective load-sharing structure, with smooth and level upper and ceiling surfaces, certain matters concerning site organisation and workmanship are especially important. Dimensional regularity and moisture stability should be considered together. Joist material with a sufficiently constant depth may be obtained either in a sawn state (by selection of a source of supply or by subsequent selection and sorting of material), or by using 'regularised' timber with top and bottom faces machined, or by obtaining CLS timber, for example, machined on all faces. In each case there may still be a tendency for some members to twist or bow along their lengths, particularly if they are exposed to damp conditions on site for more than a few days by being stored without cover or incorporated in a building that remains unroofed for a long time; both of these situations should be avoided whenever possible.

Given that there is distortion (and there will inevitably be some), the way timber is stacked on site becomes important: it can either encourage or discourage carpenters to pick out the joists and place them in the building in a favourable sequence to give, ideally, a gradual upwards camber in the middle of floor as indicated in *Figure 35.1*. This will tend to counter deflection due to the final dead weight of the floor.

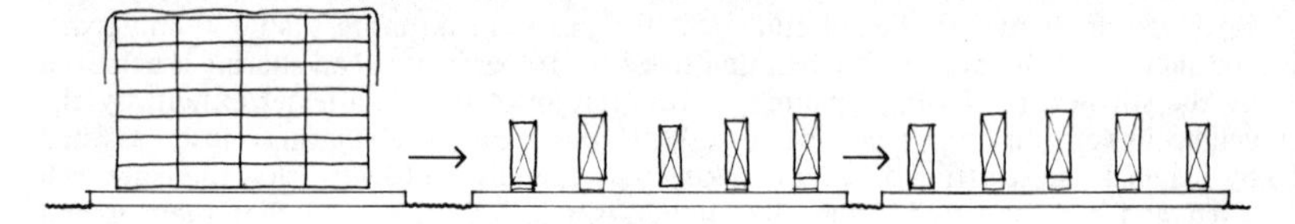

Figure 35.1 On-site sorting of floor joists

If joists are not sorted in this way, extra time and skill will be wasted in bringing upper and lower joist surfaces sufficiently into line to be able to fix reasonably true floors and ceilings. This may involve packing or straining joists when fixing herringbone strutting or solid blocking, which will hinder effective load sharing in the completed floor. *Figure 35.2* illustrates some of the practical implications of this and similar points of detail.

Moisture protection of timber on site is particularly important with deep joists (especially in houses to be centrally heated), because excessive shrinkage is likely to cause unacceptable gaps beneath skirtings; it may cause even greater problems in connection with upper floor partitions and cupboards, which are likely to have fixings also to walls that will not settle correspondingly. The protection of flooring timber from damp is of course vital to avoid subsequent shrinkage, although somewhat less critical when WPB plywood (see Section 34) is used.

Economical design of floors

The traditional method of constructing the upper floors of dwellings with softwood joists still predominates and is likely to do so in the foreseeable future, in spite of efforts to introduce alternatives based on precast concrete components. Essentially this is because of the ease of handling, trimming, fitting and fixing of timbers. However, there are ways of further improving the economy of timber joisted domestic floors, some of which are obvious but few of which are consistently practised.

The first simple rule is to use joists of the minimum practical breadth, which is another way of saying maximise the depth in relation to the cross-sectional area, which itself is an index of cost. A 38 × 225 mm joist has a cross-sectional area only marginally greater than a 50 × 175 mm joist, but has a 12 per cent greater spanning capacity.

A second simple rule is to try to reduce spans as much as possible in the planning of a dwelling, by the full use of internal loadbearing partitions and, where appropriate, by using exposed timber main beams. The potential effect of this is shown by observing, from tables, that 38 mm wide joists at 400 mm centres spanning 4000 mm involve 33 per cent greater depth (and therefore greater volume per unit area of floor supported) than ones spanning 3000 mm, and 60 per cent greater depth than ones spanning 2500 mm.

A third simple rule, with the new timber stress grades, is to consider using the higher SS or MSS grades instead of GS or MGS, particularly for the longer spans. The minimum extra spanning capacity is at least 8 per cent (as shown in tables of Schedule 6 of the Building Regulations 1976), and the extra cost of SS or MSS timber is likely to remain considerably less than 8 per cent.

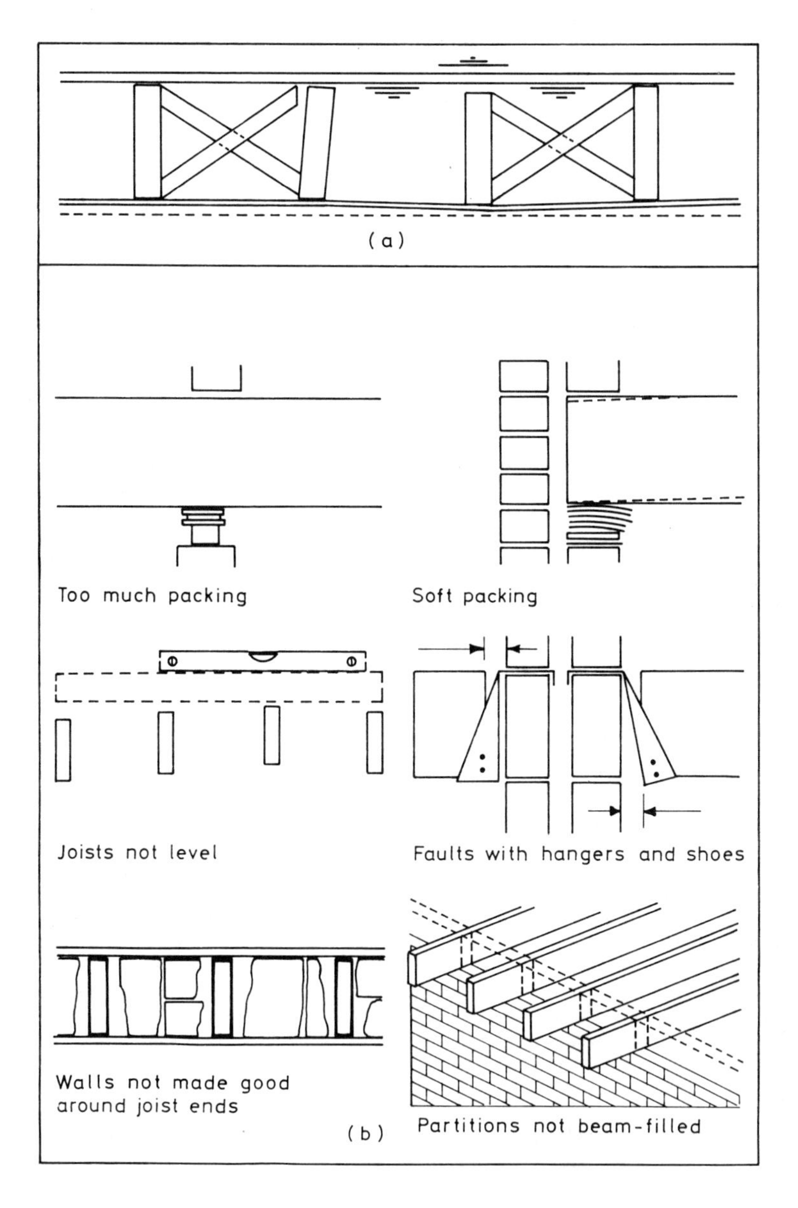

Figure 35.2 (a) Badly built-in joists may cause springy floors or wavy ceilings; (b) six most common defects of built-in joists (courtesy NHBC)

There are other variables, such as joist spacings, that can significantly affect flooring costs, especially when plywood is used. By careful selection and manipulation of all the variables, very substantial savings can be achieved compared with average practice, without any loss of strength or stiffness.

STRUCTURAL FLOOR PANELS

The idea behind the use of stressed-skin floor panels is to reduce the amount of joist material, by structurally linking it with the flooring material and possibly also the ceiling material, so that these materials are also stressed, longitudinally. By this means one can hope to reduce the joist material by about 25 per cent if plywood is used for the stressed-skin material. However, the extra cost of plywood, as an alternative to chipboard or tongue-and-groove softwood flooring (or plasterboard, in the case of a ceiling skin), would currently take up all the potential saving. Any economic advantage from adopting the stressed-skin principle is dependent on first wanting to use plywood flooring, panelised components or 'shallow' construction for other reasons. These need to be weighty ones, since the not inconsiderable costs of structurally connecting the membranes to the joists, by controlled gluing or dense nail stitching, also has to be covered.

The panelisation of complete floors may make good operational and economic sense, in large contracts involving the use of relatively light cranes, without the use of stressed-skin action, using chipboard or tongue-and-groove softwood as the flooring material. In this case there is no significant saving or extra in material costs, but there is an opportunity of significantly reducing operational and labour costs, which are likely to account for between 30 and 40 per cent of total floor-element costs in housing. Tongue-and-groove softwood flooring, which tends to be relatively labour-intensive and productive of waste on a site, could be made much less so in factory conditions.

FLOORS WITH EXPOSED BEAMS

A form of timber floor construction, having a long historical tradition and considerable potential for greater economy, involves exposed timber beams reducing joist spans to about 2–3 metres. As compared with clear spans of 4 metres this can lead to savings in joisting material of the order of 50 per cent. Further possibilities of structural economy are then opened up if one has continuous joists over two spans, or switches to SS or MSS grades giving significantly higher 'f' values when 'E' values no longer govern. In many domestic plan arrangements, such savings should more than offset the extra cost of a few beams. There are probably two main reasons why such constructions are not currently more widely used. It may not be generally appreciated that such beams are permitted by the Building Regulations to the extent of a 300 mm downstand, giving clear headroom of 2000 mm in housing, or that beams of 75 mm width or more will almost certainly provide the necessary 30 minutes fire resistance in themselves.

Trimming of stairwell openings

The NHBC Handbook gives in clause Ca 10(b) rules for the sizing of trimmers as follows:

'(b) The following limitations shall be observed:
Joist centres 600 mm maximum.

Trimmer carrying 1–4 joists – minimum of 12 mm wider than general joists.
Trimmer carrying 5–8 joists – minimum of 25 mm wider than general joists.'

The sizes of trimming joists, which normally support the trimmer at each end, are just as critical. Normally a pair of general joists will be required at the very least, and it is important that these are securely connected together, to act as one in taking the point load from the trimmer. The sizes of trimmer and trimming joists and connectors should be checked by calculation.

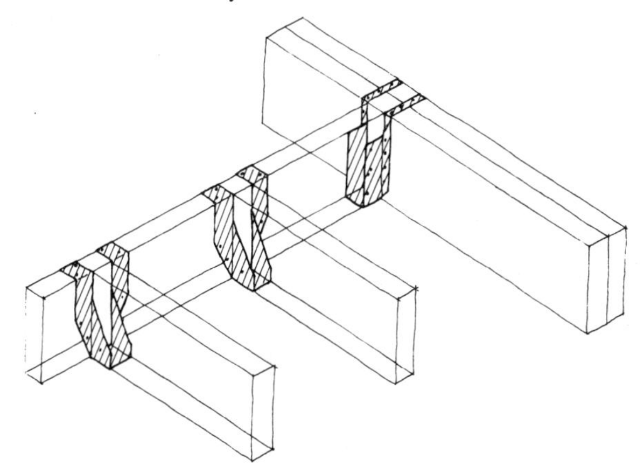

Figure 35.3 Metal shoe/hangers for trimming stairwells

Suitable carpentry joints, such as tusk tenons, between trimming members require craft skill and in practice are seldom used. Ordinary notching weakens the supporting member. The most reliable form of fixing is a galvanised steel shoe/hanger similar to the types in *Figure 35.3;* this provides for face nailing into the supporting member and in addition a top seating tab or tabs, which also should be nailed down.

Partition support

It is not advisable to support block and brickwork partitions off joisted timber floors, because joist deflection and shrinkage are likely together to cause cracking, especially in well heated houses. If it is done, supporting joists should be designed to tight deflection limits.

Timber-framed or proprietary lightweight partitions are more suitable. If they are well fixed through to the floor joists, they can often act in the manner of distributing beams in transmitting their own weight, and can generally stiffen up the structure by acting as diaphragms.

Suspended timber ground floors – ground cover

Historically and up to World War II there was generally no ground treatment beneath timber ground floors, apart from the removal of vegetable matter and topsoil, sometimes with the addition of a layer of ashes, and the provision of a few underfloor ventilators. It is strongly arguable that, if ground floor joists are impregnated with preservatives as a safeguard against future faults in damp-proof courses, blocked ventilators, or houses being unoccupied for long periods, this traditional practice becomes 'foolproof'.

Between the wars, the Model By-laws picked up a suggestion in a Concrete Code that oversite concrete beneath timber floors might be advisable on certain damp sites. Then, when the new Building Regulations came out in 1965, a deemed-to-satisfy regulation was introduced and wrongly interpreted as though it were mandatory on all sites.

On the other hand, polythene is amongst the most efficient barriers against both vapour and liquid moisture. It has been common practice to use it as ground cover in Canada and Scandinavia for many years, and it has been used for a number of houses in this country recently. It is obtainable in 4 metre wide rolls; this reduces the need for joints, which should overlap at least 300 mm and be weighted down with bands of sand or weak concrete or blocks, as should the sheet edges. It is best laid on a blinding of sand, which should be made up so that the lowest level of the polythene is not below the lowest level of the adjoining ground. It is best for the surface levels to fall towards the external walls, so that any water from floods or overflowing sinks would drain off at the edges. On a sloping site such falls would be in one direction and there would be very little fill to bring up levels, compared with the deemed-to-satisfy recipe for concrete ground cover, involving a horizontal surface corresponding with the highest level of adjoining ground.

In combination with adequate underfloor ventilation and impregnation of joists, polythene ground cover should be more reliable than the deemed-to-satisfy concrete treatment and considerably less expensive. Suspended floors have important technical and economic advantages on sloping sites. However, it is advisable to get prior confirmation of acceptability in particular cases and localities.

Party floors with sound insulation

In broad principle there are four main types of timber-based compartment floor construction that have been shown capable of meeting the deemed-to-satisfy performance standards for sound insulation in field tests; two rely on 'floating' floors and two on suspended ceilings. They are shown in *Figure 35.4,* parts (a) to (d).

The traditional method, (a), which corresponds to the deemed-to-satisfy material specification in Part G of the Building Regulations, involves the use of a floating floor laid on a quilt draped over the joists, with a ceiling construction fixed to the underside of the joists, heavily 'pugged' with dry sand. If this is used, it is important that the sand is adequately supported, carefully laid and contained. Also there must be complete avoidance of any nailing down of the floating floor. The method involves various operational difficulties, but has been used successfully in spite of these.

Method (b) has a floating floor, with additional mass incorporated in it (usually in the form of plasterboard planks) instead of 'pugging' above the ceiling. This has been used successfully on many jobs with separating or compartment walls of timber-frame construction; however, it does not provide much margin for poor workmanship or unexpected flanking transmission, in terms of the performance standard in the Building Regulations. In terms of the mandatory functional requirement of being 'adequate', there should be less doubt about its suitability since nearly all of the many floors that have been field tested have been within the ISO standard. However, with solid-wall construction it would not be suitable because of transmission down the walls. Some improved versions of this type of floor have recently been developed, but tend to be somewhat costly.

Method (c) has a 'fluid' screed of dry sand, laid on a thin plywood subfloor, with the finished floor resting on it. In this case the ceiling membrane is resiliently suspended from the joists by means of acoustic bars or timber battens fixed with neoprene

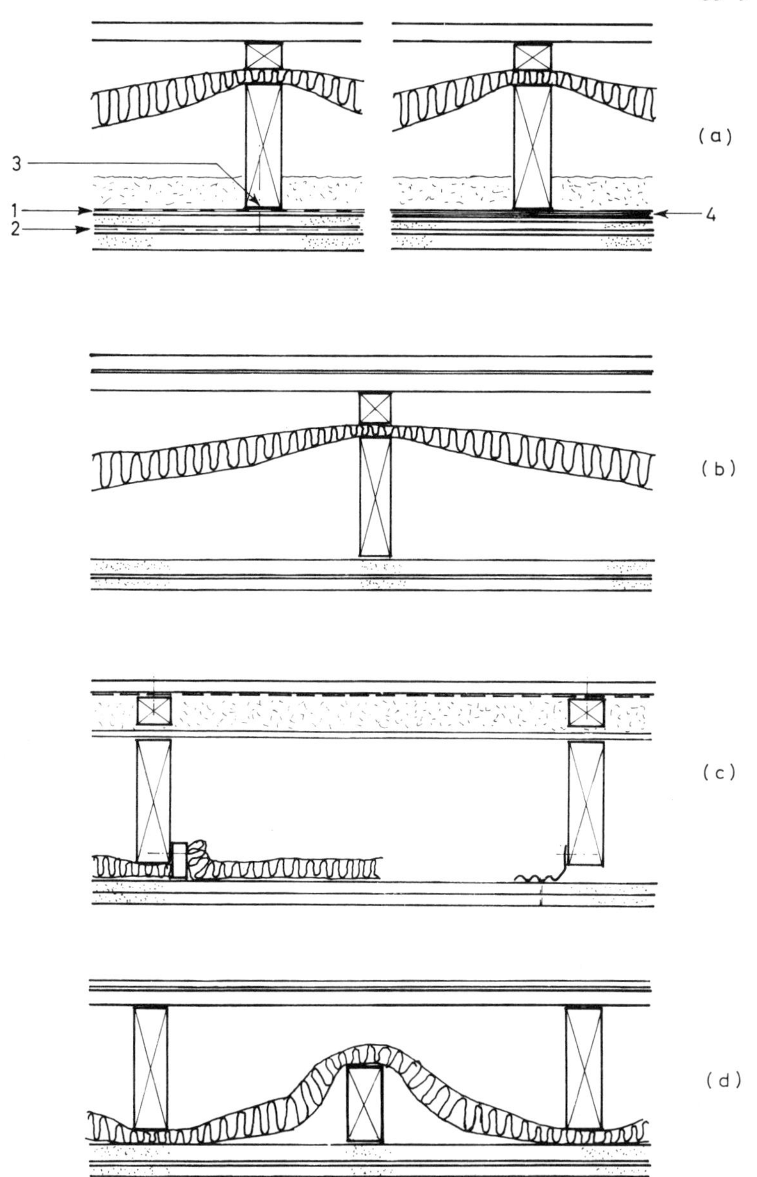

1. Polythene 2. Chicken wire 3. Nail specification based on CP 112 withdrawal values
4. Plywood or similar sand support sheet

Figure 35.4 Basic types of timber sound-resistant party floors; all ceilings are two-layer plasterboard

Description	Construction (max. joist spacing 600 mm)	*Loading* kg/m²	N/m²
Normal domestic floor ½ hour (modified) fire resistance. Sound resistance lower than grade 2.	Floor boards	12.2	120
	13 mm plasterboard	9.8–11.2	95–110
		22.0	215
Intermediate floor of maisonettes Any ½ hour floor except in houses. Sound resistance lower than in grade 2. Also wet lath and plaster.	Floor boards	12.2	120
	19 mm plasterboard	14.7–17.1	145–170
		26.9	265
Sound and fire resistant floor ½ hour fire resistance; probably one hour with chicken wire beneath mineral wool. Sound resistance possibly grade 2 with light walls, grade 1 with heavy walls	Floor boards	12.2	120
	Battens	1.5	15
	Quilt, 25 mm mineral wool	3.4	35
	Mineral wool plugging	14.7	145 min
	19 mm plasterboard	14.7–17.1	145–170
		46.5	460

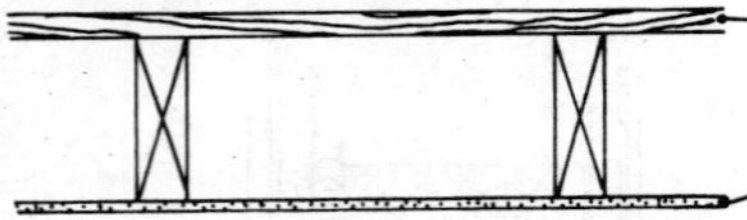

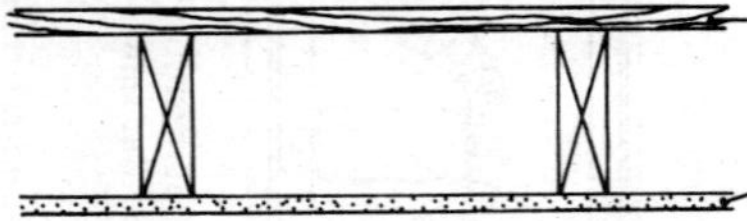

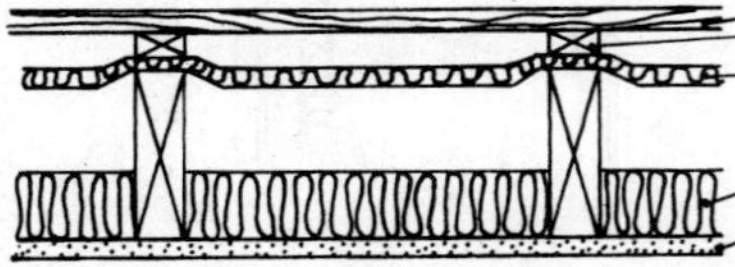

Figure 35.5 Dead loadings of typical flat-roof constructions; fire and sound performances given are probable capabilities (courtesy TRADA)

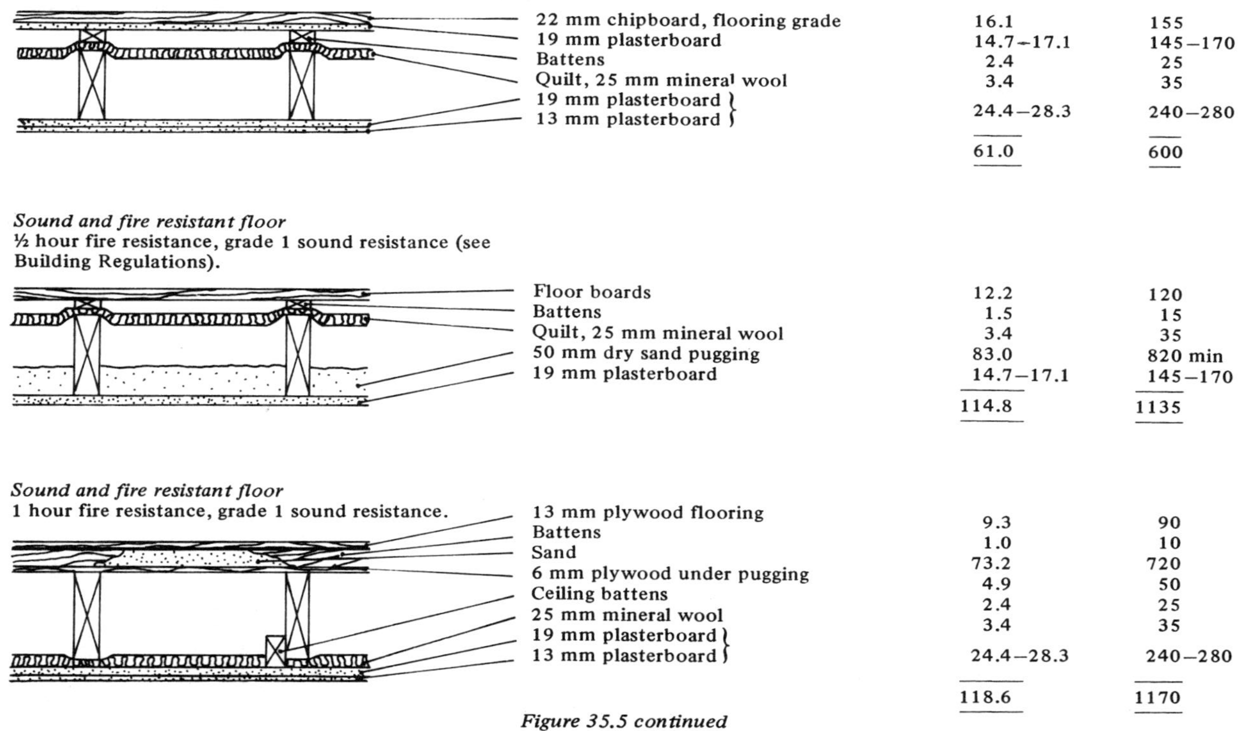

Sound and fire resistant floor
Without pugging: 1 hour fire resistance, grade 1 sound resistance (when installed in a timber-framed building).

22 mm chipboard, flooring grade	16.1	155
19 mm plasterboard	14.7–17.1	145–170
Battens	2.4	25
Quilt, 25 mm mineral wool	3.4	35
19 mm plasterboard } 13 mm plasterboard }	24.4–28.3	240–280
	61.0	600

Sound and fire resistant floor
½ hour fire resistance, grade 1 sound resistance (see Building Regulations).

Floor boards	12.2	120
Battens	1.5	15
Quilt, 25 mm mineral wool	3.4	35
50 mm dry sand pugging	83.0	820 min
19 mm plasterboard	14.7–17.1	145–170
	114.8	1135

Sound and fire resistant floor
1 hour fire resistance, grade 1 sound resistance.

13 mm plywood flooring	9.3	90
Battens	1.0	10
Sand	73.2	720
6 mm plywood under pugging	4.9	50
Ceiling battens	2.4	25
25 mm mineral wool	3.4	35
19 mm plasterboard } 13 mm plasterboard }	24.4–28.3	240–280
	118.6	1170

Figure 35.5 continued

'grommets'. Both versions were used in the Watery Lane timber-framed four-storey maisonettes designed by TRADA about eight years ago. In theory the method should be as suitable for use in conjunction with solid-wall construction as the first mentioned method is considered to be, but (as far as is known) this has not yet been established in practice in this country. It tends to be one of the more expensive types of construction, unlikely to compete with concrete floors in flats with solid load-bearing wall construction, except in special circumstances.

Method (d) is an adaptation of one commonly used in North America, with shallow independent ceiling joists and a double layer of plasterboard ceiling. The flooring, fixed direct to floor joists in the usual way, is of 22 mm chipboard with a foam-backed floor covering. It has been used successfully in timber-framed flats but is not recommended with solid-wall construction. It tends to be the least expensive of the types discussed. In all cases performance depends much on the details and circumstances, and the guidance given here should not be regarded as complete and sufficient.

TIMBER-JOISTED FLAT ROOFS

Structural considerations

Flat roof joists are, from a structural viewpoint, designed on a similar basis to floors as load-sharing structures, except that the superimposed load is considered to be related to snow and therefore of medium-term duration; permissible bending and shear stresses are therefore increased by 25 per cent. The term 'flat roof' is regarded as covering all roofs with a pitch of 10° or less to the horizontal.

Domestic flat roofs accessible only for maintenance purposes are designed for superimposed loadings of 720 N/m^2 (73.2 kg/m^2); those accessible to foot traffic are designed for superimposed loadings of 1440 N/m^2 (146.5 kg/m^2) – the same as for floors. The dead weight of a roof construction is likely to vary between 240 and 720 N/m^2, depending mainly on the roof covering (whether of bituminous felt, with or without chippings, or of mastic asphalt, for example), and span tables normally cover this range in suitable steps. *Figure 35.5,* taken from *Modular metric joist tables (redwood or whitewood) for domestic floors and roofs*, published by TRADA, indicates loadings for typical flat-roof constructions.

Operations and workmanship affecting quality

The satisfactory performance of a waterproof membrane is very dependent on it being laid on a suitable and well executed substrate. Unless a roof is designed to retain a 'sheet' of water, it should incorporate falls sufficient to overcome sag or deflection in beams and joists and to avoid the build-up of sediment, likely to cause ponding. Joists will either be shaped or laid to falls, and selective positioning of joists may be necessary to achieve true sloping planes. In order to avoid vertical joint movements beneath the roof membrane, all edges of plywood or chipboard roof sheeting should be supported or tongued and grooved; timber boarding must be tongued and grooved; plain-edge boarding is unsuitable. The heads of nails should not protrude above the surface, allowing for the possibility of subsequent nail popping. If chipboard is used it should preferably be of a prefelted type, with joints sealed or roofing laid directly after fixing. Otherwise special precautions may be necessary to avoid excessive moisture movement.

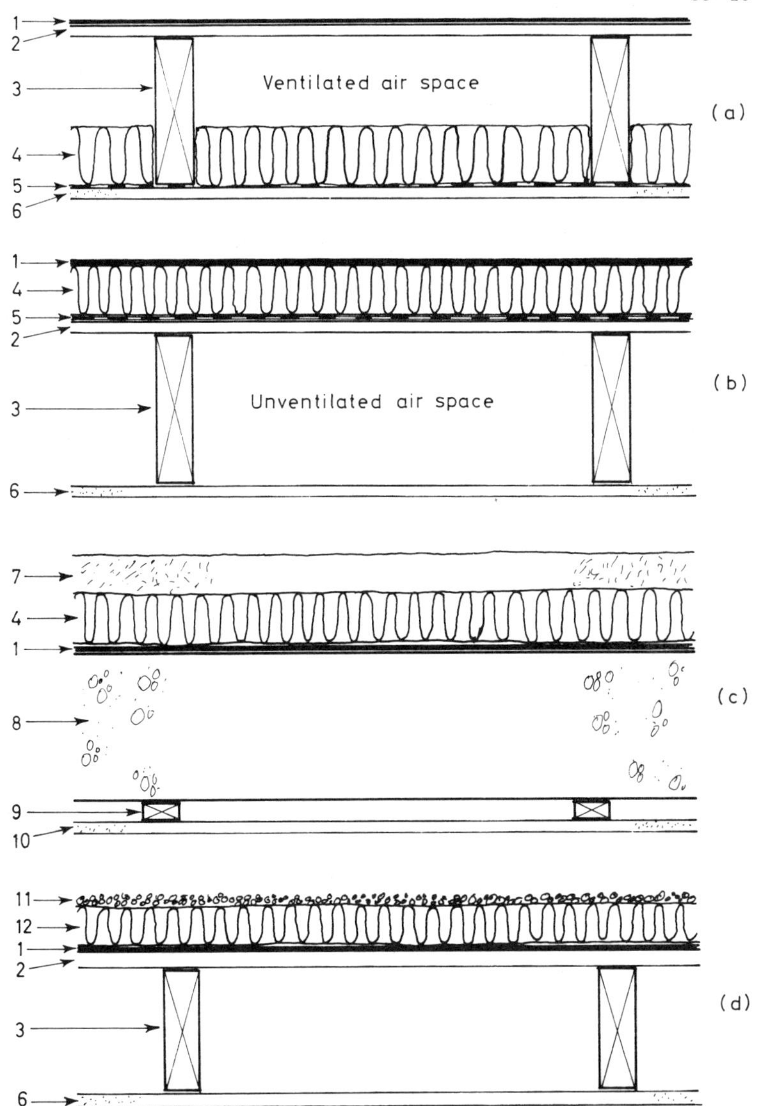

1. Waterproof covering 2. Decking 3. Joists 4. Thermal insulation 5. Vapour barrier 6. Ceiling 7. Concrete slab or ballast 8. Concrete deck 9. Battens 10. Foil-backed plasterboard 11. Gravel 12. Waterproof insulation

Figure 35.6 (a) 'Cold' roof; (b) 'warm' roof; (c) 'inverted' concrete roof; (d) 'inverted' timber roof, not recommended because of hazard of condensation beneath waterproof covering

Thermal insulation, ventilation and condensation-prevention

These functional requirements have to be considered together, because:

- condensation only occurs on cold surfaces, and barriers to vapour flow should therefore be on the warm side of the insulation;
- if condensation results in insulation materials (such as glass or mineral wool) getting wet, their performances will be much reduced;
- ventilation is the most effective means of dispersing vapour to the outside before it condenses, but excessive ventilation can cause increased heat losses especially if the void being ventilated is on the warm side of the insulation.

As far as flat roofs are concerned, there are three alternative basic solutions (shown in *Figure 35.6*) that provide satisfactory relationships in this regard: *cold roof* (insulation below air space), *warm roof* (insulation above air space), and *inverted roof* (insulation above waterproof covering). The inverted roof is not recommended with timber construction.

Cold roof, Figure 35.6(a). A vapour barrier immediately above the ceiling and on the warm side of the insulation is essential, to reduce the amount of vapour to be dispersed by ventilation; special care should be taken to ensure that the barrier is continuous and imperforate. The strategy of letting vapour breadth through the ceiling to be dispersed by free ventilation may make good sense with pitched roofs but not with flat roofs.

In practice the flat roof will have to overhang the external walls to provide a soffit condition for the fullest possible ventilation of every roof void. In many situations it will be better to provide falls by means of counter battens, as shown in *Figure 35.7*, rather than firring strips along the joists. With wide roofs it will be necessary to introduce flat-roof ventilators at the watershed, to get adequate ventilation.

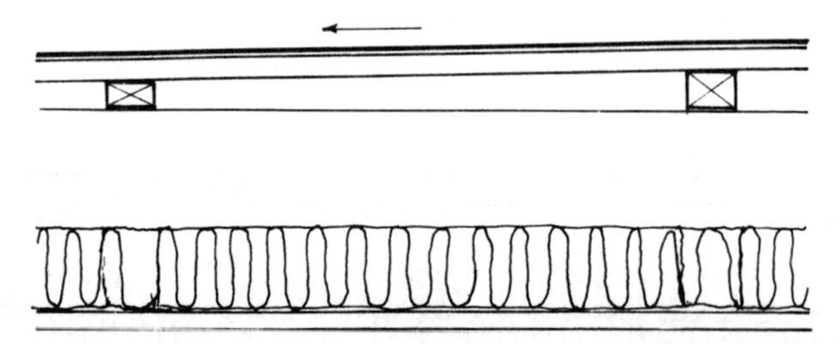

***Figure 35.7** Counter battens to provide falls and assist through ventilation*

A potential hazard with this type of solution in some regional climates, including many parts of Scotland, is what amounts to dew or hoar-frost formation on cold nights on the underside of the roof boarding from humid outside air. Such risks will be increased if thermal insulation at ceiling level is increased beyond, say, 50 mm thickness, because the underside of the boarding will receive little warmth from the building. Risks will be reduced if a small amount of insulation, in the form of 12 mm impregnated fibreboard for example, is laid above the boarding and beneath the waterproof covering. Timber boarding itself provides some useful insulation and capacity to absorb and later dispel moisture, so in many areas and building contexts the risks may not be very great. However, it is advisable for all timber in this type of roof construction to be treated with preservative.

Warm roof, Figure 35.6(b). It is important that the ceiling does not have, or amount to, a vapour barrier: the timber joists must be effectively in the warm and ventilated internal atmosphere of the building. The practical interpretation of the

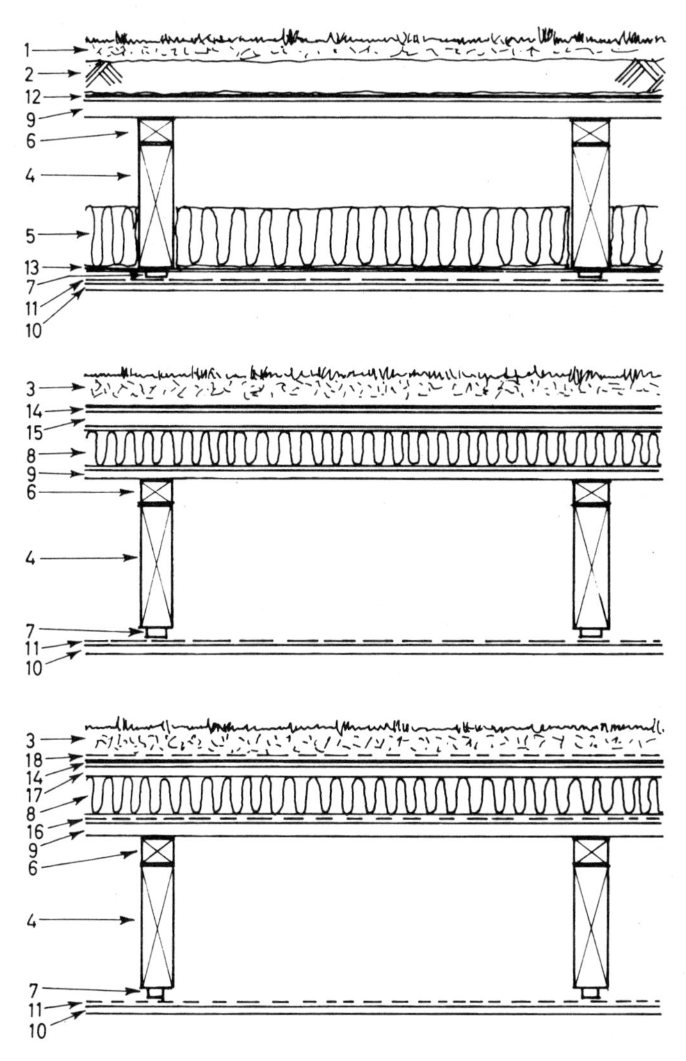

1. Local turf 2. Peat 3. Seeded meadow grass on 50/50 mix of forest bark and sifted topsoil 4. Joists 5. Mineral fibre insulation 6. Firrings 7. Softwood battens 8. Polystyrene 9. 16 mm plywood 10. 9 mm medium-density hardboard (class 1) 11. 500 g polythene 12. Plainfaced bituminous felt 13. 50 mm mesh chicken wire 14. Two-layer felt 15. 12.7 mm impregnated fibreboard 16. Vapour barrier 17. Prefelted chipboard 18. Waterproof covering

Figure 35.8 Experimental 'green' roof constructions

insulation layer above the boarding, in terms of materials and detailed specifications, was not too difficult when sufficient insulation was provided by 25 mm of bitumen-impregnated fibreboard, bonded with bitumen to a bituminous felt vapour barrier and with the roofing bonded to it to overcome uplift. However, with greater insulation standards and thicknesses, there are many difficult detailed considerations. Some of these, such as wind uplift and overheating of the waterproof covering, can be overcome by laying on top of this a 'soil carpet'. *Figure 35.8* shows some roof constructions on these lines incorporated experimentally in new office and educational buildings constructed by TRADA at its headquarters near High Wycombe.

Inverted roof. This type was developed for concrete flat roofs – *Figure 35.6(c)* – but is almost certainly unsuitable for timber flat roofs – *Figure 35.6(d)* – although it has appeared in BRE News 41 in a way that might be taken to imply that it is suitable. The serious hazard is that cold water, perhaps from melting snow, will percolate through the insulation and run over the waterproof covering, chilling it and causing condensation on its underside, which is in contact with the timber deck.

EXTERNAL SOUND INSULATION

Insulation from the sound of drumming of heavy rain or from aircraft noise can be difficult to provide with timber flat roofs. However, a 'soil carpet' can be a very effective means of achieving good sound insulation.

Fire performance

Fire resistance is not normally required of a roof element because, on balance, venting a fire through a roof helps with firefighting. However, roofs have to resist penetration by fire from neighbouring buildings but there are no great difficulties with timber roof structures with meeting specifications regarding this, such as those given in Schedule 9 of the Building Regulations 1976.

Fire-stopping of roof voids is required above any compartment wall, otherwise its effectiveness might be much impaired. This may conflict with through ventilation in some plan arrangements, and should be considered at planning stage.

Ceilings beneath roofs have to provide Class 1 surface spread-of-flame performance, except in circulation spaces where Class 0 is required, and in rooms and circulation spaces of one- and two-storey houses where Class 3 is acceptable. These requirements may be interpreted as applicable to the soffits of timber roof beams exposed beneath a ceiling.

FURTHER INFORMATION

Fire resistance of floors

Increasing the fire resistance of existing timber floors, BRE Digest 208 (1977)
BS 5268: Part 4, *The fire resistance of timber structures*
Section 4.1, *Method of calculating fire resistance of timber members* (1978)
Section 4.2, *Timber stud walls and joisted floor constructions* (in preparation)

Timber floor and flat-roof structures

Sizes of joists and beams, Schedule 6 of *Building Regulations 1976*, HMSO
Span tables for Canadian spruce-pine-fir (stress-graded timber in sawn sizes), Building Research Establishment

Span tables for M75 grade home-grown Sitka spruce (sawn sizes only), Building Research Establishment
Span tables for one-family houses on not more than three storeys (stress-graded timber in sawn and regularised sizes), Swedish Finnish Timber Council
Swedish redwood and whitewood stress-graded to BS 4978 (stress-graded timber in sawn sizes), Swedish Finnish Timber Council
Load span tables for beams of Swedish redwood and whitewood (stress-graded timber in sawn and regularised sizes), Swedish Finnish Timber Council
Floor joists in housing – comparative span tables, and *Timber frame construction – a guide to platform frame*, Council of Forest Industries of British Columbia (London)
Modular metric joist tables (redwood or whitewood) for domestic floors and roofs (stress-graded timber in sawn sizes), Timber Research and Development Association
Span charts for solid timber beams (1976 edition covering stress-graded timber), Timber Research and Development Association

Compartment floor and suspended timber ground floor constructions

Timber Review, Timber Research and Development Association (1974)

36 PITCHED ROOFS

STRUCTURAL FUNCTIONS 36–2

NON–STRUCTURAL FUNCTIONS 36–9

36 PITCHED ROOFS

J. OLLIS, AADip, RIBA
Chief Architect, Timber Research and Development Association

This section describes the various types of structural arrangement used for pitched roofs, and examines the critical non-structural subjects of thermal insulation, ventilation and condensation prevention.

STRUCTURAL FUNCTIONS

Domestic buildings

A pitched roof is composed of various members with specific structural duties, but it should act structurally as a whole and benefit from its overall geometric form. It has to be thought of, and made, very much as a load-sharing structure, beyond the fact that rafters and ceiling joists are designed on a load-sharing basis as are joisted floors and flat roofs (see Section 35).

In the words of clause Ca 21 of the NHBC Handbook: 'Members shall be triangulated or otherwise arranged so as to form a coherent structure' and 'all forces shall be resolved'. Such forces include gravitational forces causing a pitched roof to spread, which may upset the stability of supporting walls, and wind forces, which may cause a roof to deform and in so doing apply eccentric buckling forces to individual members and displace bearings on walls. One structural function of a roof is to stabilise the walls that support it, particularly flank walls, by holding the tops of them in the correct vertical position. Rules for providing such stability are given in Schedule 7 of the Building Regulations 1976.

Hipped roofs tend to resolve wind and spreading forces through the overall geometry. On an individual house approaching a square in plan, spreading forces tend to be resolved by the plates acting as ties, if these are well connected at the corners. Gable roofs, whether formed of trussed rafters or common rafters, with or without trusses and purlins, need diagonal bracing members fixed to the underside of the rafter members, as indicated in *Figure 36.1*.

ROOFS WITH COMMON RAFTERS

Schedule 6 of the Building Regulations 1976 provides a good basis for member sizes, including rafters, purlins, ceiling joists and binders, but assumes the application of carpentry skill and knowledge in connecting the members so that the roof acts as a structural whole. Tables cover a range of pitches and loadings appropriate for various roofing materials, and SS or MSS grades of timber as well as GS and MGS. As far as rafters of up to 150 mm depth are concerned, SS grades are shown as having a spanning capacity about 14 per cent greater than GS grades, and they are therefore likely to provide considerably better structural value for money in this context.

A close couple roof in its simplest form comprises pairs of rafters tied together with a ceiling joist. It should be noted that, with low pitches, the forces in the ceiling joist/tie members and in their connections with the feet of the rafters are proportionately greater. With any tie member made of two or more pieces, the joints along its length must be designed to take the full tensile load. With simple unsupported ceiling joists, the maximum economical span of a close couple roof in softwood is

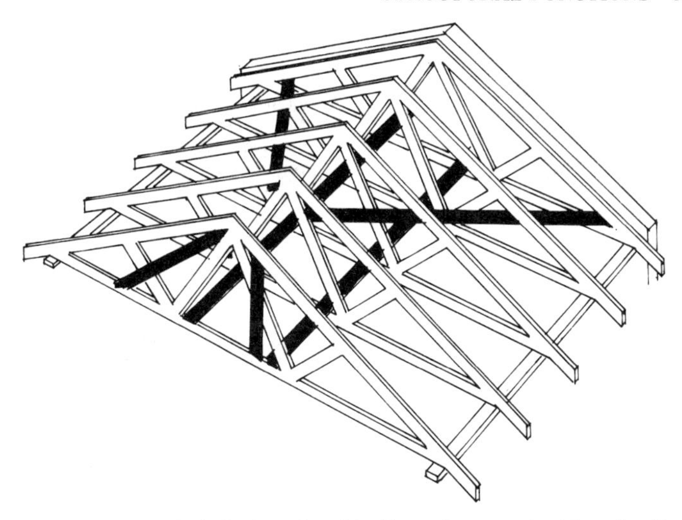

Figure 36.1 Bracing of pitched roofs, with diagonals on undersides of rafters and longitudinals above ceiling members, to resist wind on gable flank walls

about 4 metres: with ceiling joists supported by internal partitions or supported from the ridge by means of hangers and a central line of binders, spans of up to 6 metres can be economical. See *Figure 36.2.*

Roofs of common rafters supported on simple purlins have no limit on their overall front-to-back span. However, the spans of the purlins themselves are limited to about 3.5 metres when of normal softwood, unless propped from intermediate loadbearing walls, in which case overall spans of multiples of 3.5 metres can be practicable.

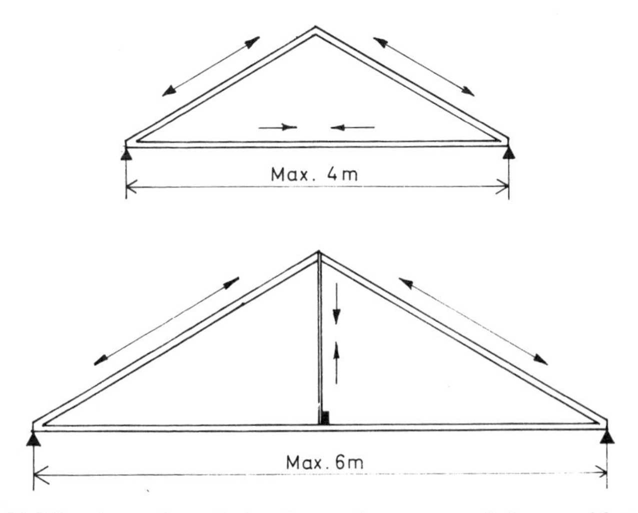

Figure 36.2 Simple couple roofs, showing maximum economical spans with softwood

(a)

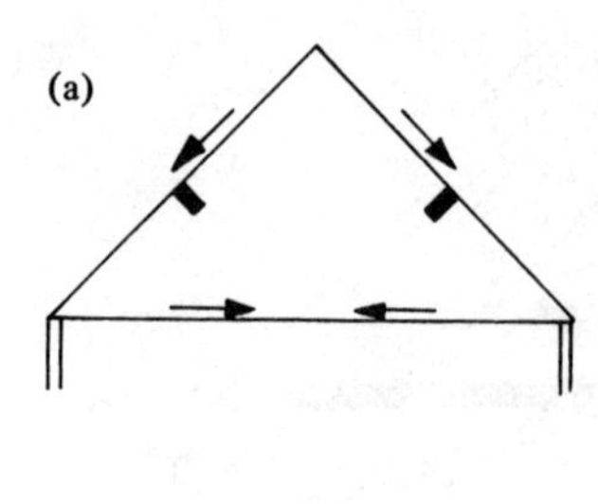

An angled purlin acts solely as a support for the rafters, and possibly also for the ceiling joists. Resistance to outward thrust is provided entirely by the ceiling ties.

(b)

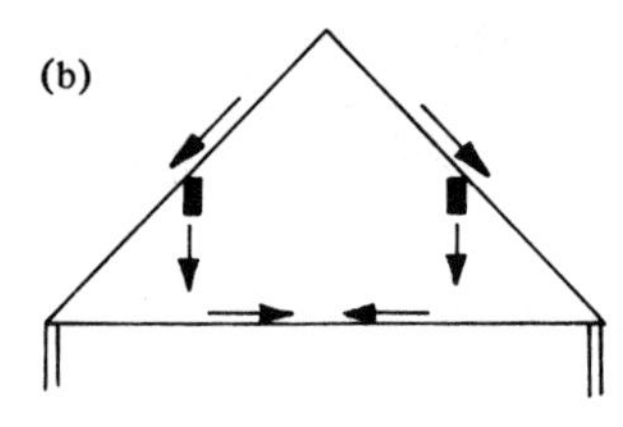

An upright purlin supports the rafters (and possibly the ceiling joists), and can partly resist outward thrust–provided the rafters are notched over the purlin or connected with metal brackets.

(c)

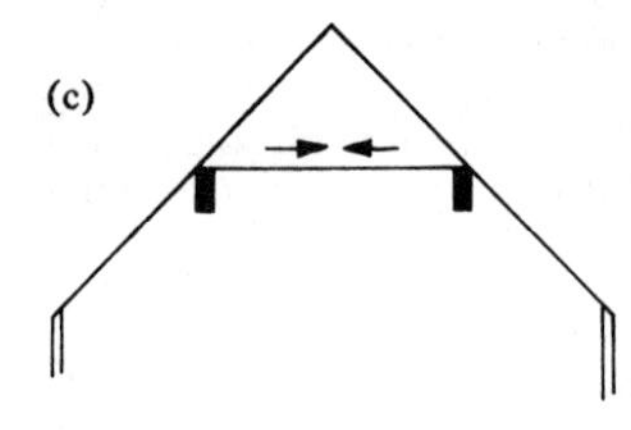

A variation of (b): the tie-members are above the purlin. As with (b), horizontal forces are resolved vertically downwards. It is not necessary to notch the rafters, or to use metal brackets, provided the tie members are securely fixed to the rafters.

(d)

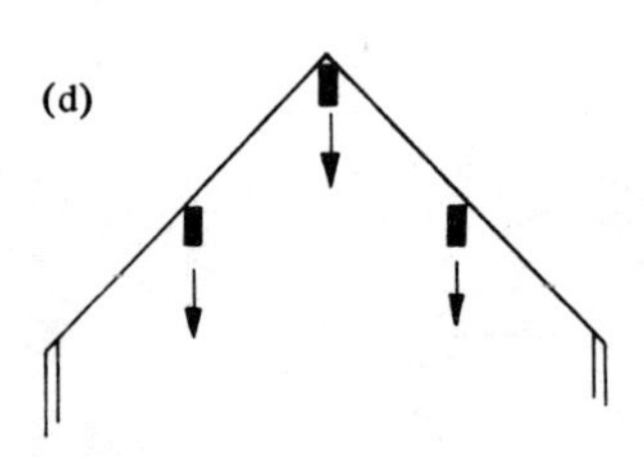

Adding a top ridge purlin eliminates outward thrust, provided the rafters are notched or metal-bracketed to the purlins. No tie members are needed. The simplest way of installing a ridge-purlin into an existing roof is to connect each pair of rafters with a batten (the same section as the rafter) seated on to the purlin (as shown below). Inserting a ridge purlin should be considered when it is necessary to remove tie members (e.g. ceiling joists).

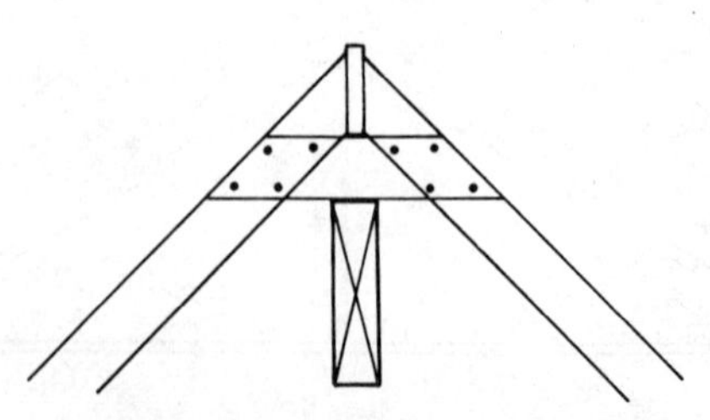

Figure 36.3 Functions of purlins (all the cases shown depend on loadbearing eaves conditions)

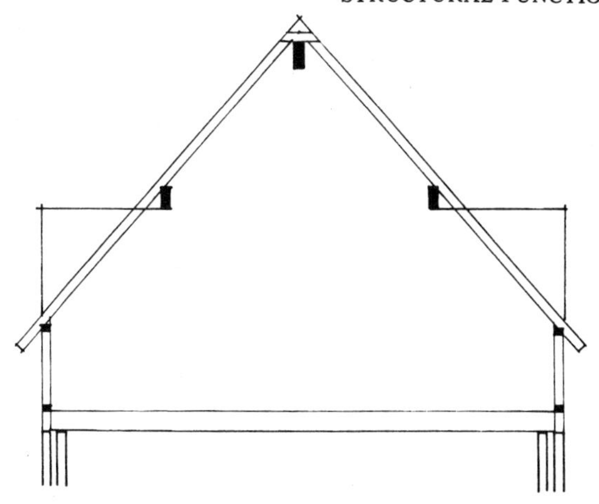

Figure 36.4 Advantages of a ridge purlin in chalet bungalow construction

With solid purlins of Canadian Hemlock or Douglas fir of Canadian or home-grown origin, all of which can be obtained with depths up to 300 mm, clear spans of up to 4.5 metres are practicable; with certain structural hardwoods, purlins can economically go up to 5.5 metre clear spans. Although such purlins are not given in Schedule 6 of the Building Regulations 1976, their sizes can be readily determined with tables obtainable from TRADA.

Depending on the way rafters are fixed to them, purlins can assist in resolving the spreading tendencies of pitched roofs, and if a ridge purlin is used it can eliminate outward thrust, as shown in *Figure 36.3.* The ridge purlin device can be particularly useful not only in conversion work but also with more steeply pitched roofs to incorporate rooms either subsequently or initially, as indicated in *Figure 36.4.*

ROOF TRUSSES

A roof truss is basically a structural framework to provide intermediate support for purlins along its top edges and, if there is a ceiling, binders along its bottom edge to reduce the spans of the ceiling joists. The points of connection of purlins and binders should be at nodal points of the truss framework to reduce adverse effects of local bending moments in the outer members. The top sloping members are called principal rafters.

Figure 36.5 shows a traditional king-post truss, a collar truss, a queen-post truss, and a 'W' truss. The king post itself is really a hanger from the firm ridge apex, providing at its bottom end a node that gives mid-span support to the bottom tie and a base for raked struts to support one purlin on each side. It is a fairly efficient form of truss for small to medium spans, but involves some obstruction of the central space. The collar truss is really a form of tied arch, the collar completing the compression line. It can be a very efficient structural form for spans up to about 8 metres, where the ceiling can be supported by intermediate partitions. The queen-post truss is basically a collar truss with two hangers to pick up ceiling binders. With pitches of about 50° and spans of about 7 metres, the clear central space makes it particularly suitable when rooms are to be provided in the roofspace. The 'W' truss is structurally similar to a king post, with the post dividing to provide two support nodes at the

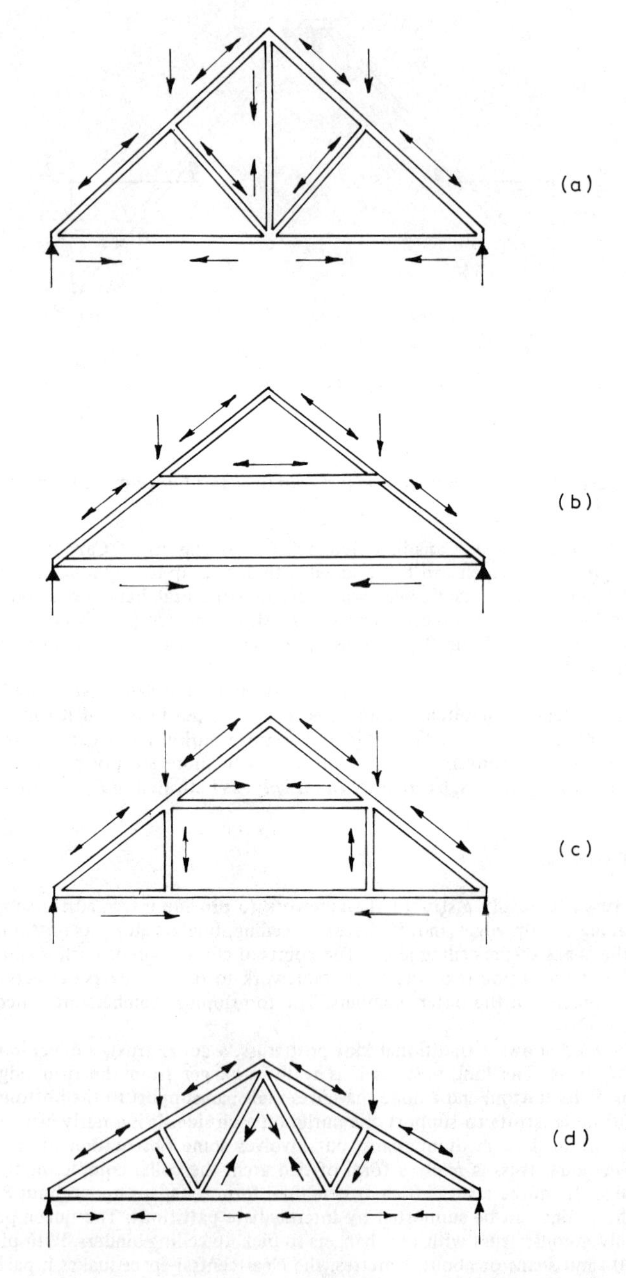

Figure 36.5 Common truss types: (a) king-post, (b) collar and tie, (c) queen-post, (d) 'W' or Fink

bottom. This form became more practicable with the introduction of modern timber connectors and has a more efficient geometry with lower pitches.

The 'W' truss form is the basis of one range of TDA (TRADA) trusses widely used in the postwar years before trussed rafters were introduced. A novel feature of these is that the principal rafter is in the same plane as and also serves as a common rafter and similarly the tie also serves as a ceiling joist. The trusses are usually 1800 mm apart. This type of truss still has its use today, lending itself better than trussed rafters to attic storage uses and the introduction of dormer windows, up to 1800 mm wide.

TRUSSED RAFTERS

The basic idea of a trussed rafter is that every pair of rafters in a roof is combined with a ceiling joist to form a truss (thus removing the need for any purlins), usually with some internal framework to reduce the effective spans of the outer members. All the members are normally single members, in one plane, held together by metal or plywood nail plates or pressed tooth plates on both faces. Apart from substantially reducing the timber content of a roof, the main advantage is the speed and simplicity of site erection of the rafter and ceiling planes in one operation.

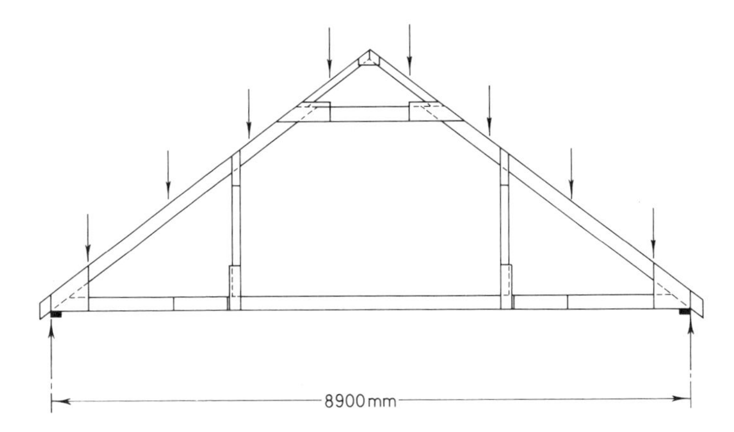

Figure 36.6 Trussed rafters designed by Council of Forest Industries of British Columbia

Trussed rafters now provide the predominant means of constructing domestic roofs of relatively low pitch, around 22½°. They usually follow the geometry of a 'W' (or Fink) truss. The suitability of this form becomes questionable with steep pitches, when the queen-post geometry can be more appropriate particularly if the roof space is to be used. *Figure 36.6* shows trussed rafters for use at 600 mm centres, designed by the Council of Forest Industries of British Columbia on these lines using nailed plywood gussets and a pitch of 36°. With a steeper pitch of 40 to 50°, perhaps with a shorter overall span, the form would be even more advantageous.

Trussed rafters are normally used with relatively heavy roofing systems of tiles or slates sufficient to counteract wind uplift forces in most exposure situations. With light roof coverings, special holding-down straps are likely to be necessary.

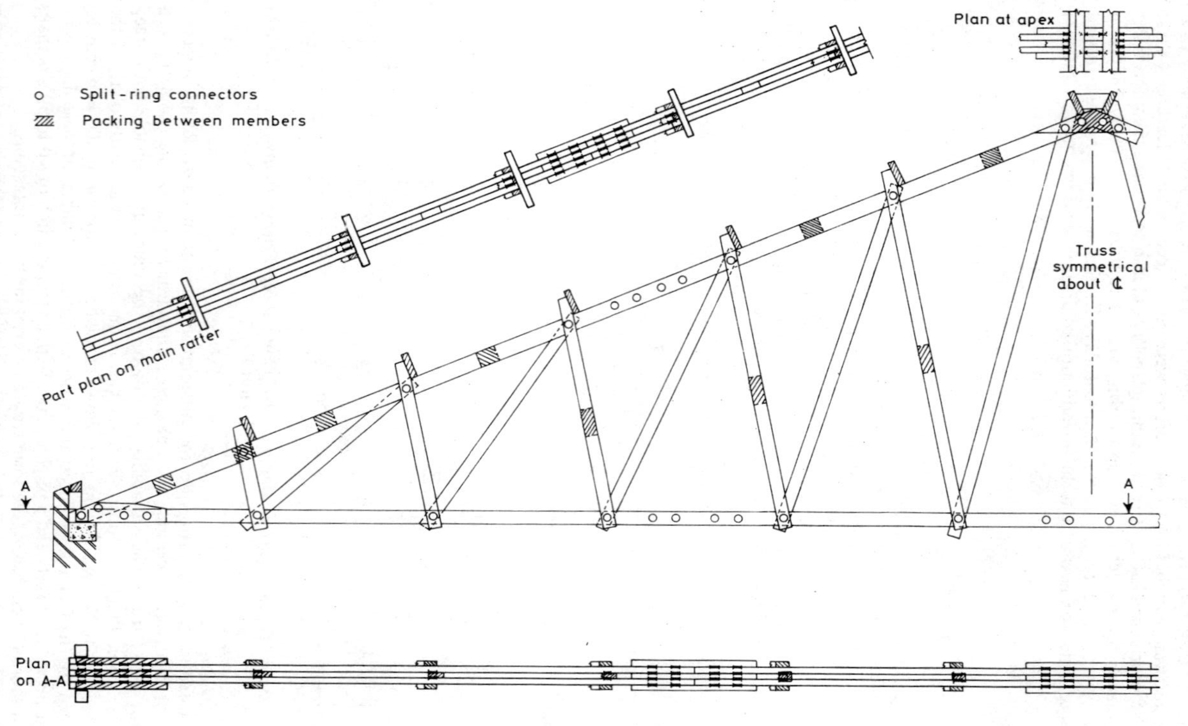

Figure 36.7 TRADA standard industrial truss

Non-residential buildings

The domestic roof structures described are equally applicable to small and medium span non-residential buildings. Trussed rafters have been used satisfactorily for spans up to about 12 metres. The lateral stability of trussed rafters then becomes even more critical in relation to wind forces, to tendencies of the top members to buckle under compression, especially if there are no tiling battens or similar members linked with diagonal wind-bracing members, to provide restraint and to help with the stability.

Many different structural forms have been applied to timber industrial and farm buildings and there is only space to mention a few. The TRADA standard industrial truss based on the use of split-ring connectors, as shown in *Figure 36.7,* caters for spans up to 20 metres with a pitch of 22½°. Similar trusses using structural hardwoods have been constructed to span up to 30 metres. Another form of truss successfully used to give such spans comprises pairs of hardwood trussed rafters bolted side by side with spacer blocks. Structural hardwoods have also been shown to provide economical solutions in a ridge portal form with nailed plywood knee and ridge gussets, for spans up to 15 metres.

NON-STRUCTURAL FUNCTIONS

When a pitched roof has a horizontal ceiling beneath it, the best place to put thermal insulation materials is immediately above the ceiling. The roof space can then be fully ventilated without significantly increasing heat losses, assuming that the insulation is about 50 mm thick or more. Except where the ceiling is over spaces that generate excessive vapour (e.g. bathrooms, washrooms, kitchens etc.) it is probably best not to include a vapour barrier in the ceiling but to let the building 'breathe' into the roof space, reducing the risks of internal condensation. It then becomes increasingly important to provide as much roof ventilation as practicable.

NHBC-approved guidance, published in the September 1977 issue of *The House Builder*, recommended eaves ventilation on both sides in the form of a continuous soffit slot 10–12 mm wide, or the equivalent area in the form of soffit grilles, and this is likely to prove sufficient in nearly every circumstance. There are two possible objections to continuous slots. In the first place, insects such as wasps are more likely to form nests in the roof space and cause a nuisance. Secondly, at least one fire has been reported where flames, bursting out of windows, have re-entered the roof space through the eaves soffit slot, causing premature ceiling collapse. With these points in mind, it is arguably better to use grilles positioned some distance from window heads, using a copper mesh small enough to restrict the entry of wasps.

Ridge ventilators provide effective assistance to eaves ventilation and are to be recommended where there is an exceptional risk of lack of ventilation (if a site is very sheltered from the wind by tree belts or by being in a hollow, for example) since they can also induce thermal convection currents. They are particularly useful for narrow frontages, for long-span low pitches (17½° or less), and for high pitches (40° or more) where eaves ventilation alone may not provide sufficient turbulence to dry out condensate in the higher peak of a roof. It is also arguable that in normal situations, if two or more ridge ventilators are provided, the area required for eaves ventilation could be halved and yet provide equivalent air flow.

In some climatic regions, there may be a risk that openness of the roof space to the external climate could lead to dew formation on the underside of the roof slopes, regardless of vapour arising from the occupied part of a dwelling. This is unlikely to matter if there is free ventilation to dry the timbers when there is a change of weather. The moisture content of pitched roof timbers is likely to fluctuate seasonally between 13 and 18 per cent moisture content, and can go up to 22 per cent for short periods without harmful effects. This represents quite a large reservoir capacity for the absorption of occasional condensation.

FURTHER INFORMATION

Basic structural design principles:
Timber roof design, Part 2 Details of rafter loading and Part 3 Domestic pitched roofs, H.J. Burgess, TRADA

Sizes of softwood pitched roof members:
Schedule 6 of *Building Regulations 1976,* HMSO

Span tables for pitched roof members:
Span tables for Canadian spruce-pine-fir (stress-graded timber in sawn sizes), Building Research Establishment
Span tables for M75 grade home-grown Sitka spruce (sawn sizes only), Building Research Establishment
Span tables for one-family houses of not more than three storeys (stress graded timber in sawn and regularised sizes), Swedish Finnish Timber Council
Swedish redwood and whitewood stress-graded to BS 4978 (stress-graded timber in sawn sizes), Swedish Finnish Timber Council
Span tables for domestic purlins (E/IB/4), TRADA

Trussed rafter design:
CP 112 : Part 3: 1973, *Trussed rafters for roofs of dwellings*

Domestic roof trusses, domestic trussed rafters with plywood gussets, and industrial roof trusses:
TRADA Standard Design Sheets

Attic conversions:
Home improvements and conversions booklet, TRADA (1975)

Trimming around dormers:
Registered house-builder's site manual, pp 79, 81, 91, NHBC (1974)

Condensation in pitched roofs:
The House Builder, September 1977 (NHBC reprint)

ACKNOWLEDGEMENT

This section is based partly on information first published in the December 1977 issue of *The Architect.*

37 TIMBER-FRAMED WALLS AND PARTITIONS

INTRODUCTION 37–2

MATERIALS 37–4

FABRICATION METHODS 37–5

ERECTION SEQUENCE 37–9

TIMBER-FRAMED WALLS AND PARTITIONS 37

37 TIMBER-FRAMED WALLS AND PARTITIONS

F.W. SUNTER, MRAIC, FRAIA, AIWSc
Architectural Consultant to the Council of Forest Industries of British Columbia

This section explains in detail how timber-framed walls and partitions, and hence timber-framed houses, should be designed and constructed to achieve structural stability, freedom from damp and condensation, and good standards of insulation and finish.

INTRODUCTION

Timber-framed walls and partitions are generally accepted as part of modern British building technology, but the principles of design and fabrication are not covered by 'deemed-to-satisfy' constructions in the current Building Regulations. The Regulations require that loadbearing timber-framed walls should be designed in accordance with CP 112: Part 2: 1971 and its amendments. Local authorities and the National House-Building Council require that the structural design calculations be done by a suitably qualified engineer. Builders should check that this requirement has been complied with, prior to commencing fabrication.

The durability and performance of timber-framed external walls depend, to a large degree, upon the correct location and specification of vapour barriers, moisture barriers and insulation. In this regard the recommendations contained in this section conform to the requirements of the National House-Building Council and are based upon principles developed over many years of experience in North America. Builders should not deviate from them.

Loadbearing timber-framed walls are designed to transmit all the vertical live and dead loads imposed on the structure through to the foundations, and at the same time to resist racking and overturning forces that may result from wind pressure and, in certain locations, seismic shock or ground settlement. In conjunction with the application of internal linings, insulation and external claddings, the walls must be weatherproof and durable with the thermal, acoustic and fire-resistant properties required by the current Building Regulations. Non-loadbearing walls and partitions vary in their function, depending upon their location, and may be built to incorporate any or all of the functions listed above.

There are several variations in the design and fabrication of timber-framed walls, but the principal method explained in this section is an integral part of the platform frame system. Each wall or partition is erected on a platform of concrete or timber and in units one storey high. This method combines flexibility, economy and ease of erection, and incorporates the cavity barriers required by the Building Regulations. Builders adopting other methods should ensure that any additional members required as cavity barriers are installed.

The constructions shown here will meet or exceed the current requirements of the Building Regulations, and can be easily modified to comply with future changes in regard to fire, thermal or acoustic properties.

Terminology

See *Figure 37.1.* The timber-framed wall consists of uniformly spaced vertical members called studs, to which are nailed top and bottom plates, usually of the same

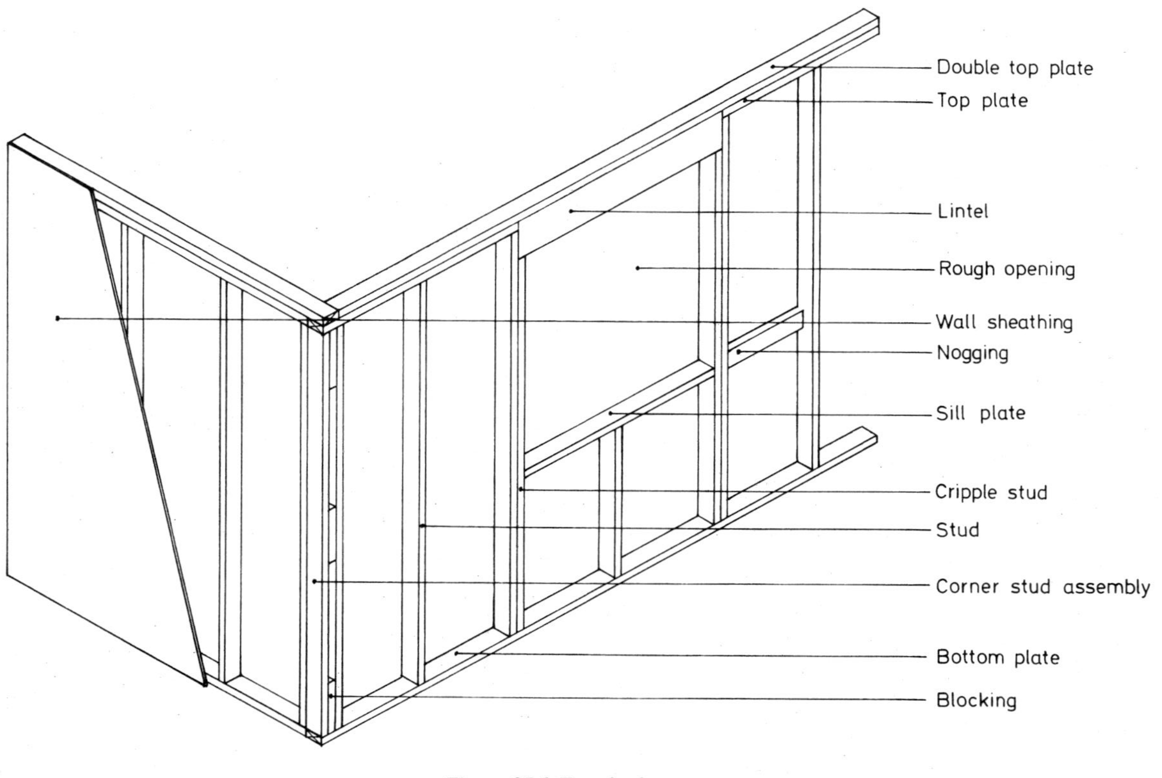

Figure 37.1 Terminology

dimension. In loadbearing walls openings for doors and windows are formed with timber lintels, which are supported on shorter studs called cripple studs. There may be either one or two cripple studs, depending upon the load carried by the beam, with full-length studs nailed alongside. To provide rigidity to the framed exterior walls, the frames must be braced. This is usually achieved by nailing a durable, weatherproof sheet material called wall sheathing to the frame. In some locations, such as party walls, the frames are braced by nailing 25 × 100 mm boards diagonally to the studs and plates inside the cavity between the walls. Horizontal members, called noggings, are nailed in between studs where required to provide fixings for wall-attached plumbing and cabinet fixtures, etc.

MATERIALS

Timber

All the timber used in wall framing must be accurately sized, either by machining one side and one edge or (in accordance with North American practice) by planing all round. Good construction cannot be achieved with rough-sawn lumber. The timber, apart from the lintels, is unlikely to be highly stressed, and the grades most commonly used are GS (to BS 4978) and Stud or Construction/Standard (graded to the Canadian Lumber Standards NLGA rules). In accordance with the Building Regulations, all timber used should carry an approved grade stamp. Engineering calculations will determine the grade and size of the members. Structural design is not the only criterion to be considered. The width of the stud may be dictated by the thickness of insulation required and by the necessity to accommodate various services within the stud cavities. The rigidity or feel of the wall, particularly if it incorporates doors, is also a factor to be considered.

Following North American practice, it is generally accepted that 38 × 89 mm for loadbearing walls and 38 × 63 mm for non-loadbearing walls should be considered as minimum acceptable structural sizes.

Sheathing

If plywood is specified, it should be a fully exterior sheathing grade with a grade stamp indicating the country of origin, the species and the quality of the glueline. Tempered hardboard, medium hardboard and bitumen-impregnated insulating board may also be specified as wall sheathing. Each material has its own characteristics in regard to weather resistance, stability, workability and vulnerability to physical damage, and must be treated accordingly by the builder. In common with other materials used in timber-frame construction, sheathing should be stored in dry conditions, well supported and clear of the ground.

Nails

Common bright wire nails are most commonly specified for assembling the wall frames. They should be long enough to penetrate the fixing by half their length; e.g. fixing a 38 mm plate to a stud requires 75 mm long nails. Longer and heavier-gauge nails are not only unnecessary but can be less effective if they cause the timber to split. Table 37.1 specifies the number of nails required for each connection.

Table 37.1 NAILING SCHEDULE

Wall plate to stud	Two nails.
Stud to stud	Nail at 600 mm centres.
Second top plate to top plate	Nail at 600 mm centres.
Bottom plate to sub-structure	Nail at 600 mm centres.
Studs to lintel	Two nails to each lintel member.
Studs to noggings	Two nails.
Sheathing to framework	Nail at 150 mm centres around the perimeter and at 300 mm centres to intermediate supports. Use 45 mm nails for sheathing up to 8 mm thick, and 50 mm nails for sheathing up to 12.7 mm thick.

Other specification points

Vapour barrier. The vapour barrier shall be an impervious sheet material such as 250 gauge polythene. A polythene or aluminium-foil vapour barrier may be incorporated as an integral part of the interior wall lining.

Moisture barrier. The moisture barrier shall be a breather-type building paper conforming to BS 4016: 1972.

Preservative treatment. CP 112: Part 5, when published, will include recommendations for the preservative treatment of structural timber in various locations. The bottom-most member or sole plate as shown in *Figures 36.2 and 36.3* should be treated. The treatment of the timber in the wall frames themselves can be considered as optional, but it is often specified to allay any doubts a potential owner may have about the durability of timber frames.

Fabrication and erection. Timber-framed walls and partitions shall be constructed in accordance with the approved drawings, butted square on bearings, closely fitted, accurately set to required lines and levels, and secured rigidly in place. On completion, check that all framing presents a flush and even plane to accept lining materials.

FABRICATION METHODS

The builder has three basic options available to him for the fabrication of timber-framed walls. The first is to buy pre-assembled panels from one of the many plants specialising in prefabrication. The size, weight, and material content may vary from one manufacturer to another, and the builder should ensure that he has the facilities to handle and to store the panels where necessary. Each panel should be identified as to its location and checked against the specification and for dimensional accuracy.

The builder's second option is to manufacture the panels in his own workshop, either on or off the site, utilising panel sizes suitable to his handling facilities. The third option is to fabricate the panels on the floor of the building and then lift them into the vertical position. This is still the most widely used method in North America. If a builder adopts it, he should either buy all his timber trimmed to precise lengths or ensure that a radial-arm power saw is available on site. For any appreciable amount of construction, consideration should be given to the use of a portable compressor and power-nailing equipment.

Whichever method is adopted, the principles of design and construction are essentially the same and there should be no difference in the final product.

Base

The builder should ensure that the base on which he fabricates the walls, whether a concrete slab or a suspended timber floor, is level and dimensionally accurate. The latter is particularly important for prefabricated panels since changes to panel dimensions are costly and time-consuming. On a rough concrete slab designed to receive a cement topping it is standard practice to use a sole plate of the same dimension as the bottom plate, e.g. 38 × 89 mm under all walls and partitions. Locate the plates accurately and set dead level on an approved damp-proof course. Where necessary a solid bed of mortar is used under the plates. The plates are fastened in position with approved mechanical fastening devices, a minimum of two to each length of plate.

For external walls the fastening devices include 12.5 mm diameter anchor bolts, 'U' shaped steel anchors, and cartridge-driven hardened-steel studs (4 mm diameter)

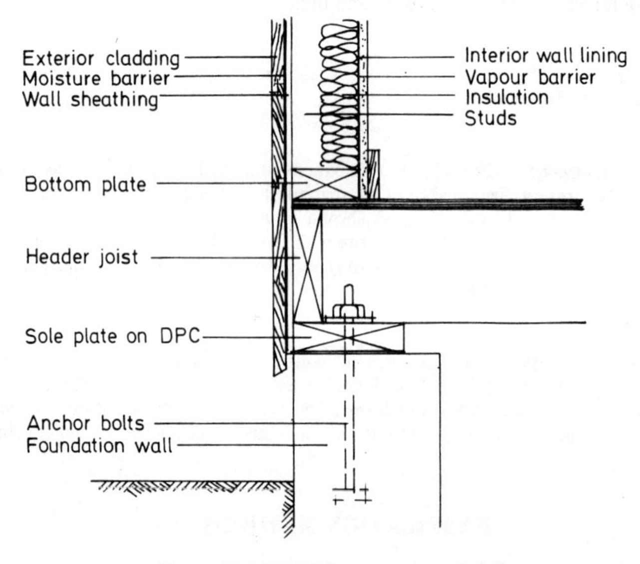

Figure 37.2 Wall frame on suspended floor

penetrating the concrete by a minimum of 25 mm. The steel studs are the most widely used device for internal plates. On suspended timber floors the positions of the walls and partitions are accurately located and marked with a chalk line or pencil. On timber floors and monolithically finished concrete slab the sole plate is not required. See *Figures 37.2 and 37.3.*

Studs and lintels

Assuming that a double top plate is specified, subtract from the overall panel height the thickness of three plates to give the standard stud length, e.g. 2400 – (3 × 38) = 2386 mm. Calculate the lengths of cripple studs and shorter-length studs under

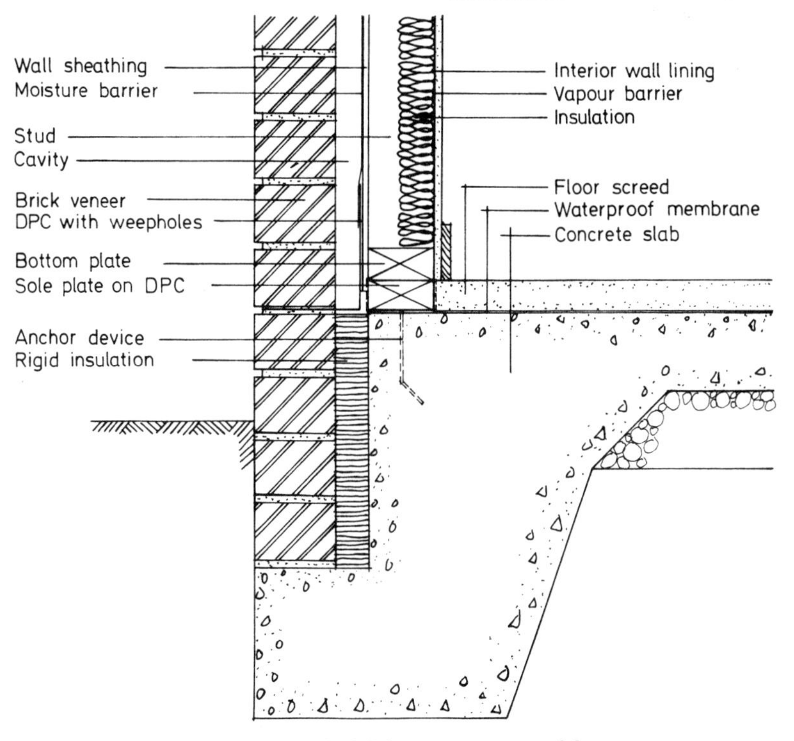

Figure 37.3 Wall frame on concrete slab

window sills, and cut all studs to length. A long bench with appropriate stops and the power saw will facilitate this operation.

Cut the lintels and sill member accurately to length. Where the lintels are made from two pieces of joist material (a standard practice to simplify material ordering), nail the lintel members together, using plywood spacers as required to bring them to the same net thickness as the width of the studs.

With on-site construction each wall is fabricated in the entire length before being erected.

Stud spacing

Interior lining and exterior sheathing materials are supplied in modular sheet sizes, usually 1200 × 2400 mm. The spacing of the studs, at 300 mm, 400 mm or 600 mm centres, is adopted to ensure that joints in the sheet materials will always occur on the centre line of the stud. Whichever spacing is chosen, it must be maintained throughout the length of the wall or partition. See *Figure 37.4*. There must also be supporting members at all corners and intersections for the interior lining and exterior sheathing materials. To achieve this, and to strengthen the framing, a combination of three studs is used at corners and intersections in walls (*Figure 37.5*). As an alternative, interior partitions may also be fixed to noggings spaced at 600 mm centres (*Figure 37.6*). Point loads from wall or roof beams require additional studs

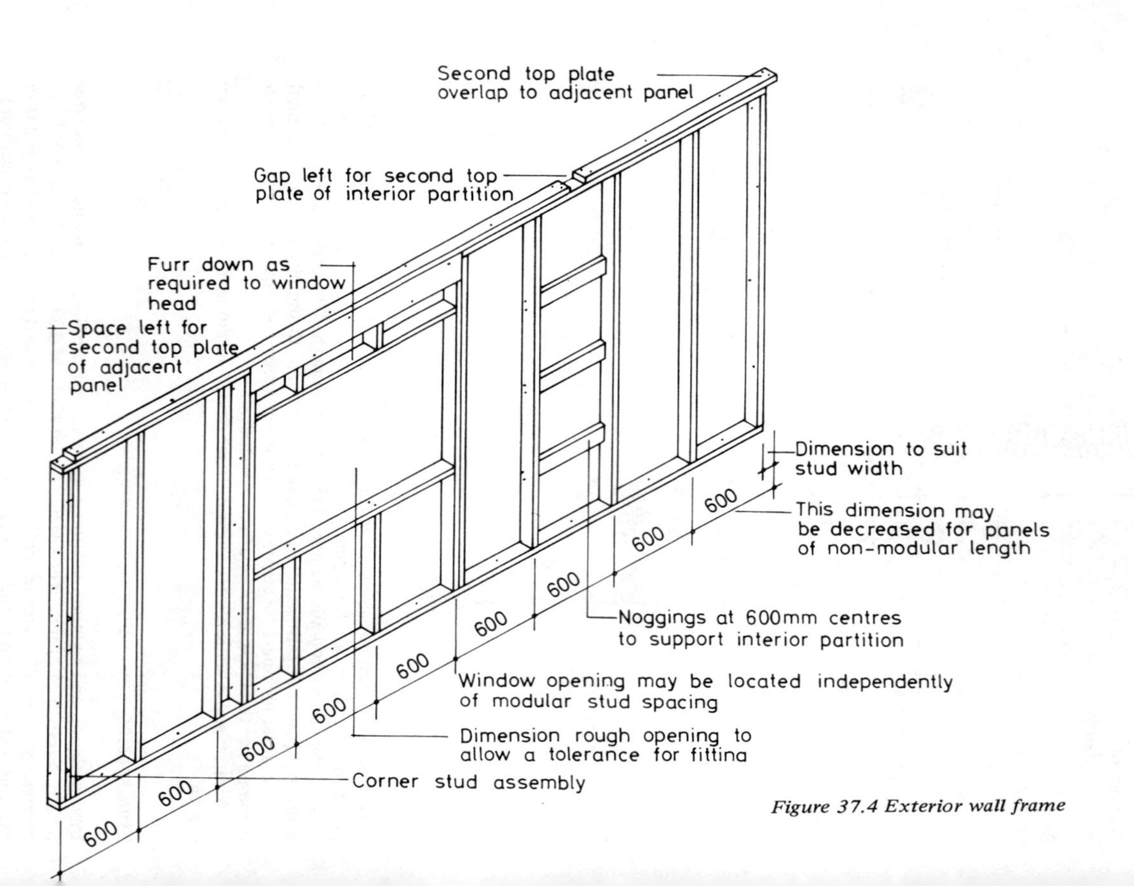

Figure 37.4 Exterior wall frame

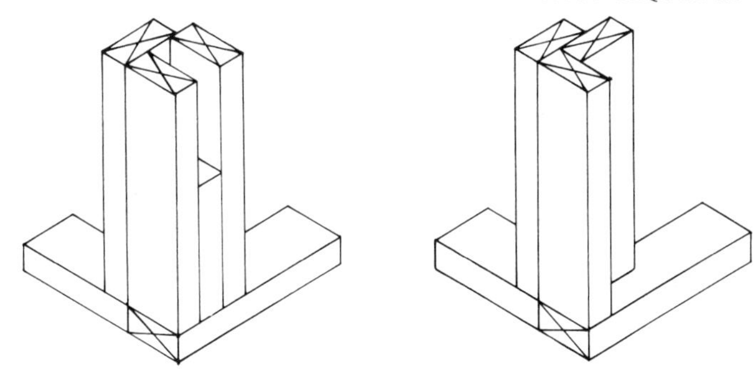

Figure 37.5 Corner stud assemblies

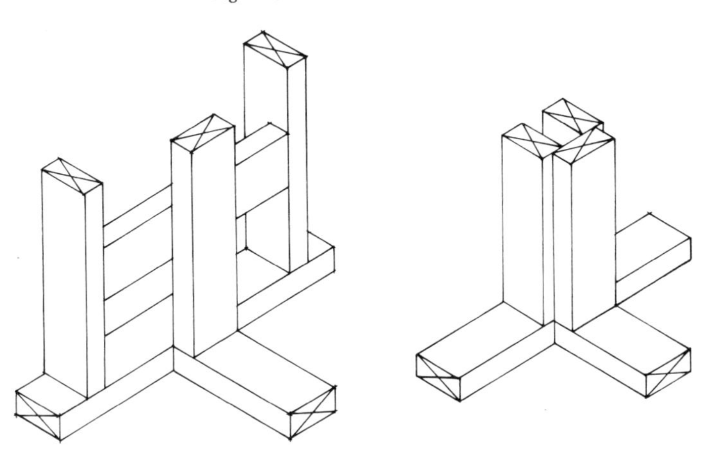

Figure 37.6 Partition intersections

for support as shown in *Figure 37.7.* Where this occurs in upper storeys, additional studs must be included in the wall framing below to transmit the load to the foundations.

ERECTION SEQUENCE

Lay the bottom and the first top plate for the first wall, accurately trimmed to length, in position on the deck. Any necessary joints should occur on the centre line of a stud. Using the modular spacing chosen, mark the stud positions on both plates. Mark any door or window openings in the wall on the plate. These are called 'rough openings' and a fitting tolerance should be added to the net dimension. The openings may be located independently of the modular stud spacing (see *Figure 37.4*).

Separate the plates and lay the studs in position. Nail the plates to the studs, using blunted nails at the ends of plates to reduce splitting. The cripple studs are nailed to the sill members before the full-length stud is fixed in position. For added rigidity in transport or erection the bottom plate is usually carried through the door openings.

A saw kerf should be made on the underside to facilitate its subsequent removal. An additional stud is installed at one corner only, incorporating 200 mm long spacers (see *Figure 37.4*). The other end of the panel is foreshortened by the width of the adjoining wall.

Square the panel up by checking that the diagonal dimensions are identical. Fasten the sheathing material, accurately trimmed around the openings, using the nailing schedule in Table 37.1. It is good practice to allow a 25 mm projection below the bottom plate. This helps in locating the panel in position and allows for a more rigid attachment to the sole plate or suspended timber floor.

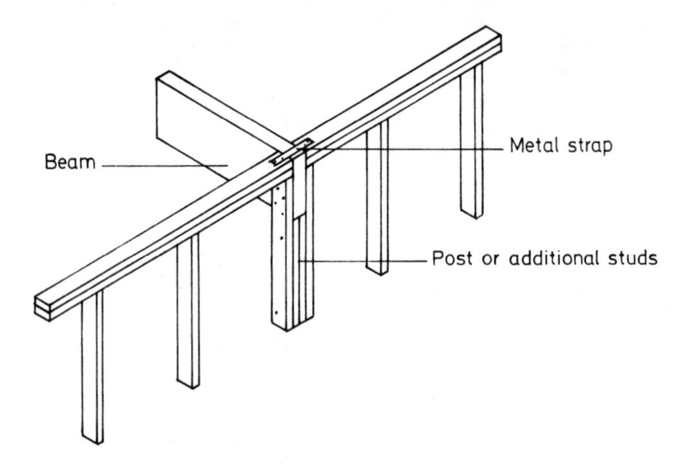

Figure 37.7 Beam support

Tilt the panel into the vertical position, plumb up with a level, and install temporary braces. Additional support will be provided by adjacent walls and partitions. Nail through the bottom plate to the sole plate, or to the joists of a suspended floor.

The remaining exterior wall panels are then fabricated and erected in sequence around the perimeter of the building.

The interior partitions are fabricated in the same fashion, with the omission of the sheathing. The requirements for the modular spacing of studs are the same as for exterior walls.

Diagonal strapping may be applied temporarily to keep walls and partitions square, but if made accurately to dimension they will automatically come into square when the whole building is plumbed up. It is essential, however, that partitions are set in a true vertical alignment before nailing in place at intersecting walls.

The second top plate can now be nailed into position. Joints in this plate should occur over studs, but should not coincide with joints in the first plate. The overlapping detail used at the intersections (*Figure 37.4*) provides a positive tie at the top of partitions and walls. The double top plate also provides sufficient support between studs to allow joists and rafters to be located independently of the stud positions. If the second top plate is omitted, 50 mm wide metal straps should be nailed on top of the plate to connect components, and joists and rafters should be located directly over the studs.

The structure should then be checked for horizontal and vertical alignment. In wall constructions supporting a pitched roof, this is done prior to fixing the roof members. Where the walls support floor or flat-roof joists, it is done after the joists are nailed in position but prior to fixing the floor or roof deck. The procedure usually followed is to check the vertical alignment at two corners and to block out a chalk line 25 mm and stretch it between the two corners at the top-plate level. The wall can then be pushed or pulled into its correct alignment.

Where interior wall or partitions run parallel to joists or trussed rafters, noggings are installed at 600 mm centres to provide fixing for the top plate and also to support the ceiling lining material (see *Figure 37.8*).

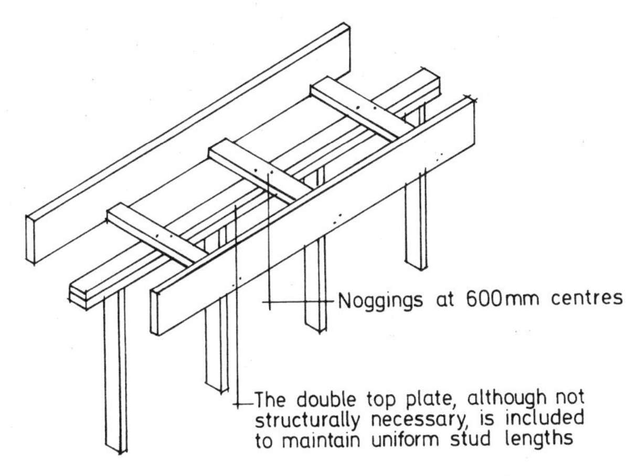

Figure 37.8 Partition parallel to joists

Gable walls may be fabricated by extending the stud lengths to the line of the roof, but it is more usual to make up the gable ends as separate components (*Figure 37.9*). The double top plate may be omitted. Accurate dimension to conform with the roof line is essential. The line of the top plate coincides with the top or bottom of the rafter, depending on whether or not a gable overhang is detailed.

Timber-framed walls and partitions on subsequent storeys are similarly fabricated and erected.

Windows and doors

Standard profiles of window and door frames can be installed in timber-framed walls. In external walls it is recommended that a positive weather barrier is incorporated into the frame. Where this is a metal or plywood strip let into the frame, as shown in *Figure 37.10*, it doubles as a fixing device for the frame. When this detail is not used, apply a 300 mm strip of breather paper around the opening, returned on to the face of the studs before installing the frames. Where brick veneer is used as a cladding, windows and doors should be supported on the timber frame and allowance made for differential movement between the frame and the brickwork; this is illustrated in *Figure 37.11*. The cavity stop shown also provides the required weatherstop. With wall finishes other than brick veneer a similar detail can be used, and by changing the dimension of the stop a variety of claddings can be accommodated. A non-corrosive flashing should be applied over openings where the distance from the head to the eaves soffit is more than one-quarter of the horizontal projections of the eaves. The flashing should extend a minimum of 50 mm up under the breather paper behind the

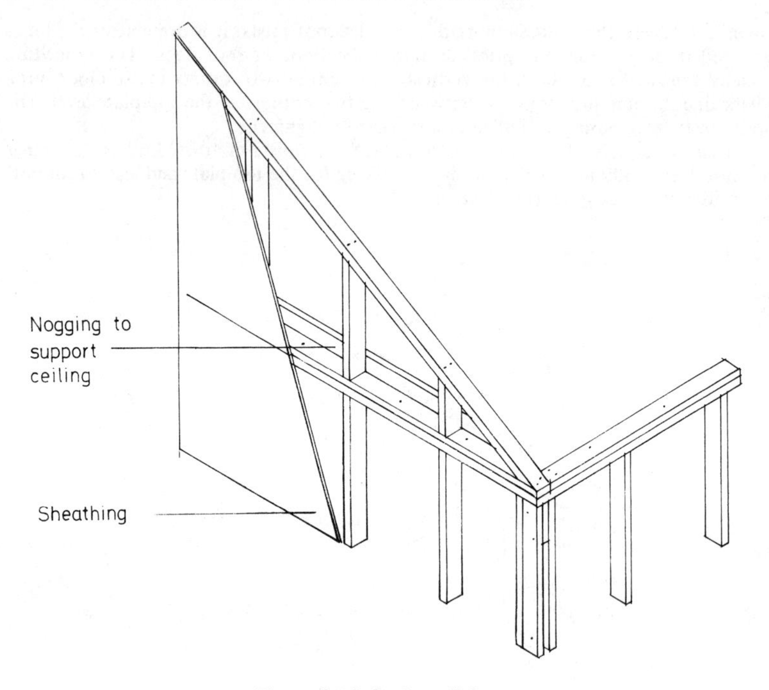

Figure 37.9 Gable wall frame

cladding. To install windows, shim the sills level where necessary and tack in place. Plumb the jambs, and nail to the cripple studs through the shims. Check the diagonals to ensure that the unit is not out of square, and tack the head to the lintel. Any gaps between the window or door units and the wall framing should be filled with insulation.

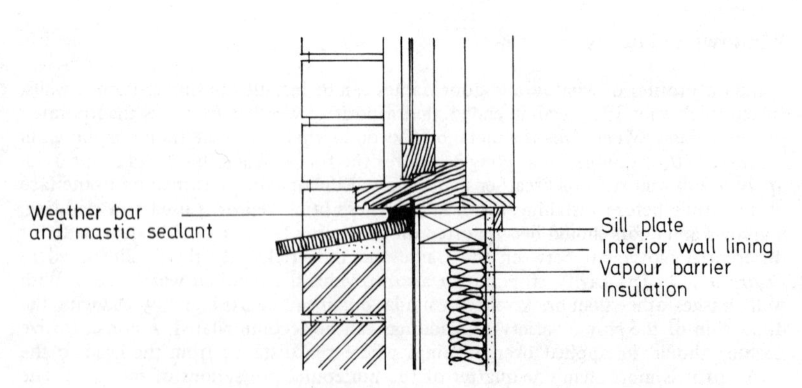

Figure 37.10 Window sill (vertical section)

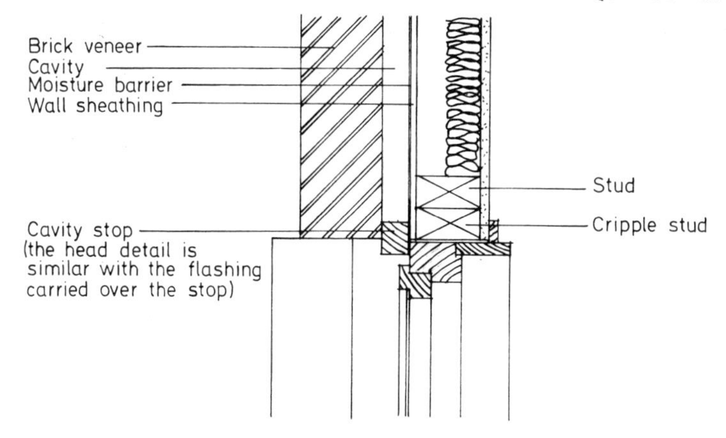

Figure 37.11 Window jamb (horizontal section)

Breather paper

This is used behind the cladding material basically as a second line of defence against the ingress of moisture. It is usually applied just prior to the installation of the cladding. It is applied horizontally over the whole of the exterior wall surface, with upper sheets lapped 100 mm over the lower sheets and nailed or stapled at 150 mm centres at laps and at 300 mm centres for the remainder. This fixing may be reduced where the cladding is fixed directly on top of the sheathing.

Exterior cladding

BRICK VENEER

Brick or masonry veneer should be supported directly on the foundations, and should be supported from the timber frame with flexible corrosion-resistant ties, using approximately four ties to each square metre. The ties are nailed through the sheathing to the studs. A minimum cavity of 38 mm is recommended, and should be kept clear of mortar droppings.

TILE HANGING

Concrete, clay or stone tiles may be applied in accordance with standard practice for this trade. The tile battens are fixed over the breather paper and nailed through the sheathing to the studs.

TIMBER CLADDINGS

Many profiles in a variety of timbers are available for horizontal, vertical or diagonal application. All softwood cladding (with the exception of Western Red Cedar and Californian Redwood) should be treated with a preservative prior to being installed.

In horizontal application, the boards are usually fixed directly over the breather paper by nailing through the sheathing into the studs. In vertical applications, 25 mm horizontal battens at 600 mm centres are nailed to the studs and the boards are nailed to the battens. All nails should be galvanised or aluminium, with only one nail used at each fixing. The ends of the boards should be sealed with paint or stain, depending upon which is specified to prevent moisture intake.

References should be made to the manufacturer's literature for the application of finishes such as aluminium and plastic claddings, cement render on metal lath, and any of the wide range of textured surface coatings now available.

Services and fittings

Electrical and plumbing services are usually installed within the cavities of timber-framed walls. The studs and plates may be notched or drilled within the following limits in order to accommodate these services. In loadbearing walls not more than one-third of the cross-section may be cut out or drilled, and in non-loadbearing walls not more than one-half. If this amount is exceeded, the stud or plate should be suitably reinforced with metal straps. Provision must also be made for back-up members in the form of noggings to support wall-hung cabinets, basins, cisterns, electrical fittings etc. Offcuts from plates and studs are useful for this purpose and should be set with the wider face vertical using two nails at each end. This permits services and insulation to be installed behind them with a minimum of cutting and drilling. As far as possible services should not be installed in party walls, but when this is unavoidable electrical outlet boxes should not be located back-to-back.

Inspection of frame

It is essential that the framing is properly inspected after the installation of services and prior to the installation of insulation, vapour barriers and internal wall linings, to ensure that no damage has occurred to the structure and that the framing is in every way ready for the subsequent operation. Damaged timber should be removed or repaired. If studs are twisted or bowed out of alignment with the wall surface they should be sawn three-quarters of the way through and a straight stud nailed alongside.

Thermal insulation

The insulation is installed in the stud cavity of all the exterior walls. Most types of insulation can be used effectively, but the most common are glass-fibre or rockwool batts with a bitumen paper bonded to one side.

The paper overlaps the insulation and is stapled to the face or to the side of the studs. The insulation should fit tightly in the cavity to prevent air leakage, and should pass on the outside of any services. Although 25 mm of this type of insulation is usually adequate to meet the requirements of the Building Regulations 1976, a recommended minimum thickness is 50 mm. This will almost double the thermal resistance of the wall.

Vapour barrier

The vapour barrier is installed between the interior wall lining and the insulation. Where a separate sheet material is used there should be a minimum of joints, and these should occur over solid fixings and be well lapped. Every care should be taken to keep openings in the vapour barrier to a minimum to prevent the ingress of moisture from the interior of the building into the stud cavities.

Although British Standards specify that structural timber should have a maximum moisture content of 22 per cent when it is incorporated into a building, it can be difficult to implement this requirement and to ensure that the moisture content does not rise during construction. The correct positioning and installation of the vapour and moisture barriers ensure that the timber will dry to an equilibrium moisture content of around 14 per cent, so there should be no danger of fungal decay. However, if the internal finishes are applied to timber with a high moisture content the subsequent shrinkage due to seasoning can cause defects in these finishes, particularly at the door and window casings and the baseboard. The vapour barrier and internal linings should not be applied until the moisture content of the wall frame is 18 per cent or less. If necessary a combination of space heating and ventilation should be employed to dry the timber.

Timber-frame construction methods recognise that some shrinkage is likely to occur during construction and that it can be accommodated by correct design and construction. Where a building incorporates loadbearing masonry or concrete walls in association with timber-framed walls, particular care is required to ensure that defects do not occur as the result of differential settlement.

Interior finishes

The most widely used interior lining material is paper-faced gypsum plasterboard, and the technique of application is commonly called 'dry-lining'. The plasterboard provides the fire resistance required by the Building Regulations, and since it is classified as having a Class 0 spread-of-flame rating its use is not restricted. In general terms 12.7 mm plasterboard gives half-hour modified fire resistance and 25.4 mm gives one hour modified fire resistance. When properly applied to well constructed wall framing it provides a sound, crack-free, completely flat surface suitable for a wide variety of wall finishes such as paint, wallpaper, ceramic tiles and wood panelling.

Sound insulation

Timber-framed wall construction provides adequate standards of insulation against the passage of sound between rooms in a dwelling. In certain locations (such as

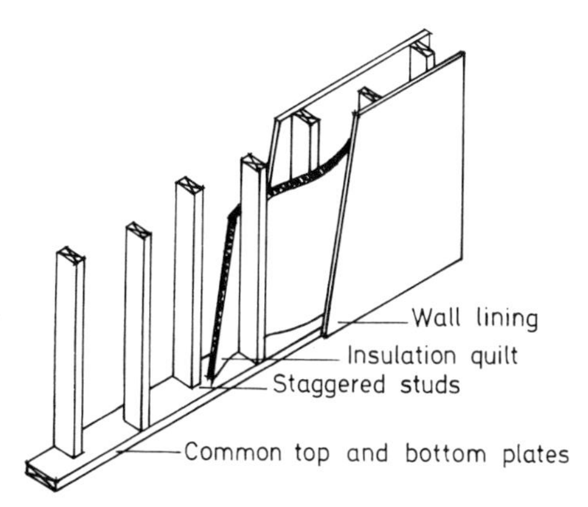

Figure 37.12 Staggered-stud wall

between a bathroom and a bedroom, or between living areas and children's sleeping areas) it may be considered desirable to increase the sound resistance. A variety of methods are used depending upon the degree of resistance required. Among those more commonly used are resilient metal bars used to partially isolate the plasterboard from the studs, a layer of insulation board between the plasterboard and the studs, and a staggered stud partition as shown in *Figure 37.12*.

The Building Regulations require a specific resistance to the passage of sound in certain locations such as party walls. The accepted construction is to separate the timber-framed walls of adjoining dwellings by a minimum of 100 mm. The walls are lined on the interior of the dwelling with 32 mm of plasterboard applied in two layers, and with 50 mm of rockwool or glass-fibre insulation in the stud cavity. This construction has been proved to exceed the requirements of the Regulations for both sound transmission and fire resistance.

38 NON-TRADITIONAL CONSTRUCTION

INTRODUCTION 38–2

CONCRETE PANEL CONSTRUCTION 38–5

IN-SITU CONCRETE CONSTRUCTION 38–6

TIMBER-FRAME CONSTRUCTION 38–6

STEEL-FRAME CONSTRUCTION 38–7

VOLUMETRIC CONSTRUCTION 38–10

NON-TRADITIONAL CONSTRUCTION 38

38 NON–TRADITIONAL CONSTRUCTION

W.W.L. CHAN, BSc, PhD, DIC, CEng, FICE, FIStructE, FIWSc, MBIM
Consulting Engineer

This section deals with construction techniques that embody assemblies of wholly or partially prefabricated components, or of site-built units using a minimum of skilled labour; these include large-panel concrete, in-situ concrete, timber-frame, steel-frame and volumetric units.

INTRODUCTION

Most non-traditional constructions were developed in the late 1950s, when the cost of labour on site rose at a faster rate than that of materials, and as skilled labour became scarcer. They have been particularly well developed for houses and flats because of the repetitive nature and the limited variety of the plan forms, and because of the comparatively large housing demand both in the public and private sectors. Similar developments, but to a lesser extent, have also taken place in non-traditional buildings for schools, hospitals and offices. For all practical purposes, 'industrialised building' and 'system building' are synonyms for non-traditional construction.

Whether the technique of construction involves off-site prefabrication or on-site rationalisation, the advantages sought from non-traditional construction include:

- maximum repetitive use of standard components;
- minimum requirement for skilled labour;
- minimum use of wet trades on site, as by using dry jointing;
- faster rate of construction;
- minimum wastage of materials on site.

To realise these advantages, capital expenditure is required both to pre-design a range of components that, when combined in various ways, form building types most commonly in demand and, in the case of prefabricated components, to set up plant and factory to produce them.

High capital investment to produce prefabricated components may be economic if there is a firm prospect of large and continuous orders. However, the pattern of building demand is more often intermittent and erratic, and techniques that are competitive on a batch-production basis are more likely to be viable commercially.

The sponsor of a non-traditional technique of construction is usually a contractor or manufacturer[1,2], but there are also systems of construction designed by professional consultants available for exploitation under licence.

Because each sponsor may base his preferred technique on different dimensional disciplines and choice of materials, the traditional form of contract, in which the employer specifies the plans and specification in every detail, is unsuitable for competitive tendering. Mutual advantage to the employer and contractor is best obtained by basing the tender on outline plans and performance requirements, leaving the sponsor to demonstrate how his system can comply. It is almost always uneconomic for a non-traditional technique of construction to be specially altered for a particular project, unless the project is exceptionally large. Thus, non-traditional constructions are usually undertaken as 'design and build' contracts, in which the builder assumes responsibility for design as well as execution.

The techniques discussed in this section refer to building superstructures. Foundations and substructures for such techniques of construction are essentially traditional, being little or no different from those for traditional building. However, where prefabricated components are used, the top of the substructure receiving the components needs to be built to very close tolerances.

Safety on site

Workmen should be trained and supervised in the handling, storage and erection of prefabricated units to avoid danger to them and damage to the construction[3].

Large prefabricated panels should be stored on edge on the site in simple but sturdy racks that prevent sideways collapse of the panels and permit the withdrawal of any panel without causing instability to the remainder.

Lifting lugs should be designed as an integral part of the construction technique. In addition, additional points of connection for temporary bracing should also be considered.

Partially completed panel constructions are often unstable, and clear site instructions for the sequence of construction and temporary bracing should be given. In the case of lightweight panels, the resistance to wind uplift and overturning of a partially completed building may be less than of the completed building, due to the full dead weight not being present and the increased wind pressure caused by the presence of large openings; these conditions should be covered in design and site instructions.

Acceptance

Non-traditional construction must satisfy all the relevant requirements of Building Regulations or the appropriate local authority building by-laws[4,5,6,7]. As its name implies, it often embodies materials and techniques for which there are neither 'deemed to satisfy' provisions nor any relevant British Standards and Codes of Practice. In these circumstances, structural calculations, and test certificates of suitability of materials and constructions, are usually required as part of the process of obtaining statutory building approval. Such documentation can usually be standardised for a large range of applications of the particular construction technique or system.

A proprietary material or component not covered by British Standards may instead have or apply for an assessment to obtain an Agrément Certificate from the Agrément Board. It certifies the suitability of the material or component for specified conditions, including the Board's opinion on the compliance with the relevant parts of the Building Regulations and local authority by-laws.

A similar form of certificate, called the Agrément Board/National Building Agency Certificate, is also issued for non-traditional systems of construction. In addition to evaluation of the suitability of materials used (which may be covered by British Standards or Agrément Certificates) the certificate covers the method of fabrication and construction and the extent to which the completed buildings meet established standards of layout. The Agrément Board/National Building Agency Certificate is not mandatory for building approval, but provides added assurance to building clients and control officers on the acceptability of the system concerned, particularly where relatively novel materials or techniques are incorporated.

For house building in the private sector covered by the National House-Building Council (NHBC), the Council has published technical criteria governing the use of non-traditional construction. Structural calculations and other important building

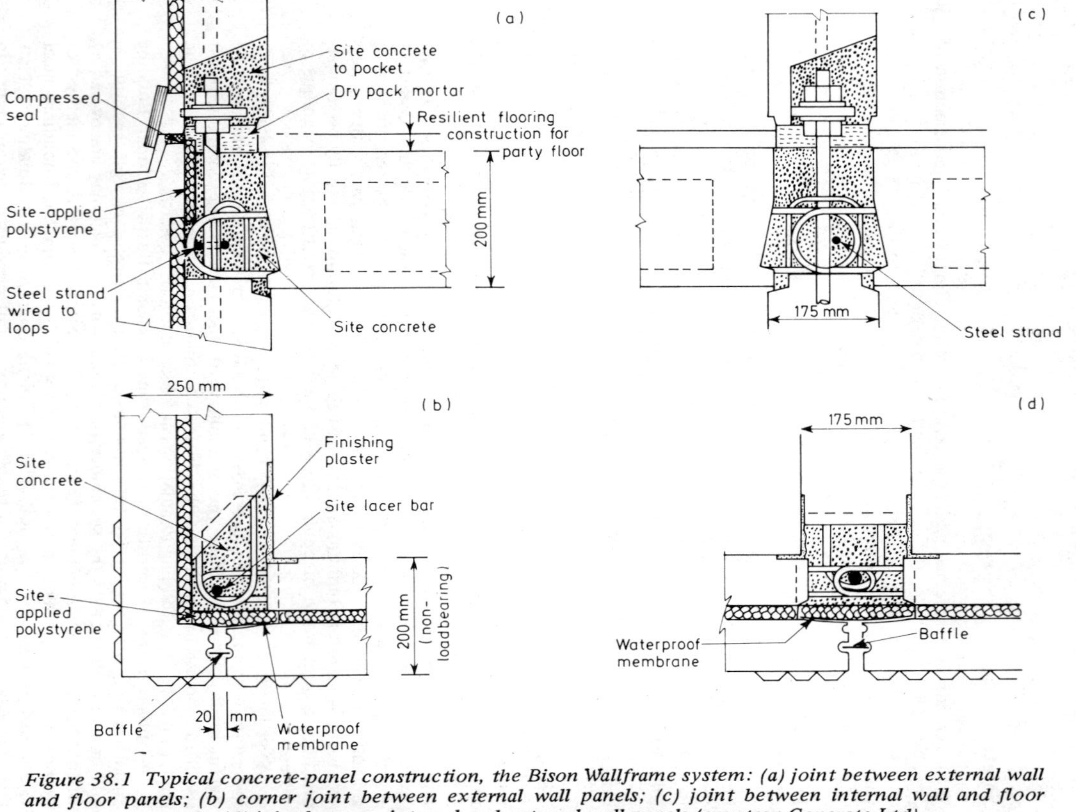

Figure 38.1 Typical concrete-panel construction, the Bison Wallframe system: (a) joint between external wall and floor panels; (b) corner joint between external wall panels; (c) joint between internal wall and floor panels; (d) joint between internal and external wall panels (courtesy Concrete Ltd)

details, prepared by a qualified designer, must be approved by the NHBC prior to each house being built. Alternatively a system may be granted 'blanket approval' from the NHBC; an appropriate system having an Agrément Board/National Building Agency Certificate may be acceptable for NHBC 'blanket approval'.

CONCRETE PANEL CONSTRUCTION

This technique employs large prefabricated panels of concrete in storey-height external and internal walls, floor slabs, stairs, and often including segments of lift shafts and refuse chutes.

Most concrete panel constructions are of room-size components, which facilitate the masking of wall and floor joints, although some systems have smaller panels, some as narrow as 1.2 m wide. For structural integrity, projecting reinforcement from adjacent panels is connected by looping and threading with site-placed reinforcement, or site welding, before the joints are filled with *in-situ* concrete (*Figure 38.1*).

Such panels may weigh from 2 tonnes to 10 tonnes and cranes are always necessary for erection. High accuracy is required in positioning the panels, usually to about ±5 mm to ensure the efficacy of the weather-proofing joints in external walls and adequate bearing of floor panels on top of wall panels. Levelling screws are usually provided at the top of wall panels to facilitate the positioning and levelling of the next wall panel over. The placing of heavy panels to such accuracy requires diligent supervision and skill. For high-rise tower blocks, tower cranes are invariably used; for medium-rise or lower-rise slab blocks, rail or track mounted cranes may be suitable.

External wall panels are usually of sandwich construction, with outer and inner concrete leaves sandwiching a foamed plastic insulation. The concrete leaves may be connected at the panel perimeter (the insulation stopping short there) or throughout the panel by wall ties; in the latter case non-ferrous ties such as silicon-bronze or stainless steel are used because of the risk of prolonged exposure to damp; care should be taken to prevent direct contact between such ties and the steel reinforcement, so as to avoid electrolytic corrosion. The horizontal joint is formed by lapping the upper over the lower panel and closed with a compressed hermetic seal. The most successful vertical joint is the 'drained joint'[8] in which driving rain first meets and runs down the face of a small baffle strip inserted between opposing grooves of adjacent panels; water that penetrates the baffle is drained away in the cavity provided behind the baffle, and the water is prevented from penetrating the solid concrete joint by a site-applied waterproofing seal.

The drained joint can accommodate construction tolerances and does not rely for its weather-tightness on mastic sealants (which frequently fail due to movement of joints) or on close-tolerance compression of sealing strips. The external faces may incorporate textured, sculptured, fluted, mosaic or exposed-aggregate finishes in different colours, brick facing with cast-in brick slips, or full-size bricks cast integrally with the panel.

Internal wall panels are of solid concrete, often fitted with levelling screws as for external walls. The wall faces are smooth finished for direct painting or wallpapering.

Floor panels are of solid or cored construction, reinforced or pre-tensioned. In blocks of flats, insulation against impact sound is usually provided by incorporating a layer of 25 mm foamed plastic under the floor screed or by providing a resilient floor covering over the screed. Care must be taken to avoid sound-bridging between the floor and walls.

Most concrete panel systems are factory-produced by large precast concrete specialist firms, using heavy steel moulds, and panels are transported a few at a time by lorries. Internal walls and solid floor slabs are often produced in vertical 'battery' moulds, but external sandwich panels and cored floors are usually cast flat.

Walls are structurally designed as masonry (unreinforced) walls, and floors as reinforced or prestressed concrete. Economy of design is achieved by simultaneously providing structural strength, fire resistance and sound insulation, particularly for blocks of flats. However, some of these requirements are over-provided for low-rise houses, where strength, fire resistance and sound insulation requirements are lower; hence concrete panel construction may be less economic for low-rise, except in mixed development of flats and houses.

IN–SITU CONCRETE CONSTRUCTION

In this type of construction, dimensionally coordinated formwork is used to produce a variety of commonly used building or house layouts. Apart from the off-site production of formwork, no other capital investment is incurred, but the construction is exposed to the risk of bad weather and is dependent on skilled site labour to a greater extent than prefabricated construction. As the formwork is usually in large heavy units, cranes (or forklift trucks for low-rise work) are required.

No-fines concrete

This technique has been successfully employed for constructing loadbearing external and internal walls in houses and, in conjunction with a cast *in-situ* reinforced concrete frame, non-loadbearing walls for medium- and high-rise flats. Because the concrete surface is naturally rough textured and the wet concrete exerts no hydrostatic pressure (lateral pressure is exerted only by the coarse aggregate), the formwork required may be of relatively low-quality materials and joints.

Only solid walls (with or without openings) can be readily cast. No-fines concrete is porous and has paths for water and damp penetration in external walls; it is therefore necessary to heavily render external surfaces and this requires good workmanship. Alternatively, a separate external lining of brick, weatherboarding or tile hanging may be used. Internal walls also require smoothing by rendering or by dry-lining with plasterboard.

Floors are usually of timber board and joist construction for houses, and cast *in-situ* reinforced concrete construction for flats. Roofs are usually of timber trussed-rafter tiled construction for houses and low-rise flats, and cast *in-situ* reinforced concrete for high-rise flats.

Cross-wall concrete

This is used for medium- or high-rise flats where a series of goalpost-shaped units of formwork are used for concreting two parallel walls, one transverse wall and the floor above, in each storey. The formwork units are so hinged and articulated that they can be struck and withdrawn from the 'tunnels' of concrete formed, through the front or the rear of the building. In some systems, electrical heating cables are incorporated in the forms for accelerated curing of the concrete. The remaining internal walls, external walls, stairs etc. are added later by traditional construction or by using prefabricated components.

TIMBER–FRAME CONSTRUCTION

Timber frame was introduced from North American and Scandinavian countries, where this is the traditional form of house building. It is a lightweight form of construction, suitable for houses and flats not exceeding four storeys, although the practical maximum height is three storeys.

In principle, loadbearing walls are formed from a series of softwood vertical studs spaced at not exceeding 600 mm centres, butting against horizontal timber plates of similar section at the top and bottom. This arrangement provides a fixing frame for lining such as plasterboard internally and for sheathing or cladding material externally. Additional cross noggings are provided between studs to support joints of lining materials or heavy fittings. The spaces between studs are partially or completely filled with lightweight insulation material in the case of external walls.

A widely used arrangement of building components is the 'platform frame', in which storey-height wall panels are used. The board and joist upper floor is supported over the top of the wall panel below, and in turn supports the wall panel above[9] (*Figure 38.2*).

Usually walls are prefabricated as a series of room-size panels, each weighing about a quarter of a tonne, complete with lintols for window and door openings. Internal linings (usually of plasterboard) are normally site-applied after the installation of wiring and piping. External wall panels are partly or wholly sheathed with structural plywood or suitable types of fibre building board, mainly to provide stiffening or resistance to racking. Because of the lightweight construction, timber-frame panels often require to be anchored down to prevent uplift under high wind exposure, and in turn the foundation should have the necessary dead weight at the points of maximum uplift.

Sound insulation across separating walls (as in terrace houses or adjacent flats) and separating floors (as between upper and lower flats) requires special consideration because of the lightweight construction. For separating walls two leaves of timber-frame wall are provided which are structurally isolated from each other thus reducing to an acceptable level the transmission of sound[10,11].

For separating floors, impact sound insulation is provided similar to concrete panel construction, by laying the floor boarding over resilient bearings placed over the joists; air-borne sound insulation is provided by added mass[12], such as sand-pugging in the joist space, or additional layers of boarding (such as plasterboard) over the joists.

Except for temporary buildings, external cladding for timber-frame building is usually in more traditional materials such as a brick skin, tile hanging (upper floors) or weather-boarding. Brick is the most usual cladding, and must be suitably wall-tied to the timber frame so that some vertical movement between brick and frame can be accommodated.

Properly detailed and constructed timber-frame buildings do not normally require preservative treatment, except in the case of areas infested with Longhorn beetle prescribed in the Building Regulations, or when exceptional precautions are considered necessary.

Timber-frame systems are offered by many contractors throughout the country; the frames are prefabricated by a large number of timber engineering component suppliers.

Economy of design is achieved by choosing the right balance between size and structural grade of timber, by keeping the number of different timber sizes to a minimum, and by providing simple and reliable connection details. Structural calculations are usually required to justify the design.

STEEL–FRAME CONSTRUCTION

The steel frame is a well-established loadbearing structure and can be used for low-rise or high-rise construction. However, in non-traditional housing, the steel frame has been developed mainly for low-rise construction. The main differences compared to concrete and timber construction are that steel, being a good heat conductor,

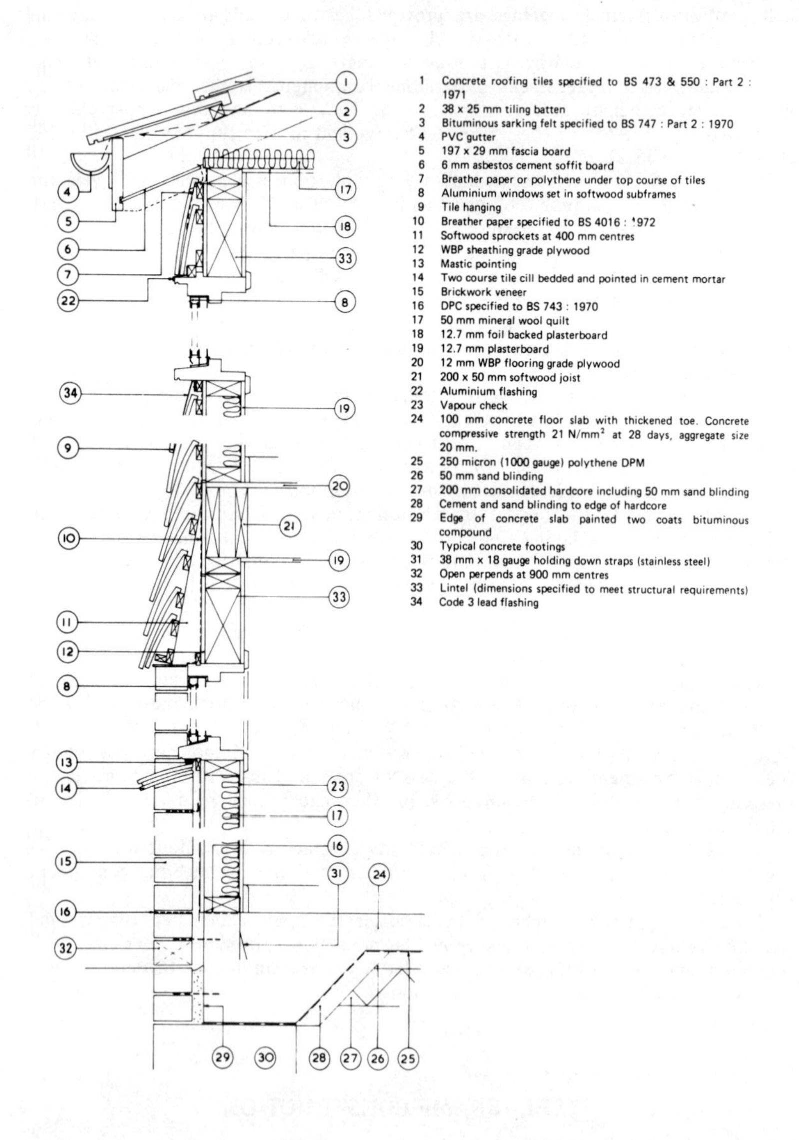

Figure 38.2 (a) Typical timber-frame construction: front and rear wall sections (from Certificate No. 76/S3, courtesy Agrément Board/National Building Agency and Lovell Housing Ltd)

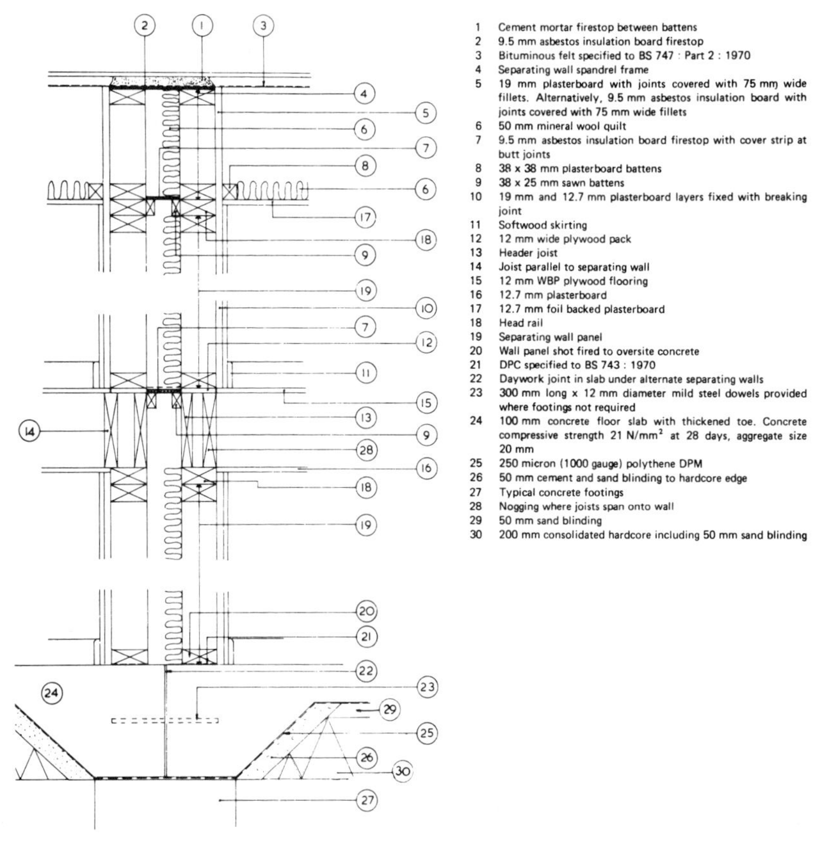

Figure 38.2 (b) Typical timber-frame construction: separating wall section (from certificate No. 76/S3, courtesy Agrément Board/National Building Agency and Lovell Housing Ltd)

requires special measures to provide fire protection and to prevent cold-bridges; protective treatment against corrosion is also required. Care should be taken in design to avoid condensation occurring on the steel surface and to eliminate water traps at the connections.

Skeletal construction

Here, the steel frame consists essentially of widely spaced steel columns at the corners of the building and intermediately as necessary, supporting steel primary beams at the perimeter and intermediately on the roof and upper floor levels. The beams support the roof and floor structure, cladding and lining. Non-loadbearing infill walls are then added, which span between storey heights, capable of resisting wind and other laterally applied loads. The roof and floor structure can be in concrete, timber or steel components.

The steel frame components are structurally connected by bolting or welding on site. The components receive corrosion protection during fabrication, with suitable paint schemes or hot-dip galvanising. Fire-resistance is provided by suitable cladding and lining materials.

The skeletal frame requires to be accurately fabricated and set out on site. It then provides a rigid and accurate template for the remainder of the construction. Another advantage is that, because the remaining construction is non-loadbearing, the sequence of construction following the steel frame becomes more flexible. The skeletal frame can also more easily accommodate variation in plan types.

Standard hot rolled joists, channels and angles are used for skeletal steel frames.

Steel wall-frame construction

In this method, the frame is arranged (as with timber-frame walls) as a series of cold-formed, light gauge steel studs, usually of channel profile with the flanges forming fixing faces for cladding or lining. As part or whole of corrosion-prevention measures, the steel strip is either pregalvanised before cold-forming, or of resistant quality such as 'Cor-ten', or a suitable paint scheme is applied.

The wall-frames may be built up on site from precast and preholed components bolted together, or prefabricated into wall panels as in timber frames. Racking resistance is conveniently provided by adding diagonal members between adjacent studs at suitable positions of each external wall elevation; by these means sheathing material is unnecessary. The wall studs may be one or two storeys in height.

The floor structure is preferably of steel construction, such as cold-formed or lattice beams, so avoiding differential movement from using dissimilar structural materials.

Cladding, lining and flooring materials are most conveniently attached to the frame by means of self-tapping screws, inserted by a light electric power drill with a special chuck attachment.

Additional mineral quilt lining may be required to augment the fire resistance afforded by the lining to the frame. In addition it may be necessary to avoid direct heat-conducting paths through the fixings to the steel frame, by using a series of isolator bars separately fixed to the frame, and then to fix the lining to the isolator bars. The mineral quilt also provides thermal insulation to external walls. It should be noted that the 'modified half-hour' fire resistance does not apply to a floor consisting of steel joists since they have no inherent fire resistance once the fire has penetrated the ceiling lining; instead the full half-hour fire resistance requirement applies even for a single-occupancy upper floor.

Sound insulation in steel-frame wall construction is provided by the principle of isolation in separating walls, and added mass for separating floors, similar to timber-frame construction.

Cold-formed steel sections are supplied to order by a number of specialist companies, who use either profile rolling machines or press-brakes for cold-forming from strip steel. Economy in design is achieved through the adoption of a minimum number of simple profiles, and a single thickness of strip. Profiles should have sufficiently wide faces to receive and fix cladding and lining, and to provide adequate structural connections between frame members.

VOLUMETRIC CONSTRUCTION

Volumetric or box construction enables the maximum degree of prefabrication to be applied to the structure, to interior decoration and fittings.

Units of houses or flats are usually made up of two or more boxes on each floor joined side by side on site. The boxes may be of precast concrete, or of timber or steel frame. The enclosed space in the units is prefinished, complete with staircase, wiring and piping, kitchen, bathroom fittings and built-in furnishings. Interior decoration has to be made good at the joints between the box units. Joints between boxes are arranged as far as possible at positions of internal partitions, to miss door and window openings, and to divide the space on each floor into a minimum number of boxes of comparable size and weight.

To provide protection from exposure to bad weather, soiling, theft and vandalism, and to provide rigidity for handling, transport and erection, the open sides or top of each box unit require secure temporary closure, bracing and weatherproofing. This often results in some duplication of wall, ceiling or floor elements between abutting floors and ceilings of contiguous box units.

Special consideration should be given to the problems of transport and erection. The maximum width of box units is restricted by the maximum permissible load-width on lorries with or without police escort during road transport; this means that usually two or three units are required to make up a flat or one storey of a house, whether the units are oriented with their side or end forming the front or rear elevations. Lifting attachments should be capable of being attached and detached above and not at the sides of the units, so as to facilitate the placing of abutting units. Large cranes with a long operating radius are required, and special lifting frames have to be provided so as to avoid inducing excessive horizontal forces on the box units during lifting. Finally, the design of the boxes, the application of linings and finishes, and the installation of fittings and furniture, require special care to ensure that the vibration, shock, twisting and warping of the units that inevitably occur during transport and erection do not cause damage, distortion and loosening; such effects cannot be fully investigated by design analysis or calculation, and are best tested by prototype trials under realistic conditions.

Volumetric units are usually loadbearing, but there are some systems that employ light non-loadbearing units supported by an open structural frame.

Any form of suitable cladding may be used with volumetric units. Brick cladding, which is often preferred for low-rise housing, can be site-applied after the units are erected.

Volumetric units are made by sponsoring companies which often buy in components supplied by others. Large indoor assembly areas are required for the units and for the storage of a large variety of materials and components. In addition, considerable areas for outdoor storage of complete units are also required.

Economy of design is achieved by maximising the amount of prefinishing and by distributing the weight of the contents of the units as evenly as possible. Clearly, with a limited number of standardised volumetric units, the flexibility in variety of plan type is somewhat restricted. Being heavy and voluminous, their economy is particularly sensitive to haulage distance.

REFERENCES

1. *British systems year book*, published annually by the System Builders Section of the National Federation of Building Trades Employers and the British Woodworking Federation
2. Deeson, A.F.L., *The comprehensive industrialised building systems annual*, Morgan-Grampian (1970)
3. *Safety guide for industrialised building*, System Builders Section of the National Federation of Building Trades Employers (1966)
4. *The Building Regulations 1976*, HMSO
5. *The Building Standards (Scotland) Regulations*, HMSO
6. *The Building Regulations (Northern Ireland)*, HMSO

7. *The London Building (Constructional) By-laws*, HMSO
8. *Drained joints in precast concrete cladding*, The National Building Agency (1967)
9. *Timber-frame housing: design guide*, Timber Research and Development Association (1966)
10. *Sound insulation and new forms of construction*, Building Research Station Digest 96, HMSO (1971)
11. *Sound insulation of lightweight dwellings*, Building Research Establishment Digest 187, HMSO (1976)
12. *Sound insulation of traditional dwellings*, Building Research Station Digests 102 and 103, HMSO (1971)

39 WATER AND SANITARY SERVICES

WATER SERVICES 39–2

SANITARY SERVICES 39–21

WATER AND SANITARY SERVICES 39

39 WATER AND SANITARY SERVICES

T.A. TOMPSON, FRSH, FIPHE
Partner, T.A. Tompson and Associates, Consulting Engineers

This section considers the Model Water By-laws, design considerations, hydrants, pumped systems, ballvalves, metering, sampling water supplies, water softeners, cold water storage and services, materials, connections to cisterns, electrolysis, bidets, sterilisation of installations, hot water storage and services, systems of sanitation, design considerations, traps, rainwater pipes, testing, and waste-disposal units.

WATER SERVICES

Model Water By-laws

It is not intended to itemise all the Model Water By-laws, but the following summary will give some idea of the content of those by-laws often encountered by the engineer. Reference should also be made to the actual by-laws in all cases of specific doubt arising as to the exact interpretation of any by-law(s).

- By-law 7 requires generally that no worn or damaged fitting be installed in such a manner as will cause or permit waste, undue consumption, misuse, erroneous measurement or contamination of water supplied by the undertakers, or reverberation in the pipes.
- By-law 7A relates to prevention of electrolytic action between dissimilar metals.
- By-law 8 requires generally that water in a service pipe shall not be connected with any other water, but permits a service pipe to discharge through a ballvalve not less than 152.4 mm above other water, not necessarily of identical purity to that in the service pipe.
- By-law 9 concerns the provision of drinking water.
- By-laws 10 and 11 concern frost and damage protection.
- By-law 12 requires pipes to be accessible for examination, repair and replacement within practical limits.
- By-law 15 specifies the minimum depth of pipes below ground.
- By-laws 18–22 concern specification of pipe materials.
- By-law 24 requires provision of a stopvalve to isolate the water supply to a building, such valve being located as soon as practicable following entry of the pipe into the building.
- By-law 25 concerns valving of distributing pipes fed by a cistern exceeding 18.2 litres capacity.
- By-law 27 concerns provision of standpipes.
- By-law 28 concerns water supplies to a drinking trough or drinking bowl for animals or poultry.
- By-law 29 requires screwdown pattern taps to comply with BS 1010, other taps to withstand a test pressure of 2068.428 kN/m^2 and if less than 50.8 mm nominal size to be constructed in a corrosion-resisting material.
- By-law 30 relates to British Standards for stopvalves and sluice valves, other stopvalves to withstand a test pressure of 2068.428 kN/m^2 and if less than 50.8 mm nominal size to be constructed in a corrosion-resisting material.
- By-law 32 concerns ballvalves.
- By-law 33 concerns accessibility of domestic storage cisterns for inspection and cleansing.

- By-law 34 concerns support and cover for domestic storage cisterns.
- By-law 35 concerns storage cisterns below ground level.
- By-laws 36 and 37 specify materials for storage cisterns.
- By-law 38 concerns capacity of domestic storage cisterns.
- By-law 39 concerns ballvalves.
- By-law 40 concerns storage cisterns of less than 4546 litres capacity.
- By-law 41 concerns storage cisterns of more than 4546 litres capacity.
- By-law 42 concerns the maximum run of hot water distributing pipe not subject to circulation (see Table 39.1).

Table 39.1 DEAD–LEGS ON HOT WATER SERVICES

Largest internal diameter of pipe	*Length* (m)
Not exceeding 19 mm	12.192
Exceeding 19 mm but not exceeding 25.4 mm	7.62
Exceeding 25.4 mm	3.048

- By-law 44 requires that no ballvalve shall be fitted for a hot water storage cistern.
- By-law 45 requires that a cold water feed pipe serving a hot water apparatus (not of the instantaneous type), a cylinder or a tank shall not serve any other fitting in addition.
- By-law 46 concerns the method of connecting any service pipe to a water heating apparatus.
- By-law 47 concerns mixing valves served by hot water distributing pipes.
- By-law 48 specifies materials for hot water distributing pipes.
- By-laws 49 and 50 specify materials for a hot water cylinder or tank.
- By-law 51 relates to capacity of hot water storage cisterns, cylinders or tanks.
- By-law 52 concerns outlets from and/or supplies to baths, wash basins, sinks or similar appliances.
- By-law 53 concerns the prevention of contamination of water supplies by means of back siphonage through any draw-off tap or other fitting. The provisions of this by-law may be broadly summarised as follows:

(a) The outlet of any tap supplying water to a sanitary fitting must be at least 12.7 mm above the lowest part of the top edge of the appliance.
(b) In the case of a tap or fitting incorporating a hand-operated hosepipe, an effective means must exist to prevent siphonage of water back through every pipe conveying water to the fitting.
(c) Where a tap discharges lower than the level mentioned in (a) above:
  (i) it must be served from a storage cistern or from a cylinder or tank having an open vent to the atmosphere;
  (ii) the connection to the storage cistern, cylinder or tank must not be less than 25.4 mm above the lowest part of the top edge of the appliance served; and
  (iii) any pipe serving such a tap must not serve any other fitting (other than a draining tap) that discharges at a lower level than that of the lowest part of the top edge of the first appliance.

- By-laws 54–58 relate to flushing cisterns.

Design considerations

In the majority of domestic installations, the incoming service pipe will commence from the Water Undertaking's main isolating valve, which is located immediately adjacent to the boundary of the premises. This valve is for use by the Water Undertaking only and should be secured in the 'open' position; a further valve should be provided for the occupier of the premises, preferably as soon as the service pipe rises from below ground level.

Choice of rising position for the service pipe to the storage cistern ballvalve at roof level will depend upon various factors, but it should be in an accessible position, although free from possible damage or attack by frost.

Pipework should be properly graded to prevent air locks and to assist draining down when required. Adequate British Standard pattern drain valves should be provided at all low points of the installation and also immediately inside the building, downstream of the isolating stopvalve.

Service pipes within a building should be properly insulated, both to prevent condensation occurring on the cold-pipe surface and to minimise temperature increase of the contents of the pipe. When installed below ground, service pipes should be covered by at least 762 mm to prevent damage to the pipe. This depth should also be sufficient to prevent frost damage during normal weather in the winter season. However, it has been known for frost to penetrate up to a depth of 1.067 m in some districts of the United Kingdom, so it is always preferable that the builder should check records of frost attack in the locality concerned.

Plug cocks or plug valves are not permitted to be used in service pipes (Model Water By-law 24(7)), and no draining tap shall be buried in the ground or so placed that its outlet is in danger of being flooded (Model Water By-law 26).

The service pipe is intended to serve a drinking-water tap at the kitchen sink, for drinking water and food preparation, and also the main cold water storage cistern, together with a feed and expansion tank when applicable as part of a central heating installation. All other cold water requirements should be served from the aforementioned cold water storage cistern in order that sanitary facilities remain in use during periods of mains failure.

Additional valves should be provided whenever economically possible, on pipes serving ballvalves or taps, to facilitate maintenance of washers.

The domestic water requirements of any premises must not be served from a joint source that also serves fire protection requirements of the premises, unless the service pipe is divided as shown in *Figure 39.1*, immediately after the Water Undertaking's main isolating valve. It is imperative that no fire protection appliance is inadvertently isolated by a valve used to isolate a domestic supply.

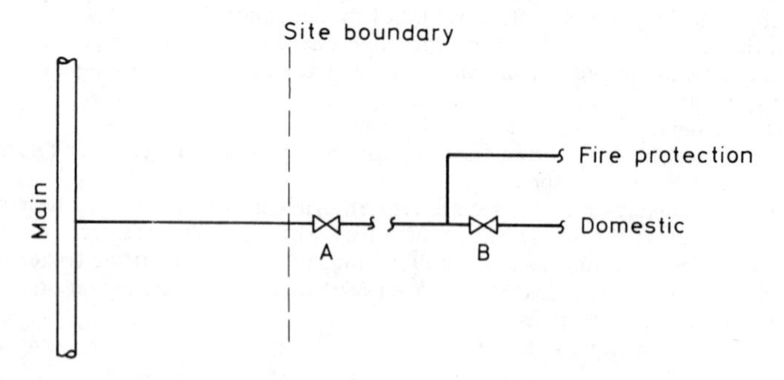

Figure 39.1 Division of service pipe (note that valve B should be inside the building)

When dealing with cold water services provided from storage, it is sometimes thought, erroneously, that the design of a cold water distribution system is simply a case of connecting up all the points that require cold water to a cistern that supplies it. In the ensuing text, it is hoped to show that many considerations are essential if the resultant installation is to be economic, efficient and practical.

It is advisable, in some cases essential, for pressure of hot and cold water supplies to a sanitary fitting to be approximately equal. Though it is possible to use brassware taps with a dual waterway, to prevent backing up of water into the lower pressure supply, it may be considered inconvenient to the user if pressures of hot and cold water supplies vary a great deal. Provision of intermediate storage levels for cold water throughout a tall building has a lot to commend it as a design principle and, whenever possible, hot water storage should be divided in a similar manner.

Thus the pressure of hot and cold water services in each section of the building is approximately equal. Intermediate storage also obviates the necessity for pressure-reducing valves, which may either prove faulty or have to be removed for maintenance, to the inconvenience of the building's occupants.

When routing distribution pipework to sanitary fittings, the engineer should give thought to the maintenance and/or continuity of supply in the installation. Vertical distribution of cold water by main pipelines, with shorter horizontal branches to fittings or groups of fittings, is considered preferable to horizontal distribution. While proper valving can greatly assist either type of system to function adequately, a vertical main, of perhaps larger diameter, can be more easily accommodated in a vertical duct than is the case with a similar pipe routed horizontally in a false ceiling. Maintenance of the pipeline in such a situation may involve disruption of many rooms or departments. In addition the effect on room heights (and possibly the height of the building) that false ceilings impose must be considered.

Hydrants

Depending upon the size of any housing development, the builder may be required to provide hydrants on the main service pipe, for fire protection purposes. Hydrants may either be of a screwdown or sluicevalve pattern and must comply with the provision of BS 750 or 336.

Hydrants are usually connected to an underground water main and are arranged so that a separate standpipe may be connected to them.

In some instances a pillar hydrant may be fixed permanently to them. Pillar hydrants should be 63.5 mm minimum bore, but where two- or three-way outlets are fitted, the bore of the standpipe should be larger, i.e. a hydrant with a double outlet should have a bore equivalent to twice the area of 63.5 mm bore pipe and a hydrant with three outlets should have a bore equivalent to three times that area.

Standpipes should be of a minimum 63.5 mm bore. Wherever possible, there should be at least two outdoor hydrants in addition to any indoor hydrants within the sites of the building concerned.

Outdoor hydrants should normally be fixed not less than approximately 6 m from the buildings they are intended to protect. Care should be used in siting these hydrants that use of them will not cause obstruction to exits or to other fire appliances used in fighting a fire. The hydrants should be so located that they provide complete protection to the premises. Where it is necessary to guard against damage by frost, an automatic frost valve should be fitted.

To prevent loss of time in locating an outside hydrant, an indicator plate complying with the fire authority's requirements should be fixed not less than 914 mm nor more than 1.83 m above ground level near to the hydrant.

Pumped systems

It may not always be possible for a water supply pressure to be sufficient to serve drinking water and cold water storage requirements for particular premises, as in the case of a block of flats.

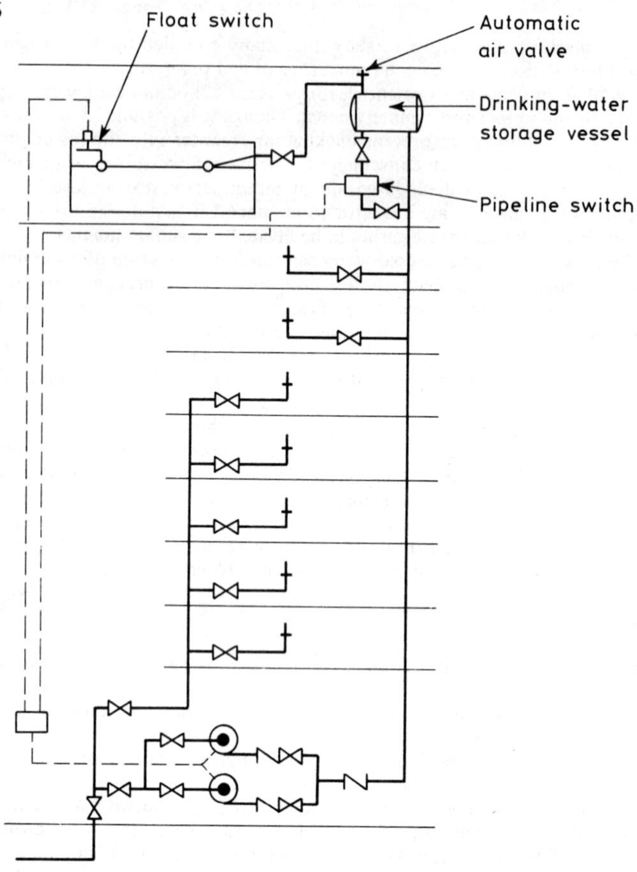

Figure 39.2 Remote-controlled pump system. Alternative: lower floors within reach of mains pressure may be served from right-hand riser, with size of drinking water storage vessel adjusted accordingly to protect all levels against mains failure

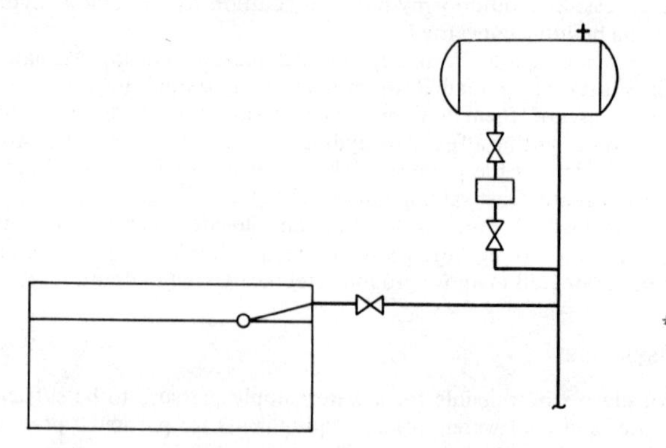

Figure 39.3 Incorrect method of serving ballvalve

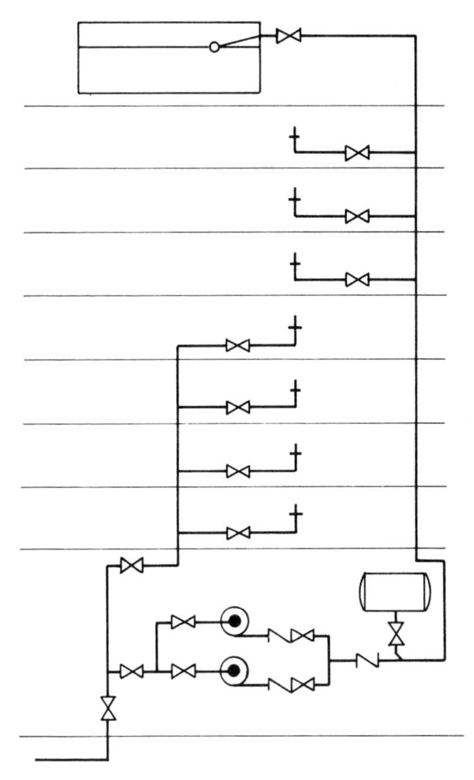

Figure 39.4 Auto-pneumatic pumped system with main storage at roof level

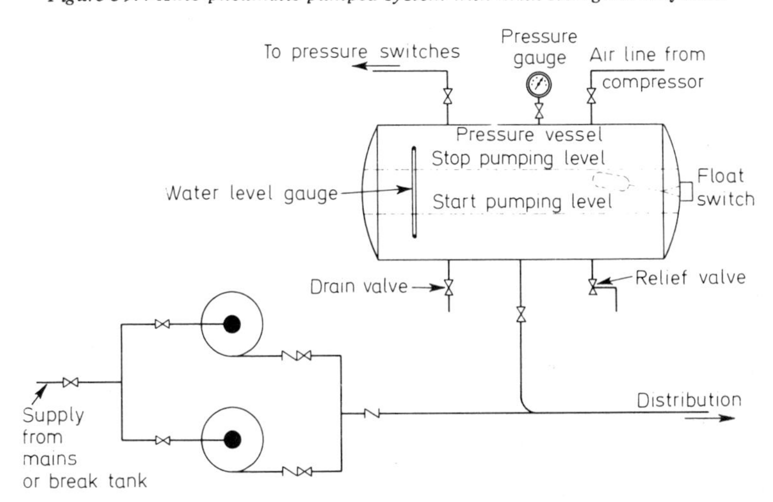

Figure 39.5 Auto-pneumatic system: schematic detail

There is a practical limit to which the mains pressure can be increased in a distribution system serving any locality where large areas of low-rise dwellings are to be served, especially when the density of such developments is to be increased beyond the original density. An increase in demand on the available water supply causes a progressive reduction in the pressure of that supply.

The detailed method of increasing the pressure of a water supply will vary according to the circumstances, but generally there are two main methods that are used to raise water in a service pipe to above its normal reach; both methods involve pumps but the control of the pumps is entirely different with each method.

The first method (*Figure 39.2*) comprises duplicate pumps (one duty, one standby, wired for auto changeover) which may be operated either by a pipeline switch located below the drinking water storage vessel, or by the float switch located within the main storage cistern.

The second method (*Figures 39.4 and 39.5*) utilises a pressure vessel in addition to the pumps, the water being lifted either by a cushion of compressed air in the pressure vessel or by the pumps. No remote controls are required as local pressure switches operate the duty pump.

Ballvalves

PORTSMOUTH PATTERN

These ballvalves (covered in BS 1212), used in conjunction with floats of various sizes, can be used for low pressure supplies up to 275.79 kN/m^2, medium pressure supplies up to 689.476 kN/m^2 or high pressure supplies up to 1378.952 kN/m^2. Reference should be made to BS 1212 for details of component assemblies for each pressure.

Portsmouth-pattern ballvalves are probably the most used ballvalves in the United Kingdom, bearing in mind all the flushing cisterns and domestic storage cisterns that incorporate them. The pattern has been tried over many years and is a reliable one for most general purposes. Portsmouth-pattern ballvalves are manufactured in nominal sizes of 9.5 mm, 12.7 mm, 19 mm, 25.4 mm, 31.8 mm, 38.1 mm and 50.8 mm with six patterns of body: I, II, III, IV, V and VI. The 9.5 mm and 12.7 mm sizes are made in two body forms, patterns I and II; 19 mm size in pattern III; 25.4 mm size in pattern IV, 31.8 mm and 38.1 mm sizes in pattern V; and 50.8 mm size in pattern VI.

When larger-diameter supply and/or high-pressure connections to storage cisterns are required, it can readily be seen that the force needed to close the valve would be far greater than the normal Portsmouth-pattern ballvalve could provide. It is therefore necessary to utilise an equilibrium (or balanced) type of ballvalve, which, by virtue of its design, counterbalances the hydraulic load on the valve element by a corresponding load in opposition.

EQUILIBRIUM PATTERN

Equilibrium-pattern ballvalves are constructed so as to provide balanced pressures surrounding the valve element, making operation of large ballvalves more effective than the Portsmouth-pattern piston action could be under similar conditions.

Equilibrium-pattern ballvalves are available in a wide range of types and sizes. Some types, for example, are manufactured in sizes up to 457.2 mm nominal bore. These valves are available for use in conjunction with working pressure of, in some cases, up to 1378.952 kN/m^2. When specifying equilibrium-pattern ballvalves, the builder should ensure that the ballvalve used for any purpose is approved by the manufacturer for that application.

DELAYED–ACTION PATTERN

Probably one of the best known of these is the 'Arclion' valve although there are other delayed-action valves available. The 'Arclion' Type A valve incorporates a ballvalve conforming to BS 1212 and is manufactured in sizes from 19 mm to 50.8 mm bore, which, at inlet pressures of from 68.95 kN/m² to 413.7 kN/m², will discharge from 0.4 litre/s to 6.7 litre/s approximately. The Type B valve incorporates a ballvalve of 'Underhayes' equilibrium pattern and is manufactured in the same sizes and discharge limits as the Type A valve. Type C is the suspended-pattern valve incorporating a ballvalve of 'Firmin' straight-through equilibrium pattern; the sizes and discharges are the same as for Types A and B. The Type D valve again is a suspended-pattern valve incorporating a ballvalve of equilibrium pattern, either angle or straight-through pattern, and is manufactured in sizes from 63.5 mm to 152.4 mm bore. At inlet pressure of from 34.47 kN/m² to 172.37 kN/m² these valves will discharge from 8.8 litre/s to 99 litre/s approximately. The 'Arclion' valve is accepted by the Metropolitan Water Board and other Water Undertakings where the use of delayed-action ballvalves has been approved.

The valve operates as follows. The unit incorporates a normal ballvalve with the float operating in a subsidiary container within the main storage cistern. A second float below the base of the container operates a quick-opening valve which provides a straight open/shut control. The main ballvalve remains open until the water level has reached the top of the subsidiary container, which fills immediately and shuts off the valve cleanly. This valve does not re-open until sufficient water has been drawn off to lower the secondary float at the bottom of the container, thus emptying the subsidiary container and allowing the ball control valve to open freely at once.

Metering

In order that fair assessment of water consumption by the occupants of any premises can be made, it is often necessary to install a suitable meter in the service pipe to record the quantity of water passing through it. An appropriate charge by the Water Undertaking is then made, following periodical reading of the meter. In practice, however, most Water Undertakings charge a flat rate to domestic consumers of water and meter only those supplies serving large-consumption premises such as factories, hotels, hospitals etc.

Where a metered supply forms the sole source of water to any premises, the Water Undertaking may permit the installation of a valved bypass to the meter. Thus the supply to the premises can be maintained during the removal of the meter for any purpose by the Water Undertaking. Apart from this instance, most Water Undertakings consider that a valved bypass is unnecessary and that there is always a possibility of the valve on the bypass being left in the open position by unauthorised persons.

It is normal practice for fire fighting or fire protection services using water to be unmetered, owing to the possible blocking of the water supply by the meter at a critical period. Accordingly, any such services should be fed from a separate main, although this may be a branch from the general-purposes supply taken prior to any meter or isolating valve, except as indicated in *Figure 39.1*.

Many types of meter are available for the varied circumstances in which they are to be used, and it is usual for the Water Undertaking to provide the appropriate meter on a rental basis for installation by the responsible contractor. The meter is then maintained by the Water Undertaking. It is probably because of this practice that domestic premises are not mechanically metered, as the resultant countless small meters would be uneconomic in view of the small quantity of water consumed.

Sampling

A builder may sometimes be required to arrange for samples of a water supply to be taken for analysis. The water supply may be from a piped main supply or from natural sources (to determine its suitability for human consumption or industrial use, or to establish the existence of contamination in an installation). Samples may also be specified as necessary by the builder concerned with a new installation of services pipework. These samples would be taken from various points in the installation following chlorination of the pipework, fittings and cisterns or other equipment containing the water in question.

It will be readily appreciated that, unless such samples are very carefully taken, the results obtained by the analyst following his tests could be rendered useless. The Ministry of Health Report No. 71, *The bacteriological examination of water supplies*, to which further reference should be made, recommends that all water samples taken should be examined as soon as possible after collection and certainly within six hours of collection. The changes in coliform and *Bact. coli* content can be delayed by packing the sample in ice during transit.

Water softeners

The term 'hardness', when applied to water supplies, actually relates to two types of hardness, temporary and permanent, which are collectively termed total hardness. Hardness is measured in terms of 'parts per million' of calcium carbonate to water (see Table 39.2).

Table 39.2 A GUIDE TO THE CLASSIFICATION OF WATER SUPPLIES IN TERMS OF HARDNESS

Classification	*Parts per million*
Soft	0–50
Moderately soft	50–100
Slightly hard	100–150
Moderately hard	150–200
Hard	200–300
Very hard	above 300

Temporary hardness is caused by the presence of calcium carbonate and magnesium carbonate, which are held in solution by combination with carbonic acid in the form of bicarbonates. Permanent hardness is caused by the presence of calcium sulphate and magnesium sulphate, and sometimes by their nitrates and chlorides.

Treatment for removal of temporary hardness in small quantities of water may be by the simple process of boiling the water. This causes the bicarbonates to break down into carbon dioxide and carbonates. The carbon dioxide is given off to the atmosphere with the steam, and the carbonates, which are virtually insoluble, are deposited as scale on the sides of the containing vessel.

Treatment for larger quantities of water may be effected by various means, but it can be stated generally here that this treatment usually consists of the lime process, with or without the addition of soda, or by ion exchange (also referred to as the base-exchange process). The treatment for permanent hardness may be effected by the addition of carbonate of soda, which converts the sulphates or chlorides into carbonates, or by ion exchange.

The generally accepted form of domestic water treatment for excessive hardness is by means of a small ion-exchange unit. These are available in a wide range of capacities, the capacity being considered as the quantity of water that can be treated by the unit before the softening material is exhausted. The material must then be regenerated by passing a brine solution through it to render it capable of absorbing a further quantity of hardness. Naturally, the capacity of the unit will depend on the classification of the water supply (i.e. the p.p.m. of calcium carbonate); the harder the water supply, the lower the capacity of the softening unit. However, as a guide, units are available for treatment from approximately 4500 litres to 22 500 litres per regeneration. Each unit may be used and regenerated about seven times before the salt in the brine tank needs replenishing. Operating pressures vary from 172.37 kN/m^2 to 517.11 kN/m^2 and the units are acceptable for connection direct in the service pipeline.

Automatic water softeners are available, which initiate their own regeneration process when required. A typical unit will treat approximately 3400 litres of water of average hardness per regeneration, and it is claimed that the only maintenance required is the addition of 50.8 kg of salt after every fifteen regenerations. This unit has been designed for incorporation into a modern kitchen layout; the top is 914.4 mm above floor level, and the cabinet is of glass-fibre construction.

Cold-water storage and services

Provision of storage facilities for cold water used for general purposes, other than drinking, is considered as normal practice in the United Kingdom. While desirable, it is, however, not always mandatory, being provided usually for the convenience of the building occupants in the event of failure of the main supply. Storage may also be required in order to reduce maximum demand on the main supply or to limit the water pressure on sanitary fittings. Some Water Undertakings will permit their service pipe to be used as a general cold water supply within the limitations of their by-laws. Failure of the main supply will render such installations useless pending resumption of the water supply.

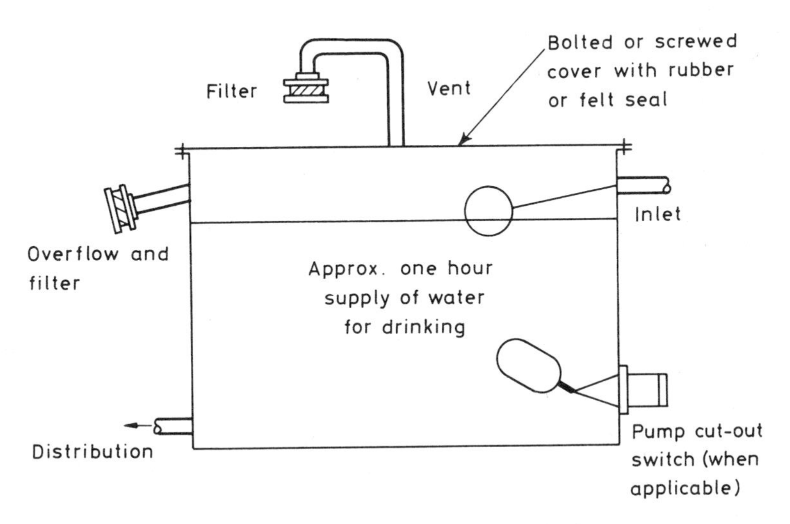

Figure 39.6 Typical drinking-water storage cistern

Once storage of cold water becomes an accepted part of the cold water services installation design, by-laws are in existence concerning standards or connections relative to such storage. In the United Kingdom it is normal practice to allow storage provision of cold water to the extent of one day's normal usage by the installation. Drinking-water storage should be as illustrated in *Figure 39.6*.

Table 39.3, based on information extracted from CP 310: 1965, *Water supply*, lists suitable storage rates for cold water other than drinking water, for various types of building. These rates apply to actual water storage, not nominal storage.

Table 39.3 PROVISION OF STORAGE TO COVER TWENTY–FOUR HOURS INTERRUPTION OF SUPPLY, BASED ON OCCUPANCY (extract from CP 310: 1965, reproduced by permission of the British Standards Institution)

Type of building		*Storage* (litres)*
Dwelling houses and flats	per resident	91
Hostels	per resident	91
Hotels	per resident	136.4
Offices without canteens	per head	36.4
Offices with canteens	per head	45.5
Restaurants	per head per meal	6.8
Day schools	per head	27.3
Boarding schools	per resident	91
Nurses' homes and medical quarters	per resident	113.6

*Conversions to metric equivalents have been made by the author.

Model Water By-law 38 requires generally that, where a storage cistern is provided in a house, i.e. premises separately occupied as a private dwelling, it shall have a storage capacity of not less than 113.65 litres if not used as a feed cistern. Where the storage cistern is used as a feed cistern it shall have a capacity of not less than 227.3 litres. When the storage is provided by means of two or more cisterns, their total capacity shall be the same as the quantities for a single storage cistern.

Consideration should always be given by the builder to the division of water storage into separate volumes, either by using separate cisterns or, in the case of sectional storage cisterns, by provision of a division plate. While this practice usually requires connections to and from each storage compartment, these can greatly assist maintenance operations by enabling part of the storage to function temporarily while the remainder is drained down and the cistern cleaned. It is common practice to connect all similar connections by means of 'header pipes' suitably valved to permit any section of storage to be isolated. It should be noted that any water storage required to serve flushing valves should be totally separate from that serving domestic distribution pipework.

Water-storage accommodation should not be subjected to frost attack and must be suitably protected. Where the storage is external to the building, proper insulation should be applied, complete with weatherproofing to the outside of the cistern(s). Where the storage is within the building, the room containing the cistern(s) should be maintained above a minimum temperature of 7.2°C.

Storage cisterns should not be installed in such a location that there is a risk of their being flooded or liable to contamination or pollution. When these requirements can be met, and a cistern is partially or wholly sunk in the ground, unless it is a concrete cistern designed and constructed in accordance with the relevant recommendations in BS 5337, space should be provided around and beneath it to facilitate maintenance and the detection of any leakage. CP 310: 1965 recommends that any cistern partially or wholly sunk in the ground should have its inlet pipe discharging into the air not less than 152.4 mm above its top edge, or alternatively it should be a closed vessel with a tightly fitting access cover bolted or screwed in position and any air inlet and overflow pipes suitably screened.

Materials

MILD STEEL

For storage cisterns of up to 4546 litres nominal capacity or 3364 litres actual capacity, it is possible to use galvanised mild steel cisterns complying with BS 417. Although this British Standard includes details of size and position of connections to the cisterns covered, it is normal practice to have the cisterns delivered to the site without any holes for connections. The holes are subsequently cut by using a proper tool designed for the purpose.

Every precaution should be taken to protect the galvanising from damage by either physical or chemical causes (e.g. electrolysis). All swarf or filings produced by the cutting of holes for connections to the cistern should be removed before the cistern is filled with water.

Galvanised mild-steel storage cisterns should not be used to contain water with a pH value of less than 7.3 and with a temporary hardness of at least 210 p.p.m. (see Table 39.4).

Table 39.4 USE OF GALVANISED METAL IN RELATION TO THE pH VALUE OF THE WATER

pH *value*	*Use of galvanised metal*
7.2 or under	Galvanised metal should not be used.
7.3	Galvanised metal can be used when the temporary hardness is greater than: 210 p.p.m.
7.4	150 p.p.m.
7.5	140 p.p.m.
7.6	110 p.p.m.
7.7	90 p.p.m.
7.8	80 p.p.m.
7.9 – 8.5	70 p.p.m.

Whether or not the cistern is to be used for drinking-water storage in the manner described earlier, it should be painted internally with two coats of non-toxic bitumastic paint before being filled with water.

ASBESTOS CEMENT

Asbestos-cement storage cisterns are manufactured in accordance with BS 2777 and are available in twelve sizes ranging from approximately 27 litres nominal capacity to 818 litres. The cistern can be provided complete with a one-piece lid having a bold lip on two sides to retain the lid in position so as to exclude dirt (as required by the Model Water By-laws) when the cistern is used to store drinking water. An asbestos-cement storage cistern is available with a nominal storage capacity of approximately 1136 litres and can be provided with a two-piece lid if required.

Asbestos-cement cold-water storage cisterns are immune from electrolytic action and are unaffected by hard or soft water. The smooth internal finish and the wall thickness combine to resist the effects of ice formation in the cistern if it is installed in an exposed position.

Drilling the walls of the cistern for pipework connections is a simple operation, but it is recommended that holes are not drilled less than 101.6 mm from the base of the cistern. In the case of the 1136 litre cistern, holes should not be drilled less than 203.2 mm from the base of the cistern. Normal unions are required with the threaded portion of sufficient length to accommodate the wall thickness and usual sealing washers. The latter are used in pairs, i.e. one washer of hard quality together with one of soft quality. One pair should be positioned on the inside and a further pair on the outside face of the cistern.

GLASS-REINFORCED POLYESTER

Glass-reinforced polyester storage cisterns are available in a range of sizes, from 32 litres nominal capacity to 341 litres. Storage cisterns manufactured in this material are accepted by the National Water Council. The smaller cisterns may be used as feed and expansion cisterns and are unaffected by hot-water discharges into them. The largest of this kind of cistern has been so designed that it can be passed through a 609 × 609 mm opening and thus can be used for replacement of an existing cold-water storage cistern or cisterns. This cistern is supplied complete with a coated metal strap, which should be fixed before the cistern is filled with water. Covers are also manufactured in this material of suitable sizes to be used in conjunction with the range of cisterns.

Glass-reinforced polyester cisterns should be supported across the whole of the base, and adequate support is required to adjacent connecting pipework to prevent strain on the walls of the cistern.

PLASTICS

Polythene cisterns are available for use with cold water and are manufactured in accordance with BS 4213. They are produced in either cubic or cylindrical form, with capacities ranging from 32 litres nominal capacity to 318 litres. Lids are also available for each size cistern.

Polythene cisterns are useful as replacements for domestic cisterns and, being pliable, may be passed through restricted openings that would prevent other materials being used. They have been tested and approved by the National Water Council. Where it is required that the cistern be used as a feed and expansion cistern, polypropylene should be used as this is more suitable for use with hot water.

Plastic cisterns should be supported across the whole of the base, and adequate support is required to adjacent connecting pipework to prevent strain on the walls of the cistern. Each cistern is normally supplied complete with a metal backplate to be fixed in conjunction with the ballvalve. The manufacturers recommend that PTFE tape is used to seal all threads. Linseed-oil based compounds, such as boss-white, must never be used with plastic cisterns.

Connections to cisterns

Connections to or from storage cisterns generally comprise the ballvalve, wash-out pipe, overflow and/or warning pipe(s) and distribution pipe(s).

BALLVALVE

Model Water By-law 39 requires that the following conditions be fulfilled:

(1) Every pipe supplying water to a cold-water storage cistern shall be fitted with a ballvalve or shall have some other not less effective device for controlling the inflow of water so designed as to prevent overflow:

Provided that where two or more cold-water storage cisterns at the same level are connected together this paragraph shall not apply to a pipe used only to connect one cistern to another.

(2) Every such pipe, whether fitted with a ballvalve or not, other than a pipe used to connect one cistern to another, shall be fitted in such a position that it discharges at a level higher than the overflowing level of the overflow pipe (or, if there is more than one overflow pipe, the highest overflow pipe) by not less than the diameter of the said overflow pipe, unless there is an effective means of preventing the siphonage of water back through the inlet.

(3) Where a ballvalve is fitted to a cistern, the size of the orifice, the size of the float and the length of the level shall be such that, when the float is immersed to an extent not exceeding half its volume, the valve is watertight against the highest pressure at which it may be required to work.

(4) Every ballvalve shall be securely and rigidly fixed to the cistern which it serves.

A ballvalve should be so sized that it will pass a sufficient quantity of water to fill the storage cistern(s) in a given period of time. This period of time can only be assessed by the builder after due consideration of all the factors involved. For example, when storage capacity is low, a filling period of, say, one hour may be acceptable. However, a similar filling period applied to a large storage capacity may require unnecessarily large supply pipework and ballvalve(s), whereas an allowance of, say, four hours would be far more practical. It may also be that the service-pipe capacity or pressure is such that the flow rate to replace storage is limited to protect the supplies to other premises.

It is not necessary for the ballvalve to be capable of passing an equal flow rate to that calculated for the distribution from the cistern. Peak flows from the cistern are likely to be of short duration only and can be met by the storage reserve in the cistern. Make-up of the water used can be at a slower rate. In certain circumstances, it may be possible to utilise a connection to the storage cistern(s), sometimes called a rapid-fill connection, which comprises an open-ended service pipe with a manually operated valve to facilitate rapid replacement of storage if required. Such a connection must be in full accordance with the Water Undertaking's by-laws.

After making due allowance for the loss in head caused by the operation of a ballvalve, a minimum pressure of at least 6.1 m head should be available in the service pipe to the ballvalve.

WASH–OUT PIPE

A wash-out pipe may be fitted to the storage cistern(s) for use on occasions when it is required to drain the water storage quickly, so that maintenance may be carried out. The pipe is terminated at a suitable point for the disposal of the water to drain, but in such a manner that it complies fully with the Water Undertaking's by-laws for the prevention of contamination.

When large quantities of water are concerned, it may be thought unduly wasteful to discharge purified water direct to the drainage system. In such circumstances, it is better to arrange the system in such a manner that part of the storage capacity may be drained by normal usage before being isolated for maintenance. The remaining storage then provides water to the system. Obviously the maintenance procedure to be adopted for the installation will also have a bearing on draining-down facilities. When maintenance staff are to be available on a permanent basis, it is practical to alternate usage of storage cisterns in the manner described above. When maintenance is intended to be carried out by staff visiting the premises for that purpose only, it must follow that the draining-down procedure is required to be as rapid as possible, e.g. by means of a wash-out pipe connection to the storage cistern(s). A wash-out connection from a storage cistern or cisterns must, by definition, be at the lowest possible level in the cistern, with minimum projection into the cistern(s).

OVERFLOW AND/OR WARNING PIPE(S)

An overflow pipe from a cistern is one that permits water to be conveyed away from the cistern to prevent submersion of the ballvalve in the event of its failure for any reason to stop the inflow of water to the cistern. This is its prime object, namely, to protect the inflowing water from contamination or back-siphonage. Indirectly, of course, it also prevents local damage by the water flowing over the top of the cistern.

When the overflow pipe is so discharged or terminated as to be readily seen to be discharging water, it is termed a 'warning pipe'. Most Water Undertakings require that such pipes should create a nuisance when discharging, to ensure immediate action in preventing undue waste of their water.

DISTRIBUTION PIPE(S)

Distribution pipes are generally of two types: those serving the cold-water distribution pipework system and those serving the feed pipework to hot-water apparatus, for domestic purposes. All distribution pipes should connect to the storage cistern as far away as is practicable from the ballvalve, to minimise stagnation within the cistern. Each connection should be made in such a position as to leave a space of 50.8 mm between the bottom of the storage cistern and any pipe drawing water from the cistern. This space is advisable to accommodate silt and any other solid matter entering the cistern, and prevents it being passed into the distribution system. When connections are made through the base of the cistern, the pipe should continue into the cistern by a similar amount.

Some difference of opinion exists as to the relative positions of connections to the cistern for cold-water distribution or cold-water feed pipework. Some engineers maintain that the feed pipe to hot-water apparatus should be connected to the cistern at a level lower than cold-water distribution pipework. They argue that, by so doing, a warning is given to the building occupants by the failure of the cold-water services, that the storage is emptying, and thus the feed to any hot-water apparatus is to some extent 'protected'. On the other hand, there is much support for the view that it is preferable for the hot-water service to fail before the cold water, especially where mixer fittings are involved. It can readily be seen that, unless this happens, possible scalding can occur to anyone using such an appliance. In the light of such variance of opinion, the builder can only seek the views of the relevant Water Undertaking and be guided by their requirements.

Electrolysis

Electrolysis may be defined as 'chemical decomposition by action of an electric current'. This action occurs whenever two dissimilar metals are placed in contact by a conducting medium (e.g. water). The speed of the action varies, and one of the best known examples is between zinc and copper. When these two metals are connected, as in the instance of a galvanised cistern and copper pipework, the zinc is decomposed and enables the steel to be attacked by rust, causing eventual failure of the cistern. Protective painting of the cistern can reduce the electrolytic action but will not guarantee its prevention, owing to the possibility of the paint being scratched at any time.

It is possible to place metals in a list, referred to as the electrochemical series, from which the relative reactions of the metals can be assessed. Each metal corrodes the one below it in the list. The further apart any two metals are in the list, the greater the corrosion.

The electrochemical series of metals

1. Gold	7. Tin
2. Platinum	8. Nickel
3. Silver	9. Iron
4. Mercury	10. Zinc
5. Copper	11. Aluminium
6. Lead	12. Magnesium

Bidets

Model Water By-law 53 requires that:

> Every draw-off tap or other fitting (other than the flushing pipe of a flushing cistern) which discharges water into a bath, wash basin, sink or similar appliance shall be fitted in such a position that it cannot discharge at a level lower than 12.7 mm above the lowest part of the top edge of the appliance:
>
> Provided that:
>
> (a) this by-law shall not require a fitting which incorporates a hand-operated hosepipe or to which such a hosepipe is attached to be so fitted that it cannot discharge through the hosepipe at a level lower than that level, if there is an effective means of preventing the siphonage of water back through every pipe conveying water to the fitting: and
>
> (b) this by-law shall not require any fitting to be so fitted that it cannot discharge at a level lower than that level, if every pipe conveying water to that fitting:
>
> (i) draws water only from a storage cistern or from a cylinder or tank having a vent open to atmosphere; and
>
> (ii) is connected to the cistern, cylinder or tank at a level not less than 25.4 mm higher than the level of the lowest part of the top edge of the appliance; and
>
> (iii) does not convey water to any draw-off tap or other fitting (other than a draining tap) which discharges water at a level lower than the last mentioned level.

Any bidet incorporating a flushing rim and ascending spray, with a diverting valve or fitment, is subject to adequate precautions to prevent siphonage of water from the bidet into the service pipework.

When all the bidets are at one level, they may be served from a storage cistern open to atmospheric pressure. The Water Undertaking may permit a suitable connection from the main storage cistern or may recommend a separate storage cistern for such appliances.

When the bidets are at different levels, a separate service from the storage cistern for each level served may be required. Should the appliances be located at different levels in a multi-storey building, it may be preferable to suitably locate a number of sub-storage cisterns at intervals as required by the distribution of the appliances, rather than a large number of distribution pipes being routed downward through the building.

It will be appreciated that, when dealing with domestic premises, where a bidet is incorporated in the bathroom this should have a separate pair of service connections and storage to those serving the other fittings.

Perhaps it should be mentioned here that the practice sometimes found of 'looping' the service connections up above the bidet, with an automatic air valve at the top of the loop, is not advised. Experience has shown, for example in hospitals, that, owing to the spitting action of the automatic air valve as it operates, staff may wrap a cloth or something else around the valve, rendering its action ineffective.

Sterilisation of installations

Sterilisation of a completed installation is an important aspect that is too frequently given insufficient attention or is even ignored. A suitable means for sterilisation of the installation should be clearly specified, and should be carried out if necessary with a representative of the Water Undertaking also in attendance.

Clause 409 of CP 310: 1965 relates to this subject and is reproduced below:

(a) All mains and services to be used for water for domestic purposes should be thoroughly and efficiently sterilised before being taken into use, and after being opened up for repairs.

(b) The sterilisation of mains should be carried out by specialists. Service pipes should, if possible, be sterilised, together with the mains.

(c) Storage cisterns and distributing pipes can be sterilised as follows:

The cisterns and pipes should first be filled with water and thoroughly flushed out. The cistern should then be filled with water again and a sterilising chemical containing chlorine added gradually while the cistern is filling, to ensure thorough mixing. Sufficient chemical should be used to give the water a dose of 50 parts of chlorine to 1 million parts of water. If ordinary 'bleaching powder' is used, the proportions will be 0.68 kg of powder to 4546 litres of water; the powder should be mixed with water to a creamy consistency before being added to the water in the cistern. If a proprietary brand of chemical is used, the proportions should be as instructed by the makers. When the cistern is full, the supply should be stopped, and all taps on the distributing pipes opened successively, working progressively away from the cistern.

Each tap should be closed when the water discharged begins to smell of chlorine. The cistern should then be 'topped up' with water from the supply pipe and with more sterilising chemical in the recommended proportions. The cistern and pipes should then remain charged for at least three hours, whereupon a test should be made for residual chlorine; if none is found, the sterilisation will have to be carried out again. Finally, the cistern and pipes should be thoroughly flushed out before any water is used for domestic purposes.

Other methods may be used, and the builder is recommended to seek the views of the Water Undertaking concerned with any particular premises.

Hot-water storage and services

In the absence of any specific requirements being given, storage of hot water may be assessed using Table 39.5, which has been based on data contained in CP 342: 1950, *Centralised domestic hot-water supply*.

Table 39.5 ASSESSMENT OF HOT–WATER DEMAND AND STORAGE (BASED ON DAY OF HEAVIEST DEMAND DURING WEEK) (extract from CP 342: 1950, reproduced by permission of the British Standards Institution)

Building	*Max. daily demand per person* (litres)*	*Storage per person* (litres)*
Dwelling houses:		
low rental	114	see note †
medium rental	114	45.5
high rental	136	45.5
Flats (blocks):		
low rental	68	22.7
medium rental	114	32
high rental	136	32
Hostels	114	32
Hotels:		
first class	136	45.5
average	114	36.4
Offices	14	4.5
Pavilions, sports (with spray-type showers)	36.4	36.4

* Conversions to metric equivalents have been made by the author.
† Storage normally a minimum of 114 litres with a 4-hour heat-up period.

For small, single-family dwellings of about 88.26 m² floor area, CP 342 requires that there should be a hot-water storage vessel of not less than 113.65 litres actual capacity, to enable 13.64 litres of hot water to be drawn off at the kitchen sink at a temperature of not less than 60°C immediately prior to two baths being taken in succession. Unless the storage vessel is accommodated within the linen cupboard, some means of warming this is also recommended. The contents of a hot-water storage vessel in small, single-family dwellings are required to be raised from 10°C to 65.5°C in four hours while the primary heat source is also meeting other loads upon it.

In the case of larger premises, consideration should be given to using a number of storage vessels to make up the total storage, in order to facilitate maintenance without interruption of hot-water supplies.

Generally, it is recommended that for normal usage hot-water should not be stored at a temperature exceeding 65.5°C. The heating-up or recovery period should normally be in the region of two to three hours, depending on factors relating to any particular premises.

While distribution of hot water follows that of cold water in many respects, it is subject to certain additional considerations, such as the requirement of an open vent from what is termed 'the secondary circuit'. This has to be terminated over the feed cistern or discharged through a three-way valve in such a manner that, when the valve is closed to the vent, the storage vessel is open to the atmosphere through the third outlet.

There are also factors concerning the maintenance of a suitable hot-water supply temperature involving additional 'secondary return' pipework, which ensures circulation of secondary water and replaces cooled water with hot water. This circulation is more or less a continuous process, in order to offset the effect of heat losses throughout the distribution system. Secondary circulation may be effected by gravity wherever possible, but in view of the pipe sizes often involved in gravity circulations, it is quite common practice to use a small pump.

CP 342: 1950 requires that no cold-feed pipe shall be less than 25.4 mm in diameter.

It is customary to locate the secondary circulation pump in a valved bypass to the main flow, with a non-return valve parallel to it. During periods when there is no draw-off, water is constantly circulated throughout the secondary circuit. When draw-off takes place, this invariably requires a greater flow than that passing through the circulating pump, so the non-return valve opens. On cessation of flow, the non-return valve is closed and circulation continues.

An advantage of locating the pump in a bypass is that it can be removed for maintenance or replacement without interruption of the hot-water services. Location of the pump in the secondary flow as it leaves the storage vessel (but after the open vent is taken off) also enables the energy of the pump to be absorbed by the secondary circuit to a large extent before the water returns to the storage vessel. There will, of course, still be some pressure available at the point of connection between the secondary return and the storage vessel, but this should be minimal if the pump has been correctly sized.

It may be thought that it would be simpler to locate the circulating pump in the secondary return as it approaches the storage vessel; if it is so positioned, however, the 'head' or energy provided to the hot water will cause it to pass up the open vent and discharge into the feed cistern. This provides an easier path for the release of energy than driving the water around the secondary circuit.

Proper attention should be paid to the need for proper grading of the pipework in order to obviate air-locks in the system and to assist draining-down operations. A recommended minimum gradient of 1 : 240 should be present in all pipes.

Due regard must also be given to the expansion and contraction that is inherent in all hot-water systems. The pipes must be adequately supported but free to move without placing strain on any branches or fittings. The notching of timber joints to enable hot-water services to be installed within the floor is not recommended and may irritate the occupants of the building because of creaking noises as the pipes move, unless wrapping is provided.

Full-way gatevalves are considered suitable for normal regulation and control purposes but are not considered suitable for positive isolation purposes.

Regulating valves should be provided on all main and branch secondary return pipes in order that the system will be balanced on completion. Where these are permitted by the Water Undertaking, lubricated plug-type valves are recommended for hot-water services as isolation valves on pipelines over 50.8 mm diameter. These should be labelled to indicate that they must be slowly operated to avoid damage by 'water hammer' and should clearly show closed and open positions. When encrustation in hot-water service pipelines is a problem, it is considered that lubricated plug-type valves have an advantage over many other types of valve, in that they are less affected by encrustation. For smaller-bore pipelines, and individual isolating valves other than main isolating valves, 'Ballofix' valves may be used.

Before any choice is made between a direct or an indirect system, full details should be obtained from the Water Undertaking concerning the cupro-solvency or plumbo-solvency characteristics of the water supply, to ensure correct choice of materials.

Direct systems should not be used when the temporary hardness of the water supply exceeds 160 parts per million, unless they can be thermostatically controlled. When the source of the local water supply is moorland, the water is likely to be soft and corrosive, and is likely to cause corrosion of cast-iron boilers when used in a direct system. Discoloration of the water may also occur, which will cause irritation to users of the water. Alternatives are to use bower barffed cast-iron boilers if the water is suitable, or a copper boiler and pipework when the installation is a small one.

For larger installations an indirect system using a copper calorifier can be used, or the water supply can be treated to render it safe from the possibility of corrosion.

'Dead-leg' is a term generally adopted to cover any pipe that is served by a hot-water service system and that does not form part of any circulating pipework in that system. Model Water By-law 42 requires that no such pipe shall exceed the lengths shown in Table 39.1.

SANITARY SERVICES

Systems of sanitation

Accepting the principle that soil and surface water should be separately piped within a building, there are three main systems of sanitation: the two-pipe system, the one-pipe system and the single-stack system. The reasons for separation of soil and surface water within a building, even though the drainage may be combined outside the building, will be explained later.

In practice, circumstances are seldom ideally suited to the application of just one of the above systems, except perhaps in housing of a low-rise type, which generally follows the requirements for the single-stack system.

The titles given to the systems are perhaps misleading in that the two-pipe system includes separate pipes for soil or waste water, each of which may have its own vent pipe, thus utilising four pipes and not two. The one-pipe system similarly involves the use of two pipes including the vent. Only the single-stack system is precisely what it implies, one pipe only. Further descriptions and illustrations given below will serve to amplify the difference between the types of system, but invariably the resultant scheme of sanitation for any premises will require slight variations or combinations of the normally defined systems to achieve a practical and economical design.

TWO–PIPE SYSTEM

In the two-pipe system of sanitation, the soil and waste discharges are kept separate, each being provided with its own vent and anti-siphonage pipework. The two discharge pipes are ultimately combined into one common drain below ground, but with the provision that the main waste pipe is trapped by a gully at the foot before connection to the drain. When the drainage system is a combined one, taking foul and surface water, any rainwater pipes are also trapped by a gully at their foot prior to being connected to the drain.

The two-pipe system of sanitation is rarely used today and has been largely replaced by a more economic design. At one time it was an accepted system for housing developments, utilising a hopper head outside the external wall of the building to collect bath and basin wastes at first-floor level. The hopper head was criticised as an offensive-smelling fitting. When a blockage of the waste pipe from the hopper head occurred, often due to freezing during winter months, waste water would run down the face of the building. Freezing of the hopper head also rendered use of the bath and basin at first-floor level impossible. The Building Regulations 1976 now restrict the installation of soil or waste pipes outside the external walls of a building, and prohibit hopper heads.

ONE–PIPE SYSTEM

In a typical one-pipe system of sanitation, all soil and waste discharges into one common pipe and all branch ventilating or anti-siphonage pipes discharge into one main ventilating pipe. The positions of access must be determined by the details of design. This system has replaced the two-pipe system and is admirably suited for use in multi-storey development.

All traps used below waste fitments should be capable of maintaining a water seal of 76 mm under working conditions. In recent years, it has been found that in practice repetitive planning at each floor level of a multi-storey building favours localised application of single-stack principles of sanitation, which reduces the total amount of ventilating pipework required. This has led to the development of the modified one-pipe system. Once again, a water seal of 76 mm is required to be maintained in all traps used below waste-water fitments.

SINGLE–STACK SYSTEM

Figures 39.7 to 39.9 illustrate the main principles of the single-stack system, and it can be readily appreciated that this is the most economical system of any considered

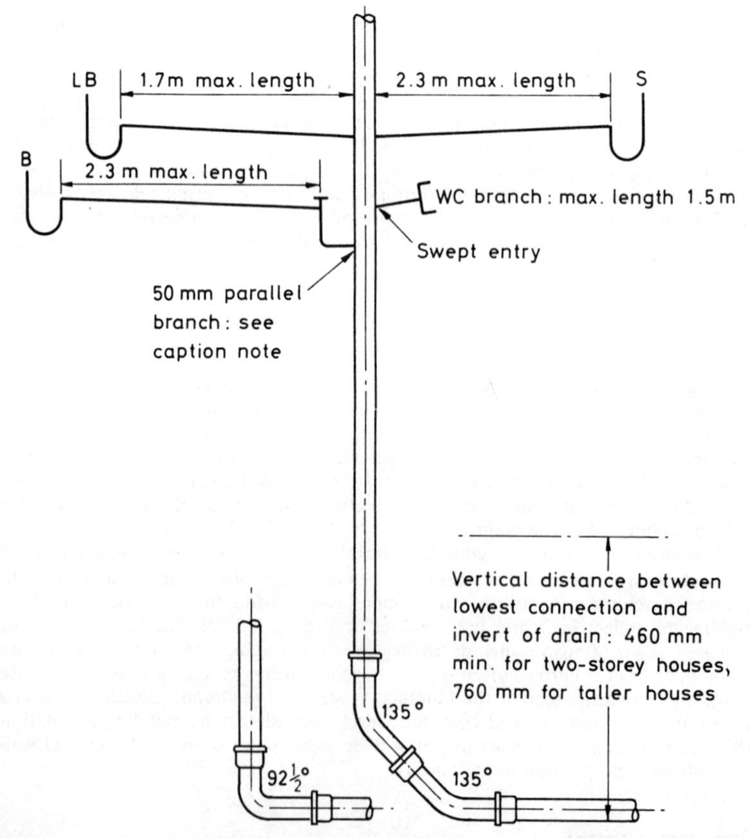

Figure 39.7 Main design features of single-stack system; no offsets are permitted below the topmost connections to the stack. Note: 50 mm parallel branch to be introduced only when bath waste would otherwise enter soil stack in area shown in Figure 39.9

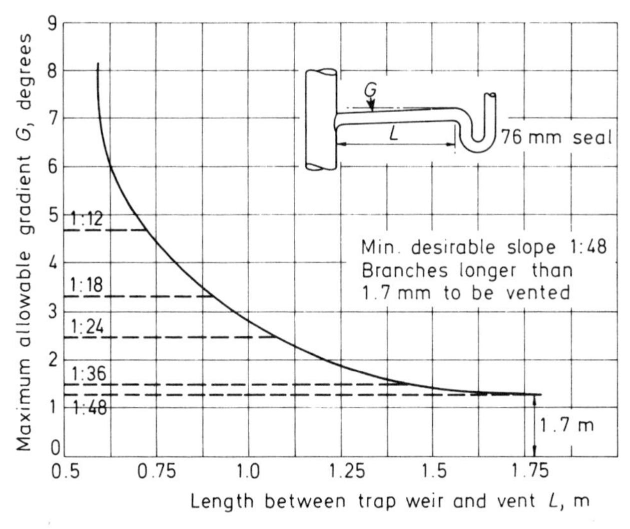

Figure 39.8 Length and fall of basin waste

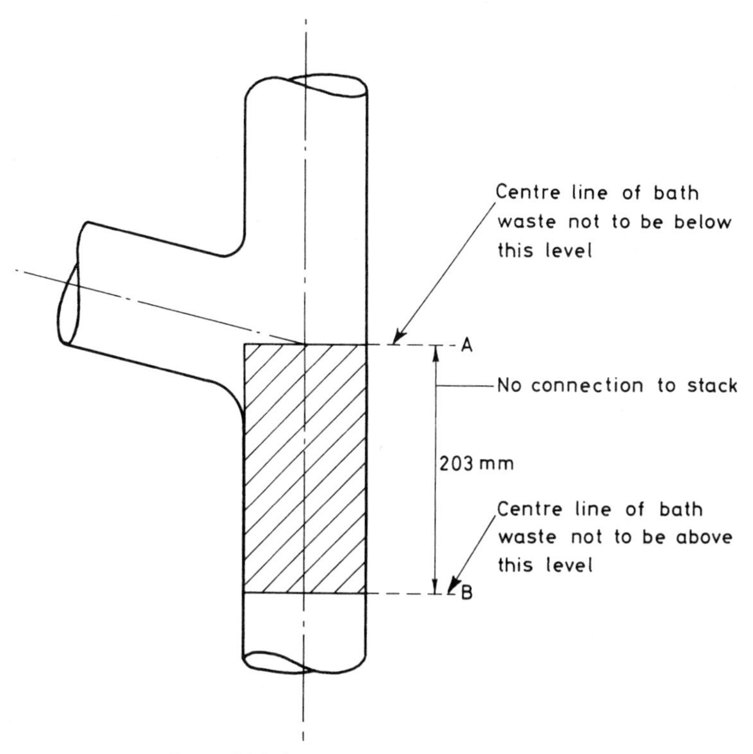

Figure 39.9 Connection of bath waste to stack

so far. It is regrettable that a system so simple to install has been subjected to so much abuse, in that insufficient attention is frequently given to the stringent design limitations. Despite the obvious economy of the single-stack system, there have been those who have sought even greater savings by permitting every fitment to discharge into one pipe, regardless of gradients, the method of connection to stack, anti-siphonage and other important design limitations.

The single-stack system is admirably suited for use in low-rise housing developments of up to about ten floors. For multi-storey buildings it may require to be partially ventilated and so becomes the modified one-pipe system previously considered. It is essential that the principles of the single-stack system of sanitation are adhered to on site. Little purpose is achieved by accurate design and sizing if the results are given insufficient attention during actual installation.

Table 39.6 provides information on the design of single branches and location of fittings. This has been drawn from CP 304: 1968, *Sanitary pipework above ground.*

Design considerations

Reference has been made previously to the separation of soil water and rainwater within a building. It is true that some local authorities will permit the discharge of rainwater into the top of a soil or waste pipe at roof level. However, while the provisions of the Public Health Act 1936 do not extend to Scotland, Northern Ireland or London unless otherwise expressly provided within the Act, Section 40 (Clause 1) of the above Act requires that 'no pipe for conveying rainwater from a roof shall be used for the purpose of conveying the soil or drainage from any sanitary convenience'. Clause 3 of the same section requires that 'no pipe for conveying surface water from any premises shall be permitted to act as a ventilating shaft to any drain or sewer conveying foul water'.

In the event of a blockage occurring within the installation, it would be possible for fittings that discharge to the affected stack to be placed temporarily out of use until the blockage had been cleared. However, when rainwater is able to enter the affected stack, the effects of the blockage are aggravated and may even cause the building to be flooded with sewage.

The practice of combining soil and surface water within the building, depending on the location of the building, is therefore either illegal or extremely inadvisable.

Water Undertakings usually require that an overflow pipe must be conspicuous when it operates, to ensure that immediate attention is given to the cause of the overflow.

The Building Regulations 1976 require that any overflow pipe connected to a waste appliance shall either discharge into a waste by the trap installed in accordance with Regulation N6, or discharge in such a way as not to cause dampness in, or damage to, any part of the building.

It would be impractical to consider sanitation as anything other than one of the many services required to enable a structure to function for a given purpose. However, whereas change of direction in many other pipelines has a marginal effect on their efficiency, an offset in a pipe carrying waste water from the building can be a perpetual cause of blockages and costly maintenance.

The Building Regulations 1976 require that if it is necessary to have a bend, the soil pipe, waste pipe or ventilating pipe shall be so constructed that the bend does not form an acute angle, but has the largest practicable radius of curvature, and that there is no change in the cross-section of the pipe throughout the bend. The Greater London Council By-laws require that a soil pipe shall not have any bend except where unavoidable, in which case the bend shall have an obtuse angle as large as possible, have the largest practicable radius of curvature, and not change in any way the cross-section of the pipe.

Table 39.6 DESIGN OF SINGLE BRANCHES AND FITTINGS (extract from CP 304: 1968, reproduced by permission of the British Standards Institution)

Component	*Design recommendations*	*Possible troubles to be guarded against*
Bend at foot of stack	Bend to be of 'large radius', i.e. 150 mm minimum root radius or, if adequate vertical distance is available, two 'large radius' 135° bends are to be preferred. Vertical distance between lowest branch connection and invert of drain to be at least 460 mm for a two-storey house and 760 mm for taller dwellings. Where this distance cannot be achieved, ground floor appliances should be connected directly to a drain (see *Figure 39.7*).	Back pressure at lowest branch, foaming of detergents.
WC branch connection to stack	WC connections should be swept in the direction of flow, with radius at the invert of not less than 50 mm. Cast-iron fittings to be BS 416. Fittings in other materials should have the same sweep as cast-iron fittings. Unvented WC branches up to 1.5 m long have been used successfully.	Induced siphonage lower in the stack when WC discharged
Basin waste 32 mm trap and 32 mm waste pipe	'P' traps to be used. The maximum fall of the waste pipe to be determined from *Figure 39.8* according to the length of the waste, any bends on plan to be not less than 76 mm radius to centre line. Waste pipes longer than the recommended maximum length should be vented. As an alternative, 38 mm diameter waste pipes longer than 3 m have been used successfully, but maintenance may be necessary to maintain bore.	Self-siphonage
Bath waste 38 mm trap and 38 mm waste pipe	'P' or 'S' traps may be used. Owing to the flat bottom of a bath, the trailing discharge normally refills the trap and the risk of self-siphonage is much reduced. Waste pipes 2.3 m long at a fall of 1¼° to 5° (1:48 to 1:12) have been used successfully. Position of entry of bath water into stack to be as in *Figure 39.9*.	Self-siphonage
Sink waste 38 mm trap and 38 mm waste pipe	'P' traps to be used. Owing to the flat bottom of a sink, the trailing discharge normally refills the trap and the risk of self-siphonage is much reduced. Fall not greater than 5° (1:12) for lengths up to 700 mm. For longer lengths up to a maximum of 2.3 m, fall not greater than 2½° (1:24).	Self-siphonage

Conversions to metric equivalents have been made by the author.

Note Where the length or fall of the discharge pipe serving a waste appliance is greater than the recommended maximum in this table, the discharge pipe should preferably be vented or a larger diameter discharge pipe could be used. This should have a maximum length of about 3 m.

Little more can be added, except to stress the importance of routes as direct as possible for all soil and waste pipes if subsequent blockages caused by abuse of the installation are to be minimised. The cost of ensuring a straight route has to be paid once only; the cost of clearing blockages is a recurring expense.

It is sometimes suggested that limitations on branch waste-pipe lengths, such as those imposed with the single-stack system, can be eased by oversizing the branch waste pipe. It is argued that, if a 32 mm trap discharges into a 38 mm waste pipe, the discharge cannot fill the waste pipe and the water seal cannot be broken. This may be so initially, depending on gradients involved, but once the fitting has been used for some time, the free area of the waste pipe will only be that which can be maintained by the flow rate. Under this condition, the water seal can be readily broken.

The function of a ventilating pipe is, as the name implies, to ventilate an installation and maintain balanced air pressure throughout. Owing to the normally dry state of these pipes, rust and debris may collect at the foot of the main vent pipe before it connects to the main soil or waste pipe. Alternatively, detergent foam may enter the base of the vent pipe and the pipe ceases to function. To avoid these problems, it is advisable to make the base of the vent pipe a 'wet vent' by connecting one or two waste fittings into it, thus keeping the pipeline clear.

A main vent pipe should connect back into the main soil or waste pipe that it serves, if practicable at least 914 mm above the highest soil or waste branch. Where the vertical distance between the lowest connection to the main soil or waste pipe and the manhole is less than 3 m, the main vent pipe may be connected into the soil or waste pipe not less than 228.6 mm and not more than 609.6 mm below its lowest branch, or connected direct into the manhole.

The provision of access for rodding purposes should be approached in a practical manner, in relation to the function of maintenance. Little purpose is served by providing an access plate for rodding purposes if it is in an inaccessible position. Rodding access on branch soil or waste pipes should never be within false ceilings because of the danger of spillage on to absorbent surfaces. Such access should be extended upwards through the floor above.

Rodding access on vertical stacks should be above the flood-level of any sanitary fittings served at that particular level. Removal of the access cover then provides less risk of sewage spilling into the duct.

It is far better, if sewage is to be spilled, that it occurs on to non-absorbent surfaces. Access to branch soil or waste pipes above the floor level should, once again, be located above the flood-level of the fittings served.

A large variety of access fittings is available for specification by the engineer. For locations with special floor finishes, Wade International (UK) Ltd manufacture access fittings with alternative surface finishes for inclusion in most floor types.

Sanitary fittings at ground-floor level should always be discharged to the drain separately from upper-floor sanitation. Some protection is thus provided against flooding of the ground floor by higher fittings, should a drain blockage occur, in that backing-up of sewage may be evident from ground-floor open gullies outside the building before any spillage occurs internally. For this reason, it is preferable that at least one outside open gully is available for the protection of any building. Some evidence exists to favour the extension of this principle to cover ground and first floors of multi-storey buildings. If a blockage then occurs at the foot of the main stack, sewage may be able to back-up sufficiently to apply enough pressure to the blockage to clear it.

Consideration should always be given to draining basement levels of buildings by gravity to a pump installation serving only the basement. While the level of the sewer may be below basement level, the latter must always run the greatest risk of flooding if backing-up occurs in the outfall drain. Similar consideration applies to underground service ducts that require drainage. It is better to install a small pump unit than to risk flooding such a duct with sewage.

Traps

The Building Regulations 1976 require the provision of 'a suitable and readily accessible trap of adequate diameter, having an adequate water seal and means of access for internal cleaning' from any soil or waste appliance, and fitted close thereto, except if the appliance itself has an integral trap (e.g. water-closets). Provision is made within these Regulations for trunk wastes or channels that receive a number of waste appliances, and that may be served by a common trap. Regulation N3 requires that 'such provision shall be made in the drainage system of a building, whether above or below the ground, as may be necessary to prevent the destruction under working conditions of the water seal in any trap in the system or in any appliance which discharges into the system'.

The Building Regulations 1976 are held to apply generally throughout England and Wales, with the exception of the inner London boroughs (the area of the former London County Council) where the London Building Acts will continue to prevail. Table 39.7, extracted from CP 304: 1968, indicates minimum internal diameters of traps. The Greater London Council By-laws differ from the recommendations of CP 304: 1968 in that the GLC require a minimum bore of 25.4 mm for traps serving drinking-water fountains.

Table 39.7 MINIMUM INTERNAL DIAMETERS OF TRAPS (extract from CP 304: 1968, reproduced by permission of the British Standards Institution)

Domestic appliances	*Minimum internal diameter* (mm)	*Non-domestic appliances*	*Minimum internal diameter* (mm)
Wash-basin	32	Drinking fountain	19
Bidet	32	Bar well	32
Sink	38	Hotel and canteen sink	38
Bath	38	Urinal (bowl)	32*
Shower bath tray	38	Urinal (stall) (1 or 2)	50
Wash tub	50	Urinal (stall) (3 or 4)	63
Kitchen waste-disposal unit (tubular trap)	38	Urinal (stall) (5 or 6)†	76
		Waste-food disposal unit (tubular trap)	50

* In hard water districts this may need to be larger.
† Where there are more than six stalls in one range, more than one outlet should be provided.

While various types of resealing traps have been manufactured for many years, not all local authorities will accept them. Before using such traps, therefore, the views of the authority concerned should be sought. Building Research Station Digest 80 refers to the use of resealing traps, but makes no recommendation beyond that each design should be judged on its merits.

The earlier types of resealing trap utilised an enlarged chamber that retained a quantity of water until the siphon was broken, whereupon the water fell and stabilised, thus resealing the trap. This pattern is still available. Some engineers argue that unless adequate maintenance is provided to these traps, the enlarged chamber fills with grease, soap etc., until the waterway through it is similar to that of the remainder of the trap.

A more recent pattern of resealing trap has been manufactured by Greenwood and Hughes Ltd. It is called the 'Grevak Monitor anti-siphon trap'; it is a bottle-type trap and is manufactured in brass or white hostalen. The Grevak Monitor trap has been tested by the Department of Scientific Research at the Building Research Station and their conclusion reads as follows:

In the siphonage tests the traps were reasonably quiet in operation and an adequate seal was maintained under conditions more severe than are likely to occur in a normal plumbing system. A seal greater than 50.8 mm was maintained when using a 32 mm diameter waste pipe up to 6.1 m long and at slopes up to 5°.

No information was obtained on the effects of age and use on performance, but the turbulent self-cleansing action within the trap during discharge was noted.

It is worth nothing that following introduction of their Monitor trap, Greenwood and Hughes Ltd have developed a normal trap called the 'Maxflo' trap for use in standard conditions. It is also a bottle-type trap, almost identical in external appearance to the Monitor and is also manufactured in brass or white hostalen. This range of traps can thus be used for varying conditions, while providing conformity in appearance.

Treatment and/or disposal of grease-laden water invariably provides many problems. Many people believe that the increasing use of detergents has removed this problem from sanitation. However, there is a growing belief in the fact that detergents merely enable the grease to be transported from the source for subsequent deposit as scum further along the pipeline. This scum eventually solidifies and may cause blockage, and may even require replacement of the pipeline.

The traditional method of removing grease from waste water from kitchens was to install grease-trap chambers at suitable locations in the pipeline. Periodically the cover would be removed from each chamber and the solidified grease removed. Experience has indicated that these units are seldom properly maintained, and eventually, owing to flooding, they cause a great deal of trouble.

A type of grease trap has been introduced in the UK by Wade International (UK) Ltd. It is called the 'Wade Actimatic hydra-filter grease interceptor'. The method of operation is quite simple, in that the grease-laden water entering the initial baffle chamber begins to separate into grease and water. Grease is retained by the interceptor and the water flows on to the drainage system. Instead of being stored for periodic removal, the grease is digested by dosing the interceptor with 'Actimatic compound', a mixture of viable organisms, bacterial enzymes and food supplements.

The manufacturers claim that, by this natural digestion process, grease is converted permanently into a water-soluble compound. Subject to regular and proper dosing of the installation, the need to clean the interceptor is virtually eliminated. These interceptors are manufactured in a large range of sizes, and are suitable for most practical purposes.

Flushing gullies are manufactured to overcome problems associated with cleansing of sterile areas or locations where cross-infection may occur. Once the floor is washed down, the contents of the gully can be flushed away and replaced with clean water. Wade International (UK) Ltd manufacture two sizes of flushing gully: 76 mm and 101.6 mm nominal bore outlet. The flushing rim is in lacquered cast iron with a large flange and clamping device to receive a waterproof membrane. The trap is lined with PVC and has a 127 mm deep seal, and the floor grating is in polished bronze. Two flushing connections are required; a 19 mm connection to the flushing rim and a 9.5 mm connection to the heel of the trap. Alternative grating finishes are available, either nickel bronze or PVC coated.

The provision of floor-draining facilities creates various problems for the engineer. Apart from instances where a slab or stall urinal is installed, recessed into the floor slab so as to function also as a means of floor drainage, floor gullies have to be installed as an independent fitting. They are usually poorly maintained and even odorous and, with modern cleaning techniques, often unnecessary. Cleaning today is generally achieved with a wet mop rather than with copious quantities of water. At one time a further disadvantage arose with floor drains in that the grating was often out of level with the finished floor level.

Wade International (UK) Ltd now provide a range of adjustable floor drains that may be set accurately at the time the floor finish is applied. It may be argued that the existence of a floor drain facilitates the discharge of overflows when sanitary fittings are remote from an external wall. While this may be true, operation of overflow pipes is infrequent and cannot be relied upon to maintain the water seal of the floor drain which, as a result, often evaporates. Once the seal is broken, a direct vent exists from the foul sewer into the building.

Rainwater pipes

At one time, it was common practice in the UK to allow an arbitrary area of roof for each unit area of cross-section of any rainwater pipe. Division of the roof into individual areas, each served by an appropriately sized rainwater pipe, would follow, and the system installed accordingly. Research during more recent years has led to much greater economy in rainwater-pipe installations and a greater understanding of the factors involved.

The function of a rainwater outlet may be defined as being to convey rainwater into the outlet with the minimum of turbulence, and to encourage an efficient discharge of the water into the down-pipe, where it gravitates to the drain. It is no longer considered essential for the rainwater pipe beneath an outlet to possess the same diameter as the nominal bore of the rainwater outlet. Water enters the rainwater outlet under conditions of very little hydraulic head, whereas within the rainwater pipe a greater hydraulic head exists, through gravity, and a smaller pipe may be used to convey the contents.

Location of rainwater outlets serving a flat roof must take into account factors such as planning at lower floors of the building and the economics of screeding the roof to proper falls. A given roof area can, in theory, be served by one outlet if no limit is placed on maximum screed thickness. The normally accepted minimum gradient for asphalt is in the region of 38 mm in 3.048 m, so discretion must be exercised as to the optimum spacing of rainwater outlets relevant to economic screeding of the roof. In this connection it is also useful to seek early guidance from the structural engineer concerned, as to the maximum permissible loading screed thickness for any roof area.

The height of the building should also be considered, in that the weight of the roof screed adds progressively to the structural costs of a tall building. Alternatively, a lower building may be less affected by a slight increase in screed thickness to reduce the number of rainwater pipes. Once the rainwater outlets have been suitably located at roof level, there is no reason why the discharge from these may not be collected below the roof into a lesser number of pipes and thence to the drainage system.

Pitched roofs may be broadly divided into two classes: those requiring smaller-size gutters to serve them, and those served by large valley gutters. Small gutters should ideally be installed with a suitable gradient to reduce their tendency to silt up. Larger valley gutters should have a minimum width of 300 mm to facilitate access along them for maintenance purposes. The sides of a valley gutter usually rise at the same gradient as the roof they serve, until the gutter has a capacity suitable for the particular circumstances. The sides should then rise vertically to provide a 75 mm freeboard above the design capacity to allow for wave action produced by wind.

Once again, rainwater pipes and their sizing should be considered in relation to dropping positions and accessibility for maintenance.

In designing an effective installation for the removal of surface water from the roof of any premises, run-off from vertical building faces above and adjacent to the roof area should also be considered. Just as the sun casts shadows, so prevailing rain-bearing winds will cast a 'dry shadow' behind a building, with a consequent uneven deposit of rainwater around that building.

With low-rise buildings the effect is unlikely to prove troublesome, but high-rise buildings can greatly influence run-off calculations for low-lying roof areas adjacent to them. Where the face of a high-rise building is smooth, rainwater may build up on it until serious problems arise with weatherproof qualities of the building. Unless periodic ledges can be incorporated in the design of the building face to shed the rainwater, hidden gutters are recommended at intervals down the building to collect the rainwater and convey it into the drainage system.

A percentage of rainwater striking the face of a tall building is reflected off and will influence the intensity of precipitation on lower flat areas. It is recommended, therefore, that the above factors should be borne in mind when dealing with buildings other than low-rise structures. It is not suggested that extensive calculations are necessary, but that some allowance should be made for any 'concentrated' run-off that may occur and inconvenience the building occupants.

It is sometimes suggested that rainwater pipes should be cast within the structure. The argument is usually given that the pipes may be thus concealed without additional ducts or boxing-in being required. This practice is not recommended for the following reasons:

- Installation of the rainwater pipes is necessary as the structure, usually concrete columns, is cast. This necessitates uneconomic attendance of plumbing tradesmen to install each length of pipe as pouring of concrete proceeds.
- Once installed the pipework cannot possibly be maintained or replaced – a serious consideration where the life of the pipework material is less than the expected life of the building.
- The choice of materials is limited as most of those available are subject to fracture, damage or collapse during the pouring of concrete.
- It is difficult to achieve the testing requirements of the local authority for the installation. Failure to meet these test requirements can involve demolition of part of the structure to remedy the defect(s).
- Water leakage from the rainwater pipes has been known to result in rusting of the structural reinforcement, with subsequent weakening of the structure. Decorations to structural surfaces may also be affected.
- Problems may arise in turning the rainwater pipe(s) out from the base of any column(s).

Testing

Regulation N4 Clause 4(e) of the Building Regulations 1976 requires that any soil pipe, waste pipe or ventilating pipe shall 'be so constructed as to be capable of withstanding a smoke or air test for a minimum period of three minutes at a pressure equivalent to a head of not less than 38 mm of water'.

The two main methods of testing used in the United Kingdom for sanitation pipework are the water test and the air test. No pipework should be covered up until satisfactory test results have been obtained. Rainwater pipes installed within buildings should be tested to the same standards as soil, waste or ventilating pipes.

The water test is generally applied to that part of the installation likely in practice to be charged with water in the event of a drain blockage. Little purpose is served by applying the water test at a higher level than the lowest sanitary fitting. Testing is commenced by inserting a test plug into the foot of the pipe, which is then filled with water. The maximum pressure applied at any point by the water test should not exceed 6.1 m.

The air test may be applied by introducing air or smoke under pressure into the sanitation pipework after properly sealing off open ends of pipework such as vent pipes. To satisfy air test requirements, an installation must support a pressure equal to 38 mm water gauge for a period not less than three minutes.

To test a waste appliance for any tendency towards self-siphonage of its trap, the appliance should be filled with water to its overflowing level and discharged in the normal manner. The seal remaining in the trap should be measured following discharge of the appliance. Water closets are flushed in the normal manner.

Apart from any tendency towards self-siphonage, appliances must also be tested for the occurrence of siphonage from any trap, induced by operation of adjacent fittings. The form of testing is generally carried out by simultaneously discharging a number of appliances and noting any loss of seal in the appropriate traps.

Experience indicates that the worst conditions occur when appliances on the upper floors are discharged. It is preferable, therefore, that one water-closet, one lavatory basin and one sink should be operated from the top of the building, and any other appliances used in the test should be distributed along the stack being tested.

Waste-disposal units

Since their introduction to the United Kingdom many years ago, kitchen waste-disposal units have found great favour as a means for the disposal of wet refuse from the kitchens of residential accommodation. Some criticism has been made concerning the problems that the resultant effluent may create at the treatment works and in pump installations, where they may produce a build-up of sludge, formed by ground solid matter. For this reason, it is preferable that effluent from waste-disposal units is kept in constant motion by being discharged direct to the sewerage system, and not via any pump installation fed by a sump. which may function as a settling tank.

Against such criticisms must be considered the advantages of disposal at source of wet refuse, which invariably causes smells and attracts flies when placed in any form of receptacle that may not be emptied for up to a week.

When it is unavoidable that the effluent from waste-disposal units has to be pumped, it is essential that regular maintenance is carried out to the pump sump. Experience indicates that otherwise, apart from settling-out of the incoming sewage, small particles of solid matter unite with emulsified fats and rise to form a curd on top of the sewage in the sump. Unless sufficient inflow is occurring to operate the pump(s) and evacuate the sump, the curd solidifies. Further inflow builds up the slab of grease and, when the pump operates, it pumps out the sewage, but leaves large slabs of solidified grease to be built up further. The electrodes or controls operating the pump may also be affected or impaired by the formation of these slabs.

It is not argued that formation of solidified grease slabs is to be found only in installations including waste-disposal units, but that they may slightly aggravate the situation by providing particles of solid matter to bond the grease into stronger slabs. One method of prohibiting the formation of slabs described above is to provide a chemical drip feed, of a grease-dissolving type, into the pump sump. Similarly, coating the electrodes or controls with silicone will inhibit the adhesion of grease to their surface.

FURTHER INFORMATION

Tompson, T.A., *A guide to sanitary engineering services,* Macdonald and Evans Ltd
The Building Regulations 1976, HMSO
Model Water By-laws (1966 Edition), HMSO
Relevant British Standards and Codes of Practice, obtainable from British Standards Institution, 2 Park Street, London WIA 2BS

ACKNOWLEDGEMENTS

The author gratefully acknowledges permission by the publishers of his original work *A guide to sanitary engineering services* (Macdonald and Evans Ltd) for him to utilise information therefrom, in the preparation of this section.

40 GAS SERVICES

INTRODUCTION 40–2

DOMESTIC GAS APPLICATIONS 40–3

BUILDERS' WORK 40–8

REGULATIONS 40–11

GAS SERVICES 40

40 GAS SERVICES

J.H. MILAN, CEng, FIGasE
Sales Officer, Products and Utilisation, British Gas

This section first describes the main applications of gas in domestic buildings, and then deals with important practical points that the builder should bear in mind when carrying out installation work; the relevant regulations are also briefly listed.

INTRODUCTION

Since gas is widely used in new buildings, both private and public, and also in remedial housing, the work of every builder is likely to be affected, in greater or lesser degree, by the particular requirements that gas installations place on builder's work.

Depending on the size and scope of a project, the builder's activities may be affected by the laying of mains and services, in the case of large, greenfield estates, or may simply have to take account of the installation of services within the premises. Involvement with gas services is therefore likely to arise at all stages of a contract and impinge at many points on the appropriate statutory regulations and standards, with the need to maintain contacts with the Gas Authorities from the first planning stages through to completion.

In addition to a ready source of reference for these regulations and how they affect the builder, it is also useful to have a good working knowledge of the main types of gas system likely to be encountered, for heating, domestic hot water and cooking, and their relation to the builder's activities, in terms of location, ventilation and flueing requirements and builders' work generally.

It is the aim of this section to present this information in a practical and concise manner, bearing in mind always that no guide can cover every eventuality but that the necessary expert advice is always available from the gas industry itself.

Moreover, as in all building services, changes are continually taking place in gas services and systems, and the builder will need to keep in touch with those aspects that are likely to affect his work. Recent years have seen a number of such changes, from the advent of wall-hung boilers to storey-height warm-air heating systems, and this pattern is bound to continue.

An awareness of the need for an understanding of the interrelation between gas services and builder's work is in the interests of the builder, the gas industry and the client, whether private individual or local authority.

Preliminary considerations

It is important to make earliest possible contact with the Housing Development Manager of the Region of British Gas concerned, for consultation on all new housing and modernisation projects. Advice and services from this source include:

- gas availability for the site and recommendations for bringing a supply on to the site;
- advice on suitable central heating systems, domestic hot water supply and other gas services for different house types;

- technical advice on Gas Safety Regulations, Building Regulations, statutory requirements or Codes of Practice relating to gas services;
- preliminary cost estimates for selected gas systems;
- detailed design drawings and specifications to meet stipulated heating standards and hot water requirements;
- estimate for gas supply and, if required, installation of the complete system;
- assistance with bulk purchasing arrangements of appliances for estate developments.

Consultation with British Gas on the design and installation of central heating systems will often indicate comparatively minor modifications to the builder's plans that will simplify the work and may lead to cost reductions. This is particularly true in the case of gas warm-air central heating systems, when considerable savings in installation and operating costs can be achieved by seeking British Gas advice on the best way to design the house around a suitable warm-air system.

By seeking advice on the selection of appliances and ancillary equipment, the best possible use can be made of the available space as well as ensuring ideal comfort conditions at minimum cost. Examples are the selection of wall-mounted boilers or combined gas-fire/back-boiler units and wise choice of radiators, warm-air grilles and other items.

Gas supply

A *gas main* is defined as a pipe that is designed to carry gas to more than two customers but that does not enter a customer's premises. Its size depends on the estimated future gas usage of the area or estate it serves, and this will be determined by the local Gas Distribution Engineer. In the case of new housing, the Region plans the provision of gas services in conjunction with the housing developer.

Installation of mains is the responsibility of the Region, and in a major new housing development the network is laid before the roads and other civil works are constructed. Consultation between the Region and those responsible for laying down the roads, pathways etc. is essential.

A *service pipe* is defined as the pipe between the gas main and a primary meter control on the premises of a consumer. In the case of buildings occupied by a number of individual households, the service is that pipe from the main into the building, and it may supply any number of consumers in the building. The service, which is of steel or polyethylene, is normally of 25 mm diameter, depending on the Distribution Engineer's estimate of gas usage.

Services are laid underground by the Region or its contractor, and enter the building either up the outside wall to a meter box installation or as near as possible to the meter location. It is recommended that the service connection should be made at an early stage of construction.

All primary meters are the property of British Gas and are installed by the Region concerned. All internal pipework should be laid to a position adjacent to the meter location.

DOMESTIC GAS APPLICATIONS

Before considering builders' work provisions and the appropriate regulations, it is first necessary to summarise the principal gas applications likely to be encountered.

Hot water heating

Small-bore heating systems served by gas-fired boilers and electric circulating pumps are the most popular domestic central heating system. The boiler, of which several types are available, provides both heating and domestic hot water, feeding an indirect hot water cylinder with an independently controlled loop to permit closing down of heating while the hot water supply continues.

One or more loops of copper piping, of 15, 22 or 28 mm o.d. to BS 2871, supplies the heat emitters (radiators, natural convectors or skirting convectors) with hot water at a flow temperature of 82°C and a drop of around 11°C for an average of 77°C at the emitters. A pump, which can be integral with the boiler, drives the water through the system.

Where a bathroom towel rail is heated, it should preferably be on the primary circuit to the domestic hot water cylinder, prior to any control valve, for all-the-year-round operation.

Controls should include a programmer or time clock, a room thermostat suitably located, or thermostatic radiator valves to provide individual room-temperature control.

Micro-bore systems are a more recent development. They are similar to small-bore systems, with three important exceptions.

- Smaller-diameter copper tubing is used, normally 6, 8, 10 or 12 mm o.d. to serve the heat emitters.
- Heat emitters on each floor are served individually by radial pipework from centrally located distribution manifolds. Pipework from boiler to manifolds is of 15, 22 or 28 mm o.d.
- A more powerful pump is needed to overcome the greater flow resistance.

Advantages of micro-bore heating include:

- Smaller holes are required in walls, floors etc.
- Copper tube supplied in flexible rolls simplifies setting out holes, permits continuous jointless pipe with fewer fittings and, in screed or solid floors, allows minimum-length runs between manifold and heat emitter.
- Micro-bore tubing can be concealed (chased into walls or run into floor screed in new building, or behind skirting in existing houses) provided all concealed piping meets water authority regulations.
- Future extensions are possible by providing spare boiler capacity.
- Reduced water content gives quicker response to thermostatic or manual control.

Boilers

It is advisable for all builders to be aware of developments in boiler design, which often affect location and other factors. Recent trends that help the builder include the space-saving low-water-content boiler and wall-hung types. Other helpful developments include pump and controls integral with the boiler and room-sealed flues.

Three boiler types are available: floor-mounted, wall-mounted and gas-fire/back-boiler types.

- Floor-mounted boilers are normally installed in the kitchen but can be located elsewhere in the house. They are supplied in styles suitable for kitchen layouts or as a basic boiler unit for installation in prepared cupboards.
- Wall-mounted boilers contain less than 1 kg of water, compared with 16–18 kg in a conventional unit of equivalent output. Weight-saving ranges between one-fifth and two-fifths of a comparable floor-mounted type. Most designs are room-sealed for installation on or adjacent to a suitable outside wall, but open-flued types are available for mounting on internal walls.

• Back boilers, combined with a gas fire, are installed in a conventional fireplace opening, providing heating and domestic hot water from the boiler while the gas fire provides radiant and convected heat into the room in which it is located. These units use existing flues, suitably lined, and thus save space.

Combination units, which incorporate boiler and hot-water storage cylinder, are also available, as a complete package, especially for modernisation projects. An even more complete package (comprising boiler, hot-water storage and cold-water cylinders, feed and expansion tanks) can be supplied, usually in pre-fabricated and pre-assembled form, and is often described as a 'heart unit'.

Heat emitters

The most widely used type is the pressed steel panel radiator, which can be manually controlled or fitted with thermostatic valves for individual automatic control. Straight, curved or angled radiators are available to suit all wall configurations.

Wall convectors can be wall-mounted or recessed into room walls or cupboards. They comprise a pressed steel casing, with louvres at top and bottom, and enclose a finned tube in which hot water circulates, warming air drawn in through the bottom louvres and discharging it through the top by natural convection.

Skirting heaters employ the same principles but are smaller and usually installed in continuous runs, at skirting level or set back into walls.

Convectors fitted with fans increase the volume of air passing over the finned tube, giving high heat outputs quickly. Thermostatic controls can switch the fan on/off to maintain a pre-determined temperature. Because a fan convector of given heat output is considerably smaller than the equivalent radiator, these units are useful where wall surfaces are not available for panel radiators or where large rooms have to be heated quickly.

Warm-air heating

Gas-fired warm-air systems provide whole house or partial house heating by ducting warm air from heater units, with recirculation, except from bathrooms and kitchens. Because of the need to instal ducting, these systems are normally restricted to new building or where major reconstruction is taking place.

A typical system comprises:

- a heater unit,
- supply ducting,
- warm-air outlets (or registers),
- return-air ducting and extract grilles,
- automatic thermostatic control and programmer/control unit.

Heater units. These are of two basic types:

• Storey-height, supplied as ceiling-height cabinets containing heater, controls, flue and rising duct, for installation in kitchens or halls.

• Units designed for installation in cupboards to complement kitchen fittings, or for siting in other suitable cupboards.

Operationally, warm-air heaters also fall into two basic types:

• Downflow or counterflow types, with the fan mounted above the heat exchanger and burner compartment, and where return air is forced downwards over the heat exchanger to low-level ducting. These types are particularly suitable for cupboard and recess locations where floor space is limited.

- Upflow types, with the fan mounted below the heat exchanger and burner compartment, so that the air is drawn upwards for high-level discharge into ducting. These types are usually located in basements.

Supply ducting. For whole house warm-air heating, three basic types of supply ducting are available:

- radial warm-air systems,
- extended plenum systems,
- stepped duct systems.

Warm-air outlets. Four basic types are available, each with advantages and disadvantages, requiring a choice to be made depending on various factors.

- Floor-perimeter outlets can be located in front of windows, especially storey-height windows or glazed doors to minimise cold radiation. They are unobtrusive and the location of return-air grilles is not critical. Disadvantages are size restrictions, to avoid unduly large floor openings and the risk of unwanted materials entering the ducting.
- Low side-wall outlets are efficient if located to avoid discomfort due to moving air currents. Outlet air temperature should not exceed 38°C, and air velocity must not be too high. The entire grille area must not extend more than 230 mm above floor level.
- Ceiling-perimeter outlets are ideal for bungalows, flats with concrete slab roofs, and two-storey houses where upper floor heating cannot be supplied from supply ducting within the upper floor. Advantages include freedom from intrusion of dust and dirt and from obstruction by curtains etc. Disadvantages: they may require some low-level outlets in main living rooms to obviate cold floor effects, and return-air grilles must also be at low level.
- High side-wall outlets have the advantage that they cannot be obstructed by curtains or furniture, but, since they have to be placed on inside walls and require high outlet velocities, this system is seldom used.

Return-air ducting. Ducting returns are not necessary from every heated room but only from main living rooms, plus the hallway at each floor level. Other rooms can have relief grilles to hallways. The return air must have a positive duct onto the heater from outside.

Return-air grilles should generally be located in stagnant areas and must not be sited either in kitchens or bathrooms.

Individual space heaters

Gas fires, most popular form of individual space heating for many years, are usually chosen and fitted by the eventual occupier. A suitable flue and gas point should therefore be provided during construction. Gas fires can either use a conventional fireplace or a specially constructed flue opening.

Room-sealed types are available for fitting on to a suitable outside wall, together with a through-the-wall room-sealed flue assembly.

More recently, individual wall convector heaters have achieved considerable popularity, with either natural or fanned convection. Models are available with either manual or automatic time/temperature controls. Installation is on an outside wall, with room-sealed or fan-assisted flues.

Domestic hot water

Where domestic hot water is not provided as an integral part of the wet central heating system, there are several other forms of gas-fired water heating appliance. One of these types will be required with warm-air heating and in other cases where it is convenient or economic to have a separate hot water supply.

Gas circulators are small boilers connected to hot-water storage cylinders, maintaining the water in the cylinder at a pre-determined temperature under thermostatic control. Circulators can be installed almost everywhere – kitchen or bathroom wall, under the sink, in a cupboard or attached to the storage cylinder. The last-named procedure is a preferred system for warm-air heating when heater and circulator use the same flue.

Note: only room-sealed circulators may be installed in bathrooms. Also, where circulators are fitted in airing cupboards, flues should be protected against combustible materials, such as clothes.

Storage water heaters incorporate the gas burner and insulated water cylinder in a self-contained unit that is connected to the domestic supply tank.

Instantaneous water heaters are connected directly to a cold water supply; this should be at constant pressure and with a minimum head, which varies with the model. If connected direct to the main supply, an automatic valve should control the gas supply should water pressure fall. A water governor should be fitted where mains water pressure varies widely.

Instantaneous heaters are of two main types. Single-point models supply only one output, such as a sink, wash-basin or bath, while multi-points normally serve kitchen sink, bath and hand-basin from the same appliance, through suitable pipework.

Room-sealed instanteous heaters should be fixed to an outside wall, or in a cupboard on an external wall.

Controls

Types of controls available with modern gas-fired heating systems include:

- Programmers and time clocks, fitted either on the appliance, inside its case or at any other convenient point. They range from simple temperature/time clock controls to programmers providing independent control of both heating and hot water over twice-daily periods.
- Room thermostats, suitably located, provide whole house temperature control through the boiler and circulating pump controls.
- Thermostatic radiator and convector valves control individual room temperature in response to air temperature changes.
- External thermostats provide a means of ensuring that radiator temperature is directly related to outside air temperatures.
- Low-temperature thermostats provide protection against cold weather by overriding the clock until the safety level is reached, when the normal programme resumes.
- Cylinder thermostats avoid the risk of the domestic hot water supply overheating by closing a control valve.
- Zone control valves can provide different temperatures in different parts of the house or switch heat from one part of the home to another at different times.

Gas kitchen appliances

This brief section is included for completeness' sake, with the recommendation that much more information be obtained from British Gas reference sources.

Two types of kitchen are involved:

- a completely fitted kitchen, pre-planned before supply pipework is installed;
- 'unplanned' kitchens, with the sink unit fitted and other gas appliances fitted after installation of gas supply network.

With the fitted kitchen, the positions of service connections are known and the layout of food storage, sink and cooker can be determined to the best advantage. For the 'unplanned' kitchen, it is only possible to make the best possible arrangements for the number and location of gas connections.

Gas appliances available include:

- free-standing gas cookers, now fitted with flexible connections, approved by British Gas;
- separate built-in hotplate, oven and grill for installation in fitted kitchens;
- gas refrigerators.

BUILDERS' WORK

Many of the implications of gas appliances for the builder will have emerged in the preceding paragraphs. Some important practical considerations are as follows.

Services into buildings

Wherever possible, services should enter on the side of the building nearest to the line of the main, terminating through an external wall at the meter box immediately on the other side, or at any other position that has been agreed with the Gas Region.

In the case of flats and high-rise buildings, services must be installed in protected shafts (Building Regulations 1976, Sections E.10 and E.14), using one of two alternative methods:

- a continuous shaft, ventilated to the outside at top and bottom, with fire stops at lateral service points; or
- a vertical protected shaft, fire-stopped at each floor and at each lateral service; ventilation to the exterior is at both high and low level on each floor.

In both cases, sealed fire-resistant panels must be installed at each floor, in the public way, to provide access to valves and for maintenance/service. Service risers must be adequately supported at the base, incorporate a dust trap, rise vertically and be supported at intervals.

Other requirements include a valve adjacent to the riser, protected against unauthorised operation, and meter controls adjacent to individual meters. Welded pipework is recommended, and compression joints with elastomeric seals should not be used.

Meter boxes

External meter boxes have been used to house the meter installation for a number of years by some Regions, but an agreed design and specification for a British Gas meter box is now available for use by all Regions.

Manufactured in glass reinforced plastic, it is installed by the builder in the cavity and outer leaf of brickwork of the cavity wall of a new house, with the meter door opening to the outside. The service pipe runs up the outside wall and terminates inside the meter box. At present the box will only accommodate the U6 credit meter and can only be installed in new dwellings, but work is continuing to widen its use.

Installation pipework

Pipework can be laid into the thickness of the wall during the construction of new dwellings, permitting a wide choice of fittings and controls with a minimum of exposed pipework.

Protected in accordance with CP 331, pipework can be installed:

- on the ground floor slab and covered by screed;
- in the first floor space with risers;
- chased into wall surfaces and covered by plaster finish.

Ventilation

Many of the most important builders' work requirements when dealing with gas installations concern ventilation and flueing.

Where appliances are fitted in purpose-built compartments, ventilation has to be provided (see CP 337, Pts. 1 and 2 – Revised 1975).

Flueless appliances may or may not require specific provision for room ventilation, such as refrigerators, towel rails, airing cupboard heaters and boiling rings. All other types of flueless appliances require at least an openable window. Any additional ventilation will depend on the rating of the appliance and room volume.

Open-flued appliances must have a permanent air vent in rooms in which they are installed. It must have direct access to the outside, either from the room itself or via another room or space, other than a hallway. All air vents should have minimum free area of 450 mm^2/kW input rating, less 3500 mm^2 adventitious allowance (i.e. air entering through cracks around windows, doors, etc.).

COMPARTMENT VENTILATION

A boiler compartment is defined as a totally enclosed space within a building containing boiler and ancillary equipment (except gas meter). The requirements are:

- It should be a fixed, rigid structure, with a floor conforming to the boiler installation instructions. Any internal surfaces of combustible material should be at least 75 mm from any part of the boiler, or lined with non-combustible material (see BS 476), or treated with fire-retardant paint (check with Gas Region).
- It should incorporate adequate air vents for ventilation and, where necessary, combustion. (For details on this and many other matters in this section, see *Gas in housing: a technical guide*, 1978, British Gas publication.)
- It should be large enough to provide access but no larger than necessary (to discourage use for storage).
- It should be fitted with a door large enough to permit withdrawal of the boiler and other equipment.

AIR VENTS

Where permanent air vents are required, the materials of which they are constructed should be corrosion resistant and dimensionally stable (see BS 493 for guidance). Size of the vent is most important, i.e. the effective area through which air can pass. Manufacturers of proprietary vents should provide this information.

Existing air vents should be taken into account when assessing requirements, and the supply needs of any other appliances in the space not burning gas should also be allowed for.

Vents should be sited so that air is introduced as near as possible to the appliance, but air must not pass through occupied areas. Ceiling-level entry is good.

Note: when considering indirect venting, with more than one vent in series, consult the Gas Region on the additional effective area required.

Where there is more than one appliance in the same space, air vent requirements should be whichever is the largest of the following:

- total flueless appliance requirements;
- total open flued requirements;
- largest individual requirement of any other appliance type.

Note that oil or solid-fuel appliances should be treated as similarly rated gas appliances, if also present.

FLUES

Provision of flues makes especially important demands on builders' work. The three basic types are:

- Rigid single- or twin-wall flue pipes, from the appliance to a suitable terminal, after completion of basic construction work;
- Precast concrete flue block systems built into the dwelling as construction proceeds;
- Flexible flue linings for brick chimney flues (e.g. in existing buildings). Since no builders' work is required, installation is usually carried out by the heating contractor.

In the case of flue block systems, these should incorporate a flue with a nominal size equal to that of the draught diverter outlet. It should be as near vertical as possible, with bends kept to a minimum. Reference should be made to appliance manufacturer's instructions and to relevant Codes of Practice. Flue runs should be inside the building whenever practicable.

Flue pipes must conform with Regulation L16 of the Building Regulations 1976. Regulation L17 details the requirements in respect of fire risk or entry of combustion products into the building.

In two-storey construction, flue pipes will normally rise to an external terminal above the roof or as a ridge tile fitting. Regulation L21 should be consulted for requirements in respect of builders' work on terminals.

PRECAST CONCRETE FLUE BLOCKS

Several systems are available as an economic alternative to conventional brick chimneys for gas appliances located in fireplace openings, e.g. gas fires, boilers behind cover plates and gas-fire/back-boilers. They can be used in external, party or internal partition walls.

Flue blocks should be British Gas approved and erected in accordance with the manufacturer's instructions. They are usually taken up to just above first floor ceiling level, with a length of flue pipe connecting to a roof or ridge terminal.

Flues should incorporate a recessed panel at the base and should run vertically for the maximum possible length. Raking and offset blocks should be avoided if possible.

If a flue pipe links the flue with the terminal, the piping should conform to Building Regulations and CP 337. It should not be at an angle less than 135° from the horizontal.

Avoid damage to flue blocks during on-site stacking and handling and use a high alumina cement-based mortar, in accordance with the block manufacturers' recommendations. Remove any parging likely to restrict the flue cross-section, and employ coring during and after construction.

ROOM SEALED APPLIANCES

These require a hole in the outer wall with dimensions specified in the manufacturer's literature. The terminal should not be sited so that combustion products can feed back through doors and windows.

Terminals should not be located immediately beneath eaves or balcony, on a re-entrant position on the face of the building, within 300 mm of ground level or within 600 mm of a corner or surface facing the terminal. A suitable guard should be fitted externally where necessary.

REGULATIONS

It is essential for builders to conform to the requirements of the Building Regulations 1976 and the Building Standards (Scotland) (Consolidated) Regulations 1971 and Amendments 1975, as well as the Gas Safety Regulations 1972, when dealing with the installation of gas appliances, their flues and ventilation requirements. The following summary is for general information only; the regulations themselves should be consulted and the advice of British Gas Regional staff sought if necessary.

Building regulations

Relevant sections of the Building Regulations and Bulding Standards (Scotland) are Part L (Chimneys, Flue Pipes, Hearths and Fireplace Recesses) Regulations 14–22, Part M (Heat Producing Appliances and Incinerators) Regulations 8–12, and Building Standards (Scotland) Part F (Chimneys, Flues, Hearths and the installation of heat producing appliances).

PART L, REGULATIONS 14–22

- Regulation L14 deals with chimneys for Class II gas appliances (i.e. those with input ratings not exceeding 45 kW), covering materials of construction and lining, with a table detailing the maximum length of certain flues.
- Regulation L14(5) covers the fitting of flexible stainless steel liners in existing flues.
- Regulation L15 deals with asbestos-cement spigot and socket type pipes.
- Regulation L16 lists types of flue pipes for Class II appliances, including the two types now mainly encountered, i.e. twin-walled metal and asbestos-cement types.
- Regulation L17 deals with shielding of flue pipes for Class II appliances.
- Regulation L18 details the sizes of flues, covering the cross-sectional area requirements. The requirements are precise but readily understood.
- Regulation L19 deals with openings into Class II appliance flues.
- Regulation L20 refers to flues communicating with more than one room or internal space, and will not normally apply to two-storey-houses.
- Regulation L21 deals with outlets for Class II appliances, of which Part 1 refers to flues serving one appliance and Part 2 to flues serving two or more appliances on different storeys. It describes types of terminals and locations.
- Regulation L22 refers to insulated metal chimneys serving Class I and Class II appliances.

PART M, REGULATIONS 8–12

These regulations deal in detail with the ventilation requirements of gas appliances.

- Regulation M8 contains details of the combustion air requirements, location and appliance ventilation of different types. It is a lengthy and important regulation, which must be closely studied in the light of relevant British Gas requirements, as indeed must all Building Regulations.
- Regulation M9 is another lengthy Regulation dealing with situations where Class II appliances can be discharged otherwise than into a flue.
- Regulation M10 deals with exceptions permitting the discharge of more than one Class II appliance into the same flue, as in multi-storey flat or maisonette construction.
- Regulation M11 describes the provisions and exceptions for Class II gas incinerators.
- Regulation M12 deals with the supply of combustion air to Class II appliances.

Building Standards (Scotland) (Consolidated) Regulations

These regulations are on similar lines to those for England and Wales. They are listed from Regulation F1 to Regulation F29. Regulations F1 and F2 deal with the scope of the Standards and describe the terminology.

Regulations F3 to F20 deal with solid fuel and oil burning appliances, and Regulations F21 to F29 apply to Class II gas appliances.

Gas Safety Regulations 1972

These supplement the Building Regulations, in which they could not be included because they were produced under separate enabling legislation. Much of the responsibility for carrying out the regulations rests with the installation contractor. The Gas Region will advise in cases of doubt. A British Gas publication, *A Guide to the Gas Safety Regulations*, is available from the local British Gas Region.

There are eight clearly defined sections in the Regulations:

- Part I consists of definitions of terms used throughout the Safety Regulations.
- Part II deals with the installation of service pipes defining the responsibilities of the Region and of the builder (in respect of builders' work).
- Part III deals with the installation of meters, and Part IV with the installation of pipework; in this connection, compliance with British Standard CP 331: Part 3, *Low pressure installation pipes*, is important.
- Part V deals with the installation of gas appliances and Part VI with the safe use of gas in appliances, including any that may have been installed prior to the introduction of these regulations.
- Part VII states that alterations and replacements to gas installations after 1 December 1972 must also conform to all the relevant regulations.
- Part VIII defines the penalty for non-compliance with the regulations.

FURTHER INFORMATION

Technical documents

STATUTORY REGULATIONS

Building Regulations 1976

Building Standards (Scotland) (Consolidated) Regulations 1971 and Amendments 1975

Gas Safety Regulations

GAS INDUSTRY INTERPRETATION OF STATUTORY REGULATIONS
A guide to gas safety regulations

GAS INDUSTRY TECHNICAL STANDARDS
List of tested and approved gas appliances

SPECIAL GAS INDUSTRY PUBLICATIONS
Material and installations specifications for domestic central heating and hot water
Gas in housing: a technical guide, 1978

Codes of practice

CP 331, *Installation of pipes and meters*
  Part 1: 1973: *Service pipes*
  Part 2: 1974: *Low pressure metering*
  Part 3: 1974: *Low pressure installation pipes*
CP 333, *Selection and installation of gas hot water supplies*
  Part 1: 1964: *Domestic premises*
CP 334, *Selection and installation of gas cooking and refrigerating appliances*
  Part 1: 1962: *Domestic cooking appliances*
CP 335, *Selection and installation of miscellaneous gas appliances*
  Part 1: 1973: *Domestic laundering and miscellaneous appliances*
BS 5376: 1976, *Code of practice for selection and installation of gas space heating*
  Part 1, *Independent domestic appliances*
  Part 2, *Central heating boilers of rated output not exceeding 60 kW*
  Part 3, *Ducted warm air systems*
BS 5440: 1976, *Code of practice for flues and air supply for gas appliances of rated output not exceeding 60 kW*
  Part 1, *Flues*
  Part 2, *Air supply*
BS 5449, *Code of practice for central heating for domestic premises*
  Part 1: 1977, *Forced circulation hot water systems*

British standards

BS 567: 1973, *Asbestos-cement flue pipes and fittings, light quality*
BS 669: 1960, *Flexible tubing and connector ends for appliances burning gas* (Amendments June 1961, October 1973)
BS 715: 1970, *Sheet metal flue pipes and accessories for gas-fired appliances*
BS 835: 1973, *Asbestos-cement flue pipes and fittings, heavy quality*
BS 1179: 1967, *Glossary of terms used in the gas industry*
BS 1387: 1967, *Steel tubes and tubulars suitable for screwing to BS 21 pipe threads*
BS 2871, *Copper and copper alloys, tubes*
  Part 1: 1971, *Copper tubes for water, gas and sanitation* (Amended November 1976)
  Part 2: 1972, *Tubes for general purposes*
BS 4814: 1976, *Specification for expansion vessels using an internal diaphragm, for sealed hot water systems*

Note Several of the above Codes and Standards are under revision at the time of going to press.

41 ELECTRICAL SERVICES

REGULATIONS AND STANDARDS 41–2

WIRING SYSTEMS 41–3

PROTECTIVE EQUIPMENT 41–6

CONSUMER'S MAIN SWITCHGEAR 41–9

WIRING ACCESSORIES 41–11

TESTING AND INSPECTION 41–14

41 ELECTRICAL SERVICES

C.D. KINLOCH, DFH, CEng, MIEE
Assistant General Manager, National Inspection Council for Electrical Installation Contracting

The IEE Wiring Regulations set out the requirements for safety from fire and shock, but are not a sufficient basis for the complete design of an installation. This section explains how, within the general rules, the designer has considerable scope for the selection and location of items of electrical equipment.

REGULATIONS AND STANDARDS

Wiring regulations

As long ago as 1882, the increasing use of electricity called for a set of rules for ensuring that this new form of energy could be used with safety from fire and shock. The Society of Telegraph Engineers (the forerunner of the Institution of Electrical Engineers – the IEE) set up a committee which published, within six weeks, *Rules and regulations for the prevention of fire risks arising from electric lighting*. The Rules were the first edition of the British national wiring rules, which since 1924 have been called *The regulations for the electrical equipment of buildings* or, more popularly, 'the IEE Wiring Regulations'.

The present status of the IEE Wiring Regulations in the UK is probably unique in the world. Most other countries have national wiring rules that are statutory, and in many of these countries only registered contractors and operatives are permitted to carry out electrical installation work.

The IEE Wiring Regulations are not themselves statutory, and there is no statutory restriction within the UK on firms or individuals who wish to install electrical equipment. These Regulations are, however, invariably accepted as providing the minimum standard of safety from fire and shock; moreover, compliance with the IEE Wiring Regulations is 'deemed to satisfy' the relevant clauses of various statutory regulations such as the Electricity Supply Regulations 1937 and the Building Standards (Scotland) Regulations 1971.

Additionally, the Scottish Building Regulations reproduce Part I of the IEE Wiring Regulations as clauses N.2 to N.7, N.9 and N.10, thus, in effect, making compliance with the IEE Wiring Regulations compulsory in Scotland both for new work and for alterations and additions to existing work.

The 14th Edition of the IEE Wiring Regulations was published in 1966 and has been amended, including metrication, on eight occasions up to April 1976. It is expected that the 15th Edition will be published in 1979 but, to allow completion of contracts and education in the new requirements, the 14th Edition will not be withdrawn until one or two years after publication of the 15th Edition.

The 14th Edition, following the pattern of earlier editions, contains Part I, which sets out in general terms the 'Requirements for Safety', and Part II, which gives 'Means of complying with Part I'. The introduction to the Regulations makes it clear that Part II does not preclude the possibility of other means of complying with Part I, but departures from the requirements of Part II should be the subject of a specification by a Chartered Electrical Engineer or other responsible person, who must satisfy himself that the proposed departure provides a standard of safety no less than that which would have been achieved by compliance with Part II.

An installation having departures from the requirements of Part II may not be described as complying with the IEE Wiring Regulations, unless the departure is effectively covered by a certificate issued by the IEE under the Assessment of New Techniques Scheme (ANTS).

ANTS Certificates

To facilitate the introduction of new products, materials or methods that were not envisaged when the IEE Wiring Regulations were framed, the IEE Assessment of New Techniques Scheme was set up to provide an authoritative assessment and, when appropriate, to issue certificates; these include a statement to the effect that the product or method, when installed in accordance with any conditions stated in the Certificate, provides a standard of safety no less than that which would have been achieved by compliance with the relevant regulations in Part II of the IEE Wiring Regulations.

Certificates are valid for three years with the intention that either the IEE Wiring Regulations will be amended or an appropriate British Standard will be issued. If, however, neither of these events has occurred and if the product or method has a satisfactory record, a new Certificate may be issued for a further period of three years. Details of the Scheme may be obtained from the Secretary, IEE, Savoy Place, London WC2R 0BL.

British Standards and Codes of Practice

British Standards exist for nearly all electrical installation materials. The IEE Wiring Regulations make specific reference to many of these Standards, but additionally there is the following general requirement: 'All apparatus, accessories and fittings shall comply with ... the relevant requirements of the applicable British Standard ... .' Accordingly, only materials complying with British Standards should be used; otherwise it is not possible to claim that the installation complies with the IEE Wiring Regulations.

British Standard Codes of Practice deal with specialised techniques that are not included in the IEE Wiring Regulations, e.g. fire alarms (CP 1019), lightning protection (CP 326) and emergency lighting (BS 5266). Although the latter document does not apply to private dwellings, it does deal with common access routes within blocks of flats.

For the complete list of British Standards and Codes of Practice, reference should be made to the BSI Year Book or to the Electrical Engineering sectional list obtainable from the British Standards Institution, Sales Branch, 101 Pentonville Road, London N1 9ND.

WIRING SYSTEMS

The IEE Wiring Regulations state the installation requirements that must be observed for the various available systems of wiring, but give little or no guidance to assist the designer in selecting the most suitable system for buildings of various kinds.

The features of some commonly used systems are described under two main headings:

- *Insulated cables,* which require additional mechanical protection.
- *Sheathed cables*, having basic insulation and a robust sheath, which generally do not require additional mechanical protection.

Insulated cables

Such cables are required to be installed in conduit or trunking, which may be made of steel or rigid plastics (e.g. rigid p.v.c.).

Steel conduit should comply with BS 4568, which provides for four classes according to the degree of protection against moisture, as follows:

Class 1 Light protection both inside and outside; e.g. priming paint.

Class 2 Medium protection both inside and outside; e.g. stoved enamel.

Class 3 Medium-heavy protection. Inside as Class 2, outside as Class 4; e.g. stoved enamel inside, sheradised outside.

Class 4 Heavy protection both inside and outside; e.g. hot-dip zinc coating or sheradising.

Steel conduits are also classified according to wall thickness as heavy gauge or light gauge. Heavy gauge steel conduit will accept a screw thread and, together with suitable conduit accessories, forms a rigid assembly that not only protects the non-sheathed cables but also acts as the earth-continuity conductor for the installation; on the other hand, light gauge steel conduit, of circular or oval section, will not accept a screw thread and is therefore used only where earth-continuity is provided by other means.

Rigid p.v.c. conduits may be used as an alternative to steel conduits in nearly all locations, the principal exceptions being for flameproof installations and for the mechanical protection of cables located immediately under floorboards. For damp or corrosive situations, p.v.c. conduits are superior to steel because, of course, they do not rust or corrode with consequent loss of mechanical strength and earth-continuity. P.V.C. conduits should comply with BS 4607.

With the exception of prefabricated systems, all conduit systems are required to be completely erected before the cables are drawn in, thus ensuring that the cables can be subsequently withdrawn and replaced.

Prefabricated systems, available from a number of manufacturers, usually consist of extruded plastic sections having the requisite number of tubes or channels each containing one single-core cable. The principal application of these systems is in repetitive building work, where a large number of identical installations permits the manufacturer to supply complete wiring kits for each dwelling.

Conduits are generally available in sizes between 16 mm and 32 mm diameter, to accommodate various sizes of cables and numbers of circuits. Where multiple circuits are required, it is often more convenient to use trunking of rectangular section with a removal lid, thus permitting cables to be laid in rather than to be drawn in as with conduit.

Trunking also facilitates the installation of additional cables at a later date, but the installer must take account of the rating factors for permissible currents in cables installed in groups, in respect of both the original installation and the additional circuits required. Steel trunking should comply with BS 4678, Part 1, which provides for a range of sizes up to 150 × 150 mm; p.v.c. trunking will be covered by a further part of BS 4678.

A range of miniature p.v.c. trunkings has been introduced in recent years. Apart from providing a downward extension to the range of available trunkings for general applications, these small sizes are useful for surface wiring for additions or alterations where it is not possible or desirable to cut chases to bury cables: being of rectangular sections and not requiring external fixings, they may be installed unobtrusively above skirting boards, adjacent to door frames or in the angles of walls, to contain either sheathed or non-sheathed cables.

Sheathed cables

Types are available for almost any application, ranging from mines and quarries down to domestic installations. Where the duty is arduous, armouring by steel strands or other means is provided to protect the cores against mechanical damage, but for many types of installation an insulating sheath without armouring is satisfactory.

The types commonly used in domestic installations are considered below.

P.V.C. insulated, p.v.c. sheathed cables (p.v.c./p.v.c.). These cables are made to BS 6004, Table 5, and are available with two or three cores with an earth-continuity conductor; single-core and two- or three-core types are also available without earth-continuity conductor (BS 6004, Table 4). They are suitable for use in the great majority of single-occupancy dwellings and many other premises where the conditions are similar.

P.V.C. insulated, p.v.c. sheathed cables with metallic covering. Two versions of these cables are available; both of them are covered by ANTS Certificates (75/8 and 76/7), which give information on current ratings and methods of installation that must be observed.

Mineral-insulated metal-sheathed cables (m.i.m.s.). These are made to BS 6207 and are available in a wide range of sizes with or without an overall covering of p.v.c. Although more rigid than some other types of sheathed cable, their small size is often an advantage for surface wiring. Because of their fire-resisting properties, they are often used for fire alarm or emergency lighting circuits; the p.v.c. covered type is suitable for installation in locations exposed to the weather and for underground wiring to outbuildings and the like.

P.V.C. insulated steel wire armoured cables (p.v.c./s.w.a./p.v.c.). Manufactured to BS 6346, Tables 6 to 12, these cables are primarily intended for industrial installations, but they are often useful for sub-main wiring in service ducts for blocks of flats or to outbuildings.

Current rating of cables

The size of cable required in any situation depends on several factors, all of which must be correctly applied to ensure that the cable does not overheat, with consequent deterioration of insulation.

The cable rating tables in the IEE Wiring Regulations show the maximum current that may safely be carried by cables forming a single circuit installed under normal conditions, either exposed directly to the air or enclosed in conduit, trunking etc. with an ambient temperature of 30°C.

If the conditions of installation differ from the norm in respect of ambient temperature or the grouping of circuits, rating factors must be applied; these reduce the effective current carrying capacity of the cables, thus necessitating a larger size for a given load current.

It is also necessary to take account of voltage drop due to the current flowing in the cable. After selecting a cable on the basis of its temperature characteristics, as indicated in the preceding paragraph, a check must be made to ascertain whether the voltage drop requirements have been met, namely a maximum of 2½ per cent of the declared voltage of the supply (i.e. 6 V for a 240 V supply) at any point on the installation. In the event of the voltage drop exceeding this figure, a larger size of cable must be used.

These calculations are particularly important for long sub-main cable runs in blocks of flats and other large buildings, but the requirements apply wherever cables are installed, although experience shows that the cable sizes commonly used for lighting and socket-outlet circuits in dwellings are adequate in respect of current rating and volt-drop. It may, however, be necessary to make the required calculations in respect of cooker or water-heater circuits, cable runs to outbuildings, and any other circuits with special characteristics.

PROTECTIVE EQUIPMENT

Every electrical installation must be adequately protected against excess currents and earth-leakage, either of which, if allowed to persist, could result in damage to the installation or risk of fire or electric shock.

Although the excess-current protective equipment may also provide the necessary degree of protection against earth-leakage, it is convenient to consider the two types of protection separately.

Excess-current protection

When an electric current flows in a cable or other conductor, heat is generated, thus raising the temperature of the conductor and its associated insulation. The cross-sectional areas of all the conductors in an installation are selected so that, under normal conditions, this heating effect will not cause an excessive temperature rise, which would lead to deterioration of the insulation. Because this effect can be determined quite accurately by reference to cable current-rating tables, there is no need to apply a factor of safety when selecting cables, but for this very reason there is a minimum of overload capacity.

Whenever an excess current occurs, it must be interrupted before it can cause damage of any kind; two types of equipment are available, namely fuses and circuit-breakers.

FUSES

Basically, a fuse consists of a short length of conductor of smaller cross-sectional area than any other conductor in the circuit that it is protecting. This small conductor, or *fuse element*, therefore heats more rapidly than the cables, and if the excess current is of sufficient magnitude it melts, thus interrupting the current and protecting the circuit.

The simplest type of fuse in use today is the *semi-enclosed rewirable fuse*. This consists of a *fuse base* having terminals for the circuit conductors and fixed contacts to receive a plug-in *fuse carrier*, which contains the fuse element consisting of a short length of *fuse wire* attached to terminals. When the carrier is inserted into the base, the element is enclosed so as to reduce the dissipation of heat, thus producing a more controllable melting rate, and to assist in extinguishing the electric arc produced as the circuit is broken.

While the rewirable fuse is a simple device, it suffers from the following disadvantages:

- It is relatively slow in operation.
- It has limited fault-interrupting capability.
- It is possible to replace the element by a larger size, thus reducing or destroying the effectiveness of the protection.
- Replacement of the element requires time and a certain amount of skill.

An improvement over the rewirable fuse is the *cartridge fuse*, which consists of a tube of ceramic or other insulating material having metal end-caps and containing the fuse element embedded in a suitable granular filling material. The cartridge fuse-link is clipped or bolted to a fuse carrier, which is inserted into the base in the usual way.

The operating characteristics of the cartridge fuse are controlled by the manufacturer and are not appreciably affected by the method of installation; accordingly they can be made to operate faster than rewirable fuses and to interrupt larger prospective fault currents.

For domestic applications, cartridge fuse-links (to BS 1361) have differing body sizes according to their current rating, so as to ensure that only a link to the correct rating is used for replacement.

CIRCUIT–BREAKERS

Circuit-breakers are, in effect, mechanical switches that are arranged to open automatically in the event of an excess current. The larger circuit-breakers have adjustable trip settings but, for final circuit protection, units are available having preset characteristics that ensure reliability of performance. These are known as moulded case circuit-breakers (m.c.c.b.) (to BS 3871, Part 2) and miniature circuit-breakers (m.c.b) (to BS 3871, Part 1); the principle is the same in both instances, the difference in nomenclature relating only to the range of current ratings.

One of the advantages of both m.c.b.s and m.c.c.b.s is the ability to reclose them simply by the use of an operating lever or pushbutton, so that no time is lost in restoring the supply once the fault has been repaired or disconnected. From the safety point of view, the principal advantage lies in the fact that the operating current is determined by the manufacturer and no subsequent alteration by the user is possible.

M.C.B.s are installed in distribution boards in a similar manner to fuses, and occupy similar space. Some designs are arranged for fixing by screws or bolts, but others are arranged for plug-in assembly; at least one manufacturer offers a range of domestic consumer units that will accommodate rewirable fuses, cartridge fuses or m.c.b.s, all of which plug-in to the standard bases according to current rating. To some extent this reduces the inherent safety of m.c.b. protection, because a low-rated m.c.b. could be replaced by the user with an over-rated rewirable fuse, with consequent loss of protection.

Earthing and bonding

To ensure safe discharge of electrical energy in the event of a fault, any exposed metalwork associated with the electrical installation must be effectively earthed.

In principle, this is achieved by means of earthing and bonding conductors, which may be referred to collectively as protective conductors, run throughout the installation and connected to all exposed metalwork of the installation and extraneous fixed metalwork. The various protective conductors are connected together to the consumer's earthing terminal, located in or adjacent to the main switchgear for the installation, and from there by means of an earthing lead to the earth electrode or other means of earthing.

Although not under statutory obligation to do so, Electricity Boards will, whenever possible, provide an earthing terminal to which the consumer's earthing lead should be connected. Where an earthing terminal is so provided, it may be connected by the Electricity Board *either* to the metal sheath of the incoming service cable *or* to the neutral conductor of the supply under the system known as protective multiple earthing (p.m.e.). Because there are somewhat different requirements for bonding within the consumer's installation, it is desirable to ascertain at an early stage which type of earthing, if any, will be provided.

Where the Electricity Board is unable to provide an earthing terminal for use by the consumer, it is necessary to provide an effective earth electrode such as an earth rod or plate. In practice, it is difficult and often expensive to provide effective earthing by this means alone, because of varying soil conditions and other factors, but the addition of an earth-leakage circuit-breaker will, in most cases, provide the necessary degree of protection for the installation.

An earth-leakage circuit-breaker (e.l.c.b.) is a switch that opens automatically when a leakage of current to earth occurs as a result of insulation failure or other cause. Two distinct types of e.l.c.b. are available, namely current-operated and voltage-operated, the former being generally preferable because of its inherent discrimination and reduced susceptibility to incorrect operation.

Current-operated e.l.c.b.s (sometimes known as residual-current or differential-current e.l.c.b.s) have operating currents in the range 1A to 30 mA or less. The maximum permitted value of operating current for a given installation is determined by the value of earth-loop impedance in accordance with the IEE Wiring Regulations, but there are advantages in selecting an operating current lower than the maximum permitted value, thus allowing for a possible subsequent increase in earth-loop impedance and providing a useful degree of protection against fire and shock.

At present, the increased cost of the more sensitive current-operated e.l.c.b.s is a deterrent to their general use, coupled with the possibility of so-called 'nuisance tripping' due to transient leakage caused by condensation on cooking appliances and water absorption by storage radiators. Nevertheless, e.l.c.b.s of 30 mA operating current do provide much enhanced safety from risk of fire and fatal shock, as compared with, say, 500 mA tripping currents.

Voltage-operated e.l.c.b.s function by detecting the rise of potential on the metalwork of an installation when a fault occurs. While this may appear to be a simple concept, in practice a number of factors combine to reduce the reliability of this method of earth-leakage protection.

Firstly, it is difficult to achieve discrimination between two or more voltage-operated e.l.c.b.s in the same premises or even in adjacent premises. It is not unknown for a fault in one house to cause e.l.c.b.s to operate in adjacent houses, to the annoyance of the occupants. Voltage-operated e.l.c.b.s are also liable to failure of operation due to the presence of so-called parallel earth paths, caused by cross-bonding to other earthed metalwork such as water pipes, thus effectively short-circuiting the trip coil of the e.l.c.b.

The current-operated e.l.c.b. is entirely free from these effects. Complete discrimination is ensured, and the presence of cross-bonding tends to assist rather than hinder the operation under fault conditions.

EARTH CONTINUITY

Having established the main earthing by one of the means described above, it is necessary to ensure that all exposed metalwork of the electrical installation is connected, by means of an earth continuity conductor (e.c.c.), to the main earthing terminal.

Every circuit of an installation is require to have an e.c.c. throughout its length, even if initially only all-insulated or double-insulated (Class II) equipment is to be installed, thus providing for subsequent alteration to metal-cased (Class I) equipment.

The e.c.c. may be a separate conductor, a core of a multi-core cable, metal conduit and trunking, or the metal sheath of a cable. Because of the importance of ensuring proper earth-continuity, the IEE Wiring Regulations indicate requirements in some detail and specify the tests required to confirm the effectiveness of earthing.

BONDING

In addition to ensuring proper earth-continuity throughout the electrical installation, it is also necessary to ensure that there are no possible voltage differences between the earthed metalwork of the installation and any extraneous fixed metalwork, such as water and gas pipes and fittings and accessible structural metalwork.

Where it is not possible to ensure complete segregation for such extraneous metalwork, it is necessary to establish a definite electrical connection by the provision of one or more protective bonding conductors. In practice, segregation is not easy to achieve and it is usually preferable to rely on the principle of effective bonding.

The IEE Wiring Regulations require a main bond between the consumer's earthing terminal and metal gas and water services to be effected as near as practicable to the point of entry of such services. As the intention of this requirement is to ensure bonding of pipework within the building, it is recommended that these bonds should be made to the consumer's pipework immediately adjacent to the water stop-cock and the gas meter.

If the pipework has metal-to-metal joints providing good electrical continuity, these main bonds are normally sufficient to meet the bonding requirements throughout the building. However, if, for example, non-metallic items such as plastic water tanks are installed, a local bond between the incoming and outgoing pipes is necessary. Similarly, if the water taps are not in good metallic contact with a metal sink, a bond must be established to an earthing terminal on the sink.

When other metallic services, such as oil feed pipes, are installed, these must also be bonded; although in this case not a specific requirement of the IEE Wiring Regulations, it is desirable to establish the bond where the pipe enters the building.

In the case of a p.m.e. supply, the bonds are referred to as 'electrical connections' to avoid confusion with the bonds required by the IEE Wiring Regulations. In many cases, however, one conductor serves the dual function and it is only necessary to ensure that the correct size of cable is installed. The p.m.e. electrical connections are required by the PME Approval to be established between the Electricity Board's earthing terminal and the point of entry of all incoming metal services (other than telecommunications) and structural steelwork that are, or are likely to be, in contact with the general mass of earth. There are no other requirements for bonding in the p.m.e. requirements, but bonding to IEE Wiring Regulation requirements must still be effected.

Electricity Boards have statutory authority to require compliance with the terms of the PME Approval before connecting a consumer's installation to a p.m.e. supply. Accordingly, the relevant Board should be consulted in any case of doubt as to the specific provisions required in any given building.

CONSUMER'S MAIN SWITCHGEAR

In deciding on the location of the main switchgear for any building, the designer has to consider various factors that may, to some extent, be conflicting:

- length of mains service cable;
- access for meter reading;
- space for Electricity Board's equipment;
- space for consumer's equipment;
- accessibility by consumer;
- minimum length of circuit cables, particularly for the cooker and other heavy load such as an instantaneous electrically heated shower unit.

Many Electricity Boards now provide outdoor meter cabinets to contain the service-cable termination and the meter. Clearly, the location of this cupboard must be agreed with the Board because it must be accessible to the meter reader, but it should also be placed in a convenient position relative to the position of the consumer's main switchgear, preferably back-to-back, so as to achieve the minimum length of the connections between the meter and the main switchgear (the meter tails).

If the use of extended meter tails is unavoidable, consultation with the Electricity Board is essential. If the Board permits, it may be convenient to install a switch-fuse in the meter cupboard, but if there is not enough space or it is not permitted, the Board may require the extended tails to be run in steel conduit or mineral-insulated metal-sheathed cables.

To comply with the basic requirements of the IEE Wiring Regulations, the consumer's main switchgear must have the following items:

- main switch;
- main excess-current protective device, i.e. a fuse or circuit-breaker;
- excess-current protective devices for each outgoing circuit.

However, subject to the use of meter tails of appropriate cross-sectional area related to the Electricity Board's service fuse, the main excess-current protective device may be omitted where the main switch and circuit excess-current protective devices are either immediately adjacent or, preferably, combined in the same assembly.

This exemption permits the use of consumer control units (c.c.u.) complying with BS 1454. These units incorporate a double-pole switch and from four to twelve fuses or miniature circuit-breakers (m.c.b.) for final circuits. Variants of the standard units are available for controlling both off-peak and normal supplies, but it is often more convenient to use two independent consumer units, each with the appropriate number of ways.

The number of ways required in the c.c.u. is determined by the arrangement of final circuits in the house. Except in the case of small flats or cottages, it is usual to provide at least two final circuits for lighting, each rated at 5 amperes (A), thus avoiding complete loss of lighting in the event of a fault on either circuit.

For larger dwellings, additional lighting circuits may be necessary to ensure that the load on each circuit does not exceed 5A, namely 1200 watts for tungsten lamps or 660 watts for fluorescent lighting. In the absence of specific information on the load requirements, each 5A circuit should supply not more than ten lighting points.

The number of ways for ring final circuits supplying 13 A socket-outlets to BS 1363 in domestic premises must be determined on the basis of at least one ring for every 100 m² of floor area or part thereof. Experience has shown that the normal load requirements within such an area will not exceed 30 A, irrespective of the number of socket-outlets or fixed appliances supplied from fused connection units; accordingly the number of outlets is unrestricted, but where more than one ring final circuit is provided the outlets should be reasonably distributed between the rings.

Where significant fixed loads are to be supplied, it is desirable to provide individual circuits to avoid undue loading of ring circuits with consequent reduction of available capacity for general purposes. For example, if electric water heating is required by means of an immersion heater or other normal tariff heater, a separate 15 A way should be provided in the consumer unit. Even if a water heater is not installed in the first instance, it is good practice to allow for possible future installation by allowing a blank way in the consumer unit. Similarly, provision should be made for a 30 A or 45 A way for an electric cooker, even if it is known that gas cooking will be used initially.

Other loads that may be required include electric shower units, which may have a rating of 6 kW; these require the provision of a separate final circuit of 30 A rating. In any case, it is desirable to allow spare ways on the consumer unit for possible future additions to the installation. These spare ways should be blanked off.

For non-domestic premises, it is necessary to make an assessment of the load requirements and to provide an appropriate number of circuits. Where the ultimate use of a building has not been determined (e.g. speculative office blocks, etc.) it is recommended that ring final circuits should be provided on the same basis as in domestic premises, namely one ring for each 100 m^2 of floor area.

WIRING ACCESSORIES

Wide ranges of matched accessories are available from British manufacturers. Almost without exception these comply with relevant British Standards, but within the dimensional limitations of the Standard there is scope for individuality in mechanical and electrical design, with varying ease of installation and customer appeal.

Most accessories are designed for flush-mounting on appropriate metal or plastics boxes that contain the terminations of the cables buried under wall finishes. Where surface-mounting accessories are required for any reason, suitable boxes are available to convert the standard flush accessories. Reference should be made to manufacturer's catalogues for full details of the wiring accessories available for domestic and other installations.

Socket-outlets

Careful thought should be given to the placing of socket-outlets, bearing in mind the probable usage of each room.

The minimum number of sockets to be installed in dwellings is set out in various documents, including the following:

- Parker-Morris Report, *Homes for Today and Tomorrow* (1961);
- National House-Building Council Handbook (1974);
- Scottish Building Regulations (1971);
- Electrical Installation Industry Liaison Committee Recommendations (1975).

The provisions set out in columns 1, 2 and 3 of Table 41.1 represent a considerable advance over the number of socket-outlets previously considered adequate, but even now they should be treated strictly as minimum numbers and should be increased where necessary, because insufficient or wrongly placed socket-outlets must lead to customer dissatisfaction coupled with an early need for additions or the use of long lengths of flexible cord or multi-way adaptors. Having taken these factors into account, column 4 represents a reasonable provision of socket-outlets for the foreseeable future.

In Table 41.1, a two-gang socket-outlet is counted as two sockets. There is a restriction on the number of two-gang sockets in columns 1, 2 and 3, but the EIILC recommendations specify that all sockets shall be two-gang; accordingly, in column 4 the actual number of points is half that shown, i.e. eight sockets are provided by four two-gang units.

The mounting height of socket-outlets should be carefully considered in relation to the anticipated usage of each room and the physical abilities of the likely occupants. The IEE Wiring Regulations recommend that the minimum height of a socket-outlet above the base floor or above a working surface should be 15 cm, so as to permit easy insertion and withdrawal of the plug without restriction by the largest size of flexible cord that may be required. For many locations, socket-outlets could, for convenience, be located well above the minimum height, e.g. in bedrooms for bedside lights, electric blankets etc., and wherever access to a low-level socket may be impeded by furniture or household equipment.

In old people's dwellings, it is very usual to mount all socket-outlets at about 1 m above the floor to avoid the need for stooping, but this must be weighed against the greater tripping hazard due to the resulting loop of flexible cord to floor-standing appliances.

Table 41.1 MINIMUM NUMBERS OF SOCKET–OUTLETS

Room	*Parker-Morris Report*	*Scottish Building Regulations*	*NHBC*	*EIILC*
Kitchen	4	3	4	8
Living room	3	4	3	12
Dining room	1	2	2	6
Double bedroom	2	2	2	8
Single bedroom	2	2	2	6
Study bedroom	2	2	2	6
Study	-	2	1	-
Landing/stairs	-	}	1	2
Hall	1	}	1	2
Garage	1	} 2	-	4
Store/workroom	1	}	-	-

Similarly, the choice between switched or non-switched socket-outlets requires consideration. For general-purpose sockets, switched units are more convenient for the user, especially where appliances do not incorporate switches, but there is no need for switches for appliances that are permanently in use, such as refrigerators and freezers. Where the eventual use for sockets cannot be determined in advance, switched sockets should be provided throughout.

Lighting switches

Switches are usually located about 1.4 m above the floor, but this may not be the optimum height because they are out of reach of children and are not particularly easy to operate when entering a room carrying a tray or other object; to avoid these disadvantages, it may be preferable to mount lighting switches at about 1 m, but this is a matter to be determined by the designer or architect rather than by the installer.

For most installations the widely available ranges of 'plate switches' are suitable, but for other applications grid-type switches offer all possible combinations of finish and technical requirements. Rocker-type switches appear to find favour with many people, although experience indicates that elderly people prefer the more positive action and clearer indication of the on and off position offered by the more traditional dolly-type switches.

Dimmer switches offering easy control of levels of illumination are often installed as replacement for standard switches and it is likely that demand for this will increase. In the past, not all dimmer switches have been satisfactory but the publication of BS 5578 for these units should result in better and more reliable products. Standard dimmers are suitable only for tungsten lighting; it is possible to dim fluorescent lighting but special equipment is necessary before this can be achieved.

Cooker controls

For most housing, it is usual to make provision for electric cooking even if the use of another fuel is intended initially.

An isolating switch must be provided alongside the cooker, in a position where it can easily be reached even when the cooker is in use; it should not be mounted directly above the cooker, where it would not be readily accessible and could interfere with the installation of a cooker hood if required.

The cooker isolating switch may be incorporated in a cooker control unit having a socket-outlet controlled by a separate switch,with or without neon indicators. The cooker control unit was originally designed to give easy provision of a socket-outlet in a kitchen, and in many early installations it was the only 'power' socket-outlet available in the house. With the ready availability of socket-outlets supplied by general-purpose ring circuits, there is no need for a socket-outlet in a cooker control unit and it is better to use a cooker control switch without a socket-outlet.

Cooker circuits are normally rated at either 30 A or 45 A; a 30 A circuit in domestic premises is permitted to supply a cooker of total loading up to 18.4 kW while a 45 A circuit will supply a load of 30.4 kW, these figures being reduced to 14.4 kW and 26.4 kW respectively if a socket-outlet is incorporated in the cooker unit.

For split-level cookers, a single circuit and switch may be used to isolate both parts of the split cooker, provided that each part is within 2 m of the switch. For more widely spaced items, separate circuits and switches would be required for each item.

For connection to a movable cooker, it is desirable to provide an outlet terminal box at about 0.75 m above the floor, thus permitting the final connection to be made by a flexible cable rather than 'twin and earth' cable, even though the latter is permitted by the IEE Wiring Regulations.

Apart from the cooker circuit, a sufficient number of socket-outlets and/or fused connection units must be provided at appropriate locations, particularly where it is expected that stationary items such as refrigerators, freezers, washing machines, waste disposal units etc. will be used.

Special requirements

BATHROOMS

Because of the special risks in rooms containing a fixed bath or shower, it is not permitted, under Regulation D.17 of the IEE Wiring Regulations, to install an ordinary socket-outlet nor to make provision for the use of a portable appliance other than an electric shaver. Fixed appliances are permitted but any switch or other means of electrical control, such as a thermostat knob, must be located out of reach of a person using the bath or shower.

When electrical equipment is installed in a bath or shower room, it is essential to ensure that there is effective electrical continuity between the exposed metalwork of the electrical installation and other fixed metalwork such as metal baths and exposed metal pipework.

Shaver outlets in bathrooms must comply with BS 3052 although, for other rooms, outlets complying with BS 4573 are acceptable. Outlets to BS 3052 have isolation from the mains and earth by means of a double-wound transformer, in addition to some form of overcurrent protection to ensure that the outlet cannot be misused for loads in excess of that normally imposed by electric shavers; those of BS 4573 have only the overcurrent protection.

Electric shower units may be installed within reach of a person using the shower, because either the knob is not an electrical control or the manufacturer has obtained an ANTS Certificate to give exemption from the requirements of Regulation D.17. There is, as yet, no British Standard for shower units, but several models are included on the supplementary list published by the British Electrotechnical Approvals Board.

There are two distinct designs of unit, namely those having metal-sheathed heating elements and those having the heating wires immersed directly in the water. The installation of the metal-sheathed type is no different in principle from, say, an ordinary immersion heater, the basic requirements being means of isolation and earthing of exposed metal parts. For the immersed-element type, additional requirements are stated in Regulations C.44 to C.50, of which the most important are the more stringent earthing and bonding requirements.

The water supply to immersed-element heaters must be through metal pipes that are in effective contact with earth by a means independent of the earth-continuity conductor to the heater. Where the water supply is by metal pipes having metal-to-metal joints, this requirement is assured by the main bond to the incoming water service, even where the latter is non-metallic, but if the cold water system is non-metallic, it must be ensured that the final pipe-run to the heater is metal (for a minimum length of, say, 50 cm) and that a separate bonding lead is run from that pipe to the main earthing terminal of the installation.

Because of the load requirements (usually 6 kW) a separate 30 A circuit is essential for electric shower units.

OUTDOOR LOCATIONS

In view of the increasing use of electrically operated garden tools, it is desirable to allow for this by the provision of a suitably placed socket-outlet, perhaps in a garage or other outbuilding.

Because of the greater risk arising from the use of long flexible cords, and hence the difficulty of maintaining reliable earth-continuity, most such tools are of double-insulated construction and therefore are not required to be earthed. Despite this, there may still be a shock hazard if the flexible cord is damaged, and it is therefore strongly recommended that any socket-outlet provided for such tools should be individually protected by a current-operated earth-leakage circuit-breaker having an operating current of not more than 30 mA.

If a socket-outlet, or indeed any other electrical equipment, is located outside a building, only weatherproof items should be installed; ordinary domestic accessories are not suitable, even if they are protected from driven rain and snow, because of deterioration caused by dampness.

TESTING AND INSPECTION

The large number of regulations that need to be observed make it essential for every electrical installation to be inspected and tested, on completion, to ascertain whether the requirements of the IEE Wiring Regulations have been met.

Under Regulation E.12, a Completion Certificate must be handed to the person who ordered the work; on this certificate, the person responsible for the work is required to give details of the installation and to state that it complies with the IEE Wiring Regulations.

If departures from the IEE Wiring Regulations have been required by specification or for other reasons, such departures should be shown on the appropriate space on the certificate, together with comments on the existing installation where the certificate relates to an extension.

The Completion Certificate is an important contractual document, and forms a safeguard both to the customer and the contractor. It should not be confused with the 'Connection Certificate' or other document required by the Electricity Board before connection of supply, although in some cases a copy of the IEE Completion Certificate may be accepted for the latter purpose.

The IEE Completion Certificate should be signed by or on behalf of a Chartered Electrical Engineer, by a member of the Electrical Contractors Association or the Electrical Contractors Association of Scotland, or by an Approved Contractor on the

Roll of the National Inspection Council for Electrical Installation Contracting. If the certificate is actually signed by another person (e.g. an employee who carried out the tests), the name and designation of the person accepting responsibility for the accuracy of the certificate must be stated.

The two Electrical Contractors Associations (ECA) referred to above, in addition to providing a full range of commercial services for their members, are also concerned to ensure that their members carry out safe and satisfactory installations. They operate guarantee of work and guarantee of completion schemes, and it is also a requirement of membership that the firm should be approved by the National Inspection Council for Electrical Installation Contracting (NICEIC).

The NICEIC is a non-profit-making consumer protection organisation set up in 1957 by all sections of the electrical industry to establish and maintain a Roll of Approved Electrical Installation Contractors. The Council has a force of Inspecting Engineers who examine samples of installation work by each contractor before acceptance on the Roll and thereafter at regular intervals.

The standard of work required is compliance with the IEE Wiring Regulations; if a contractor's work falls below the required standard, the interval between inspections is reduced, and if no improvement follows the contractor is removed from the Roll.

The NICEIC will also investigate allegations of bad workmanship and require the contractor to carry out remedial work to correct departures from the IEE Wiring Regulations; but it does not deal with commercial matters, such as the price charged for work, neither can it take any action in respect of work done by firms not included on the Roll.

42 HEATING SERVICES

COMFORT	42–2
HEATING SYSTEMS	42–2
INSULATION	42–9
VENTILATION	42–11
DESIGN	42–14
THE FUTURE	42–15

HEATING SERVICES 42

42 HEATING SERVICES

D.H. HAYES, CEng, FInstF, FIDHE, MIGasE, MCIBS
Consulting Engineer

This section analyses the object of heating systems, which is to give comfort, and discusses the main options available in fuels, appliances and systems. Information is also given on the associated subjects of ventilation, condensation and insulation. The section should be read in conjunction with Section 43.

COMFORT

The purpose of space heating is not to provide heat, but to provide environmental comfort. Our own biological control (about the most reliable mass-produced thermostat ever) adjusts to astoundingly wide variations, but compensation is uncomfortable in extreme conditions. Comfort is when biological adjustments occur unnoticed.

The commonest causes of discomfort are too high or too low air temperatures, too little or too much air movement, and chilly walls or uncurtained single glazing, which cause a chilling sensation even when measured air temperatures are high. Our forefathers' traditional open fire partly compensated for high body radiation losses into the masses of cold masonry around them, and they hung up tapestries not only to improve the decor, but to improve comfort conditions.

Comfort limits are narrowing as we become more conditioned to better comfort standards at work and at home. People live longer than they did, and old people need higher temperatures than younger ones, who keep themselves warm by their own greater physical activity and better blood circulation. Higher temperatures are also required as indoor clothing tends to become lighter.

Recent increases in fuel costs have motivated moderation in raising indoor temperatures, and sensibly so considering that lowering a thermostat by just 1°C can afford an 8 per cent reduction in room heating costs.

Still air soon offends as stale air, but air movement must be moderate to avoid draught. Even warm air, if moving too fast, can cause justified draught complaint, as householders living with a badly designed warm-air system have found to their discomfort.

Comfort criteria are therefore:

- good insulation to the outside shell to keep internal surfaces warm;
- widespread distribution of heat input, preferably with some radiation;
- automatic control to prevent overheat, wastage or underheating discomfort;
- daily heat input of sufficient time to keep structure, furniture and fitments above chill temperatures, and to reduce condensation risk;
- sufficient ventilation to reduce the probability of condensation and to maintain air freshness;
- avoidance of excessive air movement or over-ventilation.

HEATING SYSTEMS

Fuels

The Englishman's home was traditionally his castle, and to warm it he has presumed an absolute right to use whatever fuel he chooses. But the economics of house building and fuel services inevitably reduce freedom of choice, and in fact there is little evidence that nowadays occupiers have a marked preference for one form of heating or fuel over any other; their basic concern is that the home can be made comfortable at reasonable running cost.

Gas has become favourite because of its lowest heat cost and its adaptability to local appliances, and also because gas cooking is often preferred. But limited reserves of natural gas preclude a future of low-cost abundance, and logically as this premium fuel runs out so its price will increase.

There are greater reserves of oil, but depletion must motivate conserving oil for aviation and petrochemicals etc., rather than burn it for space heating. It is unlikely that oil prices will regain their pre-1973 price advantage.

Solid fuel is mostly required as a smokeless processed product, which adds to its cost. Appliances burning non-processed coal are steadily improving, and at the time of writing (1978) these 'smoke consuming' appliances offer the lowest-running-cost domestic heating. There are ample reserves of coal. Mechanisation should at least stabilise if not reduce coal prices, and it is political good sense that the price of coal will be held below that of increasingly precious gas and oil.

Electricity is now (1978) in surplus, and construction of ever larger fossil-fuel and nuclear power stations suggests a continuing abundance. But the capital costs of electricity generation are enormous, so electricity cannot provide low-cost heating.

A builder's decision must be largely affected by capital costs and availability of services. Developers encounter widely differing constraints from gas and electricity suppliers according to local policy and mains-laying costs. On balance it is probably best, both for the builder and the householder, that the lowest-capital-cost heat source is provided, preferably giving the householder the option of solid fuel if preferred, at some cost adjustment as appropriate.

Appliances

Heating appliances are heat converters, heat emitters or both. Central heating implies a central converter, e.g. a boiler or warm-air furnace serving several emitters, be they radiators, convectors or warm-air grilles. Applications can be understood by careful study of the huge amount of published commercial literature.

Boilers in the past have been floor standing, but now include wall-mounted and hearth-sited appliances – the latter to replace the traditional back boiler with an open fire before it.

It is worth mentioning that gas-fired hearth boilers can make distracting ignition and combustion noises. Solid-fuel hearth boilers impose the chores of carrying fuel and ash through living rooms, but these labours are made less by magazine-type boilers, which need only occasional stoking; also ash handling is made cleaner by purpose-designed containers.

Self-contained appliances such as gas fires or 'Hot-Line' gas convectors need little explanation, and some can now be thermostatically and time controlled to provide flexible domestic heating. Generally speaking, however, these appliances are less efficient and less satisfying than a designed central heating system.

Hot-water radiators are generally very reliable, effective and unobtrusive, and can be sized to room requirements. Single-panel pressed steel radiators emit 55 per cent of their output as comforting radiant heat and are preferred to double-panel or convector design unless space is at a premium. Siting under windows is preferred because this counteracts the cold air drop that windows cause. A radiator on the opposite side of a room to a window will, by upward air movement, intensify cold air drop discomfort. Good curtaining and double glazing will reduce this discomfort.

Skirting radiators provide unobtrusive and satisfying space heating at the expense of some wall staining, and expansion noises can be difficult to cure. Convector units, whether water or directly heated, produce either natural-draught or fan circulation.

Natural-draught convectors tend to cause high temperature gradients because as warm air is emitted so it rises vertically to the ceiling. Fan convectors effect a more even temperature distribution but at the expense of some noise, and there can be air-movement discomforts from cooler air returning to the unit for re-heat.

Composite gas-boiler/hot-water-storage units are efficient and compact. However, National House-Building Council requirements include a heated linen cupboard; this suggests siting the appliance therein, but can impose difficulties in ensuring unrestricted fresh-air supply and avoidance of fire risk.

Direct electrically heated appliances using on-peak electricity can be simple and should be inexpensive. There is no gain from complicated designs, as any electric heater emits as heat exactly the electricity load it absorbs. Thus central heating by on-peak electricity is illogical and cannot be economically justified.

As houses become better insulated and heat loads reduce accordingly, so off-peak or 'white-meter' electric heating becomes an attractive proposition. Uncontrolled 'storage radiators' are unsatisfactory because they are hottest in the morning and coolest in the evening (quite contrary to comfort requirements), neither do they reduce their output when incidental domestic heat gains from people, hot water or other domestic appliances increase the house temperature – as will rapidly happen in a 'low energy' house of minimal heat loss.

'Controlled output' off-peak heaters with a thermostatically switched output fan are an improvement on uncontrolled heaters, but the fan control does not prevent overheat if the unfanned standing heat emission of the heater is greater than the room requirement.

Builders should ensure that appliances are approved by appropriate authorities. The increasing volume and variety of imported ideas and equipment makes vigilance more important than hitherto. For instance, some continental radiators may be aesthetically more pleasing than those of British design, but the unfamiliar characteristics of aluminium and such materials 'foreign' to us should prompt some caution; indeed, this applies even more to grossly inefficient appliances such as 'gas-logs'.

Approving authorities for appliances and components in the UK are:

- The British Standards Institution, 2 Park Street, London, W1A 2BS.
- British Gas Corporation, Watson House, Peterborough Road, London SW6.
- Solid Fuel Advisory Service, Hobart House, Grosvenor Place, London SW1.
- Domestic Oil Burning Equipment Testing Association, 3 Savoy Place, London WC2.
- British Electrotechnicals Approvals Board, 153 London Road, Kingston-Upon-Thames, Surrey.
- The National Water Council, 1 Queen Anne's Gate, London SW1.

Central heating systems

Evolution is the process of the survival of the fittest, hence the popularity of domestic central heating provided by steel panel radiators served through smallbore pipes from a central boiler, this same boiler also providing hot water by an indirect cylinder.

Most contractors have now learned how to successfully design and install a wet system; user satisfaction is by proven reliability and all-over house comfort without noise. Large radiators do radiate comfort and also spread warmth by gently convected air movement.

Warm air from a central heater has the advantage of lower installation costs if the warm-air supply grilles are centralised around the unit. But 'stub-duct' systems have lost popularity because they have too often been designed down to a price rather than up to a standard of performance. The commonest complaints are localised overheating and wide-area underheating, and uncomfortably sudden increases or decreases in temperature caused by thermostatic on/off switching of the heating output. (A recent innovation of modulating both air flow and gas flame size will help overcome such discomforts: this is the 'Modaireflow' system, made by Johnson and Starley Ltd, Northampton.)

Extended ducts to supply warm air under windows or around the house perimeter would have avoided many justified complaints of discomfort, but warm air is handicapped by having a low heat-carrying capacity. This is illustrated by the fact that an average room needs only a 10 mm diameter hot-water pipe, compared with a 200 mm warm-air pipe, to provide its heat input. Also, longer ducts impose higher resistances to air flow and thus require larger (and probably noisier) air-moving fans.

Another disadvantage of warm-air heating is that, because of its low heat-carrying capacity, the whole air content of a house has to be circulated and the whole house thereby heated to an overall temperature. Differing temperatures can be maintained by siting and controlling water-heated radiators, which is more conducive to economy than high temperatures throughout a house.

The same warm-air difficulties apply to a central warm-air fanned unit heated by hot water from a boiler. This and the fact that these units require very hot water to work effectively, much hotter than would be needed at hot water taps or to make radiators usefully warm, renders such units unjustifiable except in special circumstances of replacement or special architectural requirements.

Oil-fired systems, in addition to running-cost disadvantage, are handicapped by inflexibility compared with gas in wall mounting, tuck-away siting, or the development of ever smaller output boilers to match ever better insulated houses. Also oil burners do need more expert and more regular cleaning than appliances heated by other means, and most oil burners create some noise. Combustion products from an oil-fired boiler are so unpleasant that low level discharge is not really practical. A good flue is much preferred.

Electrical heating has suffered setbacks by careless application causing high running costs. It is nonetheless clean, simple, flexible and elegant. Better house insulation reduces heat demand, so some builders are sincerely convincing their customers that the advantages of a super-insulated house more than compensate for the disadvantage of high electricity costs.

Electric ceiling heating is handicapped by necessarily using high-cost on-peak electricity; but properly applied it provides the ideal comfort of widespread gentle radiation, moderate air temperatures, no noise and no draught.

Any on-peak electrical usage for space heating can only be justified in a low-energy compact property, or for top-up in a conventional house, background heated by white-meter storage radiators. Heat storage can be in an edge-insulated solid floor, and there are also electrically self-heating wall paints and concrete awaiting application. It is reasonable to expect structural heat storage in the future, which will help banish the bad image of bulky block-storage heaters.

Gas-fired 'Hot-Line' units are in direct competition with electrical heat-storage radiators, and do not suffer heat-input time limitations; their output can exactly match heat requirements at all times. But they need to be sited on outside walls, they do not give the radiation benefit of panel radiators, they do not adapt to fine control, and their size range does not match the heat losses of individual rooms.

Outside restraints may affect system choice. An outside sited appliance can be noisy and offensive visually. Balanced-flue gas boilers can create objectionable steam pluming, and low-level products of gas combustion can be a nuisance by their characteristic odour. Oil tanks are bulky, best sited in an enclosure to enhance both appearance and weatherproofing. Solid-fuel storage requires planned accommodation, both to improve appearance and to minimise handling inconvenience.

Air conditioning and heat pumps

Air conditioning has more status than practical appeal in Britain because our temperate climate and al-fresco living habits make air conditioning generally impracticable.

Appliances so far provided are really only summer air coolers, plus perhaps a little air warming in winter. Humidification may be an added extra, and may have some advantage in an overheated house but is not generally necessary in a normal home, where temperatures are moderate, and where kitchen/bathroom moisture diffuses into the house atmosphere.

All air coolers, refrigerators and freezers discharge extracted heat, usually to waste. If the cycle is reversed so that some bulk heat source (usually outside air) is refrigerated, then the heat so extracted can be upgraded as useful heat, and this reversible machine is known as a 'heat pump'. Heat pumps are gaining some publicity, but to date only as an adaptation of an air-cooling appliance, and thus of only limited application. The advantage of a heat pump is that the electrical power put into it is augmented by the upgraded outside heat, thus producing 2–5 kW heat output for every 1 kW of power used. This ratio of advantage is known as the Coefficient of Performance, or COP. A disadvantage of electrically operated air coolers and heat pumps is that (in common with any appliance containing a fan) some noise is inevitable. A further disadvantage is that air conditioners and heat pumps need the apparatus sited outside the house, which could justify visual and noise objections.

Hot water supply

Requirements for hot water are instant availability at constant temperature, and in quantities appropriate to the usage of baths and kitchen appliances.

Constant temperature is easy to achieve by thermostatic control of the heating medium. Continuous availability means continuous input; alternatively, if input must be intermittent, large storage and minimum standing heat loss by high-grade insulation are paramount. Hot-water costs usually relate more to 'the heat you lose, rather than the heat you use'.

Heavy peak demands also obviously call for the largest practical storage capacity. Too little storage will result in 'running out of hot water'; the minimum size now suggested for an average family in the average semi is 140 litres[1].

Replacement of hot water after usage should be at least equal to that of the usual 3 kW electric immersion heater, which will replace a bathful in approximately an hour. A combined system performance is best from either a rapid-recovery cylinder with a large internal heating surface, or from a coiled primary heater and pumped primary circulation.

When the central heating as such is not required (that is, for the whole of the summer and for a surprisingly large part of the winter heating season), a large boiler usually operates at a poor load factor, rather like a double-decker bus running with only two or three passengers in it, and this can be very inefficient (see *Figure 42.1*). Avoidance of such wastage is important, and is further discussed later under Controls. The timing and temperature requirements of space heating are different to those of hot water supply, so separate and unconnected sources may be a more sensible way to provide hot water than a combined system. Alternative independent hot water sources are direct gas-fired storage heaters (as illustrated in *Figure 42.2*), a gas instantaneous heater or an electric immersion heater.

Instantaneous heaters avoid storage standing losses but are handicapped by a restricted flow rate and final temperature, particularly in cold weather when the incoming mains water is abnormally cold. Instantaneous gas heaters are more applicable for small family units or for supply to independent showers requiring only

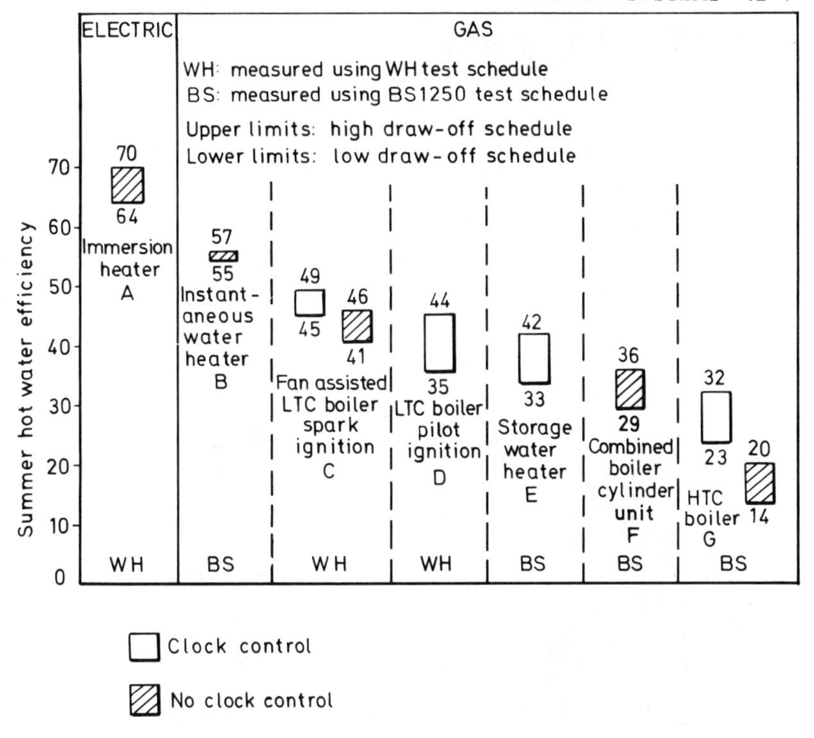

Figure 42.1 Efficiencies of different gas-fired water-heating systems (courtesy British Gas)

a spray of hot water. Electric instantaneous heaters are impracticable for more than a small shower, because of limited flow rate and high running cost.

Off-peak white-meter electricity into an overnight heated oversized storage vessel may be both cheaper to install and cheaper to run than a combined system, and will provide a reliable and satisfactory hot-water service[2].

When various authorities' approvals are established, the combined low-loss heating plus hot-water appliance (such as the one shown in *Figure 42.3*) will prove the claims of higher efficiency and lower running costs made for it.

Hot-water costs can be made unnecessarily high by careless system design. Hot water is always used most often at the kitchen sink, and thus storage is best sited close to the kitchen. The sink draw-off is best run in 15 mm pipe direct from the storage and separate from other hot-water supplies; it should also be out of contact with other pipes to prevent metal-to-metal heat transfer, and pipe insulation would not come amiss.

Towel rails are an extravagance if they continuously dissipate heat from the hot-water system. An unheated towel rail over a radiator is a more sensible bathroom arrangement.

Showers can be both uncomfortable and hazardous if other household water usage causes sudden water pressure changes. Thermostatic shower mixers eliminate such effects and are recommended if their higher cost can be justified to the householder.

Figure 42.2 Self-contained gas-fired direct water heater (courtesy Johnson and Starley Ltd)

Controls

Space-heating temperature control is the paramount economy requirement: a 1°C overheat can result in an 8 per cent running-cost increase. Individual appliance or radiator thermostats provide a more accurate and acceptable control than a central instrument, which can only measure the temperature at the point at which it is sited.

Electrical thermostats are on/off switches that cause a stop/go heating effect. Thermostatic radiator valves are less prone to stop/go complaints because they gradually alter water flow through a radiator, and because whatever hot water is entering the radiator gravitates to the top, so avoiding complaints of cold radiators, even though their heat emission may be only a fraction of full output.

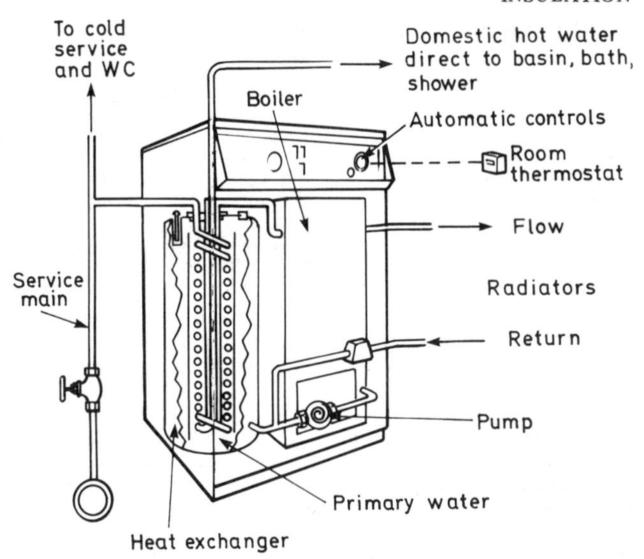

Figure 42.3 Combined heating and hot-water generator (courtesy Worcester Engineering Ltd)

Too severe timing restrictions to a heating system may not give sufficient time for the mass structure of a house to warm, causing discomfort by inverse body-heat loss into the cold surfaces around, and probably condensation thereon. These risks are less if 'set back' is arranged, whereby during off times temperatures do not fall below, say, 6–7°C below occupational temperatures; this adaptation is usually very simple.

Gas or oil appliances should not be allowed to 'short cycle', i.e. needlessly turn themselves on and off when no heating or hot water is required. All control manufacturers now offer schemes whereby this wastage can be simply avoided by proper system design.

INSULATION

The need for structural insulation has been poorly understood by both the British building trade and the public, but rising fuel costs and considerable publicity have now created new awareness of the need to reduce heat loss from houses.

Building Regulations have in the past been more concerned with safety of structure than with conservation of heat. Even Part F of the Building Regulations, upgraded in 1974, specifies insulation hardly adequate to conserve heat at today's fuel costs, and would certainly be less adequate as fuel costs rise.

Promotion of higher insulation standards has been undertaken initially by the electricity supply industry because their heating customers have most to gain. But homes heated by other means will enjoy proportionate benefits of heat-loss reduction and reduced running costs. Table 42.1 is reproduced from an Electricity Council brochure[3], as is *Figure 42.4*, which shows how insulation reduces heat demand, and reduces the length of the heating season by several weeks.

The house builder is under no obligation to improve on Building Regulations, but there are economic arguments for so doing. A well insulated house almost halves its

Table 42.1 EXISTING AND RECOMMENDED LEVELS OF THERMAL INSULATION

	Thermal insulation standards (W/m² °C)		
	Roofs	*Walls*	*Floors*
1972 Building Regulations	1.42	1.70	1.42
1974 Building Regulations	0.60	1.00	1.00
Electricity Council recommendations	0.30	0.40	0.50
	Draught-proofing of all external doors and windows		

space-heating requirements and can be effectively heated from a smaller and lower-cost system. A three-bedroom house with a heating load of 3–4 kW is now a rational reality, and better insulated houses will undoubtedly soon be demanded by discerning house purchasers.

An important advantage of structural insulation is that a house can be made comfortable in a much shorter time than if the mass structure is slow to warm up. Thus heating costs are reduced not only by a lower heat requirement at steady-state heating (when the heat input equals the heat loss), but the time taken to attain a steady state is reduced from hours to minutes.

Insulation can be difficult to apply in houses of abnormal design or structure. Flat roofs are perhaps the commonest abnormality: apart from any weatherproofing difficulties, a flat roof pointing to a clear sky on a cold night can be chilled to sub-arctic temperatures; conversely, bright sun causes cooking temperatures overhead. Thus roof insulation is more crucial than for a pitched or ventilated roof.

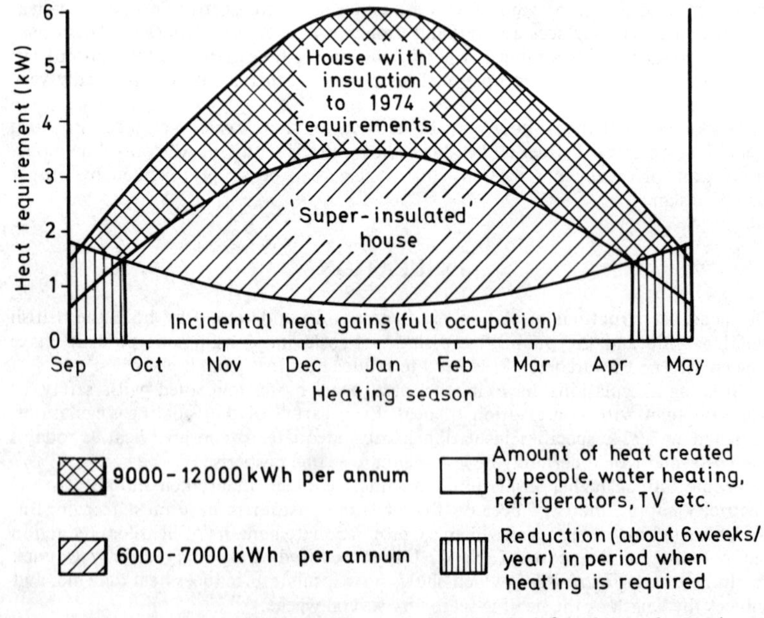

Figure 42.4 Reduced heating requirements for a typical 3-bedroom house 'super-insulated' to Electricity Supply Industry recommended U-values

Figure 42.5 Low-energy conservation house built by Wates at the Centre for Alternative Technology (courtesy Wates Ltd)

Another abnormality has been very high glass areas to living rooms. Proper use of solar (or greenhouse) gain can be beneficial, but as a rule the high heat loss in dull or windy weather is not compensated for by sunshine gains. By contrast, one nationally known builder constructed a low-heat-loss house at Machynlleth Centre for Alternative Technology; the house has prison-like tiny windows, and walls containing 450 mm of insulant-filled cavity. See *Figures 42.5 and 42.6.*

VENTILATION

As comfort standards increase and fuel costs rise, householders are more likely to apply draught proofing and keep windows closed, tending towards an unhealthy, humid (and condensation-prone) internal atmosphere.

The simple and invariably satisfying insurance against condensation problems is an open chimney, which also offers the joyous prospect of a solid-fuel fire or the efficacy and ventilation benefit of a gas fire or solid-fuel stove.

Homes built near airports or other noisy areas tend to have windows permanently closed; in such cases, mechanical ventilation can provide necessary fresh air with the minimum of noise intrusion. Most builders will have handled the small automatic toilet vents that run during electric light usage and for some time after. More complex developments of this idea are common on the continent of Europe where high-density flats would not naturally ventilate. These systems have proved very effective in avoiding condensation, ensuring a pleasant atmosphere and avoiding excessive heat loss by over-ventilation. *Figure 42.7* shows some typical equipment now coming into use in the UK.

Control of ventilation can also be incorporated within a gas-fired warm-air system, which adapts, in the summer, to control ventilation only (the 'Modaireflow' system by Johnson and Starley Ltd).

If there is any flued fuel-burning appliance in a room that depends on natural

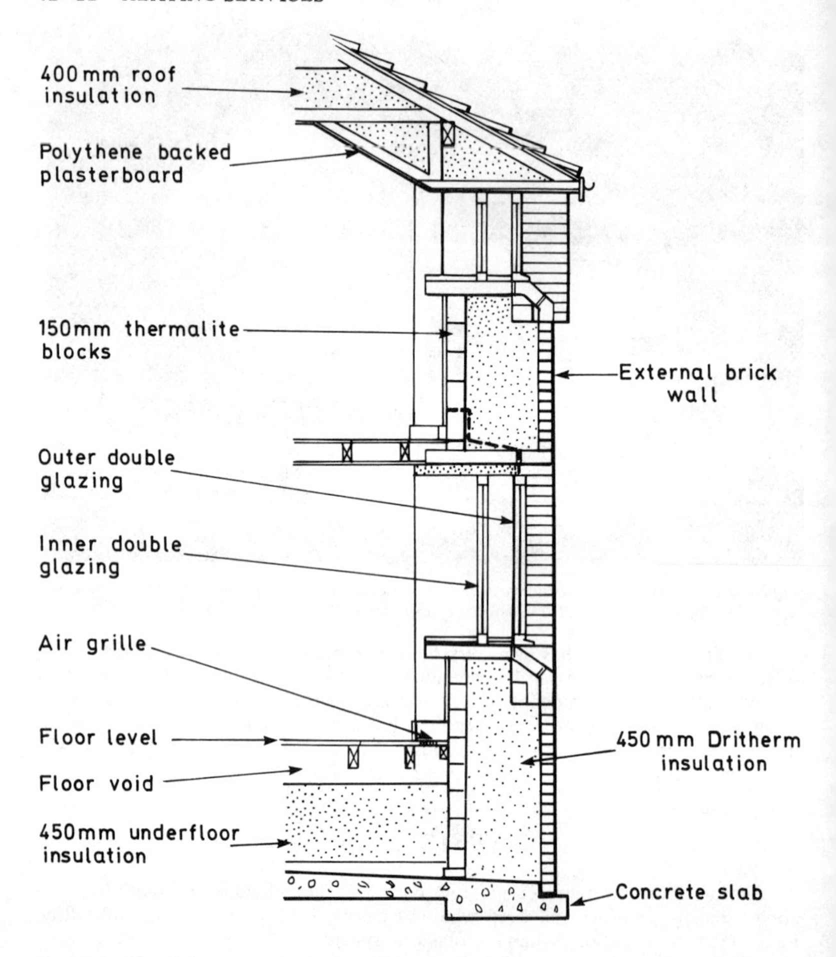

Figure 42.6 Section showing outside insulation of the low-energy conservation house (courtesy Wates Ltd)

draught to exhaust its products of combustion via a chimney, a kitchen extract fan or clothes dryer with a fanned vent to outside can exert strong suction; if this suction is greater than the flue draught, it will pull products of combustion into the room containing the appliance. This danger can be prevented by ensuring extra outside air additional to the unrestricted fresh air for combustion and flue dilution.

Ventilators are best at low level immediately behind or beside any open-flued appliance, provided they are unlikely to be blocked outside. Otherwise they are best at high level, where they are least likely to be blocked externally or covered internally because they cause draught or discomfort. An extract fan is best fitted with a shutter to avoid cold draught coming in through it when it is not running. Both Building Regulations and other Codes of Practice[4–7] give details of these requirements.

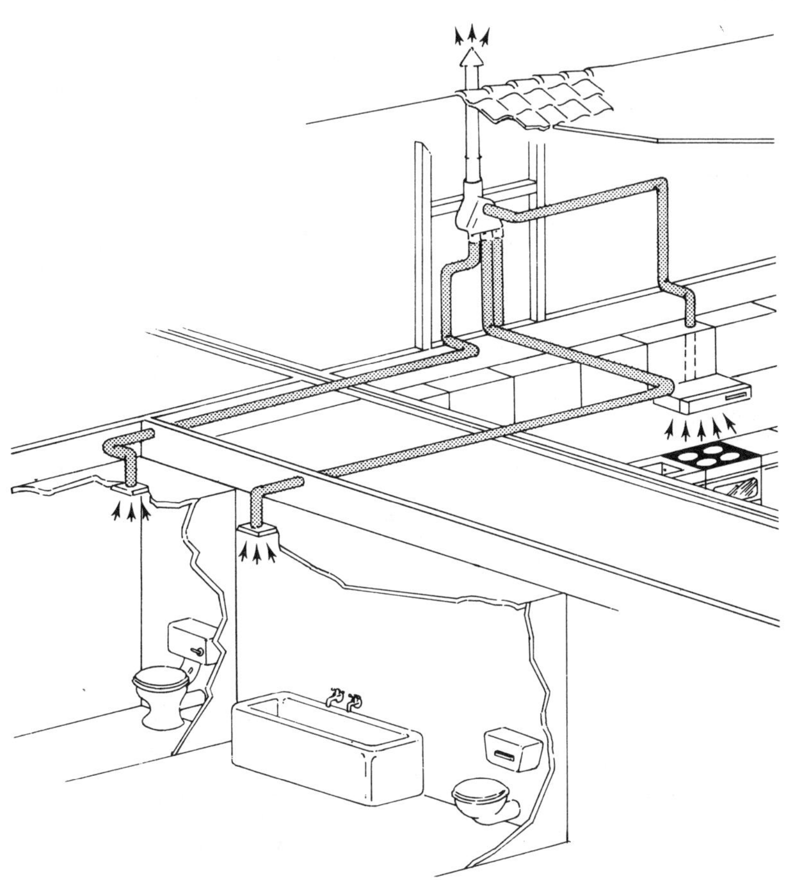

Figure 42.7 Typical controlled ventilation system (courtesy Colchester Fan Marketing Co Ltd)

Flues

The rooftops of Britain 30 years or more ago were crowned with a profusion of chimney pots of good purpose and proven fuction, and it is good that the hazards of recent flueless homes are causing the chimney to re-appear.

As houses become more airtight, so the air that any chimney requires for proper function becomes less available. Fireside ventilators from under a suspended floor, or ducted in a solid floor, ensure ample air with minimum draught discomfort.

Brick chimneys are always best located centrally and above the roof ridge to provide maximum draught reliability. Also, internal chimneys act as a thermal bank to absorb heat given to it from the appliance it serves, and then steadily return it to the householder. The Coal Research Establishment has claimed that a solid-fuel appliance can attain an overall efficiency of more than 75 per cent if the heat recovered from a central brick chimney is included in the system efficiency calculation.

Factory-made double insulated flue pipes are less generous in heat payback, but probably cost less than a lined brick chimney. Low-cost flue blocks, contained within the thickness of a wall, are satisfactory only within the limitations of their stated performance, and should not be used if there is any possibility of other usage.

Condensation

Condensation has become a dreaded nuisance, mostly in high-mass structures where the heating has been insufficiently used, and in many instances where unflued oil or gas appliances condense their products of combustion (comprising mostly water vapour) on cold surfaces.

Condensation has been a common problem in low-cost housing heated with warm air. This is because warm air has a rapid response (it gives a sensation of warmth very soon after the first switch-on), therefore the running time has tended to be too severely restricted, and to compensate for shorter timings abnormally high air temperatures tend to be used to compensate for cold-structure discomfort. The hotter air is, the more water vapour it can absorb; thus overheated air circulates more water vapour for deposit as wet patches on cold areas than is likely with other heating methods.

House construction methods to avoid condensation (i.e. vapour barriers and structural ventilation) are explained elsewhere. It is worth repeating that insulated internal surfaces will warm up quickly and so reduce condensation probability.

DESIGN

The performance of a heating system is clearly defined in the NHBC Registered House Builders' Handbook, but authorities may have different standards. Responsibility for the success or otherwise of a system design is best defined before any contracts are placed – disputes and litigation have in the past caused irreplaceable loss of goodwill, time wasted and not a few bankruptcies.

Most fuel authorities and many manufacturers offer a design service, probably providing a more efficient and satisfactory design than a sub-contractor (however good a craftsman, he is not usually a skilled design engineer) would provide.

An accurate design can save costs by avoiding oversized components, a fault more likely with designs from contractors whose natural conservatism prevents them readily accepting progressive ideas. For instance, it is now generally accepted by design engineers that the normal domestic hot-water demand does not require any additional boiler capacity over the space-heating load on which the boiler sizing can be based.

An expert design will usually be supported by a drawing, a good safeguard against the misunderstandings that frequently arise from non-specific contractor tenders. Fuel-supply authorities rightly reiterate that cost savings and inconvenience can be avoided if they are involved as early as possible in the design stage[8,9].

Inevitably there are teething troubles on new installations. Responsibility for remedial action is best defined within a heating contract or sub-contract. Approved equipment, preferably from an established manufacturer, strengthens guarantees, which may or may not include labour costs and may date from manufacture, merchant sale, or commissioning – such details very much strengthen or weaken the value of a guarantee.

The householder's instruction manual for the system as installed should be comprehensive but readable, and not overcomplicated. Maker's literature usually embraces a multitude of applications, most of which do not apply to the particular system; thus irrelevances should be eliminated or deleted, or only the relevant part included in the overall householder instructions.

Important instructions are to keep heat within the house and to keep temperatures down. Thermostat settings should be so explained that the reasons for recommended settings are understood. Facilities for all necessary services and future maintenance must also be described and explained.

To conclude, here is a digest of design recommendations prompted by the author's everyday encounters with domestic heating problems:

- Structural planning, duct provision and component access make for smoother cooperation between the trades.
- Outside metering is very sensible.
- Continuous unjointed pipe lengths in concrete floors avoid future troubles by leakage at fittings.
- Mixed emitters (for instance an assortment of radiators and convectors) tend to uneven performance and lack of effective control.
- Outside boiler siting wastes heat from the boiler when the system is running, and running time has to be greatly extended in cold weather to protect against frost damage.
- Microbore systems are more subject to site damage.
- Sealed systems are more vulnerable to troubles by imperfect workmanship or site damage.
- Wet systems must be so designed that no air is entrained, and no water is lost under all normal running conditions.

Contracting

Sub-contracting legalities are beyond this author's brief, but he can earnestly say that common-sense and cooperation are always better than litigation. Communication between the developer and his sub-contractors to ensure fullest possible understanding of each other's problems and requirements is crucial in avoidance of difficulties and disputes.

A common weakness of contract is informality of completion. A completed installation should be tested and approved between parties and the relevant authorities before it is formally handed over as complete.

A second expert opinion is often a simple stepping-stone over difficulties. Fuel interests and manufacturer's representatives can help, but within the limits of their loyalty and commercial interests. However, unbiased and uninhibited advice can be sought from professional organisations (for example, Institute of Domestic Heating Engineers, 93 High Road, South Benfleet, Essex; Chartered Institute of Building Services, 49 Cadogan Square, London SW1).

Independent consultants will also design systems and in so doing accept responsibility for their effective performance to specification. So briefed, a consultant would also ensure that the quality of materials and workmanship are to the required standard.

THE FUTURE

Houses will increasingly include more insulation and have lower heat loss. Such structures will create the probability of overheat in the summer and therefore some cooling will be necessary. It would be logical to combine the needs of controlled ventilation, winter heating and summer cooling with a heat pump. *Figure 42.8* is a diagram of the low-energy house illustrated in *Figure 42.5*.

Solar heating has suddenly become a popular idea, but at the time of writing (1978) only a few larger builders are experimenting with roof structures incorporating solar panels to preheat domestic hot water. Total solar heating requires heat-storage facilities so vast and expensive that they cannot really be considered practicable.

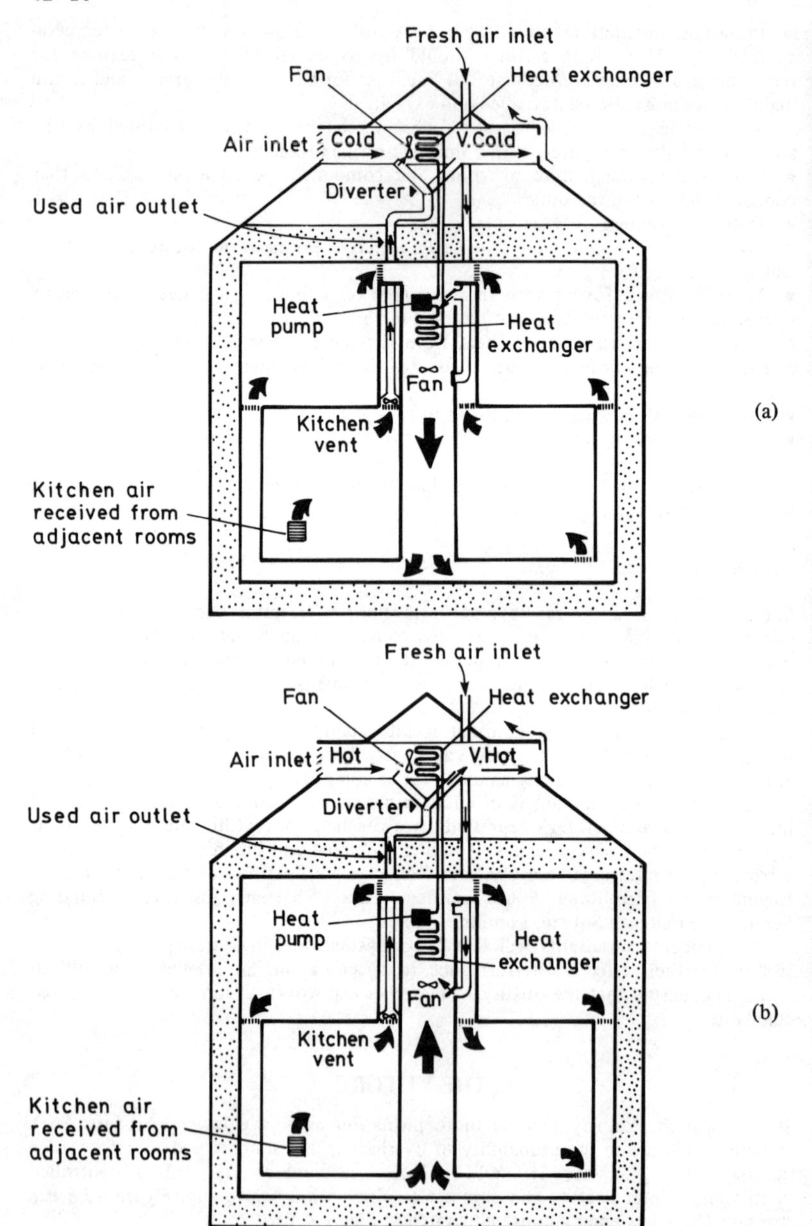

Figure 42.8 Heat pump installed for both heating and cooling in the low-energy conservation house: (a) winter, (b) summer (courtesy Wates Ltd)

Traditional house design and site planning can, and indeed should, be adapted to make the most use of available solar gain through south-facing partitioned garden rooms. Large glass areas, including double-glazed windows or doors, will lose more heat than they gain if they are in permanent year-round open communication with heated living zones.

Fortunately solar development has attracted responsible interest by Government departments and academic institutions. Reliable information and experience is mostly channelled through the International Solar Energy Society (UK: The Royal Institution, 21 Albemarle Street, London W1).

The National Coal Board are now developing appliances that radiate heat towards a living room, and are attractive from inside it, but that can be fuelled from a combined storage/stoking lobby. These appliances can consume as little as 0.35 kg of solid fuel per hour.

Future developments in gas heating could be in appliances that absorb all the heat of combustion without wasting up to 25 per cent of it (central heating appliances) or up to 60 per cent (gas fires) [10] in the flue products. 100 per cent efficiency is theoretically possible by condensing the water vapour of the flue products back to water at room temperature. Replacement of gas pilot lights with spark ignition will save gas.

Figure 42.9 Laying underfloor plastic heating pipes (courtesy Multibeton Ltd)

Electrical heating will probably make better use of house structure for off-peak heat storage. It is surely more logical and elegant to have widespread inconspicious structural heat storage than localised cumbersome appliances around the house.

Plastic pipe has proved itself for low-pressure and low-temperature applications. It is now being developed for underfloor widespread space heating, particularly on the Continent. Underfloor heating (except with thick carpets!) provides a very satisfying and unobtrusive means of house warming. See *Figure 42.9.*

Reflective insulation behind panel radiators prevents heat loss through the wall on which the radiator is fixed, but designers should know that such insulation reduces the overall radiator output.

Older housing stock with solid walls can be outside-insulated with cladding and rendering methods now being developed (*Figure 42.10*).

In most households hot-water expenditure is money not well spent, in as much as the residual heat of the water after bathing or washing all goes down the drain. Domestic launderettes now recover most of the heat from their waste water, and there is no reason why householders should not usefully regain waste heat, probably via a heat exchanger to preheat cold water entering the hot-water system.

Figure 42.10 External insulation cladding, comprising lathing on which Dritherm laminated glass-fibre and rendering is applied to greatly improve U-value and waterproofing (courtesy Grant-K-Lath Ltd)

Cold-water storage cisterns have long since been eliminated in Continental housing; thus it is probable that British homes will in future have mains-pressurised hot and cold water distribution. A more permissive water authority would also approve many of the combined heating and hot water appliances in common use in Europe.

REFERENCES

1. Billington Report, *Domestic engineering services,* Chartered Institution of Building Services
2. *The design of water heating systems in new homes*, Electricity Council
3. *The minimum resources home,* Electricity Council
4. BS 5376, *Code of practice for selection and installation of gas space heating:* Part 2, *Boilers of rated input not exceeding 60 kW* (1976)
5. BS 5440, *Code of practice for flues and air supply for gas appliances of rated input not exceeding 60 kW:* Part 2, *Air Supply* (1976)
6. BS 5449, *Code of practice for central heating for domestic premises*: Part 1, *Forced circulation hot water systems* (1977)
7. *Guide to good practice* (small-bore and microbore domestic heating), Heating and Ventilating Contractors Association
8. *Gas in housing,* British Gas Corporation
9. *Electricity – the profit builder*, Electricity Council
10. *Which?*, Consumer Association (September 1975)

43 INSULATION

THERMAL INSULATION 43–2

SOUND INSULATION 43–17

INSULATION 43

43 INSULATION

J. LAWRIE, DSC
Vice-Chairman, Structural Insulation Association

This section deals with both thermal insulation and sound insulation. In both cases the underlying theory is given, together with examples of how desirable results can be achieved. Some design data are included.

THERMAL INSULATION

Thermal requirements in buildings

Where the climate is such that buildings have to be heated for a substantial part of the year in order to maintain comfortable internal conditions, the cost of providing the necessary heat represents an important feature in the annual running cost of the building. The heat requirement will vary with the degree of comfort demanded, and with the amount of heat not usefully employed but dissipated (either through the fabric of the building or by the removal of warmed air from the building and its replacement with cold air).

Comfort conditions can be described simply as even temperatures of a level compatible with the activity or inactivity of the persons concerned, an absence of draughts, and an absence of cold surfaces towards which heat will be drawn. In housing the desired temperature can vary from about 16°C when there is some degree of activity to about 25°C when activity is minimal. These temperatures are very important, because the difference between internal and external temperature is one of the main factors controlling the loss of heat through the fabric of the building – roof, walls, windows and floors. To understand this process of heat loss it is necessary to study briefly how heat is transferred through materials or from one material to another.

Heat transfer

Heat moves from an area of higher temperature to an area of lower temperature by one of three routes:

- by *conduction* through solid materials, where the heat is passed on by particles by their contact with each other;
- by *convection* in liquids and gases, where the warmer particles expand and, becoming less dense, rise towards the top of the fluid, their places being taken by colder particles;
- or by *radiation* from a warm surface towards a colder surface as a pure energy wave passing through a gas or a vacuum without loss of heat.

In solid materials the rate of conduction or thermal conductivity is defined as the amount of heat in joules that passes in unit time (second) through unit thickness (metre) of a material of unit area (square metre) for unit difference in temperature (kelvin) between its faces. It can therefore be expressed as

$$\frac{\text{joule.metre}}{\text{second.square metre.kelvin}}$$

but since one joule/second = one watt the more usual expression would be watt.metre/square metre.kelvin. Using standard symbols this is written W m/m^2 K and contracted to W/m K or W/m °C, since one kelvin and one degree Celsius represent the same temperature difference.

Broadly speaking, the density of a solid material is a good (though not absolute) guide to its thermal conductivity; the denser the material the greater will be the rate of heat transfer through it. Thus metals tend to have high conductivities, masonry materials rather lower values, and timber products even lower. Typical examples for some common construction materials are given in Table 43.1, a study of which will show that some materials have values rather better than might be expected from their densities. One reason for this is the effect of trapped air.

Table 43.1 THERMAL CONDUCTIVITIES OF CONSTRUCTION MATERIALS

Material	*Density* kg/m^3	*Conductivity,* W/m K
Stone	2800	2.65
	2600	2.33
	2200	1.53
Glass	2500	1.05
Concrete	2350	1.44
	1850	0.65
	1550	0.47
Mortar	2000	1.14
Brick	2200	0.80
	1800	0.54
	1500	0.38
Tiles	2200	1.16
	1900	0.94
Plaster	1100	0.38
Plasterboard	960	0.17
Timber	900	0.17
	700	0.14
	500	0.11

When it is prevented from moving, air (being of very low density) has a very low thermal conductivity, but when it is in motion its rate of heat transfer is greatly increased by the effects of convection. Air entrapped within a material will therefore tend to reduce the conductivity of the material, and this effect will increase or decrease according to the ratio of entrapped air to solid material and to the size of the air cells involved, which will in turn control the amount of air movement (and hence convection) that can take place.

This principle is the basis from which insulation materials have been developed. Originally identified in natural products (for example, wood, where the air cells are finely divided but the ratio of air to cellulosic material is fairly low, or thatch, where the air cells are relatively large but the ratio of air to reed is very high), the theory could be extended as it became more clearly understood, until modern insulation materials could be produced.

An ideal insulating material would be one consisting almost entirely of air held in infinitely small cells by a matrix of solid material infinitely thin. Even better, the air could be replaced by a lighter gas of still lower thermal conductivity, or the material could be faced with reflective foil of low thermal-absorption characteristics. Such combinations can be, and indeed have been, produced to combat conduction, convection and radiation.

The importance of preserving the presence of the air or other gas must be stressed. If the air is expelled by filling the cells with water the basic conductivity will be that of water rather than that of air. This is what tends to happen when materials get wet. The thermal conductivity rises as the moisture content increases, and will not be reduced until the process is reversed. The material will only achieve its best performance when dry; hence all external structural materials that can hold water have their thermal conductivity values adjusted for the moisture contents they will experience in normal conditions. Similar adjustments are necessary for all materials in contact with transient moisture, although not on the exterior of the building, and this is particularly true of masonry materials.

Calculation of U-values

The U-value of a structure is the expression used to cover the amount of heat that is able to pass from the air on one side of the structure (comprising a variety of materials each of a different thickness) to the air on the other side, taking into account the resistances of cavities and of surfaces.

The effect of still air on the passage of heat has already been described, and it will be apparent that an air cavity in a structure will set up a resistance to heat transfer that will be governed by the degree of convection that can occur.

The ability of surfaces to absorb and emit heat has also been briefly described. Highly polished surfaces are poor absorbers, while dull black surfaces are good absorbers. Thus surfaces can be graded between these two extremes in terms of their ability to accept or reject heat. In a homogeneous material the amount of heat passed through it will be inversely proportional to its thickness but directly proportional to the other controlling factors (time, area and temperature difference). This means that in terms of overall heat transfer rate we can ignore the factors of time, area and temperature, and can concentrate on thickness. Since the proportionality is inverse it is convenient to work in the inverse ratios of the thermal conductivities of materials, which are known as *thermal resistivities*. Thus for each conductivity (k) we will have a thermal resistivity ($1/k$), which when multiplied by its thickness in metres will give its *thermal resistance*.

For each surface and each cavity the ability to resist the passage of heat can be expressed as a thermal resistance. Practical values are given in Tables 43.2 and 43.3.

The U-value of the element of structure represents the total rate of heat transfer, and its total resistance to heat flow is represented by R_u, which is the reciprocal of the U-value. However, the total resistance is also the total of the resistances of all the individual components of the element; thus it can be written:

$$R_u = R_{so} + R_1 + R_2 + R_3 \ldots + R_{au} + R_{av} + R_{si}$$

where R_{so}, R_{au}, R_{av} and R_{si} have the appropriate values from Tables 43.2 and 43.3, and R_1, R_2 etc. are the resistances of the individual components.

Table 43.2 PRACTICAL THERMAL RESISTANCES OF SURFACES

Surface type	Structural element	Direction of heat flow	Thermal resistance, m^2 K/W	
			Surface of high absorption	*Surface of low absorption*
Internal (R_{si})	Walls	Horizontal	0.12	0.30
	Ceilings, roofs or floors	Upward	0.11	0.22
	Ceilings or floors	Downward	0.15	0.56
External (R_{so})	Walls	Horizontal	0.05	0.06
	Ceilings, roofs or floors	Upward	0.04	0.05
	Floors exposed on under side	Downward	0.09	-

Table 43.3 PRACTICAL THERMAL RESISTANCES OF AIR SPACE(S)

Air space type	Description and/or dimension	Direction of heat flow	Thermal resistance, m^2 K/W	
			Surface of high absorption	*Surface of low absorption*
Unventilated (R_{au})	5 mm	Horizontal or upwards	0.11	0.18
	5 mm	Downwards	0.11	0.18
	20 mm	Horizontal or upwards	0.18	0.35
	20 mm	Downwards	0.21	1.06
Ventilated (R_{av})	20 mm or more: Loft space in tiled pitched roof (a) without sarking,	Upwards	0.11	
	(b) with felt and eaves ventilation.	Upwards	0.18	
	Flat roof air space with ventilation	Upwards	0.18	
	Air space between tiles and felt on pitched roof.	Upwards	0.12	
	Air space in cavity wall.	Horizontal	0.18	

Before proceeding to practical examples of U-value calculation it is necessary to refer again to the effect of moisture on the thermal conductivity (k) of masonry materials. In order to calculate the U-value of a structure incorporating one or more masonry materials it is necessary to know the moisture content of the masonry components as they are operating in practical conditions. This can only be done after the building has been erected and tests for moisture can be carried out based on practical sampling. This therefore can have no application in designing buildings, where the results have to be assessed in advance of building.

For this reason the concept of standard moisture contents has been developed and is applied to brickwork and blockwork in all construction.

- For the external leaf of all walls that may be exposed to driving rain and are either unrendered or rendered with untreated mortar porous to moisture, and for masonry roofs where the waterproof roof finish may entrap condensed vapour or residual moisture, the moisture content is assumed to be 5 per cent by volume.
- For the internal leaf of all walls with continuous air spaces or air spaces that have been filled with an approved insulant, and for internal partitions or walls protected by tile hanging etc., the moisture content is assumed to be 1 per cent by volume for brickwork or 3 per cent for concrete.

Table 43.4 provides conversion factors for the thermal conductivity of masonry materials at various moisture contents. These factors are applied as multipliers to the k values of dry materials, or can be used proportionately with the k value at any moisture content.

Table 43.4 CONVERSION FACTORS FOR k VALUES OF MASONRY MATERIALS

Moisture contents, %	1	1.5	2.0	2.5	3.0	4.0	5.0
Conversion factor	1.30	1.40	1.48	1.55	1.60	1.67	1.75

A simple practical example of U-value calculation is that covering the cavity wall. The basic construction taken is two leaves of 105 mm brick with a 50 mm cavity; the internal surface is 9.5 mm plasterboard bonded to the brickwork with a 20 mm air space. The basic formula can then be written:

$$R_u = R_{so} + R_1 + R_{au} + R_2 + R_{au} + R_3 + R_{si}$$

Hence:

R_{so}	External surface (Table 43.2), high absorption	= 0.05
R_1	105 mm brick at 5 per cent moisture (Table 43.4) at k of 0.48, i.e. $105 \times \dfrac{1}{0.48 \times 1.75}$	= 0.123
R_{au}	50 mm unventilated cavity (Table 43.3), high absorption	= 0.18
R_2	As R_1	= 0.123
R_{au}	20 mm unventilated cavity (Table 43.3), high absorption	= 0.18
R_3	9.5 mm plasterboard at k =- 0.158, i.e. $0.0095 \times \dfrac{1}{0.158}$	= 0.060
R_{si}	Internal surface (Table 43.2), high absorption	= 0.12

$\therefore R_u = 0.836$

$U = 1/R_u = 1.196$ W/m² K

The effect of insulating this wall can be illustrated by substituting cavity fill material in the 50 mm cavity. The basis of the calculation then becomes $R_u = 0.836$. From this $R_{au} = 0.18$ must be deducted, leaving 0.656, to which is added 50 mm of material at 0.045 (Table 43.5), i.e. $0.050 \times (1/0.045) = 1.111$, making a total for the new R_u of 1.767. Then:

$$U = \frac{1}{R_u} = 0.566 \text{ W/m}^2 \text{ K}$$

The most complex calculation for the U-value involves the pitched roof consisting of two parts, one of which is horizontal (the ceiling) and the other inclined to the horizontal (the roof). Since the areas of the two parts are different and thermal transmittance is based on unit area, some correction must be applied to the resistance of the sloping part of the roof to bring it into line with the horizontal ceiling value. The correction is based on the cosine of the angle of the roof, and the following example illustrates how it is applied.

The construction is a 30° pitched roof of tiles on battens felted, over a loft space and with timber joists and a 9.5 mm plasterboard ceiling.

R_{so}	Pitched roof, external surface (Table 43.2)	= 0.04
R_1	15 mm tiles at $k = 0.60$	
	i.e. $0.015 \times \dfrac{1}{0.60}$	= 0.025
R_{av}	Air space between felt and tiles (Table 43.3)	= 0.12
R_2	3 mm felt at $k = 0.15$	= 0.020
		0.205
Correction for slope = cos 30° × 0.205 = 0.87 × 0.205		= 0.178
R_{av}	Loft space (Table 43.3)	= 0.18
R_3	9.5 mm plasterboard at $k = 0.158$	
	i.e. $0.0095 \times \dfrac{1}{0.158}$	= 0.060
R_{si}	Internal surface (Table 43.2)	= 0.11
		$R_u = 0.528$
		$U = 1/R_u = 1.894$ W/m² K

Again to illustrate the effect of insulation we introduce 100 mm of mineral (glass or rock) wool with a k value of 0.040 W/m K. Here we do not need to omit the value of the loft cavity but merely add the resistance of the insulation, i.e. $0.1 \times (1/0.040) = 2.5$, to the existing R_u to give a new R_u of 3.028 and a U value of 0.33 W/m² K.

Thermal insulation materials

By definition, thermal insulation is the means by which a barrier can be erected against the transference of heat through the elements of the structure – roof, walls, windows and floors. It will be appreciated of course that each element of structure does in itself provide a partial barrier, but the thermal performance of most structural materials is very inferior to that of any insulating material. A very wide range of insulating materials is available and some of those most commonly encountered in housing are given in Table 43.5. Many combinations of materials are also available but for simplicity have been omitted.

From a practical viewpoint it is convenient to deal with each element separately, and the relative need can be judged from *Figure 43.1*. In an uninsulated house only 25 per cent of the heat generated will be gainfully used in maintaining temperatures at the desired levels. Of the heat lost through the fabric, 30 per cent goes through the roof, 25 per cent through the windows, 35 per cent through the walls and 10 per cent through the floor.

Table 43.5 THERMAL CONDUCTIVITIES OF INSULATING MATERIALS

Material	*Density* kg/m^3	*Conductivity* W/m K	*Equivalent thickness,* m
Boards			
Expanded polystyrene	16–32	0.040/0.035	1.6/1.4
Fibre insulating	300	0.050	2.0
Glass fibre	150	0.035	1.4
Urethane foam	32–48	0.025	1
Woodwool	600	0.090	3.6
Cavity fills			
Expanded polystyrene beads	16–32	0.045/0.040	1.8/1.6
Expanded vermiculite granules	70	0.045	1.8
Glass fibre	36	0.035	1.4
Rockwool	72	0.045	1.8
Urea formaldehyde foam	12	0.045	1.8
Mats, quilts and slabs			
Mineral fibre (glass and rock)			
mat, quilt	12–18	0.040	1.6
semi-rigid slab	16–48	0.035	1.4
rigid slab	60–120	0.035/0.030	1.4/1.2
Rigid blocks and screeds			
Cellular glass blocks	130	0.050	2.0
Insulating concrete blocks, screeds	400	0.15/0.16	6.0/6.4
	600	0.18/0.20	7.2/8.0
	800	0.23/0.26	9.2/10.4
	1000	0.34/0.38	13.6/15.2
	1200	0.45/0.50	18.0/20.0
	1400	0.50/0.55	20.0/22.0
	1600	0.55/0.60	22.0/24.0

Note 1. The values quoted above are meant to be indicative only and do not refer to specific manufactured products.
2. Two values are shown against each density of insulating concrete to show the approximate effect of increasing moisture content from 3 to 5 per cent.

In an uninsulated bungalow the ratio of roof area to wall area is much greater. The heat loss through the fabric is distributed 40 per cent through the roof, 25 per cent through the windows, 25 per cent through the walls and 10 per cent through the floor. All these figures should be regarded only as approximate, since the areas will vary with the shape of the house and there will also be variations in the U-values of the elements of the strucutre.

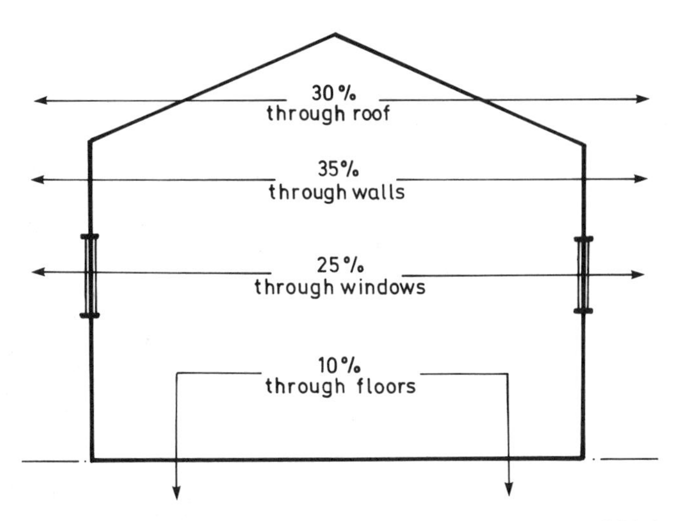

Figure 43.1 Approximate distribution of heat loss through uninsulated fabric of two-storey house

Figure 43.2 Laying 100 mm mineral wool (glass or rock) between joists in attic space (courtesy Fibreglass Ltd)

Roofs

In houses by far the most popular roof construction is the pitched roof of tiles on battens on timber rafters, with joists supporting a plasterboard ceiling. Sarking felt will probably be incorporated over the rafters. Timber trussed rafters may also be used in conjunction with joisted ceilings without affecting the thermal transmittance of the roof, which will be about 1.9 W/m² K.

By far the easiest way of insulating a pitched roof is by the application of a suitable insulation product in the attic space at ceiling level. To safeguard the ceilings if any repair work has to be carried out in the attic at a later date, the joists should not be covered and the insulation should therefore be applied between the joists.

Probably the most suitable materials for this application are mineral (glass and rock) wool mats. These are supplied to fit standard joist spacings and in easily handled rolls that can be unrolled very simply between the joists (*Figure 43.2*). Care must be taken to ensure that the eaves are not blocked and a free flow of air is allowed over the insulation. It is important to ensure that a gap of about 4 mm is left along the length of the eaves to allow air to penetrate into the attic and thereby prevent condensation occurring on the under side of the sarking felt. An alternative method of insulation is to fill the space between the joists with expanded polystyrene granules or with expanded vermiculite granules. It is necessary when using these granulated materials to ensure that the joist ends are closed and that the air penetrating through the gaps left in the eaves closure boards passes over the joist cavities and does not disturb the insulation.

It is also possible to insulate at ceiling level using combinations of plasterboard or fibre insulating boards with expanded polystyrene insulation or boards of expanded polyurethane as replacement for the plain plasterboard in the basic roof.

With all these materials the thicknesses to be used will depend on the results required. At present the Building Regulations set a standard that can be met by 50 mm of mineral (glass and rock) wool, by 80 mm of expanded vermiculite, by 50 mm of expanded polystyrene board or 60 mm of granules, or by 40 mm of expanded polyurethane in the constructions described. However, since these Regulations came into force the relative cost of fuel has risen further against the cost of insulation and the economic thicknesses are considerably greater. For example, in mineral (glass and rock) wool the economic thickness has already passed 75 mm and will very soon reach 100 mm.

In flat roofs (which, although less widely used than pitched constructions, are nevertheless quite frequently encountered) the basic construction in houses is generally timber on joists with a plasterboard ceiling, the external finish being roofing felt and granite chippings. Such roofs will have a U-value of about 1.6 W/m² K. Insulation can be provided by applying fibre insulating board, expanded polystyrene board, glass fibre board or expanded polyurethane board above the boarding and before the application of the finish. Before the insulation is applied it is important to apply a vapour barrier to the upper surface of the timber and to arrange that the roof space is adequately ventilated by the provision of air inlets, in order to ensure that dampness is not allowed to develop in the cavity. It is also good practice to include a good vapour check (e.g. polythene sheeting) above the ceiling boards.

Alternative methods are available using flexible mineral (glass or rock) wool mats, either draped over the joists before the roof boards are applied, or fixed between the joists at the back of the ceiling boards as these are being secured. The same provisions to combat the risk of dampness in the cavity should be taken as in the previous example.

Insulating boards of the types previously described can also be substituted for plain plasterboard ceilings or used in conjunction with plain plasterboard in a variety of composite ceiling treatments. Flexible materials can also be used in the same way.

Thicknesses of insulation for timber flat roofs will again vary with the product used and the position in the structure, but broadly will follow the same pattern as with pitched roofs. Building Regulations are currently met by 50 mm of mineral (glass or rock) wool or the equivalent in any of the other products, but very much greater thicknesses can be economically justified even at this time.

Concrete flat roofs are very much less often used in houses than either of the previously described constructions, but are encountered more often in flats or maisonettes. A typical 150 mm dense concrete slab roof, screeded to falls and finished with bitumen felt or asphalt, will have a thermal conductance of about 2.6 W/m^2 K, nearly 50 per cent higher than a pitched roof. Insulation is therefore even more important with this type of structure, but unfortunately the results are not quite so easily obtained.

Although any of the insulating boards already mentioned can be laid on top of the roof structure on a bitumen vapour barrier, either hot or cold applied, it is

Table 43.6 TYPICAL ROOF CONSTRUCTIONS TO MEET SPECIFIED U–VALUES

Roof construction	*Insulation thickness* (mm) *to give U of* 0.60	0.50	0.40	0.30
1. *Pitched roof of tiles on battens with sarking on timber rafters and with joists and plasterboard ceiling,* U = 1.9 W/m^2 K				
with mineral (glass and rock) wool	45	60	80	115
with expanded polystyrene pellets	55	70	90	125
with expanded vermiculite granules	55	70	90	125
with added expanded polystyrene slab	45	60	80	115
with added expanded polyurethane slab	30	40	50	70
with fibre insulating board in place of plain plasterboard	60	75	100	140
2. *Flat roof of timber boarding on joists with plasterboard ceiling and finished externally with roofing felt and granite chippings,* U = 1.6 W/m^2 K				
with fibre insulating board	55	70	95	135
with expanded polystyrene board	40	55	75	110
with glass fibre board	35	50	65	95
with expanded polyurethane board	25	35	45	70
with mineral (glass or rock) wool	40	55	75	110
3. *Dense concrete roof slab 150 mm thick screeded to falls with roofing felt or asphalt,* U = 2.6 W/m^2 K				
with fibre insulating board	65	80	110	150
with expanded polystyrene board	50	65	90	120
with glass fibre board	45	55	75	100
with expanded polyurethane board	35	40	55	75
with cellular glass slab	60	75	100	140
with lightweight concrete screed	315	395	530	720
with mineral (glass or rock) wool on suspended ceiling	45	60	90	120

Thicknesses are not intended to indicate availability of products but only the result of theoretical calculations.

difficult to achieve the desired results without using thicknesses that are somewhat more difficult to lay and hence rather more costly. With the higher application costs the economic thickness of the insulation tends to be reduced, and it is often considered better to use an application that is based on suspending the ceiling below the roof slab (and thus allowing thicker insulation to be accommodated more economically) or on bonding a thicker board product direct to the soffit. Another alternative would be to use a lightweight concrete screed, but very large thicknesses must be used and the additional weight is not usually acceptable. Examples of different treatments to meet different U-values are given in Table 43.6.

Walls

In housing, the commonest type of wall construction is that based on two leaves of masonry materials with a cavity between them. Traditionally the masonry material on which the design was founded was the 4½ inch brick, and with a 2 inch cavity the 11 inch cavity brick wall was established. Although with metrication the overall dimension of the wall is now 260 mm without external rendering or internal plaster or other finishes, its general shape and function is unaltered. Such a wall has a U-value of 1.4 W/m^2 K, and in the days of inadequate heating and cheap fuel this was considered a good enough standard for normal use. Improvements in heating standards, demands for better comfort conditions and accelerating costs for fuel, however, gradually caused a trend towards improvement in thermal conductivity.

The first move towards improvement came from the introduction of the lightweight concrete block, which was substituted for the inner leaf. Various types of standard blocks are used for this application with resulting U-values for the wall of down to 0.96 W/m^2 K, which meets the current Building Regulation requirement of 1.0. In terms of energy conservation, however, these values do not approach the economic answer, which at present lies between 0.6 and 0.4 W/m^2 K depending on the treatment employed.

In constructing such a wall the simplest approach to achieving a higher standard of insulation is to incorporate a slab of mineral (glass or rock) wool to fill the cavity as it is constructed (*Figure 43.3*). Care is necessary to ensure that mortar droppings are not allowed to fall on the edge of the slabs, since these would act as a bridge across the cavity for moisture. A similar treatment, but with a lower thermal efficiency, employs a slab of half the cavity thickness of either expanded polystyrene or mineral (glass or rock) wool held in the cavity and against the inner leaf by proprietary fixing methods.

A third approach is to blow or pump insulating materials into the cavity after the wall is constructed. For this purpose systems employing foamed urea formaldehyde, mineral (glass and rock) wool granules and expanded polystyrene pellets are available, and the choice will depend on the exposure of the building and the standing of the particular system under the Building Regulations applicable at the time. This should be discussed with the Local Authority.

Insulation can also be applied to the inside surface of the internal wall. This method would not normally be used with a cavity masonry wall, because a different design would probably be more economical, but from a practical point of view the application is suitable. There are two different methods. The first is based on battening out the wall to a suitable depth, filling the cavity with a slab insulant such as expanded polystyrene or mineral (glass and rock) wool, and finishing with plasterboard. The second employs plasterboard in composite treatments with expanded polyurethane, expanded polystyrene or mineral (glass or rock) wool, the exact method varying with the materials used. In all these applications the question of incorporating a vapour check arises. There are different opinions as to the degree of vapour barrier that should be incorporated, but it is generally recommended that at least a good-quality vapour check should be applied on the inner side of the insulation layer. This is often achieved by using aluminium-foil-faced plasterboard instead of plain plasterboard, or by incorporating a layer of polythene sheeting.

Solid walls of masonry materials are not normally used in modern construction unless the masonry material is of lightweight concrete. A large variety of such walls is available with U-values varying from 1.4 W/m² K down to about 0.90 W/m² K. To bring such walls down to the value associated with higher comfort or energy conservation, the treatments described in the previous paragraphs for application to the inner surface of cavity walls are equally applicable.

Figure 43.3 Constructing wall with 50 mm Fibreglass 'Dritherm' slab in cavity (courtesy Fibreglass Ltd)

Gaining in popularity for housing in this country is the timber-framed wall, which can be given a variety of external treatments including tile hanging, timber cladding etc. The internal lining would generally be plasterboard and the U-value would be around 1.3–1.5 W/m² K, depending on the thickness of boarding used. To meet current Building Regulations a variety of sheet materials can be used either to replace the plasterboard or in conjunction with it. In this way the same boards, expanded polystyrene, expanded polyurethane, compressed straw slab, fibre insulating board and mineral (glass or rock) wool slabs can be used in greater thickness, either over or between the studs, to considerably improve on current practice and to achieve high degrees of energy conservation.

A selection of wall treatments to achieve different U-values is given in Table 43.7.

Table 43.7 TYPICAL WALL CONSTRUCTIONS TO MEET SPECIFIED U-VALUES

Wall construction	*Insulation thickness* (mm) *to give U of* 1.00	0.80	0.60	0.40
1. *Brick/cavity/brick wall,* $U = 1.4\ W/m^2\ K$				
with inner leaf replaced by lightweight concrete				
autoclaved aerated	70	110	180	320
foamed slag	160	255	420	740
clinker block	280	445	725	1275
expanded clay aggregate	100	155	250	445
with insulation in the cavity				
mineral (glass and rock) wool slab	20	30	45	80
expanded polystyrene slab	20	30	45	80
expanded polyurethane slab	10	20	30	50
urea formaldehyde foam	15	25	40	70
mineral (glass or rock) wool granules	15	25	40	70
expanded polystyrene pellets	20	35	50	90
with insulation on inner face of the wall and faced with plasterboard				
expanded polystyrene slab	10	20	40	70
mineral (glass or rock) wool slab	10	20	40	70
expanded polyurethane slab	10	15	25	45
2. *Solid masonry wall of lightweight concrete c.250 mm,* $U = 2.00\ W/m^2\ K$				
with insulation on inner face and faced with plasterboard				
expanded polystyrene slab	20	30	45	80
expanded polyurethane slab	15	20	30	50
mineral (glass or rock) wool slab	20	25	40	70
3. *Timber framed wall with external tile hanging or timber cladding and plasterboard lining,* $U = 1.5\ W/m^2\ K$				
with insulation in the cavity				
mineral (glass or rock) wool slab	10	20	35	65
expanded polystyrene slab	15	25	40	75
expanded polyurethane slab	10	15	25	45

Thicknesses are not intended to indicate availability of products but only the result of theoretical calculations.

Windows

The straightforward answer to thermal insulation of windows is double glazing. Certainly there are arguments about the economics of replacing existing single-glazed windows with double-glazed units when the existing windows are adequate for all other purposes, and there can be little doubt that the repayment time is much too long in such a case to make the proposition very attractive. In new construction this position does not arise, however, because only the additional cost of the double glazing over the single glazing comes into the calculations and the repayment times are reduced very considerably. When the additional comfort achieved from the elimination of cold zones and the reduction of air movement in the house is added to the fuel saving, the resultant equation undoubtedly favours double glazing, which is already becoming increasingly a part of standard specifications in housing and undoubtedly will continue to do so.

Figure 43.1 shows that 25 per cent of the heat lost through the fabric of a house is lost through the windows. This arises not only because the area of windows in houses is quite large but also because the U-value of a single sheet of glass is very high – about 5.6 W/m^2 K. By double glazing, the U-value is reduced by rather more than 60 per cent and the heat loss reduced accordingly.

Floors

It is not generally considered necessary in this country to insulate the solid concrete raft floor that is so widely used in present-day houses. Nevertheless it is undoubtedly true that during the heating season the temperature of the subsoil, although considerably above freezing, is sufficiently low to cause a considerable loss of heat downwards. One way of combating this is to finish the concrete raft with a screed of lightweight concrete of a thickness not less than 50 mm. Another is to lay 50 mm of mineral (glass or rock) fibre or expanded polystyrene below a finishing screed (*Figure 43.4*). Expanded polyurethane may also be used to a thickness of 40 mm in the same way, or foamed glass 75 mm thick.

Figure 43.4 Screeding a floating floor on glass-fibre quilt (courtesy Fibreglass Ltd)

Building Regulations do require concrete floors to be insulated when the underside is exposed to the air, as occurs when houses are built on stilts with garage/assembly space below, or when a house built on a hill is projected out over a carport or similar area. Such floors are normally treated by the application of insulation underneath the concrete slab, either as permanent shuttering or secured by pinning to the soffit. Suitable materials are foamed glass, expanded polystyrene, woodwool slabs and expanded polyurethane, and a thickness of 50 mm will suffice. The insulation of suspended timber floors is important because of the possibility of a high degree of air circulation across the underside of the floor. Although the basic U-value of this type of floor is 0.7 W/m^2 K, implying that little action is necessary to control heat loss, additional insulation is easily provided during construction and would be very difficult to provide at a later date. Energy conservation is therefore much easier to achieve when erection is taking place, and the application of fibre insulating board, corkboard or mineral (glass or rock) wool mat over the joists, with the floorboards nailed through 50 mm of insulation, will give a reasonable return.

Problems with thermal insulation

Perhaps the most talked-of 'problems' in relation to thermal insulation concern condensation. While there is no direct link between the insulation of a building and the moisture in a building, there are links that can suggest wrong conclusions if the subject is not clearly understood.

The ability of air to carry moisture depends on the temperature of the air. The hotter it is the more moisture it can carry. The cooler it is the less moisture it can carry. If the air in a room is near saturation and the temperature is raised it will cease to be saturated and can carry more moisture, but if the reverse applies and the temperature is lowered the excess moisture that the air can no longer carry will condense out on the coldest surfaces available to it. Thus when a house is insulated and the advantages are taken in increased temperature rather than decreased fuel bills, there is a danger of creating conditions where condensation may occur due to the increased moisture-carrying capacity of the air and the inevitability of localised cold spots (e.g. single-glazed window surfaces, unventilated roof space, etc.).

It is essential therefore in considering house design for thermal insulation, to keep four points in mind:

- Plan to reduce moisture penetration.
- Adequately ventilate to the outside all 'wet areas', e.g. kitchens, bathrooms, drying cupboards etc., where large volumes of moisture can be produced.
- Avoid cold bridges, which pass through insulated elements and produce internal cold spots where condensation can occur.
- Ensure that all roof spaces are adequately ventilated.

It is advisable to study carefully BS 5250:1975, which gives detailed information on these points.

A second 'problem' lies with the variety of materials available for carrying out all insulation work and the danger of using the wrong material for a specific purpose. It is always better to check with the manufacturers of the material before it is used than after it has failed. Free choice between materials must be maintained, but, having chosen, check the suitability of the chosen product for the work it has to do.

One other point must be mentioned, and this concerns the effect of moisture on materials. Not many insulation materials come to any serious damage when exposed to water, but the reverse is true in terms of retention of their insulating properties when wet. Materials should therefore be stored dry and should never be installed in a wet condition. Nor should they be applied in exposed positions during wet weather or left exposed overnight when the weather happens to be dry. It is much easier to keep materials of this type dry before and during installation than to dry them out later.

Finally, when insulation is applied in attic spaces, make sure that any cold water tanks and piping are not left exposed, but that the insulation is carried up to and over the tanks, the bottom remaining exposed. All pipes should be well insulated.

Economics of thermal insulation

The actual return obtained from thermal insulation will vary very considerably with a number of factors, among which the following are the most important:

- the cost of heating equipment and its availability in different output sizes;
- the cost of the insulation materials and the labour to apply them;
- the cost of maintenance of the fabric of the building and of the heating equipment;
- the cost of fuel;
- the interest rates on borrowed capital.

With the exception of the last two, all these factors will vary according to the specific design of the building, the choice of materials and equipment used, and the operating choices of the ultimate occupants.

Whilst it is impossible in these variable circumstances to given an indication of savings that will be universally applicable, it is useful to give a general guide, based on an average two-storey detached house. In the uninsulated state such a house would lose about 75 per cent of the heat produced, only 25 per cent being usefully employed in heating the occupants. The basic cost of insulating the roof, walls and windows at current prices would be of the order of £500, and there would be a reduction in the size of the heating plant, radiators etc. of about £200. The increased capital cost of the project would therefore be £300. This sum would be part of the original mortgage and would therefore be subject to normal interest rates and to tax relief on the interest paid.

In annual terms some reduction in maintenance on the fabric of the house due to reduced pattern staining, reduced condensation etc. is taken into account, as well as lower maintenance costs on the reduced heating system. The overall average efficiency of the insulating system is calculated at 55 per cent, which means that of the 75 per cent of heat lost from the building about 41 per cent can be saved. Using current costs of labour, materials, fuel etc. throughout, and assuming tax allowance on interest only at the standard rate, the overall annual saving is about £70/year.

It is necessary to emphasise that the savings will not accrue evenly over the whole insulation system. This will depend on the shape of the building and the ratios of area of roof to wall, window to wall etc., but generally the most economic application will be in the roof (especially a pitched roof), followed by the walls, then the floors, where these need to be insulated, and finally the windows.

If more accurate information on the calculation of the economics of insulation is required, readers are recommended to study the calculation method given in Housing Development Note IV, *Thermal insulation in housing, 2: Relationship between construction and heating costs*, published by the Department of the Environment.

SOUND INSULATION

Nature of sound

Sound is an energy wave that is caused by fluctuations in air pressure and is transmitted directly from its source in all directions, its rate of progress varying according to the material through which it is passing. If its velocity through air is taken as 1, its velocity through water would be 4, through timber 10 and through steel 15. Although it moves in a direct line from its source, when it impinges on a secondary medium it travels both along and through that medium (again by direct paths) until it meets another medium.

As each medium intervenes, and as distance from the source increases, so more and more energy is lost in overcoming the resistance of the various media to the pressure variations or vibrations. The energy is dissipated as heat and the sound gradually dies out.

Thus a sound originating in one room can pass into the next room directly through the connecting wall or indirectly by the link to another wall or floor structure. It can also pass through any linking aperture or by direct or indirect links through pipes, ducts etc.

Sound transmission

When we discuss sound transmission in relation to dwellings we are inevitably discussing noise nuisance, a subjective matter the effect of which varies with the person, the time and the conditions. If we are sitting quietly in the garden and gently dropping off to sleep and the chap next door starts revving his motorcycle the noise is unbearable, but if we happen to be laying a carpet in the lounge and hammering in tacks all the time when the disturbance occurs we may not even notice it.

While we may object strongly to other people's noise we tend to be rather more tolerant with our own, and this ties in to some extent with our ability to tune out tolerable sounds that do not come to us unexpectedly. In so doing we also tune out other sounds of a similar amplitude and create what to us is an acceptable set of conditions.

In a recent illustration of this point a major road, which carried a good deal of traffic and had done so for many years, was downgraded by diversion of the traffic at the instigation of the occupants of a number of blocks of maisonettes that faced on to the road. With the reduced level of external noise the occupants became aware of noises from internal sources of which they had not previously been conscious. A new series of noise nuisances have now to be faced.

Because remedial treatment of sound problems can be very expensive indeed, especially if it has to be applied after the building is complete, the builder is well advised to take certain simple precautions:

- Design to the best available standards, bearing in mind the subjective nature of the problem.
- Having done so, do not claim specific advantage in this sphere. Above all do not claim that any building is soundproof. No reasonable design could ever claim such a standard.
- Pay particular attention to siting the property so that obvious sound paths are avoided.

Sources of noise

Noises heard within buildings are of two broad types, those originating outside the building and those coming from within.

External noise comes from many varied sources – the aircraft flying overhead, the pneumatic drill in the street, the children's voices from the nearby school, the steam hammer in the factory a few streets away, the trundling of heavy-duty traffic or the cacophony of the discotheque across the way – and is often of a type that the builder may feel that he can do little about, but this is seldom the case. Obviously with occasional happenings such as the pneumatic drill the builder cannot be expected to do very much, and the remedy may lie elsewhere, but there are often actions that the builder can and should take to ensure that the conditions are as good as the ultimate occupants will be expecting.

The problem of a busy road, for example, can be much reduced if say a block of flats is orientated so that daytime working spaces face the road and sleeping areas face away from the noise. This approach is also possible with the noisy factory: housing could be sited with gable ends facing the noise source. Another possibility is to build a wall nearer the factory so that the noise can be deflected over or away from the residential buildings, or better still it may be possible to negotiate with the factory owner to carry out remedial treatment in the factory itself.

External noise is therefore generally tackled by consideration of siting, orientation and treatments related to the source of the noise, and only in very special cases by treatments to the buildings themselves. One such special case is the house that is constructed near a major airport, where between certain hours and in certain positions in relation to the runways in use the noise levels can be very high indeed. Some reference to suitable treatments is given in Section 58, Insulation Improvements.

Internal noise also comes from many sources and is of two types, airborne and structure-borne. In the former category we have the sound of music or voices and in the latter the sound of water equipment, footsteps etc. The point at which transmitted sound becomes disagreeable noise varies with each recipient and his or her mood. Thus the music from next door may make a pleasant background or become obtrusive and annoying, or the patter of tiny feet from the upstairs flat may appear to some more like the thunder of hooves. Thus we must regard sound transmission as something to be kept as low as possible.

Flats, maisonettes, semi-detached houses and terraces

The greatest noise problems occur in these categories of dwelling, where the space occupied by one family is in direct contact with at least one other family and perhaps with as many as seventeen. It is also in contact with access corridors. *Figure 43.5* illustrates a block of eighteen flats on three stories, and shows the centre flats on the middle floor directly linked with all the other flats by direct sound paths.

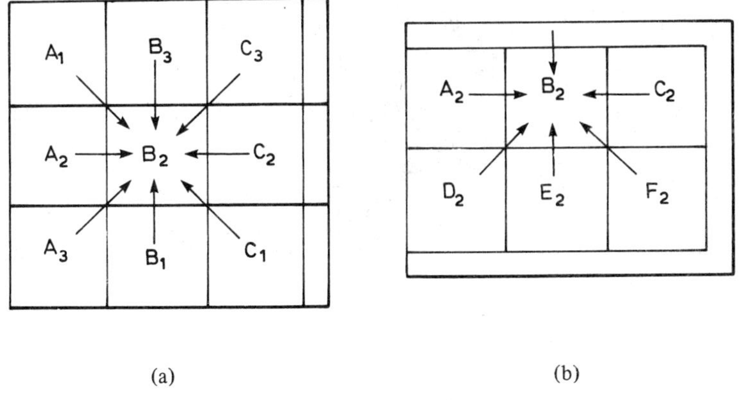

Figure 43.5 Direct sound paths in block of flats as they affect the middle flat: (a) section through three floors; (b) plan of middle floor

In addition there are indirect sound paths by which noise can be transmitted from even further afield, by passing along elements of the structure, by passing though convecting ducts, plenums or other spaces, or by travelling along water pipes, refuse chutes etc. The complexity of the problems is only too obvious and yet the solutions are not too difficult.

Primarily resistance to transmission of sound depends on the mass of the element of structure. Sound is transmitted by vibrations, and the higher the mass of the element the more difficult it will be to make it vibrate. Thus a wall or floor of high mass will resist the passage of airborne sound, which will expend its energy in trying to make the mass vibrate. The sound energy will be converted to heat energy and dissipated. Thus if we were able to build masonry walls 600 mm thick we would not experience any difficulty with airborne sound transmission through them. This of course is not a practical proposition, and we must find methods that give adequate results without involving ridiculous expense.

Fortunately mass is not the only answer, although it may be the simplest. One way to ensure that sound is not transmitted would be to arrange for its energy to be used up rather than transmitted. If a sound impinges on a construction consisting of layers of materials separated by air, the energy used in vibrating the first layer is partially used and the sound transmitted on the other side is of a lower energy level. This is in turn damped down by the resistance of the air to being compressed, so less sound is transmitted to the next layer, and so on. Thus the sound finally reaching the other side of the construction is considerably reduced from its original intensity. The arrangement can be even more effective if the airspaces are filled with a medium carrying a high volume of air, such as mineral (glass or rock) wool; this not only acts as a damping material but also allows much simpler constructions than would otherwise be necessary.

Combinations of high-mass materials, air cavities and air-containing insulation can be and are used effectively in dealing with airborne sound transmission.

The sounds associated with footfalls, tapping on pipes, running water etc. are known as impact sounds, and are also transmitted by vibration of the structure. In this case the vibration is brought about by physical force rather than by the impingement of the sound wave, and the treatment is correspondingly different.

Figure 43.6 Expanded polystyrene for sound insulation in floors, prior to screeding (courtesy EPS Association)

As a means of absorbing force, flexibility has been applied in a variety of ways. The boxer's heavy blows at the punch bag are absorbed with barely perceptible movement of the bag; the heavily swung pillow imparts much less force to the head of the victim in the pillow fight than was expected in the swing. So it is flexibility to which we turn in combating impact sound transmission.

The simplest solution would be to persuade all neighbours to use carpet slippers and to install thick carpets or linoleum or plastic floor coverings with felt backings, but this is hardly practical from the constructional viewpoint and would in any case only partially solve the problem. The real answer lies in designing the floor to carry a floating top finish that is cushioned on a layer of mineral (glass or rock) wool or expanded polystyrene materials specially developed for this application (*Figure 43.6*). It must be stressed that the grades of materials used for normal thermal-insulation applications are not suitable for sound-insulation use. *Figure 43.7*

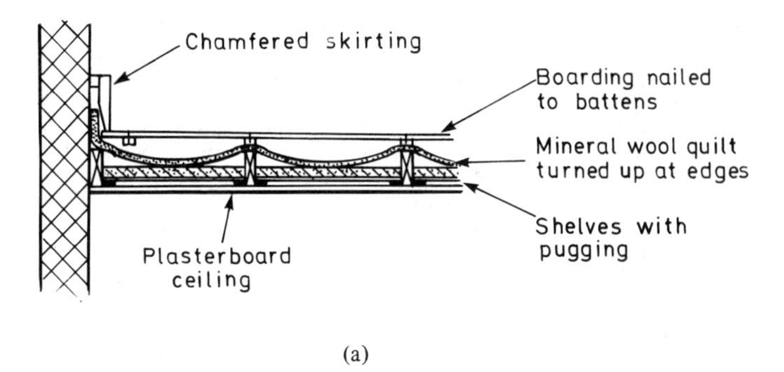

(a)

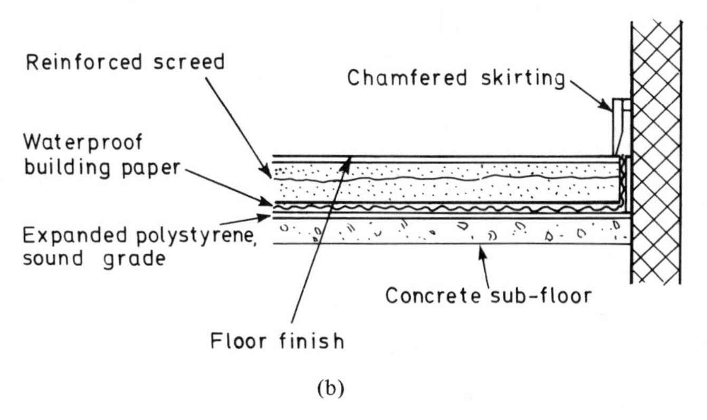

(b)

Figure 43.7 Typical floating-floor constructions: (a) timber-joist floor, (b) concrete floor

illustrates typical constructions for floating floors on timber-joist floors and on concrete.

The standards laid down by the Building Regulations for the construction of party walls and floors in dwellings are not explicit, but are couched in terms of adequacy and are backed by lists of deemed-to-satisfy constructions, which are subject to revision from time to time. No useful purpose would be served merely by repeating these constructions here, and readers are referred to the Regulations appropriate to the country in which the dwelling is to be erected for the up-to-date position at any time.

Detached houses

Since in a detached property the occupant is very largely concerned with his own or his family's noise he is much better able to control it, to contain it, or at worst to suffer it, and the specialised treatments described in the previous paragraphs will not normally be required. There may nevertheless be occasions where a high sound insulation value may be required and where the incorporation of floating floors may be advantageous. This can also apply to the provision of sound-insulated partitions, more often encountered in terms of conversion and dealt with in Section 58, Insulation Improvements.

A problem area for all dwellings can be the positioning of a WC in relation to a bedroom, or the positioning of a soil pipe in relation to either bedrooms or living rooms. It is not now considered good practice to erect uninsulated partitions between a WC and a bedroom, or to allow soil pipes to pass through living rooms or bedrooms without providing an insulated duct in which they are concealed. Some detail on these applications is also given in Section 58.

Problems with sound insulation

Many of the problems related to sound insulation are concerned with people and their reactions, and these have already been described. Here we are more concerned with practical problems.

We have discussed how sound travels in direct lines in all directions from its source, and how it can search out apertures and pass through them. It might not be evident, however, that these apertures can be very small indeed. A small keyhole in an otherwise massive door may in effect allow all the sound through. The same thing will occur with a brick wall where the mortar has been unevenly applied and voids left, or where plastering has been indifferently carried out. Any degree of porosity is potentially dangerous.

A side window that opens on to a wall may allow sound to be reflected through the window next door, or above or below, thus flanking perfectly good wall, ceiling or floor insulation.

Hard surfaces will reflect sound much more easily than soft surfaces. Hard high-gloss paints may be of great value in terms of cleanliness, but in terms of sound reflectivity soft finishes would be better. The noise level in a room equipped with hard surfaces can build up to the point that transmission to other rooms is increased. Specialised methods of treatment for reducing sound levels are discussed in Section 58.

A wall that is adequately insulated against airborne sound may have little value against impacts if they are strong enough. An example of this is the office typing desk that sits on an impact-insulated floor and is virtually unheard next door until the desk is moved so as to bring it in contact with the connecting wall. The difference can be quite alarming.

44 JOINERY

INTRODUCTION 44–2

MATERIALS 44–2

DOORS 44–4

WINDOWS 44–13

STAIRS 44–17

OTHER JOINERY 44–19

IRONMONGERY 44–20

44 JOINERY

J.K. CHAPMAN
Kent/Sussex Area Manager, National House-Building Council

This section deals with purpose-made and mass-produced joinery; it discusses the materials used (including ordering, storage and preservation), describes the construction of doors, windows, stairs etc., and lists the appropriate ironmongery.

INTRODUCTION

Joinery nowadays falls into two main categories:

- Factory mass-produced articles: standard doors, windows, stairs, cupboards, and mouldings used for skirtings, architraves and cover fillets.
- Purpose-made items of the same type, made to measure for a specific job and produced by a specialist joinery company or department of a building company with the appropriate technical and workshop facilities.

Manufacturers of standard joinery plan the operation of timber purchasing, converting, machining and assembling of the component parts as a continuous production line. This enables the competitive market to exist.

Purpose-made joinery on the one-off or small-run basis results in a far higher unit price. This is often reflected in repair and renovation work where small quantities of a skirting or a special window design are required to match existing work. The replacing of a window or door in an older building with a modern mass-produced unit often results in an unacceptable appearance, in spite of the joinery used being of acceptable quality and standard.

Manufacture of joinery has traditionally been carried out by forming accurately cut joints or intersections of the various timber members making up the completed article. In modern large-scale joinery production the traditional methods of jointing have fallen into disuse because they are basically an individual hand operation that is high in skilled labour content. The use of dowelled joints instead of mortice and tenon in door manufacture is a common example. The use of the modern synthetic glues (both cold-application and heat-treatment types) has allowed the joinery manufacturer greater scope in securing joints in timber. It may be considered that the loss of traditional methods is to be deplored, but from a cost point of view to maintain these would add a considerable burden to the industry, with its ever-declining labour force. Traditionally made joinery is still available from many companies, and is often required where the repair and renovation of older buildings is involved and the existing joinery is to be matched.

MATERIALS

Timber

Joinery is manufactured from both softwoods and hardwoods, or a combination of both, the classification being based on the tree leaf. *Hardwoods* are the broad-leaf deciduous trees that shed their leaves in winter. *Softwoods* are the needle-leaf coniferous trees. This classification can be misleading as to the strength and durability of a timber: for example, balsa, a hardwood, is very soft, whereas pitch pine, a softwood, is durable and hard.

All timbers are susceptible to the absorption and loss of moisture, either from direct contact with a heat or moisture source, or by hydroscopic means depending on the moisture content in the air.

The quality of timber can vary considerably due to knots, shakes, twist, sapwood, beetle and fungus attack etc., and considerable care must be exercised when selecting timber for joinery. This particularly applies to work that is to be treated with a clear finish. In the case of the knotty-pine type of finish for surface cladding, the frequency of knots is not necessarily a defect but an asset.

Blue stain in sapwood is acceptable in joinery that will have a fully decorated surface, but must be rejected for clear finishes. Knot sizes are restricted under BS 1186:Part 1:1971, because if these form a major part of the cross-section of a component it will be weakened considerably.

Timber dimensions are given as nominal or sawn sizes; actual or wrot sizes will be 6 mm less than nominal, the loss occurring in the planing of the surfaces.

Manufactured boards

These are used extensively for large panel areas in joinery, furniture and for surface claddings. Plywood, composite and particle boards are examples manufactured from timber. Proprietary manufactured boards tend to overcome the differential moisture absorption and movement that occur with large areas of solid timber. These boards provide a stable finish as supplied, or can be used as a backing for other decorative finishes.

Woodchip or particle boards are manufactured by mixing timber chips with synthetic resins and forming this into boards under heat and pressure processes. These boards are made in various densities, the higher the number the greater the density; e.g. 500 grade is 500 kg/m^3.

Blockboard is manufactured from strips of timber not exceeding 25 mm in width forming the core, faced on both sides with a sheet veneer.

Laminboard is of similar construction except that the inner core consists of timber strips not exceeding 6.5 mm in width. It is considered superior to blockboard.

Plywoods are made by gluing laminates of timber together under pressure, the grain of the face sheets running longitudinally, with inner sheets alternately at right angles. This results in any plywood being made of an odd number of laminates, giving the common name of three-ply.

Fibreboards are manufactured to varying densities, and consist of fibrous woody materials compressed by a rolling process that uses the natural adhesive quality of the material, along with added bonding agents. Grades are medium or semi-hardboard, standard, and tempered or super hardboard, with densities ranging from 500 kg/m^3 for the semi-hardboard to 960 kg/m^3 for the super hardboard. Surfaced hardboards are of the standard or super grades, moulded or embossed on the face in varying patterns; flutes, linenfold and reeded are but a few. These boards are also available finished with plastic or enamelled face, which makes these sheet products a versatile cladding for decorative treatment at a reasonable cost.

Perforated hardboard or pegboard can be used for many purposes, but should not be confused with purpose-made perforated sheet materials designed specifically for acoustic work.

Insulation fibreboard is seldom used in joinery work except as infilling panels where sound insulation is required. Due to the soft face of such panels they are usually covered with hard sheet cladding.

Ordering and storage

This quite critical phase of the construction industry is often a poorly organised and programmed operation and can result in high extra costs due to damage, theft and problems of site storage. Points to remember are:

- All internal joinery must be stored under cover (NHBC Jo 8).
- External joinery must be stored off the ground and covered (NHBC Jo 8).
- Order joinery in sequence and to dates required. Delivery of stairs at first lift of brickwork is not good organisation, just the opposite.
- Large orders to manufacturers may give unit price reduction, but if usage rate is slow the cost of the capital outlay, site storage, possible damage, theft and extra handling required can exceed the original saving.
- Strict site control of the joinery-storage area is required to avoid wrong frames being placed and later abandoned; site control also involves correct stacking to prevent twisting of frames, and correct cutting (and preservation) of projecting horns to heads and sills.
- Allocation of internal joinery per house or job unit can prevent the excessive waste seen on various building projects of all sizes and quality.

Preservation

Preservation of external joinery is mandatory (NHBC Jo 2 and 3). Care should be taken to specify a preservative treatment that is compatible with the final decorative finish.

For one-off or small-run joinery, the supplier should be informed of any preservative requirement at the time of ordering. The large manufacturers supply joinery that has a preservative treatment as part of their production programme, and this is usually noted on the trade invoices and catalogues.

Cut ends of frames must be re-treated on site with two flood coats of preservative (NHBC Jo 12(e)).

Shrinkage and movement

This is an age-old problem in joinery, often increased by today's general requirement of higher heating standards in all types of buildings. It is necessary to consider at the design stage not only the joinery and type of timber to be used but the environment it will be subjected to in the completed and occupied building.

Inadequate fixings and slight section of timber can contribute to maintenance problems and costs at a later date. Moisture content of joinery timber is listed in BS 1186 and NHBC Jo 7.

Provision of 1½ pairs of butts or hinges for doors close to heat sources is desirable, in addition to airing-cupboard doors and external doors (NHBC Jo 11(a)). Heating units placed under stairs are another potential source of trouble from shrinkage, and should not be sited in this position if at all possible, but with the architect or designer under pressure to use all space this has at times to be accepted.

Different types of timber and manufactured boards that have a different shrinkage potential and are used together in the construction of joinery can result in movement problems. This could have been avoided by careful consideration at the design stage, with detailing that allows movement to take place under cover mouldings or at joints with a bead or moulding and a wide tongue to keep the joint light-tight.

DOORS

Ledged and braced. The maximum practical width for this type of door (*Figure 44.1*) is 900 mm. It is constructed of vertical battens either rebated or tongued-and-grooved together, fixed to *ledges* (the horizontal members) 150 mm wide. *Braces* 75

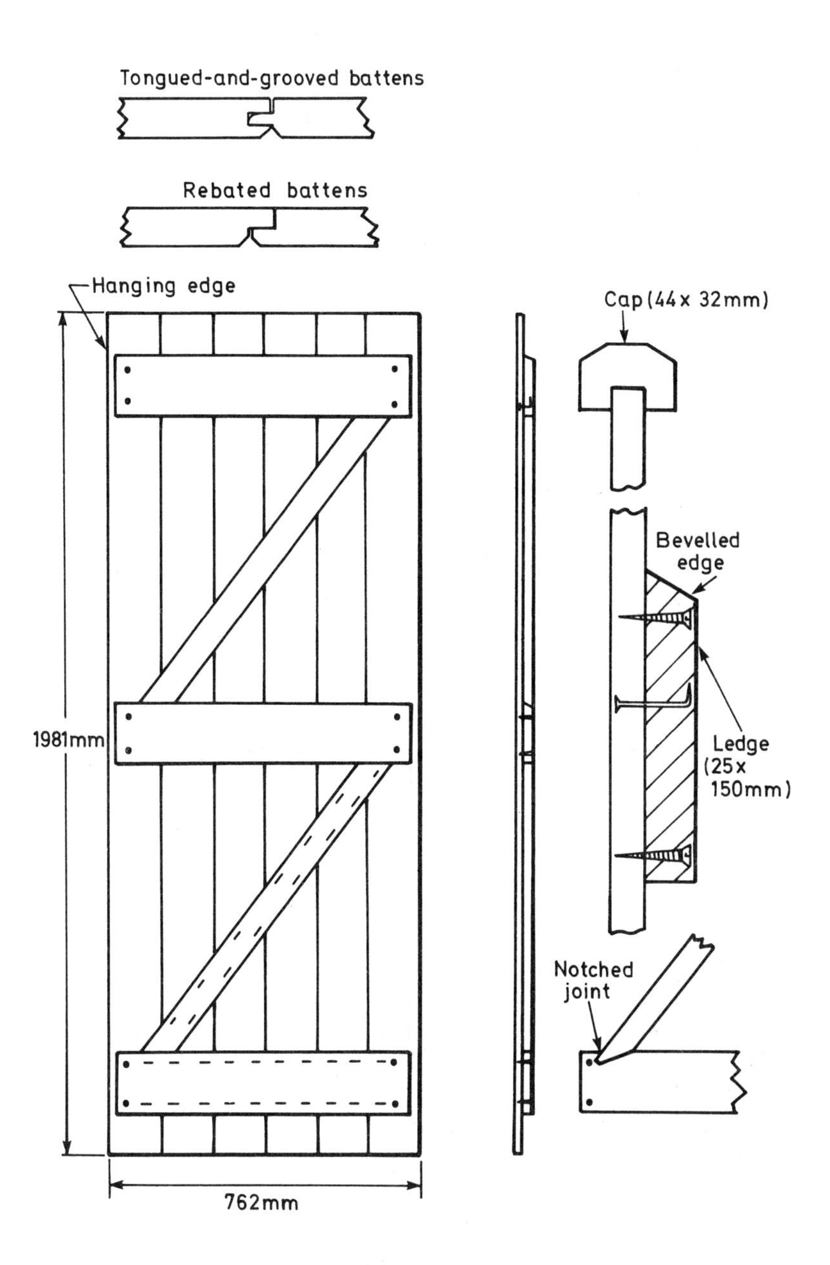

Figure 44.1 Ledged and braced door

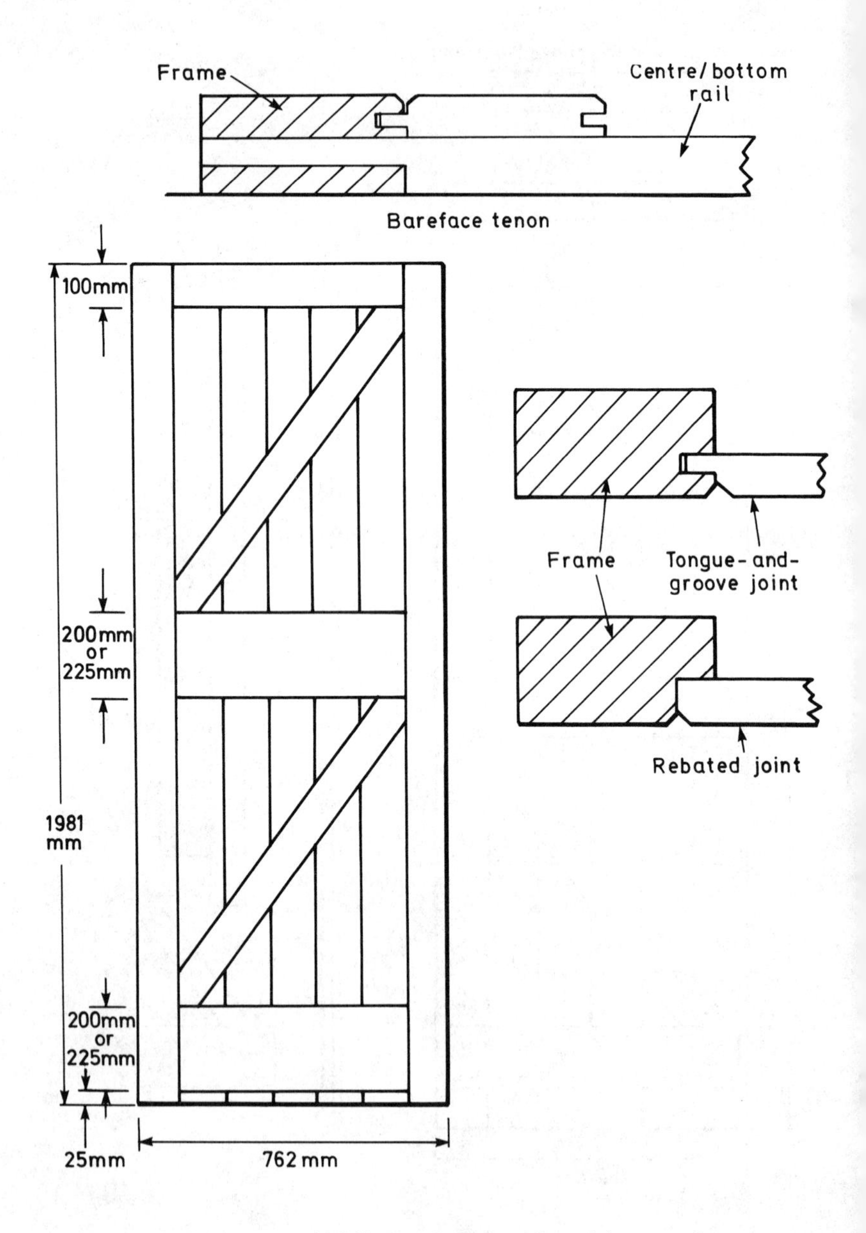

Figure 44.2 Framed, ledged and braced door

mm wide are fitted between the ledges at an angle of not less than 45 degrees, and must thrust upwards away from the hanging edge. Braces can be notched into the ledges (bottom right of the figure) or can abut the ledges. Where exposed to the weather the top edge of the ledge should be bevelled, and if used as an external door or gate it is advisable to provide a capping to prevent entry of water into the top edge. Ledges are twice screwed at each end to the battens from the face of the ledge, and the battens are nailed from the batten face and clenched over into the ledge.

Framed, ledged and braced. Doors of this type are constructed from a morticed and tenoned frame with a matchboard infill panel (*Figure 44.2*), the boarding being rebated or tongued into the frame. The centre and bottom rails are thinner to allow the boarding to finish flush with the frame, these rails having a bareface tenon. This type of door was commonly used for garage doors before the popularity of the up-and-over type.

Stable doors. The name indicates the use of these doors, which are made in two parts, enabling the top section to be opened for ventilation etc. They are constructed as ledged and braced doors, with the top ledge of the lower door projecting to form a stop and weather check when both are closed, *Figure 44.3(a)*. A framed type of this door, *Figure 44.3(b)*, is used in housing, usually in conversion work on older cottage

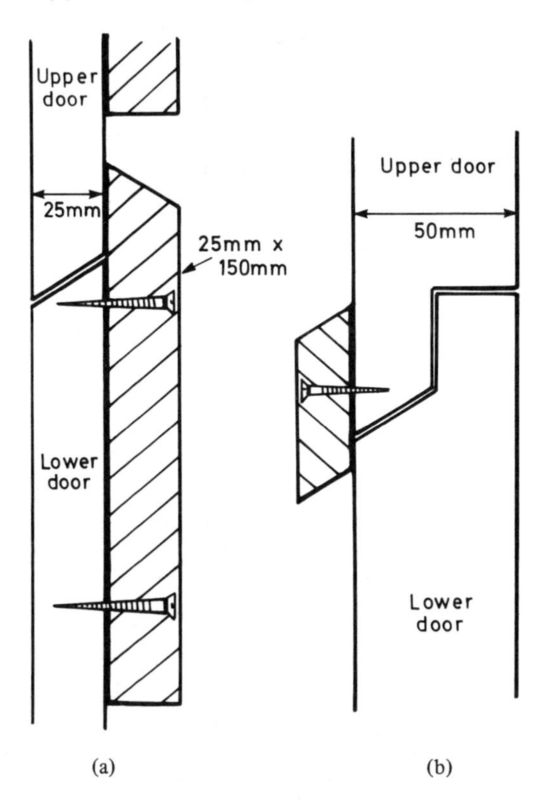

Figure 44.3 Stable door details: (a) ledged and braced; (b) framed, ledged and braced

property, but provision must be made to connect both parts with locks or bolts to enable the door to be used as one unit and provide adequate security.

Panelled doors. These are made in many variations of numbers and types of panel, both in timber and glazing. The most common type is shown in *Figure 44.4*. Mouldings are either formed on the panels or stiles in the solid, or planted on the face of the panel. *Figure 44.5* shows some panel details. Bead and butt panelled doors are perhaps the most common, where a bead is formed on the edge of the panel or stile,

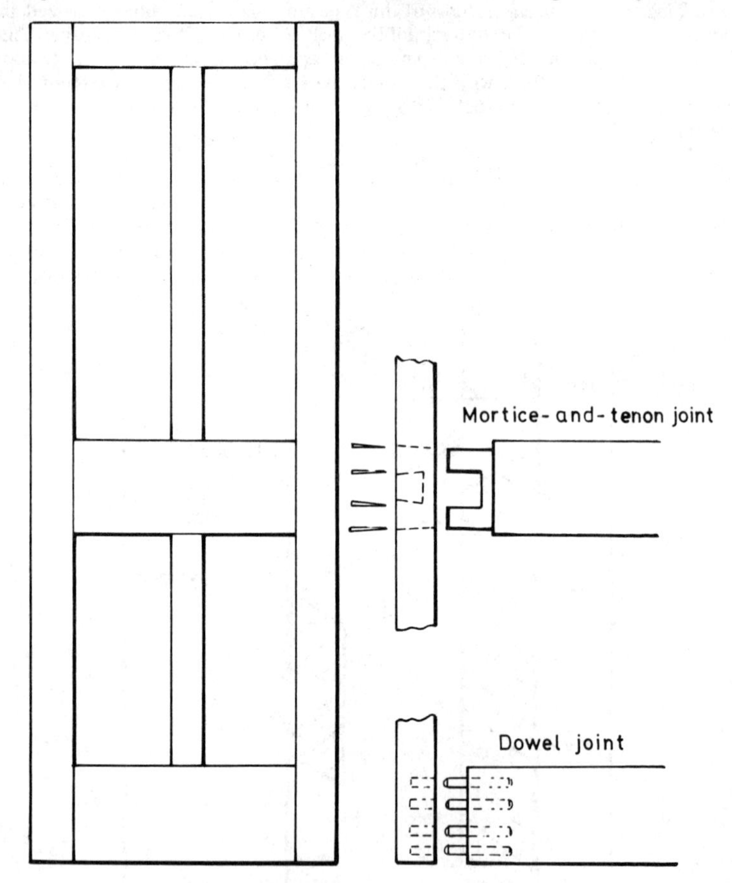

Figure 44.4 Panelled door

with a butt joint to the rails; see (a) and (b) of the figure. Door stiles and rails are jointed together by mortice-and-tenon or dowel joints, as shown at the right of *Figure 44.4*.

Fully glazed or partly glazed doors again offer many alternative arrangements of the panel. A fully glazed door often only has one pane of glass, which is secured into the rebates by glazing beads, moulded or square, secured by screws and cups or panel pins. Screw-fixed beads are always used on first-class work, and this method should always be used on clear-finish or hardwood joinery.

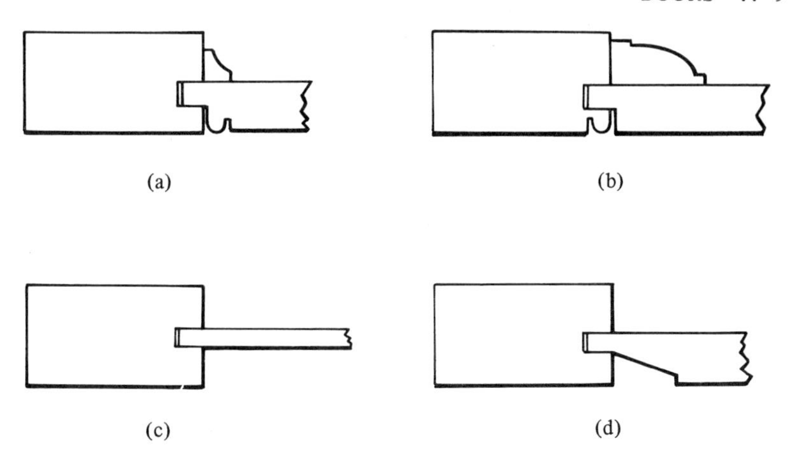

Figure 44.5 Door-panel joint details

Diminished-stile or gunstock-stile doors. The stiles are reduced in width above the centre rail, giving a more pleasant appearance (see *Figure 44.6*); a detail of the glazing bead and panel is shown on the right of the figure.

Flush doors. Various types of flush door, both internal and external, comprise the greater part of all door production. The use of internal flush doors for all new housing development is the rule rather than the exception in the UK at the present time. Variations of flush doors range from the light 'egg box' type core with a slight timber frame covered with hardboard panels, to the solid timber core covered with hardwood-faced plywood and lipped on all edges with matching hardwood. Although the standard internal flush door would appear to be of a light construction compared with traditional doors, the fact that large numbers have been used without problems must indicate the value of this type. Fixing blocks are provided for locks and door furniture in the hollow type of flush door. Hollow flush doors must have grooves or holes across the internal framing to allow the air pressure to equalise within the door to that of its environment. *Figure 44.7* gives typical details of sections through flush doors, (b) showing a fire-check construction. Fire-check flush doors are cored with plasterboard and/or asbestos-based board rebated into the framing under the plywood face panel. The thickness of fire-retarding sheet varies with the amount of resistance specified. Manufacturers supply half-hour fire-check doors from stock. Where a longer fire resistance is required a solid timber core is often the best type to use.

Door linings and frames

Door linings. These are usually made from 32 mm thick timber, 30 mm wider than the wall to which it is to be fixed; this allows a 15 mm projection on each side of the wall to accommodate the thickness of plaster or drylining to the walls. *Figure 44.8* shows a solid lining 50 mm thick, with a rebate 20 mm deep on the rear of each leg. The width of the lining depends on the partition being built: in this case, 75 mm blockwork requires 105 mm finished-size lining. Linings can be made with the legs either terminating at the top of the head, or extending to the full height of the room;

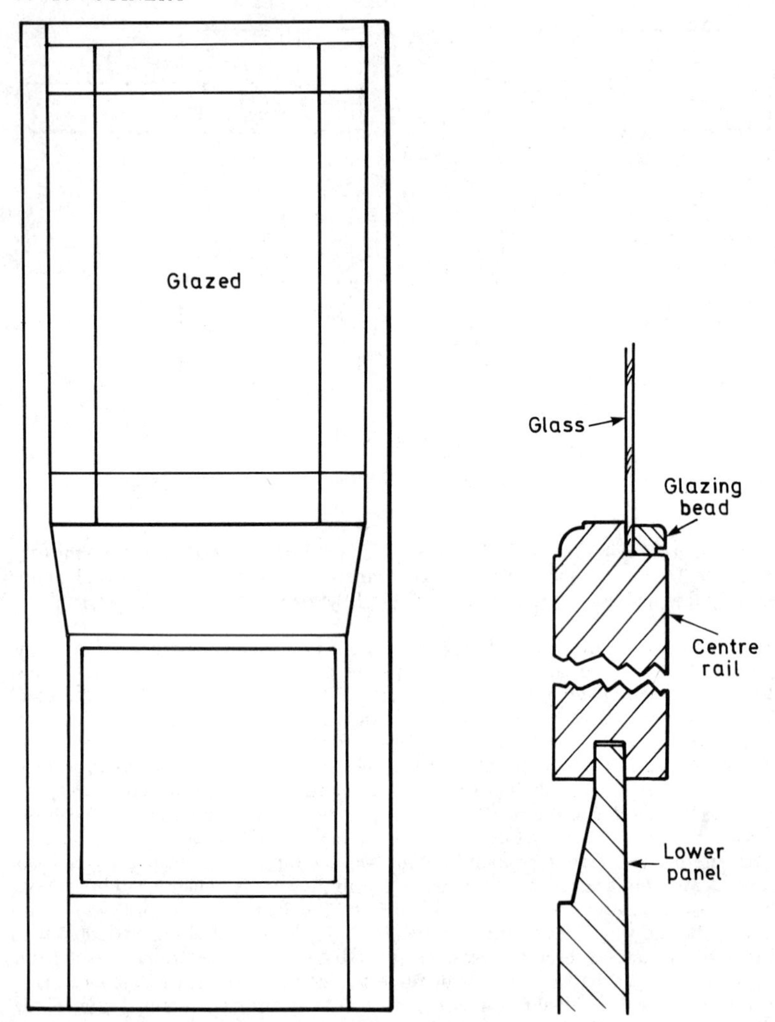

Figure 44.6 Diminished-stile or gunstock-stile door

the latter method provides a more positive fixing into joists or noggins. Storey or full-height linings can be fitted with glazing or a solid panel above the door; on the other hand, if the blockwork is carried over the doors, the width of the lining should be reduced to the width of the blockwork above the head of the lining. Rebates for the doors are formed by planting stops.

Figure 44.9 shows a door lining for a stud partition with plasterboard lining to the wall.

With walls exceeding 150 mm thick and where a full-width timber lining is required, it is necessary to form a skeleton frame from wrot timber and to secure this to the wall by means of plugs and screws, then cover with a plywood or solid

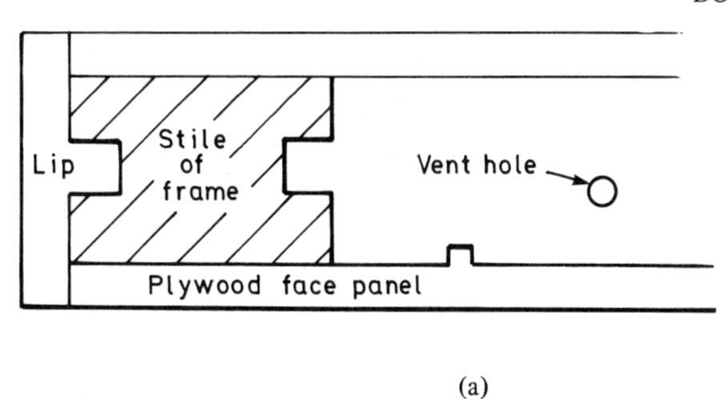

(a)

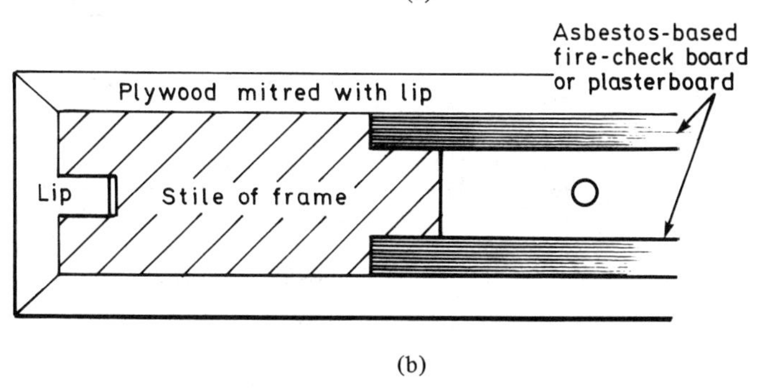

(b)

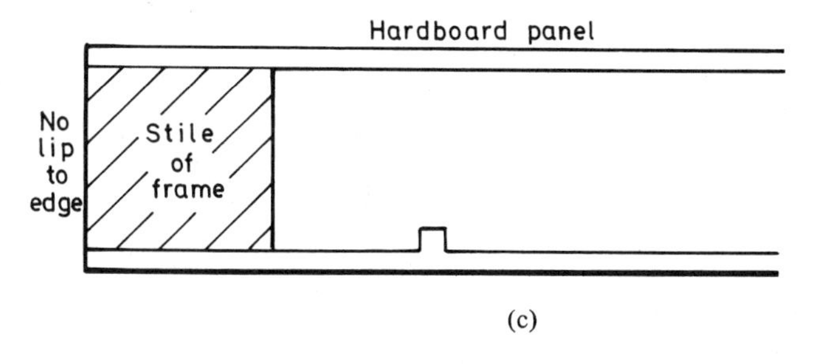

(c)

Figure 44.7 Typical sections through flush doors

timber panel set back to form the rebate for door hanging as required; see *Figure 44.10(a)*. The alternative is to construct a fully framed, panelled and rebated lining – *Figure 44.10(b)* – and then fix it to full-length battens or grounds that have been secured upright to the wall.

It is essential that all door linings and grounds for door linings are fixed with all vertical work plumb, and horizontal members level; also, the linings must not twist or wind across the face of a leg or head, otherwise the problems at the time of hanging the door will be many and will result in a sub-standard finish.

Where planted stops are used to form the rebate, they are fixed after hanging the door; architraves are fixed to leave an equal margin of the lining showing around the door opening, and to cover the joint between the lining and the plaster or wall lining.

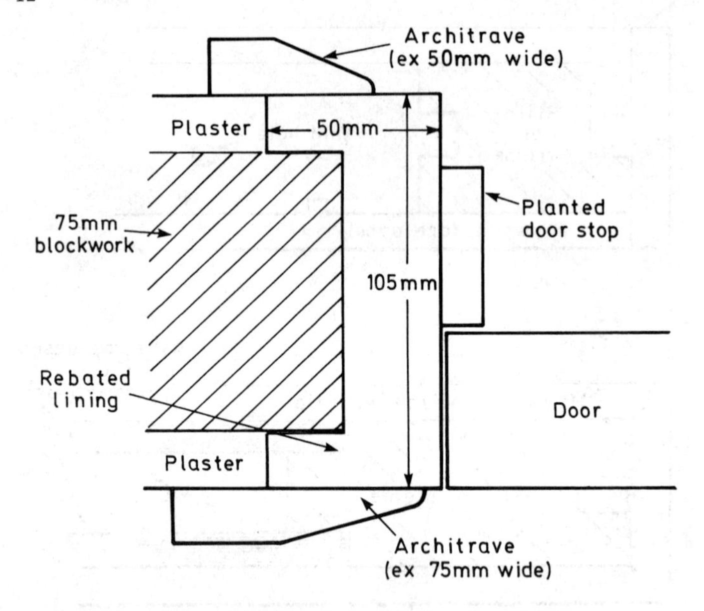

Figure 44.8 Door lining: blockwork partition

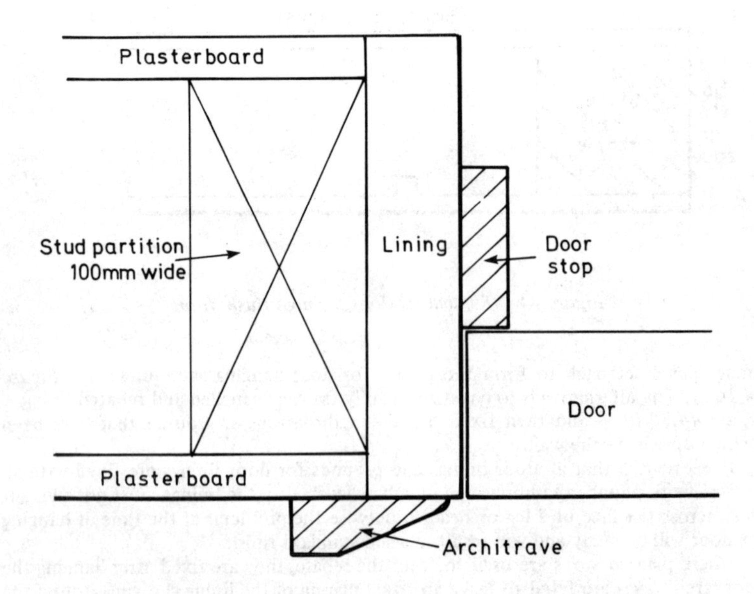

Figure 44.9 Door lining: stud partition

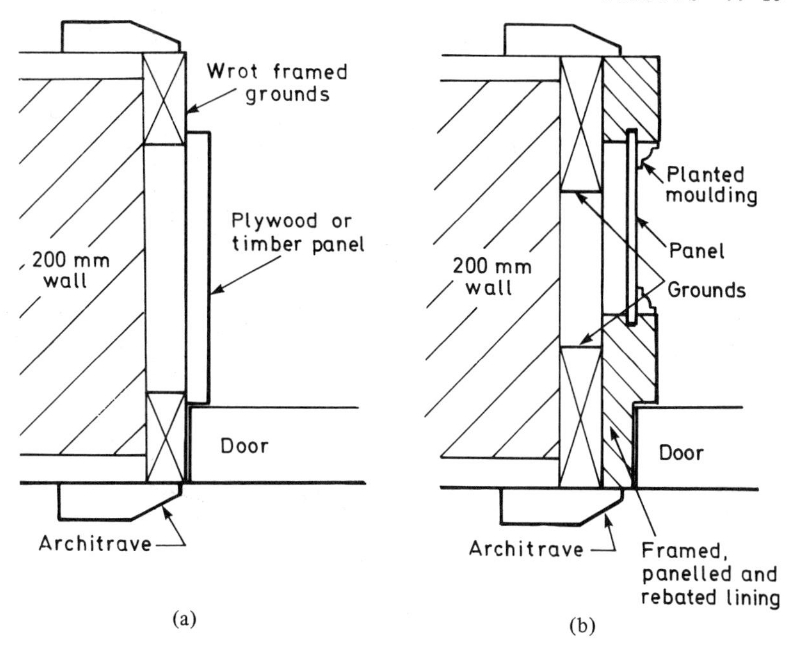

Figure 44.10 Door lining: walls exceeding 150 mm thick

Door frames. Door frames constructed out of solid timber can be used both in internal and external walls, but NHBC Jo 12(c) requires solid-timber frames on all external doors. These are usually of 50 × 100 mm or 63 × 100 mm timber, and are built-in as work proceeds by means of frame clamps at a maximum of 600 mm centres, including fixings at not more than 150 mm from both top and bottom of the frame. It is essential that both vertical and horizontal damp-proof courses are fitted to prevent the entry of moisture at the junction of joinery and external walls – this applies to all door and window frames (NHBC Br 15). In some types of good-class work, door and window frames are fixed after the completion of the structure, and it is essential to ensure that accurate profiles are used when the openings are formed, so that the joinery can be fixed without the need for excessive packing or cutting-out of work. Methods of fixing vary, depending on hardwood, softwood, joinery finish and wall construction. When this type of work is planned it is necessary to decide the fixing details before manufacture, so that any special requirements can be incorporated.

WINDOWS

Casement windows. These fall into two basic categories: the standard square-edge sash fitted into a solid rebated frame of varying design, and the stormproof casement window favoured by the large joinery manufacturers (who usually produce a modified version of the type specified by BS 644). The sashes of stormproof windows are rebated over the face of the frame, with a capillary groove in the edge of the sash and frame rebates to prevent water penetration; *Figure 44.11* shows a

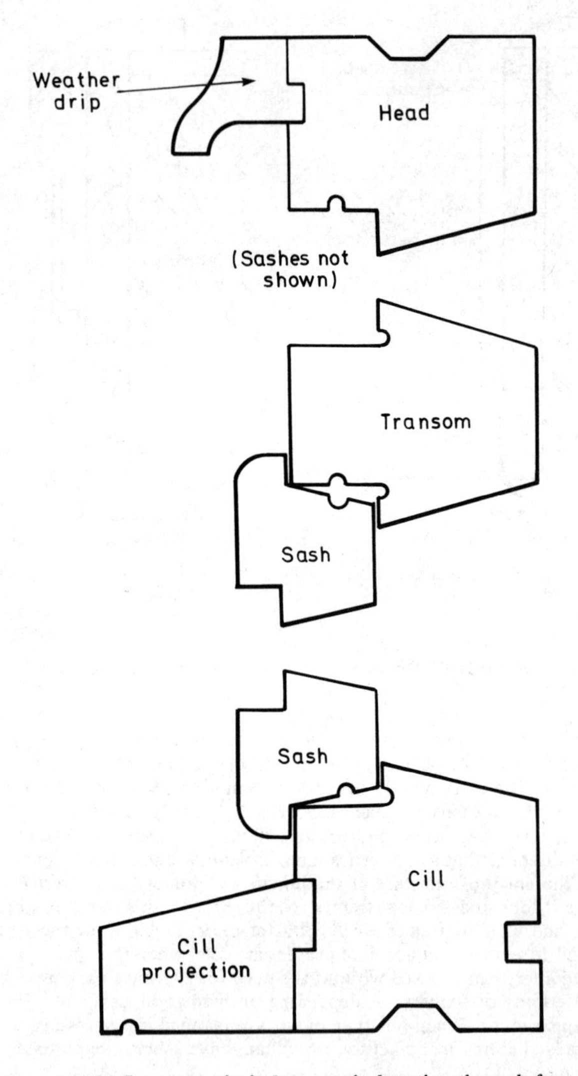

Figure 44.11 Stormproof window: vertical section through frame

section through the frame. Since the tops of the sashes are rebated against the frame these require a drip or weather check to prevent water running down into the top joint. Drips and cill projections are rebated into the frame, because to produce these sections from solid timber would not be economical.

Bay windows. Square and splayed bay or projecting windows are made from standard sashes fitted to a frame with mitred cills and heads and solid double-rebated corner posts. Circular bay windows consist of square sashes fitted between splayed posts to form facets around a curve on plan, the outside line of the cill and the brickwork under forming a curve that gives the impression of a curved bay window. The window boards usually follow the line of the sashes. A true curved bay window would have frame, sashes and window board all curved on plan, and would also require the glass to be curved. To carry out this type of construction would be very expensive, since all the rails would have to be cut out of solid timber to the correct radius, and then individually machined on a spindle moulder fitted with a ring attachment, the final finishing being carried out by hand with a compass plane.

Steel casement windows. When fitted to wooden surrounds, these require a double-rebated frame to allow the steel frame to be bedded into the timber surround. When using steel sashes for opening lights, but glazing direct into the rebate of the other

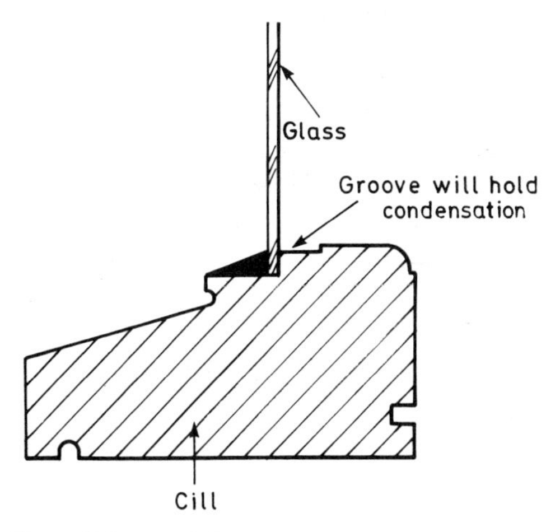

Figure 44.12 When cill intended for steel casement is direct glazed, groove formed by unused rebate will hold condensation

lights, the latter will create a problem in the form of a 3 mm groove in the top of the cill that will hold condensation (*Figure 44.12*). As with all frames, it is most essential that those used as surrounds for steel casements are not allowed to twist, or difficulty will be experienced in getting the sashes to close properly.

Double-hung sash windows. This type of window consists of two counter-balanced windows sliding vertically between grooves formed by the jambs and planted beads. Sashes are hung on cords or wires, connected on each side to a weight that is concealed in the hollow formation of the jamb: see *Figures 44.13 and 44.14.* Another method of supporting double-hung sashes is by the use of 'Unique' spiral balances, which can be grooved either into the solid frame or the edge of the sash. This method reduces both timber and labour content in manufacture, and eliminates the replacement of sash cords.

Venetian windows. These are double-hung sash windows with additional hung or fixed lights to the sides of the centre window, the side lights being one-third or half the width of the centre sash. Where side lights are required to open, the boxed mullions have to accommodate two sets of weights, resulting in a cumbersome appearance. Where side lights are fixed, the weights for the centre sash can be hung in the outside jamb by running the cords over a pulley in the solid mullion, through a

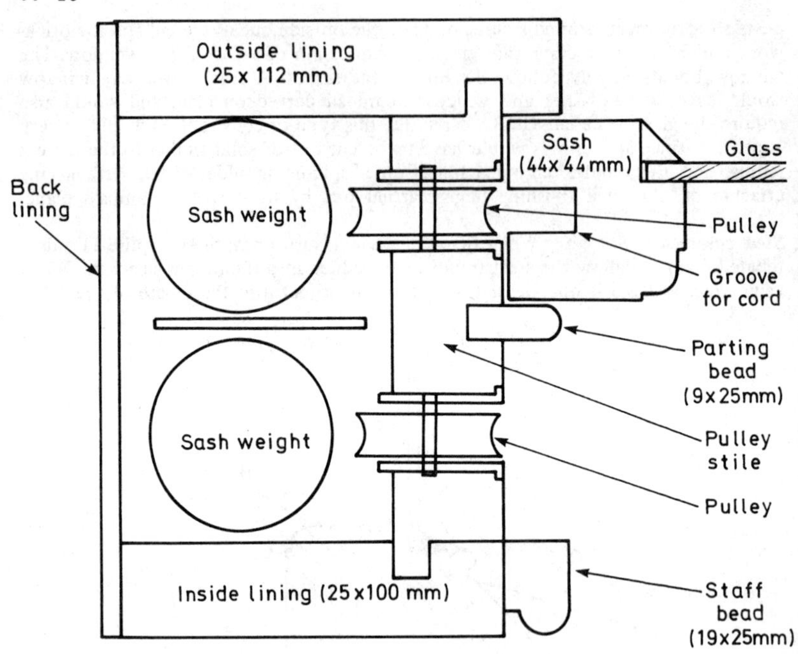

Figure 44.13 Double-hung sash window: section through jamb

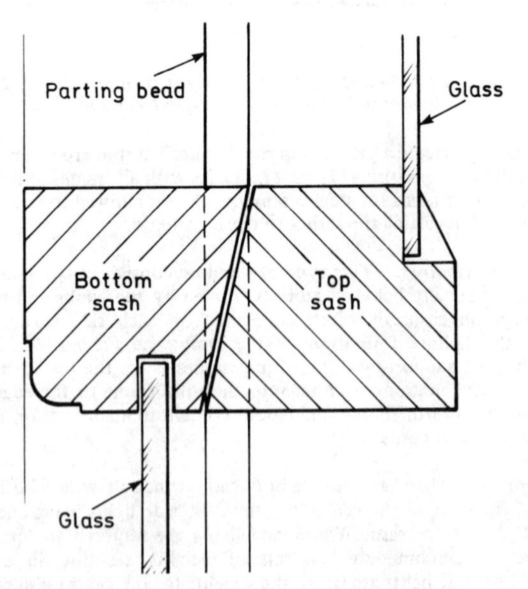

Figure 44.14 Double-hung sash window: section through meeting rails

groove in the top sash. A cord cover is fixed in the head of the frame in the bottom sash groove.

Pivot-hung windows. These can be either the conventional type (where the sash pivots on a pin protruding from the frame that engages with a grooved plate on the edge of the sash) or the type that uses a proprietary hinge incorporating an adjustable friction boss to enable the sash to remain open without the need of stay bars. This does allow a variable amount of ventilation. Back-flap type hinges fixed externally to operate pivot-hung windows can be questioned as a security risk if easily accessible, since the removal of the hinges will allow the sash to be lifted out. Friction-boss hinges are usually semi-concealed, being fitted into the frame rebate and the edge of the sash.

STAIRS

Stairs provide a means of access between differing floor levels within a building. At times stairs are designed to be a central focal point of graceful proportions and elegance, but with present-day pressures on costs of land, labour and materials they have become rather utilitarian. A complete stair consists of strings, treads, risers, landings, newels, balustrade and handrails, with many variations as to the finished details.

The basic principle of stair design is that the amount of rise and going of each step must be equal within each staircase. Production of stairs is again a choice between the mass-produced factory article or an individual item produced specifically for a particular job. The factory-produced stairs are competitive in price, and because of greater purchasing potential manufacturers can often obtain a superior selection of timber that perhaps is not always available to the individual. Part H of the Building Regulations 1976 gives certain definitions as to the interpretation of the components and the adjoining construction of stairways, landings, ramps etc. Regulation H3 sets out specific requirements for stairways, relating them to the Purpose Group the building comes within, as defined by the Table to Regulation E2. This gives the descriptive title of the groups of buildings or compartments within a building, 'small residential' and 'other residential' being Groups 1 and 3 respectively. Regulations H3 (4) (j) and H6(2)(d) both restrict openings between treads or in balustrades to a size that will not allow a 100 mm sphere to pass through. This applies to Purpose Groups 1 and 3 and also to Group 2 where persons of under five years of age are liable to use the stairways.

The main points from Building Regulations, for domestic housing in Groups 1 and 3, are:

- Minimum stair width is to be 800 mm, unless serving only a bathroom or closet or one room not a kitchen or living room, when it can be 600 mm.
- Pitch of flight is not to exceed 42 degrees.
- Number of risers is not to exceed sixteen.
- Height of riser is to be not less than 75 mm or more than 220 mm.
- Going of step is to be not less than 220 mm.
- Total of the going plus twice the riser is to be not less than 550 mm or more than 700 mm.
- Landings are to be not less than the width of the stairs.
- Tapered tread or winder is to be not less than 75 mm at any part of the going. The going is to be not less than 220 mm; this point is determined at the centre if the stairs are less than 1 metre wide, or at 270 mm from each side if over 1 metre wide.
- Handrails are required to any flight of stairs with a rise of more than 600 mm.
- Handrails are required on both sides if the width exceeds 1 metre.
- Handrails are to be securely fixed at a height of not less than 840 mm or more than 1 metre, measured vertically above the pitch line.

- Balustrade guarding a flight of stairs is to be a minimum height of 840 mm.
- Balustrade guarding a landing is to be a minimum height of 900 mm.
- Headroom is to be at least 2 metres vertically from the pitch line or the surface of a landing or ramp.

See *Figures 44.15 and 44.16.* The pitch line is the notional line drawn along the front edge of the nosings. When designing open-rise stairs it is advisable to remember that the support to the treads of the riser is lost, so the thickness of the tread should be increased to a minimum of 38 mm. With open-rise stairs the tread above must overhang the lower one by at least 15 mm. A hardwood rod or bar screwed to the underside of the treads with a spacer can meet the requirement of reducing this opening.

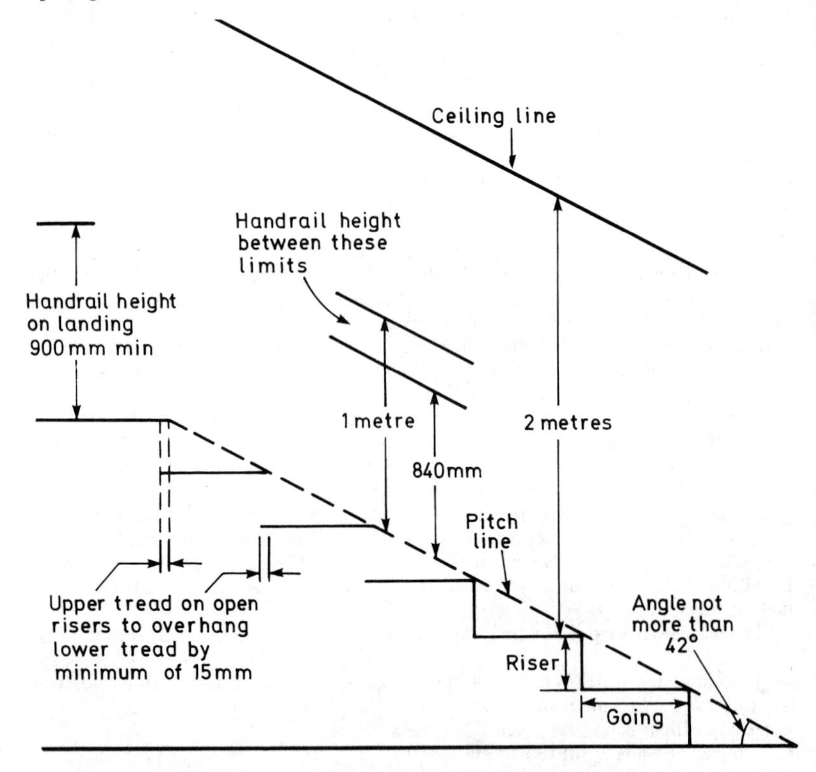

Figure 44.15 Stairway regulations for domestic buildings

On-site problems that arise with stairs are:

- Lack of adequate storage and protection, with stairs often delivered too early.
- Loss and damage to components that are supplied loose.
- Damage after fixing, lack of protection strips on treads etc.
- Incorrect trimming to openings of floor joists: either too narrow, often resulting in hacking away part of trimmer, or too wide, being made out with excessive packing.
- Variation in floor-to-floor heights, resulting in varying depth to top or bottom risers, or stairs being fixed with the going out of level.
- Excessive easing of tenons between strings and newels, resulting in movement of the newels when in use.

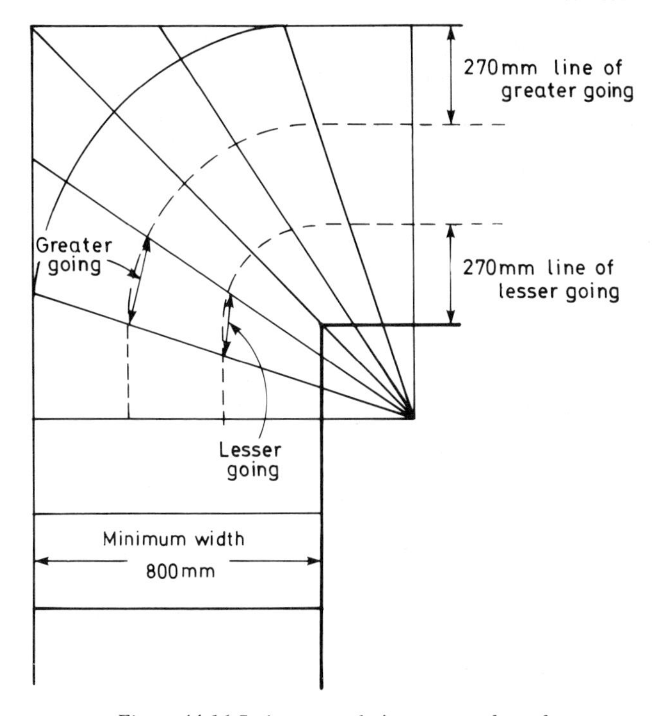

Figure 44.16 Stairway regulations: tapered treads

- Failure to drawbore the newel/string tenon and drive home dowels, and lack of fixing where newels abut joists.
- Failure to screw the bottom edges of risers to winders, and/or to glue the joints in this on-site operation.
- Failure to secure winders with full-length wedges fully glued.
- Failure to fix glue blocks to riser tread angle.

All these items are a result of inadequate site supervision and workmanship.

Riser and going formula. The Building Regulations allow a variation of 150 mm in the basic formula: 'total of two risers plus one going = not less than 550 mm or more than 700 mm'. Another similar formula is 'twice the riser plus one going equals 580 mm', e.g. two risers of 175 mm = 350 mm plus one going of 230 mm = 580 mm.

OTHER JOINERY

Trim and mouldings

These take the form of architraves, skirtings, cover mouldings etc. They vary widely in shape and size, although they have become standardised under mass production, as against each local timber merchant producing his own variation of moulding profile. Chamfered and round-edge skirting and architrave are the most used today in speculative and local-authority housing developments. The width of skirtings is usually 75 mm or 100 mm, and 38 mm or 50 mm for architraves. In renovation work skirtings fixed to grounds and formed of several sections of solid timber are still encountered, as are single-piece skirtings up to 200 mm wide, often finished with a lamb's-tongue or torus moulding. Picture rails and dado or chair rails are seldom used these days.

Skirtings and similar mouldings are fixed direct to the plastered walls in most work; the use of grounds consisting of timber battens fixed prior to plastering has fallen into disuse except in some high-class work. Intersections at external angles of mouldings should be mitred; for internal angles one member should be scribed over the face of the adjoining piece. Architraves are fixed to the door frames or linings showing an equal margin all round, and must also be fixed through the back edge into the wall.

Kitchen cupboards

Today kitchen cupboards are often purchased from a merchant or manufacturer in a knocked-down form for assembly on site. These fittings are obtainable in a wide range of quality, both in structure and finish, ranging from unpainted softwood to plastic laminates. Specialist firms will provide fitted kitchens including cookers, hob units, sink tops etc., with worktops and cupboards finished with laminates to the purchaser's choice. It should be remembered that, once ordered, these special items have to be paid for, so beware of the client who is indecisive. When making up worktops or fittings to be covered with laminates, the opposite surface must also be covered with a similar material to balance the board used, otherwise warping can occur. Joinery manufacturers offer a wide range of floor and wall units that will meet the needs of most domestic housing developments.

IRONMONGERY

This covers the whole range of door, window and cupboard fittings used for hanging and opening doors and sashes, and providing means of security for these.

Butt hinges for hanging doors are supplied in cast iron, steel, brass and plastic. Butts are sunk into both frame and edge of the door. 1½ pairs of 100 mm butts are required to external doors, 1½ pairs of 75 mm butts to airing cupboards (NHBC Jo 11). This provides additional restraint not given by the normal one pair of butts.

Rising butt: the faces of the two leaves are formed as helical curved surfaces that cause the door to rise when opened, in order to clear a carpet or similar obstruction. It is necessary to bevel the top back edge of the door to allow it to rise when the butts are fitted.

Back-flap hinge: used for thin doors and shutters where it is not possible to fit a hinge on the edge.

Cross-garnet: steel hinge with a long leg screwed to the face of a matchboard door and the short vertical leg to the frame.

Strap hinge (or cup and ride): heavy steel hinge bolted and screwed to the face of a door, the vertical pin being supported in a cast-iron cup fixed to the frame; used on framed garage doors and similar.

Rim lock: surface-fixed lock with the tongue engaging in a box staple fixed to the frame.

Mortice lock: as the name indicates, fitted into a mortice sunk into the edge of the door. Available in many sizes including rebated edges.

Security mortice lock: supplied by specialist manufacturers, with steel reinforced tongues of extra length, and multiple fixed striking plates.

Security bolt: small mortice bolt fixed into a bored hole in a sash or door, with a striking plate sunk into the frame, and operated by a splined key.

Rim-lock furniture: usually round or oval type as handle is away from frame, and has only one plate as lock is fixed to face of door.

Mortice-lock furniture: usually lever type unless a lock of at least 150 mm deep is used, because shallow locks with knob type furniture tend to foul the door frame.

Cylinder-rim night latch: the most popular type of door lock, available in many finishes and forms.

Casement stay: available in varying legnths and finishes, supplied with two pins to allow the sash to be secured in both open and closed positions.

Casement fastener or *cockspur:* available in mortice or face-fixed wedge type, to secure sash when closed.

Ball catch: spring-loaded, sunk into the edge of a door, available in many sizes.

Fibre track: for sliding doors on cupboards etc., supplied both for surface or rebated fitting, with sliders being fitted to bottom of doors. Manufactured from high-density plastic.

Sliding door gear: available for doors of many widths and weights. It is advisable to check weight and size of door and location to be used before ordering. Track for heavy multiple-leafed doors is available from specialist suppliers, who issue detailed ordering sheets.

Locking door set: incorporates the locking device within the handle that secures the mortice bolt.

Door furniture can be supplied in prepacked door sets applicable to the door type and position to be fixed. This method does help with stock control, and reduces the need for counting and issuing many individual items from bulk stores.

FURTHER INFORMATION

BS 455: 1957, *Schedule of sizes for locks and latches for doors in buildings*
BS 459, *Doors*
  Part 1: 1954, *Panelled and glazed wood doors*
  Part 2: 1962, *Flush doors*
  Part 3: 1951, *Fire-check flush doors and wood and metal frames*
  Part 4: 1965, *Matchboarded doors*
BS 584: 1967, *Wood trim (softwood)*
BS 585: 1972, *Wood stairs*
BS 644: *Wood windows*
  Part 1: 1951, *Wood casement windows*
  Part 2: 1958, *Wood double-hung sash windows*
  Part 3: 1951, *Wood double-hung sash and case windows (Scottish type)*
BS 1088 & 4079: 1966, *Plywood for marine craft*
BS 1142, *Fibre building boards*
  Part 1: 1971, *Methods of test*
  Part 2: 1971, *Medium board and hardboard*
  Part 3: 1972, *Insulating board (softboard)*
BS 1186, *Quality of timber and workmanship in joinery*
  Part 1: 1971, *Quality of timber*
  Part 2: 1971, *Quality of workmanship*
BS 1195, *Kitchen fitments and equipment*
  Part 2: 1972, *Metric units*
BS 1227: Part 1A: 1967, *Hinges for general building purposes*
BS 1331: 1954, *Builder's hardware for housing*
BS 1455: 1972, *Plywood manufactured from tropical hardwoods*
BS 1567: 1953, *Wood door frames and linings*
BS 3444: 1972, *Blockboard and laminboard*

BS 4011: 1966, *Recommendations for the coordination of dimensions in building. Coordinating sizes for building components and assemblies*
BS 4471, *Dimensions for softwood*
Part 1: 1969, *Basic sections*
Part 2: 1971, *Small resawn sections*

Registered house-builder's handbook, National House-Building Council, revised 1974 (Joinery, pages 61–65)

Building Regulations 1976, HMSO (Regulation E2 and table, pages 58–59; Part H, Stairways, ramps, balustrades and vehicle barriers, pages 104–119)

45 GLAZING

PERFORMANCE REQUIREMENTS AND SAFETY	45–2
MECHANICAL CHARACTERISTICS	45–6
MATERIALS	45–8
CONTRACTS	45–10
WORK ON SITE	45–10

45 GLAZING

D.A.S. GOODMAN, DipArch, FRIBA
Partner, Briars Goodman Associates, Chartered Architects and Industrial Designers

This section deals with all the characteristics governing the choice and installation of glazing in building. Although the builder's main responsibility may be work on site, a full understanding of all aspects and materials is essential to ensure correct installation, even when the responsibility for the design and specification may rest with others.

PERFORMANCE REQUIREMENTS AND SAFETY

Glass in building is used in three ways:

- the glazing of external walls, where it has to provide light, allow ventilation, and prevent ingress by rain and wind;
- internal glazing of screens etc., where the main function is to allow light and sometimes vision between different internal areas;
- the guarding of changes in floor level, balconies and stairs, where the glazing has to provide a physical barrier.

Aspects of performance requirements and safety are closely related to these uses. The main requirements are set out below in some detail.

Daylighting. Glazing must allow light to enter without wind and rain. The use of new methods of construction and materials allowing the formation of window walls has tended to increase the window area of modern buildings. This tendency, however, may be reversed by the requirements of the Building Regulations 1976 in terms of insulation.

A window area of not less than one tenth of the floor area is usually required, and Part K of the Building Regulations sets out minimum open space outside windows. More accurate assessments of daylighting requirements in large areas such as schools, offices, factories, etc. involve the calculation of daylight factors.

Ventilation. Part K of the Building Regulations 1976 requires ventilation openings of one twentieth of the floor area for habitable rooms. These can open direct into a conservatory or enclosed verandah, provided they have opening windows of one twentieth of the combined floor area of room and verandah.

Windows that are side hung, horizontally sliding and vertical pivoting allow rain to penetrate when open, and for night-time ventilation top-hung lights or horizontal pivots can give minimum ventilation without rain penetration. Hit-and-miss grilles in the frame or horizontal inserts fitting into the top of the glazing rebate can also provide ventilation without rain penetration.

Vision – out and in. Apart from the obvious function in housing of allowing views over gardens, windows can provide a visual link with outside activity and movement in the street, essential with old people, and for mothers wishing to keep an eye on children playing. Good detailing of full-height windows and doors can avoid the barrier that the traditional window and french door arrangements give between garden and house.

In large factory areas windows at a viewing height give those working inside a vital link with the outside environment, even if only overlooking an internal court. Though shops provide the best example of 'vision in' as a requirement for the display of goods and processes to those outside, glazing can enclose an industrial process, boiler house, etc. so that the function can be viewed or monitored from outside.

Access. Means of access are often incorporated in glazing. On the domestic scale patio doors give access to gardens and form the visual link referred to earlier. In commercial and shop premises, doors are provide in large glazed areas. Full consideration must, however, be given to the safety aspects, which are described later.

Weather resistance. Glazing forming a barrier between the external and internal environment must provide resistance to penetration by water and air. Such penetration can take place in three areas:

- between the window frame and the wall opening in which it is fitted:
- between the glass and frame in which it is bedded;
- between an opening frame and a fixed frame.

Tests, which are generally related to BS Draft for Development 4 of 1971, are directed at the third category. However, serious leaks can occur with the first two categories, which are associated more with failure of workmanship on site than with faulty design.

Water and air penetration can pose serious problems on exposed sites on coasts and hilly areas. Modern windows incorporating draughtproofing devices are far more resistant to penetration than older windows, but good weather sealing, with little or no air leaks, can lead to increased condensation on the glass, causing problems with run-off, often mistaken for water entering from outside. Many draughtproofing materials have a limited life in use and may need replacing within the lifetime of the window. Fixings must be accessible without the removal of complete window frames.

Thermal insulation. Part F of the Building Regulations 1976 lays down insulation requirements for walls, which can limit the glazing areas. Unfortunately these requirements take no account of orientation, so solar heat gain from a southerly aspect is ignored.

Double glazing of windows halves the heat loss through the glass, but the loss is still considerably more than through solid walls. On existing buildings the energy savings achieved by insulating walls and roofs, and by draughtproofing windows, give far greater value for money than does double glazing on its own. Two types of double glazing are commonly used:

- Sealed double-glazed units, formed of two panes of glass with an airspace hermetically sealed either by an edge spacer and channel, or by physically joining the glass edges. These are glazed into the glazing rebate in the normal way, although larger panes will need a deeper rebate. Certain pane sizes are mass-produced to suit standard windows, giving considerable economic advantage.
- Separate glazing systems mounted internally, consisting usually of combinations of fixed and sliding units in aluminium frames. These are largely used for double glazing existing windows because they are independent of the existing sashes, making them simple to use on such windows. Because the air space is not hermetically sealed, condensation still can take place on the inside surface of the outer pane. There is no great advantage in thermal insulation in increasing the air space between the glazed panes.

Acoustic insulation. Double glazing provides acoustic insulation similar to thermal insulation, with the exception that a greater air space improves the performance acoustically. For this reason double-glazed panes are not as good as a separate glazing system mounted on the inside of a reveal, with maximum air space, and where

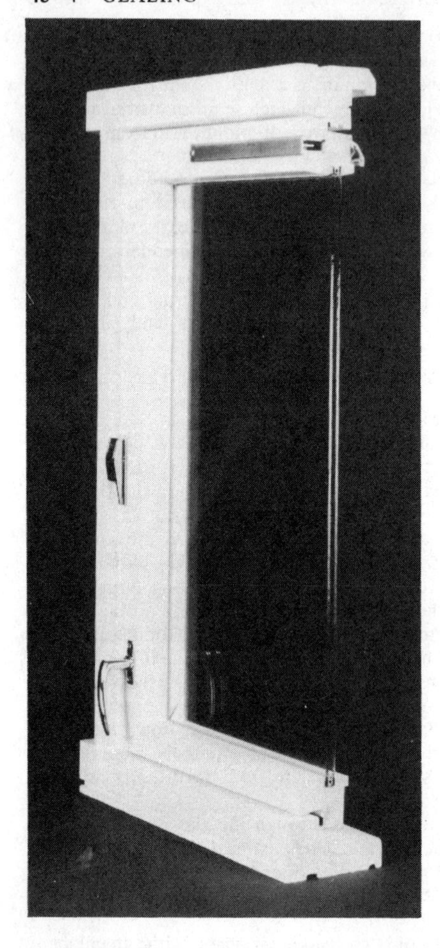

Figure 45.1 Cutaway section of timber-framed, double-glazed 'High Performance' window (courtesy Boulton and Paul Ltd)

possible sound absorbing material fitted to the soffits and reveals between the glazing.

Window cleaning. Access given by opening lights for the cleaning of adjoining fixed lights from the inside must be considered above first floor level. Side-hung windows with offset hinges and all types of pivot windows can be cleaned from the inside and give access to side lights. Vertical and horizontal sliding windows can be difficult to clean completely from the inside.

Modular coordination. Windows in external walls are largely available in sizes enabling them to be coupled together in modular increments, based on 100 mm horizontally and vertically. The coordinating sizes in which windows are normally stocked are in increments of 300 mm horizontally and 200 mm vertically. All windows are produced with a minus tolerance under the coordinating size, the actual size of the unit being known as the 'work size'.

Safety

Safety must be assessed in terms of the type of hazard to enable the most appropriate solution to be selected. Hazards fall into four broad categories:

Occupational: where the process or user produces a hazard, as in schools, gymnasia, swimming pools and high pedestrian traffic zones.
Visual: where full-height glazing down to floor level does not have sufficient manifestation to be seen in certain conditions of light, with the danger of people 'walking through', e.g. patio doors, shop doors, side lights, etc.
Barriers: where glass is used as a screen separating different levels, such as with balconies, stairs and landings.
Fire prevention: where glass is part of a barrier with a time period of resistance to the passage of fire and smoke.

To meet such hazards, glass must either *break safely* so that the risk of injury from fragments of glass is eliminated, or *remain in position* after impact or when broken, so that containment is not impaired.

Figure 45.2 'High Performance' windows in a dwelling at Wetherby (courtesy Boulton and Paul Ltd)

The following glazing materials are generally recognised as breaking safely:

- toughened glass, which breaks into small blunt fragments, similar to that used in a car windscreen;
- laminated glass, which consists of two panes of glass with a single clear plastic film that holds the glass in position if broken;
- plastic sheet, generally of acrylic or rigid PVC.

Those that remain in position when broken are as follows:

- wired glass containing a 12 mm galvanised wire mesh within the thickness of the glass;
- laminated glass as described above.

It is important to note that no form of glass can be proof against penetration by very severe impact, although the forces needed to penetrate the above glasses are many times greater than for plain glass.

The following should be considered when selecting suitable glass for use against the four main hazards:

Occupational hazard. Wired glass is commonly used, though where the hazard is from personal impact at low level there is a danger from the sharp edges. Laminated, toughened and plastic: all are suitable, although cost could be the most important factor, and the use of plastic might be affected by fire resistance requirements.

Visual hazard. The choice should be a glass that breaks safely, or one with sufficient thickness to withstand reasonable impact. The Code of Practice on glazing, CP 152, allows the use of single sheets of plain annealed glass of 6 mm thickness in areas not greater than 2.5 m^2, although such glass is not acceptable in the USA and certain European countries, and the specifier should consider each application on its own merits.

Toughened or laminated glass in such locations provides the best 'break safely' characteristics. Wired glass is not recommended in such areas, owing to the sharp fragments that remain. The possibility of accidental 'walking through' large glass panes can be considerably reduced by the introduction of a visible manifestation, such as a waist rail, marking on the glass, etc. Manifestation does not avoid the risk of danger from falling, tripping etc., and also is generally sited above the eye-level of young children.

Barriers. Where glass is used as a barrier between different levels containment is a most critical factor; since all glass can be penetrated if the impact is hard enough, the energy level of potential impact must be considered according to the application. CP 152 recommends toughened or laminated glass, although wired glass has been widely used on the basis that it has certain powers of containment.

Wired and laminated glasses that break but hold together may deflect considerably from the impact, and it is important to provide sufficient depth of rebate and strong enough fixing to avoid the pane falling out of its seating. Choice of pane sizes can be critical in such circumstances.

Fire. Wired glass is the only glazing material accepted to give a period of fire resistance – usually when fitted as viewing panels in doors that themselves have to provide a period of fire resistance. Sizes of glass panels are strictly limited, and the correct material must be used for the glazing rebate and fixing beads.

MECHANICAL CHARACTERISTICS

Method of operation

Opening lights in windows can be operated in four main ways:

- hinged on one edge,
- sliding horizontally or vertically,
- pivoting horizontally or vertically,
- louvres.

Windows hinged on one side are either top- or side-hung, although some bottom-hung opening in lights are provided for special purposes. Metal side- and top-hung windows hinged on one edge are often fitted with offset or projection-type hinges, allowing both access for cleaning and extra ventilation. Timber side- and top-hung windows normally have plain hinges.

Vertical sliding windows provide ventilation at both high and low levels, and have been traditionally used in the past. Horizontal sliding windows and doors are now commonly used on a domestic scale. Some horizontal sliding units have a top-hung vent for night-time ventilation.

Pivot windows can pivot vertically or horizontally. The pivot is usually central but can be slightly offset with vertical pivots. Pivot windows can interfere with internal nets, blinds and curtains, even where these are fitted on the inside of the reveal. Cleaning of both sides of pivot windows is possible from the inside, particularly if the unit is fully reversible.

Glass louvres are usually made up of 150 mm blades held at either end and spanning up to approximately 1 m. They provide one hundred per cent ventilation with very little movement and the minimum of projection both internally and externally.

Figure 45.3 Timber-framed casement windows at Great Hockham (courtesy Boulton and Paul Ltd)

Earlier types of louvres tended to leak at the contact points of the blade holders, but the introduction of plastic injection-moulded blade holders allows far greater accuracy of movement and sealing. When closed there is a glass-to-glass contact and overlap of approximately 10–12 mm, sufficient for normal weather conditions. Louvres are a potential security hazard, as the glass blade can often be removed from the outside.

Outward opening windows at ground level can be a hazard to pedestrians where a pathway directly adjoins the side of a building.

Hardware

This has three functions, providing:

- the means of window movement, i.e. hinges, slides etc.;
- the means of operating the window;
- the method of securing the window.

Movement. Side- and top-hung windows are provided with butts; these should operate freely for the lifetime of the window, can be fitted with friction plates to hold the window in an open position, and can also have the pivot point of the hinge offset externally to allow the window to open leaving a gap on the hinged side.

Pivots fitted to pivoting windows normally have friction pads to hold the window in position, although these will not hold it securely in a high wind.

Traditional timber vertical sliding windows operating with weights and sash cords need the cords renewed at intervals, and such renewals should always take place before redecoration. Tubular spring balances are now often provided in the sash itself.

Horizontal sliding windows are mounted on bottom rollers or nylon sliders. All horizontal sliding windows can provide maintenance problems if the tracks are not kept completely clear of leaves and dirt, and their life in use may not be as long as the windows.

Operation. Small windows can be opened and closed by handling the frame, although larger units may require separate handles for moving them, and high units may require poles or remote-control gear. Windows without friction butts will require peg or friction stays to hold them in the open position, and such stays are often used for locking the windows closed.

Security. Normally a locking lever knob or handle, operated from the inside, is provided to maintain the window closed, and is fitted so that it cannot be operated from the outside. Such handles are not usually accepted for insurance purposes, and greater security is obtained by a deadlock fitted with a removable key. Horizontally sliding windows must have head stops to prevent the sash being lifted out of its frame when in the closed position.

MATERIALS

Four groups of materials are involved in glazing:

- those forming the structural surround;
- those forming the frame in which the glass is held;
- the glass itself;
- the bedding.

The *structural surround* in the external walls in which the glazing frame is fitted can be brick, block, concrete, timber, plastic, steel, or profiled cladding sheets. All materials involve different details between the surround and the frame, but the main requirement is always to prevent the ingress of air and water between the structural surround and frame.

Window frames are most commonly available in timber, steel and aluminium, although there are now on the market plastic frames with a steel core, GRP frames and stainless steel, the latter sometimes being used to encase aluminium or steel. All materials for frames require initial protection and most require regular maintenance, such maintenance being essential for timber and steel. Aluminium can be supplied mill finished, i.e. as extruded, but it is better when anodised, avoiding white oxide pitting on the surface. Plastic windows need no maintenance but their use in hot climates should be checked for possible degradation by ultraviolet light.

Glass is provided in a number of different forms:

- Clear annealed glasses supplied in sheet form, or as float, which is the equivalent of plate.

- Obscured glasses with one surface figured or frosted to give obscurity, or cast glass. Obscured glasses should be checked when specifying, since the amount of obscurity varies considerably with the different types.
- Safety glasses including those that break safely, i.e. are less likely to cause injury, and those with powers of containment, i.e. remaining in place when broken, such as toughened, laminated, wired, etc. as described earlier.
- Double-glazed units formed of two sheets of glass with edges physically joined and sealed.
- Solar control glasses with tint formed in the body of the glass itself, on one surface only, or with two sheets of glass on either side of a tinted film; in the latter case the film often has mirror-like properties. All these glasses have different powers of reflection of solar heat, and of admission of daylight.

Figure 45.4 Aluminium vertical sliding sash (courtesy Crittall Windows Ltd)

Bedding, i.e. joining glass to frame, is carried out by three broad groups of materials:

- bedding compounds,
- rigid beads,
- preformed gaskets.

Bedding compounds include linseed-oil putties used on timber windows, metal casement putties used for steel windows, and a variety of gun-applied and preformed mastics. The latter have varying lives, but linseed oil and metal casement putties must be painted immediately.

Bead glazing is usually carried out in conjunction with a bedding for both seating the glass and providing the waterproof backing. The glass itself is physically held in place in its rebate by the bead.

Preformed extruded gaskets are often used on special applications, and combine the seating of the glass with the fixing for holding it in place. They are usually preformed of rubber or plastic extrusions, with a 'zip in' insert that locks the glass in place. Gaskets can be designed to fit into a groove in the frame, or in turn clamp round a nib. Wherever mastics or gaskets are used, the advice of the window, glass and bedding manufacturers must be obtained; their recommendations will be influenced by the glass pane size, frame materials, degree of exposure, access for future maintenance and glass replacement, etc.

CONTRACTS

On large contracts window manufacturers will be invited to tender by the architect on the basis of being appointed a nominated supplier, or nominated sub-contractor, under the RIBA Form of Contract, and the appropriate sub-contract forms should be used. Such invitations to tender are generally in advance of the appointment of the main contractor. Tenders will usually be based on drawings and specifications.

It is most important for the conditions of tender to be clearly stated, particularly in respect of attendance by main contractor, use of scaffolding, whether the work is to be carried out in one or more visits to the site, etc.

On small contracts windows may be specified, or billed in detail, leaving the main contractor to obtain them direct from his own suppliers. Delivery time may be critical, particularly if special units are involved, and it is important for this to be confirmed at the time of tender. Most manufacturers state that their delivery time commences from approval of their drawings, but do not specify how long they take to prepare these.

WORK ON SITE

Handling and storage. Nominated suppliers and sub-contractors will normally require the main contractor to unload, and it is necessary for storage space to be prepared before delivery and for unloading facilities to be readily available. Time of delivery must be agreed in advance, as materials arriving on site in the late afternoon or at weekends may be dumped without appropriate protection, etc.

After delivery, windows and glazing materials must be stored so that they are fully protected from the weather and ground moisture, and stacked in such a way as to avoid distortion. If enclosed space is not available, external storage must be on timber palettes, or bearers on ground covered by tarpaulins or polythene. All materials should then be sheeted over.

Stacking areas should not be adjacent to site roads carrying a lot of traffic. When windows are being hoisted on site by crane, they should be lifted in such a way as to avoid twisting of the frames and local damage to finishes.

Erection. Windows can either be erected as walling proceeds, i.e. with the masonry built-up around the window, or inserted into prepared openings after the walling has been completed. Building the window in as work proceeds avoids any problems of tolerance, but does subject the window to damage during construction, and it is not recommended for windows supplied ready glazed.

Building in prepared openings allows windows to be inserted after the carcass of the structure is complete, and reduces dangers of damage considerably. However, prepared openings must have sufficient tolerance to allow the window to be physically inserted and to be squared and levelled. In this case tolerances on the diagonals and levels of the opening are just as important as the main dimensions.

Figure 45.5 Steel-framed windows, factory-coated in polyester panels, being fitted in renovation work (courtesy Crittall Windows Ltd)

Windows with prefinished frames, such as aluminium etc., are usually supplied with protective tape over the frame, and this should not be removed until the bedding of the window and internal plastering is complete, to avoid damage to the frame and glass by mortar droppings.

Where metal windows are supplied in a timber subframe, it is often possible to build in the subframe as work proceeds, leaving the metal window to be inserted later. This avoids the tolerance problems, although the subframe must be maintained square and vertical. On long runs of glazing a careful check should be maintained to avoid dimensional creep due to accumulation of + or– tolerances. Use of modular coordinated window sizes makes such checking quite simple.

Glazing. Glazing a window on site must be carried out with the material specified; where compound beads and gaskets are used, the work must be strictly in accordance with the drawings and specification. Conditions for glazing must be dry, and glazing compounds must be fresh.

Cleaning. Final cleaning of windows, both glass and frames, may have to take care of the removal of dirt, mud, mortar, paint etc., and must be carefully carried out so as to avoid damage to surface finishes and the glass.

FURTHER INFORMATION

Trade associations

Aluminium Window Association, 26 Store Street, London WC1E 7EL
Steel Window Association, 26 Store Street, London WC1E 7BT
Timber Research and Development Association, Hughenden Valley, High Wycombe, Bucks HP14 4ND
Glass and Glazing Federation, 6 Mount Row, London W1Y 6DY
Flat Glass Manufacturers Association, Prescot Road, St Helens, Lancs
Patent Glazing Conference, 13 Upper High Street, Epsom, Surrey
Sealant Manufacturers Conference, 15 Tooks Court, London EC4A 1LA
National Association of Shopfitters, 411 Lumpsfield Road, The Green, Warlingham, Surrey, CR3 9HA
Insulation Glazing Association, 6 Mount Row, London W1Y 6DY

British Standards

DD4:1971, *Recommendations for the grading of windows*
CP 145:Part I:1969, *Patent glazing*
CP 152:1972, *Glazing and fixing of glass for buildings*
CP 153, *Windows and rooflights*
  Part 1:1969, *Cleaning and safety*
  Part 2:1970, *Durability and maintenance*
  Part 3:1972, *Sound insulation*
  Part 4:1972, *Fire hazards associated with glazing in buildings*
BS 544:1969, *Linseed oil putty for use in wooden frames*
BS 644, *Wood windows*
  Part 1:1951, *Wood casement windows*
  Part 2:1958, *Wood double hung sash windows*
  Part 3:1951, *Wood double hung sash and case windows (Scottish type)*
BS 952:1964, *Classification of glass for glazing and terminology for work on glass*
BS 990, *Steel windows generally for domestic and similar buildings*
  Part 1:1967, *Imperial Units*
  Part 2:1972, *Metric Units*
BS 1787:1951, *Steel windows for industrial buildings*
BS 4254:1967, *Two-part polysulphide-based sealants for the building industry*
BS 4255, *Preformed rubber gaskets for weather exclusion from buildings*
  Part 1:1967, *Non-cellular gaskets*
  Part 2:1975, *Cellular gaskets*
BS 4873:1972, *Aluminium alloy windows*

Other sources of information

BRS Digest 122, *Security* (October 1970)
BRS Digest 68, *Window design and solar heat control* (Revised 1971)
BRS Digest 140, *Double glazing and double windows* (April 1972)
BRS Digests 128 and 129, *Insulation against external noise* (April and May 1971)

46 ROOF TILING

INTRODUCTION 46–2

BASIC ROOF TILING 46–3

ROOF FINISHES AND DETAILS 46–7

46 ROOF TILING

R.H. YOUNG
Redland Roof Tiles Ltd

This section concentrates on recommendations for fitting single-lap, interlocking concrete roof tiles, which are now used on 90 per cent of new roofs and form the cheapest pitched roofing material. Recommendations for straightforward double-lap plain tiling are included but space does not allow coverage of special features in this type of tiling.

INTRODUCTION

The pitched, tiled roof is the most effective means of protecting the structure and finish of a building from the weather. Aside from this functional role the pitched roof can add character, colour and beauty to a building – or demonstrate only too visibly a sad lack of craftsmanship.

A good roof is one that channels moisture efficiently to the system of guttering. It is fixed according to a specification appropriate to the tile profile and local wind-load conditions. All roof finishes should be weatherproof, and all perpendicular and diagonal lines created by the tiling should be straight.

Roof tiling is a potentially hazardous trade and great caution should be exercised when moving and working on a roof. Readers are referred to the requirements of the Health and Safety at Work etc. Act 1974. It is always advisable to work in pairs, and many parts of the work will in any case require two pairs of hands.

The first part of this section concerns basic roof tiling of a straightforward, gable-to-gable roof; the second, the more skilled tasks of making finishes weatherproof and constructing roof details on more complicated roofs, such as those that have valleys, hips etc.

List of tools

The following tools will be found in most roof tilers' tool bags:

Tiler's hammer/axe
Slater's hammer
Slater's ripper
Square
Chalk line and red ochre paste
Trowels for bedding and pointing
Pincers
Tenon saw for sawing battens
Tile cropper
Heavy-duty electric saw with carborundum cutting discs
Goniometer (inclinometer)

BASIC ROOF TILING

Before starting work the tiler should check the architect's drawing and specification. This should be fully understood and any variation from normal roofing practice should be queried.

Roof structure

The tiler will then check the roof structure. If he works over a fault he may be said to have condoned it and be held responsible for stripping off the tiling so that the defect can be put right. The following aspects of the structure should receive attention.

Timbering Check that the timbering is complete. Diagonal bracing should be fixed where trussed rafters have been used. Are all the rafters on a level plane? Check the batten section against rafter spacing. (See Table 3 of BS 5534: Part 1: 1978.)

Pitch Using a goniometer (inclinometer), the tiler should check that the minimum pitch on the roof is not less than the minimum recommended by the tile manufacturer for the profile being used. Pitch variation on either side of a valley should not be greater than 5°, preferably not more than 2½° if valley troughs are being used.

Support for felt at eaves Continuous support at the eaves should be provided to ensure that roofing felt does not sag behind the fascia, or between the rafters, allowing moisture to collect. This is usually feather-boarding, a wedge-shaped piece of wood, or alternatively 9 mm marine ply can be used nailed at the rafters.

In double-lap plain tiling only, a 'bell' eave effect may be created by the use of sprockets. The angle of belling should not be less than the minimum plain-tile pitch of 35°.

It should always be remembered that the lower portion of a roof carries the most water.

Felting

The purpose of roofing felt is two-fold:

- to keep the tiles securely on the roof by protecting the tile layer from air pressure from within the building;
- to conduct any moisture, e.g. condensation that has formed beneath the tiles, down into the guttering.

Special roofing felts can also be used to insulate the roof.

SPECIFICATION

Extent of felted area Felt must extend a minimum of 150 mm over the apex of the roof; it must extend 50 mm over the fascia board at the eaves to carry water into the gutter. Felt should be cut off at the verge, to allow a sufficient area of solid bedding on the outer leaf.

Laps of felt at joints Horizontal laps at joints in felt on an unboarded roof should not be less than:

100 mm for pitches of 35° and above
150 mm for pitches between 15° and 34°,
225 mm for pitches between 12½° and 14°.

On a boarded roof the lap at horizontal joints may be reduced to not less than:
75 mm for 35° pitches and above,
100 mm for pitches between 15° and 34°,
150 mm for pitches between 12½° and 14°.
Where a horizontal lap in the felt occurs between battens it should be held down by an extra batten in order to prevent it opening under wind uplift.

Vertical laps should not be less than 100 mm, and should be nailed not less than 50 mm from the edge of the felt. (See BS 5534, Table 7.)

METHOD

Felting and battening usually proceed together, and the horizontal laps in the felt are set out and marked at the same time as the gauge for the battens. (See 'Setting out' below.) Handling a heavy roll of roofing felt alone on a roof is not easy and is more safely accomplished by two people.

Joints Vertical joints in the felt should always be made at, and be supported by, a rafter. Many tilers will lap the felt the full distance between adjacent rafters, or pleat the two ends together to ensure a moisture-proof joint.

On reaching the ridge the felt should be left proud until the top batten is fixed, in order to prevent ripping. Felt on the second side of the roof will be tucked under the felt on the first side not less than the minimum lap.

Nailing Nail holes in the felt should be kept to a minimum, and only extra-large 3 mm diameter clout nails (copper, aluminium or galvanised) should be used. Runs of felt may be temporarily held in position at the verges by clout nails. (The risk of tearing felt can be reduced by nailing through a small block of wood.)

Setting out

Time spent in setting out is always worth while. In addition to ensuring a professional appearance for the roof it can minimise the amount of tiles used, and the number of tiles that have to be cut at abutments, protrusions or openings in the roof.

Setting out is providing a grid of horizontal and vertical guide marks to which the battens and thus the tiles will be fixed. The aim is to equalise the gauge (the depth of tile exposed) at each course and, by determining the amount of shunt to be taken up in the side locks of interlocking tiles, to ensure that all perpendicular and diagonal lines created by the tiles are true.

There are three fixed points on a roof:

- the eaves course of tiles should overhang the middle of the gutter;
- The tiles and undercloak should overhang at both verges a minimum of 38 mm, and when unsupported a maximum of 50 mm;
- ridge and hip tiles should overlap the tile courses a minimum of 75 mm on either side of the roof.

There is one fixed point on a tile: the minimum headlap recommended by the roof-tile manufacturer. For single-lap tiles, gauge = length of tile – headlap. For double-lap (plain) tiles, gauge = length of tile – lap (65 mm minimum) ÷ 2.

METHOD

There are seven basic steps for setting out and battening a roof.

Fix position of eaves-course batten Holding a tile hard on to a piece of battening at the right-hand verge, mark the felt at the top edge of the batten at a point where it allows the tail of the tile to hang over the centre of the gutter. Repeat at the left-hand verge.

With a chalk line soaked in red ochre, tied to a nail at the mark on the left-hand verge, snap a horizontal line on to the felt. Nail the eaves-course batten at each rafter, ensuring that if it is bowed it does not deviate from the line marked.

By marking the rafters at both verges it is possible to ensure that the battens, and thus the tiling, will be parallel to the gable ends even if the fascia board itself is not true.

Fix position of top-course batten The top-course batten must be fixed at a point where it will allow the ridge tile to lap the top course of tiles by a minimum of 75 mm. Measure the length of the rafters at the left-hand and right-hand verges, from the top of the eaves-course batten, to the top of the top-course batten.

Calculate number of courses Calculate the number of tiles needed by dividing the distance between the top edge of the eaves-course batten to the top edge of the top-course batten by the maximum gauge permitted by the roof-tile manufacturer. Any odd millimetres over should be divided by the number of courses and added on to the headlap, thus reducing the gauge, to establish the gauge for the particular roof. Measure and mark the rafters at both verges, and strike horizontals as described. At no time should the gauge encroach on the minimum headlap recommended by the manufacturer.

Once the appropriate gauge for a given rafter length has been calculated this must be strictly maintained. Any variation in the gauging of the battens when using a high-profile tile will result in a deviation in the diagonals, and bring the maintenance of the minimum headlap under suspicion.

Batten out the roof • Keeping the battens hard on to the red line when passing over the rafters will ensure that all warp in the batten is lost and that the battens are parallel with the elevation. • Battens should be sawn, not cut with an axe, to provide good butt joints. • Joints in battens should be staggered (local building regulations may call for every joint to be staggered, or perhaps only every third course). • Joints should occur in the middle of a rafter. • Two nails should be driven in at a slant to ensure both nail points are firmly bedded in the rafter. • Battens should be fixed with any wane occurring at the bottom edge to keep tiling level and secure. • Nails at verge rafters should be 'spiked' and not driven home, to allow for withdrawal to enable the undercloak to be fixed. The nails should be sufficiently secure, however, to allow the tiler to use the battens as a ladder to reach the upper courses.

Lay in full course to eaves batten To set out perpendiculars, first lay in a full course of tiles at the eaves, allowing the correct overhang at both verges (38 to 50 mm). Depending on the width of the roof, it may be found possible to gain or lose sufficient space to avoid tile cutting. (Obviously, the smaller the roof the less opportunity there will be to make adjustments.)

Lay in full course to top batten As above.

Strike perpendiculars Remove every fourth tile and strike perpendiculars down the left-hand edge of every third tile, using a wet red line to mark each batten from eaves to top course. Horizontals and perpendiculars should be square. (In the case of broken-bonded tiling, the tile will line up vertically every other course and special left- or right-hand verge half-tiles are used.)

An alternative method of establishing perpendiculars is by the use of a gauge rod marked with the width of three tiles + half the shunt.

Special note: setting out dropped eaves, etc. Some roof designs may present the tiler with more than three fixed points, i.e. verge, ridge and eaves, having two or more ridge or eaves heights. In calculating his batten gauge the tiler should firstly establish which is the longest rafter length and then relate this to the eaves/ridges. If he does not he may find it impossible to maintain the same gauge throughout. Not only would this look unprofessional, it might also indicate that the tiler has been tempted to sacrifice the tile manufacturer's minimum headlap requirement.

Tile fixing

The number of tiles that should be nailed or clipped (or both) will be determined by:
- the exposure of the site;
- the effective height of the building;
- the roof pitch;
- the higher windloadings encountered at eaves, ridge and verges.

(See CP 3: Chapter V: Part 2: 1972, *Wind loads,* and BS 5534: Part 1: 1978.)

SINGLE-LAP TILING

Single lap, interlocking roof tiles designed for low pitches are usually clipped. (Check manufacturer's appropriate technical data, and see Clause 17 of BS 5534: Part 1: 1978.)

All tiles at verges, on either side of valleys and hips, and where the roof meets an abutment along an inclined line should be mechanically fixed. Top and eaves courses should be mechanically fixed.

Pitches over 45° Each tile should be nailed with one nail where only one nail hole is provided and with two nails if two nail holes are provided.

Pitches over 55° At pitches over 55° the lower edge of the tile should in addition be secured with a wire hook or by other suitable means.

DOUBLE-LAP PLAIN TILING

Each tile should be nailed twice with 38 mm nails at least every fifth course; depending on degree of exposure, tiles may need more frequent nailing. Tiles at verges, abutments and on either side of valleys and hips should be nailed. At eaves and top courses, two courses of tiles should be nailed.

METHOD

Loading out Correct distribution of the required number of tiles on the roof speeds laying. It also helps maintain an even loading on the roof. Both sides of the roof should be loaded equally before commencing laying. From the grid of horizontal and perpendicular lines it is easy to calculate the number of tiles required to fill a square.

Tiling in All single-lap tiles have their side lock channels on the left-hand side. Work on each side of the roof therefore should be started from the right-hand verge working to the left. Tiles are placed on the roof keeping the nibs hard on to the battens. Adjust the shunt every third tile to conform with the perpendiculars.

Every third and fourth tile at the left-hand verge may be left out to provide a ladder to the upper courses. These are slipped into place as the tiler descends the roof for the last time.

Eaves-course single-lap tiling The eaves course is fixed first. Before this is laid, where required, an eaves filler unit is nailed to the fascia. No mortar or under-eave course should be used. The eaves course of tiles should be on the same plane as all other courses of tiles.

Eaves-course double-lap tiling In the case of plain tiles a double course of tiles should be laid to the eaves, both courses being nailed. The course of under-eaves tiles must be broken-bonded to the course of full-eave tiles.

A special, shorter, undercourse tile is produced by manufacturers for use at both eaves and ridge.

ROOF FINISHES AND DETAILS

The second half of this section deals with the more advanced aspect of the roof tiler's work, making the roof finishes weatherproof and constructing details.

Mortar bedding

Good practice in mortar bedding is applicable generally, and the following points should be noted.

Soaking All units to be bedded (verge, ridge and hip tiles etc.) should be first soaked to reduce the possibility of shrinkage and cracking of bedding mortar.

Mortar All mortar used in a roof should be compounded of one part of cement to three parts fine aggregate. Mortar may be tinted by the addition of not more than 1 part in 60 of approved pigments (see BS 1014). A colour match between the roof tiles and bedding mortar is difficult to achieve. It is recommended, therefore, that mortar be tinted black to provide a contrast, or red, according to tile colour.

Any mass of bedding mortar should be reduced by thinning out with broken tiles or dentil slips to prevent cracking and shrinkage.

Bedding and pointing Pointing should not be carried out as a separate operation; bedding mortar should be squeezed down and struck off to give a smooth face.

Running a verge

Running a verge may be defined as fitting a tile, slate or asbestos-cement strip undercloak to enable verge tiles to be bedded on to either brick or timber.

Fixing asbestos strip undercloak A piece of timber is first fixed to the outer skin of the cavity wall (or to the barge board). This supports the undercloak until the mortar sets and provides a straight edge from which to work. A 75 × 50 mm timber, 50 mm side uppermost, is then tacked the full length of the verge, fractionally below the top surface of the wall to give the bedded verge a slight outward tilt, to deflect water from the building.

At a barge board the tiler, lifting the ends of the spiked battens, slides in the asbestos strip roughside up, and when the battens are nailed to the barge rafter they secure the asbestos undercloak in position:

- if laid on timber the strip should be nailed at no greater than 300 mm centres;
- if laid on brick the undercloak should be bedded in mortar and struck off flush with the external face of the wall.

The undercloak should have the same overhang at the verge as the verge tiles (between 38 and 50 mm). Joints in the strip should be close butted. The eaves support board should be cut away at the eaves corner to accommodate the strip and avoid an unsightly finish.

Bedding the verge The tiles at the verge should all be bedded at the same time as the tiling takes place. Sufficient mortar is laid on the undercloak, and the verge tiles are pressed down on to this bedding. It may be necessary to remove the outer nib to get the verge tile to sit correctly.

As the verge tiles are pressed down the mortar will spread to the full width, and any excess should be struck off at the outer edge, forming a sloping face. This assists drainage and makes the mortar appear much thinner.

The combination of a flat tile profile and a low pitch is particularly vulnerable to water penetration at the verge, and consideration should be given to the use of a PVC dry verge system.

Fixing plain tile undercloak The method is essentially as above, and on a plain tiled roof a tile undercloak is considered preferable. The following points should be noted.

- The undercloak tiles should be used in their width, not lengthwise.
- If laid on boarding each tile forming the undercloak should be nailed twice.
- Brickwork must be kept a minimum of 25 mm below the top surface of the rafters.
- Plain tiles used for undercloak should be laid granule face down.

Valleys

In every roof containing a valley, the valley must be the starting point of operations.

Notching fascia board The tiler, using the valley trough tile as a guide, marks the fascia board where it will need to be notched to allow the valley trough to sit flush with the top of the fascia. Saw cuts are made and the material is removed with a chisel.

Felting A width of roofing felt is run down the valley and tacked into position, with the nails being placed well away from the centre of the valley. The main roof is then felted, and the ends of the strips lapping into the valleys are cut off on the mitre so that they lap 150 mm beyond the valley rafter from both sides.

Positioning valley battens A valley trough is set into the notched fascia and another laid at the top of the valley. A full length of batten is nailed to the rafters on either side of the troughs.

It is good practice to use *firring battens* on either side of valleys. Firring battens, diminishing thicknesses of battens parallel to the valley rafter, allow the main battening to rise gradually up to the valley batten and prevent the tiles cocking. It should be remembered that the roof surfaces on either side of the valley will act as a funnel directing wind and water into the valley, so this area is particularly vulnerable.

On roofs of 30° pitch and over, the battens on the main roof area are mitred up to the valley battens and skew nailed to them. For pitches below 30°, the tile battens are carried up on top of the valley battens and skew nailed. This enables the cut tiles in each course to be laid satisfactorily over the double thickness where the valley troughs lap.

Laying valley troughs The rest of the troughs are laid up the valley, so that their edges are supported by, and are flush with, the inside edges of the valley battens, making sure that they are properly lapped and there is no sideways rock.

Before tiling is started it is necessary to cut away a V portion from the tail edge of the bottom trough to carry the line of the gutter to prevent water overshooting.

Guide timber A length of 125 × 50 mm timber is laid up the centre of the valley with the 125 mm face uppermost as a guide to the finished exposed width of the valley. This is secured by short pieces of batten nailed into the battens on the left-hand side of the valley. The timber also acts as a wedge to hold bedding mortar in place until it has set.

Tiling The roof slope to the right of the valley is tiled in first. When the bedding is set the batten pieces are removed and tiling to the left of the valley is completed.

The tiler strikes a perpendicular from the bottom left-hand edge of the left-hand side of the valley from eaves to ridge. On the right side of this line he will work from left to right to obtain the cut into the valley for each course.

Where the tiles are being clipped the tiler follows this procedure to obtain the cut, but the tiles are then lifted and relaid right to left as normal to fix the clips. The timber guide is used to indicate where tiles should be scribed to obtain the correct angle for cutting.

Bedding tiles at valleys The tiles to be cut are taken off the roof, cut and the nibs removed where necessary to allow them to seat properly. Bedding mortar should be thrown hard up against the 50 mm edge of the guide timber. It is essential that the mortar should be kept clear of the two tapered bars on either side of the trough, otherwise water will dam over the obstruction and cause it to leak into the roof.

The guide board is removed when the mortar has taken on its initial set and the exposed edges of the mortar are cleanly pointed with a trowel. The valley should finally be swept down to remove any debris.

Saddles It is recommended that a lead saddle should be placed where the valley joins the ridge. This is placed over the head of the shouldered valley trough at the ridge. It is not advisable to rely on the mortar bedding of the ridge tile to make the top of the valley watertight.

Purpose-made valley tiles In plain tiling, special valley tiles may be used so that the courses of tiles on the main roof continue through the valley. Purpose-made valley tiles do not need nailing and should not be bedded.

Other types of preformed valleys are to be fixed according to the manufacturer's instructions.

Metal valleys

Metal valleys can be used:

- if the variation in the degree of pitch on either side of the valley is more than 5° (in practice, because of the angle of tilt, it will be difficult to lay tiles smoothly up to the higher side of the trough or other preformed valley where the variation is more than 2½°);
- if the rafter pitch is less than 22½° or more than 44°.

Metal valleys (i.e. lead), unlike troughs or preformed valleys, require layboards as support.

Felting and layboards A continuous width of felt should be laid up the valley underneath the layboards. Layboards should not be more than 13 mm thick or else the valley rafter must be recessed to receive them. The boards should be good-quality ply.

A *triangular fillet* is laid down the angle formed by the layboards to avoid a sharp angle in the lead.

Undercloak A strip of natural slate or asbestos cement 115 mm wide is placed between the lead and the mortar bedding of the tiles at the valley, in order to allow the lead to expand and contract with changes in temperature.

Hips

In single-lap tiling, third-round or angle ridge tiles, or sometimes specially designed hip tiles, are used for covering the hip angle. In double-lap plain tiling, hip ridges or special bonnet hip tiles are used.

SINGLE-LAP TILING

Felting and hip iron Hips are felted after the main area of the roof; a strip of roofing felt not less than 600 mm wide is run down the hip. A hip iron is screwed to the base of the hip rafter to support the weight of the hip ridges throughout the life of the roof.

Cutting tiles to hips The main roof tiles are scribed on the roof, removed and cut away from the roof to avoid damaging the roofing felt.

Bedding hip tiles The bottom hip tile is placed on the roof, marked, cut to follow the eaves line on both sides, and bedded into position. The top hip tile is then placed in position and a string line is fixed on one side from top to bottom of the hip to give a straight line for bedding.

The hip tiles are then laid to the string line working up the roof, each tile being continuously edge bedded and solid bedded at butt joints. Pointing is carried out as part of the same operation. Hip tiles should be placed in position and gently tapped down into the mortar already placed.

Where a bold profile tile is being used the troughs should be packed out with dentil slips to avoid deep mortar bedding, which when dry would probably shrink and crack.

BONNET HIPS (Plain tiling)

The method for ordinary ridge tiling a hip on a plain-tiled roof is as described above. When bonnet hips are being used the following points should be noted.

- Each bonnet tile must be nailed with the nail penetrating at least 25 mm into the hip rafter, or batten fixed to the hip rafter.
- Tiles are scribed and cut as described above. In plain tiling it will sometimes be necessary to support a small piece of cut tile by notching and driving a long nail into a batten or the hip rafter. This prevents the piece from slumping when the hip tile is pressed home into the bedding.

• Mortar bedding of the tail of the bonnet must be applied during fixing, and the mortar must be either struck off smoothly at the lower edge of the bonnet or kept slightly back and pointed to give a shadow effect.

Ridges

The method of building a ridge is similar to that used for hips. The following points should be noted.

• As a matter of safety any cut ridge tiles should not be placed at the ends of the ridge.

• It is advisable to dry lay the ridge tiles first, marking the top course tiles on both sides as a guide.

• Ridges must be straight and level, and the mortar bedding must be sufficiently thick throughout to maintain the correct level.

• Ridge tiles should be edge bedded, and packed out with tile pieces and solid bedded at butt joints. Dentil slips should be used to pack out the troughs of the top courses on either side of the ridge in bold profile tiling.

• A mitre occurs wherever a hip joins the ridge. It is recommended that a lead saddle be used under the converging ridges to ensure complete watertightness.

Gas-flue vent ridge It may be found necessary when fixing a gas-flue ridge to remove the section of tiling between the rafters at this point to allow the fitting to be installed.

It is often necessary to remove part of the top course of tiles to accommodate the vent ridge. A new nail hole will be drilled, and the part tile fixed to a secondary batten. This will allow a continuous line through the tails of all the top-course tiles.

Abutments

Where tiling meets an abutment there is a potential point of ingress for water. This is prevented by the provision of flashings. In the case of profiled single-lap tiling, a single lead overflashing dressed into the brickwork is acceptable. Special abutment flashing units are manufactured for flat-profiled tiles.

The roofing felt should have an upstand of 100 mm at abutments; this is turned up above the level of the tile layer. The tiles should be cut as close up to the abutment as possible.

Extent of flashing The lead flashing should extend over the first profile of the tile and be dressed hard against the brickwork and cut on the step. The lead is then tucked into the mortar joint, which is previously raked out. The lead is secured with lead plugs.

Secret gutters Where a secret gutter is formed at an abutment a small gap must be left, between 25 mm and 38 mm wide. This will allow the gutter to be cleared of debris during maintenance, and is particularly important where the roof is near trees.

Where the gutter does not empty directly into the rainwater gutter at the eaves, the tiler should ensure that it will discharge safely on to the main roof. It may be necessary to provide a timber fillet to give sufficient rise in the gutter to achieve this.

Forming gutter The lead is dressed down the wall over the sole plate and rafter and up over the end of the batten. Care must be taken to ensure that the sole plate is sufficiently below the top of the rafter so as to give adequate depth for the flow of water. The return edge of the lead must not cause cocking of the tiles at this point, otherwise it may be necessary to fir-out the tiling battens with reducing counter battens as at the valley.

Double-lap tiling In the case of double-lap tiling, each individual tile has its own soaker and the soakers are fitted as the tiling is laid. The soakers are the length of the tile plus 25 mm. The soaker is bent at 90°, one edge being placed against the wall, the other against the tile. The extra 25 mm is bent over the top of the tile to prevent slipping.

Compatibility of materials It should be noted that either zinc soakers and zinc flashings should be used together, or lead with lead.

Chimneys/dormers Lead aprons should be used beneath chimneys/dormers, and with a bold profile tile it will be found necessary to fill the troughs with dentil slips to enable the lead to be dressed flat.

Monopitch roofs On a monopitch extension to a dual-pitch roof, the outer leaf of brickwork must be taken as the centre line of the roof. When fixing monopitch ridge tiles it is recommended that a horizontal batten be fixed to the brickwork, so as to support the weight of the tile by the bottom edge of the vertical leg resting on this batten. When the mortar has set the batten can be removed.

FURTHER INFORMATION

Health and Safety at Work etc. Act 1974
BS 5534: Part 1: 1978, *Slating and tiling design* (first section of revision of CP 142)
CP 3: Chapter V: Part 2: 1972, *Wind loads*
BS 5534: Part 2 (not yet available)
Tile manufacturers' technical manuals

47 ASPHALT ROOFING

MASTIC ASPHALT 47–2

DESIGN PRINCIPLES 47–3

DESIGN AND SPECIFICATION 47–10

SITE PRACTICE 47–15

ASPHALT ROOFING 47

47 ASPHALT ROOFING

A.F. CONSTANTINE, CEng, MRAeS, HonMIAT
Director, Mastic Asphalt Council and Employers Federation Ltd

This section discusses the general characteristics of mastic asphalt, and the design principles used with this material; it then describes good asphalt roof design in more detail, and outlines the necessary site facilities and procedures.

MASTIC ASPHALT

General characteristics

The asphalt used for waterproofing and paving within the confines of buildings is known as mastic asphalt; the term refers to a type of asphalt that is laid hot without compaction, and is defined in British Standards as:

> 'a type of asphalt composed of suitably graded mineral matter and asphalt cement in such proportions as to form a coherent voidless impermeable mass, solid or semi-solid under normal temperature conditions, but sufficiently fluid, when brought to a suitable temperature, to be spread by means of hand tools'.

Mastic asphalt is a reversible thermoplastic, i.e. it is capable of being resoftened by the further application of heat. In consequence, by using the correct techniques local areas of mastic asphalt can be removed or repaired and a completely homogeneous joint to the remaining material achieved. Although the material is capable of considerable plastic deformation it has no elastic properties, and in consequence once elongated it can only wrinkle if subsequently compressed due to contraction of the base on which it is laid.

It should also be remembered that mastic asphalt, particularly when laid on a relatively soft insulant, as is now relatively common, may fracture when subjected to impact loads; the hairline fractures so caused are not readily apparent but will subsequently open up and permit the entry of water. The importance of protecting the finished mastic asphalt waterproofing from subsequent damage and abuse cannot, therefore, be over-emphasised.

Following a survey of flat roofs on over three hundred Crown buildings, the Building Research Establishment in Digest 144[1] expressed the view that: 'mastic asphalt roofing properly designed and laid should prove capable of lasting 50–60 years'.

Specifications

There are two British Standards for roofing grade mastic asphalt: BS 988[2] (limestone aggregate) and BS 1162[3] (natural rock asphalt aggregate).

The BS 988 type is manufactured utilising petroleum bitumen limestone powder and limestone aggregate (BS 988/B) or with 25 per cent of Trinidad Lake Asphalt incorporated in the asphaltic cement (BS 988/T). The BS 1162 type is manufactured utilising natural rock, normally from Swiss or French sources, plus additions of petroleum bitumen and/or Trinidad Lake Asphalt and siliceous coarse aggregate (BS 1162, Table 3, column 1 or 2 according to the amount of Trinidad Lake Asphalt incorporated in the asphaltic cement).

Generally speaking nowadays, due to its high cost, the use of natural rock mastic asphalt is confined to roofs of prestige and similar buildings, and for most applications material to BS 988 is specified. The Code of Practice governing the use of mastic asphalt for waterproofing roofs is CP 144 Part 4[4], and the specifications for roofing felts used for vapour barriers and for sheathing felt used as a separating membrane under mastic asphalt are contained in BS 747[5].

DESIGN PRINCIPLES

Basic roof types

Generally speaking, flat roof constructions can be categorised in two ways: firstly in terms of the type of construction, i.e. rigid or flexible, and secondly in relation to the location of the insulation, which can be:

- beneath the primary roof structure (cold roof), or
- above the primary roof structure but beneath the waterproofing membrane (warm roof), or
- less frequently, above the waterproofing membrane (externally insulated, 'inverted' or 'upside down' roof).

Rigid roofs

Generally speaking, rigid roofs constructed of an *in-situ* cast concrete slab with brick parapets have the lowest thermal and moisture movement characteristics and provide the most stable base for mastic asphalt waterproofing. Such constructions permit the mastic asphalt to be carried straight off the horizontal to the vertical upstand without the interposition of a free-standing kerb or similar detail. This remains true whether or not a cementitious or bituminous lightweight screed or a board-type insulant is interposed between the slab and the waterproofing. Typical edge upstand details utilised with this type of construction are shown in *Figures 47.1 and 47.2.* An exception to this rule, and a type of construction requiring careful detailing, is where a precast concrete parapet unit or ring beam is used in conjunction with an *in-situ* slab; in such cases a detail that takes account of the differential movement that will in all probability occur at the junction is essential.

Systems comprising prestressed concrete slabs or panels laid to precast concrete beams, when covered with a sand/cement screed to falls, can generally be treated as a rigid roof as far as detailing the mastic asphalt is concerned, except where local differential movement of the structure can occur, such as vertical cladding panel joints to which the asphalt is directly applied.

It is seldom that rigid roofs are designed as 'cold roofs', because amongst other things their construction does not lend itself to locating an effective vapour barrier and insulation at the soffit of the ceiling, and also it is generally considered preferable to protect the structure from extremes of temperature.

Externally insulated (inverted) roof

The basic difference between the externally insulated roof system (also known as the 'inverted' or 'inside-out' roof system) and traditional flat roof construction is that in the former the insulation is located above the waterproofing membrane.

Locating the insulant above the waterproofing membrane obviates the necessity for the incorporation of a separate vapour barrier and minimises the risk of condensation or trapped-water conditions occurring, because the waterproof

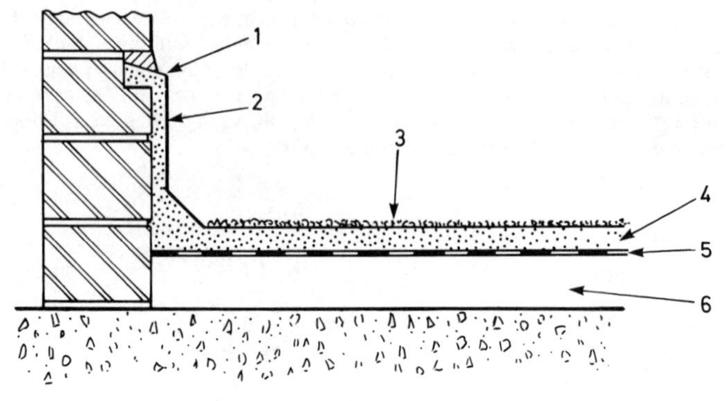

1. 25 mm × 25 mm chase in brickwork; asphalt tuck-in pointed up with cement mortar, preferably containing PVA additive.
2. 12 mm two-coat skirting in roofing-grade mastic asphalt, with two-coat fillet at junction to horizontal.
3. Solar-reflective treatment: chippings bedded in bitumen adhesive or suitable paint on horizontal, paint on vertical.
4. 20 mm two-coat horizontal roofing-grade mastic asphalt laid breaking joint on separating membrane.
5. Separating membrane: BS 747 Type 4A(i) sheathing felt.
6. Screed to falls.

Figure 47.1 Typical rigid roof design

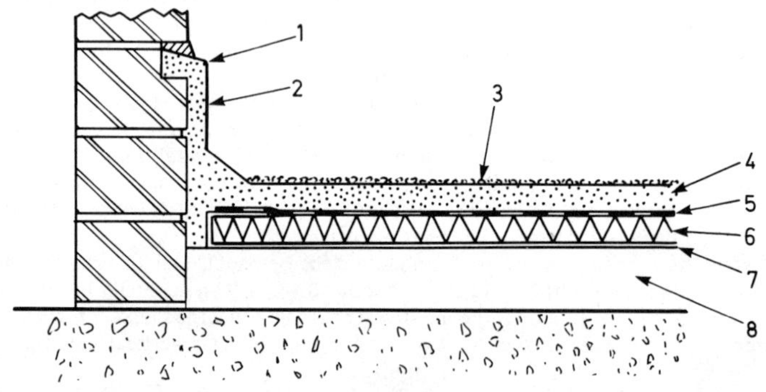

1. 25 mm × 25 mm chase in brickwork; asphalt tuck-in pointed up with cement mortar, preferably containing PVA additive.
2. 12 mm two-coat skirting in roofing-grade mastic asphalt, with two-coat fillet at junction to horizontal.
3. Solar-reflective treatment: chippings bedded in bitumen adhesive or suitable paint on horizontal, paint on vertical.
4. 20 mm two-coat horizontal roofing-grade mastic asphalt laid breaking joint on separating membrane.
5. Separating membrane: BS 747 Type 4A(i) sheathing felt.
6. Board-type insulant bedded in hot bitumen on vapour barrier kept back 15–20 mm from upstand.
7. Vapour barrier, not less than BS 747 Type 1B roofing felt, lapped and bonded with hot bitumen.
8. Screed to falls.

Figure 47.2 Rigid roof with insulant

membrane is kept warm and thus no major horizontal cold face on which moisture vapour can condense exists within the roof construction. Other advantages of the system are that the waterproof membrane is protected from both damage and ultra-violet degradation and, provided the pavings are located on spacers, extremely quick drainage of the upper surface is achieved.

The roof construction should incorporate falls of 1 : 80, in which case a normal two-coat mastic asphalt specification is utilised. It is important to provide adequate skirting height, as the thickness of the insulant and pavings can be 100 mm or more and the skirtings should rise above the paving level by a further 150 mm. If, exceptionally, a 'no-falls' roof is constructed, this should be regarded in a similar

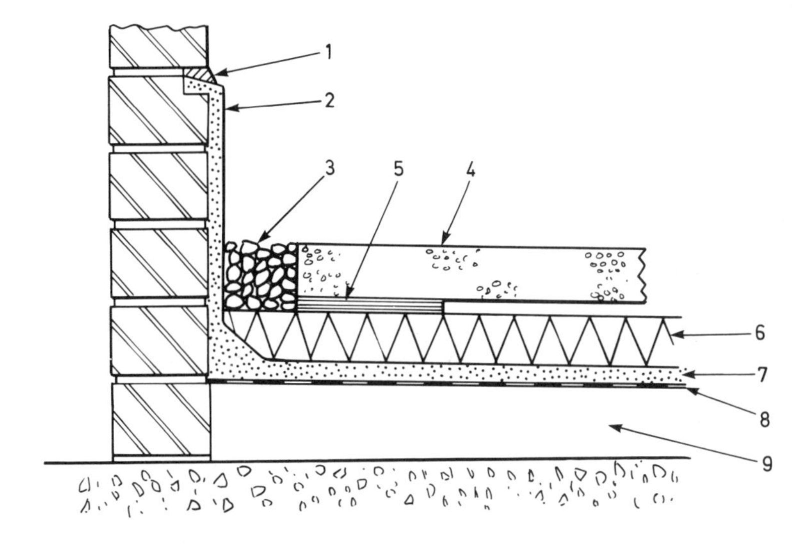

1. 25 mm × 25 mm chase in brickwork; asphalt tuck-in pointed up with cement mortar, preferably containing PVA additive.
2. 12 mm two-coat skirting in roofing-grade mastic asphalt, with two-coat fillet at junction to horizontal.
3. Optional gravel between pavings and skirting.
4. Paving slabs laid loose on levelling pads.
5. Multi-layer roofing-felt levelling pads or equivalent proprietary pads.
6. Suitable non-water-absorbent insulant.
7. 20 mm two-coat horizontal roofing-grade mastic asphalt laid breaking joint on separating membrane.
8. Separating membrane: BS 747 Type 4A(i) sheathing felt.
9. Screed to falls.

Figure 47.3 Externally insulated (inverted) roof

light to a water reservoir roof and a three-coat mastic asphalt specification utilised. It is extremely important that the insulant and pavings or gravel should be laid by the asphalt contractor immediately the waterproofing has been completed, to prevent trafficking and damage of the waterproof membrane.

Mastic asphalt is a particularly suitable waterproof membrane for this type of roof, as it is seamless and amongst other things has no joints to upset the levels of the insulant when it is laid. A typical construction is illustrated in *Figure 47.3*.

Flexible roofs

Flexible joisted roof constructions (including tongued-and-grooved boarding, marine-quality plywood or wood-wool slab decks laid on timber joists, and metal decking laid on metal joists) are treated in a similar manner to rigid roofs, except that they require special edge treatment to isolate the mastic asphalt membrane from the differential vertical, horizontal and angular movements that can occur at perimeter upstands. Roofs of timber construction are often designed as cold roofs, with a vapour check and insulating quilt located immediately above the ceiling level. With such a design it is essential to ventilate the roof space adequately to the outside; CP 144 Part 4[4] recommends not less than 300 mm² of opening per 300 mm run of roof on two opposite sides. Typical constructions are illustrated in *Figures 47.4 to 47.6*.

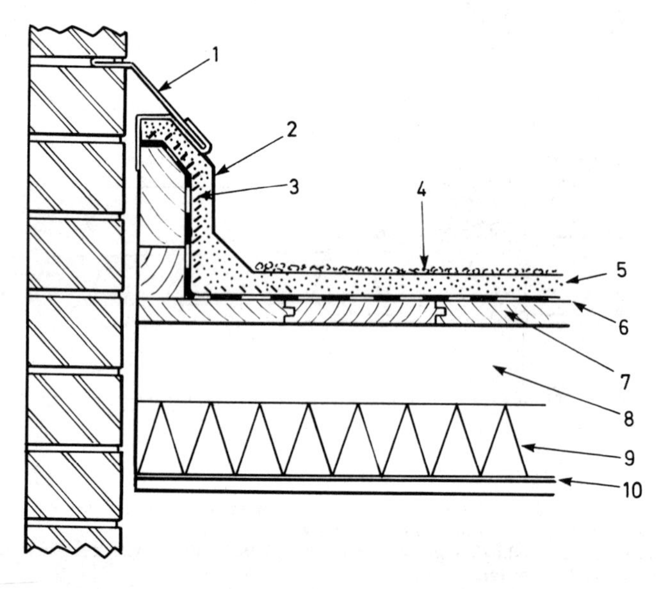

1. Code 4 lead flashing or equivalent.
2. 20 mm three-coat skirting in roofing-grade mastic asphalt, with two-coat fillet at junction to horizontal.
3. Expanded metal lathing, clout-nailed at 150 mm centres over separating membrane to free-standing kerb; e.m.l. continued 75–100 mm on horizontal.
4. Solar-reflective treatment: chippings bedded in bitumen adhesive or suitable paint on horizontal, paint on vertical.
5. 20 mm two-coat horizontal roofing-grade mastic asphalt laid breaking joint on separating membrane.
6. Separating membrane: BS 747 Type 4A(i) sheathing felt.
7. Tongued-and-grooved board deck laid to falls of not less than 1 : 80 by use of firrings.
8. Joist space ventilated: not less than 300 mm² per 300 mm run in each of two opposite walls or equivalent.
9. Insulant laid between joists.
10. Foil-backed plasterboard or equivalent vapour check above ceiling finish.

Figure 47.4 Timber joisted (flexible) roof: cold roof design

Free-standing kerbs

As previously mentioned the incorporation of free-standing kerbs to isolate the mastic asphalt from movement at perimeter upstands of flexible roofs is essential, and typical methods of providing such kerbs are also shown in *Figures 47.4 to 47.6.*

Falls

It is desirable to design flat roofs such that adequate falls are provided to drain rain-water. If no falls are incorporated at the design stage, ponding will occur, since it is impossible to construct a truly flat roof and adverse tolerances are bound to be present. In addition, because there is little flow of water, such a roof will become dirtier and the probability of moss and other vegetable growths occurring is greatly

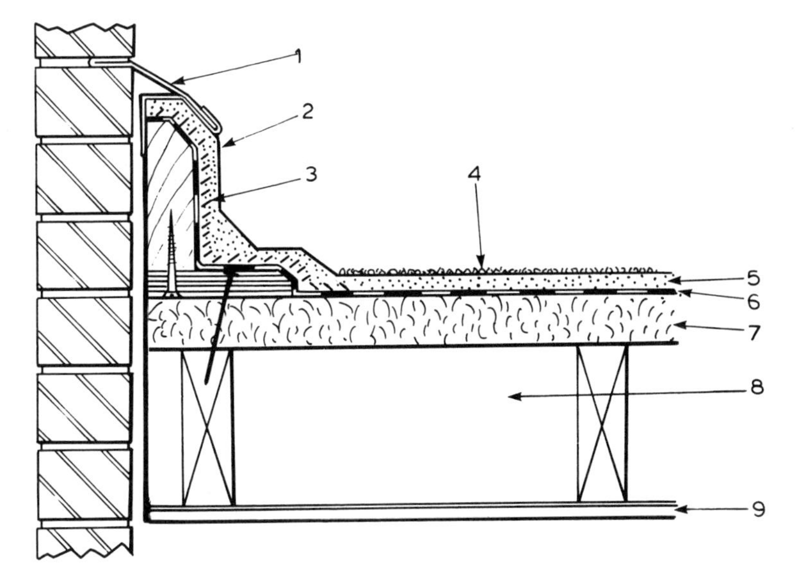

1. Code 4 lead flashing or equivalent.
2. 20 mm three-coat skirting in roofing-grade mastic asphalt, with two-coat fillet at junction to horizontal.
3. Expanded metal lathing, clout-nailed at 150 mm centres over separating membrane to free-standing kerb; e.m.l. continued 75–100 mm on horizontal.
4. Solar reflective treatment: chippings bedded in bitumen adhesive or suitable paint on horizontal, paint on vertical.
5. 20 mm two-coat horizontal roofing grade mastic asphalt laid breaking joint on separating membrane.
6. Separating membrane: BS 747 Type 4A(i) sheathing felt.
7. Wood-wool slab deck prescreeded with joints punned and taped or, if pre-felted, with a single layer of building paper interposed between the felt and separating membrane.
8. Joist space, not ventilated to outside.
9. Ceiling finish.

Figure 47.5 Timber joisted (flexible) wood-wool slab decked roof: warm roof design

increased. However, BRE Digest 144[1] makes it clear that there is no evidence indicating that lack of falls has any significant effect on the life of mastic asphalt.

CP 144 Part 4 recommends the use of minimum falls of 1 : 80. It is now generally agreed that utilising a design fall of 1 : 80 will not eliminate all ponding; where it is essential that no ponding shall occur, e.g. on access balconies or play areas, it is necessary to incorporate a design fall of between 1 : 40 and 1 : 60, to take account of contra-falls created by material and constructional tolerances, if a finished fall of 1 : 80 is to be achieved and all ponding eliminated.

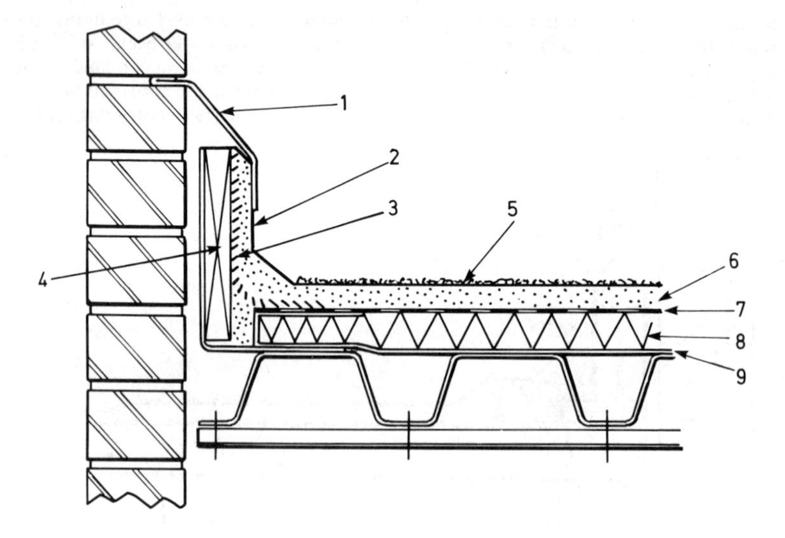

1. Code 4 lead flashing or equivalent.
2. 20 mm three-coat skirting in roofing-grade mastic asphalt, with two-coat fillet at junction to horizontal.
3. Expanded metal lathing, clout-nailed at 150 mm centres over separating membrane to free-standing kerb; e.m.l. continued 75–100 mm on horizontal.
4. 25 mm thick timber fixed to metal upstand.
5. Solar-reflective treatment: chippings bedded in bitumen adhesive or suitable paint on horizontal, paint on vertical.
6. 20 mm two-coat horizontal roofing-grade mastic asphalt laid breaking joint on separating membrane.
7. Separating membrane: BS 747 Type 4A(i) sheathing felt.
8. Board-type insulant bedded in hot bitumen on vapour barrier.
9. Vapour barrier not less than BS 747 Type 1B roofing felt, lapped and bonded with hot bitumen.

Figure 47.6 Metal decked (flexible) roof: warm roof design

Mastic asphalt has to be laid to a constant thickness, and it is therefore necessary that any falls required are provided in the structure. In the case of timber joisted roofs the deck should be laid to falls by the use of firrings, by sloping the joists or similar means. In the case of concrete roofs the falls should be provided by screeding, either using a cementitious or bitumen-bound screed, laid to a plane surface and free from ridges or hollows. Alternatively, in suitable cases the *in-situ* roof slab can be laid to falls, thus eliminating the need for a separate screeding operation.

Vapour barriers and vapour checks

In both cold-roof and warm-roof designs, it is necessary to incorporate a membrane located beneath the insulant to prevent the migration of moisture vapour into the roof construction and its subsequent condensation at cold faces, e.g. the underside of the mastic asphalt waterproofing or metallic pipes penetrating the roof. Many, but not all, insulants lose their insulating qualities with the increase of moisture content, in addition to which any organic insulants and organic components of the roof construction will bio-deteriorate in the presence of a high moisture content.

In the case of a cold-roof design the impermeable membrane is located at ceiling level immediately below the insulant. Very often it is provided by the use of a foil-backed plasterboard or the interposition of a polythene membrane between the ceiling and the joists. Such a membrane is inevitably pierced by services, and in consequence can only be regarded as a vapour check. However, provided the roof cavity is adequately ventilated in the case of cold-roof designs, the provision of such a check generally proves adequate.

The location and provision of a vapour barrier in warm-roof designs involves complex design considerations. However, if a vapour barrier is specified it should always be located on the warm side of the insulation and should comprise not less than a BS 747 Type 1B roofing felt, lapped and bonded with bitumen. In the case of lightweight cementitious screeds, however, the incorporation of a vapour barrier over the concrete slab and beneath the screed has been shown to be an unsatisfactory practice because it isolates the screed from its base; this permits the screed to crack and curl, with disastrous effects on the waterproof membrane. These considerations do not apply in the case of bitumen-bound screeds. In all cases where a vapour barrier is incorporated beneath a board-type insulant, it should be lapped around the edge of the board at perimeters to maintain continuity and connection to the mastic asphalt waterproofing at these locations; see *Figure 47.2.* The Department of Environment Advisory Leaflet No. 79[6] provides further information on this subject.

Insulants

A wide variety of insulants are suitable for use in conjunction with and directly beneath mastic asphalt. These include corkboard, fibre boards, polystyrene, cellular glass, lightweight cementitious screeds, bitumen-bound screeds and foamed plastics boards of the polyisocyanurate types.

Corkboard was probably the first insulant used in conjunction with mastic asphalt. It provides a stable base for mastic asphalt and does not collapse in the presence of water. With the advent of more efficient insulants, coupled with the increased insulation now required for buildings, it is now less frequently used than formerly.

Fibre insulation board is probably the most widely used and economical insulant. Being an organic material, however, it is susceptible to collapse in the presence of moisture. In consequence, careful storage and protection from inclement weather during laying is essential. Improved types, having a higher compressive strength and containing perlite to improve their insulating properties, are now available.

Cellular glass is a rigid, inert and extremely stable insulant that provides a good base for mastic asphalt. The brittle nature of the material requires careful storage and installation of the material, including full bedding in hot bitumen and the incorporation of a double layer of building paper between the insulant and the sheathing felt underlay to prevent undue adhesion.

Plastics insulants. Polystyrene and polyurethane boards require the interposition of a fibre insulating board between them and the sheathing felt, as they are unable to directly withstand the heat of the mastic asphalt at the time of laying. Polyisocyanurate boards do not need to be protected in this manner, and to date are the only plastics material capable of accepting, without collapse, directly overlaid mastic asphalt.

Lightweight cementitious screeds can be used to provide both falls and a degree of insulation. However, to obtain the degree of insulation now required for domestic and office roofs, a board-type insulant requires to be laid over the screed if an undue thickness of screed is to be avoided.

Lightweight bitumen-bound screeds. Screeds of the bitumen-bound type have now been available for more than a decade. They have the advantage of not introducing quantities of water into the roof structure at the installation stage, compared with cementitious screeds, with which they share the advantage of being able to provide both falls and insulation. Their insulation characteristics can be supplemented by the interposition of polystyrene or similar board between the vapour barrier and the screed. A number of proprietary types are available, generally based on either a leca or perlite aggregate.

DESIGN AND SPECIFICATION

Specifications for trafficked roofs

Mastic asphalt is widely used as a combined waterproofing and paving for roofs subjected to traffic. In this connection it should be noted that paving grades of mastic asphalt to BS 1447 should not be used on their own as a roof waterproof membrane, and for such applications should always be laid over at least one coat of roofing-grade mastic asphalt. In addition, flooring grades of mastic asphalt to BS 1076 are not suitable for external trafficked applications. Some typical specifications are as follows.

Access balconies and private terraces: 25 mm two-coat mastic asphalt to BS 988.
- First coat: 10 mm thick laid on sheathing felt.
- Second coat: 15 mm thick, having 5–10 per cent additional grit added at the re-melting stage.

This specification will accept a fair degree of pedestrian traffic, but not point loads such as chair and table legs, without indenting.

Private balconies: 20 mm two-coat mastic asphalt laid on sheathing felt overlaid with concrete or asbestos tiles. This specification will accept point loadings.

Access balconies (general duty) and public areas: 38 mm two-coat work.
- First coat: 13 mm BS 988 laid on sheathing felt.
- Second coat: 25 mm BS 1447 paving grade.

Alternatively, a normal two-coat roofing specification can be overlaid with pavings bedded in cement mortar or laid dry on terrace supports of adequate size.

Rooftop car parks: 55 mm three-coat work.
- 20 mm two-coat BS 988 laid on sheathing felt.
- 35 mm single-coat BS 1447.

Heavier-duty specifications are available suitable for service, standing and loading areas for commercial vehicles.

Thicknesses and number of coats

The thickness and number of coats of mastic asphalt used for roof waterproofing are dependent on both the pitch and nature of the surface to which it is applied. For this purpose CP 144 Part 4[4] defines horizontal, sloping and vertical work as follows:

'*horizontal work:* asphalt laid on surfaces up to and including 10° pitch;

sloping work: asphalt applied to surfaces over 10° up to and including 45° pitch with the horizontal;

vertical work: asphalt applied to surfaces having a pitch greater than 45°'

For roofs subject only to maintenance traffic the thickness of mastic asphalt and number of coats that should be applied are as follows.

Horizontal work An average total thickness of 20 mm laid breaking joints in two coats on an underlay, normally of BS 747 Type 4A(i) sheathing felt laid loose over the base. (*Note:* other thicknesses can be laid, e.g. three-coat work 30 mm thick for rooftop gardens or reservoirs.)

Sloping work. On sloping concrete or similar solid bases, an average total thickness of 20 mm laid in three coats breaking joint directly to the base, i.e. without the interposition of an underlay. On timber slopes an average total thickness of 20 mm laid in three coats breaking joint on expanded metal lathing (complying with BS 1369) fixed by nailing or stapling at 150 mm centres over a BS 747 Type 4A (i) sheathing felt underlay.

Vertical work (other than skirtings). On concrete, brick or other solid surfaces, an average total thickness of 20 mm laid in three coats breaking joint without an underlay. Normally one of the preparatory treatments referred to on page 47–16 is utilised, or, optionally, expanded metal lathing fixed by shot-firing or other means at 150 mm centres can be used on concrete or brick, but in this case sheathing felt is not interposed between the expanded metal lathing and the concrete or brick. On timber, an average total thickness of 20 mm laid in three coats breaking joint on expanded metal lathing (complying with BS 1369) fixed by nailing or stapling over a sheathing felt underlay.

Skirtings. On new work, skirtings normally comprise an average total thickness of 13 mm laid in two coats, minimum height 150 mm, with a two-coat fillet at all internal angles. The asphalt shall be tucked into a 25 mm × 25 mm chase at the top edge, provided by and pointed by the main contractor.

- Skirtings over 300 mm high are regarded as vertical work.
- Skirtings on timber require to be three-coat work on expanded metal lathing, as for vertical work.
- Skirtings on old concrete or brickwork may require an additional 'dubbing out' coat, in which case the total thickness will be approximately 20 mm.

Details

The importance of correct design and execution of details such as eaves, verges, and collars to pipes and balustrade supports cannot be over-emphasised. Typical details are illustrated in *Figures 47.7 to 47.9.*

Movement joints

Movement joints should always be located at the top of falls and should be of the twin upstand kerb type (*Figure 47.10*) unless the use of a flush expansion joint is

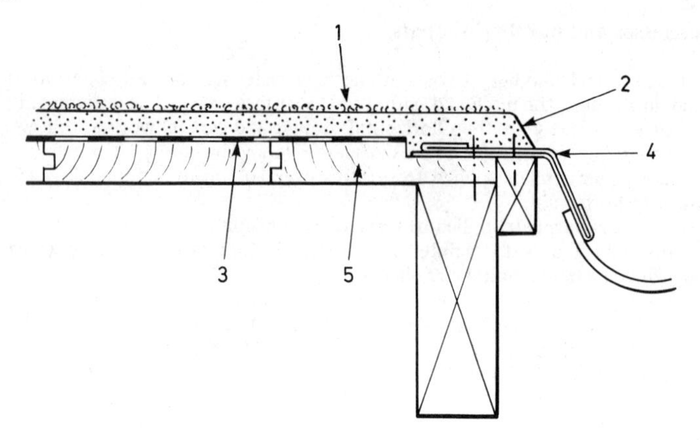

1. Solar reflective treatment: chippings bedded in bitumen adhesive or suitable paint on horizontal, paint on vertical.
2. 20 mm two-coat horizontal roofing-grade mastic asphalt laid breaking joint on separating membrane.
3. Separating membrane: BS 747 Type 4A(i) sheathing felt.
4. Code 4 lead flashing or equivalent, securely fixed in rebate to permit full thickness of asphalt to be maintained at edge of roof.
5. Tongued and grooved board deck, laid to falls of not less than 1 : 80 by use of firrings.

Figure 47.7 Gutter detail

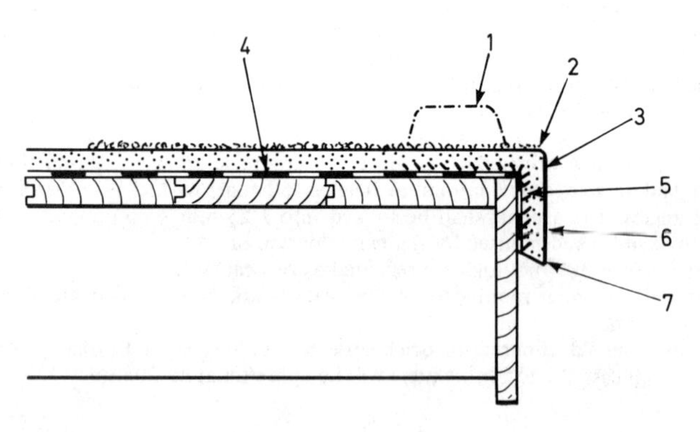

1. Optional solid mastic asphalt water check.
2. Solar-reflective treatment: chippings bedded in bitumen adhesive or suitable paint on horizontal, paint on vertical.
3. 20 mm two-coat horizontal roofing-grade mastic asphalt laid breaking joint on separating membrane.
4. Separating membrane: BS 747 Type 4A(i) sheathing felt.
5. Expanded metal lathing, clout nailed at 150 mm centres extending to 100 mm on to flat.
6. 20 mm three-coat roofing-grade mastic asphalt apron.
7. Undercut drip edge.

Figure 47.8 Roof verge drip detail

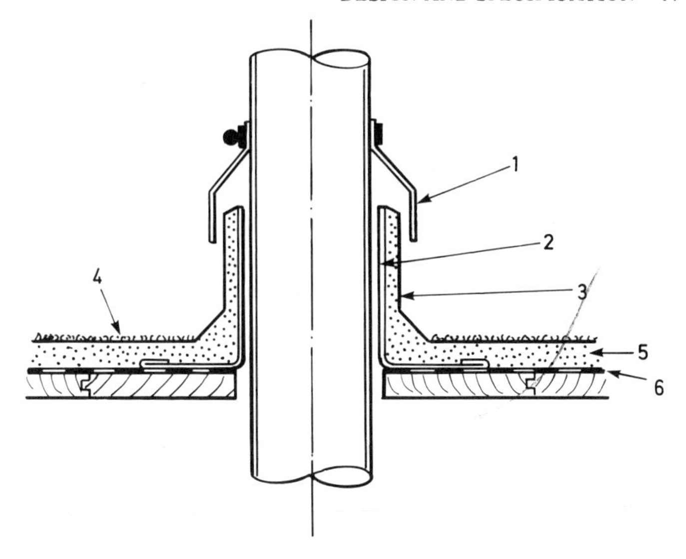

1. Metal collar clipped to pipe.
2. Non-ferrous metal sleeve with welted slate, to be wire brushed and treated with suitable rubberised bitumen emulsion.
3. 12 mm two-coat vertical roofing-grade mastic asphalt, with two-coat fillet to horizontal mastic asphalt.
4. Solar-reflective treatment: chippings bedded in bitumen adhesive or suitable paint on horizontal, paint on vertical.
5. 20 mm two-coat horizontal roofing-grade mastic asphalt.
6. Separating membrane: BS 747 Type 4A(i) sheathing felt.

Figure 47.9 Collaring of pipe through roof

essential. Particular care should be exercised in selecting the type of flush movement joint to ensure that it will accept the expected type of traffic and movement to be expected, that its materials are compatible with mastic asphalt, and that a good joint can be made between it and the waterproof membrane.

Solar reflective treatments

The use of a solar-reflective treatment on mastic asphalt roofs is recommended, particularly where the mastic asphalt is laid directly over an efficient insulant.

For horizontal surfaces either light-coloured chippings bedded in a bitumen-based adhesive compound or a solar-reflective paint may be used; the latter is the only treatment that can be used in skirtings and vertical work.

Generally the first choice for the horizontal surface should be chippings, but in some circumstances (e.g. if the roof has been damaged and the possibility exists that all such damage has not been discovered) it may be preferable not to use chippings as they will hide leaks or other defects.

It is important that the solar-reflective treatment is applied as soon as practicable after the mastic asphalt has been laid, because it is in the initial period, when maximum movement and settlement in the building will occur and the asphalt is

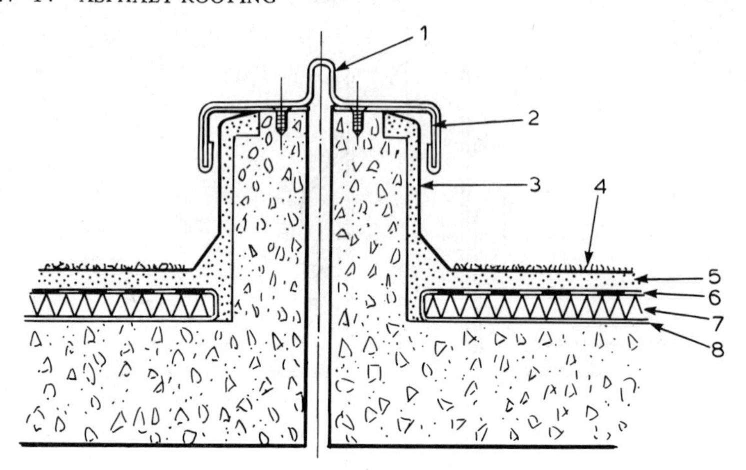

1. Non-ferrous metal capping.
2. Non-ferrous cleats fixed to concrete to hold capping in position.
3. 12 mm two-coat skirting in roofing-grade mastic asphalt, with two-coat fillet at junction to horizontal.
4. Solar-reflective treatment: chippings bedded in bitumen adhesive or suitable paint on horizontal, paint on vertical.
5. 20 mm two-coat horizontal roofing-grade mastic asphalt laid breaking joint on separating membrane.
6. Separating membrane: BS 747 Type 4A(i) sheathing felt.
7. Board-insulant, bedded in hot bitumen on vapour barrier, kept back 15–20 mm from upstand.
8. Vapour barrier: not less than BS 747 Type 1B roofing felt, lapped and bonded with hot bitumen.

Figure 47.10 Kerb-type expansion joint

black, that the reduction in solar heat gain (and consequential thermal movement) is most required.

Paints used as a solar-reflective treatment on mastic asphalt must be suitable for the purpose and should not have an unduly high film strength. Generally they are either of the pigmented acrylic emulsion type or comprise a bitumen-based binder pigmented with flake aluminium. Paints having unduly high shrinkage characteristics and/or film strength, or incorrectly applied paints, can cause unsightly crazing of mastic asphalt.

Surface finishes

Where pedestrian or vehicular traffic is expected, mastic asphalt can be overlaid with asbestos or concrete tiles, or with pavings or *in-situ* concrete.

Asbestos or concrete tiles can be bedded in a bitumen adhesive direct to the mastic asphalt. The tiles should be set back 75 mm at all perimeter upstands.

Pavings can be bedded in a sand/cement screed on the mastic asphalt, but it is essential, firstly, that a separating membrane of building paper or two layers of polythene sheet should be interposed between the mastic asphalt and the screed, secondly that the mastic asphalt is not damaged during the laying process, and thirdly that perimeter gaps and movement joints as specified for tiles are incorporated.

In-situ concrete can also be laid over mastic asphalt, subject to the same three provisions that apply to overlaying with pavings.

SITE PRACTICE

Site facilities and services required by the mastic asphalter

Site for plant The location of the asphalt mixer or cauldron and the asphalt materials should be as close as practicable to the point of laying, in order to avoid undue cooling of the material during transportation.

Storage of materials Dry, safe storage for insulants, roofing felts, underlays, expanded metal lathing, edge trims and similar materials is required. In the case of gas-fired equipment, a suitable compound having a fence not less than two metres high for the storage of gas bottles is also required.

Cooperation by main contractor It is normal practice for the main contractor to provide adequate scaffolding, edge protection and hoisting facilities for the asphalt contractor. It is also usual for the main contractor to arrange the preparation of the various surfaces, the cutting of chases, the fixing of metal work such as flashings, drips and outlets, and the pointing up of chases after the mastic asphalt work has been completed.

Re-melting and laying procedures

Re-melting The mastic asphalt is normally delivered to site in the form of blocks, where it is re-melted either in a mechanically agitated mixer or in a cauldron. The temperature of the material during re-melting should not exceed 230°C. Buckets should be used for conveying the material to the laying point. To prevent sticking, the inside of the bucket should be coated with a fine inert dust; ashes or oil must not be used for this purpose.

Laying The sheathing-felt separating membrane should be laid loose with 50 mm lapped joints. Timber battens should be used to determine the thickness and bay width of each coat of horizontal mastic asphalt. Bay joints in successive coats should be staggered by not less than 75 mm. The final coat of horizontal mastic asphalt should be sand rubbed using a float and clean sharp sand.

It should be noted that, in the case of vertical and slope work, the thickness of successive coats will not necessarily be equal, because the first coat is primarily an adhesive layer and in consequence may be thinner than succeeding layers.

Fillets The junction between the vertical and horizontal asphalt should be thoroughly warmed and cleaned with hot mastic asphalt and a substantial two-coat fillet formed at the internal angle.

Priming of metal Metal surfaces should be cleaned and primed to provide adhesion.

Preparation of other vertical surfaces Where vertical or sloping concrete is very smooth one of the following treatments may be necessary:
- removal of surface laitance by wire brushing;
- stippled application of a proprietary sand/cement plastic emulsion (normally used as a key for plaster);
- application of a proprietary bitumen/rubber emulsion.

Mastic asphalt cannot be applied directly to lightweight concretes, because excessive blowing would occur. One solution to the problem is to use expanded metal lathing fixed to the lightweight concrete over a sheathing felt as for timber. Excessive use of mould oil will result in a loss of keying and damage to the mastic asphalt.

Brickwork should be clean and the horizontal bed joints lightly raked. Weak lime/sand mortars are not a suitable base for mastic asphalt.

Protection of mastic asphalt roofing

It is essential that the mastic asphalt is protected from damage by following trades, by storage of materials or plant on the roof, by spillage of paint, gas or diesel oil or white spirit or other solvents, or by droppings of concrete or mortar.

Impact blows may fracture mastic asphalt, and the full extent of the hairline cracks so caused will not be readily apparent; they will, however, subsequently open up. It is therefore particularly important that the mastic asphalt is protected from such damage if ingress of water into the roof structure is to be avoided.

REFERENCES

1. *Asphalt and built-up felt roofings durability,* Building Research Establishment Digest 144
2. BS 988, 1076, 1097, 1451, *Mastic asphalt for building (limestone aggregate),* British Standards Institution (1973)
3. BS 1162, 1418, 1410, *Mastic asphalt for building (natural rock asphalt aggregate),* British Standards Institution (1973)
4. CP 144, *Roof coverings,* Part 4, *Mastic asphalt,* British Standards Institution (1970)
5. BS 747, *Specification for roofing felts,* British Standards Institution (1977)
6. *Vapour barriers and vapour checks,* DOE Advisory Leaflet No. 79, HMSO

48 BUILT-UP FELT ROOFING

REGULATIONS AND SPECIFICATIONS	48–2
SUBSTRUCTURE	48–3
SELECTION AND SURFACE FINISH OF BUILT–UP FELT	48–6
WEATHERING DETAILS	48–10
MAINTENANCE AND REPAIR	48–14

BUILT-UP FELT ROOFING 48

48 BUILT–UP FELT ROOFING

I. RANKIN, BSc(Hons), MPhil, MIOB
South East Regional Director, National House-Building Council

This section concentrates largely on the selection of materials, procedures and details of laying built-up felt roofing. Consideration is given to the overall roof structure, with a list of regulations and requirements applicable to flat roofs. The important subject of substructure work is also discussed.

REGULATIONS AND SPECIFICATIONS

Building Regulations

The statutory requirements of Building Regulations cover not only the roof finish but also the total roof structure; consideration must therefore be given to structural loadings, thermal insulation, fire precautions and other aspects, in addition to the provision of a built-up felt finish that is weather resistant. Table 48.1 lists the main points to consider in relation to the Building Regulations in England and Wales, and Scotland. (The Northern Ireland Regulations and London Building Acts should also be considered where appropriate.)

Table 48.1 SUMMARY OF BUILDING REGULATIONS RELEVANT TO FLAT ROOFS

Building Regulations 1976		*Building Standards (Scotland) Consolidated Regulations 1976*	
B1	Fitness of materials	RA8A	Classification of roofs
C10	Weather resistance of roofs	RA12A	Deemed-to-satisfy specification and schedule D
D	Structural stability (dead, imposed and wind loads)	B	Materials and durability
E	Structural fire precautions and schedule 9, part IV	RE15	Internal linings
		F	Chimneys, flues etc.
F	Thermal insulation and schedule 11, parts II & III	G8	Resistance to moisture from rain and snow
L	Chimneys and flue pipes	J3	Resistance to transmission of heat
		J6	Control of interstitial condensation

The Building Regulations consider materials complying with the specification of a British Standard, or used in accordance with a British Standard Code of Practice as being 'deemed to satisfy' the general or performance requirements of the Regulations. Hence felts complying with BS 747, *Specification for roofing felts,* and used in accordance with CP 144: Part 3, *Built-up bitumen felt,* will generally satisfy the requirements for fitness of materials and weather-resistance of roofs. Details of these points are discussed later in this section.

NHBC specifications

The NHBC has detailed specifications dealing with flat roofs, which again closely follow British Standard publications. The main sections are summarised in Table 48.2,

Table 48.2 SUMMARY OF NHBC REQUIREMENTS FOR FLAT ROOFS

Concrete superstructure
Co 8 to Co 14 deal with minimum falls to concrete roofs; the laying of screeds; and precautions against condensation.

Carpentry
Ca 2(b) calls for preservative treatment of joists to flat roofs; Ca 9 to Ca 20 deal with design objectives; the laying of softwood, plywood and chipboard decking; finished falls and precautions against condensation.

Roof coverings
RF 16 to RF 23 deal with the detailed laying procedure for built-up roofing.

but it should be noted that two important items are more onerous than required by Building Regulations or CP 144. The first requires the provision of a *finished fall not less than 1 in 60*, and the second requires that *joists to flat roofs should be treated with a preservative* where the roof is over habitable accommodation.

SUBSTRUCTURE

Built-up felt may be laid on most traditional types of roof structure, provided sufficient care is taken to minimise the effects of movement between the substructure and felt finish. In housing, the most common substructure used is timber joists supporting some form of decking, e.g. softwood boarding, chipboard, plywood, blockboard or woodwool slabs.

Such substructures with the incorporation of many joints are relatively flexible and subject to differential movements and shrinkage or expansion. This must be allowed for, if the resulting stresses are not to fracture or buckle the felt, causing failure. The inevitable movement between roof structure and felt finish can generally be accommodated by following the correct details for laying the felt and weatherproofing at abutments or protrusions, as outliined later in this section (e.g. nailing or partial bonding of the first layer of felt; provision of upstand and separate flashing at abutments).

However, to minimise further adverse effects and excessive movements between substructure and finish, consideration must be given to preservative treatment, provision of adequate falls, protection of materials during construction, and protection against condensation moisture.

Preservative treatment

BS 5268: Part 5, which deals with the structural use of timber and preservative treatments, recognises that there is a particular hazard when dealing with flat roof timbers, defining such timbers as a hazard of category C, quoted as being:

> 'where experience has shown that there is an unacceptable risk of decay, whether due to the nature of the design or the standard of workmanship and maintenance, or where there is a substantial risk of decay or insect attack which, if it occurs, would be difficult and expensive to remedy.'

Clearly it is vital, when considering both initial cost and the degree of risk and cost involved with failures, to ensure that timber joists and firrings are treated with a suitable preservative. This is of particular importance when one realises that damp conditions may not be evident in the flat roof until substantial damage has been caused, resulting in disproportionate remedial costs.

The NHBC Handbook requires treatment of all flat roof joists over habitable accommodation. Simple immersion in a tank or brush-applied treatment does not provide the necessary degree of protection, and it will be necessary for the builder to purchase suitably treated timber from a stockist.

Treatments considered suitable are pressure treatment with copper/chrome/arsenic (CCA) salts, the boron diffusion process, and the double vacuum process. In all cases a statement certifying treatment should be obtained.

Provision of adequate falls and drainage

Roofs having a slope or fall of under 10° (1 in 6) are regarded as flat roofs, although this would normally be considered as a particularly steep slope.

One of the main reasons for past failures in flat roofs is that water has been allowed to pond on the roof surface and has been given time to seek out and penetrate any small cracks or irregularities in the felt surface, which can cause immense damage. Again, comparing the initial cost with the very high cost of remedial work should failure occur, it is obviously better to allow for as large a fall as possible, to discharge any water efficiently.

As a minimum, 1 in 60 (1°) is regarded as a reasonably safe fall, and NHBC requirements do in fact quote this level. It must be emphasised however, that 1 in 60 should be the *finished* fall, after account is taken of any deflection, shrinkage or movement in the substructure. In practice it is desirable to design for a steeper slope, say of 1 in 40 (1½°), to allow for these movements and for inaccuracies in setting out or workmanship. Table 48.3 can be used as a guide to slopes and dimensions, the steeper slopes being more important over larger spans, where maximum deflection is to be expected.

Table 48.3 SLOPES, ANGLES AND WORKING DIMENSIONS

Slope	*Angle*	*Working dimensions*
1 in 60	1°	Allow 18 mm per metre run
1 in 40	1½°	Allow 25 mm per metre run
1 in 30	2°	Allow 35 mm per metre run
1 in 10	5°	Allow 100 mm per metre run
1 in 6	10°	Allow 166 mm per metre run

The falls to timber-joisted flat roofs are normally provided by firrings cut to fall and fixed on top of the joists. The minimum dimension of the firring should be 25 mm. Alternatively, cross battens of varying thickness may be used, with the added advantage of providing free circulation of air. Unscreeded woodwool slabs and concrete decks, which require a screed, can incorporate the required fall in the laying of the screed.

The size of gutters and downpipes required for disposal of rainwater should be based on a rate of rainfall of 75 mm per hour (unless localised conditions indicate particularly wet areas) and the effective roof area. In the calculation it can be assumed that 1 mm of rainfall on an area of 1 m^2 gives 1 litre of water. BRE Digests 188 and 189 give further information and charts on the sizing of gutters, downpipes and outlets or receivers. Manufacturers' literature should also give helpful guidance.

Protection during construction

It is essential to ensure that all materials used in built-up felt roofing are protected during construction, both from physical damage and also from excessive moisture, which could result in problems at a later date.

The protection must start with the adequate storage of materials when they arrive on site. Many builders recognise the need to store insulating materials such as glass fibre under cover (and indeed under lock and key to prevent theft), but fail to take the same precautions with other materials like softwood boarding, chipboard, block-board or woodwool slabs. A great deal of unnecessary damage and waste is therefore caused to these moisture-susceptible materials. Adequate cover should be provided by storing materials off the ground in garages, or under cover of tarpaulins or polythene sheeting securely held down.

Similarly when roof decks of these materials are laid they are often left exposed for many days before the felt is applied, causing unnecessary damage and waste. All roof decks of moisture-susceptible materials must either be covered with the first layer of felt as the boarding proceeds or be protected by tarpaulins or polythene sheet overnight. Care should be taken to ensure that temporary protection is securely fixed and not damaged.

An alternative protection available is to order prefelted materials such as prefelted chipboard or woodwool slabs, which (provided the joints are taped) will offer immediate protection to the decking. Such prefelting layers should not, however, be counted as one of the layers in the built-up roofing system.

Protection against condensation moisture

An average family creates, through breathing, cooking and washing, about 12 litres of water vapour a day in their home. This water vapour can pass easily and quickly through most building materials, and when the moist vapour is cooled to a temperature known as the dew point it begins to condense. It is therefore necessary to protect the materials in the roof from this moisture by providing an impermeable barrier that will stop the movement of moist air from reaching the point where it will cool and condense to water. Such impermeable barriers are referred to as vapour checks or vapour barriers, and usually consist of polythene sheets or felt.

The vapour barrier or check should be positioned on the warm side of the insulation material. The two basic types of flat roof (illustrated in *Figure 48.1*) are the 'warm roof' and 'cold roof' principles, which have the position of thermal insulation and vapour barrier fixed as standard. It is recommended that one of these systems be adopted.

The *warm roof* has the thermal insulation placed on top of the roof decking, with the vapour barrier immediately below. The roof void is unventilated. Common insulation materials used are: polystyrene grade HD with top surfaces prefelted; polyurethane to British Reinforced Urea Formaldehyde Manufacturers Association (BRUFMA) standard with both surfaces glass-faced; glass-fibre, mineral-wool or perlite roof-grade boards; foamed-glass slabs or baked-cork or wood-fibre roof-grade boards.

The *cold roof* has the thermal insulation placed in direct contact with the ceiling, the vapour check again immediately below the insulation. This time the roof void is ventilated, to dispel any small amounts of moist vapour that may infiltrate through. Ventilation should be provided at a minimum rate of 1000 mm^2 of opening per metre run of eaves, on both sides. This may sound extensive but merely involves a gap of 1 mm along the eaves or equivalent, in the form of standard air bricks or the slim proprietary vents that fit into perpend joints. On large roof areas it may be necessary to vent through the roof structure to achieve adequate circulation (e.g. where the span between openings exceeds about 12 m).

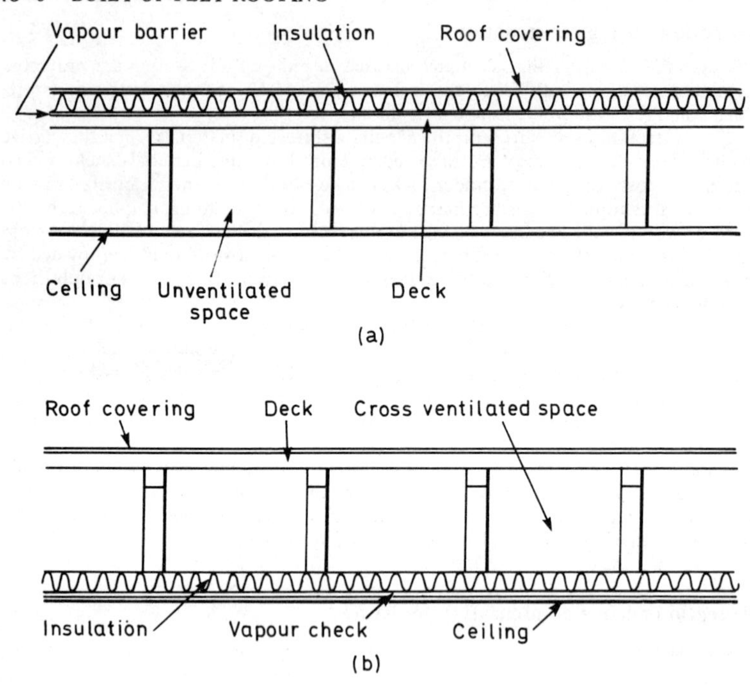

Figure 48.1 (a) Warm roof principle; (b) cold roof principle

Any increase in thermal insulation standards could affect the choice of materials involved, and it may be more practical at times to combine the warm and cold roof principles. If such a combination is made, serious thought must be given to the position of vapour barriers and checks, to ensure that moist vapour is not allowed to reach the dew point.

SELECTION AND SURFACE FINISH OF BUILT–UP FELT

Bitumen roofing felts

Roofing felts for built-up roofing should comply with the relevant grades specified in BS 747. Felts in accordance with this standard are manufactured by impregnating interwoven fibre sheets with bitumen, which provides waterproofing. By varying the type of fibre used, the spacing of the strands, the number of sheets and the surface finish, felts of wide-ranging strength, weight, thickness and performance properties can be obtained. The felts suitable for built-up roofing are specified in three main classes according to the type of fibre predominant:

Class 1 Fibre based (cotton, jute, flax, wood pulp, hair, wool)
Class 2 Asbestos based (usually incorporating a percentage of fibre base)
Class 3 Glass-fibre based

The asbestos and glass-fibre based felts have the advantage of being rotproof and increasing the fire resistance.

Table 48.4 SUMMARY OF FELTS TO BS 747 AND COMMON USES

Class of felt:	*Fibre base*	*Asbestos base*	*Glass-fibre base*	*Nominal length of roll* (m)	*Common use*
Colour code:	*White*	*Green*	*Red*		
Fine granule surfaced felts	1B	2B	3B	10 or 20 10	Lower layers of built-up roofing; or top layer when finished with chippings in bitumen or other suitable finish
Mineral surfaced felt	1E	2E	3E	10	Vertical surfaces, flashings, skirtings etc.; possible surface finish on steep roofs over 5° (1 in 10 fall)
Reinforced felt	1F	–	–	15	Not for built-up roofing except as a vapour barrier below insulation; mainly used under tiles or slates on pitched roofs
Venting base layer	–	–	3G	10	First layer of built-up roofing where partial
Venting base layer	–	–	3H	20	bonding or venting of the first layer is required

The three main classes are subdivided into categories, B,E,F,G and H according to the surface finish or surfacing material (A,C and D categories have been discontinued).

A colour coding system has been introduced by BS 747 to enable the three main classes to be identified, and a new type of venting base layer felt has been added. The colouring should take the form of a continuous band up to 50 mm wide along the roll. Although the colour code is not applicable to reinforced felts (type 1F) and the venting base layers (types 3G and 3H), these felts should be readily distinguishable; type 1F felts have the pronounced hessian showing on the underside and the venting base types are perforated.

All felts should be supplied in standard 1 metre wide rolls with lengths of 10, 15 or 20 metres. In addition the British Standard requires the packages to be marked with the number BS 747, the type of felt, and the nominal length, width and mass of the roll. Table 48.4 gives the various grades, colour codes, length of roll and common use of the felts specified in BS 747.

Selection of felt for substructure

Built-up felt roofing should be specified in three layers, with the first layer being specially chosen and fixed to suit the particular substructure upon which it is laid. Table 48.5 gives the commonly recommended details using in housing work. For details of nailing and bonding techniques see below.

In recent years materials have been developed by several manufacturers that give results higher than BS classification, especially for fatigue resistance (ability to withstand repeated cycle of loading). These should be seriously considered when using insulation materials with a high sensitivity to temperature change.

Table 48.5 SELECTION OF FELTS FOR SUBSTRUCTURE

Roof deck	*Substrate directly below 1st layer of felt*	*Method of fixing 1st layer*	*Recommended types of roofing felt*		
			1st layer	*2nd layer*	*3rd layer or capsheet*
1. Timber	Timber	Nailing	2B	B	B
2. Plywood	Plywood	Partial bond	3G or 2B	B	B
3. Woodwool	Cement based screeds or pre-felted surface	Partial bond	3G or 2B	B	B
4. Concrete	Concrete or cement based screeds	Partial bond	3G or 2B	B	B
5. Any of the above listed 1 to 4 (generally for the warm roof design principle, with a vapour barrier between the roof deck and the insulation slabs or boards)	(a) Polystyrene grade HD with top surface pre-felted *or* polyurethane to BRUFMA with both surfaces glass-faced	Full bond on fibreboard underlay *or* partial bond	3G	B	B *or* proprietary roofing not to BS but with extra fatigue resistance should be considered
	(b) Glass-fibre, mineral-wool or perlite roof-grade boards	Full bond	B *or* extra fatigue resistant roofing	B	B
	(c) Foamed-glass slabs or baked cork or wood-fibre roof-grade boards	Full bond	B	B	B

The vapour barrier positioned under the insulation, with the warm roof principle, should project about 225 mm beyond the insulation at the perimeter to allow this to be turned up and folded back on top of the insulation to provide all-round protection.

On large roof areas it may be necessary to form expansion joints through the roof structure, especially where the roof is shaped in anything other than a rectangular form (e.g. L or T shaped).

The inherent movement between the substructure and the built-up felt finish means that it is of crucial importance to isolate the built-up felt, as far as possible, from the effects of such movements. This is achieved by nailing or partial bonding the first layer of felt. Subsequent layers should be fully bonded.

In the UK such fixing details have proved adequate to resist wind uplift, but under exceptionally exposed conditions additional holding-down devices should be used.

Nailing

On timber structures such as softwood boarding or chipboard the first layer of felt should be asbestos type (2B) and this should be nailed. The nails should be galvanised, round, extra large headed clout nails not less than 19 mm long (to BS 1202). The felt should be unrolled and laid with subsequent rolls in the direction of slope, lapping each layer by 50 mm (or 75 mm at the ends of each length of felt).

Nails should be positioned at 50 mm centres along the centre of the lap, with an additional row of staggered nailing midway between laps at 150 mm centres. As stated above, second and third layers should be fully bonded.

Partial bonding

On cement screeds the first layer can be asbestos-type felt (2B) partially bonded, or preferably a venting base layer type (3G), which will help release the pressure that may build up between the roof deck and the bonding compound. Partial bonding is achieved by spot or strip sticking the felt with bitumen to the deck, or in the case of the venting base layer (which usually has 25 mm diameter perforations positioned at 75 mm pitch throughout) with an edge margin to facilitate lapping. Venting channels 150 mm wide should be left around the perimeter of the roof and at abutments to allow for movement. The remainder of the perimeter should be fully bonded for a width of about 450 mm to ensure adequate fixing.

Additional protection can be provided against movement by forming minor movement joints in the substructure or deck, and laying a strip of felt or building paper 300 mm wide over these joints and bonded at the edge only. The first layer of felt is then applied as normal.

Full bonding

As the name implies, full bonding involves spreading hot bitumen over the complete surface and laying the felt on top. Two methods are generally used for this technique, referred to as either pouring and rolling or mopping. With both methods the layer of felt should be laid in position and then rolled back halfway. The bitumen bonding compound is then poured from a container or spread out in front of the roll by mops dipped in a bucket; the process is repeated with the other half of the roll.

Table 48.6 SURFACE FINISHES

Roof slope and use	*Surface finish*
Flat roofs up to 10° pitch where access required for maintenance only	• Bitumen dressing compound and 13 mm aggregate (limestone, granite gravel, calcinated flint) • Bitumen dressing compound and 6 mm aggregate (calcite or felspar)
Promenade roofs, subject to continuous foot traffic, not above 2° pitch (1 in 30)	• Bitumen macadam, not less than 13 mm consolidated thickness • Concrete or asbestos tiles, bedded in hot bitumen compound • Cement screeded layer cut into squares not exceeding 600 × 600 mm with *vee* joints for the full depth of the screed, afterwards filled with bitumen compound
Roofs over 5° pitch (1 in 10) with no possibility of localised ponding	• Where the flat roof has a relatively steep slope, mineral surfaced felts type 2E and 3E may be used

Surface finishes

A surface finish is required for built-up felt flat roofs, to provide protection against physical damage, to reduce solar heat gain and also to provide increased fire resistance. The common treatments used are summarised in Table 48.6.

WEATHERING DETAILS

Many of the procedures associated with built-up felt roofs necessitate strict compliance with details and good workmanship. The procedures associated with weathering and the provision of flashing are perhaps the most important in their need for strict control, as they contribute very greatly to the prevention of failures.

The following points should be noted and adhered to if trouble-free felt roofing is required.

Eaves details

The eaves detail should be finished with a welted drip or a preformed metal drip, capable of shedding the surface water into the gutter.

The welted drip shown in *Figure 48.2* is formed with a separate strip of felt, usually type 3E, and turned down into the gutter. The strip should be at least 250 mm wide and is first nailed to a fixing batten, which provides the overhang into the gutter, then folded back to form an envelope and returned at least 100 mm on to the roof. Hot bitumen is poured into the envelope and along the roof to secure. The depth of apron should not be less than 50 mm. Subsequent layers oversail this return. Alternatively a metal drip may be used; this is fixed on top of the first layer of felt, usually with screws, and taken over the fixing batten to ensure adequate disposal into the gutter. Care should be taken to allow for expansion gaps longitudinally, when using metal drips.

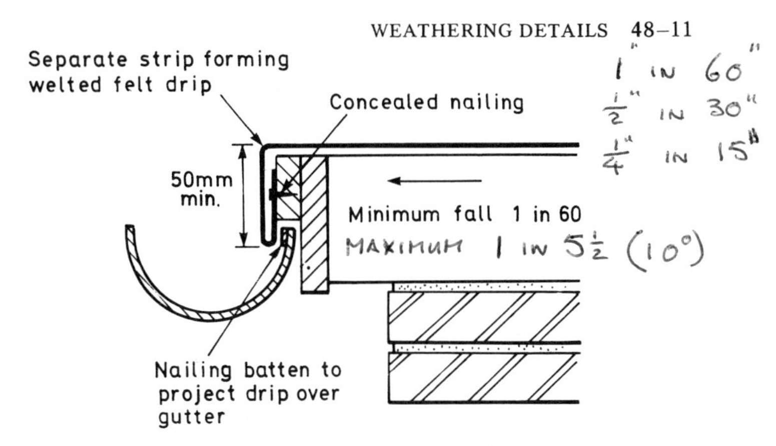

Figure 48.2 Eaves detail finished with a welted drip

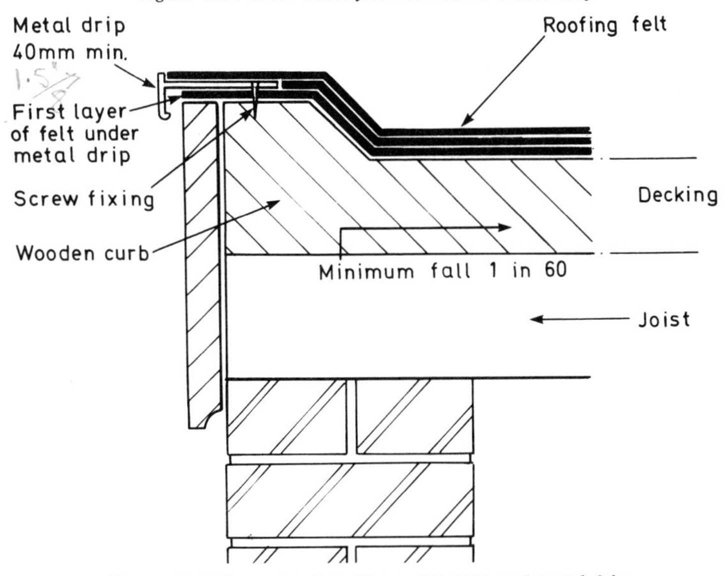

Figure 48.3 Verge detail finished with curb and metal drip

Verge details

The verge can be finished with a welted drip or metal drip as described above, but preferably a curb or fillet should be formed as shown in *Figure 48.3*. The three layers of built-up felt are taken up over the curb, and the separate layer of felt forming the drip should oversail these layers to ensure weathertightness.

Abutments with vertical walls

At all abutments of the flat roof to walls, the three layers of felt should be taken up over an angle fillet (30 × 30 mm) at least 150 mm and stuck to the wall. It is important that the felt is not tucked into the wall but merely finishes vertically. A separate flashing is then wedged and pointed into a brick joint or chase and dressed down over the upstand. This method will impose the least strain on the felt and ensure weathertightness. The chase should not be less than 25 mm deep, and the flashing should be tucked immediately under any cavity tray or dpc that is incorporated in the wall. See *Figure 48.4*.

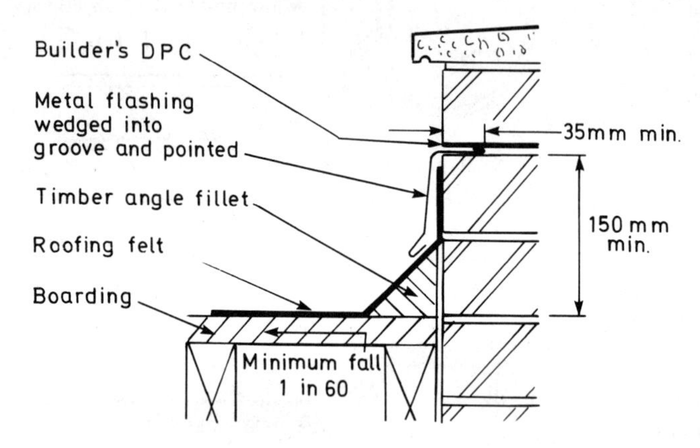

Figure 48.4 Abutment with vertical wall

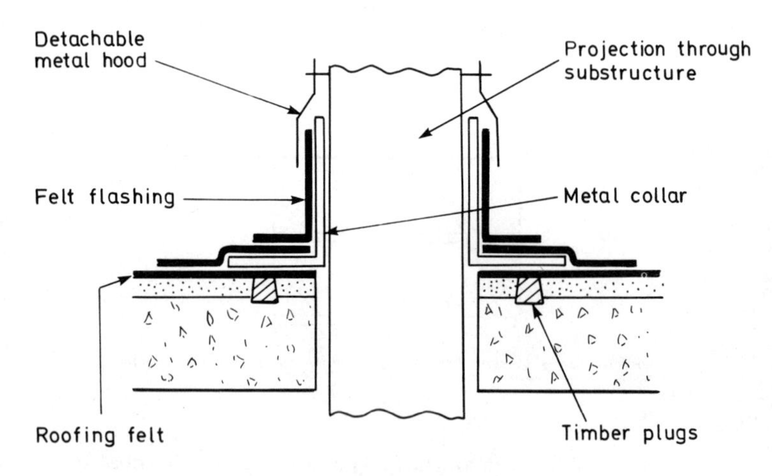

Figure 48.5 Projection through substructure

Mineral-surfaced felt may be used for the flashing (bonded to the wall with hot bitumen) but a more reliable and weathertight finish will be achieved if a lead flashing is used. The lead should be in lengths of not more than 2 metres, wedged firmly with lead into the chase, and lapped at least 100 mm at each joint. A 13 mm check welt should be formed on upper and lower ends of the lead, and the chase pointed in with mortar to complete.

Projections through roof surface

The best way to ensure weathertightness where pipes and ventilators pass through the built-up felt and substructure is to completely isolate the projection from the felt finish by using a metal collar. See *Figure 48.5.*

The metal collar, which should be flanged, is secured to the roof deck and projects at least 150 mm above the felt finish. A separate flashing should then be fixed to the collar and lapped on to the roof covering. Finally a separate metal cap or hood should be fixed around the projection to oversail the flashing and shed water.

A common omission with projections is to ignore handrails that pass through the felt finish. It is equally important in such instances to ensure weathertightness, and this can be easily done by turning the roof covering up the handrail and again finishing with a metal cap or hood.

Rooflights

Manufacturers' details for rooflights or domes should be followed closely, but in general they should be mounted on curbs 150 mm above the roof, with a separate flashing taken over the curb and bonded to the built-up felt finish before the lights are fixed.

Box gutters and outlets

Drainage of felt flat roofs should preferably be taken by the general slope of the roof to the eaves, and discharged into a gutter fixed on the fascia. Where box gutters

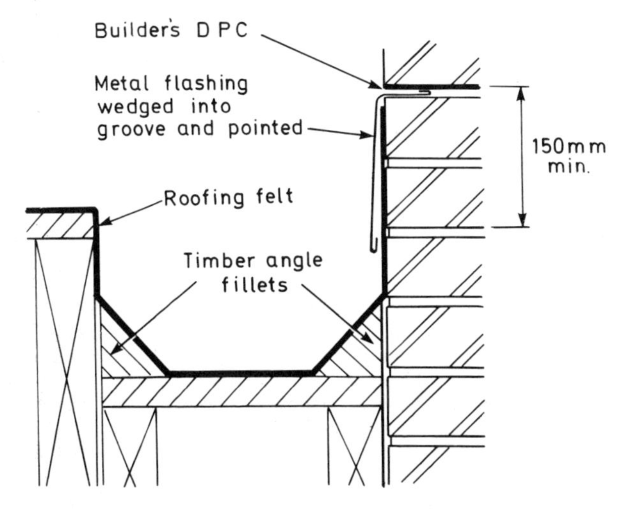

Figure 48.6 Box gutter treatment

are necessary, special care must be taken in forming these, especially with timber or lightweight roof decks, which will be subject to considerable movement.

Three layers of asbestos or glass-fibre based, fine-granule-surfaced felt may be used (mineral-surfaced felt is not to be used) as outlined in *Figure 48.6*. It is,

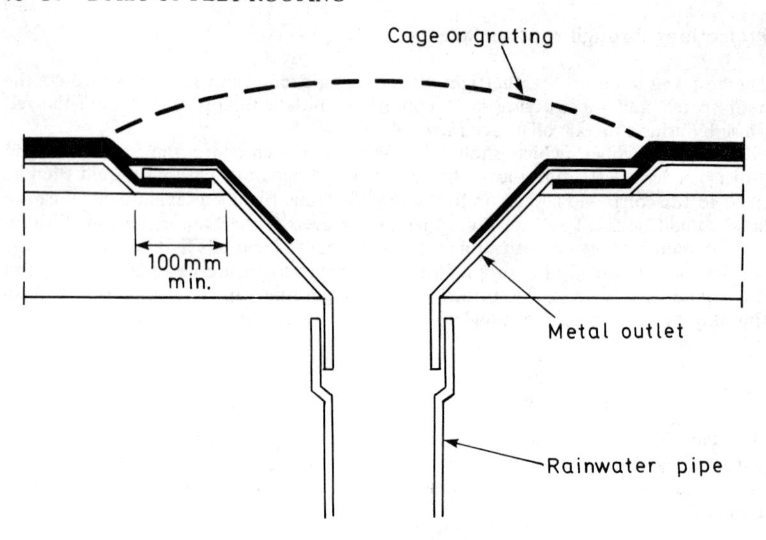

Figure 48.7 Detail at drainage outlet

however, worth considering some of the proprietary materials available that have greater strength or resistance to movement. Outlets and apron chutes should be formed with metal, well below the felt finish and suitably flanged to allow a bonded lap of at least 100 mm with the felt. A cage should be fitted over the outlet to stop possible blockage by leaves etc. See *Figure 48.7*.

MAINTENANCE AND REPAIR

The code of practice for built-up roofing does not envisage that any routine maintenance will be necessary if the recommendations of the code were followed. Obviously, however, routine inspection of the flat roof is advisable, in case something does go wrong; it makes sound economic sense to identify any failure as early as possible before too much damage results.

If failure areas are spotted early enough the actual repair to the membrane itself will be relatively easy and cheap. To this end a full check should be made on the built-up felt finish at the end of the laying, to ensure that the finished falls are correct and that there are no obvious cracks, blistering etc.

A check should also be made yearly by the occupier to clear any obstacles preventing proper drainage, and to check for failures. As it is in the builder's interest to be informed as early as possible of any likely repairs before serious damage occurs, it is recommended that on roofs of any great area the occupier be advised accordingly, or else the builder should arrange his own inspection during the general maintenance period.

On blocks of flats it is also advisable to ensure that access to the roof can be obtained from within the property. This need not involve direct access through a door, which would tempt continuous foot traffic. A simple trap door accessible from a ladder should provide sufficient access for inspection and carrying out minor repairs, without too much expense.

FURTHER INFORMATION

Built-up roofing manual, Felt Roofing Contractors Advisory Board, Maxwelton House, Boltro Road, Haywards Heath, Sussex

Building Research Establishment Digests, available from Her Majesty's Stationery Offices:

Digest 180, *Condensation in roofs*

Digest 8, *Built-up felt roofs*

Digests 188 and 189, *Roof drainage*

Digest 108, *Standard U-values*

Digest 144, *Asphalt and built-up-felt roofings durability*

British Standards:

BS 747: 1977, *Specification for roofing felts*

BS 5250: 1975, *Code of basic data for the design of buildings: the control of condensation in dwellings*

CP 144: Part 3: 1970, *Built-up bitumen felt*

49 PLASTERING AND RENDERING

INTERNAL PLASTERING 49–2

EXTERNAL RENDERING 49–9

PLASTERING AND RENDERING 49

49 PLASTERING AND RENDERING

J.F. RYDER, BSc
Formerly Head of Limes and Plasters Section, Building Research Establishment

This section deals with internal plastering, with mixes based on lime, cement or the various types of gypsum plaster, and with external rendering with cement-based mixes. The properties of these materials, and the characteristics of different backgrounds, are described, and guidance is given on the choice of specifications for particular backgrounds and performance requirements.

INTERNAL PLASTERING

The main purpose of plastering the internal walls and ceilings of a house is to correct any unevenness in the background and to provide a finish that is smooth, crack-free, hygienic, resistant to damage and suitable for painting or papering. A plaster finish may also be required to improve fire resistance, sound insulation or thermal insulation, to reduce condensation, or to provide a specified standard of fire protection to structural steelwork.

Organisation

Although a large proportion of plastering work is sub-contracted, it is important that the builder should fully understand the problems and requirements of the plasterer and the interrelation of plastering and other building operations. The use of a particular type of partition block and failure to protect stacked blocks during wet weather, for example, may considerably delay plastering or lead to problems with cracking, loss of adhesion or efflorescence. Time and money can be saved if the plasterer does not have to wait for other trades, or make good after them. Careful planning is even more important if full advantage is to be taken of mechanical plastering or projection plastering. Plastering should generally be done after the carpenter's first fixings are completed, floors have been screeded and the first stages of plumbing and electrical services have been installed. The builder's time schedule should allow sufficient time for walls to dry out after the house has been made weathertight, so that plastering work may then proceed without subsequent damage to it or its decoration.

Because he is producing a surface finish, the plasterer needs a higher general level of lighting than most other building trades. For plasterwork that will eventually be illuminated by shallow angle lighting, as from recessed trough lights or close-wall lights, it is essential that similar side-lighting that shows up minor irregularities in the finish should be used by the plasterer. A 1200 mm 40 W tubular fluorescent lamp, housed in a suitable reflector and placed close to and parallel with the surface being worked, has been proved suitable for this purpose.

Basic principles

There are a number of main groups and subdivisions of plastering materials, which will be described later, and many different backgrounds to which these may have to be applied. Some factors, however, are common to all plasterwork, and an understanding of them will help in selecting the appropriate plaster for a particular background and in avoiding defects and failures.

All plasterwork consists of successive layers of different materials, and with changes in temperature or moisture content these layers, and the background, tend to expand or contract to a different extent. The differential movement of a well-bonded undercoat is normally restrained by the background, and relative movement of a final coat is restrained by the undercoat. These restraints, however, result in the build-up of shear stresses at the interfaces between the layers, and these stresses can eventually cause a failure in adhesion of poorly-bonded plaster. *Figure 49.1(b) and (c)* show that expansion of the plaster coat, or shrinkage of the background, produces compressive

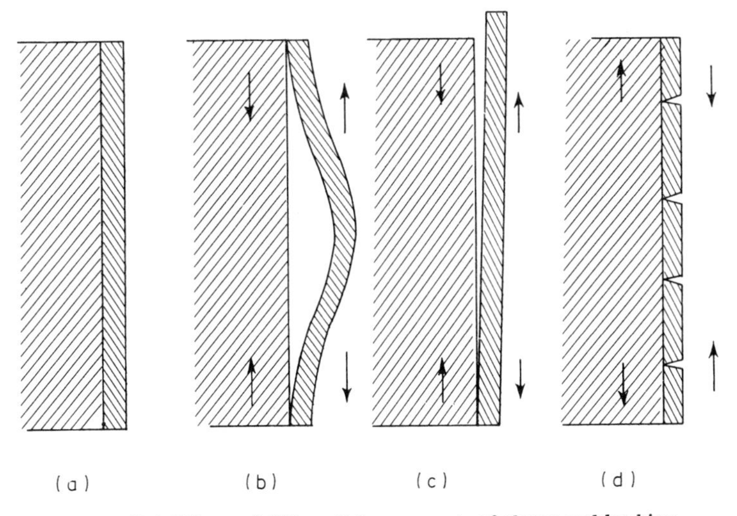

Figure 49.1 Effects of differential movements of plaster and backing

stresses in the plaster layer that tend to cause it to blister or shell off. Such failures can be prevented by ensuring that the plaster is not too strong for the background and is well bonded to it. Any expansion or shrinkage of the plaster can then be restrained by the background, but if the plaster is too strong it will shear the surface off a weak background.

The four basic principles governing the design and specification of successful plasterwork, which are the basis of many of the recommendations of BS 5492: 1977, *Code of practice for internal plastering,* are as follows:

- Differential movement between plaster and background, or between successive coats of plaster, must be minimised.
- Good adhesion between plaster and background, or between successive coats, must be ensured by provision of an adequate mechanical key and by control of suction.
- The undercoat plaster must not be too strong to be restrained by the background, nor must it be too weak in relation to the final coat. In *Figure 49.2,* for example, the final coat must not be stronger than the undercoat, nor must the latter be stronger than the background. A progressive decrease in strength from background through undercoat to final coat is, however, permissible and indeed advantageous. The exception to this rule is that it is permissible to use a strong gypsum plaster, which

has virtually no drying shrinkage movement, on a weak background such as wood-wool slabs; the reason for this is that the strong plaster tends to restrain the drying shrinkage of the weak backing, and thereby prevents cracking at the joints.

- Plastering coats must not be unduly thick, that is they should not appreciably exceed the thicknesses recommended in BS 5492 (Clause 36). The importance of plaster thickness can be seen in the two-coat plastering system illustrated in *Figure 49.2*. Any tendency of the plaster to expand in relation to the background produces

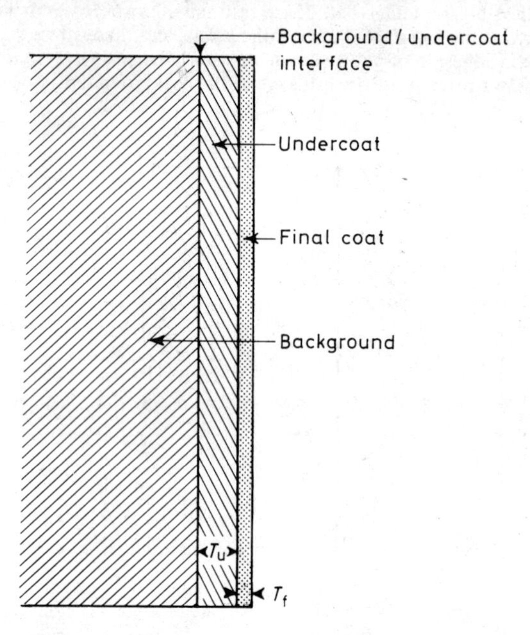

Figure 49.2 Effect of thickness and elasticity of plasters on shear stress at interface of background and undercoat

a compressive strain in the plaster. The resultant shear stress at the background/undercoat interface is then proportional to $(T_uE_u + T_fE_f)$, where E_u and E_f are the elastic moduli of the undercoat plaster and final-coat plaster, and T_u and T_f are their respective thicknesses.

Plastering materials

The three main groups of plastering materials are those based on lime, on Portland cement and on gypsum plasters; but there are sub-divisions of these main groups, and lime may be used in combination with cement or gypsum plaster.

LIME MIXES

Lime was the traditional plastering material in Britain, but lime plasters have two serious disadvantages; they shrink considerably on drying, and therefore tend to crack and craze, and they harden only very slowly by carbonation. Lime plastering mixes are therefore now gauged with cement or gypsum plaster, to provide early strength and reduce the time interval necessary between successive coats. Lime may also be used to replace a proportion of the cement in cement/sand mixes, or for gauging certain types of gypsum final-coat plasters.

The plasterer's main requirement of lime is that it shall provide a mix that has good working properties, hangs well on the trowel, spreads easily and does not stiffen too quickly on a high-suction backing. These properties are most readily obtained by using lime putty. This can be made on site by soaking a high-calcium hydrated lime for at least 16 hours before use, or it can be obtained ready for use from suppliers of ready-mixed mortars.

CEMENT MIXES

For interior work, cement-based plastering mixes are normally only used as undercoats, although they may also be used as a final coat when resistance to damp conditions or to hard knocks is required, as in a wash-house or garage. They should only be given a wood-float finish, as trowelling is likely to cause surface cracking. Although a 1 : 3 cement/sand mix is the undercoat recommended for a Keene's plaster final coat, such a mix is only suitable for a strong background with a good mechanical key, such as brickwork or no-fines concrete, since its shrinkage on drying and its high early strength produce stresses that weaker backgrounds are unable to restrain. These stresses are reduced if the proportion of sand to cement is increased, and the leaner mix is plasticised with lime or an air-entraining additive. The type III mixes (1 : 1 : 6 or equivalent) of Table 49.1 are strong enough to be finished with a neat class B or class C gypsum-plaster final coat, yet are not too strong for most backgrounds. The weaker type IV mixes (1 : 2 : 9 or equivalent) are preferable for non-rigid backgrounds such as lathing or wood-wool slabs, but should not be used in combination with a strong final coat.

Since cement-based undercoats shrink on drying and tend to crack, it is advisable not to apply the final coat before the initial shrinkage has taken place, but difficulties may then arise from the high suction of dry cement/lime/sand undercoats. The equivalent air-entrained mixes of Table 49.1 develop less suction on drying, and may

Table 49.1 CEMENT-BASED MIXES FOR PLASTERING AND RENDERING

Mix type	*Cement/lime/sand*	*Cement/ready-mixed lime/sand*		*Cement/sand (using plasticiser)*	*Masonry cement/sand*
		Ready-mixed lime/sand	*Cement/ready-mixed material*		
I	1 : ¼ : 3	1 : 12	1 : 3	–	–
II	1 : ½ : 4 to 4½	1 : 8 to 9	1 : 4 to 4½	1 : 3 to 4	1 : 2½ to 3½
III	1 : 1 : 5 to 6	1 : 6	1 : 5 to 6	1 : 5 to 6	1 : 4 to 5
IV	1 : 2 : 8 to 9	1 : 4½	1 : 8 to 9	1 : 7 to 8	1 : 5½ to 6½

therefore be preferable for summer use. However, if these aerated mixes are applied in cold and humid conditions they may be slow to develop sufficient suction to ensure satisfactory adhesion of the final coat. Although all the cement-based undercoats are more effective than gypsum plaster undercoats in preventing soluble salts from wet walls migrating to the drying surface of the plaster and forming efflorescence, the aerated mixes are an even more effective barrier than the corresponding cement/lime/sand mixes. They are therefore to be preferred when plastering wet, and possibly salty, brickwork or blockwork.

Premixed lightweight cement plaster is a proprietary backing plaster based on cement and lime with a lightweight perlite aggregate. It has the convenience and controlled composition of a ready-mix, and its light weight makes for ease of application and provides enhanced fire resistance and thermal insulation. In other respects it is similar to a cement/lime/sand mix.

GYPSUM PLASTERS

Except for gypsum projection plaster, for which there is at present no British Standard, gypsum plasters should satisfy the requirements of the relevant part of BS 1191, *Gypsum building plasters,* and their properties can most readily be considered in relation to their classification in this Standard.

BS 1191: Part 1: 1973, *Gypsum building plasters – excluding premixed lightweight plasters,* covers classes A to D as follows.

- Class A – Plaster of Paris. This form of hemihydrate plaster sets too quickly for most plastering purposes, but can be used in small batches for gauging lime final coats or for small repairs or patching work.
- Class B – Retarded hemihydrate gypsum plaster. These plasters contain a retarder to lengthen the setting time. They are available in two undercoat grades for use with sand, browning plaster (type a1) and metal lathing plaster (type a2), and two final-coat grades. The type b1 finish plaster can be used neat or gauged with not more than a quarter of its volume of lime putty, and is for use on gypsum-plaster/sand undercoats. It can also be used on cement-based undercoats that are not weaker than the type III mixes of Table 49.1. The type b2 board-finish plaster is mainly used for single-coat work on suitable building-board backgrounds, and must never be gauged with lime for this purpose. It can also be used for single-coat work on other level backgrounds of uniform suction, and particularly on surfaces that have been treated with a bonding agent.
- Class C – Anhydrous gypsum plaster. Plasters of this class are intended for use only as final coats on sanded undercoats of low to moderate suction and of sufficient strength. They should not be applied direct to plasterboards or other building boards. Their gradual and progressive setting characteristics enable a high standard of finish to be obtained more easily than with the quicker-setting class B plasters, but these characteristics can also lead to failures due to delayed hydration if these plasters dry too quickly on suction backings or in hot weather. In addition to providing a smooth surface when used neat, or if desired gauged with up to a quarter volume of lime putty, they can provide a textured finish when used neat or mixed with sand.
- Class D – Keene's plaster. This is an anhydrous final-coat plaster that can provide a very hard and smooth surface, but is now mainly used for specialist plasterwork, such as squash courts and cold-storage rooms, rather than for housing work.

Premixed lightweight gypsum plasters. These plasters conform to Part 2 of BS 1191. They are mixtures of retarded hemihydrate gypsum plaster with suitable lightweight aggregates, and require only the addition of clean water to prepare them for use. Four undercoat grades are available, and it is essential that the correct grade for a particular type of background is used. They are as follows.

- Browning – type a1 of BS 1191: Part 2. This has an expanded perlite aggregate, and is for undercoats on solid backgrounds of normal suction, such as brickwork and blockwork.
- Browning HSB – type a1. This contains a water-retaining additive, and is for undercoats on solid backgrounds of very high suction, such as certain kinds of aerated concrete.
- Bonding coat – type a3. This has an exfoliated vermiculite aggregate, and is for

undercoats on plasterboard, low-suction backgrounds such as dense concrete or surfaces treated with a bonding agent, and composite surfaces.

- Metal lathing – type a2. Both expanded perlite and exfoliated vermiculite aggregates are used in this plaster, which also contains a rust inhibitor. It is for undercoats on expanded metal lathing, Grant-K-lath, Newtonite lath, wood-wool slabs and expanded polystyrene.

The finishing grade (type b of BS 1191: Part 2) of lightweight gypsum plaster has an exfoliated vermiculite aggregate, and is suitable for final coats on any of the above undercoats.

These lightweight gypsum plasters are similar to the class B hemihydrate plasters in general properties. They are quick-setting and no long drying periods are required between coats. The final drying-out, however, may take longer, due to the water absorbed by the aggregate. They provide better fire resistance, on account of their better thermal insulation, which also helps to prevent surface condensation by providing a surface that warms up more quickly in conditions of intermittent occupation. The more elastic nature of these lightweight plasters allows them to provide greater resistance to cracking due to structural movement.

Gypsum projection plasters. These are a blend of hemihydrate and anhydrous gypsum plasters, formulated for use with mechanical plastering machines that mix and spray the plasters as a direct one-coat application. They are then levelled and trowelled by conventional methods, although powered rotary floats may also be used. For walls other than concrete the thickness of the single coat may be up to 13 mm. For concrete walls and soffits the thickness should be kept to a minimum, but should not be less than 5 mm or exceed 10 mm. The ordinary grade is only for use on low-suction backgrounds; the multi-purpose grade is for general use on most internal backgrounds of high or low suction. Certain types of background, however, are not suitable for these plasters and the advice of the manufacturer should be sought.

The economic benefits of projection plastering are best realised on sites that allow continuity of operations, so that the cost of buying or hiring the machines can be offset by increased productivity.

Thin-coat plasters. These are gypsum-based plasters of high water retention for final coats and single-coat work on level backgrounds of high suction, such as some types of aerated concrete panels. They are applied and worked by normal hand plastering techniques.

Veneer-coat plasters. These plasters have an organic binder, and harden only by drying. They are normally supplied as a wet mix, are applied by spraying or with a broad spatula, and are often left with a textured finish produced by spraying. They are best suited for use on large areas of level backgrounds, where little masking around openings is necessary, and are therefore not commonly used in housing work.

Backgrounds

For success in plastering, and particularly to prevent failures in adhesion, the specification for plasterwork must be selected to suit the characteristics of the background to be plastered. The properties that are of importance are as follows.

STRENGTH

The strength of the background is of particular importance when cement-based undercoats are to be used, since the background must be strong enough to restrain their shrinkage on drying. Gypsum plaster undercoats do not shrink appreciably on drying, so when they are used the strength of the background is of less importance.

SUCTION

The suction of the background needs to be considered in relation to both the strength and the adhesion of plasters. The recommended ratio by volume of gypsum browning plaster to sand ranges from 1 : 3 for backgrounds of moderate to high suction, such as ordinary clay brickwork, to 1 : 1½ for low-suction engineering brickwork. The higher sand content mixes require more water to make the mix workable, and the 1 : 3 mix would therefore give too weak a plaster unless the background removes some of the water before the plaster sets. By absorbing some of the water, the high-suction background reduces the final water/plaster ratio of the 1 : 3 mix to about the same level as that of a 1 : 1½ mix on a low-suction background, so that the final strengths of the two undercoats are also similar.

The proportions of gypsum plaster to the lightweight aggregate in ready-mixed gypsum plasters have also been designed to match the suction of their recommended backgrounds.

Adhesion failures of plaster coats can result from either unduly high suction of a backing, or from insufficient suction. A background of very high suction, such as aerated concrete, may absorb so much water from the plaster that the contact layer is weakened and failure results; alternatively, a gypsum-plaster final coat that is applied to a cement-based undercoat before the latter has dried sufficiently to develop some suction will be only weakly bonded and is likely to fail in adhesion as the undercoat shrinks on drying.

KEY OR BOND

Backgrounds such as metal lathing, wood-wool or no-fines concrete offer a good mechanical key, which should be fully utilised by working the plaster well into the interstices with the trowel. The key offered by some other backgrounds can be improved, by raking the joints of brickwork, for example, or by the use of special shuttering for concrete. Bonding agents of the PVAC emulsion type can be used to improve the adhesion of plaster to smooth surfaces, and are also useful in controlling the suction of backgrounds or dry undercoats. When bonding agents are used for the latter purpose, however, it is essential to use plasters that are appropriate to low-suction backings.

DRYING SHRINKAGE AND THERMAL MOVEMENT

All concrete products shrink as they dry, and the shrinkage of walls built of concrete blocks or bricks can cause adhesion failures of plaster. It is therefore necessary to allow them to dry out as completely as possible before plastering. Although wood-wool slabs have a high drying shrinkage, this causes little trouble when the slabs are fixed dry and are restrained by their fixing. If used as permanent shuttering, however, their shrinkage can cause serious cracking of the plaster unless this is strong enough to restrain the movement.

Thermal movements of backgrounds are rarely large enough to cause serious defects of plasterwork, although the movements of heated concrete floors or of flat concrete roofs can crack and disrupt plaster at their junctions with walls. The thermal expansion of gypsum plaster is appreciably greater than that of most backgrounds, and on smooth surfaced backgrounds having particularly small coefficients of expansion, such as concrete made with a limestone aggregate, differential thermal movements can cause failures in adhesion of poorly-bonded plaster. Such failures can be prevented by the use of a PVAC bonding agent.

EFFLORESCENCE

Any soluble salts that may be present in a background tend to be brought forward to the surface of the plaster as the walls dry out. These salts generally form as a fairly soft efflorescence that can be brushed off before decoration, when the drying-out process is complete. Certain salts form a smooth glass-like efflorescence, and others may cause erosion of the surface of the plaster. Soluble salts may be present in some clay bricks or blocks, or in concrete products. Even a very small proportion of salts in the background may be sufficient to produce efflorescence, as it becomes concentrated at the drying surface.

Ideally, backgrounds should be dry before plastering, or at least before decorating. If backgrounds likely to contain efflorescent salts must be plastered before they are dry, it is advisable to use an aerated cement/sand undercoat that will present a barrier to the passage of these salts.

Plastering systems for various backgrounds

In Table 49.2 the various backgrounds for plastering are categorised according to their strength, and other properties that are relevant to the selection of a plastering system are tabulated. Suitable alternative undercoat plasters and their appropriate final coats are shown for each type of background. For many backgrounds there is a choice between several cement-based undercoats, a sanded gypsum browning plaster, and a lightweight gypsum plaster. This choice will be influenced by various factors, such as the time schedule, the state of the background, namely whether it is unduly wet or salty, and the required decoration. To guide this choice, Table 49.3 summarises the advantages and disadvantages of the different plastering systems, and indicates some precautions that are necessary when using them.

EXTERNAL RENDERING

Basic principles

Like internal plasterwork, external renderings consist of successive layers of possibly different materials applied to a background; to avoid failures in adhesion due to the differential movements shown in *Figure 49.1*, it is similarly necessary to ensure that the final coat is not unduly strong for the undercoat, and that this is not too strong for its drying shrinkage to be restrained by the background. The same background characteristics (strength, suction, key or bond, drying shrinkage and thermal movement, and possible salt content) must also be considered when selecting a rendering system. However, the choice must also be influenced by the particular degree of exposure to wind, rain and atmospheric pollution if the rendering is to effectively exclude rain and maintain an acceptable appearance.

Types of finish

There is a wide choice of finishes for renderings, ranging in texture from a smooth floated finish to the coarsest roughcast or dry-dash. Experience has proved that the

Table 49.2 SUITABLE PLASTERING SYSTEMS FOR VARIOUS BACKGROUNDS (Crown copyright: reproduced from Digest 213 by permission of the Director, Building Research Establishment)

	Background				
Class	*Type*	*Suction*	*Key or bond*	*Drying*	*Type of finish*
Solid	Normal clay brickwork and blockwork	Moderate to high	Good if joints well raked or bricks keyed	Negligible	I Gypsum/lime II Weak lime III Lightweight IV Cement V Projection
	Dense clay brickwork (other than engineering brickwork) and calcium silicate or concrete blockwork or brickwork	Low to moderate	Variable: bonding treatments may be required for cement based undercoats	Low to high for materials other than clay brickwork	I Gypsum/lime II Weak lime III Lightweight IV Cement V Projection
	Clay engineering brickwork, dense concrete	Low	Hacking, spatterdash or bonding treatment may be needed	Negligible for clay brickwork, low to high for dense concrete	I Gypsum/lime II Weak lime III Lightweight IV Cement V Projection
	No-fines concrete	Low	Good	Low to moderate	I Gypsum/lime II Weak lime III Lightweight IV Cement V Projection
	Close textured lightweight aggregate concrete blocks	Low	Poor: bonding treatments may be necessary	Moderate to high	I Gypsum/lime II Weak lime III Lightweight IV Cement V Projection
	Open textured lightweight aggregate concrete blocks	Low to moderate	Good	Moderate to high	I Gypsum/lime II Weak lime III Lightweight IV Cement V Projection

Suitable alternative undercoat plasters				*Remarks*	*Single-coat work (excluding projection plaster)*
Premixed lightweight plasters		*Sanded browning plasters (proportions by volume)*	*Undercoats based on cement (proportions by volume)*		
Gypsum	*Cement*				
–	–	1:3	1:1:6	Background should be dry to minimise efflorescence	Not suitable for this type of background
–	Backing	1:3	1:1:6 or 1:2:9		
Browning	Backing	–	–		
–	–	–	1:¼:3, 1:1:6 or 1:2:9		
–	–	–	–		
–	–	1:2	1:1:6	Background should be dry to minimise shrinkage except for dense clay brickwork	Not suitable for this type of background
–	Backing	1:2	1:1:6 or 1:2:9		
Browning or bonding	Backing	–	–		
–	–	–	1:¼:3, 1:1:6 or 1:2:9		
–	–	–	–		
–		1:1½	1:1:6		Sufficiently level concrete surfaces may be plastered with thin-wall plasters or board finish plasters: a bonding treatment may be necessary
–		1:1½	1:1:6 or 1:2:9		
Bonding		–	–		
–		–	1:¼:3, 1:1:6 or 1:2:9		
–		–	–		
–	–	1:2	1:1:6		Not suitable for this type of background
–	Backing	1:2	1:1:6 or 1:2:9		
Browning	Backing	–	–		
–	–	–	1:¼:3, 1:1:6 or 1:2:9		
–	–	–	–		
–	–	1:1½	Not normally suitable for this type of background	Background should be dry to minimise shrinkage movement	Sufficient level surfaces may be plastered with thin-wall plasters or board finish plasters: a bonding treatment may be necessary
–	Backing	1:1½			
Bonding	Backing	–			
–	–	–			
–	–	–			
–	–	1:2	1:1:6	Background should be dry to minimise shrinkage movement	Not suitable for this type of background
–	Backing	1:2	1:1:6 or 1:2:9		
Browning	Backing	–	–		
–	–	–	Not suitable		
–	–	–	–		

Table 49.2 *continued*

	Background				
Class	*Type*	*Suction*	*Key or bond*	*Drying shrinkage*	*Type of finish*
	Aerated concrete slabs and blockwork	Moderate to very high	Poor: bonding treatments may be necessary	Moderate to high	I Gypsum/lim II Weak lime III Lightweight IV Cement V Projection
	Any surface treated with a bonding agent: for example glazed bricks, tiled or sound painted surfaces all treated with bonding agents	Low	Poor: bonding treatments necessary	Usually negligible	I Gypsum/lim II Weak lime III Lightweight IV Cement V Projection
Slab	Wood-wool slabs	Low	Good	Usually fixed dry, but may be high when used as permanent shuttering	I Gypsum/lim II Weak lime III Lightweight IV Cement V Projection
Board	Plasterboard	Low	Adequate with suitable plasters	Fixed dry	I Gypsum/lime II Weak lime III Lightweight IV Cement V Projection
	Expanded plastics boards	Low	Adequate with suitable plasters	Fixed dry	I Gypsum/lime II Weak lime III Lightweight IV Cement V Projection
Lathing	Metal lathing	Low	Good	None	I Gypsum/lime II Weak lime III Lightweight IV Cement V Projection

Suitable alternative undercoat plasters				Remarks	Single-coat work (excluding projection plaster)
remixed lightweight lasters		Sanded browning plasters (proportions by volume)	Undercoats based on cement (proportions by volume)		
ypsum	Cement				
-	–	1:2	1:1:6	Some blocks of very high suction may require special plasters of high water retentivity or sealing with a bonding agent	Sufficiently level surfaces may be plastered with thin-wall plasters or board finish plasters: a bonding treatment may be necessary
-	Backing	1:2	1:1:6 or 1:2:9		
rowning r HSB rowning	Backing	–	–		
-	–	–	1:1:6 or 1:2:9		
-	–	–	–		
-		1:1½	Not suitable for this background		Sufficiently level surfaces may be plastered with thin-wall plasters: a bonding treatment may be necessary
-		1:1½			
onding		–			
-		–			
-		–			
-	–	1:2	–		Not suitable for this type of background
-	Backing	1:2	1:2:9		
letal athing	Backing	–	–		
-	–	–	1:2:9		
-	–	–			
-		1:1½	Not suitable this background		Sufficiently level surfaces may be plastered with board finish plasters
-		1:1½			
onding		–			
-		–			
-		–			
-		1:1½	Not suitable for this background	Not suitable for paper faced laminates	If the boards are fully bonded to a firm background sufficiently level surfaces may be plastered with board finish plasters
-		1:1½			
onding r metal athing		–			
-		–			
-		–			
-	–	1:1½ followed by 1:2	1:1:6	Three coats required. Metal lathing should be well braced. Mixes for first undercoat should contain hair or fibre. Lathing to receive spray application available	Not suitable for this type of background
-	Backing	–	1:1:6 or 1:2:9		
letal athing	Backing	–	–		
-	–	–	1:1:6 or 1:2:9		
-	–	–	–		

Table 49.3 PROPERTIES OF VARIOUS PLASTERING SYSTEMS, AND INDICATIONS FOR AND AGAINST THEIR USE

Undercoat	*Final coat*	*Advantages*	*Disadvantages*	*Precautions*	*Comments*
Lime/sand, gauged with cement or gypsum plaster	Weak lime	Good workability	Weak and liable to shrinkage cracking	Do not overwork and kill the set of the gauging	Gauging with gypsum plaster helps to reduce the shrinkage
Cement-based mixes of Table 49.1	As undercoat	Resistant to water and hard knocks	Liable to shrinkage cracking	Finish only with a wood float, not a steel trowel	The air-entrained mixes are a more effective barrier to efflorescence, and develop less suction on drying
	Class B or class C gypsum plaster*	Can act as a barrier to efflorescent salts	Liable to shrinkage cracking	The undercoat must not be too strong for the background and must be allowed to shrink and develop suction before the final coat is applied	Use of these systems on wet blockwork can lead to shelling of the final coat. See Building Research Advisory Leaflet TIL 14
Premixed lightweight cement	As recommended by manufacturer of u/c	Convenient and controlled mix, easy to apply			
Premixed lightweight gypsum	Lightweight gypsum	Convenient and controlled mix, easy to apply. Hardens quickly, negligible shrinkage	Not a barrier to efflorescent salts	Use the undercoat appropriate to the background. Never apply browning over bonding	Gypsum plastering systems save time in completion work. They need only a short interval between coats, can be allowed to dry quickly*, and can then be decorated with any permeable finish
Class B gypsum browning and sand mixes	Class B or class C gypsum plaster*	Hardens quickly, negligible shrinkage	Not a barrier to efflorescent salts	Use the higher proportions of sand only on suction backgrounds	
Single-coat systems					
Gypsum projection plaster		Hardens quickly, negligible shrinkage. Speed of application by machine	Not a barrier to efflorescent salts	Use the ordinary grade only on low-suction backgrounds	Offers the advantages of a two-coat gypsum plaster system on suitable backgrounds
Thin-coat plaster Veneer-coat plaster		Quick drying	Only suitable for very level backgrounds of uniform suction		Dries very quickly if applied to dry backgrounds, and then allows a free choice of decoration

*The class C final-coat plasters have a more gradual set than those of class B, and can therefore be more readily worked to a smooth finish. However, because their hydration is slow, too rapid drying may lead to delayed hydration and blistering or loss of adhesion.

smooth finishes are less suitable for the more severe conditions, since they tend to craze and allow rain penetration, whilst the flow of rainwater over their surface tends to become channelled and therefore produces uneven and unsightly discoloration. Scraped or textured finishes are generally less liable to crack or craze, and small cracks are less noticeable. Although the rougher texture tends to pick up dirt more readily, it is more evenly distributed and is therefore less apparent. The flow of rainwater over the surface is also more evenly distributed, so that penetration and irregular discoloration are less likely. The coarser roughcast or dry-dash finishes have the same advantages as the textured finishes and are more suitable for severe conditions of exposure, since their weatherproofness, durability and resistance to cracking are superior. However, these coarse-textured finishes should preferably not be applied to undercoats weaker than those of type II of Table 49.1, and cannot therefore be used on relatively weak backgrounds, such as lightweight concrete blockwork.

Textured finishes of various types can also be applied by machine, often using proprietary materials and processes that may require specialist operators. On backgrounds that are not themselves resistant to water penetration, two undercoats are needed for this type of finish. If a proprietary undercoat is not supplied with the finishing material, a type III (1 : 1 : 6 cement/lime/sand) mix can be used.

Mixes and materials

The proportions by volume of rendering mixes of strength categories I to IV are shown in Table 49.1. For each category there is a choice of a cement/lime/sand mix, a cement/sand mix aerated by means of a mortar plasticiser, or a masonry cement/ sand mix. The quality of the sand is important for all rendering mixes, and a well-graded sand complying with BS 1199 is desirable, but the performance of an aerated rendering mix is especially sensitive to sand grading, particularly on suction backgrounds. Working such a mix after it has lost water to the background tends to break down the bubble structure, unless the sand contains sufficient fine material to stabilise the bubbles and maintain the workability of the mix. This breakdown of the bubble structure can lead to surface cracking of the rendering and to rain penetration.

Rendering mixes for backgrounds likely to contain soluble sulphate salts should be based on sulphate-resisting Portland cement.

Recommendations

The rendering mixes that are recommended in the Code of Practice for various backgrounds, in relation to the type of finish required and the exposure conditions, are shown in Table 49.4. The column headings of *severe, moderate* and *sheltered* correspond to the exposure gradings described in Building Research Digest 127, which are based on driving-rain indices as modified by local conditions and building heights. For severe conditions of exposure to prolonged driving rain, the first coat, at least, of the rendering should be as rich in cement as the strength of the background will allow. A continuous spatterdash coat, of a 1 : 1½ to 2 cement/sand mix, used as the first of three coats, will satisfy this requirement. The bagged rendering mixes that include alkali-resistant glass fibre to prevent the formation of shrinkage cracks have also been found effective for these conditions. Since these renderings are normally applied as a thin (5–8 mm) coat, their moisture movements are more easily restrained than those of normal cement-rich renderings, even by weak backgrounds.

Table 48.4 RECOMMENDED MIXES FOR EXTERNAL RENDERINGS IN RELATION TO BACKGROUND MATERIALS, EXPOSURE CONDITIONS AND FINISH REQUIRED (Crown copyright: reproduced from Digest 196 by permission of the Director, Building Research Establishment)

Note The type of mix shown in bold type is to be preferred.

Background material	*Type of finish*	*First and subsequent undercoats*			*Final coat*		
		Exposure			*Exposure*		
		Severe	*Moderate*	*Sheltered*	*Severe*	*Moderate*	*Sheltered*
Dense, strong, smooth	Wood float	II or III	II or III	II or III	III	III or IV	III or IV
	Scraped or textured	II or III	II or III	II or III	III	III or IV	III or IV
	Roughcast	I or II	I or II	I or II	II	II	II
	Dry dash	I or II	I or II	I or II	II	II	II
Moderately strong, porous	Wood float	II or III	III or IV	III or IV	III	III or IV	III or IV
	Scraped or textured	III	III or IV	III or IV	III	III or IV	III or IV
	Roughcast	II	II	II	as undercoats		
	Dry dash	II	II	II	as undercoats		
Moderately weak, porous*	Wood float	III	III or IV	III or IV	as undercoats		
	Scraped or textured	III	III or IV	III or IV	as undercoats		
	Dry dash	III	III	III	as undercoats		
No fines concrete†	Wood float	II or III	II, III or IV	II, III or IV	II or III	III or IV	III or IV
	Scraped or textured	II or III	II, III or IV	II, III or IV	III	III or IV	III or IV
	Roughcast	I or II	I or II	I or II	II	II	II
	Dry dash	I or II	I or II	I or II	II	II	II
Woodwool slabs*+	Wood float	III or IV	III or IV	III or IV	IV	IV	IV
	Scraped or textured	III or IV	III or IV	III or IV	IV	IV	IV
Metal lathing	Wood float	I, II or III	I, II or III	I, II or III	II or III	II or III	II or III
	Scraped or textured	I, II or III	I, II or III	I, II or III	III	III	III
	Roughcast	I or II	I or II	I or II	II	II	II
	Dry dash	I or II	I or II	I or II	II	II	II

*Finishes such as roughcast and dry dash require strong mixes and hence are not advisable on weak backgrounds.
†If proprietary lightweight aggregates are used, it may be desirable to use the mix weaker than the recommended type.
+Three-coat work is recommended, the first undercoat being thrown on like a spatterdash coat.

FURTHER INFORMATION

Internal plastering

BS 5492: 1977, *Code of practice for internal plastering,* gives detailed guidance on the choice of plastering systems for various backgrounds, on design considerations, application and maintenance. It has appendices on projection plastering and on lighting for plastering, and includes a useful table showing the classification of proprietary plasters. Obtainable from British Standards Institution. Building Research Establishment Digest 213 is a convenient summary of this Code of Practice. Obtainable from HMSO.

Building Research Advisory Service Information Sheet TIL14, *Shelling of plaster finishing coats,* deals with adhesion failures of gypsum final coats on cement-based undercoats.

Internal plastering on dense concrete cast in situ, National Federation of Plastering Contractors, 82 New Cavendish Street, London, W1.

The following Department of the Environment Advisory Leaflets are obtainable from HMSO:

No. 2, *Gypsum plasters*
No. 9, *Plaster mixes*
No. 15, *Sands for plasters, mortars and renderings*
No. 21, *Plastering on building boards*

External rendering

BS 5262: 1976, *Code of practice for external rendered finishes,* gives guidance on the choice of renderings for various backgrounds, in relation to the type of finish required and the particular exposure conditions. It includes sections on design and detailing, application, maintenance and repair of renderings. Obtainable from British Standards Institution. Building Research Establishment Digest 196 is a convenient summary of this Code of Practice, and also gives some information on textured finishes based on organic binders, and on some recently developed renderings. Obtainable from HMSO.

External rendering, a brochure obtainable from the Cement and Concrete Association, includes illustrations of various finishes and advice on techniques.

Department of the Environment Advisory Leaflet 27, *Rendering outside walls,* is a very concise summary of the recommendations in the Code of Practice. Obtainable from HMSO.

FURTHER INFORMATION

Internal plastering

[illegible] for internal plastering [illegible] detailed guidance on [illegible] of plastering [illegible] for various [illegible] and [illegible] considerations [illegible] plasters [illegible] and includes a [illegible] classification of [illegible] plasters [illegible] Building [illegible] Code of Practice [illegible]

[illegible]

[illegible]

50 PLASTERBOARD DRY LININGS

INTRODUCTION 50–2

TYPES OF PLASTERBOARD 50–3

PERFORMANCE 50–7

DESIGN AND PLANNING CONSIDERATIONS 50–11

SITE ORGANISATION 50–12

SITE WORK 50–12

PLASTERBOARD DRY LININGS 50

50 PLASTERBOARD DRY LININGS

R.W. SMITH, MSAAT
Technical Literature Officer, British Gypsum Ltd

This section deals with the use of gypsum plasterboard for dry internal linings to walls, ceilings and roofs. It discusses plasterboard types, tools and accessories, performance, fixing systems and joint treatment. For full advantage to be obtained from using dry linings, the design and site information should be carefully considered.

INTRODUCTION

Plasterboard dry linings are becoming increasingly popular because they are an alternative to plaster finishes, and hence offer several distinct advantages, which can be summarised as follows:

- Little or no water is introduced into the structure at a late stage of the building operations. The drying-out time of a building is considerably reduced, and earlier completions and occupations are therefore possible. Freedom from dampness is also enjoyed when the building is first occupied.
- Decoration can start as soon as the joint treatment of the lining has dried. As there is no risk of the decoration being damaged (e.g. due to efflorescent salts leaching out), permanent, instead of temporary, decoration can be selected if required. This provides a much greater freedom of choice in the type of first decoration.
- Services can be accommodated in the cavity between the plasterboard and the wall, or chasing is reduced to a minimum.
- Sites are tidier, with the minimum of cleaning-up required.
- Dry-lining operations are less arduous than wet plastering.
- Maintenance is reduced because cracking is virtually eliminated.
- Plasterboard linings can provide good standards of thermal insulation to external walls, depending on the grade of plasterboard and the fixing system selected. In intermittently heated buildings the wall linings respond to heating more quickly than plasterwork, so desired room temperatures are obtained more quickly and consequently with less fuel being used.
- By improving the standard of thermal insulation on the inside of the walls the linings can also help to reduce the risk of surface condensation. In order to reduce the risk of interstitial condensation occurring within the structure behind the lining, vapour-check grades of plasterboard can be used, or a vapour-barrier treatment can be provided to the exposed surface of the boards.

With careful consideration given during the design stage, and with proper site organisation and workmanship, plasterboard dry lining can be comparable in cost with good quality two-coat plaster. Correct fixing and jointing of the grades of plasterboard available with tapered edges can produce smooth surfaces that compare with traditional plastering.

Limitations

Plasterboard dry linings should not be considered as a means of isolating dampness, a purpose for which they are entirely unsuitable. They should be kept dry, and are not suitable for use in any area subject to continuously damp or humid conditions.

TYPES OF PLASTERBOARD

In the UK, plasterboard is manufactured by British Gypsum Ltd. Plasterboard for dry lining is different from the plastering grades because it has an ivory-coloured surface for receiving most forms of decoration. The following types of plasterboard are available for dry-lining purposes.

Gypsum wallboard (trade name: Gyproc wallboard). The board has an ivory-coloured exposed surface and a grey surface on the back. The board edges can be supplied tapered, square or bevelled, depending on the joint treatment required. The types of joint treatment are illustrated in *Figure 50.1*. The board is available in an

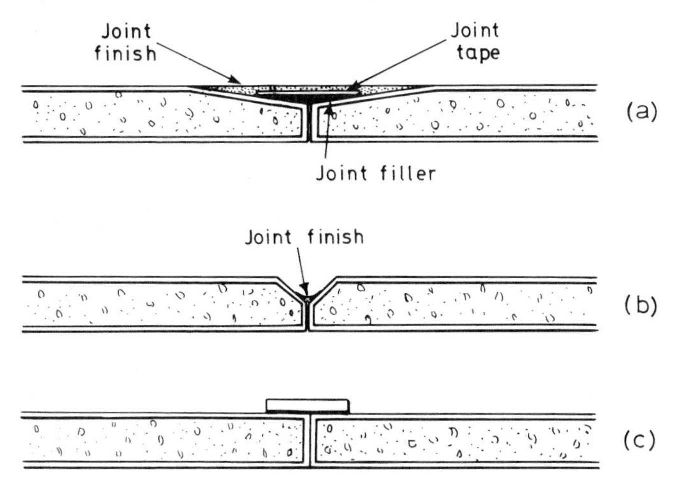

Figure 50.1 Types of jointing for plasterboard dry lining: (a) tapered edge for smooth, seamless surface; (b) bevelled edge for featured V-jointing; (c) square edges

insulating grade, which has a backing of polished aluminium foil. The foil acts as a reflective thermal insulator when the board is fixed facing a 20 mm minimum airspace.

The board is also available in a vapour-check grade, which has a backing of water-vapour-resistant blue polyethylene film. It is available with tapered or square edges. The polyethylene film reduces the passage of water vapour from inside the building into the structure, and can therefore help to reduce the risk of interstitial condensation. The board can be upgraded to form a vapour barrier by suitable treatments.

Gypsum fireline board (trade name: Gyproc fireline board). Similar to gypsum wallboard but only available in a 12.7 mm thickness and tapered edges. The gypsum core incorporates glass fibre and vermiculite to improve its fire-protection performance.

Gypsum plank (trade name: Gyproc plank). This is similar to gypsum wallboard but thicker, and is normally supplied with tapered edges for dry-lining purposes. A square-edge plank with grey surfaces on both sides is used as the first layer in certain two-layer lining systems. Also available in an insulating grade with a polished aluminium foil backing.

Gypsum wallboard/expanded polystyrene laminate (trade name: Gyproc thermal board). This is a laminate composed of gypsum wallboard bonded to a backing of expanded polystyrene. Available with tapered or square edges. The board provides a good standard of thermal insulation.

It is also available in a vapour-check grade, which incorporates a water-vapour-resistant membrane at the interface between the wallboard and the expanded polystyrene. The membrane reduces the passage of water vapour from inside the building into the structure, and can therefore help to reduce the risk of interstitital condensation. If a complete vapour barrier is required the standard grade of board should be used with a suitable water-vapour-resistant treatment applied to the entire exposed surface of the board, e.g. a proprietary water-vapour-resistant primer.

Prefabricated gypsum wallboard panel (trade name: Paramount panels). The panels are formed from two gypsum wallboards (with ivory-coloured exposed faces) separated by, and bonded to, a core of cellular construction. Available in a 38 mm thickness and with tapered or square edges. Used as an internal lining to external walls. Also available in an insulating grade with a backing of polished aluminium foil.

Gypsum plasterboard cove (trade name: Gyproc cove). Although the ceiling/wall angle in a room may be finished in the same way as any other joint, plasterboard cove makes this operation unnecessary and provides a very attractive finish.

British Standards

Gypsum plasterboard conforms to BS 1230: 1970, *Gypsum plasterboard*. The expanded polystyrene backing of the Gyproc thermal boards described in this section conforms to BS 3837: 1977, *Specification for expanded polystyrene boards* (Type A). The prefabricated gypsum wallboard panels (for use as an internal lining to external walls) as described above conform to BS 4022: 1970, *Prefabricated gypsum wallboard panels.*

Description

Composition. Gypsum plasterboard consists of an aerated gypsum core encased in and firmly bonded to paper liners.

Tools and accessories. Plasterboard products are complemented by a comprehensive range of drylining tools and accessories for the fixing, jointing and finishing operations. The tools are briefly described below. (Information on the accessories available, including the approximate requirements, is available from British Gypsum.)
Gyp-C jack board lifter lifts any size board to ceiling; low height 1829 mm, raises to 2743 mm, or 3048 mm with extension piece.
Foot lifter (shown in use in *Figure 50.3*) jacks wallboard into position tight against ceiling as the wearer steps down, leaving hands free for fixing.
All-steel drywall hammer for nailing plasterboards, with rounded and chequered head that will not bruise the board or break the paper.
Wallboard trimmer trims up to 112 mm: steel wheels perforate both sides of the board so that it can be snapped off cleanly.
Wallboard saw with 400 mm blade.
Utility saw for cutting outlet and service holes in wallboard: 150 mm blade with sharp point penetrates easily and eliminates drilling.

Table 50.1 GYPSUM WALLBOARD

Width (mm)	*Edge and thickness*	*Length* (mm) 1800	1829 (6 ft)	2286 (7.5 ft)	2350	2400	2438 (8 ft)	2700	3000	3300	3600†
600	Tapered 9.5 or 12.7 mm	*	—	*	*	*	*	*	*	—	—
	Square 9.5 or 12.7 mm	*	*	*	*	*	*	*	*	—	—
900	Tapered 9.5 or 12.7 mm	*	—	*	*	*	*	*	*	—	—
	Square 9.5 or 12.7 mm	*	*	*	*	*	*	*	*	—	—
	Bevelled 12.7 mm only	*	—	—	*	*	—	*	*	—	—
1200	Tapered 9.5 or 12.7 mm	*	—	*	*	*	*	*	*	*	*
	Square 9.5 or 12.7 mm	*	—	*	*	*	*	*	*	—	—
	Bevelled 12.7 mm only	*	—	—	*	*	—	*	*	—	—

† 12.7 mm thickness only
The vapour-check grade of wallboard is available in the same sizes as above but in 900 mm and 1200 mm widths only.

Table 50.2 GYPSUM FIRELINE BOARD

Width (mm)	*Edge and thickness*	*Length (mm)* 1800	2400	2700	3000	3600
900	Tapered 12.7 mm	*	*			
1200	Tapered 12.7 mm		*	*	*	*

Screwdriver clutch attachment converts standard electric drill to power screwdriver for fixing boards to any Gyproc metal-framed system.
Mixing tool for Gyproc bonding compound, drywall or multi-purpose adhesive.
200 mm applicator with rigid blade for applying Gyproc joint filler and joint finish.
Narrow taping knife with 50 mm blade for pressing back joint tapes.
Jointing sponge, circular with wood handle, for cleaning off surplus jointing material, feathering edges and giving final finish.
Long-handled broad knife with 175 mm blade to flatten joint tape and remove excess cement. Replacement blades available.
Bladed cement pan, 350 mm long, with removable steel blade for wiping broad knife clean.
Universal sander, long-handled with universal joint at head, enables both ceiling and wall joints to be sanded from the floor. Sanding papers available.

Table 50.3 GYPSUM PLANK

Width (mm)	*Edge and thickness*	*Length* (mm) 2350	2400	2700	3000	3200
600	Tapered 19 mm	*	*	*	*	*

Table 50.4 GYPSUM WALLBOARD/EXPANDED-POLYSTYRENE LAMINATE (AND VAPOUR CHECK GRADE)

Width (mm)	*Thickness* †	*Length* (mm)
1200	22 mm (9.5 mm wallboard/12.7 mm polystyrene) 25 mm (12.7 mm wallboard/12.7 mm polystyrene) 28 mm (9.5 mm wallboard/19 mm polystyrene) 32 mm (12.7 mm wallboard/19 mm polystyrene)	1800, 2400, 2438 (8 ft)

†Non-standard thicknesses available to order.

Table 50.5 PREFABRICATED GYPSUM WALLBOARD PANELS

Width (mm)	*Edge type*	*Thickness* (mm) 38	*Length* (mm) 1800	2286 (7.5 ft)	2350	2400	2438 (8 ft)
900	Tapered	*	*	*	*	*	*
	Square	*	*	*	*	*	*
1200	Tapered	*	*	*	*	*	*
	Square	*	*	*	*	*	*

Size. The sizes available for gypsum wallboard, gypsum plank, gypsum wallboard/expanded-polystyrene laminate and prefabricated gypsum wallboard panels are indicated in Tables 50.1 to 50.5. (Imperial lengths will be withdrawn as demand declines.)

Weight. Approximate weights of boards are:

Gypsum wallboard 9.5 mm thick, 6.5 kg/m^2 to 8.5 kg/m^2
12.7 mm thick, 9.5 kg/m^2 to 12.0 kg/m^2

Gypsum plank 19 mm thick, 14.0 kg/m^2 to 17.5 kg/m^2

Gypsum wallboard/expanded polystyrene laminate

22 mm thick, 6.5 kg/m^2 to 8.5 kg/m^2
25 mm thick, 9.5 kg/m^2 to 12.0 kg/m^2
28 mm thick, 6.5 kg/m^2 to 8.5 kg/m^2
32 mm thick, 9.5 kg/m^2 to 12.0 kg/m^2

PERFORMANCE

Fire protection

Due to the unique behaviour of the non-combustible gypsum core, when subjected to high temperatures plasterboard linings provide good fire protection.

Although plasterboard is combustible (when tested to BS 476: Part 4: 1970), the ivory (exposed) and grey (backing) surfaces of standard Gyproc wallboard, fireline board, plank and aluminium-foil backed (insulating grade) surfaces are designated Class 0 in accordance with the requirements of the Building Regulations. Class 0 is the highest standard required for restricting the spread of flame over wall and ceiling surfaces.

Periods of fire resistance for various constructions that incorporate gypsum plasterboard dry linings are given in the 'deemed-to-satisfy' schedules of the Building Regulations. Information on the fire resistances of other constructions (including the advantages of using gypsum fireline board to contribute to the fire resistance of certain structures) is available from British Gypsum Ltd.

Sound insulation

Information on the sound-insulation values (average sound-reduction indices) for various constructions incorporating plasterboard dry linings is available from British Gypsum Ltd. For maximum sound insulation to be obtained for a building element on site, all air paths such as perimeter cracks and gaps should be sealed. This can be achieved, for example, by the use of acoustical sealants around perimeters, or gypsum cove at the wall and ceiling angles. Ideally, a building element should be imperforate for optimum sound insulation.

Thermal insulation

Plasterboard, due to its aerated gypsum core, is a low-thermal-capacity lining material. Plasterboard dry-lining systems can improve the thermal insulation of a building element and therefore reduce heat loss. When used in intermittently heated buildings, a plasterboard dry lining responds to heating more quickly than denser plaster finishes and allows rooms to warm up more rapidly. This reduces the heating requirement and consequently saves fuel. Combined with efficient thermostats, plasterboard lining systems can provide greater heating efficiency.

Although legislation does not apply to existing housing, a large number of the existing houses in this country have masonry walls of low thermal insulation. A most effective method of upgrading these walls to present thermal insulation standards is to apply an internal lining of gypsum wallboard/expanded-polystyrene laminate boards. The boards can easily be fixed to pressure-impregnated timber battens, or (where the background is suitable) can be fixed with a special adhesive. Further information on these fixing systems is included later, under 'Site work'. The requirement for the vapour-check grade of board to be used or a vapour barrier treatment should be carefully considered. See 'Condensation' below for guidance.

Table 50.6 U–VALUES (W/m²°C) FOR TYPICAL EXTERNAL WALL CONSTRUCTIONS INCORPORATING PLASTERBOARD DRY LININGS

Construction	*Dot & dab lined*				*Metal-channel lined or fixed to timber battens*					
	Gypsum wallboard		*Gypsum wallboard/ expanded-polystyrene laminate*		*Gypsum wallboard*	*Gypsum insulating wallboard (20 mm min. airspace)*	*Gypsum wallboard/expanded-polystyrene laminate*			
	9.5 mm	12.7 mm	25 mm	32 mm	12.7 mm	12.7 mm	22 mm	25 mm	28 mm	32 mm
Brick/cavity/brick	1.26	1.23	0.87	0.75	1.09	0.92	0.81	0.80	0.71	0.70
Brick/cavity/concrete block *k* of block = 0.19 W/m °C†	0.87	0.85	0.66	0.59	0.79	0.69	0.63	0.62	0.57	0.56
Brick/cavity/concrete block *k* of block = 0.26 W/m °C†	1.0	0.98	0.73	0.65	0.89	0.77	0.69	0.68	0.62	0.61
Brick/cavity/concrete block *k* of block = 0.32 W/m °C†	1.07	1.05	0.77	0.68	0.95	0.82	0.73	0.72	0.65	0.64
Timber-framed wall clad externally with 22 mm thick weatherboarding on battens	–	–	–	–	–	–	–	0.95*	–	0.82*
Twin-leaf wall having a cavity not less than 50 mm wide with brick outer leaf and an inner leaf of timber framing	–	–	–	–	–	–	–	0.92*	–	0.80*

Existing 220 mm solid brick wall plastered	—	—	—	—	1.39	1.12	0.96	0.94	0.83	0.81
							Bonded direct with suitable foam polystyrene adhesive			
							1.16	1.14	0.97	0.95

*Fixed direct to timber framing (25 mm or 32 mm thicknesses of gypsum laminate are used where maximum contribution is required from the plasterboard element of the lining to the fire resistance of the wall).
†At a moisture content of 3 per cent by volume.

The 25 mm and 32 mm thicknesses of gypsum laminate are the most suitable for use with metal channel and the modified dot and dab system.

For the purpose of the calculations, the thickness of the brick outer and inner leaves has been taken as 105 mm, the density of the brickwork as 1700 kg/m^3, the thickness of the concrete blocks as 100 mm, the resistance of the cavity formed by the metal channel system as 0.18 $m^2\,°C/W$ (0.35 $m^2\,°C/W$ with insulating wallboard) and that formed by the dot and dab system as 0.08 $m^2\,°C/W$.

In order to reduce the risk of interstitial condensation, the use of the vapour-check grade of lining material or vapour barrier treatment and other precautions should be carefully considered for all constructions.

Thermal conductivity:
Gypsum wallboard and plank, $k = 0.16$ W/m °C.

Thermal resistance:
9.5 mm thick gypsum wallboard, $R = 0.06$ m^2 °C/W.
12.7 mm thick gypsum wallboard, $R = 0.08$ m^2 °C/W.
19 mm thick gypsum plank, $R = 0.12$ m^2 °C/W.

The above resistances are increased by 0.35 m^2 °C/W if the insulating grades of plasterboard (i.e. with polished aluminium-foil backings) are fixed facing an unventilated airspace of not less than 20 mm width.

22 mm thick gypsum wallboard/expanded polystyrene laminate, $R = 0.40$ m^2 °C/W
25 mm thick gypsum wallboard/expanded polystyrene laminate, $R = 0.42$ m^2 °C/W
28 mm thick gypsum wallboard/expanded polystyrene laminate, $R = 0.57$ m^2 °C/W
32 mm thick gypsum wallboard/expanded polystyrene laminate, $R = 0.59$ m^2 °C/W

Thermal transmittance (U-value):
See Table 50.6 for U-values of typical external wall constructions that incorporate plasterboard dry linings.

Effect of high and low temperatures

Gypsum plasterboards are unsuitable for use in temperatures above 49 °C. They can be subjected to freezing temperatures without risk of damage. Low temperatures may, however, cause the polyethylene backing of gypsum vapour-check wallboard to become brittle.

Condensation

To prevent condensation occurring, adequate heating and ventilation are necessary in addition to good thermal insulation. The absence of any of these requirements can result in condensation and mould growth on internal surfaces (surface condensation) or harmful condensation within the structure (interstitial condensation).

In intermittently heated buildings, plasterboard dry linings respond quickly to heating, and surface temperatures soon rise above the dew point. Consequently, the risk of surface condensation is reduced. This advantage improves as the thermal insulation of the dry-lining system is increased.

However, the use of an internal thermal-insulating lining may increase the risk of interstitial condensation by making other parts of the structure colder. In this situation a vapour-check or vapour-barrier treatment, and possibly other precautions, should be carefully considered.

In the case of external walls, if the wall is permeable the use of the vapour-check grade of plasterboard dry lining should be satisfactory. However, if the wall incorporates any materials of low permeability, such as dense concrete, very dense rendering, oil paint finishes or tiles, a vapour-barrier treatment must be provided. Wherever practical, however, the materials of low permeability should be removed or the permeability of the material improved.

Gypsum vapour-check wallboard can be upgraded to a vapour barrier by the use of one of the following treatments applied to continuous timber framing.

- Preformed water-vapour-resistant sealing strips (e.g. Denso tape as manufactured by Winn and Coales Ltd).
- Water-vapour-resistant gun-applied mastic sealant (compatible with polyethylene film of vapour-check wallboard).
- A thick application of chlorinated rubber paint.

The above treatments must be applied to the timber framing so as to ensure a complete and continuous contact with the polyethylene film at all ends and edges of the boards.

The vapour-check grade of gypsum wallboard/expanded-polystyrene laminate is only recommended for use in vapour-check applications. However, where this laminate is required to perform the function of a vapour barrier, the standard grade of laminate should be used and a suitable water-vapour-resistant treatment applied over the whole of its ivory-coloured surface (e.g. a proprietary water-vapour-resistant primer).

In tiled and slated pitched roofs without a vapour check, water vapour can pass through the plasterboard dry-lined ceiling and condense on the underside of the roofing felt or boarding. A vapour check at ceiling level can reduce this, but it should be used in conjunction with insulation otherwise the water vapour could condense on the vapour check itself, which is in immediate contact with the cold air of the roof space. It is also generally recommended that ventilation of roof spaces should be provided by means of suitable eaves ventilation. In addition, the roof timbers should be treated with a preservative.

For sources of further information on condensation control and the provision of vapour barriers, vapour checks and other precautionary measures, see the end of the section.

DESIGN AND PLANNING CONSIDERATIONS

Whilst plasterboard dry linings can be used in almost any type of building it is recommended that the following design and planning factors are taken into account.

- Ideally, plasterboard dry linings should be used throughout the building and not mixed with traditional plastering. However, this recommendation may mean that other design (and workmanship) factors may have to be considered: for example, to ensure that building elements of dwellings that are dry lined will provide a satisfactory standard of sound insulation as required by the Building Regulations.
- The total thickness of a dry-lined wall system will be a combination of the thickness of plasterboard used plus the cavity formed by the fixing method. This total thickness should be considered when determining the sizes of door frames, window sills, cupboard units, etc. Dry linings usually continue into window and door reveals, and with certain types of system the thickness of door and window frames may have to be increased.
- The cost of fixing and jointing is more expensive for cut boards than for uncut boards. Narrow reveals and other small areas involving the cutting of boards should therefore be kept to a minimum.
- Certain joinery work such as staircases and cupboards should be designed to be fixed to prepared grounds after completion of dry lining.
- The inclusion in the design of plasterboard lightweight partitions offers planning, organisational and cost advantages, since they employ a common material. (See Section 51.)
- If timber floors are specified the flooring should be butted to the walling.
- Where tapered-edge plasterboards are specified to provide smooth surfaces, jointing of the dry linings can be carried out by hand or by the Ames mechanical jointing system. The Ames system is particularly suitable for larger contracts. Comprehensive information on this jointing system, including the hiring of tools and training of operatives, is given in British Gypsum's literature.
- Fixings that would normally be made to a plastered wall can be made to plasterboard dry-lined walls.
- A dry-lined wall fixing system allows the inclusion of services behind the lining or minimises the depth of chasing required.

SITE ORGANISATION

Valuable time, labour and materials can be saved by careful site organisation to ensure that delivery times, quantities received and unloading points fit in with work requirements. It is very important that the dry-lining operations are carefully planned and prepared for, and that the correct type and amount of materials are ordered well in advance. Delivery costs for small loads are high, and it is both frustrating and uneconomical to wait for some of the materials to finish a job.

Access and storage areas. Ideally, gypsum materials should be kept in a dry, protected store or building. Hard access lanes and turning areas should be provided for vehicles (and be clearly signposted on larger contracts). Gypsum materials are normally delivered on rigid or articulated flat-bed lorries when supplied direct from works. Hard accesses should also be provided where distances are involved between the storage and work areas.

Off-loading. To avoid delay and congestion, materials should be off-loaded at one central point and taken to work areas as needed. Gypsum materials should not be moved from a protected store or building to a new area during wet weather without adequate protection.

Phasing of work. The times between storage and use should be kept as short as possible in order to reduce the risk of damage to the materials. Plasterboards can be taken into the area where they are to be used immediately after the roof has been completed and the building made watertight. The fixing of plasterboards can usually be carried out after internal loadbearing walls have been built, joists, structural floors and joiner's first fixings completed, and the first stage of installation of services carried out. The ceiling linings should be in position before dry lining external walls and partitions. It is desirable that glazing should be completed before drylining commences. The operations that follow are usually joiner's second fixings, service connections and decoration. Decoration should follow as soon as possible after board fixing and joint treatment.

By adopting the above sequence of operations the possibility of damage to the linings is reduced and therefore any repair work is kept to a minimum.

SITE WORK

Handling

Waste can be minimised and delays avoided by careful handling and storage of materials.

Plasterboards and prefabricated panels should always be carried on edge by two men and not dropped on corners or edges. They should not be carried with their surfaces horizontal. When a board or panel is stacked, or removed from a stack, its long edge should be placed down first before being lifted. They should not be slid over each other as this may scuff the surfaces.

Care should be taken in handling all dry-lining jointing materials and other accessories. Paper bagged materials can be easily punctured unless proper precautions are taken.

Storage

Boards and panels must be stacked flat on a level surface in a dry place, preferably inside a building, and protected from rising damp and inclement weather. They should be neatly aligned, and the height of the stack should not exceed 900 mm for plasterboards or 54 panels for 38 mm thick panels. The stacks should be situated away from areas where they are liable to impact or abuse, etc.

All dry-lining jointing materials and other accessories should be stored in dry, clean conditions and protected from rising damp – particularly for systems involving adhesives of any kind.

Cutting

Cutting of the boards and panels should be carried out with a fine-toothed saw. Alternatively, gypsum wallboard and plank can be cut by scoring with a sharp knife, snapping the core over a straight edge and then cutting though the paper on the opposite side. Plasterboard with an ivory-coloured surface should always be cut with the ivory side up.

Fixing plasterboard dry linings

The systems available for fixing plasterboard dry linings are briefly described as follows.

Fixing to timber framing. British Gypsum's literature gives recommendations on fixing to timber framing, including the requirements for nailing, spacing of timber supports and noggings, etc. DOE Advisory Leaflet No. 64, *Plasterboard dry linings,* provides additional information on fixing to timber battens.

Dot and dab fixing (trade name: Thistlebond). Standard tapered-edge gypsum wallboard can be applied to most dry internal brick, block or concrete walls, using small, perforated, bitumen-impregnated, fibreboard pads (or dots) aligned to set the levels. Dabs of an appropriate gypsum-plaster adhesive secure the boards to the wall, and special double-headed nails driven through the boards into the pads hold the boards in position until the dabs have set; the nails are subsequently removed. See *Figures 50.2 and 50.3.* This system cannot be used to fix insulating or vapour-check grades of gypsum wallboard. Using a modified technique, backgrounds suitable for the dot and dab fixing method can also be lined with 25 mm or 32 mm thicknesses of gypsum wallboard/expanded-polystyrene laminates to provide an internal lining and additional thermal insulation in one fixing operation. A gypsum plaster multi-purpose adhesive must be used in this application. The boards are skew-nailed into the set dabs of adhesive to maintain the fixing in the event of fire. Consult British Gypsum Ltd for fuller fixing details.

Metal channel fixing (trade name: Gyproc metal furring system). This is an advanced method of fixing tapered-edge plasterboards to most dry, internal brick, block or concrete walls. The wall is 'straightened' by applying and aligning a series of vertical proprietary lightweight metal channels bedded to the wall with an appropriate gypsum plaster adhesive. When all the channels on a run of wall have been applied and lined up, short lengths of channel are bedded horizontally between the channels at the top and bottom of the wall. When the adhesive has set, the plasterboards are screwed to all the channels using a power screwdriver. This system can be used to fix standard, insulating or vapour-check grades of gypsum wallboard or gypsum wallboard/expanded-polystyrene laminates. Consult British Gypsum Ltd for fuller fixing details.

Bonding to existing plastered walls. Gypsum wallboard/expanded-polystyrene laminates can be fixed direct to suitable, dry, plastered external walls using an

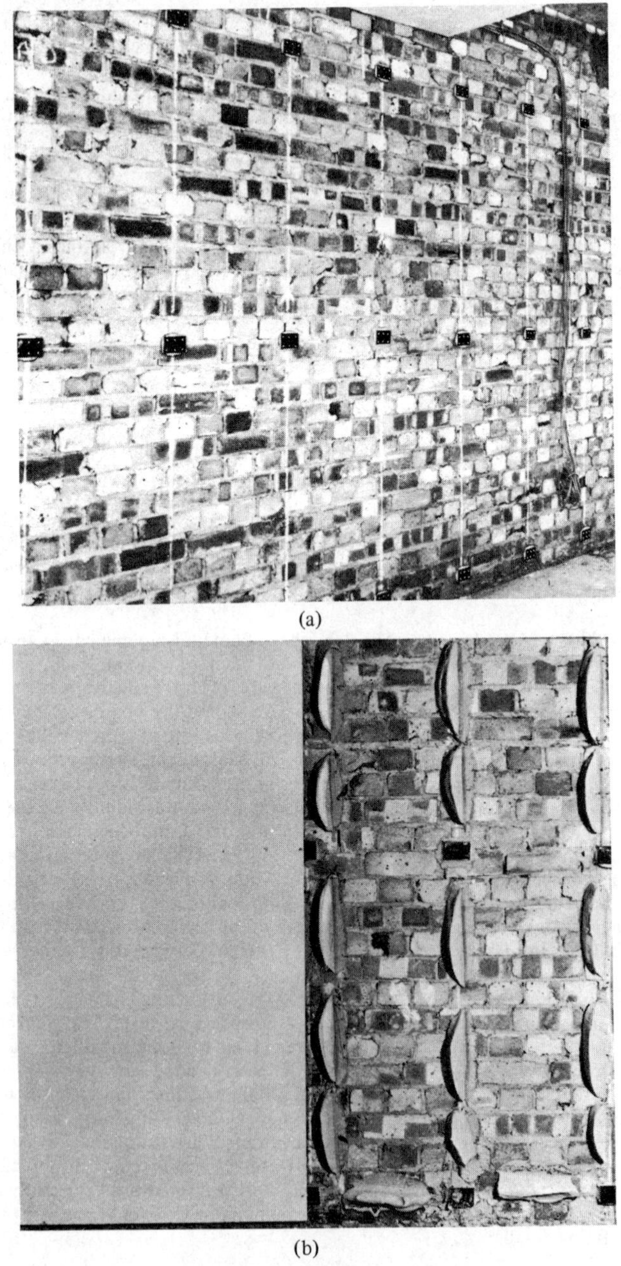

(a)

(b)

Figure 50.2 Thistlebond method of dry lining: (a) align and fix pads; (b) apply plaster dabs (courtesy British Gypsum Ltd)

Figure 50.3 Thistlebond method of dry lining: fix wallboard (courtesy British Gypsum Ltd)

appropriate foam polystyrene adhesive. A partial mechanical fixing is also required to maintain the fixing in the event of fire. Consult British Gypsum Ltd for fuller fixing details.

Manual jointing

Gypsum plasterboards used for dry linings are jointed according to the type of edge detail of the boards' long edges. See *Figure 50.1.*

Tapered edge. The abutting long edges of the boards form a depression, which is filled with joint filler and reinforced with joint tape. The joint is finished with joint finish, and then the whole surface of the board is given a final application of a thin slurry of joint finish. This treatment provides smooth, seamless surfaces ready for receiving most forms of direct decoration. Approximate requirements of hand-jointing materials are as follows:

Joint filler 22 kg/100 m^2
Joint tape 128–146 m/100 m^2
Joint finish 33 kg/100 m^2

The tools required are a 200 mm jointing applicator, a 50 mm taping knife and a jointing sponge.

The work is divided into five stages as follows:

1. Using the applicator, apply a continuous thin band of joint filler to the taper at the board joints. Cut the required length of joint tape and press it into the band of filler using the taping knife. Make sure the tape is firmly embedded and free from air bubbles, but with sufficient filler left under it to ensure good adhesion.
2. Using the applicator again, follow immediately with another coat of joint filler over the tape to fill the taper level with the surface of the boards. Before the filler stiffens, moisten the sponge with water and wipe off surplus material from the edges of the joint, taking care not to disturb the main joint filling. Rinse the sponge.
3. When the filler has set but not necessarily dried (about an hour), apply a thin layer of joint finish over the joint, using the applicator. Feather out the extreme edges with the dampened sponge.
4. When the first coat of joint finish has dried, apply another coat in a broad band over the joint and feather out as before.
5. When the joints are dry, even up the difference in surface texture between the board surface and the joint by distributing a slurry of joint finish over the whole wall or ceiling surface with the jointing sponge to leave a light, even texture.

Cut edge joints. Cut edges should ideally occur only at internal angles, but where they are unavoidable they can be taped as follows. First sandpaper the edges to remove paper burrs, then fill the gap between the boards flush with the board surface with joint filler. After the filler has set, make a thin application of joint finish over the joint and bed the tape in this as tightly as possible. Two more coats of joint finish should then be applied as described in steps three and four above, the first being allowed to dry before the second is applied.

Nail heads. Nail or screw heads should be 'spotted', i.e. the depression over the heads should be filled flush with the board surface. This is done in two thin coats, the first of joint filler, and the second with joint finish. This can be done while the main joints are setting or drying between stages.

Internal angles (including wall/ceiling angle). Fill any gap between the boards at angles with joint filler. Cut the tape to the required length and crease it firmly down the middle. Apply a thin band of joint finish to either side of the angle and press the tape into the angle ensuring that air bubbles are eliminated but with a thin layer of finish under it to ensure good adhesion. Apply a thin layer of joint finish over the tape, and feather out the edges with the damp sponge. When this has dried, make a further application of joint finish to both sides of the angle, and feather the edges again.

External angles. External angles can be reinforced with corner tape to give a clean sharp edge. Drywall angle bead is used where maximum protection is required.

Bevelled edge. The abutting long edges of the boards form a V-joint, which provides featured joints after bridging the base of the joint with joint finish.

Square edge. Used to provide featured or cover-strip joints.

Decoration

After joint treatment and nail or screw head spotting has dried, decoration, including any decorator's preparatory work, should follow with the minimum of delay. The manufacturers of paint or applied finishing products should be consulted regarding their recommendations for their products on gypsum plasterboards for dry lining. Gyproc wallboard primer has been specially formulated for use on the ivory-coloured surface of plasterboard. A single coat seals the surface and allows greater coverage and more uniform finish to be obtained from the paint used. Wallboard primer must be used before paperhanging to facilitate the subsequent removal of the paper. Where textured finishes are to be used, or where decoration is delayed, it is strongly recommended that wallboard primer is used as part of the decorative treatment.

Fixtures to plasterboard dry-lined walls

The recommendations given here are principally concerned with fixtures at the time of erection. Subsequent fixtures in use are rather more awkward, particularly where heavier fixtures are required, since steps must be taken to prevent deflection of the lining, for example.

With dry-lined walls there is normally a cavity to be bridged between the boards and the background; the fixing device should be long enough to allow for this and to penetrate well into the solid wall or background.

It is extremely important that the drilling of the board and fixing of the fixing device is carried out as carefully and accurately as possible, otherwise the performance of the fixing can be considerably affected. The manufacturer's recommendation as to drill size for any device is particularly important, and holes should always be drilled and not punched.

Table 50.7 gives a range of fixing methods and devices for installation of fixtures to plasterboard linings. It should be noted that the references to Fischer and Rawlplug devices are not exhaustive and other similar devices are commercially available, but in all cases manufacturers' recommendations should be closely followed.

Where heavy to medium weight fixtures are to be made to walls with plasterboard linings, and the loads are to be applied to the plasterboard face through narrow or small-area bearings such as wall brackets or coat hooks, then timber face plates should be used.

Table 50.7 FIXING METHODS AND DEVICES FOR USE WITH PLASTERBOARD DRY LINING

Construction	*Heavy (e.g. heating units, water heaters)*	*Medium to heavy (e.g. radiators, wall cupboards)*	*Medium (e.g. floor cupboards, light fittings)*	*Medium to light (e.g. wall mirrors)*	*Light (e.g. small pictures)*
Thistlebond system of dry lining	Extra plaster when fixing boards to give solid area at fixing point Plugs and screws can be used in the above condition provided plug is in background. If no solid area has been provided, use Fischer stud screw and plug with reversed-type BU flange nuts. Rawlbolts with spacers. Fischer RS plugs*.	Provision as with heavy fixing. Use plugs and screws. If not solid, plugs and screws with spacers.	Rawlnut 316. Fischer NA 8 × 40 rivet anchor.	Rawlnut 316. Fischer NA 8 × 40 rivet anchor.	Steel pin and hook.
Gyproc metal furring system	Rawlbolts and spacers. Fischer stud screws and plugs with reversed-type BU flange nuts.	Fischer RS plugs*. Plugs and screws with spacers.	Rawlnut 316. Fischer NA 8 × 40 rivet anchor	Rawlnut 316. Fischer NA 8 × 40 rivet anchor. PK screws to metal furrings.	Steel pin and hook
Thistlebond (thermal board) system	As for metal furring.	As for metal furring.	As for metal furring.	As for metal furring	As for metal furring system.
Gyproc metal furring system with thermal board	As for metal furring	As for metal furring	As for metal furring	PK screws to metal furrings using spacer.	As for metal furring system.

Health and safety

Precautions to be observed in the use and fixing of plasterboard dry linings, systems and accessories should be obtained from British Gypsum's relevant literature.

The installation of electrical services should be carried out in accordance with the recommendations of the Institution of Electrical Engineers. The cables should be protected by conduit, or other suitable precautions must be taken, to prevent abrasion where these pass through metal framing.

FURTHER INFORMATION

BS 5250: 1975, *Code of basic data for the design of buildings: the control of condensation in dwellings*

DOE advisory leaflets:
No. 34, *Thermal insulation*
No. 45, *Warmth without waste*
No. 61, *Condensation*
No. 64, *Plasterboard dry linings*
No. 79, *Vapour barriers and vapour checks*

Performance details of plasterboard dry linings are also given in British Gypsum's relevant literature.

51 PLASTERBOARD LIGHTWEIGHT PARTITIONS AND SEPARATING WALLS

INTRODUCTION 52–2

TYPES OF LIGHTWEIGHT PARTITION AND SEPARATING WALL 51–3

PERFORMANCE 51–9

SITE WORK 51–13

PLASTERBOARD LIGHTWEIGHT PARTITIONS AND SEPARATING WALLS 51

51 PLASTERBOARD LIGHTWEIGHT PARTITIONS AND SEPARATING WALLS

R.W. SMITH, MSAAT
Technical Literature Officer, British Gypsum Ltd

This section (which should be read in conjunction with Section 50) describes the plasterboard lightweight partitions and separating walls that are in everyday use. The sound-insulation and fire-resistance performances of these constructions are given, with information on the Building Regulations applicable to sound insulation of dwellings.

INTRODUCTION

When compared with masonry constructions, the plasterboard lightweight partitions and separating-wall systems described in this section offer several distinct advantages, which can be summarised as follows:

- The constructions are very lightweight, and this can result in substantial savings on the structural design. The approximate weights of the partitions start as low as 23 kg/m² and the separating walls at 58 kg/m². The range of weights for both types of construction is much less than for masonry constructions of comparable thickness.
- Additional floor space can be achieved by the selection of 50 mm, 57 mm and 63 mm thick partitions. They can also prove to be very beneficial in rehabilitation work where floor space is limited.
- The constructions are erected quickly and simply.
- They can provide economical constructions. The partitions are of minimal thickness and this can save on the cost of door frames.
- A demountable partition system is available. The erection sequence is simple and the partition can be easily dismantled and, where suitable, re-used.
- The installation of services is generally very simple.
- Although all the proprietary constructions described in this section are non-loadbearing, a plasterboard dry-lined timber-framed construction can be used where there is a loadbearing requirement.
- The constructions provide good standards of sound insulation. The range of sound-insulation performances (average sound reduction indices, 100–3150 Hz) varies between 29 dB and 52 dB. This means that most sound-insulation requirements can be satisfied in virtually all types of building – with the additional advantages of using a plasterboard dry-lined construction.
- In addition to their other advantages, the constructions provide a range of fire-resistance performances between ½ hour and 2 hours. The demountable partition system provides standards of fire resistance not usually achieved by other systems.
- For work in the Inner London Area, the strength and stability requirements for lightweight separating walls require that they provide an adequate resistance to pressure. An explosion test has shown that the 211 mm Gyproc jumbo metal stud separating wall provided considerably more resistance than a 215 mm thick solid brick wall.

The constructions described in this section provide a versatile range of systems that can be used equally in simple home improvements or in modern building construction.

Limitations

The constructions are not suitable for use in areas subject to continuously damp or humid conditions.

TYPES OF LIGHTWEIGHT PARTITION AND SEPARATING WALL

Four types of plasterboard lightweight partition are in everyday use.

Prefabricated gypsum wallboard cellular-core partition (trade name: Paramount dry partition). Non-loadbearing, this is constructed on site from prefabricated panels that incorporate two Gyproc wallboards separated by, and bonded to, a core of

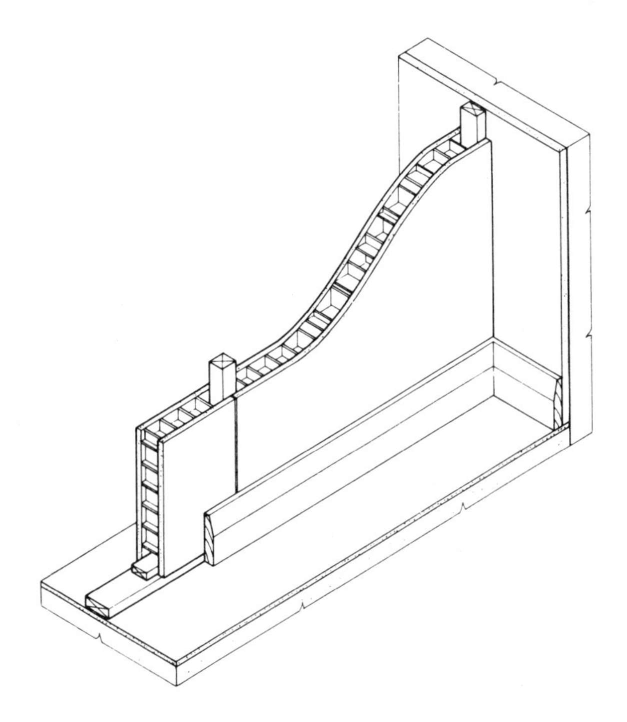

Figure 51.1 Paramount dry partition

cellular construction; see *Figure 51.1.* The panels are fixed to timber core battens at the perimeter and joints. Three types of panel are available:

- Ivory-coloured faces with tapered edges, which when correctly jointed give smooth, seamless surfaces ready to receive most forms of decoration.
- Ivory-coloured faces with square edges for featured joints.
- Grey faces with square edges for gypsum plastering.

The partition can be erected as either a single- or double-leaf construction, depending on the performance requirements. The panels are manufactured to BS 4022:1970, *Prefabricated gypsum wallboard panels.*

Laminated gypsum plasterboard partition (trade name: Gyproc laminated partition). This non-loadbearing partition consists of three layers of plasterboard bonded together on site with a suitable gypsum-based adhesive (Gyproc bonding compound or Gyproc multi-purpose adhesive). Alternatively, two layers of plasterboard (Gyproc plank) may be bonded and fixed on each side of a timber perimeter frame with a glass wool blanket provided in the cavity – *Figure 51.2.* Boards with ivory-coloured faces and tapered edges are used for the outer layers which when correctly jointed give smooth, seamless surfaces ready to receive most forms of decoration.

Metal-stud partitions (trade names: Gyproc metal-stud partition and Gyproc jumbo metal-stud partition). These consist of lightweight, non-loadbearing metal framing incorporating metal studs at 600 mm centres. Either one or two layers of plasterboard are screw-fixed to each side of the framework. Gyproc self-drilling and self-tapping screws are used to increase the ease and speed of erection, particularly as

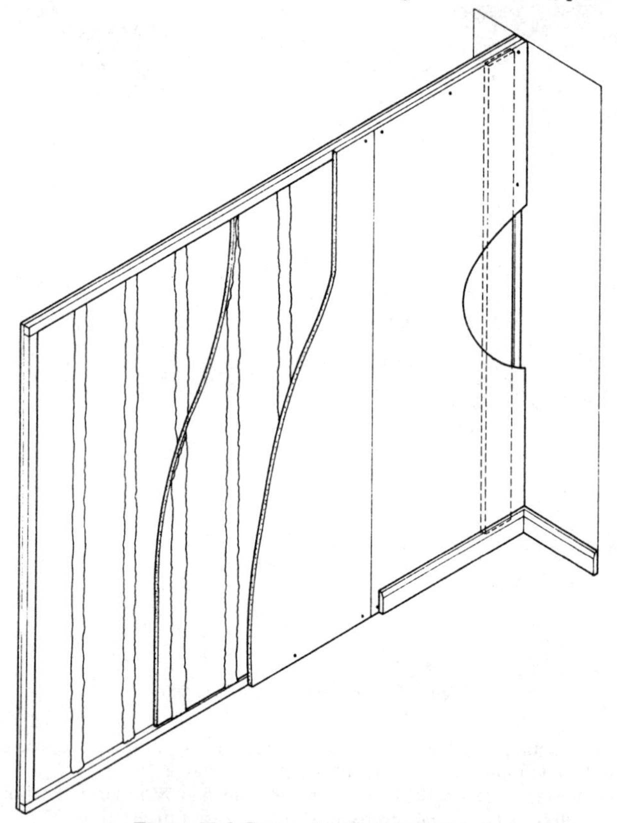

Figure 51.2 Gyproc laminated partition

screws can be inserted with an electric screwdriver having a magnetic bit with adjustable screw-depth control.

See Table 51.1 for the improvements in the contribution to fire resistance that the use of gypsum fireline board can make to metal-stud partitions.

A demountable partition system based on the metal-stud partition components is also available (trade name: Gyproc demountable partitions).

Timber-stud partitions. Loadbearing or non-loadbearing timber-stud partitions that incorporate single- or double-layer plasterboard linings nail-fixed to the timber framework are widely used.

Separating walls

The following non-loadbearing plasterboard lightweight separating wall constructions have the capability of satisfying the sound-insulation performance requirements of Building Regulations for dwellings, subject to height limitations and the suitability of their associated structures.

Metal-stud separating wall (trade name: Gyproc metal-stud separating wall). An inner layer of one 19 mm Gyproc plank is screw-fixed horizontally and an outer layer of one 12.7 mm Gyproc wallboard is screw-fixed vertically to the outside faces of two independent frames incorporating 48 mm Gyproc metal studs. A 60 mm glass wool blanket (12 kg/m^3) is provided in the cavity. See *Figure 51.3*.

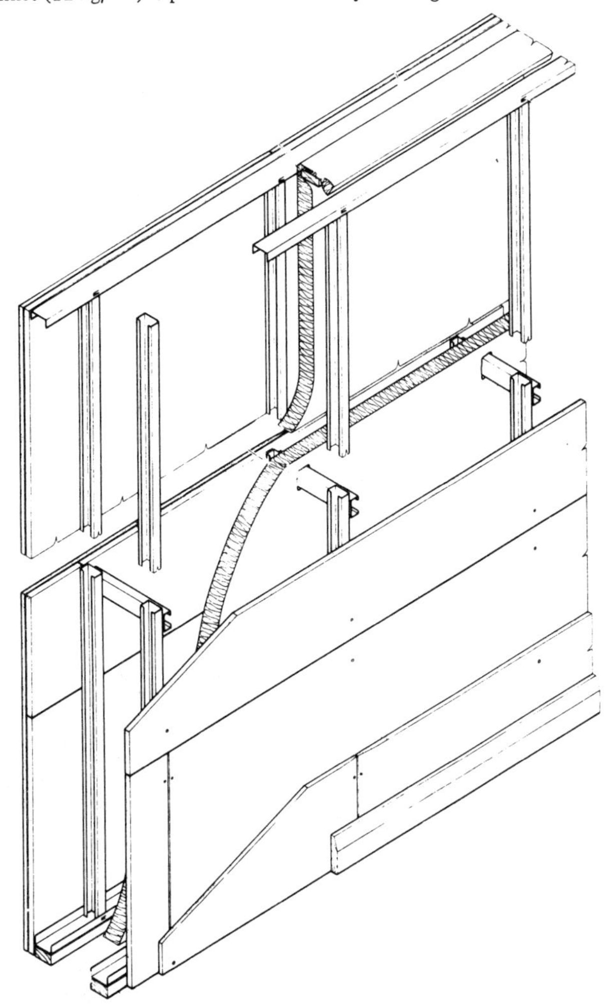

Figure 51.3 Gyproc metal-stud separating wall

Table 51.1 PARTITION AND SEPARATING–WALL CONSTRUCTIONS

Detail	*Construction*	*Dimensions:* overall thickness	maximum height	length	*Approx. weight* (kg/m²)	*Fire resistance** (h)	*Average sound-reduction index** 100-3150 Hz (dB)
PARTITIONS							
Paramount dry partitions	50 mm thick	50 mm	2400 mm	no limit	23.0	½	29
	57 mm thick	57 mm	2700 mm	”	23.0	½	29
	63 mm thick	63 mm	3600 mm	”	27.0	½	30
	Double-leaf partitions:						
	Two 50 mm in contact	100 mm	2400 mm	”	46.0	1	33
	(a) Two 57 mm with 50 mm cavity	164 mm	2700 mm	”	46.0	1	39
	(b) Two 63 mm with 50 mm cavity	176 mm	3600 mm	”	54.5	1	42
	As (a) above, but with 25 mm minimum glass wool blanket in cavity.	164 mm	2700 mm	”	46.5	1	45
	As (b) above, but with 25 mm minimum glass wool blanket in cavity.	176 mm	3600 mm	”	55.0	1	47
Gyproc laminated partitions	12.7 mm Gyproc wallboard each side of 19 mm Gyproc plank	50 mm	2600 mm	”	43.5	1½	32
	Three layers of 19 mm Gyproc plank	65 mm	2800 mm	”	54.5	2	31
			3200 mm	7300 mm†			
	Two layers of 19 mm Gyproc plank each side of 75 × 50 mm timber perimeter framing, with 25 mm minimum glass wool blanket in cavity	150 mm	2600 mm	6000 mm†	73.0	1½	48
			2800 mm	3700 mm†			
Gyproc metal stud partitions	12.7 mm Gyproc wallboard each side of 48 mm studs	75 mm	2600 mm	no limit	24.0	½	35
	As above, but with 25 mm minimum glass wool blanket in cavity				24.5	½	41
	12.7 mm Gyproc fireline board each side of 48 mm studs, with 25 mm minimum glass wool blanket in cavity	75 mm	2600 mm	”	25.0	1	41
	Two layers of 12.7 mm Gyproc wall-board each side of 48 mm studs	100 mm	3600 mm	”	46.0	1	43

	Two layers of 12.7 mm Gyproc fire-line board each side of 48 mm studs	100 mm	3600 mm	,,	46.5	2	43
	As above, but with 25 mm minimum glass wool blanket in cavity				46.5	1	44
	12.7 mm Gyproc wallboard each side of 70 mm studs	98 mm	3000 mm	,,	24.5	½	35
	As above, but with 25 mm minimum glass wool blanket in cavity				25.0	½	41
	12.7 mm Gyproc fireline board each side of 70 mm studs, with 25 mm minimum glass wool blanket in cavity	98 mm	3000 mm	,,	25.5	1	41
	Two layers of 12.7 mm Gyproc wall-board each side of 70 mm studs	124 mm	3600 mm	,,	46.5	1	44
Gyproc jumbo metal stud partitions	Two layers of 12.7 mm Gyproc wall-board each side of 146 mm studs	200 mm	4200 mm+	,,	47.5	1½	47
	As above, but with 25 mm glass wool blanket in cavity				48.0	1½	48
	Two layers of 12.7 mm Gyproc fire-line board each side of 146 mm studs	200 mm	4200 mm	,,	48.0	2	47
Gyproc demountable metal stud partitions	12.7 mm Gyproc wallboard each side of 70 mm studs (alternate studs boxed)	98 mm	2700 mm	,,	24.5	½	36
	Two layers of 12.7 mm Gyproc wall-board each side of 48 mm studs (alternate studs boxed)	100 mm	3000 mm	,,	46.0	1	43
Timber stud partitions (75 × 50 mm non-loadbearing studs at 600 mm centres)	Lined both sides with: 12.7 mm Gyproc wallboard	100 mm			27.0	½	34
	As above, but with 25 mm minimum glass wool blanket in cavity	100 mm			27.5	½	37
	19 mm Gyproc plank	113 mm			38.0	1	35
	Two layers of 12.7 mm Gyproc wall-board	126 mm			49.0	1½	39
	As above, but with 25 mm minimum glass wool blanket in cavity	126 mm			49.5	1½	41

Table 51.1 *continued*

Detail	*Construction*	*Dimensions:* *overall thickness*	*maximum height*	*length*	*Approx. weight* (kg/m²)	*Fire resistance** (h)	*Average sound-reduction index** *100–3150 Hz* (dB)
SEPARATING WALLS							
Gyproc metal stud separating wall	As described in text	200 mm or 300 mm	2700 mm	,,	60.0	1	51
Gyproc jumbo metal stud separating walls	As described in text	211 mm	4200 mm‡	,,	58.0	1½	52
	As described in text	224 mm	4200 mm‡	,,	68.5	2	50
Timber framed separating walls	Two separate timber frames consisting of 75 X 50 mm timber studs at 600 mm centres. Three layers of 12.7 mm Gyproc wallboard on each side. Timber frames spaced 70 mm apart.	296 mm			70	1	51 (field test)
	As above, but with one layer of Gyproc plank and one layer of 12.7 mm Gyproc wallboard on each side; 25 mm minimum glass wool blanket (density 12 kg/m³ in cavity. Timber frames spaced 80 mm apart.	293 mm			63	1	53 (field test)

*The fire resistances and average sound-reduction indices are applicable to imperforate constructions using tapered-edge Gyproc wallboard, Gyproc fireline board or Gyproc plank as the outer layers, the joints of which are taped and filled in the recommended manner (not applicable to Gyproc demountable metal-stud partitions). All plasterboard joints in double- or multi-layer lining systems should be staggered.

†The maximum lengths can be exceeded if vertical stiffeners are inserted.

‡British Gypsum's advice should be sought if this height is to be exceeded.

Jumbo metal-stud separating walls (trade name: Gyproc jumbo metal-stud separating walls). An inner layer of one 19 mm Gyproc plank is screw-fixed horizontally and an outer layer of one 12.7 mm Gyproc wallboard is screw-fixed vertically to each side of a metal frame incorporating 146 mm Gyproc jumbo metal studs at 600 mm centres. A 25 mm minimum thickness glass wool blanket (12 kg/m^3) is provided in the cavity.

A second version of the above wall is also available, but with three layers of 12.7 mm Gyproc wallboard screwed to each side of the metal frame. No cavity infill is required.

Masonry and timber framed separating walls. Field test evidence, in the form of acoustic test reports, is available for a wide range of masonry and timber-framed separating walls incorporating plasterboard dry linings that have provided standards of sound insulation satisfying Building Regulation requirements. It is essential that all the mortar joints of masonry walls that incorporate plasterboard dry linings are completely filled, and that bricks or blocks of adequate density and correct acoustic composition are selected.

Tools and accessories

Plasterboard products are complemented by a comprehensive range of dry-lining tools and accessories for the fixing, jointing and finishing operations. The tools that are particularly applicable to this section are briefly described below. (Information on the accessories available is available from British Gypsum.)
Stud interlocking tool punches through stud to lock two lengths of steel together when boxed or extended studs are needed.
Stud shear cuts 48 mm, 60 mm and 70 mm metal studs, reducing cutting time by 75 per cent; drilled base for bench mounting.
Crimping tool punches and locks metal stud and channel sections together; used when crimping is required at base and head.
Applicator brush, in tough nylon, for applying bands of Gyproc bonding compound or Gyproc multi-purpose adhesive when building laminated partitions.
Mastic gun, 1 litre capacity, for use with Gyproc resilient wallboard adhesive and Gyproc acoustical sealant.

Dimensions and weights

The overall thicknesses, maximum heights and approximate weights of the partition and separating-wall constructions are given in Table 51.1.

The panel sizes and edge types available for the erection of Paramount dry partitions are indicated in Table 51.2 by an asterisk. (Imperial lengths will be withdrawn as demand declines.) The sizes available for Gyproc wallboard and Gyproc plank, and the approximate weights of the boards, are given in Section 50.

PERFORMANCE

Fire protection

Due to the unique behaviour of the non-combustible gypsum core when subjected to high temperatures, plasterboard linings provide good fire protection.

Although plasterboard is combustible (when tested to BS 476: Part 4: 1970), the ivory (exposed) and grey (backing) surfaces of standard Gyproc wallboard, fire-line board and plank are designated Class 0 in accordance with the requirements of the Building Regulations. Class 0 is the highest standard required for restricting the spread of flame over wall and ceiling surfaces.

Table 51.2 PARAMOUNT DRY PARTITION

Width (mm)	*Edge type*	*Thickness* (mm) 57	63	*Length* (mm) 1800	2286 (7.5 ft)	2350	2400	2438 (8 ft)	2700	3000†	3300†	3600†
600	Tapered	*	*	*	*	*	*	*	*	*	–	–
	Square	*	*	*	*	*	*	*	*	*	–	–
900	Tapered	*	*	*	*	*	*	*	*	*	–	–
	Square	*	*	*	*	*	*	*	*	*	–	–
1200	Tapered	*	*	*	*	*	*	*	*	*	*	*
	Square	*	*	*	*	*	*	*	*	*	–	–

50 mm thickness is available in 900 mm width, 2350 mm and 2400 mm lengths, with tapered or square edges.

† 63 mm thickness only.

Types. For direct decoration, with ivory faces and tapered or square edges. (For plastering, with grey faces and square edges.)

The fire resistances of the partitions and separating walls as given in Table 51.1 are the results of tests carried out on imperforate constructions to BS 476:Part 1: 1953 or BS 476:Part 8:1972, or assessments as provided by appropriate authorities. The advantages of using Gyproc fireline board to contribute to the fire resistances of metal-stud partitions should be particularly noted.

Sound insulation

The average sound-reduction indices (100–3150 Hz) as given in Table 51.1 are the results of laboratory tests carried out on imperforate constructions to BS 2750:1956. The separating wall constructions have the capability of satisfying the sound-insulation performance standards of the Building Regulations, subject to the suitability of their associated structures.

Where necessary, the sound insulation of new or existing plasterboard-lined partitions and separating walls can be improved by resiliently fixing an additional layer of 9.5 mm or 12.7 mm thick Gyproc wallboard to one or both sides. The average sound-reduction indices for the imperforate partition constructions given in Table 51.3 are the results of laboratory tests to BS 2750:1956.

For maximum sound insulation to be obtained for a building element on site, all air paths such as perimeter cracks and gaps should be sealed – for example, by the use of an acoustical sealant (e.g. Gyproc acoustical sealant) around perimeters, or gypsum cove at the wall and ceiling angles. Ideally, a building element should be imperforate for optimum sound insulation.

Sound insulation legislation

The Building Regulations 1976 and the Building Regulations (Northern Ireland) 1977. Part G of these regulations gives specific requirements to limit sound transmission into dwellings, either from other adjoining buildings or from other parts of the same building that do not form part of the dwelling. Separating walls and floors are required, in conjunction with their associated structures, to provide adequate resistance to the transmission of airborne sound. Separating floors are also required to resist the transmission of impact sound, with the exception of certain floors described in Regulation G3 (2).

Two deemed-to-satisfy provisions are given in the regulations for ensuring that separating walls and floors provide adequate standards of sound insulation. Regulations G2(1), G4(1) and G5(1) define performance standards against which the field performances of the walls and floors can be compared. The standard for walls is 'House party wall grade' and 'Grade 1' for floors (as defined in CP 3 : Chapter III : 1972). In order to allow for small variations of performance, a total adverse deviation of up to and including 23 dB is permitted. As alternatives to the performance standards, certain traditional separating-wall and floor constructions are also deemed-to-satisfy the sound-insulation requirements. The performance standards of the regulations are given in Table 51.4.

The Building Standards (Scotland) (Consolidation) Regulations 1971. The sound-insulation requirements as defined in Part H are similar to those for England, Wales and Northern Ireland (see above) with a few exceptions, including that a lower level of airborne sound insulation is permitted for the separating walls between flats, i.e. Grade 1 (as defined in CP 3 : Chapter III : 1972). The regulations define mandatory performance standards against which the field performances of separating walls and floors must be compared. Alternatively, certain traditional wall and floor constructions are deemed-to-satisfy the sound insulation requirements. The performance standards of the regulations are given in Table 51.4.

Table 51.3 AVERAGE SOUND REDUCTION INDICES (100–3150 Hz)

Partition construction	*Standard construction*	*9.5 mm Gyproc wallboard fixed using Gyproc resilient wallboard adhesive*		*12.7 mm Gyproc wallboard fixing using Gyproc resilient wallboard adhesive*	
		1 layer on one side of partition	*1 layer on each side of partition*	*1 layer on one side of partition*	*1 layer on each side of partition*
50 mm Paramount dry partition	29 dB	32 dB	36 dB	32 dB	36 dB
57 mm Paramount dry partition	29 dB	32 dB	36 dB	32 dB	36 dB
63 mm Paramount dry partition	30 dB	32 dB	35 dB	33 dB	37 dB
50 mm Gyproc laminated partition	32 dB			36 dB	
65 mm Gyproc laminated partition	31 dB			37 dB	
48 mm Gyproc metal studs at 600 mm centres, with a single layer of 12.7 mm Gyproc wallboard screwed to each side	35 dB			39 dB	43 dB
70 mm Gyproc metal studs at 600 mm centres, with a single layer of 12.7 mm Gyproc wallboard screwed to each side	35 dB			42 dB	45 dB
75 mm × 50 mm timber studs at 600 mm centres, with a single layer of 12.7 mm Gyproc wallboard nailed to each side	34 dB			40 dB	42 dB

All the above partitions were tested with Gyproc tapered-edge wallboard as the additional layer, the joints of which were taped and filled in the recommended manner. The vertical joints of the additional layer were staggered with those of the base layer.

Table 51.4 LEVELS OF SOUND INSULATION IN DWELLINGS, SEPARATING WALLS AND FLOORS

Frequency (Hz)	*Airborne sound*		*Impact sound*
	Sound reduction (dB)		*Octave band sound pressure level* (dB)
	(a) *England, Wales, Northern Ireland and Scotland: separating walls.* See column (b) for separating walls between flats in Scotland.	(b) *Scotland: separating walls and floors between flats.* *England, Wales and Northern Ireland: separating floors.*	*England, Wales, Northern Ireland and Scotland: separating floors.*
100	40	36	63
125	41	38	64
160	43	39	65
200	44	41	66
250	45	43	66
315	47	44	66
400	48	46	66
500	49	48	66
630	51	49	65
800	52	51	64
1000	53	53	63
1250	55	54	61
1600	56	56	59
2000	56	56	57
2500	56	56	55
3150	56	56	53

London Building (Constructional) By-Laws 1972. Although the by-laws exercise no control over sound insulation, the Department of Architecture and Civic Design of the Greater London Council recommend that the requirements of the Building Regulations should be adopted. The GLC Development and Materials Bulletin No. 15 (second series) dated May 1968 refers to this subject. It is thought that the provisions of Part G (and Part F) of the Building Regulations 1976 will be eventually applied to work in the inner London area.

SITE WORK

For information on the handling, storage and cutting of plasterboards and certain accessories, and also the decoration of plasterboard dry linings, refer to Section 50.

Brief descriptions of the partition and separating-wall systems are given earlier, but comprehensive installation details, including information on the fixing of plasterboard dry linings to timber-framed constructions, should be obtained from British Gypsum's current literature.

Plasterboards used for dry linings are jointed according to the type of edge detail of the boards' long edges; see Section 50.

Table 51.5 FIXING METHODS AND DEVICES FOR USE WITH LIGHTWEIGHT PARTITIONS

Construction	*Heavy (e.g. heating units, water heaters)*	*Medium to heavy (e.g. radiators, wall cupboards)*	*Medium (e.g. floor cupboards, light fittings)*	*Medium to light (e.g. wall mirrors)*	*Light (e.g. small pictures)*
Gyproc laminated partition		Timber batten inserted at erection. Fischer Pi10 mushroom plugs.	Fischer Pi6 mushroom plugs.	Fischer type S-RS plugs.	Steel pin and hook.
Paramount dry partition		Timber plugs inserted at erection.	Rawlnut 316. Fischer NA 8 × 40 rivet anchor.	Rawlnut 316. Fischer NA rivet anchors.	Steel pin and hook.
Gyproc metal stud partition	Metal and timber framing at erection stage.	Gyproc fixing channel and/or timber noggings at erection stage. PK screws to studs.	PK screws to studs. Fischer NA 10 × 55 rivet anchor (double boarded). Fischer NA 8 × 40 rivet anchor (single boarded). Rawlnut 316L double. Rawlnut 316 single.	PK screws to studs, as with medium.	Steel pin and hook.

Fixtures to plasterboard dry-lined lightweight partitions

The recommendations given here are principally concerned with fixtures at the time of erection. Subsequent fixtures in use are rather more awkward, particularly where heavier fixtures are required, as this will often require the identification of the basic frame in hollow partitions.

When timber- or metal-framed partitions are used, fixings can be made into the timber or metal studs, or to timber noggings or metal fixing channels if medium to heavy weight fixtures are required between the studs.

It is extremely important that the drilling of the board and fixing of the fixing device is carried out as carefully and accurately as possible, otherwise the performance of the fixing can be considerably affected. The manufacturer's recommendations on the drill size for any device are particularly important, and holes should always be drilled and not punched.

Table 51.5 gives a range of fixing methods and devices for installation of fixtures to lightweight partitions. It should be noted that the references to Fischer and Rawlplug devices are not exhaustive and other similar devices are commercially available, but in all cases manufacturers' recommendations should be closely followed.

Where heavy to medium weight fixtures are to be made to lightweight partitions, and the loads are applied to the plasterboard face through narrow or small area bearings such as wall brackets or coat hooks, then timber face plates should be used.

Information is available from British Gypsum on the following partitions' abilities to receive and hold fixings for lightweight and heavyweight fixtures:

- 57 mm thick Paramount dry partition;
- 75 mm thick Gyproc metal stud partitions incorporating 48 mm wide studs with a single-layer 12.7 mm Gyproc wallboard lining on each side;
- 50 mm thick Gyproc laminated partition.

Health and safety

Precautions to be observed in the use and fixing of plasterboard dry linings, systems and accessories should be obtained from British Gypsum's relevant literature.

With metal-stud partition and separating-wall constructions the installation of electrical services should be carried out in accordance with the recommendations of the Institution of Electrical Engineers. The cables should be protected by conduit, or other suitable precautions must be taken, to prevent abrasion where these pass through the metal framing.

52 FLOOR AND ROOF SCREEDS

INTRODUCTION 52–2

LAYING A SCREED 52–2

PROBLEMS AND DEFECTS 52–4

FLOOR AND
ROOF SCREEDS 52

52 FLOOR AND ROOF SCREEDS

P.H. PERKINS, CEng, FASCE, FIMunE, FIArb, FIPHE, MIWE
Senior Advisory Engineer, Cement and Concrete Association

This section discusses in detail how screeds should be laid to give a trouble-free life. Consideration is given to defects that can occur, and to the special problems of roof screeds.

INTRODUCTION

A screed is defined in BS 4049, *Glossary of terms applicable to internal plastering, external rendering and floor screeding,* as 'a layer of mortar or other material applied to a sub-floor and brought to a defined level'.

The definition is sufficiently wide to cover almost any type of floor finish, and consequently has given rise to a number of misunderstandings, which unfortunately sometimes result in what may be termed 'floor failures'. In fact, the normal cement/sand screed as laid by a sub-contractor in building work in the UK is quite unsuitable to be used as the final wearing surface to the floor. Nor should it form the basis for high-quality cement-based toppings such as granolithic. If a granolithic topping is laid on a cement/sand screed, the granolithic will curl and de-bond itself from the screed below, with all the trouble that this will cause.

The author therefore prefers to define a screed as a layer of mortar that is laid on a structural floor/roof slab, but that does not form the final wearing surface. Materials such as concrete and resin mortars that are laid on a structural slab to form the wearing surface should be referred to as 'toppings'. It is true, but unfortunate, that floors and roofs give rise to more trouble than any other parts of a building.

The normal purpose of a screed is to bring the surface of a concrete floor/roof slab to a sufficiently well regulated level and surface finish so that the final covering can be laid on it. This 'final covering' may consist of a carpet, sheet vinyl, vinyl tiles, ceramic tiles etc. The precision required for finishing the surface of a screed depends on the type of floor covering that will be laid on it. It is the normal practice to use a skim coat (underlayment) in order to provide the screed with the smooth, closed finish needed to receive thin sheet and tile floor covering, and to a lesser degree carpet. For ceramic and concrete tiles a skim coat should not be needed.

If the designer/specifier of the floor requires the contractor to work to specific tolerances, these must be written into the specification and the tenderers' attention drawn to them. The relevant Code of Practice is CP 204, *In situ floor finishes,* and clause 120 recommends 'an appropriate limit for localised variations in level for a nominal flat floor' as ±3 mm under a 3 m straight edge.

LAYING A SCREED

Probably the most satisfactory way of laying a screed is to lay it monolithically with the base concrete; that is, it should be laid within 2½–3 hours of the completion of the compaction and finishing of the structural floor slab. See *Figure 52.1.* However, this is hardly ever done in practice, because subsequent building operations are likely to damage the screed, and so it is usual for the screed to be laid quite late on in the building programme when the concrete slab has hardened. See *Figure 52.2.*

The following are the basic requirements necessary to obtain a satisfactory cement/sand screed:

- The cement mortar (cement and fine aggregate) should be batched by weight and not by volume.
- The fine aggregate should be well-graded concreting sand complying with zone 1 or 2 of BS 882.
- The water/cement ratio (which should include an allowance for moisture in the sand) should not exceed 0.5, i.e. 25 litres of water to each 50 kg of cement. If this does not give the necessary workability for proper compaction, a plasticiser can be used.
- The cement, sand and water should be properly mixed in a mechanical mixer.
- The mix proportions should be in the range 1 : 3 to 1 : 4 by weight.

The screed material should be laid between screeding battens and brought to a level surface and well compacted. Light beam tampers should be used for this. The usual method of just patting the screed down with a piece of batten is not good enough. It should be remembered that cement/sand mortar is similar to concrete in as much that, however high the quality of the material when it comes from the mixer, it will not have strength and durability and bond to the concrete base slab unless it is properly compacted in the position in which it is laid. This is well known in the case of concrete, but many contractors do not seem to appreciate that the same applies to screeds.

On completion of compaction and trowelling the screed must be properly cured. This means that the evaporation of water from the screed should be prevented, as far as practicable, until the mortar has reached a reasonable strength. In normal weather conditions in the UK four days should be adequate for this. A good practical method of curing is to cover the screed with polythene sheets well lapped, and held down around the edges by scaffold boards. A sprayed-on resin-based membrane, similar to that used for concrete roads, will be satisfactory from the point of view of curing the screed. However, the presence of this membrane on the surface of the screed will interfere with the bond of any subsequent coverings or layers that may be applied to the screed.

Thickness

The thickness of the screed will depend largely on the purpose for which the screed is provided, and on the degree of bond with the base concrete. If electric conduits are buried in the screed, it should be 65–75 mm thick; but if it is only required to provide a smooth even surface on which to lay vinyl tiles or carpet or similar, it can be appreciably thinner. The presence of conduits in the screed will introduce planes of weakness, and some cracking along the line of the conduits should be expected if the screed is thinner than 65 mm.

If the screed is laid monolithically there is no need for it to be thicker than 20–25 mm (see *Figure 52.1*). However, if it is laid separately, i.e. after the base concrete is hardened, then the minimum recommended thickness is 40 mm (see *Figure 52.2*), but a thickness greater than 75 mm is not recommended. If for any reason, due to a change or error in the levels, a thickness greater than 75 mm has to be provided, it is better to use a concrete topping (1 : 1½ : 2½) using 10 mm aggregate, and to lay the screed monolithically on that.

In the case of solid concrete ground-floor slabs, it is generally necessary to lay a vapour-resistant membrane in some part of the floor thickness. Sometimes this is laid under the structural slab, sometimes on top. If it is laid on the surface of the floor slab, this will prevent or reduce bond between the screed and the base concrete (see below). In these circumstances it is advisable for the screed to be not less than 50 mm thick. This thickness should be adequate if the vapour-resistant membrane is

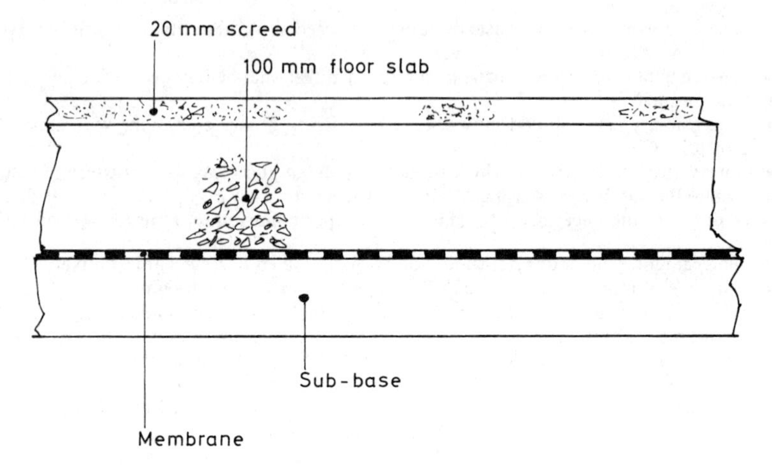

Figure 52.1 Monolithic screed on concrete ground-floor slab

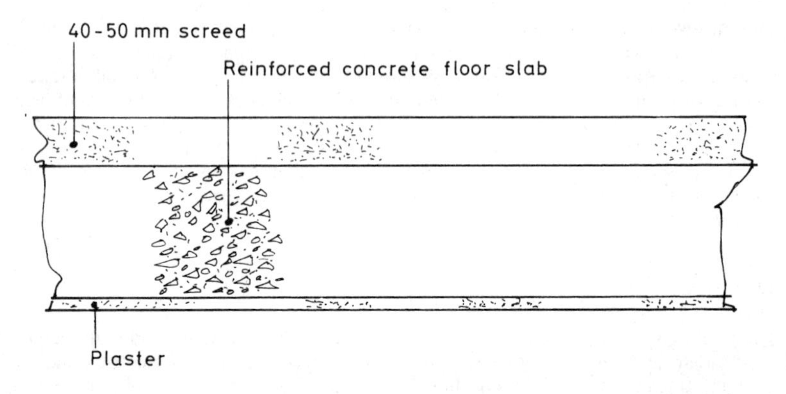

Figure 52.2 Semi-bonded screed on suspended reinforced-concrete slab

finished with a smooth surface, such as polythene sheets that have a rubber-bitumen adhesive on the lower side; this material does not get into rucks and wrinkles, which would substantially reduce the thickness of the screed at these points.

PROBLEMS AND DEFECTS

Bonding

An important problem, but one that is not easy to solve in a practical way, is the preparation that should be given to the base concrete to receive the screed in order to secure the necessary degree of bond. This depends largely on the use to which the floor will be put, and also on the thickness of the screed. In domestic buildings and offices, a screed 40 mm and more in thickness can be laid on the concrete slab after it has been well brushed (to remove all dirt, loose material and weak laitance) and dampened down the day before.

When a floor will be subjected to heavy wear, as in a department store, a high-strength screed (mix proportions 1 : 3 by weight) is essential unless the floor finish is ceramic tiles, thick carpet or similar. Thin sheet or tiles will transmit impact straight through to the screed, which then requires the strengthening effect of good bond to the base concrete, or else must be of adequate thickness.

When a screed is laid on a membrane, it is advisable for the water content to be as low as possible, if possible 0.4, and to use a plasticiser to provide the necessary workability for proper compaction. The mix proportions can be between 1 : 3.5 and 1 : 4 by weight. The reason for these restrictions is that the membrane prevents bond with the base concrete, and a cement-rich 'wet' screed material would be liable to curl badly at bay joints.

Regarding the actual preparation of the base slab for maximum bond, there are differences of opinion on this. The orthodox method, and one which is recommended by such organisations as the Cement and Concrete Association and the Building Research Establishment, is to mechanically hack the surface of the concrete so as to expose the coarse aggregate. This exposure of the coarse aggregate can be done by other means, such as grit blasting and high-velocity water jets. The latter is probably the best if the circumstances on the site allow it to be used, because it does not create any noise, vibration or dust; there is the problem of getting rid of the water, however.

The other method to secure good bond is to thoroughly brush the surface of the concrete with a wire brush so as to remove all dirt, loose particles and weak laitance etc., and then to apply a bonding coat and to lay the screed on the bonding coat within 20 minutes of its application. This limit of 20 minutes is essential: if the bonding coat has already started to set before the screed is laid, it will have the opposite effect of what is intended, and will act as a de-bonding layer. Effective bonding agents on floors not subject to dampness are polyvinyl acetate (PVA) adhesives, but on floors where there may be some dampness it is better to use a styrene butadiene based latex emulsion. The SBR resins, as they are called, are best used as a grout mixed with two parts ordinary Portland cement to one part emulsion (25 litres of the emulsion to 50 kilos of cement). The bonding agent should be well brushed in to clean the surface of the concrete, and as stated above the screed should follow immediately.

Bays

The next problem relates to whether the screed should be laid in bays and if so, what the dimensions of the bays should be. There is a difference of views on this matter: Building Research Establishment Digest 104 suggests that there is no need for the screed to be laid in bays, whereas the publication on floor screeds by the Cement and Concrete Association recommends that the screeds should be laid between screeding battens. The author is of the opinion that it is better to lay the screed in bays of limited width. This will enable the screed to be properly compacted, provided the screeding battens are not further apart than 3 m.

There is the question of how long the bays should be. This again is a matter of opinion, and some authorities feel that the length is not important because the screed is likely to develop a transverse crack about every 4–5 m when it has a bay width of 3 m. There is likely to be less curling and lipping at a transverse crack than at the end of the bay. However, it is advisable to provide plain butt joints in a screed to coincide with joints in precast concrete units when these form the structural slab. Also, all full and partial movement joints in the base slab should be carried through the screed (see *Figure 52.3*).

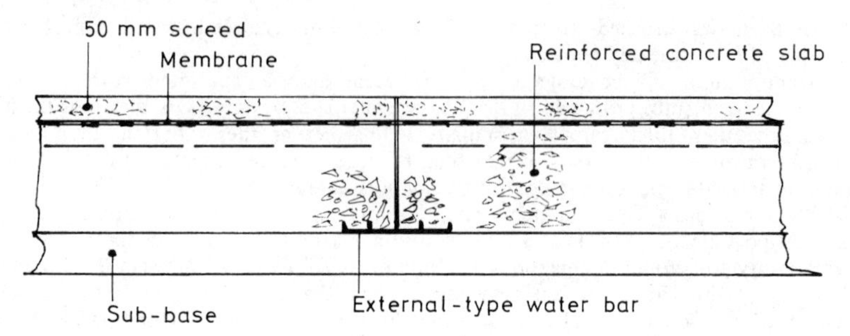

Figure 52.3 Control joint in concrete ground-floor slab and screed

Screeds laid on thermal insulation

Cement/sand screeds are sometimes specified to be laid on various types of thermal insulation (*Figure 52.4*). This insulation can vary from compressible glass-fibre mats to almost incompressible materials such as extruded polyurethane or cork. There is a serious physical incompatibility between laying a relatively thin, rigid cementitious screed on a flexible compressible base. In the author's experience, it is very seldom that a satisfactory cement/sand screed can be laid on a compressible layer. Therefore, if thermal insulation is required in the floor or roof slab, it is better for it to take the form of either a lightweight aggregate or thermal insulating screed, or for the thermal insulation to be of the rigid, virtually incompressible type mentioned above. The screed can be made satisfactorily of various types of artificial lightweight aggregate such as expanded clay, sintered PFA, vermiculite, or similar materials. In the case of

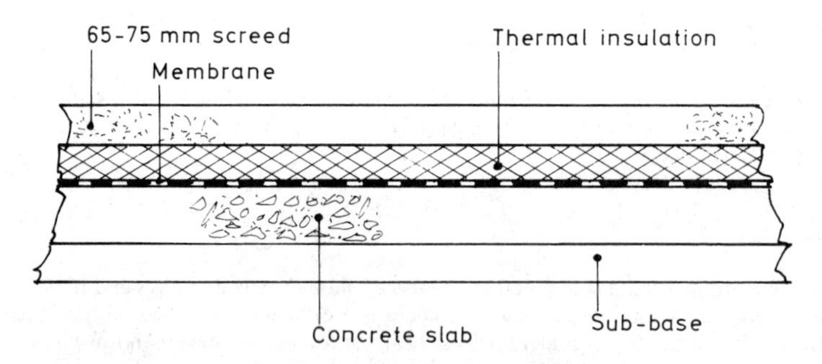

Figure 52.4 Screed on thermal insulation

vermiculite, it is normally recommended that the screed should be finished with at least 20 mm of cement/sand mortar so as to give a harder and stronger surface.

The thickness of screeds laid on thermal insulation should not be less than 60 mm. If the insulation is compressible 75 mm is better, but an entirely satisfactory screed may not be obtained.

Screeds laid on precast units

There are definite problems and hazards in laying cement/sand screeds on precast floor units. The main one is difficulty in obtaining bond, since the precast units should not be mechanically hacked and the concrete is usually very dense and impervious. Precast prestressed planks may move slightly relative to one another, and this may cause cracking in the screed.

Defects in floor screeds

The principal defects in floor screeds are:
- curling of the screed at bay joints and at wide cracks, and lipping at bay joints;
- hollowness, indicating lack of bond with substrate;
- excessive dusting on the surface of the screed;
- general wear and abrasion of the surface;
- random cracking.

It is very seldom that a screed of any size is laid without any visible defects at all. However, some of the 'defects' may have no significant long-term effect on the satisfactory performance of the screed.

The use to which the floor is put will have an important influence on the decision as to what defects can be accepted and what cannot. For example, slight curling and lipping at bay joints, and a few small areas of hollowness can generally be accepted where the floor is to be finished with ceramic tiles or carpet that has a thick underlay. However, it may well be that such defects may not be acceptable for a similar floor that is to be finished with thin sheet or vinyl tiles and that is likely to have heavy wear. The same applies to the actual quality of the screed and the extent to which it bonded to the base concrete. The importance of bond to the base concrete has already been mentioned.

If, during the period between the completion of the screed and the laying of the floor covering, serious surface wear has taken place without any apparent reason, this indicates a poor-quality screed; the right decision would probably be to take it up and lay a new one. Borderline cases are always the most difficult to deal with.

Hollow areas can often be adequately dealt with by the injection of a low-viscosity resin, which should flow into the very narrow space between the underside of the screed and the surface of the concrete slab. It is this very narrow void that causes the hollow sound when the screed is tapped. However, it should not be expected that this injection technique will completely cure all hollowness. The objection to hollowness from a durability point of view is that under heavy impact the screed may shatter over the hollow area; the injection of the resin may prevent this even though it does not completely eliminate the hollow sound.

Special problems with roof screeds

A screed is provided on a flat concrete roof for one or both of the following reason:
- to provide falls on the roof, discharging to rainwater outlets;
- as thermal insulation.

All screeds can be troublesome, and roof screeds are no exception. Therefore, if they can be eliminated so much the better. This can be done by providing the necessary falls in the structural slab itself. There is no special problem with *in-situ* concrete for roof slabs; this and precast units are discussed in Section 31. Generally a gradient of between 1 in 80 and 1 in 100 should be adequate. A roof requires a waterproof membrane, and the position of this in relation to the structural slab and screed will influence the details and specification for the screed.

The principles involved in providing a good-quality roof screed are essentially the same as for a floor screed. It is important to take into account the magnitude and type of live load that the roof will have to bear. Generally roofs are more lightly loaded than floors, but this is not always the case.

Most flat concrete roofs are provided with thermal insulation, and this can be provided in the form of a special screed. Suitable materials are pumice, artificial lightweight aggregates and vermiculite. There is one important factor to be remembered with roof screeds of all types, but especially with thermal insulating screeds: saturation, partial or complete, will greatly reduce the thermal properties of the screed. In addition, the absorbed water cannot readily be removed, and unless there is a waterproofed membrane below the screed this water is likely to slowly penetrate the slab. This problem is discussed in more detail in Section 31.

FURTHER INFORMATION

CP 204:1970, *In-situ floor finishes*

Floor screeds, Building Research Establishment Digest 104, HMSO (1969)

Perkins, P.H., *Floor screeds,* Cement and Concrete Association (1971)

Perkins, P.H., *Floors – construction and finishes,* Cement and Concrete Association (1973)

Warlow, W.S., and Pye, P.W., *The rippling of thin flooring over discontinuities in screeds,* Current Paper 94/74, Building Research Establishment (1974)

Guidance note on floor screeds and finishes, National Federation of Plastering Contractors (1972)

Developments in roofing, Building Research Establishment Digest 51, HMSO (1968)

Laing, J., *Lytag for floor and roof screeds,* Technical Bulletin No. 4 (1960)

The vermiculite handbook, Mandoval Ltd (1961)

Cable, J., 'The use of lightweight cellular screeds', *Insulation* (May-June 1960)

Condensation in dwellings, Part 1, *A design guide,* HMSO (1970)

53 FLOOR FINISHES

INTRODUCTION 53–2

IN–SITU FLOOR FINISHES 53–4

PREFORMED FLOOR FINISHES 53–5

53 FLOOR FINISHES

A.A. THOMPSON, MSAAT, LIOB
Lecturer in Construction Technology, Vauxhall College of Building and Further Education

This section is concerned with finishes providing the top wearing and covering surfaces to normal domestic floors, and therefore excludes less common finishes like marble, slate and brick, and in-situ finishes such as granolithic and terrazzo.

INTRODUCTION

Regulations and standards

Mandatory requirements for floor finishes are limited. Part C5 of the Building Regulations 1976 influences the construction of timber finishes on solid ground floors. Part F3 deals with thermal insulation of floors between a dwelling and the external air of ventilated spaces, and Part G3 similarly governs the sound insulation of floors, but these factors are mainly affected by the structure and insulation rather than the finish.

Despite the interchange of flooring products between different countries, there are currently no international standards for floor coverings. It is likely that the efforts of the ISO will result in a measure of standardisation, as committees and working parties are already making progress in this direction.

Various British Standards, listed at the end of this section, define minimum acceptable standards for floor finishes, and since these will frequently be mentioned in contract documents it is important for builders to be familiar with their content and to adhere to their standards.

The Agrément system of certification will often have relevance where unfamiliar flooring products are specified. Whilst innovation should not be discouraged, when confronted with novel materials the builder will be wise to seek some kind of assurance as to performance. Other useful sources of information are BRE Digests and the publications of the various trade associations.

Factors influencing the choice of floor finishes

The comments in CP 203 that 'appearance should not be the primary consideration' and that 'the logical approach is to eliminate all those materials which are unsuitable for reasons of performance and cost and only then to choose on the basis of taste' are the best advice for the builder to follow.

Wear resistance is probably the single most important factor, and in making a selection account should be taken of expected traffic. In areas such as entrance halls subjected to heavy wear accelerated by transfer of outside grit, the extra cost of especially durable materials of good appearance such as vitreous tiles, or a high-vinyl-content PVC finish may be justified. In bathrooms, where water resistance is important, rubber may be selected, and in kitchens clay or polyester products may be chosen for their oil and grease resistance.

The need for safety in the home will encourage the selection of soft resilient finishes, which tend to provide non-slip surfaces, in critical areas, finished with polishes that are not intrinsically slippery. Combustibility is also a safety factor, and although there is no requirement in the Building Regulations regarding restriction of spread of flame over floor surfaces, finishes should be incombustible or with a low flame spread, particularly in kitchens.

Comfort conditions are particularly relevant in living and sleeping areas. Carpets and other soft finishes give an impression of warmth. They also improve the acoustics by lowering the reverberation time, as well as reducing the nuisance of impact sound passing through floors. Resilient floors improve comfort, as they are less tiring to walk upon.

Regarding appearance, where possible an attempt should be made to visualise the material in its actual setting by viewing similar finishes installed in existing buildings under comparable conditions. Builders should particularly consider the degree of permanence of the colours and the ability of the floor to disguise marking. Consideration should be given to factors that control the long-term appearance of the floor, such as the tendency to retain dirt, and ease of cleaning, maintenance and repair. Wherever possible, advice should be given to the customer as to the limitations of each material, together with maintenance advice.

In practice, cost is often the governing factor, but account should be taken of preparatory work as well as the finishing, and the client should be encouraged to consider the total cost of the floor over a long period, including maintenance, repair and renewal, and not just the initial cost.

Organisation

The various Codes of Practice stress the need for skilled operatives and efficient supervision. It is also vital to consult manufacturers about their products in respect of matters such as material storage, laying conditions, adhesives, sealing and maintenance.

If the work is to be sub-contracted, clarification must be obtained on items such as unloading, attendance, power, lighting, cutting and waste, and cleaning.

The work must be realistically programmed to permit sequencing with other trades, and access arrangements are vital. Integration is necessary with other work such as installing services beneath the floor finish. Where possible, 'wet' and 'heavy' trades should be completed before the floor finishes, but where this is not possible adequate protection must be provided. Sheeting should be used to protect the finish during painting work.

Associated work

The sub-floor must be free from loose areas, and must be strong and rigid enough to support the finish. Alternatively, if some movement is expected, the finish must be flexible enough to avoid cracking resulting from such movement. CP 202 and CP 204 provide guidance on level tolerances for hard tiles and *in-situ* finishes. Generally the base must be level enough to permit the finish to be applied or laid to a consistent finish to arrive at the required level, with permissible tolerances dependent to some extent on the laying method. See *Figure 53.1*.

The sub-floor must be appropriate to the finish in respect of compatibility and smoothness. If an *in-situ* finish or cement bed is intended, a degree of roughness may be needed to provide a good mechanical key, but thin tiles or sheets need a smooth surface. Regardless of Building Regulation requirements, some finishes such as carpets and flexible PVC require a damp-proof membrane on the concrete to prevent rising damp. Manufacturers should be consulted as to requirements for their products.

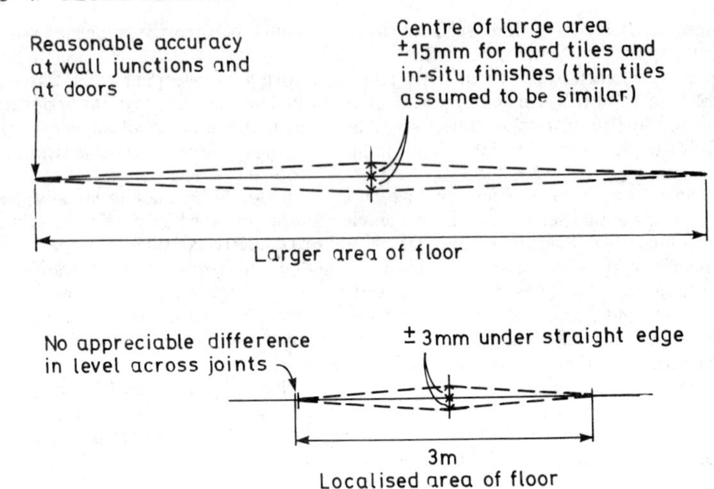

Figure 53.1 Permissible tolerances on level of floor finishes given in CP 202 and 204

IN-SITU FLOOR FINISHES

Asphalt and pitchmastic

The basic materials used for asphalt finishes are natural rock asphalt or limestone aggregate and asphaltic bitumen, and the finish is available in black or pigmented colours of dark brown and red. A single coat 15–20 mm thick is appropriate for domestic work, and coved skirtings can be formed in the same material but laid in two coats.

The asphalt is spread hot on an underlay of black sheathing felt, and hand floated to a uniformly level surface. Space will be required for the mixing cauldron, near to the laying area, to minimise cooling of the molten asphalt. Traffic must not be permitted on the finished surface until the material has cooled to room temperature, and then protection is required from things such as builder's equipment, cement and paint. The surface needs to be kept clean, particularly from oils and grease, and can be finished with a water-based seal to preserve the appearance.

Domestic-grade asphalt is a durable floor, but heavy point loads cause indentation. It is impervious to moisture, but not to oils and grease, and provides a dustless, clean, slip-proof surface. Care needs to be taken if asphalt is used in conjunction with underfloor heating, as it softens at certain temperatures. The finish is fairly low priced.

Pitch mastic is similar to asphalt in respect of laying techniques, quality etc., but is based on silica aggregate bound with coal-tar pitch. It has superior resistance to mineral oils and grease, but inferior resistance to alkalis.

Magnesium oxychloride

This is composed of calcined magnesite, fillers such as sawdust and powdered asbestos, together with a solution of magnesium chloride, and is available in a variety of colours and mottled effects, generally rather dull. For domestic work it will typically be laid in a single coat 15 mm thick, but two-coat work is necessary for a mottled finish.

The mix should be thoroughly compacted and trowelled to a smooth finish, and it will set in a few hours. Skirtings may be formed in a similar material. A final coat of chloride solution should be applied after 12 hours, and curing procedures used for the first 24 hours. It is best to seal the finish soon after laying with a solvent-based clear seal.

This finish is sometimes considered as a substitute for asphalt. Its appearance may be preferred, but it is more expensive, not suitable for damp, humid conditions, and a damp-proof membrane is essential for solid floor construction. Otherwise, if properly laid, it is reasonably durable, dustproof, oil and grease resistant, and fairly warm and resilient if wood-based fillers are used.

Cement rubber latex

This finish is a mixture of cement, different aggregates, fillers and pigments, gauged on site with a rubber latex emulsion. It is available in various colours and commonly laid 6 mm thick.

The material adheres well to most surfaces, but wood bases should be reinforced with wire netting. Base surfaces must be primed with a thin latex-cement trowelled coat, with the finish laid immediately the priming coat has set, spread with a straight edge and steel trowelled. Traffic must not be permitted for three days. The surface can be sealed with a water-based seal if required.

Mixes can be selected to achieve most functional requirements. If wearing qualities and water resistance are of prime importance, a hard stone aggregate with a high latex content in the mix can be used, whereas for a quieter, resilient finish a cork aggregate will be chosen. A damp-proof membrane is necessary for solid ground floors.

Plastic

These finishes use synthetic resins such as polyester or polyurethane, and are essentially proprietary finishes laid to a variety of thicknesses by different methods. Polyester-based floors are typically screeded about 5 mm thick, laid and trowelled similarly to a rubber latex floor. Polyurethane floors are generally self-levelling, about 1.5 mm thick, with the materials more liquid and allowed to flow into place. They may consist of two or three coats on which are scattered tiny flakes or chips of plastic or stone, finished with several coats of clear polyurethane sealer.

Although a fairly recent development, plastic finishes are suitable for many conditions if laid correctly, and provided precautions are taken to combat the high shrinkage during curing. Various mixes are available to produce tough, chemical-resistant finishes, easy to clean and of attractive appearance.

PREFORMED FLOOR FINISHES

Timber

Despite many alternative finishes available, wood retains its popularity due to its appearance, warmth and resilience. It must be used so as to resist fungal attack and avoid becoming a fire hazard.

Softwood and hardwood boards are over 100 mm wide, fixed together by tongued-and-grooved joints along their edges, and to every floor joist, or to fillets attached to solid concrete floors. A damp-proof membrane is required in solid ground floor construction. See *Figure 53.2.*

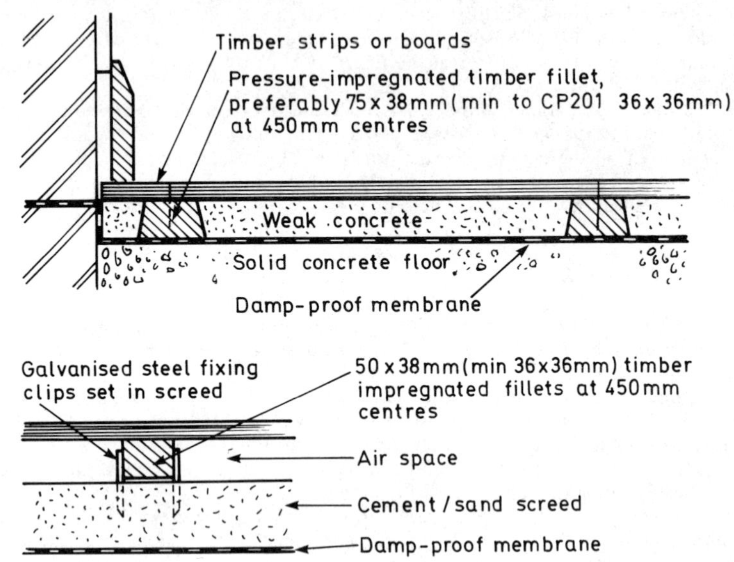

Figure 53.2 Strip or boarded ground floor on solid concrete slab

Hardwood boards are often secret nailed, and wider boards are screwed. Softwood boards have their ends butt-jointed centrally over supports and staggered, and ends of hardwood boards are matched (i.e. tongued-and-grooved). Strip flooring, up to 100 mm wide, is commonly in hardwood, secret nailed. See *Figure 53.3.*

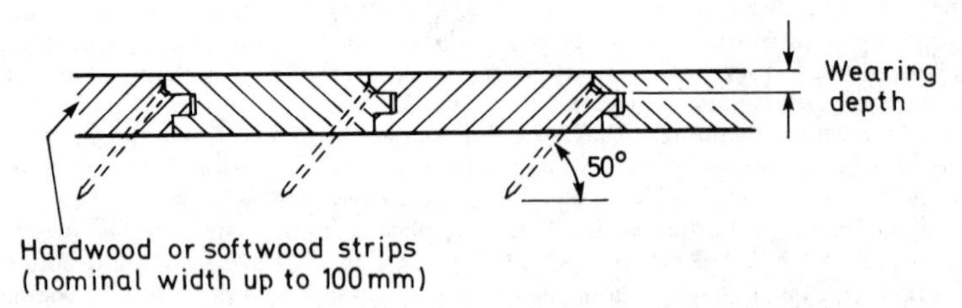

Figure 53.3 Wood strip floor secret-nailed (to CP 201: Part 2) at each intersection with joists or battens; plain-shanked nails (not less than 2½ times thickness of part of board through which nail is driven) are driven in just above tongue

Wood blocks are usually hardwood, up to 89 mm wide, 150–300 mm long and 19–38 mm thick, laid to various patterns, generally on a screed. See *Figure 53.4.*

In ground floor construction the adhesive used must be laid so as to serve also as a damp-proof membrane as required under the Building Regulations. If hot bitumen is used as an adhesive, the screed is treated with a black varnish primer and the bottoms of the blocks are dipped in hot bitumen or coal tar. More commonly a cold latex bitumen emulsion is used in accordance with the supplier's instructions.

Setting out is from the centre of the room to achieve symmetry, and expansion joints are required around the perimeter, filled with cork or similar material or hidden by the skirting.

Parquet floors are used on account of their appearance, being made from specially selected hardwoods, of various sizes, generally 6.5–9.5 mm thick, with square edges, and laid to several designs. They are commonly fixed to a timber sub-floor with adhesives or panel pins, or secured to a backing sheet and laid in panels 300–600 mm square. Parquet strip may also be used, 50–75 mm wide and 9.5 mm thick with edges tongued-and-grooved and fixed by secret nailing. Wood mosaic is similar to parquet,

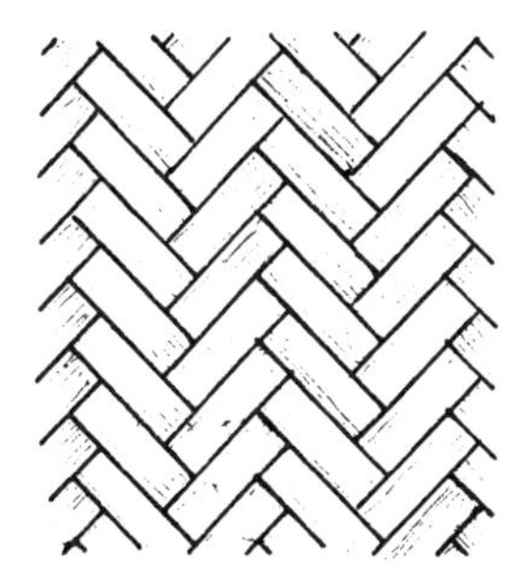
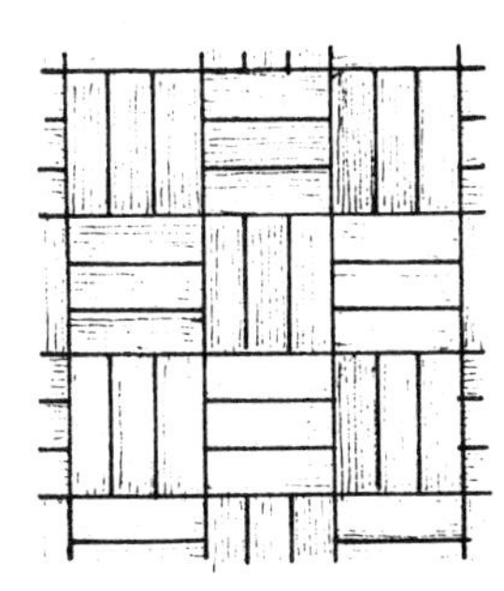

Figure 53.4 Wood-block flooring patterns: left, herringbone; right, basket

and the term is applied to fingers about 114 × 25 × 9.5 mm thick arranged in squares of 114 mm. Parquet and mosaic floors are economical compared to other hardwood floors, and suitable for light domestic wear in water-free areas.

Timber floors need correct maintenance, because unsealed wood will absorb dirt and water. They should be sanded (except parquet and mosaic floors, which are too thin for this treatment) and then sealed with a solvent-based wax.

Timber-based

Plywood decking is becoming increasingly popular as an alternative to softwood boarding, being suited to a fast programme and providing a good finish for carpets. Panels 12–31 mm thick, tongued-and-grooved and fixed at each joist position, are supplied in overlay quality for covering, or clear finishing quality.

18 or 22 mm flooring-grade chipboard is used in lieu of plywood or softwood boarding, principally due to cost considerations. Boards are supplied with either square or tongued-and-grooved edges, and should be fixed at each joist position, and around all four edges if sheets are square-edged. It is advisable to check that chipboard flooring is acceptable to the local authority. Chipboard with tongued joints is sometimes laid loose over insulation to form a floating floor in conjunction with a concrete slab.

Plywood parquet formed of tiles, typically 228–914 mm square, is available as a cheaper alternative to traditional parquet. Similarly, composition blocks composed of a mixture of cement, wood, gypsum, linseed oil and pigments may be considered as an alternative to wood blocks.

Thin resilient tiles and sheets

A smooth and level base is vital for this group of materials, particularly the thinner products. Fixing will invariably be by adhesive, used according to the floor supplier's recommendations. Certain materials, as advised by the supplier, require a damp-proof membrane under them, and in no case should an adhesive be considered to perform this function. The covering should be set out from the centre of the room to achieve symmetry. Sheet materials are advisable when water spillage is likely. There is a wide range of sizes and thicknesses, and those given below are indicative. Matching skirting is available for most coverings. Generally the life expectancy of the floor will relate to the thickness.

Cork finishes are composed of compressed granulated cork of various densities, supplied in sheets 1829 mm wide and 3–6 mm thick, and tiles 305 mm square and 3–7 mm thick. It is normal to fix the tiles with headless steel pins, in conjunction with the adhesive. The finish requires sealing with a solvent-based clear seal to prevent dirt absorption. Cork finishes are not suitable for point loads, but otherwise are hard-wearing (particularly the denser coverings), quiet, warm and slip-resistant except when wax polished.

Rubber is another traditional finish, manufactured from natural or synthetic rubber and filling compounds to give a variety of attractive colours and textures. Sheets are 914–1829 mm wide and 3.2–6.4 mm thick, sometimes with a foamed-rubber backing for extra quietness and resilience. Tiles, 152–1219 mm square, are cut from the sheets. The finish is hard-wearing (particularly the thicker grades), quiet, and of attractive appearance; also, being moisture-resistant, hygienic and non-slip, it is suitable for bathrooms. In areas where fat and grease are likely to be encountered, the synthetic rubber-based product is advisable. Rubber is a comparatively expensive finish, but has a long life expectancy.

Linoleum is produced from a base of drying oils with resin, cork or wood flour, mineral fillers and pigments, pressed on a jute-canvas backing. It is supplied in a wide range of colours and patterns, plain, printed and inlaid. Sheets are 1829 mm wide and 1.4–6.7 mm thick, and tiles are 305 mm square and 2–4 mm thick. Linoleum is reasonably quiet, resistant to water, oils and grease, has good wearing qualities and is easy to maintain.

Thermoplastic or asphalt tiles are composed of asphalt and resins, fibres, fillers and pigments. They are available in limited colours, and in tile form only, 229 mm square and 2–4.5 mm thick. They are low priced, hard-wearing and water-resistant, but rather brittle and noisy.

Flexible PVC is produced from a mixture of PVC resins, mineral fillers and pigments, in a wide range of colours and patterns. Sheets, which can be welded at the joints, are up to 2430 mm wide and 1.5–3 mm thick, and tiles are 305 mm square. The material is tough and hard-wearing, resistant to water, oil and grease, but tends to show sub-floor irregularities and is easily damaged by cigarette burns.

Vinyl asbestos consists of PVC resins, asbestos fibres, fillers, plasticisers, stabilisers and pigments, in a wide range of plain or mottled colours, and is produced in tile form, 229 mm square and 2.5 or 3 mm thick. The tiles are water, oil and grease resistant, can be used in conjunction with underfloor heating, and do not require a damp-proof membrane, although this may be required under the Building Regulations. They are cheaper, but less flexible than flexible PVC.

With the exception of cork, which requires a solvent-based wax, the above floor surfaces can be finished with a water-emulsion floor wax.

Carpets

Carpets are manufactured from natural materials such as wool, animal hair, mineral fibres such as jute, and man-made nylon, acrylic, and rayon. There is a wide choice of qualities, patterns and sizes, but most are liable to be damaged by water, and a solid floor needs the protection of a damp-proof membrane. They should be laid on a felt or latex underlay, and fixed by perimeter nailing, adhesives or special grip-fast fixings. Carpet tiles, 600 mm square, may be secured with an adhesive, but the thicker tiles remain in place due to interlocking of the edge fibres.

Hard tiles

Fixing techniques are generally similar for the various hard tiles described below. There are three bedding methods; see *Figure 53.5*. The first aims at isolating the bedded finish from the base by a 'separating layer' of bitumen felt, building paper or polythene, which is laid on top of the sub-floor to receive the tiles bedded on a 1 : 3 mortar mix, about 15 mm thick. As well as permitting differential movement

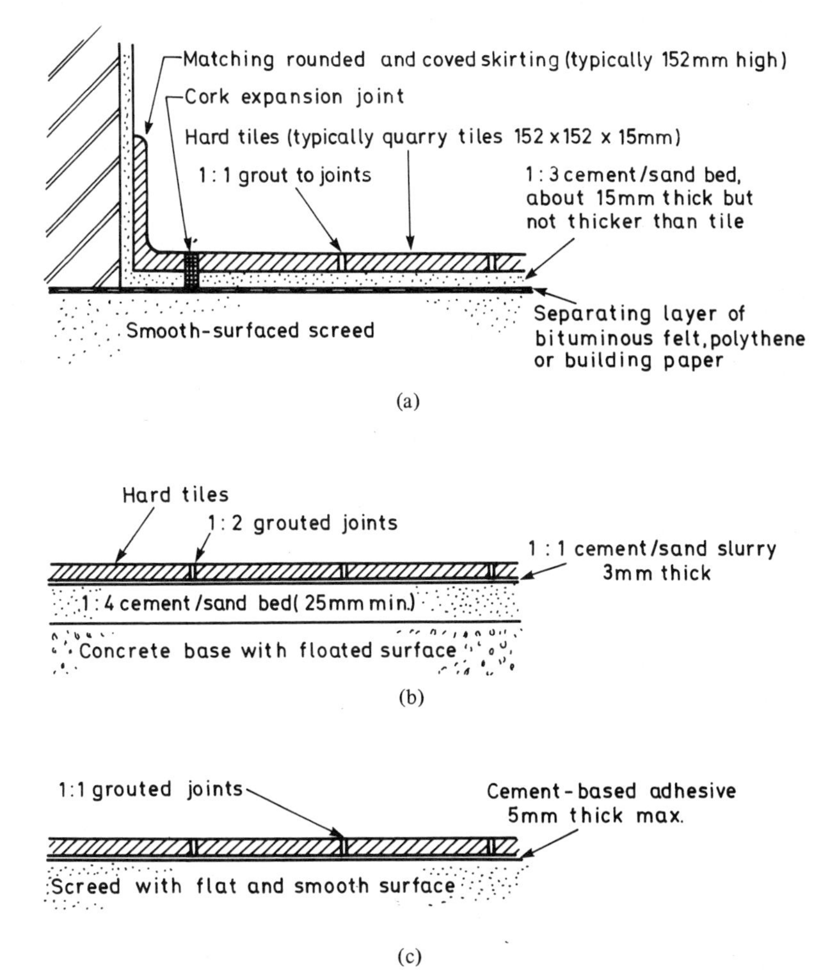

Figure 53.5 Typical fixing methods for hard tiles (generally in accordance with CP 202): (a) 'separating layer' method; (b) 'thick bed' method; (c) 'thin bed' method

between finish and base, the separating layer can serve as a damp-proof membrane. In the second method a 'thick bed' is used of semi-dry 1 : 4 cement/sand, minimum thickness 25 mm (40 mm typical), with the top surface finished with a 1 : 1 cement/sand slurry 3 mm thick before placing the tiles in position. No separating layer is needed as the mix is too dry to bond with the base, and sub-floor tolerances are less critical. The third method uses a thin bed, not exceeding 5 mm, of cement-based adhesive, and requires an accurate sub-floor as it is not possible to adjust discrepancies in the bedding. The adhesives must be used in accordance with the manufacturer's instructions.

Movement joints are required around the perimeter of each floor area. A full range of matching skirtings and special fittings is available for most rigid tiles.

Concrete tiles, typical size 305 mm square and 16 mm thick, are made with coloured cements mixed with hard aggregates on a concrete backing. They are resistant to moisture but not to oil, often competitively priced, but (like all hard, rigid tiles) cold and noisy.

Terrazzo tiles, made from crushed marble aggregate embedded in a white or coloured cement matrix, in sizes from 152 mm square and 19 mm thick, are hard-wearing, water-resistant and likely to be used where appearance is of primary importance and cost is not the first consideration.

Quarry tiles are produced from plain unrefined clay, in colours of red, brown and buff, with a common size for domestic work of 152 mm square by 15 mm thick. Quarries are hard-wearing, hygienic, with good water resistance, but may be slightly uneven, and care must be taken that their appearance is not spoiled by cement stains.

Plain clay tiles are more expensive than quarries, being manufactured from refined clay; typical sizes are 75, 102 and 152 mm square and 13 mm thick. They have similar qualities to quarries and are available in a wide range of colours. Due to their superior, regular, smoother appearance, they are sometimes used in areas such as entrance halls.

Vitreous tiles incorporate calcined flint and felspar and are available in similar sizes to plain clay tiles. They are impervious to water, oil and grease resistant, and generally the best and most expensive of the clay group of tiles.

Concrete tiles can be sealed with a clear polyurethane sealer. It is not usual to seal clay tiles, but a water-based seal may be used if required.

FURTHER INFORMATION

The Building Regulations 1976, HMSO (Parts C5, F3, F4, G3, G4, G5)
BS 776, *Materials for magnesium oxychloride (magnesite) flooring*
Part 2:1972, *Metric units*
BS 810: 1966, *Sheet linoleum (calendered types), cork carpet and linoleum tiles*
BS 1187: 1959, *Wood blocks for floors*
BS 1197, *Concrete flooring tiles and fittings*
Part 2: 1973, *Metric units*
BS 1286: 1974, *Clay tiles for flooring*
BS 1450: 1963, *Black pitch mastic flooring*
BS 1711: 1951, *Solid rubber flooring*
BS 1763: 1975, *Thin PVC sheeting (calendered, flexible, unsupported)*
BS 1863: 1952, *Felt backed linoleum*
BS 2592: 1973, *Thermoplastic flooring tiles*
BS 2739: 1975, *Thick PVC sheeting (calendered, flexible, unsupported)*
BS 3260: 1969, *PVC (vinyl) asbestos floor tiles*
BS 3261, *Unbacked flexible PVC flooring*
Part 1: 1973, *Homogeneous flooring*

BS 3672: 1963, *Coloured pitch mastic flooring*
BS 4050: 1966, *Wood mosaic flooring*
BS 4131: 1973, *Terrazzo tiles*
CP 201, *Flooring of wood and wood products*
Part 2: 1972, *Metric units*
CP 202: 1972, *Tile flooring and slab flooring*
CP 203, *Sheet and tile flooring (cork, linoleum, plastics and rubber)*
Part 2:1972, *Metric units*
CP 204, *In-situ floor finishes*
Part 2: 1970, *Metric units*
CP 209, *Care and maintenance of floor surfaces*
Part 1: 1963, *Wooden flooring*
Design of timber floors to prevent decay, Building Research Establishment Digest 18, HMSO
Sheet and tile flooring made from thermoplastic binders, Building Research Establishment Digest 33, HMSO
Granolithic concrete, concrete tiles and terrazzo flooring, Building Research Establishment Digest 47, HMSO
Clay tile flooring, Building Research Establishment Digest 79, HMSO
Technical specifications for ceramic floor tiling, British Ceramic Tile Council
Finply flooring, Technical publication 7, Finish Plywood Development Association
Flooring with mastic asphalt, Mastic Asphalt Council and Employers Federation
Specification, The Architectural Press

54 PAINTING AND DECORATING

INTRODUCTION	54–2
PREPARATION AND PRIMING	54–2
PAINTING	54–7
MAINTENANCE PAINTING	54–11
PAINTING DEFECTS	54–14
PAPERHANGING	54–14
COLOUR	54–18

PAINTING AND DECORATING 54

54 PAINTING AND DECORATING

J.L. REYNOLDS
Lecturer in Painting and Decorating, Chelmer Institute of Higher Education

This section describes good practice for painting on timber, plaster, metal and other common materials. It gives information on the main types of material in common use, paper-hanging and the selection of colour schemes. Maintenance painting and painting defects are also considered.

INTRODUCTION

There are comparatively few buildings that do not depend to some extent on painting to protect them from deterioration or decay. To perform this function adequately neither the specification nor the work should be skimped, but should be to the best possible standard that circumstances permit. All too often, a finished job can look well, but fairly quickly fail due to insufficient coats of paint or inadequate preparation. The range and quality of paints available has never been better. There is no need for such failures if these materials are correctly applied to well prepared surfaces as intended by the manufacturers.

Because of the wide variety of surfaces painted and the conditions to which they are exposed, no single paint system can possibly meet all these demands. Instead, manufacturers offer an almost equally wide variety of priming and finishing paints, each particularly suited to different circumstances. It is in the wise selection of these materials for a particular job, together with correct preparation, that the specifier or painting contractor will show his ability.

Another factor that can influence the effectiveness of the paint is good structural design and detailing of the building. Very few paint systems, however well applied, are completely impermeable. Where the design of any unit allows moisture to collect and have no escape route, it will eventually be absorbed, with subsequent rotting or corrosion. Wooden cills insufficiently weathered or water trapped by capillarity are particular examples of this. The painter can have no control over these matters; nevertheless they are still relevant to the effectiveness of his work.

For the purposes of painting, and particularly priming, it is convenient to group surfaces as:

- absorbent or porous – timber and timber products, weathered plaster or cement surfaces;
- non-porous – metals, glass, ceramics;
- chemically active, usually alkaline – new cement, lime plaster, asbestos sheeting etc.

Porous surfaces require a primer that will satisfy the porosity. Non-porous surfaces require a primer that will adhere to the surface and, where necessary, help prevent corrosion. Chemically active surfaces require a paint system unaffected by the chemical activity, and often sufficiently porous to allow any surface moisture to dry out.

PREPARATION AND PRIMING

Timber

For the purposes of painting, timber is best grouped as softwood, hardwood, oily and resinous wood. Timber products such as plywood, laminboard and blockboard can be considered as either softwood or hardwood, according to the face veneer used.

The timber should be dry with a moisture content of below 14 per cent for general purposes, but where it is used in centrally heated areas the moisture content should be around 7 per cent. Apart from any consideration of rot setting in, the paint will not key efficiently to the surface if moisture is present, and will either blister or flake off. The timber should be free from grease, large knots or other major defects. Where resinous areas are present, these should be cut out and plugged or filled with sound timber.

GENERAL PREPARATION

Softwoods should be lightly rubbed down across the grain with medium-grade sandpaper, and all arris edges lightly rounded off at the same time. Taking these sharp edges off will allow a better build-up of paint. All knots and any resin pockets should then receive two thin coats of patent knotting. This is best stippled on to give the primer a better key.

Hardwoods can be prepared in the same manner; usually there are very few knots present. Oak, which may still be used for cills, does not always offer a good surface for paint, which may flake off after a comparatively short time. To overcome this, the old-fashioned method of sealing with a coat of red lead and gold size has much to commend it. Alternatively use calcium plumbate primer.

Oily woods such as cedar or teak are not usually painted, and are used often because of their natural resistance to weather or decay. However, there are occasions when painting is necessary. The problem is to remove the natural oil from the surface pores of the wood to enable the priming paint to be absorbed slightly and thus obtain a key. This can be done by following the preparation for softwood and then, immediately before priming, washing over the surface with a strong solvent such as zylol. Evaporation of the solvent with some of the natural oil will be quick, and there need be no delay before priming.

Resinous timbers such as Columbian or Oregon pine are difficult to paint satisfactorily because resin will exude from the timber whenever the temperature rises to any degree, pushing the paint off with it. There is no way of preventing this exudation, but nevertheless it may be overcome reasonably satisfactorily. After sanding down, apply a thick coat of water paint if available. Thinned water-based filler will make a suitable substitute. This coat will perform a similar function to blotting paper, and will tend to absorb any exuded resin. After light sanding down, an aluminium wood primer can be applied.

PRIMING PAINTS FOR TIMBER

In general, the purpose of a priming paint on wood is to partially satisfy the porosity of the timber and provide a key for the following coat of paint. It is not essential that it should obscure the surface. Brush application is best, as this allows the paint to be forced into the grain, cracks and nail holes. The primer should be well brushed out and not too thickly applied.

Once primed, timber should not be left for any great length of time before proceeding with the next coat of paint. The reason for this is that the timber can still absorb moisture through the priming and can become greasy from handling; both are potential causes of early paint failure.

Lead paints were the standard primers for timber over many years. They proved to be satisfactory in almost every respect. However, because of health hazards, they have been withdrawn from the market over the last year or two. There is no law forbidding their use, but they would now be rather difficult to obtain in any quantity. Nevertheless, anybody with a source of supply is entitled to use these paints if he so wishes, bearing in mind the health hazard and provided he complies with the Lead Paint Regulations.

Lead-free primers have now replaced the old lead primers. They have similar properties but are non-toxic. For the general priming of softwoods and hardwoods, these are the best paints to use. The manufacturer's instructions should be followed, particularly as regards the thinning of these paints. It is not wise to dilute with linseed oil and turps substitute, as was common practice, unless the manufacturer specifically recommends this.

Acrylic primer/undercoats are now widely used and are suitable for both hardwoods and softwoods. They are similar to the emulsion paints in general use, being water-thinned, but have been especially formulated as primers. Normal emulsion paints should not be substituted. Acrylic primers are particularly useful when time between coats is limited. Brushmarks are somewhat coarse, and there is a tendency to raise the grain of wood. Being water-thinned they are rust-stimulative, so any metal fixings must be spot primed with a metal primer before using the acrylic. Always make sure adequate time is allowed for moisture to dry out before overpainting with oil paints. Trapped moisture could result in blistering or flaking. The use of knotting is not always necessary beneath acrylic primers, but it is best to use it should there be any doubts.

Aluminium wood primers are among the more waterproof of paints, and therefore should not be used if moisture penetration from behind is possible. (Such penetration can occur, for example, if skirtings and frames are not back-primed before fixing.) Using the primer under these circumstances could result in flaking of the whole paint system. These primers are particularly useful for oily and resinous woods, being more adhesive than the lead-free primers. Their darker colour may cause some problems with certain finishing colours, but only if the minimum number of coats is being applied. They are also useful for painting over old creosoted surfaces required to be finished in oil paint; their structure prevents the bleeding through that would normally occur.

Universal primers are now being introduced to the market. These are usually of silver-grey colour, and are based on the pigment zinc phosphate. Apart from their function as wood primers they are also primers for plaster and metal. This makes them very suitable for composite construction – for example, metal window frames with wooden sub-frames.

Factory primers are available for large-scale priming of joinery off the site. These are highly specialised paints, not normally suitable for brush application. Special applicators or coaters are available. Drying time is such that after about ten minutes the joinery units can be stacked. If this method of priming is contemplated one would be well advised to consult a reliable supplier, who would advise on the equipment and primer most suitable to meet a particular need.

Builder's primer is usually much cheaper than the primers so far mentioned. Tests have shown it to be inferior, and it is frequently the cause of early breakdown of paint systems on new wood.

HARDBOARDS

There is little preparation needed before priming these materials. Tempered hardboards may have traces of oil on the surfaces, which should first be washed off with white spirit. The same primers as for softwoods are suitable, together with hardboard primers offered by the paint suppliers. In some situations, it may be necessary to use hardboards that have been impregnated with fire-retardant salts. These should never be coated initially with water-thinned materials, even if an emulsion finish is required, as this may result in the leaching out of the salts and possible efflorescence. They should always be coated with an alkali-resistant primer.

INSULATING BOARDS

These are not usually finished in oil paints, but when this is required they should be primed with acrylic primer applied with a roller. As with hardboards, they may have been treated with fire-retardant salts, in which case an alkali-resistant primer will have to be used.

Plasters, rendering, asbestos sheeting and brickwork

Gypsum plasters and lightweight plasters can safely be painted with oil paint when thoroughly dry, provided no lime was added to the mix. Cement rendering and lime plaster are safe to paint after weathering for six to twelve months, after which time the alkaline salts should have leached out. Asbestos sheeting can be painted when fixed, provided it is dry and there is no chance of moisture reaching any unpainted areas. If this is not possible, then it too should be allowed to weather before painting.

Any grease should be washed off with solvent, nibs scraped off and the whole area dry brushed to remove dirt and any efflorescence. Large cracks should be raked out and primed before filling. Small cracks and minor defects may be filled with a proprietary filler. When dry these should be sanded down to a level surface. Two coats of alkali-resistant primer can then be applied or universal primer used.

New cement rendering, brickwork, damp asbestos, lime plaster and gypsum plaster containing lime. All these surfaces are highly alkaline when new. They should be allowed to dry out for a period of six to twelve months according to drying conditions. Any attempt to coat them with oil paints in their early stages will lead to saponification (i.e. chemical breakdown of the paint, the oils in the paint combining with the alkalis to form a soap). Apart from this, there is also the possibility of upsetting the set or hardening of the underlying surface.

When an early form of decoration is required, oil-free coatings such as emulsion paint should be used. It is advisable to use the minimum number of coats possible. This will allow the substrate to dry out through the paint film. Left uncoated to weather, the surfaces may then be treated as for the weathered surfaces described above. If these surfaces have been temporarily emulsion-painted, see the sub-section on maintenance painting later.

Ferrous metals

There is comparatively little exposed iron or steelwork used in domestic dwellings these days. Where it is used, thorough preparation and coating (before fixing if possible) will ensure easy maintenance and prolong the interval between maintenance. For best results it is essential that the metal is free of all rust and mill-scale, dry and free from grease. According to the scope of the work and practical situation, various methods are available for de-rusting and scaling. The most effective of these is shot or grit blasting. It is also the most expensive. Two methods are available. 'Open blast', where the grit is not recovered, can be a health hazard if safeguards are not taken. 'Vacuum blast' recovers the grit and there is no hazard. Flame cleaning, using especially designed oxy-acetylene equipment, is another effective method. The equipment is somewhat bulky and fairly expensive to run. As well as de-rusting and scaling, this method dehydrates the metal and leaves it warm for priming. Chemical treatment is possible, but needs strict control and is really only suitable for off-site preparation where the necessary control is possible. Needle guns, chipping hammers and mechanical wire brushes powered either by electricity or compressed air will remove medium to fine rusting, but have variable effect on millscale. Finally there is hand chipping and wire brushing, which are often the only practicable methods. These are ineffective for the removal of millscale; their efficiency in removing rust will depend on the condition of the metal and the persistence of the operator. The importance of the operation cannot be overstressed, because it is upon this, together with subsequent paint coatings, that the efficiency of the primer will depend.

Zinc-rich and metallic-lead primers are the most expensive, and the most effective provided they are applied to virtually clean metal. They are intolerant of poor surface conditions and will fail if so used. Zinc chromate, zinc phosphate, red lead and calcium plumbate are all rust-inhibitive paints and will actively resist corrosion. Zinc chromate is a pale greenish yellow in colour, and is also widely used for painting non-ferrous metals. Zinc phosphate is a pale grey and useful when light finishing colours are required; this too may be used for non-ferrous metals. Red lead is the traditional primer; it may be slow drying, and its strong orange colour may cause problems in using light finishing paints, but it is the most tolerant of surface preparation. Calcium plumbate is available in a limited colour range from stone to dark grey. It is similar to red lead in many ways, but shows remarkable adhesive properties to surfaces, as noted elsewhere. These primers are best applied by brush, as soon as possible after preparation. Where there are a number of sharp edges involved it will be beneficial to allow two coats of primer, even if one coat is confined to the edges only. This will give a better build up of paint at these vulnerable points.

Non-ferrous metals

Unlike ferrous metals, these are usually painted for decoration and not protection, particularly in a domestic situation. The problem is one of getting good adhesion between paint and surface. It is sufficient, in most cases, to degrease and abrade the surface with emery cloth or steel wool, but do not use steel wool for aluminium. Any steel left embedded in the aluminium will give rise to detrimental electrolytic action. After preparation as described, proceed as follows:

- *Copper* – apply zinc chromate primer.
- *Copper hot-water pipes* – no primer, but use finishing paint direct.
- *Zinc, galvanised or sheradised surfaces* – apply calcium plumbate primer. The adhesion of this is far superior to other primers.
- *Aluminium* – apply zinc chromate; lead must not be used.
- *Lead* – apply finishing paint direct.

When protection of the metal is of importance, in areas of acidity, an alternative method is far more effective. Use an etch or wash primer and follow with zinc chromate primer. These etch primers are water thin and contain an acid, usually phosphoric. As the name implies they etch the surface and give good adhesion for the normal paint system. Note that lead does not respond to this treatment. No mention has been made of aluminium paint or red oxide for use on metal. By themselves they are ineffective as primers. If the surface is damaged then corrosion is rapid. However they are used in combination with other materials to form proprietary primers.

Miscellaneous surfaces

From time to time it is necessary to use an oil paint on fairly unusual surfaces. The problems associated with the surface vary, but the more usual ones are as follows.

Vinyl guttering and the like. Although it is not necessary to paint this for protection, a colour change is required from time to time. There will usually be traces of release agent on the surfaces; this must be washed off with either detergent or white spirit. After etching the surface with steel wool or sandpaper, apply gloss paint direct.

Glass and ceramic tiles. These surfaces cannot be etched. They must be thoroughly cleaned to remove all traces of grease and then a strong adhesive coating applied. Gold size is suitable for this purpose. If possible, add a small amount of mica to assist adhesion.

Expanded polystyrene. This will dissolve if coated with oil paints. If an oil paint finish is required, first coat in with emulsion paint, making sure there are no misses.

Fabrics. Occasions arise when wall hessians are to be painted. Oil paints tend to make these brittle and they are best coated in with emulsion paints.

PAINTING

Filling

After priming, timber and plaster surfaces will require nail holes and other surface defects filled to an even surface. The quality of the finish required will dictate the extent of the filling operations.

For external painting, it is usually sufficient to use putty for this purpose. External grades of proprietary cellulose filler may also be used. These are particularly useful for any fine shrinkage cracks in rendering.

Internal painting is usually more concerned with decorative appeal, therefore more attention should be given to filling. There are two basic types of filler, water-thinned and oil-based. The water-thinned are the more widely used, and may be ordinary gypsum plaster for plaster walls or proprietary cellulose-type filler. This may be used on either wood or plaster, before or after priming. Coatings of not more than 3 mm thickness should be applied, the surface overfilled slightly and then rubbed down with a dry abrasive to a smooth finish. Proprietary forms of alabaster are also used in similar fashion; the surface should be wetted in before application. When using these fillers a thinned coat of varnish or gold size should be applied to the filled areas. This evens up porosity and prevents loss of gloss or colour change later.

Spatchel fillers, rather like a thick oil paint, are becoming more used for general surface filling. They can easily be applied with a squeegee, over either bare or primed wood. If correctly applied, little rubbing down is required.

Oil-based fillers are used when a really good finish is called for, in which case they should be specified. They are applied with a filling knife to primed surfaces. Again the surface should be overfilled and, when hard enough, rubbed down with a water-proof abrasive. This particular operation is fairly slow and consequently expensive, but can give a glass-smooth surface.

Choice of paint

There is now available a very wide range of oil paints, some glossy, some matt, some thixotropic (jelly-like), some requiring undercoats, some not.

Paints may be described as alkyds or polyurethanes, or by some other name intended to indicate their content. They are highly complex products, but in simple terms they have two components, a fine coloured powder, the *pigment,* dispersed in a liquid called the *medium.* The pigment serves to give colour to the paint, and to reflect the damaging ultraviolet rays in sunlight. It may also serve other purposes. Generally speaking, it is common to all types of paints. The medium is the part that gives a paint its main characteristics, and consists basically of a resin, which may be combined with a drying oil. So, when we talk about an alkyd paint or an epoxy paint, we are talking about a paint based on those resins. Natural resins, the sticky substance that exudes from trees, were used, but nowadays practically all are man-made.

As a general guide, most paints used on dwellings are alkyds. These are modified in manufacture to be glossy, matt or semi-matt, to be of normal consistency or thixotropic. The alkyd may be blended with other resins to give harder or more water-resistant paint films (for example, polyurethanes and silicones that can be thinned with white spirit).

Only gloss paints should be used externally for finishing coats on timber and metal. The matt and semi-matt oil paints are not sufficiently weather-resistant and are for internal use only. Two coats of paint are the very minimum that should follow the primer. An additional coat would offer greater protection externally and opportunity to produce a better finish internally. Where normal alkyds are used the extra coat could be an undercoat inside and gloss outside.

Generally, thixotropic paints give thicker coatings and only two coats will be required. The manufacturer's instructions should be followed closely as there is some variation between them. The practice with semi-matt finishes also varies: some are applied over a standard undercoat, others are applied coat on coat. There may be those who consider that undercoats should always be used, but some paints are formulated in such a manner that the undercoat can be replaced by an extra coat of finishing. These paints are in no way inferior. Provided paints are used and applied according to the maker's instructions, no problems should arise.

WATER–THINNED MATERIALS

All water-thinned paints have certain similarities. They are all rust-stimulative, require storage in frost-free conditions, and cannot be successfully applied in frosty weather. Apart from oil-bound distemper, all are oil-free and can be applied over alkaline plaster or cement surfaces just as soon as these are dry.

Emulsion paints are the most widely used and vary considerably in quality. When used externally PVC or copolymer types should be used; the vinyl emulsions are for internal use. They are most successfully applied to slightly porous surfaces, two coats usually being sufficient for most purposes, although certain colours will require an extra coat. These colours are usually indicated by the maker on the colour card, and usually are in the yellow range. In addition to the standard matt finish, semi-matt or eggshell finishes are available as well as a limited range of glossy finishes.

Distempers are now very rarely used and not easily obtainable. Their use is sometimes called for when a dead matt finish is required.

Masonry paints are not necessarily water-thinnable, but many are. They are tough materials incorporating a bonding agent such as mica or shredded nylon, and are intended for external painting of rendering, brickwork etc. Two coats are usually sufficient. Colour range is good.

Cement paints use white Portland cement as a base. They are for use in similar situations to masonry paints, and are usually supplied in powder form. Colour range is fairly limited. Make sure the surface is not too dry when applying; if it is dry it is quite safe to hose over the surface first with clean water. These materials tend to powder with age, sometimes needing a sealer before redecorating.

Texture relief paints, also known as plastic paints, may be supplied in powder form or ready for use. The application of these materials in some areas is left to sub-contractors, but it is a reasonably straightforward job for the decorator. Mixing and surface-preparation instructions should be followed closely. Texturing can be achieved by stippling with rubber stipplers or wood floats. It may be combed or dragged. Whatever method is used, it should be one that can consistently be repeated.

CLEAR WOOD FINISHING

This is always in steady demand for timber that is to be left in its natural state, both externally and internally. When this is required it is essential that the timber is left free of glue or hammer marks, and that the surface is not sanded across the grain. Any such defects will be exaggerated by clear finish.

Varnishes are available as gloss or semi-matt finish. The latter is for interior use only. They are now rarely described under the old names of carriage varnish or church-oak varnish (although boat or yacht varnish is), but usually by the resin used – e.g. alkyd gloss, polyurethane eggshell etc. Special varnishing systems have been developed for use where a class 1 flame-spread rating is required; these need to be applied exactly as stipulated by the manufacturer. There are also a variety of floor varnishes, which may be oleo-resinous, epoxide or polyurethane, single-pack or two-pack. Again, application should follow the manufacturer's directions. When none are given, for general varnishing proceed as follows. Sand down in direction of grain. Whether a gloss or matt finish is required, apply a first coat of gloss varnish thinned by approximately 30 per cent white spirit. Nail holes or cracks can be filled with coloured putty or a proprietary wood filler. Apply a full coat of gloss varnish, and then apply a finishing coat of gloss or semi-matt as required. Lightly sand down between coats. An additional varnish coat is recommended for exterior work.

Boat varnishes, despite a tendency to yellow, appear to give the best results for exterior work. They are both hard and flexible enough for most situations. *Polyurethanes* are good outside, but very quickly flake off if moisture can penetrate from behind.

A superior finish is from time to time required for interior surfaces. To obtain such a finish, rub in a grain filler, use polyurethane gloss varnish and coat in as described. After allowing the final coat of varnish to thoroughly harden off, carefully abrade with fine steel wool and wax polish, finally polishing with a soft cloth.

It should be remembered that, however well applied, external varnished areas are attacked by the ultraviolet rays in sunlight and cannot be guaranteed to last for more than eighteen months. To overcome this problem, varnishes are being introduced incorporating ultraviolet reflectors to give them a similar life to normal paints. Apart from the comparatively short life of exterior varnish, there is often the need to strip existing coatings before revarnishing. This can be avoided by using non-film forming materials in the first place. These are available from various manufacturers and are not varnishes but waxlike coatings, usually with a resin to give a degree of hardness.

SPECIAL PAINTS

For general domestic painting the alkyds or modified alkyds are normally adequate. In certain circumstances a more resistant or decorative finish will be required, however.

Chlorinated rubber paints are particularly suitable for application to concrete swimming pools and surrounding areas. They are very water-resistant and unaffected by chlorine that may be added to the water. Colour range is limited. Application is by brush and roller. The second coat is liable to lift the first, so this is best applied by roller, even if the first coat is brushed.

Epoxy paints are very tough materials, able to resist chemicals and abrasion. They may be single-pack or two-pack. In the latter case the material is supplied in two cans, which are intermixed immediately before use. Once mixed the usable life of the paint is about sixteen hours. The two-pack type is about twice the price of ordinary paints. A limited colour range is available for use as floor paints; clear finishes are also produced for the same purpose. All the above require their own particular thinners and brush cleaner and cannot be applied over standard paints.

Multicolour paints make possible the application of base colour and speckled finish as one operation. Application is by spray only. There are three basic types available, which may conveniently be described as high odour, medium odour and low odour. Colour range is quite wide. The material provides a tough decorative finish, very suitable for the walls of common staircases. Air pressures used for spraying should be closely watched, since they affect the spot size. The recommended pressures vary between different brands. The thinners used between types will also vary. According to type, it may or may not be possible to apply multicolour paint over existing paintwork without the use of a buffer coat. Shelf life is usually limited to six months.

Anti-condensation paints are sometimes required where mild condensation is present. They are ineffective in cases of severe condensation. The paint contains shredded cork or similar insulator and gives a lightly textured finish. Application is by short-pile roller or brush and stipple. The paint is usually only available as white, but can be over-painted.

Flame-retardant paints are available as gloss paints, eggshell paints, varnishes and emulsions. Correctly applied, they are able to raise a class 4 rating of flame spread to class 1. There are two general types in use: *intumescent,* which foams up and insulates the surface in the presence of fire, and *antimony chloride,* which insulates the surface with an inert gas in the presence of fire. Application is as for standard paints. Shelf life is usually limited to six months. Colour range is good. The spreading rate should be as stipulated by the maker.

Methods of application

Brush application is often the most convenient and the most widely used method, but neither roller nor spray gun should be ignored: each is far superior to a brush on occasions.

Rollers are available in a wide range of shapes and coverings, and are not confined to flat surfaces. They are ideally suited to the application of emulsion and semi-matt finishes, and are particularly useful over textured surfaces or very porous surfaces such as brickwork. In these conditions they apply a more even coat of paint much quicker than a brush. It is accepted that rollers use slightly more paint than brushing, but this is more than balanced by the higher application rate, which at the very least is twice as fast as by brush. For textured surfaces use lambswool rollers. The deeper the texture the deeper the pile. Mohair or short-pile rollers are best for oil paints and smooth surfaces. They will show only a light orange-peel texture, which if found objectionable can be laid off with a paint brush. Gloss paints can also be applied with short-pile rollers, but the results are variable according to the characteristics of the particular paint used.

Spray application is by far the quickest over all types of surface and with all types of paint. The main objection to its use is the need for masking off and the 'fog' that may occur. Both these objections can be overcome to some extent, the first by the use of templates where practical and the second by using the minimum air pressure necessary to obtain satisfactory atomisation of the paint. Manufacturers' literature will show the wide variety of equipment available. Hire firms, operating in most areas, can supply much of this.

Spray equipment will consist of a compressor, spray gun or guns, possibly pressure pots, and the necessary air and fluid hoses. Compressors may be electric, petrol or diesel powered. They are graded in size according to the *amount* of air delivered rather than the *pressure* of air; this is because the amount of air delivered is directly related to the amount of paint applied by the gun per minute. The pressure among other considerations controls fineness of the finish. Outputs from compressors are pretty standard between the different manufacturers, 1.20 litre/s (2.5 ft^3/min), 3.45 litre/s (7 ft^3/min), 7.15 litre/s (15 ft^3/min), 11.8 litre/s (25 ft^3/min), being common; the maximum pressures will vary, however. A point to note concerning electric-powered compressors is that those delivering 1.20 litre/s will run from a 3 amp 250 volt supply, and the 3.45 litre/s type will run from a 13 amp supply. Those above this will require a greater amperage at 250 volts, but may be obtainable at 410 volts. For site work compressors are available at 110 volts, or of course a transformer can be used.

Spray guns will also vary in the amount of air required as well as the types of material they can efficiently apply. Compressor and gun (or guns, in the case of the larger compressors) should therefore be matched so that the gun never requires more air than the compressor can supply.

Figure 54.1 shows four types of spray painting set-up. Note that with the constant-bleed type no air is stored, and constant-bleed guns *must* be used. A 1.2 litre/s compressor will feed one constant-bleed or one pressure-feed gun, a 3.45 litre/s compressor one suction-feed or one pressure-feed gun, and a 7.15 litre/s compressor two suction-feed or two pressure-feed guns. With the electric airless set-up usually one gun is used, but larger units can operate more guns.

Airless spray units have entirely overcome the problem of 'fog', With this system, no air actually mixes with the paint, but instead the paint is forced at high pressure (200 bar) through the narrow orifice of the gun. The equipment is not at all bulky. Paint is applied far more quickly than with a conventional spray. Because there is no overspray, it is possible under certain circumstances to cut edges in with a brush and then spray to within 25 mm of the edge.

Apart from large areas, spraying is also a very convenient method for painting such items as radiators, particularly the tubular type. Also, as mentioned elsewhere, multicolour paints must be sprayed on. Normal spray equipment can be used, but several manufacturers offer a compact unit for this purpose.

Painter's mitts and pads are often disregarded as methods of applying paint. The mitts are often a quick and convenient way of painting plain iron railings. Pads are sometimes the best way of applying very thin materials without splashing – for example, certain flame-retardant clear finishes.

MAINTENANCE PAINTING

External

It is generally true that little external repainting is carried out unless it is obviously necessary. When paint ages it becomes porous, allowing moisture through to the underlying surface. This must be allowed to dry out thoroughly before repainting; springtime following a wet winter is therefore not the best painting time; the job should be done later, when the warm sunshine has had a chance to dry out the moisture.

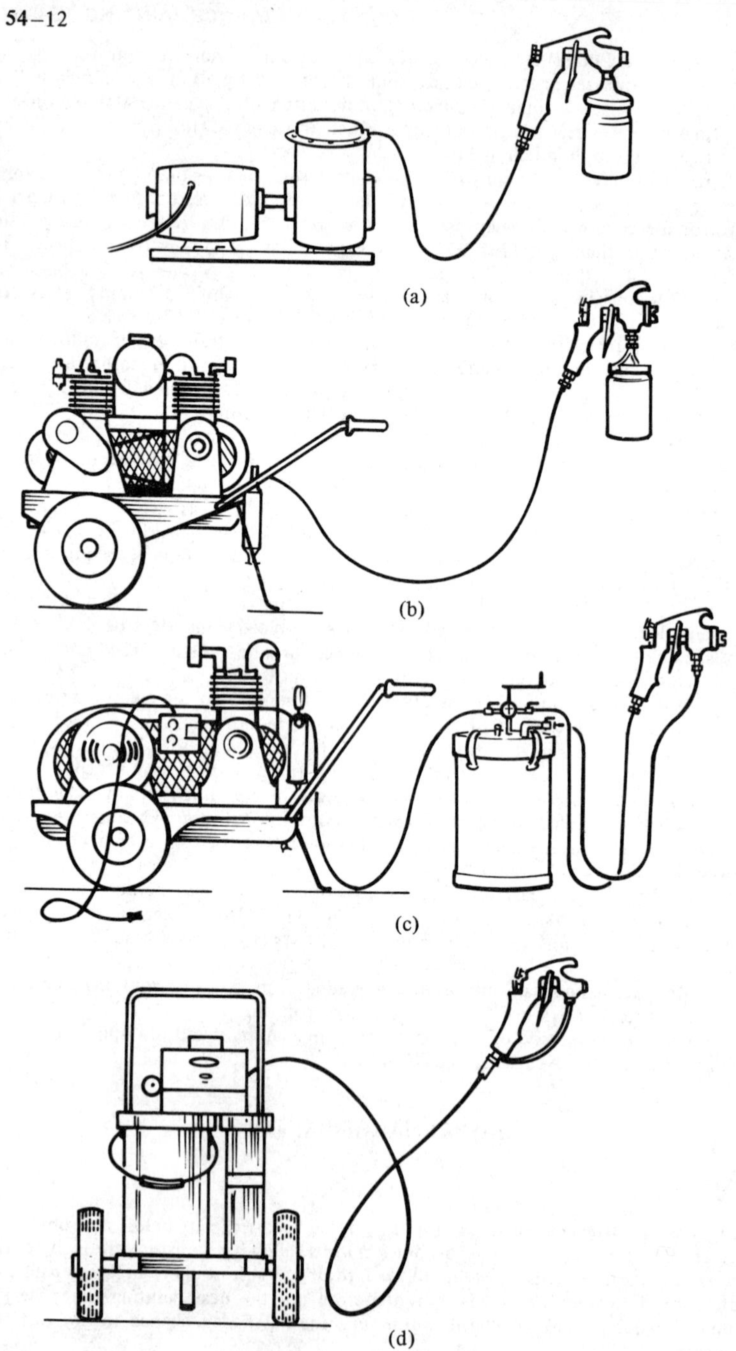

Figure 54.1 Types of spray equipment: (a) constant bleed; (b) suction feed; (c) pressure feed; (d) electric airless

Preparing and repainting woodwork. There is no point in removing paint in sound condition. It will require abrading, and the accumulated dirt or atmospheric grease must be washed off with sugar soap. Some paints will powder towards the end of their useful life; these too may be abraded, but if they are porous wash them with white spirit. Surfaces upon which the paint has flaked, blistered or is badly cracked must be stripped. Gas torches are the best for this purpose. Note, however, that acrylic primer cannot be burnt off; use solvent paint remover instead. After removal of old paint, the surface should be treated as new softwood.

In the general preparation of timber, glazing putties should be raked out if loose and the exposed surface, together with any cracks or nail holes, spot primed before puttying. It is essential that putty has paint beneath it, or else it will lose its oil to the surface, shrink and fall out. After preparation, one undercoat and one finishing coat will suffice, although an additional coat of undercoat or gloss would prolong maintenance periods, even if this extra coat were confined to cills and exposed areas.

Preparing and repainting renderings. For satisfactory repainting it is usually best to use the same type of finish as previously. Surfaces should be well scraped and brushed down with a stiff brush, and any flaking material removed to obtain a sound surface. Making good can be carried out using cement and sand or external cellulose filler. Do not use cement if oil paints are to be applied. Any lichen or mould must be treated either with a strong bleach solution or fungicidal wash and scraped off. Powdery surfaces require coating in with a binder-type sealer or stabilising solution. After preparaton, two coats of the selected material can be applied. Unless there is a radical colour change there is no point in extra coats.

Preparing and repainting steelwork. The main preparation is concerned with cleaning off grease and grime and abrading. Any bare or rusty spots should be well abraded and spot primed; insides of steel gutters can be cleaned out and given two coats of bituminous paint. Provided the paint is in reasonably good condition one undercoat and one gloss coat will be sufficient. Exterior painting of any sort should not be undertaken in direct hot sunshine. Neither, of course, should it be undertaken in inclement weather. There is no virtue in assiduously drying off the moisture with one hand and painting with the other. The paint will be blistering or flaking off within six months, and the painter's reputation with it.

Job procedure. Generally, it is best to get all the dirty cleaning-down finished before any painting, although on extensive buildings this is not always possible. Burning off and priming should follow: gutters should be cleaned out and under-coated, together with soffits, gables etc. Any painting of stucco work should be ahead of the painting of windows and doors, and should be completed before finishing any paintwork. Glass should be cleaned before final coating and any unwanted paint splashes removed. Care of shrubs, flowerbeds and lawns should always be taken.

Internal

Rooms should be as clear as possible, with carpets lifted and any furniture protected by dust sheets. The same basic rules apply as for exterior repainting. Wash and abrade surfaces, only removing paint that is in poor condition. Generally, work from top to bottom and get all the dust and mess away before any painting commences. Whenever possible, always strip existing wall hangings before hanging new ones. There is always a strong chance of edges lifting if this is not done. When washing surfaces not to be repainted, use a mild solution of sugar soap or detergent and wet in from bottom upwards. This will avoid streaking.

PAINTING DEFECTS

The majority of complaints concerning defective paintwork can usually be traced to one of three basic causes. These are bad preparation, moisture, and wrong use of materials. Although often blamed, the paints, if of reputable brands, are rarely at fault. The more common failings are listed below.

Defect	*Cause*
Flaking	Long interval between coats, allowing moisture or grease to be absorbed. Failure to back prime. Painting moist timber.
Loss of gloss	Insufficient paint beneath gloss. Gloss paint adulterated, e.g. thinned with paraffin. Dampness on drying paint. Slow drying.
Blistering	Moisture or solvents trapped in surface.
Bittiness	Poor preparation of surface. Dirty brushes. Unstrained paint. Using very old paint. Mixing odd paints together.
Sheeriness (paint looks patchy)	Wrong colour undercoat. Paint over-thinned. Dirty brushes.
Chalking	Natural long-term aging of some paints. Insufficient paint or diluted finishing paint.
Failure to dry or slow drying	Low temperature. Failure to neutralise paint remover. Grease or nicotine tar on surface. Lack of light.
Bleeding	Pigment in existing paint dissolved by new paint.
Cracking, crocodiling, crazing	Bad system of coatings, brittle over elastic. Using old paint. Paint laid off too heavily.
Fading	Slight fading is normal with some colours. Undercoating wrong colour for finishing. Thinned finishing paint.

PAPERHANGING

Preparation

For best results wallpaper should be hung to smooth surfaces that are slightly but evenly porous. The surfaces should be dry, dimensionally stable, and free from grease, efflorescence, mould growth, cracks and any other surface imperfections. Preparation should be aimed at meeting these requirements.

- *Grease* should be washed off with detergent.
- *Efflorescence* should be dry-brushed off and moisture kept to a minimum.
- *Mould growth* should be coated in with strong bleach (Parazone, Domestos or

similar) or a proprietary fungicidal wash. Surrounding areas must be included. Scrape off and collect the growth, leave for two weeks; if growth reappears repeat.

- *Cracks* should all be filled with cellulose plaster-type filler or similar. If cracks are left the paper will cover them, but it will ridge up over the cracks later and show.

New plaster walls are ideal grounds for wallpaper once they have dried out. Preparation is straightforward: lightly rub down, fill cracks, and apply a coat of suitable paperhanger's size to satisfy the excessive porosity. Lightweight plasters may show some mild efflorescence; brush it off.

Dry lining, plasterboard, hardboard, strawboard: make sure all nails are lightly punched in and fill. It is advisable to apply two coats of alkali-resistant primer over the surface. The purpose is to ensure that the wallpaper can be subsequently stripped without damage to the surface.

Size-bound distempered surfaces are still found in redecoration work. They should be washed off with warm water back to bare plaster. If this is not possible, apply a binder-type sealer, then proceed as for new plaster.

Emulsion-painted plaster: if it is in poor condition, scrape and scrub off to bare plaster, then proceed as for new plaster. If it is in sound condition, rub down well with coarse sandpaper, fill cracks, then size with a handful of plaster added to the size. This will provide better key for the paper.

Gloss or eggshell painted plaster: treat exactly as above, for preference rubbing down with coarse caustic block. After applying size with plaster added, cross-line the walls. The purpose is to provide something into which the paste from the top paper can be absorbed. If this is not done, the paste will saturate the paper and cause staining.

Surfaces liable to movement (e.g. blockboard, tongued-and-grooved boarding): stop, prime, and fill nail holes and joints. Size with a fairly strong mixture, then cross-line with reinforced or linen-backed lining. This will take up the movement between boards and prevent the face paper wrinkling or cracking.

Foundation papers

Lining paper. This has two purposes: to provide a superior ground over which to hang wallpapers, and to provide a suitable ground to receive emulsion or oil paint. Nominal size 0.55 × 10 metres as single rolls; also obtainable as double or quadruple length rolls. Graded by weight per ream: 300, 480, 600, 800. Although 300 is the cheapest it takes longer to hang; 480 is only slightly dearer and is quicker to hang. The quadruple rolls give least waste. Buying in quantity makes a saving in cost.

Reinforced lining or linen-backed lining. This has a backing of scrim, and should be hung scrim face to the wall. Its purpose is to take up movement.

Chip paper or wood chip or plain ingrain. Incorporate wood chips, giving a textured base paper. Available as coarse, medium or fine. Nominal size 0.55 × 10 metres. The purpose is to give a textured surface, particularly if the existing surface is rough.

Anaglypta and supaglypta. Supaglypta is the heavier paper of these two. They represent textured plaster finishes of a wide variety, widely used on ceilings. Nominal size 0.525 × 10 metres.

Expanded polystyrene. Available in rolls 3 mm thick, 0.55 m wide and 10 metres long, also in panels 9 mm thick and usually 1200 × 1200 mm. Its purpose is to provide some thermal insulation. Must be hung with the manufacturer's recommended adhesive. Normal papers can be hung over this.

Pitch paper and metal foils. The purpose of these is to cover damp patches or soluble stains. Wherever possible, the cause of dampness should be treated at source since these materials only cover the damp. The pitch paper should be hung pitch face to the wall and lapped, using a fungicidal paste. The rolls of metal foil, which are the more effective, are hung metal face to the wall, being sure to use the recommended adhesive.

Pastes and adhesives

Pastes used for standard wallpapers are either starch pastes, which look like flour, or cellulose pastes. Wallpaper manufacturers normally recommend a number of brands of paste for their products, from which the decorator can choose. The solids content of the paste is a good guide to the properties one may expect it to have. This can roughly be obtained by comparing the amounts of powder needed to mix the pastes to unit volume for a specific purpose. Cellulose pastes have lower solids contents than starch pastes.

COMPARISONS OF PROPERTIES

Property	*High solids*	*Low solids*
Staining resistance	poor	good
Wet adhesion	good	poor
Open time	good	poor
Mould inhibited	not usually	yes
Mixed storage	poor	good
Slip	good	fair

There are several ready-mixed proprietary brands, which are intended for vinyls or specialised hangings.

Size and sizing

The purpose of sizing walls is to even up the porosity over the surface. It also provides better slip for the paper hanging and lessens the chances of edges lifting. It is usual to use diluted paste as size when using cellulose paste. Fungicidal sizes made from animal glue are also widely used. These require mixing with either hot or cold water.

Wallpapers and hangings

It is convenient to group these materials as wallpapers, vinyls and specialised hangings.

Wallpapers. These are usually subdivided into several groups, but a practical division is: duplex, simplex, smooth or polished face, and washable. *Duplex* papers are embossed and consist of two papers pressed together. They should not be over-soaked; ten minutes is the usual maximum time. It is important that they are hung free from blisters; any that subsequently arise should be left, and they will dry out. *Simplex* papers may be of varying thickness, embossed or otherwise. As a general rule, the thicker the paper, the thicker the paste used and the longer the soaking time. *Smooth or polished* papers are a form of simplex. Walls must be prepared to a really smooth finish, since any defects will be exaggerated by this type of paper. *Washable* papers may be of any type, and the general notes above apply. In addition, a mould inhibited paste must be used for the paper and for any lining beneath it, as well as an inhibited size.

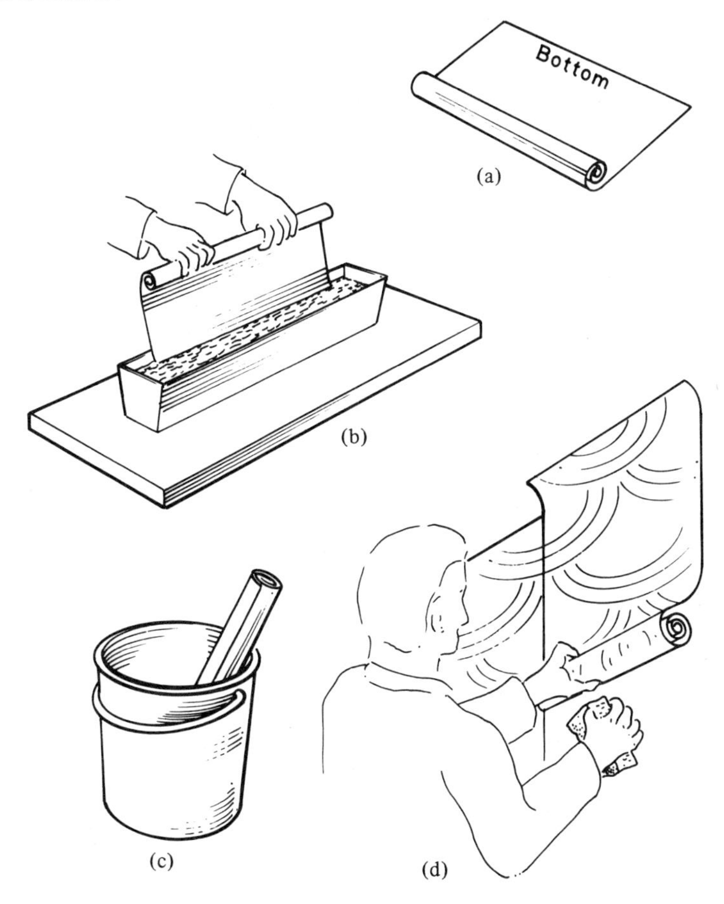

Figure 54.2 Hanging pre-pasted wallpaper; (a) cut paper to length and roll up from top; (b) roll back through water trough; (c) stand in empty bucket to drain; (d) unroll in position on wall

Vinyls. These are either normal or pre-pasted. With the normal types, use the recommended paste, but note that overlaps will not usually stick, although they appear to at the time. There are two-pack pastes available that overcome this problem. Also adhesives such as Copydex can be used on any overlapping edges.

Prepasted vinyls are often hung other than as intended; the wall is sometimes pasted or the back of the paper coated in with thin paste or water. All these methods are time-consuming and unnecessary. A better way is as follows (see *Figure 54.2*). Cut paper to length; roll up each length face outwards, starting from the top; place in the tray and roll through the water so that the top of the paper is now on the outside. Stand in a bucket to drain. Then hang by rolling downwards on the wall. This method ensures all paste is wetted, provides a convenient roll for hanging, and keeps the tray away from the work area.

Specialised hangings. Grouped together here are hessians, silks, grasscloths, felts, corks (paper-backed or tiles), and flocks both vinyl and traditional. Preparation of the surfaces should be thorough, and cross-lining is recommended to give the ideal porosity. Trimming of any selvedge edges should be by knife and straightedge. Use the paste recommended. If the material is unbacked (e.g. hessian), paste the wall and roll the material down. Materials that are backed should be pasted in the normal way.

Measuring for wallpaper

There are a number of different ways of doing this. The use of ready-reckoner tables is the most simple, and is reasonably accurate.

A widely used method is to find the area to be papered (in square metres) and divide by five, adding one extra roll in seven to allow for waste. This method is of variable accuracy.

Although more involved, the following method is the best. Divide the height of the area to be papered (in metres) into 10. Ignore any fraction. This gives lengths per roll.

Thus: height = 2.40 m; divide into 10 = 4

Measure the total length of the area to be papered (in metres) and multiply by two (go to the next whole number). This gives the number of lengths required.

Thus: length round room = 19.40 m; multiply by 2 = 38.80 m; go to next whole number = 39

Divide the second answer by the first, again going to the next whole number in the answer.

Thus: 39 divided by 4 = 9.75, i.e. 10 rolls required

Allow one roll in seven for waste, but note that no deduction has been made for openings. So if there are a number of doors or windows, no extra for waste will be required.

The amount of paper required for ceilings can vary according to the direction in which the paper is hung. Normally one can expect to hang parallel to the window wall, but with a good paperhanger this is not essential.

COLOUR

BS 4800

Builders and decorators will most often apply the colours specified by the client or architect. There are, however, occasions when they will be asked to advise on colour selection. Such advice must be given on a sound basis and not from guesswork or any imagined flair for colour sense. Fortunately, BS 4800, *Paint colours for building purposes,* is a useful instrument for the purpose. This standard sets out a range of 86 colours plus black and white, each colour with its own unique code of identification. Paint makers can supply all these colours, but for practical reasons usually limit the number available in any particular range of material. At the same time they include some of their own non-standard colours. To give a general idea, the trade colour range of gloss paints consists of those colours that are popularly used and available over the counter at any merchants. The remainder of the BS 4800 colours will be made, but held at area depots. Because emulsion paints are normally used on wall areas, the colours offered will be those considered suitable for this purpose. Semi-matt paints will also normally be limited because of their use, and also because of manufacturing problems with some colours. When choosing colours, therefore, one must always check their availability in the required material.

BS 4800 does not set out to name colours, only to give them a code. Paint makers put their own names to colours, so that the 'Oasis' offered by one maker is not necessarily the same as 'Oasis' offered by another. Regardless of any name, any two different brands of paint with the same colour code will be the same colour.

The code consists of an even number from 0 to 24, a letter A to E, and an odd number 01 to 55. Each colour is therefore indicated by a three-group coding, e.g. 02D45, 12E51 etc. The even number indicates the basic hue from which the colour is made; all colours with the same hue number are therefore derived from the same basic hue. The letter indicates greyness; thus A indicates maximum greyness and E nil greyness. All colours with the same letter will have the same greyness. The final odd number indicates weight. Again, all colours with the same number have the same weight.

Colour scheming

Warm colours are those containing red. As the name indicates they give the feeling of warmth, and also, according to their strength, have a closing-in effect. *Cool* colours are those containing blue. They have opposite properties to warm colours. *Neutral* colours are those that have no obvious tendency towards a particular hue, and sensibly can include those that act as a foil to strong colours, whether true neutrals or not.

When selecting colours, the following points should be kept in mind:

- the wishes of the client, and any existing colours, e.g. carpets or furnishings;
- the amount of sunlight the room receives;
- the use of the room;
- the shape or style of the room.

Beware of too much or conflicting pattern: usually one third of all areas, including the floor, is about the maximum. Avoid the use of too many colours. If strong colour is used, it is best confined to small areas; if a strong colour is used on a wall of any size, however, have plenty of quiet colours on other surfaces. The eyes will need a rest.

Bold patterns tend to make walls look smaller. Small patterns may be lost on large areas. Cool colours can be used to advantage in sunny rooms. Warm colours will cheer rooms that receive little or no sunlight. Yellow, correctly used, can give the impression of sunlight. Try and lose unpleasant shapes such as unsightly pipes by painting them in the background colour.

Warm colours are those containing red, and can include pure yellow tints. On the BS 4800 range they include hue numbers 02, 04, 06, 08, 10 and 24. The remainder make up the cool colours.

A range of colours selected from any single hue range will produce a satisfactory monochromatic scheme. This type of scheme may lack interest, however, which can be overcome by introducing small amounts of the opposite colour. Opposite or near-opposite hues will be found by adding or subtracting twelve to the BS 4800 hue number. For example, if the scheme has been selected from the 10 hue range a suitable colour will be found in the 22 hue range. It would also be satisfactory to use a colour from the adjoining range, in this case 20 or 24. Other schemes can be selected from closely related hue numbers, perhaps 04, 06 and 08. Another variation is to choose one hue, and also the two hues adjoining its opposite.

Finally, however, the test of any colour scheme is the answer to the question: can you live with it?

FURTHER INFORMATION

Paints and their uses are being constantly improved. To try to ensure that their products are correctly used, manufacturers produce comprehensive manuals covering their range and use. These make useful and up-to-date reference books. Other useful reading includes:

CP 231: 1966, *Painting of buildings*
CP 98: 1964, *Preservative treatments for constructional timber*

DoE advisory leaflets:
55, 56 and 57, *Painting walls*
70, *Painting ferrous metal*
71, *Painting non-ferrous metal*
73 and 106, *Painting timber*

A. Fulcher and others, *Painting and decorating, an information manual,* Crosby Lockwood Staples (1975)

55 DEFECTS IN PRIVATE HOUSE BUILDING

INTRODUCTION 55–2

STRUCTURAL DEFECTS 55–2

NON-STRUCTURAL DEFECTS 55–8

CRITERIA FOR DEFECTIVENESS 55–20

DEFECTS IN PRIVATE HOUSE BUILDING 55

55 DEFECTS IN PRIVATE HOUSE BUILDING

M.J.V. POWELL, MSc, MIOB
Research Manager for Building, Construction Industry Research and Information Association

This section (based on studies by the National House-Building Council) analyses the incidence, causes and prevention of structural and non-structural defects in private house-building, and makes suggestions for the quantitative definition of what does and does not constitute defective work.

INTRODUCTION

The present ten-year protection scheme for the purchasers of new private dwellings was introduced by the National House-Building Council in 1965. Thus, with two million houses in or having passed through the scheme, there is now available over twelve years of feedback information on defects that have occurred. Priorities for the improvement of private building quality and standards are clear, and the purpose of this section is to discuss them in detail. Regrettably, there is no comparable national source of data relating to public-sector housing as distinct from other types of building. Technological innovation has without doubt been greater in the public sector and there has been press reporting of some major failures, due apparently to the misuse of innovatory materials and techniques. One of the benefits of the private sector's comparative conservatism is that it has not experienced disasters in superstructure work. It is impossible to say whether the public sector has experienced defects in substructure work and general quality of building to a greater or less extent than the private sector.

STRUCTURAL DEFECTS

Analysis

Under the NHBC protection scheme, if a major structural defect occurs in the third to tenth years of the warranty period, the cost of remedial work is met by NHBC. Such claims give a nationally comprehensive indication of the types of structural defect that occur.

Table 55.1 shows an analysis by both number and cost of major structural defect claims met by NHBC in the two-year period ending June 1976. The table is in three sections: substructure – infill (i.e. oversite infilling to support concrete ground floors); substructure – defects due to hazardous ground and other causes; superstructure – all work above dpc. This grouping is chosen because 51.1 per cent by number and 54.6 per cent by cost of claims are attributable to infill, and 25.1 per cent by number and 35.8 per cent by cost are the other substructure defects. Thus substructure as a whole accounts for 76.2 per cent by number and 90.4 per cent by cost, and superstructure accounts for 24.6 per cent by number and only 9.1 per cent by cost.

Infill

Part A of the table, infill, includes the 'textbook' failures that are said to arise from poor infill material, particularly A15 (action caused by gypsum, i.e. old plaster, in

Table 55.1 MAJOR STRUCTURAL DEFECT CLAIMS PAID IN TWO–YEAR PERIOD ENDING JUNE 1976

Subject of claim		Frequency		Cost	
		No.	%	£	%
A	*Substructure–infill*				
A1	Building off slab	5	0.5	9 044	0.6
A2	Soft ground beneath	22	2.0	45 695	2.9
A3	Use of chemically active shale	110	10.0	226 990	15.4
A4	Sand, gravel, hoggin	44	4.0	52 491	3.4
A5	Fine ash	5	0.5	10 644	0.7
A6	Soil as fill	16	1.5	33 760	2.1
A7	Loss of fines	8	0.7	10 499	0.6
A8	Consolidation over 600 mm	55	5.0	81 378	5.1
A9	Consolidation below 600 mm and unrecorded depths	247	23.0	309 356	20.7
A10	Lack of fines	2	0.2	506	–
A11	Subsoil shrinkage	24	2.2	36 468	2.3
A12	Washout–water table	4	0.4	5 137	0.3
A13	Washout–leak	5	0.5	3 643	0.2
A14	Inadequate oversite	1	0.1	210	–
A15	Gypsum action	3	0.3	1 112	0.1
A16	Wet chalk	1	0.1	1 981	0.1
A17	Dry rot action	1	0.1	1 343	0.1
		553	51.1	830 257	54.6
B	*Substructure–hazardous ground and other causes*				
	Hazardous ground				
B1	Clay (usually with tree problem)	110	10.0	266 319	17.8
B2	Peat, silt and soft ground	13	1.2	40 023	2.7
B3	Sand	3	0.3	5 735	0.4
B4	Made ground	15	1.4	26 870	1.8
B5	Differential settlement	15	1.4	9 670	0.6
B6	Subsidence–unclassified	36	3.4	73 056	4.8
B7	Split-level dwelling	1	0.1	15 000	1.0
B8	Land slip	4	0.4	5 047	0.3
B9	Swallow hole	1	0.1	15 500	1.0
	Other				
B10	Retaining wall failures	2	0.2	557	–
B11	Damp-proof membranes and DPCs	14	1.3	9 820	0.6
B12	Workmanship–various	13	1.2	7 164	0.5
B13	Problems caused by drains	18	1.7	25 677	1.7
B14	Pile failures	1	0.1	13 540	0.9
B15	Water on oversite	9	0.9	7 741	0.5
B16	Sulphate–brickwork	11	1.0	15 848	1.0
B17	Raft failures	4	0.4	3 406	0.2
		270	25.1	540 973	35.8
C	*Superstructure*				
C1	Defective bricks	4	0.4	3 341	0.2
C2	Defective mortar	12	1.1	9 109	0.6
C3	Lintels	40	3.8	12 125	0.8
C4	Chimneys and flues	6	0.5	1 150	0.1
C5	Brickwork–various	14	1.3	8 535	0.5
C6	Wet rot	26	2.5	8 022	0.5
C7	Dry rot	16	1.5	11 919	0.8

continued overleaf

Table 55.1 *continued*

	Subject of claim	Frequency No.	Frequency %	Cost £	Cost %
C8	Flat roofs	61	5.8	33 175	2.2
C9	Pitched roofs	45	4.3	20 314	1.4
C10	Joisted floors	17	1.6	20 875	1.4
C11	Staircases	1	0.1	86	–
C12	Various	4	0.4	851	–
C13	Trussed rafters	10	1.0	6 890	0.5
C14	Balconies	3	0.3	1 657	0.1
		259	24.6	138 049	9.1
	Summary				
A	Substructure–infill	553	51.1	830 257	54.6
B	Substructure–hazardous ground and other	270	25.1	540 973	35.8
C	Superstructure	259	24.6	138 049	9.1
		1 082	100.8	1 509 279	99.5

the infill) and A17 (dry rot caused by the presence in the infill of old timber infected with dry rot spores).

Defect A3 (use of chemically active shale) is also a 'textbook' fault but its value here is extremely high, 10 per cent by number 15.4 per cent by cost. The majority of defects in Section A are not textbook defects. Generally they are covered by the poor compaction of infill (A8 and A9) and the use of inappropriate materials (A4, A5, A6, A7, A10, A12, A16). Also significant are categories A2 (soft ground) and A11 (subsoil shrinkage). With these, the ground underlying the infill has proved incapable of supporting infill, floor and any part of the house structure carried by it. Category A1 (building off the oversite slab causing subsidence thereof) is an example of careless workmanship and supervision on site.

It had become apparent by the late 1960s that settlement of ground floor slabs, partitions and fittings was the most frequent and costly major type of defect. Slab settlement occurred with all depths of infill from 300 mm to 1.5 m and more. The remedial costs were highest with the greater depths. Costings made in 1973 by builders showed that, where infill exceeded 600 mm deep, it would be cheaper to provide reinforced concrete suspended floors with rough infill as formwork than good infill consolidated layer by layer. Therefore it was made mandatory, where infill exceeded 600 mm, to use suspended floors of *in-situ* reinforced concrete, precast concrete or timber joists without sleeper walls. NHBC published *in-situ* slab design tables for the assistance of builders. Since the tables inevitably contained an element of over-design, builders were given the alternative of commissioning floor designs for their particular houses from professional engineers.

Category A3 (use of chemically active shale) includes as would be expected some cases of sulphate attack. These tend to occur in industrial areas where red shale and industrial brick rubble are common. The major unpredicted and catastrophic problem has been in the Cleveland area. This has affected good and bad builders alike. Briefly, what happened was that the houses were built using local sources of shale for fill beneath the ground floor slabs. Subsequently lifting of floors and interior walls and pushing out of external walls was experienced, movements of up to 100 mm occurring in some cases.

A study by the Building Research Establishment concluded that the floor heave arose primarily from a volume increase in the shale, associated with oxidation of pyrite and the resulting crystallisation of gypsum. Sulphate attack on concrete in contact with the slab may have been a contributory factor. On subsequent investigation the BRE found that the phenomenon had been experienced before in Canada, but that UK antennae had not picked up the very faint warning signals. The consequence of NHBC's experience with these problems has been that a virtual ban is operated on shale as infill, and all industrial waste materials are viewed with extreme scepticism.

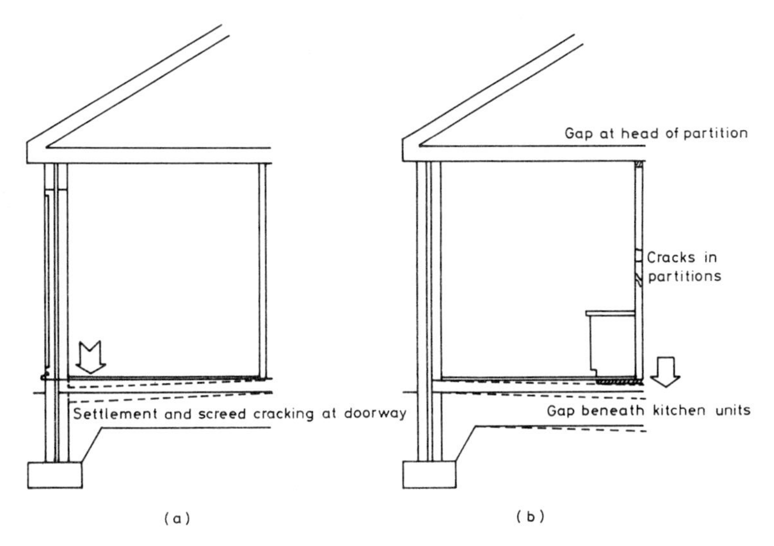

Figure 55.1 Floor slab settlement: (a) fill consolidated under load; (b) ground consolidated under load

Figure 55.1 shows the difference between the distortion resulting from consolidation within the infill and that resulting from consolidation due to poor ground.

Hazardous ground

Table 55.1 shows that, after infill, hazardous ground is the main cause of major structural defects. Progressively since the 1939–45 war, and particularly in the last decade, the community at large has expressed the will that development should not spread regardless into agricultural areas and the countryside, and should not necessarily be allocated the best available ground. Infill developments and the redevelopment of previously used sites have been common. More controversially, there is a widespread view that local authorities have kept better land for their own house building and allocated the more hazardous ground for private development. The claims experience of NHBC has emphasised that, given this situation, the industry must increase its technical capacity to meet the problems posed by building on hazardous ground, and must budget realistically for the abnormal costs that arise from the foundations necessary to meet hazardous conditions. High prices cannot be paid for hazardous sites: the money is required for the foundations.

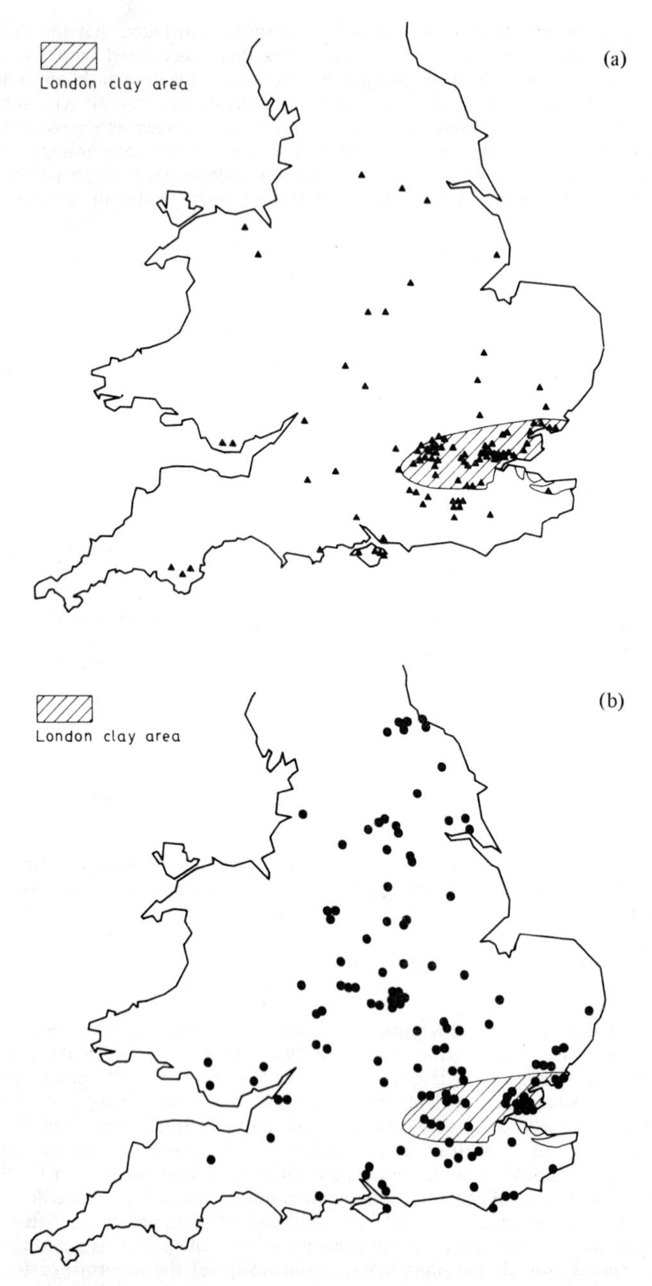

Figure 55.2 Distribution of defects arising from clay/tree problems: (a) June 1974 – May 1976; (b) December 1975 – November 1976

Apart from the clay and tree problems discussed below, the problems arising from hazardous ground are mixed and include examples of most that might be expected. There is some tentative evidence that about one failure in five is in an engineered foundation, either piling, raft or deep strip. These failures are thought to arise mainly from inadequate site appreciation and ground investigation and insufficient control on site, rather than from design errors *per se*.

One of the most common problems arising on small clay redevelopment and infill sites was what to do about the trees on the site itself or on adjacent sites. Table 55.1 has shown that this is the largest single cause of major structural defects on hazardous ground. In 1969 NHBC first published Practice Note 3, whose main object was to give guidance on deep strip foundations. Builders were unable to keep houses clear of trees by one to one-and-a-half times the height of the tree and use shallow foundations, and they believed that the alternative of piling was prohibitively expensive on the small or slower-moving site.

The 1975–76 drought increased the incidence of clay and clay/tree problems. It is interesting to note how widespread the clay problem became in the drought period (see *Figure 55.2*). Part of the reason may be that, outside the dry areas of south-east England, the clays do not normally become critically dry.

Superstructure

Part C of Table 55.1 shows that the greatest costs for superstructure defects are attributable to C8 (flat roofs), C9 (framed roofs) and C10 (joisted floors). In private house building, flat roofs can be either the large roofs to blocks of flats built mainly in London or on the coast, or the simple dormer-type roofs used widely on chalet houses. Often with the former, and almost always with the latter, the faults arise from failure to observe the four simple crucial rules for flat roofs shown in *Figure 55.3*.

Where the failure of the flat roof of a block of flats is not attributable to one or more of the four basic faults, it is usually caused by poor design decision-making and the unsuitability of the design for the situation. Confusion of warm-roof construction with cold-roof construction can result in the rapid breakdown of the roof structure. See *Figure 55.4*. It is held by some technical authorities that the use, in cold-roof construction, of degradable deck materials, such as the lower grades of chipboard, is not appropriate for a major element such as the roof of a block of flats. The flat roof is architecturally attractive, but the use of low-cost specifications has done no good to its cause.

The major defects that occur in pitched roofs often illustrate the fact that the centuries-old rapport between architect and craftsmen is often not present nowadays. There was a time when an architect could detail an unusual roof in the fullest detail and the carpenter could readily appreciate that what was being asked for was practical and sensible, subject possibly to the clarification of some minor points on site. Alternatively, the architect could give a broad outline only, knowing that the carpenter would use his skill and judgement to produce an adequate structure, again subject to some discussion and clarification of detail. The problems now arise with roofs of slightly unusual shape. Neither the architect nor the carpenter has sufficient detailed knowledge and skill; each relies on the other, with disastrous results. This is not a criticism, although it is possibly a cause for regret. We live in different times and we have to accept new realities. The almost total eclipse of traditional roofing by the trussed-rafter roof means that, within a generation, skilled roofing carpenters will not be available for general house construction.

Failure in joisted floors often occurs in narrow-fronted terraced or semi-detached houses, and is often due to a combination of circumstances, basically in design. Deep floor joists are used to span from party wall to party wall. These are subject to

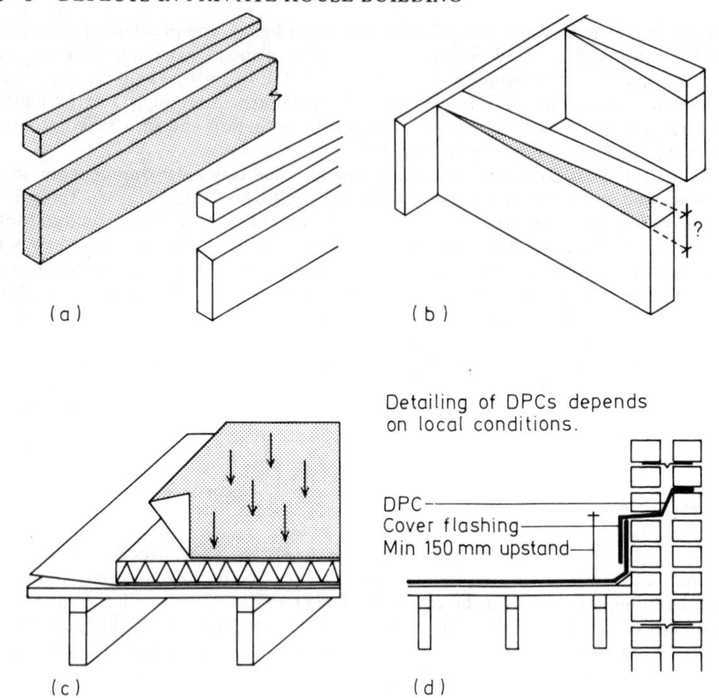

Figure 55.3 Four rules for flat roofs: (a) joists and firrings must be preserved; (b) fall must not be less than 1 in 60; (c) deck must be kept dry with temporary protection until roof covering is laid; (d) upstands must be detailed to prevent moisture penetration

deflection and shrinkage. The airing cupboard and hot water cylinder are so located that their load has to be taken by the critical joints between the trimming and trimmer joists (*Figure 55.5*). The heat from the cylinder causes exceptional shrinkage. Supplementary causes of the defect are likely to be failure to keep structural timbers reasonably dry during construction, thereby adding to shrinkage, and inadequate workmanship in framing the floor, which makes the joints less strong than they should be.

NON-STRUCTURAL DEFECTS

Analysis

Under the NHBC protection scheme, if a non-structural defect occurs in the first two years of the warranty period, and the builder and purchaser are unable to reach agreement on whether or not it should be rectified, NHBC provides a conciliation service. 1028 conciliation cases were analysed and were found to involve 5372 adjudged defects. The defects were of 375 different types. Table 55.2 gives details of the incidence, likely causes and means of preventing the twenty most frequent defects. These accounted for 18.3 per cent of all defects. Table 55.3 gives the incidence, likely causes and means of preventing defects that, on the basis of

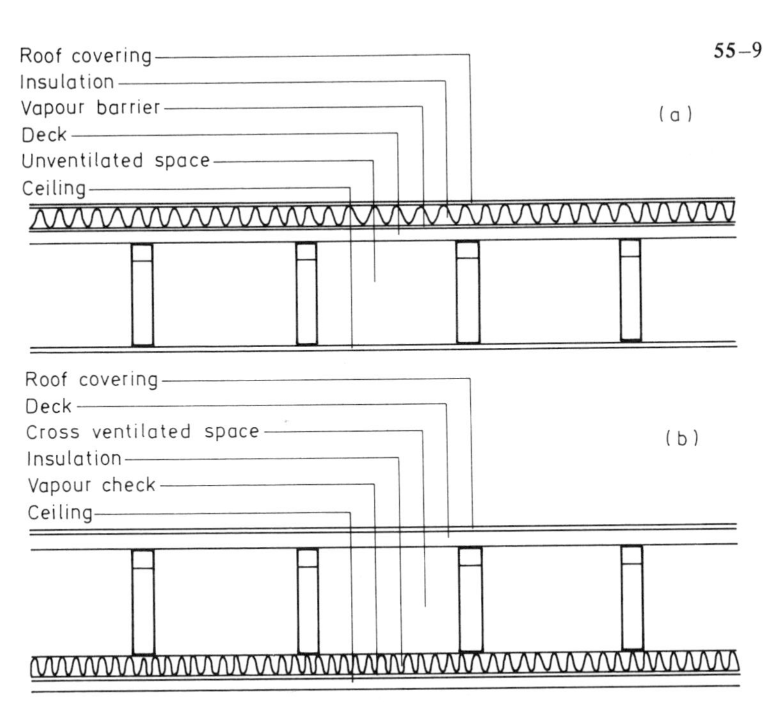

Figure 55.4 (a) Warm roof principle; (b) cold roof principle

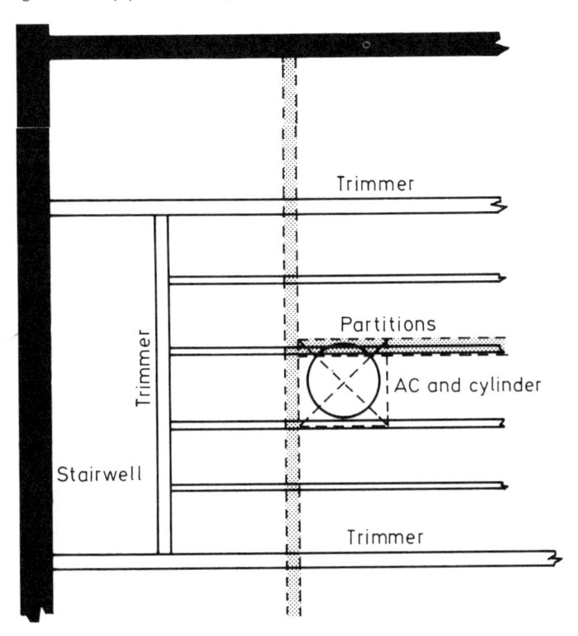

Figure 55.5 Joisted floor failure

frequency multiplied by a points allocation for seriousness, emerge as important. Following normal quality-control practice, the points allocation used was:

- 50 points demerit for serious defects, including damp entry and any defect affecting health or safety;
- 15 points demerit for non-serious defects that could not reasonably be lived with, e.g. loose or damaged fittings;
- 5 points demerit for non-serious defects that could be lived with by most people, e.g. rough paintwork.

The five most frequent defects

Rough internal paintwork A standard of painting that is tolerable outside the house may not be acceptable inside, where it has to be looked at daily and dusted weekly. A building site is a dirty, dusty or muddy place, and often the dust and dirt gets ingrained in the internal woodwork, which must be cleaned down before it is painted. In addition to this, there is the poor machining of some bought-in joinery. One

Table 55.2 MOST FREQUENT NON–STRUCTURAL DEFECTS

Defect	*Occurrences*	*Likely causes*	*Prevention*
Rough internal paintwork	67	Lack of sanding and filling. Lack of skill and care in paint application. Rough finish left by joinery manufacturer.	Greater care. Better supervision
Incomplete external paintwork	65	Lack of skill and care in workmanship and checking.	Use of completion check-list.
Missing mastic pointing to windows	63	Lack of skill and care in brickwork. Failure to check house before handover.	Use of completion check-list.
Incomplete internal paintwork	56	Lack of care in workmanship and checking.	Use of completion check-list.
Shrinkage gap between bath and wall tiling	53	Settlement due to the weight of a filled bath. Joist shrinkage.	Storage of timber off the ground. Use of a flexible joint between bath and tiling.
Ponding on drives	53	Lack of skill and care. Inadequate consolidation of ground after backfilling of trenches before laying drive or path.	Better supervision.
Ill-fitting internal doors	51	Lack of skill and care. Shrinkage due to excessive moisture in door.	Greater care in workmanship and storage.
External door requiring easing	49	Swelling in winter. Hygrothermal stress due to differences in internal and external humidities.	Use of more stable and robust door. Minimise movement by careful storage.
Cracks in wall plaster	48	Shrinkage in wall behind plaster. Differential shrinkage between rendering and setting coats.	Protection of blocks against damp before fixing.

Table 55.2 *continued*

Defect	*Occur-rences*	*Likely causes*	*Prevention*
Incomplete plaster making good	48	Failure to return after other trades. Lack of care in checking.	Use of completion check-list.
Flaking plaster	46	Insufficient thickness of plaster over conduits and cable sheathings. Shrinkage in plaster or wall behind.	Chasing in. Protection of materials against damp.
Breakdown of external paintwork	46	Painting on wet timber. Insufficient body in the paint system due to watered-down or low-quality material.	Better supervision.
Faulty pumps, switches, valves and thermostats	45	Normally a manufacturing deficiency in the component.	
Loose wall tiles	45	Use of wrong or insuffīcent adhesive. Differential shrinkage between wall, plaster and tiles.	Following of the manufacturer's fixing instructions. Protection of materials against damp.
Loose or cracked floor tiles	45	Insufficient adhesive. Damp from condensation, pipe leak or rising damp. Tiles laid on uneven surface. Movement in screed.	Greater skill and care. Clean and level surface before laying tiles. Refuse to lay tiles on unacceptable surfaces.
Loose glass or putties	43	Glazing wet frames. Lack of skill and care.	Greater skill and care.
Leaks in pipes, radiators or boilers	43	Bad workmanship. Failure to test adequately. Entry of oxygen into a closed system. Bi-metallic action. Chemical action on the exterior of the pipe by cementitious materials. Misuse of new types of pipe.	Greater skill and care.
Roof leaks around chimneys	42	Failure of the chimney/roof flashing system to keep out damp.	On sites of normal exposure, greater care in execution of normal flashing details. On exposed sites, the use of special details such as a lead tray or a second DPC.
Movement of window sashes	42	Seasonal swelling and shrinkage. Hygrothermal stress due to difference in internal and external humidities.	Use of more stable and robust sashes. Minimise movement by careful storage.
Rough plaster/ uneven wall areas	42	Inadequate use of rules. General lack of skill and care. Scarring of finished surfaces with trowels.	Greater skill and care in plastering. Greater skill and care in rubbing down and filling before decoration.

Table 55.3 MOST SIGNIFICANT NON–STRUCTURAL DEFECTS

Defect	*Occur-rences*	*Likely causes*	*Prevention*
Leaks at gutter joints	36	Failure to make the joint in accordance with the manufacturer's instructions. Slightly loosening joint causing grit from roof tiles to enter and force it open still further.	Greater skill and care
Loose pipes/water hammer	36	Oscillations in the pipe.	Use of more pipe clips. A maximum distance of 1.2 metres between clips is normally recommended.
Damp entry at jambs of openings	35	Failure to isolate the internal skin of the cavity wall from the external by the use of either vertical DPCs, rebated reveal construction, or complete isolation of the two skins.	Use of a 150 mm wide vertical DPC and not 112 mm. Use of rebated reveals. Non-closure of cavity in very exposed areas.
Non-alignment or overflow of gutters	34	Fall of the gutter away from instead of towards the RWP. Dip in the run of gutter preventing water from reaching the RWP. Occasionally, undersizing.	Greater skill and care. Use of more brackets. Greater design care.
Leaks in plumbing pipes	34	Lack of skill and care and failure to test adequately. A bi-metallic problem. Chemical action on the exterior of the pipe by ground or adjoining materials. Misuse of new types of pipe.	Greater skill and care. Knowledge of chemical composition and likely actions of local water.
Wire balloon missing (soil pipe)	33	Lack of care in checking.	Use of completion check-list.
Damp entry between door and frame	29	Excessive gap between door and frame.	Deeper rebate sections and greater care in the fitting of doors. Use of weather strip.
Broken or loose vertical tiles	27	Failure to nail each tile with two nails. Damage by ladders.	Greater skill and care.
Warped external door requiring replacement	26	Swelling in winter. Hygro-thermal stress due to differences in internal and external humidities.	Use of more stable and robust door. Minimise movement by careful storage before fixing.
Cavity bridged	25	Cavity blocked with mortar or mortar standing on a DPC tray or trunking. Wall ties sloping inwards or spots of mortar on wall ties.	Greater skill and care in brickwork.

Table 55.3 *continued*

Defect	*Occurrences*	*Likely causes*	*Prevention*
Weather board missing from door	25	Non-provision of a weather board in conditions where one is necessary to keep rain out.	Provision of weather board.
Leak at joint of WC with soil stack	25	Lack of skill and care in the making of the joint. Differential movement between the floor (if timber, subject to shrinkage), the pan and the soil pipe.	Greater skill and care.
Defective manholes	20	Failure to check levels before building manhole or to adjust completed manhole when found wrong. Lack of skill and care.	Greater care in checking levels at completion.

Note The following defects, listed in Table 55.2 as being frequent, were also regarded as important on a frequency × demerit rating basis: roof leaks around chimneys, faulty pumps and other heating accessories, leaks in heating pipes, loose or cracked floor tiles, ponding on drives, missing mastic around windows.

sometimes hears on site that it is not in the painter's price to clean-down after other trades, or do extra work to poor-quality joinery. This is fair comment, as far as the painter is concerned, if it was not included in his specification, but it is the builder's job to see that all necessary preparation and cleaning down is in someone's specification, and it is the agent's job to see that it is in fact done by that person. It is no good turning a blind eye to poor preparation until the decorations are complete, and then deciding that it is too expensive or too late to do anything.

Incomplete external paintwork It easily happens. The painter goes up the ladder and does not start quite where he left off, and there is a gloss-coat gap. He has to do the job piecemeal, half today and half next week, because someone else has not finished something. This is the way small things, like a window casement or the garage fascia, get forgotten. To a certain extent this is inevitable, but what is not inevitable is that things get left and nobody goes round to check that everything has been done; people get paid for incomplete work, and then no power on earth, apart from conscience, will bring them back to finish off. Like so many problems, this one boils down to careful, firm supervision. One person must be responsible for snagging finished houses before occupation.

Shrinkage gap between the bath and wall tiling Joists shrink for natural reasons, and they shrink more if they are allowed to get excessively damp before building in. They deflect under load, especially heavy point loads such as the feet of baths. Hence, because of shrinkage and deflection, after a few months an unsightly gap often appears between the edge of the bath and the bottom edge of the wall tiling. Some of the grout sticks to the tiles, some to the bath, and some falls out. A flexible joint between the bath and the tiling was made an NHBC requirement in 1975.

Ponding on drives Service trenches must sometimes run across drive areas. They may not be back-filled until late in the job, and today 'return, fill and ram in layers not exceeding 150 mm' sounds rather quaint. Inevitably, the hardcore base of the drive may consolidate within itself or sink a little into soft ground. Any of these contingencies may lead to ponding on the surface of the finished drive. Wading through puddles to get to the car annoys people. The defect can be prevented if sufficient time is allowed between backfilling trenches, laying hardcore and drive surfacing, and if the supervision of these operations, particularly the laying of the hardcore, is good. A crossfall or cambering also helps to shed water. If all else fails, *speedy service* will prevent the purchaser from finding more minor defects in the drive, the boundary wall, the garden, the garage, the hall . . .

Incomplete plaster making good The plasterer forgot to make good after the plumber, after the electrician, after the heating fitter, etc. Whether the plasterer is to blame or not, the site agent most certainly is. The agent must remember to look into the airing cupboard, the cupboard under the sink, around socket plates, around the chimney breast and in the cupboard under the stairs. These are some of the places where incomplete plasterer's work is found.

Responsibility

During the two years up to June 1976, complaints about small builders were proportionally more frequent than complaints about large builders. Small builders were found to build 31 per cent of new houses, but accounted for 66 per cent of complaint cases. The largest builders built 16 per cent of the houses but were responsible for only 5 per cent of the complaint cases. A detailed appraisal of the characteristics of the small firms prominent for defects showed them as a class to have neither local roots nor long building traditions. See Table 55.4.

Table 55.4 DISTRIBUTION OF DEFECTS CASES BY BUILDER SIZE GROUPS

Builder size group (number of houses built in 1969)	*% of houses built by size group*	*% of defect cases*
Over 500	16.0	5.0
100–500	27.0	15.0
50–100	13.0	9.0
30– 50	10.0	5.0
Under 30	34.0	66.0
Total	100.0	100.0

Defects were analysed into responsibility areas as shown in Table 55.5. *Design* defects are self-evident as a class. *Site supervision* was considered primarily responsible for not adequately checking a small number of operations known to be critical; for not ensuring that materials were properly protected to minimise shrinkage; for not condemning work which was obviously very rough; and for letting houses be handed over to purchasers with large numbers of minor defects and omissions clearly visible. *Workmanship* was considered responsible for fixing and jointing and similar craft processes. Failure in *materials* is self-evident as a class.

From Table 55.5 it can be seen that site supervision was considered to be responsible for more defects than any other interest, with workmanship second. When only those defects costing over £100.00 to remedy were considered, the main responsibility seemed to lie with site supervision and designers. Overall, therefore, site supervision was judged to be the vital area.

Table 55.5 ANALYSIS OF DEFECTS BY MEMBER OF THE BUILDING TEAM JUDGED TO BE RESPONSIBLE

Responsibility group	*% of all defects*		*% of defects costing over £100.00 to remedy*	
Defects primarily attributable to design				
Selection of site and design of appropriate foundations, groundworks and drains	1.4		19.4	
Design faults in superstructure	1.6		4.9	
General joinery detailing	1.2		–	
Failure to design for shrinkage	5.9		8.7	
Design faults in plumbing and engineering services	0.5			
Total		10.6		33.0
Defects primarily attributable to site supervision				
Failure in control of critical operations	3.9		27.1	
Shrinkage due to failure to keep materials dry	8.8		4.0	
Failure to condemn rough work	4.4		3.9	
Failure to check the house systematically before handover	18.7		1.9	
Total		35.8		36.9
Defects primarily attributable to workmanship				
Bricklayer	3.5		4.9	
Carpenter and joiner	9.1		–	
Roof tiler	4.2		1.0	
Plumber and engineer	3.9		2.9	
Plasterer and pavior	4.8		5.9	
Painter and decorator	3.8		–	
General	1.3		3.0	
Total		30.6		17.7
Defects primarily attributable to material behaviour and manufactured goods				
Bricks, plaster and glass	1.5		1.0	
Timber and manufactured joinery	4.6		–	
Plumber's and engineer's goods	3.0		1.9	
Total		9.1		2.9
Defects belonging to more than one group		13.9		9.5
Total		100.0		100.0

It was considered that the greatest single action that can be taken to reduce numbers of defects is the prevention of minor defects (including damaged items and incomplete work) by competent pre-handover inspection of the near-finished house by the builder's site supervision staff. Such snagging defects accounted for 18.7 per cent of the total. It was estimated that 30 per cent of all the defects analysed were clearly visible at the time the house was handed over to the purchaser. An outline pre-handover checklist, based on the defects identified in the analysis, is given in Table 55.6. How such a check can be used is illustrated in *Figures 55.6 and 55.7*.

Table 55.6 CHECKLIST FOR FINAL INSPECTION

1. External elevations
Missing, loose or broken roof tiles.
Loose or broken cladding tiles.
No chippings on flat roofs.
Missing lengths of gutter or downpipe.
Broken gutter or downpipe.
Missing wire balloon (soil and rainwater pipes).
Incomplete pointing.
Mortar splashes on brickwork.
No making good around waste pipes or to putlog holes.
Chipped or split window frames.
No mastic around frames and large gaps left.
Damaged sills or drip moulds.
Damaged doors.
Incomplete paintwork or top coat missed (especially at high level and on soffits of sills).

2. Around the dwelling
Drives or paths not completed; incomplete making good around drains, gullies, etc.
Manholes, gullies and stopcock boxes at wrong level.
Missing gully grids and kerbs.
Loose manhole covers.
Soil banked around house.
Rubble not cleared from site.

3. In the roof space
Brickwork or blockwork not made good around ends of purlins, binders, etc.
Brickwork or blockwork not made good up to roof line (especially party walls).
Torn or loose underlay requiring patching or refixing.
Roof insulation incomplete between joists or rafters.
Tank cover not provided.
Walk-boarding not provided.
Pipes and tanks not lagged where water liable to freeze.

4. All rooms
Chipped or split window frames.
Damaged doors.
Incomplete skirting, chipped skirting.
Floor boards not made good around pipes.
Insufficient pipe clips.
Damaged socket plates or light switches.
Missed coats of paint both on visible surfaces and under window boards, etc.
Paintwork not touched up or untidily touched up.
Missing floor tiles – especially around cupboards, fittings, doorways.
Paint splashes on doors or floors.
Loose balustrades and newels.

5. Kitchen and bathroom
Insufficient clips to plumbing and heating pipes.
Damaged sanitaryware.
Loose bath panel.
Incomplete plaster making good around pipes (ceiling and walls).
Cracked or missing wall tiles.
Missing or loose plaster vents over air-bricks.
Inadequately fixed kitchen fittings.

Geographical distribution

In England and Wales NHBC operates in six regions, which contain approximately equal numbers of newly built houses. Table 55.7 shows that, in the period examined, there were more defects in the south than in the north and more in the west than in the east.

The difference between the west and the east is due in part to the higher incidence of damp penetration defects in the west; see Table 55.8. This is logical and is due in part to climatic conditions. However, damp penetration did not account for the whole difference. It was considered that a possible cause of the further difference was that the west has more concentrated industrial areas than the east, and over the last century is likely to have experienced greater competition for good labour between building and other industries.

Previous unpublished research carried out by NHBC had shown that, taking a very broad general view, building standards were probably higher in the south as a whole than in the north as a whole. This was also found to be the subjective view of people with a national knowledge of the private house building industry. At first sight,

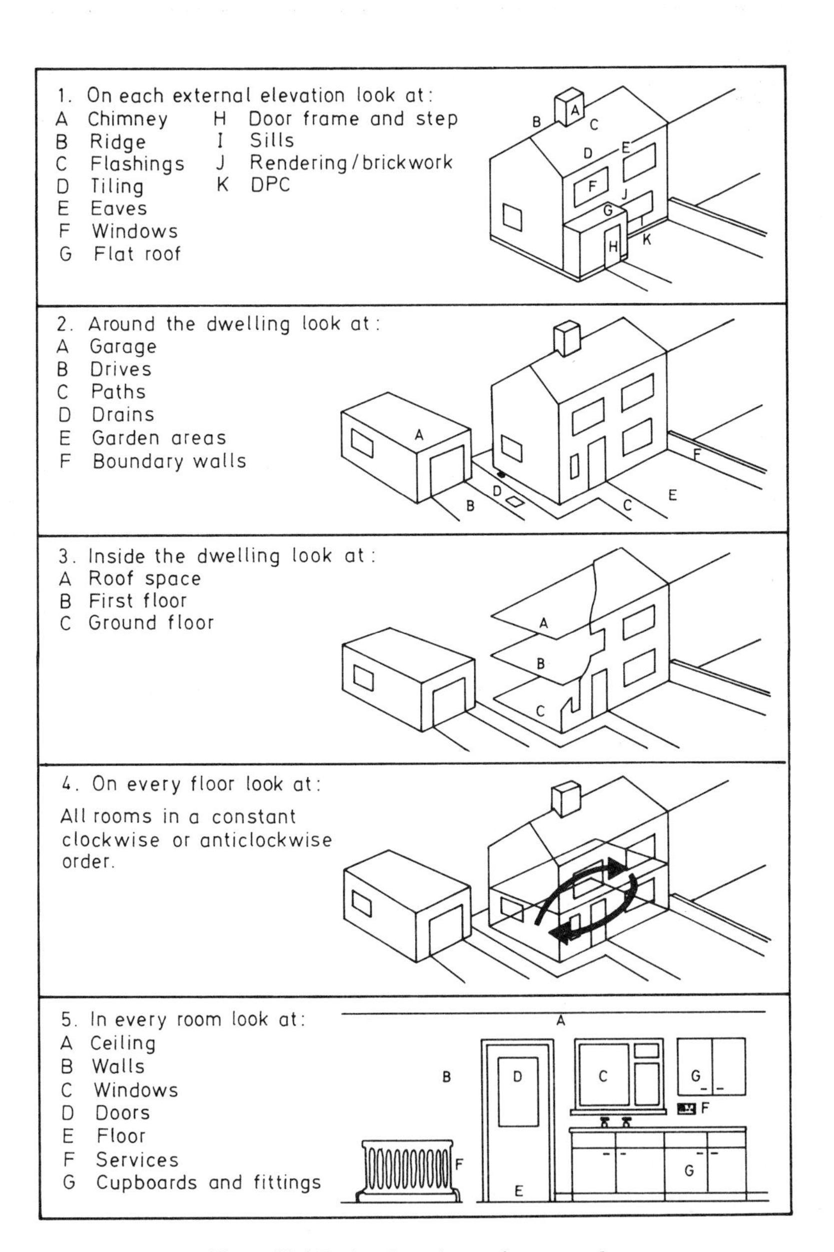

Figure 55.6 Pre-handover inspection: procedure

Snagging Inspection

Property

NHBC HB 72

Inspected by.. Date ....

Inspect the property systematically.

As you inspect, note all defects and outstanding work on this form.

After completing this form:—

1. Note the defects/outstanding work you have found on Trade Defects Sheets and hand top copies to trade foremen or sub-contractors.
2. Chase up work not done on time.
3. Check that work has been done satisfactorily.

Location	Defects to be rectified/outstanding work to be completed
Front elevation	
Left elevation	
Back elevation	
Right elevation	
Garage	
Drives and paths	
Garden area	

Snagging Inspection

Trade Defects Sheet

Property

NHBC HB 73

Work to be done by ....(trade)

Please rectify/complete items listed below by...(date)

Signed.. Date..

Location	Defects to be rectified/outstanding work to be completed

Figure 55.7 Pre-handover inspection: forms

Table 55.7 ANALYSIS OF DEFECTS BY NHBC REGIONS

	West	*East*	*Total*
North	997	528	1525
Midland	920	775	1695
South	1121	1031	2152
Total	3038	2334	5372

Table 55.8 ANALYSIS OF DAMP PENETRATION DEFECTS BY NHBC REGIONS

	West	*East*	*Total*
North	120	46	166
Midland	83	47	130
South	95	41	136
Total	298	134	432

therefore, it was surprising that the number of defects increased from north to south. If the standard of work was the sole criterion for complaint, the reverse should have occurred. It was deduced that there must be another factor causing the unexpected result.

Surveys of new purchasers carried out by National Opinion Polls for the Housing Research Foundation had given information on the social classes of purchasers in the west midland and south-east regions. It is reasonable to suppose from Table 55.9 that in the south-east region 72 per cent of new house buyers were in the AB/C1 social classes, compared with only 47 per cent in the west midlands. It seemed

Table 55.9 SOCIAL CLASSES OF NEW HOUSE PURCHASERS IN THE SOUTH–EAST AND WEST MIDLAND REGIONS

Social class	*% of new house purchasers in social class*			
	West midlands		*South-east*	
AB	16	47 (AB + C1)	31	72 (AB + C1)
C1	31		41	
C2	44		24	
DE	9		3	

reasonable to infer that purchasers of high social class, paying high south-east prices for their houses, had higher standards of expectation as regards quality of building. No comparable data is available for a northern region, but common sense suggests that the consistent trend towards a lower social class leading to lower standards of expectation continues northwards. The conclusion was tentatively drawn that social class and price of house caused the greater incidence of complaint in the south, even though there was evidence that building standards were higher.

CRITERIA FOR DEFECTIVENESS

Philosophy

The preceding sections on structural defects and non-structural defects indicate the main technological priorities for the improvement of house building in the private sector. Perfection, however, will never be achieved and distortion will always occur as a result of the natural behaviour of buildings and materials. Consequently it is necessary to know what degrees of imperfection ought reasonably to be accepted. This section is devoted to a detailed consideration of that question.

Structural distortion

Many buildings move to a greater or lesser extent. Often the movement is small, uniform and unnoticed. Sometimes, however, it is sufficiently great or differential in its effect for damage to occur to the structure, principally cracking and tilting. Damage can be aesthetic, including cracking of walls, sloping of floors and bulging and tilting of walls; more serious are cases where cracking is not weatherproof, doors and windows jam and window panes break; and finally, damage can affect stability, i.e. there is a probability of part of the structure collapsing. The Building Research Establishment has developed a five-point classification scale for structural damage, as shown in Table 55.10. It will be seen that various descriptions indicate the type of remedial work that is likely to be appropriate in most circumstances.

All beam-type structures such as floors or joists deflect. For every material and circumstance there is a calculable permissible deflection for which allowance is made For example, in Building Regulation tables for the selection of joists for particular spans and loads, normally this would be 0.003 times the permissible span. Any deflection not exceeding that is not normally justifiable as a defect.

As discussed above, and as shown in *Figure 55.1,* the settlement of ground floor slabs laid on infill is one of the most significant structural defects in new houses. NHBC normally regards a maximum deformation of 18 mm as being a minor defect to be tolerated or at most to be masked with a skirting quadrant. When the deformation exceeds 18 mm, relaying of the floor or in the worst cases total removal of the floor slab and infill may be required. However, even if the deformation is less, remedial work is necessary if there is damage to partitions, door openings, kitchen fittings or the like.

Distortion due to drying-out of timber

Shrinkage is not a new problem, but it has become more acute with the greater use of central heating and other continuous heating units (all-night coal fires, for example) in modern houses, and with the better insulation that houses have against loss of heat. Warmer, better insulated houses mean that the timber loses more moisture than it used to in the days when most houses were cold and draughty. This increases the risk of shrinkage.

The amount of shrinkage depends on a number of things: the design of the house, the moisture content of the timber when the builder bought it, and whether it was protected from the rain and the damp ground when it was stored on site. NHBC strongly recommends that timber used in house building should be dried to a certain moisture content before it is used. See Table 55.11.

NHBC requires that timber should be adequately protected on site; carcassing timber must be stored off the ground on bearers, and manufactured components and joinery must be stored off the ground on bearers and under cover. A correct sequence

Table 55.10 CLASSIFICATION OF VISIBLE DAMAGE TO WALLS, WITH PARTICULAR REFERENCE TO EASE OF REPAIR OF PLASTER AND BRICKWORK OR MASONRY

Category of damage	*Degree of damage**	*Description of typical damage (ease of repair is indicated by italic type)**	*Approximate crack width†* (mm)
		Hairline cracks of less than about 0.1 mm width are classed as negligible.	$\ngtr 0.1$
1	Very slight	*Fine cracks that can easily be treated during normal decoration.* Perhaps isolated slight fracturing in building. Cracks rarely visible in external brickwork.	$\ngtr 1.0$
2	Slight	*Cracks easily filled. Redecoration probably required. Recurrent cracks can be masked by suitable linings.* Cracks not necessarily visible externally; *some external repointing may be required to ensure weathertightness.* Doors and windows may stick slightly.	$\ngtr 5.0$
3.	Moderate	*The cracks require some opening up and can be patched by a mason. Repointing of external brickwork and possibly a small amount of brickwork to be replaced.* Doors and windows sticking. Service pipes may fracture. Weather-tightness often impaired.	5–15 (or a number of cracks $\geqslant 3.0$)
4	Severe	*Extensive repair work involving breaking-out and replacing sections of walls, especially over doors and windows.* Window and door frames distorted, floor sloping noticeably.‡ Walls leaning‡ or bulging noticeably, some loss of bearing in beams. Service pipes disrupted.	15–25 but also depends on number of cracks
5	Very severe	*This requires a major repair job involving partial or complete rebuilding.* Beams lose bearings, walls lean badly and require shoring. Windows broken with distortion. Danger of instability.	usually >25 but depends on number of cracks

* It must be emphasised that, in assessing the degree of damage, account must be taken of the location in the building or structure where it occurs, and also of the function of the building or structure.

† Crack width is one factor in assessing degree of damage and should not be used on its own as direct measure of it.

‡ Local deviations of slope, from the horizontal or vertical, of more than 1/100 will normally be clearly visible. Overall deviations in excess of 1/150 are undesirable.

Table 55.11 MOISTURE CONTENTS FOR TIMBER

Element	*Moisture content (%)*
Joists, rafters and general carcassing	22
Specialist pre-fabricated work	17
Windows and frames	17
Internal joinery:	
intermittent heating	15
continuous heating	12
close proximity to heat source	9

of building is vital to ensure that glazing or alternative methods of protection are available before floor boarding and internal joinery are fixed. In a non-centrally heated environment, the maximum permissible shrinkage is normally 1 mm in 50 mm. Where central heating is provided, 1 mm in 40 mm is the maximum. In either case, if the shrinkage is such that the element does not perform its function, then, regardless of any calculation, a defect exists.

Workmanship tolerances for internal finishes

Traditionally judgements about internal finish in new houses have been based on the vague, subjective opinions of builders, architects, sub-contractors and purchasers. Inevitably there have been conflicts. A report by Rankin (see 'Further Information' at the end of the section) seeks, for the first time, to give objective guidance. The tolerances suggested are not based on theory but on measurements of what was actually achieved in typical private-sector housing sites. It is suggested that the tolerances can be used both for guiding builders on what is a reasonable standard of work to be aimed at in production, and for settling disputes on whether or not work is of an acceptable standard.

Twenty lower- and middle-price sites were chosen for convenience from Essex/Hertfordshire and South Wales. From each site dwellings were selected that were ready for occupation but not yet heated or fully dried out. Detailed measurements were taken on 125 dwellings. The measurements showed that the majority of houses achieved relatively small (i.e. good) tolerances, but that 5 per cent of houses accounted for much larger (i.e. poor) tolerances. It was accordingly decided to recommend builder guideline tolerances that were equal to the statistical median (the

Table 55.12 TOLERANCES FOR INTERNAL FINISHES

Item	*Builder guideline tolerance (median)*	*Unacceptable work tolerance*
Joint between floor and skirting	2 mm	6 mm
Level of floor, general	3 mm in 2.0 m	10 mm in 2.0 m
Level of floor at doorway	3 mm in 1.0 m	8 mm in 1.0 m
Flatness of wall	3 mm from 2.0 m straightedge	
Plumb of wall	5 mm in 1.0 m (For practical purposes can be extended to 10 mm in floor-ceiling height.)	12 mm in 1.0 m
Width of reveal	3 mm in full height	8 mm in full height
Level of reveal soffit	1 mm in 1.0 m	5 mm in 1.0 m
Plumb of reveal	2 mm in 1.0 m	7 mm in 1.0 m
Reveal–soffit angle	2 mm in 500 mm	7 mm in 500 mm
Vertical junction of wall	5 mm in 500 mm	10 mm in 500 mm
Flatness of ceiling at wall–ceiling line	6 mm from 2.0 m straight edge, provided there is no abrupt deviation of more than 5 mm in 300 mm.	
Square of door frame	1 mm in 500 mm	3 mm in 500 mm
Door twist or bow	5 mm	10 mm
Fit of door	Shutting edge 3 mm Hinged edge 2 mm	7 mm 5 mm
Level of window sill	1 mm in 1.0 m	6 mm in 1.0 m
Plumb of window frame	2 mm in 1.0 m	6 mm in 1.0 m

tolerance achieved in the 62.5th house) and unacceptable work tolerances that were based on the level achieved in the worst 5 per cent of houses. The latter could be used in disputes to decide when work should be remedied. Some examples of these tolerances are shown in Table 55.12.

Whole house quality

No house is perfect. Each will contain a variety of minor defects and imperfections, but provided the total number of these is not excessive the house should be regarded as of reasonable, acceptable quality. Not all imperfections or degrees of imperfection have the same 'values'. Therefore a system of weighting is necessary similar to that described above for the identification of 'important' non-structural defects. In his

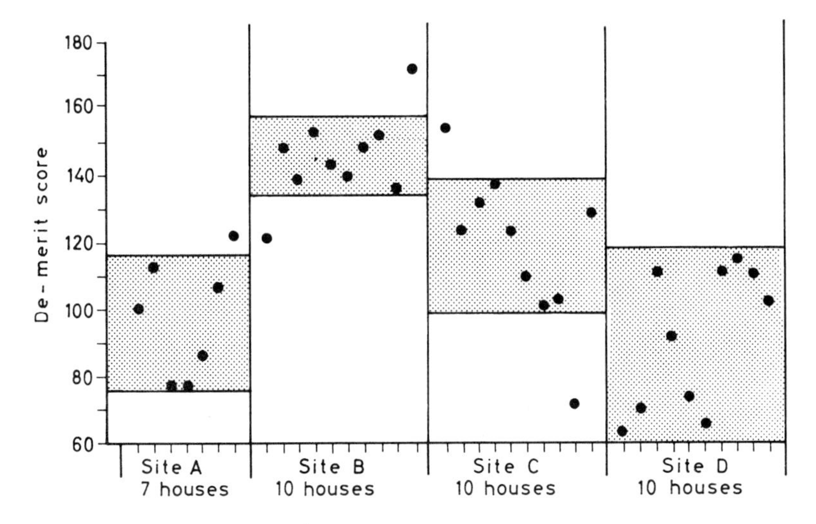

Figure 55.8 Comparative quality of internal finish

work on tolerances for internal finishes, Rankin allocated a demerit rating of 5 points or 2 points according to his opinion of each item's seriousness. Demerit points were allocated to a house if the item did not reach the guideline tolerance level (median). *Figure 55.8* shows the demerit scores achieved on the houses on four sites. The relative quality levels of the sites are clearly apparent. The achievement of high scores would indicate to the builder that improvement was needed on future houses.

FURTHER INFORMATION

Chapman, D.A., Dyce, R.E., and Powell, M.J.V., 'NHBC structural requirements for housing', *The Structural Engineer,* **56(a)** No.1, 3–10 (Jan 1978)

Driscoll, R., *Foundations for low-rise housing – assessment of damage,* Building Research Establishment Notes, Reference B 492/77

Powell, M.J.V., *The incidence, causes and prevention of defects in the construction of new houses,* Housing Research Foundation (1971)

Powell, M.J.V., *Quality control in speculative house building,* Occasional Paper 11, The Institute of Building (1976)

Rankin, I., *An investigation into tolerances for the internal finish of new private dwellings,* Housing Research Foundation (1978)

ACKNOWLEDGEMENTS

The author wishes to thank his former employers, the National House-Building Council, for permission to quote extensively from their published research results, and Ian Rankin for permission to quote from his work on tolerances for internal finishes.

Table 55.10 is Crown Copyright and is reproduced by permission of the Director of the Building Research Establishment.

56 DIAGNOSIS OF STRUCTURAL FAILURE

BASIS OF DIAGNOSIS 56–2

SITE INVESTIGATION 56–2

CONSTRUCTION DRAWINGS AND CALCULATIONS 56–3

EFFECT ON THE STRUCTURE 56–3

CONCLUSION 56–11

DIAGNOSIS OF STRUCTURAL FAILURE 56

56 DIAGNOSIS OF STRUCTURAL FAILURE

J. COTTINGTON, CEng, FICE, MIMunE, FGS, FIArb, MConsE
J.C. CORMACK, BSc (AppSc), CEng, MICE
Cottington Phillips and Associates, Consulting Engineers

This section offers a systematic and thorough approach to the diagnosis of structural failure. It describes various cracking patterns and types of damage, with examples from the authors' own experience.

BASIS OF DIAGNOSIS

Diagnosis is directed by the four most common forms of failure, which may occur singly or in combination. They are:

- failure to consider the ground on which the structure is founded, and that for some depth below, as an integral part of the structure;
- inadequate design;
- the use of sub-standard materials, or failure to observe particular requirements and limitations related to proprietary materials;
- poor workmanship.

There is frequently a tendency to form an immediate conclusion related to one of these but, following detailed examination, it is seldom that only a single cause of failure is found. Examples of particular failures are discussed later, and it has been found in the ninety or so cases on which this section is based that the first form is the predominant or primary cause, followed by the second. The two latter are usually linked, and directly connected with the others in only twenty per cent of the cases.

The diagnosis of failure must follow the stages and sequence of design presented in other sections, and detailed notes should be made of each stage in order to determine the significant features, which may be related to subsequent stages. A form of checklist will be found useful. The form of this can be based on the NHBC Foundations Manual[1].

SITE INVESTIGATION

In the last decade, for many reasons, land that previously would have been considered unsuitable for building has been purchased without preliminary investigation to ascertain the feasibility of economic development. Development has then proceeded without detailed investigation to determine the engineering properties of the ground, and thus without data for acceptable layout, design, sequence of construction and temporary works, in most cases with predictably dire consequences. Potentially unsuitable land may be classed in five groups:

- landslip and coastal cliff areas;
- marshy areas, whether permanent or seasonal;
- alluvial plains and estuarine basins;
- reclaimed land, including tip areas;
- land affected by recent and adjacent construction.

It is necessary to check, by examination of the site investigation reports, whether cognisance has been taken of these possible features. Are the reports fully comprehensive in relation to all the factors concerned? Were there sufficient borings? Were they deep enough? What tests, *in situ* or laboratory, were applied? Finally, what recommendations were made, and were these followed? Sadly, again within the last decade, inadequate reports have been presented, possibly because of limitation of finance for the work or incorrect interpretation of the factors revealed by the investigation and tests. Given proper investigation and qualified interpretation, most land that is on the face of it 'unsuitable' can be used for building. Such land should not be used, however, if there are tight financial constraints on construction, since the danger of subsequent failure lies with inadequate design in order to reduce costs.

CONSTRUCTION DRAWINGS AND CALCULATIONS

These must be examined and checked in relation to the site investigation report. Starting with the general layout of the site and the type of dwellings, points to consider are:

- Does the layout pattern conform to the natural drainage, both surface and sub-surface, of the site? Have soakaways endangered stability of the site as a whole or of individual units?
- On a sloping site, have there been cut-and-fill operations to an extent that would impair the overall stability of the site? In particular units, has the 'fill' area been fully compacted under controlled conditions? Is the soil suitable as 'fill'?
- Have terraced units been sited on ground of dubious or variable support? If so, are these sub-divided to permit differential settlement without structural failure?
- Where piled foundations have been used, are these of suitable type in consideration of permissible deviation from verticality, possible 'necking', or attack by corrosive ground conditions? Has possible horizontal thrust by sub-surface movement been considered?
- Does the foundation design allow for long-term integrity such as must be considered with ground subject to landslip or mining subsidence? Following this, is the possible movement of the superstructure in consequence of overall foundation movement considered in terms of reinforced brickwork or tied beams at floor and roof level?
- Does the foundation design preclude 'short-term' failure such as differential settlement on ground of considerable local variation? This may be found also where construction has taken place over previous foundations, cellars, and 'live' but disused drainage systems.
- Are the foundations of lighter or heavier wings, garages or other extremities designed to give a loading equal to that of the main structure, thus avoiding fracture at joints?
- In the use of proprietary structural units such as precast beams, floor and roof members, blockwork, sealing materials and the like, it is necessary to obtain proof calculations with test data and to ensure that these relate to the particular loadings and situation of the structure as a whole.

The examples that follow are factual cases selected as an indication of the need to observe an analytical approach to each problem.

EFFECT ON THE STRUCTURE

Structural failure is nearly always first suspected from the presence of hitherto unobserved cracks, or deformation of parts of the structure. Some typical failure symptoms, and their associated causes, are discussed below.

It is important to remember, as previously stated, that there may be a multiple cause for the damage visible to a structure; for example, the vertical movement of a wall due to foundation subsidence may cause an inadequate roof structure to spread, thus applying a lateral thrust to the wall at eaves level. Detailed observation, together with testing of soils and construction materials, is essential in providing valuable clues to the probable causes of failure. Equally useful is research into the history of the site, and the testimony of occupiers and nearby residents who can often assist in dating crack formations and other unusual occurrences. Place names such as 'Giddylake', 'Spring Meadows', etc., are strong pointers to a dubious situation.

The crack patterns discussed below often occur in both leaves of an external cavity wall. However, because of the significant difference in the construction of each leaf, the pattern and width of cracking on the outer and inner faces of a cavity wall may vary.

The external brickwork, particularly if clay bricks are used, possesses strength properties similar to that of the mortar in the joints. In addition, the brick outer leaf has three times as many bed joints as a blockwork inner leaf. It is thus a relatively flexible structure able to withstand minor vertical movements. By contrast, the blocks of which the inner leaf are composed are constructed more with thermal-insulation than with strength characteristics in mind. The mortar used to bond them

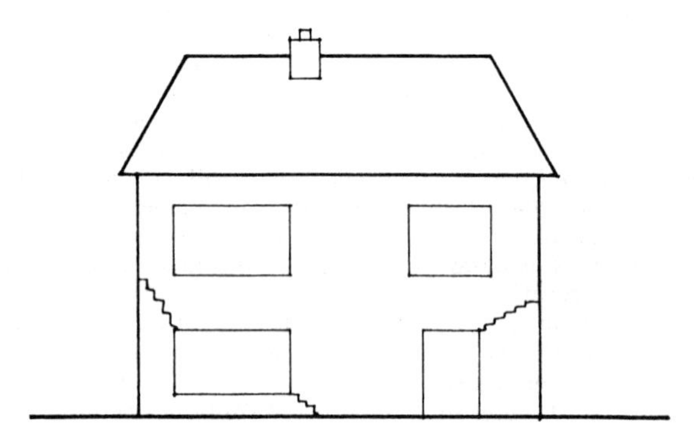

Figure 56.1 Diagonal cracks in external walls

together has usually a much stronger cement/lime content than the block manufacturers specify. The result is a comparatively inflexible inner leaf of varying strength characteristics. When such a wall is called upon to withstand minor vertical movements, brittle failure can occur, with cracks following a fairly straight line, from corners of openings. Because of the brittleness of the blocks, failure is more sudden and tends to produce wider cracks that in the outer leaf.

These differences in the behaviour of the leaves of a cavity wall go some way towards explaining why, upon inspection of a cracked structure, a marked difference may often be observed between the degree of visible damage suffered internally and externally.

A further cause may well be the variations in loading sustained by the two leaves. Normally, the outer leaf carries only its own self-weight, plus a portion of the roof load. By contrast, the inner leaf, in addition to various dead loads, has to sustain the variations of live load applied to the roof and floors.

Diagonal cracks – external walls

Such cracks normally emanate from areas of high stress concentration such as the corners of doors and window openings, as shown in *Figure 56.1*. The route of the cracks is generally a zig-zag one through mortar perpends and bed joints, or, in more severe cases, passing through the bricks themselves. Failures of this type usually imply differential settlement of one or more parts of the structure relative to the remainder, caused by the failure of the foundations to withstand non-uniform bearing conditions beneath the structure.

The non-uniformity may be due to the variable soil or rock conditions encountered at the time of excavation for the foundations and subsequently not catered for in the design of the structure. Alternatively, the presence of flowing water beneath the structure, perhaps from a leaking land drain or sewer, can assist in the leaching out of fine material from certain soils, with a corresponding significant reduction in their bearing capacity.

The prolonged spell of dry weather experienced in 1976 highlighted the vulnerability of structures built on certain clay soils, mostly encountered in south and south-east England. These clays can be broadly categorised as shrinkable, because of their tendency to decrease in volume when the water they contain (which is an inherent part of their physical and chemical structure) is removed. This removal may be by pumping from a well, or by nearby vegetation attempting to obtain the moisture it is not receiving in the form of rain. Because of the structure of the clay, the volume decrease takes place not only vertically, but also (to a lesser extent) laterally, resulting in a downward and sideways movement of the foundations. Thus, when a failure is attributable to a shrinkable clay subsoil, in addition to crack formation as in *Figure 56.1,* some lateral movement of walls may be observed, often as a sliding movement across the damp-proof course. The extreme lengths (literally) to which the roots of vegetation will go to obtain water mean that interior walls and ground-floor slabs may also be affected by sub-soil shrinkage.

Heave, the process whereby a clay expands in volume, produces a similar crack pattern to a shrinkage failure, and is usually precipitated by one or two causes. In certain types of clay, the clay structure has been precompressed by the weight of an overburden that has varied through several geological periods. Excavation in such a clay can release this precompression, the excavated bottom heaving up from its previous level. Such reactions, however, tend to occur mainly in relatively deep excavations, such as for basements. A more common type of heave is gradual in its effect, and therefore may not be noticed until construction is well advanced. This type is due to shrinkable clay taking up moisture to which it was previously denied access. Such a situation could occur if trees were cleared on a shrinkable clay site. The uppermost three metres or so of soil on the site would have a relatively low moisture content due to the removal of water by tree roots. Felling of the trees causes the equilibrium moisture content of the soil to rise, with a corresponding increase in volume. Measurements taken by the Building Research Station on an office block constructed in such circumstances revealed an upward movement of about six millimetres a year over a six-year period. Like the shrinkage type of failure, some degree of lateral thrust, as well as vertical, is exerted upon the structure.

In the design of remedial works to counteract the effects of clay shrinkage and heave, the main objective is to lower the level at which foundation loads are transmitted to the subsoil. This is normally done by the introduction of mass concrete piers, or alternatively piles, beneath the structure, with reinforced-concrete linking beams carrying the walls. The effects of clay movement on the piers or piles is safeguarded by surrounding piers with a granular material and piles with a slip membrane. In heave cases, the installation of a suspended ground floor is also advisable. Major wall cracks should be repaired by 'stitching' new brickwork or blockwork across the crack, bedded in epoxy mortar.

Vertical cracks (uniform width) – external walls

A typical occurrence of this type of cracking is shown in *Figure 56.2.* It is usually associated with continuous lengths of brick walling such as may be found in terraced housing or long blocks of offices or flats. Short returns in the external walls render the structure particularly vulnerable, since they tend to act as hinges.

The cause is usually thermal movement due to expansion and contraction of the walls with changes of temperature. For this reason this type of failure is often observed on north- or south-facing elevations of structures, as these normally experience a wider range of temperatures.

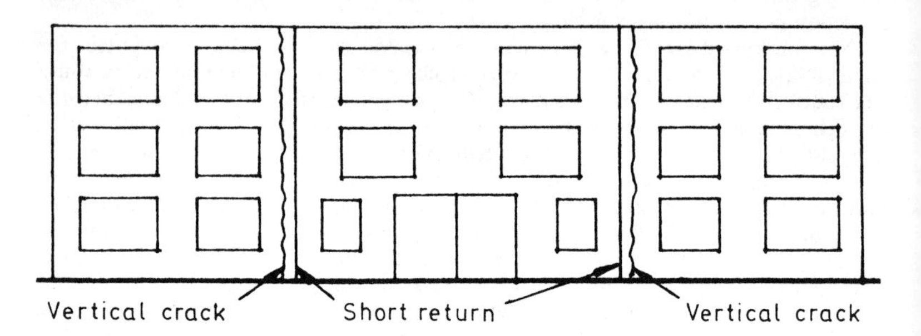

Figure 56.2 Vertical cracks (uniform width) in external walls

The cracking is due to the lack of provision of sufficient (or indeed any) vertical movement joints, and is accentuated by dark-coloured bricks and the presence of cavity-fill insulation. To permit crack formation, sliding usually occurs along the damp-proof course, so cracks normally terminate at this level, rather than continuing down to the foundations.

Remedial works involve the careful removal of bricks or blocks on either side of the crack, and reinstatement of the wall incorporating straight vertical movement joints, with careful attention to weatherproofing details. To ensure stability of the structure, it may be necessary to incorporate dowel bars across the joints in each storey.

A type of failure that produces very similar crack patterns is that due to moisture expansion of brickwork, which can occur in new buildings. However, the risk is not high unless the bricks had an abnormally low moisture content at the time of construction. Under normal British weather conditions, this type of failure is unlikely to manifest itself when more than six months have elapsed since construction.

Vertical cracks (varying width) – external walls

The presence of cracks following an approximately straight line but of varying width in a wall is often an indication of rotation of some part of the structure.

The point in the wall where the crack tapers to its minimum width gives some clue as to the centre of rotation, i.e. the point about which pivoting has taken place. Such a case is the building illustrated in *Figure 56.3.* This building was constructed of well-bedded sandstone blocks and was approximately 150 years old. Although it had no foundations, the walls were corbelled out below ground level to act as quasi-foundations. The extension on the gable end was constructed on shallow strip foundations on a shrinkable clay close to the site boundary, beyond which was an orchard. During the drought of 1976, the roots of the nearest apple trees penetrated beneath the extension in an attempt to find sufficient water. This caused settlement of the extension, which unfortunately had been securely bonded to the main building by toothing in the brickwork with the sandstone courses. The result was the crack pattern shown in *Figure 56.3,* which was repeated on the rear elevation. The end walls of the extension were of solid 225 mm brickwork with no openings, and were functioning as deep beams cantilevering off the gable wall of the main building. This caused the gable wall to rotate, creating tapering cracks as shown. It is interesting to note that the crack did not occur where the gable walls joined the front and rear walls, because of the strength of this part of the structure due to the bonded quoin blocks. Instead, the failure occurred at the first weak point along the wall from the corner, where the edge of a door and window opening in the same vertical line provided a ready-made failure plane. The degree of movement of the gable wall was approximately 60 mm at the level of the tiling battens.

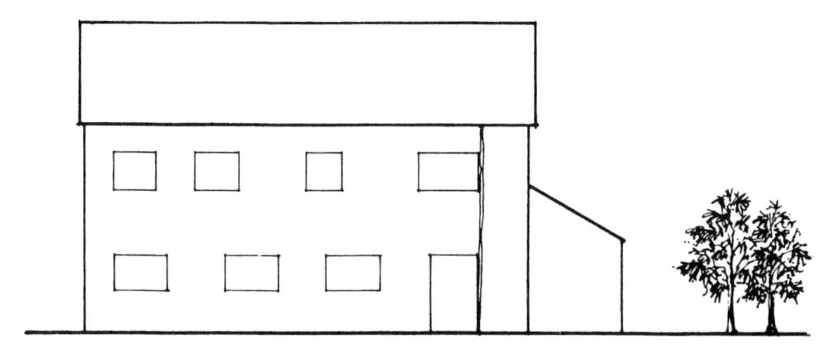

Figure 56.3 Vertical cracks (varying width) in external walls

Remedial works included underpinning of the extension, and the toothed joint replaced by a straight vertical one. Stitching of new stonework across the line of the rotational cracks was carried out on each elevation, and there was selective felling of those apple trees nearest to the extension.

In a structure that has been constructed on traditional strip footings, the presence of such a rotational crack may well reveal a physical break through the concrete footing. In the design and construction of remedial works to the structure, particular care would have to be given to this region.

Horizontal cracks – external walls

The presence of horizontal cracks in the walls of a structure usually indicates failure along one or more bed joints in the brickwork, caused by rotation or sliding of part of the wall. This implies the application of some force other than a vertically-applied one, which the wall was not designed to withstand.

The situation illustrated in *Figure 56.4* is a typical example. The wall exhibiting the failure cracks was directly supporting a concrete roof, which, because of poor mixing and placing of the concrete, was allowing the passage of rainwater through to the soffit finish. In an attempt to cure this a black bituminous-based waterproofing compound was applied to the roof surface, but no reflective finish such as granite chippings was applied. Subsequently, a prolonged spell of warm sunny weather occurred, and the solar radiation absorbed by the roof caused expansion of the concrete slab, with a consequent displacement of the supporting wall. Irrespective of the colour, when large areas of concrete roof are supported on peripheral walls, consideration should be given to the problem of dissipating forces induced in this way.

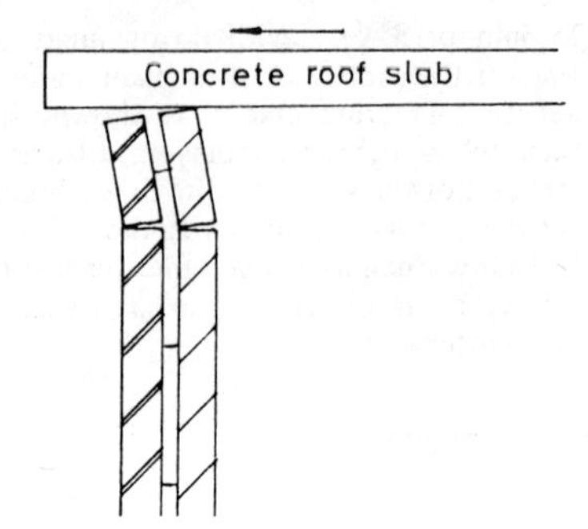

Figure 56.4 Horizontal cracks in external walls

Deformation of external walls – older properties

The following example, illustrated in *Figure 56.5*, is an indication of the behaviour of a brickwork wall under the influences of a gradually-applied and increasing lateral load.

The wall structure was of double-leaf cavity brickwork, bridged at eaves level by a wall plate supporting a roof structure that proved to be inadequate. Tie members linking the feet of pairs of rafters were provided only infrequently, and the roof thus had a tendency to spread at eaves level, causing a lateral thrust against the top of the supporting walls. As is often the case with roof structures that are inadequate but not

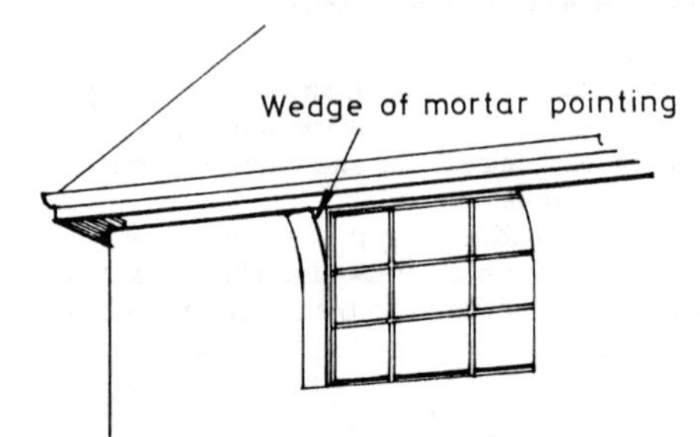

Figure 56.5 Deformation of external walls

disastrously so, movement was gradual, taking place over several decades. The resulting deformation of the external walls is shown in *Figure 56.5* with the tell-tale 'wedge' of mortar pointing in the window reveals. Examination of the wedge showed it to be composed of a number of small wedges each of different mortar, indicating the long-term nature of the problem. Just before major remedial works to the structure were carried out (which included underpinning) a horizontal crack in the plaster of the inner leaf occurred at mid-storey height, which implied that a more serious failure of the wall was imminent.

Because of the condition of the timber, complete reconstruction of the roof was necessary in this case, together with the rebuilding of the wall from first floor to eaves level.

Vertical cracks (uniform width) – inner leaf, external walls

In many reported cases of material failure in the construction industry, the root cause is found to be not an inherent defect in the material, but ignorance, or wilful disregard, of the manufacturer's recommendations concerning its use. The failure illustrated in *Figure 56.6* gives an example of this. It shows the inner leaf, constructed in 100 mm concrete blockwork, of a cavity wall. The blocks were manufacturerd by a reputable company, and finished with the usual 12 mm plaster coat.

A large central heating radiator was installed some 50 mm from the wall, and brought into use when the house was occupied, some three months after completion. Within a month the cracking illustrated was observed in several similar locations within the building. Subsequently, samples of the blockwork were removed from the

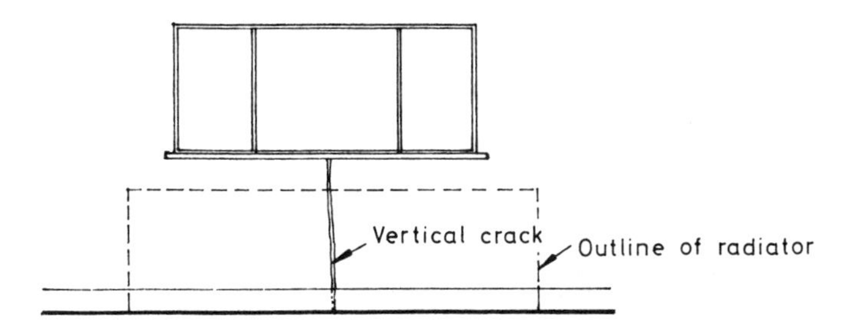

Figure 56.6 Vertical crack in inner leaf

wall and, upon testing, proved to comply in every respect with the manufacturer's specification. Analysis of the mortar in the joints, however, revealed a high cement content, the original mortar mix being approximately 1 : 2½ cement/sand. This was a significant deviation from the manufacturer's requirements of 1 : 6 cement/sand. In addition, workmen employed on the construction of the house testified that no attempt was made to protect the blocks from heavy rainfall either before or after construction – again a contravention of the manufacturer's requirements.

The cracking was thus attributable to shrinkage failure, exacerbated by several factors. The degree of drying shrinkage to be expected in a cement-based product such as mortar or concrete is directly related to its cement content. Assisted by the proximity of the radiator, the drying shrinkage of the mortar and the removal of moisture from the block set up horizontal tensile forces within the wall. Concrete, a material not noted for its tensile strength, possesses even less when the concrete is in an aerated form, as in this type of block. Whether the failure would have occurred without the aggravating effect of the radiator is a debatable point, but this example typifies many householders' heavy-handed treatment of their new properties, and their ignorance of the vulnerability of 'still-green' construction.

As the cracks did not impair the stability of the property, they were stopped with epoxy mortar and an insulation medium was introduced between the radiator and the wall.

Manifestations of sulphate attack – walls and floors

Figure 56.7 illustrates typical failure patterns in walls and floors that can occur when Portland cement-based products such as mortar and concrete are attacked by sulphate, a prime source of which are products containing gypsum. The effect of the sulphate is to cause an increase in volume of the attacked material. Thus a wall render containing sulphates can, when applied to a backing wall, cause expansion of the mortar in the wall joints, leading to the type of cracking shown in *Figure 56.7(a)*.

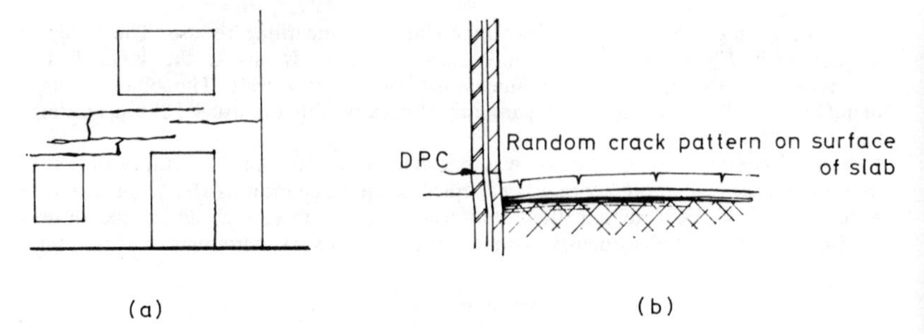

Figure 56.7 Effect of sulphate attack on (a) walls, (b) floor

Remedial works in such a situation are impossible – all affected walls must be demolished and, ideally, none of the bricks or blocks should be re-used as they may be contaminated; nor should they be broken up and used as hardcore under a concrete floor slab, as this could result in the floor slab failing as in *Figure 56.7(b)*. Here, expansion of the slab has caused it to crack and arch upwards from the hardcore beneath. This expansion puts a considerable lateral thrust on the peripheral walls to the slab. Lateral movement of the walls is particularly likely because of the proximity of the damp-proof course. Similar floor failures have occurred when colliery shales, another source of sulphates, were used as fill material.

Cracking and movement of internal walls on ground floor slab

Because of the varieties of wall construction and finish used for internal walls, it is not possible to generalise on typical failure patterns. However, the presence of cracks other than those from minor shrinkage in such a wall can indicate movement, usually downwards, of the concrete floor slab. In non-loadbearing walls (partitions) this is often accompanied by the formation of a gap at the junction of the wall with the ceiling. Many reasons are possible for such slab movement, the more common being listed below.

Inadequate site preparation. When preparing the ground to receive the hardcore beneath the slab, failure to remove all friable topsoil and vegetable matter leaves a compressible layer in the ground. Peaty heathland soils are a particular hazard in this respect.

Shrinkage of clay subsoil. As discussed earlier, clay subsoil shrinkage primarily affects walls. Because of the sheltering effect of the structure floors within it are less likely to be affected, but failure can occur, for example, where root growth has penetrated beneath the structure to a considerable extent.

Incorrect fill material and poor compaction. Fill material beneath a ground floor slab must be clean and free from organic or inorganic matter likely to cause settlement or attack the concrete. It is also important that it be well graded in order that a dense void-free layer can be achieved by suitable compaction.

Inadequate sub-grade beneath hardcore. If the virgin ground, after topsoil strip, has a low bearing capacity, due either to the nature of the original soil, or perhaps the presence of unsuitable fill, the benefits of a suitable hardcore layer beneath the slab can be negated if, under load, this layer penetrates the sub-grade. In such situations suspended floors should be installed, bearing in mind the deflection possibility outlined in the next paragraph.

Deflection of suspended floors. The type of internal wall movement described above can also occur when suspended floors, be they concrete or timber, are designed without due regard to the possibility of excessive deflection.

Lack of foundations. It is essential that all internal walls sustaining loads other than their own weight be carried down to separate foundations, passing through the ground-floor slab in such a way as to transfer no load to it.

CONCLUSION

Seldom is it found that structural failures are identical, similar though they may be, but the details of each are dictated by many factors. It follows that the remedial works will vary in each case, not only in type but also in extent when related to the age and the designed longevity of the structure.

Recommendations for remedial works follow the diagnosis of failure, and it should be borne in mind that inaccurate design of the remedial works could lead to structural failures of a different nature in the future.

REFERENCE

1. *Registered house-builder's foundations manual,* National House-Building Council (1977)

57 DIAGNOSIS OF DAMP CONDITIONS

SOURCES OF DAMPNESS 57–2

DIAGNOSIS 57–7

DIAGNOSIS OF DAMP CONDITIONS 57

57 DIAGNOSIS OF DAMP CONDITIONS

E. THOMPSON, BSc
Director of Technical Services, National Federation of Building Trades Employers

This section describes the main types of dampness that can occur in newly built and older dwellings, and explains how they can be diagnosed in order that the correct remedial work can be put in hand.

SOURCES OF DAMPNESS

In a survey carried out by the Building Research Station over the period 1970–74, 510 building defects were investigated in depth. Of these, half were found to be the direct result of dampness of one form or another (BRE Digest 176). The defects attributed to dampness varied widely, ranging from small isolated damp patches on a wall, or blistering of paintwork, to extensive dry rot of structural timber. Naturally, whenever a defect shows itself it is always desirable, and often vital, that it should be rectified as soon as possible. To do this it is essential to diagnose correctly the basic cause of the defect; without this, remedial work can be both expensive and abortive.

Rain penetration

When considering the possibility of rain penetration it is essential to appreciate that much of the moisture that finds its way through external walling does so via the mortar joints, or through the fine capillaries between the joints and the units. It follows, therefore, that it is an advantage – from the point of view of resistance to rain penetration – for the units comprising the wall to have some absorptive capacity. Conversely, the walls most susceptible to rain penetration are those that consist of low absorptive units such as dense bricks or cast stone, and where the mortar joints are dense and recessed. With such walls almost all the water that falls on the face will run down and feed into the mortar joints, and will eventually find its way to the inner face of the wall.

The introduction of cavity construction in place of solid walling meant that any moisture that found its way through the outer leaf would normally run down the inner face of this leaf and would make no further progress to the inside of the building. Cavity construction, not unnaturally, therefore, has reduced considerably the incidence of rain penetration. Only in some very exposed areas, for example on the north-west coast of Scotland (see BRE Digest 127, *An index of exposure to driving rain*), is it likely that wind-driven rain will be blown across the cavity to be absorbed by the inner leaf. Under normal conditions, direct rain penetration through a cavity wall can usually be attributed to faulty detailing and/or defective workmanship.

Regrettably, careless bricklaying still results in appreciable mortar droppings accumulating on the wall ties, which are then readily able to transmit moisture across the cavity. The appearance of isolated damp patches along a wall, usually facing the prevailing wind, is a fairly sure indication of the presence of dirty, or sometimes inward sloping, wall ties.

Inadequate detailing and poor workmanship in installing cavity trays, damp-proof courses around windows and doors, etc., are common causes of dampness, the precise location of which is usually a reliable pointer to the deficient detail. Particular care should be taken when installing long cavity trays in severely exposed walls where appreciable rain penetration of the outer leaf may be expected. The material used for such long cavity trays should preferably be sealed and not just lapped at the joints. Water dripping from lintols over window openings, or damp patches at skirting level where concrete floors bridge the cavity, are indications of a faulty cavity tray. The formation of cavity trays is frequently a difficult operation, for example at changes in direction of the brickwork, stop ends, etc.; since this is a site operation, subject to the vagaries of the weather, it is surprising that cavity trays do not prove to be defective more often. Pre-formed cavity trays are now available, and their use in awkward situations merits serious consideration. A common defect, less excusable, is the omission of weep holes, which means that water collected by the tray is unable to drain away.

Recent demand for higher thermal insulation has led to the filling of cavities of external walls with insulating material, usually a urea formaldehyde foam, and there is evidence that the incidence of dampness problems has been increased by this process. Many of the defects reported have related to very exposed sites, but occasionally well sheltered walls have given trouble. In these latter instances the rain penetration has been attributed to inferior foam material and/or foaming techniques. The risk of dampness problems as a result of filling the cavity is now widely recognised, and there is a greater awareness of the need to be more selective when deciding which properties can be insulated by means of a cavity-foam method. Many cavity insulation materials are now the subject of Agrement Certificates, and BS 5617: 1978, *Urea-formaldehyde (UF) foam for thermal insulation of cavity walls,* covers the manufacture of urea-formaldehyde foam. The associated Code of Practice, BS 5618, *Thermal insulation of cavity walls (with masonry inner and outer leaves) by filling with urea-formaldehyde (UF) foam,* indicates how a contractor should decide at the outset whether a building is suitable for foam filling, and lists suitable forms and materials of construction. BS 5618 also sets out various checks that should be made on the building before, during and after filling, and explains the principles to be followed to ensure a satisfactory end result.

Defective roof coverings (including tiles, felt and asphalt) can obviously lead to rain penetration into the building, and an inspection of the roof usually enables the point of entry to be ascertained without much difficulty. Poorly designed parapet walls are often a source of moisture penetration and it is not unknown for this to be wrongly diagnosed as a roof leakage. The design of parapet walls to exclude rain penetration is dealt with in BRE Digest 77; an even more effective way of removing this source of rain penetration is to eliminate parapet walls at the design stage!

Rising dampness

In the years before damp-proof courses became mandatory, a common defect in dwellings was rising ground moisture, the effects of which can readily be seen in many old buildings and are familiar to most people. Rising dampness is characterised by a line of staining roughly parallel to the floor, rising to a slightly higher level in the corners of the room, or behind furniture, where the rate of evaporation is somewhat lower. Below this line the wall is often persistently wet, particularly during the winter months when the water table in the ground is generally at its highest. Another feature of rising dampness, particularly where the defect is of long standing, is that the severity of the dampness can change with changes in humidity of the atmosphere. This feature can be attributed to the presence of deliquescent or hygroscopic salts. usually chlorides or nitrates, derived from the soil and carried in solution and deposited in the wall. Such salts, which are capable of absorbing moisture from the atmosphere under humid conditions, tend to concentrate at the surface of the plaster or wallpaper at the upper limit to which the moisture has risen. Hence, in the renovation of old property, after providing a suitable damp-proof membrane in the wall it is essential to remove the salt-contaminated plaster if the dampness is to be eradicated. Even then, if the brickwork itself is heavily contaminated with salts, there is the possibility that over a number of years these may migrate through the new plaster to give rise to dampness once again.

In new buildings with cavity walls, provided that the damp-proof courses are of a suitably durable material and have been correctly positioned and installed, rising ground moisture should rarely be a problem. Where it does occur it can be attributed to bridging of the damp-course, for example by excessive mortar droppings or brick bats in the cavity, or by an adjacent floor screed. In these cases the location and extent of the dampness usually provide a clear pointer to the offending bridging. Although dampness that is confined to the lower portion of a ground floor wall is strongly indicative of rising ground moisture, where this is confined to external walls built off a solid ground floor the possibility of condensation should not be overlooked.

Now that it is standard practice to incorporate a damp-proof membrane at some point in the flooring system, it is rare for solid floors to be affected by rising dampness. Where it does occur it is mainly caused by perforation of the membrane before being covered up. Rising dampness in a floor can usually be confirmed by the effect on various finishes: wood block or strip flooring will tend to swell; thermoplastic tiles may disintegrate at their edges through salt action; and sheet lino and PVC may become loose and expand. It must not be overlooked, however, that residual construction moisture, which is discussed in the following paragraphs, can have a similar effect on floor finishes; in view of the expensive remedial measures called for in the case of rising moisture through a solid floor, it is particularly important for the diagnosis to be accurate.

Residual construction moisture

It is often not appreciated how much water is used in the traditional construction process. Even under favourable conditions, where the materials remain substantially dry during the building process, several tonnes of water are introduced by way of the wet trades such as bricklaying, plastering, concreting and screeding. Premature covering of floor screeds with impervious finishes can give rise to the defects referred to in the previous paragraphs; as a general rule of thumb, the drying-out period of a concrete floor slab under favourable conditions is about one month for every 25 mm of concrete or screed. Often, particularly during the winter months, an appreciably longer drying period is necessary if subsequent defects are to be avoided.

The flaking of paintwork and the appearance of salts on the internal surfaces during the early life of the building is often caused by drying-out of the construction moisture, and unless there are other positive indications that rain penetration is occurring it would be unwise to attempt any remedial action until the building has been occupied for at least one heating season.

Lightweight roof screeds can also contain large amounts of moisture, which become sealed in once the impervious felt or asphalt roof covering has been applied. Blistering and possible splitting of the roof covering when it warms up in the sun is usually an indication of entrapped moisture. Once the roof covering has split, of course, rain penetration can occur. Because of this the original cause of the defect, i.e. entrapped moisture, is often overlooked. A somewhat dramatic indication of entrapped screed moisture is the dripping of a dark brown liquid from light fittings on the upstairs ceilings; the brown colour is bituminous material that has been leached out of the sheathing felt by the alkaline screed moisture. Entrapped moisture in roof insulation, by reducing the thermal insulation value, may also contribute to the dampness problem by increasing the risk of condensation on the underside of the roof slab.

Condensation

According to the BRE survey previously referred to, apart from rain penetration by its various sources, condensation is by far the most common cause of dampness. Unfortunately the basic principles underlying condensation are not widely understood, and there is little doubt that many dampness problems are incorrectly diagnosed as rising ground moisture or rain penetration, where condensation is the real source of the offending moisture.

Condensation occurs when warm moist air comes into contact with a cold surface; the presence of beads of moisture on the cold water tap when having a bath is a good reminder of this basic principle. The more moisture there is in the atmosphere (usually indicated in terms of relative humidity), the higher the temperature (the dew point) at which condensation will occur. Thus it will be realised that in a warm house where the relative humidity is maintained at a low level the risk of condensation will be slight. Contrary to general belief, the air inside an occupied dwelling usually contains more moisture than the outside air; consequently ventilation to the outside air generally serves to reduce the humidity inside the dwelling, which in turn reduces the condensation risk. The older type of dwelling with chimneys, and ill-fitting windows and doors, may well have been cold and uncomfortable through draughts but it rarely suffered from condensation. The recent practice of building houses without chimneys and with increased emphasis on weather-stripping of windows and doors has meant that the dwellings have little fortuitous ventilation, which previously served to keep down the moisture levels of the inside air. Thus ventilation must be achieved consciously by opening windows, a practice that many occupants, not unnaturally, are reluctant to follow during the cold winter months. Modern window design has not helped either; many windows have no small opening top lights, which have the merit of providing some ventilation without introducing an undue draught and without causing a great deal of heat loss.

It is not uncommon on an estate consisting of identical dwellings for isolated properties to be subject to severe condensation. This can generally be attributed to the living pattern of the occupants and the effect this has on the heating and ventilation levels within the dwelling. The mere presence of occupants means that the relative humidity inside the dwelling will increase; it is estimated that the breathing alone of a single occupant produces about a litre of moisture per day. Having a bath introduces moisture into the atmosphere at the rate of about 1½ litres per hour, and doing the washing can produce as much as 3 litres per hour. Heating by unflued paraffin heaters causes about the same amount of moisture to be discharged as paraffin that is burned, and cooking, drying clothes, etc., also obviously put additional moisture into the air. It follows, therefore, that where the family includes one or more small children, with the attendant extra washing, bathing, etc., the possibility of condensation occurring is considerably enhanced. It has also been found that dwellings occupied by elderly people are particularly prone to condensation; often these occupants are unable to afford to heat their dwellings adequately and, being particularly sensitive to draughts, are less inclined to open their windows.

Unless it is very severe and prolonged – which is most unusual – condensation does not normally produce visible dampness except on impervious surfaces such as glass and tiles. However, the moisture produced by condensation invariably leads to mould growth on wall and, sometimes, floor surfaces, and it is the appearance of these moulds that usually prompts complaints of dampness. Housewives are particularly sensitive to the presence of moulds as they feel that it reflects badly upon their cleanliness. Moulds require moisture for their growth, and their presence on cold bridges such as concrete lintols over windows and along the bottom of external walls (where the ground floor is of solid construction) is a good indication that condensation is occurring at these points. Unless suitably insulated, the perimeter of a solid ground floor will be at a lower temperature than the remainder, and in turn this means that the base of the wall in contact with the floor around the perimeter will also be slightly colder than the rest of the wall. The latter pattern of mould growth is frequently incorrectly attributed to rising ground moisture.

The presence of moulds on clothes and shoes in wardrobes, often of the built-in type on an external wall, is a positive pointer to a high internal relative humidity. Moulds do not necessarily require liquid water for their growth; the maintenance of a relative humidity above about 75 per cent for long periods will suffice. With humidities of this magnitude the temperature of internal wall and floor surface will have to be maintained at a fairly high level if surface condensation is to be avoided. For example, if the relative humidity of the air is 75 per cent at a temperature of 20°C (a comfortable room temperature), deposition of moisture will occur on any surface whose temperature drops below about 15°C. At a lower, more normal humidity level of 50 per cent at 20°C the corresponding surface temperature at which condensation will occur would be 10°C. It is not difficult to envisage the temperature dropping to either of these levels during a winter night when the heating is turned off.

Single-glazed windows with their low thermal insulation are invariably the coldest surfaces within a room and this is where condensation will first occur, a fact that is clearly borne out by the moisture deposition visible on bedroom windows in the mornings. This particular problem can, of course, be eliminated by installing double glazing. However, although double glazing will reduce the overall heat loss, which will tend to make the internal wall surfaces slightly warmer, there is the possibility that by replacing large areas of single glazing – which serve as a condensing plate to remove moisture from the air within the dwelling – the incidence of condensation on other wall surfaces may be increased.

Describing the results of an extensive survey carried out by the Department of the Environment among local authorities in England and Wales, Humphreys[1] reported that the incidence of condensation was increased in:

- dwellings left unoccupied for long periods;
- unheated bedrooms at the same level as living rooms;
- unheated areas projecting from the main part of the building;
- dwellings at tops and ends of blocks and terraces;
- *in-situ* and precast concrete structures;
- north-facing walls;
- dwellings where the standards of thermal insulation were at or below Building Regulations minima;
- dwellings where the standard of heating was below Parker Morris minimum.

The modern emphasis on heat conservation by increasing the thermal insulation of dwellings, notably at upper ceiling level, has meant that the loft space in pitched roofs is colder. Moreover, it is now standard practice to provide sarking felt below the tiles, thus creating a virtually sealed roof space. These two factors have resulted in increased condensation on the undersurface of the sarking felt; the moisture has then dripped on to the insulation, eventually causing damp patches to appear on the bedroom ceilings. A similar pattern of dampness is often encountered with flat roof construction, and it is not unknown for the blame for such dampness to be attributed to a defective roof covering.

Deliquescent salts

As previously mentioned when discussing rising dampness, deliquescent salts are able to absorb moisture from the atmosphere during humid conditions and often give rise to what, on the face of it, appear to be inexplicable areas of dampness. Intermittent damp patches, often brown-stained, on or adjacent to upstairs chimney breasts are not uncommonly mis-diagnosed, resulting in chimney stacks being taken down and rebuilt, tiles replaced around the stack and flashings removed in abortive remedial attempts. Although direct rain penetration through the roof cannot be ignored, this particular form and location of dampness is invariably caused by the presence of deliquescent salts derived from the condensate of the flue gases, which over the years has slowly permeated through the brickwork and plaster. The associated brownish staining serves to confirm this diagnosis, but if further confirmation is required simple chemical tests can be carried out to determine the presence or otherwise of chlorides or nitrates, the common deliquescent salts. In modern construction, where the chimneys are required to be lined, it is rare for this form of dampness to occur.

DIAGNOSIS

The essential requirement, when attempting to diagnose any building defect including dampness, is to keep an open mind until the full range of data have been obtained and assessed. These data are normally derived from four basic sources:

- background information about the pattern of the dampness – time of its initial appearance, relation to weather conditions, etc.;
- study of constructional details, e.g. position of damp-proof courses;
- observations on site, e.g. presence of moulds, staining, weather-stripping of doors and windows;
- results of moisture and humidity measurements and, occasionally, chemical tests.

Often much valuable information can be obtained from talking to the occupants of the dwelling and thereby ascertaining their general pattern of living, e.g. is the house unoccupied for long periods in the day? At what level is the heating system operated? To what extent and when are windows opened?

Although nothing can take the place of visual observations, relatively simple and inexpensive instruments are now available that can be a useful tool in the diagnosis of

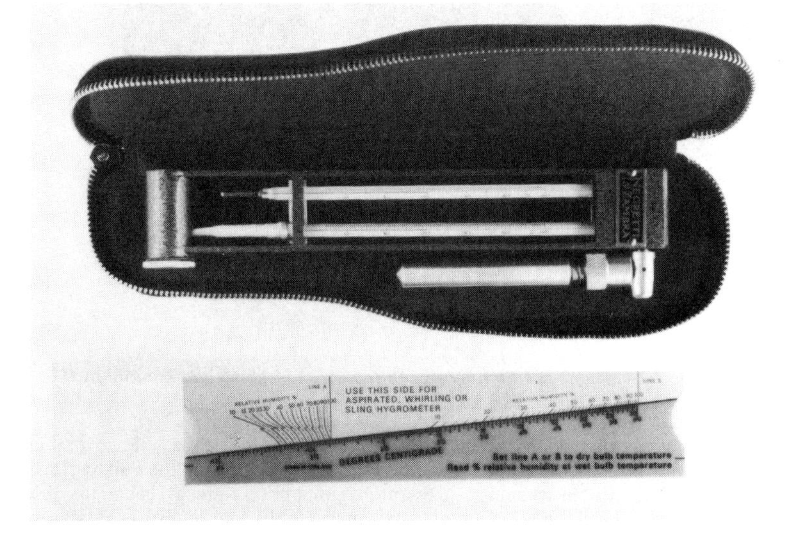

Figure 57.1 Whirling hygrometer (courtesy Negretti and Zambra Ltd)

dampness. However, moisture meters need to be used – and their results interpreted – with discretion. Most moisture meters work on the principle of electrical conductivity; whilst they can be calibrated to measure the dampness of timber in terms of percentage moisture content, they cannot be used with such precision on materials such as plaster and brickwork. Although some moisture must be present for the meter to register, the presence of soluble salts, which are always present to some extent in plaster and brickwork, can have a significant effect on the readings. Nevertheless, moisture meters can help considerably in plotting the extent of any dampness. In the case of timber, they are particularly valuable in being able to measure whether it has reached a moisture level at which it is likely to be attached by the highly damaging dry-rot fungus.

In addition to measuring dampness, some of these meters have suitable attachments for measuring surface temperatures. This is a particularly valuable facility when potential condensation problems are being investigated.

Measurements of the humidity in the atmosphere can be made by means of a simple instrument known as a whirling hygrometer or psychrometer (*Figure 57.1*). This consists of 'wet' and 'dry' thermometers that are whirled round like a football supporter's rattle; with the aid of suitable tables, the thermometer readings enable the relative humidity to be calculated.

Figure 57.2 Portable thermohygrograph (courtesy Negretti and Zambra Ltd)

When condensation is suspected it is often very useful to be able to obtain continuous readings of the temperature and humidity conditions in the building, and relatively inexpensive instruments – thermohygrographs – are available for this purpose (*Figure 57.2*). These instruments, which are fairly robust, give a continuous recording of temperature and relative humidity over a period of seven days.

REFERENCE

1. Humphreys, W.E., 'Condensation and remedial measures', *Building Technology and Management* (June 1972)

FURTHER INFORMATION

All the following publications are obtainable from HMSO:
BRE Digest 27, *Rising damp in walls*
BRE Digest 110, *Condensation*
BRE Digest 163, *Drying out buildings*
BRE Digest 176, *Failure pattern and implications*

DoE Advisory Leaflet 47, *Dampness in buildings*
DoE Advisory Leaflet 61, *Condensation*

Condensation in buildings Part 1 – *A design guide*
Part 2 – *Remedial measures*

58 INSULATION IMPROVEMENTS

THERMAL INSULATION 58–2

SOUND INSULATION 58–5

INSULATION
IMPROVEMENTS 58

58 INSULATION IMPROVEMENTS

J. LAWRIE, DSC
Vice-Chairman, Structural Insulation Association

This section is concerned with the problems that may be encountered in existing property in terms of providing improved insulation performance. Since the basic principles are the same as in new construction, it should be read in conjunction with Section 43.

THERMAL INSULATION

Pitched roofs with attic space

The application is identical to that in new construction. If no insulation has been used, the maximum recommendations in Table 43.6 would apply, but if some insulation is already present some reduction of thicknesses may be possible. It should be noted, however, that the labour cost will be a substantial part of the total cost regardless of the thickness of material used, and the most economic application in the long run may be to use the maximum thicknesses.

Difficulty may be experienced in gaining access to some roofs and it may be necessary to cut an access trap. Where the pitch is too low to allow entry to the roof space it will be necessary to erect a suspended ceiling to carry the insulation, but this is not always practical due to lack of headroom. In such cases remedial treatment is expensive, and involves the removal of the ceiling and the re-erection of ceiling and insulation together.

Whenever the insulation is applied in the attic space care must be taken to ensure that the space is adequately ventilated, and where possible the incorporation of a vapour check immediately above the ceiling is advisable, especially with low pitch roofs where ventilation may be difficult.

Pitched roofs with no attic space

In older property where the roof rafters are exposed on the underside it is possible to apply the insulation between the rafters and face out with a variety of finishes, e.g. plasterboard, hardboard, timber etc. The insulation best suited to this type of application is mineral (glass or rock) wool slabs, which can be pressed between the rafters and retained there while the facings are applied. Facings may either enclose the rafters or be fitted between, depending on individual choice. It is also practical to use the expanded polystyrene or polyurethane boards either between or over the rafters. In all cases a vapour check should be applied immediately under the insulation.

Flat roofs of timber

Improved insulation is achieved by the same methods described in Section 43, but it is necessary to remove the roof finish before work can be carried out on top of the roof, or remove the ceiling if the work is to be done below. If there is sufficient headroom it is also possible to construct a false ceiling, using cross-battens of appropriate depth, and apply the insulation between before finishing in plasterboard. In some situations combinations of insulation materials and facings may also be used, e.g. plasterboard/expanded-polystyrene laminate. In all these methods the need to provide the vapour check and to ventilate the roof cavity must be emphasised.

Flat roofs of concrete

Once again improved insulation follows the pattern of new construction, but with the roof finish being removed before insulation is applied. Similarly, false ceilings can be used or certain boards (mainly those of expanded plastics) can be bonded direct to the concrete or secured by machine-fired fixing systems. The need for adequate vapour barriers must again be stressed.

Cavity walls

By far the simplest way of insulating a cavity wall is to fill the cavity, and for this three materials are available: urea formaldehyde foam, mineral (glass or rock) wool granules, and expanded polystyrene granules. The materials are applied by specialist contractors, and the choice will depend on the exposure of the building. Under the Building Regulations, Type Relaxation is necessary for certain types and it is advisable to check with the local authority to ensure that no breach of regulations is committed. Mineral wool types can generally be used in any climatic zone without restriction.

Another approach to the problem is by dry lining using a proprietary glass-fibre product, DP Wall Liner by Fibreglass Ltd. This product is provided with regular holes over its whole surface and is applied to the inside of the inner leaf and plaster dabs placed on the wall through the holes. Plasterboard is secured by pressure in to the plaster dabs and in intimate contact with the DP Wall Liner. Before application of the liner the walls should be cleared of any protrusions so that good contact can be achieved. Jointings and angle finishing follow best established practices, but full details are available from British Gypsum Ltd.

Cavity walls can also be battened out on the inner face and flexible, semi-rigid or rigid insulation materials applied between the battens and plasterboard secured over. The application is more fully detailed in Section 43, where information on materials and results is also given.

Where it is not possible to use cavity insulation or to apply the insulation inside the house (which has the disadvantage of reducing the size of the rooms concerned), there are available a number of proprietary systems for insulation externally. One of these involves the erection of an additional cavity, using battens and rendered expanded metal, and filling the cavity with urea formaldehyde foam. Full details are available through the National Cavity Insulation Association. Another utilises glass-fibre mat or slab fitted between battens fixed externally, and the whole rendered in sand/cement on a suitable base. Full details are again available from the manufacturers.

Solid walls

Improved insulation of solid walls is governed to a considerable extent by available space. Ideally the insulation will be applied internally and will follow the same principles as for cavity walls, i.e. one of the dry-lining methods, but if space is limited external insulation is the only answer, again following the same practices as with cavity walls.

Timber-framed walls

Usually timber-framed walls are insulated to a reasonable standard when erected, but older properties will not be up to modern requirements. Treatment is inevitably expensive and will involve gaining access to the cavity, probably by the removal of tiles and boarding, which can be achieved without disturbing the interior of the building.

A vapour check should be inserted (if one is not already provided) and insulation placed in the cavity behind the boarding, which is then replaced and the tiles rehung. Existing insulation can often be re-used in conjunction with the new, which can be of the mineral wool type or any of the expanded plastics.

Timber floors

Insulation of existing timber floors follows the same general pattern as that for new floors, but whereas it is simple to apply insulation before the floor is laid it is more complex and much more expensive if the floor has to be completely lifted before the insulation work can be carried out.

In some dwellings it will be possible to get under the floors and fit rigid insulation between the joists and supported on timber fillets. Semi-rigid slabs can also be inserted between the joists and retained by light laths sprung into place against the slabs and wedged between the joists. Many buildings, however, will not be provided with sufficient space under the floors so that a person can gain access, and there is then no practical alternative to raising the floors and tackling the problem from above.

Full details of the materials available and the thicknesses to use are given in Section 43.

Concrete floors

There is no simple way of insulating an existing concrete floor, and work of this type would be contemplated only when a major reconstruction was in hand and considerable change of floor levels could be accommodated. The application would then take on the character of a new construction, and the method used would be the same as given in Section 43.

Where some change of level can be allowed, some benefit is obtained by setting fibre building board in bitumen on the concrete floor and finishing in hardboard with the joints staggered. The hardboard should be adhered to the fibreboard, as pinning is not satisfactory in this situation. The thickness of board used will depend on how much the level can be raised, but should not be less than 15 mm if any practical advantage is to be obtained.

It should be noted that this type of floor should not be subjected to point loading. Furniture legs etc. should be located in cups so as to distribute the load.

Cylinders, tanks and pipes

All hot-water cylinders or tanks should be insulated with a minimum of 75 mm good-quality insulation. Proprietary cylinder jackets are freely available in the standard sizes, and these can be adapted to suit non-standard cylinders. For tanks, slabs of expanded polystyrene, compressed straw or mineral (glass or rock) wool should be used, and where tanks are housed in airing cupboards it is advisable to cover the materials with a washable protective coating, which as well as retaining fibres will also allow surfaces to be kept clean. Another method is to construct a surrounding box to the tank, leaving a space at least 75 mm deep, and fill the space with expanded vermiculite or expanded polystyrene granules. The box sides should be constructed somewhat higher than the tank so that a top cavity is also available to take the insulation. The top should be protected with boarding, fibreboard or plasterboard, again covered with a washable surface such as polythene.

Hot-water pipes should also be insulated (unless the surface of the pipes is an important element in the heating system), particularly where the piping is under floors or in ducts, chases etc. If the piping is easily accessible but out of sight, flexible pipe-wrapping materials such as mineral wool or polyurethane can be used to a thickness of about 50 mm, but this will not be possible in all cases. Often it will be necessary to construct timber or chipboard chases to surround the pipes, and to pack these with any of the usual loose materials. This same method is followed for existing ducts and chases.

Cold-water tanks, cisterns and piping should only be insulated if they can be exposed to air temperatures near and below freezing. This is especially likely in attic spaces that have been insulated, and we have already described how this should be dealt with. Two further points should be made in this connection. Firstly it is very important to see that the insulation is not laid carelessly and will not allow cold draughts to enter under the insulation in the vicinity of tanks or pipes. Freezing is most commonly found at joints that have been covered poorly or not at all. Secondly it is important to ensure that a good cover is provided to tanks before the top insulation is applied, and that access for overflow pipes is provided through the insulation and the cover. Most if not all water authorities will insist on this arrangement.

SOUND INSULATION

Partitions

In many situations in existing dwellings there will be the need to provide partitions that give some protection against sound transmission from an adjacent room. There are many ways of constructing such partitions, and the results will vary not only with the type of partition but also with the type of surrounding structure. In Section 43 the principles are covered in some detail, and the following constructions are examples of good practice in this field.

If the structure will stand the superimposed loading, a brickwork partition will probably give the best results against airborne noises, provided it is constructed without spaces in the mortar, it is well jointed into the adjacent walls with all gaps sealed, and its whole surface is plastered on both sides. Where such partitions are constructed on wooden floors the partition must follow the line of an existing joist, which should in turn be reinforced with timber to the full depth of the joist and along its full length.

Another method employs compressed-straw slab faced with plasterboard on both sides. The method of fixing will vary with the conditions and the slab manufacturer should be consulted for details. Again it is very important that all joints are adequately sealed.

Also in general use are many types of timber-framed partition that can be constructed on site and that employ insulation quilts in the cavities under the plasterboard facings. There are two basic methods in use.

In the first method the studs are erected in parallel, and the mineral (glass and rock) wool sound-deadening quilt is applied across the face of the studs and secured to them with broad nails and light washers. Plasterboard (wallboard grade) is then secured to the studs and the whole surface skimcoated. Best results are obtained with a second layer of quilt similarly applied and finished on the other side of the stud, but the layer of quilt on the quieter side may be omitted for economy reasons, although with some loss of efficiency. In both cases care must be taken to ensure that top and bottom plates are properly secured to the surrounding structure, and that studs and boards at the sides in addition to being secure are also sealed, preferably with a mastic composition to ensure that there are no gaps and that variations due to timber movement can be taken up.

The other method, known as the *staggered stud* system, employs studs set vertically parallel but horizontally offset. Every alternate stud is offset by a distance equivalent to the thickness of the sound-deadening quilt; the quilt is applied to the face of the recessed stud then passed behind the next stud, to which it is secured on

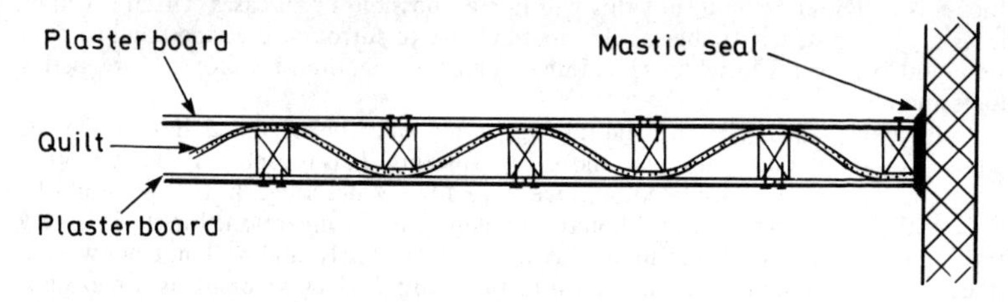

Figure 58.1 Staggered stud partition

the other side, and so on. The plasterboard in turn is secured only to the exposed studs on each side (see *Figure 58.1*). Both sides of the partition are skimcoated with plaster, and the same precautions as before must be followed regarding sealing.

Partitions of the type described are suitable for use in separating WCs from bedroom accommodation.

Timber floors

Where a property is being converted from some other use to that of dwelling, it will be necessary to provide for sound insulation of intermediate floors if the dwellings are to be flats. There is no way of doing this without some increase of floor level, but the minimum increase is possible if the following method is followed (*Figure 58.2*).

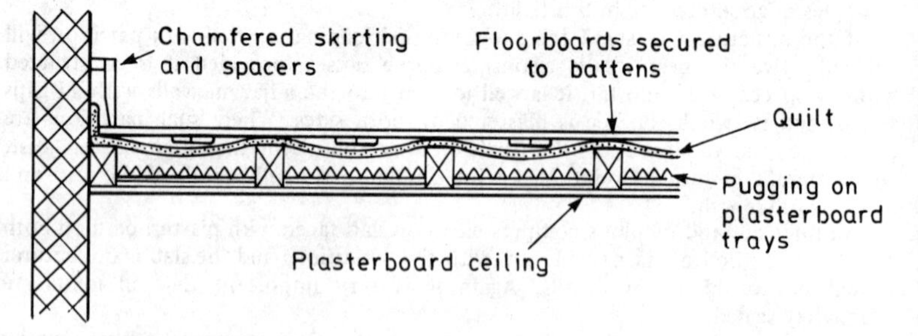

Figure 58.2 Offset floating timber floor on timber joist sub-floor

The skirting boards and floor boards are removed, thus exposing the joists and the ceiling of the floor below. Timber fillets are nailed to both sides of the joists and about 15 mm from the ceiling. Plasterboard shelves are cut to fit between the joists, and are placed on the fillets so that a continuous shelf is constructed. Dry sand and ashes pugging is loaded on the shelves to a loading of 120 kg/m², including the weight of the shelves and the ceiling.

Mineral (glass or rock) wool sound-deadening quilt is then laid across the joists and secured with very occasional nails hammered well home so as not to protrude, and using fibrous washers to ensure that the nail does not cut through. The edges are turned up all round the room. Sections of floorboards are then prepared with the boards screwed to battens positioned so that they will come between the joists, and with the battens offset 150 mm so that the sections can be screwed to each other on assembly. Sections are placed progressively, ensuring that the edges of the quilt are above the edges of the boards and that each section is screwed to the protruding battens of its neighbour. Battens on adjacent sections must be offset so as to allow the protrusions to fit underneath. It may also be necessary to trim the ends of the boards and the final edge in order to accommodate the turned-up quilt.

Before refixing the skirting boards over the quilt, the lower edge should be chamfered to a point so that contact between skirting and floor is minimised. It may also be necessary to insert a spacing piece under the top edge of the skirting board to create a space for the quilt and to similarly seal off the skirting at doors. If the floor loading required in this application is not permissible, sufficiently good results may be achieved by using mineral (glass or rock) wool loose fibre on the shelves. Packing should be at about 30 to 40 kg/m^2 excluding the weight of shelves and ceiling, and it should be carefully distributed to ensure that there are no voids.

Windows

In many houses a good deal of external noise is found to come through windows. If the noise level is not too high a satisfactory solution may be to reglaze the windows using a heavier glass. Normally window glass will be about 3 mm thick, and this can be replaced with heavier substances up to 10 mm without any alteration of the window structure. If a high level of noise is experienced it will be necessary to adopt double glazing. Normal double-glazing units are not effective against high levels of noise, although they may be satisfactory against moderate levels. For best results it is necessary to construct double windows with a space between the independent panes of not less than 100 mm.

The reveals within the cavity should be lined with mineral (glass or rock) wool sound-absorbent strips, faced out with perforated metal or hardboard with free space of not less than 3 per cent. The best results will be obtained if the glass used for each side is of a different weight, but both should be of the heavier types.

As with all sound problems, care should be exercised to ensure that the frames are sealed into the openings using a mastic sealing compound.

Special problems

Aircraft. One of the most difficult problems with external noise relates to aircraft, and around airports the levels of noise experienced under certain conditions can be so high as to cause physical pain to sensitive ears. Fortunately such situations seldom occur, but quite often occupants can find conditions quite intolerable. There are four actions that can be taken practically but at considerable cost.

- First, the windows can be replaced with the 100 mm cavity double-glazed type described above.
- Secondly, all apertures must be sealed and mechanical ventilation with acoustic baffles provided. This includes the sealing of all windows not double glazed (e.g. those in WCs) and providing ventilation at all appropriate points.
- Thirdly, all cavity walls should be insulated in the cavity, or external wall insulation applied.
- Finally, pitched roofs should be insulated with a minimum of 150 mm of mineral

wool; the whole upper surface of the insulation should be covered with heavy-grade plasterboard carefully fitted over the joists with good butt joints, all joints being stripped in the normal way.

When such an application is carried out is is necessary to check that the ceiling structure is capable of taking the additional weight. Walkways must also be provided above the plasterboard if access to the attic is necessary, and it is advisable to relocate water tanks and piping below the ceiling level if this is practical. Alternatively a housing of the same specification should be built over the tanks and piping, taking care that a good fit is achieved between the housing and the remainder of the insulation.

Soil pipes. Where soil pipes pass through living rooms or bedrooms, or are housed in the structure adjacent to rooms of these types, it is often necessary and always advisable to ensure that adequate insulation is provided. The insulation is provided by wrapping the pipe in a minimum of 25 mm flexible insulating material of reasonably high density (the higher density mineral wool products are suitable) and encasing the pipe in a duct constructed of 25 mm timber on framing. The duct including the framing should be entirely dissociated from the pipe.

Where the pipes are already housed in a duct within the structure, the best results possible in the circumstances will probably be obtained by packing the duct as tightly as possible with loose mineral wool, after carefully fitting a 25 mm mineral wool rigid preformed section around the pipe. If there is insufficient space to accommodate 25 mm of the rigid section it may be possible to alter the duct, or it will be necessary to accept a reduction of performance by trimming the section in thickness.

In all applications where pipes or tanks are encased in insulation, or run in chases, ducts etc., it should always be remembered that access to the pipe or tank may be required at a later date to effect repairs, and provision should be made for this.

WC fittings. Badly sited WCs will often cause problems in relation to bedrooms and living rooms. Suitable treatments for partitions and soil pipes have already been described, but often the problem may be concerned with the connection to the waste pipe or to the floor. Flexible connectors are now available, and their use also allows resilient floor mountings to be used. These arrangements will considerably reduce the transmission of impact sound, and additional improvement is possible if pugging can be introduced to the floors immediately underneath and adjacent.

Reducing sound levels

In areas where high noise levels are causing annoyance to other people, it is often advisable to carry out some treatment at the source of the noise. This involves the incorporation of areas of sound-absorbing treatment, usually on the walls or ceilings of the source room. The subject is very complicated and should not be contemplated without first receiving detailed advice from the material or system manufacturer.

General

Throughout the discussion of remedial sound treatments, emphasis has often been placed on identifying free air paths and sealing them off. Too much emphasis cannot be placed on this requirement, lack of attention to which can often destroy the effect of very expensive treatments.

59 ALTERATIONS AND CONVERSIONS (1)

INTRODUCTION 59–2

DEALING WITH AN ENQUIRY 59–2

MARKETS 59–4

PRACTICAL CONSIDERATIONS 59–6

COMPANY POLICY 59–10

ALTERATIONS AND CONVERSIONS (1) 59

59 ALTERATIONS AND CONVERSIONS (1)

L. HILL, FIOB, MBIM
Chairman, R. Mansell Ltd

This section gives guidance on what to consider at enquiry and tender stages. It comments on the nature of the domestic and commercial markets, and gives practical guidance on the execution of work.

INTRODUCTION

Alteration and conversion work is widespread and, with some exceptions, is usually the market for builders in the categories of 'small' to 'large medium'. Major contractors are sometimes employed on this work when time is the essence of the contract, or when their usual market for new building work is weakened by economic pressures. A number of the largest contractors have specialist sections engaged in this market, often trading under a different name.

DEALING WITH AN ENQUIRY

Assessing the enquiry

Any alteration to a building must be worthwhile, and some careful questioning can enable the builder to ascertain how essential the work may be and thereby to assess the likelihood of it going ahead. If doubt arises, an approximate price should be given, indicating willingness to go into greater detail if the client's interest in the scheme is sustained.

With any building work, the builder ought to decide for himself whether the result will justify his client's expenditure. If not, then from the outset it should be clear that the builder may be wasting time, and interest and expenditure should appropriately diminish.

Giving an estimate directly to a client is a matter for shrewd appraisal. The builder needs to satisfy himself that the client's intentions are serious and that they are not satisfying a whim to find out how much some work might cost. To quote for work to an architect or surveyor indicates that the clients have already committed themselves to some expenditure and will usually have been given an indication of cost levels.

A useful yardstick for either domestic or commercial and industrial work is to consider whether it would be cheaper to sell off the existing property and buy another, more suitable, property for less than the cost of alteration. If this appears to be the case, no harm can result in putting forward this theory; it may well bring out the client's reasoning, and help the builder to decide if the scheme is likely to go ahead.

Estimating and tendering

Always remember that the production of an estimate is time-consuming and highly expensive. Estimating as an occupation is only a means of securing work, and once the quotation is produced and submitted every effort must be made to 'sell' the job.

To our infinite regret, the building industry committed itself to giving estimates free, but it is worth remembering that the estimate belongs to the builder, who has a substantial investment in it in terms of money and time, and no builder should allow this fact to escape him. It is therefore essential to put maximum effort into securing work for which the estimate has been produced before going on to deal with the next enquiry. When dealing with the client directly, it is good practice to present an estimate personally and try hard to get a commitment for the work to proceed at that time.

When writing a specification, it is in the builder's interest to be precise. This eliminates problems arising from misunderstanding of intention in scope and quality. A carefully written specification also serves as a check on the estimate and removes the dangers of omitting vital costs.

One of the biggest pitfalls awaiting the builder is a gradual raising of quality during the couse of the work, best exemplified by the fact that substitution of hardwood for softwood involves far greater costing than the value of the timber. Often these problems do not manifest themselves until the final account for the work is presented, and by then the builder cannot negotiate from a position of strength.

When the specification or Bill of Quantities is provided by the architect or surveyor, a quality appraisal is essential, coupled with a realistic analysis of the builder's ability to work to the limits required. Here again it is the cost of producing estimates that must be considered, and if the work required is above or below the quality or style that the builder can accommodate it is probably advisable to return the enquiry without wasting time and money on it.

Speed of work, too, can be equally onerous, and a job calling for greater resources than the builder has available in terms of men or money should be abandoned. Equally, a careful appraisal must be carried out of the rate of going of a contract, which can be too slow to warrant the skills and time of the foreman or supervisor. This often applies when only parts of a building can be made available, and the work must proceed in stages, each completed before another starts. On-cost and profit ratios should be considerably increased to compensate for the lack of earnings potential of the supervision tied up in this way. The following considerations must therefore be borne in mind when submitting an estimate.

Imagine that the builder's requirement to pay for overheads and profit in the current trading year is £120 000, and that he employs six foremen of equal skill and stature. Then each foreman must earn £20 000 per annum for the company.

Therefore, if the builder's estimate is for work totalling £40 000 to be executed in six months, and his normal percentage addition for overheads and profit is 15 per cent, he would show a gross income of £6000 from this work. Reference to the earning capacity of his foreman shows:

Value of contract to be carried out in 6 months	=	£40 000		
6 months for 1 foreman at £20 000 per annum	=	£10 000	=	25 per cent

Therefore to add less than 25 per cent will, in actuality, show a loss.

It is essential strategy in carrying out alteration work that this calculation be made on every estimate submitted, and if the work cannot produce appropriate on-costs the builder must wait for a faster-moving job to arrive.

Naturally, not all foremen or types of work will carry the same earning capacity, but a strict understanding of this rule is essential. For a small builder who is preparing his own estimates and supervising his own work, this consideration of the earning capacity of work, when related to his own involvement, can be the deciding factor before he starts on the preparation of an estimate, or indeed any involvement at all in the project. The larger builder can make similar calculations; therefore, related to the figures used in the example above, a yardstick could be prepared, using the following concept, but with the obvious unequal ability that really exists:

	Individual earning cap. per annum	*Total per annum*
The builder's most efficient and skilled foreman	£25 000	£25 000
2 good foremen	15 000	30 000
3 working foremen	12 500	37 500
3 trade foremen supervising work in their own trades (e.g. plumbers)	5 000	15 000
4 decorators with appropriate gangs on domestic and commercial work	6 000	24 000
Total desired earnings		£131 500

Some margin over the desired overhead and profit figures must be built in to allow for voids between contracts.

It is not possible to give a guide to the real earning rates because the building market varies so much throughout the country. However, by starting with the total desired earnings figure, accurate enough guides can be obtained.

Never overlook the rate of earnings, and never assume that a higher percentage addition will suffice. Obviously, the converse can apply and a high throughput can equate to a lower addition for on-costs, although if the work warrants two of the foremen to get it done, then the extra earning capacity of both must not be overlooked.

In setting up the supervision of an alteration job, it is advisable to abstract the time allowed for the various trades as a means of checking time spent against work done. If it is possible to complete an analysis of this time, before the estimate is submitted, it may prove of even greater value. To illustrate this point, it may be patently clear that a carpenter/joiner will be needed on site for the whole running time of a 30 week contract, whereas the time allowed task by task in the estimate could be much shorter.

MARKETS

Domestic market

The domestic market is a much more difficult area to work in than it appears. One must remember that the work is being executed in the client's home, and the construction will frequently be undertaken in full view of the person paying the bill. Building is a comparatively simple operation, and the builders will find that the client will be able to follow the progress of the work and be critical of it.

From acceptance of this fact it will be realised that an early grasp of the client's character is essential, and the builder's estimate must reflect the level to which he will be made to perform. This will not necessarily be equal to the level of the construction of the existing house, but can perhaps be more quickly appreciated by observing the way in which the house is furnished and kept.

Separate the construction work from the client by physical barriers as far as possible, and make arrangements for the workmen to mess and work independently of the household. These remarks apply whether or not an architect has designed the work.

Before giving an estimate, some detailed discussion with the client or architect is essential. The builder needs to know to what extent he may occupy the site and entrance way and drive. Who will be responsible for making good lawns, paving, hedgerows, gardens and even plants?

Before commencing the work, there should be some briefing by the builder to the client indicating the extent of disturbance to be expected; whether or not water, power, gas, drainage and telephone will be interrupted, and for how long. The client should understand that power and telephone may have been assumed by the builder to be provided for his use. If the building is empty and no architect is employed or contract signed, then the question of insurance of the building must be settled and the client's insurer notified of the extra risk while work is being carried out. Vandalism must be considered and appropriate precautions taken.

Most alterations involve some work to the roofs. No form of temporary covering (other than a complete temporary roof) is proof against heavy rain or gale-force winds. Therefore the builder needs to point out the risks and agree who is accepting them. At all times a responsible workman or the foreman must be briefed to ensure that water supplies are turned off and no form of fire is alight when the building is left at night or weekends. Other forms of heating need to be considered in the winter period and the costs included.

Strangely, the builder is usually insured against negligence by his employees, but when some catastrophe occurs one can rely on the workpeople denying wholeheartedly that they were in any way negligent, thereby weakening the claim.

The builder must be extremely wary about the cost of additional work ordered by the client and all instructions should be confirmed, preferably quoting firm additional costs or indicating the likely increase. It is dangerous to quote upper and lower figures. An extra cost of £400 to £500 implies not less than £400 or more than £500, although this may not be what the builder intends to convey. It is better to give a single figure as a guide.

Credit-checking a private client is difficult but necessary, and it may be advisable to enquire bluntly where one can obtain a reference for the amount in question and never to assume that a client can meet the bill. Make sure the client knows that payments on account will be requested, and ensure that these payments are made and honoured at the times outlined. If a payment is not received on time, stop work.

Be careful of materials to be supplied by the client, or to be re-used. Sometimes unit furniture has its instructions in a foreign language, which can be expensive to interpret, and re-used fitments can cost more to adapt than to replace. It is much better to voice one's dislike of work on the owner's fittings than to be involved in heavy costs that are difficult to justify later.

Commercial market

There are essential differences in tackling this market. It is more likely that the client has an expert to guard his interests, but if an architect or surveyor is not employed it would be wise to ascertain the powers of the person dealing with the work, and the extent to which he may commit his company to any given degree of expenditure.

Read carefully any official order given for work to be done, as the terms may suit the client's trade or profession and be inappropriate for building work. When preparing the quotation, make sure how much work needs to be carried out in overtime, night-time or weekends, and clearly state what has been allowed, or make it clear that one has assumed that the work will be carried out in normal working hours.

Try not to be made unduly responsible for noise or dust levels, and get a clear brief indicating how much space you will be allocated for work and storage at any time. Even the most sophisticated clients often overlook how much their demands, in terms of minimising discomfort for their staff, can cost. If the brief demands high levels of security, low noise and dust levels, and generally reduces the level of activity of the builder, it is wise to indicate how much this is costing above normal rate.

Commercial clients have a habit of demanding that all work be stopped at certain times for specific reasons of their own. They must be made aware from the outset that each time the men stand idle at their request, the total cost will be added to their bill.

Commercial work is usually in the city centre, and one should check with the local police to find out how long they will allow for loading and unloading. Sometimes the restrictions are from 7 a.m. to 7 p.m., which can be expensive to overcome. Alternatively, if the builder wants and assumes that he will use the client's loading bay or parking facilities, this must be made clear and confirmed from the outset.

PRACTICAL CONSIDERATIONS

Shoring, supporting and demolition

It is becoming increasingly difficult to find men able to create timber shoring, and mostly this work is now carried out in tubes by scaffolding contractors. Ensure that the shoring work is carried out before access scaffolding is erected. The method and technique can be left to a competent scaffolding firm, who will also carry out the design, but the builder will find he must consider the place and style of the shoring.

It is necessary, throughout the process of alteration work, to give the shoring constant reconsideration. If there is a complete drawing of the building, sketch on it where the shoring is placed, and look again and again at the holes that have been cut to ensure that support is in the right place.

If load-bearing walls are being cut and steels placed above the openings, it is usually advisable to get the beams into position at the foot of the wall before the needling or support is erected. Often the shores themselves will make it difficult or impossible to get the beams into position. If the opening is on an upper floor, serious consideration must be given to shoring on the same line on each floor below, not forgetting the basement if there is one. If the opening is really large and the load considerable, then the plaster on ceilings must be cut down and head trees wedged directly to joists.

In house reconstruction, the shoring on the ground floor should be taken through a wooden suspended floor and based on the oversite concrete.

It must be clear that these precautions are relative, and hinge entirely on the stability of the building and the load being carried. In no circumstances, during the course of the work, should an opening being formed be left without support. One or more puncheons wedged into an opening overnight can save a minor catastrophe, and certainly in most buildings extra care should be taken to provide support over the weekend. When giving consideration to these points, it will be realised that some extra adjustable props of varying sizes on site are a wise investment; it is difficult to foresee every eventuality. Do not overlook a careful survey of the wiring, plumbing and heating in a building before commencing on the demolition of any part; have electricity and telephone cut off before work commences, and isolate pipe runs.

Beware of leaving isolated lengths of external wall unpropped. Even a high wind can overturn them. A short length of 50 mm flat steel cut into a mortar joint is enough to give lateral support to a raking adjustable prop set on one or both sides.

Go back to the original scheme for support work again and again, and see how it relates to the work in hand and add vertical or horizontal support as required.

Certainly water should be used to keep down dust when demolishing or when removing rubbish. A chute is an excellent idea, but provision for it must be made in the scaffolding and a flexible tail provided to prevent dust flying from the skip or lorry. If the fall of the debris is more than 6 metres, the chute must be cranked, preferably with an access door on the upper side of the crank for clearing obstructions, and a hose must be played over the rubbish as it goes into the skip or lorry.

During a constant period of demolition and preparation work, it is difficult to keep floor areas clear of debris. It is worthwhile using the system employed in engineering works, i.e. marking lines on the floor for throughways, thereby keeping obstructions out of the way.

Dust screens may be obligatory during this stage of the work, but it is worthwhile preparing the occupants of the building for fine dust, which percolates for several floors up and down from the places of work and cannot be obviated.

It is during demolition work of this type and of any size that builders must give some thought to the environment. Drills or breakers of sufficient size must be employed for the shortest possible time, and the removal of rubbish must be considered from the outset. Ensure that foremen know enough about the regulations concerning use of skips, in order that they do not cause a contravention.

Bricklaying

The major initial consideration in this trade must be the problem of marrying new to old work. Constant thought must be given to bonding satisfactorily so that the ties of the building remain. If it is not possible to obtain a bond when filling openings, use expanded metal firmly pinned to the reveal and built into the joints of the new work.

Pinning up between new brick or block and existing work must be carefully and continuously carried out. If slate is unobtainable tiles should be used, but driven in so that the full thickness of the wall is pinned.

If it is possible to argue that even greater care is needed, then it must be when pinning up between a new beam and the work over. If possible, sling a substantial weight from the centre of the beam to ensure maximum deflection before pinning up. On application, the steel supplier can provide the weight most suitable.

Externally If possible, arrange a break between new and old face brickwork. Even 50 mm will do to avoid the problem of matching. Avoid at all costs the use of common bricks against facing bricks, and consider a contrasting brick where a match is not possible. Strangely enough, it is the jointing or pointing of new brickwork in a different colour or style that often shows to the greatest disadvantage, and it is incumbent on any good builder to experiment with some mortar to find a mix that, when dry, has a similar property to the existing work. Be especially wary of the problems of joining new hollow walls to existing solid walls in older buildings, and devise means of overcoming the bridging effect of a solid wall passing dampness across a cavity.

Internally When building up from timber floors, ensure that there is a bearer joist to take the weight of a new partition and pin up the top to some timbers or a plate rather than to a soft plastered ceiling.

Plastering

In alteration work this trade can create many problems, because the results of the plasterers' work get close attention from the client.

The use of the appropriate soft plaster can avoid many of the shrinkage problems that can occur in making good, but ensuring that the existing work is dampened and that new material is not too rapidly dried out by sunlight will help. When conducting negotiations with one's clients it is advisable to warn that making good usually shows, especially if subjected to top light on a wall or side light at 180° from windows.

Danger besets the builder who puts one or more newly plastered walls into an existing room of an old building. It shows the original walls to such disadvantage that the client is prompted to argue that they are not correctly made good. Remember to pre-empt this with some warnings. Plain or regency-type wallpaper also highlights old walls or those with making good, particularly if wall-bracket lighting is used.

Where the builder has to work by artificial light in dark spaces, it is wise to find out what form of permanent lighting will be used, and to sling a temporary strip-light or bulb in a similar way to that ultimately proposed whilst the men are at work.

In this trade, thickness still tends to produce quality. If the specification limits the work to two coats of plaster it is unreasonable to expect superb finish, and the client or architect should be made aware of this before the work is started.

Excessive sun or wind drying will craze the plaster, and the builder should also ensure that all rainwater pipes are fixed back or repaired before plastering commences, especially in old buildings where the walls are usually solid.

Plumbing and drainage

When tracing existing drainage runs, dig the discovery trench at right angles to the expected line of the drain. When working on older buildings it is dangerous to make any assumptions about pipe runs; they must all be traced out to their ultimate conclusions.

Particularly if the building has its own septic system, it is paramount to ensure that the arrangement of the tanks has been located before making any connections. Increasingly it is concluded in law that the builder is considered to be an expert and can be made liable for any subsequent problems relating to the work done. Consequently, to slavishly follow the drawings that are provided is inadequate when making new connections to old drainage systems.

Great care must be taken to guard open drainage excavations around an existing building. The contractor is often found liable to damages for personal injury to people who may appear to be improperly on the site.

Supporting trench walls is far more important when working around buildings in alteration work than on an open site. Heavy rain or frost can cause expensive damage to nearby buildings or to paving and fencing.

Notify the client immediately if the sanitary inspector requires work done beyond the extent of the estimate submitted. It is not a clear responsibility of the client to pay for extra work done to comply with statutory requirements.

When preparing an estimate for plumbing work in alterations, one should beware of the risk of re-using existing equipment. The expression 'take out and set aside and later refix bath, basins and WCs, traps, flushing cisterns etc.' should not be accepted as a 'risk' item. Damage frequently results, for which the builder may be held responsible.

Making good or replacing equipment 'to match existing' is another expression written into specifications which, when applied to sanitary ware or fittings, may be unduly risky.

Too often, clients will insist on personally making purchases of major items such as sanitary or kitchen fitments, mistakenly believing that they are gaining a substantial discount. This can lead to astonishing problems in the event of damage during or after fixing. It is advisable to point out that this risk must be accepted by the clients if they make their own purchases.

When carrying out alteration work to plumbing and heating, great care must be exercised. Never fill up a system at the end of the day after work has been carried out, and in a severe cold spell the question of responsibility for the building and its plumbing should be settled with the owner beforehand. It is by no means certain that the problem of the result of frost damage rests with the builder if working in an existing building.

Carpentry and joinery

The appraisal of costs and skills outlined earlier in this section has particular application in this trade. The availability of the correct skills is essential and usually overlooked at foreman level. In alteration and reconstruction work the ability of the contractor to perform the work must be a first consideration, but subsequently it will be found that aptitude in carpentry and joinery is not wide. The cost of getting a joiner to fix a small bay of shuttering or erect a few panels of fencing can be astounding, and the butchery resulting from a good shuttering man's work on joinery has to be seen to be believed. Work containing small series of items of different categories of carpentry and joinery therefore needs premium costing rates.

As in some of the other trades, lovely easily written phrases appear in specifications. 'Overhaul and repair or renew sashes and frames as necessary' can mean almost anything, and there appears to be no sound reason why the builder should accept the risk of such items. It is arguable that if the person writing the specification cannot be precise, then equally the builder's estimator cannot quantify the work required.

Repairs carried out on a daywork basis to certain types of buildings are an historic problem in this industry, notably in work on greenhouses and conservatories and historic buildings. It is wise in such cases to agree a number of days or weeks of work with the architect or client, and then conduct a review. When the account for this type of work is submitted, it is unacceptable to all parties to be faced with the problem that it would have been cheaper to renew.

Roofing

Perhaps the predominant matter in this trade is what is delightfully called 'a Builder's Risk Item'. This is often dismissed in the specification provided for the building estimate as 'overhaul and repair roof surfaces as necessary'. Sometimes the roofs can only be inspected by erecting scaffolds or maybe by aerial survey. Almost certainly the problem arises when quoting for alterations to existing buildings.

Unless the extent of the repairs can be quantified, this item should be negotiated out of the builder's problems, either by limiting the amount of work for which one has allowed or by including a provisional sum. This should state clearly whether the scaffolding required is included elsewhere in the tender or is specifically required for this work.

Temporary covering whilst work is in progress means what it says, and is not proof against heavy rain or storm. The risks must belong to the client, who must choose the ratio of cost to preparedness. There is no satisfactory way of covering flat roofs, and when work on them is in progress emergency measures must be ready all day and proper cover available at night and weekends.

Never overlook the delay to roofers in periods of heavy rain. Three weeks of standstill for rain can mean that the arrival of roofing trades on your site will be set back by a like amount.

There is a classic quandary awaiting every builder estimating for work to roofs. It often appears to be quite ridiculous to allow for an access scaffold for a small amount of work. The estimator therefore will allow an amount of money in the estimate for gaining access. Usually this is too much for ladder access and far too little for scaffolding. The moral is that if one cannot solve the problem in one's mind when actually looking at the difficulty, include for scaffolding as it will certainly be the only solution.

Decorating

In the building industry this trade is generally afforded a lower level of consideration, and frequently the decorators say that by the time they arrive on a building site the time and the money have all been used. In fact it is the decorators' work that is actually seen, and if it is well done it can provide the final lustre to some well-executed earlier work. Of course the reverse applies.

As in so many of the items of work in building, it is not possible to get high-quality finish unless the appropriate number of coats of paint are applied and the surfaces to be decorated are carefully prepared. This must be made clear to the architect or client before the work is commenced. It is intolerable for the builder to be forced to apply additional coats at his own expense to overcome blemishes that insufficient paint cannot conceal.

As mentioned earlier, the raising of quality during the course of the work is a great danger to the builder when decorating. The subtle raising of quality of the finishings of a space cannot be paid for by increasing the cost of the wallpaper.

Externally, when dealing with alterations and repair work, quality is the greatest problem. In almost every specification appear those words of doom: 'burn off where necessary and paint two additional coats etc.'. In some cases, concealed elsewhere in the preamble to the specification, one finds that where any part of a surface is to be burnt off, then the whole area shall be similarly prepared. In many cases the knowledgeable builder can come to some conclusion, but if the decision-making is too borderline it is wise to return the enquiry and not waste valuable time taking part in a specualtion. Alternatively it may be advisable to take some more sweeping decision and assume all woodwork on south and west elevations needs the extra work outlined.

Never be inveigled into starting external decoration too early or too late in the year. The costs of men waiting to start or finish work when the weather is frosty or unsettled fall upon the builder, as does the cost of rectification if the work is spoiled.

COMPANY POLICY

The operation of a building company working in the field of domestic and smaller industrial and commercial work is a matter for science and finesse. The senior people must combine the skills of management, sales, public relations, labour relations, technical expertise, haulage management, accountancy, buying and so on. The real importance of understanding these many functions is to realise that they exist, and that each must be given the right amount of time and priority in each day.

The ability to make notes adequately on note pads of adequate size and quality is most important. Keep systems simple. Value the top person's time properly, and ensure that he does not become the most highly paid delivery man in the town.

Buying plant and equipment is a problem that can be solved by applying a simple rule. Buy lorries, ladders, concrete mixers, tractors, dumpers and so on of the size you most often need. It costs comparatively little more to buy the next size up, but the temptation should be resisted because this raises costs enormously. When you need equipment larger or smaller than average, hire it for the shortest possible time.

If you own, say, three mixers, then two more should be on hire before you buy an additional one. Keep equipment that you own in use all the time, and tend to place wholly owned plant on the longest running jobs. Hire from outside to expand plant stock for short periods.

From these basic thoughts a pattern will emerge. If some work needs very long ladders, it can be made to pay for them by hiring. If the cost of hiring loses the contract, that will be the correct result.

The correct theme that will allow a building company to prosper is to do *more* of what one does *well.* Every departure from this pattern should be considered carefully and the work avoided if it does not fit into this rule.

This attitude may seem restricting, but it will then become a conscious decision to take work involving new techniques. Some pitfalls for the ambitious builder who has no previous experience in alteration and conversion work are: basement construction, reinforced concrete frames, work from cradles, city-centre contracts, multi-storey buildings, earthworks and pipelaying. Commercial or industrial building, domestic work, and work for local authorities and government agencies, are all quite different from each other.

Provided that the builder is aware that he is addressing a new field of operation and organises his entry into it with awareness, it may be considered progressive. Similarly, the size of contract undertaken has to be carefully calculated. The amounts are relative to the size of the enterprise, but experience of two contracts of £50 000 does not prepare a builder for the problems of one contract of £100 000, which demands a totally different management technique, as the amounts increase, so the changes are greater.

Finally, be aware that all builders are operating in a market probably peculiar to their own locality. It changes constantly, and contractors change their styles reluctantly.

60 ALTERATIONS AND CONVERSIONS (2)

PLANNING THE CONTRACT **60**–2

CARRYING OUT THE CONTRACT **60**–4

COMPLETION OF CONTRACT **60**–16

ALTERATIONS AND CONVERSIONS (2) 60

60 ALTERATIONS AND CONVERSIONS (2)

N.J. RAGSDALE, LIOB, FFB, AMBIM
Managing Director, H.A. Marks Ltd, Building Contractors

This section discusses management of alteration and conversion work, i.e. the coordination of labour, materials, plant and specialist services, with appropriate payments at the correct stages of a contract.

PLANNING THE CONTRACT

Receipt and interpretation of instructions

Although it is not advisable to make any positive start to work until official instructions are received, it is as well for the builder and his team to consider the workload and financial budgeting requirements that a prospective contract will impose on his organisation.

On receipt of official instructions – and no instructions are official until confirmation is in writing or a contract has been exchanged or signed – the builder should examine thoroughly the contract drawings and specifications. Normally architects and surveyors place the onus on contractors for checking the accuracy of their drawings and specifications. The builder and/or his staff should pay high regard to this point, since agreements to variations are sometimes difficult to establish if left too long after a start on site is made, particularly if an operative has removed, demolished or altered the very item upon which a claim may be based.

Although an estimator should have taken into consideration any requirements called for in the preliminaries section of a Bill of Quantities or specification, it is as well that the builder should be prepared to implement these requirements in order to avert a claim by the employer for breach of contract. Where there is no need to conform to specific requirements, the builder can consider the sums allowed as a reserve against other sections of the work on which more money may need to be spent than has been budgeted for.

A Bill of Quantities is a price in itself and should present little problem in respect of identification and price of a particular item. A builder should arrange for his estimate to be broken down into what are called builders' quantities, the advantage being that a rate can be established for the job. This serves to protect the contractor when an omission or an addition is required; this will be referred to later. Most surveyors require a priced specification, which follows the same principle. However, the builder should be on his guard against giving too much detailed price information.

Programming

There have been various views expressed as to the advantage of compiling a programme, particularly with alteration and improvement work. For instance, discovering or tracing the extent of dry rot can have the effect of substantially altering a programme. The advantage of having a programme can be stated briefly: 'it is better to have some sort of plan of operations than none'. In the building industry, flexibility is one of the most common occurrences encountered. Here again it is not uncommon for the client's agent to require a programme as a condition of the contract.

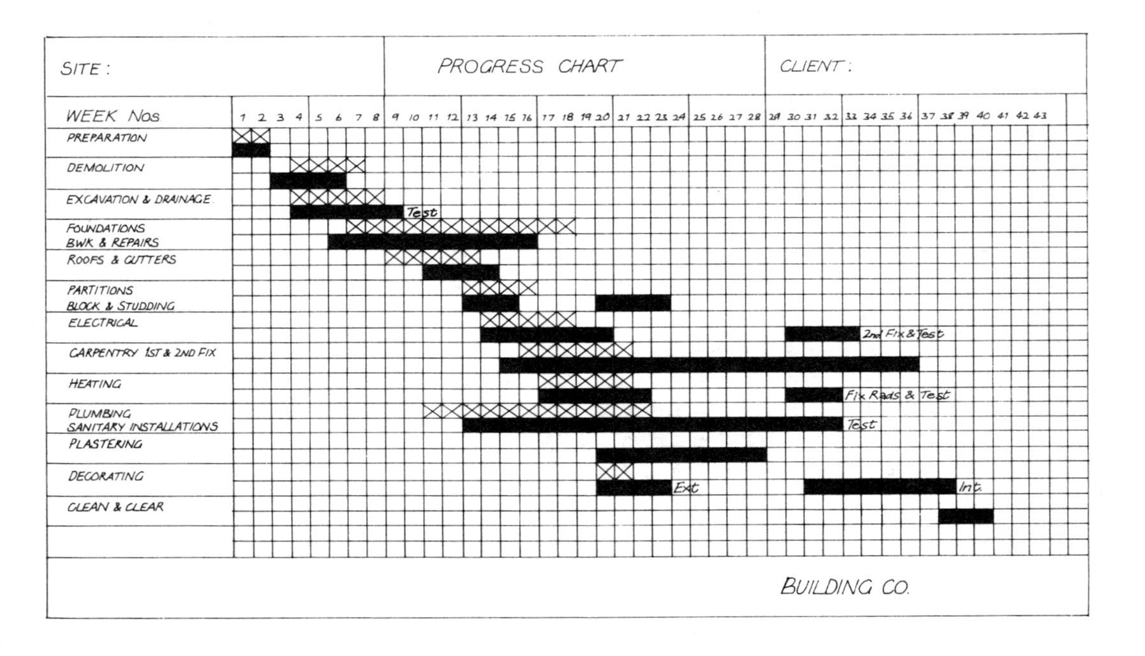

Figure 60.1 Progress chart

A programme sufficient for alteration and improvement work can take the form of a single bar chart or progress chart as illustrated in *Figure 60.1.* A successful builder would already have indicated his estimated duration of work. A progress chart has the advantage of enabling him to plan the sequence of all trades, particularly the specialist operations such as heating and electrical work. The builder has the option of stating when he requires the nominated sub-contractors on site; if they are unable to meet the builder's reasonable requirement the responsibility is placed on the architect, surveyor or employer (whoever is responsible) for extending the programme as necessary. This obviously has the advantage of protecting the builder from incurring a penalty under the 'liquidated and ascertained damages' clause; at the same time it allows an extension of Preliminaries, which represents money to the builder in compensation for maintaining his site through delay. The builder must equally be on his guard and must at all times ensure that his own work-force are keeping to a timetable. At the first indication that delays are likely through reasons beyond the builder's control, it is essential that he inform the architect/surveyor to establish the fact and protect himself.

A typical conversion contract could be programmed as illustrated in *Figure 60.1,* a job value of say £40 000–50 000 scheduled for a work duration of nine months. It involves an old property large enough to lend itself to conversion into four two-bedroom units. A specification would embrace the renewal of the roof, structural repairs to brickwork such as settlement, window arches, kitchen and toilet extensions. The progress chart illustrates the planned progress midway through a contract (say week 21), and the actual progress. The spacing for each trade and section is divided into two, the lower half representing the programmed period and the upper half the actual site progress.

Visual aids are very much easier to comprehend and appreciate than dates on paper or letters in files. By continually examining a chart the builder and his site staff can see at a glance where a delay is likely to occur, so that remedial action can be taken in advance.

If one examines the chart, a cause for concern in this hypothetical case is that the second phase of partitioning has not yet started nor has any work started in the plastering trade. Plastering is the 'halfway house' in a building contract: certain work can only be done before the plasterer comes on the job, and no work can be completed until after he is finished. It is essential therefore to try to work to a plan. Remember that there is a duty to the specialists and nominated sub-contractors, who also have to plan their respective organisations and need to budget for their future labour and material requirements.

CARRYING OUT THE CONTRACT

Variations (extra work)

There is hardly an occasion when in the course of a contract a variation does not occur. Extra work comes about mainly for the following reasons:
(a) change of mind by the client;
(b) necessity to alter or increase the work as a result of opening up;
(c) additional work considered to be economical whilst a contract is in progress;
(d) insufficient information in tender documents;
(e) errors and omissions observed as the work proceeds.
There are few builders who in the course of carrying out alterations, conversions or new work have not at some time experienced a client who either cannot make up his mind or is continually wanting to alter or increase a scheme. Since a builder must be prepared to accept work and instructions from the general public there are occasions when he must be prepared and on his guard when dealing with a difficult client,

subject of course to the reasonable assumption that he will be allowed to perform his duties and be properly paid. Where professional assistance has not been sought – particularly with smaller works – before a contractor has been appointed or invited to submit an estimate, he is more likely to come up against indecision and mind changing. If a builder feels it to be worthwhile and has the opportunity to advise a prospective client, then it is in his interest to do so. After all, the builder is a professional and can use his expertise to the mutual benefit of his company and client. Therefore a builder should not allow himself to be rushed into starting a job until he has satisfied himself that his client has thoroughly considered the work and made up his mind as far as is possible.

Category (b) causes are more uncontrollable, for the reason that design and decisions cannot always be finalised until a structure has been opened up for examination. It is not unusual for older property that is available for alteration and conversion to be subject to attack by dry rot or woodworm. Diseases such as these have to be traced to their limit and beyond, and one can appreciate that the extent of such work involved in eradication cannot be accurately forecast, hence these sorts of operation become variations.

Category (c) work needs little description, and is often the result of a change of circumstances affecting the client or building owner and the proposed use of a building. Alternatively, the client may have been notified to comply with certain statutory regulations and it is sensible to incorporate the extra work while the builder and his organisation are on site.

Categories (d) and (e) are very sensitive areas, and a competent builder will not look upon the variations as a 'pot of gold'. The builder must take every precaution to safeguard his interests, to ensure that circumstances do not place him in a position of accepting responsibility not properly his. Immediately an error or an omission is discovered, the contractor should notify the architect or surveyor pointing this out; at the same time it will be helpful if the builder can make some suggestion as to how the problems should be overcome. If it is possible to be able to offer a firm or budget figure, or even advise that further costs will be involved, this should be done.

~~SITE INSTRUCTION~~ / VARIATION ORDER

To: Surveyor

Serial № 801 Date

Dear Sir,

re: CONTRACT London JOB No. 1000

We confirm ~~our~~/your ~~phone~~/verbal instruction of (date) Recent date to

Take out and renew rotted joists and wall plates.

Replace with new

- 2nd Floor Front Large Room

Yours faithfully,

BUILDING Co.

Figure 60.2 Variation order

An effective method of advising a client or his surveyor of the necessity to vary the work is for the builder to issue his own confirmation of instructions as shown in *Figure 60.2*. This form is from a site instruction/variation order pad serialised with

BUILDING CO.

EXTRA WORKS.

*Name of Client*__________ *Week Ending*__________

Address of Job London *Job No* 1000

Description of Work

Variation Order No. 801 dated ....

Take out rotted joists and wall
plate and replace with new - 2nd
Floor Front Room.

PLEASE ENTER DETAILS OF WORK
FINISHED AND RETURN THIS SHEET
TO OFFICE.

NAME	M	TUE	W	TH.	FRI	SAT	SUN	TOTAL HRS	RATE

Plant & Materials Used

*Work authorized by (Signature)*__________

Figure 60.3 Extra work sheet

coloured copies. It should be in quadruplicate, the first copy going to the client or architect, the second to the quantity surveyor (if one is appointed), the third to the site foreman or whoever is in charge, and the fourth copy retained in the builder's office.

The issuing of one of these statements thus establishes the fact that a variation has occurred. Even if the particular item is too complicated to enable it to be priced, at least the builder has established the fact, and if the statement is undisputed it serves as a permanent contract record. On larger work, and some months later when the accounts are being finalised, the builder will readily be able to collate all the essential information, particularly if he has kept an up-to-date record of all instructions and variations. Coupled with this information the builder can send a duplicate of his extra work job sheet attached to the foreman's site copy of the variation order; this enables the documents to be properly matched. A typical example of an extra work sheet is illustrated in *Figure 60.3*. The prompt issue of such a sheet to the foreman renders it less likely for the details of the time and material to be wrongly recorded, lost or forgotten. The sheet when completed can be returned to the office, checked against the duplicate so that it is not listed as missing, attached to the site instruction/variation order, and kept in hand for valuations and final account procedure.

SITE INSTRUCTION / ~~VARIATION ORDER~~

To: Plastering Co.

Serial № 802 Date

Dear Sir,

re: CONTRACT London JOB No. 1000

We confirm our/~~your~~ ~~phone~~/verbal instruction of (date) .... to

Take down faulty section of cornice at main door and replace.

Yours faithfully,

BUILDING Co.

Figure 60.4 Site instruction

The quadruplicate form can be used as an instruction or variation simply by deleting what is not applicable. It also serves to confirm the issue of a variation or an instruction to the sub-contractors, whether or not the builder's own specialists are nominated. The examples given in *Figures 60.2 and 60.4* illustrate the general type of use.

Dayworks

Daywork is defined as a system of building operation work that is accounted for on a time and materials basis. Unless the proper procedure is carried out, the result can cause controversy between the builder and his client. Perhaps it is understandable that professional advisers are not keen to involve their clients in daywork charges, since it can be considered by some that the builder is being given an 'open cheque'. Certain operations, particularly those in conversion and alteration work, can only be carried out on a 'time and materials' or 'cost of work plus' basis – for example, exploratory work in connection with tracing the extent of dry rot, the repair of collapsed brickwork in the vicinity of a new door opening, or the exposure of steelwork to ascertain the extent of damage through rust and lamination. In such cases, the contractor would be extremely ill-advised to give an estimate for work that he was not sure could be fully ascertained or allowed for. These and similar cases amply justify the use of daywork as an effective means of accounting. However, if, as has been mentioned earlier, a builder is not to become involved in a dispute over the amount of work done and its cost under this system, the following guidelines should assist.

The contractor in the course of running his business should already have compiled a schedule of daywork rates listing his charges for labour, materials and miscellaneous expenses. The costs of labour need to be revised with every nationally agreed wages fluctuation or statutory requirement. A typical schedule of daywork charges is illustrated in *Figure 60.5*. For labour calculations, it may be considered easier and

BUILDING COMPANY

Schedule of daywork charges as from . . . .

Section 'A' The cost of labour as listed below is deemed to cover for all payments to be made in respect of:

(a) The cost of wages to be paid to operatives
(b) Holidays with pay, public holidays
(c) National Insurance/pension
(d) Sickness insurance scheme
(e) Obligations under Redundancy Payments Act
(f) Industrial training levy
(g) Head office overheads
(h) Supervision
(i) Employer's liability insurance
(j) Non-chargeable plant
(k) Profit

(A)	Foreman carpenter/Painter/Plumber/Bricklayer	3.90 per hour
(B)	Shop joiner/Machinist	4.10 per hour
(C)	Plumber/Plasterer/Carpenter/Roofer/Bricklayer/Painter/ Truck driver	3.31 per hour
(D)	Labourer	3.05 per hour

Section 'B'	Materials	Add 15 per cent
Section 'C'	Chargeable plant Truck hire @ £1.20 per hour Men's fares etc.	Add 10 per cent
Section 'D'	Attendance and profit on Nominated Sub-contractors	Add 5 per cent

Figure 60.5 Schedule of daywork charges

more practical to give a gross figure rather than become involved with adding percentages to a net labour figure. The gross figure for labour should therefore reflect the builder's costs of his employees' wages and statutory contributions (i.e. holiday pay, sickness pay, training levy appropriation, etc.), his establishment (i.e. rent, rates, lighting, office staff, etc.), and his percentage addition for profit. These of course will vary from one contractor to another depending on the type and size of the business. It is not unusual for materials to receive the addition of 15 per cent for profit and handling and 10 per cent for expenses. The generally accepted percentage

JOB No. 1000 | **DAYWORK SHEET** | BUILDING CO.

CONTRACT LONDON | D/W SHEET No.: 6 | WEEK ENDING – – –

DESCRIPTION OF WORK	OPERATIVE	TRADE	HOURS WORKED M	T	W	T	F	S	S	T	R	WAGES £	p	FARES & ALLOWANCES £	p
Rear Ground Floor. Removing flooring. Hacking off plaster and leaving ready for surveyor to trace extent of Dry Rot. Instructions: Surveyor	A. Brown	F/MAN				2	1			3	3.90	11	70		35
	C. Davis	LAB				6	7			13	3.05	39	65	1	20
			TOTAL TO SUMMARY									51	35	1	55

MATERIALS	R	£	p
NIL			
TOTAL TO SUMMARY	£		

SUMMARY	NETT COST £	p	% ADDED	VALUE OF % ADDED £	p	TOTAL £	p
WAGES	—	—	—	—	—	51	35
FARES & ALLOWANCES	1	55	10		16	1	71
MATERIALS	—	—	—	—	—	—	—
PLANT HIRE ETC.	5	08	10		51	5	59
TOTAL VALUE						58	65

PLANT HIRE TRANSPORT ETC.	R	£	p
2 No 10FT Trestles @ £1·70		3	40
2 No 14FT Boards @ 0·84		1	68
TOTAL TO SUMMARY	£	5	08

Signed on behalf of Client

Figure 60.6 Daywork sheet

addition for specialist sub-contractors is 5 per cent, but this can vary according to the amount of work to be supervised and the control the builder must exercise. This is referred to later when dealing with sub-contractors.

In many tender documents, the builder is required to state the percentage he will require to be added to his labour, based on the National Joint Council's latest agreement for wages. He should ensure that in these circumstances his percentages are correct, for there will certainly be a considerable difference between the percentage addition he requires on his own labour rates and the percentage addition he must have on the NJC agreed wages structure.

The sample daywork sheet shown in *Figure 60.6* should again be in quadruplicate, each copy being printed in a different colour for easy identification. The daywork operation having been defined and authorised, the foreman or whoever is responsible for making up the sheet should extract his information from the time sheets or log-book for hours and the expenses of each man employed in the work, and should record this information on the daywork sheet, together with the materials ordered and the plant allocated to the job with a brief description of the works.

At this point, the sheet can be either fully priced out or retained, as desired. However, the important point is that daywork sheets should be dispatched to the client or his representative as soon as possible after the end of the week to which the sheet relates. The method of despatch should be to send the first and second copy to the client, requesting an authorising signature and return of the latter.

It may be that the builder's staff may prefer to price out the sheets at a later date, perhaps when actual costs of materials and plant hire have come in. The important thing is that the signed sheet should never leave the builder's office. It is his proof of authority, not necessarily that the work is correct or acceptable, but as confirmation of the labour content, material quantities and plant.

Valuations

No builder can exist without stage or progress payments during the course of a contract. It is an accepted principle of trading in the industry, and a builder should make ample provision in his estimate or note the detail of an appendix to a tender document to ensure that the facility for drawing payment on account is allowed for.

Valuations for work carried out to a certain date normally take place monthly during the progress of a contract. With much emphasis being placed on cash flow, it is not uncommon for a builder to arrange for fortnightly valuations. On small works of a private nature, the builder will most probably arrange for one or two progress payments, this being a somewhat informal arrangement between the builder and his client.

The accepted methods of valuing for payment of works in progress are: (a) if a quantity surveyor is appointed, he either carries out a valuation from an inspection of the work without assistance from the contractor, or else both he and the contractor's surveyor or representative mutually agree the valuation of works; otherwise (b) the builder prepares his own valuation and presents this to the client, architect or surveyor.

Referring to earlier notes on Bills of Quantities or priced specifications, one of the principal advantages of having these is that the work can easily be evaluated and the amount allocated to the total payment. The usual method is for the work completed to date to be taken as a percentage of the total amount of work to be undertaken in a particular section. A percentage of the builder's site cost (i.e. the sums allocated in the Preliminaries) is also included. An important factor not to be disregarded is the value of unfixed materials on site. Since the builder will generally have to meet his monthly account charges, it is important that these sums are carried into a valuation. It will also be possible for a builder to obtain a payment in respect of goods and materials not on the site – for example, work being carried out on the builder's own premises, perhaps the manufacture of joinery, special-size windows or a staircase. Under these circumstances, if he is going to request payment, he must make his premises available for the work to be inspected if requested, in order to verify his claim. The valuation is usually subject to an agreed retention amount. VAT is now also to be taken into account, but the complexities of this are beyond the scope of this section.

It is important and of practical and financial advantage to the builder that he be properly represented at a valuation meeting by being present himself or arranging for his representative to be present. The site foreman should also be available, as there are many queries that arise at a valuation meeting when the benefit of the foreman's intimate knowledge of the work enables these to be resolved.

A builder will find it to his financial advantage to arrange for the greatest allocation of preliminaries in a valuation towards his site overhead costs in the early stages of a job. This will enable him to recover the cost of installing plant and scaffolding, organising a site office and mess-room facilities, etc. Payments towards these expenses early in a contract considerably assist the builder's cash flow. Examining the programme, again it can be seen how vital it is to rely on a supply of money to feed the job, particularly in the third, fourth and fifth month of the contract when site labour and activity is at its greatest.

The builder should bear in mind that his own specialists and the nominated sub-contractors also have their own cash-flow budgets to meet; therefore arrangements should be made for their applications for payment and valuations to coincide with that of the main contractor. When dealing with his own specialist contractors the builder follows the same procedure as between surveyor and main contractor, i.e. originally a specialist's estimate should be either quantified or compiled on a measured basis. All too often the builder accepts a lump sum figure. If the type of work only requires the latter, or if the builder is dealing with a specialist who can only produce an overall figure, then he should insist that items are separated or broken down for ease of identification and valuation for payment.

The builder should not be put to disadvantage with the sub-contractor by having terms and conditions of payment less favourable than his own. It is usual for the contract to carry a retention clause, and this until practical completion is 5 per cent of a contract sum. On large contracts the amount of retention has a limit, and after reaching the limit by virtue of certified payments no further deductions take place under the heading of retention.

However, on small works the agreement to deduct retention is a matter between the architect or surveyor and the builder. It may be that the builder, when working directly for a client, will allow a percentage retention as an expression of good faith, or as a guarantee or undertaking to attend to defects after an agreed period of time. As has been stated, whatever arrangement the builder makes or is required to observe, the specialist contractor should come under the same conditions.

Staffing a contract

The importance of having correct and reliable labour on a job cannot be underrated. It is the one factor that has the greatest influence on the success or failure of a contract. Without doubt the most important figure on a job is the team leader, whether he be an agent (if the contract is of such a size as to warrant this standard of supervision), a general foreman or a working foreman.

For most works of a conversion and alteration nature a good working foreman is quite capable of supervising the overall work from commencement to completion. Men placed in a position of job responsibility should preferably have had a sound basic training in one of the established building trades, preferably carpentry and joinery or brickwork. The foreman in carpentry has the advantage of seeing a job through to the decorating stage.

Too much emphasis cannot be placed on the choice of the right man for the job. The foreman is the 'site anchor'; he is the one his employer will rely on, the one whom the architect or surveyor will look to for help and opinion, the one whom the client will automatically rely on. It should be remembered that the foreman is the employer's public relations representative, and although he will work and act within the limits of his own status and capabilities, nevertheless he should be held in high esteem by those around him. A most important aspect is the fact that the man on the job is in charge of a large part of his firm's investment.

Since the building industry has become increasingly bound by legislation in recent years, it is essential that the foreman should be reasonably literate; he should also be capable of inspiring those of whom he is in charge, since the pace of the leader sets the pace of the team. However, the responsibility of a job does not rest solely on the foreman's shoulders. To give of his best, he must receive proper direction and encouragement from management. As has been stated, the choice of a foreman is very important in any size of job, and the prime factor is salary. The employer must be prepared to reward adequately by means of wages, benefits and perhaps some form of target bonus, as he considers suitable.

Regarding the labour that a foreman will be required to supervise, it is the employer's duty and in his interest to staff a job with reliable tradesmen. Although this may be difficult at times, owing to holiday periods or sickness, well paid tradesmen will prove of greater benefit to the job and the employer in the long run. Reliable staff can be trusted, which is of great importance in a conversion or alteration, particularly where there is client or tenant occupation, or property to be protected and safeguarded.

Small alteration work requires more team work amongst the work force than other work, simply because the size of the job does not permit other than a minimum number of operatives. There is no room for demarcation argument; all must assist each other in the common aim of bringing a job to a speedy and successful conclusion.

Site organisation and management

Time, money and frustration can be saved by efficient site organisation. The builder, in consultation with his foreman, will decide the most expedient methods of carrying out the proposed works in accordance with the programme or agreed timing.

Bearing in mind the sequence of the work and the size of the job, a room nearest the entrance to the site – and one that has the least structural upheaval – is the most suitable for use as a site office. In cases where a building has suffered fire damage or delapidation to the extent that it is virtually open to the elements, the builder will be best advised to have a site hut erected either within the building or outside. However, it is essential to maintain records and drawings in a safe and secure location. The office will also need to be used for receiving visitors to the job.

A further requisite to the efficient running of the work is a site phone, which allows the foreman or any duly authorised person to communicate outside. To reduce the risk of unauthorised calls being made, key-operated dial locks are available.

Delivery of materials and acceptance of hire plant play as important a part in alteration and conversion work as in any other contract. A builder can lose money and time if there is not a close check-system operating. The foreman, his deputy or someone who takes charge in his absence should check all goods coming to the site for condition and quantity. Delivery notes, although sometimes discarded, are extremely important documents particularly when there is a dispute over quantity.

Delivery notes should be kept and returned to the office weekly. When a shortage is discovered, the immediate response from the supplier is that the goods and quantity were signed for. Continuing shortages should make the management query the honesty of the site staff. Quality and condition are of prime importance, and the foreman should instruct that certain articles used in finishings (for example doors, baths, formica and even prepared timber) should be examined before being unloaded or in the presence of the drivers. After two or three men have carried a bath up three storeys, it is extremely annoying to find a surface blemish or chip, sometimes three or four days after the delivery has been made. There is nothing that can be done but to repeat the order – at the builder's expense.

The same care must be exercised with the delivery of plant being hired to the contract. If plant is delivered damaged but usable this should be noted on the driver's docket and the builder's copy. When plant is returned, collected by the hirer's own transport or returned by the builder's transport, great care should be taken to ensure that the builder or his staff obtain a signature for the returned article. On larger work the foreman will no doubt deputise a reliable labourer with the duty of safeguarding the plant and ensuring none gets cleared with rubbish. It is not unusual for the builder to be charged for the repair of a damaged item of plant when it was delivered in that condition. Proper inspection at the time of delivery will prevent this.

The care and custody of tools and power tools on a site is important. The foreman should be issued with a strong lockable site box that can be either screwed or bolted to the floor if possible. Tools should be kept sharp and in proper condition, thus ensuring that when required they can be used to their full advantage and time is not wasted relying on equipment that is faulty. Tools that are in need of repair should be returned to the builder's office so that the stock of equipment is kept in prime condition and always ready for issue. Tools should only be issued to sites on a signature. There is little an employer can do if an operative loses a tool, but it does establish a moral responsibility.

Coupled with the care of plant and materials is the security of premises, particularly if unoccupied outside the working hours. The foreman or any person he nominates should be responsible for ensuring the premises are left in a secure condition to prevent ingress by squatters, vandals or thieves. Unfortunately little can be done to dissuade the determined thief. Ladders should be taken in at night or weekends. Windows not fully glazed and locked should be boarded securely. These precautions, considered a chore by site staff, should be insisted upon by the builder or management. Prevention of damage or loss is only a fraction of the cost of repair or replacement.

A builder can minimise the risk of the expense of theft, wilful damage and break-in by consulting with his insurers and negotiating a 'Contractors All Risks' policy (see Section 5). For a premium proportionate to the contract sum, he can arrange cover for most eventualities likely to occur on a building site.

Health and safety at work plays an important role in the construction industry. Regulations impose a responsibility on both employer and employee to observe current legislation and to act within the building firm's Statement of Policy in respect of the Act. It is the employer's duty to issue protective clothing against both the elements and danger from building operations. Particularly on alteration work, defective flooring is often removed and requires denailing, therefore the supervisor or foreman should order the wearing of gloves. In fact gloves, eye protection and helmets should be the basic safety equipment issued to all site staff. In addition to this an operative should not be allowed on site unless he is wearing correct footwear. All visitors should be encouraged to wear safety helmets. The greater the encouragement given to safety as a result of the example set by the builder and his team, the easier it becomes for men to accept the maintenance of safety standards as part of everyday building life. As a matter of health and welfare, ample provision must be made to provide a mess room and an area for drying out wet clothing. One of the first requisites in setting up the site in the provision of temporary toilet accommodation if these facilities are not available.

It is a statutory requirement that the following notices must be exhibited in a prominent position:

- the company's valid certificate of Employer's Liability insurance;
- Electricity (Factories Act) Special Regulations 1908 and 1944, if appropriate;
- the Construction (Lifting Operations) Regulations 1961;
- the Construction (General Provisions) Regulations 1961;
- the Construction (Health and Welfare) Regulations 1961;
- the Construction (Working Place) Regulations 1961;
- all amendments to the above mentioned Regulations;
- the Factories Act 1961 and all amendments thereto;
- the Health and Safety at Work Act 1974 and all amendments thereto.

In addition an accident book must be kept on the site and any accident, however minor, must be duly recorded. In cases where the employer considers a claim could materialise he will be well advised to consult his insurance company. The first-aid equipment must be maintained in complete order and replenished as required.

Site records play a vital role in the smooth running of a contract, however small. The most important item is the diary. A large desk-type diary suffices for all general purposes. All visitors to site should be recorded, and also the names of sub-contractors, their trades and numbers. A brief description of the day's work, and the state of the weather, should be entered. Lastly, if the job carries a relatively small workforce the same diary will serve as a time book. At the completion of the contract the book should be removed for safekeeping. The continual maintenance of such a record is one of the foreman's most valuable contributions. There are many occasions at the final account stage when the builder is negotiating with the client or his representative when reference to a diary can prove or disprove an argument.

Regular site meetings serve to keep all informed of the progress of the work and bring to light any delays or problems with delivery of materials and coordination of trades. These meetings should always be properly recorded and copies of the record sent to all those attending and interested parties. It is usual for the architect or surveyor to produce minutes of a meeting. However, it is in the builder's own interest to produce his own notes, and to publish them if no minutes are forthcoming.

Spot estimating

As has been mentioned earlier, variations inevitably occur with alteration and conversion work. There is much to be said for being able to advise a cost at the time of instruction, and in certain circumstances the 'spot item' method of estimating can be employed. This system can be entrusted to a reliable foreman if the matter is urgent. The method employed is to assess the hours required on a gross daywork basis, and the cost of materials required, then make a percentage addition for profit. The following example shows the method.

Take off 6′6″ × 2′6″ × 1³/₈″ softwood flush door and replace with half-hour fire resisting door. Remove existing stops and replace with 25 mm × 25 mm screwed to frame. Re-use existing furniture and lock. Fix set aside door closer.

Labour

Remove door and clear, ditto stops	½ hour	
Fit new door, fix furniture and closer	3 hours	
Cut and fix stops	½ hour	
	4 hours at say £3.60 =	£14.40

Material		
6′6″ × 2′6″ × $1\frac{3}{8}$″FR flush door	£17.50	
1 pair 4″ CI butts	1.01	
5 m 25 mm × 25 mm SW	1.15	
Sundries etc.	0.60	
	£20.26	
Add 15 per cent	3.04	
		£14.40
	£23.30	£23.30
		£37.70

Say £39.00 to £40.00

Thus it is possible, using the principle of the above example, to advise an item price.

Specialist sub-contractors

Sub-contractors fall generally into two categories: those providing a complete labour and material service, and those known as self-employed or providing a 'labour only' service. The former are frequently nominated by the supervising officer, and on the exchange of letters or the receipt of an order become employed by the main contractor. On a major contract, or any contract where the builder has signed an RIBA form of contract, it is advisable that he enters into a contract with a specialist sub-contractor. The sub-contractor is then bound by the same conditions of contract laid down in the appendix as the main contractor. It must always be remembered that, although the sub-contractor may be nominated by the client or architect, the builder is responsible for his performance and standard of work. The builder has the right to refuse a nomination, but such grounds for refusal must be based on fact and past experience. It is the builder's duty to provide a nominated sub-contractor with all site facilities, such as messing, use of plant and scaffolding, water and electricity use. For these services the builder will be entitled to a percentage addition on the sub-contractor's charges for attendance. Profit is a separate percentage addition irrespective of any attendance.

Self-employed or 'labour only' sub-contractors are usually dealt with in the same way as the traditional sub-contract trades that the builder employs direct, such as plasterers, tilers and roofers etc.

Notices to local authorities

All construction work comes under the jurisdiction and supervision of the local authority surveyor, and it is the responsibility of the builder to notify the surveyor or building inspector of the intended work. During the course of a job the inspector will continue to visit a site. His services and advice are particularly valuable in alteration and conversion work where loads and stresses have to be regarded – for example, when extra floors, extensions or openings are being formed.

There are few works of alteration and conversion that do not require alteration to the drainage system. It is a requirement of the builder to give notice to the Environmental Health Department in whose area the work is situated to obtain approval for the scheme. The same authority will be responsible for health and hygiene standards.

Under present legislation, if a contract is scheduled to run for more than six weeks the local office of the Factory Inspectorate must be notified on the appropriate form. Factory Inspectors make unspecified visits to construction sites, and the builder should always ensure that his site is a safe one and that his operatives and staff are observing the safety regulations.

COMPLETION OF CONTRACT

The completion of a building contract, irrespective of size and nature, is usually in two parts – practical completion and final completion. Practical completion is defined as that stage of completion which allows the building owner the use of the premises or structure as intended from the resulting work. Generally, small items of finishing-off carry on after the work has been practically completed. It is essential for the builder to agree the date of practical completion with the architect/surveyor or, if working direct, with the client. The period between practical and final completion is in ordinary terms referred to as the maintenance period, and can be of any duration agreed by both parties. It is usual for such periods to be either three or six months. During this period the builder must maintain the work he has contracted to undertake, and must also remedy any defects at an inspection when the date of practical completion has been established. Therefore it is important to agree this date, because it limits the time the builder must be responsible for defects. This period also allows for the preparation of final accounts, and ideally accounts should be agreed before final completion.

At the agreed date of final completion a further inspection usually takes place and any further defects are noted such as shrinkages, paint irregularities, defective equipment etc. On the satisfactory completion of these outstanding items, the final certificate can be issued or the final payment made and the contract fully completed.

61 MANAGEMENT OF MAINTENANCE

SCOPE OF MAINTENANCE WORK 61–2

FINANCIAL MANAGEMENT AND BUDGETARY CONTROL 61–7

HUMAN ASPECTS OF MAINTENANCE MANAGEMENT 61–11

WORK PLANNING AND CONTROL 61–16

MANAGEMENT OF MAINTENANCE 61

61 MANAGEMENT OF MAINTENANCE

H. GRAHAM, FIMBM, FFB, MIOB
Formerly Assistant Director of Housing (Maintenance), Greater London Council

This section illustrates the application of management disciplines to the building maintenance organisation. Although the examples used are from a very large organisation, the principles can be applied to one of any size.

SCOPE OF MAINTENANCE WORK

Planned maintenance service

A maintenance organisation is concerned primarily with the cyclical work of redecoration and the like, the operation of services, and carrying out day-to-day running repairs to both building structures and engineering services. Table 61.1 shows typical maintenance service standards of a large organisation.

Particular attention is drawn to the response times shown for item 4 of the table (day-to-day repairs and servicing). In the GLC maintenance department, for example, response times were calculated by grouping together jobs reported by tenants and others, to ascertain the economical grouping of work in the various trades. It was found, after plotting these jobs on a location plan, that for an economical bricklayer's workload it took about five weeks for an adequate number of jobs to accumulate, for carpenter's work about four weeks, for plumber's work three weeks, for glazing jobs two weeks, for electrical work one to two weeks. It was accepted that these reponse times were reasonable, on condition that special arrangements were made for emergency work that could be undertaken immediately by an operative in a small van. Experience proved these to be acceptable to tenants, and produced best performances from the operatives at the lowest possible cost at the time of introduction.

Rectification of defects

Quite distinct from the above aspects, but also part of the maintenance organisation's work, is the rectification of problems found to result from incorrect detailing, poor specifications and choice of materials, defective construction techniques and workmanship, and failures of inadequately tested innovatory constructional methods requiring urgent attention to protect the structure. Typical examples of major defects that arise are:

Roofs Breakdown of asphalt surfaces over party walls etc., owing to differential movement of components, thermal movement, inadequate sheathing felt, failures of substrate, and movement owing to inadequate size of roof members. Premature breakdown of roof insulation owing to absence of vapour barrier and interstitial condensation. Breakdown of strawboard roofing materials owing to condensation.

Rain penetration of structure Fracturing of brickwork owing to a combination of concrete frame 'creep' and expansion of brickwork, exacerbated by poor workmanship and inadequate ties. Failure of bond between finishing surface materials and the structure owing to chemical reaction between different materials or thermal movement, e.g. brick-slips, defective mosaic. Rain penetration through concrete panel

Operation	Frequency	Relationship to other operations or time allowed to complete operation
1. Repairs prior to painting (external painting)	–	Inspection 1½ years in advance of painting. All major work to be included in Estimates, and executed 1 year ahead of external painting; minor repairs three months before painting. All external hinges, tee bands to be lubricated every 5 years.
2. *Painting and decorating (external)*		
Normal external painting of all property	5 years	Including common staircases to flatted blocks with lifts.
New property external painting	3–4 years	After inspection if found necessary.
Short-life property external painting.	5 years	To be programmed and painted if life is two years or more and it has not been painted for five years.
3. *Painting and decorating (internal)*		
Main entrances to all blocks of flats, including hallways and internal corridors together with staircases to blocks of flats without lifts.	2½ years (or earlier on inspection if in poor condition)	To be included with external painting programme as wet-weather work and additional treatment to fit programme. Surfaces not painted to be cleaned down. All main entrances to flatted blocks to be inspected annually and treated immediately if in a poor condition.
Normal programme work (limited liability), if finance available, to the following:		All existing metal lift-off hinges (particularly rising butts) to be re-lubricated with grease at least every 5 years or during cyclical redecoration of the premises. Cylinder locks/latches to be re-lubricated with light oil.
(a) house or maisonette	5 years	Hall, landing or staircase only.
(b) flat or bungalow	5 years	Hall, and any landing and staircase together with one bedroom.
(c) sashes and frames	5 years	Throughout the building if necessary to preserve the fabric and prevent wet rot or decay.
(d) special cases	5 years	Restricted to OAPs, one-parent families or disabled persons, and covering minimum re-decoration to all occupied rooms.
Empties – estate properties at change of tenancy	3 weeks (maximum time allowed)	Repair all work necessary to rooms being redecorated. Check window safety catches, oil or grease all hinges. Check all rising butts and oil and treat as agreed. Renew where burring has occurred on splayed bearing surfaces. Work to be completed within 3 weeks of receipt of keys.
Empties – general properties	3 weeks	Repair all work necessary to rooms being redecorated. Check window catches, safety catches, oil or grease all hinges. Treat rising butts as above.
Common staircases to flatted dwellings without lifts	2½ years (approximately)	To be included with external painting programme as wet-weather work. Non-painted surfaces to be washed down.
Laundries	2 years	On average, including machinery.

Table 61.1 *continued*

Operation	*Frequency*	*Relationship to other operations or time allowed to complete operation*
4. *Day-to-day repairs and servicing*		
Emergencies (safety, security, health reasons)		Complete within 24 hours of notification to management, or make safe if a long job. (Progress must be checked daily by management until completed.)
Bricklaying repairs within	5 weeks (average)	5 weeks average programme, but no repairs to be outstanding more than eight weeks.
Carpentry repairs within	4 weeks (average)	4 weeks average programme, but no repairs to be outstanding more than six weeks.
Plumbing repairs within	3 weeks (average)	3 weeks average programme, but no repairs to be outstanding more than five weeks.
Glazing renewals within	2 weeks	2 weeks average programme, but no repairs to be outstanding more than four weeks.
Electrical repairs	1–2 weeks	Average programme, but no repairs to be outstanding more than three weeks.
Space heating	3 days or ASAP	Average programme.
Miscellaneous repairs	5 weeks	5 weeks programme
Fencing repairs	5 weeks	5 weeks programme
5. *Servicing of applicances and equipment etc.*		
Gas appliance including boilers	Annually	Only specially trained staff to be engaged on this work. Central heating units to be serviced during summer period, hot water heaters all year round. Special attention to be given to ventilation.
Electric heating (all types)	Annually	To be treated as emergency repairs and dealt with immediately.
Immersion heaters	Annually	To be treated as emergency repairs and dealt with immediately.
Vent Axia fans	Annually	To be cleaned, serviced and tested.
Door closers	Annually	To cover complete range of door closing equipment.
Soil stacks	Annually	To be thoroughly cleaned. Back surges to be reported to management for attention to drains.
Scale reducers	Annually	Or more frequent where required.
Cleaning of ventilation ducts and individual extractor fans	Annually	During summer period.
Gulley and drainage channel cleaning	Annually	To include all public and private balconies.
Petrol and oil interceptors	Annually	To be inspected and cleaned out where necessary.

Table 61.1 *continued*

Operation	*Frequency*	*Relationship to other operations or time allowed to complete operation*
6. *Inspections and tests*		
Roof storage tanks	Annually	Special attention to be given to covers, overflows and trap.
(a) Multistorey blocks externally and common parts internally	5 years	To be thoroughly inspected 18 months ahead of external painting and followed by preparation of programme of works to meet requirement of item 1. Any Defective Premises Act or Health and Safety at Work Act matters to be reported and dealt with immediately. Blocks over 6 storeys in height to be inspected from cradles.
(b) All other property externally		
Power-tool inspection and tests (including portable hand tools)	(a) Daily inspection when in use	Immediate attention to be given to any defects. Defective tools shall not be used.
	(b) Monthly and annually	To be inspected by the manufacturers or their agents at intervals not exceeding six months, and any defect corrected immediately. Unserviceable machines not to be used under any circumstances.
Scaffold and plant	Monthly	Plant for external painting serviced in winter period and for internal painting to be serviced in summer period. All to be inspected monthly and appropriate action taken to ensure that only scaffold and plant in good condition is available for use.
7. *Refuse*		
Cleaning of hoppers on staircases and access balconies	3 times each year	Check to ensure hopper seal working correctly, check ventilation and that the chute lining is in a sanitary condition. Any grease or other dirt to be removed by approved treatment.
Clean dust chute chambers	3 times each year	Paint internally or wash down as specified. Any defects to be reported to management for action.
Clean chute cut offs	3 times each year.	Repair immediately any defects or renew cut off.
8. *Horticultural work*		
Grass cutting	Every 9 working days in normal growing periods	Reduce to 7 working days or less in peak growing periods, or increase to 10 working days in slow growth periods of drought.
Hedge cutting	Twice a year	April/May and August/September. Cut only once in the year following splitting.
Hedge splitting to 750 mm high × 250 mm wide.	1/5 each year	5 year programme. Between October and March.
Winter programme	Yearly	

Table 61.1 *continued*

Operation	*Frequency*	*Relationship to other operations or time allowed to complete operation*
9. *Inspections and servicing of mechanical plant, equipment and laundry equipment to comply with Health and Safety at Work Act 1974 and other appropriate legislation*		
Laundries and drying rooms		
(a) Check mechanism of washing machines, hydro extractors, spin dryers, and tumbler dryers	Quarterly	Resulting work to be programmed for immediate attention. Defective equipment must be made inoperable by withdrawing the fuses, locking the room, turning off gas or other services, disconnecting or capping off if necessary.
(b) Examine and test electrical circuits of washing machines, hydro extractors, spin dryers, and tumbler dryers.	Quarterly	Any resulting work to receive immediate attention.
(c) Examine and test electrical circuits of electrically operated drying cabinets	Monthly	Any resulting work to receive immediate attention.
(d) Examine and test the operation of gas-fired drying cabinets, and ensure adequate ventilation and satisfactory working of safety devices	Monthly	Any resulting work to receive immediate attention.
Woodworking machinery (including timber treatment plant, compressed air plant and equipment, dust extractor plant) and all other machinery	(a) Daily inspection (b) Not exceeding 6 months period	Give immediate attention to any defect or servicing required. To be inspected by the manufacturer or his agents at intervals not exceeding six months. Correct any defect immediately. Unserviceable machines not to be used under any circumstances.
10. *Playground equipment*		
All playground equipment	Monthly	All playground equipment to be tested monthly and action taken immediately on any defects.
Exhaustive condition tests on playground equipment	5 yearly	Equipment found to be defective must be immediately immobilised and made safe, followed by immediate action to repair.
11. *Dry risers and all fire-fighting equipment*	Twice per year	Check dry-risers to ensure that all valves and fittings are in position and undamaged. Check all hose reels and fittings and all fire extinguishers to ensure that they are fully charged and securely fixed. Resulting work to be programmed immediately. Arrange for fire prevention officer to check at the appropriate intervals.
12. *Electrical re-wiring (renewals)*	20–30 years	
13. *Renewal of gas water heaters*	25 years	

joints in industrialised construction methods. Failure of concrete owing to inadequate cover of reinforcement or use of excessive amounts of calcium chloride causing surface spalling. Wet rot to timber in windows, door frames and sashes etc. owing to incorrect construction, poor design of joints, insufficient sizes of components, and internal condensation on glass causing wet rot to sashes. Rain penetration through windows, curtain walling, doors, and the structure generally.

Services Extensive work required to drains owing to failures of workmanship; replacement of drains laid in made-up or poor soil, resulting in settlement and failure. Excessive corrosion of water mains owing to aggressive soil conditions. Excessive and continual furring up of hot water services in hard water areas owing to design of system (use of direct instead of indirect system). Sludging up of heating systems owing to gassing and chemical action from different metals. Poor design of warm-air heating units. Defects in electrical services owing to sub-standard workmanship. Estate lighting schemes having no regard to possible vandalism.

Materials Many recent problems have been caused by the use of high alumina cement, polyurethane insulation, asbestos products (blue asbestos), sealants, adhesives, mould growth, calcium chloride, marine aggregates and glass reinforced plastic (GRP) materials.

What options are available to the building team to minimise these defects and to ensure that value for money is obtained by producing buildings that are long-lasting, with minimum maintenance costs? Probably the best answer to this increasingly expensive remedial problem is to direct the awareness of the design team to the effects of ill-considered and irresponsible design on the long-term costs in use of their buildings; these costs are much increased by design and constructional defects. Mid-career training courses are necessary for architects and technicians to familiarise design teams with the problems of maintenance and to re-dedicate them to reducing maintenance costs to the minimum.

In addition, quality control in manufacture and on site must be improved to achieve an acceptable performance standard and to increase the life-cycles of building components. More care is required from all members of the building team to eradicate defects and improve construction standards, to ensure that in future our buildings require the minimum maintenance.

FINANCIAL MANAGEMENT AND BUDGETARY CONTROL

It is important in any maintenance organisation that the repair and maintenance requirements of the estate, its property and equipment (including any necessary servicing arrangements to meet the agreed maintenance service standards and the carrying out of the work to an accepted quality standard) should be adequately assessed and procedures laid down for correct financial management, estimating procedures and budgetary control for the whole organisation.

Many financial officers in municipal authorities and commercial organisations calculate the annual maintenance estimates on the basis of the previous year's approved financial budget plus a factor for inflation and growth. Since budgets for the coming year are based on historic expenditure levels not necessarily complying with the required standards, this could result in an over- or under-assessment of the true financial requirement. This method of budgeting for repairs, maintenance and servicing is unsatisfactory and economically unsound, if the property is to be adequately maintained to preserve the capital asset or the property investment. It does not take account of the evaluated workload and servicing arrangements needed to properly maintain the stock of property, and is bad management.

It is therefore necessary for managers to lay down an adequate annual estimate preparation system to assess the actual maintenance requirement and acceptable levels of operative performance. Work can be carried out more economically by directly employed staff, if properly managed; the author's experience shows that the use of work study, production and cost control, management by objectives and budgetary control procedures makes this a practicable and economic system in calculating the funding requirements for maintenance and in controlling actual expenditure against the approved financial budget.

Manpower budget

Using work-study or work-measurement techniques, manpower budgets can be calculated by trade or other requirements in advance of the preparation of annual estimates. The manpower budgets should be subsequently reviewed every six months to adjust them in the light of the latest work trends prior to operation for the financial year. Records should be maintained of all incoming repair requests and estimated workloads; after a period of time these records give a reliable indication of the trends and provide a basis for forecasts. Greater weighting is given to more recent results of operative performances, and any sizeable changes are carefully investigated. Increased performances will result in a reduction of labour requirements, and vice versa.

In calculating the numbers of directly employed operatives and supervisory staff required to carry out the total workload, provision must be made for absence due to sickness, holidays and other acceptable reasons. The manpower budget for each trade is calculated from the following formula:

$$\text{Number of men required} = J \times WC \times \frac{100 + SL}{EP} \div C$$

where J = forecast number of jobs per year in the particular trade
WC = forecast work content per job, in Productive Standard Hours in the trade
EP = forecast effective performance
SL = percentage for sickness and leave
C = conversion to man-years
(40 hours/man-week × 52 weeks/man-year = 2080 hours/man-year)

Example: J = 8320 jobs per year
WC = 2½ hours
EP = 80
SL = 20 per cent

$$8320 \times 2\tfrac{1}{2} \times \frac{120}{80} \div 2080 = 15 \text{ operatives}$$

In addition to the main trades, budget provision has to be made for indirect staff, i.e. first-line supervision, mates, general labourers, stores, etc. A detailed main-trades manpower budget is established for each department, and indirect staff are added either as a ratio to main tradesmen or according to the number of depots, stores etc., and the general geographic size of the organisation area. Where any doubt exists, work measurement should be carried out to obtain factual manning information. Experienced managers can give an accurate estimate for indirect staff.

Workload

Generally the number of jobs is directly proportional to the number of properties or services to be maintained, although the types of properties and needs of the estate must also be taken into account.

There will be weekly variations in the number of incoming repair requests from tenants and other sources, owing to seasonal fluctuations and emergency situations. When the number of repair requests is low, this can be offset by feeding programmed routine maintenance and preventive maintenance into the system for execution. Peaks of work that cannot be economically programmed for a fixed labour force can be covered separately by using term contractors. Budget forecasts should therefore include a provision for the use of outside contractors.

Technical staff should undertake a thorough survey of property in the year preceding that for which external painting is programmed. The survey is concerned mainly with the fabric of the building and with identifying the repairs required to bring the property up to the approved standard before painting is carried out. The work to be carried out is detailed on job tickets, which are fed into the programme as workload permits, but all work must be completed before commencement of the external painting programme. The budgeted number of jobs takes account of the known requirements of the pre-painting repairs.

The maintenance service standards (Table 61.1) set out a number of routine programmes to be completed each year. Managers should maintain records of the work detail covered by each programme.

A work-study based incentive bonus scheme that measures the work content of all jobs will show the average content per job for each trade in terms of Productive Standard Hours per job. Using the records and a careful examination of anticipated trends, an assessment of the work content of both repair and routine programme work and servicing operations can be made.

Multiplication of the forecast number of repair jobs for each trade and property, plus the number of routine programmed jobs, by the forecast average work content per job produces the total workload in terms of Productive Standard Hours for the directly employed labour force. For convenience this is calculated as a weekly figure.

Multiplication of the forecast number of properties in each of the programmes for external painting and internal redecoration work by the forecast average work content per property to be treated produces the total workload in terms of Productive Standard Hours. For convenience this is calculated as a weekly figure.

To convert the total workloads (Productive Standard Hours) into the numbers of directly employed operatives and supervisory staffs, it is necessary to establish the number of Productive Standard Hours of working that can be produced per man-hour worked. The detailed management controls that result from the workstudy schemes, and that are produced weekly from each level of management, provide the information required for this calculation.

$$\text{Effective performance} = \frac{\text{Productive Standard Hours}}{\text{Gross hours worked}} \times 100$$

The effective performance is calculated weekly for each main trade. The work measurement unit maintains records of this information, and from these an accurate performance level can be forecast for the budget year. The forecast will take the following into consideration:

- actual results achieved during previous years;
- a workstudy-based budget effective performance, which represents a 'standard' level of efficiency;

- an objective target, which calls for the highest level of managerial ability and against which each maintenance manager is measured.

In practice, in well-managed maintenance organisations most performance targets are achieved and the budget is set at this level, thereby ensuring that financial provision is calculated to cover a high value-for-money performance.

Financial budget

Having determined the manpower budget required to carry out the estimated workload for the following financial year, the cost of this labour force is then calculated. The following factors are taken into consideration:

- basic wage plus bonus, including performance trends of trade teams;
- overhead charges and all other on-costs;
- establishment and staff on-costs;
- materials costs;
- element of profit;
- purchase of new plant and equipment.

The current weekly wage, including average bonus earnings, for all operative trades (with separate rates for chargehands, tradesmen, apprentices/trainees, leading hands and labourers and other staff) is calculated. The cost of indirect support staff, i.e. transport, chargehands, clerks, pool drivers and mates, scaffolders and stores staff, is added to each trade in proportion to the support rendered. To obtain the basic annual cost of each class of operative, the weekly cost is then multiplied by the estimated number of weeks of productive work (i.e. excluding annual and sick leave and public holidays) in the year.

Overhead charges are added to the labour costs, as a percentage addition calculated to cover the following items (all indirect charges must be included):

- holidays and sick pay scheme;
- National Insurance and superannuation, death grants;
- rents and rates of works accommodation;
- purchase, lease or hire of vehicles and plant, including maintenance and repairs;
- establishment expenses and charges by other departments;
- gratuities for non-pensionable staff;
- return on capital employed.

These various components of the percentage addition are monitored monthly from costing information provided, and are adjusted to ensure an adequate recovery rate.

In addition to costs included in overhead charges above, a further percentage addition is made to cover the salaries of administrative, professional, technical, executive, financial and clerical staff employed on maintenance-management activities.

The estimated cost of materials is calculated, in conjunction with materials-management operations, on the basis of current expenditure. For budgeting, the total cost is spread over the various trades in proportion to the estimated volume of material usage per trade or activity.

The basic cost per year of each trade, with the addition of overheads, on-costs and materials costs, is multiplied by the numbers of staff in the manpower budget to produce the total estimated cost of the labour force. Part of the total cost is allocated, on the basis of the estimated workloads, to all activities of estate and property maintenance (e.g. painting, repairs, special maintenance, upkeep of gardens and sports grounds, ancillary improvement work, repairs and improvements to all properties, car parks, roads, drains, sewage works etc.).

In many efficient organisations, because of fluctuations or peaks and troughs in the total workload the direct labour force is responsible for about 80 per cent of the overall programme, the remainder is put out to contractors. Contract work can be divided into three main categories:

- trade support to cover peak workloads or shortages of operatives;
- local contracts for specialist work (e.g. asphalt roofs);
- major contracts for large works for which there is no labour allocation.

The first two categories are mainly covered by term contracts based on a schedule of rates, for which competitive tenders are invited annually and budgeted for accordingly. Major contracts, the third category, are generally prepared for all major maintenance jobs, remedial and improvement work, and estimated costs from technical information are prepared for budgeting.

In cases where the financial resources to be made available for the ensuing year are below those calculated to comply with the maintenance service standards, it will be necessary to amend the standards to avoid overspending.

Budgetary control and value for money

It is essential for the maintenance manager to keep a constant check on actual expenditure against budget. Computerisation of financial accountancy is generally accepted, giving the following:

- budgetary computer-control sheets, showing a running total of all expenditure against the various quotas for direct-labour activities;
- expenditure on all types of contract work shown against allocation of funds.

The maintenance manager is responsible for ensuring that his budget is not overspent. Orders are placed to comply with the programme at the appropriate time in the financial year, and a record of the resultant committed expenditure is maintained. Financial accountancy procedures give detailed materials costs, to ensure efficient materials management.

All contract work must be effectively supervised for quality control and to ensure value for money. Omissions and additions to contracts must be fully documented. Variations must be covered by the issue of a variation order at the time it is found necessary. The rates in the contract provide the basis for adjustment. Where items are not covered by the contract rates, a quotation must be obtained from the contractor for agreement by management.

It is of the utmost importance in both direct-labour and contract work to obtain value for money, to monitor all costs and to achieve cost targets set. One of the methods of establishing the cost-effectiveness of a direct-labour organisation is to compare actual direct-labour costs with the costs of comparable competitive tender work carried out by contractors. By this method one large municipal undertaking has recently shown that it would cost that authority an additional £8 million per year if maintenance work now carried out by directly employed labour were to be placed with term contractors at the lowest competitive tender rates; this indicates that value for money is being achieved.

HUMAN ASPECTS OF MAINTENANCE MANAGEMENT

A manager must make things happen; the art is, of course, to make the right things happen. The academic often contends that a good manager has no regular duties to perform and that he should delegate down the line for others to execute the work, since they should be well acquainted with their duties and objectives and should be left alone to get on with the job.

Personal experience has shown that one can delegate authority, but that responsibility is still expected to be vested in top management. In maintenance management the responsibilities are so complex and critical that top management must be involved to ensure that the five basic principles of management are applied:

1. *Plan* – examine the future and draw up a plan of action.
2. *Organise* – build up the material and human structure of the undertaking.
3. *Command* – maintain activity among the personnel.
4. *Coordinate* – bind together, unify and harmonise all activity and effort.
5. *Control* – see that everything occurs in conformity with the established rule and expressed command.

A manager's time is most valuable and must be used wisely. In considering whether he is giving sufficient time to things that really matter, he has the following alternatives for each activity:

- Don't do it at all – it's surprising what can be done without.
- Let someone else do it – this involves delegation.
- Do it himself – some things cannot be delegated.

The principle of 'management by exception' (in which top management is involved only when objectives are not being achieved and when programmes are not being met) is now widely used. This is satisfactory in many circumstances as long as all levels of management are able to achieve the plan or objectives, but experience indicates that shortfalls are often deliberately hidden by staff; consequently effective monitoring must be introduced to guard against this possibility.

'Management by crisis' should not be encouraged, but in maintenance and repair operations emergencies will occur and programmes must be flexible to give immediate attention to a crisis situation (e.g. fires in property, fractures occurring in a structure, the receipt of a dangerous structure notice). Programmes must be capable of absorbing this type of work.

In crisis situations in organisations dealing only with repairs, allowances for emergencies can be planned by having emergency teams to deal with choked drains and sewers, dangerous cases of broken glass (particularly in high-rise buildings), securing of property, defective sanitary fittings, electrical defects etc.

Management by objectives (MBO)

Work-measurement and work-study based incentive bonus schemes are adequate methods of motivation for operative staff, given that good labour relations and working conditions prevail. The men will generally respond and produce good performances to earn bonus, as take home pay is important to them. Supervisory staff can generally be motivated by similar methods and they usually align themselves closely with the thinking of the men. Managers, however, are not similarly motivated and require different methods to produce the required results.

The discipline of management by objectives gives a procedure whereby long-term motivation can be achieved and key targets set for the different levels of management staff. Job descriptions showing clearly a manager's duties and responsibilities, and performance plans giving critical objectives to be worked to, are prepared. These enable decision-making to become more effective, and allow delegation with precision. When setting up MBO, attitudes have often to be changed because encouraging people to think in terms of results is quite difficult.

For example, in the writer's own organisation it was decided to employ experienced consultants to assist in introducing MBO. They held introductory meetings with the various managers involved, the objects of the programme were explained, and job questionnaire forms were issued to ascertain the nature of each particular job. A management training programme was designed for senior management; sessions were highly participative, and course subjects and speakers were chosen involving both the staff and the consultants. Subjects covered a thorough explanation of MBO procedures, budgetary control, estimating, communications, O & M procedures, network analysis, the need for job descriptions, performance plans, objectives of the organisation, monitoring of performances, man management, labour relations and motivation.

The next step was to agree job descriptions and performance plans, with key target areas (KTA). The important function of preparing the performance plans was carried out between the chief manager and the senior managers individually. The critical key target areas relevant in each case were agreed as follows:

- budgetary control;
- planning the workload;
- controlling the effective working of the labour force;
- quality control;
- industrial relations and communications.

Performance targets to achieve the key target areas were then agreed and recorded on the performance plan. There may, of course, be several sub-headings included in the performance target to assist and guide the manager to achieve his targets.

When the system is in operation, the manager's performance against targets is checked and recorded, and results are entered weekly by the manager on the performance record. If the manager achieves his target a green square is recorded; if not, a red square. After several weeks the trends indicated by the colours are obvious and the manager motivates himself to achieve the targets. The records are checked by the chief manager with managers individually at monthly intervals, and guidance is given on action to be taken to obtain a more successful performance. Letting managers know how well they are doing at these meetings on performance plans is a psychological motivation and must be undertaken at all levels of management.

Having completed the introduction of management development to the top managers, the scheme must be introduced down the line to foreman level, covering surveyors, work-study officers, maintenance superintendents and foremen. Job descriptions must be agreed with all staff. The same pattern should be adopted for each level of management as was carried out at the top level, the performance plan with key target areas, performance targets and methods of checking being agreed between each member of staff and his immediate superior.

The interval between performance reviews can vary, depending upon the development of the person being reviewed. In the writer's experience, periods of between four and eight weeks are advisable to encourage enthusiasm to achieve the key targets and to kindle motivation, with a six months period between new performance plans; the latter time will, however, be influenced by the organisation's policy patterns.

Experience has shown that the skilled use of work study will result in a marked reduction in labour costs, and the level of savings will increase, as momentum continues, by the introduction of better methods of operation. Response times will be improved and maintained within stringent maintenance service standards, but with an improvement in quality, and operating costs will be reduced. Additionally, improved gang performances will result in increased bonus earnings to the operatives.

Among the many changes that the introduction of MBO should achieve is a great improvement in communications and staff relations up and down the management structure. With regular meetings, performance plans and key targets are freely discussed, ideas encouraged and the aims and objectives of the organisation constantly aired, with senior staff taken into confidence to contribute to policy decisions.

The labour force

Maintenance operatives must be not only competent craftsmen, in order to carry out repairs effectively, but also capable of remedying building defects that are the result of poor building or poor design. Consequently they require expert supervision and advice, and often additional training.

Current examples of this are that painters (brushhands) may require training in paperhanging or glazing work, and plumbers in repairs or servicing to heating equipment such as gas-fired boilers, hot-air units or other gas appliances. Electricians may require additional training in diagnostic work to identify defects in modern electrical controls or electronics, and may also be required to work in servicing teams with other trades and arrange their work to maximise the output of these teams.

Carpenters and joiners are often required to remedy rain penetration through wood windows and curtain walling, and in many cases this cannot be achieved without partly redesigning the windows. They are required to know how to make improvements that have a reasonable chance of success (in order to avoid the alternative of expensive renewal), and also to give advice to clients or owner occupiers on remedial measures.

Bricklayers, in addition to their normal duties, are required to execute remedial works to slated and tiled roofs, flat roofs, jointing or sealants to concrete panels in industrialised buildings, anti-damp measures, drainage work, minor plastering defects and the application of insulants.

To acquire these skills, maintenance operatives require careful training. This could be achieved by on-the-job training by experienced supervisors, or by special training courses organised by the employer or by the Local Education Authority. A considerable reduction in costs could be achieved if, with trade union cooperation, operatives were encouraged to cover the skills of more than one craft to become general-purpose technicians, who could be rewarded by increased wages.

Productivity and incentives

Maintenance and repairs are generally labour intensive. The work is often fragmented and scattered, with management having great difficulty in properly controlling and supervising all the work. Much has to be left to operatives, and good production performances are unlikely unless some form of incentive (such as work measurement coupled with output targets and bonus for improved production) is introduced.

Adequate financial reward for good performance and good standards will offset the difficulties that craftsmen and management have to contend with in executing repairs and remedying building defects.

In the manufacturing field and in new building construction, incentive schemes and the measurement of work are quite simple. More elements of the manufacturing process produced, or more bricks laid, concrete placed, joinery, plumbing or electrical work completed, can be measured accurately and incentive schemes subsequently prepared.

In building maintenance and servicing, however, where repetitive jobs can vary considerably in work content, work measurement is much more difficult. Nevertheless, by work study the elements of work can be measured and tabulated, and will form an accurate basis for a productivity scheme.

Historic time values obtained from past records are sometimes used but this method is inaccurate, resulting in operatives having little confidence in the time values because they have not been involved in their compilation and because the base from which they were derived is not usually open to inspection and adjustment. Time values are often loose and exploited by the operatives, with costly results.

The work-study approach is to be preferred. Here, the operatives or their stewards are involved in obtaining the time values and assessing operative rating, and this provides the confidence required in any work-study scheme. Standard times are measured and accepted between operatives and management, since men are timed doing actual elements of work on the site or in the workshops.

As the similar work elements in repairs can vary quite considerably in time values, five or six time studies should be taken to provide an average acceptable time in standard minutes. Because the measurements are factual and are recorded in a logical sequence they are open to inspection and rechecking, and this builds confidence between management and men.

Additional allowances should be:

- a daily allowance to cover getting tools from locker and packing away at the end of the day, making out time sheets, sharpening tools, washing hands;
- a job allowance to cover gaining entry to the work place, attending to tools, arranging materials;
- allowance for wasted calls, to cover cases of no access to the property, writing out and leaving an appointment card, pay arrangements;
- travelling time, whether by walking or pushing a handcart, by minivan or lorry, or by public transport.

Improvement in the manner in which work is organised, as well as its execution on site, is the major source of increased labour productivity. There is always a better way of doing things, and good work-study will identify incorrect or poor methods being used; management should take action to correct and improve any method deficiencies. The use of mechanical aids and labour-saving devices must be examined and introduced where practicable.

It is essential that men are conveyed to, or report direct to, the work site as early as possible and that the maximum possible time is spent on productive work. It is also essential that the required materials, tools and plant are on site or conveyed by vehicle with the men, so that no time is lost due to lack of materials or awaiting instructions on site. Arrangements must be made for access to the work place at all times, ensuring that no delays or waste of time ensues. Work measurement will provide the yardsticks against which management can also judge its own efficiency in all these respects.

Supervisory staff must ensure that detailed instructions and advice when required are given to all operative staff, that adequate quality control is exercised, and that jobs are approved or passed on completion to provide client or customer satisfaction. Work measurement, because it includes an element-by-element build-up of work, provides an ideal source for standard method instructions and quality standards.

All these factors are of paramount importance to achieve a positive increase in labour productivity. Many authorities, council departments and private firms have introduced work study with success. Some have doubled productivity and achieved a considerable saving in financial and manpower resources, but there is certainly much room for improvement in both the private and public sectors.

The use of work study is strongly recommended for any maintenance organisation employing twenty operatives or more. It is often advisable to employ experienced consultants to study the firm or organisation: a feasibility study will indicate the savings to be made against the additional outlay required. The consultant if engaged will recommend the most economical method of operation to be adopted, will advise on and supervise the time and method studies, will train supervision and management on the controls required to operate successfully, and will advise on budgetary control preparation of estimates for labour requirements and any other matters that the directors or principals include in their brief.

Table 61.2 shows the increase in productivity achieved in the GLC for operatives engaged on housing maintenance work where a work-study based incentive bonus scheme was introduced; a historic-time-value scheme obtained from past records was previously used, and management had considered that the organisation was working productively. Similar results can be achieved for any private firm, and are probably necessary when one considers the large number of bankruptcies each year in this important section of the building industry.

Table 61.2 PERFORMANCE INCREASE IN GLC HOUSING BRANCH

Trade	*Effective performance (EP)* *Reference period 1965–1966 (Traditional bonus scheme)*	*December 1977 (WS bonus scheme)*	*Increase*
Painters	56	107	91%
Plumbers	38	104	174%
Bricklayers	25	106	324%
Carpenters	26	98	277%
Electricians	37	107	189%

	Reference period 1965–1966	*Annual level as at December 1977*	*Increase*
Productive Standard Hours (PSHs), all trades	4 022 000	5 032 000	25% workload increase by introduction of stringent maintenance service standards.

	Reference period 1965–1966	*December 1977*	*Decrease*
Labour force	6250	3900	38%
Comparative number of dwellings	240 000	220 000	8%

Non-financial incentives

Anyone who has managed men soon learns that money is without doubt a great motivator of all levels of staff, but it is not the only motivator. If it were, production engineering would be simple, and the firm that paid well would automatically achieve economic performances and recruit adequate staff.

To encourage motivation it is important that operatives receive praise for good performance. Direct them towards a goal and help them to achieve that goal; if you ignore them you will not receive co-operation. Talk to the staff as responsible people; criticism must be constructive, and good working conditions and relationships are essential. Supervision and management must be keen and well trained, and must work with the men as a team to achieve the objectives of the organisation.

WORK PLANNING AND CONTROL

Planning for the client's requirements

In planning work it is necessary to thoroughly understand the requirements of the various clients or customers for which the maintenance service is to be carried out. To industry and commerce, reliability, speed of operation, competitive prices and value for money are all essential factors. In London it is most difficult for owner occupiers to obtain the services of good maintenance firms, and delays of many months are quite common after the initial contact with a firm to completion of the job. Owner occupiers are forced by desperation into a 'do it yourself' solution, which may be acceptable for small simple jobs but causes great problems for skilled maintenance work. House agents also find great difficulty in obtaining the services of reliable maintenance firms, and consequently planning workloads becomes difficult.

Different policies are required for the various types of maintenance client. A few examples are given below.

- All work in factories must be pre-planned with the management so that production lines are not disturbed. Agreed commitments must be complied with as delays cost money and loss of production; this often requires the maintenance staff to work out of normal hours, weekends and often night shifts.
- Trading establishments such as shops, stores or supermarkets are customer conscious and will not wish maintenance work to be undertaken during business hours, which might upset their customers. Evening work and weekend working is therefore common. Rubbish accumulating from such work must be removed immediately from the site, generally by clearance skips, and care must be taken to keep the premises clean and tidy to create a pleasant environment at all times. Plant and scaffolding left on site may be a danger to the public and must not be left unattended.
- Planning of maintenance work for local authorities will also vary depending upon the department for which the service is to be undertaken. Work in schools will require extra care to prevent noise, and prevent accidents to children. Head teachers will require immediate attention to any defect, and discretion is required when working in girls' schools. Much work can be executed in holiday periods, but completion dates are important so that opening dates of schools following holidays are not disrupted. Work for the police, fire service or ambulance service needs special care so that the operational efficiency of these services is not adversely affected; planning of work must be approved with the senior officer in each case. Public Health Notices issued by Environmental Health Departments must receive urgent attention, since the abatement requires resolving within 28 days. Tenants can arrange for a summons to be served on landlords neglecting property, and immediate attention to repairs is required in such cases. Repairs to council property require attention within the maintenance service standards approved by the council, and repairs programmes should be accordingly prepared.
- Hospital maintenance requires its own particular planning arrangements. Work in operating theatres and central sterile supply departments requires special protective clothing for infection reasons. Differing priorities and methods of operation are necessary for work in hospital wards, intensive care units, X-ray departments, outpatients and special treatments departments, offices, nurses' and doctors' homes etc., and work plans must be agreed with the hospital management and the chief nursing officer.

The Defective Premises Act 1972 and the Health and Safety at Work etc. Act 1974 define responsibility in cases of defects or accidents, and particular attention is necessary to ensure that the various clauses of this legislation are complied with. Working with special materials such as blue asbestos (crocidolite) requires special precautions to comply with appropriate Acts, and the Health and Safety Executive must be consulted on these matters. Ignorance of the law will not excuse non-compliance.

Other key considerations in planning

The wide variety of maintenance activities makes work planning essential. Maintenance is often difficult to organise because of its fragmentation, but it is unacceptable to leave operatives to arrange for their own work planning, as they must be completely involved in the execution of the various detailed tasks allocated to them. To obtain the best possible results with cost effectiveness, the work should be programmed in separate trades and jobs grouped together so that there is the least travel between each job.

Plant, tools, equipment and materials must be available at the work place for immediate use when required. The operative must be able to spend the maximum amount of time during his working day on productive activities and the work must be organised accordingly. The line manager or supervisor must be available to give assistance and advice to the operative when required, and should not be responsible for more operatives than he can adequately control to obtain the best results.

The high cost of labour makes it imperative that transport be made available to convey the operatives to the work place or site with the minimum of delay. A successful method where the jobs are in isolated locations is for a small van to be used for this purpose, stocked with adequate materials and driven by the operative to the place of work. Where an adequate workload is available for several operatives for a day's work or more, a team bus can be provided with space for materials, plant, tools and equipment, the bus being driven by the supervisor. This is a highly efficient method of working. Alternatively, if a mix of trade work can be pre-planned a mixed gang may be economic, but care must be taken to ensure a full workload for the various operatives allocated to the jobs, and a continuous work flow is essential for mixed-trade gangs.

An experienced supervisor knows his men, their likes, dislikes and varying aptitudes, and on a 'horses for courses' principle he allocates jobs to workmen from whom he can expect the best results. Difficult jobs should be pre-inspected by the supervisor so that he will be acquainted with all details of requirements, and can ensure that materials and special equipment are available on the transport or delivered to the site in advance of the men, e.g. baths, bricks, joinery, special pre-formed plumbing etc. It will be necessary to post-inspect all jobs on completion for quality control, customer satisfaction, and possibly a materials check for invoicing. It may also be necessary to arrange for follow-up trades to piece up brickwork after a plumber, or paint new joinery.

Daily planning

To achieve the performance required for the completion of maintenance and repair work, it is quite impossible to effectively control the large volume and high turnover of job tickets unless there is a system of daily work planning and programming. This daily job-ticket planning is necessary at all levels of works management and supervision, although its form will vary with each level. The objective is to ensure:

- that each job receives quick and adequate attention;
- that work is progressed in date and urgency sequence, and completed within the maintenance service standards;
- that problem jobs are identified, and that action is taken to overcome the problem and also to ensure work completion within the objective times.

Daily planning is a discipline that must be observed to achieve stringent service standards. It will not be easy and requires a conscious and continued effort to achieve, but the rewards are considerable in terms of job satisfaction, performance improvement and reduced response times, and improved tenant or client relationships.

Any review of old job tickets will show that the majority of these tickets have become old because they represent an unpleasant, disagreeable or problem job. Any delay for this reason is unacceptable. The management and supervisory staffs have sufficient experience, knowhow and facilities to overcome every one of these problems, and the best way to tackle them is by continual cooperation between all levels of staff.

Receipt of job tickets The cornerstone of good daily planning is to scrutinise all incoming jobs (on day one) and to immediately identify any that may cause problems or require special attention (including diagnostic investigatory work, special joinery etc.). These should be passed immediately to the appropriate supervisor, most frequently the trade chargehand, for inspection on the following day; he should report back to the senior supervisor each evening on special materials or plant required, and should prepare a sketch with dimensions for special joinery or special plumbing etc.

Preparation of programmes Programmes must be realistic, and because situations are constantly changing the programmes must therefore be reviewed daily by the senior supervisor and amended to take account of the latest information. An out-of-date or impossible programme given to a chargehand will get the justice it deserves and bring all programmes into dispute. Programmes must be prepared in conjunction with chargehands in such a way that they can be accepted as practical and achievable, but they will include some additional jobs to cover the cases of no access.

Daily feedback on results The foundation of good daily planning is the end-of-day discussion between the chargehand and the senior supervisor when the results are reviewed. Every job ticket for action on the programme for that day which has not been completed *must* be jointly discussed to ensure immediate follow-up action to complete the job. The supervisor must help the chargehand to overcome any problem jobs. This may require a joint site visit during the following day, and the provision of special tools, equipment or materials or technical advice. Whatever policy action is agreed upon, the senior supervisor must check that the job has been carried out when he meets the chargehand at the end of the working day.

Daily work planning by the chargehand Although care is taken by the works office on programme preparation, it can easily be ruined by poor planning of jobs issued to tradesmen. The chargehand knows his men well, their strengths and weaknesses. He will need to make a careful distribution of job tickets so as to give a well-balanced workload of interest and variety to achieve job satisfaction, yet within each man's capacity. He must also keep a record of the jobs distributed to each man so that he can make site visits to help and advise and check quality. At the end of each day the operative must return every job ticket to the chargehand so that he is always in a position to discuss the uncompleted work with his senior supervisor for immediate progress or re-programming.

Control meetings

To carry out his duties efficiently a manager must exercise adequate control of the manpower and the work and cost elements for which is responsible. It is important that he constantly reviews the performance achieved against the targets and budgets set, and makes decisions on any action to be taken to correct adverse trends. To exercise this control it is recommended that a weekly control meeting should be held of all appropriate management and supervisory staff responsible for the relevant work programmes to be reviewed. The manager should chair the meeting, show intense interest, and ensure that an action sheet is prepared of agreed decisions for supervisory staff to follow in order to improve any performances that are below target and to set achievable dates for performances to be back on target.

Table 61.3 SAMPLE STANDARD AGENDA OF WEEKLY MANAGEMENT CONTROL MEETING (GLC HOUSING MAINTENANCE)

Item	*Special features in addition to performance cost, quality and workload*
Performance	
1. Gardening	Summer and winter programmes, machine availability.
2. Painting	Internal/external programmes, empty property, pre-painting repairs, quality.
3. Glaziers	Availability of glass.
4. Bricklayers	Programme work, mates, sealant work.
5. Carpenters	Joinery requirements, fencing, boarding up.
6. Plumbers	Gas servicing, welding.
7. Electricians	Emergency work.
8. Transport	Availability of vehicles, reduction of hiring, pool bonus, vehicle utilisation, skip clearance service.
9. Miscellaneous	Dust and drains, empty clearance, scaffolding, sector labourers, heating plant attendants.
Special programmes	
10. Technical inspections	Five-yearly multi-storey, anti-condensation and playground inspections. Fire inspection of some specially designed maisonettes.
11. Work checking and disciplinary cases	
12. Programme work	All programmes as per service standards.
13. Old job tickets	Daily planning, cancellations, no access, materials.
14. Out-of-hours emergency services	
15. Contract work	Progress, work quality, accounts etc.
16. Budgetary control	Expenditure against budget, action to adjust adverse trends.
Staff and welfare	
17. Safety Committees	
18. Stewards' meetings	
19. Annual leave	
20. Accommodation	
21. Staff vacancies	
Any other special items	

It will not be necessary to deal at length with every item listed in the fixed agenda (see Table 61.3), but it should be used as an 'aide memoire'. Of course the manager will prepare his own agenda pattern to suit his sphere of operation.

Particular reference must be made to the job-ticket position. The meeting must review incoming, outgoing and in-hand numbers of job tickets from the records maintained in the works office. The job ticket analysis print-out of information provides evidence of the results achieved, but it cannot be used as an action document; as it is an historic document, forward planning must at all times be undertaken. Decisions must be made to ensure that *all* the job-ticket standards will be met during the coming week. Individual jobs that will not be completed within standards during the following weeks must be discussed, to ensure they are all completed immediately. The forward plan must ensure that all job tickets over four weeks old are either complete or in an advanced state of progress.

Part of the agenda should be concerned with staffing and welfare matters, which are so essential to the smooth operation of the maintenance organisation. The manager must be satisfied that problems or potential problems are brought to light or resolved.

Control of performance

When monitoring performances and costs, the following guidelines should be considered.

1. *Labour costs.* A high labour cost indicates inefficiency in one or more respects and must be investigated further. A low labour cost, whilst indicating a generally good trend, may hide particular areas of inefficiency that require attention if an adverse trend is to be avoided. Managers must therefore always examine low-labour-cost results in this light, and not blindly accept that a low labour cost in itself cannot be improved.

2. *Trade transport cost,* i.e. the cost of all the vehicles allocated to that trade. The retention of surplus vehicles will increase the trade transport costs. Managers should ensure that there is only sufficient trade transport to meet the average requirement, bearing in mind levels of sickness and holidays, and not to cover every tradesman on the books; surplus vehicles should be returned for use elsewhere. Vehicle planning on a daily basis to obtain maximum efficiency is essential, and transport supervisors must ensure that this is done efficiently.

3. *Individual pay performance.* Under a normal incentive bonus scheme, a low pay performance indicates that some (or all) of the operatives are not producing at expected levels. This may be due to poor supervision, lack of job planning, poor programming, or a lack of motivation in individuals. The manager should ask for the names of the low performers and arrange for each operative to be interviewed to establish the reasons for the poor results. Action must be taken immediately to overcome whatever problems are identified at this interview. It is not necessary to use a big stick to operatives, but motiviation must be achieved by persuasion, providing good conditions of work, assisting operatives with problems, and achieving good industrial relationships. On the other hand, a high pay performance may be indicative of high personal performances and good motivation, but managers must be satisfied that the work is being done in the correct way, to high quality standards, and that work is being booked properly for bonus calculation purposes. Careful checks are necessary, particularly where average pay performances are extremely high (overbooking of work is not acceptable, and deliberate overbooking is fraud).

4. *Gang or trade group performance.* This increases as individual pay performance increases, but decreases with lost time, unmeasured work and non-productive work. A low *group* performance coupled with a high *individual* pay performance indicates poor management efficiency and lack of job planning and control. This is the worst possible effect, as the men are earning high bonus for a low productive output – resulting in high cost. The manager must identify the gang or trade group, and the causes of the low group performance, and take immediate action to increase efficiency. It is most likely that he will need to review the performance of the chargehand or chargehands involved, including giving personal instructions and arranging on-job training of the systems for work planning and control. When a low *group* performance is coupled with a low *individual* pay performance, the manager must follow the guidelines in (3) above to improve the performance of individual operatives, as well as taking action to reduce management/supervision inefficiency.

5. *Cost effectiveness.* In examining the reasons for poor cost, productivity, and efficiency performances, the following information must be used to identify causes:

   Work content – increasing work content must be checked to ensure overbooking is not present.

   Unmeasured work – must not exceed 2 per cent on average.

   Lost time – must be avoided; explanations for lost time must be sought, and action taken to overcome problems.

Non-productive work – indicates planning and supervision inefficiency and is a key area for reducing costs.
Material costs – jobs with an excessive material cost must be identified for work checking.

6. *Work programming.* The manager must review each programme, establishing what results have been achieved during the previous week against objectives, and agreeing and minuting the lines of action to be taken during the coming week. Efforts should be increased where work is behind programme.

Planning and control by computer

Some maintenance organisations operate computer schemes for job-ticket analysis and control of wages, bonus payments, management control and standard costing procedures. When computerisation was first introduced it would have been inadvisable to include work-planning, programming and servicing data. It was generally found that the clerical work in preparation of the data made this operation uneconomical, but it is pertinent to reconsider this matter now that technology has improved.

Work planning by computer is now practical; the job tickets or works orders with the work programme can also be printed by the computer. This would indicate the numbers of operatives required to complete the programme in a given timescale, and the job details would be entered on the computer order.

OBJECTIVES

The basic objectives would be as follows.

- To accurately and quickly record new jobs via a direct computer access system, with the facility to check that the job was not already held on the files (this would prevent costly duplications and speed up programming).
- To note access arrangements and give firm appointments to tenants or clients when jobs were recorded, if required (this would reduce operating costs).
- To separately identify different types of work, e.g. tenants' repairs (emergency and normal), routine maintenance, preventive maintenance, vandalism, environmental repairs, repairs to empties, work for other departments or firms etc.
- To prepare work programmes for each trade group and to produce job tickets or schedules according to these programmes.
- To monitor job completion against programme, and highlight any jobs not done on programme for immediate action.
- To regularly monitor jobs outstanding and to print out full details, including address identification of jobs not completed within the service standards or programme.
- To produce data on input, output, in-hand balance and a weekly profile of response times.
- To accept, via the direct access system, details of jobs completed for bonus calculation, and for the costing scheme (this would reduce the cost of bonusing and job costing).
- To prepare all management control information.

Each day the computer would be informed of the current and immediate future manpower availability so that programmes could be adjusted as necessary. At the end of each day details of programme completions would be indicated to the computer, and this information would be taken into account for future programme adjustment. The objective would be to complete all jobs in programme sequence. This should result in the reduction of response times for tenants' repairs, and help to prevent queries or complaints.

A computer scheme would be programmed to produce daily, weekly and monthly statements of the job-ticket and programme positions. Information would include number and type of jobs received, number of jobs completed, number of jobs in hand, average response times, a profile showing the range of response times within the average, and details of any job outstanding over the agreed service standards. It would also produce individual performance records for operatives, with gang, trade and district summaries.

Statistics in respect of job-completion performance, mix of work, trends in incoming tenants' complaints, levels of environmental repairs or vandalism etc. could be produced as required. It should be possible to computerise an MBO discipline for all management staff from line managers to top management.

BENEFITS

The introduction of a computer scheme would have the following benefits:

- Clearer job ticket description by the use of standard work codes; this would simplify and produce more accurate job planning from more accurate job ticket information.
- Elimination of job-ticket duplication and resultant time-wasting.
- Automatic appointment system, with a reduction in wasted visits and non-productive time.
- Greater flexibility in identification of job types.
- Immediate entry of job-ticket details on the computer file.
- Automatic job programming, reducing the clerical and planning staff workload.
- Pre-inspection planning when job descriptions were inadequate.
- Printed job tickets and associated programme sheets, reducing clerical and supervisory time.
- Automatic programming of routine maintenance work (painting, gas appliance servicing etc.).
- All work given the required priority rating for programming.
- Immediate entry of job ticket, schedule or programme-completion details on to the computer file.
- Reduction in bonus staffing, as standard minutes would be taken directly off the job ticket.
- Completion of entries for bonus, costing and job-ticket analysis with a single input, thus considerably reducing the workload.
- Greater flexibility in monitoring the performances in respect of job tickets, trade performances, programme completions etc., thus enabling speedier maintenance-management decisions on performance and cost control.
- Improved sub-divisions of types of work, resulting in better manpower planning, budgetary control, preparation of annual estimates, long-term forecasting, selection of policy options.
- Information received earlier by management, so that quicker action results.

A computer scheme used to plan, control and cost maintenance work would provide many benefits to management, both financial and in terms of increased effectiveness. It would above all provide the flexibility in scope that is so essential in a works organisation, where policies, procedures and problem areas are constantly changing.

FURTHER INFORMATION

Building maintenance, Research and Design Bulletin, Department of the Environment, HMSO

An introduction to incentive schemes in building maintenance, Research and Design Paper, Department of the Environment, HMSO

Maintenance by design, report of a seminar, Ministry of Public Building and Works (1969)

Maintenance management, Institute of Building

Maintenance manuals for buildings, Research and Design Bulletin, Department of the Environment, HMSO

Geary, R., *Works study applied to building,* Godwin (1970)

Rogers, J.M., *Management and maintenance techniques with particular reference to local government,* Joint Committee of Student Societies of the Institute of Municipal Treasurers and Accountants

62 SCOTTISH PRACTICE

INTRODUCTION	62–2
SCOTTISH BUILDING LEGISLATION	62–2
OTHER RELEVANT LEGISLATION	62–5
OTHER CONSTRAINTS	62–8
TRADITIONAL PRACTICE	62–8
BUILDING CONTRACTS	62–9

SCOTTISH PRACTICE 62

62 SCOTTISH PRACTICE

M.R. MILLER, BArch (L'pool), MPhil, RIBA
Scottish Development Department

This section summarises the mandatory controls that apply to new building and to the improvement of existing buildings in Scotland. It comments on other controls, and also on those traditional design features and trade practices that still remain in common use in Scotland.

INTRODUCTION

Whilst the information contained in earlier sections of this volume is generally valid for the builder working in Scotland, there are a number of differences, mainly in the legislation, of which he must be aware. This section does not attempt to give details of the legal aspects of the purchase and registration of land, but does draw attention to the forms of contract documents used in Scotland.

Although this section may appear to be over-concerned with legislation, it is important to appreciate that many of the differences in practice between Scotland and England are a result of the seperate legal systems in the two countries.

SCOTTISH BUILDING LEGISLATION

Prior to 1963 building in Scotland was controlled by By-laws made under powers in the Burgh Police (Scotland) Act of 1892, the Public Health (Scotland) Act of 1897 and a multitude of Local Acts. Model By-laws were first issued in Scotland in 1932 but, as in England and Wales, they were not mandatory and some local authorities still chose to operate the Building Rules contained in the 1892 Burgh Police (Scotland) Act. These Rules had one great merit, brevity, but were framed with special regard to the erection of high tenements and other heavy stone buildings and were not satisfactory when applied to other forms and methods of construction.

After the war it was generally accepted that the system of control required a thorough overhaul to allow the use of new materials and methods of construction. Owing to the rather restricted terms of reference of the working party, which was set up in 1952, the changes following the review still did not provide a comprehensive modern building code. Therefore in 1954 a new committee was appointed with the remit 'to examine the existing law for the general regulation of building in Scotland so far as it relates to the control of building standards by the Secretary of State and by local authorities; to consider the extent of the jurisdiction exercised by the Dean of Guild Courts as regards building standards; and to recommend what changes may be necessary to secure a control of building standards which would operate as uniformly as possible in Counties and Burghs and be flexible enough to take account of the development of new techniques and materials.'

The Report of the Committee on Building Legislation in Scotland set out the basic philosophy for building control in Scotland. Published in 1957 and better known as the Guest Report, it is well worth close study by anyone who is interested in building control. Amongst the recommendations made in the Report were:

- that there should be uniformity of building requirements throughout Scotland;
- that these requirements should be contained in Regulations made by the Secretary of State for Scotland;
- that as far as possible flexibility should be encouraged by laying down requirements in the form of performance standards;

- that housing standards should continue to be laid down along with the more general requirements in the building code.

The recommendations of the report were accepted and embodied in the Building (Scotland) Act of 1959. The fourth schedule of this Act lists the matters in regard to which building standards regulations may be made by the Secretary of State. Other sections of this Act particularly relevant to this discussion are as follows.

- Section 4 allows the Secretary of State to relax the requirements of any regulation. It should be noted that buildings authorities cannot relax any regulation where a new building is concerned. They may relax certain regulations if the application is in connection with a building erected prior to 1964.
- Section 6 requires a warrant for the construction of the building to be obtained from the buildings authority before building work can commence.
- Section 9 requires the buildings authority to issue a certificate of completion if, but only if, the building complies with the conditions on which the warrant was granted. The building cannot be occupied before a completion certificate has been issued.
- Section 11 allows a buildings authority to require existing buildings to comply with the building standards regulations when it is considered necessary in order to secure the health, safety or convenience of the persons who will use the building. This power is not available for use with all Parts of the Standards. The first regulation in each Part states whether or not the provisions of that Part may be required under Section 11 procedure.
- Section 18 gives the buildings authority power to inspect buildings in the course of construction and to have materials tested.
- Section 20 allows the buildings authority to charge such fees as may be prescribed.

Amendments to this basic Act are contained in the Building (Scotland) Act 1970, the Local Government (Scotland) Act 1973 and the Health and Safety at Work etc. Act 1974, but the powers outlined above are not changed.

Unfortunately it has still not proved possible to include in a single piece of legislation all the controls on all aspects of all buildings. Other relevant legislation includes the Housing (Scotland) Act 1974, the Sewerage (Scotland) Act 1968, the Fire Precautions Act 1971 and the Fire Certificates (Special Premises) Regulations 1976 made under the Health and Safety at Work etc. Act. However it is true to say that the Building Standards (Scotland) Regulations is the basic document controlling standards of construction in Scotland.

The current edition of the Buildings Standards (Scotland) Regulations, referred to as the Standards for the remainder of this section, was issued in 1971 and amended in 1973 and 1975. Even a cursory examination of these documents will reveal differences, both in form and content, from the Building Regulations applying in England and Wales. Housing standards are set out in considerable detail, mainly in Parts Q and L and the schedules to these Parts. Deemed-to-satisfy specifications are gathered together in Schedule 10 and do not appear in the body of the Standards because they are not to be confused with the Standards. It is unfortunate that there is no index printed in the Standards document, although there is one included in the Explanatory Memorandum. The most up-to-date official index is that in the 'Amendments to 1973 Supplement to the Explanatory Memorandum'.

Since 1971 the Standards have allowed the designer a choice between the older, less flexible, standards for the size of rooms, detailed in Table 17 of Schedule 10, and the basically Parker Morris space standards, detailed in Table 18. The standards in Table 18 are explained in greater detail in 'Metric Space Standards' published by the Scottish Development Department in 1968. The majority of private developers still prefer to use the standards in Table 17, mainly in order to avoid any problems of interpretation that may arise when the standards in Table 18 are used.

The Secretary of State does issue from time to time Class Relaxations. These are relaxations of particular regulations not related to a specific building but to apply in general when the circumstances specified in the Class Relaxation are complied with. Details of Class Relaxations are in future to be circulated by the Scottish Office to selected technical journals when they are introduced, and copies may be obtained from the Scottish Development Department. A Class Relaxation may be considered to be the Scottish equivalent of a Type Relaxation now used in England and Wales. Each Class Relaxation is usually valid for three years, unless withdrawn, and the subjects so far covered range from shower units to timber-framed houses. There is no index of those Class Relaxations currently in force.

The Scottish Development Department has issued two series of Explanatory Memoranda, the current series in 1972 and 1973. Unfortunately these Memoranda are not up to date and therefore cannot be relied upon for detailed guidance. They do, however, contain much useful information and the illustrations do show generally acceptable construction. A section of particular interest to the builder of houses and other small buildings is section 3 in the Explanatory Memorandum for Parts A, B and C. This section not only explains ways of complying with Part C for the small building but also gives joist sizes and foundation dimensions for various conditions.

A most useful series of checklists and an index are contained in *The Building Regulations Scotland checklists and index*, published by House Information Services.

Four other documents supplement the Standards. The Building (Procedure) (Scotland) Regulations, the Building (Forms) (Scotland) Regulations, the Building Operations (Scotland) Regulations and the Building Standards (Relaxation by Local Authorities) (Scotland) Regulations.

The Building (Procedure) (Scotland) Regulations 1975 and the amendment of 1977 amplify those sections of the Building (Scotland) Acts that enable the Secretary of State to make procedural requirements. The main points are as follows.

- Regulation 8 requires the buildings authority to issue the certificate of completion, or notify the applicant of their refusal to do so, within 14 days.
- Regulation 9 and schedule 2 prescribe fees that the buildings authority may charge for building warrants. The revised scales detailed in the 1977 Building (Procedure) (Scotland) Amendment Regulations came into force on 1 February 1978.
- Regulation 12 and its associated schedules detail the procedure for applying for a building warrant.
- Regulation 13 requires the applicant to inform every affected proprietor, defined in regulation 5, of the details of the application.
- Regulation 32 requires buildings authorities to come to a decision on an application for a relaxation of a regulation within two months, otherwise the applicant may appeal to the Secretary of State as if the application had been refused by the buildings authority.

The Building (Forms) (Scotland) Regulations 1975 set out in detail the information required on the various applications and notices necessitated by the Building (Scotland) Acts and the Procedure Regulations.

The Building Operations (Scotland) Regulations 1975 give details of the work the builder is required to do in order to protect the public.

The Building Standards (Relaxation by Local Authorities) (Scotland) Regulations 1975 give details of those provisions of the Standards that the buildings authority has the power to relax. This power is only valid for buildings erected or approved for erection prior to 15 June 1964.

The Standards apply to all building work, whether entirely new construction or extension or alteration of an existing building, by virtue of the exceptionally wide definition of 'building' contained in Regulation A3 in the Standards. This regulation defines 'building' as 'any structure or erection of what kind or nature so ever, whether temporary or permanent, and every part thereof, including any fixture affixed thereto...'. There are some exclusions related to roads, bridges, sewers, water mains, aerodrome runways, railway lines etc. However, although the Standards have such a wide application, there are exemptions allowed by Regulation A9(1). These are listed in Schedule 3 of the Standards. Despite the detailed requirements in these various documents some slight differences in procedure still occur, particularly in authorities that have a long history of building control.

OTHER RELEVANT LEGISLATION

The Housing (Scotland) Act

The Housing (Scotland) Act 1974 is the main piece of legislation concerning existing houses. Under this Act Local Authorities are given the power to make improvement grants. The sections that are particularly relevant are as follows.

- Section 5 gives details of the amounts of grant available for improvement and repair.
- Section 9 sets out the conditions attached to improvement grants.
- Section 11 gives details of the increased grants available for houses situated in housing action areas, which are described in sections 16 and 17.

Orders made under the 1974 Act have raised the maximum of approved expense on which improvement grants are based. The latest relevant statutory instruments are the Housing (Amounts of Approved Expense) (Scotland) Order 1977 and the Housing (Standard Amenities Approved Expense) (Scotland) Order 1977.

The Scottish Development Department has specified the standard to be achieved in a house that has been improved with the aid of a discretionary improvement grant. This document, Circular No. 72/1975, gives internal planning standards to be achieved as well as the minimum number of fittings. It also makes a clear statement of the relationship between the standards required in a house improved with a grant and the requirements in the Standards. Any work of construction or alteration must comply with the Standards unless a relaxation has been granted by the appropriate authority; those parts of the house that are not being altered may not be required to comply with the Standards unless the new work would cause the unchanged parts to fail to conform to the Standards by an even greater degree. If change of use is involved, all the work must comply with the Standards.

The whole question of housing finance is currently under review. At the same time that *Housing policy: a consultative document* was issued in England, the Scottish Office issued *Scottish housing: a consultative document.* Amongst other proposals the document suggested several changes in the financing of improvement and rehabilitation.

A Bill that seeks to implement many of the proposals in the consultative document, the Housing (Financial Provisions) Bill, is mainly intended to introduce changes in housing grants and subsidies, but the opportunity has been taken to introduce certain new provisions affecting house improvement and repair. Perhaps the point of greatest interest to the builder is the amendment proposed to allow grants towards the expenses of repairing houses. Clause 8 of the Bill introduces a new section 10A in the Housing (Scotland) Act 1974 to empower local authorities to make repairs grants in cases of financial hardship. The existing power of local authorities to make repairs grants relates only to houses in housing action areas. These housing action area powers will be unchanged, but under the new section 10A grants will be available to owners of houses in need of repair broadly in the same way as house improvement grants are available to owners of houses in need of improvement. The repairs grant will only be available in cases of financial hardship. The intention is that provision of repairs grants should be combined with a more positive use by local authorities of their powers to compel house owners to deal with conditions of serious disrepair. The repairs notice procedure is amended by Schedule

2 of the Bill. Shops and other non-residential premises in tenements or in other buildings containing houses will be within the scope of repairs notices.

The Town and Country Planning (Scotland) Act

Planning control in Scotland differs less from the practice in England and Wales than does building control. The basic legislation is the Town and Country Planning (Scotland) Act of 1972. Part III, 'General Planning Control', contains the sections under which development control is operated. The main items are as follows.

- Section 20 gives details of developments that require planning permission.
- Section 21 gives the Secretary of State powers to provide for the granting of planning permission.
- Section 22 requires applications to local authorities for planning permission to be made 'in such a manner as may be prescribed by regulations under this Act'.
- Sections 23–25 require publication and notification of applications.
- Section 26 requires local authorities to refuse or grant, 'either unconditionally or subject to such conditions as they think fit', planning permission.
- Section 27 gives grounds for some conditions that might be imposed in a conditional approval.
- Section 28 enables the Secretary of State to regulate the method of dealing with planning applications.
- Section 31 requires the planning authority to keep registers of all applications, which must be available for inspection by the public at all reasonable times.
- Sections 32–35 give details of the Secretary of State's poweıs to deal with appeals against a decision by a local authority.
- Sections 38–41 limit the duration of a planning permission in most cases to five years, and state procedures to be followed in a termination of a planning permission by reference to the time limit.

If the Town and Country Planning (Scotland) Act is considered to equate to the Building (Scotland) Act, then the Town and Country Planning (General Development) (Scotland) Order 1975, as amended in 1976, is the equivalent of the Standards. There is one major difference: planning officers are expected to use discretion in applying planning 'standards', normally termed policies, whereas the building control officer, in theory at least, must apply the Standards. The main points in the General Development Orders are:

- Articles 3 and 4 and schedule 1 give details of developments that are permitted without any application to any planning authority. These permitted developments differ from the buildings, listed in Schedule 3 of the Standards, for which no building warrant is required, and also from the permitted developments in England and Wales.
- Article 5 gives the planning authority power to require not more than four copies of the application form, drawings, plans and descriptive material necessary for the authority to determine the application.
- Article 7 requires the planning authority to notify the applicant of its decision, in writing, in normal cases within two months and in cases affecting trunk or special roads within three months.
- Article 11 gives details of the consultations that should be made by the planning authority before determining any application.
- Article 14 gives details of the appeals procedure open to an applicant who is not satisfied with the decision of the planning authority; such an appeal must be made in most cases within six months of the decision. There is no time limit for the consideration of the appeal by the Secretary of State.

In April 1977 the Scottish Development Department issued a consultative paper, *The review of the management of planning.* Amongst many other proposals this document suggested amendments to the list of permitted developments in Schedule 1. The proposed increase in the size of an extension to an existing building permitted without obtaining planning permission will be the main point of interest to most builders.

The development control policies applied by the planning authority are not determined by any directives from the Scottish Development Department. Unlike the Department of the Environment, the Scottish Development Department does not publish any Planning Bulletins nor issue Development Control Notes; however, it does issue Circulars, but these are not available through the Stationery Office. It is not uncommon to find planning authorities in Scotland using the Bulletins issued by the Department of the Environment.

The Sewerage (Scotland) Act

The Sewerage (Scotland) Act 1968 superseded the sewerage sections of the Burgh Police (Scotland) Act 1892 and the Public Health (Scotland) Act 1897. It is important to note that the local authority with respect to this Act is the regional council.

The main provisions of the Act that are of interest to the builder are as follows.

- Section 1 places a duty on the regional councils to provide public sewers for draining their area of domestic sewage, surface water and trade effluent; and to take the public sewers to such a point or points as will enable the owners of premises that are to be served by the sewers to connect their drains or private sewers to the public sewers at a reasonable cost. However, regional councils are not to be required to do anything that is not practicable at a reasonable cost. Section 1(4) provides for determination by the Secretary of State of questions arising as to reasonable cost.
- Section 2 requires the regional council to inspect, maintain, repair, cleanse, empty, ventilate and where appropriate renew all sewers, sewerage treatment works and other works vested in them.
- Section 8 allows the regional council to make an agreement with a developer, but only in those cases where no duty under section 1 exists. Such an agreement may specify the terms and conditions on which the work is to be carried out.
- Section 12 details the rights of connection to the public sewer. The regional council must, within 28 days of the receipt of an application for a connection, either refuse permission or grant approval 'subject to such conditions as they think fit'. Section 12(5) allows an applicant to appeal to the Secretary of State against the decision, or any conditions, of the regional council.
- Section 14 allows the regional council to require a different method of construction if they consider that the proposed drain, sewer or works is, or is likely to be, needed to form part of a general sewerage system. If the regional council's requirements involve the developer in increased costs, the additional cost is to be borne by the regional council. Section 14(3) allows the right of appeal to the Secretary of State against any order by the regional council under 14(1).
- Section 17 allows the regional council to take over private sewage treatment works, either by agreement or by notice. Section 17(3) again allows the right of appeal to the Secretary of State.

Part II of the Act deals with trade effluents and is not discussed further here.

Since the Act came into force some developers have experienced problems, due in some cases to the interactions between the provisions of the Act and the requirements of Part M of the Standards. In order to avoid any problems that might arise, developers are strongly advised to discuss their proposals at the earliest possible stage with the regional council.

OTHER CONSTRAINTS

The design of roads

The requirements in *Roads in urban areas* still form the basis of many local authorities' standards. Many of the standards in this document were seen to be unnecessarily wasteful of land in the normal housing layout. Therefore a fully revised issue of Scottish Housing Handbook No. 2 was published in 1977. Now titled *Housing development layout, roads and services* and renumbered 3, the document is much wider in scope than its predecessor. It proposes a hierarchy of roads from local distributor roads down to minor access roads, and suggests design criteria for each. It also discusses the implications of various house types and housing layouts. The bibliographies are extensive and the checklist should prove to be useful to the designer.

Design guides

As yet few planning authorities in Scotland have issued the Design Guides that seem to be a growing part of development control in England and Wales.

The National House-Building Council

Until 1974 the requirements for Scotland were published in a separate volume, but now the requirements for the entire area covered by the NHBC are printed in the one volume. A suffix against each requirement makes clear which specifications are relevant in Scotland. Requirement number 1 is that 'every dwelling shall comply with all relevant building legislation and the regulations of statutory undertakings and other proper authorities'. One important difference is the refusal of the NHBC in Scotland to accept the use of certain types of material as insulation in the cavity at the time of construction.

TRADITIONAL PRACTICE

Many people connected with the building industry in Scotland consider that Scottish construction is superior to that in England and Wales. Whilst this may still be true the differences between the two countries are disappearing. The influence of the large national building firms and the growth in the use of standard components, often manufactured in England, are tending to kill the older trade practices. One could well argue that most of the remaining differences are a result of the greater incidence of more highly exposed sites being used for housing in Scotland. On such sites a higher standard of detailing and workmanship is necessary.

In 1970 the Stationery Office published a report on *The cost of private house building in Scotland.* The report had been commissioned from Professor Sidwell to investigate the fact that private houses in Scotland seemed to be more expensive than comparable houses in England. The report therefore compared several aspects of the private house building market in the two countries with particular reference to costs. One aspect of the study involved an enquiry into trade practices. Sidwell suggested that historical contracting methods had been a strong factor in the development of the different practices. The two main areas identified in the report as those in which traditional Scottish practice still persisted were the use of scaffolding and general carpentry and joinery. In general each trade provides the scaffolding to suit its own needs, with little or no consideration of the needs of any other trade. Virtually all the bricklaying and joinery work is done from inside the building using planks, trestles and ladders as necessary. Frequently the only operation that requires scaffolding is the application of the external render or harling. Even then it is often of such a flimsy nature as to scarcely qualify as scaffolding.

The traditional single trade contracting system is no longer used, but some of the other practices that it encouraged still persist, possibly because the resulting construction works well.

The Scottish joiner tended to prefer heavier sections than his counterpart in England, but with the increase in the cost of timber and the growth in the use of prefabricated components this practice is now much less common. The most obvious remaining traditional joinery practice is the use of roof sarking. Sarking in traditional construction consisted of one inch thick timber boards butt-jointed. Sidwell suggested a number of possible reasons for this method of construction:

- to provide a continuous flat surface on which to nail the slates;
- to brace the roof frame;
- to provide additional support for the heavier Scottish slate;
- to seal the roof space against the entry of wind and wind-driven snow.

Traditionally the slates were nailed directly to the sarking, but with the almost total replacement of slate by concrete tiles as the roof finish it is now common practice to use battens and counterbattens over felt on the sarking. Today the sarking is often insulating board rather than timber, as the deemed-to-satisfy specifications for the thermal insulation of roofs recognise the acceptability of insulating boards. It is worth noting that the Standards do not require sarking to be provided.

The single trade contracting system combined with climatic conditions are presumably the reasons behind the standard details at door and window openings. Frames are usually fitted in the plane of the inner leaf of the wall, traditionally after the brickwork is completed, although the practice of building the frames in as work proceeds is growing.

Possibly due to the lack of relatively flat building sites, the ground floor construction in low-rise housing generally consists of suspended timber joists with tongued-and-grooved boards fixed by 'secret' nailing. This method of fixing is obviously more expensive than the face nailing usual in England and Wales but it does produce a less unsightly finish. The growing tendency to use an unreinforced concrete slab over hardcore and fill is discouraged by the NHBC on sites that require more than 600 mm of fill, due mainly to failures following inadequate consolidation of the fill.

A traditional term that is in common use in connection with the preparation of the site is 'solum', a term for which there is no precise equivalent in English building terminology. It is defined by Pride in the *Glossary of Scottish building* as 'the area within walls of building after removal of top soil, the subsoil'. A number of specifications for the treatment of the solum are included in the deemed-to-satisfy clauses in the Standards.

BUILDING CONTRACTS

The traditional system of contracting in Scotland involved separate contracts between the building owner and each of the building trades. This system has been almost entirely superseded by the use of a suitably modified version of the commonly termed RIBA contract. Owing to the differences between English and Scottish law, the main contract document and its related sub-contract documents all require amendments before they are acceptable for use in Scotland. These amended documents are issued by the Scottish Building Contract Committee and should be used for any contract in Scotland.

For other legal aspects of building in Scotland the builder should refer to the *AJ legal handbook,* which contains sections relating to Scotland as well as a comprehensive bibliography.

FURTHER INFORMATION

HMSO publications

Building (Scotland) Act 1959, 7 & 8 Eliz. Ch. 24
Building (Scotland) Act 1970, Ch. 38
The Building Standards (Scotland) (Consolidation) Regulations 1971, SI No. 2052
The Building Standards (Scotland) Amendment Regulations 1973, SI No. 794
The Building Standards (Scotland) Amendment Regulations 1975, SI No. 404
The Building (Forms) (Scotland) Regulations 1975, SI No. 548
The Building (Procedure) (Scotland) Regulations 1975, SI No. 550
The Building (Procedure) (Scotland) Amendment Regulations 1977, SI No. 2189
The Building Operations (Scotland) Regulations 1975, SI No. 549
The Building Standards (Relaxation by Local Authorities) Regulations 1975, SI No. 547
Sidwell, N., *The cost of private house building in Scotland,* a Report for the Scottish Housing Advisory Committee (1970)
Explanatory Memoranda to the Building Standards (Scotland) Regulations, various dates 1972–1974
Fire Certificates (Special Premises) Regulations 1976, SI No. 2003
Fire Precautions Act 1971, Ch. 40
Health and Safety at Work etc. Act 1974, Ch. 37
Housing (Amounts of Approved Expense) (Scotland) Order 1977, SI No. 1687
Housing (Financial Provisions) (Scotland) Bill 1977
Housing policy: a consultative document, Cmnd. 6851 (1977)
Housing (Scotland) Act 1974, Ch. 45
Housing (Standard Amenities Approved Expense) (Scotland) Order 1977, SI No. 2074
Local Government (Scotland) Act 1973, Ch. 65
The New Scottish Housing Handbook, Bulletin 1, *Metric space standards* (1968)
Report of the Committee on Building Legislation in Scotland, Cmnd. 269 (1957)
Roads in Urban Areas (1973)
Scottish housing: a consultative document, Cmnd. 6852 (1977)
Scottish Housing Handbook, Bulletin 3, *Housing development, roads and services* (1977)
Sewerage (Scotland) Act 1968, Ch. 47
Town and Country Planning (Scotland) Act 1972, Ch. 52
Town and Country Planning (General Development) (Scotland) Order 1975, SI No. 679
Town and Country Planning (General Development) (Scotland) Amendment Order 1976, SI No. 693
Town and Country Planning (General Development) (Scotland) Amendment No. 2 Order 1976, SI No. 1307

Other official publications

Housing (Scotland) Act 1974: requirements to be met by houses improved with the aid of grant, SDD Circular 72/1975, Scottish Development Department, Edinburgh (1975)

Review of the management of planning, Scottish Development Department, Edinburgh (1977)

Other publications

Davey, P., and Freeth, E. (editors), *AJ legal handbook,* Architectural Press, London (1977)
The Building Regulations Scotland checklists and index, House Information Services, London (1975)
Pride, G.L., *Glossary of Scottish building,* distributed by the Scottish Civic Trust, Glasgow (1975)
Registered house-builder's handbook, NHBC, London (1974)
External environment: drainage, Scottish Local Authorities Special Housing Research Unit, Edinburgh (1975)

AUTHOR'S NOTE

The views expressed in this section are those of the author and do not necessarily represent the views of the Scottish Development Department.

63 THE METRIC SYSTEM

PRINCIPLES OF THE METRIC SYSTEM 63–2

THE METRIC SYSTEM IN THE CONSTRUCTION INDUSTRY 63–5

EXAMPLES 63–5

THE METRIC SYSTEM 63

63 THE METRIC SYSTEM

J.D. HEALY, BSc, PhD
Senior Lecturer in Science, Vauxhall College of Building and Further Education

This section begins with the concept of fundamental units and the relationship between fundamental units and derived units. The development of the metric system, and some advantages of the metric system over the Imperial system, are considered. The nomenclature and conventions used in the SI metric system are reviewed, and the section concludes with some worked examples and a series of problems with answers.

PRINCIPLES OF THE METRIC SYSTEM

Basic and derived units

All systems of measurement contain two types of units, i.e. basic and derived units. The basic or fundamental units, as their name suggests, represent the measurement units of basic quantities. These fundamental units supply the base from which the derived units may be developed. Examples of quantities that have fundamental units are given in Table 63.1.

Table 63.1 BASIC UNITS

Physical quantity	*Unit*	*Symbol*
Length	metre	m
Mass	kilogram	kg
Time	second	s

Table 63.2 DERIVED UNITS

Physical quantity	*Basic unit combination*	*Symbol*
Area	length × length	m^2
Volume	length × length × length	m^3
Velocity	length/time	m/s or $m\ s^{-1}$
Acceleration	length/time × time	m/s^2 or $m\ s^{-2}$
Density	$mass/length^3$	kg/m^3 or $kg\ m^{-3}$

Derived units are related to basic units by simple combinations of appropriate basic units, normally involving multiplication or division of the basic units. For example, the derived unit for area is found by multiplying a length by a length. Examples of quantities that have derived units, and the appropriate basic unit combination, are given in Table 63.2.

Development of the metric system

The first metric system appeared in France during the 1790s. The system depended on the recognition of constant standards for basic or fundamental units. During the development of the metric system a variety of standards for fundamental units have been used. Two forms of the metric system popular with scientists were the so called cgs and MKS systems, where the fundamental units of length, mass and time were expressed in centimetres, grams, seconds and metres, kilograms, seconds respectively. The availability of multiple variations of the metric system posed difficulties in communication between the advocates of each variation and necessitated conversion from one system to the other. The latest internationally agreed rationalisation is the Systême International d'Unitēs, generally referred to as SI units. Under the SI system all the basic units are internationallly agreed and all countries are now encouraged to use the full SI system. As an example of the accuracy of the definition of a basic unit, the SI definition of the metre is in terms of the wavelength of the radiation due to a specific electronic transition in one of the isotopes of krypton; this definition is absolute and hence not open to interpretation.

Advantages of the metric system

Unlike the Imperial system, where the magnitude of the basic units depended on a variety of traditional and arbitrary factors, all the derivatives of the metric system are decimal. Therefore any calculations are significantly simpler, because all the units are multiples or divisions of ten. Under the older Imperial system the relationship between smaller and larger units was non-decimal; for example, there were sixteen ounces in one pound and fourteen pounds in one stone.

Furthermore the various multiple units in the metric systems are recognisably related, while the Imperial units are not. For example, it is obvious that the gram and the kilogram are related, while it is not obvious what relationship exists between the mile and the furlong.

Using the metric system, decimal fractions may be regarded as vulgar fractions that have 10, 100, 1000 etc. as their denominators. Thus 5/10, 73/100, 627/1000 are all fractions that may be directly converted to decimals by removing the denominators and placing decimal points in front of the numerators.

Thus: 5/10 = 0.5, 73/100 = 0.73, 627/1000 = 0.627

Those fractions that make lengthy decimals when converted need only be expressed to the degree of accuracy required by the work concerned.

Nomenclature and conventions

The SI system is completely decimal, all multiples and sub-multiples (decimal fractions) of units are powers of ten, and the names given to the various multiples and sub-multiples are used for all units. Generally, the choice of which unit to use should be confined to powers of ten that are multiples to the power of plus or minus three. Examples of the recommended names and symbols of the commonly used multiples and sub-divisions are given in Table 63.3. Consequently the name for a particular unit is established by selecting the appropriate basic unit and adding the correct prefix.

It is now accepted that when indicating a decimal point a full point on a line level with the foot of the numerals is used. Commas are no longer inserted to mark off in three the digits of large numbers; instead it is recommended that a space be left. Examples of correct and incorrect usage are:

Correct	*Incorrect*
3 694.5	3,694.5
3.924 691	3.924,691

Table 63.3 MULTIPLES AND SUB–DIVISIONS OF SI UNITS

Multiplying factor	*Prefix*	*Symbol*
10^{12}	tera	T
10^{9}	giga	G
10^{6}	mega	M
10^{3}	kilo	k
10^{-3}	milli	m
10^{-6}	micro	μ
10^{-9}	nano	n
10^{-12}	pico	p
10^{-15}	femto	f
10^{-18}	atto	a

Numbers less than one require a zero placed before the full point; for example 0.621 and 0.0045.

A cross should be used to indicate multiplication between numbers, but it may be omitted between letters. For example 2.09×3.84 and zy are both correct when zy means multiply quantity y by quantity z. Division of numbers may be satisfactorily indicated by any of the following conventions:

$$\frac{94}{107} \qquad 94/107 \qquad 94 \times 107^{-1}$$

The last of the conventions given above is the preferred one.

SI fundamental and derived units

The fundamental and derived units are given in Tables 63.4 and 63.5 respectively. Table 63.5 also indicates the relationship between the derived unit and other SI units.

Table 63.4 BASIC SI UNITS

Quantity	*Unit*	*Symbol*
Length	metre	m
Mass	kilogram	kg
Time	second	s
Electric current	ampere	A
Thermodynamic temperature	kelvin	K
Luminous intensity	candela	cd

Table 63.5 DERIVED SI UNITS

Quantity	*Special unit (if any)*	*Symbol of unit, and connection with SI units*
Velocity		m/s or m s^{-1}
Acceleration		m/s^2 or m s^{-2}
Force	newton	N = kg m/s^2 or kg m s^{-2}
Work, energy, quantity of heat	joule	J = m N
Power	watt	W = J/s or J s^{-1}
Torque		N m
Area		m^2

Table 63.6 SENSIBLE UNITS

Quantity	*Unit*	*Applications*
Length	km	large distances
Length	m	common uses
Length	mm	detailed measurement
Area	km^2	land measurement
Area	m^2	interior work
Area	mm^2	close measurement
Volume	m^3	materials ordering
Volume	mm^3	materials evaluation
Mass	kg	common uses
Mass	g	materials evaluation

It is essential that sensible selection of units is made, ensuring that the quantities represented are recognisable. Table 63.6 suggests sensible sizes of common quantities often met in the construction industry.

THE METRIC SYSTEM IN THE CONSTRUCTION INDUSTRY

Traditionally the construction industry has used Imperial units, and problems are sometimes encountered when metric systems are introduced. This is because of the necessity of having to convert from an established and familiar system into a new and preferred system. Table 63.7 indicates a variety of Imperial units and their metric equivalents.

Table 63.7 METRIC EQUIVALENTS

Quantity	*Imperial unit*	*Metric equivalent*
Length	1 in	25.4 mm
Length	1 mile	1.609 km
Area	1 $yard^2$	0.836 m^2
Area	1 acre	4050 m^2
Mass	1 pound	0.454 kg

The traditional use in the construction industry of Imperial units is being increasingly eroded. Although the industry is conservative and includes a large number of small companies, two factors that operate against innovations, the use of metric units is accelerating. Larger companies within the industry are insisting on metric measurement, and materials and machinery from the EEC are quantified and calibrated in metric units. Furthermore, any development work deriving from the scientific community is invariably presented in metric units.

However, the overriding factor in favour of adoption of a metric system is economic. It is far more efficient to standardise both within a single company and between the various crafts by using a metric system. Metrication is an extremely standardised system and if applied properly will increase efficiency and reduce costs.

EXAMPLES

Worked examples

Converting fractions to decimals This is accomplished simply by dividing the numerator by the denominator. The division is continued until the required degree of accuracy is reached.

Convert 5/9 to decimals.

Dividing 5 by 9 gives: $9\overline{)5.000\,0}$

$0.555\,5$

Since the last figure is 5, the answer correct to three decimal places is given as 0.556. Had the figure been 4 or less the answer would have been 0.555, i.e. the last figure would have been ignored.

Expressing values to significant figures The same procedure is adopted as that given above, counting all the figures including noughts in the answer.

Express 6.089 26 to four significant figures.

The number contains six figures in total. Since the last figure is greater than 5 the preceding figure is increased to 3; since the fifth figure is now 3 and is less than five it is deleted, leaving the fourth figure as 9. The answer to four significant figures is then 6.089.

Addition This is a simple operation, provided it is remembered that for neatness and accuracy the decimal points are kept directly beneath each other.

Sum the following: 6.089 23, 4.076, 3.974.

$$\begin{array}{r} 6.089\,23 \\ +4.076 \\ +3.974 \\ \hline 14.139\,23 \end{array}$$

Subtraction The same procedure as that adopted for addition is followed.

Subtract 4.098 21 from 8.353 64.

$$\begin{array}{r} 8.353\,64 \\ -4.098\,21 \\ \hline 4.255\,43 \end{array}$$

Multiplication To ensure that the decimal point is placed in the correct position after the multiplication, it should be remembered that the number of decimal point places is always the sum of the number of decimal places in the two numbers that have been multiplied.

Multiply 13.74 by 5.45.

$$\begin{array}{rl} 13.74 & \\ 5.45 & \\ \hline 68.70 & \text{found by multiplying by 5} \\ 5.496 & \text{found by multiplying by 0.4} \\ 0.687\,0 & \text{found by multiplying by 0.05} \\ \hline 74.883\,0 & \end{array}$$

Check that the decimal point is in the correct position by counting the number of decimal places in the two numbers to be multiplied: 2 + 2 = 4, so count four from right to left in the answer. Remember that *all* the figures including the noughts in both the answer and the multiplying numbers must be considered.

Division Various techniques are available for rationalising division of decimals. One of the simpler methods is to move the decimal point of the divisor to the right until the divisor becomes a positive integer. To ensure that the correct answer is obtained it then becomes necessary to move the decimal point of the numerator or dividend an equal number of places in the same direction. The division then follows normal arithmetic methods.

(a) Divide 18.75 by 0.25.

The problem is first written in the accepted form, which is:

$$\frac{18.75}{0.25}$$

Moving the decimal point two places to the right, the problem becomes:

$$\frac{1875}{25}$$

```
     75
25)1875
   175
   ---
    125
    125
    ---
    000
```

(b) $\frac{8.464}{2.64}$ to three significant figures.

The problem becomes:

$$\frac{846.4}{264}$$

```
      3.202
264)846.4
    792
    ---
     54.4
     53.8
     ----
      0.600
```

The answer is therefore 3.20, to three significant figures.

Problems

1. Express the following in decimals:
   (a) 9/10 (b) 47/100 (c) 9/100
   (d) 28/100 (e) 672/100 (f) 942/1000

2. Express in decimal form, correct to three significant figures:
   (a) 17/42 (b) 18/63 (c) 5/8
   (d) 27/49 (e) 16/17 (f) 34/608

3. Find the value of the following:
   (a) $6.94 + 8.21 + 19.64$
   (b) $8.37 + 16.91 + 8.32 - 6.41$
   (c) $443.62 - 321.86$

4. Find, correct to four significant figures:
   (a) 22.19×17.62
   (b) 284.3×1.4
   (c) $539.4 \div 18.67$
   (d) $23.216 \div 6.94$

5. Calculate the number of m^2 in an area 0.62 km × 0.32 km.

6. The inside dimensions of a water tank are 0.61 m × 0.71 m × 0.82 m. Calculate the volume of water in the tank when it is filled to 0.73 of its total capacity.

Answers

1. (a) 0.9 (b) 0.47 (c) 0.09
   (d) 0.028 (e) 0.672 (f) 0.942

2. (a) 0.405 (b) 0.286 (c) 0.625
   (d) 0.551 (e) 0.941 (f) 0.056

3. (a) 34.79 (b) 26.19 (c) 121.76

4. (a) 390.9 (b) 398.0 (c) 28.62 (d) 3.345

5. 198 400 m^2

6. 0.259 m^3

64 SOURCES OF BUILDING INFORMATION

INTRODUCTION 64–2

TRADE ASSOCIATIONS 64–2

INFORMATION AND ADVISORY SERVICES 64–6

PROFESSIONAL SOCIETIES 64–10

EMPLOYERS' ASSOCIATIONS AND TRADES UNIONS 64–10

GOVERNMENT DEPARTMENTS AND ORGANISATIONS 64–11

PLANNING AND CONSERVATION 64–13

PUBLICATIONS AND BOOKSHOPS 64–13

TECHNICAL PRESS 64–14

SOURCES OF BUILDING INFORMATION 64

64 SOURCES OF BUILDING INFORMATION

J.F. GEORGE, BSc, *Chief Executive*
A.F. WILKINS, *Information Officer*
Mrs J. HUNTING, *Librarian*
The Building Centre Group

This section provides information on the best sources for the wide variety of information that a building contractor may require in the course of his activities.

INTRODUCTION

This section is intended for use as a day-to-day reference source. Builders and subcontractors will very often be operating with known products in familiar situations, but there will be occasions when information on technical aspects of a problem are not available from the more regularly used published material or from their usual commercial contacts.

The listings are intended to be comprehensive, but initially some omissions and changes will come to notice during the life of the book. It is hoped that it will be of value not only to builders but to architects, surveyors, builders' merchants and others engaged in the construction industry.

Almost 300 organisations and services are listed; many restrict their service to members, others provide a service for a fee, some for specifiers only, also for a fee, and only a proportion offer a free information service. All will, however, direct enquirers to an alternative and appropriate source if they cannot provide the information desired.

TRADE ASSOCIATIONS

Products

The names listed cover associations concerned with promotion of goods and materials allied to the various trades concerned in the manufacture of building materials. In almost all cases the associations are supported by the industry they represent, and in the main give technical information and supply technical literature, which can be obtained by letter or telephone.

Aluminium Federation Ltd,
Broadway House, Calthorpe Road, Birmingham B15 1TN — 021–455 0311

Aluminium Window Association,
26 Store Street, London WC1E 7EL — 01–637 3578

Asbestos Association,
6th Floor, Royal Exchange, Manchester M2 7FB — 061–832 3017

Asphalt and Coated Macadam Association,
25 Lower Belgrave Street, London SW1W 0LS — 01–730 0761

Association of British Plywood and Veneer Manufacturers,
Epworth House, 25 City Road, London EC1Y 1AR — 01–628 5801

Association of British Roofing Felt Manufacturers,
69 Cannon Street, London EC4N 5AB — 01–248 4444

Association of Building Component Manufacturers,
26 Store Street, London WC1E 7BT — 01–580 9083

British Adhesive Manufacturers Association,
20 Pylewell Road, Hythe, Southampton SO4 6YW (0703) 843765
British Blind and Shutter Association,
251 Brompton Road, London SW3 01–584 5552
British Carpet Industry Technical Association,
Aykroyd House, Hoo Road, Kidderminster, Worcestershire DY10 1NB (0562) 4053
British Ceramic Tile Council,
Federation House, Station Road, Stoke-on-Trent ST4 2RU (0782) 45147
British Constructional Steel Work Association,
Silvertown House, 1/2 Vincent Square, London W1 01–834 1713
British Electrical and Allied Manufacturers Association,
Leicester House, 8 Leicester Street, London WC2H 7BN 01–437 0678
British Lead Manufacturers Association,
Firlands Lodge, Firlands Avenue, Camberley, Surrey GU5 2HY (0276) 23196
British Lock Manufacturers Association,
91 Tettenhall Road, Wolverhampton WV3 9PE (0902) 26726
British Plastics Federation,
5 Belgrave Square, London SW1 01–235 9483
British Pre-cast Concrete Federation,
60 Charles Street, Leicester LE1 1FB (0533) 28627
British Pump Manufacturers Association,
37 Castle Street, Guildford, Surrey (0483) 37997
British Ready Mix Concrete Association,
Shepperton House, Green Lane, Shepperton, Middlesex TW17 8DN (093 22) 43232
British Rubber Manufacturers Association,
90/91 Tottenham Court Road, London W1P 0BR 01–580 2794
British Wood Preserving Association,
150 Southampton Row, London WC1 01–837 8217
British Woodworking Federation,
82 New Cavendish Street, London, W1M 8AD 01–580 4041
Builders Merchants Federation,
15 Soho Square, London W1 01–439 1753
Cement Admixtures Association,
Dickens House, 15 Tooks Court, London EC4A 1LA 01–831 7581
Chipboard Promotion Association,
7a Church Street, Esher, Surrey KT10 8Q5 (0372) 66468
Clay Pipe Development Association,
Drayton House, 30 Gordon Street, London WC1H 0AN 01–388 0025
Cold Rolled Section Association,
King Edward House, New Street, Birmingham B2 4QP 021–643 5494
Concrete Block Association,
60 Charles Street, Leicester LE1 1FB (0533) 28627
Concrete Pipe Association of Great Britain,
Binley, Andover, Hampshire (026 473) 317
Contractors Plant Association,
28 Eccleston Street, London SW1W 9PY 01–730 7117
Copper Development Association,
Orchard House, Mutton Lane, Potters Bar, Hertfordshire EN6 3AP (0707) 50711
Copper Tube Fittings Manufacturers Association,
Crest House, 7 Highfield Road, Birmingham B15 3ED 021–454 7766
Decorative Lighting Association,
Llanerchymedd, Anglesey (024 876) 383

Dry Lining and Partitions Association Ltd,
15 South Street, Lancing, Sussex BN6 8AE (079 18) 5700
Expanded Polystyrene Product Manufacturers Association,
PO Box 103, Haywards Heath, Sussex (0444) 57871
Federation of Stone Industries,
Admin House, Market Square (North Side), Leighton Buzzard, Bedfordshire (052 53) 75252
Felt Roofing Contractors Advisory Board,
Maxwelton House, 41/43 Baltro Road, Haywards Heath, Sussex (0444) 51835
Fibre Building Board Development Organisation,
6 Buckingham Street, London WC2N 6BZ 01–839 1122
Finnish Plywood Development Organisation,
Broadmead House, 21 Panton Street, London SW1Y 4DR 01–930 3238
Floor Quarry Association,
Federation House, Station Road, Stoke-on-Trent ST4 2RU (0782) 45147
Galvanizers Association,
34 Berkeley Square, London W1X 6AJ 01–499 6636
Glass Manufacturers Federation,
19 Portland Place, London W1N 4BH 01–580 6952
Gypsum Products Development Association,
15 Marylebone Road, London NW1 5JE 01–486 4011
Hardwood Flooring Manufacturers Association,
Clareville House, Oxendon Street, London SW1Y 4EL 01–839 3381
Institute of Plumbing,
Scottish Mutual House, North Street, Hornchurch, Essex RM11 1RU (040 24) 51236
Iron and Steel Trades Confederation,
324 Grays Inn Road, London WC1X 8DD 01–837 6691
Kitchen Furniture Manufacturers Association,
Turret House, Station Road, Amersham, Bucks HP7 0AB (024 03) 2143
Laminated Plastics Fabricators Association,
c/o W.H. Foster and Sons Ltd, Cardale Street, Rowley Regis, Warley, West Midlands B65 0LX 021–559 2028
Lead Development Association,
34 Berkeley Square, London W1X 6AJ 01–499 8422
Lighting Industry Federation,
25 Bedford Square, London WC1B 3HH 01–636 0766
Manufacturers Association of Radiators and Convectors,
24 Ormond Road, Richmond, Surrey TW10 6TH 01–948 4151
Metal Sink Manufacturers Association,
Fleming House, Renfrew Street, Glasgow 041–332 0826
Mortar Producers Association Ltd,
274 High Street, Slough SL1 1NB (0753)36936
National Building and Allied Hardware Manufacturers Federation,
5 Greenfield Crescent, Edgbaston, Birmingham B15 3BE 021–424 2177
National Cavity Insulation Association,
178 Great Portland Street, London W1 01–637 7481
National Council of Building Materials Producers,
26 Store Street, London WC1E 7BT 01–580 3344
National Federation of Clay Industries,
3rd Floor, Weston House, West Bar Green, Sheffield S1 2AD (0742) 730261
Oil Appliance Manufacturers Association,
Chamber of Commerce House, PO Box 360, 75 Harborne Road, Birmingham B15 3DH 021–454 6171

Paintmakers Association of Great Britain,
Alembic House, 93 Albert Embankment, London SE1 7TY 01–582 1185
Patent Glazing Conference,
13 Upper High Street, Epsom, Surrey KT17 4QY (037 27) 29191
Pitch Fibre Pipe Association of Great Britain,
35 New Bridge Street, London EC4V 6BH 01–248 5271
Plastic Pipe Manufacturers Society,
82 Cornwall Street, Birmingham B3 3B7 021–236 1866
Plastic Tanks Cistern Manufacturers Association,
c/o Peat Marwick Mitchell and Co, Glen House, Stag Place,
London SW1E 5AT 01–834 9823
Prefabricated Building Manufacturers Association of Great Britain,
4 High Street, Epsom, Surrey KT19 8AF (037 27) 24481
Sand and Gravel Association,
48 Park Street, London W1Y 4HE 01–499 8967
Sealant Manufacturers Association,
Dickens House, 15 Tooks Court, London EC4A 1LA 01–831 7581
Stainless Steel Fabricators Association of Great Britain,
39 Hillside Gardens, Brockham, Betchworth, Surrey (073 784) 2068
Steel Castings Research and Trade Association,
East Bank Road, Sheffield S2 3PT (0742) 28647
Steel Window Association,
26 Store Street, London WC1E 7BT 01–637 3571
Timber Trade Federation of the United Kingdom,
Clareville House, Whitcomb Street, London WC2H 7DL 01–839 1891
Vitreous Enamel Development Council,
New House, Ticehurst, Wadhurst, Sussex TN5 7AL (0580) 8252
Zinc Development Association,
34 Berkeley Square, London W1X 6AJ 01–499 6636

Services

Associated Master Plumbers and Domestic Engineers,
Longwood, Kenley Lane, Kenley, Surrey CR2 5DR 01–660 3336
British Decorators Association,
6 Haywra Street, Harrogate, Yorks HG1 5BL (0423) 67292
Contract Flooring Association,
23 Chippenham Mews, London W9 2AN 01–286 4499
Council of Registered Gas Installers (CORGI),
140 Tottenham Court Road, London WC1 01–387 9185
Regional Offices
Northern: Stephenson House, Northumbria Way, Killingworth,
Newcastle upon Tyne NE99 1GB (0632) 683833
North-Western: Victoria House, Victoria Road, Hale, Cheshire
WA15 9AF 061–941 2548
North-Eastern: Negas, New York Road, Leeds LS2 7PE (0532) 445909
West Midlands: 5 Wharf Road, Solihull, West Midlands B91 2JP 021–704 4101
East Midlands: PO Box 145, De Montfort Street, Leicester
LE1 9DB (0533) 50510

Eastern: Star House, Mutton Lane, Potters Bar, Hertfordshire EN6 2PD (0707) 50793
Scottish: Granton House, West Granton Road, Edinburgh EH5 1YB 031–552 6960
Wales: Snelling House, Bute Terrace, Cardiff CF1 2UF (0222) 395398
South-Western: Pixash Lane, Keynsham, Bristol BS18 1TP (027 56) 61051
North Thames: 346/8 High Street, Chiswick, London W4 5TA 01–995 9110
South-Eastern: 1 Katherine Street, Croydon CR9 1JU 01–688 7564
Southern: 223 Portswood Road, Southampton SO2 1NF (0703) 553835

Electrical Contractors Association
Esca House, Palace Court, London W2 01–229 1266

Federation of Civil Engineering Contractors,
Romney House, Tufton Street, London SW1P 3DU 01–222 2533

Federation of Master Builders,
33 John Street, London WC1N 2BB 01–242 7583

Federation of Piling Specialists,
Dickens House, 15 Tooks Court, London EC4A 1LA 01–831 7581

Glass and Glazing Federation,
6 Mount Row, London W1Y 6DY 01–629 8334

Heating and Ventilating Contractors Association,
Esca House, 34 Palace Court, London W2 4JG 01–229 2488

Heating, Ventilating and Air Conditioning Manufacturers Association Ltd,
Unit 3, Phoenix House, Phoenix Way, Heston, Middlesex TW5 9ND 01–897 2848

Mastic Asphalt Advisory Council,
Construction House, Paddock Hall Road, Haywards Heath, Sussex (0444) 57786

National Federation of Painting and Decorating Contractors,
82 New Cavendish Street, London W1M 8AD 01–580 4041

National Federation of Plastering Contractors,
82 New Cavendish Street, London W1M 8AD 01–580 4041

National Federation of Roofing Contractors,
Soho Square, London W1V 5FB 01–439 1753

National Federation of Terrazzo-Mosaic Specialists,
1st Floor, 251 Brompton Rd, London SW3 2EZ 01–584 5552

National House-Building Council,
58 Portland Place, London W1N 4BU 01–387 7201

National Inspection Council for Electrical Installation Contracting (NICEIC),
Alembic House, 93 Albert Embankment, London SE1 7TB 01–582 7746

North Wales Slate Quarries Association,
Bryn, Llanllechid, Bangor, Gwynedd LL57 3LG (0248) 600656

Swedish Finnish Timber Council,
21 Carolgate, Retford, Nottinghamshire DN 22 6BZ (0777) 706616

INFORMATION AND ADVISORY SERVICES

Information services

This section covers many organisations that provide information and advice. Some provide a service free of charge to the enquirer (including the public), some of the services are available only to the building trade and professions, and some charge a fee.

Asbestos Information Committee,
2 Old Burlington Street, London W1X 2LH 01–734 0081

Association of Special Libraries and Information Bureaux (ASLIB),
3 Belgrave Square, London SW1X 8PL — 01–235 5050

Barbour Index,
New Lodge, Drift Road, Windsor, Berkshire SL4 4RQ — (034 47) 4121

Building and Allied Trades Enquiry Bureau,
Zodiac House, 163 London Road, Croydon CR9 2RP — 01–686 5644

Building Centres:

Birmingham: Engineering and Building Centre,
Broad Street, Birmingham B1 2DB — 021–643 1914

Bristol: The Building Centre Bristol,
Colston Avenue, The Centre, Bristol BS1 4TW — Admin. (0272) 22953
Info. (0272) 27002

Cambridge: The Building Centre Cambridge,
15–16 Trumpington Street, Cambridge CB2 1QD — (0223) 59625

Coventry: Coventry Building Information Centre,
Department of Architecture and Planning, Tower Block, New Council Offices,
Coventry CV1 5RT — (0203) 25555 ext. 2512

Durham: Northern Counties Building Information Centre,
Green Lane, Durham City DH1 3JY — (0385) 62611

Glasgow: The Building Centre Scotland,
6 Newton Terrace, Glasgow G3 7PF — 041–248 6212

Liverpool: Liverpool Building and Design Centre,
Hope Street, Liverpool L1 9BR — Admin. 051–709 8566
Info. 051–709 8484

London: The Building Centre,
26 Store Street, London WC1E 7BT — Admin. 01–637 1022
Info. 01–637 8361

Manchester: The Building Centre Manchester,
113/115 Portland Street, Manchester M1 6FB — Admin. 061–236 9802
Info. 061–236 6933

Nottingham: Nottingham Building Centre Ltd,
17/19 Goosegate, Nottingham NG1 1FE — (0602) 581931

Southampton: The Building Centre Southampton,
18–20 Cumberland Place, Southampton SO1 2BD — (0703) 27350

Stoke-on-Trent: The Building Information Centre,
College of Building and Commerce, Stoke Road, Shelton,
Stoke-on-Trent ST4 2DG — (0782) 24651

Building Research Establishment,
Garston, Watford, Hertfordshire, WD2 7JR — (092 73) 74040

Building Statistical Services,
14 Great College Street, London SW1P 3RZ — 01–222 8686

Cement and Concrete Association (Research Station and Advisory Service),
Wexham Springs, Fulmer, Slough SL3 6PL — (028 16) 2727

Data Express,
26 Store Street, London WC1E 7BT — 01–637 8361

Farm Buildings Information Centre Ltd,
National Agricultural Centre, Kenilworth CV8 2LG — (0203) 22345

Fire Protection Association,
Aldermary House, Queen Street, London EC4N 1TJ — 01–248 5222

National Building Agency,
NBA House, 7 Arundel Street, London WC2R 3DZ — 01–836 4488

National Building Specification,
6 Higham Place, Newcastle-upon-Tyne NE1 8AF — (0632) 29594

Plastic Bath Information Bureau,
143 Fleet Street, London EC4A 2BP — 01–353 3541

Timber Decay Advice Bureau,
16 Gander Green Lane, Sutton, Surrey SM1 2EJ 01–643 2248

Research organisations

These organisations exist basically to finance and promote research and to assist manufacturers in the marketing of their products.

Agrément Board,
PO Box 195, Bucknalls Lane, Garston, Hertfordshire (092 73) 70844

Albury Laboratories Site Investigation,
Albury, Guildford, GU5 9AZ (048 641) 2041

British Cast Iron Research Association (BCIRA),
Alvechurch, Birmingham B48 7QB (0527) 66414

British Fire Service Association,
86 London Road, Leicester LE2 0QR (0533) 542879

British Glass Industry Research Association,
Northumberland Road, Sheffield S10 2AU (0742) 686201

British Standards Institution,
2 Park Street, London W1A 2BS 01–629 9000

Building Research Establishment,
Garston, Watford, Hertfordshire WD2 7JR (092 73) 74040

Building Services Research and Information Association,
Old Bracknell Lane, Bracknell, Berkshire RG12 4AH (0344) 25071

Concrete Society,
Terminal House, Grosvenor Gardens, London SW1W 0AJ 01–730 8252

Construction Industry Research and Information Association,
6 Storey's Gate, London SW1P 3AU 01–839 6881

Electrical Research Organisation,
Cleeve Road, Leatherhead, Surrey KT22 7SA (037 23) 74151

Institute of Asphalt Technology,
24 Grosvenor Gardens, London SW1 0DH 01–730 7175

Paint Research Association,
Waldegrave Road, Teddington, Middlesex TW11 8LD 01–977 4427

Rubber and Plastics Research Association,
Shawbury, Shrewsbury SY4 4NR (093 94) 383

Timber Research and Development Association,
Stocking Lane, Hughenden Valley, High Wycombe, Bucks HP14 4ND (024 024) 3091

Tin Research Institute,
Fraser Road, Perivale, Greenford, Middlesex UB6 7AG 01–997 4254

Water Research Centre,
45 Station Road, Henley-on-Thames, Oxfordshire (049 12) 5931

Advisory services

Building Research Establishment,
Advisory Service, Bucknalls Lane, Garston, Watford, Hertfordshire WD2 7JR (092 73) 74040

Cement and Concrete Association,
52 Grosvenor Gardens, London SW1W 0AQ 01–235 6661
Research Station and Advisory Service, Wexham Springs, Fulmer, Slough SL3 6PL (028 16) 2727
Training Centre, Fulmer Grange, Fulmer, Slough SL2 4QS (028 16) 2727

Scotland:
Peter Russell, BSc, FICE, FIStructE, FIHE, FIOB
Cement and Concrete Association,
2 Rutland Square, Edinburgh EH1 2AS 031–229 5085

North:
R. Colin Deacon, BSc(Eng), FICE, FIHE
Cement and Concrete Association,
18 Appleby House, Town Centre, Thornaby,
Stockton-on-Tees, Cleveland TS17 9EY (0642) 65603

North-west (Lancashire, Greater Manchester, Merseyside, Cheshire, Clwyd, Gwynedd, Isle of Man):
Alan Pink, BSc(Tech), MICE, MIHE
Cement and Concrete Association,
550 Mauldeth Road West, Chorlton-cum-Hardy, Manchester
M21 28J 061–881 5394

North midlands (West Yorkshire, South Yorkshire, Derbyshire, Nottinghamshire, Leicestershire):
J. Martin Dykes, BSc, MICE, MINES
Cement and Concrete Association,
7 Lindrum Terrace, Lincoln LN2 5RP (0522) 25876

Midlands (Staffordshire, Leicestershire, Shropshire, West Midlands, Powys, Hereford and Worcester, Warwickshire, Gloucestershire):
Michael E. Fitzgibbon, BA, BAI, MICE, FIHE
Cement and Concrete Association,
Engineering and Building Centre, Broad Street, Birmingham B1 2BD
021–643 1914

East (Norfolk, Suffolk, Northamptonshire, Cambridgeshire, Oxfordshire, Buckinghamshire, Bedfordshire, Hertfordshire, Essex):
J. Gregory-Cullen, MA, MICE, FIHE
Cement and Concrete Association,
25 Mill Street, Bedford MK40 3EU (0234) 52486

South-west (Dyfed, Gwent, West Glamorgan, Mid Glamorgan, South Glamorgan, Avon, Somerset, Cornwall, Devon, Dorset, Scilly Isles):
Richard T.L. Allen, MA, FICE, FIHE, FIOB
Cement and Concrete Association,
67 Park Place, Cardiff CF1 3AS (0222) 40840

South-east (Berkshire, Wiltshire, Surrey, Kent, Hampshire, West Sussex, East Sussex, IoW, Channel Islands):
Geo. F. Blackledge, BSc(Tech), MIStructE, MIHE
Cement and Concrete Association,
29 High Street, Guildford GU1 3DY (0483) 76206

National Federation of Building Trades Employers,
Technical Advisory Service, 82 New Cavendish Street, London
W1M 8AD 01–637 4771

Timber Research and Development Association,
Stocking Lane, Hughenden Valley, High Wycombe, Bucks
HP 14 4ND (024 024) 3091
Scotland: 6 Newton Terrace, Glasgow G3 7PF 041–221 0719
North: 48 Dudley Court, East Square, Cramlington,
Northumberland NE23 6QW (0670) 713238
North-west: 115 Portland Street, Manchester M1 6DW 061–236 3740
North-east: 18 Park Row, Leeds LS1 5JA (0532) 457256
Midlands: Engineering and Building Centre, Broad Street,
Birmingham B1 2BD 021–643 1914
East: 16 Trumpington Street, Cambridge CB2 2QA (0223) 66287

South-west: Building and Design Centre, Colston Avenue, The Centre, Bristol BS1 4TW (0272) 23692
South-east: The Building Centre, 26 Store Street, London WC1E 7BU 01–636 8761
West: Pearl Assurance House, High Street, Exeter EX4 3NS (0392) 54034

PROFESSIONAL SOCIETIES

Architectural Association Incorporated,
34 Bedford Square, London WC1B 3ES 01–636 0974
Artworkers Guild,
6 Queen Square, London WC1N 3AR 01–837 3474
Chartered Institution of Building Services,
49 Cadogan Square, London SW1 01–235 7671
Contractors Mechanical Plant Engineers,
48 Coburn Avenue, Hatch End, Pinner, Middlesex HA5 4PF 01–428 4795
Faculty of Building,
10 Manor Way, Boreham Wood, Hertfordshire WD6 1QQ 01–953 7053
Guild of Architectural Ironmongers,
15 Soho Square, London W1V 5FB 01–439 1753
Guild of Surveyors,
13 Lubbock Road, Chislehurst, Kent BR7 5JG 01–467 9415
Institute of Building,
Englemere, King's Ride, Ascot, Berkshire S15 8BJ (0990) 23355
Institute of Clerks of Works of Great Britain,
41 The Mall, London W5 01–579 2917
Institute of Landscape Architects,
12 Carlton House Terrace, London SW1Y 5AH 01–839 4044
Institute of Quantity Surveyors,
98 Gloucester Place, London W1H 4AT 01–935 1859
Institution of Heating and Ventilating Engineers,
now Chartered Institution of Building Services
Institution of Industrial Safety Officers,
222 Uppingham Road, Leicester LE5 0QG (0533) 768424
Institution of Municipal Engineers,
25 Eccleston Square, London SW1V 1NX 01–834 5082
Institution of Structural Engineers,
11 Upper Belgrave Street, London SW1X 8BH 01–235 4535
Plastics and Rubber Institute,
11 Hobart Place, London SW1W 0HL 01–245 9555
Royal Institute of British Architects,
66 Portland Place, London W1N 4AD 01–580 5533
Royal Institution of Chartered Surveyors,
12 Great George Street, London SW1P 3AD 01–222 7000
Royal Town Planning Institute,
26 Portland Place, London W1N 4BE 01–636 9107
Society of Architectural and Associated Technicians,
1–4 Argyll Street, London W1V 1AD 01–437 0976

EMPLOYERS' ASSOCIATIONS AND TRADES UNIONS

Employers' Associations

Association of Painting and Decorating Employers,
26 Orphanage Road, Erdington, Birmingham B24 9HT 021–350 5209

Confederation of British Industry,
21 Tothill Street, London SW1H 9LP 01–930 6711
Institute of Directors,
10 Belgrave Square, London SW1 01–234 3601
National Federation of Building Trades Employers,
82 New Cavendish Street, London W1M 8AD 01–580 4041
National Joint Council for the Building Industry,
11 Weymouth Street, London W1N 3FG 01–580 1740

Trades unions

Amalgamated Union of Engineering Workers, Construction Section,
Construction House, 190 Cedars Road, London SW4 0PP 01–622 4451
Association of Official Architects,
66 Portland Place, London W1N 4AD 01–580 5533
Electrical, Electronic, Telecommunications and Plumbing Union,
Hayes Court, West Common Road, Bromley, Kent BR2 7AU 01–462 7755
Furniture, Timber and Allied Trades Union,
Fairfields, Roe Green, Kingsbury, London NW9 0PT 01–204 0273
National and Local Government Officers Association,
8 Harewood Road, London NW1 01–262 8030
Transport and General Workers Union, Building Crafts Section,
Transport House, Smith Square, London SW1 01–828 7788
Union of Construction, Allied Trades and Technicians,
UCATT House, 177 Abbeyville Road, London SW4 9RL 01–622 2442
Association of Building Technicians (now STAMP section of UCATT),
UCATT House, 177 Abbeyville Road, London SW4 9RL 01–622 2442
Wall Paper Workers Union,
223 Bury New Road, Whitefield, Manchester 061–766 3645

GOVERNMENT DEPARTMENTS AND ORGANISATIONS

Also includes government-sponsored councils etc., and nationalised industries.

British Gas Corporation,
59 Bryanston Street, London W1A 2AZ 01–723 7030
British Steel Corporation,
PO Box 403, 33 Grosvenor Place, London SW1X 7JG 01–235 1212
Central Electricity Generating Board,
Sudbury House, 15 Newgate Street, London EC1A 7AU 01–248 1202
Central Office of Information,
Hercules Road, London SE1 7DU 01–928 2345
COI regional offices
Andrews House, Gallowgate, Newcastle-upon-Tyne NE1 4TB (0632) 27575
10th Floor, City House, New Station Street, Leeds LS1 4JG (0532) 38232
Cranbrook House, Cranbrook Street, Nottingham NG1 1EW (0602) 46121
Block D, Government Buildings, Brooklands Avenue,
Cambridge CB2 2DF (0223) 58911
St. Christopher House Annexe, Sumner Street, London SE1 0TB 01–928 2371
Market Place House, Reading RG1 2EH (0734) 581261
The Pithay, Bristol BS1 2NF (0272) 291071

Five Ways House, Islington Row, Middle Way, Birmingham B15 1SH — 021–643 8191
22nd Floor, Sunley Buildings, Piccadilly Plaza, Manchester M1 4BD — 061–832 9111

Department of the Environment,
2 Marsham Street, London SW1P 3EB — 01–212 3434

Department of Trade,
1 Victoria Street, London SW1H 0ET — 01–215 7877

Design Council,
28 Haymarket, London SW1Y 4SU — 01–839 8000

Electricity Council,
30 Millbank, London SW1P 4RD — 01–834 2333

Forestry Commission,
231 Corstorphine Road, Edinburgh EH12 — 031–334 0303

HM Land Registry,
Lincoln's Inn Fields, London WC2A 3PH — 01–405 3488

Her Majesty's Stationery Office,
Atlantic House, Holborn Viaduct, London EC1P 1BN — 01–248 9876

Housing Corporation,
Maple House, 149 Tottenham Court Road, London W1P 0BN — 01–387 9466

Metrication Board,
22 Kingsway, London WC2B 6LE — 01–242 6828

National Coal Board,
Hobart House, Grosvenor Place, London SW1X 7AE — 01–235 2020

National Water Council,
1 Queen Anne's Gate, London SW1H 9BT — 01–930 3100

New Towns Association,
Glen House, Stag Place, London SW1E 5AJ — 01–828 1103

Local government

Association of County Councils,
Eaton House, 66a Eaton Square, London SW1W 9BH — 01–235 5173

Association of District Councils,
25 Buckingham Gate, London SW1E 6LE — 01–828 7931

Association of Metropolitan Authorities,
36 Old Queen Street, London SW1H 9JE — 01–930 9861

Convention of Scottish Local Authorities,
3 Forres Street, Edinburgh EH3 6BL — 031–225 1626

Joint Advisory Committee on Local Authority Purchasing,
36 Old Queen Street, London SW1H 9JE — 01–930 9861

National Association of Local Councils,
100 Great Russell Street, London WC1B 3LD — 01–636 4066

National Federation of Housing Associations,
86 Strand, London WC2 0EG — 01–836 2741

National Housing and Town Planning Council,
34 Junction Road, London N19 — 01–272 8448

National Society for Clean Air,
136 North Street, Brighton, Sussex BN1 1RG — (0273) 26313

Scottish National Housing and Town Planning Council,
19 Bentinck Drive, Troon, Ayrshire — (0292) 314552

PLANNING AND CONSERVATION

Planning

For methods *not* directly involving specific proposals (where it would be appropriate to consult the local authority concerned):

Department of the Environment,
Becket House, 1 Lambeth Bridge Road, London SE1 7ER 01–928 7855
Greater London Council (Building Regulations Division)
Department of Architecture and Civic Design, Middlesex House, Vauxhall Bridge Road, London SW1V 2SB 01–633 3827
Royal Town Planning Institute,
26 Portland Place, London W1N 4BE 01–636 9107

Conservation

Building Conservation Association,
26 Store Street, London WC1E 7BT 01–580 9098
Civic Trust,
17 Carlton House Terrace, London SW1 01–930 0914
Historic Buildings Council for Scotland,
25 Drumsherrgh Gardens, Edinburgh EH3 7RN 031–226 36114
Historic Buildings Council for Wales,
Pearl Assurance Buildings, Greyfriars Road, Cardiff CF1 3RT (0222) 44151
Institute of Advanced Architectural Studies,
Kings Manor, York YO1 2EP (0904) 24919
Society for the Protection of Ancient Buildings,
55 Great Ormond Street, London WC1N 3JA 01–405 2646

PUBLICATIONS AND BOOKSHOPS

Building Research Digests and Current Papers
Published by: Building Research Establishment, Garston, Watford, Hertfordshire WD2 7JR (092 73) 74040
Obtainable from:
(a) Her Majesty's Stationery Office:
49 High Holborn, London WC1V 6MB
13a Castle Street, Edinburgh EH2 3AR
41 The Hayes, Cardiff CF1 1JW
80 Chichester Street, Belfast BT1 4JY
Brazenose Street, Manchester M60 8AS
258 Broad Street, Birmingham B1 2HE
Southey House, Wine Street, Bristol BS1 2BD
(b) The Building Bookshop, The Building Centre, 26 Store Street, London WC1E 7BT 01–637 3151
(c) The Building Bookshop, The Building Centre, 15–16 Trumpington Street, Cambridge CB2 1QD (0223) 59625
(d) The Building Bookshop, The Building Centre, Colston Avenue, The Centre, Bristol BS1 4TW (0272) 22953
(e) The Building Bookshop, The Building Centre, 113–115 Portland Street, Manchester M6 6FB 061–236 6933

Publications can also be obtained through bookshops who act as agents for HMSO.

Transport and Road Research publications
Issued by the Transport and Road Research Laboratory of the Department of the Environment. Obtainable from: HMSO and Building Bookshops, and through bookshops who act as agents for HMSO.

Advisory Leaflets
Published by: Department of the Environment. Obtainable from: HMSO and Building Bookshops.

Commodity File – a service for specifiers
Obtainable from: National Building Commodity Centre, 26 Store Street, London WC1E 7BT — 01–637 1022

Architects' Standard Catalogues
Standard Catalogue Information Services Ltd, Medway Wharf Road, Tonbridge, Kent TN9 1QR — (0732) 365251

Barbour Index
Barbour Index Ltd, New Lodge, Drift Road, Windsor, Berkshire SL4 4RQ — (034 47) 4121

British Standards
Counter service: BSI, 2 Park Street, London W1 — 01–629 9000
Post: BSI Sales Dept, 101 Pentonville Road, London N1 — 01–837 8801

Agrément Certificates
Agrément Board, PO Box 195, Bucknalls Lane, Garston, Watford, Hertfordshire — (092 73) 70844

RIBA publications
RIBA, 66 Portland Place, London W1N 4AD — 01–580 5533

National Federation of Building Trades Employers publications
NFBTE Publications Department, 82 New Cavendish Street, London W1M 8AD — 01–580 4041

TECHNICAL PRESS

Provides a variety of general and specialist subjects and items of trade news to architects, builders, and those allied to the building trade. With few exceptions most are published by commercial publishers, some by trade associations and others by house journals.

The Architect
4 Addison Bridge Place, London W14 — 01–603 4567

The Architect's Journal
9/13 Queen Anne's Gate, London SW1H 9BY — 01–930 0611

Builder and Decorator
243 Caledonian Road, London N1 1ED — 01–278 6949

Builders Merchants Journal
Sovereign Way, Tonbridge, Kent TR9 1RW — (073 22) 64422

Building
1–3 Pemberton Row, London EC4P 4HL — 01–353 2300

Building Design
30 Calderwood Street, London SE18 6QM — 01–855 7777

Building Maintenance and Service
886 High Road, London N12 9SB — 01–446 2411

Building Specification
43 High Street, Croydon CR0 1YP — 01–686 8564

Building Trades Journal	
Theba House, 49/50 Hatton Garden, London EC1N 8X5	01–278 6571
Chartered Surveyor	
29 Lincoln's Inn Fields, London WC2 3ED	01–242 6451
Construction News	
Elm House, 10 Elm Street, London WC1X 0BP	01–278 2345
Contract Journal	
Surrey House, 1 Throwley Way, Sutton, Surrey SM1 4QQ	01–643 8040
Heating and Ventilating Engineer, and Journal of Air Conditioning	
886 High Road, London N12 9SB	01–446 2411
Heating and Ventilating News	
Faversham House, 111 St. James's Road, Croydon CR9 2TH	01–684 4082
House Builder	
82 New Cavendish Street, London W1M 8AD	01–636 6575
Insulation	
39 High Street, Wheathampstead, Hertfordshire AL4 8DG	(058 283) 2771
Master Builder	
33 John Street, London WC1N 2BB	
National Builder	
82 New Cavendish Street, London W1M 8AD	01–580 4041
Painting and Decorating Journal	
2 Woburn Street, Ampthill, Bedfordshire	(0525) 404328
Plumbing and Heating Equipment News	
Peterson House, Northbank, Berrygill Industrial Estate, Droitwich, Worcestershire WR9 9BL	(090 57) 5564
RIBA Journal	
66 Portland Place, London W1N 4AD	01–580 5533
Surveyor	
Surrey House, 1 Throwley Way, Sutton, Surrey SM1 4QQ	01–643 8040

INDEX

Accidents, 17–2, 17–3
  see also Safety
Account books, 1–4, 1–5, 1–6
Accounts, 1–4
  adjustments to, 1–5
  analysis of receipts and payments, 1–5
  legal necessity for, 1–4
  period and summary of, 1–7
  'set of accounts', 1–7
Accruals accounting, 1–6
'Acrow' telescopic prop, 30–15
Active pressure, 21–8, 21–9
Administration, 1–33
Admixtures, concrete, testing, 27–6
Advance Payments Code, 20–2
Advertising, 4–8
Advisory, Conciliation and Arbitration
    Service (ACAS), 13–4
Advisory services, 64–8
Agents, 3–6
Aggregates, 27–2, 27–4, 32–12
  impurities, 27–4
  storage, 27–5
  testing, 27–6
Agrément certificate, 38–3
Air conditioning, 42–6
Air photographs, 24–4
Air vents for gas services, 40–9
Aircraft, sound insulation, 58–7
Airing cupboard, 16–26, 16–31
Alterations and conversions,
  Bills of Quantities, 59–3, 60–2, 60–10
  bricklaying, 59–7
  carpentry, 59–8
  commercial market, 59–5
  completion of contract, 60–16
  contract planning, 60–2
  credit-checking, 59–5
  daywork, 60–7, 60–8, 60–10
  decorating, 59–10
  demolition, 59–6
  domestic market, 59–4
  drainage, 59–8, 60–15
  dust screens, 59–7
  enquiries, 59–2
  estimating and tendering, 59–2, 60–14
  extra work sheet, 60–7
  foreman role, 60–12
  health and safety, 60–13
  insurance, 59–5, 60–13
  joinery, 59–8
  labour requirements, 60–11
  legislation, 60–15
  local authorities, 60–15
  plant and equipment, 59–10, 60–13
  plastering, 59–7
  plumbing, 59–8
  programming, 60–2
Alterations and conversions (*cont.*)
  progress chart, 60–4
  records, 60–14
  roofs, 59–5, 59–8
  security, 60–13
  shoring, 59–6
  site instruction, 60–7
  site organisation, 60–12
  site overhead costs, 60–11
  specifications, 59–3
  statutory requirements, 60–14
  sub-contractors, 60–11, 60–15
  supporting procedures, 59–6
  valuations of works in progress, 60–10
  variations (extra work), 60–4, 60–5
Angle of repose, 19–3
ANTS Certificates, 41–3
Appropriation account, 1–8
Approval schemes, 12–5
Arbitration, 6–16, 6–17, 8–10, 8–11
Architect's approvals, 6–7
Architect's instructions, 6–5
Articles of Association, 1–40
Asbestos-cement storage cisterns,
    39–13, 39–14
Asbestos sheet, painting, 54–5
Asbestos tiles, 47–14
Asphalt floor finishes, 53–4
Asphalt roofing, 31–13, 47–2
  basic roof types, 47–3
  cold-roof design, 47–9
  design principles, 47–3
  details, 47–11
  falls, 47–7
  fillets, 47–15
  flexible roofs, 47–6
  free-standing kerbs, 47–7
  insulants, 47–9
  'inverted' roof, 47–3
  laying, 47–15
  movement joints, 47–11
  protection of, 47–16
  re-melting, 47–15
  rigid roofs, 47–3
  site practice, 47–15
  solar reflective treatments, 47–13
  specifications, 47–10
  surface finishes, 47–14
  vapour barriers and checks, 47–9
  warm-roof design, 47–9
  see also Mastic asphalt
Atterberg limits, 24–12
Augers, 24–6
Authorised capital, 1–12

Balance sheet, 1–9, 1–10, 1–40
Balconies, 47–10
Ballvalves, 39–8, 39–9, 39–15

Balustrade, 44–18
Bank reconciliation, 1–5
Bank statement, 1–5
Bar bender and cutters, 9–16
Bar chart, 11–4
Basements,
design, 26–32
drainage, 39–26
excavation, 26–33
foundations, 26–30, 26–32
piled, 26–36
Batching plants, 9–7
Bathrooms,
design, 16–27, 17–12
electrical services, 41–13
internal, 16–31
layouts, 16–28
planning, 16–18
Beams, 29–2, 29–4, 29–8
reinforced concrete, 29–16
steel, 29–15
timber, 29–12
Bearing capacity of soils, 23–4, 23–15
Bearing pressures for footings, 23–16
Bending force in beams, 29–5
negative-type, 29–7
positive-type, 29–6
stress due to, 29–8
Bending moment, 29–5, 29–6, 29–7, 29–13, 29–16
Bidet, 16–30, 39–17
Bills of Quantities, 1–14, 1–15, 1–17, 2–14, 6–17, 7–3, 7–4, 7–5, 7–9, 7–11, 59–3, 60–2, 60–10
Bitumen roofing felts, 48–6
Blocks, *see* Concrete blocks
Boilers, 39–20, 42–3
gas, 40–4, 40–7, 40–9
Bolts, 34–15
Bonding, 41–9
Bookkeeping, 'double-entry', 1–7
Boreholes, 24–6, 24–8, 24–11
Borrowing, 1–34
Box construction, *see* Volumetric construction
Breach of contract, 6–16
Breakers, 9–15, 9–16
Breather paper, 37–13
Brick veneer, 37–13
Bricklaying,
alterations and conversions, 59–7
safety, 10–12
Bricks,
calcium silicate, 25–16, 32–9, 32–28, 32–33
British Standard, 32–9, 32–10
compressive strength, 32–10
drying shrinkage, 32–11
limits of size, 32–10
manufacture, 32–9
modular, 32–11
walls, 32–34
Bricks (*cont.*)
clay, 32–2, 32–21, 32–26
British Standard, 32–4, 32–7, 32–8
classification, 32–4
efflorescence, 32–9
extruded or wire cut, 32–3
Internal Quality, 32–21
limits of size, 32–6
manufacture, 32–2
modular, 32–7
moisture expansion, 32–9
Ordinary Quality, 32–5, 32–9, 32–16, 32–21
pressed, 32–3
soft mud, 32–3
Special Quality, 32–5, 32–9, 32–21, 32–26, 32–29
strength, 32–8
walls, 32–34
concrete, 25–16, 32–11, 32–12, 32–28
coping, 32–37
engineering, 25–17, 32–4, 32–5, 32–8
facing, 32–4, 32–10
see also Brickwork; Mortars
Brickwork,
carbonation shrinkage, 32–25
cracking, 32–24, 32–34, 56–4
damp-proof courses, 32–29, 32–34
defects, 55–3
deformation due to loads, 32–23
degrees of exposure, 32–18
dimensional stability, 32–22
durability, 32–17, 32–19
expansion and contraction, 32–33
foundation movement, 32–24
frost action, 32–20
moisture expansion, 32–37
movement, 32–23, 32–24, 32–35
movement joints, 32–33
narrow piers, 32–31
painting, 54–5
problems, 32–16
semi-detached houses, 32–26
structural performance, 32–23
sulphate attack, 32–20, 32–24
terraced housing, 32–37, 32–39
thin plates, 32–23
wall functions, 32–26
British Ready Mixed Concrete Association, 12–5
British Standards,
calcium silicate bricks, 32–9, 32–10
clay bricks, 32–4, 32–7, 32–8
concrete, 27–12
concrete blocks, 32–12
concrete bricks, 32–11, 32–12
dimmer switches, 41–12
electrical services, 41–3
felt roofing, 48–6
floor finishes, 53–2
mastic asphalt, 47–2, 47–10
paint colours, 54–18

British Standards (*cont.*)
plasterboard dry linings, 50–4
plastering, 49–3
reinforcement, 28–4, 28–11, 28–13
rendering, 49–17
timber, 34–6, 48–3
Budgeting, 1–34, 7–7, 7–8
Builders' merchants, 3–5
Building contracts, 1–2, 6–2, 13–5, 30–2, 42–15
alterations and conversions, 60–2
arbitration, 6–16, 6–17, 8–10, 8–11
architect's instructions and approvals, 6–5
breach, 6–16
completion, 6–10, 6–16, 60–16
determination, 6–11
disturbance and damage to works, 6–11
final certificate, 6–16
financial matters and settlement, 6–13
glazing, 45–10
injury to or in connection with works, 6–12
nomination, 6–8
principles, 6–2
programming, 6–9, 11–3
Scottish, 6–18, 62–9
variants, 6–2
work by others, 6–7
Building Regulations, 16–12, 16–15, 17–3, 17–5
concrete floors, 43–15
fire resistances, 35–2, 35–16, 37–15
flat roofs, 48–2
floor finishes, 53–2
floors, 31–2, 31–4, 35–8
flues, 40–10
foundation depths, 25–9
gas services, 40–11, 40–12
insulation, 42–9
non-traditional construction, 38–3
roofs, 31–12, 36–2
sanitary services, 39–24, 39–30
Scotland, 62–2
soil stacks, 16–31
sound insulation, 37–16, 51–11
stairs, 44–17
strip footings, 23–15
thermal insulation, 45–3
timber-framed walls, 37–2
traps, 39–27
ventilation, 45–2
Building Research Establishment, 16–12, 16–31

Cables, electrical, 41–3
California Bearing Ratio (CBR) test, 20–10
Capital allowances, 1–37
Capital Gains Tax, 1–36, 1–37, 1–39
Capital items, 1–5
Capital reserve, 1–12
Capital Transfer Tax, 1–36, 1–37, 1–38
Car parking, 15–9, 16–13
rooftop, 47–10
Carpentry for alterations and conversions, 59–8
Carpets, 53–8
Cartridge hammers, 9–17
Cash Book, 1–4, 1–5
Cash-flow, 1–35, 1–36, 7–7, 7–10
Cavity insulation, 33–16
Cavity ties, 33–10
Cavity walls, 58–3
Cellular glass, 47–9
Cement, 27–2, 27–3
delivery and storage, 27–3
for mortars, 32–15
mixes, 49–5
Portland, *see* Portland cement
sulphate-resisting, 27–3, 32–22, 32–28
testing, 27–6
Cement paints, 54–9
Cement rubber latex floor finishes, 53–5
Chemically active shale, 55–4
Chimney stacks, 32–31
damp-proofing, 33–15
defects, 55–3
Chipboard, 34–11
Cills, damp-proofing, 33–10
Circuit breakers, 41–7, 41–8, 41–14
Claims, 7–10
Clay, *see* Soils
Clay bricks, *see* Bricks
Clay problems, 55–7
Clay tiles, 53–10
Close companies, 1–36
Closed-shop agreements, 13–7
Codes of Practice, 13–4, 16–15
electrical services, 41–3
gas installations, 17–12
screeds, 52–2
see also British Standards
Colour, 54–18, 54–19
Common law, 30–3
Communication, 13–11, 13–12
Community Land Act, 1–2, 1–36
Compaction equipment, 9–6
Compressibility modulus, 23–8
Computer control of maintenance, 61–22, 61–23
Concrete, 27–2
admixtures, 27–5
British Standards, 27–12
buried, 25–15, 25–16
chemical attack, 23–21, 25–14, 25–16, 26–14
colour faults, 30–14
compaction, 27–3
cross-wall, 38–6
curing, 27–3, 27–7
cutting, 9–17
dimensional errors, 30–14
durability, 27–10

Concrete (*cont.*)
formwork, 30–8
fresh, 27–2, 27–6, 27–7
frost protection, 27–7
handling and placing, 9–8, 27–3
hardened, 27–2, 27–7
in situ, 38–6, 47–15
materials, 27–3
mix proportions, 27–11
no-fines, 38–6
precast flue boxes, 40–10
ready-mixed, 12–5
reinforced, 26–15, 28–15, 29–4, 29–16
retarders, 27–5
safety, 27–12
specifications, 27–2, 27–10, 27–11
strength measurement, 27–10
sulphate-resistant, 27–10
testing, 27–2, 27–6, 27–7, 27–10, 31–9
workability, 27–2
workmanship, 27–3
see also Floors; Reinforcements; Roofs
Concrete beam foundations, 25–8
Concrete blocks, 25–16, 32–12, 32–28, 32–35
British Standard, 32–12
drying shrinkage, 32–14
moisture movement, 32–14
walls, 32–34
Concrete bricks, *see* Bricks
Concrete panel construction, 38–5
Concrete pumps, 9–8
Concrete tiles, 47–14, 53–10
Concreting plant, 9–7
Condensation, 31–10, 35–14, 42–11, 42–14, 48–5, 50–10, 54–10, 57–4, 57–8
Conservation organisations, 64–13
Construction costs, summary of, 1–18
Consultation, 13–20
Contracts, *see* Building contracts
Control of Pollution Act 1974, 10–8
Convector heaters, 40–6
Conversions, *see* Alterations and conversions
Cookers, 16–25, 17–9, 41–12
Copings, 32–28, 32–37
Cork floor finishes, 53–8
Corkboard, 47–9
Corporate planning, 4–2
Corporation Tax, 1–36
Cost adjustment sheet, 1–32
Cost books, 1–7
Cost controls, 1–13
Cost increases, 2–19
Cost/value comparisons, 1–22, 1–30, 7–9, 7–11
Costing, cost of, 1–3
Cove, gypsum plasterboard, 50–4
Cracks in structure, 32–25, 56–3 *et seq.*
Cranes, 9–8, 9–11
Credit-checking, 59–5
Creosote, 34–13
Critical Path Networks, 11–4
Cupboard fittings, 44–20
Cupboards, 16–24, 17–11
Current liabilities, 1–12
Cylinders, thermal insulation, 58–4

Damp, sources of, 33–2, 57–2
Damp-proof courses, 32–29, 32–34, 57–3, 57–4
Damp-proof membrane, 33–3, 33–6
Damp-proofing,
chimney stacks, 33–15
cills, 33–10
floors, 33–2
jambs, 33–12
lintols, 33–15
parapets, 33–19
pile foundations, 33–8
raft foundations, 33–7
roofs, 33–16
threshold, 33–10
walls, 32–29, 33–9
Daylight, 16–15, 45–2
Death-watch beetle, 34–12
Decorating, *see* Colour; Painting; Paperhanging
Deep freeze, 16–24
Defective Premises Act 1972, 8–15
Defects,
criteria for, 55–20
design, 55–14
geographical distribution, 55–16
most frequent, 55–10
non-structural, 55–8
rectification, 61–2
responsibility for, 55–14
structural, 55–2
see also under specific types of defect
Delegation, 13–12
Deliquescent salts, 57–6
Demolition for alterations and conversions, 59–6
Department of the Environment, 16–12, 16–16, 17–2
Design guides, 62–8
Design load, 29–4
Determination of contract, 6–11
Development design, 15–2
see also Convenience; Safety
Development Gains Tax, 1–38
Development Land Tax, 1–2, 1–36, 1–38
Deviator stress, 24–15
Dewatering, 9–15
Diaphragm wall construction, 26–34
Diminution principle, 1–38
Direct labour, 1–20
Disabled persons, 17–2
Discipline, 13–16
Dismissal, 13–16, 13–17, 13–18
arising from redundancy, 13–19
arising from sickness or injury, 13–18

Dismissal (*cont.*)
  checklist, 13–20
Distempers, 54–8
Door fittings, 44–20
Door frames, 37–11, 44–13
Door linings, 44–9
Doors, 17–9, 44–4
  design requirements, 17–7
  diminished-stile or gunstock-stile, 44–9
  flush, 44–9
  framed, ledged and braced, 44–7
  glazed, 44–8, 45–3
  ledged and braced, 44–4
  panelled, 44–8
  placing of, 17–7
  stable, 44–7
Drainage, 16–31, 20–14, 20–15, 21–2, 29–21
  alterations and conversions, 59–8, 60–15
  basement, 39–26
  floor, 39–28
  land, 33–9
  roof, 39–29, 47–7, 48–13
  see also Surface water drainage
Drilling, 9–15
Drills, electric, 9–16
Drives, 20–15, 55–13
Dry rot, 34–13, 55–3
Dumpers and dumptrucks, 9–9
Dust screens, 59–7
Dustbins, 59–7
Dwelling design,
  access, 16–13
  advisory information, 16–12
  appearance, 16–33
  basic form, 16–6
  circulation space, 16–17
  common core, 16–7
  constraints, 16–3, 16–12
  flexibility, 16–7
  internal planning, 16–17
  mandatory regulations, 16–12
  market considerations, 16–2
  orientation, aspect and view, 16–15
  repetition, 16–6
  'shell', 16–9
  types and mix, 16–3

Earth-retaining structures, 19–4
Earthing, 41–7, 41–8
Earthworks, 19–2
  temporary, 19–5
Eating zone, 16–25
Electrical Contractors Association, 41–15
Electrical heating, 42–17
Electrical installation, 17–12
Electrical services,
  bathrooms, 41–13
  bonding, 41–9
  British Standards, 41–3
  cables, 41–3

Electrical services (*cont.*)
  circuit-breakers, 41–7, 41–14
  Codes of Practice, 41–3
  Completion Certificate, 41–14
  consumer control units, 41–10
  cooker controls, 41–12
  earthing, 41–7, 41–8
  excess-current protection, 41–6
  fuses, 41–6
  mains, 20–14
  outdoor locations, 41–14
  protective equipment, 41–6
  socket-outlets, 17–11, 41–11
  switches (lighting), 41–12
  switchgear (main), 41–9
  testing and inspection, 41–14
  wiring accessories, 41–11
  Wiring Regulations, 17–12, 41–2, 41–10, 41–11, 41–13, 41–14
  wiring systems, 41–3
Electricity Council, 17–7
Electricity generation, 42–3
Electrochemical series of metals, 39–17
Electrolysis, 39–16
Employers' associations, 64–10
Enquiry form, 3–7
Entrances, design requirements, 17–4
Estate roads, *see* Roads
Estimating, 2–2
  alterations and conversions, 59–2
  analytical, 2–18
  basic information sources, 2–3
  computing costs, 2–8
  feedback, 2–24, 7–14
  net cost, 2–14
  programme, 2–6
  'spot item' method, 60–14
  training, 2–24
Excavations, 10–11, 10–12, 19–4, 30–16
Excavators, 9–5
Expansion joints in concrete, 31–6
Extendible homes, 16–5, 16–12

Falsework, 30–15
Family business, 1–36
Fans, extract, 42–12
Felt roofing, 31–13, 48–2
  abutments, 48–11
  bonding, 48–9
  British Standards, 48–6
  Building Regulations, 48–2
  cold roof, 48–5
  condensation, 48–5
  details, 48–10, 48–11
  drainage, 48–4, 48–13
  maintenance and repair, 48–14
  nailing, 48–9
  NHBC specifications, 48–2
  projections through roof surface, 48–13
  protection during construction, 48–5
  substructure, 48–3
  surface finishes, 48–10

Felt roofing (*cont.*)
warm roof, 48–5
see also Roofs, flat
Fibre building board, 34–11, 34–16
Fibre insulation board, 47–9
Fill materials, collapse of, 23–20
Financial management, 1–2
Financial ratios, 1–12
Fire insurance, 5–7
Fire resistance, 17–3
glazing, 45–5, 45–6
paints, 54–10
plasterboard dry linings, 37–15, 50–7
plasterboard lightweight partitions, 51–9
roofs, 31–12, 35–16
separating-wall systems, 51–9
steel-frame construction, 38–9, 38–10
timber floors, 35–2
Fireguard, 17–7
First aid, 10–14
Flame-retardant treatments, 34–14
Floor areas, minimum, 16–13
Floor drainage, 39–28
Floor finishes, 17–7, 53–2
asphalt, 53–4
British Standards, 53–2
Building Regulations, 53–2
cement rubber latex, 53–5
Codes of Practice, 53–3
hard tiles, 53–9
in situ, 53–4
magnesium oxychloride, 53–4
permissible tolerances, 53–3
pitchmastic, 53–4
plastic, 53–5, 53–8
preformed, 53–5
sub-floor requirements, 53–3
thin tiles and sheets, 53–7
timber, 53–5, 53–7
Floor heave, 55–5
Floor screeds, *see* Screeds
Floor slabs, 31–5, 31–6, 55–5
Floor tiles, *see* Floor finishes
Floors,
Building Regulations, 31–2, 31–4, 31–9, 35–8, 43–15
Codes of Practice, 31–9
concrete, 31–2, 33–5, 58–4
damp-proofing, 33–2
defects, 55–7
economical design, 35–4
floating, 43–15, 43–21
hollow vented, 33–4
lightly loaded, 31–2
medium and heavily loaded, 31–5
precast prestressed, 31–7
reinforcement, 31–5
sound insulation, 35–8, 58–6
structural panels, 35–6
suspended, 31–7, 35–7
thermal insulation, 31–5, 43–15, 58–4
timber, 33–4, 35–2, 58–4, 58–6
Floors (*cont.*)
U-value, 31–5
uniformly supported on ground, 31–2
with exposed beams, 35–6
Flues, 40–10, 42–13
Food preparation, 16–24
Footings,
bearing pressure on, 23–4, 23–16
construction of, 23–17
design charts, 23–13
failure of soil beneath, 23–2
settlement of, 23–6, 23–15
stress-settlement curve, 23–3
strip, 23–15
Footpaths, 1–15, 20–15
Footwear, 10–10
Foreman role, 60–12
Forklift trucks, 9–8, 9–10
Formwork, 30–8
basic materials, 30–9
dimensionally coordinated, 38–6
proprietary equipment, 30–9
rectangular-column, 30–11
small-beam, 30–11
stripping, 30–14
surface finish, 30–8
Foundation stress, 23–2
Foundations,
allowable bearing pressures, 25–10
basement, 26–30, 26–32, 26–36
boring depths, 24–8
carrying capacity, 26–23
chemical attack, 25–14
concrete beam, 25–8
continuous beam, 26–29
depth requirements, 25–9, 25–12
materials, 25–14
movement, 32–24, 32–33
multi-storey buildings, 26–28
narrow strip, 25–6, 25–12
pad, 26–29
pile caps, 26–25
piled, 26–16
basements, 26–36
continuous, 26–34
damp-proofing, 33–8
durability, 26–24
end-bearing, 26–16
friction, 26–16
ground beam, 26–28
groups, 26–23
H-section soldier, 26–33
negative skin friction, 26–23
precast concrete, 26–16
secant, 26–34
small-displacement, 26–20
steel, 26–20, 26–25
testing, 26–25
timber, 26–24
raft, 26–2
angular distortion, 26–12
cellular, 26–7

Foundations (*cont.*)
raft (*cont.*)
damp-proofing, 33–7
on sloping ground, 26–14
plain slab, 26–2
slab and beam, 26–6, 26–12
stiffened-edge, 26–4
structural design, 26–11
relative rotation, 26–12
seasonal swell and shrink, 23–17
shrinkable clays, 25–11
strapped, 26–30
strip, 25–2, 25–16
minimum width, 23–14
on made ground, 25–13
timber sheeting, 26–33
underpinning, 30–18
Frost heave, 23–17
Fuels, 42–2, 42–3
storage, 16–32
Fund raising, 1–34
Furniture beetle, 34–12
Furniture sizes, 16–16, 16–17
Fuses, 41–6

Garage, 16–7
Gas fires and convectors, 40–6, 42–3
Gas mains, 20–14, 40–3
Gas service pipe, 40–3, 40–8
Gas services,
air vents for, 40–9
builders' work, 40–8
Building Regulations, 40–11, 40–12
central heating, 40–3, 40–4
flueless appliances, 40–9
flues, 40–10
future developments, 42–17
installation, 40–8
kitchen appliances, 40–7
meter boxes, 17–7, 40–8
pipework, 40–9
room sealed appliances, 40–11
safety regulations, 17–12, 40–12
space heaters, 40–6
terminals, 40–11
ventilation requirements, 40–9
warm-air heating, 40–5
water-heating, 40–7
Generators, 9–17
Geological maps and guides, 19–3, 24–3
Glass-reinforced plastic, 30–9
storage cisterns, 39–14
Glazing, 45–2
bedding, 45–9
contracts, 45–10
double, 45–3
fire resistance, 45–5, 45–6
handling and storage, 45–10
materials, 45–8
safety requirements, 45–5
structural surround, 45–8
thermal insulation, 45–3
Glazing (*cont.*)
types of glass, 45–8, 45–9
weather resistance, 45–3
work on site, 45–10, 45–12
see also Doors; Windows
Glues, 34–15, 34–16
Goodwill, 1–12, 1–39, 1–40
Government departments and organisations, 64–11
Graders, 9–6
Grevak Monitor anti-siphon trap, 39–27
Ground-pressure calculation, 21–9
Ground strength assessment, 20–10
Groundwater, 29–21
aggressive substances in, 26–14
sulphate-bearing, 25–17
Groundwater levels, 24–11, 24–12
Gutters, 39–29
Gyproc metal furring, 50–13
Gyproc partitions, 51–3, 51–4
Gypsum action, defects due to, 55–2
Gypsum fireline board, 50–3
Gypsum plank, 50–3, 50–6
Gypsum plasterboard cove, 50–4
Gypsum plasters, 49–6
Gypsum thermalboard, 50–4
Gypsum vapour-check wallboard, 50–10
Gypsum wallboard, 50–3, 50–5
Gypsum wallboard/expanded polystyrene laminate, 50–4, 50–7, 50–13
Gypsum wallboard panel, prefabricated, 50–4, 50–6

Hand vane tester, 24–10
Handrail, 17–4, 17–5, 44–17
Hardboards, painting, 54–4
Hardwoods, *see* Timber
Haulage equipment, 9–9
Hazardous ground defects, 55–3, 55–5
Health and Safety at Work etc. Act 1974, 5–7, 9–4, 13–5, 13–9, 17–3, 30–3, 46–2, 60–13, 60–14
Heat converters, 42–3
Heat emitters, 40–5, 42–3
Heat losses, 43–8
Heat pumps, 42–6, 42–15
Heat transfer, 43–2
Heating services, 17–11
central, 40–3, 40–4, 42–4, 42–6
comfort criteria, 42–2
controls, 40–7, 42–8
design, 42–14
electric, 42–5, 42–17
electric instantaneous, 42–7
electric underfloor, 33–8
future developments, 42–15
gas, 40–3, 40–4, 40–6, 42–5, 42–17
kitchen, 16–24
off-peak electric, 42–7
oil-fired, 42–5
warm-air, 40–5, 42–5

Heave, clay, 56–5
Highways Act 1959, 20–2
Hoists, 9–14
Home extension, 16–5
Housing (Scotland) Act 1974, 62–5
Humidity measurements, 57–8
Hydrants, 39–5
Hydrostatic pressure, 29–21

IEE Assessment of New Techniques Scheme, 41–3
IEE Completion Certificate, 41–14
IEE Wiring Regulations, 17–12, 41–2, 41–10, 41–11, 41–13, 41–14
Incentive schemes, 7–6, 13–15, 13–16
Industrial and Commercial Finance Corporation (ICFC), 1–34
Industrial relations, 13–2
Industrialised building, 38–2
Infill defects, 55–2, 55–4
Inflation, 2–19
Information services, 64–6
Inland Revenue, 1–39, 1–40
Insolvency, 6–11
Inspection procedure, 12–6, 55–15, 60–16
Institution of Civil Engineers, 20–15
Insulating boards, painting, 54–5
Insulation, *see* Sound insulation; Thermal insulation
Insurance, 5–2, 6–13, 6–18, 59–5, 60–13
  'Contractors All Risks', 5–3, 5–5
  employer's-liability, 5–6, 6–12
  engineering, 5–7
  fire, 5–7
  of money, 5–7
  personal accident and/or sickness, 5–7
  professional indemnity, 5–7
  public liability, 5–5
  third-party, 6–12
  work in progress, 5–3
Interim payments, 6–9, 6–16, 7–10
Ironmongery, 44–20

Jambs, 37–13
  damp-proofing, 33–12
Joinery,
  alterations and conversions, 59–8
  grading, 34–7
  mass-produced, 44–2
  ordering and storage, 44–3
  preservation, 44–4
  purpose-made, 44–2
  shrinkage and movement, 44–4
  trim and mouldings, 44–19
  see also Doors; Stairs; Timber; Windows
Joint Contracts Tribunal, 6–2, 6–3, 6–17, 6–18

k values, conversion factors, 43–6
Kitchen, 17–8
  design, 16–19, 17–8
Kitchen (*cont.*)
  fitments and equipment, 16–24
  food preparation, 16–21, 16–24
  future design trends, 16–27
  heating, 16–24
  lighting, 16–22
  services, 16–22, 16–25
Kitchen appliances, gas-fired, 40–7
Kitchen cupboards, 44–20

Labour,
  annual worked hours, 2–10
  built-up rates, 2–8
  'standard hourly base rate', 2–12
  prime cost of daywork, 2–9
Labour allocation, 1–21, 7–6
Labour overhead checklist, 2–9
Labour variances schedules, 1–21
Laminated plasterboard partition, 51–3
Land costs per unit, 1–15
Land drainage, 33–9
Land Stock Account, 1–15
Landscaping, 15–10
Laundering, 16–20, 16–26, 17–11
Lay-offs, 13–18
Legislation, 13–4, 13–22, 30–3
  alterations and conversions, 60–15
  Scotland, 62–2
  sound insulation, 51–11
Lifting gear, 9–11, 30–19
Lighting, 17–11
  kitchen, 16–22
  stairways, 17–5
  switches, 17–7
  working surfaces, 16–22
Lime, lime mixes, 32–15, 49–4
Limited liability company, 1–33
Linoleum, 53–8
Lintols, 37–6
  damp-proofing, 33–15
  defects, 55–3
Loading, structural, 29–2, 29–4
Local authorities, 60–15
London Building (Constructional) By-Laws 1972, 51–13
Long-term liabilities, 1–12
Lorries, 9–10

Magnesium oxychloride floor finishes, 53–4
Maintenance,
  client's requirements, 61–16
  computer control, 61–22
  control meetings, 61–19
  cost effectiveness, 61–21
  daily planning, 61–18
  financial management, 61–7, 61–10
  incentives, 61–14
  job tickets, 61–18
  labour costs, 61–21
  labour management, 61–11
  labour requirements, 61–8, 61–13
  management by objectives, 61–12

Maintenance (*cont.*)
- materials, 61–7
- performance control, 61–21
- plant, 9–3
- productivity, 61–14
- programmes, 61–19
- services, 61–7
- standards, 61–2
- transport cost, 61–21
- work planning and control, 61–16
- workload, 61–9
- *see also* Defects

Majority holding, 1–38
Management, function and role of, 1–2, 1–3
Manholes, 21–7, 22–8
Marketing, 4–2
Mastic asphalt,
- British Standards, 47–2, 47–10
- falls, 47–8
- thickness, 47–11
- *see also* Asphalt roofing

Materials,
- checking, 10–7
- costs, 2–9
- defects, 55–14
- quality control, 12–4
- scheduling, 7–5

Materials-received sheet, 1–28
Materials take-off schedules, 1–16, 1–18
Materials variance analysis, 1–18
Mechanical plant, 2–13
Medicine cabinets, 17–12
Merchant banks, 1–34
Mess hut, siting of, 10–5
Metals, painting, 54–5, 54–6
Meters, 16–14, 17–7, 40–8
- water consumption, 39–9

Method study, 2–7
Metric system, 63–2
Mines, abandoned, 24–3
Mining subsidence, 23–20, 26–9
Minor building works, 6–17
Minority holding, 1–38
Mixers, 9–7
Mixing-points, 10–8
Moisture barrier, 37–5
Moisture, residual construction, 57–4
Monorail, 9–11
Monthly valuations, 1–22
Mortars, 32–14
- 'balling', 32–16
- defective, 55–3
- materials, 32–15
- mixes, 32–16
- proportioning, 32–16
- roof tiling, 46–7
- selection, 32–16
- sulphate attack, 32–21, 32–24

Motivation, 13–2, 13–12
Mouldings, 44–19
Multi-storey buildings, 26–28, 29–3

Nails, 34–14, 37–4
National Building Agency, 16–5
National Federation of Building Trades Employers, 6–18, 7–7, 13–5
National House-Building Council, 6–2, 8–2, 16–17, 16–27, 16–31, 48–2
- *see also* NHBC scheme

National Inspection Council for Electrical Installation Contracting (NICEIC), 41–15
National Working Agreement, 13–18
Needling, 30–18
Net cost estimates, 2–14
NHBC scheme, 8–2, 55–2, 55–8
- application, 8–4
- arbitration, 8–10
- builders' obligations to purchasers, 8–7
- certification, 8–6, 8–8
- changes in law affecting builders' liabilities, 8–15
- conditions, 8–4
- discipline, 8–5
- future development, 8–14
- inspection, 8–6
- major structural defect liability, 8–12
- register, 8–4
- Scotland, 62–8
- successors in title, 8–9
- technical requirements, 8–6
- Top Up Cover, 8–13, 8–14
- two-year warranty, 8–8
- undertakings to purchasers, 8–11

Noise problems, 16–31, 43–18, 43–19
- *see also* Sound insulation

Nominal ledger, 1–7
Nomination principle, 6–8
Non-traditional construction, 38–2

Occupier's Liability Act 1957, 13–9
Oedometer test, 24–15
Oil fuels, 42–3
Old people, 17–2
Open fire, 17–7
Order forms, 3–7
Order schedules, 1–16
Ordinary Shares, 1–12
Ordnance Survey, 18–2, 18–7
Overheads, 1–5, 1–34, 1–36
- allocation and recovery, 7–7
- company, 2–18
- site, 1–15, 1–25, 2–13

Painting,
- alterations and conversions, 59–10
- anti-condensation paints, 54–10
- asbestos sheeting, 54–5
- boat varnishes, 54–9
- brickwork, 54–5
- brush application, 54–10
- cement paints, 54–9
- chlorinated rubber paints, 54–10

Painting (*cont.*)
choice of paint, 54–8
clear wood finishing, 54–9
colour choice, 54–18, 54–19
defects, 54–14, 55–10, 55–13
emulsion paints, 54–8
epoxy paints, 54–10
filling, 54–7
flame-retardant paints, 54–10
hardboards, 54–4
insulating boards, 54–5
lead paints, 54–3
maintenance, 54–11, 54–13
masonry paints, 54–8
metals, 54–5, 54–6, 54–13
multicolour paints, 54–10
plasters, 54–5
polyurethanes, 54–9
preparation, 54–2
priming, 54–3, 54–4
rendering, 54–5, 54–13
roller application, 54–10
spray application, 54–10
steelwork, 54–13
texture relief paints, 54–9
timber, 54–2, 54–3, 54–13
varnishes, 54–9
water-thinned materials, 54–8
Paperhanging, 54–14
colour scheming, 54–19
measuring for, 54–18
Paramount panels, 50–4
Parapets, damp-proofing, 33–19
Pareto's law, 3–6
Parquet floors, 53–7
Particle board, 34–11, 34–16
Partitions, internal, 16–9
sound insulation, 58–5
supports, 35–7
timber-framed, 37–2
see also Plasterboard lightweight partitions; Separating-wall systems
Partnership, 1–34
Paths, 17–4, 20–15, 20–16
Pavings, 47–14
Payments, analysis of, 1–5
Personnel selection and engagement, 13–5
Perth penetrometer, 24–10
Petty Cash Book, 1–4
Phased completion, 6–10
Photography, 14–5
aerial surveys, 18–6
Piled foundations, *see* Foundations
Pipe laying, drainage, 21–10, 22–7
Pipes, flow capacity of, 21–3, 22–5, 22–6
Pipework, gas services, 40–9
Pitchmastic floor finishes, 53–4
Planning and building control, 11–6, 14–3, 16–33
Planning organisations, 64–12
Plant, 2–13, 9–2
allocation, 7–5
electric power supply, 9–17
hire, 9–3
safety, 9–4
Plasterboard dry linings, 50–2
British Standards, 50–4
condensation, 50–10
cutting, 50–13
decoration, 50–17
design and planning, 50–11
fire protection, 50–7
fixing methods, 50–13, 50–17
handling, 50–12
health and safety, 50–19
jointing, 50–3, 50–16
limitations, 50–2, 51–3
performance, 50–7
site organisation, 50–12
sound insulation, 50–7
storage, 50–12
temperature effects, 50–10
thermal insulation, 50–7
tools and accessories, 50–4
U-values, 50–10
Plasterboard lightweight partitions, 51–2
dimensions and weights, 51–9
fire protection, 51–9
fixing methods, 51–15
health and safety, 51–15
performance, 51–9
site work, 51–13
sound insulation, 51–11
tools and accessories, 51–9
Plastering, 49–2
alterations and conversions, 59–7
backgrounds, 49–7, 49–8
British Standards, 49–3
cement mixes, 49–5
defects, 55–14
design and specification, 49–3
efflorescence of background, 49–9
gypsum plasters, 49–6
lime mixes, 49–4
painting, 54–5
thicknesses, 49–4
thin-coat, 49–7
veneer-coat, 49–7
Plastic floor finishes, 53–5, 53–8
Plasticity index, 24–12
Play spaces, 17–4, 17–10
Plumbing, *see* Water services
Plywoods, 34–8, 34–16
decking, 53–7
sheathing, 37–4
Pocket penetrometer, 24–9
Polythene membrane, 25–17
Polythene storage cisterns, 39–14
Portland cement, 27–2, 54–9
ordinary, 27–3, 31–4, 31–5
rapid-hardening, 27–3
sulphate-resisting, 27–3, 31–4

Practical completion, 6–10, 6–16
Prefabrication, 38–2, 38–10
Preference Shares, 1–12, 1–40
Price earnings (P/E) ratio, 1–13
Pricing, types of, 4–7
Prime cost contracts, 6–18
Priming, *see* Painting
Privacy, 15–8, 15–9, 16–13
Private development, 1–2, 11–3
Production control, 11–6
Production planning, 11–2
Products, trade associations, 64–2
Profit, 2–19, 4–3, 12–2
Profit and Loss Account, 1–8
Profit statement, 1–29, 1–33
Psychrometer, 57–8
Public utility services, 20–14
Publications, 64–13
Publicity, 4–8
Pumps, 9–14, 9–15
Purchase order form, 1–29
Purchases Journal, 1–6
Purchases Ledger, 1–6
Purchasing, 3–2
  delivery conditions, 3–5
  source selection, 3–6
Purlins, 36–3, 36–5

Quality assurance, 3–3, 12–2
Quality control, 12–2
  authorisation schemes, 12–5
  components, 12–5
  correction procedures, 12–6
  costs, 12–7
  feedback, 12–6
  materials, 12–4
  sampling, 12–5
  training, 12–7
  workmanship, 12–6
Quantity surveying,
  documents, 7–3
  functions, 7–2, 7–5
  training, 7–14
Quantity Variance Schedule, 1–18
Quarry tiles, 53–10

Radiators, 40–5, 42–3
Raft foundations, *see* Foundations
Railway, 9–11
Rain penetration, 57–2, 61–2
Rainfall intensity, 22–2, 22–5
Rainwater pipes, 33–17, 39–29
Reaction force, 29–5, 29–6, 29–13, 29–16
Receipts, analysis of, 1–5
Redundancy, 13–19, 13–20
Reflection factors, 16–15
Refrigerator, 16–24
Refuse storage and collection, 16–32
Reinforcement, 28–2
  bar, 28–8, 28–16
  bending tests, 28–6
  bond characteristics, 28–6
Reinforcement (*cont.*)
  British Standards, 28–4
  chairs, 28–19
  characteristic strength, 28–4
  detailing, 28–15
  ductility, 28–6
  elongation, 28–6
  fabric, 28–13, 28–16
  fibre, 31–7
  fixing, 28–17
  floor slabs, 31–5
  labelling, 28–18
  location, 28–16, 29–18
  ordering, 28–17
  specification, 28–2
  tying, 28–19
  ultimate strength and margin, 28–4
  welding, 28–8, 28–19
Rendering, 49–9
  British Standards, 49–17
  mixes and materials, 49–15
  painting, 54–5, 54–13
Research organisations, 64–8
Risk assessment, 2–19
Roads, 1–15, 20–2
  alignment, 20–4, 20–7
  design, 15–2, 15–6, 20–4, 20–9, 62–8
  flexible construction, 20–11
  ground conditions, 20–10
  lateral fall, 20–9
  rigid construction, 20–13
  sub-base thickness, 20–11, 20–13
  traffic estimation, 20–10
  width, 20–9
Roof screeds, *see* Screeds
Roof slabs, waterproofing, 31–13
Roof spaces, 16–7, 16–11, 17–6
Roof structure, 46–3
Roof tiling, 46–2
  abutments, 46–11
  battening, 46–5, 46–8
  double-lap, 46–6, 46–7, 46–10, 46–12
  felting, 46–3, 46–8
  fixing, 46–6
  flashing, 46–11
  hips, 46–10
  mortar bedding, 46–7
  ridges, 46–11
  setting out, 46–4
  single-lap, 46–6, 46–7, 46–10
  tools, 46–2
  valley, 46–8, 46–9
  verge, 46–7
Roof trusses, 36–5, 36–9
Rooflights, 48–13
Roofs, flat,
  alterations and conversions, 59–9
  Building Regulations, 31–9, 31–12, 35–16
  Codes of Practice, 31–9
  cold, 35–14, 55–7
  concrete, 31–9

Roofs, flat (*cont.*)
condensation, 31–10, 35–14
defects, 55–7, 61–2
drainage, 39–29
insulation, 31–10, 35–14, 35–16, 43–10, 58–2, 58–3
inverted, 31–12, 35–16
timber, 35–12
ventilation, 35 14
warm, 35–14, 55–7
waterproofing, 31–10, 31–13, 33–16, 33–19
see also Asphalt roofing; Felt roofing
Roofs, pitched,
alterations and conversions, 59–9
Building Regulations, 36–2
common rafters, 36–2
defects, 55–7
drainage, 39–29
insulation, 36–9, 43–10, 58–2
trusses, 36–5, 36–7
ventilation, 36–9
waterproofing, 33–16
Room sizes, 16–17
Rubber floor finishes, 53–8

Safety committees, 13–20
Safety factor, structural, 23–2, 29–4, 29–16
Safety in the home, 17–2
gas services, 40–12
glazing, 45–5
Safety on site,
attitudes, 10–9
checklist, 10–11, 10–14
plant, 9–4
prefabricated units, 38–3
training, 10–10
see also Health and Safety at Work etc. Act
Sales Journal, 1–6
Sales Ledger, 1–6
Sales promotion, 4–8
Sales turnover, 4–3
Sands for mortars, 32–15
Sandwich construction, 38–5
Sanitary services, 39–21
alterations and conversions, 59–8
Building Regulations, 39–24, 39–30
by-laws, 39–2, 39–24
design considerations, 39–24
distribution pipework, 39–5
fittings, 16–27
flushing gullies, 39–28
grease traps, 39–28
one-pipe system, 39–21
rainwater pipes, 39–29
rodding access, 39–26
single-stack system, 39–22
testing, 39–30
traps, 39–22, 39–27, 39–30
two-pipe system, 39–21
Sanitary services (*cont.*)
ventilating pipe, 39–26
waste-disposal units, 39–31
water pressure, 39–11
WC positioning, 43–22
Sap stain, 34–13
Saws, 9–16
Scaffolding, 30–3
fittings, 30–8
loads on platforms, 30–5
regulations, 30–8
safety, 10–12, 30–3
sub-contracting, 30–8
Scotland,
building legislation, 62–2
Building Regulations, 62–2
contracts, 6–18, 62–9
costs, 62–8
mandatory controls, 62–2
NHBC, 62–8
traditional practice, 62–8
Scrapers, 9–6
Screeds, 52–2
bonding methods, 52–4
Code of Practice, 52–2
defects, 52–4, 52–7
floor, 31–6, 52–7
on precast floor units, 52–6
on thermal insulation, 52–6, 52–8
roof, 31–10, 31–11, 52–7, 57–4
thickness, 52–3
Screws, 34–15
Security on site, 2–4, 10–8, 10–14
Selling price of dwellings, 1–13
Separating-wall systems, 51–2, 51–4
dimensions and weights, 51–9
fire protection, 51–9
sound insulation, 51–11
Services,
bathroom, 16–27
kitchen, 16–22, 16–25
laundering, 16–26
mains, 14–4, 20–14
safety, 17–7
Serving area, 16–25
Setting out, 18–2, 18–8
Settlement cracks, 32–25
Settlement, soil, 23–5, 23–15, 26–11, 55–5
Sewage outfalls, 21–12
Sewage treatment, 21–12
Sewerage (Scotland) Act 1968, 62–7
Sewerage systems, 21–2
Code of Practice, 21–3
costs, 1–15
flow calculation, 21–2
ground pressure calculation, 21–8, 21–9
house connections, 21–11, 21–12
joints, 21–7, 21–10
layout, 20–15, 21–4
materials, 21–6
pipe laying, 21–10

Sewerage systems (*cont.*)
setting out, 21–7
testing, 21–11
see also Surface water drainage
Shale, chemically active, 55–4
Share transfers, 1–40
Share valuation, 1–39
Shaver outlets, 41–13
Shearing force in beams, 29–5, 29–9, 29–10
Shoring, 30–17
alterations and conversions, 59–6
Shower units, 16–30, 41–13, 42–7
Shrinkage gap between bath and wall, 55–13
Shrinkage of timber, 55–20
Shuttering panels, 30–11
SI units, 63–2
Sickness or injury, dismissal arising from, 13–18
Site accommodation, 10–3
Site investigation, 14–2, 18–2, 24–2
access and circulation, 14–4, 15–6
checklist, 2–3, 14–4
earthworks, 19–2
features, 14–4
historic data, 24–4
in structural failure diagnosis, 56–2
laboratory tests, 24–12
legal and statutory aspects, 14–5
maps and guides, 24–3
neighbouring properties, 14–4
photography, 14–5, 24–2
planning considerations, 14–3
preliminary investigation, 24–2
services, 14–4
site visit, 24–4
subsurface investigation, 24–5
underground hazards, 14–4
Site meetings, 10–7
Site office, 10–6
Site organisation, 10–2
alterations and conversions, 60–12
bulk material location, 10–8
compound, 10–8
incoming materials, 10–7
production control, 11–6
roads, 10–3
safety, 10–9, 10–10, 10–14
security, 10–8, 10–14
vehicles entering and leaving, 10–13
Site selection, 14–2
Site work costs, 1–16
Skirtings, 44–20
Slope instability, 19–3, 23–21
Slump test, 27–2, 27–7
Soakaways, 22–9
Social classes of house purchasers, 55–19
Socket-outlets, 17–11, 41–11
Soft ground defects, 55–4
Softwoods, *see* Timber
Soil mechanics, 19–3, 23–2
Soil pipes, sound insulation, 58–8
Soil stacks, 16–31
Soils, 23–2
acidic, 25–14, 26–14
bearing capacity, 23–2, 23–4, 23–15
chalk, 23–17
chemical tests, 24–12
clay, 23–4, 23–10, 23–16, 23–18, 24–9
cohesive, 23–2, 23–4, 23–10, 26–23
compressibility, 23–8, 24–15
description method, 24–9
frost action, 23–17
granular, 23–5, 23–11, 23–17, 24–10
hillside creep, 23–21
non-cohesive, 23–2, 29–19
particle diameters, 24–11
plasticity tests, 24–12
relative density, 24–10, 24–11
settlement, 23–5, 23–15, 26–11, 55–5
sheer strength, undrained, 24–10, 24–12, 24–15
shrinkable clay, 23–17, 23–18, 25–11
softening, 23–17
stiffness, 23–8
sulphate-bearing, 23–21, 24–12, 25–14, 25–16, 26–14
swelling, 23–17, 23–18, 25–12
variability problems, 23–16
Solar energy, 42–15, 42–17
Solar gain, 42–11, 42–17
Solar reflective treatments, 47–13
Sole traders, 1–34
Solid fuels, 42–3
Sound insulation, 38–7, 43–17
aircraft, 58–7
Building Regulations, 37–16, 51–11
detached houses, 43–21
flats, maisonettes, semi-detached houses and terraces, 43–19
floors, 35–8, 58–6
improvements, 58–5
legislation, 51–11
plasterboard dry linings, 50–7
plasterboard partitions, 51–11
problems, 43–22
roofs, 35–16
separating-wall systems, 51–11
soil pipes, 58–8
timber-framed walls, 37–15
WC fittings, 58–8
windows, 45–3, 58–7
Sound levels, reducing, 58–8
Sound transmission, 43–18, 43–19
Spatchel fillers, 54–7
Speculative developments, 5–3, 5–4
Staggered-stud partition, 37–16
Stairs, 44–17
Building Regulations, 44–17
design requirements, 17–5
on-site problems, 44–18
riser and going formula, 44–19
sloping soffit shutter, 30–11
Stairwell openings, trimming, 35–6

Standard Method of Measurement (SMM), 7–3, 7–4
Standard Penetration Test, 24–11
Standpipes, 39–5
Starter homes, 16–6, 16–12
Steel-frame construction, 38–7
Steel piles, 26–20, 26–25
Steelwork, preparing and repainting, 54–13
Sterilisation of water-supply installations, 39–18
Stock relief, 1–36, 1–37
Storage cisterns, 39–11 *et seq.*
  overflow and/or warning pipe, 39–16
  sterilisation, 39–18
  wash-out pipe, 39–15
Storage cupboards, 16–24
Storage, design requirements, 16–31
Storage units, 17–11
Stormwater overflows, 22–8
Stress,
  bending, 29–8
  in beam sections, 29–8
  permissible, 29–4
  shear, 29–9, 29–10
Stress distribution beneath footings, 23–5, 23–9
Stress-distribution diagram, 29–17
Stress grading, timber, 34–6
Structural defects, 55–2
  due to trees, 55–7
Structural distortion, 55–20
Structural elements,
  design, 29–3
  handling, 30–18
Structural failure, diagnosis, 56–2
Structures, theory and design, 29–2
Sub-contractors, 3–2, 5–4, 6–8, 6–9, 7–7, 13–4, 30–3, 30–8, 42–15, 60–11, 60–15
  analysis form, 1–31
  costs, 1–22
  ledger card, 1–23
  nominated, 7–7, 13–4
  payment certificate, 1–24
  selection and negotiation, 7–7
Sub-contractors Journal, 1–6
Sub-contractors Ledger, 1–6, 1–23
Subsidence, 55–4
  mining, 23–20, 26–9
Substructure defects, 55–3
Subsurface investigation, 19–3, 24–5
Sulphate attack,
  on concrete, 25–15, 26–15, 56–9
  on mortars, 32–20, 32–24, 56–9
Sulphate content of soil, 24–13
Sulphate-resisting cement, 27–3, 32–22, 32–28
Sulphate-resisting concrete, 27–10
Sunlight, 16–15
Superstructure defects, 55–3, 55–7
Supervision, 13–13
Suppliers, 3–6, 3–9
Surface water drainage, 21–2, 22–2
  flow calculations, 22–2
  permeability and area, 22–4
  pipe laying and testing, 22–7
Surface water outfalls, 22–7
Survey stations, 18–4
Surveying, 18–2
  aerial, 18–6
  levelling method, 18–7
  recording and plotting, 18–7
Swallow holes, 23–21
Switches (lighting), 41–12
Switchgear (main), 41–9
System building, 38–2

Tax Deduction Schemes, 1–22
Taxation, 1–36
Telephones, 20–14
Telescopic props and centers, 30–15
Temporary works, 30–2
Tendering, 1–2, 2–2, 2–18, 6–17
  alterations and conversions, 59–2
  finalising, 2–22
  plant, 9–2
  pre-tender planning, 2–6
  programme, 2–6
  risk sources, 2–22
  specimen summary sheet, 2–23
Terrazzo tiles, 53–10
Thefts, 10–15, 10–16
Theodolite, 18–5, 18–6, 18–8
Thermal conductivity, 43–3, 43–6, 43–8
Thermal insulation, 43–2, 57–3
  Building Regulations, 45–3
  cylinders, 58–4
  economics, 43–16
  flat roofs, 31–10, 35–14, 43–10, 58–2, 58–3
  floors, 31–5, 43–15, 58–4
  glazing, 45–3
  improvements, 58–2
  k values, 43–6
  materials, 43–7
  pipes, 58–4
  pitched roofs, 36–9, 43–10, 58–2
  plasterboard dry linings, 50–7
  problems, 43–16
  screeds, 52–6, 52–8
  U-values, 43–4, 43–11, 43–15, 50–10
  walls, 33–10, 37–14, 43–12, 58–3
  windows, 43–14
Thermal resistance, 43–4, 43–5
Thermal resistivity, 43–4
Thermal roofs, 43–10
Thermohydrograph, 57–8
Thermostats, 42–8
Thistlebond method of dry lining, 50–13
Threshold, damp-proofing, 33–10
Tile hanging, 37–13
Tiling, *see* Roof tiling
Timber, 34–2
  Canadian, 34–7

Timber (*cont.*)
costing, 34–5
distortion, 55–20
dryness, 34–4
fixings, 34–14
fungal attack, 34–12
grading, 34–5
hardwoods, 34–8, 44–2, 53–6
insect attack, 34–12
joinery, 44–2
joinery grading, 34–7
lengths, 34–4
manufactured boards, 34–11
moisture content, 34–4, 55–21
moisture movement, 34–16
non-joinery uses, 34–7
painting, 54–2, 54–3, 54–13
preservation, 34–13, 37–5, 48–3
priming, 54–3
resawing, 34–4
sheet materials, 34–9
shrinkage, 55–20
sizes, 34–2
softwoods, 34–2, 44–2, 53–6
storage, 34–17, 35–4
stress grading, 34–6
see also Plywood
Timber-based floor finishes, 53–7
Timber claddings, 37–13
Timber floor finishes, 53–5
Timber floors, 33–4, 35–2, 58–4
Timber-frame construction, 38–6, 38–7
Timber-framed walls and partitions, 37–2, 58–4
base, 37–6
erection sequence, 37–9
exterior cladding, 37–13
fabrication methods, 37–5
inspection of frame, 37–14
interior finishes, 37–15
loadbearing, 37–2
materials, 37–4
services and fittings, 37–14
sound insulation, 37–15
stud spacing, 37–7
studs and lintols, 37–6
thermal insulation, 37–14
Tippers, 9–10
Tolerances for internal finishes, 55–22, 55–23
Tooth-plate connector units, 34–15
Top Up Cover, 8–13, 8–14
Town and Country Planning (General Development) (Scotland) Order 1975, 62–6
Town and Country Planning (Scotland) Act 1972, 62–6
Tractor dozers, 9–5
Tractors, 9–11
Trade associations, 12–5
products, 64–2
services, 64–5
Trade press, 64–14
Trade unions, 13–7, 64–11
Trading account, 1–7, 1–8
Trailers, 9–11
Transport and Road Research Laboratory, 20–9
Traps, 39–27, 39–30
Tree problems, 23–19, 24–12, 25–12, 55–7, 56–5
Trenchers, 9–5
Trenches and trench supports, 30–16
Trial balance, 1–7
Trial pits, 24–6, 24–11, 24–12
Triangulation, 18–3, 18–5, 18–9
Trimming joists, 35–7

U-values, 43–4, 43–11, 43–15, 50–10
Undercloaks, 33–17
Underground hazards, 14–4
Underpinning, 30–18
Under-recovery of increased costs, 2–20

Valuations,
progress chart, 7–12
unfixed materials on site, 1–26
work done, 1–27
work in progress, 60–10
Value of a business, 1–39
Vapour barriers, 37–5, 37–14
Varnishes, 54–9
Vegetation effects in clay soils, 23–18
Ventilation, 16–26, 17–11, 42–11, 45–2
boiler compartment, 40–9
Building Regulations, 45–2
control, 42–11
gas-service requirements, 40–9
roof, 35–14, 36–9
Vitreous tiles, 53–10
Volumetric construction, 38–10

Walkways, 10–11
Wall cupboards, 16–24, 17–11
Wallboard, *see* Gypsum wallboard
Wallpaper, *see* Paperhanging
Walls,
cavity width, 33–9
cracks, 55–21, 56–4 *et seq.*
damp penetration, 33–3
damp-proofing, 32–29, 33–9
deformation, 56–8
diaphragm, 29–23
external, 32–26
gable, 37–11
gravity retaining, 29–18
no-fines concrete, 38–6
reinforced retaining, 29–25
sandwich construction, 38–5
solid, 58–3
thermal insulation, 33–10, 37–14, 43–12, 58–3
see also Bricks; Brickwork; Separating wall systems; Timber-framed walls and partitions

Washing-up, 16–26
Waste disposal, 17–2, 39–31
Waste pipes, 16–31
Water closets, 16–27, 16–30, 43–22, 58–8
Water hammer, 39–20
Water heating,
gas, 40–4, 40–7
immersed-element, 41–14
Water services, 17–11, 20–14, 39–2
air-locks, 39–20
alterations and conversions, 59–8
analysis, 39–10
and fire protection, 39–4, 39–9
ballvalves, 39–8, 39–9, 39–15
bidets, 16–30, 39–17
by-laws, 39–2, 39–12, 39–13, 39–15, 39–17, 39–21
choice of materials, 39–20
cold water, 39–4, 42–18
distribution pipework, 39–16
storage and services, 39–11, 39–12
design considerations, 39–3
discoloration of water, 39–20
distribution pipes, 39–16
electrolytic action, 39–16
fire protection, 39–4, 39–5, 39–9
frost damage, 39–4
hardness, 39–10, 39–11, 39–20
hot water, 42–6, 42–9, 42–18
dead-legs, 39–3, 29–21
demand and storage assessment, 39–19
feed pipework, 39–16
storage and services, 39–18
hydrants, 39–5
immersed-element heaters, 41–14
mains pressure, 39–8
metering, 39–9
pH value, 39–13
piping, 39–4
pumped systems, 39–5
regulating valves, 39–20
sampling, 39–10
sanitary fittings, 39–5
secondary circuit, 39–19, 39–20
service pipes, 39–4
Water services (*cont.*)
standpipes, 39–5
sterilisation of installations, 39–18
storage cisterns, 39–11 *et seq.*
Water softeners, 39–10
Welding plants, 9–17
Wet rot, 34–12, 55–3
Whirling hygrometer, 57–8
Wind load, 29–3
Window cills, 32–31, 37–12
Window cleaning, 45–4, 45–12
Window fittings, 44–20
Window frames, 45–8, 45–11
Windows, 16–16, 17–5, 37–11, 44–13
bay, 44–14
casement, 44–13, 44–15
design requirements, 17–7, 17–8
double-hung sash, 44–15
hardware, 45–7
high performance, 45–5
louvres, 45–7
mechanical characteristics, 45–6
modular coordination, 45–4
pivot, 44–17, 45–7
security, 45–8
sound insulation, 45–3, 58–7
steel casement, 44–15
stormproof, 44–13
thermal insulation, 43–14
Venetian, 44–15
see also Glazing
Wiring accessories, 41–11
Wiring Regulations, 17–12, 41–2, 41–10, 41–11, 41–13, 41–14
Wiring systems, 41–3
Wood blocks, 53–6
Woodworm, 34–12
Work-in-progress, 1–36
insurance, 5–3
Working capital, 1–12
Working conditions, 13–13
Working surface, 16–22, 16–24
Workmanship defects, 55–14
Works Committees, 13–20
Worktops, 17–9